The Migration Ecology of Birds

The Migration Ecology of Birds

Ian Newton
Monks Wood Research Station,
Cambridgeshire, UK

Illustrations by Keith Brockie

AMSTERDAM • BOSTON • HEIDELBERG • LONDON • NEW YORK • OXFORD
PARIS • SAN DIEGO • SAN FRANCISCO • SINGAPORE • SYDNEY • TOKYO
Academic Press is an imprint of Elsevier

Academic Press is an imprint of Elsevier
84 Theobald's Road, London WC1X 8RR, UK
30 Corporate Drive, Suite 400, Burlington, MA 01803, USA
525 B Street, Suite 1900, San Diego, CA 92101-4495, USA

First edition 2008

Notice
No responsibility is assumed by the publisher for any injury and/or damage to persons
or property as a matter of products liability, negligence or otherwise, or from any use
or operation of any methods, products, instructions or ideas contained in the material
herein. Because of rapid advances in the medical sciences, in particular, independent
verification of diagnoses and drug dosages should be made

British Library Cataloguing in Publication Data
A catalogue record for this book is available from the British Library

Library of Congress Cataloguing in Publication Data
A catalogue record for this book is available from the Library of Congress

ISBN: 978-0-12-517367-4

For information on all Academic Press publications
visit our web site at books.elsevier.com

Typeset by Charon Tec Ltd (A Macmillan Company), Chennai, India
www.charontec.com
Printed and bound in Great Britain

08 09 10 11 12   10 9 8 7 6 5 4 3 2 1

Contents

Preface

From ancient days the migration of birds has excited the wonder of thoughtful observers. (J. A. Thomson 1913.)

The phenomenon of bird migration has long fascinated its human observers, who have been continually impressed by the sheer scale and regularity of the movements. It has repeatedly prompted familiar questions about birds, such as where do they go or come from, how do they know when and where to travel, and how do they find their way? For more than a century now, bird movements have been subjected to scientific study, and by increasingly sophisticated methodology. In the past 25 years, hardly a year has gone by without the publication of a new book or symposium volume dealing with some aspect of bird migration, and each year dozens of papers have appeared in the scientific journals. In this book, I hope to provide an up-to-date synthesis of much of this information, taking account of both older and newer findings. However, the emphasis throughout is on ecological aspects: on the different types of bird movements, how they relate to food supplies and other external conditions, and how they might have evolved. It is mainly in the weight of attention devoted to ecological aspects – which have received scant attention in previous reviews – that this book differs from earlier ones. It is also in these aspects that, with my own background, I feel most at home with the subject matter.

After a brief introduction and survey of methodology, the book is divided into five main sections. The first deals with the journeys themselves: with the constraints and limitations of bird flight, the influence of weather, fuelling needs, migration strategies, travel speeds, the problems of navigation, and vagrancy. The second section is concerned with the annual cycles of birds, with how migration relates to breeding and moult, and with the physiological control of these various processes. The third section describes geographical patterns in bird movements across the globe, and the various types of bird movements, such as dispersal, irruption and nomadism, emphasising the ecological factors that underpin them. The fourth section is concerned with the evolution of migration and other movement patterns of birds, with the role of glacial history in influencing current migration patterns, and with recent changes in migration related to climate change and other human influence. The fifth section discusses how the population ecology of migratory birds differs from that of sedentary ones, and the influence of migration events on the population levels of birds. In particular, it considers the extent to which migratory bird numbers are limited by conditions in breeding, migration or wintering areas. This section is followed by a glossary, references and index.

Although the book is intended mainly for research students, I have tried to write simply, in the hope that the text will appeal to anyone with an interest in this fascinating subject, including the many bird-watchers and ringers who have contributed so much over the years to its development. To keep the book within bounds, I could not mention all recent work on bird migration, and have sought to cite examples rather than every study. Nevertheless, the reference list (up to and including 2006) relates to more than 2500 scientific papers and more than 50 books. It is inevitable in a book of this type that the same topics recur in different chapters, as they are relevant to more than one aspect of the subject, but I have tried to reduce this repetition to a minimum, and cross-refer between chapters. Nevertheless, each chapter is intended as a stand-alone read. So much of the book is concerned with geography that, while I have tried to provide some helpful maps in the text, some parts would be better read with an atlas, or preferably a globe, close at hand.

For permission to reproduce diagrams and other material from scientific journals, I thank the various publishers, ornithological societies and individuals involved, and for providing electronic copies of particular diagrams, I thank John Croxall, Thord Fransson, Mark Fuller, Sidney Gauthreaux, Yossi Leshem and Richard Phillips.

I owe a great deal to the many colleagues in the field who have discussed various aspects of the subject with me over the years, and to several friends for helpful comments on particular chapters, namely Bill Bourne (Chapter 4), Bill Clarke (Chapter 7), Alistair Dawson (Chapters 11 and 12), Barbara Helm (Chapters 11, 12, 20 and the Glossary), Lukas Jenni (Chapters 5 and 6), Peter Jones (Chapters 22, 24 and 25), Mick Marquiss (Chapters 15 and 18) and Tim Sparks (Chapter 21). Other colleagues, in their capacity as referees, commented helpfully on certain papers which preceded the book. I owe a particular debt to David Jenkins, who read the whole book in draft (some parts more than once), and offered many constructive suggestions for improvement. Finally, my wife, Halina, supported me through the writing process, and commented helpfully on the penultimate draft.

Ian Newton

Common Cranes *Grus grus* on migration

Chapter 1
Introduction

That strange and mysterious phenomenon in the life of birds, their migratory journeys, repeated at fixed intervals, and with unerring exactness, has for thousands of years called forth the astonishment and admiration of mankind. (H. Gätke 1895.)

The most obvious feature of birds is that they can fly. This facility gives them great mobility and control over their movements. Many species can travel quickly and economically over long distances – up to thousands of kilometres, if necessary crossing seas, deserts or other inhospitable areas. They also have great orientation and navigational skills, and are able to remember and re-find remote places they have previously visited. Birds can thereby occupy widely separated areas at different seasons, returning repeatedly to the same localities from year to year, and adopting an itinerant lifestyle of a kind not open to less mobile creatures.

Although migration is evident in other animal groups, including insects, mammals, pelagic turtles and fish, in none is it as widely and well developed as in birds. The collective travel routes of birds span almost the entire planet. As a result of migration, bird distributions are continually changing – on regular seasonal patterns, and on local, regional or global scales. Movements are most marked in spring and autumn, but can occur in every month of the year in one part of the world or another. These facts raise questions about the ecological factors that underlie the movements and distributions of birds that simply do not arise with more sedentary organisms.

Birds are also pre-adapted for long-distance migration in ways that other animals are not. One of the main advantages of flight is its speed, which is much faster than the alternatives of walking, running or swimming. Flight requires more energy per unit time, but because of the greater distance covered, it is also the cheapest mode of transport overall. One type of flight, by soaring–gliding, is cheaper still, but is practised mainly by larger species, such as albatrosses, which can travel the Southern Ocean with little more energy expenditure than sitting still (Chapter 3). Long-distance flight also allows birds to cross hostile areas that would otherwise act as barriers to their movements. Nevertheless, while most birds migrate by flying, penguins and some other seabirds migrate by swimming, and some landbirds by walking for part or all of their journeys.

Most birds are of a size that enables them to become airborne, and have wing shapes that ensure efficient flight. The wings are powered by massive breast muscles, the pectoralis and supra-coracoideus, which are responsible for downward and upward strokes, respectively. The two pectoralis muscles, one on each side of the breast, are by far the largest muscles in the body of flying birds, forming more than one-third of the total body mass of some species. They are well supplied with blood vessels, and consist of fast-contracting fibres (red fibres), which in many species can beat the wings continuously for hours or days on end.

Compared with other animals, birds are not only homiothermic (warm-blooded), but they also have exceptionally efficient respiratory, cardiovascular and metabolic systems. Together these systems ensure that the specialised wing muscles are kept well supplied with oxygen and energy-rich fuel, and that waste products are swiftly removed, preventing the muscle pain and fatigue so familiar to human athletes. The breathing mechanism of birds also results in much more efficient gas exchange than that in mammals. A bird's lung is connected by an array of tubes to a system of thin-walled air-sacs. Air is continuously directed through the lungs during both inspiration and expiration, thereby increasing the efficiency of oxygen extraction. By possessing all these various traits, birds are pre-adapted for the development of long-range movement patterns. Compared with resident bird species, migrants have these same features more highly developed as specialised adaptations for long-distance migration. It is this combination of features that enables some species of birds to perform some of the most remarkable migrations in the animal world.

TYPES OF BIRD MOVEMENTS

The terms resident and sedentary are usually applied to birds that occupy the same general areas year-round, and to populations that make no obvious large-scale movements resulting in changes in geographical distribution. The term migration is less easily defined because it means different things to different people. Ornithologists tend to use the word only for return movements between breeding and non-breeding areas, but biologists working with other organisms often use the term more widely. For purposes of convenience in this book, I shall divide bird movements into six main types:

- First, there are the **everyday routine movements** centred on the place of residence, which occur in all birds, whether classed as resident or migratory. Typically, they include the flights from nesting or roosting sites to feeding sites, or from one feeding site to another, and can occur in any direction. In most birds these movements are short and localised, restricted to a circumscribed home range, and extend over distances of metres or kilometres. But in other species (not-ably pelagic birds) regular foraging movements can extend over hundreds of kilometres out from the nesting colony.
- Second, there are one-way **dispersal movements**. In both sedentary and migratory bird species, after becoming independent of their parents, the young disperse in various directions from their natal sites. Individual young seem to have no specific inherent directional preferences, so within a population, dispersal movements seem to occur randomly in all directions. In most bird species, dispersal distances can be measured in metres, kilometres or tens of kilometres, but in a few species (notably pelagic birds), such distances can be much greater (Chapter 20). Post-fledging dispersal of this type does not usually involve a return journey (see below), but in any case most surviving young subsequently settle to breed at some distance from their hatch-sites (called natal dispersal). In addition, some adults may change their nesting locations from year to year (breeding dispersal), or their non-breeding locations from year to year (here called non-breeding or wintering dispersal).
- Third, there is **migration,** in which individuals make regular return movements, at about the same times each year, often to specific destinations. Compared with the above movements, migration usually involves a longer journey over tens, hundreds or thousands of kilometres and in much more restricted and fixed directions. Most birds spend their annual non-breeding period at lower latitudes than their breeding period, but some migrate to similar latitudes in the opposite hemisphere where the seasons are reversed. Such migration occurs primarily in association with seasonal changes in food availability, resulting from the alternation of warm and cold seasons at high latitudes, or of wet and dry seasons in the tropics. Overall, directional migration causes a massive movement of birds twice each year between regular breeding and wintering ranges, and a general shift of populations from higher to lower latitudes for the non-breeding season.
- Fourth, there is another category of migration, which I have called **dispersive migration,** in which post-breeding movements can occur in any direction from the breeding site (like dispersal), but still involve a return journey (like other migration). Although these movements occur seasonally between breeding and non-breeding areas, they do not necessarily involve any change in the latitudinal distribution of the population, or any change in its centre of gravity. They are evident in some landbird species usually regarded as 'resident' (Chapter 17), and include altitudinal movements in which montane birds shift in various directions from higher to lower ground for the non-breeding season. In addition, many seabirds can disperse long distances in various directions from

their nesting colonies to over-winter in distant areas rich in food, returning to the colonies the following spring.

- Fifth, there are **irruptions (or invasion migrations),** which are like other seasonal migrations, except that the proportions of birds that leave the breeding range, and the distances they travel, vary greatly from year to year (the directions are roughly the same but often more variable between individuals than in regular migration). Such movements are usually towards lower latitudes, and occur in association with annual, as well as with seasonal, fluctuations in food supplies. In consequence, populations may concentrate in different parts of their non-breeding ranges in different years. Examples include some boreal finches that depend on sporadic tree-seed crops and some owls that specialise on cyclic rodent populations (Chapters 18 and 19).
- Sixth, there is **nomadism,** in which birds range from one area to another, residing for a time wherever food is temporarily plentiful, and breeding if possible. The areas successively occupied may lie in various directions from one another. No one area is necessarily used every year, and some areas may be used only at intervals of several years, but for months or years at a time, whenever conditions permit. The population may thus be concentrated in largely different areas in different years. This kind of movement occurs among some rodent-eating owls and raptors of tundra, boreal and arid regions, and among many birds that live in desert regions, where infrequent and sporadic rainfall leads to local changes in habitats and food supplies (Chapter 16). Because these changes are unpredictable from year to year, individual birds do not necessarily return to areas they have used previously, and may breed in widely separated areas in different years.

These different kinds of movements intergrade, and all have variants, but in any bird population, one or two kinds usually prevail. Almost all bird species show post-fledging dispersal movements, in addition to any other types of movement shown at other times of year, and some species show both nomadic and irruptive movements (Chapters 18 and 19). Through migration, irruption and nomadism, birds exploit the resources of mainly different regions at different times. The birds thereby achieve greater survival and reproductive success (and hence greater numbers) than if they remained permanently in the same place, and adopted a sedentary (resident) lifestyle.

The main variables in these different types of bird movements include: (1) the directions or spread of directions; (2) the distances or spread of distances; (3) the calendar dates or spread of dates; and (4) whether or not they involve a return journey. They also differ in whether they occur in direct response to prevailing conditions, or in an 'anticipatory' manner, in adaptation to conditions that can be expected to occur in the coming weeks, and leading birds to leave areas before their local survival would be compromised or arrive in other areas in time to breed when conditions there are suitable. Each of these aspects of bird behaviour can be independently influenced by natural selection (Chapter 20), giving overall the great diversity of movement patterns found among birds, related to the different circumstances in which birds live.

This book is concerned with all these types of bird movements, but the emphasis is on the seasonal return movements of migration and irruption, which are by far the most spectacular and extreme. Migration itself varies greatly between species, as well as between populations, sex and age groups, in respect of distances travelled, routes taken, timing of journeys and behaviour en route. It is often useful to distinguish between 'short-distance' migrants that make mostly overland journeys within continents, and 'long-distance' migrants that make longer journeys between continents, often involving substantial sea-crossings. There is, of course, no clear division between the two categories, but a continuum of variation in the distances travelled and terrain crossed. Similarly, in terms of timing, some birds can complete their migrations in less than a day each way, while others may take more than three months each way, and may therefore be on the move for more than half of each year – most of the time they are not breeding.

In theory, some birds might benefit from remaining on the move at all times of year, for they could then take advantage of rich food supplies wherever and whenever they occurred. It is mainly the needs of breeding that tie birds to fixed localities for part of the year, because individuals need to remain at their nests, or visit their nests frequently, in order to feed their young. However, in some species, notably some seabirds, one parent can be away for long periods (often days, sometimes weeks at a time), while the other remains at the nest. This enables parents to collect food hundreds or even thousands of kilometres away from their nesting places. As their single chick grows, it may be able to survive on its own for long periods, enabling both parents to be away foraging at the same time. Some of the foraging flights of albatrosses undertaken while breeding can cover up to 15 000 km, a distance far greater than the total annual migrations of the vast majority of landbirds.

In many bird species, individuals do not breed until they are two or more years old. The immature, non-breeders of such species are not locality-tied in the same way as breeders, and are free to feed away from nesting areas throughout the year. It is not unusual in these species for adults and immatures to concentrate in different places in the breeding season, and in some such species the young remain in 'winter quarters' year-round, returning to nesting areas only when they are approaching breeding age (Chapter 15). This holds for many kinds of seabirds, shorebirds, large raptors and others.

ADAPTATIONS FOR MIGRATION

One of the most amazing aspects of migration is how birds find their way over long distances. Many species are capable of migrating between exactly the same breeding and wintering places year after year, even if these places lie thousands of kilometres apart on different continents. Young birds migrating alone can find their own way to the usual wintering areas for their species, and back to their natal areas the following spring. Some pelagic seabirds wander widely over the oceans, yet each year return unfailingly to their own particular nesting islands. Great Shearwaters *Puffinus gravis*, for example, nest on the isolated Tristan da Cunha islands, lying at 40°S in the South Atlantic and more than 2000 km from Africa, the nearest continent. In the non-breeding season these birds migrate northward in their millions, ranging over large parts of the North Atlantic. But they return each year with pinpoint accuracy to their tiny breeding islands, which are spread over only 45 km of ocean, and individuals occupy the same nest burrows from year to year, often lying within a metre of those of other individuals. These and other seabirds that migrate long, overwater distances to small oceanic islands must surely be among the greatest of animal navigators, possessing extremely accurate orientation mechanisms.

Like human navigators, birds and other animals can find their way over long distances only with the aid of a reliable reference system by which to navigate. Research has confirmed that birds use at least two types of system, based on geomagnetic and celestial cues (the sun by day and the stars at night) respectively. However, a compass is of little value to a migratory bird unless it 'knows' beforehand – either by inheritance or experience – what course it needs to take. The mechanisms of bird orientation and navigation are discussed in Chapter 9.

The timing of bird migration is equally remarkable. Many long-distance bird migrants arrive at their nesting or wintering places every year at around the same date. This implies the existence in the birds of precise timing mechanisms that, in response to external stimuli, trigger migration at about the same dates each year and maintain it for long enough to allow the bird to cover the distance required. Such mechanisms ensure that individuals arrive in their nesting areas as conditions become suitable for breeding and leave before conditions deteriorate and affect survival. The relatively small variations in timing that occur from year to year are mainly associated with variations in prevailing weather or food supplies (Chapter 12).

A third adaptation that facilitates seasonal migration is the ability of birds at appropriate times of year to accumulate large body reserves (mostly fat) to fuel the flights (Chapter 5). Small birds that cross large areas of sea or desert in which they cannot feed are able to double their usual weight beforehand through fuel deposition, and some species also reduce the mass of other body organs not directly concerned with migration, thus reducing the overall energy needs of the journey. The seasonal changes in body composition that occur in migratory birds are some of the most extreme of the animal world. Birds are also unusual in the speed and efficiency with which they can convert the fatty acids in fuel reserves to the energy needed to power the wings.

The migratory lifestyle requires that periods of movements are integrated with other events in the birds' annual cycle, especially breeding or moult. In most bird species, these events normally occur at different times of year, with minimal overlap between them. Because the act of breeding requires that birds remain within restricted localities, it is obvious that individuals cannot breed and migrate at the same time. And because feather replacement can temporarily reduce flight efficiency, it is also desirable that moult and migration are separated as much as possible. Studies of the annual cycles of birds, and the physiological control of migration within these cycles, are discussed in Chapters 11 and 12.

An interesting aspect of bird migration concerns the extent to which individuals are pre-programmed by inheritance to do the right things at the right times of year. Without innate programming, an individual would have little sense of when to migrate, in which direction to fly or for how long. Nor would it know when on its journey to do specific things, such as change direction or accumulate extra body reserves in preparation for a long sea-crossing. All these aspects require an endogenous schedule which promotes particular kinds of behaviour at appropriate times of year or stages in a journey. This inherent component of some bird movements adds an additional fascination to study of the controlling mechanisms (Chapter 12).

Yet despite being partly under genetic control, migration patterns among birds show great flexibility and facility for rapid change (Chapter 20). Many bird families contain both migratory and non-migratory populations, showing little phylogenetic constraint on the development of migratory behaviour. Within species, changes in migratory patterns are presumed to have occurred repeatedly through the Pleistocene glacial cycles and, more strikingly, even in recent decades, as particular populations have become more sedentary, or shortened their migrations, in apparent response to climate warming (Chapter 21). Further understanding of the evolution of migration systems can be inferred from present distribution and movement patterns, as well as from palaeontological and molecular evidence (Chapters 22 and 23).

To accommodate a long-distance migratory lifestyle, participants must be able to live in two or more different parts of the world, often on different continents. They must often occupy somewhat different habitats and climatic regimes, deal with different foods, and exist within different communities, filling distinct niches in both their summer and winter homes. Such split lives have consequences that a sedentary lifestyle does not. In particular, the population levels of migratory birds can be influenced by conditions in breeding, migration and wintering areas, and conditions experienced in wintering or migration areas can affect subsequent survival and breeding success (Chapter 26). Recent widescale declines in the numbers of many migratory species, from both the Eurasian–African and the North American–South American bird migration systems, have stimulated research into what limits the population sizes of migrants, and whether the limitation occurs primarily in wintering, breeding or migration areas (Chapters 26–28).

THE DIVERSITY OF MIGRATION

Migration occurs to some degree in most bird species that live in seasonal environments, from arctic tundras to tropical savannahs and grasslands. It is in strongly seasonal environments that food supplies vary most markedly through the year, fluctuating between abundance and scarcity in each 12-month period. Generally speaking, birds time their migrations so as to be present during the periods of abundance and absent during the periods of scarcity. Only in the relatively stable conditions of tropical lowland rainforest, where food supplies remain fairly constant year-round, do the majority of bird species that breed there remain all year, but even these forest areas receive a seasonal influx of wintering migrants from higher latitudes. Worldwide, in response to seasonal changes in food supplies, more than 50 billion birds are thought to migrate every year on return journeys between breeding and non-breeding areas (Berthold 1993).

Because almost all migratory birds travel to milder climes for the non-breeding period, they move mainly on a north–south axis. However, many populations also have an easterly or westerly component in their movements, especially those that breed in the central parts of the northern landmasses and move to the warmer edges for the winter. Thus, the predominant autumn migration direction of intra-continental migrants in western Europe is southwestward, but the further east they breed within Europe, the stronger the westerly component in their autumn journeys. Western Europe is warmer in winter than equivalent latitudes anywhere else on the Eurasian landmass, so acts as a major wintering area for Eurasian migrants, including up to two million waterfowl. Nevertheless, some birds from eastern Europe move southeast to winter in the Middle East, East Africa or India. Similarly, in much of North America, many birds move southeastward in autumn, towards the warm southeastern States, or onward to the Caribbean Islands or South America (most of which lies in longitudes east of North America).

Some bird species move almost directly east–west on their migrations. For example, the Pochards *Aythya ferina* which breed in Siberia move up to 4000 km in autumn to winter in western Europe, in the process crossing up to 80° of longitude (M. Kershaw, in Wernham *et al.* 2002). Many species in southern Africa move from the arid west in summer to the wetter east in winter (Brooke 1994). Many seabirds, shorebirds and waterfowl of high latitudes fly east or west in spring along the northern edge of the continents before moving inland to nest on the open tundra to the south. In the autumn, they retrace their journeys along the northern coastline, until they reach the continental edges when they veer southwards towards their wintering areas (Alerstam & Gudmundsson 1999).

The Bald Eagles *Haliaeetus leucocephalus* that breed in southern North America show an unexpected pattern. The young are raised in winter or early spring, then move generally northwards for up to 2200 km, spending from May to September in Canada and Alaska, where they feed largely on salmon which fill the rivers at that time (Broley 1947). The young eagles therefore travel north in spring and south in autumn with the conventional migrants but, unlike them, they have been reared in the south before doing so. Many adult eagles also leave the south in spring, but it is not clear from ringing whether they travel as far as the juveniles. This migration was first established from ringing nestling eagles in Florida (Broley 1947), but more recently it has been confirmed in radio-tracked young from California (Hunt *et al.* 1992), and in colour-marked young from Texas (Mabie *et al.* 1994). This last study also showed that young returned to their natal areas to breed.

The young of several species of herons, raised in winter in the southern USA, disperse in various directions but mainly northward, again presumably to exploit the presence of fish in shallow water in the northern spring and avoid the effects of drought in more southern areas (Lincoln 1935a). Similar but less marked summer movements have been recorded among herons in Europe, and in the southern hemisphere some heron species in Australia also migrate to higher latitudes after breeding (Maddock 2000), as do flamingos in South America (Sick 1968b). In addition, the non-breeders of some seabird species, including the Little Auk *Alle alle* and several skua species, spread up to several hundred kilometres beyond their natal colonies in summer, exploiting the summer flush of food at higher latitudes, and some winter-breeding petrels and shearwaters also move to higher latitudes after breeding (see later).

Difficult journeys

Bird migrations may vary from a few tens to many thousands of kilometres, but it is the long and difficult journeys that best reveal the capabilities of migratory birds. Among landbirds, spectacularly long journeys are made by those species that fly regularly between northern Eurasia and southern Africa or Australasia,

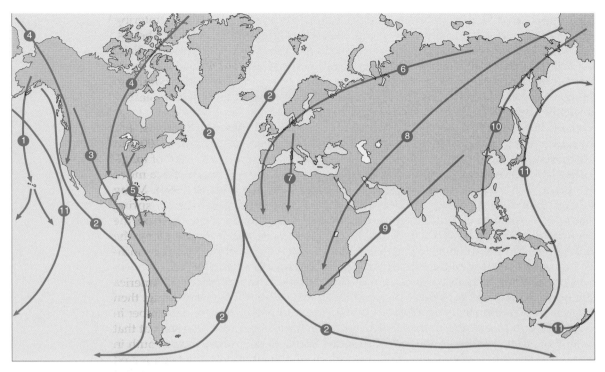

Figure 1.1   Some long-distance migrations of birds. 1. Alaskan population of Pacific Golden Plover *Pluvialis dominica*; 2. Arctic Tern *Sterna paradisaea*; 3. Swainson's Hawk *Buteo swainsoni*; 4. Snow Goose *Chen caerulescens*; 5. Many North American breeding species that cross the Gulf of Mexico; 6. Ruff *Philomachus pugnax*; 7. Many European breeding species that cross the Mediterranean Sea and Sahara Desert; 8. Northern Wheatear *Oenanthe oenanthe*; 9. Amur Falcon *Falco amurensis*; 10. Arctic Warbler *Phylloscopus borealis*; 11. Short-tailed Shearwater *Puffinus tenuirostris*. Partly after Berthold (1993).

or between northern North America and southern South America or Australasia (**Figure 1.1**). Such long movements are performed each year by many shorebirds, and some seabirds, passerines and others. Even on the shortest routes, this entails some individuals flying more than 25 000 km on return migration each year. Some of the participants are small enough to be held comfortably in the palm of your hand. The major advantage in migrating so far between the northern and southern hemispheres derives from the fact that the seasons are reversed. The species involved thus pass both breeding and non-breeding seasons in summer conditions when food is plentiful, although no such birds are known to breed regularly at both ends of their migration route (Chapter 13).

Most birds that migrate overland have plenty of places to stop and feed. They can therefore migrate, rest and feed almost every day, accomplishing their journeys by a series of short flights. Other birds cross mainly hostile areas, where they cannot stop and feed. They therefore have to accumulate larger body reserves, and make long flights between widely spaced stopping places (Chapter 5). For example, shorebirds typically complete their migrations in 2–4 long stages, refueling before each stage, and often travelling 1000–4000 km between suitable estuaries, even when mainly following coastlines. The flights themselves comprise long periods of muscular work without food or water, at great heights over inhospitable terrain, and usually require pinpoint navigation to widely separated refuelling areas. Flight paths and stopping sites of some shorebirds have been worked out in some detail from synchronised counts at different estuaries, from ring recoveries, and in some regions also from radar observations and studies of body weights.

To elaborate with one example, those Bar-tailed Godwits *Limosa lapponica* that winter in West Africa north of the equator have to face a journey of 10 000 km to their breeding grounds in Siberia (Piersma 1994a). Taking off from Guinea-Bissau at dusk, a bird could reach the next major mud flat, the Banc d'Arguin in Mauritania 1000 km away, by mid-day. From there, another 16 hours and 1000 km of flight would get the bird to the next suitable estuaries in Morocco and another 16 hours and 1000 km to the estuaries of the Loire and Gironde in western France, and yet another 10 hours and 600 km of flight to the Wadden Sea which fringes the northern Netherlands, Germany and Denmark. After a long period of refuelling on the Wadden Sea coast, most godwits seem to make the rest of their journey to Siberia in a single flight of 4000 km. In fact, most godwits also seem to fly from Banc d'Arguin to the Wadden Sea in one flight, as the Moroccan and French estuaries are used only by a small proportion of the population. Travelling at 60 km per hour (without the benefit of a tailwind), the entire journey translates to 167 hours of airtime, equivalent to a solid seven days and nights of flight, excluding breaks for refuelling (Piersma 1994a).

Some shorebird species that breed across the arctic show an astonishing array of migration routes. In the Ruddy Turnstone *Arenaria interpres*, for example, Alaskan birds migrate down the entire western seaboard of the Americas to winter as far south as Chile, while most of the Canadian birds head for the coasts of the Caribbean and beyond. The Greenland and eastern Canadian birds move to Britain and Ireland, and Scandinavian ones to West Africa. The central Siberian birds move south to the Middle East, the shores of the Indian Ocean and on to southern Africa, while the east Siberian/west Alaskan birds winter in southeast Australasia and Pacific Islands. Except at high ice-bound latitudes, few rocky shorelines anywhere in the world do not support wintering Ruddy Turnstones from one part of the breeding range or another.

Landbirds that migrate over oceans provide some of the most extreme examples of endurance flight and precise navigation. They travel without opportunity to feed, drink or rest, over vast stretches of open water devoid of helpful landmarks. They cannot stop, as birds do overland, when the weather turns against them. Yet millions of landbirds regularly cross the Mediterranean Sea and Gulf of Mexico at their widest points (about 1200 km), and smaller numbers regularly cross longer stretches, such as the western Atlantic between northeastern North America and northeastern South America (2400–3700 km), or the northern Pacific between Alaska and Hawaii and other central Pacific Islands (5000 km). However, the most impressive of all overwater migrations by a landbird is undertaken by the Bar-tailed Godwits *Limosa lapponica* from eastern Siberia and Alaska, which in autumn apparently accomplish an astonishing 175-hour non-stop 10400 km flight to New Zealand (Chapter 6). Apart from the length of the journey, imagine the navigational precision required. From the departure point in Siberia or Alaska, the target area of New Zealand subtends an angle of only 5°, extending over a relatively tiny part of the southern Pacific. To judge from their normal flight speeds, landbirds would take more than 100 hours of non-stop flight in still air to accomplish the longer of their overwater journeys, but by taking advantage of favourable winds, they can shorten their flight times, sometimes by as much as one half. Participants include many passerines and shorebirds, but also waterfowl which, unlike the others, can rest on the sea if need be.

Some overland journeys are also difficult. Long desert crossings are made by the many species (including passerines) that travel between Eurasia and tropical Africa. Most west European species cross at least 1500 km of the Sahara Desert immediately after crossing the Mediterranean Sea, an overwater journey of up to 1200 km. In autumn some species may make this Mediterranean–Saharan flight without a break, a total journey of 1500–2500 km, depending on the route taken (Chapter 6). Other birds from further east cross the central Asian deserts, and then another 1700 km of southern Arabia and its bordering gulfs, before reaching East Africa. In Australia, some waders cross the central desert in moving between southern and northern coasts, a journey of more than 2000 km.

Yet other birds cross high mountain ranges, including the Himalayas and Tibetan plateau. One such species is the Bar-headed Goose *Anser indicus* which in the process can rise to more than 8 km above sea level, where the air is thin and very cold (Chapter 6). Other species cross extensive areas of pack-ice that lie in spring between Siberia and Alaska or between Norway and Svalbard. A few species cross 2000 km of the 2-km-high Greenland ice cap on journeys between northeastern Canada and western Europe. No landbirds regularly cross the Southern Ocean to Antarctica (which holds only seabirds), and none is known to cross the North Pole, even though the tundras on either side are only 2000–3000 km apart (Gudmundsson & Alerstam 1998). There would probably be no advantage in trans-polar movements, which in any case might also present navigational problems (Chapter 9).

The various migrations mentioned above are among the longest and most impressive undertaken by landbirds. Most long journeys involve movement from one hemisphere to another, but this is not true of all. Brent Geese *Branta bernicla* are restricted to northern latitudes, yet some of their breeding and wintering areas can be as much as 5000 km apart, involving substantial east–west shifts, as well as north–south ones. The journey from northeast Canada to Ireland involves these geese crossing polar seas, the Greenland ice cap and the North Atlantic.

At the other end of the spectrum, the shortest seasonal migrations by landbirds are undertaken by some altitudinal migrants, which move only a few kilometres from mountains to valleys for the winter, and by various gallinaceous birds which typically move over short distances (up to several tens of kilometres) between their breeding and wintering sites (Chapter 17). Such migrations often occur in various directions and intergrade with purely local movements.

Seabird movements are just as varied as landbird movements. Their study is hampered by the obvious improbability of getting ring recoveries from pelagic regions, but observations from ships and the satellite-based tracking of radio-marked individuals have helped to fill out the picture (Chapter 2). Some species, as well as moving north–south, cross from one side of an ocean to another. In the Atlantic, Manx Shearwaters *Puffinus puffinus* move from western Europe to eastern South America after breeding, and Arctic Terns *Sterna paradisaea* from eastern Canada cross to West Africa, before continuing southward to the Southern Ocean. In addition, Sooty Terns *S. fuscata* cross the Atlantic from breeding colonies in the Caribbean to non-breeding areas off West Africa, but in the process make little or no latitudinal shift.

Like landbirds, many seabirds perform exceptionally long migrations, which are perhaps less demanding than the transoceanic flights of landbirds. This is partly because most seabirds are larger and more robust than the majority of landbirds, but also because many can more readily rest on the sea surface or feed en route. In moving between the Arctic and Southern Oceans, the Arctic Tern *Sterna paradisaea* may perform the longest migration of any tern, entailing a round trip of 30 000–50 000 km each year (this species is not known to rest on the sea). The birds move down the western coasts of South America and Africa, and on reaching the Southern Ocean, travel eastwards on the winds, some passing south of Australia and New Zealand. Many juveniles may continue eastwards, circling the Antarctic, before heading north again on their return flight, two or more years later (Salomonsen 1967b). These movements have been revealed by observations and by some striking ring recoveries, showing the presence in winter of North American birds off South Africa and of European birds off South Africa and Australia. Because some Arctic Terns may reach an age of 25 years, they might cover more than a million kilometres on migration during their lifetimes, about three times the distance between the earth and the moon. Some Common Terns *Sterna hirundo* from northern Europe have also been found off Australia, but this species apparently does not extend south towards Antarctica.

Conversely, some seabirds breeding in the southern oceans spend their non-breeding season in the northern hemisphere. Examples include the Sooty Shearwater *Puffinus griseus* which migrates between the South and North Atlantic, and between the South and North Pacific. Seventeen individuals from breeding colonies in New Zealand were tracked on migration using miniature archival tags to record geographical position, dive depth and ambient temperature (Shaffer *et al.* 2006). The tags revealed that these birds flew across the Pacific in a figure-eight pattern while travelling an average of 64 037 ± 9779 km round-trip. They took 198 ± 17 days over the journey, and reached speeds up to 910 ± 186 km per day. Each shearwater made prolonged stops in one of three discrete productive regions – off Japan, Alaska or California – before returning to New Zealand through a relatively narrow corridor in the central Pacific. The birds obtained food from the surface or down to depths of 68 m. Similar figure-eight migrations had previously been inferred from ring recoveries for the Short-tailed Shearwater *P. tenuirostris* which breeds on islands off southeast Australia and 'winters' in the north Pacific (**Figure 1.1**).

Several Antarctic seabirds, such as immature Southern Giant Petrels *Macronectes giganteus*, perform circumpolar migrations, flying eastward around the world in the Southern Ocean. Radio-tracking results from albatrosses have revealed the extraordinary distances travelled by some species in short time periods. For example, a Northern Royal Albatross *Diomedia epomophora* was found to fly up to

1800 km in 24 hours, and a Grey-headed Albatross *D. chrysostoma* circled the globe in just 46 days (Croxall *et al.* 2005). Albatrosses can cover long distances on routine foraging flights from their nesting islands, as well as on migration. Thirteen young Wandering Albatrosses *Diomedia exulans*, radio-tracked by satellite from their natal areas on the Crozet Islands, flew eastward in the southern Indian Ocean, where they foraged back and forth along a 2000 km strip of ocean, just north of the sub-tropical front. In their first year of life, they covered an average of 184 000 km (range 127 000–267 000 km), corresponding to a distance of 4.6 times round the earth at its widest part (Åkesson & Weimerskirch 2005).

Not all seabirds migrate to lower latitudes in winter. Some species that breed in winter migrate to higher latitudes after breeding, like the Bald Eagles mentioned earlier. In the northern hemisphere, they include the Black-vented Shearwater *Puffinus opisthomelas* and Brown Pelican *Pelecanus occidentalis* off western North America and Bonin Petrel *Pterodroma hypoleuca* of eastern Asia, and in the southern hemisphere they include the Kerguelen Petrel *Lugensa brevirostris* and the King Penguin *Aptenodytes patagonicus*. For those penguins nesting on the Crozet Islands, this involves swimming about 1600 km (over 8° of latitude) to exploit the rich food supplies available at the edge of the winter pack-ice (Bost *et al.* 2004). Many other seabird species disperse eastward or westward after breeding to spend the winter away from their nesting colonies, concentrating at upwellings and other areas of abundant food, but remaining within the same oceanic zones year-round. Examples include the Northern Fulmar *Fulmarus glacialis* and Kittiwake *Rissa tridactyla* in the northern hemisphere, and various albatrosses and petrels in the southern hemisphere. Those albatross species whose breeding cycle lasts more than a year have an off-year after each successful breeding attempt. This applies to Wandering Albatrosses *Diomedia exulans* nesting on the Crozet Islands in the southern Indian Ocean. The adults leave the foraging areas that they frequent while breeding and spend their sabbatical years in sea areas 1500–8500 km away, elsewhere in the southern Indian Ocean or in the southwest Pacific. The ocean areas they occupy then range from tropical–subtropical waters (females) to sub-Antarctic–Antarctic waters (males), and individuals probably visit the same areas in each sabbatical (Weimerskirch & Wilson 2000). These albatrosses do not therefore perform a seasonal north–south movement like other birds but, as different sectors of the population breed in different years, by moving far from their breeding areas the sabbatical birds avoid competing with others that are breeding that year and thus operating from the colony. In this sense, their movements are again food-related.

While many landbirds face the hazards of ocean crossings, some seabirds make long overland journeys. The birds are seldom seen because they fly too high for human vision, and the evidence for their overland movements is based on unusual events and ring recoveries. In both Eurasia and North America, part of the migrations of the marine Sabine's Gull *Xema sabini*, Long-tailed Skua *Stercorarius longicaudus*, Arctic Skua *S. parasiticus*, Arctic Tern *Sterna paradisaea* and several other seabirds may take place overland to and from their tundra nesting areas. Arctic Terns *Sterna paradisaea* may even cross the central parts of the Eurasian landmass en route between the Indian Ocean and the Siberian tundra. Occasional ringed birds were found dead more than 1000 km inland from any sea-coast (Bourne & Casement 1996).

The above examples give some idea of the variety of migration patterns found among birds from around the world, and of some of the more spectacular journeys.

Migration routes

For reasons of habitat and weather, birds do not always take the most direct (great circle) routes between their breeding and wintering areas; nor do they necessarily take the same routes in autumn and spring (Chapter 22). Thus, sea-ducks often migrate long distances around coastlines rather than taking over-land shortcuts between breeding and moulting or wintering areas. For example, most Eiders *Somateria mollissima* breeding on islands far up the St Lawrence River in eastern Canada fly a coastal route of 2250 km to reach a point on the coast of Maine scarcely 640 km distant from their nesting islands, a shortcut which is taken by only a minority of birds (Reed 1975). Such detours may offer several benefits to migrating birds, such as continuous suitable habitat in which they can stop and feed, or reduced risk from adverse weather or predators. Mainly because of wind conditions, some roundabout routes may also offer reduced energy costs, despite the longer journey (Chapter 3; Alerstam 2001). The trade-off between time-saving or energy-saving direct routes on the one hand and risk-reducing detours on the other may have favoured different patterns in different species, according to their flight capabilities and refuelling needs, as well as the habitat and seasonal weather patterns encountered en route.

Most bird species set off on a broad front between their breeding and winter-ing areas, but during the journey they may be concentrated to some extent along mountain ranges, sea-coasts and other 'leading lines'. Such streaming is particu-larly marked in waterbirds, which often migrate between specific sites, following river valleys or other routes offering wetland areas where they can rest and feed. It is also marked in raptors and other soaring birds that favour routes where ther-mals and other updrafts develop, and minimize any necessary water crossings. Because these features depend on geography and topography, such species tend to follow the same narrow traditional routes year upon year.

In taking roundabout routes, many birds divide their journey into distinct stages, each with a different main orientation. For example, most of the European passerines that migrate southwest to Iberia must then turn south or southeast if they are to reach West African wintering areas. Similarly, those that migrate from Europe southeast to the Middle East must then turn southwest to reach East African wintering areas. Their journeys are thus undertaken as two distinct stages, with different headings, although neither stage is necessarily accom-plished non-stop. Many Eurasian–Afrotropical migrants make their return north-ward flight somewhat to the east of their autumn southward flight, probably in response to prevailing wind or feeding conditions (Chapter 22).

The tendency to take different routes in autumn and spring is especially marked in those species, such as the American Golden Plover *Pluvialis domin-ica*, that migrate between northeastern North America and southeastern South America (**Figure 1.2**). In autumn, when winds are favourable, these birds take the shortest route over the Atlantic, but in spring when winds over the Atlantic would be against them, they take the longer overland route through Central and North America. Such loop migrations are performed by many species in many

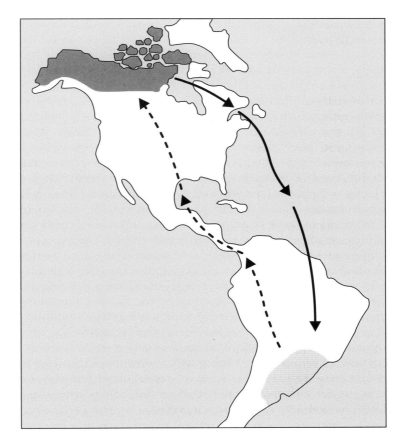

Figure 1.2 Loop migration of the American Golden Plover *Pluvialis dominica* between breeding areas in North America and wintering areas in South America. The outward route occurs largely over the Atlantic Ocean and the return route largely overland. Partly from Byrkjedal & Thompson (1998).

different parts of the world, illustrating the effects of seasonal local conditions on the development of migration routes (Chapter 22).

Costs of migration

While migration allows participants to exploit the resources of different regions at different times of year, the travel involved is not without costs. In addition to the energy required for the journey, migrants must travel through unfamiliar and sometimes hostile terrain, adjust to atypical habitats, face unfavourable weather and intense competition for the limited resources at staging sites, and suffer the consequences of navigation errors (Chapter 10). As explained above, some species perform long and hazardous journeys on which they can neither feed nor rest. Although the mortality associated with such journeys is hard to estimate, it may often be substantial, and storms have killed thousands or even millions of birds at a time (Chapter 28).

In addition, at some stopover sites, where birds replenish their fuel reserves, large numbers of individuals must often gather at one time. Local food supplies can then be greatly depleted, leading to intense competition, to the detriment of many. Avian predators often concentrate at the same places, giving a high ratio of predators to prey, which increases mortality and disrupts the feeding of many individuals (Chapter 27). Moreover, in passing through a wide range of areas, migratory birds are likely to encounter a greater range of parasites and pathogens than are resident birds that remain in the same areas year-round. The high densities that migrants experience at some stopover sites favour the transmission of certain kinds of parasites and pathogens, which in turn can compromise the migratory performance, survival and breeding success of infected individuals. Compared to residents, some migrants have larger immune defence organs, such as spleen and bursa of Fabricius (Møller & Erritzoe 1998). The bursa is found only in sexually immature birds, but it is relatively larger in migrants, even before their first migration, than it is in residents. Disease agents may therefore play a greater role in the lives and deaths of migratory birds than in resident ones, and impose greater costs in terms of immunoprotection. In the process, migratory birds can also transport parasites and pathogens over long distances, as in recent years with the spread of the H5N1 strain of 'avian flu'. These are aspects of bird migration that have so far received little attention, and on which more research is needed. Overall, however, the benefits of migration to the participants, whether they accrue through increased survival, increased reproductive success or both, must on balance be greater than the mortality costs of the journey, for otherwise migration could not have evolved and could not persist.

SEDENTARY POPULATIONS

At the opposite end of the spectrum from migratory populations are sedentary (or resident) ones. A sedentary bird population can be defined as one whose distribution and centre of gravity remain more or less the same all year round, and from year to year. Individuals of sedentary populations typically show no directional bias in their movements at any time of year (unless imposed by local topography), and generally move over much shorter distances than migrants. In Britain, as elsewhere, large numbers of many resident bird species have been ringed as chicks and adults, and the subsequent recoveries of birds found dead and reported by members of the public have given some idea of their overall movement patterns. Typically, most birds of non-migratory species were found (up to several years later) near where they were ringed, in all directions, but with progressively fewer at increasing distances. In many resident songbird populations, the median distance moved between ringing and recovery sites was less than 1 km, but some individuals had reached more than 20 km. This pattern held for such sedentary passerine species as House Sparrow *Passer domesticus* and Carrion Crow *Corvus corone*, and for non-passerines such as Moorhen *Gallinula chloropus* and Grey Partridge *Perdix perdix*. All these birds are likely to have made their longest movements in the immediate post-fledging period, soon after becoming free of parental care. Of course, other 'sedentary' species make longer movements, yet in all such species the population as a whole retains the same broad-scale distribution year-round.

The everyday movements that birds make to obtain their daily sustenance vary enormously between species, depending on their particular lifestyles. In species that live year-round in territories, individuals may spend their whole adult life in a confined area, varying from less than one hectare in some small songbirds up to a few hundred square kilometres in some large eagles (Newton 1979). In species in which breeding and feeding occur in different areas, everyday movements may be longer. For example, some radio-tagged female Brown-headed Cowbirds *Molothrus ater* studied in California travelled 9–16 km every day between their breeding and foraging areas, and even further to a communal roost (Curson *et al.* 2000). Other flock-feeding species, which base themselves in nesting colonies or communal roosts, forage over even larger areas. For example, radio-tagged Common Grackles *Quiscalus quiscula* studied in Oklahoma commuted an average of 24 km between roost and feeding area, and ranged during the weeks of study over an average of 325 km^2 (Bray *et al.* 1979). Among colonial seabirds, individuals may forage over very much larger areas, extending over many thousands of square kilometres, as mentioned earlier. In all such species, the same individuals may maintain the same distribution pattern year-round, or at least over large parts of each year.

HIBERNATION

While many birds alleviate seasonal food shortages by migrating elsewhere, many other animals cope with seasonally difficult periods by hibernating, remaining dormant for up to several months at a time. They survive at much reduced metabolic rate on body reserves, and emerge when conditions improve. At one time, the disappearance of most birds from high latitudes for the winter was attributed to hibernation rather than migration. In fact, at least one species of bird does hibernate in winter. This was discovered in 1946, when a Common Poorwill *Phalaenoptilus nuttallii* (a sort of nightjar) was found in a torpid state in a rock crevice in a California desert (Jaeger 1949). The bird was inert, its respiration and heart rate were barely detectable, and its body temperature was 18–20°C, about half the usual level for birds. The individual was ringed, and in subsequent winters it was found hibernating again in the same crevice. Since then other poorwills have been found in similar sites in the same condition, and their physiology has been studied in laboratory conditions (Withers 1977). The energy consumption of torpid birds was so low that they could live off their body fat for more than three months. Other kinds of birds can also become torpid but remain so only overnight (hummingbirds) or for at most a few days at a time (swifts and colies). Evidently, long-term hibernation is at best extremely rare among birds, most escaping difficult conditions by migration instead.

SUMMARY

The large-scale movements of birds can conveniently be divided into dispersal, dispersive-migration, migration, irruption and nomadism, although these different types of movements intergrade with one another, and the same populations may show more than one type of movement at different stages of their annual cycle. This book is concerned with all these types of movements, but chiefly with migration, defined as a seasonal return movement in fixed directions between separate breeding and wintering ranges. Migration occurs to some degree in most species of birds that live in seasonal environments. It leads to massive twice-yearly changes in the distributions of birds over the earth's surface.

Some migratory birds travel relatively short distances of a few tens of kilometres between their breeding and wintering areas, but others travel hundreds or thousands of kilometres, sometimes crossing long stretches of sea or desert or high mountain ranges, where they cannot rest or feed. They accumulate large reserves of body fat for the journey. Such birds show impressive navigational skills which enable individuals to return to the same breeding and wintering sites year after year. Although migration occurs mainly on a north–south axis, many species have a strong east–west component in their journeys, especially those that move from the seasonally hostile centre to the milder edges of the northern land masses. Individuals in so-called sedentary populations mostly move over short distances of at most a few tens of kilometres, and show no directional preferences, so that the population occupies essentially the same range year-round. Only one bird species is known to hibernate through the unfavourable season.

Ringing a Curlew Sandpiper *Calidris ferruginea*

Chapter 2
Methodology

The study of living birds by the banding method, whereby great numbers of individuals are marked with numbered aluminium leg rings, has come to be recognised as a most accurate means of ornithological research. (Frederick C. Lincoln 1935.)

The development of any area of science depends heavily on the methodology available. Migration studies began in the simplest possible way, by observation, and have developed over the past 120 years by the addition of progressively more sophisticated methodology, including in recent years the use of miniature radio-transmitters to track individual birds on their journeys. At no stage, however, has any method been dropped from the arsenal, and systematic observations can be just as revealing now as they were 120 years ago. In this chapter, different study techniques are described, highlighting their pros and cons. Two hundred years ago practically no evidence was available that birds migrated, apart from their seasonal appearance and disappearance in particular areas. Hibernation was often thought to be responsible for the disappearance of many species from high latitudes in winter, a notion that gained support even from the scientific community, but again without evidence.

An early indication that individual birds could actually travel long distances to winter elsewhere was provided by a White Stork *Ciconia ciconia* which was seen in Germany in 1822 flying around with a spear stuck through its body. When the bird was shot it was found that the spear could be attributed from its design to a part of West Africa. This probably provided the first firm indication from Europe of a long-distance movement by an individual bird. Since that time, more than two dozen other storks have been recovered in Europe in similar circumstances. In more recent times, bird migrations have been studied by observations (made directly or with radar), by bird counts made in particular places at different dates, by widescale surveys of bird distributions at different seasons, by use of ring recoveries, or in recent years by the use of radio-transmitters and other devices fixed to individual birds which can then be followed on their journeys.

OBSERVATIONS OF BIRDS ON MIGRATION

It is widely known that, at particular localities, some bird species appear only in the breeding season, and others in the non-breeding season or at times of passage. Watching birds on migration has become a favourite pastime for thousands of bird watchers, and in many countries the concentration points (such as coastal promontories, offshore islets and mountain passes) are now well known. Hawk Mountain in Pennsylvania, which is famous as a viewing site for raptor migration, attracts about 20000 raptors each autumn, but more than 100000 human observers. Large numbers of people also visit Cape May in New Jersey each year, the Santa Ana Wildlife Refuge in southwest Texas, and Falsterbo in Sweden. At some sites, diurnal migration has been observed systematically over many years, giving information on the numbers, directions and passage periods of different species, on prevailing weather effects on migration, and on long-term changes in numbers and migration timing (Chapter 7).

For most bird species, however, counts of birds seen on the ground or flying over represent only a small and variable proportion of those passing overhead. This is because most migrating birds fly much too high to be seen with the naked eye or even with binoculars, and in any case many species migrate mainly at night. It is chiefly when they encounter headwinds that birds fly low enough to be easily seen. Migrants come to ground mainly to rest or refuel, or after they have drifted off course because of side winds, or been forced down by headwinds, mist or rain. Hence, visual counts of migrants cannot usually reflect the true volume of migration, or the weather conditions that most favour it (for a critique of observational methods of study, see Kerlinger 1989). On the other hand, any birds seen can usually be identified to species by their appearance or calls. Large-scale studies of visible bird migration, with observers posted at different localities, formed some of the earliest cooperative projects involving bird-watchers.

Watching seabirds on migration is most profitable when onshore winds cause birds to fly closer to land, and often thousands per hour can be seen streaming past headlands. Such counts depend on wind conditions bringing migrating birds within view. In other conditions, migration would occur too far out to be visible to a land-based observer. Also, as seabird species forage at long distances from their nests, it is impossible at certain times of year to distinguish migration from foraging flights.

At night, ground-based observers are much more limited, but on clear nights low-flying birds can be seen as they cross the lit surface of the moon (moon-watching). The drawbacks of the method are that it can only be used near full moon in clear weather and the observation cone has a relatively small angle (on average 0.52°), covering only a tiny portion of the night sky. By adventurous calculations involving the moon's bearing and elevation, counts of birds crossing the face of the moon can be transformed into estimates of the numbers passing over, their direction of movement and even their height and speed (Nisbet 1959a). Using a telescope with 40× magnification, it was estimated that about 50% of the birds flying at 1.5 km distance from the observer were detected, reducing to zero at 3.5 km, based on comparison with radar and infrared observations (Liechti *et al.* 1995). The moon-watching method is also hard on the eyes, and ideally needs several observers taking turns. The most impressive large-scale count programme ever undertaken on the basis of moon-watching was in central Asia where, at the time it was done, no other study methods were available there (**Box 2.1**).

Other observers have used a strong spotlight directed skywards to count the birds passing through the beam. The best device for this purpose is a ceilometer, which is normally used at airports for measuring cloud height. In warm weather, the lower part of the beam tends to be full of insects, but birds seen flying through the upper part can be recorded in the same way as for moon-watching, but with limitations on distance as the beam typically extends only to a few hundred metres. Hebrard (1971) used a horizontally directed portable ceilometer placed on a tower to illuminate birds at they took off from the tree canopy at night.

Other evidence of nocturnal migration can be obtained by listening for the calls of birds as they pass invisibly overhead. The unaided human ear cannot pick up the normal flight calls of birds beyond about 400 m, but use of a parabolic reflector and amplifier can extend the range to 3000 m or more. Birds call more during mist and poor visibility than in clear skies, and some species seem not to call at all, so the numbers of calls heard are only broadly related to the number of birds passing (Farnsworth *et al.* 2004). Nevertheless, the opportunity that listening affords for identifying species makes it a useful accessory to other methods.

Early indication of the numbers and species of birds migrating at night was provided by 'kills' of low-flying birds attracted to lighthouses and illuminated communication masts (Chapter 28; Gätke 1895, Clarke 1912). Spectacular slaughter has sometimes been recorded at particular sites, such as the 50 000 birds of 53 species killed at one site in Georgia in one night (Johnston & Haines 1957). Some species, such as Common Snipe *Gallingo gallingo*, Water Rail *Rallus aquaticus* and Common Grasshopper Warbler *Locustella naevia* in Europe, seem notoriously prone to such accidents. Mortality occurs mainly on overcast or foggy nights, and the resulting corpses have provided information on the migration seasons, body weights and condition of different species.

Indications that landbirds cross the sea have been obtained from coastal observations of birds flying out to sea, or from ships or oilrigs of birds passing over or stopping by, and from radar observations. The recent series of oil and gas

Box 2.1 The use of 'moon-watching' to assess migratory bird numbers in central Asia.

The most extensive series of counts ever made by moon-watching was obtained by a team of Russian observers stationed at various points across a 2200-km zone stretching from the Caspian Sea to the eastern part of the Tien Shan Mountains in central Asia (Dolnik & Bolshakov 1985). At each point, birds were counted in spring as they crossed the lit surface of the moon, using $30\times$ telescopes. The totals were converted to numbers of birds crossing 1 km of a latitude line, and then extrapolated to the whole 2200 km. Although skies were generally clear throughout the migration season, the observers could not count around times of a new moon, when too little of the moon surface was lit. For those darker nights they assumed that the same numbers of birds passed as on nights around full moon, when counts could be made. All the main types of birds expected were seen, but passerines formed 73–89% of the totals at different localities. Overall, an estimated 731 million birds crossed the 42°N parallel on spring migration. From the directions taken, 85% of these birds came from winter quarters in southern Asia (heading mainly north-northwest), and the rest (110 million) probably mainly from Africa (heading mainly northeast). The volume of migration was not uniform across the whole 2200-km stretch, perhaps because some birds were diverted by topographical features, including the Tien Shan Mountains.

In autumn, the number of migrants would be expected to be at least twice the number counted in spring, owing to reproduction. This was equivalent to three times the autumn estimate made for North Africa which spans approximately double the length of latitude as the Asian survey. The implication was that about six times more migrants left this part of Asia in autumn than entered Africa. The difference may be due partly to the different methods of assessment, each with their own errors; to greater bird densities in central Asia where habitats have been less disturbed than in Europe; and to the fact that many of the Asian birds may have wintered south of 42°N, but at latitudes north of the northern latitude of Africa. The 220 million birds that could have been heading for Africa across this 2200-km front would form nearly 4.5% of the 5000 million Eurasian migrants estimated by Moreau (1972) to enter Africa each autumn.

drilling platforms in the North Sea has provided additional information on the movements of birds between Britain and continental Europe (Bourne *et al.* 1979, Anderson 1990).

RADAR AND OTHER DEVICES

The use of radar for the systematic recording of bird migration began in the 1950s. A radar emits short pulses of radio-waves and records their echoes from targets, whether birds or aeroplanes. Because radio-waves travel at the constant speed of light, the distance between the radar and the target can be calculated from the time lapse between pulse emission and echo reception. The use of radar revolutionised the study of bird migration because it made observations almost independent of flight altitudes and weather, totally independent of light conditions, and hence fully comparable by day and night. It has taught us much about unseen migration and about the influence of weather on bird movements (Chapter 4). It has provided reliable information on the seasonal and diurnal timing of migration, and on the speed, direction and altitude of flight (for reviews, see Eastwood 1967, Bruderer 1997a, 1997b, Gauthreaux *et al.* 2003). Radar also swiftly disposed of the idea that migration occurred only in spring and autumn. Birds of one species or another could be seen migrating somewhere on earth at almost any time of year.

Individual birds can be followed by radar over enough of their journeys to reveal how they orientate during migration and react to different weather conditions, and hence how their flight behaviour is shaped by prevailing atmospheric conditions. The density of birds on a radar screen cannot be precisely related to the true number of birds flying over (because several birds flying close together may appear as a single echo-spot), but it provides a relative measure of abundance that can be used by day and night.

The most obvious disadvantage of radar work is the cost: the equipment itself is expensive, and trained personnel are needed to maintain and operate it. For the most part, it is available only at a limited number of fixed installations (although mobile units are also available). The main operational drawback is that the identities of the species are usually unknown, apart from broad categories distinguished by body size, flight speed, or wing-beat patterns. The radar echoes often show rhythmic fluctuations that can be recorded and used to estimate the wing-beat frequency. This procedure enables waders and waterfowl (continuous wing-beats) to be distinguished from passerines (wing-beats broken by pauses), and perhaps two size classes in each group. Other drawbacks are that birds flying close to the ground below the radar horizon are usually missed, and backscatter from the ground can sometimes blur the image. Surveillance radars, like those used for traffic control at airports, have a fan-beam of wide vertical angle (10–30°) and narrow horizontal angle (up to 2°). By rotating the radar antenna, a wide swathe of sky can be scanned for echoes with a high horizontal resolution, but no altitude resolution. Spanning an area of more than 100 km across, surveillance radars are therefore good for studies of migration intensity, speed and general direction. On some modern radar sets, small songbirds can be detected to beyond 100 km, and larger birds to more than 500 km, providing they are high enough (Bruderer 1999). With most radar sets, the displays can be easily recorded on film for subsequent playback and analysis. A useful way of recording the slow-moving echoes of birds is with time-lapse photography, the radar screen with a clock beside it being photographed with a cine camera every 1–2 minutes. Projected at normal speed, a whole night's migration can then be viewed in a few minutes. The combination of records from many different surveillance radars at different locations has been used to provide a broad picture of bird migration on particular dates over large regions including much of North America (**Figure 2.1**; Lowery & Newman 1966, Gauthreaux *et al.* 2003). The latter study used a new system of WSR-88D weather surveillance radars. Nocturnal data on such a huge geographical scale could not have been obtained in any other way.

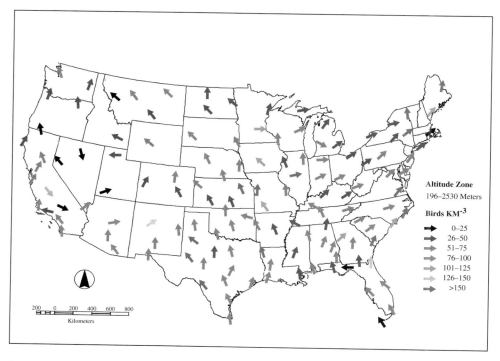

Figure 2.1   Map depicting bird migration over the United States within the altitude zone 196–2530 m on the nights of 10–11 May 2002. Arrows reflect the positions of weather surveillance radars, and show the directions and volumes of migration overhead. Map provided by S. Gauthreaux. For further details, see Gauthreaux *et al.* (2003).

In contrast to surveillance radar, a tracking radar emits a narrow 'pencil beam' by which individual birds or flocks can be tracked. When operated in automatic tracking mode, the radar locks onto a particular target bird (or flock) and records repeat measurements of distance, elevation and azimuth angles, from which the speed and direction of the tracked migrants can be calculated, and their flight trajectories plotted in three-dimensional space (Bruderer *et al.* 1995). Alternatively, the beam can be used in a conical scanning mode to provide information on the spatial distribution of migrants (although calculations of bird numbers from conical scanning present problems). Wind profiles can be obtained by using radar to track ascending weather balloons carrying aluminium foil for maximum reflectance. The heading and airspeed of the birds can then be calculated from the tracking data against the wind data. Another radar technique involves a vertical set designed to quantify the amount of migration taking place and the heights at which the birds are flying. This gives similar results to those obtained using a ceilometer, except that the radar can look through clouds and detect the birds at all heights.

Another method of remote detection involves the use of an infrared sensor to pick up the heat radiated from birds flying overhead. By pointing a thermal imaging device of 1.45° opening angle to the sky, migrating birds can be detected from 300 m up to 3000 m (Zehnder *et al.* 2001). Flight tracks are recorded on video, and targets are grouped into size classes to estimate flight altitudes. Infrared sensors work best at night under clear skies, so are not good for assessing weather effects on migration. If such an instrument is combined with a distance measure, quantitative information on migration can be obtained.

The most comprehensive picture of migration is obtained by a combination of radar and visual observations by ground-based observers in the same area. However, in their studies of migration of soaring birds in Israel, Leshem & Yom-Tov (1996a) also used a motorised glider in which they could actually accompany flocks of large soaring birds on part of their journeys. This enabled these researchers to record in detail the ups and downs of the birds' flight, as they climbed in each thermal and glided, losing height, to the next.

DISTRIBUTION STUDIES

For many years museum collections formed our main source of information on bird distributions, especially of the wintering areas of those northern hemisphere birds that migrate to the tropics for the non-breeding season. The aim of skin collectors, operating mainly in the nineteenth century, was to preserve representative samples of all the species occurring in different areas. These specimens still provide invaluable information on the tropical wintering areas of many migrants which have yielded few ring recoveries or observational records. Among European breeding birds, for example, the winter distribution of the Common Cuckoo *Cuculus canorus* in Africa is still better known from museum skins than from ring recoveries.

Over much of the world, however, increasing information is becoming available on the breeding and non-breeding distributions of birds through the collective efforts of bird-watchers (Chapter 13). For some parts of the world these distributions have been depicted at relatively fine scale in recent 'atlas' projects. However, in most tropical regions, where many high-latitude breeding species spend the non-breeding season, bird distributions are still poorly mapped, despite greater travel by bird-watchers. The main value of such distributional data in our present context, however, is in showing where the same species occur at different times of year; in other words, in revealing breeding and non-breeding ranges, as well as migration routes.

RINGING

Around the end of the nineteenth century, research on bird migration received a major boost with the start of scientific bird ringing, which is still the mainstay of migration studies around the world. This activity began with the efforts of a school master, Hans Christian C. Mortensen, in Denmark in 1899, but it quickly spread to other places in Europe, North America and elsewhere. A ring (or band) is a light but tough metal band which can be placed loosely around the leg of a nestling or adult bird, with different sizes for different species. The British scheme currently uses rings of 20 sizes, with internal diameters of 2–26 mm. Each ring carries a unique engraved number, identifying the individual bird, and an address to which a recovery can be reported. The bird can be identified unequivocally, and its whereabouts are thus known at least twice in its life – at ringing and recovery. In general, birds ringed as nestlings are of most value because their precise natal locality is known, whereas birds ringed as adults may be of less certain provenance; depending on when and where they were caught, they may have been local breeders, winter visitors or passage migrants. Some recoveries of ringed birds are provided by other ringers who trap the birds alive and release them again, while other recoveries are provided by hunters or by other members of the public who may report the birds dead or injured.

The recovery rates of ringed birds are generally low: in many small species less than 0.1% of ringed individuals are ever reported again, but in larger species, especially those that are hunted, the proportion can rise above 20%. Of course, for ringers operating repeatedly in the same place, local recapture rates can be very much higher, rising to nearly 100% in some species, but such local records reveal little about bird movements. In general, therefore, getting useful information about migration in this way depends on ringing very large numbers of individual birds, from which varying proportions may be subsequently reported from elsewhere. Moreover, because nestlings suffer higher mortality rates than older birds, many more nestlings than adults must be ringed to provide a given number of recoveries.

Another problem is that recovery rates can vary enormously along migration routes, according largely to the density and literacy of the local human population. For example, of nearly 300 000 House Martins *Delichon urbica* ringed in Britain, just over 1000 (0.4%) have been recovered. More than 90% of these reports were from within Britain and Ireland, and so were of little help in indicating migration routes, while only one came from Nigeria, within the presumed wintering range (I. A. Hill, in Wernham *et al.* 2002).

In 1903, following the pioneering work of Heinricke Gätke on Heligoland Island in the southern North Sea (**Box 2.2**), a modern-style bird observatory and ringing station was established at Rossitten (now Rybachi) on the Courland Spit in the southern Baltic, a site where migrant birds are concentrated. Subsequently, many other bird observatories were established at other sites in Europe and North America and most are still in operation. Together, they provide a network of well-placed sites, where migrants can be observed and, more importantly, trapped and ringed in large numbers. During the early twentieth century, many countries came to operate their own institutionalised ringing schemes, in most of which ringing was carried out largely by amateurs operating in their home areas, but also making ringing expeditions to more remote areas. Nowadays in Europe, the various national ringing schemes are linked by EURING, which coordinates techniques and the electronic handling of data, unifies standards and formats, and stimulates projects and analyses on a pan-European basis. All ringers are trained, tested and licensed before they can operate alone.

Many of the techniques used to trap birds are developments of ancient methods used to catch birds for food. One important development was a giant funnel trap, big enough to enclose bushes, known as the Heligoland trap, because it was first constructed on Heligoland Island. At the end of the funnel is a glass-fronted catching box into which birds are driven (**Figure 2.2**). However, the numbers ringed increased greatly in the 1950s with the development of more efficient trapping methods, including mist nets and cannon nets, which increased the range of species that could be caught in large numbers. Mist nets are essentially walls of fine, almost invisible netting, each up to 20 m long and up to 2 m high. Each net is erected on

Box 2.2   Heligoland Bird Observatory

The first bird observatory, of very different style from those of today, was established on the island of Heligoland (German Helgoland) in the southeastern North Sea, about 60 km west of Denmark and about 80 km north of the German town of Wilhelmshaven. The observatory became famous mainly through the work of one man, Heinrich Gätke, who spent more than 50 years on the island, observing and shooting birds. The skins were sold to museums and private collectors, providing a useful supplement to the income of Gätke and his local collaborators. In the process, Gätke amassed a great deal of information on the timing and volume of bird migration, and on the occurrence of vagrants on the island. The business of skin collecting meant that particular emphasis was paid to rarities, as in much of modern bird-watching. His famous book, *Heligoland as a Bird Observatory*, was translated into English and published in 1895. Until the spring of that year, he had recorded 398 different bird species on the island. The book is full of fascinating information, and most of his ideas and interpretations have stood the test of time, although in the absence of proper measuring devices, he greatly overestimated the speed and altitude of bird migration.

The bird observatory still survives on Heligoland, but like other modern observatories, it has become a centre for ringing and scientific study. It is the original home of the so-called Heligoland trap, a large horizontally placed wire-netting funnel, big enough to enclose many bushes, and through which birds can be driven and caught in a glass-fronted box at the end.

Figure 2.2   Drawing of a Heligoland bird trap, a large funnel through which birds can be driven and caught in a glass-fronted box at the end.

poles, and set against a background of trees and shrubs to ensure that the net does not show against the sky. Any small bird that hits the net slides into a pocket of net formed by one of three or four shelf strings, which are threaded horizontally at different levels through the length of the net.

Figure 2.3   Cannon-netting of Oyster-catchers *Haematopus ostralegus*.

A different method was developed for catching waders, waterfowl or others that gather in large concentrations on the ground. A cannon- or rocket-propelled net is placed furled on the ground near where birds assemble (a roost or baited feeding area). The several rockets, or projectiles from cannons, are then fired simultaneously, pulling the large net rapidly over the unsuspecting birds.

By the end of the twentieth century, using a variety of trapping methods, more than 200 million birds had been individually ringed worldwide, giving hundreds of thousands of recoveries, revealing the movement patterns of different populations. Over the years, several 'atlases' of bird movements, based on ringing data, have been published (e.g. Schüz & Weigold 1931, Zink 1973–85, Wernham *et al.* 2002, Bakken *et al.* 2003).

Ringing activities tend to be concentrated in particular regions, where opportunities and interest levels are high. Although many of the ringed birds then move on, the subsequent recoveries are probably biased, as mentioned above, towards areas with high-density, literate human populations. Even in so well studied a region as western Europe, spatial variation in the reporting of ring recoveries could give an unrepresentative idea of migration patterns. Over this whole area, much migration occurs on a northeast–southwest axis, as amply confirmed by ring recoveries; but almost certainly ring recoveries greatly underestimate the amount of migration that occurs on a northwest–southeast axis. This is because the chances of getting ring recoveries from southeast Europe are much lower than from elsewhere on the continent. Yet other information from migrants caught in central and eastern Europe indicates that many species show predominantly southeast directional preferences in autumn, their movements not being picked up to any significant degree by subsequent ring recoveries (Busse 2000, 2001).

Care is therefore needed in the interpretation of ring recoveries, although they can still be useful in defining the flyways and wintering areas of particular breeding populations, the annual and seasonal timing of movements, and any sex and age differences in movements that might occur within species (Chapter 15). Some of the most geographically complete information on migration relates to North American waterfowl. It results from a planned, geographically dispersed ringing effort over many years, and subsequent recoveries provided from all parts of the continent by millions of hunters.

Overall, ring recoveries comprise our main source of information on bird movements. Taken together, they have revealed a network of bird migration routes that encompass all habitable parts of the globe, and that are travelled annually by millions of migrating birds. It has sometimes been possible to set up coordinated collaborative projects in a wide range of localities along a migration flyway, in which many observers collect data on the same species in a standardised way. The EURING projects on Barn Swallow *Hirundo rustica* and other European–African songbird migrants provide examples.

One drawback of ringing is that the ring can only be re-read if the bird is in the hand, alive or dead. Not surprisingly, therefore, researchers have been keen to develop methods that enable the re-sighting of marked birds without the need to trap them. Marking has been achieved in many different ways depending partly on the species, such as colour rings on the legs, large rings bearing numbers or letters that can be read through a telescope, coloured or numbered neck collars or wing tags. Such colour-marking schemes have greatly increased the rate of information gain for some species, especially waterfowl and waders, and have often yielded multiple records of the same individuals at different places. They have given more accurate information than ring recoveries on the speeds of migration and the duration of stopovers. The information yield from such schemes is, of course, greatly increased if observers along the potential migration route are alerted to look out for tagged birds. For example, in the Black-tailed Godwit *Limosa limosa* in Britain only 2.5% of ringed birds were ever recovered, but following the introduction of a colour-marking programme and additional observer input, more than 80% of marked birds were subsequently reported, many at several different places on the migration route. In the same way, the use of colour leg-tagging has greatly increased our knowledge of shorebird migration in Eastern Asia–Australasia and in North–South America, providing information on the timing and speed of migration, and of the locations of important stopover sites (Minton 2003).

Birds trapped for ringing can be sexed and aged, enabling differences in timing and other aspects of migration between sex and age groups to be identified. Individuals in the hand can also be measured and weighed, providing information on weight gain and fat deposition in different species, which can be related to the types of journeys undertaken. Laboratory analyses of carcasses have revealed that substantial changes in body composition accompany migration, not simply the gain and loss of fat (Chapter 5). In some species caught on migration, measurements (mostly wing and bill) can give some idea of provenance, enabling the passage periods of different populations to be assessed. Nowadays this type of information can often be augmented from ring recoveries or studies on DNA or isotope markers (see later). Blood samples can provide information on the levels of specific hormones, metabolites or red blood cells, all of which can help in understanding migration. Details of plumage development also reveal how moult is fitted into the annual cycle of different species, along with breeding and migration (Chapter 11).

From early in the twentieth century, bird ringing led to the experimental manipulation of birds in order to learn more about their navigation abilities. Large-scale displacement experiments, in which birds were caught in one locality and released in another far away, were done to see whether birds could re-find their home areas, or how translocation affected their migrations, using subsequent ring recoveries to provide the necessary information. No less than 24 such large-scale experiments, involving a wide range of species from Barn Swallows *Hirundo rustica* to White Storks *Ciconia ciconia*, were done in the first half of the twentieth century at Vogelwarte Rossitten on the Courland Spit (Schüz *et al.* 1971), followed by others in other parts of Europe and in North America (Chapter 9). Recoveries of birds trapped, ringed and released without displacement provided the control comparisons. Such experiments revealed much about the orientation and navigational abilities of birds, and about differences in behaviour between young and older individuals. In more recent years, much work on orientation and navigation has involved homing pigeons, which are easy to keep and handle. Or it has involved wild birds which were trapped at migration times and their directional preferences assessed in 'orientation cages', after which they were released to continue their journeys (see later). Such captive birds were sometimes subjected to simulated displacements by adjusting the celestial or magnetic cues to which they were exposed, and their directional preferences then re-assessed.

RADIO-TRACKING

In recent decades an additional study method has become available, namely the day-by-day tracking of radio-tagged birds on their journeys. Initially, aircraft were used to follow the tagged birds (Chapter 8; for various thrushes see Cochran *et al.* 1967; for raptors see Hunt *et al.* 1992; for cranes see Kuyt 1992). But since the mid-1980s, however, tracking has been made much easier by use of satellite-based receivers (Chapter 8). The transmitters, called platform transmitter terminals (PTTs), can be tracked individually and automatically by the Argos satellite system. This is a joint French–American venture, originally designed to locate objects on earth such as floating weather stations and buoys. It is based on satellites that continually circle the globe over the poles, and is capable of detecting signals from anywhere on earth, with an accuracy of 150–3000 m depending on the angle of the satellite pass and the quality of the PTT signal. The satellites then transmit the information to a ground station. By measuring the Doppler shift of the emitted signal, the system can measure the exact distance between the transmitter and the satellite, and knowing the parameters of the satellite orbit, the system can also calculate the exact coordinates of the transmitter. The method is expensive, but the data provided are some of the best available on the movements of migratory birds. As the accuracy of each reading is known, the less reliable ones can be discarded if necessary.

The use of PTTs enables large birds to be monitored day by day on their journeys, and to be followed all the way from their breeding grounds to their winter quarters, and back again, regardless of where in the world they move (Chapter 8). With this new method, fieldwork on bird migration is advancing in new directions, providing information on migration routes and progress, stop-over location and durations, flight speeds, wind and weather effects and orientation abilities.

In one of the earliest radio-tracking studies, six male Wandering Albatrosses *Diomedea exulans* had satellite transmitters attached to them at their nests on the Island of Crozét, midway between South Africa and Antarctica (Jouventin & Weimerskirch 1990). Four of the six birds were followed for about a month as they wandered around the ocean looking for food. One albatross covered a distance of 10 427 km over 27 days. The satellite located it 314 times and its maximum flight velocity between location points was 63 km per hour. Another albatross flew a total of 15 200 km during 33 days. The satellite located it 385 times, and its maximum flight velocity between location points was 81 km per hour. On one day this albatross covered a total of 936 km. For some years, satellite tracking provided the only way to obtain such information, especially in birds that cover such huge distances over the open sea. Additional information gained from the satellite-based tracking of albatrosses and other birds is discussed in Chapters 8 and 17.

Because of their weight, PTTs could until recently be carried safely only by birds weighing at least 1 kg (since reduced to 600 g), which excludes the majority of bird species (ideally the transmitters should not exceed 3–4% of the bird's weight). Most studies have involved swans and geese, raptors, cranes, storks, pelicans and albatrosses. Each transmitter provides information for a period of months or years, until the battery or transmitter fails. It gives immediate information on the daily movements of individuals and, if necessary, also on aspects of the physiology of the wearer and on the conditions of the environment through which the bird passes. Various devices, such as intermittent transmission, can be used to lengthen the life of a battery-powered transmitter, but as yet few battery-operated PTTs have lasted longer than a year. However, from 1995 solar-powered transmitters became available which, unlike battery-powered ones, could in theory last for many years. The current world record holder is a White Stork *Ciconia ciconia*, so far tracked (with periodic transmitter changes) over a 10-year period on six outward and six return journeys between its nesting place in Germany and its wintering places in different parts of Africa. In some years, this adult female wintered at localities within a few degrees of the equator in East Africa (about 7000 km from its nesting place) and in other years it wintered at places in southern Africa (about 11 000 km from its nesting places) (Chapter 8; Berthold *et al.* 2004). Only radio-tracking has so far revealed that individual storks have wintered in widely separated places in different years. For other species, satellite-based radio-tracking has also revealed previously unknown breeding or wintering areas (Meyburg *et al.* 1998, Ueta *et al.* 2002).

Other kinds of electronic and data-storage tags can now be used to track migrating birds on a worldwide scale. Geolocation systems (GLS) are based on continual measurements by photosensors of the ambient light intensity to record the geographical coordinates (latitude from daylength and longitude from absolute times of dawn and dusk), while global positioning systems (GPS) receive data from satellites for calculating the position of the bird. When first introduced, both systems required the recapture of the birds to recover the tags (attached to leg rings) and accumulated data. This was not difficult with seabirds, for example, returning annually to the same nest sites (e.g. Croxall *et al.* 2005). However, recent developments to link GLS and GPS to satellite transmitters now allow the data stored on the bird to be retrieved without the need for recapture, and some current solar-powered models can operate over periods exceeding ten years, providing that light levels are sufficient to generate the necessary power. GPS operate through a network of satellites launched by the United States Department of Defense. The bird is equipped with a GPS receiver, which collects locations at pre-set intervals (say every hour) from the GPS satellite network. These data can then be relayed to ground-based Argos processing centres, again at pre-set intervals (say every few days) (Seegar *et al.* 1999). Because the locations determined in this way are accurate to within 20 m, the method can be used to gain precise assessments of a bird's home range at different seasons, as well as its migration routes as often as required. Used in conjunction with high-power satellite images or aerial photographs of the ground, a bird can be placed accurately within a landscape situated thousands of kilometres from the observer who is seated comfortably at home in front of a PC. Other sensors can be added to a PTT in order to measure other environmental variables, such as altitude of flight or ambient temperature, but they also add weight.

To be of most value, radio-tags and sensors should ideally have no effect on the flight or other behaviour of the wearer. In practice, the capture and handling of birds (for whatever purpose) is likely to cause some temporary stress, and there could be an energy cost to carrying any extra load. However, birds are normally able to compensate for such effects, so that they are not obvious to the observer, and the weight of the attachment can be trivial compared to the weight of a meal or internal body reserves. Attempts to test the effects of tags on the flight performance of birds have either used experimental approaches (such as comparing the flight or energy consumption of tagged and untagged birds flown in wind tunnels, e.g. Holliday *et al.* 1988), or have compared the migratory timing and progress of radio-tagged birds with those of untagged birds, where this was measured (Beekman *et al.* 2002, Igual *et al.* 2005). So far as I am aware, no such study has revealed significant impacts of radio-tags or sensors on the flight performance of migrants, providing these items weighed less than 3–4% of the birds themselves. Michener & Walcott (1966) could detect no differences in the flight performance of homing pigeons *Columba livia* carrying transmitters weighing as much as 15% of pigeon body weight, although some effects were detected in a later study of homing pigeons by Gessaman & Nagy (1988). Hence, although we can generally assume that the data on migrations obtained by radio-tagging have not been significantly distorted by the effect of the tag on the wearer, this may not necessarily have been true of all such studies (for further discussion see Kenward 2001, Phillips *et al.* 2003).

ISOTOPES AND OTHER MARKERS

The use of markers present within the tissues of migratory birds to analyse broad-scale movement patterns offers an alternative approach in species that yield few or no useful ring recoveries, and are too small to carry PTTs. In particular, analyses of stable isotopes (different forms of the same element) in bird tissues can provide information on the broad provenance of trapped migrants. Stable isotopes of several abundant elements, including hydrogen (H), carbon (C), strontium (Sr) and others, have spatial distributions that vary consistently either across broad geographical regions or between bird habitats and food types (Hobson 1999). For example, in North America, the ratio of hydrogen to its isotope deuterium (δD) in precipitation varies across the continent, from deuterium-enriched in the southeast to deuterium-depleted in the northwest. These patterns are transferred through food webs from plants to higher organisms. Birds absorb isotopes from their food and deposit them in body tissues, giving isotope signatures which reflect either the region where the food was eaten, or the habitat and type of food. Birds that move between regions or food webs can retain information of previous feeding locations for periods that depend on the turnover rates of particular isotopes in their body tissues. Keratinous tissues, such as feathers, are metabolically inert following synthesis, and maintain an isotopic signature reflecting the food eaten at the time and place of their formation. Other tissues are metabolically active, and retain their signatures for periods ranging from a few days (in the case of liver or blood plasma) to several weeks (in the case of muscle or whole blood), to the lifetime of the individual in the case of bone collagen (Hobson 2003). Isotope ratios also change as the chemicals concerned pass from prey to predator, upwards through food webs, because of differential loss through excretion and respiration, but these changes are known and can be allowed for.

In practical terms, by catching a bird and pulling a single feather, analysing by mass spectrometry its isotope signature, and comparing this with known geographical patterns in isotope ratios, it is possible to find (in very broad terms) where the bird grew its feathers. Even without a baseline reference, one can tell whether different breeding populations have their own distinct wintering areas, or whether wintering populations have their own distinct breeding areas. It is thus not necessary to re-capture birds, and the method is equally applicable to museum specimens.

Many birds moult in their breeding areas, and retain the feathers grown there for up to a year. In migration studies, linkages between breeding and wintering sites have been established using this approach. For example, in Black-throated Blue Warblers *Dendroica caerulescens*, δ^{13}C, δD and δ^{87}Sr values in feathers varied systematically across the breeding range, while equivalent values from wintering sites in the Caribbean region indicated that the birds there had been drawn from northern parts of the breeding range. The δD and δ^{13}C values among individuals from local wintering sites showed greater variation than those among individuals from local breeding sites, which implied that migrants sampled at each wintering locality were drawn from more than one part of the breeding range (Chamberlain *et al.* 1997). Even more strikingly, δD signatures for five species of Neotropical migrants sampled at a single locality in Guatemala represented individuals from across the breeding ranges of these species (Hobson & Wassenaar 1997); and δD signatures for Northern Bullfinches *Pyrrhula p. pyrrhula* obtained in Scotland in 2004 indicated that these irruptive migrants derived from a large part of the European boreal region, as far east as the Urals (Newton *et al.* 2006). In contrast, material from Wilson's Warblers *Wilsonia pusilla* showed isotopic evidence for 'leap-frog migration', where more northern birds migrate to wintering areas further south than those of southern breeders (Kelly *et al.* 2002).

In contrast to the above species, Eurasian Willow Warblers *Phylloscopus trochilus* moult in summer and winter, and analysis of carbon and nitrogen isotopes of winter-grown feathers plucked on European breeding areas confirmed that the two subspecies (*Phylloscopus t. trochilus* and *P. t. acredula*) found breeding in different parts of Sweden winter in different (west and east) parts of Africa (see **Figure 22.2**; Chamberlain *et al.* 2000). Similarly, stable isotope analysis of Barn Swallow *Hirundo rustica* feathers (grown in winter) has indicated that Swiss and English breeding birds probably winter in different parts of Africa (the δ^{13}C signatures being more depleted in Swiss birds, which indicates wintering in more wooded areas than the

English birds, Evans *et al.* 2003a). More surprisingly, analyses of Barn Swallows breeding in Denmark revealed that isotopes in feathers grown in winter quarters had a bimodal distribution, suggesting two different wintering areas (Møller & Hobson 2004). The two types of birds also differed in phenotype, as did their off-spring. Depending on the season of moult, different species are suited for summer or winter population differentiation, while a twice-yearly moult makes some species useful for study in both summer and winter areas. The same is true for species with split moults, which grow some of their feathers in breeding areas and others in wintering areas (Chapter 11).

Sometimes migrants from northern areas winter within the range of more southern birds which are resident. It is then hard to tell the relative proportions of migrants and residents in the same wintering area. Using measurements of deuterium (δD) and δ^{13}C values in feathers of Loggerhead Shrikes *Lanius ludovicianus*, it was established that northern breeders made up about 10% of the Florida population in winter, 4% of the Texas population and 8% of the Mexican population (Hobson & Wassenaar 2001). The differences in proportions between States were not statistically significant, but the figures showed that northern migrants formed only a small part of these lower-latitude wintering populations. This would have been difficult to establish reliably by ringing.

Although useful in identifying the regional origins of migrants, and filling gaps in other information, the method of isotope analysis cannot provide anything near the geographical resolution that is possible with other approaches, such as ringing or radio-tracking. The levels of deuterium (δD), which follow patterns in rainfall, are perhaps the most useful in studies of migratory birds: they give good latitudinal precision, but less good longitudinal precision (Hobson 2005). However, δ^{15}N and δ^{13}C values are of less value in this respect because drought conditions can enrich both, and natural regional variations in both are increasingly modified by human activities, such as fertilizer use and atmospheric pollution, reducing their value as geographical markers. Nevertheless, the method of isotope analysis provides better-than-nothing information for species with low recovery rates in the regions concerned, especially where analysis of several elements rather than one can give greater discrimination power.

Isotope analyses of soft tissues have also been used to address other questions, such as: (1) whether eggs were formed from food eaten in the immediate breeding area or from food imported to the breeding areas as body reserves accumulated in migration or wintering areas (Chapter 5; Hobson *et al.* 2000, Klaassen *et al.* 2001); (2) whether birds breeding in one region had accumulated body reserves on the same or different stopover sites (Atwell 2000, cited by Hobson 2003); and (3) whether particular individuals examined in a breeding area had spent the winter in good or poor habitat (knowledge which can then be related to migration and breeding performance, Marra *et al.* 1998). In addition, analyses of muscle have provided information on the changes in feeding areas and diets that occur during the course of a single migration (see Minami *et al.* 1995 for shearwaters migrating from the Southern Ocean to the North Pacific).

Analyses of trace elements in feathers have also been used to indicate the broad geographical origins of birds. This is possible because the proportions of different elements in feathers vary from region to region, according to geological substrate. Sand Martins *Riparia riparia* breeding in different parts of Europe differed markedly in the elemental composition of their tail feathers, indicating that the birds from different breeding areas had moulted in different parts of Africa. Moreover, tail feathers from the same individuals in different years were similar in elemental composition, implying that individuals were consistent in their moulting areas from year to year (Szép *et al.* 2003). Other studies of this type have involved Peregrine Falcons *Falco peregrinus* (Parrish *et al.* 1983), and various species of geese (Hanson & Jones 1976).

Some other bird populations drawn from the same species can sometimes be distinguished by their DNA. Using as a reference DNA samples from different breeding areas, birds sampled on migration or in wintering areas could be assigned to one or more of these areas (for Dunlin *Calidris alpina* see Tiedemann 1999, Wennerberg 2001, for passerines see Smith *et al.* 2005 for review see Wink 2006). However, the use of DNA markers is limited among birds because of the weak genetic differentiation found in many populations. This in itself results partly from the mobility of birds, and the consequent genetic exchange between populations in different breeding areas, and partly from their recent (post-glacial) colonisation of northern regions, which has given too short a time for spatial genetic variation to arise (Chapter 22). In addition, DNA-based methods tend to be expensive and time-consuming.

CONNECTIVITY

All these various methods, from ringing and radio-tracking to isotope ratios and other internal markers, provide information on 'connectivity' – the geographical linking of populations at different times of year between specific breeding, migration and wintering areas. Connectivity can be classed as strong or weak (diffuse) depending on the degree to which individuals from different breeding areas mix in their non-breeding areas, or vice versa (Boulet & Norris 2006). It can be classed as strong if all individuals from a limited breeding area migrate to the same limited wintering area, and as weak if such individuals scatter among many wintering areas, intermixing with individuals from elsewhere. The importance of understanding connectivity is that conditions experienced by individuals in one part of the world can affect their subsequent performance in another part, hundreds or thousands of kilometres away (Chapter 26). These carry-over effects can be studied only if the birds from particular breeding areas can be linked to specific wintering and migration areas. They can occur at the population level (as with density-dependent effects on reproduction and mortality), or at the individual level (as when body condition at one time of year influences performance at another). Understanding such geographical linkages is also relevant to questions of the ecology, disease transmission and genetic structure of populations, and for effective conservation (Webster & Marra 2005).

STOPOVER ECOLOGY

Some birds perform their entire migration with a single bout of flying. This is true of some short-distance migrants which can cover the whole distance within a single day or night and of shorebirds and others whose migration routes lie entirely over water or other unsuitable habitat, and may take up to several days and nights of non-stop flight (Chapter 6). However, the majority of birds break their journeys for hours, days or weeks at a time in places where the food situation can influence their fattening rates, and subsequent breeding or survival (Chapter 27). The term stopover ecology has been increasingly used for studies at stopover sites, where birds are often crowded, present for only a short time, and feeding hard in order to replenish their depleted body reserves and continue their journeys. Attempts are made to estimate the periods that individual birds are present, and by repeated trapping and weighing, to determine their rates of fuel deposition.

Stopover periods are generally hard to measure because one can seldom be sure exactly when particular birds arrive or leave. However, precise measures have been obtained for marked individuals of conspicuous species that can be seen throughout their stay (such as cranes and swans), radio-tagged birds that can be monitored daily, and other birds that visit such small sites that they can be caught within hours of their arrival, and then seen throughout their stay, and where there is no danger of confusing migrants with local residents, as at the garden of St Catherine's monastery in the Sinai desert (Lavée *et al.* 1991). Otherwise birds are often present in an area before they are seen or captured, and may also stay some time after the last observation or capture. Hence, in most studies stopovers have been estimated as minimum values between first and last sighting or first and last capture. Such estimates can be improved in various ways, for example by use of various statistical methods (Kaiser 1999, Schaub *et al.* 2001).

LABORATORY RESEARCH ON PHYSIOLOGY, MIGRATORY RESTLESSNESS AND DIRECTIONAL PREFERENCES

Laboratory research on bird migration began in the mid-1920s and rapidly gained ground. Hundreds of experiments on migratory physiology, orientation and other aspects have now revealed most of the relevant physiological processes and controlling mechanisms, at least in broad terms (Chapters 11 and 12). An important discovery was that, at migration times, migratory birds in captivity developed migratory restlessness (*Zugunruhe* in German), in which they hop and flutter around their cages, an activity that can be registered automatically by use of electronic trips under perches (Chapter 12). Migratory restlessness in captive birds occurs either by day (in diurnal migrants) or at night (in nocturnal ones), and has been regarded by some as the laboratory equivalent of migration itself (see **Box 12.1**). It appears chiefly in birds from migratory populations and much less so, or not at all, in birds from resident populations. The number of days on which migratory restlessness is shown has been found to correlate with the natural duration of migration (and hence distance travelled) in the population concerned. Migratory restlessness therefore provides a useful means of comparing the migration seasons of captive birds from different populations, and of testing the influence of various factors on migration timing (Chapter 12).

In particular, the role of daylength in influencing migration timing has been examined by manipulating the artificial daylengths (photoperiods) to which captive birds are exposed, and then recording their condition and behaviour. Metabolic rates, food consumption, fat deposition, body weights and migratory activity can all be studied at the same time.

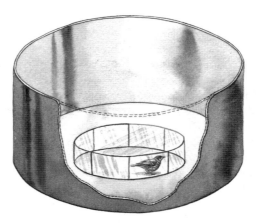

Figure 2.4 Equipment for measuring directional preferences of birds in field conditions: protective non-transparent wall around, and test cage within. The cage is shaped like a round cake. It is made of two circles of wire, connected by eight vertical wires. The top is covered with wire-netting through which the test bird can see the sky. The sidewall is covered by transparent foil (kitchen wrap or cling film) on which pecks and scratches are made by the bird in its attempts to escape the cage. The cage is placed in the centre of a circular fence of uniformly coloured solid plastic that prevents the bird from seeing any landmarks other than the sky. After a standard time (say 10 minutes) in the cage, the bird is removed, as is the transparent foil, and the pecks and scratches are counted in each sector of the foil. A new piece of foil is then attached in preparation for the next bird. With two cages available, up to six birds can be tested each hour by a single observer. From Busse (1995, 2000).

Another discovery was that captive birds also developed strong directional preferences at migration times. Such preferences could be measured in individuals using circular 'orientation cages', which typically have solid sides and wire tops affording a view of the sky. In one early type, the cage was shaped like a vertical funnel, with an inkpad on the floor. The pattern of footprints up the sides of the funnel (lined with white filter paper) indicated the directional preferences of the occupant (Emlen & Emlen 1966). Automatic registration was achieved in circular cages equipped with radially-arranged perches fitted with micro-switches or other devices to record directional activity (e.g. Wiltschko 1968). Such apparatus has been used to study the orientation and navigation behaviour of laboratory birds, or of wild ones trapped and tested in field conditions during the migration season (Busse 2000). The simplest of all such cages was designed by Busse (1995) specifically for use in the field on birds trapped on migration (**Figure 2.4**). Using only two cages, up to six birds can be tested each hour by a single observer.

Orientation cages can provide unbiased information on the directional preferences of birds caught at migration times at particular localities (Ozarowska *et al.* 2004). Most methods of analysing directional preferences (so-called circular statistics) start with the assumption that the birds sampled show only one main directional preference. Yet many individual birds that have been tested have shown more than one migratory axis which is best expressed as a bi-vector individual pattern (see Busse & Troćinska 1999 for analytical method). In addition, when samples of migrants are caught and tested at particular localities, two or more migration axes frequently emerge, as birds migrating through the same locality but from different areas take somewhat different directions (Busse 2000). This is another reason for using a statistical method that can pick out the different directional preferences, rather than calculate a single amalgamated mean from all birds tested. While in theory directional information could be obtained from ring recoveries, in practice many years of data are needed, and geographical bias in ring reporting can distort the picture.

Another major benefit of studying migratory orientation in caged birds is that the external information received by the bird can be manipulated. For example, the perceived position of the sun can be altered by use of mirrors, star patterns can be modified in a planetarium, or the geomagnetic field can be altered using large magnetic coils (Wiltschko & Wiltschko 1995). These procedures facilitate study of the external cues that might be used by birds to determine their migratory direction (Chapter 9).

Wind tunnels

Recently developed wind tunnels have revealed much about the mechanics and energy needs of bird flight (Pennycuick et al. 1997). A wind tunnel creates a smooth (laminar) air flow in a test section where birds are trained to fly. The artificial wind speed can be adjusted so that, when the bird flies against the wind, it maintains a constant position in an observation section. Low turbulence is important in order to generate a natural situation reflecting flight through non-turbulent air but, if desirable, turbulence can be created by inserting nets or other objects upstream from the test section. Wind tunnels have been used to test flight mechanical theory (Chapter 3), to measure the metabolic costs of flight and to study flight style using high-speed video cameras. To yield meaningful results, especially on energy consumption, birds must be trained in the wind tunnel beforehand, so that they 'feel at home' there, and fly steadily, maintaining constant position against the wind for long periods.

In consequence of work with captive birds, we now have some understanding of how birds orientate on migration, of the energy costs of flight, and of the physiological preparation for migration that occurs at appropriate times of year, including the deposition of internal body reserves to fuel the flights.

BREEDING PROGRAMMES

Large-scale breeding programmes for captive migratory birds have revealed much about the genetic control and inheritance of different aspects of migration behaviour, whether timing, duration or directional preferences (Berthold 1996). The most convincing results have come from cross-breeding individuals of the same species but drawn from populations with different migratory behaviour (Chapter 20). In general, the resulting offspring showed migratory behaviour that was intermediate in timing, direction and duration between their two parents. By selecting and breeding only from the most migratory individuals in a population, migratory behaviour could be enhanced over several generations, and similarly by selecting the least migratory individuals, populations became increasingly non-migratory. These experiments, conducted mainly on Blackcaps *Sylvia atricapilla*, confirmed that all major aspects of migratory behaviour were genetically controlled, and could be altered by selection (Chapter 20).

MATHEMATICAL MODELS

It is not enough in biology to know what animals do and how they do it. It is also important to understand the adaptive significance of morphological and behavioural features, and hence why animals have evolved to look and behave the way they do. Evolution can be regarded as a process of improvement (or optimisation) as animals become better adapted to what they have to do in order to survive and reproduce in contemporary environments. So-called optimisation analysis is a powerful approach to the study of adaptation, and has been increasingly used to test hypotheses about bird migration (Alerstam & Lindström 1990, Alerstam & Hedenstrom 1998a). Models are used to predict behavioural and other patterns that 'should' be observed if individuals follow one strategy or another, and the predictions are compared to field observations or experimental findings to see how well they fit, and hence to infer the likely strategy being followed by the bird. Optimal behavioural decisions during migration may involve matters of habitat selection, flight speed and altitude, whether to fly now or later, whether by day or by night, by flapping or soaring flight, which direction to head, how much account to take of winds, and so on. The theory of bird flight yields quite specific predictions on the speed and altitude of flight and how it is expected to vary with wind or fuel loads, all of which can be tested with field data.

Although the combination of modelling and field observation comprises a potentially powerful method for studying migratory adaptations, such models are heavily dependent on the assumptions on which they are based. These assumptions may be unrealistic and in any case are dependent on current knowledge which may be inadequate. Such models are nonetheless important in directing research, through defining more precise questions and the types of data that need to be collected. So far in migration research, models have proved especially useful in understanding flight behaviour, patterns of fattening, the timing and duration of the individual flights and stopovers that comprise migration, and the responses of birds to wind conditions (see papers in Alerstam & Hedenstrom 1998b). At present, there are more models than critical tests of their assumptions and predictions, giving plenty of scope for further research. Like any other ideas, however, formal models (often couched in mathematical terms) must be continually tested against experiments and field observations. Progress is often most rapid when predictions fail or are not supported by new data, showing that seemingly plausible ideas are probably wrong.

Optimality models have another potential pitfall. Optimality does not require that all individuals in a population behave identically. It requires only that individuals make decisions that maximise their own fitness, including making the best of a bad job. Different individuals may therefore pursue different tactics, dependent on their own physical condition at the time, and on the prevailing environment as it affects them. Considerable variation in individual behaviour might then occur. The population might appear to be following no particular strategy, when in reality all individuals are behaving optimally for their own particular circumstances. One cannot interpret the significance of variability without information on the environmental conditions that affect the state and performance of the birds themselves at the time. Factors that influence individual variation are increasingly being incorporated into optimality modelling. In this book, I shall not dwell on the mathematical details of the various models (which are under continual revision), but with the physiological and ecological understanding that has emerged from their use.

As the study of migration has progressed over the years, rigorous statistical techniques have been developed to analyse the resulting data. Examples include the estimation of migration routes (Perdeck & Clason 1983), rates of movement (Nichols & Kaiser 1999), stopover durations and turnover rates of migrants at particular sites (Kaiser 1999), as well as the application of so-called circular statistics to the analysis of directional information (Batschelet 1981, Busse & Trocinska 1999).

CONCLUDING REMARKS

The scientific study of bird migration has developed over a period of about 120 years, beginning with the observations conducted at bird migration hotspots, such as Heligoland in the southern North Sea (Gätke 1895) and Fair Isle off northern Scotland (Clarke 1912). Early studies were mainly observational, but were frequently augmented by use of a gun to aid identification. The ringing and release of live birds began around the turn of the nineteenth century, and rapidly expanded as a scientific and recreational pursuit throughout the twentieth century, bringing an end to shotgun ornithology in migration studies. Observation and bird ringing were the major methodologies used through the first half of the twentieth century, but bird ringing also facilitated large-scale transplantation experiments designed to study the homing and navigational skills of birds. Throughout the latter half of the twentieth century, newer and increasingly more sophisticated methodologies were continually added, gradually spreading migration research to an increasing range of species, and enabling previously intractable questions to be addressed. Most of these methodologies, including radar and radio-tracking, were developed for very different purposes, but soon proved of value in studies of bird migration. Different approaches are being increasingly applied through collaborations between different specialists, integrating approaches which combine theory, field observations and laboratory studies, and linking physics, physiology and ecology with behaviour.

Migration research has been heavily dependent on the complementary contributions of amateurs and professionals. Amateur participation has greatly increased the numbers of active investigators and the geographical spread of studies. This has been especially evident in studies that depend on large-scale ringing and recovery (see the recent migration atlases, such as Wernham *et al.* 2002). Moreover, unlike many aspects of science, old methods (such as observation and ringing) are still contributing greatly to the growth in understanding. The cheapness of these methods means that they can be used effectively by anyone with an interest in birds, however impecunious, and even the most ardent rarity-hunter has added usefully to our understanding of bird migration (Chapter 10).

Looking to the future, further breakthrough is likely once tracking devices have been miniaturised to such an extent that they can be used without ill-effect on smaller birds, when effective migration studies have spread to parts of the world where there has so far been little or no interest, and when better means are available for measuring the physiological condition of individual birds throughout their journeys. Intellectual breakthrough is also needed in several aspects, notably in the study of bird navigation.

SUMMARY

Bird movements have been studied by observations (made directly or with radar), by bird counts at particular localities in different seasons, by widescale distribution surveys, by use of ring recoveries to elucidate routes, and in recent years by the use of radio-transmitters or position locators fixed to individual birds which can then be tracked day by day on their journeys.

Most bird migration occurs at heights too great to be seen only with binoculars, and many species travel by night, so counts of migrating birds seen from the ground represent a small and variable proportion of those passing overhead. For most species, visual counts cannot, therefore, reveal the true volume of migration, or the weather conditions that favour it. At night, birds can be seen through binoculars or a telescope as they cross the lit surface of the moon, or through a powerful upwards-directed light beam. Night migrants can also be heard passing overhead, especially with a parabolic reflector and amplifier. The best measures of the volume of bird migration are made using radar, which can be used day or night in all weathers, but can seldom give precise identification of species.

Ringing is the main means by which the migration routes and wintering areas of breeding birds have been studied, together with any age or sex differences within populations. Ringing identifies individuals unequivocally, but tends to be concentrated in particular regions with high human interest. Similarly, recoveries tend to be biased towards areas with high human populations and literacy. Live birds in the hand can also be measured and weighed, providing information on weight gain and fat deposition; they can also be tested in orientation cages for directional preferences. Colour rings and other conspicuous tags enable birds to be identified at a distance without their being recaptured or killed.

Satellite-based radio-tracking can be used to follow individuals precisely on their journeys, wherever in the world they may go but, because of their weight, PTTs can be used safely only on large birds. Analyses of isotope or trace element signatures in bird feathers or other tissues have provided additional insights, linking birds from particular breeding areas with particular wintering areas, or vice versa.

Laboratory work has revealed relevant physiological processes and controlling mechanisms, and wind tunnels have been used to study various aspects of bird flight, including energy consumption. Studies of migratory restlessness and migratory orientation on captive birds have provided details of migratory timing and directional preferences in particular populations. Such studies are being increasingly extended to free-living birds.

Part One
The Migratory Process

King Eiders *Somateria spectabilis* on migration

Chapter 3
Migratory flight

Early in the morning I was apprised by my servant that an extraor-dinary flock of birds was passing over, such as he had never seen before. Hurrying out and ascending the grassy ramparts, I was perfectly amazed to behold the air filled and the sun obscured by millions of pigeons, not hovering about, but darting onwards in a straight line with arrowy flight, in vast mass a mile or more in breadth, and stretching before and behind as far as the eye could reach. (Major King, Ontario, late nineteenth century, writing of the migration of the Passenger Pigeon.)

No one can fail to be impressed by the migrations undertaken by birds. Some journeys extend over distances of more than 10 000 km, and may involve the crossing of seas and other inhospitable areas. Such migrations require not only extraordinary navigational skills, but massive body reserves to fuel the flights, coupled with sustained non-stop effort for tens of hours at a time. Migration differs from ordinary day-to-day flight, not only in the much greater length of journey, but in the greater altitude at which it usually occurs, with most small birds flying at heights well beyond the range of human vision. Once underway, therefore, birds are usually exposed to a cooler and thinner atmosphere, with reduced buoyancy and oxygen levels. While en route, migrants must stop and feed to replenish depleted body reserves, often in unfamiliar places; they must respond appropriately to prevailing weather, and correct for any off-course drift. Moreover, areas of favourable habitat, where a migrant can feed rapidly and safely, may be limited and widely spaced. Little wonder that mortality rates are often high at migration times (Chapter 28).

Astonishingly, some species, such as grebes, rails and gallinules, may have hardly flown for months before they set off on migration, having moved around mainly by walking or swimming. Yet at the appropriate time, and having accumulated the necessary fuel reserves, they suddenly ascend into the night sky and fly for hundreds of kilometres non-stop. In Barnacle Geese *Branta leucopsis*, telemetrically measured heartbeats revealed that in the weeks before autumn departure the birds flew for no more than a few minutes per day. Yet on migration, they flew non-stop for up to 13 hours at a time, with only occasional breaks in their 2500–3000 km journeys (Butler *et al.* 2000). Evidently, the amount of practice needed by such birds is minimal, compared with that needed by a human athlete to perform for much shorter periods.

One of the main advantages of flight is its speed. Whether by flapping or gliding, flight is fast compared with walking, running or swimming. It thereby facilitates long-distance travel, and allows migration to be accommodated as a twice-yearly event within the annual cycle. Nevertheless, the lengths and types of journeys that birds can undertake are greatly influenced by the body size, wing shape, flight powers and other features of the bird. These various features constrain the speed and mode of flight and the amount of fuel that can be carried as body reserves. This chapter is concerned with: (1) the relationships between body weight and fuel reserves, flight speed and duration; (2) the type of flight, whether mainly by flapping or soaring–gliding; and (3) migration by walking or swimming. These different aspects involve costs and benefits, and the interest is in finding where the balance is drawn in different species according to their sizes and shapes, and the particular conditions and constraints under which they operate.

BODY WEIGHT, SPEED AND FLIGHT MODE

The importance of flight speed for migration is obvious: for a given time airborne, a faster bird can cover a greater distance. The flight speeds of birds have often been measured using a car or aeroplane travelling alongside, or by using radar to track the movements of particular flocks or individuals (Bruderer & Boldt 2001). Measures taken from a vehicle or airplane cannot be corrected for wind effects, and are often of doubtful accuracy, so are of limited value. Radar measures can be obtained specifically for birds on migration and can be corrected to allow for wind speed, but do not always provide a reliable identification of species. Other values used for comparative purposes are the theoretical flight speeds calculated from aerodynamic principles on the basis of body mass, wingspan and wing area (Pennycuick 1969). The main findings to emerge from all these various sources of information are that, while individual birds can vary their flight speeds according to circumstance, larger birds generally fly faster than small ones, although body and wing structure also have a major influence.

Flight is clearly not a uniform activity. Many species can vary their speed and wing-beat frequency according to prevailing conditions and intent, and some can switch between different types of flight, such as flapping and gliding. In consequence, some species can change their still-air flight speeds by more than threefold (Bruderer & Boldt 2001), with wind effects adding yet further variation. And as birds change their speed and type of flight, their power requirements change accordingly.

Within species, the relationship between power requirement and flight speed in flapping flight is not linear, but U-shaped (**Figure 3.1**). There is a particular speed at which the power required for flight is minimal (V_{mp}), and flying either slower or faster than V_{mp} is more costly (Pennycuick 1969, 1975). This may explain why birds on migration do not usually fly at the maximum speed of which they are capable: full speed demands too much energy. Birds are likely to fly either at the 'minimal power speed' (V_{mp}), which gives minimal rate of fuel use and hence maximum time airborne, or at the somewhat faster 'maximum range speed' (V_{mr}), which gives the longest distance on a given amount of fuel. These two speeds span the most economical range of flight speeds for converting fuel to distance. On migration, birds are likely to fly faster than V_{mr} only in special circumstances, such as in countering a headwind or falcon attack, or in attempting to reach land before dark (V_{mt} – flight on minimal time or 'full speed').[1]

Comparing different species, the theoretical minimum power speed (V_{mp}) and the theoretical maximum range speed (V_{mr}) vary in direct proportion to the square root of body weight. On average, 10-g birds have a theoretical maximum range speed in flapping flight of around 22 km per hour, 20-g birds of around 32 km per hour, 100-g birds of around 55 km per hour, 1000-g birds of around 85 km per hour and 10-kg birds of around 90 km per hour (Pennycuick 1969). As a rule of thumb, the theoretical maximum range speed of birds roughly doubles for every 100-fold increase in body mass up to around 15–20 kg, the approximate weight limit for flying birds. These are mean theoretical values, however, and species of similar weight would be expected to vary somewhat in their actual flight speeds, according to body and wing shape and other features, which vary from one type of bird to another. Hummingbirds, pigeons, ducks and auks fly faster than expected from their body weight, while terns, harriers and owls fly slower.

The relationship between the actual flight speeds of migrating birds and their mean body weights is shown in **Figure 3.2**, based on radar studies by Bruderer & Boldt (2001). Such measured flight speeds have given broad agreement with V_{mr} values predicted from body weight (Rayner 1985, Kerlinger 1989, Alerstam & Lindström 1990, Welham 1994). However, the actual slope of the relationship found in field studies of this type varies somewhat depending on the species included (Rayner 1988, Bruderer & Boldt 2001). This would be expected because

---

[1]*For more information on the theory of bird flight, see Pennycuick (1969, 1975, 1978, 1989, 1998), Rayner (1990), Hedenström (1993), Videler (2005).*

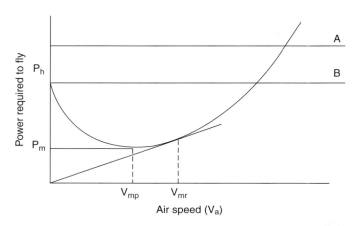

Figure 3.1   The energy costs of flight within species, as expressed by the U-shaped power curve, resulting from the relationship between the power needed to keep aloft and the power needed for forward thrust at different flight speeds. The power needed to keep aloft declines with rising flight speed (as aerodynamic lift takes over), while the power needed to overcome drag increases. At very slow or very fast speeds, the energy required for flight is greater than at intermediate speeds. Only where the combined power curve is less than the bird's maximum possible power output can flapping flight occur. For species that can hover, the minimal flight speed is zero.

Four attributes of flight are used in discussions of bird migration: (1) V_{mp} is the speed that requires minimum power, which minimises energy cost per unit time; (2) V_{mr} is the speed that gives the maximum range, which minimises the energy cost per unit distance covered. It is found by drawing a tangent from the zero point to the power curve, and could be adopted by birds maximising the distance flown in a migratory flight with a certain amount of fuel; (3) the minimum power requirement for flight (P_m); and (4) the power required for true hovering flight (P_h). Lines A and B indicate two species with different maximum power available for flight. Modified from Pennycuick (1969) and Kerlinger (1989).

The power requirement on the y axis can be expressed in either of two ways. A mechanical power curve shows the rate at which the flight muscles must do mechanical work in steady level flight as a function of air speed, while a chemical power curve shows the rate at which chemical fuel energy is consumed. The mechanical power can be calculated directly from the mechanics of flight, and the chemical power is then calculated from the mechanical power (Pennycuick 2006). Calculation of the chemical power is less well-based than the mechanical power, because it involves additional assumptions of a physiological nature, for which no underlying theory has yet been developed.

Hovering, in which a bird can remain suspended in one place in still air, is so energetically expensive (except for hummingbirds) that even small birds use it for only brief periods. Raptors such as kestrels do not hover, although they can fly very slowly (4–6 m per second, Videler *et al.* 1983). Instead of flying at zero air speed, they fly at zero ground speed, while maintaining a positive air speed. In effect, they face into the wind, and fly forward at the same speed that the wind would otherwise blow them backwards. At very slow air speeds, kestrels flap continuously, whereas at faster speeds (in stronger wind) they incorporate bouts of gliding, in which they appear to hang on the wind.

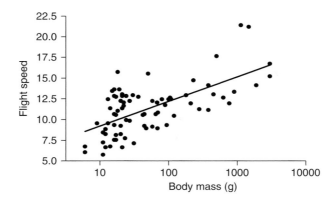

Figure 3.2   Relationship between flight speed and body mass in birds that fly primarily by flapping flight. Flight speed measured from visually identified individuals in level flight, followed on migration with high-precision tracking radar. Body weights were taken as likely averages for each species. Flight speeds were corrected to allow for wind effects (according to radio-tracked wind-measuring balloons) and altitude, all measures being corrected to expected still-air sea level equivalents. Speed $= -5.1 + 2.97 \times$ body mass, $r^2 = 0.415$, $P < 0.001$, $N = 77$. Data from Bruderer & Boldt (2001).

different kinds of birds differ in wing shape, as well as in body weight. In general, small birds fly slightly faster than expected on theoretical grounds, while large birds fly somewhat slower (Welham 1994). Another problem with published data on flight speeds is that they are not always corrected for altitude, but as air density declines with increasing altitude, the flight speeds of birds would be expected to rise with increasing altitude (see later).

Wing shape

Wing design varies greatly between different kinds of birds – long-and-thin or short-and-broad – according to the balance of selection pressures that act upon it, whether provided by habitat, foraging habits and predator evasion, or by migration (Rayner 1985). Bird wing size is usually described in terms of wing-loading (body weight divided by wing area), while wing shape is described by aspect ratio (wing span squared divided by wing area), which varies from about 5 in grouse to more than 20 in albatrosses. Both measures depend on wing area, which is taken as the area of both wings, including the body between them, projected on a flat surface. Both V_{mp} and V_{mr} increase with both body mass and wing-loading, because higher speeds become necessary if lift is to be sufficient to support body mass, while both speeds decrease slightly as aspect ratio rises, because longer wings are more aerodynamically efficient. Those birds that fly slowly or hover while foraging typically have long, high-aspect-ratio wings, which reduces V_{mp} and V_{mr}, and also the power needed to fly at those speeds, as well as the maximum speed possible. In contrast, divers, ducks, geese and auks have relatively short, high-aspect-ratio wings, which allow high speeds without unduly high power. Many small passerines have short, rounded wings of large area and low aspect ratio. Their low wing-loading permits a large increase in weight prior to migration without flight becoming impossible. More extreme wing designs, as in swifts or hummingbirds, are associated with specialised lifestyles, and may limit flight flexibility in other respects.

Power requirements in relation to body weight

Using standard aerodynamic models, theoretical relationships between power requirement and flight speed have been calculated for birds over a wide range of body weights. The power (P) required for flight at maximum range speed increases with body weight (W) roughly according to the proportion $P = W^{1.17}$ (Pennycuick 1975). However, because larger birds have lower metabolic rates (in the proportion $Mass^{0.75}$), flight at V_{mr} is disproportionately more costly for larger birds (up to the weight of obligatory flightlessness). In other words, the chemical power required to fly at V_{mr} (or V_{mp}) is a larger multiple of basal metabolic rate (BMR) in larger than in smaller birds. (*Note*: Basal metabolic rate is the rate of energy consumption by an inactive bird, not requiring extra energy to move around, digest food, or thermoregulate; that is, the lowest rate at which an inactive living bird consumes energy.)

One consequence of this body-weight–power relationship is that small birds have more power available to them than do large birds, relative to that required to fly (Pennycuick 1969). This in turn means that the smaller the species the relatively more extra fuel they can carry for migration (Hedenström 1993). Larger birds become progressively more restricted in the proportion of extra weight that they can carry which, despite their greater speed, reduces their potential non-stop flight range. Thus the proportionate weight of fuel that can be carried by a bird at maximum range speed (V_{mr}) has been estimated to decrease linearly with increasing body size, down to none at about 6 kg. Birds with even greater fat-free weight than 6 kg, and carrying fuel reserves, would have to fly at speeds less than V_{mr}. For example, the observed flight speed of migrating swans is around 64–72 km per hour, which is substantially lower than their V_{mr} (Klaassen *et al.* 2004). In contrast, species with a fat-free body mass less than 750 g can theoretically double their weight through fuel deposition and still have sufficient power to fly at maximum range speed.

Effects of migratory fattening

Theoretically, the relationships between body mass and maximum range speed (V_{mr}) derived from comparisons between species should also hold within individual birds as they change weight through fuel deposition and use. The addition of body reserves for migration must automatically increase V_{mr} and V_{mp}, which vary in direct proportion to the square root of body mass (Pennycuick 1969). Hence, maximal fuel load in long-distance migrants not only provides the necessary energy, but also obliges the birds to fly at higher speed if they are to achieve maximum range. This effect is not trivial. A bird doubling its lean body mass through fuel deposition (as some do) should theoretically increase its V_{mr} at the start of the flight by about 1.4 times. However, to achieve this, power output must be increased about 2.8 times at the start of the flight, according to calculations by Pennycuick (1969). Muscle growth is one way in which increased power output could be achieved, but there are clearly limits to this.

During a long flight, in which mass declines through fuel consumption, theoretical V_{mr} (and V_{mp}) also decline, so the bird would be expected progressively to reduce its cruising speed (and power output) if it is to achieve the maximum possible range. This raises the possibility that individual birds might reduce their flight speed as they lose weight during the course of a long journey, in order to fly at V_{mr} for as much of the flight as possible. Reduction of flight speed during migration has been confirmed in radio-tracked Brent Geese *Branta bernicla* (Green & Alerstam 2000). However, maximum range could be achieved only if birds flew at V_{mr} throughout the flight. They are unlikely to manage this with a heavy fuel load, because this would require exceptionally hard work, and excessive demands on the heart and lungs early in the flight.

In addition, all theoretical flight speeds (including V_{mp} and V_{mr}) are expected to increase with increasing altitude, owing to reduced air density. A bird the size of a thrush has been calculated to have a V_{mp} of 35.3 km per hour at sea level, which would increase to 45.4 km per hour at 5000 m (although it is not known for certain that thrushes ever do fly at such high altitude) (Pennycuick 2006). Actually the bird must fly faster at higher altitude if it is to fly at all. It has to generate the same forces as before to support its weight and propel itself forward, but because the air is less dense at higher altitude, the bird is forced to fly faster to compensate.

Because the power output required to fly declines during long flights, as a result of fuel consumption, one might expect either that muscle mass would decline during a long flight or that reduced power output would be expressed as reduced wing flapping. Several studies have confirmed a reduction in muscle mass during flight (Chapter 5), and others have shown decreasing metabolic rate during flight (reflecting declining power output). For example, Barnacle Geese *Branta leucopsis* were tracked between Svalbard and Scotland in autumn using satellite transmitters and data-loggers recording heartbeat as an index of metabolic rate (Butler *et al.* 1998). Migration distance was about 2500 km. During the journey, heartbeats declined from 315 to 225 per minute, reflecting a decline in estimated metabolic rate of 104 to 74 W, and paralleling a loss in mean body mass from 2.30 kg at the start of migration to 1.83 kg at the end. I know of no studies that have checked for a reduced rate of wing flapping during a migratory flight.

The problem of climbing to high altitudes

With increasing body weight, the climb rate and manoeuvrability of the bird are also reduced (Chapter 5). In birds tracked from take-off by radar, climb rate was inversely correlated to body size, from more than 2 m per second in a 50-g Dunlin *Calidris alpina* to 0.32 m per second in a 10-kg Mute Swan *Cygnus olor* (Hedenström & Alerstam 1992). This constraint of reduced climb rate in large birds could be important as they climb to reach their flying altitude or to cross mountain ranges. The high cost of climbing in large birds that use flapping flight may be one reason why swans usually fly at relatively low altitudes, even though winds may often be more favourable higher up. For Bewick's Swans *Cygnus c. bewickii* tracked while migrating mainly overland from Denmark to northern Russia, mean flight altitude was only 165 m (maximum 759 m, Klaassen *et al.* 2004); and for Whooper Swans *Cygnus cygnus* migrating over water between Iceland and the British Isles, mean flight altitude was 228 m (range 68–387 m, although one individual reached 1856 m, Pennycuick *et al.* 1999). These altitudes are low compared with those attained by smaller birds (Chapter 4). Moreover, after taking off over land, Whooper Swans did not usually exceed 500 m altitude until they had flown for more than 50 km, indicating an extremely slow rate of climb (Pennycuick *et al.* 1999). This gave further indication that their power requirement in level cruising flight near the start of migration is close to the maximum possible, and cannot be increased much further. Greater effort over a long period may be physically unattainable or involve risks to muscle and tendons, as well as higher respiratory rates which result in greater water loss (for estimates in swans, see Klaassen *et al.* 2004). Considering their energy expenditure and associated water loss, it is not surprising that swans (and geese) are seen to drink copiously on arrival from migration.

Effect of wind conditions

For any flying bird, wind conditions can have a big effect on travel speed, and hence on flight times and associated energy expenditure. Wind can be visualised as a large body of air moving at some specific speed (V_w) in a particular direction. The bird is embedded in this moving air, and flies at a speed relative to the air: its air speed (V_a). The bird's speed and direction with respect to the ground represent the vector sum of its still-air flight speed and the prevailing air speed. If the wind blows in the same direction as the bird is heading (tailwind), the bird's theoretical ground speed is $V_a + V_w$, whereas if the wind is in the opposite direction (headwind), the bird's theoretical ground speed is $V_a - V_w$. If the wind does not flow parallel to the bird's flight-path, but at some angle to it, the bird could be blown off course (**Figure 3.3**) or alternatively it could allow for this, notably by adjusting its heading in such a way that the net effect of bird direction and wind direction compensate for one another and thereby keep the bird on track with respect to the ground (Chapter 4). Headwinds of a given speed have much more effect on the flight range of slow-flying (= small) birds than of fast-flying (= large) ones. Against a headwind of 10 m per second (36 km per hour), small passerines would make no significant progress, while larger migrants could make some progress, but have their still-air flight range reduced to less than half.

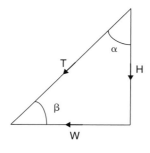

Figure 3.3   Diagram showing relationships between heading (H, the direction the bird faces), the wind direction (W), and the track (T, direction the bird travels). The track vector (T) is the sum of the heading (H) and wind (W) vectors. The angle α between track and heading depicts the amount of drift with constant heading, while the angle β shows the wind direction in relation to the track direction.

In general, then, the smaller (and slower) the bird, the more it can be affected adversely by unfavourable winds, and the greater the safety margin of fuel it is likely to need on migration. Loss of flight range through flying against a headwind cannot be prevented, but it can be reduced by flying somewhat faster than V_{mr} if sufficient power is available. Small birds are better off than large ones in this respect, on account of their greater power margin, which gives greater endurance. This is of special importance for landbirds blown out to sea, when they may be better off slowing down to V_{mp} to conserve energy and thereby remain airborne as long as possible on the chance of reaching land. In fact, the best strategy for a bird attempting to achieve maximum range would be to fly slower with a tailwind and faster against a headwind (**Figure 3.4**). Such adjustments have been shown to occur. For example, the ground speed of night migrants tracked by radar increased with a tailwind, but less than expected (3–4 knots for a 10-knot

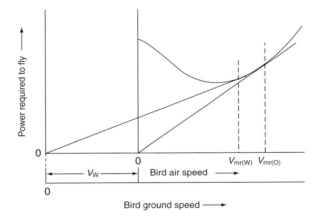

Figure 3.4   Flight speed in relation to power requirement in different wind conditions. The 'power curve' is U-shaped, so that the maximum range on a given fuel reserve will come near the bottom of the power curve (**Figure 3.1**). It will not come at the very bottom because the slight increase in power needed to fly faster is compensated by the extra length of journey achieved. The best possible speed for cruising is given by the longest tangent from the power curve which passes through the origin ($V_{m(0)}$) in still air). The graph shows the situation in still air. However, with a wind blowing, the graph would need to be altered to take into account the real distance over the ground achieved by the bird. This can be done by shifting the origin of the graph to the left if the bird is being helped by a tailwind (V_W) and to the right if the bird is being hindered by a headwind. The resulting tangents, giving the best speed for the maximum range, show that the bird should fly slower with a tailwind ($V_{mr(W)}$) and faster against a headwind, compared to the maximum range speed in still air ($V_{mr(0)}$). In practice, the amount by which the best speed is varied turns out to be much less than the wind speed encountered. The effect of headwinds and tailwinds will be much greater on slower-flying than faster-flying species which in general have high maximum range speeds. From Pennycuick (1975).

increase in tailwind) (Bellrose 1967). In this study, the birds were exploiting wind assistance, but not to the maximum extent possible at V_{mr}. They were instead minimising energy expenditure, and at the same time making a gain in flight range of 30–40% (see also Chapter 4).

Cutting the costs of flight

In their migratory flights, birds are often found to perform better than predicted by simple aerodynamic models. This is mainly because birds have various ways of reducing the costs of flight. The bodies of birds are much more flexible than those of airplanes, around which aerodynamic theory was developed. Birds can also change the shape of their wings and tail, altering their body outline and lift-to-drag ratio to suit the circumstances; and some can also switch between powered flapping flight and still-wing gliding. In addition, many small birds do not flap continuously during migration but adopt a slightly undulating flight, rising while flapping, and dipping on closed or half-closed wings to reduce the drag. Combined with residual lift, bounding flight gives greater distance for a given fuel cost, and allows many small birds to fly faster than they otherwise could (Bruderer & Boldt 2001). It is also physiologically advantageous, as it gives brief rest times between bouts of flapping (Rayner 1990). Bounding flight is shown by many passerines, woodpeckers and small owls, among others, and the dipping part of the flight becomes relatively longer the larger the bird. Energy savings of 10–15% have been estimated.

Another way in which some large birds cut their fuel costs is by flying in line or V-formation. This is usual among geese, swans, gulls, cranes, pelicans, cormorants and others. Each individual flies behind and to the side of the one in front, benefiting from its slipstream, gaining lift and reduced drag (Rayner 1979). Each bird sheds vortices from its wing-tips from which the bird behind gains lift. Estimates of individual energy savings lie in the range 12–20% for birds flying in V-formation, compared with birds flying alone (Hummel & Beukenberg 1989, Alerstam 1990a). The individual at the apex of a formation has no such advantage in power saving, and frequently changes position, pulling out and joining the line further back. Other birds migrate in flocks of other shapes, but again can save on energy costs. Probably for this reason, Red Knots *Calidris canutus* and Dunlins *C. alpina* followed by radar flew about 5 km per hour faster in flocks than solitarily (Alerstam 1990a).

A further means by which some large birds reduce their fuel costs is by passive soaring and gliding on outstretched wings, making use of updrafts to climb and remain aloft, and thereby saving energy. To judge from measurements and calculations of energy costs to flying birds, soaring–gliding requires only about 5–25% of the energy required for flapping flight (**Table 3.1**). Updrafts are mainly produced by thermals (columns of rising air caused by uneven heating of the ground) or by horizontal winds being deflected upwards by slopes and cliffs ('orographic lift', Chapter 7). The flight mode adopted by different bird species is influenced by their wing-loading; that is, their weight relative to wing area. The heavier the bird, the more difficulty it has in creating lift for flapping flight by muscle power alone, and the relatively greater is the energy saving from switching from flapping to soaring. This may be one reason why soaring is seen mainly in large birds, such as eagles and pelicans (Pennycuick 1975), and

Table 3.1 Estimates of the costs of flight in birds in relation to basal metabolic rate[a]

Species	Mass (g)	Method[b]	× BMR[a]
Powered flight			
Eleonora's Falcon *Falco eleonorae*	350	PM	8–14.8
Common Kestrel *Falco tinnunculus*	213	DLW	16.2
Laughing Gull *Larus atricilla*	322	WT	11.9
Ring-billed Gull *Larus delawarensis*	427	WT	7.5
Bar-tailed Godwit *Limosa lapponica*	282	MC	8.1
Domestic Pigeon *Columba livia*	384	DLW	8.0
	412	DLW	17.5
Brunnich Guillemot (Thick-billed Murre) *Uria lomvia*	900	PM	12.5
Black-necked Grebe *Podiceps nigricollis*	389	MC	25.2–33.3
	374	PM	17.9–23.7
Gliding			
Herring Gull *Larus argentatus*	950	WT	1.5–2.2

Data mainly from Jehl *et al.* (2003), in which the original references may be found; also Butler & Woakes (1990) for Domestic Pigeon, Baudinette & Schmidt-Nielsen (1974) for Herring Gull. Other estimates of the costs of flapping flight are summarised by Norberg (1996), but not in terms of BMR.

[a]Basal metabolic rate (BMR) is the minimum rate of energy expenditure of an animal at rest, in a post-absorptive state. Note that the flight costs in this table are not directly comparable between species, because BMR values themselves vary with body weight, and in larger birds the chemical power required to fly at V_{mr} (say) is a larger multiple of BMR than in smaller birds. For their size, the grebes are exceptional in the power required to fly.

[b]DLW, doubly-labelled water; MC, mass change; PM, Pennycuick's theoretical model; WT, wind tunnel.

it is perhaps surprising that some other large species, such as geese and swans, migrate entirely by flapping (except in unusual circumstances, Klaassen *et al.* 2004). However, these large waterfowl migrate mainly in regions where thermals are weak at migration times, and their flock formations provide other ways of reducing energy costs. Even small birds, such as warblers, could save energy by soaring if they had the right wing structure, but they would then travel only very slowly (less than half their flapping speed). Moreover, the thermal convection that allows soaring is usually available for only part of each day, and by restricting the daily migration time for soaring flight, assuming time is important, the body size favouring soaring shifts even more towards larger sizes (Hedenström 1993).

Any columns of rising air must be embedded in wider areas of more slowly sinking air, which replace the rising air. This means that, regardless of body size and flight mode, any migrating bird could gain energy (in the form of net lift) if it flew more slowly through updrafts than between them, the resulting energy gain being subsequently translated into increased cross-country speed or flight range (Nisbet 1962). On this basis, rising air currents could benefit all birds, not just soaring species, and may be of special value to species that travel long distances over hot deserts. It is not, of course, necessary for an air current to be strictly vertical to permit soaring, so long as it has a vertical component that exceeds the bird's own rate of sink.

FLAPPING AND SOARING FLIGHT

In steady flapping flight, a bird must generate the forces which support its weight against gravity and which provide the forward thrust necessary to overcome the friction and other forces that make up drag. The power for both the lift and forward thrust is supplied by the breast muscles, while directional control is provided mainly by the tail. Given sufficient fuel reserves, some birds that migrate by continuous flapping flight can travel for hours or days on end. They can cross water or other hostile terrain, and can fly by night as well as by day. In moving between their breeding and wintering places, therefore, such birds often travel directly, taking the shortest routes. As populations, they migrate mostly on a broad front, but concentrate to some extent through mountain passes or along coasts or other 'leading-lines' that deviate little from their main direction. Because flapping flight is expensive, however, such species must normally lay down substantial body reserves, especially for travelling over large areas of sea or other inhospitable substrate where they cannot feed. Sustained flapping also produces heat, which may enable birds to fly at high latitudes and altitudes without having to burn extra fuel to keep warm. In hot conditions, however, heat production can result in the need for evaporative cooling (panting), which increases water loss and dehydration risk.

The situation differs somewhat in birds that migrate mainly by soaring–gliding flight, notably the broad-winged raptors, storks and pelicans, which gain most of the energy they need for flight from the ambient atmosphere. Typically, these birds circle upwards in a thermal, then glide with loss of height to the next thermal and rise again; they repeat this process again and again along the migration route, and over long distances in ideal conditions they seldom need to flap their wings (**Figure 3.5**; Chapter 7). Because the lift comes largely from rising air currents and the forward motion partly from gravity, this still-wing flight mode requires much less internally generated energy than continuous flapping (typically $1.5–2 \times$ BMR versus around $8–30 \times$ BMR, **Table 3.1**). Many soaring species use a mix of gliding and powered flight, with intermediate costs, but seek to maximise the contribution from gliding, and resort increasingly to flapping as thermal conditions deteriorate. Flapping during a glide ('power gliding') can provide additional lift and speed, but at a cost in fuel use. The extreme soaring species thus depend for their movements mainly on a source of energy external to their own bodies and, unlike flapping birds, soaring species can continually correct for the effects of crosswinds without wasting energy.

Soaring–gliding flight has other consequences. Because soaring birds travel mostly on still, outwardly stretched wings, they produce less heat than species that continuously flap their wings. In cold climates this could increase the energy needs for heating the body, but in hot climates it reduces the likelihood of overheating, and the resulting water needs for cooling. Moreover, because of their dependence on updrafts, soaring landbird species must migrate mainly over land, favouring routes where appropriate conditions develop. They are also constrained to travel by day when the sun heats the land surface, creating rising air currents. Their migration typically reaches its peak, and moves most rapidly, in the middle part of each day when thermal activity is greatest (Chapter 7). It is then that the birds achieve the greatest heights, and can make the longest and

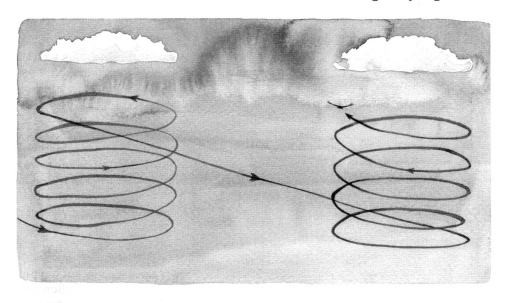

Figure 3.5 Soaring–gliding bird migration, indicating soaring within thermals, gaining height, and gliding between thermals, losing height. Thermals are often topped with cumulus clouds.

fastest glides across country (Spaar 1997, Spaar & Bruderer 1996). Soaring landbirds also tend to concentrate along narrow land bridges (such as Panama), or at narrow sea crossings (such as Gibraltar or the Bosphorus), and thereby avoid spending long periods over water where thermal soaring is seldom feasible.

In this way, soaring landbirds often take long roundabout routes between breeding and wintering areas in order to make as much of the journey as possible over land, and minimise the distance travelled by expensive flapping flight. Despite the greater distances, their total energy consumption is thereby greatly reduced.

Soaring species produce spectacular concentrations, as birds from large parts of a breeding or wintering range funnel through well-known bottlenecks, including those just mentioned. Moreover, because their travel routes are determined by geography and topography, they tend to take the same traditional narrow-front 'corridor' routes year after year (Chapter 7).

Migration by thermal soaring is analogous to powered flight with stopovers, except that the bird replenishes its potential energy by climbing in thermals, rather than by feeding to accumulate fuel during stopovers. The faster the rate of climb in thermals, the less time is lost in regaining height, and the higher the average cross-country speed. The rate of climb that can be achieved in thermals depends mainly on the strength of the thermals, with smaller birds having a minor advantage in most conditions. An average rate of climb of 0.5 m per second would represent weak soaring conditions, 2–3 m per second would be typical of good conditions, while 5 m per second is not unusual in individual thermals, but would seldom be sustained as an average.

Body weight has interesting effects on overland soaring–gliding migration (Pennycuick 1975). Compared with small soaring species, large ones (with greater wing-loading, which influences climb rate but not sink rate) have to start later in the day, when thermals are strong enough to lift them; they also tend to rise more slowly, mainly because their turning radius is greater, so they spend more time near the edges of the thermals where air currents are weaker than in the centre. But having reached the top of a thermal, they then glide more rapidly across country than smaller species, their greater weight adding speed along a given gliding angle. The starting height determines the distance that a bird can glide, depending on its glide angle, but regardless of its mass. The overall cross-country speed of soaring–gliding birds therefore depends partly on their body weight (or more strictly wing-loading), partly on aerodynamic design, and partly on the strength and height of thermals. In general, smaller species rise faster within thermals, but travel more slowly between them than larger species, and being able to make use of weaker thermals, small species can travel for longer each day. In general, observations support theoretical predictions that soaring flight gives higher migration speeds than flapping flight in bigger species (Hedenström 1993, Hedenström & Alerstam 1998). This is largely because soaring species consume less energy (relative to body mass), so have to spend less time replenishing their body reserves during the journey.

Soaring–gliding flight is not confined to landbirds. Many seabirds make use of up-currents formed either as the wind is deflected off waves (equivalent to slope soaring), or as a wave of 'swell' displaces air upwards as it moves over the sea surface. This method is effective even at times with little wind, providing there are waves or swell to displace the air upwards. Some seabirds also use 'dynamic soaring', which depends partly on wind speed being slowed by the sea surface, an effect which is lessened with height up to about 16 m. The bird first climbs into the wind, then makes a high leeward turn, gaining distance by gliding with the wind whilst losing height. After making a low turn in the trough of a wave, it starts the cycle again. A bird could also make use of discontinuities in wind flow near the sea surface, as it flies first behind a wave crest and then emerges for a time into the unobstructed wind. At this moment, the bird tilts its body so that the temporary gust strikes its ventral surface, providing lift, enabling further onward gliding flight (Pennycuick 2002). Over most oceanic areas, soaring seabirds are thus normally constrained to fly low over the sea surface, where conditions are most favourable.

The flight of the Wandering Albatross *Diomedia exulans* has been studied with miniaturised external heart-rate recorders in conjunction with satellite transmitters and activity recorders (Weimerskirch *et al.* 2000). Heart rate was taken as an instantaneous index of energy expenditure. When cruising with favourable tail- or sidewinds, these albatrosses achieved high flight speeds while expending little more energy than when resting on land. In contrast, as headwinds increased, heart rate also increased, and flight speed decreased. Heart rates were greatest when the birds took off or landed on water. In order to experience favourable winds, albatrosses flew from nesting to feeding areas on a large loop track. When heading north from the breeding island they flew on an anticlockwise loop, and when heading south they flew on a clockwise loop. Albatrosses are among the fastest of migrating birds, and can achieve average distances of 750–950 km per day, much greater than those achieved as an average by large, flapping flyers, such as swans and geese (Chapter 8).

Over tropical seas, in regions affected by tradewinds, the warm surface water heats the cool wind, leading to thermal formation day and night. Frigatebirds can soar effectively in these weak thermals because they have exceptionally light wing-loading (the lowest of all birds) yet high aspect ratio. Weimerskirch *et al.* (2003) attached satellite transmitters and altimeters to eight Magnificent Frigatebirds *Fregata magnificens*, and found that, when away from their nests, these birds remained continuously on the wing. Like soaring birds overland, they travelled in a series of climbs and descents, soaring on thermals to heights up to 2500 m and gliding down in the direction of travel. From the heights achieved, they were presumed to spot their prey on the sea surface. Their climb rates in these weak thermals were very slow, at an average of only 0.4 m per second, and they travelled forwards extremely slowly, averaging only about 10 km per hour. Their foraging trips lasted 8.4–93.8 hours, and extended 27–261 km from the nesting colony. Two individuals tracked on migration behaved in the same way. Their reliance on weak thermals may explain why these birds are restricted to tradewind zones, where soaring over the sea is possible throughout the year. This radio-tracking study also supported an earlier supposition that frigatebirds are unable to settle and take off from the water surface.

Among soaring species, flight performance can be analysed in a manner analogous to that for flapping flight, using the relationship between the rate of sink and forward gliding speed to predict optimal speeds under various conditions (Pennycuick 1975, 1989, Alerstam 1991b). The graph of sinking speed (V_z) against forward speed (V_a) is called the glide polar (**Figure 3.6**). It can be viewed as a type of power curve, in which the power is equal to the sinking speed multiplied by the body weight. This power comes from the bird's potential energy, rather than being generated by the flight muscles. During the glide the bird may close its wings to some extent, thereby reducing drag and increasing speed. In general, a bird sinks faster while travelling at higher air speeds, but the form of the relationship differs between species. As field measurements have shown, in each species the maximum glide-ratio (distance covered/height lost) is obtained at a particular air speed, which again differs between species. The maximum glide-ratio of various raptors is around 9–15:1, and for albatrosses around 23:1, while the equivalent ratio for sailplanes is typically around 38:1. In other words, from a given starting altitude, sailplanes can reach in a single glide around twice the distance of that achieved by the best gliding birds. This is one of the few instances where, in terms of efficiency, a man-made machine out-performs a bird, but the bird has much greater flexibility in its flight behaviour than a sailplane.

Like birds that migrate by flapping flight, soaring species can reduce energy costs by various means. In particular, hawks can adjust their wing and tail shapes to provide more or less surface area. This allows them to exploit a wider range of air speeds and glide ratios than would be possible if these features were constant, and reduce the drag during glides (Kerlinger 1989). Holding the wings in the level outstretched position would normally consume some energy, but some soaring species have anatomical adaptations that allow them to 'lock' their wings in place, freeing them from using muscles for this purpose, and thereby saving fuel. In albatrosses and others, a tendon sheet associated with the pectoralis

(a)

(b)

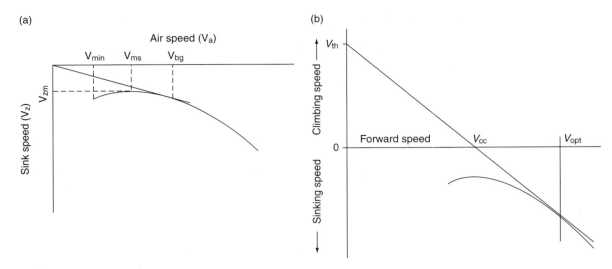

Figure 3.6   (a) The glide polar of a gliding bird or sailplane, showing minimum sink speed (V_{zm}), minimum air speed (V_{min}), air speed at best glide ratio (distance covered/loss of height, V_{bg}), and air speed at minimum sink (V_{ms}). All values are with respect to the air through which the bird flies, and not to the ground. The glide ratio (angle) at any point on the aerodynamic performance curve is equal to a ratio of V_a to V_z. The best glide ratio on the glide polar (the maximum ratio of V_a to V_z) is determined by drawing a tangent to the curve from the origin. From Pennycuick (1975). (b) Cross-country speed in thermal soaring. V_{th} is the achieved rate of climb in thermals and is plotted upward from zero on the same axis as the sinking speed. The optimum speed V_{opt} at which to glide between thermals is found by drawing a tangent to the polar from V_{th}. The tangent cuts the speed axis at the average cross-country speed V_{cc}. The cross-country speed corresponding to some inter-thermal speed other than V_{opt} can be found by the same construction, but will be less than V_{cc}. From Pennycuick (1975).

muscle prevents the wings from being raised above the horizontal, with no use of power. In addition, other soaring species typically have part of the pectoralis muscle adapted as a tonic muscle, able to hold the wings out at minimal energy cost (Pennycuick 1975). Birds that do not soar do not have this divided pectoralis muscle.

To summarise this section, migrating birds that travel by flapping flight can normally: (a) migrate day or night; (b) use direct and short routes; (c) cross large areas of inhospitable substrate without stopping; (d) use prior fat stores to migrate long distances without feeding and, if necessary, without resting; (e) migrate on a broad front rather than on a well-defined, narrow route; and (f) travel at high altitudes. In contrast, landbird migrants that travel mainly by soaring and gliding have greatly reduced fuel needs, but normally: (a) migrate only in the warmest part of the day and not at night; (b) avoid large water bodies, and often take roundabout routes with appropriate wind and thermal conditions, thereby increasing the total distance covered; (c) migrate along well-defined and relatively constant routes, some of which are taken year after year by most of the population from large areas; and (d) are usually limited in flight altitude by the height reached by thermals (seldom more than 1.5 km at noon overland). Because soaring migrants require less internal energy per distance covered, they could in theory migrate with smaller fat reserves than flapping migrants, and this is supported by what little information is available on the fat levels of soaring species (Chapter 7). More details of soaring–gliding flight are given in Chapter 7.

THE CONCEPT OF ENERGY HEIGHT

The concept of 'energy height', developed by Pennycuick (2003), measures a bird's fuel reserves in terms of what the bird can do with the fuel, rather than its energy content. For a bird that travels by flapping flight, the fuel energy height is defined as the height which a bird would reach if all of its stored fuel energy were converted into work by the flight muscles, and used to lift the bird against gravity. A soaring bird, starting from a given height above ground, can glide a certain distance in exchange for using up its initial store of potential energy. Likewise, a store of energy fuel in the form of fat corresponds to a virtual energy height from which the bird comes down at a virtual angle that depends on its aerodynamic efficiency. This virtual angle is much the same as the gliding angle. The notion of energy height thus provides a way to compare species that migrate by powered flight, replenishing their fuel energy height at stopovers, with those that migrate by soaring, replenishing their energy height by being lifted by air currents. Rates of climb in thermals are typically higher than the rates achieved through powered flapping flight, but the available height band for flight in thermals is at least one order of magnitude smaller, and the intervals at which energy replenishment is needed are correspondingly shorter (Pennycuick 2003). Albatrosses replenish their kinetic energy by exploiting discontinuities in wind flow and wave action, requiring replenishment at intervals of tens of seconds, a further two orders of magnitude shorter than in thermal soaring. In such a system, however, a bird could in theory migrate over long distances with little more daily energy expenditure than it would use at other times, a prediction borne out by the study of albatrosses mentioned above (Weimerskirch *et al.* 2000).

CONCLUSIONS ON THE ROLE OF BODY SIZE IN BIRD MIGRATION

It will be evident by now that body size (and the associated features of metabolic rate, power requirement and flight speed) have a major influence on the flight modes and migration capabilities of birds, increasing body size bringing increasing restrictions. The power available for flight declines rapidly with increase in body size, and many large birds may have insufficient muscle power or aerobic capacity to fly at V_{mr}, so are constrained to fly at some slower speed, such as V_{mp}. Among birds that migrate by flapping flight, the load-carrying capacity decreases with increasing body mass, including the fuel load (Hedenström & Alerstam 1992). This constraint in load-carrying is expected to limit the distance that can be flown non-stop by large flapping birds over terrain in which feeding is not possible, although large birds compensate for this to some extent by relatively long wings and by flying more slowly than maximum range speed

(Rayner 1988). Other large birds, such as albatrosses, vultures and eagles, can travel long distances by gliding, which requires much less fuel, but as explained above, this flight mode in landbirds can restrict their times and routes of travel.

Another major advantage of large body size is that it is in most species associated with greater flight speed. This mainly translates to reduced journey times, but it also enables large birds to migrate in more adverse wind conditions than small ones. Headwinds that would merely slow large birds might be sufficient to stop small ones altogether.

The range a bird can attain on a full load of fuel is of special significance. Large birds are not the best performers in this respect, because they are limited in the fuel they can carry, but nor are small birds, which fly too slowly. Rather some medium-sized birds, such as some shorebirds, can undertake the longest non-stop flights, because they can carry large fuel loads and at the same time fly fast. Both passerines and shorebirds can sustain flapping flight for up to a hundred hours (or more), but because shorebirds fly faster, they can cover greater distances in this time (Chapter 1). These same features may also enable shorebirds to perform better under adverse winds than slower species.

The longest apparent non-stop flights recorded for swans reached around 1700 km, for geese and passerines around 3000 km, and for shorebirds around 4000–7500 km, apart from the 10400-km flight of Bar-tailed Godwits *Limosa lapponica* from Alaska to New Zealand (Chapters 1, 6 and 8). The longest overall migrations (including stopovers) for swans reached around 3000 km, for geese around 5000 km, and for passerines and waders 12000 km or more, the latter covering the distance between the northern parts of the northern continents and the southern parts of the southern ones. Because soaring birds do not need to spend long periods refuelling, they can make longer overall migrations within the same time limits as flapping species of similar body weight. Some Steppe Buzzards *Buteo b. vulpinus* migrating between Russia and South Africa make journeys of 12000 km or more, as do Swainson's Hawks *Buteo swainsoni* travelling between North and South America. Equally remarkable, the Ruby-throated Hummingbird *Archilochus colubris* weighs about 4.8 g, yet regularly crosses the Gulf of Mexico (1100 km) in a non-stop flight of about 18 hours, requiring an estimated 3.2 million wing-beats (Nachtigall 1993).

The range of body sizes found among flying birds spans four orders of magnitude, from the smallest hummingbirds of about 1.5 g to the largest flying species of around 15 kg, including bustards, pelicans, swans, condors, vultures and albatrosses. Flightless birds can be much heavier, reaching 35 kg in penguins and 150 kg in Ostriches *Struthio camelus*, while some extinct flightless birds probably reached 400 kg. Large birds become increasingly likely to adopt less strenuous flight modes, such as gliding or formation flying, until at some undefined weight, flight becomes impossible.

THE NEED FOR REST

Different bird species also vary greatly in the distances they can fly without needing to rest. Near one extreme are certain gallinaceous birds that can fly less than 1 km before having to land (Palmer 1962). Near the other extreme is the Common

Swift *Apus apus* which outside the breeding season normally spends both day and night on the wing. As roosts are unknown from winter quarters, it is possible that, after leaving the nest in August, young swifts remain on the wing until they re-enter nest-sites as pre-breeders in the following or a later year (C. M. Perrins, in Wernham *et al.* 2002). Frigatebirds also remain on the wing continuously when away from their nests, including their pre-breeding years, but they may come to roost on land (Weimerskirch *et al.* 2003).

Consideration of the non-stop flight capabilities of migrating birds has usually centred on the need for fuel or water, rather than on the need for rest. Such flights are clearly limited by the body reserves available to fuel them, but well inside this limit, some birds may have to come down periodically in order to rest. This is difficult to prove, because migrants come to ground for various reasons, including adverse flying conditions (winds, rain, mist or heat), and not necessarily merely to sleep. However, many birds appear tired on arrival from migration, and may break their journeys for several hours without apparently feeding. Individuals that arrive low over the sea are sometimes seen to collapse onto the beach or into the first patch of vegetation available. Those that are caught are often found to have substantial body reserves, so have not run out of fuel (Chapter 6).

In many birds just after landing from a long flight, the need for sleep may take precedence over most other activities (Schwilch *et al.* 2002b). Some Pink-footed Geese *Anser brachyrhynchus* arriving at a lake in eastern Scotland after a long flight from Iceland first drank a good deal, but then spent many hours sleeping, standing on one leg in shallow water. At times they also waded into deeper water, splashed and preened for long periods. On the first day after arrival, most birds probably took no food at all, and not until the next morning did they fly to feed in nearby fields (Newton & Campbell 1970). Other observations describe shore-birds after long flights falling asleep within minutes after arrival on coastal mud-flats, and sleeping for several hours before starting to feed; or passerines after arrival from long overwater flights resting under coastal bushes for much or all of the following day before moving on, apparently without feeding (Schwilch *et al.* 2002b).

Many radio-tagged birds have now been tracked on their journeys, but because individuals are located from satellites only at intervals of several to many hours, it is not often possible to tell whether they have settled for brief rest periods. However, some Barnacle Geese *Branta leucopois* that were tracked on migration also had their heart rates monitored continuously during the journey. Heart rates were faster when the birds were flying than at other times. Records showed that birds settled about every 11 hours or so (maximum non-stop flight 13 hours) during the journey, even though this often entailed the birds resting on the sea (Butler *et al.* 2000). Similarly, some Brent Geese *B. bernicla* travelling between the Netherlands and the Taimyr Peninsula of Siberia were found to stop about every 6.5–7 hours, on average, which gave flight lengths of 450–500 km. However, some non-stop flights by these birds lasted 14–19 hours, and covered up to 1300 km. From satellite-based records of Brent Geese migrating from Europe to northern Canada, it was also surmised that birds settled to rest on the Greenland ice cap, at least on the upward part of the journey (Chapter 4; Gudmundsson *et al.* 1995). Long rests on the sea were detected by satellite-based tracking of Whooper Swans *Cygnus cygnus*, but these stops may have been in response to poor weather or complete darkness rather than to tiredness (Pennycuick *et al.* 1999). Shorebirds may also set-tle on the sea from time to time, at least under adverse flying conditions (Piersma *et al.* 2002). During stops on the sea and ice, and some of those on land, the need for feeding can be ruled out, as in many cases can the need for drinking or sit-ting out bad weather, leaving the need for rest as the most likely explanation. Of course, most landbirds could not settle on the sea and, despite the many records of birds resting on ships and offshore structures, most landbirds are assumed to fly even their longest overwater journeys non-stop. Nevertheless, the possibility that the need for sleep may help to shape the migratory behaviour of birds, and influence the routes they take, warrants further study.

MIGRATION BY WALKING OR SWIMMING

While most birds migrate by flight, others migrate by walking or swimming. These include not only flightless birds, but also some birds which are able to fly, but in some circumstances opt to walk or swim, for at least part of their journey. For example, Prill (1931) described a pedestrian migration of American Coots *Fulica americana* in the Warner Valley of Oregon during May 1929. At least 10 000 individuals were seen walking northward over a period of four days. They did not swim or fly (unless alarmed), but followed the shore, 6–25 abreast. They may have been engaged in a moult migration, and some may not have flown because their flight feathers were loosened or already shed. In western North America, the Blue Grouse *Dendragapus obscurus* performs an altitudinal migration, moving several hundred metres up and down mountainsides between the breeding and non-breeding areas (Cade & Hoffman 1993). Although the bird can fly, the radio-tracking of individuals revealed that this journey is often undertaken mainly on foot, which is perhaps not surprising in a bird that spends most of its time walk-ing. In flightless landbirds, such as ratites, all movement is inevitably by walk-ing or running. Emus *Dromaius novaehollandiae* in central Australia have been found to cover hundreds of kilometres at times of drought, with a mean speed of 13.5 km per day (Marchant & Higgins 1990).

Most species of penguins perform regular migrations by swimming, some of more than 1000 km. Having lost the power of flight, penguins have developed flippers instead of wings, enabling them to 'fly under water'. They are well streamlined and can travel under water at high speeds, larger species faster than smaller ones. Large King Penguins *Aptenodytes patagonicus* were filmed in an aquarium swimming at 3.4 m per second (more than 12 km per hour). Although penguins can float and swim at the surface, they apparently undertake long sea journeys mainly under water, where the drag on the body is less than on the sur-face, but they have to come up frequently for air. As they approach the surface, the drag on the body increases, and it has been suggested that leaping clear of the water (porpoising) is energetically less costly than surfacing, although this may hold only at speeds over about 2.5 m per second (9 km per hour).

Some other seabirds that are normally able to fly, such as auks, may migrate entirely or partly by swimming, remaining in suitable habitat throughout. Young auks have been found to travel by paddling at 40 km per day (Gaston 1983). The extinct Great Auk *Pinquinus impennis*, which nested at high latitudes, was probably a long-distance swimming migrant, wintering in the western Atlantic as far south as Florida and in the east Atlantic south to Spain, probably involving journeys of more than 1500 km (Brown 1985).

CONCLUDING REMARKS

The main conclusion from this chapter concerns the ways in which bird features (such as body weight, wing shape, flight mode and fuel load) interact with one another, and with environmental features (such as wind conditions) to influence bird migratory capability. Such interactions probably account for some of the big differences in migratory performance (speeds, routes, altitudes and distances) that occur between species, and also within species, according to their current fuel loads and ambient conditions. Theory based on aerodynamic principles is increasingly supported by field data. For example, it provides an aerodynamic rationale for the observation that long-range small bird migrants accumulate relatively greater fuel loads for migration than large birds that also travel by flapping flight, and for why many large bird species migrate by soaring–gliding flight, rather than by flapping. Large waterfowl, which travel entirely by flapping flight, are of special interest because of the limited power available to them, and their weight-related restrictions on fuel reserves.

SUMMARY

Individual birds can vary their flight speed. In species that fly by flapping flight (as opposed to soaring–gliding flight), the energy need is greater at slow and fast speeds than at intermediate speeds. The energy costs of powered (flapping) flight are 8–30 times greater than basal metabolic rate, and of gliding flight about 1.5–2 times greater. While these energy needs depend on speed in flapping flight, they are independent of speed in soaring–gliding flight, in which most energy for flight comes from the ambient atmosphere.

Body size sets constraints on migration in birds. With increasing body mass, flight costs increase, as does flight efficiency (energy cost per unit weight) and flight speed, but the amount of fuel (relative to body weight) that can be carried declines, reducing the maximum possible non-stop flight range. Wing design has additional influence on flight speed and efficiency, but seems influenced as much by the needs of everyday life as by those of migration. With favourable winds, swans (some of the heaviest of flying birds) can make non-stop flights covering up to 1700 km and geese up to 3000 km. Passerines and shorebirds can make non-stop flights of 80–100 hours, enabling passerines to cover distances of 1500–3000 km, and the faster shorebirds distances of 4000–7500 km (with one extreme flight exceeding 10000 km).

Most bird species migrate by flapping flight, but some large species migrate mainly by soaring and gliding. This method requires much less energy per unit time than flapping flight but, being dependent largely on thermals or other rising air currents, soaring landbird species often have to take roundabout routes avoiding long water crossings. They can also travel only by day, while many other (flapping) birds can travel at night.

While fuel and water could clearly limit the length of non-stop flights by birds, some species may need to rest periodically, well before they reach the limit set by fuel reserves. Species that migrate by walking or swimming travel more slowly and generally shorter distances than species that migrate by flight, but migrations exceeding 1000 km have been recorded from some penguin species.

White Pelicans *Pelecanus onocrotalus* soaring in an updraft

Chapter 4
Weather effects and other aspects

They [birds] are usually able to choose a period of mild and favouring winds. North winds [in autumn] either lateral or from the rear are favourable, and they wait for them with the same sagacity that sailors exhibit when at sea. (Frederick II of Hohenstaufen, 1244–1248; from an English translation of 1943.)

Weather has obvious effects on bird migration. It influences the times when birds can travel, the energy costs and risks of the journey, and the visibility of any celestial or ground-based cues that birds might use for navigation. This chapter is concerned with how birds behave in different weather conditions, with the altitudes at which they fly, with day–night patterns of migration, and with the influence of social factors, such as the sizes and formations of flocks. In all these aspects of migration, the interest is in seeing how birds of different kinds adjust their behaviour to prevailing conditions so as to minimise the costs and risks of long-distance travel.

MIGRATION AND WEATHER

Assessing the effects of weather on the volume of bird migration is not straight-forward. This is partly because migration depends less on the prevailing weather than on the intrinsic migratory state of the birds themselves, as is evident from several types of observations (Lack 1960b). First, weather conditions that in spring or autumn would be associated with migration have no effect at other times of year. Second, after a hold-up due to unsettled weather in the normal migration season, the onset of favourable weather may result in unusually heavy passage. Third, after a long hold-up, or late in the migration season, birds may take off in conditions that, at other times, would stimulate little or no migration. Fourth, in addition to such broad effects, the individuals in particular localities do not all set off together, but over periods of days or weeks. Such variation can only be attributed to variation in their internal states and the dates at which individual birds are physiologically prepared (with adequate body reserves) for departure. The underlying controlling factor is thus the migratory state of the individual, which interacts with weather and other external conditions to influence depart-ure dates. If the internal changes are well advanced, migration may occur under apparently unfavourable conditions, but if the migratory state is low, departure will occur only under especially favourable conditions or not at all.

The numbers of birds migrating on particular days thus depend not just on the prevailing weather, but on the weather over preceding days, the date in the season, and the number of birds ready to leave at the time. Towards the end of the migration season there may be few birds left to migrate, however suitable the weather. The association between weather and the volume of migration on particular days is therefore not constant, making analysis more complicated. In addition, because species differ in body size, flight mode and other aspects, they are affected by adverse weather to different extents, some species being able to migrate in conditions that would ground others (Lack 1960b, Alerstam 1978a, Elkins 2005). Ideally, therefore, the effects of weather should be examined at par-ticular sites in relation to the proportion of birds in a migratory state that take off each day, but this proportion is seldom measurable.

Another pitfall in assessing weather effects on bird migration is that differ-ent weather factors tend to be associated with one another, with some occurring under cyclonic and others under anticyclonic conditions (Lack 1960b, Richardson 1978, 1990). Even with the help of multivariate statistics, it is often hard to tell which factors are critical and which are coincidental.[1]

---

[1]*Given sufficient data collected day to day over several years some of the difficulties can be allowed for in analysis (e.g. Alerstam 1978a). For example, the effect of date in season on numbers migrat-ing can be allowed for by calculating how much the measure for a particular day deviated from the long-term average for that date. Weather variables whose typical values change greatly over the season (such as temperature) can also be expressed as deviations from seasonal norms, enabling temperature and date effects to be more effectively separated. The effect of weather-induced delay on migration volume can be allowed for by including the duration of delay as a variable in a multivari-ate analysis. Despite the increasing sophistication of analytical procedures, however, multivariate analyses still cannot reliably separate causal from coincidental relationships, or reveal whether birds respond directly to strongly interrelated variables, such as pressure, temperature and humidity.*

Box 4.1   Weather effects: the bigger picture. Based mainly on Richardson (1990)

To a large extent, it is possible in high latitude regions to predict the likelihood of strong bird migration on particular days within the migration seasons from examination of synoptic weather maps, showing fronts and pressure systems over wide areas. The atmosphere at such latitudes is organised into high and low pressure systems which move approximately eastward. In the northern hemisphere, winds blow clockwise around highs and counter-clockwise around lows. Fronts separate air masses in these systems: warm fronts occur where advancing warm air is replacing cold air, usually east or southeast of an approaching low; and cold fronts occur where cold air is replacing warmer air, usually south or southwest of a low. Precipitation and thick cloud occur most commonly near lows and fronts.

Given the importance of wind direction and clear skies, migration timing can be related to these large-scale atmospheric features, even if birds sense only their local manifestations. In the northern temperate region, peak southward migration tends to occur with cool northerly tailwinds as a 'low' moves away to the east, or a 'high' approaches from the west, or both. Conversely, peak northward migration tends to occur with warm southerly tailwinds, as a 'high' moves away to the east, or a 'low' approaches from the west, or both. At both seasons, much migration also occurs with light winds near the centre of a high.

When birds are concentrated by coasts, ridges or valleys, the numbers passing a given point may be quite different from what the above weather patterns would suggest. Also, the association between synoptic weather patterns and migration volume is likely to decline as birds get increasingly far from their departure points. In the southern hemisphere, different (but related) trends are expected, because winds there blow in the opposite directions around lows and highs.

Almost certainly, migrants do not react to the general weather situation as such, but to one or more components of it, such as wind and rain. Nevertheless, for the human observer, the synoptic weather situation of fronts and pressure systems gives a good indication of how much migration is likely to occur at different places on particular days (**Box 4.1**).

Yet another problem relates to the use of visual observations alone to assess weather effects. Visual records miss any birds flying too high to be seen by day, and provide little or no information on nocturnal migration. Yet radar has revealed that most overland migration of birds that fly by flapping flight occurs above visual range. In fact, the proportion of birds flying within sight, and the proportion that come to ground, tend to be greatest in conditions that are unfavourable for flight (Lack 1960b). Thus, migrants tend to fly low in opposing rather than following winds, and to settle whenever they encounter strong opposing winds, mist or rain, or reach coastlines or islands. The observer equipped only with binoculars would conclude that these were the very conditions that favoured migration, a once firmly held view but the opposite to reality. Birds also tend to fly low along coasts in these conditions, reluctant to strike out over water. It is therefore important to distinguish the influence of weather in promoting migration from its influence in making migration conspicuous (Lack 1960b, Alerstam 1978a).

The advent of radar greatly clarified the situation, because it enabled migrants to be detected at almost all heights (missing only those below the radar horizon), day and night, and in all weathers. From radar-based studies, consensus has now emerged that, within the appropriate seasons, migration is favoured by fine anti-cyclonic conditions with favourable tailwinds, and also by rising temperatures in spring and by falling temperatures in autumn. In effect, at both seasons the birds prefer to migrate under clear skies with following or light winds. Clear skies assist navigation, especially at night, by making celestial cues more visible, while following winds reduce the time and energy spent on the journey, and the risk of being blown off course. In contrast, birds seldom take off to migrate in strong opposing winds, dense cloud, mist and rain. Opposing winds make progress difficult or impossible, cloud hampers navigation, while mist or rain can soak many kinds of birds and force them down. Revealingly, the same weather conditions that are associated with heavy migration in free-living birds are often associated with strong migratory restlessness in captive ones (Schindler *et al.* 1981, Viehmann 1982, Gwinner *et al.* 1992).

The above conclusions on weather effects were drawn from a large number of short-term studies in various parts of the world (Richardson 1990). However, an almost complete picture of autumn migration was recorded near Nuremberg in Germany by use of a conically scanning pencil-beam radar (Erni *et al.* 2002). At this site, bird migration increased in volume from early August, reached high levels during September to mid-October, and then declined. Allowing for this seasonal trend, about two-thirds of the variation in daily migration volume was explained by wind and rain, the two variables likely to have most influence on flight. But because of the association between different weather factors, migration volume was also correlated with change in temperature, pressure and cloud cover. It was more closely correlated with the duration of rain than with the overall amount per night. After short heavy showers, birds continued migration, whereas on nights with continuous light drizzle, migrant densities remained low.

The proportions of days that are favourable for migration clearly vary greatly from region to region, from year to year, and from autumn to spring in the same region. The wettest regions on earth offer only a few days in each season that are ideal for migration. In some other regions, birds hardly ever experience favourable winds near the ground, but they take off and climb to higher altitudes, where wind conditions are more often favourable (see later).

The importance of temperature to migration is uncertain. In spring, warmth occurs in association with other conditions favourable to flight, as does cold in autumn. But temperature may have direct effects through influencing the energy balance of the birds, and more importantly through influencing food supplies, because all vegetative growth, insect activity, and ice melt are temperature-dependent. It is therefore of obvious advantage for migrants to adjust their migratory schedules to year-to-year variations in temperature, and they are clearly deterred in spring by extreme cold and snow (Chapter 14). It is also uncertain to what extent (if at all) migratory birds can predict the weather before it arrives, say by a change in barometric pressure. Bird movements in autumn and spring have been correlated with changes in pressure, associated with an approaching front (Richardson 1990, Dau 1992). But whether the birds were reacting to pressure as such or to associated weather variables is hard to say. More study of apparent pressure responses is needed before firm conclusions on this point can be drawn (see also Chapter 9).

Observations of birds about to depart are consistent with the findings from radar studies mentioned above. Take-off occurs most often on nights with good visibility, bright stars, no overcast, no rain, and with light or following winds, these conditions being more influential in long-distance than in short-distance migrants (e.g. Cochran & Kjos 1985, Bolshakov & Bulyuk 1999). In some species, individuals usually take off singly without preliminary activity (Hebrard 1971). Passerines leaving from trees and bushes at dusk at first flutter uncertainly upwards, in ones and twos or small parties, and then, climbing all the time, proceed speedily and confidently in the direction of travel, rapidly disappearing into the gloom. In other species, departure can be noisy and impressive, as flock after flock takes off and heads into the distance, climbing till out of sight (Piersma et al. 1990b).

Regardless of the weather at take-off, migrants can encounter poor conditions en route. If they meet low cloud and unfavourable wind, birds may be forced low and, if over the land, they can settle and wait for conditions to improve. Over the sea, as radar has revealed, landbird migrants that enter cloud or mist banks usually become disorientated, milling in all directions and gradually drifting downwind, or actively flying downwind which gives a good chance of reaching clearer weather (Williamson 1955, Lack 1960b, Richardson 1978, Bourne 1981). If cloud persists, migrants over the sea are sometimes attracted in large numbers to lighted ships or oilrigs (Bourne 1979, 1983). Although heavy overcast appears inimical for migration, some birds seem to maintain more or less straight courses with complete cloud cover. Below the cloud they can see the ground, and above it the sun or stars (although flying above the cloud seems infrequent).

Many journeys do not go smoothly, and the survival of landbird migrants over the sea is likely to depend largely on their abilities to cope with adverse weather and off-route displacement. On any one journey, depending on weather, birds may advance, then retreat, veer from a direct course, or be borne far off route by side-winds, returning to their regular course along unfamiliar routes (Williams 1950). They may sometimes perish in large numbers (Chapter 28). When over land, in contrast, most landbirds seem to settle when they encounter a rain front, and simply wait for the weather to clear. When over the sea at not too great a distance, they may turn around and fly back to land. The important point is that, once en route, the best course of action for a bird is likely to depend on prevailing circumstances, and particularly whether the bird is fat or lean, and over favourable or unfavourable terrain. Drift by crosswinds is, of course, one of the commonest ways in which migrants turn up as vagrants in places off their normal routes (Chapter 10).

Importance of wind

In general, wind speeds are stronger at mid-day than at night or early morning, and increase from the ground, where friction slows the wind, up to several thousand metres. In addition, the air mass in which birds migrate is continually changing in speed and direction, and birds must continually adjust their behaviour if they are to migrate to a predetermined destination in the most energy-efficient way. That birds respond to wind is shown by the frequent observations that: (1) they tend to depart only in favourable (following) winds; (2) they often select flight altitudes where winds are favourable; and (3) they compensate for wind drift, at least to some extent (providing that they can see the ground below).

The long-standing observational evidence that birds mostly set off with following winds has been confirmed for individual radio-tagged birds. On days when radio-tagged passerines departed from south Sweden, the tailwind component was significantly greater than on days when birds were present but did not leave (Åkesson & Hedenström 2000). Similarly, radio-tagged Brent Geese *Branta bernicla* en route from the Netherlands to arctic Siberia selected for each stage of their journey days with wind assistance (Green *et al.* 2002), and the same held for Bewick's Swans *Cygnus c. bewickii* migrating from Denmark to northern Russia (Klaassen *et al.* 2004). The reluctance of birds to fly against headwinds is clearly adaptive, but can sometimes result in considerable delay. For example, in 1994 when winds were favourable, four radio-tagged Barnacle Geese *Branta leucopsis* took 5–15 days to migrate between Svalbard and Scotland; but in 1995 when winds were unfavourable for much of the time, radio-tagged birds took 9–36 days, with five individuals arriving 10 days (three birds) and 24 days (two birds) later than the majority that year (Butler *et al.* 2003).

Wind assistance is like food or fat reserves: it is a resource that fuels migration. If a bird with a given flying speed was blessed with a tailwind of the same direction and speed, it could in theory fly twice as far on the same fuel and in the same time, or the same distance on half the fuel and in half the time. Yet a bird flying into a headwind of the same speed could make little or no progress, however great its fuel reserve. In practice, it seems from radar studies that birds flying overland do not behave in quite this way, but fly slower than usual with a tailwind, and faster than usual against a headwind (Chapter 3; Bellrose 1967, Alerstam & Gudmundsson 1999, Hedenström *et al.* 2005, Liechti 2006). It is as though they conserve energy when conditions allow, and expend more than usual when necessary. The net effect is that birds make slower progress than expected in a tailwind, and faster progress than expected in a headwind, but further study is needed to find how widespread this behaviour is. In any case, following winds reduce the energy cost of migration, which is important to many migrants. But where following winds exceed the birds' own flight speed, they also bring risks, especially over inhospitable terrain. Birds are then unable to maintain their preferred direction if the wind direction changes; and if their body reserves are exhausted or conditions deteriorate, they cannot easily return. Small-bodied, slow-flying species obviously face the greatest risks from adverse winds.

If the wind deviates to some extent from the birds' intended track, the bird can in theory correct for this by adjusting its heading so as to remain on track with respect to the ground (**Figure 4.1**). Birds do not then progress in the direction they are heading, but at some angle to it, which is closer to the intended track. The greater the crosswind component for a given flight speed, the greater this compensating angle must be (aircraft pilots refer to the angle between heading and track as the drift angle, α). This angle can be reduced if the bird flies faster.

Powered flight

1. No wind, track = heading

2. Full drift

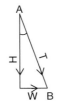

3. Full compensation

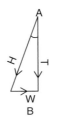

4. Partial compensation

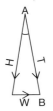

Soaring flight

5. No wind, no displacement

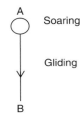

6. Passive displacement with drift

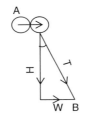

7. Passive displacement with full compensation

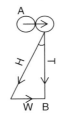

8. Passive displacement with partial compensation

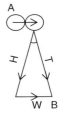

Figure 4.1 Effects of side-winds (W) on the migratory tracks (T) of birds, depending on adjustments of heading (H), leading to full drift (no compensation, 2, 6), partial drift (partial compensation, 4, 8) and no drift (full compensation, 3, 7). Heading and track are the same when there is no side-wind (1, 5). Upper diagrams relate to powered flight, and the equivalent lower diagrams to soaring flight. If a bird is to continue soaring in a thermal, it will inevitably be transported passively over the ground in whichever direction the thermal is being blown (laterally in the diagrams). The bird may compensate for this during the next gliding phase of the flight. Modified from Kerlinger (1989).

The point at which a bird is no longer able to compensate for lateral drift is thus a function of wind speed and direction, as well as the maximum flight speed that the bird itself is able to maintain (called the threshold for drift, Evans 1966b). Several types of drift can thus be distinguished in migratory birds (**Figure 4.1**). One type occurs when a bird flies along a fixed heading towards its destination, and is drifted from that heading by lateral winds (called full lateral drift). A second type occurs when the bird compensates for lateral wind, but only partly (called partial drift). A third type occurs when a bird attempts no compensation, so flies downwind, getting progressively further off course (called downwind drift). All three types of drift are commonly seen among migrating birds, as is complete compensation.

There are other ways in which birds could minimise the problems created by crosswinds (Alerstam 1979). For example, birds might allow themselves to be drifted partially at high altitudes and then correct for this by overcompensation

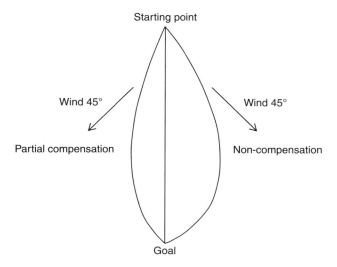

Starting point

Wind 45°   Wind 45°

Partial compensation   Non-compensation

Goal

Figure 4.2   Optimal use of a crosswind by a migrant. The bird is drifted by the wind during the initial stage of migration and later overcompensates to reach its goal. By invoking different altitude use and varying prevailing winds, variations of this model can be used to test predictions on real birds in the field (see text). From Alerstam (1979).

at low altitudes, as they approach the end of their flight. By flying high at the start they gain the advantage of faster winds at those altitudes, even though their direction is not ideal, and make the correction later in the flight under weaker winds that prevail at lower altitudes (**Figure 4.2**). Alternatively, if winds shifted direction predictably along a migration route, as is common at latitudes 25–35°N, birds could allow themselves to be drifted in one direction at the start of a flight and in the opposite direction towards the end, as seems to occur in some overseas flights (see later).

Transoceanic migrants have no benefit from obvious landmarks, making it difficult or impossible to correct for drift. At take-off, such birds are clearly sensitive to local wind conditions, but once over the ocean they usually show little or no compensation for wind drift. While birds over land might settle and re-assess their position every few hours, transoceanic migrants may sometimes fly for more than 100 hours without apparent reference to local conditions. Their orientation, as followed by radar, may be largely explained by the maintenance of a simple compass heading, giving much more uniformity in directions than is usual among migrants over land (Williams & Williams 1990). However, as they descend towards the end of their flight, they begin to respond again to local weather and topography. Selection of appropriate weather cues before departure, together with simple vector migration, normally appear sufficient to allow such transoceanic migrants to make landfall on distant coasts or island chains.

Radar observations have been used to explore not only the extent to which migrating birds can compensate for drift by crosswinds, but also how much the prevailing circumstances (wind strength, visibility, day or night, over land or over sea, and cloud cover) affect the response. Patterns of compensation, full lateral drift, partial drift and downwind drift have all been observed, depending on conditions. Birds can presumably 'know' they are being drifted off course only when they have reference to some stable feature, such as the ground below; drift would have to be substantial before it could be detected from celestial or magnetic cues. These considerations may explain why compensation seems usual in low-altitude (<1 km) diurnal migrants, but is much less frequent or complete at high altitudes, and why it is more frequent in light winds (up to 5 m per second) than in stronger ones (Bruderer & Jenni 1990). In any case, it is rarely complete in nocturnal migrants and over large water bodies (Richardson 1990). Whether birds compensate for wind may reflect not only their ability to do so, but also their remaining body reserves, which influence the relative advantages or disadvantages of doing so. In some circumstances, it may be more economical for landbirds over water to drift downwind than to use scarce fuel reserves to fight against an unfavourable wind in order to remain on track. These considerations help to explain the variations in bird behaviour found with respect to wind (for behaviour of a radio-tagged Peregrine *Falco peregrinus* in a hurricane see McGrady *et al.* 2006).

Despite the effects of crosswinds, some birds can maintain remarkably straight tracks on migration. During a radar study of migrating waterfowl in Denmark, individual flight trajectories of bird flocks showed zig-zag patterns rather than straight lines (Desholm 2003). Analysis of these continual small changes in flight directions, which were too small to be detected by satellite, showed that geese and Common Eiders *Somateria mollissima* were flying, on average, only 0.7% and 1.6% longer distances than if they had flown along exact straight lines. The frequent small changes in flight direction could have resulted from birds repeatedly compensating for wind drift as they travelled towards their goal.

But what are the overall advantages in birds paying so much attention to wind? In conditions prevailing in central Europe in autumn, a bird migrating only on nights with favourable wind can, on average, increase its flight speed by 30% compared with an individual that disregards wind (Liechti & Bruderer 1998). Selecting the most profitable flight altitude may result in an additional 40% gain in flight speed. In other words, by responding appropriately to wind conditions, a bird can greatly increase its flight speed and greatly reduce its energy use during a journey through this region. The time needed for refuelling decreases accordingly, or the safety margins provided by the body reserves are extended. The bird is also at less risk of being blown far off course. Clearly, there are great benefits to a migrant in responding appropriately to wind.

For birds undertaking long non-stop flights over the sea, selection of favourable wind conditions at departure could be even more important than for birds over land. For example, many birds fly from northeastern North America directly over the Atlantic to Caribbean Islands or South America (Williams & Williams 1990). In autumn, the passage of a cold front brings favourable south-southeast winds and triggers departure. Once over water, as radar has shown, the birds continue flying south-southeast. When they encounter the trade winds near 25°N, their tracks are drifted to the south and south-southwest as they pass over the Caribbean. This drift is essential if birds are to reach South America, unless they redirect their heading en route. By not compensating fully for wind drift on the first part of their journey, the birds realise a faster and more energy-efficient migration to latitudes where wind direction changes, which then allows them to compensate for drift in an energy-efficient manner. They are able to follow the pattern in **Figure 4.2,** making no obvious compensation because crosswinds change predictably from one direction to the opposite direction along the route. At the time of spring migration, the winds are in the same direction as in autumn, yet the birds must fly in the opposite direction. They therefore make their northward journey to the west overland so that their entire two-way migration follows a clockwise loop.

Likewise, many birds in western Europe make their southward journey through Iberia into Africa, but make the return northward journey further east through Italy, so that in this case the entire return journey follows an anticlockwise loop. Some birds may have evolved elliptical (loop) migrations partly to minimise the effects of adverse winds, including latitudinal changes in wind direction. This is one of several hypotheses on the evolution of loop migrations, applicable to seabirds as well as to landbirds (Chapter 22).

With respect to wind, soaring birds provide a partial exception to the usual patterns. Updraughts strongly reduce the energy cost of migration for such birds, which often fly in side-winds or light opposing winds if updrafts are present. Thermals develop in calm or light wind conditions, but not in strong winds, which suppresses raptor migration in some regions, regardless of wind direction. Soaring birds also show no particular tendency to migrate on cold days in autumn, as do many other birds, probably because the necessary thermals develop best on warm days (Alerstam 1978a, Kerlinger 1989). An example of the effects of autumn weather on the migrations of different types of raptors is given in **Figure 4.3**. At the locality concerned, raptor migration occurred almost entirely within visual range, and so could be recorded accurately. Nevertheless, different species favoured somewhat different conditions for migration, depending on their particular flight modes.

Wind conditions are clearly crucial to successful long-distance migration, and the reactions of migrants to weather systems can be viewed as adaptive, ensuring a more energy-efficient and safer journey. The huge day-to-day fluctuations in the volume of bird migration reflect the continual adjustment of bird behaviour to prevailing conditions (Alerstam 1981). Some long non-stop flights of birds may be accomplished only with the aid of a following wind. This was the conclusion of several studies in which the known energy reserves of migrants were compared with their estimated needs on migration (for passerines see Wood 1982, Izhaki & Maitav 1998a; for shorebirds see Stoddard *et al.* 1983, Piersma & Jukema 1990, Marks & Redmond 1994, Tulp *et al.* 1994, Butler *et al.* 1997). However, these findings may be modified as more information becomes available on the energy needs of migrating birds (Chapter 5: Appendix 5.1). In some species, wind assistance has been found necessary for birds to achieve their migrations in the time observed (Butler *et al.* 1997), and in other species, arrival dates have been related to the proportion of days with following winds during the normal migration period (for Bewick's Swans *Cygnus columbianus* see Rees 1982, for Lesser Snow Geese *Chen caerulescens* see Ball 1983, for Barnacle geese *Branta leucophrys* see Butler *et al.* 2003). In order to better understand the variable relationships between behaviour and wind conditions in migrants, more research is needed, aimed particularly at testing predictions from mathematical models, such as those of Alerstam (1979).

Most of our understanding of weather effects on migration is based on radar studies which can provide less biased information on mass migration than any

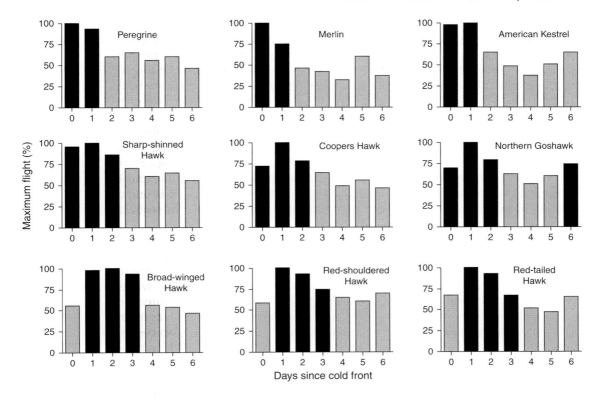

Figure 4.3 Passage rates of nine species of raptor at Hawk Mountain in Pennsylvania, 1934–91, in relation to the passage (NW–SE) of cold fronts. The height of each bar is expressed as a percentage of the highest passage rate for each species. Passage rates on days with shaded bars are significantly higher than those on other days (Tukey's Studentized range test in association with one-way analysis of variance for species in which ANOVA probability was <0.05).

Using pooled data from a 55-year period of observations, cold fronts passed this area, on average, every 4–5 days in autumn, and raptor migration increased in the first three days after the passage of each front. Three basic patterns emerged, related to the flight mode of the species involved. Falcons (top three) had their highest rates of passage on the same day of frontal passage; accipiters (middle three) had their highest rates on the first day after the passage; and buteonine hawks (bottom three) on the first three days after the passage. The falcons migrated mainly by direct flapping flight, a mode suited by the high-speed winds that typically occurred soon after frontal passage. Accipiters flap less and soar more, so are better suited by the lighter updraft-producing northwesterly winds and light thermals that begin to form within a day after frontal passage. Finally, buteos soar more and fly higher on migration than do most other raptors, making them suited by the weather conditions that occur 2–3 days after frontal passage.

During the 55-year study period, 10–20 cold fronts passed Hawk Mountain each year between 1 September and 23 November (the peak migration period). In years with few fronts, more raptors passed after each one, reflecting the greater build-up of birds between fronts. The number of cold fronts did not affect the overall numbers passing each year. From Allen *et al.* (1996).

other method. Nevertheless, observations show that some low-level flight (below the radar horizon) occurs under headwinds and other conditions that inhibit higher level flight. In these conditions, many small birds migrate by 'bush-hopping' or 'tree hopping', in which they flit from bush to bush or from tree to tree, feeding as they go, but travelling continually in the same direction. Warblers, which normally migrate at night, sometimes pass in daytime through extensive bushy or reedy areas in this way, as do tits and other canopy-feeding passerines as they travel through wooded terrain. Many small birds also fly low for short distances in the gaps between showers. Such low-flying birds take more account than high-flying birds of topographical features, being more deflected by hills and water bodies, and in strong winds taking shelter provided by ground contours or forest edges. In these ways, small birds can maintain some progress on days when they might not otherwise travel because of strong headwinds or insufficient body reserves. By their combination of flying and feeding, these small passerines resemble swifts and hirundines that pick off insects as they go. But as headwinds increase, fewer and fewer birds take part in the migration, as species after species drops out.

ALTITUDE OF MIGRATION

Study of the altitude at which birds migrate is of interest not only to ornithologists, but is also of practical relevance in conservation (because birds can kill themselves against masts and other tall structures) and in aircraft safety (because birds can get caught in jet engines and bring down planes). Over low land, to judge from radar studies, most migration takes place within 1.5 km of the ground, with decreasing numbers of birds at higher altitudes up to 3 km or more (Bruderer 1999). When birds need to cross high mountains or find favourable airstreams, they sometimes fly much higher, occasionally reaching more than 5 km (**Figure 4.4**). Radar studies

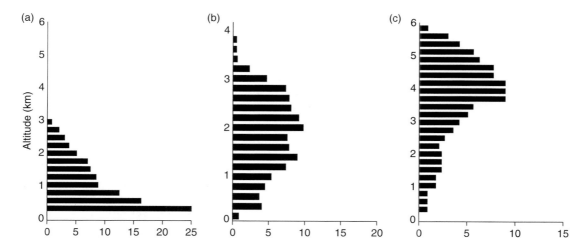

Figure 4.4 Altitudinal distribution of migrants as registered by radar studies in: (a) Lowlands of central Europe, spring (Bruderer 1971); (b) South Sweden, spring, mainly waterfowl and shorebirds (Green 2004); (c) Antigua, Caribbean, autumn (Williams et al. 1977).

in central Europe have also revealed that migration generally occurs at higher elevation in spring than in autumn and by night than by day. However, different patterns of flight altitude occur in other parts of the world, where seasonal wind patterns differ (see Bellrose & Graber 1963 for Illinois, see Klaassen & Biebach 2000 for Sahara Desert).

In some regions, the height of bird migration seems to vary relatively little from night to night, but where wind conditions are highly variable, migration height may change accordingly, with birds flying higher than usual if winds are unfavourable lower down. This was evident, for example, at various sites in the southeastern USA, when migrants were heading towards or from the Gulf of Mexico (Gauthreaux 1991). In this region, migration on many nights was concentrated at less than 500 m above ground, but when winds were unfavourable in this zone, the birds flew higher, giving over a number of nights a clear correlation between the altitude with the greatest density of migrants and the altitude with the most favourable winds (**Figure 4.5**).

Apart from the needs to find favourable winds or to cross mountains, flights at several kilometres high are likely to occur mainly on long non-stop journeys, such as over oceans. So much energy is consumed in climbing to such high altitudes that this might not be worthwhile on short flights. Migrants from northeastern

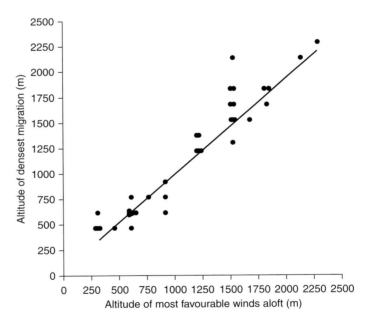

Figure 4.5 Correlation between the altitude of densest nocturnal bird migration and the altitude of most favourable wind, as measured by radar in the southeastern USA. The most favourable wind was defined as a wind blowing toward the north-northeast in spring or south-southwest in autumn and occurring at the lowest possible altitude. Regression relationship: $y = 58.6 + 0.987(x)$, $r = 0.96$, $t = 22.9$, $P < 0.01$, $N = 38$. From Gauthreaux (1991).

North America rose during the course of their autumn flight over the western Atlantic to reach heights of 4–6 km over Antigua in the Caribbean, and later dropped abruptly as they approached South America (**Figure 4.4**; Williams *et al.* 1977). But radar has also revealed that birds crossing smaller stretches of water, such as the North Sea or the Great Lakes, lose height during the night, often flying within 100 m above the water surface, but rise again at dawn (the so-called dawn ascent, Myres 1964, Diehl *et al.* 2003). Thus, overwater echoes on the radar screen disappear late in the night, as birds descend below the radar horizon, and reappear towards dawn, an appropriate distance along the route. One view is that birds descend during the night in order to find a place to settle. If they find themselves still over water, they ascend at dawn probably to avoid predation by gulls, which become active around daybreak, and also to gain a better view (Bourne 1980). They may change direction to head towards land if they are within about 30 km (and can presumably see land).

Other altitude estimates were made by 'moon-watching' in some desert, foothill and mountain areas of central Asia (Dolnik & Bolshakov 1985, Dolnik 1990). The average flight altitudes were estimated at around 1 km above sea level over low desert, 1.5 km above sea level over foothills, 2–3 km over the Caspian, and 3.2 km above sea level over the Tien Shan and Pamir Mountains. In the latter areas, some birds reached 9.7 km above sea level, but flight altitudes at night varied greatly between different types of birds, with median values above sea level of 5.0 km in passerines, 5.8 km in shorebirds and 6.9 km in ducks (equivalent to 1.3, 2.1 and 3.2 km above ground). In general, these figures are in broad agreement with those found by radar in central Europe, and confirm that most movement occurs within 1.5 km of the ground, even though in mountain areas this takes the birds to heights of several kilometres above sea level.

Several species migrate regularly over the Himalayas, and at least one species over the highest parts, namely the Bar-headed Goose *Anser indicus*, which has been recorded at more than 8 km above sea level (Chapter 6). Demoiselle Cranes *Grus virgo* can reach altitudes of 7.5 km in crossing the Hindu Kush on migrations between their nesting areas in central Asia and wintering areas in northwest India. Clearly, such birds must be pre-equipped physiologically to avoid altitude sickness (hypoxia). Not all birds have such flexibility in their flight altitudes and, whatever the advantages of high-altitude flight, many species seem confined to migrate at low elevations, presumably for physiological or energy-based reasons, as in the swans discussed in Chapter 3.

Soaring birds are limited in flight altitude by the height reached by thermals, which is greatest around noon over land, and in most conditions seldom exceeds 1.5 km. During their migrations, soaring birds are continually rising and falling, as they climb in successive thermals and lose height between them (**Figure 7.3**; Leshem & Yom-Tov 1996a, Spaar & Bruderer 1996). Unlike many other birds, therefore, they cannot maintain constant altitude over long distances. Much higher altitudes have occasionally been recorded in raptors, such as a Bald Eagle *Haliaetus leucocephalus* which reached 4.5 km above ground in North America, but it can be assumed that such birds were travelling largely by flapping flight (Harmata 2002). Soaring seabirds, such as albatrosses, seldom reach more than 30 m above the waves. Clearly, not all birds are high fliers.

Changes in conditions with altitude

As they ascend from ground level, migrating birds face progressively changing conditions. Wind speed increases, while air density, oxygen availability and temperature decline. Once birds break through the cloud layer, however, they can escape from mist, rain and snow. Wind has a major influence on the height above ground at which birds fly, for it can change in direction as well as strength, affecting the choice of cruising height. This is most strikingly demonstrated by the trade winds, which give way higher up to the anti-trades blowing in the opposite direction. This situation allows birds to find following winds in both autumn and spring, providing they fly at an appropriate height. In southern Israel, for example, the wind shear associated with the shift from trade to anti-trade winds fluctuates roughly around 1.5 km above sea level. Correspondingly, birds studied by radar flew mainly below this level in autumn, and mainly above it in spring, when some reached heights of 5–9 km above sea level. In the low-level jet streams of this region, birds also achieved ground speeds of up to 180 km per hour (Liechti & Shaller 1999).

Waders leaving the West African coast for Europe in spring must climb rapidly to altitudes greater than 3 km. It is only at such high altitudes that tailwinds keep the flight costs within reasonable bounds, thereby enabling the birds to complete their 4300-km non-stop flight to the Dutch Wadden Sea coast (Piersma 1990). Similarly, radio-tagged thrushes tracked on migration in North America ascended as much as about 3 km each evening until they found suitable winds (Cochran & Kjos 1985). With weak winds, they flew lower and accepted some lateral drift. If winds at altitudes above 75 m had unfavourable head or side components, the birds returned to the ground and did not migrate. Their migration therefore involved adoption of a constant heading, and mitigation of wind drift mainly by altitudinal adjustment rather than by lateral compensation. In most regions, where little altitudinal choice in wind direction is available, birds can only wait for favourable conditions or change their route between autumn and spring.

Apart from wind, the main weather factor influencing the altitude of migration is the cloud base. Most birds fly below the clouds where they can see the ground. The height of the cloud base therefore limits the vertical spread of migration, and if the cloud descends, it compresses the stream downwards, so the average flight altitude decreases. In some situations, however, birds fly above the clouds, presumably relying entirely on celestial or magnetic cues for navigation, with no visual reference to the ground below. Migration streams can also be compressed laterally as they are funnelled through mountain passes, or concentrate along coasts, river valleys or other 'leading lines', when birds may fly at low levels even without cloud cover.

Consequences of high-altitude flight

Air densities and associated oxygen levels decline by around 10% for each 1000 m rise in altitude, which roughly translates to a 27% decline at 3 km, a 40% decline at 5 km and a 65% decline at 10 km, with minor temporal and regional variations. The decline in air density means that, in theoretical still air, birds have to work harder to keep aloft at high altitude, but meet less resistance to forward flight. The maximum range speed increases by an estimated 5% for each 1000 m rise in altitude (Alerstam 1990a), but so does the power required to fly at this speed, which in turn entails a corresponding increase in energy and oxygen consumption. Although there may be no gain in overall range, it is still advantageous to fly high, because the increased cruising speed shortens the flight time. In addition, stronger winds and reduced turbulence at high altitudes may reduce the energy costs and flight times even further. Other advantages of high-altitude flight include: (a) a wider view of the ground (which may help birds to stay on course, and maintain a straighter track); (b) avoidance of obstacles and deflection by mountains; and (c) a reduced chance of predation (because birds get above the zone where falcons and other aerial predators normally hunt).

Compared with mammals, birds can breathe much more efficiently. Their various air-sacs, which lie within the thoracic and abdominal cavities, as well as between the integument and body walls and within the bones, are connected to the lungs, and enable birds to extract oxygen from thin air more effectively than could lungs alone (Chapter 1). Respiratory movements act mainly on the air-sacs, causing a continual stream of air to pass back and forth through the lungs. The lungs themselves do not hold a substantial residue of used air, as in mammals. In addition, the air within the air-sacs reduces the specific weight of the bird and, being warmer (and hence thinner) than ambient air, may help to provide lift. As flight speed increases with height, while oxygen availability declines, the optimal flying altitude might be regarded as that where the bird could extract just enough oxygen to maintain its maximum range speed. In practice, however, wind conditions seem to have much more influence on flight altitudes, often overriding other factors of potential importance.

One cost of high-altitude flight is the climb involved, which is a major effort for heavy species such as swans (see earlier), although some of this cost may be compensated towards the end of the flight, as the birds gain distance while losing height. Climbing capacity becomes critical to Brent Geese *Branta bernicla* during spring as they attempt to cross the 2500-m-high Greenland ice cap (Gudmundsson *et al.* 1995). These birds leave Iceland with enough fuel to reach their breeding areas in northeast Canada 2600–3500 km away. Five radio-marked individuals were tracked by satellites to west Greenland. Their climbing rates were extremely low, only 0.01–0.06 m per second, reflecting the severe limits to their flight power when carrying large body reserves. Their average movement up the ice slope was so slow (on average only 2.5–14.0 km per hour) that the geese must have stopped frequently on the way up. Their downward journey was of normal speed, 29.5–54.7 km per hour, indicating a more or less uninterrupted descending flight. Not all birds crossing high-altitude areas would have the option of resting en route, as suitable habitat may not be available. Climbing to high altitudes may be more costly for migrants making only short flights than for those making long flights, because a greater proportion of the journey is spent on the climb.

Air temperature falls by about 7°C for every 1000 m rise above sea level (or 2°C for every 1000 feet). Over much of the temperate zone, a typical night temperature at ground level in autumn might be 5°C, so that at 1-km altitude, a bird experiences an ambient temperature of about −2°C, while at 3 km it experiences −16°C and at 5 km about −30°C. To this must be added a chill-factor dependent on wind. To some extent, the heat generated by flight could compensate for the ambient heat loss caused by low air temperature, but at the high altitudes at which birds sometimes fly, heat generation could clearly impose an additional energy drain. In contrast, over hot deserts the heat generated by flight could lead to overheating which the bird could counter only by evaporative water loss through panting. In hot environments, therefore, high-altitude flight at lower temperatures could greatly reduce the dehydration risks, at least up to the point where the air becomes so thin that water loss is raised through increased respiratory panting (Chapter 6). Air temperatures are also lower at night than by day, and by flying at night, migrants realise at least the same temperature difference as they would experience between sea level and 1000 m during daytime.

The heat produced by working flight muscles is not trivial: only about one-fifth of the energy they generate is mechanical, while the remaining four-fifths is heat, the surplus of which must be dissipated by respiration or convection. The lower the ambient temperature, the more heat can be lost by convection, and the less water is required for cooling. In one early experiment, a budgerigar flying in a wind tunnel at an air temperature of 18–20°C dissipated by evaporation only about 15% of the waste heat generated in the flight muscles, whereas at 36–37°C some 47% of the heat was dissipated in this way, entailing much greater water loss (Tucker 1968). Probably most of the remaining heat at both temperatures was lost by convection from the thinly feathered under-surface of the wings.

Air humidity tends to decline with increase in altitude, especially above the cloud layer. The extreme cold at high altitudes could also enhance water loss, for cold air is relatively dry when it is breathed in, but saturated when it is exhaled. In addition, if oxygen extraction by the lungs is to remain unchanged at high altitudes, the ventilated volume of air must increase (Carmi et al. 1992). This need has been calculated to increase from sea level to 5 km altitude by as much as 175% per unit distance flown and by 254% per unit time flown in a swan (although swans are unlikely to normally reach this altitude, Klaassen et al 2004). For this reason, too, migration at very high altitudes brings not further reduction in dehydration risk, as in hot deserts, but increased dehydration risk.

Clearly, the costs and benefits of flight at different altitudes vary with circumstances, as well as between species, and the physiological constraints under which they operate. Taking all these external factors into consideration, along with the variations in body mass, shape and flight mode between species, it is not surprising that different types of birds seem to fly at different altitudes: sea-ducks and quail just above the waves, songbirds mostly up to 1.5 km, and shorebirds often at more than 3 km. Even on the same night, geese migrating over southern Sweden flew at 100–800 m, while shorebirds over the same site flew at up to 3.7 km above sea level (Green 2004).

Whatever the chosen altitude, individual migrants travelling by flapping flight have been found by height-finding radar to maintain remarkably constant heights on migration (Eastwood & Rider 1965). Although birds have no known sense organ which could detect altitude, this and other evidence suggests that they have some sort of pressure sense (Chapter 9).

The main advantages of low flying (by flapping flight) are that: (1) little energy is expended on climbing; and (2) the ground below is more clearly visible. The main disadvantages are the greater risks from overheating and dehydration, and from predation. Species that migrate by day often fly fairly low (within visual range), and react to the presence of topographical features, such as mountains or coastlines. They may form into streams, as they fly along shorelines or river valleys, or become funnelled through mountain passes. But species that migrate by night at high altitude often seem from their radar tracks to be little influenced by topography below; they usually fly on broad fronts and often cross mountains and coastlines without deviation. In such species, the need to minimise the expenditure of time or energy may have promoted the evolution of behaviour which led to birds taking the shortest routes across barriers, even at the cost of increased risk (such as severe weather). The compromise between time-saving or energy-saving direct routes on the one hand and long risk-reducing detours on the other may be drawn differently in different species, according to their flight capabilities and refuelling needs, the length of their journeys, and the habitat and weather expected en route.

In conclusion, there is little doubt that the altitude of migratory flight is related to prevailing atmospheric conditions, especially wind speed and direction, but also to cloud thickness and height, topography and other factors, as well as features of the birds themselves. We are in need of models which predict the optimal flight behaviour for a range of different species, taking all these variables into account.

Descending from migration

When landing from migration, birds do not always lose height steadily, like an aeroplane, but instead drop almost vertically from high altitude into habitat below, as described long ago on Heligoland Island by Gätke (1895). Similarly, on Cyprus small nocturnal migrants were seen to 'drop like stones from the sky at first light and immediately dive into cover' (Bourne 1959); while on the Louisiana coast in spring, migrant passerines were seen to dive nearly vertically down from more than 1 km, producing 'a whizzing sound as they pulled out of the dive just above the trees' (Gauthreaux 1972). Such 'fallout' occurs in other circumstances, as when migrating birds are confronted by a sudden headwind or downpour, and literally drop from the sky to seek refuge. Thousands of birds seem to come from nowhere, ladening trees and bushes or carpeting the ground, presenting an amazing spectacle to any onlooker. One well-known fallout site comprises 5 ha of trees at the small town of High Island, situated in the open coastal plain of Texas, which attracts hundreds of bird-watchers every spring. Geese have a particular form of flight, known as whiffling, through which they lose height rapidly when over their destination, and this behaviour is seen not only on migration but also as geese return to their winter roosts at night.

Apart from emergency landings, little attention has centred on the question of when is the optimal time of day or night for migrating birds to land. While dusk provides the latest time that a diurnal migrant is likely to land, and dawn for a strictly nocturnal migrant, both groups extend their migrations if they find themselves over water. However, radar studies have shown that many diurnal migrants stop in the middle of the day, and many nocturnal migrants stop well before the night is ended, showing that other factors influence when they settle. In both groups, the time of landing and the flight duration are highly variable, probably depending on the internal state of the bird with respect to energy reserves and tiredness, or the influence of other individuals, as well as on external factors, such as weather and habitat. Birds tend to land as they approach a coastline or other barrier late in the day, or late in the night, and generally land rather than risk flying in bad weather. As freshly arrived nocturnal migrants are found in suitable habitats at dawn, they must be able to recognise suitable habitats at night while flying overhead. Apart from the limitations imposed by overall energy stores, the need for water or sleep may also limit flight duration. What is uncertain is the extent to which an endogenous (internal) rhythm of flight and rest may influence the birds' behaviour.

DIURNAL AND NOCTURNAL FLIGHT

Some birds migrate mainly by day and others mainly by night. Nocturnal species such as owls and nightjars, or optional diurnal–nocturnal species such as shore-birds, might be expected to migrate under cover of darkness. What is surprising is that many normally diurnal species also travel at night. To judge from their eye structure, diurnal birds may have no better vision at night than do humans, but this would still enable them to fly safely through the open skies, and recognise star patterns and landscape features that might help them find their way.

Apart from soaring landbirds, which depend on daytime thermals, it is not immediately obvious why particular species migrate at one time rather than another. Among passerines: crows, finches, pipits, larks, wagtails, tits, hirundines and others migrate primarily by day; while warblers, flycatchers, thrushes, chats and others migrate primarily by night (**Table 4.1**). Among non-passerines: pigeons, raptors, cranes, herons and egrets migrate by day; while cuckoos, shore-birds, rails, and grebes migrate mainly by night. Comparing different families, there is no obvious and consistent connection between migration times and difficulty of journey, habitat, diet or other aspects of ecology. However, among closely related families, some striking differences occur, as in the passerines just mentioned, and also among waders, in which plovers (Charadriidae) migrate more by day than sandpipers (Scolopacidae). Although most species within a family seem

Table 4.1   Diurnal and nocturnal migrants among Holarctic birds.

Mainly diurnal	Divers, pelicans, gannets and cormorants, raptors, storks, herons, ibises, spoonbills, flamingos, cranes, bustards, pratincoles, raptors, grouse, skuas, gulls and terns, alcids, pigeons and doves, bee-eaters, woodpeckers, swallows and martins, jays and crows, chickadees and titmice, creepers, accentors, larks, pipits and wagtails, starlings, sparrows, cardueline finches, fringilline finches, buntings
Mainly nocturnal	Grebes, sea-ducks, bitterns, quail, rails and coots, waders, stone curlews, cuckoos, wrynecks, owls, nightjars, wryneck, hummingbirds, orioles, flycatchers, nuthatches and creepers, wrens, thrashers, chats and thrushes, bluebirds, kinglets (goldcrests), shrikes, vireos, warblers, icterids, orioles, tanagers, New World sparrows
Both diurnal and nocturnal	Albatrosses and petrels (Procellariiformes), swans and geese (Anseridae), dabbling ducks, gulls and terns, shorebirds, swifts

consistent in their migratory behaviour, occasional revealing exceptions occur, with the tendency to nocturnal migration increasing with migration distance (Dorka 1966). For example, most species of *Emberiza* buntings in Europe migrate by day over short distances, but the Ortolan Bunting *E. hortulana* migrates by night over long distances, being the only species that winters in Africa south of the Sahara. Similarly, most pigeons migrate by day over short distances within Europe, but the Turtle Dove *Streptopelia turtur* migrates partly by night over long distances to Africa. In addition, Tree Pipits *Anthus trivialis* are more nocturnal than Meadow Pipits *A. pratensis*, Willow Warblers *Phylloscopus trochilis* than Chiffchaffs *P. collybita*, and Common Redstarts *Phoenicurus phoenicurus* than Black Redstarts *P. ochropus* (**Figure 4.6**). In all these species pairs, the first mentioned species migrates further than the second. Nevertheless, there are still some puzzling exceptions: for example, European Robins *Erithacus rubecula* and Firecrests *Regulus ignacius* are short-distance migrants, but still travel mainly at night.

The division between day and night migrants is most obvious from take-off times, with diurnal migrants leaving in the morning and nocturnal ones in the evening. However, whether day or night, landbirds of both groups must continue flying if they find themselves over water, as must waders and waterfowl over dry land. This explains the appearance of typical day migrants, such as Eurasian Skylarks *Alauda arvensis*, at lighthouses at night, or of typical night migrants, such as chats and warblers, arriving on the coast around mid-day in bright sunshine. Someone watching thrushes and warblers fly ashore from the Gulf of Mexico in spring might understandably class these birds as diurnal migrants, when in fact they set off after sunset one evening, and took more than 12 hours to complete their non-stop flight. Barn Swallows *Hirundo rustica* are viewed as typical daytime

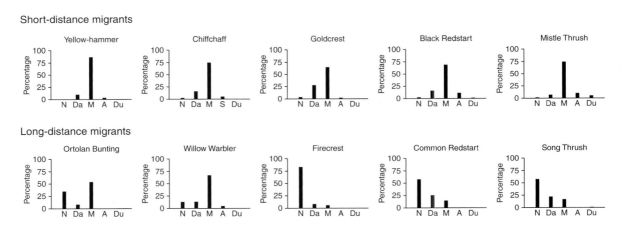

Figure 4.6   Diurnal variation in the numbers of migrants seen (or caught in mist nets) when crossing the Col de Bretolet in the Swiss Alps. N = night, Da = dawn, 1 hour on each side of sunrise; M = morning, dawn to noon; A = afternoon, noon to dusk; Du = dusk, 1 hour on each side of sunset. In each species pair (upper and lower), the longer-distance one undertakes more migration at night than the short-distance one. From Dorka (1966).

migrants, feeding on the wing as they go and spending their nights in large roosts; but they may occasionally turn up at watch points at night and, when crossing the Sahara, they appear commonly to travel in the cool of the night. Moreover, beyond the Arctic Circle, all bird movements in summer inevitably occur in daylight. Thus, whether a species is perceived as a day-migrant or a night-migrant depends partly on the locality, whether the species is seen near the start or end of a flight, and whether there are other reasons for flying at a particular time. It is not impossible that birds could become increasingly nocturnal in migratory behaviour along a gradient from facultative short-distance migrants to obligate long-distance migrants, producing differences between closely related species and also between different populations of the same species. This is another aspect of migration that warrants more research.

For similar reasons, some species appear to pass particular localities at different clock-times in autumn and spring. For example, the Long-tailed Ducks *Clangula clangula*, Common Scoters *Melanitta nigra*, Barnacle Geese *Branta leucopsis* and Brent Geese *Branta bernicla* that cross southern Finland between the Baltic and White Seas appear mainly by night in spring and mainly by day in autumn, as shown by radar (Bergmann 1977). However, this could depend on distance from departure point, the autumn birds having been longer on the wing. In addition, birds migrating later in the season for their species may be more prone to travel at night than conspecifics migrating earlier (Clarke 1912). Clearly, the division between diurnal and nocturnal in most birds cannot be taken as hard and fast. Of 147 species (from 16 bird orders) surveyed at British Bird Observatories, no species fell into an exclusively nocturnal or diurnal category, but 75% of species had been recorded at night at some time, including such usual day migrants as Chaffinch *Fringilla coelebs* and Barn Swallow *Hirundo rustica* (Martin 1990).

Nocturnal migration has been known for a long time. Not only can birds be heard at night as they fly unseen overhead, but they can be watched with a telescope as they fly across the lit face of the moon (Chapter 2). Some species can even be attracted down by high-powered lure tapes, or are killed during the night at lighthouses and other illuminated structures (Chapter 28). Moreover, even the most casual observer can see that, at migration seasons, places devoid of birds in the evening can be full of birds the following morning, as a result of an overnight fall. In recent decades, however, radar has confirmed that huge numbers of birds migrate at night, in most places far more than travel by day. By flying at night, birds lose sleep, but at those times of year they are physiologically adapted to this, and the effects seem minimal. Even captive birds reduce their sleep during migration periods, apparently without ill-effect, although they often have a marked rest period in the late afternoon before they would normally depart (Berthold 1990a). After a really long journey, birds such as waders and waterfowl, settling in safe places, sleep soon after landing, whether day or night (Chapter 3).

The main supposed advantages of nocturnal migration are that: (1) more time is left for feeding during the day, the only time that most birds can feed, so the entire journey can be accomplished more quickly; (2) temperatures are lower at night than in the day which could help to prevent overheating and dehydration in warm regions; (3) humidities are usually higher at night and early morning, which could further reduce dehydration risk; (4) energy demands are lower, because it costs less to fly in cooler denser night air than in warmer daytime air;

(5) wind speeds are generally lower at night, thus reducing the effects of head-winds or crosswinds, and vertical turbulence is less, further reducing the total energy cost of flight; (6) the use of stars for navigating is possible; and (7) the likelihood of predation during flight is much reduced. The main threat to flying migrants is from falcons or eagles during the daytime (plus gulls over water), but a wide range of other raptors take migrants when they are on the ground. Owls do not normally fly high enough to encounter migrants and in any case seldom catch prey on the wing.

The advantages of nocturnal migration can thus be summarised as energy-saving, time-saving and risk-saving. In particular, the reduced turbulence at night must greatly lessen the energy costs of flight, especially in small birds (Kerlinger & Moore 1989). Even diurnal migrants concentrate their flight into the first and last few hours of the daylight period, when turbulence is much less than in the warmer middle part of the day. Among small birds, only swallows, which are adapted to fly throughout the day regardless of turbulence, seem to migrate at similar intensity throughout the daily light period. The various advantages of night-flying are so obvious that it is hard to imagine why some birds migrate primarily by day (apart from those dependent on thermals). Excluding the soaring species, diurnal migrants mostly travel short distances overall, and restrict their flight to brief favourable periods. In flying by day, they also gain a clear view of the terrain below, which may help in recognising familiar landmarks and finding areas of favourable habitat or food supplies.

As found from radar studies, diurnal migration usually starts up to an hour before sunrise (when colour vision returns), builds to a peak in the 2 hours after sunrise and then declines from late morning to early afternoon, but occasionally with a slight resurgence in the late afternoon (Dorka 1966, Bruderer 1999). Hummingbirds are unusual in migrating largely during the mid-day period, leaving time for feeding in the morning and evening, but those that cross the Gulf of Mexico fly at night. These tiny birds are more constrained than most other birds in obtaining enough energy per day, and also respond to the diurnal pattern of nectar production on which they depend.

In general, waterfowl seem more variable in their take-off times than most other birds, starting at any time of day or night, but mostly around sunset. Night-migrating shorebirds mostly set off in the 2–3 hours before sunset, depending to some extent on the stage of the tide, and night-migrating passerines mostly in the hour after sunset (when colour vision goes). Overland migrants studied by radar reach peak densities and altitudes 1–2 hours later, well before midnight. The volume of migration remains high during the first half of the night, then declines from around 2 a.m. to mid-morning, as increasing numbers of birds settle. In addition, large numbers of migrants that normally migrate at night sometimes undertake 'morning flights', which start 1–1.5 hours after sunrise, and continue for up to 2 hours. While some morning flights by nocturnal migrants may represent a continuation of migration, others do not follow the normal migration direction, so have been interpreted as attempts by arriving birds to find better feeding areas or to compensate for overnight drift (Alerstam 1978b, Gauthreaux 1978b, Åkesson & Sandberg 1994, Sandberg et al. 2002).

As found by radar, the behaviour of nocturnal migrants changes during the course of a night, seemingly as their motivation wanes. Directional preferences are much stronger during the early part of a flight, around dusk, than towards its end during the following morning (Bruderer 2001). Night migrants tracked in a coastal area became less likely to strike out over the sea and more likely to veer along the coastline as the night progressed; they also flew at lower altitudes and lower speeds, and increasing proportions flew in the reverse direction. All these changes may have been responses to diminishing body reserves, or shifting compromises between straight flight and risk avoidance, depending on prevailing conditions (Bruderer & Liechti 1998). The frequent finding that flight tracks are less variable over the sea than over land may be explained if only the most highly motivated straight-flying birds continue out to sea, while others settle, fly along the coast or turn back inland (Casement 1966, Richardson 1978).

SOCIAL FACTORS

Many bird species migrate in flocks. Individuals take off together and probably largely remain together during the flight, although they do not necessarily stay together for the next stage of the journey. Frequent calling, especially at night or in overcast conditions, may help to maintain flock cohesion and direction. It may also induce grounded birds in appropriate condition to join the passing stream (**Box 4.2**). In caged Bobolinks *Dolichonyx oryzivorus*, nocturnal call notes, recorded and played back, increased the nocturnal restlessness of birds in migratory condition (Hamilton 1962b). Such social influences may thus help to synchronise migration between individuals. Migrating in flocks may improve navigation through the collective efforts of individuals. It may also improve food-finding and foraging efficiency during stopovers, and give more effective predator detection and evasion.

Another common advantage of migrating in flocks is the energy-saving resulting from specific flight formations, such as the V-formation usual in geese, gulls, and others, or the dense flocks of some passerines and shorebirds. As described earlier, individuals gain lift by flying in the upward-directed air flow near another bird's wings, and thus save energy (Chapter 3). Some usually-solitary raptor species are also seen in flocks on migration. While this may result as much from the sharing of narrow migration corridors as from social attraction, individuals are clearly influenced by the behaviour of others in moving from one thermal to another (Chapter 7). In effect, each bird watches and follows others along the migration route.

As is evident to any bird-watcher, typical flock sizes, densities and flight formations differ greatly between species. Among passerines, species that travel in level flight, such as Starlings *Sturnus vulgaris*, usually migrate in dense flocks, whereas in species which migrate by undulating flight, individuals keep further apart. Most birds seem to migrate in single species flocks, but some often travel in flocks with other species of similar flight speed. Various thrush species often travel together, and so do various tits. Among nocturnal migratory passerines, individuals usually travel singly, or in loose aggregations, with individuals more than 50m apart, maintaining contact by calls. In most species, flocks seldom exceed a few tens of individuals, but in some species flocks comprise hundreds or thousands of individuals (Gatter 2000). Moreover, at times of peak migration, flocks in many species follow in such quick succession that they almost run into

Box 4.2   Social influences and migration waves among Chaffinches

At the Courland Spit on the Baltic coast, Chaffinches *Fringilla coelebs* were collected at different stages of a migratory wave and their carcasses were examined to determine the fat content and the amount of food in the gut (Dolnik & Blyumental 1967). These results were then related to the volume of migration, and a remarkable and consistent pattern emerged.

Throughout the migration season, wave after wave of migrants passed through the area. On the first day of each wave only very fat birds flew: they began their movement around sunrise, without feeding beforehand, and continued for about 4 hours. There was then a pause of 1–3 hours, after which the movement was resumed and continued for 2 hours in the evening. On the second day, the volume of migration reached a peak; again the movement began at sunrise and at first only fat birds flew. As the day progressed, however, the migrating flocks contained increasing numbers of lean birds which, unlike the fat ones, usually had fresh food in their stomachs. This implied that lean birds began their flight later in the day than fat ones and after feeding. During the day the lean birds stopped to feed again: at the same time they attracted down some fat birds, although on this day the latter did not normally feed. By the afternoon, all the lean birds had stopped to feed, and in the evening fat birds were the only ones left flying. On the last day of the wave the migration did not begin at sunrise but only after the birds had fed. Fewer birds participated; they flew with frequent stops and at lower altitude. On this day, almost all the flying birds contained little fat, some feeding occurred throughout the day, and the movement did not reach a minor peak in the evening. Some flocks flew in the reverse direction. Each migratory wave usually lasted three days, but varied from one to seven. After it was over, the pause usually lasted three days, but varied from one to eight, depending partly on the weather. During each migratory wave, birds were estimated to cover up to 500 km, in which time individuals expended 2–3 g of fat. Together with stopping time, this amounted to about 500 km per six days, which was consistent with the migratory progress of Chaffinches recorded from ring recoveries (Chapter 8). No sex or age differences were noted among the birds caught on different days of a wave.

In their explanation of this pattern, Dolnik & Blyumental (1967) attached great importance to the pull that flying birds had on others which at that time were physiologically less ready to migrate. Since the first birds to fly were the very fat ones, it was presumably the presence of many fat birds, which started to fly under a common stimulus (such as favourable weather), that began each wave. Once started, however, the stream of flying birds stimulated others to join, and the larger the stream of flying birds the greater the pull. The expenditure of fat by the fat birds, and the frequent stops by the lean ones, explained the picture observed in succeeding days of the wave. When most of the fat birds had depleted their reserves, their stopping pulled the rest down and the wave was brought to a standstill. Movement was resumed after the birds had built up their reserves again, which presumably depended partly on feeding conditions

(Chapter 27), and inclement weather could further delay departure. This type of pattern may partly account for the greatly varying fat levels found in migrants caught together on migration at the same place.

The frequency with which the migratory waves appeared thus depended primarily on the time needed for spent birds to replace their fat, modified by variations in the weather which also affected the urge to fly. It was the stimulus of movement by the fat birds on others less fat that caused a large part of the population to move together and produce the wave-like pattern. Although birds moved at various stages of fatness, the amount of fat carried by a bird affected the timing and duration of its flights; and in general the fattest birds made the fastest progress.

The birds in this study were making a diurnal overland journey (along the coast) in a region where short-term variations in weather were less extreme than further west in Europe. Possibly in some other regions, the more variable weather has so much influence on migration as to obscure any underlying pattern in the behaviour and physiology of the birds themselves. This work does, however, help to explain why movement does not occur on all days when conditions are apparently ideal, and why it sometimes occurs on days when the weather is less good. The value to the birds in this behaviour probably lies in the advantages of travelling together, including predator avoidance and judgment pooling, as described in the text (see also Chapter 9).

one another. This is true even of large species, such as geese and swans, which sometimes appear as an almost continuous parade of flocks of 10–100 individuals. Despite the differences between species, the causal factors behind specific flock sizes and formations are poorly understood, apart from the obvious points mentioned above that flocks in general provide greater protection from predation, and in some cases improved navigation and reduced individual energy costs.

The benefits of migrating in flocks may explain why many birds do not start migration from their nesting places, which are scattered over a wide area, but first assemble at particular staging sites, often used year after year, and from which birds depart over a period of days or weeks. This behaviour is especially obvious in shorebirds and waterfowl, but occurs in many others, including cranes, gulls, terns and shearwaters. Although related to weather, the departures and arrivals of individuals are partly under social influence. Through social interactions, including vocal activity, individuals of flocking species seem to communicate their readiness to migrate, and thereby synchronise their departures. Some species display pre-flight intention movements, and intense calling, with repeated take-offs and landings, before finally setting off. Among shorebirds, some of these preliminary flights occur in highly structured formations, as do the departures themselves (Piersma et al. 1990b). It is presumably the advantages of migrating in flocks that encourage such synchronisation behaviour.

Some large bird species, such as swans, geese and cranes, travel in pairs or families within the flocks. Other waterfowl, including many species of ducks, form pairs in winter quarters, and the male then accompanies the female back to her breeding area (Chapter 17). Somewhat unexpectedly, however, some migrant passerines also occur in pairs, either on autumn and spring stopover sites or in winter: that is, male–female combinations occur much more often than expected by chance, and the partners behave as mated pairs (for examples see Greenberg & Gradwohl 1980). The Bearded Tit (Parrotbill) *Panurus biarmicus* has provided many instances of birds apparently migrating as pairs (D. Pearson, in Wernham *et al.* 2002); and in some other migratory species, male–female pairs defend territories in winter quarters (for White Wagtail *Motacilla alba*, see Zahavi 1971, for Stonechat *Saxicola torquata*, see Rödl 1994). Whether such liaisons persist into the breeding season or have reproductive consequences is unknown. Despite such examples, we can assume that reproductive pairing before arrival on breeding areas is not common among birds, because in most species that have been studied, the two sexes behave independently when away from their breeding areas, and males arrive, on the average, at least several days before females (Chapter 15). Only in relatively few species, including some waterfowl and cranes, do birds arrive already paired.

These various observations confirm that, in many bird species, individuals do not necessarily behave independently of one another on migration, but can be influenced to varying degrees by other individuals. The role of social influence, and its relationship with body condition, has been studied in particular detail in Chaffinches *Fringilla coelebs* (**Box 4.2**), and in field experiments on other species social influence has been found sufficient to override inherent migratory and directional tendencies (Chapter 9). On the other hand, individuals of some species, such as the Cuckoo *Cuculus canorus*, seem always to migrate alone.

REVERSE MIGRATION

Migratory flights in directions opposite to those expected occur commonly in both spring and autumn, and have already been mentioned above. They have been explained as responses to adverse weather (with birds turning back when conditions ahead are bad), as orientation errors, or as attempts to correct previous orientation errors or wind drift or overshooting (with birds back-tracking to a point en route, for example see Pennycuick *et al.* 1996). When reverse movements occur at coasts, they have also been interpreted as attempts by birds with small fuel reserves to feed inland and increase their reserves before setting out over water (Alerstam 1978b). By moving inland, the argument goes, the birds avoid the high competition, depleted food supplies or predation risk caused by the build-up of birds near the coast. Inland, birds can accumulate body reserves more rapidly and safely. To judge from recoveries of 20 passerine species ringed in southern Sweden in autumn, reverse movements varied between 9 and 65 km, and species with small fat reserves were more likely to perform reverse movements than were species with larger reserves (Åkesson *et al.* 1996). Moreover, among Chaffinches *Fringilla coelebs* and Bramblings *F. montifringilla* in the same area, the peak in reverse movements occurred about 3.5 hours after the early morning departure in the normal direction; and the average weights of reverse migrants and of birds lingering at the coast were lower than those of birds of the same species that proceeded in the normal direction (Lindström & Alerstam 1986).

In spring, if birds encounter snow or other bad weather en route or after arrival in breeding areas, they often retreat for some distance in the direction of their wintering areas, providing that body reserves permit (e.g. Gätke 1895, Williams 1950, Svärdson 1953, Gauthreaux & LeGrand 1975). Many thousands of birds can be involved in such movements. The birds can advance again with the next warm front, and at times of alternating mild and cold periods, back-and-forth shuttle movements sometimes ensue until birds can eventually settle in their nesting areas. The same occurs among birds that breed on high mountains, which having settled on their breeding areas in spring, move down-slope during spells of bad weather, and back again when conditions improve. In northern Europe, the Lapwing *Vanellus vanellus* is one of the earliest species to return each spring, and is well known for reverse migration. The first individuals to arrive on the coast of Finland in spring often turn back on the same day if they encounter cold and snow. One spring, more than 10000 individuals per day were counted moving southwest across one 10-km stretch of coast (Vepsäläinen 1968). In autumn, when birds generally migrate towards warmer climes and better feeding conditions, adverse weather is likely to play a smaller role, with journeys less disrupted. Although reverse movements often occur on the same days as normal movements, they are often at different altitude, with birds selecting wind conditions appropriate to their flight direction.

SUMMARY

The numbers of diurnal migrants seen on the move or on the ground by day constitute a variable proportion of the total participants. This is because most migration occurs at night or too high by day to be seen through binoculars. Radar is therefore the best method for studying its day-to-day volume, and relationship with weather.

As found mainly by radar studies: (1) the intensity of migration is influenced positively by clear skies and following winds, and negatively by mist, rain and opposing winds; (2) to some extent most migrants adjust their flight altitudes to prevailing winds; (3) birds can compensate for weak lateral winds, and remain on course, but they drift increasingly off course with stronger winds or at high flight altitudes, and more by night than by day; (4) birds that migrate by flapping flight usually travel on a broad front, although low-flying migrants may be temporarily deflected by topographical features into apparent streams; and (5) among nocturnal migrants adjustments to coastlines increase during the night, as landbirds become more reluctant to strike out over water. Drifted birds can re-orientate after skies clear and wind becomes more favourable. Raptors and other landbirds that migrate mainly by soaring flight provide exceptions to some of these generalisations, because of their dependence on thermals and other updrafts.

In spring at temperate latitudes in the northern hemisphere, more migration occurs on warm days (when winds blow from the south), and in autumn on cold days (when winds blow from the north), but temperature may be a correlate of winds and other conditions that favour migration, rather than an immediate causal factor. Nevertheless, temperature can have a direct effect on bird energy needs and food supplies, with potential repercussions on migration timing.

Within their normal limits, birds seem to fly at altitudes that offer the best wind conditions for their progress, although some large species – apparently for energetic or physiological reasons – seem restricted to relatively low altitudes (<1 km). With increasing height above ground, travel speeds increase and predation risks decline, but the atmosphere thins and temperature plummets. The height of bird migration varies with geographical location, topographical situation, weather, day or night, type of flight (flapping or gliding), and species. Some species have been recorded at heights of several kilometres above sea level, and at least one at more than 9 km.

Some types of birds migrate primarily by day and others primarily by night. In general, all species in particular bird families (if they migrate at all) behave the same way in this respect, but some closely related families behave differently from one another, and within some passerine genera, most nocturnal flight occurs among species that make the longest journeys. By migrating at night, birds can avoid predators and loss of feeding time by day, and travel in less turbulent (less energy-demanding) conditions.

Some bird species migrate singly and others in flocks, the structure of which varies consistently between different types of birds. The flight times and behaviour of birds are clearly influenced by those of other individuals, leading to waves in the passage of birds, characteristic flock sizes and formations. Migration in a temporarily reversed direction can occur for various reasons, both in autumn and spring.

Garden Warbler *Sylvia borin* fattening on elder berries

Chapter 5
Fuelling the flights

All creatures are fatter in migrating. (Aristotle, writing 2300 years ago.)

Some of those birds are very succulent and sanguine, and so may have their provisions laid up in their very bodies for the journey. (Written by 'a person of learning and piety', 1703.)

At times of migration, many birds put on extra body fat and other reserves for use as fuel during the journey. Typically, they divide their migration into periods of flight, during which reserves are depleted, and stopovers, when reserves can be replenished by feeding. Species that migrate over favourable terrain tend to migrate in short flights, each lasting up to several hours, broken by periods of rest and foraging, when they can replace the relatively small amounts of fuel used on each flight. Given suitable weather, migratory flight can, in theory, occur for part of every day until the journey is completed. However, birds that migrate over large inhospitable areas have to sustain much longer fasts during flights of up to several days, which are preceded by several days of feeding during which much larger reserves are accumulated. The extreme is shown by landbirds that cross large stretches of ocean, requiring up to several days and nights of non-stop flight. Passerines typically take 1–3 weeks to accumulate the fuel reserves necessary for such long journeys, and before departure some may have doubled their normal weights.

The alternating activities of fuelling and flight put different demands on the physiology of migrating birds. During fuelling, the bird should be an efficient eating machine, with a large digestive system, able to process larger than normal quantities of food rapidly for conversion to stored energy. But during flight, it must be an efficient exercise machine, with large muscles, well-functioning heart and circulatory system. It should carry enough energy-rich fuel for the journey, but a minimum of other body structures that merely add unwanted weight. One of the most interesting discoveries of recent years is that some long-distance bird migrants can drastically change not only their physiology, but also their internal body structure over periods of a few days, as they switch from fuelling mode to flight mode and back again. All muscles and body organs are energetically costly to maintain, so an ability to change their relative sizes rapidly, according to the needs of the time, can be regarded as an important adaptation, not only for migration but also for other events in the annual cycle. It enables disparate activities to be performed more efficiently than would be possible on fixed metabolic and body structures (Piersma & Lindström 1997). So-called phenotypic flexibility is apparent in other animals too, but migratory birds show some of the most extreme examples of large-scale rapid changes in body weight and structure.

In addition to fuel deposition, preparation for migration in many birds involves enlargement of the breast muscles, heart and blood vessels, and shrinkage of other organs less important in migratory flight (Piersma *et al.* 1999). It also involves the activation of enzyme systems for the storage and rapid mobilisation of fat, an increase in the erythrocyte (haematocrit) content of the blood to enhance oxygen transport during long flights (Thapliyal *et al.* 1982, Jenni-Eiermann & Jenni 1991), and the modification of different aspects of behaviour, including the diurnal rhythm of activity to permit nocturnal flights in some otherwise diurnal species (Chapter 4). This chapter is concerned with the types and amounts of fuel accumulated by birds, with the process of fuelling and other changes in body composition, and with how these features vary with the types of journeys undertaken. The full cost of migration consists of the energy needed to fuel the flight plus the energy needed to maintain the bird over the whole migration period, including stopovers. This latter cost is increased by any weather-induced delays.

ENERGY NEEDS AND BODY COMPOSITION

Per unit weight, fat provides much more energy than any other storable biochemical fuel available. The use of 1 g of fat will yield around 9.2 kilocalories (or 38 kilojoules) of energy, compared with only about 1.3 kcal (5.3 kJ) from 1 g of protein or 1.0 kcal (4.0 kJ) from 1 g of carbohydrate (**Table 5.1**). Weight for weight, fat therefore contains 7–9 times more energy than alternative fuels, and thus provides the maximum energy storage for the minimum weight gain. Fat is an even more efficient fuel than high-octane vehicle fuel, and also has the advantage for birds that its oxidation yields an equal weight of water, thus contributing to another of the bird's needs during long-distance flights. Not only can fat be

Table 5.1 Energy and water yield of the three main fuel types in birds

	Lipids in adipose tissue	Protein in skeletal muscle or digestive organs	Glycogen
Energy density (kJ g^{-1}) in dry mass	39.6	17.8	17.5
Energy density (kJ g^{-1}) in wet mass	37.6	5.3	3.5–4.4
Water content (%)	5	70	75–80
Metabolic water production (g water g^{-1} dry matter)	1.05	0.39	0.56
Total water production (g water g^{-1} wet tissue)	1.10	0.82	0.89–0.91
Water produced (g water kJ^{-1} expended from wet mass)	0.03	0.16	0.21–0.25

Note. One kilojoule (kJ) is equivalent to 0.239 kilocalaries, and a kilocalorie is popularly called a calorie in human dieting.
Modified from Jenni & Jenni-Eiermann (1998).

stored without water or protein, it can also be digested efficiently with less loss of heat and no effect on body glucose. The main recognised drawback of fat is that its metabolism requires the breakdown of small amounts of protein to provide enzymes for the chemical processes involved (the citric acid cycle). In addition, while most tissues in the body can oxidise fatty acids to release energy, some tissues cannot, and rely instead on carbohydrate or ketone bodies (a reduced form of fatty acids) for energy. Such tissues include the brain and nervous system, red and white blood cells and the kidney medulla.

Fat is laid down as adipose tissue (called fat bodies) in various parts of the bird's body, especially under the skin, and in well-defined deposits within the wishbone (tracheal pit) and around the gut. At least 15 distinct fat depots have been described in passerines (King & Farner 1965). Just before departure, the subcutaneous fat layer in some long-distance passerine migrants can be so extensive that most of the body appears to be clad in a thick layer of pale-yellow fat, only the central part of the breast muscle remaining uncovered. This subcutaneous fat is relatively soft, even at body temperature; and in the hand the bird appears strangely soft and spongy. The precise composition of the fat varies to some extent with the diet of the bird, and with the part of the body where it is stored; but in those species studied (mostly passerines) it consists largely of unsaturated fatty acids, especially oleic, linoleic and palmitic acids, mostly stored in the form of triglycerides (McWilliams et al. 2004, Pierse & McWilliams 2005).

Although mainly passerines, this consistency across species is surprising given the diverse food habits of the species involved, suggesting that birds may be selective in their diets and in the fatty acids they lay down as fuel. Trials on captive birds have shown preferences for unsaturated oversaturated acids, and for short-chain over long-chain ones (McWilliams et al. 2004). Nevertheless, the composition of the diet has some influence on the nature of the fat and other body stores accumulated, particularly the ratio of fat to protein, and hence on the flight range (Prop & Black 1998, Jenni-Eiermann & Jenni 2003).

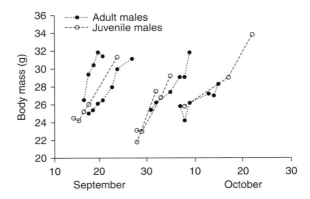

Figure 5.1   Body mass changes, reflecting pre-migratory fuel deposition, in male White-crowned Sparrows *Zonotrichia leucophrys gambelii* caught repeatedly in autumn in California. The maximum rate recorded was from a bird which went from 26.5 g to 30.4 g in 22.5 hours, a 14.7% increase. From Morton (2002).

The pectoral muscles of birds may form more than a third of total body mass, and consist predominantly of fast oxidative glycolyic fibres, which are able to beat the wings continuously at high frequencies for hours or days on end. The flight muscles of many birds are especially adapted for utilising fatty acids for energy. These muscles are highly vascularised, with greater capillary-to-fibre ratio than other muscles (Butler & Woakes 1990), and are well supplied with mitochondria and aerobic enzymes for the oxidation of fatty acids (for review, see Ramenofsky 1990). In fact, the smallest fibres and greatest capillary densities are found in the flight muscles of birds which migrate the longest distances (Lundgren & Kiessling 1988). Fatty acids are transported from the adipose tissue to the flight muscles by the bloodstream, bound either to albumin or to lipoproteins. During long flights, power is produced almost entirely from aerobic metabolism, the respiratory and cardiovascular systems supplying the necessary fuels and oxygen, and also removing the various metabolic end products (such as carbon dioxide and heat).

An idea of the amount of fuel deposited by migratory birds can be gained from their body weights (comparing individuals at different stages of fattening; **Figure 5.1**), and from their 'fat scores' based on the yellowish fat that can be seen through the skin of a live bird when the feathers are blown aside. Most such studies record the fat in the furculum (wishbone), which appears as a V-shaped hollow at the base of the neck on the underside. Such scores are useful but do not bear a linear relationship to the total fat in the bird's body; nor are they comparable between species. Their value is that they can be recorded without harm from live birds, and can be used for comparative purposes within species. More detailed studies of fuel deposition have involved analyses of bird carcasses in order to find the relative proportions of fat, water and lean dry material (the latter comprising mainly body protein, feathers and skeleton) (**Tables 5.2 and 5.3**). Such studies have shown how the body composition of particular species changes during the course of migration, and how these changes vary between species, according to

Table 5.2   Mean body composition of migrants killed at Bahig on the Egyptian coast in autumn. The birds had recently arrived at the coast after crossing the Mediterranean Sea, and within each species, individuals showed a range of weights and body composition

Species	N	Body mass	Fat mass	% fat in body	Ratio fat:lean dry mass	% water in body	Ratio water:lean dry mass
Common Quail *Coturnix coturnix*	33	90.7	14.6	16.1	0.48	50.4	1.50
Barn Swallow *Hirundo rustica*	8	16.1	2.1	13.0	0.45	59.0	2.02
Red-backed Shrike *Lanius collurio*	40	28.0	4.7	16.8	0.51	49.6	1.50
Thrush Nightingale *Luscinia luscinia*	33	24.4	5.2	21.3	0.71	49.2	1.64
Nightingale *Luscinia megarhynchos*	24	22.6	5.3	23.5	0.84	48.7	1.75
Spotted Flycatcher *Muscicapa striata*	39	16.5	3.4	20.6	0.68	49.1	1.62
Eurasian Golden Oriole *Oriolus oriolus*	51	79.0	18.7	23.7	0.90	49.9	1.89
Common Redstart *Phoenicurus phoenicurus*	60	16.9	4.4	26.0	1.00	47.9	1.84
Willow Warbler *Phylloscopus trochilus*	43	9.2	2.1	22.8	0.81	48.9	1.73
Greater Whitethroat *Sylvia communis*	31	15.8	4.4	27.8	1.10	46.8	1.85
Lesser Whitethroat *Sylvia curruca*	48	12.6	3.1	24.6	0.89	48.4	1.74

Figures are mean values. From Moreau & Dolp (1970).

the types of journeys undertaken. Some studies have also examined the composition of different components of the body separately, such as the various muscles and digestive organs (see later).[1]

The bodies of resident bird species, or of migrants outside the migration season, typically contain fuel amounting to 3–5% of their lean body mass. Some

---

[1]*The following terms are commonly used by researchers studying migratory fattening: live weight (of the living bird), fresh weight (of a carcass preserved without evaporative water loss, and therefore equivalent to live weight), dry weight (fresh weight minus the water component of the body), lean (or fat-free) weight (fresh weight minus the entire fat component of the body), lean dry weight (fresh weight minus the lipid and water components), and fat (lipid) weight (of the lipid component). In this field of study, the terms fat and lipid are usually used interchangeably (as here), as are the terms weight and mass, and body stores and body reserves. In the recent literature, the term mass is used much more often than weight.*

Table 5.3 Mean body composition of migrants killed at a Television Tower at Tallahassee, Florida, and presumably about to cross the Gulf of Mexico in autumn

Species	N	Body mass	Fat mass	% fat in body	Ratio fat:lean dry mass	% water in body	Ratio water:lean dry mass
Bobolink *Dolichonyx oryzivorus*							
Early	8	34.1	10.2	29.7	1.23	45.7	1.88
Late	19	41.9	17.5	42.0	2.13	38.7	1.98
Scarlet Tanager *Piranga olivacea*	29	41.3	17.6	42.6	2.12	37.3	1.85
Summer Tanager *Piranga rubra*	44	40.7	16.6	40.7	1.98	38.6	1.89
Red-eyed Vireo *Vireo olivaceus*							
Early	41	22.2	5.4	24.3	0.88	37.9	1.75
Late	59	20.8	5.6	25.9	1.04	47.1	1.80

Calculated from data in Odum (1960) and Odum *et al.* (1961).

migrants apparently travel with reserves no greater than this. However, most regular passerine migrants depart with fuel loads amounting to 10–30% of their lean body mass, and those making especially long flights accumulate fuel loads between 40 and 70% of their lean mass, approaching 100% in a few species (Fry *et al.* 1970, Moreau & Dolp 1970, Alerstam & Lindström 1990). Similarly, some shorebirds attain very large fuel loads, as high as 50–90% of lean body mass, but with a maximum of around 100% in those embarking on the longest non-stop flights. They may then lose up to half their body mass during their flights over the next few days.

The amount of fuel deposited by migratory birds thus varies according to the length of their non-stop flights. This variation is evident even between different populations of the same species. For example, three races of *Quelea quelea*, inhabiting eastern, western and southern Africa respectively, each migrate at the start of the wet season, but for greatly differing distances. The amount of pre-migratory fat and protein accumulated differs significantly between the three populations, and correlates with the lengths of their respective journeys (Ward & Jones 1977). Similarly, among Barn Swallows *Hirundo rustica* studied at different localities in the Mediterranean region, the pre-migratory fuel stores matched the distances to be covered across the sea and the Sahara Desert (Rubolini *et al.* 2002). Among the Northern Wheatears *Oenanthe oenanthe* that pause in spring on Heligoland Island, birds of the nominate race, which breed over much of Europe, mostly stop and feed for up to a day before moving on, whereas birds of the larger Greenland race usually stop for 10–17 days, building up larger reserves for their long oversea journey (Delingat & Dierschke 2000). In contrast to individuals in migratory populations, those in resident populations show no obvious extra fat deposition in either autumn or spring, except perhaps in females in association with egg-laying.

Pre-migratory weight increase involves not only the deposition of fat, but also of body protein. Fuel should therefore be regarded as a combination of the two, but not necessarily in consistent proportions. In the most extreme species, protein contents increase prior to migration by less than two-fold, whereas fat contents may increase by more than 10-fold. Nevertheless, the protein and fat levels usually increase in step with one another during migratory fuelling, so appear to be closely correlated (e.g. Johnson *et al.* 1989). Some species, such as Sandhill Crane *Grus canadensis*, add protein and fat at migration times in the approximate ratio of 1:10 (Krapu *et al.* 1985), whereas other migratory birds lay down approximately equal weights of protein and fat in the ratio of 1:1 (see later). This difference may result simply from differences in the diets of different species, or in their metabolism, but it may also represent an adaptation to the different types of journey they undertake. Populations that are obliged to make long non-stop flights are presumably under greatest pressure to maximise dependence on fat, rather than carbohydrate or protein.

The ratio of fat to protein deposited as body reserve can also depend on the needs of the time, and the same species can alter this ratio between seasons. For example, the 40–50-g mass gain by 200-g European Golden Plovers *Pluvialis apricaria* during autumn stopovers consists almost entirely of fat, but a similar mass gain in spring consists of proteinaceous tissue (mainly muscle). This difference may be because Golden Plovers face different needs at the two seasons. They face energy deficits on autumn migration and in winter when they eat mainly protein-rich earthworms, but in spring they risk protein deficits, when after arrival in arctic breeding areas they eat mainly berries but must soon produce eggs (Piersma & Jukema 2002). Another indication that birds can adjust their body reserves to oncoming needs comes from King Penguins *Aptenodytes patagonicus*, which double their body mass before long fasts on land, but with reserves consisting of about 14% protein before incubation, and 29% protein before feather moult (Cherel 1995).

At least 5–10% of the total energy released during a migratory flight must come from protein, in order to satisfy requirements in the breakdown of fat (Jenni & Jenni-Eiermann 1998). Hence, the amount of protein consumed depends partly on the amount of fat consumed. The protein is derived not only from the flight muscles, but also from other organs, including the gut, body and leg muscles (Piersma & Gill 1998). However, because the fat fraction contributes so much of the total energy needs, it has more influence than protein on the overall flight range.

Although protein breakdown is essential for fat metabolism, there are at least two reasons why birds should minimise the use of protein as fuel. First, as mentioned above, wet protein holds only about one-seventh as much energy as the same weight of adipose tissue (**Table 5.1**; Jenni & Jenni-Eiermann 1999), so per unit weight it is a much less efficient fuel. This is clearly a more important consideration in birds migrating by long non-stop flights than in those migrating by short flights. Second, the metabolism of protein is more complex and inefficient than that of fat and carbohydrate, and also results in toxic by-products. If protein is deposited as a major fuel, therefore, there must presumably be some other reason. Among other things, it supplies important precursors and intermediates for other physiological processes (McWilliams *et al.* 2004), but particularly in spring it contributes to egg formation in some species (see later). The breakdown of wet protein also yields about five times as much water as an equivalent weight of fat.

Carbohydrate is present in the form of glycogen in the liver and muscle tissue, but occurs in such small quantities at migration times that it is of relatively minor importance as an energy source. The highest glycogen values reported from birds amount to no more than about 3.0% of liver mass and about 0.5% of total body mass (Marsh 1983, Blem 1990). As they begin to prepare for migration, some small passerines change from a metabolism based mainly on carbohydrate (glycogen) to a metabolism based mainly on fat (Dolnik & Blyumental 1967, Jenni-Eierman & Jenni 1996). The weight of the liver then diminishes, because of the reduction in glycogen reserves, and shows less diurnal fluctuation in size. It becomes increasingly involved in lipogenesis, lipids being obtained directly from food or synthesised in the liver from carbohydrates. They are then transported as lipoproteins in the bloodstream to the adipose tissue. The lipid is hydrolysed and stored 'dry'. This is in contrast to carbohydrate and protein, the storage of each gram of which requires 3–5 g of extra water (**Table 5.1**; Blem 1990). This is a substantial weight burden, but because the water is released when carbohydrate or protein is metabolised, it may help to counter dehydration on long journeys through hot regions. So while prior to migration, glycogen is a major source of stored energy, as departure approaches, fat becomes by far the main source.

Because very little of the necessary fuel can be stored within the working muscles, their metabolic needs are met chiefly by continuous input of fuel materials via the blood system from the adipose tissues and elsewhere. Of the three fuel types, carbohydrate (glycogen) is the most readily mobilised. Based on evidence from pigeons, the three types of fuel are not used in similar ratio throughout a flight. Carbohydrates are mainly used at the start for the initial take-off and climb, while fatty acids from adipose tissues reach their steady state contribution after 1–2 hours of flight, and amino acids from tissue protein after 4–5 hours (Nachtigall 1990, Jenni-Eiermann & Jenni 2003). In birds with a fat content of more than 25%, only about 5% of the energy is derived from protein, but this proportion increases during a journey as fuel is consumed, and once the fat content falls below 5%, about 20% of flight energy derives from protein. This trend has been observed within species, as well as in comparisons between species (Jenni & Jenni-Eiermann 1998). Eventually, as the fat reserves dwindle, the bird switches even more to protein, and starvation sets in (Schwilch et al. 2002a).

The use of different types of fuel by birds has other consequences. Use of adipose tissue has little adverse effect other than to reduce energy stores and, to a small extent, body protein. But the use of too much protein could result in some functional or structural loss, because protein has no special storage form. For example, a reduction in the digestive organs (consisting mainly of muscle protein) could result in reduced ability to process food rapidly and in a lower refuelling rate during the first days of stopover (Biebach 1998, Piersma 1998). The complete loss of glycogen stores would render sudden fast flights during stopover impossible, making the bird vulnerable to predator attack or unable to chase mobile prey. To cope with emergencies, the migrant would therefore benefit from conserving some glycogen stores, or from reconstituting them soon after landing (Jenni-Eiermann & Jenni 2001, 2003).

The notion that species differ in the proportions of different fuel types used during migration is supported by findings from blood analyses of passerine birds caught directly on migratory flight, as at the high mountain pass of Col de Bretolet in Switzerland (Jenni-Eiermann & Jenni 2003). The presence of uric acid in blood plasma was taken as indicating protein metabolism, and triglycerides as indicating fat metabolism. In the blood plasma of migrants, uric acid levels were lower in five highly frugivorous species than in 13 other species that feed mainly on arthropods. Conversely, the frugivorous species showed very high levels of plasma triglycerides, indicating a heavy dependence on fat metabolism. Similarly, in birds arriving in spring at an Italian island (Ventotene) after crossing the Mediterranean Sea from North Africa (a non-stop flight of at least 500 km), plasma uric acid levels of two frugivorous and nectarivorous species were significantly lower than in seven insectivorous species (Jenni et al. 2000). Similar results were obtained for birds migrating through Israel (Gannes 2001). The implication is that different species use different ratios of fat and protein during migration, and because this is related to their diets, it gives further indication that they accumulate fat and protein as fuel in different ratios.

Unanswered questions, however, concern the extent to which migratory birds, with extremely rapid energy deposition, can influence the composition of their stores irrespective of food composition, and how much the composition of stores is controlled by metabolic or nutritional constraints. The composition of migratory fat varies to some extent with diet, but some fatty acids can apparently be synthesised, or absorbed and stored selectively (Egeler et al. 2003). Another important question is how birds 'know' they have accumulated sufficient body reserve for a migratory flight. They can clearly assess their own body condition, as can mammals, but the mechanism is unknown.

Costs and benefits of body reserves

As the body reserves of a bird increase, so does its flight and migration speed and its potential flight range, but not in direct proportion. This is mainly because the extra fuel itself requires energy to synthesise, maintain and transport, so as body reserves increase, so do the flight costs per unit travel distance (Pennycuick 1989, Lindström & Alerstam 1992, Witter & Cuthill 1993). The costs of migratory fuel are reflected in metabolic rates which rise and fall in line with body weight. For example, in a captive Thrush Nightingale *Luscinia luscinia*, basal metabolic rate (BMR[2]) increased in almost direct proportion to body mass. Over 48 hours, BMR increased by 22.7%, in parallel with an increase in body mass of 24.3% (Lindström et al. 1999). Likewise, some Great Knots *Calidris tenuirostris* on their 5400 km non-stop flight from Australia to China in spring lost about 40% of body weight during the four-day journey. BMRs measured in birds just before and just after their flight were found to have fallen by an average of 42% in association with the reduction in body mass (Battley et al. 2001). In Barnacle Geese *Branta leucopsis* travelling from Svalbard to Scotland, heart rate (reflecting metabolic rate)

---

[2]BMR = *basal metabolic rate, the rate of energy consumption of a bird at rest, in a post-absorptive state.*

was 29% lower just after the flight than before, again reflecting the associated decline in body weight. Resting metabolic rates decline with loss of body mass presumably because this loss involves some metabolically active tissue, such as muscle (Butler *et al.* 1998). Clearly, BMR varies considerably within individuals, in association with the rapid changes in their body mass, and deposition of extra fuel entails a considerable 'holding' cost.

In addition to its maintenance and transport costs, extra fuel makes a small bird less agile and more vulnerable to predation. This is another reason for a bird not to accumulate larger body reserves than necessary. Even the slight weight increase shown by small birds during the course of a normal day greatly reduces their lift-off speed and manoeuvrability, and birds accumulating migratory fat suffer much greater impediment (Witter *et al.* 1994, Metcalfe & Ure 1995, Lee *et al.* 1996, Kullberg *et al.* 1996, 2000, Lind *et al.* 1999). For example, when captive Blackcaps *Sylvia atricapilla* were exposed to simulated predator attacks, individuals carrying a fuel load equivalent to 60% of lean body mass (the maximum recorded in this species) were calculated to suffer reduction of 32% in angle of ascent and 17% in velocity, compared with lean Blackcaps (Kullberg *et al.* 1996). This degree of difference could put fat birds at substantially greater risk (Lind *et al.* 1999, Burns & Ydenberg 2002). These considerations should favour a migration strategy of short flights, frequent fuelling and low fuel loads wherever possible, with the alternative of long flights, infrequent fuelling and heavy fuel loads resorted to only when necessary. Avoidance of predation may also be one reason why many small birds migrate at night, when diurnal birds of prey are usually roosting (Chapter 4).

Water balance and thermoregulation

Birds migrating long distances over seas or deserts cannot drink on their journeys. This could give rise to dehydration, especially in hot conditions, where the birds must pant in order to remain cool. By panting, birds lose heat through the evaporation of water from the damp inside surfaces of the mouth and nostrils. The potential problem of dehydration has raised interest in the water content of migratory birds. Because fat is stored anhydrously, body water as a fraction of body weight declines as fat is deposited. For this reason, water content is best expressed in relation to lean dry mass. The ratio between the two is highly variable, but in healthy passerines is usually in the range 2:1–2.4:1 (that is, on a percentage scale water forms 200–240% of lean dry mass). In a study of Eurasian Reed Warblers *Acrocephalus scirpaceus*, Fogden (1972a) took water levels lower than 2:1 as indicative of dehydration, and found that 11 out of 80 individuals caught at spring migration time in Uganda had water contents below this level. He also calculated that, of 409 birds of 11 species obtained on the Egyptian coast after crossing the Mediterranean Sea in autumn (and analysed by Moreau & Dolp 1970), 78% had water indices lower than 2:1, while as many as 12% had ratios lower than 1.4:1. On the same basis, reduced water levels were also apparent among migrants that had crossed the Gulf of Mexico in spring (Odum 1960). While these findings would seem to indicate severe dehydration after long flights, Fogden (1972a) suggested an alternative explanation, namely that in some conditions migrants lower their water levels before departure in order to reduce total body weight, enabling them to cover greater distance on their body fat. This view was later reiterated by Johnson *et al.* (1989), on the basis of the pre-migratory changes in body composition found among Lesser Golden Plovers *Pluvialis fulva* on the Hawaiian Islands, preparing for their flight to Alaska. The question whether low water levels reflect dehydration or adaptation to long flights remains open, but birds with such low water levels would surely have little leeway to counter any overheating by panting (Chapter 6).

The body temperatures of birds measured during active flight are generally greater than 41°C, up to 4°C higher than normal (Gessaman 1990). This may improve muscle efficiency and increase maximum power output (Butler & Woakes 1990). But during normal sustained flight, birds must dissipate more than eight times as much heat as during rest in order not to become overheated. Overheating is a potential problem for migrants during continuous flights at high ambient temperatures, as occur by day in deserts (Chapter 6). Flying birds lose some heat through convection and radiation, especially through their underwings and unfeathered legs and feet, but the amount of heat that can be lost in this way is limited by simple physical processes. So at higher ambient temperatures, evaporative mechanisms (mainly through panting) play an increasing role. The resulting risk of dehydration is offset either by metabolic water production (from fat and protein catabolism), by ascent to altitudes where the air is cool enough to keep evaporative heat loss at the required level, or by flight at night when temperatures are lower than during the day (up to about 10°C). The bird might also switch to other energy sources: per unit of energy released, protein yields about five times as much water as fat, and glycogen at least seven times as much (**Table 5.1**).

MIGRATION MODE

For many birds it may be advantageous to migrate as rapidly as possible, and thereby minimise the time spent on the journey (the 'time minimisation model' of Alerstam & Lindström 1990). This gives migrants the longest possible time on their breeding, wintering or moulting sites, but requires large fuel stores to permit long, non-stop flights. Other birds may have food available throughout the migration route, so that they can stop and feed almost anywhere. Because heavy fat loads mean greater transport costs, as mentioned above, one way to save energy is to keep fat loads small and fly only short distances at a time, refuelling as necessary (the 'energy minimisation' model). Moreover, because any extra weight also reduces flight performance (notably climb rate and agility), minimising fuel loads can also reduce predation risks (the 'predation minimisation' model). The second and third options may thus be combined as the 'load-minimising' strategy. The particular migration mode adopted by any population might be a compromise between any of these different options, depending partly on the type of terrain over which populations travel, the distribution of potential feeding places, and the risks of predation. Moreover, any 'ideal strategy' that the bird might have is likely often to be compromised by external conditions, such as adverse weather and poor food supplies.

Where birds follow a 'stepping stone' migration strategy, this may occur in different forms (called hop, skip and jump by Piersma 1987), depending mainly on the distances between successive feeding sites. For a load minimiser, the best strategy would be to use a large proportion of potential refuelling sites along a migratory route (hopping). But a time minimiser would do better to put on a large fuel load at a high-quality stopover site, in order to bypass a poor-quality stopover site (skipping). It could then migrate more quickly (Gudmundsson *et al.* 1991). For example, Bewick Swans *Cygnus columbianus* stop at the White Sea in spring but mostly bypass this site in autumn when their fuel reserves enable them to travel to a more distant site along the migration route (Beekman *et al.* 2002). A bird that has to cross a large stretch of inhospitable terrain can only adopt a 'jump' strategy, flying a long distance without feeding. Studies of the duration of stopovers, rates of weight gain and departure weights of migrants making overland journeys give some idea of the strategy pursued, bearing in mind that other factors also influence the behaviour of migrants.

Seabirds migrating entirely over the sea would seem to have plenty of opportunity to pick up food en route. But this is not always the case. Many species breeding at high latitudes migrate over the equator, and tropical seas are notoriously poor in food. In any case, foods such as fish tend to be concentrated in particular localities, which may be few and far between. Evidence is accumulating that, like some landbirds, some seabirds refuel at traditional staging areas before continuing migration. For example, after breeding in western Europe, Black Terns *Chlidonias niger* assemble at one major feeding area, the IJsselmeer on the Dutch coast. Here they increase in body mass by 25–30% within 2–3 weeks, which would then enable a non-stop flight of more than 3600 km to West Africa. The birds ascend in the evening to high altitudes (>500 m) and start migrating at night. Although Black Terns are seen at localities en route, no important stopover site is known between the IJsselmeer and West Africa (van der Winden 2002). In Namibia, a similar increase in body mass of these terns was noticed just before spring migration. Other terns may also make long flights between regular rich feeding areas, rather than hunting as they travel. Arctic Terns *Sterna paradisaea* staging in Norway and Britain in autumn would need to accumulate fuel equivalent to 30–40% of body weight before conducting a direct flight of 3000–5000 km to West Africa (Alerstam 1985). These and other seabirds (mentioned later) make their journeys so quickly that they can spend little (if any) time feeding en route.

Alternative strategies

Most studies of migratory fuelling in birds have been concerned with fuel deposition before departure, and its replenishment at staging areas. However, some birds appear to depart at normal weight without prior fuel deposition, lose weight during the flight, and make it up after arrival in a staging area – an extreme load-minimising strategy. Without special reserves, small birds could not survive much more than a day without food, but they could travel in this way on short flights lasting a few hours, followed by feeding. However, large birds, such as swans, geese and eagles, can normally survive for many days without food, so in theory they could travel for longer periods without prior fuel deposition. This is especially so for soaring birds, which expend little more energy on migration than on normal daily life. Most soaring species that have been studied accumulate relatively small amounts of fuel for overland flight (for raptors see Chapter 7), and no pre-migratory fattening is apparent in White Storks *Ciconia ciconia* at either season, despite migrations of up to 10 000 km (Berthold *et al.* 2001b). These birds feed as they go, mainly in the mornings and evenings, and travel in the middle part of the day, when the thermals that permit soaring–gliding flight are best developed. But they face large stretches of the journey, notably through deserts, when they could not expect to feed for several days. In contrast, cranes also travel partly by soaring–gliding flight but accumulate substantial body reserves (up to one-third of body weight), during the several weeks they spend at favoured stopover sites (for Sandhill Crane *Grus canadensis* see Krapu *et al.* 1985).

An absence of pre-migratory fattening might also be expected in those overland species that have to leave their breeding areas as soon as parental commitments permit, in order to get out before environmental conditions deteriorate. It is not hard to imagine, therefore, a continuum of variation in which some birds migrate entirely on prior reserves, others entirely on a continually replenished deficit, and yet others on both. But it is the species that accumulate reserves before migration that have received most attention from researchers.

Different strategies for similar journeys

Closely related species migrating between the same areas sometimes show different patterns of fuelling and flight, depending partly on their food supplies. For example, Sedge Warblers *Acrocephalus schoenobaenus* travelling from Britain to Africa in autumn attain higher rates of fat deposition close to their breeding areas than further south in the Mediterranean region. Their main prey (plum reed aphids) reach peak abundance in Britain and northern France at migration time, but they have already passed their peak in southern Europe by the time the warblers arrive. This may be why Sedge Warblers typically deposit very large fat loads in southern England and northern France, from which areas many may then make a single long flight to their wintering areas south of the Sahara, a journey exceeding 3000 km (Gladwin 1963, Bibby & Green 1981). In contrast, Reed Warblers *Acrocephalus scirpaceus* eat a wider variety of insects, and can fatten at a wider range of sites until later in the season. Ring recoveries show that they usually migrate through Europe in shorter stages, stopping at various localities on their southward journeys, and accumulating large body reserves only in North Africa, just before the desert crossing. Consequently, while both these warblers make the same journey, the location of refuelling stopovers, and patterns of fattening and flight lengths differ between them. For some bird species, including the Sedge Warbler, much otherwise suitable habitat may contain insufficient food at migration time to allow efficient fuelling, leading to a high fuelling/long flight strategy even over apparently favourable terrain.

Analyses of the weights of six species of trans-Saharan migrants caught at 34 trapping stations located in widely scattered parts of Europe and North Africa revealed four types of fattening patterns on the southward autumn journey (Schaub & Jenni 2000):

1. Birds accumulate large fuel stores well before they reach the northern edge of the desert, and then fly to sub-Saharan Africa without refuelling. This pattern is shown by western populations of Sedge Warblers *Acrocephalus schoenobaenus* (as just mentioned), by western populations of Pied Flycatchers *Ficedula hypoleuca*, and possibly also by some individuals of other species.

2. Birds accumulate more fuel at each stopover than is needed to fly to the next, leading to a progressive increase in body mass southwards through Europe toward the desert. This strategy is shown by Garden Warblers *Sylvia borin* and more eastern populations of Pied Flycatchers *Ficedula hypoleuca*.
3. Birds migrate in short stages, and accumulate only enough fuel at each stopover site to fly to the next, with especially large amounts just before the desert crossing. This strategy depends on finding good feeding sites in the southern Mediterranean region, and is adopted by Eurasian Reed Warblers *Acrocephalus scirpaceus* (as mentioned above) and possibly also by Greater Whitethroats *Sylvia communis*.
4. The same strategy as (3), except that birds put on only moderate reserves before the desert crossing, relying instead on finding food at desert oases or on catching migrant insects. Various hirundines and Spotted Flycatchers *Muscicapa striata* seem to adopt this strategy, and the same may be true for shrikes which can also kill and eat their fellow migrants. However, Barn Swallows *Hirunda rustica* that migrate from Italy across the Mediterranean and Sahara accumulate up to 40% fat before the journey, which is much more than those that take the shorter sea crossing at Gibraltar (Rubolini *et al.* 2002).

Similar patterns occur in spring (Curry-Lindahl 1963, Ward 1963, Fry *et al.* 1972, Wood 1992). Thus, while most northward-bound passerine populations fatten well to the south of the Sahara, and undertake a flight much longer than the desert crossing itself, others proceed in stages to the southern edge of the desert and fatten there. In East Africa, Sedge Warblers *Acrocephalus schoenobaenus* and Great Reed Warblers *Acrocephalus arundinaceus* fatten well south of the Sahara (in Kenya–Uganda) and fly 2500 km direct to the Middle East; but most other passerine species probably fatten in Somalia much nearer to the Sahara (Pearson 1990). In West Africa, Garden Warblers are among those that fatten far south of the desert, in the Guinea Zone (Ottosson *et al.* 2005), while Sand Martins *Riparia riparia* and others fatten around Lake Chad in the Sahel Zone (Fry *et al.* 1972). The migratory stage that involves crossing the Sahara is relatively rapid (less than one week), and later stages through Europe are slower (three or more weeks).

In North America, Helms & Smythe (1969) recognised similar broad categories with respect to fuel reserves and migration itineraries:

1. Intra-continental migrants vary from (a) those that depart on autumn migration with scant fuel reserves, move relatively slowly, and may (e.g. Dark-eyed Junco *Junco hyemalis*, Savannah Sparrow *Passercualus sandwichensis*) or may not (e.g. Tree Sparrow *Spizella arborea*) add reserves as migration progresses, to (b) those that accrue moderate reserves immediately before departure and migrate fairly rapidly (e.g. White-throated Sparrow *Zonotrichia albicollis* in some areas).
2. Inter-continental migrants that behave like intra-continental migrants on the first part of their journey through favourable habitat, but accrue much larger reserves in the southern States as they approach a sea crossing (e.g. Scarlet Tanager *Piranga olivacea* and Bobolink *Dolichonyx oryzivorus*, **Table 5.3**).

In both types, breeding populations taking different routes may have different strategies, presumably developed according to conditions en route, including the spatial and temporal distributions of their habitats and food supplies. Again, the same populations may also adopt different strategies at the two seasons, and adults may differ from juveniles, but it is not known whether individuals can switch easily from one strategy to another as circumstances dictate.

At particular sites, birds sometimes stayed longer, fattened more rapidly and to higher levels in autumn than in spring (Dolnik & Blyumental 1967, Morris *et al.* 1994), and in other species in spring than in autumn (King *et al.* 1963, 1965, Butler *et al.* 1987). For example, some waders migrating between eastern Canada and South America accumulate more body fat in autumn, when they make a single long flight over the sea to northern South America, than in spring when they migrate by a series of shorter coastal flights, broken by feeding stops (McNeil & Cadieux 1972). Similarly, birds crossing the Sahara and fattening close to its edges tend to stay there longer and accumulate more body fuel in autumn, before the cross-ing, than in spring, just after it (**Table 5.4**). The fattening strategy at each season is evidently adapted to the route taken at each season and the numbers and spa-cing of potential fuelling sites.

Table 5.4 Mean estimated stopover periods in days of various passerines at an oasis in the Sinai desert. The average stopover period in most species was longer in autumn before crossing the Sahara than in spring after crossing it

	Autumn	Spring
Species present only in autumn		
Willow Warbler *Phylloscopus trochilus*	10.8 ± 3.9	–
Orphean Warbler *Sylvia hortensis*	5.6 ± 0.6	–
Red-backed Shrike *Lanius collurio*	3.7 ± 0.5	–
Greater Whitethroat *Sylvia communis*	6.6 ± 2.3	–
Yellow Wagtail *Motacilla flava*	4.5 ± 1.1	–
Bluethroat *Luscinia svecica*	13.6 ± 7.6	–
Eurasian Reed Warbler *Acrocephalus scirpaceus*	3.0 ± 0.4	–
Whinchat *Saxicola rubetra*	2.3 ± 0.5	–
European Pied Flycatcher *Ficedula hypoleuca*	3.1 ± 1.1	–
Species more numerous in autumn than spring		
Lesser Whitethroat *Sylvia curruca*[a]	5.0 ± 0.5	1.9 ± 0.6
Eurasian Chiff-chaff *Phylloscopus collybita*	8.1 ± 3.4	2.4 ± 0.2
Spotted Flycatcher *Muscicapa striata*	3.5 ± 0.7	3.2 ± 0.7
Tree Pipit *Anthus trivialis*[a]	3.6 ± 0.2	2.2 ± 0.3
Species more numerous in spring than autumn		
Blackcap *Sylvia atricapilla*	6.7 ± 1.8	3.8 ± 0.8
Redstart *Phoenicurus phoenicurus*	2.8 ± 0.6	3.3 ± 0.3
Garden Warbler *Sylvia borin*	1.7 ± 0.7	1.5 ± 0.2

[a]Difference between autumn and spring, $P<0.001$. From Lavée *et al.* (1991).

This site was so small (a small garden (0.02 km^2) at St Catherine's Monastery surrounded by barren desert) that individual birds could be easily caught on arrival, and the length of stay of a sizeable proportion of birds could be estimated reliably to within nar-row limits. It was situated 300 km from the desert's northern edge and about 2000 km north of the southern edge of the Sahara. In general, the longest stopovers were shown by birds on long migrations for which the stopover site was far from the point of initiation, while the shortest were in species for which the stopover site was close to the point of initiation in autumn.

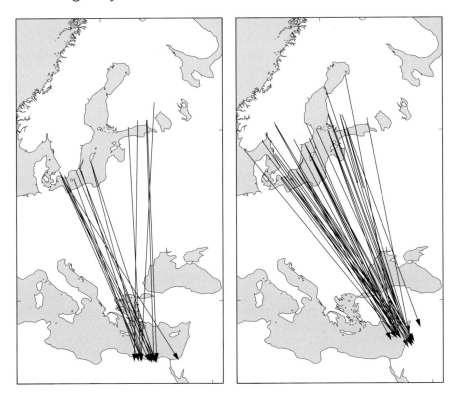

Figure 5.2   Recoveries of birds ringed in northern Europe that were found in the eastern Mediterranean region. Lines connect the ringing site with the recovery site for each individual. Left: Thrush Nightingale *Luscinia luscinia*. Right: Blackcap *Sylvia atricapilla*. These are two out of seven passerine species which migrate from northern Europe round the eastern Mediterranean in autumn for which there are sufficient ring recoveries for analysis. From Fransson *et al.* (2005).

Important stopover areas

The above inferences on migration strategies were derived from fattening patterns, but are supported by ring recoveries (Wernham *et al.* 2002). Some summer migrant species ringed in Britain, such as Eurasian Reed Warbler *Acrocephalus scirpaceus*, Greater Whitethroat *Sylvia communis* and Sand Martin *Riparia riparia*, yield recoveries widely scattered through southern France and Iberia, implying that they stop almost anywhere on their journey towards Africa. In other species, however, the recoveries are heavily concentrated in particular regions, with few or none elsewhere: Pied Flycatchers *Ficedula hypoleuca* and Tree Pipits *Anthus trivialis* in western Iberia, Lesser Whitethroats *Sylvia curruca* in the Alps region of northern Italy, and Wood Warblers *Phylloscopus sibilatrix* in peninsular Italy. These latter patterns suggest that these species migrate in a single long flight from their breeding areas to specific regions in the southern half of Europe, and from there they may move without other long pauses to sub-Saharan Africa. Their whole journey

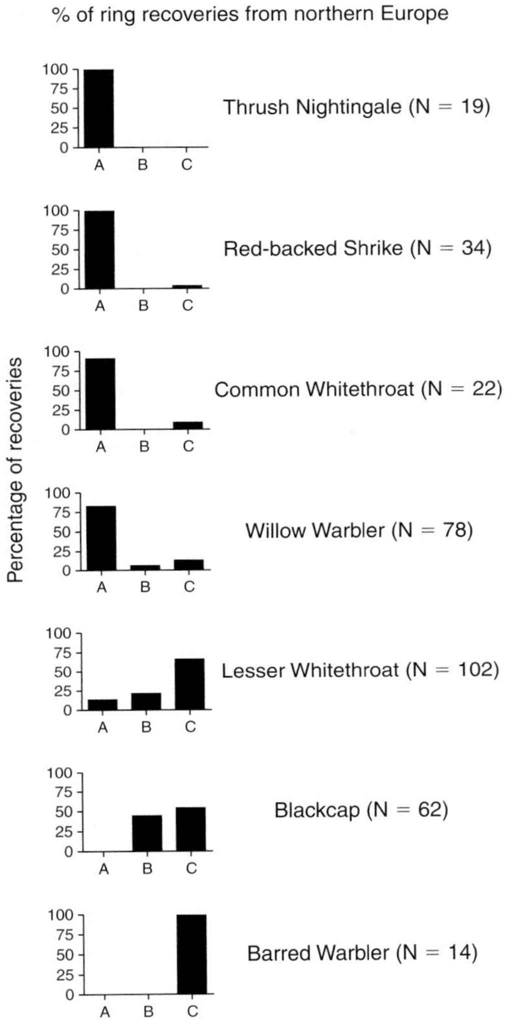

% of ring recoveries from northern Europe

Figure 5.3 Percentages of recoveries of birds ringed in northern European breeding areas that were obtained in three different parts of the eastern Mediterranean region where they were presumed to refuel (see **Figure 5.2**). A. Northern Libya–Egypt, B. Cyprus, C. Israel–Jordan–Lebanon and Turkey. From Fransson *et al.* (2005).

could thus be completed in two long stages, separated by a single long stopover for replenishment of reserves. This view is generally supported by other data from Britain, such as their body weights on departure, or their appearance or lack of appearance at coastal bird observatories. Some of the same species have different stopping areas on the return journey, and individuals are seen (and recovered) in much greater numbers in North Africa in spring than in autumn (Chapter 6).

Similar patterns have been described from trans-Saharan migrants breeding further north in Europe (Denmark, Norway, Sweden and Finland), as illustrated in **Figures 5.2 and 5.3** (Fransson *et al.* 2005). From these breeding areas, Thrush

Nightingales *Luscinia luscinia*, Common Whitethroats *Sylvia communis* and Red-backed Shrikes *Lanius collurio* are recovered almost entirely on the Egyptian coast during autumn migration, whereas Barred Warblers *S. nisoria* and Blackcaps *S. atricapilla* are recovered mainly in a region embracing Israel, Lebanon, and parts of Jordan, Syria and Turkey. Willow Warblers *Phylloscopus trochilus* and Lesser Whitethroats *S. curruca* are recovered in both regions. Trapping activities by local people who supply the recoveries take place throughout all these regions, and the different recovery patterns imply that different species have different major refuelling areas en route to winter quarters. Localised fuelling has long been known in species with restricted habitat areas, such as some waterfowl and shorebirds, but is more surprising in small birds that seem to have lots of suitable habitat distributed along their migration routes. Perhaps again, at this time of year not all this habitat may provide food abundant enough for fuel deposition. However, these staging areas are large, covering thousands of square kilometres, and birds from different parts of the breeding range may use different regions for stopover, as shown, for example, by Lesser Whitethroats from Britain (stopping in northern Italy) and from northern Europe (stopping in the Middle East).

MECHANISMS OF FUEL DEPOSITION

Three types of limitation could influence the fuelling rates of birds. The most obvious is the rate of food intake – the amount that can be consumed per 24 hours. This rate is affected mainly by external conditions: by daylength, food availability through the day, the feeding efficiency of the bird itself, and the constraints on feeding rate imposed by competitors and predators (Chapter 27). The bird might increase its intake rate in various ways, such as feeding more rapidly or for longer than usual each day, or by selecting from potential foods the most calorific and easily digestible items. Some waterfowl and shorebirds can feed both by day and by night, and can thus achieve higher rates of food throughput than other birds, and correspondingly higher rates of fuel deposition (Zwarts *et al.* 1990a).

The second type of limitation is imposed by digestive efficiency which, regardless of the rate of intake, limits the amount of food that can be processed per unit time (Diamond *et al.* 1986, Klaassen *et al.* 1997). Again, birds might maximise throughput in various ways, such as ensuring that food is always present in the crop, ready for passage down the gut as soon as space becomes available, or increasing the throughput rate so as to digest food less thoroughly but in greater quantity than usual, or by modifying gut structure so as to digest food more rapidly.

The third potential constraint comes from crop capacity. Many birds, especially herbivorous species, normally fill the crop before going to roost, and digest that food during the night. If crop capacity is small relative to night length, then crop capacity could be said to limit daily intake. To my knowledge, the extent to which crop capacity can be increased at migration seasons has not been studied.

The effects of various mechanisms in raising energy intake may be enhanced by reduction in other energy-demanding activities, such as moving around. However, all these mechanisms have costs which affect the bird adversely in other ways, so they normally occur only at migration seasons or at other times when demands are high. They are apparently under endogenous control, and normally appear only at appropriate times of year, as revealed in captive birds (Berthold 1976).

The relative contributions of several of these mechanisms to migratory fattening have been studied in certain species (Bairlein 1985c, Bairlein & Simons 1995). For example, the Garden Warbler *Sylvia borin* at migration time eats up to 40% more per day (in terms of convertible energy), and switches from a mainly insect to a mainly fruit diet (often Elder *Sambucus niger* at northern latitudes and figs *Ficus* at Mediterranean latitudes). Increase in digestive and assimilation efficiency accounts for another 20% increase in metabolised energy during pre-migratory fattening (Bairlein 1985c, 1991a). This increased digestive efficiency involves an increase in gut weight, and in the rate of synthesis of fatty acids in the liver. In contrast, no such change in utilisation efficiency was found in seed-eaters, such as White-crowned Sparrows *Zonotrichia leucophrys* and Bobolinks *Dolichonyx oryzivorus*, the extra calories for fuel deposition being obtained entirely by longer feeding periods (King 1961, 1972, Gifford & Odum 1965). The various mechanisms to increase daily energy assimilation are discussed in more detail below.

Increased feeding rates and feeding times

Hyperphagia (eating more than needed to maintain a stable body weight) is evident especially in captive passerine migrants, which at appropriate seasons suddenly begin to eat around 25–30% more per day than usual (range 10–50%). This promotes mean weight gains of up to 10% per day (**Figure 5.4**). On a 20–30-g bird, with a fattening rate of 1.0–1.5 g per day, fattening may take 4–10 days. Higher rates of weight gain have been recorded at 20% per day in White-crowned Sparrows *Zonotrichia leucophrys* (King 1972); at 25–30% in Bobolinks *Dolichonyx oryzivorus* (Gifford & Odum 1965), and at 40% in Garden Warblers *Sylvia borin* (Bairlein 1990). In studies of captive shorebirds, which could feed for 23 of the 24 hours per day under artificial light, maximum daily energy intakes reached 300–500% above existence levels (Lindström & Kvist 1995, Kvist & Lindström 2003).

Before spring migration, White-crowned Sparrows *Zonotrichia leucophrys* showed two peaks in foraging, one in the morning and the other before sunset, as observed in both wild and captive birds (Morton 1967, Ramenofsky *et al.* 2003). Toward migration time this pattern changed, as birds began to feed throughout the daylight hours, gaining in body mass and fat content. Once migration activity started, some birds ceased feeding in the late afternoon, enabling them to empty their guts, reducing excess weight before the flight, which would normally begin after dark (Morton 1967, Brensing 1989). Shorebirds in West Africa increased their foraging periods from 6–10 hours per day in winter to 12 hours per day during spring fuelling. This change involved more feeding at night, but the total time spent foraging was still limited by the tidal regime (Zwarts *et al.* 1990a). Some birds may achieve hyperphagia at the expense of vigilance, spending less time scanning for predators, as noted in Ruddy Turnstones *Arenaria interpres* (Metcalfe & Furness 1984).

The most obvious way in which a diurnal bird could conserve feeding time is to migrate at night. While this may not increase feeding time over what is usually available, it at least prevents potential feeding time being reduced by flight time. Given this advantage, together with reduced predation risk, it is hard to

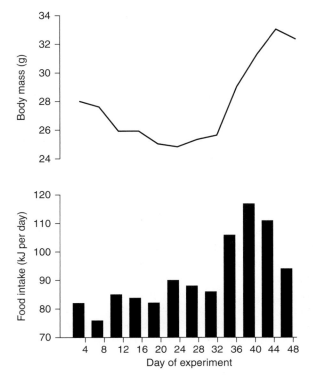

Figure 5.4 Changes in food intake and body mass in a captive White-crowned Sparrow *Zonotrichia leucophrys* during the spring migration period. After King (1961).

understand why any birds migrate by day, apart from overland soaring species (Chapter 4).

Changes in gut structure and digestive capacity

More food can be processed per unit time by increasing the size of the digestive tract, or adjusting its structure and activity to better deal with particular types of food. Changes in gut structure and action have been documented in several migratory species from passerines to geese, and usually take several days to enact (McLandress & Raveling 1983, Jordano 1987, Afik & Karasov 1995, Bairlein 1996a, McWilliams & Karasov 2005). Studies on captive Blackcaps *Sylvia atricapilla* revealed that the rate of energy assimilation under ad lib food was proportional to the size of the intestinal tract and liver (Karasov & Pinshow 2000). However, it is uncertain whether such gut changes anticipate increased food intake or occur in response to it. The latter seems most likely, but in any case such massive gut changes are presumably not without costs (Karasov 1996, Hume & Biebach 1996, Piersma *et al.* 1999). For one thing, they increase overall body mass, affecting agility and vulnerability to predation.

Change of diet

Another way in which birds can increase their intake is to concentrate on easily digestible, energy-rich food items. In both the Old and New Worlds, many passerines eat fruit at the time of migratory fattening. These include not only regular fruit-eaters, such as *Sylvia* warblers, but also others not normally considered as fruit-eaters, such as the Pied Flycatcher *Ficedula hypoleuca* and Yellow Wagtail *Motacilla flava*. This change of diet does not result simply because fruit is more available than insects in late summer, because captive birds, given a choice of insects or fruit, also select fruit then. In the Mediterranean region, at the time of autumn migration, frugivorous warblers increased in body mass about twice as rapidly as purely insectivorous ones, even though the fruit-eaters had to eat more than their own body mass in fruit per day (Ferns 1975, Thomas 1979, Izhaki & Safriel 1989). Likewise, pre-migratory weight gain in European Robins *Erithacus rubecula* showed a close relationship to fruit consumption, and was greatest from fruits relatively rich in lipids (Herrera 1981). Nonetheless, most birds take some insects at the same time as fruit, presumably to supply their protein needs, and captive Garden Warblers *Sylvia borin*, Red-eyed Vireos *Vireo olivaceus*, Hermit Thrushes *Catharus guttatus* and others gained weight more swiftly from a mixed diet than from fruit or insects alone (Bairlein 1998, Parrish 2000, Long & Stouffer 2003). For obvious reasons, most fruit-eating occurs in autumn, but many passerines also select fruit before the spring migration if it happens to be available. Some of the passerine species that spend the northern winter south of the Sahara eat many fruits in spring, especially those of *Salvadora persica*, on which they can also fatten more rapidly than on insects (Fry *et al.* 1970, Moreau 1972, Stoate & Moreby 1995).

Fruits are generally low in protein (most types 1–7% of dry mass), with a few exceptions such as Olive *Oleo europaea* (8%) and Elder *Sambucus nigra* (12–18%), both of which are favoured by birds (Jenni-Eiermann & Jenni 2003). However, most fruits contain high proportions of carbohydrate and unsaturated fatty acids that can be easily metabolised to produce body fat. Unsaturated fatty acids in fruit proved ideal for migratory fat formation in captive birds, while other individuals fed on diets containing only saturated fatty acids showed lower rates of fattening (Bairlein & Simons 1995).

The lipid content of fleshy fruits varies greatly between plant species. In temperate areas, the percentage of digestible lipid in most fruits is lower than in insects. In Mediterranean regions, where many warblers fatten in autumn before crossing the Sahara, fruits tend to be richer in lipids than those in the temperate zone; and in tropical regions, where fruits tend to be even richer in lipids, some bird species seem able to subsist on fruit alone for almost the whole year, some even raising their young on fruit (Snow & Snow 1988). Captive birds provided simultaneously with two diets identical in energy content but differing in lipid content, showed clear preferences for lipid-rich foods (Borowitz 1988, Bairlein 1990), confirming observations on wild birds (Borowitz 1988, Snow & Snow 1988). In contrast, American Robins *Turdus migratorius* preferred sugar-rich to lipid-rich fruits, and had higher absorption efficiency for sugars, although assimilation of lipids increased from summer to autumn (Lepczyk *et al.* 2000).

Another advantage of feeding on fruit is that it can be obtained more easily than insects; being concentrated, predictable and conspicuous, it requires the minimum expenditure of time and energy to obtain. Its low fibre content also makes fruit easy and quick to digest, and its high water content may reduce the need to drink. The main problem with some fruits is that they contain tannins or other toxic compounds which birds have to deal with in various ways.

Many passerines that remain insectivorous during migration prefer insects that are rich in fats (such as caterpillars) or carbohydrates (such as aphids), and can at times show fattening rates as high as those of fruit-eaters (for high rates of fuelling in Sedge Warblers *Acrocephalus schoenobaenus* eating aphids see Bibby & Green 1981). Other warblers on spring migration prefer nectar over insects, presumably for the same reason. Likewise, the Yellow-faced Honeyeater *Lichenostomus chrysops* in eastern Australia was found to increase the ratio of nectar to insects in its diet during both autumn and spring migrations (Munro 2003).

Most diet studies relevant to migration refer to small insectivorous–frugivorous passerines. It is unknown to what extent other birds change their diets at migration times, but fruit-eating is known then in such unlikely candidates as shorebirds, gulls and cranes (Glutz von Blotzheim *et al.* 1975, 1977). Also, Pink-footed Geese *Anser brachyrhychus* turned from grass to newly sown grain at migration time, more than doubling their daily energy intake (Madsen 1985), while Canada Geese *Branta canadensis* changed from a diet of corn to a mixture of corn and meadow grass in spring, the grain providing carbohydrate (which can be converted to fat) and the grass protein (McLandress & Raveling 1981). Many species of ducks turn from plant leaves and seeds to invertebrates at the time of spring migration, presumably to acquire more protein (Arzel *et al.* 2006). Raptors that eat migrant birds presumably have diets richer in lipids at migration times, when the fat contents of prey are higher than at other times.

Digestive limitations

The rate at which birds can process and digest food could also constrain rates of fuel deposition. Hence, fuelling rates could be described as 'food-limited' or 'feeding-time limited' at one level, or 'metabolically limited' at a higher level (for a likely example of a metabolically limited fuelling rate see Dierschke *et al.* 2003). Early studies undertaken mainly outside the migration seasons suggested an upper limit to the daily metabolisable energy intake (DME_{max}) of birds at around 4–5 × BMR, imposed by digestive physiology (and theoretically equivalent to a maximum rate of 2200 kJ per $kg^{0.72}$, Kirkwood 1983). This in turn implied an upper limit on the rate of fuel deposition (FDR_{max}).

During migration time, however, owing to the steps birds take to improve their food-processing capacity, they can also achieve higher rates of fuel deposition. Twelve out of 22 species examined at migration times had rates of energy intake exceeding the theoretical limit (Lindström & Kvist 1995). Studies of captive birds in a migratory state, notably shorebirds, have shown that, given sufficient food, individuals can fatten at much higher rates than normal – up to 6 × BMR, but reaching up to 10 × BMR in shorebirds given access to food for 24 hours per day (Kvist & Lindström 2003). While wild birds would not normally have free access to food all day and night, these findings show what some migratory birds can achieve under near-optimum conditions.

Reducing expenditure

As well as increasing energy intake during migratory fuelling, birds can also reduce expenditure, as mentioned above, by reducing locomotory activity. The switch from insects to berries, for example, may lead to reduction in the energy spent on foraging movements. Birds may also lower their metabolic rates and body temperatures when sleeping at night, as a means of conserving energy, and migrant hummingbirds usually become torpid at night, even when very fat for migration (Carpenter & Hixon 1988). Barnacle Geese *Branta leucopsis* show a progressive reduction in mean daily abdominal temperature (down to 4.4°C below usual), which begins just before the birds embark on their migration and continues, on average, for about three weeks (Butler *et al.* 2003).

DAILY RATES OF WEIGHT GAIN

Care is needed in assessing the rates of weight gain in birds because weight varies with time of day, and often drops (or increases less rapidly) for some hours after capture, as the bird may react to handling (e.g. Schwilch & Jenni 2001). Nevertheless, the repeated trapping of migrants in the days before they set off on migration, or at stopover sites en route, has provided information on their individual rates of weight gain (e.g. **Figure 5.1**), from which average and maximum rates for different populations have been calculated (Alerstam & Lindström 1990, Lindström 2003). Expressed as the daily (24-hour) gain in mass relative to lean body mass, average rates of pre-migratory weight gain (mostly fat), as measured in 58 populations, ranged from less than 1% to more than 7% (maximum 13%). Exceptionally high rates were recorded on particular days, but not sustained over a longer period. In captive passerines, weight gain was often greater on the second than on the first day of fattening, a difference attributed to growth of the digestive tract to facilitate more efficient food processing (Alerstam & Lindström 1990, Hume & Biebach 1996, Klaassen *et al.* 1997).

Comparing species, mean rates of weight gain decline disproportionately with increasing body size, so that large species generally accumulate reserves more slowly than small ones, and depart with relatively smaller reserves (**Figure 5.5**). The smallest birds have daily weight increases that, in relation to their lean weights, are five or more times greater than those of large birds. This is because the maximum limit to the daily metabolisable energy intake is proportional to basal metabolic rate (which declines with body mass), rather than to body mass itself (Lindström 1991). In field studies of various bird species, maximum rates of daily fuel deposition were rarely above 10% of lean body mass in the smallest birds studied (3-g hummingbirds) and rarely above 2% in the largest (3-kg geese). The three main groups of birds examined (passerines, shorebirds and other non-passerines) showed no significant differences in mass-specific rates of daily fuel deposition (Lindström 2003). Combining all species, the relationship between daily fuel deposition and body mass (M) for populations was $1.16 M^{-0.35}$ ($r^2 = 0.66$, $P < 0.001$), and for individuals $2.17 M^{-0.34}$ ($r^2 = 0.54$, $P < 0.001$). The slopes of these relationships were not significantly different from the maximum rate of fat deposition predicted on theoretical grounds as around $M^{-0.27}$, where M is lean body mass (Lindström 1991, 2003).

These various findings all go to illustrate the enormous variation in fuelling patterns found among birds, some of which may result from nutritional constraints on fattening and others from adaptive differences between populations,

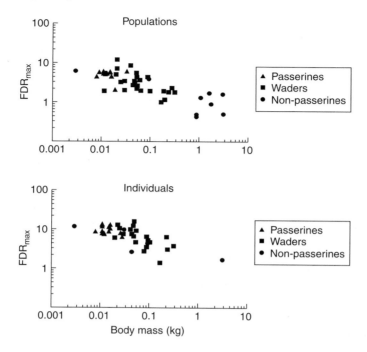

Figure 5.5   Maximum fuel deposition rates (FDR, expressed as % lean body mass per day) for populations (upper) and individuals (lower) of free-living migratory birds of different body weights. Data are based on changes in body mass over time (minimum two days), either in individuals or in populations. Only the highest value for a species is included. Maximum FDRs were negatively correlated with body mass, both for individuals and for populations. No significant differences in mass-specific daily FDRs were found between passerines, waders and other non-passerines. Combining all species, the relationship between maximum recorded daily fuel deposition and body mass (BM) for populations is $1.16\,\text{BM}^{-0.35}$ ($r^2 = 0.66$, $P < 0.001$, 95% confidence interval (CI) of slope -0.27 to -0.42) and for individuals $2.17\,\text{BM}^{-0.34}$ ($r = 0.54$, $P < 0.001$, 95% CI of slope -0.23 to -0.44). From Lindström (2003), in which the original references may be found.

sex and age groups, according to the journeys they undertake. In this section, I have been concerned primarily with mean rates of weight gain in different populations, and how they vary between species in relation to body size. But within species, rates of weight gain can be enormously variable, as can the fuel loads at departure. These variations, which are due largely to variations in food acquisition, and their influence on the speed of migration, are discussed in Chapter 27.

EXAMPLES OF CHANGES IN BODY COMPOSITION

In addition to the storage and depletion of fat that occurs over the migration period, the muscles and internal organs can undergo considerable changes in size

(Piersma 1998). Such changes occur even in short-distance migrants making frequent stops but, as expected, they are much greater in birds making long uninterrupted flights (**Table 5.5**). They serve to adapt the bird for the journey, providing necessary fuel but also reducing unnecessary weight.

Body changes during migration have been studied in particular detail in the Garden Warbler *Sylvia borin* (Bairlein 1991a, 1991b, 1998, Biebach 1998, Biebach & Bauchinger 2003). This trans-Saharan migrant can increase its body mass from 18 g in summer or winter to a maximum of about 37 g shortly before setting out over the desert in autumn or spring (Bairlein 2003). This doubling in body mass is due largely to fat deposition, but also to increase in protein and water content. In autumn in northwest Africa, preparation takes 10–14 days, with mean rates of weight gain of 0.7–1.0 g per day, and maximum rates of 1.5 g per day (10% of lean body mass), depending on the food available.

By comparing Garden Warblers *Sylvia borin* caught in autumn in Turkey (just before a southward Mediterranean–Saharan flight) with others caught in spring in Sinai (just after a northward Saharan flight), Biebach (1998) concluded that about 70% of the loss in body mass during migration comprised fat, and the rest protein and associated water. The protein came partly from the breast and leg muscles which were reduced by 19%, but mostly from the digestive tract which was reduced by 39% (Biebach 1998). Overall, 2.2 g protein and 5.1 g fat were used on this trans-Saharan flight, in an approximate ratio of 1:2.3. Assuming a flight of 2200 km and a mean weight loss of 7.3 g, this gave a mean weight loss of 3.3 g per 1000 km flown, which was not very different from the 3.6 g per 1000 km calculated for this species at sites elsewhere by Bairlein (1991b). After the trans-Saharan flight, it took 1–2 days before gut function and metabolic intake returned to pre-flight level, and 2–3 days for the digestive tract to recover its size (Biebach 1998). This was consistent with the general finding that, after arrival at a stopover site, long-distance migrants often do not gain weight for 1–3 days, and may even lose weight in this period (mean weight loss in 11 passerine species = 4.4%, range 0–13%, Alerstam & Lindström 1990).

In another study of Garden Warblers *Sylvia borin* migrating north over the Sahara in spring, Biebach & Bauchinger (2003) estimated a generally lower rate of weight loss of 1.8 g per 1000 km, but obtained more detailed information of the loss from particular organs (**Figure 5.6**). The most pronounced weight reduction took place in the liver (57%) and gastrointestinal tract (50%), followed by the flight muscles (26%), leg muscles (14%) and heart (24%). It could not be determined whether some of these changes (such as gut reduction) occurred immediately before take-off, or whether they occurred during the flight itself, but the latter seemed more likely (from the general correlation between body mass and gut mass). By reducing the size of organs before or during a flight, Garden Warblers were estimated to save around one-fifth of the energy needed for the same flight made without organ reduction. Protein catabolism made up about 34% of the overall saving, reduced maintenance costs 22% and reduced flight costs 43%. Estimated savings were roughly the same whether birds flew continuously or intermittently (flying at night and resting by day). The reduction in energy costs has presumably been a major driving force in the evolution of organ flexibility in these birds, extending the maximum possible flight range. A similar ranking of organ reductions to the Garden Warbler occurred in three other

Table 5.5 Examples of phenotypic flexibility of the exercise and the nutritional organs of birds in relation to long-distance migration

Species	Ecological context	Flexibility in exercise organs	Flexibility in nutritional organs	Source
Chaffinch *Fringilla coelebs*	Stopover during southward migration	No information	40–50% mass loss of liver and intestine with increasing fat loads	Dolnik & Blyumental (1967)
Gray Catbird *Dumetella carolinensis*	Stopover during southward migration	Increase by ca. 35% of pectoral muscle mass with increase in fat load	No information	Marsh (1984)
Garden Warbler *Sylvia borin*	At autumn departure in Turkey and spring arrival in Egypt	Decrease in pectoral, leg and heart muscle mass by 15–20% during migration	Decrease in mass of digestive tract and liver by up to 50% during migration	Hume & Biebach (1996), Biebach & Bauchinger (2003)
Dusky Thrush *Turdus naumanni*	Spring departure from Japan	Pectoral muscle, lungs and heart increase in mass before migration	No information	Kuroda (1964)
Pied Flycatcher *Ficedula hypoleuca* Willow Warbler *Phylloscopus trochilus* Barn Swallow *Hirundo rustica*	Autumn migration during short flights compared with spring after a long flight	Decrease in pectoral, leg and heart muscle mass of 15–20% during migration	Decrease in liver and gastrointestinal tract of up to 50% before or during migration	Schwilch et al. (2002a)
Yellow Wagtail *Motacilla flava*	Spring departure from West Africa	Pectoral muscle increases in mass before migration	No information	Fry et al. (1972)

Species	Situation	Flight muscles	Nutritional organs	Source
Red Knot *Calidris canutus canutus/islandica*	Northward migration along West Africa/Europe seaboard	No information	Decrease in stomach mass before northward departure	Piersma *et al.* (1993)
Red Knot *Calidris canutus rogersi*	Before northward migration from New Zealand wintering area	Increase in heart and pectoral muscle mass with readiness to migrate	Decrease in stomach and intestine mass with readiness to migrate	Battley & Piersma (1997)
Semipalmated Sandpiper *Calidris pusilla*	Stopover before transoceanic southward flight	Increase in pectoral muscle and heart mass before departure	No information	Driedzic *et al.* (1993)
Bar-tailed Godwit *Limosa lapponica*	At departure on trans-Pacific southward flight	Probable increase in heart and pectoral muscle mass before departure	Very small gizzard, liver, kidneys and intestine at departure	Piersma & Gill (1998)
Greater Snow Goose *Chen caerulescens atlanticus*	Staging during northward migration	No information	Decrease in gizzard mass during spring stopover	Gauthier *et al.* (1984)
Eared Grebe *Podiceps nigricollis*	Before southward departure from moulting area	Increase in pectoral muscle and heart mass before departure	Liver, stomach and intestine shrink by more than 50% before departure	Gaunt *et al.* (1990), Jehl (1997)

Data from body composition analyses. Mainly from Piersma (1998).
In addition to the species listed, enlargement of breast muscles (as measured by fat-free dry weight) before migration has been found in a wide range of passerine migrants (see Wingfield et al. 1990) and in various shorebirds (Davidson & Evans 1988, Evans et al. 1992). In contrast, no obvious muscle growth in association with autumn migration was found in Wood Thrush *Catharus mustelinus*, Veery *Catharus fuscescens* and juvenile Willow Warbler *Phylloscopus trochilus* (Hicks 1967, Baggott 1975).

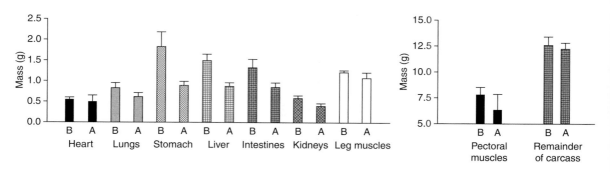

Figure 5.6 Comparison of the mass of different organs (lean dry mass (95% confidence interval)) of Garden Warblers *Sylvia borin* before (B, Ethiopia) and after (A, Egypt) crossing the Sahara Desert. From Biebach & Bauchinger (2003).

trans-Saharan migrants, namely the European Pied Flycatcher *Ficedula hypoleuca*, Willow Warbler *Phylloscopus trochilus* and Barn Swallow *Hirundo rustica* (Schwilch *et al.* 2002a).

Massive changes in body composition also occur in some shorebirds, which include some of the most impressive long-distance bird migrants, crossing some of the world's largest stretches of ocean or desert non-stop. Not only are they able rapidly to store and metabolise large amounts of fat, they also undergo many other physiological changes, affecting skeletal muscles and various internal organs (Piersma & Lindström 1997, Battley *et al.* 2000). Extreme changes were found in Bar-tailed Godwits *Limosa lapponica baueri* collected in Alaska as they hit a radio-tower, just after take-off on a presumed trans-Pacific flight of at least 10400 km to New Zealand. The majority of individuals in autumn are thought to make this flight non-stop, but some are seen to pause on various Pacific Islands (Chapter 6). The Alaskan birds had some of the highest fat contents recorded in birds, amounting to 55% of total body mass. They also had relatively large breast muscles and heart (= exercise organs), but very small gizzard, liver, kidneys and gut (= digestive organs). Upon departure in autumn, these long-distance migrants apparently dispensed with parts of their metabolic machinery that were not directly necessary during flight, presumably converting them to other tissue. They rebuilt them upon arrival at the migratory destination, in order that they could again feed at maximum efficiency (Piersma & Gill 1998). This temporarily reduced digestive function may have been more than compensated by savings on transport costs.

Attaining as much as 55% of fat in total body mass has other consequences. Not only must all of the bird's structure and systems be contained in 45% of the body mass at take-off, but consuming this huge proportion of fat requires heavy inroads into the remaining protein during flight. According to Pennycuick & Battley (2003), the journey to New Zealand (10400 km) would entail flying continuously for a week, but the fuel load was sufficient to fly much further, or alternatively to provide a reserve for adverse winds and navigation errors.

A second sample from the same population of godwits, obtained in spring before departure from New Zealand, would have run out of fat before reaching Alaska, but could have reached the Yellow Sea area, where these birds stage

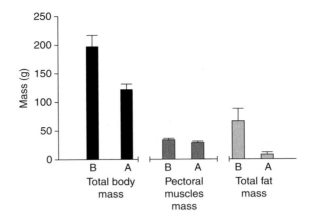

Figure 5.7 Comparison of different components of body mass in different samples of Great Knots *Calidris tenuirostris* caught just before and just after a 5420 km flight between northwest Australia and Chongming Island near Shanghai in China. From Battley *et al.* (2001).

on their northbound journey. The higher flight-range estimate for the Alaskan birds was not due mainly to their higher fat mass (only 5% higher), but to the higher proportion that fat formed of total body mass, which they had achieved in autumn by reducing the mass of other organs before departure. The 'fat fraction' was more important than the fat mass in influencing flight range (Pennycuick & Battley 2003).

Turning to another shorebird, samples of Great Knots *Calidris tenuirostris* were collected before and after a 5420 km flight from northwest Australia to Chongming Island near Shanghai in China (believed to be flown non-stop), again in order to determine the mass of fat consumed and also of protein withdrawn from flight muscles and other organs. This journey takes four days (Pennycuick & Battley 2003). Not only did these birds show reduction in fat content (but not to zero), they also showed loss of lean tissue mass, with statistically significant reductions in six organs (pectoral muscle, skin, salt glands, intestine, liver, kidney), and non-significant reductions in several others (**Figure 5.7**). The reduction in functional components between the start and the end of the flight was reflected in a lowering of BMR by 42%, one of the fastest rates of BMR change recorded in birds. Only the brain and lungs had not changed over this journey (Battley *et al.* 2000). Other Great Knots caught in northwest Australia were flown to the Netherlands, and kept without food until their body mass had declined to the level found in arriving birds on Chongming Island. Organ reductions were broadly similar but, compared with the fasted birds, those that had migrated had conserved more protein (Battley *et al.* 2001).

Another well-documented example of seasonal change in body composition concerns the Eared Grebe *Podiceps nigricollis* in North America (Jehl 1997, Jehl *et al.* 2003). Incapable of flight for months at a time, this species has the longest non-flying period of any northern hemisphere bird, totalling 9–10 months over the course of a year. In practical terms, the bird flies only to migrate and spends the rest of its life on water, in association with which it undergoes several cycles per year of expansion and contraction of particular body parts. Yet its existence depends on its ability to fly as much as 6000 km each year to reach high-yield seasonal environments that are exploitable by very few other species. After breeding mainly on prairie wetlands, the species migrates to assemble in huge numbers on a small number of hypersaline lakes, especially the Great Salt Lake in Utah and Mono Lake in California, each of which can hold more than a million grebes in autumn. The birds feed on the huge numbers of brine shrimps *Artemia* and alkali flies *Ephydra* available at this time.

The grebes arrive mainly in September (but from mid-July to early November), moult their feathers and remain until food supplies have dwindled (usually late November–December). They then migrate to wintering areas in the Gulf of California. They begin the return journey from January, travelling via the Salton Sea (in south California) where they remain for at least two months, again becoming flightless, and then in late March–April they continue to the Great Salt Lake, and thence to the prairies in April–June.

When birds leave their prairie nesting ponds in late summer they weigh about 420–450 g, but on arrival on the moulting sites after 2–3 nights of flight, they weigh as little as 250 g. Their moult leaves them flightless for 35 days, but they gradually increase in body mass to reach more than 600 g by mid-October. During this period they have accumulated massive fat stores, and their body organs have undergone large changes, involving increases in the size of the digestive organs and leg muscles, and a reduction of 50% or more in breast muscle, to below the size needed for flight. However, at the end of this autumn staging period, when the flight feathers have been replaced, the body changes are reversed. In the 2–3 weeks before leaving southward, the birds lose as much as one-third of their fat reserves, and reduce the size of nutritional organs by up to 75% while building their pectoral muscles and heart (Jehl 1997). These huge changes in body structure result in a net one-third loss of body mass to 420–450 g. In this way the grebes optimise flight efficiency by reducing weight and wing loading, and increasing flight range. No other bird species is known to reach migratory condition by losing so much mass before departure (but for nestlings of some species, see later). Most of the remaining fat reserves are used during the flight to wintering areas. Juveniles undergo similar but less extreme changes.

Additional cycles that involve less marked fattening are repeated at breeding and wintering areas and in some birds also at spring staging areas, so that body weight and composition are in continual flux. The extreme reductions in leg muscle mass that occur before every migration reduce flight costs. If the bird maintained the body structure best suited for swimming and feeding, it would need far greater energy reserves for its long-distance migrations. This continual modification of behaviour and body structure as the bird shifts between swimming and flight modes, sedentary and migratory phases, with minimal waste of resources, may have helped the Eared Grebe to become the most abundant grebe on earth.

It is clear from these various examples that a bird refuelling for long-distance migration is not like a plane landing, refuelling and taking off again. Unlike the plane, the bodies of long-distance migrant birds have to be partly reconstructed at each stopover and modified again before take-off. We can envisage that the sizes of organs carried at take-off by different populations represent evolutionary compromises between their functions during the pre-departure, flight and post-arrival phases of migration (Piersma 1998). In all populations, other tissue is invariably deposited along with fat, but in proportions that vary greatly between species, and between different populations of the same species, according to the journeys they make (**Table 5.5**). Before departure on long journeys, exercise organs (pectoral muscle and heart) tend to enlarge and nutritional organs (stomach, intestine and liver) tend to shrink. This makes sense on long flights where weight reduction is at a premium. In other species, the digestive tract is apparently reduced during the flight itself, rather than beforehand, contributing to the fuel and water needs of the migrant on its journey. Prior reductions in nutritional organs appear most pronounced in populations about to over-fly oceans that offer few or no opportunities for emergency landings, let alone feeding.

The adaptive role of muscle growth and shrinkage is not entirely clear, as it may serve more than one function. As explained already, muscle breakdown is necessary for efficient fat metabolism, providing intermediates for the citric acid cycle. It may also provide metabolites such as glucose for proper functioning of the nervous system, and uric acid which is an antioxidant that can de-toxify the free radicals produced when tissues consume oxygen (Dohn 1986, Klaassen 1996). At the same time, as muscles provide the power needed for migratory flights, their shrinkage during a flight may be an adaptation to the reducing weight (and hence reducing power needs) of the bird during its journey (Pennycuick 1975). On sustained long flights, loss of protein is thus unavoidable; but it also happens to be strategically convenient (Jenni & Jenni-Eierman 1998, Battley *et al.* 2000). However, populations differ in the ratio of fat to protein they accumulate as reserves, as do the same populations at different seasons. These facts cast further doubt on the idea that shrinking muscle size during flight serves primarily to reduce power output as the bird loses weight. This may be one function, but as flight muscle mass is not consistently related to overall body mass, it cannot be the whole story (Bauchinger & Biebach 2005).

Because protein catabolism results in a higher metabolic water yield per unit energy than lipid, it is of additional value on long, non-stop flights (**Table 5.1**; Klaassen 1996). Net water availability during continuous flight could therefore be altered by changes in the relative proportions of the different fuel types used. In Bar-tailed Godwits *Limosa lapponica* and other shorebirds on spring migration, fat and protein are deposited in almost equal amounts (fresh mass basis), but because of its greater energy content, fat provides about 90% of the energy required for the flight. The same ratio (roughly half and half) may apply to a much wider taxonomic array of birds, including Chaffinch *Fringilla coelebs* and Garden Warbler *Sylvia borin*, as well as various waders. This is far removed from an earlier assumption that all weight increase in migrant birds was due to fat, although the fat-to-protein ratio may differ greatly between species, and also between autumn and spring in the same species, as mentioned above. In some shorebirds, about two-thirds of the increase in flight muscle mass resulted from increases in myofibril mass, and about one-quarter from additional mitochondrial mass, while sarcoplasm increased very little (Evans & Davidson 1990). Most data refer to the autumn migration, and the situation may differ in spring when different routes are often taken, and when birds must ideally arrive on breeding areas with a surplus of fat and protein reserves left over for breeding (see later).

The fact that nutritional organs are not fully functional on arrival after a long flight may explain why some migrants, on reaching a stopping site, do not appear to feed. They simply rest, and depart later in the day or at night (e.g. Rappole & Warner 1976). Typically, they are less specific in their habitat needs than feeding birds, and individuals of normally territorial species make no attempt to establish a territory but sit around in groups. They are in flight mode rather than in feeding mode, and it may be more energy-efficient to move on while reserves last rather than reconstruct digestive machinery and start to feed again before this becomes necessary. An example of stopover periods recorded at a desert oasis are given in **Figure 5.8**. Most of the birds were recorded on only one day, and may have passed on without feeding, but others stayed for more than 10 days, replenishing their body reserves (for other records from another site, giving generally longer periods, see **Table 5.4**).

Body reserves for survival and breeding

In most species that have been studied, individuals were found to arrive in breeding or wintering areas with residual body reserves not totally used on the journey (e.g. Sandberg 1996, Fransson & Jakobsson 1998, Farmer & Wiens 1999, Widmer & Biebach 2001, Morrison 2006, Krapu *et al.* 2006). For example, American Redstarts *Setophaga ruticilla* arrived in Michigan breeding areas with enough body fat for at least another 1000 km of further flight, while Bar-tailed Godwits *Limosa lapponica* from Alaska arrived in New Zealand with enough fat for another 5000–6000 km of flight (Pennycuick & Battley 2003). However, remaining fuel levels varied greatly between individuals, and from year to year in the same population.

Considering the energy cost of transporting such reserves, the question arises why birds carry so much more than they need? Extra reserves might act as insurance against food shortage or bad weather en route; they might enable migrants to fly faster than the maximum range speed on flights where they are likely to encounter unfavourable weather; or they may help to ensure survival after arrival in new areas, especially in spring when local food supplies are still scarce. In spring, they might also enable newly arrived males to concentrate on fighting and territory acquisition, or females to produce eggs earlier than they might otherwise do (Chapter 27). Clearly, these various possibilities are not mutually exclusive.

If the main function of surplus reserves is to promote survival, the first arriving sex (in most species the males) would be expected to arrive with the largest reserves; and if the main function is to aid egg production, the females should arrive with the largest reserves. In the sub-arctic environment of Swedish Lapland, long-distance passerine migrants arrived with enough fat to fly an estimated further 242–500 km, depending on species, while short-distance migrants arrived with lesser amounts (Sandberg 1996). Fat reserves on arrival were thus related to migration distance. They were also related to feeding habits which were assumed to influence the amount of reserve needed during the transition period between arrival and breeding. However, in seven out of nine species in which both sexes arrived together, females had significantly larger reserves than males, favouring the egg production hypothesis. In the two species in which the sexes had significantly different mean arrival dates, males and females showed

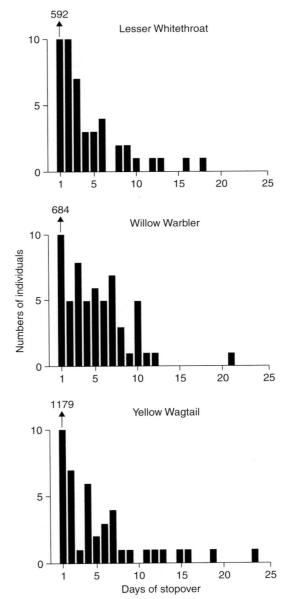

Figure 5.8 Frequency distribution of minimum stopover periods at the oasis of Sadat Farm, in the Libyan Desert of Egypt. Sample sizes for birds trapped and ringed on one day only are shown above the columns. The remaining records are based on retraps of ringed birds. From Biebach et al. (1986).

no difference in mean fat load on arrival, so the later arriving females could have had more of their reserve left than males by the time of egg-laying.

In some waterfowl, notably geese, reserves accumulated before arrival, and especially at the last stopover site, help in egg formation and survival through incubation, when little food is eaten (Newton 1977, Ebbinge 1985). Such geese often arrive before vegetation growth has begun in their breeding areas, when little food is available. They are described as 'capital breeders', because they reproduce largely on the strength of existing body reserves, and contrast with 'income breeders' which breed on the strength of food eaten at the time. In different goose populations, weight gains of 25–53% have been recorded before the birds set off on spring migration (McLandress & Raveling 1981). This is a lot for birds of this size. Females accumulate more weight than males, in association with the needs of egg production and incubation (in which males do not participate, except in nest guarding).

Among Canada Geese *Branta canadensis* wintering in Minnesota, body weights of males and females increased by 26% and 36% respectively in the month before spring migration. Weight gain of males comprised 47% lipid, 13% protein and 35% water, while that of females comprised 61% lipid, 10% protein and 21% water. Initial weight gain was predominantly protein (and accompanying water), probably contributing to gut enlargement, while most of the later weight gain was due to fat. Lipid and protein storage in this population was judged sufficient to cover maintenance after arrival in the breeding areas, egg-laying and incubation by females, and territorial defence by males (McLandress & Raveling 1981). Although large body reserves on arrival are a pre-requisite to successful reproduction in some populations of geese, they do not of course guarantee it. Nevertheless, females with larger body reserves often showed better reproductive success than individuals with smaller reserves (Chapter 27).

At least some species of high arctic geese, such as the Lesser Snow Goose *Chen c. caerulescens*, Ross's Goose *A. rossii* and Brent Goose *Branta bernicla*, can start egg-laying 2–5 days after arriving in breeding areas, before plant growth has begun, and seem to rely entirely on body reserves. Others feed and regain some weight after arrival, but still depend partly on body reserves accumulated further south (Bromley & Jarvis 1993). In Lesser Snow Geese, Ankney & MacInnes (1978) studied the relationship between body reserves and reproductive output from females collected at various stages of breeding in the Northwest Territories of Canada. They determined the potential clutch size of pre-laying females by counting the number of large vascularised follicles in the ovary, and found that females with larger body reserves had, on average, larger potential clutches. Other females collected post laying revealed that body reserves had been used partly during laying, but that after laying the mean weights of remaining reserves from females that laid clutches of different sizes were not significantly different. The authors concluded that clutch size in Lesser Snow Geese was determined by the size of nutrient reserves. Breeding females used most of their remaining fat and protein reserves during incubation (85% and 24% respectively, Ankney & MacInnes 1978). Late in incubation when females had depleted their body reserves, some left their nests to feed, while others were found dead on their nests from starvation. Hence, to reproduce successfully in this area, a female Lesser Snow Goose had to accumulate beforehand enough reserves to support the last stage of migration, egg production and maintenance during the four weeks of incubation. Only after hatch was she able to feed intensively again, and build up body condition for the return migration to wintering areas.

Some duck species also depend for reproduction at least partly on body reserves accumulated in wintering or migration areas (for Mallard *Anas platyrhynchos* see Krapu 1981, for Lesser Scaup *Aythya affinis* see Anteau & Afton 2004), but many other ducks accumulate most of the necessary reserve after arrival in breeding areas. Among Northern Pintails *Anas acuta* nesting in Alaska, individuals that arrived with the largest residual fat reserves could start nesting earlier and produce larger clutches than individuals with smaller reserves (Esler & Grand 1994). The latter produced small clutches later in the season when local food supplies had developed, and probably bred less successfully. At least one arid-zone duck, the White-faced Whistling Duck *Dendrocygna viduata,* is also a capital breeder in South Africa, producing eggs mainly from internal body reserves (Petrie & Rogers 2004).

Sandhill Cranes *Grus canadensis* nesting in arctic North America likewise accumulate substantial body reserves at spring stopover sites, increasing in body weight by about one-third before travelling to breeding areas (Krapu *et al.* 1985). By the time they arrive, about half the reserve still remains. The main difference from waterfowl is the lack of any obvious difference in reserves between the sexes (in cranes, reserves are not greater in females), and the relatively small proportion allocated to egg production (in cranes eggs and clutches are relatively smaller). The clutch in cranes weighs less than one-tenth of total body mass, compared with 20–40% in geese.

Stable isotope analysis of eggs has provided an additional means of telling whether eggs are produced from food eaten in breeding areas or from food eaten in migration and wintering areas. The method depends on the fact that foods eaten in different habitats or different regions have different isotope ratios, and that these differences are reflected for a time in the body tissues of the consumer (Chapter 2). Such analyses confirmed that Lesser Snow Geese *Chen caerulescens* were largely capital breeders, with reserves accumulated mainly south of breeding areas (Klaassen 2000). Isotope signatures differed markedly between natal down (derived from materials laid down in the egg) and juvenile feathers produced by the same individuals later in the season (derived from nutrients ingested locally). The difference between these two samples confirmed that the parent collected the nutrients for the egg from other than local sources, presumably in wintering or migration areas further south. In contrast, Greater Snow Geese *Chen c. atlanticus* nesting on Bylot Island, Canada, seemed to obtain most of their protein and fat requirements for egg production after arrival in the breeding area (Choinière & Gauthier 1995). Among different populations of arctic-nesting geese, considerable variation may exist in the extent of reliance for reproduction on post-migration residual body stores. Much of this variation may depend on the distance between the last staging area and the breeding area, and on the feeding time available between arrival and egg-laying in the locality concerned. Where food becomes available through ice-melt relatively early in the season, there is less need for geese to depend so heavily on imported body reserves.

Invertebrate food items from tundra and estuarine habitats have distinctly different carbon and nitrogen isotope ratios, which are also expressed in bird tissues, including eggs. Carbon isotope analyses of the eggs of several shorebird species in arctic Canada and Greenland showed that these eggs were made from inland food eaten in the tundra breeding areas rather than from coastal food obtained in wintering and migration areas (Klaassen *et al.* 2001). The isotopic signatures of eggs and natal down resembled the signatures of juvenile feathers grown locally, but differed markedly from those of adult feathers grown in migration or wintering areas. On the other hand, in another study, eggs in the earlier clutches of Red Knots *Calidris canutus* and Turnstones *Arenaria interpres* in the northeastern

Canadian Arctic were rich in $\delta^{13}C$ and $\delta^{15}N$, which suggested that some residual marine nutrients were used in their production (Morrison & Hobson 2004). These eggs were evidently produced partly on body reserves accumulated before the birds reached their inland nesting areas. This occurred even though *Calidris* sandpipers typically produce clutches equal to 80–120% of female body mass. Even if not used in egg production, post-migratory residual body stores in shorebirds may still contribute to breeding success by reducing energy needs on arrival, the situation varying from year to year, and probably also between fat and protein. In another study, eggs were analysed from five gull, four tern and one jaeger species nesting at Great Slave Lake in the Northwest Territories of Canada (Hobson *et al.* 2000). The eggs of most species were formed from local food supplies, but those of Caspian Terns *Sterna caspia* and Common Terns *S. hirundo* may have been produced partly from body reserves accumulated earlier while the birds were on migration.

Capital and income breeding evidently represent opposite ends of a continuum of variation found among birds. Species and populations may vary in the contribution that body reserves make to reproduction, as may individuals within the same population. Populations also vary in where on the migration route between wintering and breeding areas they acquire the reserves for breeding. Larger species, in which clutch weight forms a small proportion of total body weight, are much more likely to rely heavily on body stores for egg formation than are small species in which the clutch weighs as much or more than the female herself. Thus, the body reserves that small passerines and shorebirds carry to the breeding areas can provide at most only a minor contribution to the total protein and energy costs associated with clutch formation, but they may help with initial survival and reproductive activities, enabling earlier egg-laying than otherwise (Chapter 27).

At the least, these various studies serve to emphasise the great phenotypic flexibility of birds, the varying extents to which birds depend on previously accumulated body reserves for breeding, and the varying stages of the journey at which such reserves are accumulated. It is remarkable that some birds accumulate reserves for migration and breeding in the same places at the same time, and that they transport reserves hundreds, or even thousands, of kilometres from wintering to breeding areas. It is also remarkable that birds can shift protein from one body organ to another, as the muscles and digestive organs change their relative sizes before and after flights. Although less research has been done on the condition of birds when they arrive in wintering areas, studies indicate that residual body reserves are usual there too, and can influence subsequent survival (Chapter 27).

FAT CHICKS

In most bird species, fattening for the first migration does not begin until after the young are fully grown and able to feed themselves efficiently. In some bird species, however, the young fatten while in the nest, courtesy of their parents, and migrate independently within a few days after fledging. They include many seabirds, notably various petrels (*Procellariiformes*), in which the nestlings reach peak weights that greatly exceed the adult weight. Thereafter, they receive little or no additional food from their parents and lose weight until they leave the nest. The young continue to grow their feathers, but are still extremely fat at the time of their first flight. Although this fattening of seabird nestlings has been known for many years, it has usually been regarded as an insurance to tide the bird over difficult times until it learns to feed itself. However, in many such species the young set off within a few days after fledging on long journeys for which fuel reserves are almost certainly needed.

Take the Manx Shearwater *Puffinus puffinus* as an example. At their peak, the young weigh around 20% more than the adults, and are extremely fat, from which point they are fed at a much reduced rate. After about 10 further days, the young leave for the sea and thereafter have no further contact with their parents (Harris 1966). Their fat reserves may not only support them during times of hardship and migration, but may also allow their parents to divert food to building their own body reserves and so to migrate earlier than otherwise possible. Some 36 hours after having been ringed on Skokholm Island off Wales, some young shearwaters were recovered several hundred kilometres to the south, in the Bay of Biscay. Within six weeks, many others were recovered in the seas off Brazil, more than 9000 km across the Atlantic Ocean. One was found, an estimated three days after it had died, on the coast of Brazil 16 days after being ringed at the nest. Even if it had left immediately, it made the 9600 km journey at a rate of at least 740 km per day (Perrins *et al.* 1973). Another young bird was recovered in the Canary Islands, close to the Great Circle route from Skokholm to Brazil, six days after it was ringed on Skokholm, an average of at least 460 km per day. These flights almost certainly required substantial fuel reserves, for they left little time for feeding by such inexperienced birds. The young would gain obvious advantage in being able to fly directly to the wintering area without having to spend time in a perhaps fruitless search for food on the way. Judging from the estimated fat contents of the young, and the likely energy costs of their migration, 'it seems just possible that fat birds could make the 9,000 to 10,000 km flight on their reserves' (Perrins *et al.* 1973). This is not to imply that the birds would ignore any potential food encountered, but merely that the birds leave as fat as possible in order to make the journey if necessary without further supplies.

All species of petrels that have been studied achieve peak weights that are higher than adult weights, and lose some weight to fledging, but many still leave at above the usual adult weight (Warham 1990). This applies to non-migratory as well as migratory species, but in both groups the young typically travel long distances from the colonies soon after fledging. Their substantial body reserves presumably help to fuel these movements, whether classed as migration or dispersal. Fat reserves would in any case add buoyancy, so could presumably hinder any ability of the young to obtain food by diving for many days following fledging (in those species that feed by diving). In Wilson's Storm Petrel *Oceanites oceanicus*, chicks in the weight recession phase were found to lose body fat and water (Obst & Nagy 1993), but in the Northern Fulmar *Fulmarus glacialis* and Cape Gannet *Morus capensis* the weight loss was apparently due to water alone (Navarro 1992, Phillips & Hamer 1999). In the Northern Fulmar, fledglings had enough fat to stave off starvation for at least 18 days (Phillips & Hamer 1999).

Like petrels, young Northern Gannets *Morus bassanus* become very obese before fledging, reaching around 30% heavier than adults, and carrying around 1 kg of fat. The young leave the nest before their flight feathers are fully grown, but can glide down to the sea, reaching distances of 3–5 km from the colony (Nelson 1978, Wanless & Okill 1994). Once on the water, they seem unable to take off again, at least for several days, by which time they have lost further weight, and their flight feathers are better grown. They swim long distances in this period; and once they become airborne they can travel even longer distances, and at some later stage begin to hunt for themselves, using the aerial diving technique characteristic of the species. Six young gannets that were caught on the sea in the flightless period were recovered 10–16 days later at distances of 394–2483 km (Wanless & Okill 1994). They had mean rates of travel of 15.1–155.2 km per day, similar to those of many landbirds on migration (Chapter 8). Again, the birds were unlikely to have covered the longer of these distances without substantial fuel reserves.

The young of some landbird species may also migrate immediately they leave the nest, and without further parental support. An example is the Common Swift *Apus apus* in which the young also fly at weights that often exceed those of adults (Lack 1956). One individual was recovered in Madrid three days after it left the nest in Oxford, a journey of 1300 km in three days (C. M. Perrins, in Wernham *et al.* 2002). Immediate post-fledging migration could occur in other landbird species, as well as in other seabirds, but further research is needed. It is clear, however, that migratory fat deposition occurs not just in full-grown birds, but in some is an integral component of nestling growth.

CONCLUDING REMARKS

One of the most striking features of birds is their ability to fuel high-intensity endurance exercise with fatty acids that are stored in adipose tissue and delivered continuously to the working muscles by the circulatory system. This facility makes birds exceptional among vertebrate animals (McWilliams *et al.* 2004). Yet studies have increasingly revealed other amazing ways in which migratory birds have overcome physiological constraints in energy storage and migratory flight. For fuelling, these include various ways of temporarily increasing calorie intake rates, including the extension of feeding times (sometimes into night-time), night-time migration to leave the day for feeding, the switching of metabolic emphasis from carbohydrate to fat, change of diet, and increase in the size and effectiveness of the digestive tract. For flight, they include reduction in the size of digestive organs immediately before or during the flight, and catabolism of protein to supply metabolites needed for the utilisation of fat and other purposes. These are all measures which the bird can take for short periods only, as they also have costs, and conflict with other activities or with requirements at other stages of the annual cycle.

Growth in the gut and liver facilitates rapid fat deposition, but in the days before departure by long-distance migrants these organs may shrink, while the pectoral muscles and heart continue to grow, providing the increased metabolic power necessary for sustained flight or for heat production at high altitudes (Piersma *et al.* 1999). Improving intake and assimilation, or cutting expenditure, are all ways in which birds can increase the efficiency of migration (Piersma 2002), enabling some species to perform extreme journeys otherwise impossible, or to carry body reserves from one region in order to facilitate reproduction in another far away. What astonishing creatures birds are!

SUMMARY

Migration normally requires body reserves (mostly fat) accumulated at appropriate times of year. Depending mainly on the terrain to be crossed, birds migrate on a small fuel reserve/short flight system or on a large fuel reserve/long flight system, the latter being necessary over seas, deserts or other inhospitable areas where refuelling is not possible. Birds may vary their migration mode at different stages of their journey, depending on feeding opportunities in different parts of the route. The more fuel a bird carries, the more energy the bird uses in transporting it, which increases the energy cost per unit distance flown. Extra fuel also reduces the bird's agility in evading attacks by predators. These considerations should favour a migration strategy of short flights, frequent fuelling and low fuel loads wherever possible, with the alternative of long flights, infrequent fuelling and heavy fuel loads resorted to only when necessary.

The predominant fuel is fat which has five main advantages: (1) it provides the highest concentration of metabolic energy per unit weight; (2) it can be stored dry without accompanying water or protein; (3) it can be metabolised (by the citric acid cycle) more efficiently than protein or carbohydrate; (4) it can be oxidised efficiently and completely by most body tissues, including the all-important flight muscles; (5) muscle fibres relying on fatty acids can work for long periods without tiring. However, carbohydrate and protein are also used as fuel on migratory flights, and metabolism of fat is accompanied by metabolism of protein, which yields intermediates for the citric acid cycle, and other important metabolites. The ratio of fat to protein accumulated at migration times varies greatly between species, and even between populations of the same species, and between outward and return journeys.

In small passerine migrants, the composition of the diet seems to influence to some extent the composition of the fuel stores laid down. These stores in turn influence the relative composition of the fuel types used during migration, and (via their energy density) the flight range. The composition of body stores, notably the ratio of fat to protein, may also vary according to the nature of the journey and the terrain to be crossed, and the subsequent needs of the bird, whether for breeding or survival.

Fuel deposition results in rates of daily weight gain up to about 10% (occasionally 13% or more) of lean body mass, with slower rates in larger species, and great variation among individuals. Among passerines and shorebirds, short-distance migrants usually increase in weight by 10–30% before departure, and long-distance migrants by 70–100%, effectively doubling their body mass. Some species may accumulate no obvious reserves for migration, but travel for only parts of each day, losing weight but replenishing it on a day-to-day basis. In some seabirds, the young accumulate fat stores while in the nest, and embark on a long migration independently of their parents soon after fledging.

Rapid weight gain for migration mainly results from increased food consumption (feeding time), but in some species also from dietary change and improved digestive efficiency (involving increase in the size of the digestive tract). Rates of weight gain may be environmentally or metabolically limited. Some migrants increase muscle and heart mass before departure, and reduce the size of other internal organs (digestive tract and liver). On arrival, the digestive organs are rebuilt before normal food assimilation can occur again. These changes in body composition are much more marked in species making long-distance non-stop flights lasting several days than in short-distance migrants that can feed and drink every day during their journeys.

Many species typically arrive on spring breeding areas with a surplus of body reserves, including both protein and fat. In some species, notably arctic-nesting geese, the reserve is used for both egg production and incubation, at a time when little food is available locally. Body reserves imported to breeding areas may thus have been accumulated on wintering and migration areas at lower latitudes. For these reasons, the various seasonal changes in the body mass and composition of migratory birds are often rapid and substantial. Nevertheless, populations vary in their level of dependence on internal body reserves for breeding.

APPENDIX 5.1 CALCULATION OF FLIGHT RANGES

Much interest has centred on estimating the non-stop flight ranges of birds from their weights or fuel contents on departure. Such estimates require information, not only on the departure weight (or fuel content) of the birds concerned, but also on the rate at which weight is lost (or fuel is used) during a journey. This rate has sometimes been estimated by comparing the weights (or body composition) of samples of birds obtained at different points on a journey. Weights taken just before and just after a non-stop flight over a sea or desert give the most reliable estimates of flight costs, for only then can one be sure that the birds had not fed and replenished their body reserves en route (Nisbet 1963b, Fry et al. 1972, Biebach 1998, Bauchinger & Biebach 2003). These studies provide estimates of weight loss (or fuel use) per unit time or per unit distance covered. However, they are usually based on mean values from samples of different birds and not on measurements from the same individuals, and they are valid only for the particular wind and other conditions prevailing at the time. Nevertheless, to estimate roughly how far a bird of known weight could fly, these rates of weight loss could be used, together with knowledge (or assumptions) of the lowest weight a migrating bird could reach and still remain active (usually taken as the fat-free weight). If the rates of weight loss are measured per unit time, additional information is needed on the flight speed of the bird, in order to convert hours flown to distance covered.

Most published estimates of flight range using this procedure are based on the assumption that the total weight loss during flight is due entirely to catabolism of fat. Where part of the weight loss is due to protein, as seems usual in migratory birds, this can greatly reduce the estimate of energy expended, and thus the estimated flight costs. The 24-g Garden Warblers *Sylvia borin* mentioned earlier lost about 7.3 g body weight on a 2200-km trans-Saharan flight, about

3.3 g per 1000 km flown, which involved fat and protein in the ratio 2.3:1 (Biebach 1998). At the other extreme of body size, Barnacle Geese *Branta leucopsis* migrating 1000 km from Svalbard to Scotland weighed an average of 2.30 kg at the start of migration (including 432 g of fat) and 1.82 kg at the end, although some may have fed en route. The amounts of fat and tissue protein required during the 60 hours of flying were estimated at 415 g and 126 g respectively, a ratio of 3.3:1, so with no major hold-ups, at least some of the geese could have flown the whole route without feeding (Butler *et al.* 2003). Other findings on weight loss during migration have been expressed on the basis of loss per unit time – for example the 7.4 g per hour weight loss of Eared Grebes *Podiceps nigricollis,* which was equivalent to less than 2% of total body mass per hour (Jehl *et al.* 2003).

A second type of procedure for estimating flight range involves knowledge of the departure weight of the bird, from which its flight cost (rate of fuel use) can be calculated from standard equations depicting the relationship between body weight and BMR (Lasieski & Dawson 1967). BMR is higher for passerines than for non-passerines, so different equations are used for the two groups. Flight cost is taken as some multiple of BMR (usually $12 \times$ BMR). Modifications to the original model of Raveling & Lefebvre (1967) allow for weight loss during flight resulting from fuel consumption (Summers & Waltner 1979), use actual measurements rather than estimates of flight costs (Davidson 1984), and allow for variations in the shape of the birds themselves (notably wing-length) (Castro & Myers 1989). The advantage of this latter correction is that it does not assume that the cost of flight is a consistent multiple of BMR, but varies in a way that depends on the aerodynamics of the bird, therefore diverging from the classic 0.7 exponent of metabolic costs versus body mass. It acknowledges that two birds of the same mass can have very different flight costs depending on their aerodynamic design (Castro & Myers 1989). Only the latter model predicted decreasing costs of flight with decreasing body mass, as also expected on theoretical grounds. This method is thus readily applicable to birds of a wide weight range, providing that information is available on body mass, fuel load and flight speed.

The current preferred model for estimating the flight range of migratory birds is that of Pennycuick (1989), again derived mainly from aerodynamic theory, but with later modifications (current model, Pennycuick 2006, see Pennycuick & Battley 2003). This sophisticated model can be applied to birds of any size providing that appropriate measures are available for the species concerned. It is by far the best model yet devised, enabling the researcher to test the effects of altering any of the key variables involved. Further refinements to models are likely to occur in the future, as additional information becomes available.

All these indirect methods of estimating flight range make no allowance for the wind conditions in which particular birds migrate, for any effects of flight formations, or for other external variables that influence energy consumption. Flight range estimates could therefore be in considerable error. Nevertheless, particular models have useful comparative value for closely related species of similar shape and flight mode (Gudmundsson *et al.* 1991). Such models also provide a useful check on our understanding: if estimates of flight ranges do not match what migrants on independent evidence are known to achieve, then our knowledge is probably deficient in some way.

Rather than estimating flight costs, some researchers have attempted to measure them directly for experimental birds flying in a wind tunnel (e.g. Nachtigall 1990). This has been done from measurements of individual oxygen use or carbon dioxide production, made either directly or by the use of doubly labelled water.[3] However, all these methods involve either fitting apparatus to the bird, subjecting it to some artificial procedure or flying it in atypical surroundings. Not surprisingly, then, flight costs measured in these ways may differ considerably from those obtained in other ways from free-living birds (review Norberg 1996).

Birds often migrate more cheaply in natural than in experimental situations. The average heart rate (reflecting metabolic rate) of four naturally migrating Barnacle Geese *Branta leucopsis* over the whole journey was 253 beats per minute, which is only half the mean rate recorded from Barnacle Geese trained to fly behind a truck, and 65% of that recorded from Barnacle Geese trained to fly in a wind generator (Butler *et al.* 2003). Moreover, in the naturally migrating birds, heartbeat (as an index of metabolic rate) declined during the course of a journey from 315 beats per minute to 225 beats per minute, paralleling the loss in body weight as fuel was used (Butler *et al.* 2003).

Another study of wild birds involved Swainson's Thrushes *Catharus ustulus* and Hermit Thrushes *C. guttatus* in the United States (Wikelski *et al.* 2003, Cochran & Wikelski 2005). Using doubly-labelled water and heart rate measurements, the daily energy costs of individuals that were not migrating at the time averaged 88 ± 5 kJ per day, while flight costs in migrating birds added another 15.5 kJ per hour of flight (=4.3 watts). The longest migration recorded on any one day in tagged birds was 8 hours, giving a flight cost on that day of about 124 kJ. The entire northward migration of 4800 km over 42 days was estimated to cost about 4500 kJ or 0.9 kJ per km flown, of which the flight costs formed only 29%, the rest being spent on maintenance and fuel acquisition. This overall cost could be provided by about 120 g of fat, many times more than the maximum carried by these small birds at any one time.

In one experimental study, flight costs emerged as lower than expected. Using the doubly-labelled water technique, Kvist *et al.* (2001) measured the total metabolic power output of a long-distance migrant, the Red Knot *Calidris canutus*, flying for 6–10 hours in a wind tunnel. The total metabolic output increased with fuel load, but proportionally less than the predicted mechanical power output from the flight muscles. The authors suggested that the efficiency with which metabolic power input is converted into mechanical output by the flight muscles increases with fuel load. This is not accounted for in existing models, and (together with wind effects) may explain why some birds routinely fly further than would be predicted by these models. Existing models make assumptions about the ratio of chemical to mechanical power output, and also assume that

---

[3]*In this technique, the production of CO_2 during flight can be determined by loading the bird with $^2H_2^{18}O$ (doubly-labelled water) and measuring the levels of the isotopes in the blood before and after the flight. The method is based on the observation that the oxygen in respiratory CO_2 is in isotopic equilibrium with the oxygen in body water. Thus, the hydrogen of body water is lost primarily as water, whereas the oxygen is lost both as water and as CO_2. Carbon dioxide production can be estimated from the difference in turnover rates of the ^{18}O and $^2H_2$.*

this ratio remains constant during a flight. The discrepancies between predictions from aerodynamic models and empirical findings have implications for other models, such as those concerning stopover behaviour, optimal staging time, optimal range and optimal fuel load (e.g. Alerstam & Lindström 1990, Weber & Houston 1997a, 1997b).

On the basis of prevailing information on energy reserves and flight costs, several estimates have been made of the flight ranges of particular birds. Even the tiny Ruby-throated Hummingbird *Archilochus colubris* has an estimated flight range in still air exceeding 1000 km, easily enough to cross the Gulf of Mexico. For a passerine with a fat-free body mass of 20 g and moderate fat deposits of 10 g (yielding 39 kJ per g), with an assumed flight speed of more than 30 km per hour, theoretical flight duration amounts to more than 40 hours, and the potential flight distance in still air to more than 1200 km, and with a following wind much further, enough to cross the Sahara Desert. On the same basis, Sedge Warblers *Acrocephalus schoenobaenus* leaving southern Britain at 21 g could reach sub-Saharan Africa without further refuelling, and Reed Warblers *Acrocephalus scirpaceus* at 15.5 g could reach Iberia. For birds that have large fat deposits, faster and more efficient flight, such as shorebirds, theoretical flight distances reach values up to 10 000 km or more (Pennycuick & Battley 2003). Some migrants thus have non-stop flight ranges similar to those of large aeroplanes, and it is easy to see how such species can cross large stretches of sea (Berthold 1975, Williams & Williams 1990). Hence, one reassuring finding from these estimates is that birds seem fully capable on physiological grounds of achieving the long non-stop flights indicated by evidence from observations and ring recoveries.

Bar-headed Geese *Anser indicus* crossing over the Himalayas

Chapter 6
Incredible journeys

Lofty mountain ranges and wide belts of desert are traversed, and lesser or vaster expanses of sea are crossed. (William Eagle Clark 1912.)

Some birds make what seem to us astounding journeys over seas, deserts, high mountains, or other hostile terrain. This chapter focuses on some of these difficult journeys, and the various adaptations that enable birds to complete them successfully. The fact that some birds can make such journeys does not imply that all could do so. Much depends on features of the bird itself, such as its size and physiology, and also on the habitat to which it is adapted. Oceans are inhospitable for landbirds, continents for pelagic seabirds, open country for forest birds, forests for open country birds and barren deserts for almost all birds. And on some overland routes, waterfowl and waders may encounter few sites where they can rest and feed.

Most flying birds can migrate over small areas of hostile habitat that they can cross in a few hours. The difficulties come mainly on longer journeys which require more than 24 hours of non-stop flight, or entail physiologically harsh conditions, such as temperature and humidity extremes, or greatly reduced oxygen levels.

SEA-CROSSINGS

Landbirds that migrate over oceans provide some of the most extreme examples of endurance flight and precise navigation when they travel, without opportunity to feed, drink or rest, over vast areas of open water devoid of helpful landmarks. Nor can they stop and take shelter, as birds do over land, when the weather turns against them. While some landbirds on overwater flights take whatever opportunities for rest are available, stopping on ships, oil-rigs and other installations, or even on mats of floating vegetation or other flotsam, these individuals probably represent an exhausted minority, and almost certainly most landbirds make water-crossings non-stop.

In all long sea-crossings, the behaviour of the birds with respect to weather is paramount. Birds often accumulate in coastal areas, sometimes waiting for days at a time for conditions to improve. Even then, they often seem reluctant to leave, flying along the coast or turning back inland. To accomplish a long sea-crossing successfully, birds need a good tailwind, and an adequate supply of body fuel, sufficient not only for the journey, but also for any emergencies that might arise. Even with favourable weather at take-off, birds have no means of predicting how it might change on a flight lasting up to several days, on a journey of up to several thousand kilometres. Any long-distance migrant would therefore benefit from having sufficient reserves to cope with adverse conditions which a human pilot, benefiting from weather forecasts, could anticipate and avoid. Landbirds making long overwater flights therefore need generous fuel reserves, and it is little wonder that such species accumulate before they set off some of the largest reserves of body fat recorded among birds.

A second problem for landbirds making overwater flights concerns navigation. They have no ground features below to help keep them on course, but must rely entirely on celestial or magnetic cues. If they encounter mist or rain, they often become disoriented, and mill around for hours or drift far off course, as radar observations confirm. Nor can they shelter from rain, for if they come down on the water, they usually perish. In sea-crossings, natural selection acts harshly against any kind of mistake or mishap, and abundant evidence points to heavy mortality of migrating landbirds over water (Chapter 28). One way round these problems is to fly high enough to avoid mist and rain, and on long overwater flights birds have often been detected by radar at heights exceeding 4 km above sea level, but high-altitude flights bring other problems (see later).

Despite the difficulties, all the major seas and oceans of the world are crossed regularly by some landbird species (**Figure 6.1** and **Table 6.1**). Each year, millions of birds of a wide range of species cross on a broad front the Mediterranean Sea in the Old World or the Gulf of Mexico in the New World (Moreau 1961, Gauthreaux 1999). At their widest points both these crossings involve flights of more than 1000 km. One of the species that crosses the Gulf is the tiny Ruby-throated Hummingbird *Archilochus colubris*, which doubles its normal 3.5 g body weight with extra fat in order to do so. Frequent storms make the trans-Gulf crossing one of the most hazardous in the world, especially in the autumn hurricane season. It is unlikely that any small birds could withstand a hurricane at sea. Smaller numbers of other birds also cross all the major inland waters, such as Caspian and Baikal in the Old World and the Great Lakes in the New World.

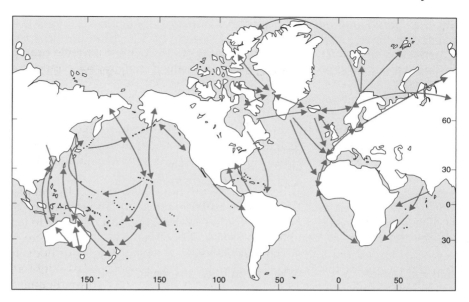

Figure 6.1 Some major sea-crossings undertaken by landbirds. Some journeys are made only in one direction, while the return is made over a more overland route. Mainly from Williams & Williams (1990).

Crossing these waters by direct flight saves considerable energy over the alternative of flying around them.

Smaller numbers of species also make long flights over parts of the Atlantic, Pacific and Indian Oceans. These journeys run up to several thousands of kilometres, and involve an estimated 50–100 hours of non-stop flight, which can be lengthened in adverse conditions. They include the most extreme long-distance endurance flights performed by birds.

Some species make the journey from eastern North America, across the Atlantic direct to South America. Judging by their appearance on Bermuda, which lies 900 km east of the North American coast, more than 80 species make this journey each autumn, and others occasionally (e.g. Wingate 1973). Ships have reported that most birds appear within 600 km from the coast, but others appear at more than 2000 km, mainly warblers and shorebirds (McClintock *et al.* 1978). Small birds flying from Nova Scotia over the Atlantic face a non-stop 2400 km flight to the West Indies, or 3700 km to South America. Small Blackpoll Warblers *Dendroica striata*, with a fat-free weight of about 11 g, can double their total body weight before departure as they accumulate sufficient fuel (Nisbet 1963b, Nisbet *et al.* 1963). Some shorebirds that breed on the northern tundras set off from staging areas in Newfoundland and Nova Scotia and strike out over the Atlantic to South America. They make the longer journey of up to 3700 km, but can fly twice as fast as the Blackpoll Warbler (60–70 km per hour versus 30–35 km per hour), so accomplish the flight in around half the time (Nisbet 1970, McClintock *et al.* 1978). They include the American Golden Plover *Pluvialis dominica*, Hudsonian Godwit *Limosa haemastica* and White-rumped

Table 6.1 Some long overwater journeys of migratory landbirds. Flight durations are calculated assuming still air, and for flight speeds of 30–35 km per hour (typical of small songbirds) and of 60–70 km per hour (typical of small shorebirds). In practice, most overseas migration occurs with the benefit of a tailwind, which reduces the flight times

Journey	Species	Distance (km)	Flight duration (hours) at	
			30–35 km per hour	60–70 km per hour
Siberia to Australia, across the Pacific	Bar-tailed Godwit, *Limosa lapponica*, Great Knot *Calidris tenuirostris*	8000	–	114–133
Alaska to Hawaii and other American Pacific Islands[a]	Pacific Golden Plover *Pluvialis fulva*, Bristle-thighed Curlew *Numenius tahitiensis* and other shorebirds	5000	–	71–83
India to southern Africa, across the Indian Ocean	Amur Falcon *Falco amurensis*	4000	–	57–67
Greenland to southwest Europe, across the Atlantic	Northern Wheatear *Oenanthe oenanthe*	2000–3000	57–100	–
China to Australia across the Pacific	Red Knot *Calidris canutus* and other waders	4000	–	57–67
Northeast North America to South America across the Atlantic[b]	Various waders some passerines (including 13 parulid warblers)	2400–3600	69–120	34–60
North America to South America, across the Gulf of Mexico	Many and various species	Up to 1300	37–43	19–22
Europe to Africa, across the Mediterranean	Many and various species	Up to 1100	31–37	16–18
Iceland to Britain, across the Atlantic	Various species	1100	31–37	16–18

[a] Include Lesser Sand Plover *Charadrius mongolus*, Whimbrel *Numenius phaeopus*, Pacific Golden Plover *Pluvialis fulva*, Bar-tailed Godwit *Limosa lapponica*, Lesser Yellowlegs *Tringa flavipes*, Wandering Tattler *Heteroscelus incanus*, Grey-tailed Tattler *H. brevipes*, Long-billed Dowitcher *Limnodromus scolopaceus*, Ruddy Turnstone *Arenaria interpres*, Sanderling *Calidris alba*, Least Sandpiper *C. minutilla* and ducks such as Northern Pintail *Anas acuta* and Northern Shoveller *A. clypeata*, although the bulk of their populations winter elsewhere. Only the Bristle-thighed Curlew *Numenius tahitiensis* winters entirely on Hawaii and other Pacific Islands. For the Bar-tailed Godwits that migrate from Alaska or Siberia to New Zealand see text.
[b] At least eight shorebird species: White-rumped Sandpiper *Calidris fuscicollis*, Least Sandpiper *C. minutilla*, Lesser Yellowlegs *Tringa flavipes*, Hudsonian Godwit *Limosa haemastica*, Red Knot *Calidris canutus*, Short-billed Dowitcher *Limnodromus griseus*, Semipalmated Sandpiper *Calidris pusilla* (McNeil & Cadieux 1972). At least 13 parulid warblers, notably Blackpoll Warbler *Dendroica striata* (Nisbet *et al.* 1963). Other warbler species make this Atlantic flight occasionally, for at least two dozen species turn up frequently in autumn in Bermuda which lies 900 km east of the North America coast at that latitude (Scholander 1955, Wingate 1973). Studies using a network of radar sites agree with the estimates that small passerines take more than 80 hours to accomplish the 3700-km non-stop flight between northeast North America and northern South America (Williams *et al.* 1978).

Sandpiper *Calidris fuscicollis*. Shorebirds and passerines on this journey have been found by radar to reach heights of 4–6 km (Williams *et al.* 1977)), taking advantage of the northwest winds that follow a front. Flight speeds measured by radar gave estimates of the flight times of passerines at 18 hours to Bermuda, 64–70 hours to the Caribbean and 80–90 hours to South America (Williams *et al.* 1978). Waders could presumably cover these distances in half the time. This overwater 'shortcut' saves over 1000 km on the more roundabout overland coastal route. In the spring, however, when winds are against them, these same species return by the overland route, northward through eastern North America.

Other long sea-crossings by North American birds are undertaken by shorebirds and waterfowl that migrate direct from Alaska over the eastern Pacific to make landfall at various sites between southern Canada (2500 km) and Baja California (up to 5000 km), depending on species. Again, some species make this flight only in autumn when winds are favourable and return in spring by the longer land-ward route. Other long overwater journeys are flown by shorebirds and ducks that migrate between Alaska and Hawaii (>4000 km) or between Alaska and south Pacific islands without stopping on Hawaii (6000–9000 km) (Thompson 1973, Johnson *et al.* 1989, 1997, Williams & Williams 1990, 1999, Marks & Redmond 1994). These journeys are remarkable, not only for the distances involved, but because of the great precision of navigation required to find such tiny wintering areas in the vastness of the Pacific. Participants include the Bristle-thighed Curlew *Numenius tahitiensis*, which is the only migratory bird species that breeds on a continent and winters entirely on Pacific Islands.

The most impressive of all overwater migrations by a landbird, however, is undertaken by the Bar-tailed Godwits *Limosa lapponica baueri* from eastern Siberia and western Alaska, which apparently fly an astonishing non-stop 10 400 km to eastern Australia and New Zealand. The shortest (Great Circle) distance from coastal Alaska to the North Island of New Zealand is 10 260 km, requiring in typical winds an estimated 175 hours (7.3 days) of flight. This flight requires enormous fat reserves, and considerable shrinkage of other body organs to reduce weight. When the birds depart, they are about twice their normal weights (Chapter 5). Astonishingly, by the time they reach New Zealand, birds have already flown more than 10 000 km without feeding, but some birds killed soon after arrival still had sufficient fat to fly another 5000 km, enough to reach the South Pole (Pennycuick & Battley 2003). The birds evidently travel with a substantial safety margin, in case they are forced down, or blown off course by adverse winds. The main advantage of this oversea migration is the saving in distance, for if the birds migrated round the coasts of eastern Asia (as they do in spring), they would have to travel about 16 000 km one way, a journey almost 40% longer than the direct trans-Pacific route. The single flight may also be safer (fewer predators) and healthier (fewer patho-gens, Piersma 1997), as well as being accomplished in a much shorter period of about seven days in favourable winds. The reason that birds take the longer route in spring is that the winds are unfavourable for a northern transoceanic flight.

This is such a remarkable migration that it is worth examining the evidence that the majority of individuals do indeed fly non-stop (Gill *et al.* 2005). The god-wits involved are a distinct race *L. l. baueri*, and many have been colour-marked, so that they can be identified from a distance without being captured. Yet virtu-ally none has been reported from along the eastern Asia coasts during the autumn migration period (a time when other marked shorebirds are commonly reported from that region). Second, the peak southward departure from Alaska is followed about a week later by the peak arrival in New Zealand. Third, considering that 150 000 *L. l. baueri* godwits make this journey every year, very few are sighted on Pacific Islands where they might be expected to stop. Such sightings occur mainly in the September–November migration period, on islands that lie on a direct route between Alaska and eastern Australia/New Zealand, and especially near journey's end where most fallout would be expected. Fourth, the bird appears energetically and mechanically able to complete such a flight non-stop. Flight simulation models, the bird's extreme fat loads, and the selection of suitable wind conditions from Alaska, all support the notion of a direct flight. With the benefit of a tailwind, godwits can travel at more than 100 km per hour, and by analogy with other shorebirds, they should be capable of remaining airborne for the required period. Lastly, known departures from Alaska occur on winds favourable for a southerly flight, rather than on winds that would take them on a southwesterly course round the coast of Asia. All these lines of evidence point to a direct non-stop flight.

On their northward migration, *baueri* godwits are thought to undertake a two-step journey, the first flight of 8000–10 000 km from New Zealand and eastern Australia to staging areas around the Korean peninsula, Japan or the north coast of the Yellow Sea, where they refuel before the shorter second stage which takes them to Siberia or Alaska. Birds that leave New Zealand in spring have fat reserves not much smaller than those that leave Alaska in autumn, but the autumn birds have shrunk their internal body organs to a much greater degree, giving them a much larger proportion of fat in the body. This greater 'fat fraction' enables them to make a much longer journey than with the same weight of fat in a heavier body. Other Bar-tailed Godwits perform long-distance migrations elsewhere in the world, but none as long as the *baueri* birds.

Other long sea-crossings of note are made by the Amur Falcons *Falco amurensis* which in autumn fly the 4000 km direct from India to southern Africa. These are the longest overwater flights regularly performed by a raptor. Similarly, Northern Wheatears *Oenanthe oenanthe* in autumn cross 2000–3000 km of the North Atlantic between Greenland and Europe before moving on to Africa. These are some of the longest overwater flights performed by a passerine. Both species take somewhat different routes on their return journeys when winds are less favourable for the most direct route. In this respect they resemble many of the species mentioned above.

Examples of long overwater migrations by southern hemisphere landbirds are provided by three species that breed in New Zealand, namely the Double-banded Plover *Charadrius bicinctus*, which crosses 2000 km of sea to Australia, the small Shining Bronze Cuckoo *Chrysococcyx lucidus*, which crosses 2500 km of sea to the Bismarck and Solomon Islands, and the much larger Long-tailed Cuckoo *Eudynamys taitensis*, which crosses 2700 km to Samoa and Fiji and other islands of the central Pacific east to the Caroline, Line and Phoenix Islands. Even within the tropics, some landbird species move up to 3000 km between breeding and non-breeding areas, part of which may involve a sea-crossing: for example, the Broad-billed Roller *Eurystomus glaucurus* migrates between Madagascar and the Congo, the Roller (Dollarbird) *Eurystomus orientalis* migrates between northern Thailand and Borneo-Java, and the Pied Cuckoo *Clamator jacobinus* and others between India and East Africa.

DESERT CROSSINGS

Overall, one-third of the earth's land area is classed as desert (**Figure 6.2**). The term covers a range of arid landscapes, from the sparsely vegetated regions of western North America to totally barren rock and sand typical of much of the Sahara. Providing there is some vegetation, some specialist bird species can survive and breed in such arid habitats, but most birds cannot find sufficient food or water, or remain active at such high temperatures. Outside the tropics, most deserts are at their best in spring, after winter rains have promoted plant growth and flowering. They are much less use in autumn, when plants have withered under the desiccating heat of summer. The problems of desert flights include the long distances without food or water, searing daytime temperatures and low humidities, which can lead to risks of overheating and dehydration. Birds can avoid the extremes of heat by migrating at night or at high altitudes, but crossing the biggest deserts requires more than 12 hours of flight, even by the fastest birds.

The main sources of water in deserts are rivers flowing through from wetter regions, which provide green bands of riparian vegetation in otherwise parched landscapes, and the scattered oases which form wherever groundwater breaks the surface. These features are visible from a long distance, being marked by taller green bushes and trees than the surrounding desert, and provide habitat, food

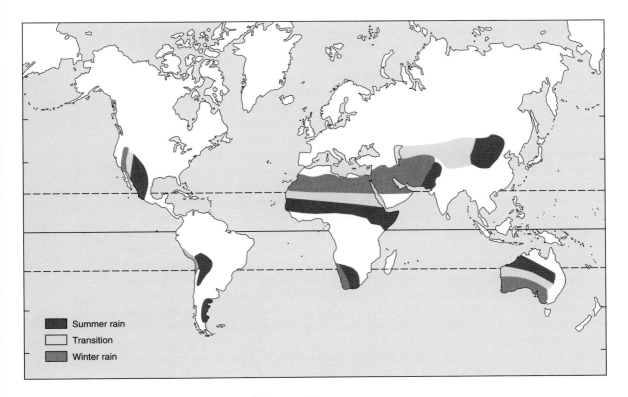

Summary
- Summer rain
- Transition
- Winter rain

Figure 6.2 The main desert areas of the world.

and water for a range of birds. However, oases and their associated vegetation typically cover small areas, of up to a few hectares, and are usually few and far between, up to several hundreds of kilometres apart. Although at migration times they can seem packed with birds, they are likely to be visited by only a tiny proportion of those passing over. Moreover, oases are normally used only by those species adapted to the habitats they provide.

The largest and most severe desert in the world sits right across one of the main bird migration routes, between western Eurasia and the Afrotropics. The Sahara stretches from the Atlantic coast in the west to the Saudi Arabian peninsula in the east, extends 1000–1500 km from north to south, and covers an area of 12 million km², about the same as the USA. Much of it is totally bereft of vegetation, lacking even the sparsest of desert scrub. Daily maximum temperatures during the autumn and spring migration periods average 25–38°C, but sand surface temperatures during the day may reach 70°C. Relative humidity can drop below 10%. The desert varies greatly in severity in different longitudes. The Atlantic coastal strip in the west provides a continuum of scrubby vegetation running north to south. In places, this coastal strip is more than 200 km wide, beyond which the scrub becomes progressively sparser into the desert. In the east, the Nile Valley provides a narrow green corridor for migrants, the only route that is abundantly supplied with both food and water. Over the rest of the desert, vegetation is mostly sparse to non-existent, apart from the few isolated oases. Yet despite its inhospitability, this desert is crossed and re-crossed at all longitudes by an estimated 5000 million birds of 186 species every year (Moreau 1972). Revised estimates, made 40 years later and based on radar observations at sites across the Mediterranean region, suggest figures of 3500–4500 million birds (B. Bruderer, personal communication), but these recent estimates (unlike the earlier one) exclude birds that enter Africa via Arabia and follow several decades of population decline among many migratory species in Europe (Chapter 24). Taking these facts into account, the two estimates, found in different ways, are remarkably consistent.

Birds that follow the Nile Valley in the east, with its narrow border of swamps and cultivated land, or the Atlantic coast in the west with its broad belt of sparse desert scrub, could in theory migrate with short flights and frequent stops, as they could find food at various points en route. The concentrations of migrants are greater at the west and east ends of the desert than in the middle section, at least in autumn, as revealed by radar, but concentrations seem unexceptional in the Nile Valley itself, apart from waterbirds (Moreau 1961, Biebach 1990, Bruderer & Liechti 1999). The option of flight from oasis to oasis is possible only along a row of mountains and oases that stretches southeast from Morocco. Elsewhere, oases are too few and far between. Furthermore, most of the Saharan oases are no longer in a natural state, the native vegetation having been largely removed or replaced by date palms, which offer little sustenance for birds. Most species apparently migrate on a broad front over the most hostile parts of the desert, but the same species may behave differently in spring and autumn, or at different points in the crossing (Moreau 1961).

The greater density of migrants at the west and east ends of the desert in autumn may be as much a consequence of the funnelling effect of land areas to the north as of the desert conditions themselves. Most of the west European migrants cross the Mediterranean at its narrowest, around Gibraltar, from which they reach the western edge of the desert, while the east European and Asian migrants cross the desert in its eastern sector or cross Arabia instead. Moreover, throughout its breadth, the desert would have offered better conditions a few thousand years ago than it does today. The difficulty of this journey is increasing all the time, as the desert expands north and south under human influence. The annual rate of desertification near the southern edge has been calculated at about 0.5%, which corresponds to an area of at least $60\,000\,km^2$ added each year by degradation. In these times of climate change, other areas on the migration route are also becoming drier, including parts of Spain, much of North Africa and the Sahel zone to the south. These various changes mean that, over the centuries, birds have had to cross ever widening areas without access to food or water, while suitable staging areas have become smaller and more widely spaced. These trends are likely to continue into the future.

The scorching daytime temperatures, combined with lack of food and water, make crossing the Sahara one of the most arduous journeys undertaken by migratory birds. For some birds, it also follows (in autumn) or precedes (in spring) a crossing of the Mediterranean Sea at its widest points. The west and east parts of the North African coastal region support abundant green habitat extending hundreds of kilometres to the south, but in the central part little vegetated habitat is available between the sea and the desert, offering only limited opportunities for refuelling. This central coastal strip over most of its length is typically less than 20 km from north to south, consisting of sparse perennial scrub, with the addition in spring of ephemeral herbage dependent on winter rain. Beyond this to the south lies barren desert.

Many species cross the Sahara by accumulating the necessary body reserves beforehand (**Figure 6.3**; Chapter 5). Most passerines increase their body weight by more than 50%, and some by around 100%, effectively doubling their weights before departure (Ward 1963, Fry *et al.* 1970). For example, the Garden Warbler *Sylvia borin* weighs about 18 g in the breeding and winter seasons, but increases its body mass up to 37 g before crossing the Sahara, whether on outward or return journeys (Bairlein 1991b).

In some bird species, at least some individuals seem to cross both the Mediterranean Sea and Sahara Desert in a single non-stop flight, while others break their journey for refuelling in North Africa (Moreau 1961, Casement 1966, Bairlein 1992, Biebach 1998). The sea-plus-desert crossing would require a flight of 1500–2500 km, depending on the route taken, and the desert crossing alone would require a flight of 1000–1500 km. The Sahel zone, to the south of the desert, is at its driest in spring, which adds another 500 km to the desert crossing. This is offset to some extent by the greening of about 300 km of the northern Sahara as a result of spring rains in the Mediterranean region (see below). The net effect of these seasonal changes is therefore to increase the desert crossing by about 200 km in spring compared with autumn. Moreover, these distances refer to north–south crossings, and any bird traversing the desert diagonally could face an even longer flight.

A combined sea and desert crossing in autumn has been claimed for various waterfowl and waders and for some slower flying species, such as Sedge Warbler *Acrocephalus schoenobaenus* (Gladwin 1963), Yellow Wagtail *Motacilla flava* (Wood 1992) and Common Cuckoo *Cuculus canorus* (Moreau 1972). Evidence stems not

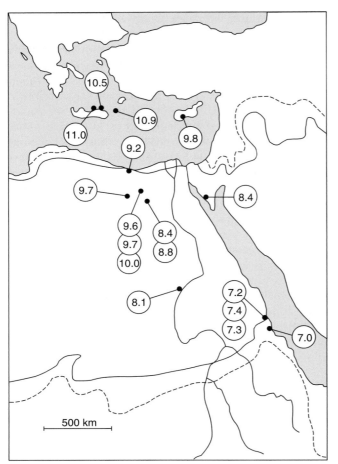

Figure 6.3 Body weights of Willow Warblers *Phylloscopus trochilus* at different stages of their autumn migration over the Mediterranean Sea and Sahara Desert. Of all the species that cross the Sahara, the Willow Warbler is most numerous. The fat-free body weight of this tiny bird is about 7 g. The mean weights of trapped samples at various locations on the autumn migration route reveal weights of 10–11 g on Cyprus and other islands in the north Mediterranean, decreasing gradually through localities en route to 7–7.4 g in the south of the Sahara. Body weights taken from the same places in different years showed little variation; but weights of birds at different sites in the same general region were often markedly different, depending on habitat. From Biebach (1990).

only from the relative scarcity of these species on the ground in North Africa in autumn, but also from the body reserves accumulated, which are calculated to be sufficient for a journey of this length. A combined sea and desert crossing may also be performed by some hirundines, swifts and bee-eaters, which could pick up food on the way, at least over vegetated areas.

While some species accumulate most of the body reserve necessary for the journey close to the desert, others accumulate most of their reserve beforehand, up to several hundred kilometres north of the desert in autumn, or several hundred kilometres to the south of it in spring (Chapter 5). This further increases the distance that must by flown without substantial further fuelling. For example, Pied Flycatchers *Ficedula hypoleuca* fatten mainly in northern Iberia in autumn, and may then migrate without feeding until they reach the south of the Sahara. Sedge Warblers *Acrocephalus schoenobaenus* are even more extreme, accumulating sufficient body reserves as far north as Britain to get them to the south of the Sahara. In the spring, Blackcaps *Sylvia atricapilla*, Garden Warblers *S. borin*, Whitethroats

S. communis and other species fatten well south of the Sahara, in the more mesic Guinea zone (Hjort *et al.* 1996, Ottosson *et al.* 2005), while Bar-tailed Godwits *Limosa lapponica* migrate directly from West Africa to the Dutch Waddensea, a distance of 5000 km (Drent & Piersma 1990). In effect, the Saharan crossing is much longer for forest-dwelling passerines than for open-country ones which can find foraging areas at the northern and southern edges of the desert. In one study, forest-dwelling species, such as the Golden Oriole *Oriolus oriolus*, Wood Warbler *Phylloscopus sibilatrix* and various flycatchers, were found to arrive in spring on Ventotene Island (off southern Italy) almost devoid of fat reserves, whereas open-country species still had enough fat, on average, for more than 300 km of further flight (Pilastro & Spina 1997). It is clear, therefore, that some species migrate without refuelling over substantially longer distances than are needed to cross only the Mediterranean and Sahara. Different species adopt different fuelling patterns, depending on the routes taken and locations of suitable feeding areas (Chapter 5).

Large birds flying non-stop north–south at 80 km per hour (say) in still air would require around 19–31 hours for the sea and desert journey, and around 13–18 hours for the desert alone; while small birds flying non-stop at half this speed would require twice as long (e.g. up to about 62 hours). The most numerous species to make this journey is the Willow Warbler *Phylloscopus trochilus*, which is also one of the smallest (fat-free weight about 7 g) and slowest fliers (still-air speed on migration 34 km per hour). This species would take 44–74 hours for the sea and desert journey, and 29–44 hours for the desert alone. In all such species, however, these times could be substantially shortened with the help of tailwinds, or lengthened by headwinds or stops, or by crossing the desert diagonally. In the daytime, frequent thermals could also provide lift and help to speed the flights of any species active then at low enough altitude. Nevertheless, this seems a staggering achievement for birds weighing as little as 12–40 g at take-off.

Conditions for trans-Saharan migration differ greatly between autumn and spring. When the migrants move south in autumn, the Sahel zone immediately to the south is near the end of its wet season. Vegetation is green, flood pools are common and insects are plentiful. This reduces the distance that must be flown without feeding, and favourable southward winds on most days help to speed the birds on their journeys. But in spring, when the migrants move north, the Sahel zone is reaching the end of the dry season, when conditions are by no means ideal for refuelling, and when the same southward winds prevail, hampering northward progress. According to Moreau (1961), birds crossing the Sahara in spring must expect to meet headwinds unless they are travelling diagonally northeast at altitudes greater than 2 km. At these higher altitudes, they meet the anti-trade winds blowing in the opposite direction to those below, and it seems likely that most birds take advantage of them (radar having revealed higher flight altitudes in spring than in autumn). Otherwise, the only way to reduce the effect of headwinds would be to fly low, say within 500 m of the ground.

So, while in autumn many small birds are apparently able to cross the Mediterranean and Sahara without replenishing their reserves, they are less likely to do so in spring. This seasonal difference in conditions may explain why many more migrants are seen in North Africa and around Mediterranean coasts in spring than in autumn, even though overall population levels are lower in spring (Nelson 1973, Shirihai 1996). As indicated above, conditions in the Mediterranean region are much better in spring than in autumn. As a result of winter rains, the migrants in spring encounter a flush of fresh vegetation and insects, together with areas of standing water. The implication is that many more birds make a combined Mediterranean–Saharan crossing in autumn than in spring, and that many more stop to refuel in North Africa in spring than in autumn.

In crossing the desert, some small passerines may make a single non-stop flight, while others have been found to break their journeys in the daytime, but mostly without feeding. At least some individuals of many passerines species (notably Garden Warbler *Sylvia borin*, Lesser Whitethroat *Sylvia curruca* and Willow Warbler *Phylloscopus trochilus*) descend to spend the day sheltering motionless near oases, or even in the open desert in the shade of rocks, continuing their journey in the cool of night (Bairlein 1987, 1988, 1992, Biebach 1990, 1992, Biebach *et al.* 1986, Bruderer 1994). In this way, they could cross the desert in a few nocturnal flights. The majority of trapped migrants at such sites showed high body mass and fat loading, with sufficient reserves for onward flight, so had not been grounded through lack of fuel. In resting by day and flying by night, such birds travelled in cooler conditions, and may thus have reduced their dehydration risks, but they would have added at least one or two nights to their journey time (giving 2–4 nights for the total journey across the desert). Yet other individuals of these same species stopped at oases where they fed and drank, usually remaining for 2–4 days, but occasionally for up to three weeks at a time, if they needed to rebuild their body reserves. However, observations at oases indicate that the birds stopping are only a tiny proportion of the numbers that must be passing. Strictly diurnal migrants, such as broad-winged raptors dependent on daytime thermals, may stop every night (as shown by radio-tracking of several species, Chapter 8), but would normally obtain no food then.

Further studies are needed to find the relative frequencies of non-stop Mediterranean–Saharan flights, two-step flights with a stop in North Africa, or multi-stage flights, stopping in North Africa and at points in the Sahara (with use of oases or of stopping points along the Nile Valley). The relative frequencies of these different patterns clearly differ between autumn and spring, linked with different winds, ground conditions and time constraints. From autumn radar studies at sites in the Egyptian Sahara, Biebach *et al.* (2000) concluded that about 20% of all migrants detected were involved in non-stop migration and 80% in intermittent migration, with stopover at the coast (70%) or in the desert (10%). Other studies have suggested that in Mauritania an even greater proportion of individuals descend and rest by day (B. Bruderer, personal communication).

Physiological constraints

On the face of it, stopovers in the desert without any possibility of refuelling would seem to waste time and energy, but they may help to conserve water, because the bird can seek shade and remain inactive, so as to minimise internal heat production. The possibility that water rather than energy might restrict the flight lengths of birds has long been appreciated, because of early findings that, at laboratory temperatures, the water loss of birds exceeds their metabolic water production during daily activities (Hart & Berger 1972). The need to retain water could therefore be a more important constraint than fuel for small birds crossing the Sahara. Only about

15–25% of the energy expended by muscle is converted to mechanical power, and the rest to heat. The body temperatures of birds measured during flight are generally greater than 41°C, up to 4°C higher than normal. To avoid overheating, the bird must pant, which leads to water loss and eventually to dehydration (Chapter 4). In the absence of drinking water, this risk can be offset by metabolic water production, or by flying at high altitudes where the air is cooler, or at night when temperatures are lower than during the day.

The metabolism of fat, carbohydrate or protein results in the production of 1.1, 0.90 and 0.82 g of water per g of wet tissue expended, or 0.03, 0.23 and 0.16 g of water per kJ of energy released (see **Table 5.1**). For flying birds, metabolic water derived in this way has been estimated as sufficient to offset the danger of overheating in ambient temperatures up to around 10°C (Biebach 1990, Nachtigall 1990). Small birds flying at air temperatures up to this level should therefore have no additional water loss, even on long flights; but above 10°C further evaporative cooling would be needed, rapidly leading to dehydration. In the Sahara in autumn, if one assumes an air temperature at ground level of 30°C by day and 8°C by night, and a decline of 7°C with each 1000 m rise in elevation, an air temperature lower than 10°C would be found by day at more than 3000 m above ground, and by night at more than 1000 m above ground. In spring, when the ground would typically be cooler than in autumn, an air temperature of 10°C would be found at an altitude greater than 1750 m by day and 500 m by night. At the time of writing, however, doubt hangs over the 10°C estimate, as subsequent work (still in progress) suggests it may be higher, which would alter the altitude estimates. At very high altitudes, reduced humidity could raise water loss, with the thinner air leading to increased ventilation, and hence to further increase in respiratory water loss (Carmi *et al.* 1992, Carmi & Pinshow 1995, Klaassen 1995, Klaassen *et al.* 1999). Overall, the best option for a small bird may be to fly at night at moderate altitude (say 500–2000 m) and rest in the shade by day.

Birds are unlikely to fix their flight altitudes in terms of water conservation alone, however, for they must also take account of other factors, notably wind conditions, which affect the energy costs of flight. Precise radar measurements of the height distributions of migrants over southern Israel in autumn and spring, combined with simultaneous altitudinal recordings of weather variables, offered an opportunity to examine the presumed importance of wind conditions, energy and water constraints on flight range (Liechti *et al.* 2000). Predictions were made of the optimal flight altitude, in the conditions prevailing, if birds took account only of tailwinds, only of energy conservation, or of both energy and water conservation, in achieving maximum flight range. These predicted flight altitudes were then compared with the actual height distributions recorded in the field. The authors concluded that wind profiles, and thus energy rather than water limitations, governed the altitudinal distribution of nocturnal migrants. However, this conclusion, drawn from large samples of birds spread over a wide range of altitudes, does not exclude the possibility that individuals may adjust their altitude to whatever was most limiting for them at the time, including water if they were becoming seriously dehydrated.

Regarding fuel type, calculations for Willow Warblers *Phylloscopus trochilus* and Eurasian Golden Orioles *Oriolus oriolus* revealed that, under most flying conditions, if the birds used fat alone as fuel, water could impose the main limitation to their total flight range, but if they used 70% fat and 30% muscle (which is two-thirds water), then energy rather than water could become the main constraint (Klaassen & Biebach 2000). The bird would then gain only 74% as much energy as from fat alone, but 26% more water. This situation is probably closer to reality, and it would also allow birds to fly at relatively very high altitudes (in cooler but thinner air) than if they used fat alone without incurring a water debt.

Water loss could also be reduced by increasing the oxygen extraction efficiency, which would allow reduction in the throughput of air, and hence in water lost during respiration.

Further insight can be gained from studies of body composition. The proportion of water within a bird's body decreases with increase in fat content, because fat is stored in anhydrous state. However, the fat-free component of a bird's body usually contains around 67% water, which can reduce by about one-third to around 55% before the bird expires from dehydration (Haas & Beck 1979). For Willow Warblers *Phylloscopus trochilus*, Biebach (1990) calculated reduction in body water over one night from 67.5 to 62.9%, which is well within the range of dehydration tolerance, and over two rest days, further reduction to 57.9%, which is still above the value at which death becomes likely. Moreover, the water contents of birds caught in the Sahara Desert were in general around 67–69% of the fat-free weight (equivalent to a water:lean dry mass ratio of around 2:1), which is within the usual range of healthy birds (Chapter 5). In addition, the levels of Na^+ ions and urea in the bloodstream, measured as further indicators of the state of body hydration, were normal in most birds, and high only in lean ones (Biebach 1990, 1991, Bairlein & Totzke 1992). Furthermore, Willow Warblers and other small passerines found dying in the Libyan Desert showed normal water levels, but had run out of fat, so in these birds, fuel rather than water seemed to have been limiting (Biebach 1991).

Most of the evidence cited for lack of water stress in migrant birds derives from the absence or scarcity of birds found with lowered water contents, even near the end of a long migration (Biebach 1990, 1991, Gorney & Yom-Tov 1994, Landys *et al.* 2000). This is not true of all samples, however, for 78% of 409 birds caught on the Egyptian coast in autumn, after crossing the Mediterranean Sea, had water:lean dry mass ratios less that 2:1, including 12% with ratios less than 1.4:1 (Fogden 1972a, Chapter 5). The latter were close to the level at which death would be expected to occur. To some extent, as emphasised already, birds could offset increasing water loss by catabolising body protein to yield water. The finding of migrants in the Sahara Desert with high fat but low protein reserves might reflect effects of dehydration.

Migrants clearly show behaviour that could be construed as anti-dehydration, such as shade-seeking, remaining immobile during the hottest part of the day, and drinking heavily on arrival at stopover sites (Biebach 1990, Klaassen 2004). At such a site in the Negev Desert, Blackcaps *Sylvia atricapilla* put on weight more rapidly when they were provided with water than when not (Sapir *et al.* 2004). This could be because, in the absence of drinking water, they had to catabolise some food to provide for body water needs, reducing the amount of food that could be stored as fuel. In contrast, water availability had no effect on fuel deposition of Lesser Whitethroats *Sylvia curruca* in the same place. Differences between these species in their adaptation to arid conditions may explain their differential response to water provision.

To conclude, these various considerations suggest that crossing the Sahara even by non-stop flight is practicable with a balanced water budget if the bird flies at an appropriate altitude. This is difficult during the day in autumn because of the need for tailwinds which occur mainly below 1000 m. This may be why many birds in autumn, in order to conserve water, fly by night and rest by day (see also Carmi *et al.* 1992). Of course, birds resting in the desert by day are also confronted with high temperatures and low humidity, but they can seek shade and remain inactive, so as not to generate additional body heat. Given sufficient fat, small warblers could easily endure three nights of flight and two stopover days without feeding or drinking, as calculated by Biebach (1991). It seems, therefore, that with sufficient body reserves, small migrants are able to avoid both overheating and dehydration in the hot desert mainly by suitable behaviour. Nevertheless, evidence from the body composition of birds obtained on the Mediterranean–Saharan crossing suggests that, while fuel energy could be limiting in some, water could be limiting in others (Chapter 5).

Asian-Afrotropical migrants

Migrants travelling between eastern Asia and the Afrotropics also face a difficult and even longer journey than those migrating between Europe and the Afrotropics. They include eastern populations of the Willow Warbler *Phylloscopus trochilus*, Sedge Warbler *Acrocephalus schoenobaenus*, Garden Warbler *Sylvia borin*, Common Redstart *Phoenicurus phoenicurus* and Northern Wheatear *Oenanthe oenanthe*, among others. Birds from these populations have overall journeys 1.5–2.0 times longer than their European equivalents. They have to cross a difficult succession of hostile areas, including the deserts and semi-deserts of western Central Asia, the almost treeless areas of the Iranian highlands, and the deserts of the Arabian peninsula. Some also have to cross the world's highest ground in western Asia, including the Tien Shan, Pamiro-Alay and the Himalayas. For forest passerines, inhospitable areas in Central Asia alone span 2000–3000 km, and for eastern Siberian populations of some species (Willow Warbler and Northern Wheatear), the total one-way migration distance may exceed 15000 km.

The Arabian peninsula offers an alternative route from the Sahara into the more genial parts of Africa (Bourne 1959). Not only is this desert less wide than the Sahara, but the hills and mountains around its borders are vegetated. It supports a number of oases, and is generally better vegetated than the Sahara, because it receives more rain. Not surprisingly, therefore, the Arabian peninsula is crossed not only by the migrants from Asia heading to Africa, but also by many birds from central and eastern Europe (including Blackcap *Sylvia atricapilla*, Lesser Whitethroat *S. curruca* and Red-backed Shrike *Lanius collurio*), which head southeast on the first part of their migration (Chapter 22).

Migration in the Asian deserts and mountains was studied over 12 years by a large-scale programme of observation and trapping at more than 20 different sites scattered between the Caspian Sea in the west and the Hindu Kush, Pamir and Tien Shan mountains in the east (**Figure 6.4**; Dolnik 1990). These sites together spanned the region from 37–48°N and 53–78°E. This region represents a crossroads of migration routes involving two main wintering areas, so that two main axes of migration were detected: northwest–southeast for Siberian birds wintering in southern Asia, mainly India, and northeast–southwest for Siberian birds wintering in the Afro-tropics. Hence, many of the birds that cross the Asian deserts also cross the eastern Sahara or Arabian Peninsula on the same migration, for which greater fat levels are needed. Of birds captured, 25 species (75.6% of all captures) winter in Africa, and 23 species (9.5% of captures) in Asia, mostly in India, while eight species (14.9% of captures) winter in both Africa and India.

The deserts of Central Asia present a much less arduous journey than the Sahara, because they support more rivers and lakes and generally more vegetation, especially in spring after winter precipitation. An average of 1.5 billion birds (85% passerines) are estimated to cross these deserts on a broad front each autumn and about 0.75 billion each spring, flying mainly at night (Dolnik 1990). The oases, desert lakes and rivers do not provide adequate foraging sites for all these birds and, because of the intense competition at such places, the rate of fat deposition there is low – about the same as in the open desert. Nonetheless, in general, the basic strategy used by passerines migrating across this arid region centres more on effective refuelling during daily stopovers than on maximal fuel storage before starting the journey. The birds thus maintain a moderate rate of travel across the desert, limited by the opportunities for refuelling, but with no net fat loss over the journey.

Many birds of forest and other habitats have little option but to feed in unfamiliar arid habitats with little vegetation. On average, a passerine migrant foraging in the Asian deserts accumulates during one day enough fat to support an estimated 1.4 hours of migratory flight in spring and 0.6 hour in autumn. At both seasons, the average fatness of small passerine migrants initiating nocturnal flight in Asian deserts was 20% of fat-free mass, and in those trapped just at the end of their nocturnal flights it was around 10%, a loss expected over at least 4 hours of night flight (Dolnik 1990). In the Asian deserts, as in the Sahara, birds found by day in shady places that offer little food tended to have high fat levels, whereas those found at potential foraging sites near water tended to have low fat levels (Dolnik 1990). Evidently, lean birds were more specific in their choice of resting sites than fat ones.

The numbers of birds seen and caught in the western deserts were much higher in spring than in autumn, about 2.6 times higher according to data obtained by 'moon-watching' (Chapter 2). This was partly because birds took mainly different routes at the two seasons, but also because many birds stopped in spring but flew over in autumn (Bolshakov 2003). In spring, practically all nocturnal passerine migrants avoided crossing the highlands of Central Asia, which at that time were still snow-covered, and took the desert route instead, whereas in autumn, mountain crossing was much commoner, especially among species wintering in India (**Figure 6.4**).

In addition, captures at stopovers and long-distance recoveries of ringed birds suggested that, in autumn, most transit passerines from the forest zone of central and eastern Siberia made a detour round the north and northwest edges of the deserts. By heading westward towards Europe, before veering southward to Africa, they migrated through more hospitable areas but lengthened the journey by another 2000 km, bringing the total to 8000–10 000 km. It seems, therefore, that, in spring when the deserts are at their most propitious, many migrants cross them, feeding en route, but in autumn birds largely avoid the deserts by migrating over the mountains to the east, or by heading westward for 2000 km or more, before veering south towards Africa. By taking somewhat different routes in autumn and spring, the birds make the best of prevailing conditions at both seasons.

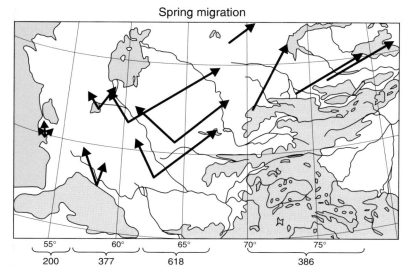

Spring migration

55°   60°   65°   70°   75°

200   377   618   386

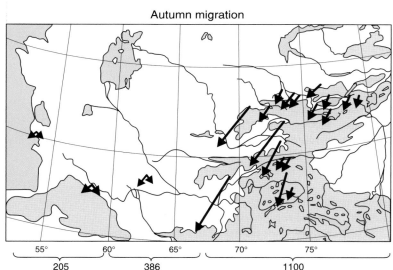

Autumn migration

55°   60°   65°   70°   75°

205   386   1100

Figure 6.4 Density (length of arrow) and main direction (azimuth of arrow) of nocturnal bird migration above moon-watch sites indicated by the origins of the arrows. The numbers beneath the maps show the total number of migrants (in millions) that passed in each season through the fronts delimited by the brackets. Large water bodies are indicated in blue, the largest being the eastern part of the Caspian Sea at the left and the Aral Sea at longitude 60°E. Mountain ranges are shaded grey; the Elburz range is at the lower left; the Hindu Kush, the Pamirs and the Tien Shan respectively are eastward from about 65°E. In autumn, birds mostly crossed the mountain ranges (then at their best), avoiding the dry deserts to the west; but in spring they mostly crossed the deserts (then at their most lush) and avoided the mountains (then snow-covered). Redrawn from Dolnik (1990).

The North American deserts

The western North American deserts are easier to cross than the Eurasian–African ones, because they are generally smaller, better vegetated, and better endowed with oases and riparian vegetation where birds can feed. Some of the desert towns, with their well-watered gardens, also offer good feeding for some kinds of birds, but many former rivers are now dry for most of the year, through excess water abstraction. Nevertheless, some oases attract many passing migrants, and hotspots for bird-watching can be found scattered through all the main desert areas. In southern California, such hotspots include the Forty-nine Springs Oasis in Joshua National Park and the Big Morongo Canyon Preserve (which includes many hectares of riparian cottonwoods and marsh) in Morongo Valley, each attracting more than 100 regular passage species at migration times. Again, however, it is likely that most migrating birds fly over without stopping.

One of the most interesting birds to cross the western American deserts each year is the Eared Grebe *Podiceps nigricollis*, as it travels from the Great Salt Lake in Utah to the Gulf of California (Jehl *et al.* 2003). The birds develop flight seasonally, especially for migration, losing body weight but building pectoral muscles in order to do so (Chapter 5). The body shape of these birds, which is adapted to foot-pro- pelled diving, is not ideal for flight, and the energetic costs are the highest recorded for any bird species, at around $25 \times BMR$ (basal metabolic rate). The birds take off from water, after a long, foot-pattering taxi. They fly only at night (perhaps chiefly to avoid predation), and have to cross more than 1000 km of mainly desert in non- stop flight, potential stopping places being almost non-existent. To judge from radar-based observations, the flight is direct and fast, taking about 17 hours, and ideally completed within a single night. By storing massive body reserves on the Great Salt Lake in autumn, grebes can postpone their migration beyond the date in autumn when their local food supplies collapse, waiting until nights are longer and thereby benefiting from safer flying time. The grebes thus migrate to their winter- ing areas later than any other North American bird, and on body reserves accumu- lated weeks previously (Chapter 5).

The grebes depart in periods of stable air or light tailwinds, avoiding the prevail- ing northwesterlies. Not every year do conditions remain favourable throughout the flight, however, and in some years thousands of birds are drifted off course or crash-land in desert snow storms (Chapter 28). If the birds encounter rain or head- winds within an hour or so after take-off, they can return to the staging lakes, as confirmed by radar observations, but they then suffer further depletion of body reserves until conditions again become suitable, for at that time of year spent reserves cannot be replenished locally. Behaviour during departure also falls short of ideal. The birds assemble in large dense flocks, and on appropriate nights, tens of thousands or hundreds of thousands take off en masse, shortly after sunset. This early departure makes best use of the available dark period, but the num- bers involved produce mid-air collisions. On some nights hundreds crash into one another, come skittling down and die within moments of taking off (Jehl *et al.* 2003). The return journey is somewhat shorter, as birds can stop and feed en route at the Salton Sea, which in spring offers abundant food.

HIGH MOUNTAINS

As mentioned already, migratory birds regularly cross high mountain ranges, often at extremely high altitudes. The high valleys may be snow-covered and frozen in spring, when they offer little or nothing for passing birds, but in autumn many birds can stop and feed in whatever vegetation is available there. The main problems are the considerable costs of climbing to high altitudes (greatest in large birds), and the low temperatures, thin air and low oxygen levels often encoun- tered (Chapter 4).

The highest mountain range in the world, the Himalayas, sits right across a major north–south bird migration route, as do most of the European ranges, such as the Pyrenees, Alps and Carpathians. Depending on their primary direction, birds may cross high mountains or fly around them. Radar studies suggest that many birds cross mountain ranges in non-stop flights, but visual observations confirm that others are funnelled through valleys and passes, where they may stop and feed. For example, at the high pass of Col de Bretolet in the Swiss Alps, about 75 species have been found regularly to rest and feed, and many more occasionally. In effect, this amounts to almost all the small species that take this route.

Several species migrate regularly each year over the Himalayas, flying at more than 5 km above sea level, and at least one species over the highest parts, namely the Bar-headed Goose *Anser indicus*, which nests on high-altitude lakes on the Tibetan plateau, and winters in the lowlands of India. This bird has been recorded at more than 8 km above sea level, where temperatures sink below −50°C,[1] and oxygen pressures below one-third of those at sea level. Taking advantage of the jet stream, these birds can achieve speeds in excess of 150 km per hour, and complete the 700–1000 km journey within a day. The species is physiologically adapted to high-altitude flight, having a large heart which beats unusually rapidly, and haemoglobin with an exceptionally high affinity for oxygen. It shows no increase in haematocrit (packed cell volume) or in haemoglobin concentration when exposed to simulated high altitude, and therefore avoids any increase in blood viscosity which could impair circulation (Black & Tenney 1980). Such birds can thus achieve feats of high-altitude performance that are shown by very few, if any, other animals, and can do so without time to acclimatise. For instance, Bar-headed Geese *Anser indicus* may begin their spring migration near sea level, and reach altitudes exceeding 7 km in a few hours.

Tests have shown that Bar-headed Geese can remain conscious and stand erect in hypobaric chambers under simulated high-altitude conditions of slightly over 12 km (Black & Tenney 1980). This altitude far exceeds the tolerance limits of most mammals. Without acclimatisation, a person is in trouble at 4 km above sea level, but with training can reach more than 5 km without becoming breathless.

Clearly, high-flying birds must be pre-equipped with mechanisms which prevent altitude sickness (hypoxia) caused by oxygen deficiency in thin air. All flying birds may have this facility to some extent (Novoa *et al.* 1991), but species living or migrating at high altitudes need special adaptations which maintain oxygen delivery to the brain, gas exchange efficiency in the lung, oxygen binding by haemoglobin, cardiovascular performance and cellular function (see review in Berthold 1996). Some birds have an interesting adaptation in their blood, as they have at least two forms of haemoglobin side by side. These forms differ in amino acid composition, and in their oxygen carrying and releasing properties. One form acts as 'normal haemoglobin' for low altitudes, and the other as 'high elevation haemoglobin', which reaches greatest concentration in the blood of species that fly at especially high altitudes, thus ensuring adequate oxygen throughout (Heibl & Braunitzer 1988). In general among birds, lung ventilation is especially

---

[1] *Temperatures fall to −56.5°C at 11 km above sea level, and remain fairly constant with further elevation, even though air pressure and density continue to decline. This altitude marks the tropopause, the boundary between the troposphere below and the stratosphere above. All migration occurs within the troposphere, and almost all within the lower half of the troposphere.*

effective because of the air-sacs, and resulting continual flow of air through the lungs. Needless to add, not all birds have such flexibility in their flight altitudes, and many species seem confined to low elevations for physiological or energy-based reasons (Chapter 4).

It is not only the altitude that presents problems in mountains, for strong adverse winds are often funnelled through the valleys. One well-known migration route in Nepal includes the Kali Gandaki Valley, which provides a useful corridor through the Himalayas, with passes reaching up to 6.5 km. This route is used each autumn by up to 50 000 Demoiselle Cranes *Grus virgo*, flocks of which start their daily migration as soon as thermals develop in the mornings, soaring to heights that enable them to cross the passes. However, time is against them, for by mid-morning immensely strong headwinds develop, against which the cranes can make only slow progress. They also experience attacks from Golden Eagles *Aquila chrysaetos*, spaced every few kilometres along the valley. Compared to other cranes, demoiselles have one of the most difficult migrations, as they pass through desert and high mountains which offer very few feeding or safe roosting sites. As revealed by radio-tracking, they accomplish their migration relatively quickly, crossing both desert and mountains within one week (Kanai *et al.* 2000).

Not surprisingly then, many birds fly around mountains rather than taking a more direct route over the tops (Bruderer 1999). Most short-distance migrants, flying northeast–southwest in autumn, tend to circumvent the Alps. However, longer distance migrants en route to Africa fly more directly north–south, at higher altitude, and mostly cross the Alps. However, both groups fly lower in headwinds, smaller species being more affected than larger ones, and hence being more often deflected by mountains from their most direct route.

Migrants captured in high alpine passes often have more fat and longer wings than conspecifics caught in nearby lowland, implying that the two groups derive from different populations, one breeding further north than the other. This view is supported by ring recoveries. For example, Robins *Erithacus rubecula* occurring in the Alps during autumn migration originate from further north (Scandinavia) than those occurring in the nearby lowlands, which originate mainly from Germany and Czechoslovakia. Similarly, northern populations of Garden Warblers *Sylvia borin* (identified by their greater wing lengths) are more likely to cross the Alps than mid-latitude birds (Bruderer & Jenni 1990). Such differences presumably arise because longer distance migrants from higher latitudes migrate at higher altitude, or have been longer on the wing, giving more time to attain high altitude by the time they reach the Alps. Among trapped samples of Garden Warblers, the proportion of long-winged individuals increases on days with headwinds which induce lower flight than on other days.

Wetland and soaring species tend to avoid the higher passes. Raptors which migrate mainly by flapping flight, such as falcons, appear in the high Alpine passes much more often than those that depend on soaring, such as kites and buzzards. The few Common Buzzards *Buteo buteo* flying in the high Alps occur early in the migration season, when soaring is most readily possible. The average climbing rate of soaring birds decreases by one-third over the autumn migration season because of decreasing lift in thermals, as temperatures cool towards autumn (Bruderer & Jenni 1990). Hence, any tendency of birds to cross the Alps varies with the innate directions, flight capabilities and physiological state of the migrants, and with local weather conditions at the time. These findings from the central European Alps may well apply also to other mountain areas at similar latitude. In the New World, by contrast, most of the mountain ranges, particularly the Rockies and the Andes, run approximately north–south, so do not need to be crossed or circumvented by most lower ground migrants travelling north–south.

OTHER BARRIERS

Some birds on their migrations also cross substantial stretches of sea-ice. For example, the many shorebirds that fly between northern Siberia and northern North America en route to wintering areas in South America have to cross 1800–3000 km of the Arctic Ocean which is almost totally ice-covered (Alerstam & Gudmundsson 1999). The main participants include Pectoral Sandpipers *Calidris melanotos* and Grey Phalaropes *Phalaropus fulicaria*. Nevertheless, such flights would seem to offer less of a challenge than overwater flights of similar length, because the birds can at least stop and rest in emergencies. Many arctic and Antarctic seabirds cross more than 100 km of sea-ice in order to reach their breeding areas in spring, including some penguin species which have to walk. A bigger challenge is presented by the Greenland ice cap, which is up to 1000 km across, and reaches more than 2.5 km above sea level at its highest points. It is crossed regularly by shorebirds and water-fowl on their migrations between breeding areas in northeast Canada or west Greenland and their wintering areas in Europe or Africa. Participants include Brent Geese *Branta bernicla* and White-fronted Geese *Anser albifrons*, which have been shown by radio-tracking to stop periodically on the way up (Gudmundsson *et al.* 1995, Fox *et al.* 2003), and apparently also Ruddy Turnstone *Arenaria interpres*, Red Knot *Calidris canutus*, Ringed Plover *Charadrius hiaticula*, Purple Sandpiper *Calidris maritima* and Northern Wheatear *Oenanthe oenanthe*. Most other birds that breed in western Greenland migrate to North America via the Davis Strait rather than to Europe.

CONCLUDING REMARKS

The various migrations described in this chapter indicate what some birds are capable of achieving on their annual travels. The journeys involved may be assumed to push birds to the limits of their endurance: long non-stop flights without sleep, rest, food or water. Most involve high fuel deposition beforehand, and special adaptations for high-altitude flight, extreme heat or cold, dehydration, low air densities and oxygen levels – conditions in which few animals could survive. The fact that some birds can achieve these high endurance feats without prior training or acclimatisation is all the more remarkable.

Many of the findings reported in this chapter have come to light in the last 25 years, but more research is needed to better understand the physiological mechanisms involved. For example, humidity has so far been considered only in terms of dry conditions which lead to excessive water loss, but the humid conditions over tropical rainforests could bring the opposite problem, and reduce the effectiveness of evaporative cooling. This could greatly raise the optimal altitude for flight.

It is uncertain how often birds need to rest while on migration, or how far they could fly without resting (Chapter 3). Resting is of course not normally possible for landbirds migrating over water, and continuous wing-beating for periods up to 70 hours or more seems incredible for small birds. Many birds seem 'tired' after overwater flights, in that they flop down as soon as they reach a place to stop, fly only with reluctance if disturbed, and can be seen to sleep in the day-time. Typically, on arrival, they drink and then sleep (Schwilch *et al.* 2002b). Many other birds stop for periods of a few hours, apparently without feeding, even where food is available (for further discussion, see Chapter 5). Large waterfowl are known from radio-tracking to sit on the sea, hundreds of kilometres from land. Such birds may be resting or simply sitting out periods of unfavourable weather. The same is true for seabirds which migrate by dynamic soaring, but which in calm conditions sit on the sea and wait for the wind to get up.

Depending on their journeys, birds need fuel and water in different propor-tions, and on non-stop flights when they can neither feed nor drink, birds can address these different needs physiologically or behaviourally. They may adjust the ratios of stored and metabolised fat, protein and carbohydrate, so that body reserves yield different ratios of energy and water. Alternatively, birds might adjust the times of day (day or night), or the altitudes at which they fly, both of which influence rates of heat and water loss. Birds can also reduce their energy needs by selecting times or altitude zones with favourable winds (Chapter 4). One of the most striking findings from studies of difficult journeys is how much birds use behavioural means to overcome difficulties, involving, for example, choice of wind conditions, flight altitudes and flight times, as well as stopover frequency and duration, and whether to feed or rest during stopovers. Some types of behaviour are appropriate in some conditions but not in others. For example, the burning of body fuel for flight releases heat, which helps to main-tain body temperature in cold conditions (high latitudes and altitudes), but could lead to overheating (and hence water loss) in warm conditions (low latitudes and altitudes). Hence, in order to maintain a favourable energy and water balance, a bird must continually adjust its flight times, flight altitudes and other behaviour according to prevailing conditions. In these various ways, a bird can not merely survive, but can maximise the time and distance it can fly on a given fuel load.

SUMMARY

Landbirds migrating over the sea must often make long non-stop flights, without rest, food or water. Many birds fly non-stop for periods exceeding 60 hours. Such flights take passerines over distances up to 3000 km, and shorebirds over 4000–6000 km. The longest non-stop flight known from any landbird species involves one population of Bar-tailed Godwits *Limosa lapponica baueri*, which travels for more than 175 hours non-stop in autumn, covering more than 10 400 km over the Pacific Ocean from Alaska or Siberia to New Zealand. Many other shore-birds make journeys of up to 6000 km over water, and hundreds of species cross shorter stretches of up to 1500 km, including those that regularly migrate over the Mediterranean Sea or the Gulf of Mexico.

The most arduous desert journeys are undertaken by the many species that cross the most barren parts of the Sahara Desert each spring and autumn, where no food or water are taken en route. Some species are thought to cross the Mediterranean Sea and Sahara Desert together in a single non-stop flight of 1500–2500 km; others break their journey to refuel in North Africa; while yet others migrate only at night, and stop in the desert during the day, remaining motionless in shady places, again without feeding or drinking. Individuals of the same species may show more than one of these different strategies. Only small proportions of birds stop at oases, but more migrate down the west and east sides of the Sahara, where conditions are better than across the central sector. Analyses of the body composition of migrants caught or found dead on the ground suggest that some may be limited on their desert crossing by energy needs, and others by water needs.

While some birds migrate around mountain ranges, others cross even the highest ranges, including the Himalayas. The main problems are the climb to high altitudes (especially for large birds), and the extreme cold, low air densities and oxygen levels found there. In spring, mountains are also often snow-covered, affording no opportunity to feed. Extreme high-altitude flights (more than 7 km above sea level) have been recorded from Bar-headed Geese *Anser indicus* that cross the Himalayas. These birds have special adaptations enabling them to extract and utilise efficiently the reduced oxygen levels at high altitudes.

Extreme migrations often involve extreme adaptations of high fuel deposition, water conservation and respiratory physiology, as well as behavioural adjustments in flight times, flight altitudes, stopover frequency and duration, and resting or feeding behaviour.

POSTSCRIPT

As this book goes to press, news is emerging of some radio-tagged Bar-tailed Godwits *Limosa lapponica* tracked on spring migration from New Zeland. Six birds accomplished the journey from New Zeland to the north end of the Yellow Sea in China in a single non-stop flight, covering more than 10 000 km in about seven days. (A seventh bird broke its journey in Micronesia and Japan.) They then refuelled for the second stage of their journey to breeding areas in eastern Siberia and Alaska (web-based information from the Western Ecological Research Center, Sacrimento, California).

Honey Buzzards *Pernis apivorus* thermal-soaring on migration

Chapter 7
Raptors and other soaring birds

if they had continued for a fortnight in the same strength as on that day, we could surely have said that they were in greater number than all the men living on the earth. . . . they are seen to pass in this way as thick as ants, and so to continue for many days. (Belon (1555), writing of the migration of Black Kites *Milvus migrans* on the Black Sea coast.)

In some large bird species, the spread wings span a large area relative to body weight, and provide good lift in rising air currents. This is true of broad-winged raptors, pelicans, storks and anhingas, and to some extent of cranes. Such species therefore depend more heavily upon soaring–gliding flight (as opposed to flapping flight) than do most other birds (Kerlinger 1989, Hedenström 1993). They migrate primarily by day, when thermals are best developed, and most avoid long water crossings (Chapter 3). They usually travel low enough to be seen with the naked eye, and in certain places, as determined by geography and topography, they form predictable migration streams. They can therefore be studied and counted at such sites by ground-based observers equipped only with binoculars or telescopes in ways that other, higher flying or night-flying birds cannot.

The seasonal timing of their migrations can be assessed accurately and day-to-day passage can be related to weather and other conditions. In some species, the different age groups can be distinguished, enabling their movements to be examined separately. Study has been helped by the publication of special field guides showing the birds from below, and facilitating the identification of species, age groups and sometimes sexes, as they pass overhead (Porter *et al.* 1974, Forsman 1999, Clark 1999, Clark & Wheeler 2001). Some aspects of soaring bird migration result from the mode of travel itself, and do not entirely apply to other types of birds, but other aspects hold more generally. Hence, some aspects of bird migration have been better studied in soaring species than in many others (see also Chapters 3 and 4).

No sharp division separates species that travel by soaring–gliding flight from those that travel by continued flapping. Different species form a continuum of variation between the two extremes, depending on their body size and wing shape, and in all soaring species the ratio of flapping to gliding varies with air conditions at the time. Among the birds of prey, vultures and eagles are most dependent on soaring–gliding, followed in descending order by *Buteo* hawks, *Milvus* kites, *Accipiter* hawks and *Pernis* honey buzzards, and then by *Circus* harriers and *Pandion* ospreys. Falcons are more active fliers, less dependent on updrafts, but making use of them when available. This order of listing broadly follows the sequence of wing-loading, from lowest to highest, and the variation in wing shape from long and broad, with slotted primary feathers, to narrow and pointed, with little or no slotting. It also reflects the dependence of these various species on updrafts, and hence the extents to which they form into concentrated migration streams and avoid long sea-crossings. It is chiefly the falcons that regularly make long (<100 km) overwater flights, but other species (*Pandion*, *Pernis*, *Butastur* and others) do so in some parts of the world. The most extreme is the Amur Falcon *Falco amurensis*, which each autumn is thought to cross the Indian Ocean between India and East Africa on a journey exceeding 4000 km. This crosswater flight occurs in late November or early December, and is assisted by prevailing winds. On the return spring migration, winds are less favourable, so migration is largely over land, running west and north of the outbound passage, up the east side of Africa into Asia.

Soaring landbirds use routes where topography favours the development of thermals and other updrafts. Well-known observation points for watching soaring migrants include Hawk Mountain in Pennsylvania, Cape May Point in New Jersey, Veracruz on the Gulf coast of Mexico, Panama in Central America, Falsterbo in Sweden, Gibraltar and the Bosphorus at either end of the Mediterranean Sea, the Black Sea coast in northeast Turkey, various localities in the Rift Valley in Israel, Suez in Egypt, Chumphon in Thailand and Kenting on the southern tip of Taiwan (**Figure 7.1**). At these points, large numbers of raptors and other soaring species pass in spring or autumn, with total numbers typically varying between tens of thousands and hundreds of thousands, even millions, depending on site.

Because they depend on geography and topography, birds take the same routes each year, but many take somewhat different routes in spring and autumn, depending on wind and other conditions. Thus the passage at Hawk Mountain in Pennsylvania is marked in autumn, but barely noticeable in spring, while at Eilat in Israel the spring passage is much bigger than the autumn one. Daily counts

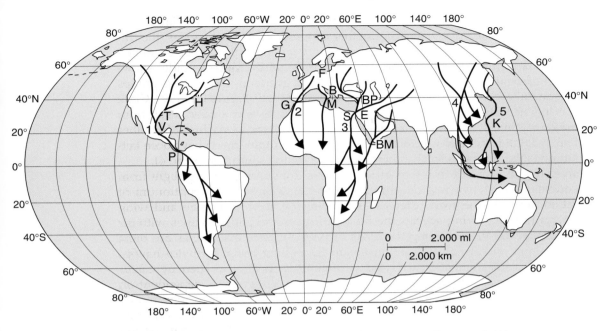

Figure 7.1   Main flyways used by soaring birds. 1. Trans-American Flyway; 2. Western European–West African Flyway; 3. Eurasian–East African Flyway; 4. East Asian Continental Flyway; 5. East Asian Oceanic Flyway. Note that no major flyways for northern hemisphere raptors extend into New Guinea and Australia. Watch sites mentioned in the text: B – Bosphorus, BM – Bab el Mandeb, BP – Belen Pass, E – Eilat, F – Falsterbo, G – Gibraltar, H – Hawk Mountain, K – Kenting, M – Messina Strait, P – Panama, S – Suez, T – Corpus Cristi, Texas, V – Veracruz, Mexico. Modified from Zalles & Bildstein (2000).

throughout the autumn or spring migration seasons have been made at several of these sites, and at some such sites counts have been made repeatedly over several to many years (for Hawk Mountain see Bednarz *et al.* 1990; for Cape May Point in New Jersey see Dunne & Clarke 1977; for various sites in western North America see Hoffman & Smith 2003, for Falsterbo in southern Sweden see Kjellén & Roos 2000; for Israel and elsewhere see Shirihai *et al.* 2000). Observers at well-watched sites now publish count data on the web, updated each year as fresh data become available.

For parts of their migrations, most soaring species travel on a broad front (Bednarz & Kerlinger 1989), but for other parts, they form into concentrated streams. Such streams typically occur along landscape features that favour soaring–gliding flight, such as mountain chains, narrow valleys and coastal plains, but also on the east or west sides of large water bodies, as birds hug the shoreline rather than crossing the water. Worldwide, most migration corridors coalesce into one of five principal flyways, two of which extend to the southern parts of the southern continents (**Figure 7.1;** Zalles & Bildstein 2000). All flyways tend to converge on narrow land bridges (such as Panama or Suez) or on short sea-crossings (such as the Straits of Gibraltar or Bosphorus). At these points, the migration streams may be only a few tens of kilometres across (or less).

1. **The Trans-American Flyway**: Each autumn, more than six million raptors travel along one or other part of this 10 000 km overland system of corridors that stretches from boreal Canada to central Argentina (**Figure 7.1**). At least 32 species, including eight buteos, migrate along the flyway's central land corridor that connects North and South America, reaching its narrowest point at Panama. Once the birds reach South America, the flight-line turns south, and many of the migrants follow the Magdalena Valley south through Colombia and onwards. The central part of the route is dominated by Turkey Vultures *Cathartes aura* (>two million), Broad-winged Hawks *Buteo platypterus* (>one million) and Swainson's Hawks *B. swainsoni* (>one million) (**Table 7.1**, Bildstein 2004). The biggest numbers of birds are seen near Corpus Christi (Texas, 840 000 birds in autumn), Veracruz (Mexico, more than six million birds in autumn) and Panama City (more than 2.5 million birds, both seasons). In some parts, the route also carries substantial numbers of other species, including Ospreys *Pandion haliaetus* (5000), Mississippi Kites *Ictinia mississippiensis* (200 000), Peregrines *Falco peregrinus* (5000) and others.

2. **The Western European–West African Flyway**: Each autumn, at least 200 000 raptors travel along one or other part of the 5000 km overland system of corridors that stretches from northern Europe to West Africa, via the short (<14 km) sea-crossing at the Straits of Gibraltar (**Figure 7.1**). At least 22 species use this flyway, which for most of its length is dominated by European Honey Buzzards *Pernis apivorus* (117 000) and Black Kites *Milvus migrans* (39 000). The biggest concentrations of birds are seen in both spring and autumn at the Straits of Gibraltar, where some 195 000 birds were counted in 1972 (Bernis *et al.* 1975; **Table 7.2**).

3. **The Eurasian–East African Flyway**: More than 1.5 million raptors travel along this 10 000 km system of largely overland corridors that extends from northeastern Europe and western Siberia through the Middle East into southern Africa (**Figure 7.1**). At least 35 species use this flyway that for much of its course follows the Great Rift Valley, and that includes narrow water crossings at the Bosphorus, Suez or Bab el Mandeb Straits. Two main known routes converge on Africa. In the western route, birds pass in autumn, east or west of the Black Sea, over the Bosphorus and on to cross Jordan and Israel, and then Sinai, entering Africa at the northern end of the Red Sea at Suez. In the eastern Caspian–Arabia route birds pass either side of the Caspian Sea, move on south through Arabia and cross into Africa via the Bab el Mandeb Straits at the southern end of the Red Sea. Six raptor species make up the bulk of the flight through the Middle East, namely the Western Honey Buzzard *Pernis apivorus* (up to 852 000), Black Kite *Milvus migrans* (37 000), Levant Sparrowhawk *Accipiter brevipes* (60 000), Eurasian (Steppe) Buzzard *Buteo buteo vulpinus* (466 000), Lesser Spotted Eagle *Aquila pomarina* (142 000) and Steppe Eagle *A. nipalensis* (75 000), along with White Storks *Ciconia ciconia* (530 000), Black Storks *C. nigra* (17 000) and Great White Pelicans *Pelecanus onocrotalus* (66 000). Major concentrations of birds are seen at several sites in Israel (see later), and at the various water crossings mentioned above, and on the east and west sides

Table 7.1   Autumn counts (August–November) of soaring birds at Veracruz, Mexico, 2002–2005

Species	2002	2003	2004[a]	2005
Turkey Vulture *Cathartes aura*	2 677 355	2 028 633	1 404 964	2 398 841
Golden Eagle *Aquila chrysaetos*	3	3	5	5
Mississippi Kite *Ictinia mississippiensis*	306 274	210 105	177 088	171 059
Swallow-tailed Kite *Elanoides forficatus*	272	202	141	198
Black-shouldered Kite *Elanus axillaris*	2	0	0	260
Snail Kite *Rostrhamus sociabilis*	5			
Hook-billed Kite *Chondrohierax uncinatus*	165	118	146	104
Plumbeous Kite *Ictinia plumbea*	2	1	0	1
Northern Harrier *Circus cyaneus*	269	130	33	228
Sharp-shinned Hawk *Accipiter striatus*	3 152	2 595	1 931	2 167
Cooper's Hawk *Accipiter cooperii*	2 030	1 899	1 294	2 220
Northern Goshawk *Accipiter gentilis*	1	1	2	
Red-shouldered Hawk *Buteo lineatus*	7	6	7	9
Broad-winged Hawk *Buteo platypterus*	2 386 232	1 745 351	2 069 336	1 807 571
Grey (Grey-lined) Hawk *Asturina nitida*	326	108	95	1 220
Red-tailed Hawk *Buteo jamaicensis*	146	122	38	95
Swainson's Hawk *Buteo swainsoni*	1 009 648	1 216 153	980 494	1 200 928
Ferruginous Hawk *Buteo regalis*	2	1	2	1
White-tailed Hawk *Buteo albicaudatus*	0	1	0	0
Common Black Hawk *Buteogallus anthracinus*	7	1	6	3
Harris's Hawk *Parabuteo unicinctus*	4	12	5	9
Zone-tailed hawk *Buteo albonotatus*	142	137	152	238
Short-tailed hawk *Buteo brachyurus*	3			
Osprey *Pandion haliaetus*	2 694	3 002	2 098	2 712
Crested Caracara *Caracara plancus*	2	1	1	
American Kestrel *Falco sparverius*	4 097	4 296	2 977	3 184
Merlin *Falco columbarius*	117	106	188	206
Peregrine Falcon *Falco peregrinus*	714	860	44 020	450
Roadside Hawk *Buteo magnirostris*	1	1	0	1
Unidentified raptor	177 347	52 635	57 948	57 940
Total raptors	6 571 019	5 266 480	4 742 971	5 649 660

[a]No count for November 2004, so overall total reduced.

From the Hawk Migration Association of North American (HMANA) website at www.hawkcount.org.

of the Black Sea, but at some sites the numbers differ greatly between autumn and spring (**Tables 7.2, 7.3** and **7.4**).

4. **The East Asian Continental Flyway**: More than a million raptors travel along one or other part of this 7000 km mostly overland system of corridors that stretches from eastern Siberia to Southeast Asia and the Indonesian Archipelago, and that includes sea-crossings of 10–60 km at the Straits of Malacca, Sunda, Bali and Lombok (**Figure 7.1**). Strangely, no raptor species is known to reach Australia in numbers, the flyway being curtailed further north

Table 7.2 Maximum annual counts of raptors (and year) at various migration sites on the Eurasian–African flyways

Species	Gibraltar Autumn	Messina, Italy Spring	Malta Autumn	Bosphorus, Turkey Autumn	Bosphorus, Turkey Spring	Eastern Pontics, Turkey Autumn	Eastern Pontics, Turkey Spring	Belen Pass, S. Turkey Autumn
Osprey *Pandion haliaetus*		25 (1997)	23 (2002)	10 (1971)	+	24 (1976)		
European Honey Buzzard *Pernis apivorus*	117175 (1972)	27297 (2000)	1317 (2002)	25745 (1971)	1211 (1978)	138000 (1976)	29937 (1993)	15967 (1976)
Black Kite *Milvus migrans*	39099 (1972)	1008 (2000)	27 (2003)	2617 (1971)	287 (1937)	5775 (1976)	9096 (1994)	506 (1976)
Red Kite *Milvus milvus*	66 (1972)	8 (1998)	1 (2002)	10 (1966)				
Egyptian Vulture *Neophron percnopterus*	3768 (1972)	12 (1999)	1 (1999)	554 (1971)	17 (1937)			874 (1976)
Eurasian Griffon *Gyps fulvus*		4 (1998)		166 (1931)	4 (1937)	30 (1977)		125 (1965)
Short-toed (Snake) Eagle *Circaetus gallicus*	8797 (1972)		5 (2001)	2517 (1937)	240 (1937)	243 (1976)	395 (1994)	728 (1976)
Western Marsh Harrier *Circus aeruginosus*	359 (1972)	3074 (2000)	2056 (2002)			385 (1976)	251 (1994)	
Hen (Northern) Harrier *Circus cyaneus*	10 (1972)	84 (1997)	1 (2002)	6 (1980)	4 (1965)	41 (1977)	34 (1994)	1 (1976)
Pallid Harrier *Circus macrourus*		83 (2000)	4 (2001)			133 (1976)	11 (1994)	
Montagu's Harrier *Circus pygargus*	1727 (1972)	866 (2000)	49 (2002)			124 (1976)	63 (1993)	

Species								
Northern Goshawk *Accipiter gentilis*	6 (1972)	2 (2000)			6 (1965)	53 (1977)	12 (1994)	
Eurasian Sparrowhawk *Accipiter nisus*	925 (1972)	14 (1999)	14 (2002)	428 (1969)	56 (1937)	1057 (1977)	3966 (1994)	
Levant Sparrowhawk *Accipiter brevipes*				6516 (1978)	253 (1965)	290 (1976)	1945 (1993)	2951 (1976)
Common Buzzard *Buteo buteo*	2647 (1972)	93 (1999)	13 (2000)	32895 (1969)	160 (1937)	205000 (1976)	136327 (1994)	
Long-legged Buzzard *Buteo rufinus*		12 (1997)		11 (1973)				
Lesser Spotted Eagle *Aquila pomarina*		5 (1997)	5 (2001)	32228 (1988)	1745 (1978)	736 (1976)	277 (1993)	1299 (1976)
Greater Spotted Eagle *Aquila clanga*		2 (1996)		20 (1966)	+	21 (1976)	24 (1994)	
Steppe Eagle *Aquila nipalensis*						434 (1977)	107 (1994)	
Imperial Eagle *Aquila heliaca*	2 (1972)	1 (2000)		22 (1931)		29 (1977)	20 (1994)	
Golden Eagle *Aquila chrysaetos*	1 (1972)	6 (1998)	3 (2002)	+				6 (1976)
Booted Eagle *Hieraaetus pennatus*	14451 (1972)	19 (1999)		523 (1971)	37 (1965)	473 (1976)		588 (1976)
Bonelli's Eagle *Hieraaetus fasciatus*		2 (2000)			15 (1937)	8 (1994)		3 (1976)

(Continued)

Table 7.2 (Continued)

Species	Gibraltar Autumn	Messina, Italy Spring	Malta Autumn	Bosphorus, Turkey Autumn	Bosphorus, Turkey Spring	Eastern Pontics, Turkey Autumn	Eastern Pontics, Turkey Spring	Belen Pass, S. Turkey Autumn
Lesser Kestrel *Falco naumanni*	545 (1972)	46 (134)	58 (2002)	+	23 (1937)	+		+
Common Kestrel *Falco tinnunculus*	1198 (1972)	934 (1997)	214 (2001)		12 (1965)	450 (1977)	63 (1994)	
Red-footed Falcon *Falco vespertinus*		1012 (2000)		391 (1972)				
Eurasian Hobby *Falco subbuteo*	217 (1972)	276 (1999)	184 (2001)	168 (1972)	17 (1937)	189 (1976)	41 (1994)	
Eleanora's Falcon *Falco eleonorae*		28 (2000)	16 (2000)	+				+
Saker Falcon *Falco cherrug*		2 (2000)	3 (2001)	13 (1971)	+	12 (1976)		
Peregrine Falcon *Falco peregrinus*		25 (2000)	2 (2002)	5 (1981)		23 (1976)		
Maximum annual counts of raptors	195187 (1972)	35197 (2000)	3905 (2002)	75176 (1971)	3991 (1978)	370000 (1976)	205131 (1994)	28641 (1976)[a]

Species seen in only very small numbers (<10) are omitted.

[a] Recently 149000 birds counted, largely White Storks *Ciconia ciconia*, and total passage estimated at 500000 birds (O. Cam 2004, SEEN Workshop, Istanbul).

Data mainly from Bernis (1975) for Gibraltar, Corso (2001) for Messina, Sammut & Bonavia (2004) for Malta, and Shirihai et al. (2000) and sources therein for the remaining sites.

Table 7.3 Mean (and maximum) counts of various migrating raptors made at different places in Israel over a period of years

	Northern Valleys Autumn 1990–1999	Kafr Qasim Autumn 1982–1987	Eilat, Red Sea Autumn 1980, 1986–1987	Eilat, Red Sea Spring 1977–1997[a]
Osprey Pandion haliaetus	72 (127)	64 (79)	3 (7)	84 (130)
European Honey Buzzard Pernis apivorus	322 727 (544 215)	337 218 (419 164)	235 (2400)	360 184 (851 598)
Black Kite Milvus migrans	1701 (2695)	676 (1195)	43 (236)	26 117 (36 690)
Egyptian Vulture Neophron percnopterus	116 (219)	322 (474)	16 (45)	363 (802)
Eurasian Griffon Gyps fulvus	42 (83)	40 (73)	17 (45)	10 (22)
Short-toed Eagle Circaetus gallicus	3224 (4548)	7429 (8045)	159 (338)	167 (345)
Western Marsh Harrier Circus aeruginosus	1227 (1881)	848 (1237)	53 (128)	177 (371)
Northern Harrier Circus cyaneus	4 (9)	14 (21)	+ (2)	+
Pallid Harrier Circus macrourus	45 (129)[b]	30 (57)[b]	4 (12)	56 (113)
Montagu's Harrier Circus pygargus	132 (252)[b]	24 (51)[b]	6 (20)	18 (55)
Northern Goshawk Accipiter gentilis	2 (4)	+ (+)	+ (6)	+ (4)
Eurasian Sparrowhawk Accipiter nisus	557 (1131)	897 (1761)	51 (117)	183 (456)
Levant Sparrowhawk Accipiter brevipes	45 353 (60 390)	31 302 (44 653)	304 (1500)	20 452 (49 836)
Common (Steppe) Buzzard Buteo buteo	1306 (3619)	281 (?)	904 (4700)	348 656 (465 827)
Long-legged Buzzard Buteo rufinus	37 (95)	27 (57)	11 (32)	56 (105)
Lesser Spotted Eagle Aquila pomarina	68 744 (83 701)	81 652 (141 868)	6 (17)	63 (211)
Greater Spotted Eagle Aquila clanga	44 (87)	52 (85)	5 (13)	8 (22)
Steppe Eagle Aquila nipalensis	142 (267)	338 (456)	9093 (24 246)	24 338 (75 053)
Imperial Eagle Aquila heliaca	9 (25)	24 (37)	18 (64)	47 (95)
Golden Eagle Aquila chrysaetos	1 (4)	–	–	–
Booted Eagle Hieraaetus pennatus	653 (1006)	1391 (1973)	55 (83)	147 (210)

(Continued)

Table 7.3 (Continued)

	Northern Valleys Autumn 1990–1999	Kafr Qasim Autumn 1982–1987	Eilat, Red Sea Autumn 1980, 1986–1987	Eilat, Red Sea Spring 1977–1997[a]
Bonelli's Eagle *Hieraaetus fasciatus*	3 (9)	+ (4)	+	+ (12)
Lesser Kestrel *Falco naumanni*	14 (26)	?	8 (107)	29 (84)
Kestrel *Falco tinnunculus*	41 (144)	?	132 (298)	69 (300)
Red-footed Falcon *Falco vespertinus*	3231 (10877)	2275 (5752)	10 (21)	9 (24)
Eurasian Hobby *Falco subbuteo*	32 (79)	45 (77)	6 (21)	23 (54)
Eleonora's Falcon *Falco eleonorae*	7 (18)	28 (44)	+ (12)	+ (33)
Saker Falcon *Falco cherrug*	+ (2)	+	+ (6)	+ (4)
Lanner Falcon *Falco biarmicus*	+ (1)	+		4 (7)
Peregrine Falcon *Falco peregrinus*	7 (17)	12 (20)	+	+
Maxima annual counts of raptors	691 761 (1997)[c]	604 000 (1986)	25 998 (1980)	1 193 229 (1985)
Great White Pelican *Pelecanus onocrotalus*	36 923 (57 042)	65 569[d]	?	–
White Stork *Ciconia ciconia*	257 442 (530 301)	202 935[d]	?	265 954 (301 048)[e]
Black Stork *Ciconia nigra*	6 226 (16 898)			

Raptors from Shirihai et al. (2000), Alon et al. (2004); storks and pelicans from Leshem & Yom-Tov (1996), Alon et al. (2004).
+ indicates seen in small numbers in some years. In autumn, approximately 20 groups of 2–4 observers were spaced some 2–3 km apart on a line extending across the path of migration from the Mediterranean Sea to the Jordan (Rift) valley, initially on the Kafr Qasim survey (1978–1987) and latterly about 100 km further north on the Northern Valleys survey (1988–1997) and in spring fewer groups of observers were spaced further south at the north end of the Red Sea near Eilat.
[a]Eight years within this period.
[b]Males only, because female Pallid and Montagu's Harriers could not be reliably distinguished.
[c]Four most numerous species reached 683 553 in 1997 (Alon et al. 2004).
[d]One year only (1987) (Leshem & Yom-Tov 1996a).
[e]Four years only (1987–1990) (Leshem & Yom-Tov 1996a).

Table 7.4   Maximum annual counts of raptors (and year) at two points of entry to Africa

Species	Suez, Egypt		Bab el Mandeb, French Somaliland[a]
	Autumn	Spring	Autumn
Osprey *Pandion haliaetus*	9 (1981)	7 (1982)	3 (1987)
European Honey Buzzard *Pernis apivorus*	79 (1981)	630 (1982)	
Black Kite *Milvus migrans*	106 (1981)	3861 (1982)	579 (1987)
Red Kite *Milvus milvus*	9 (1981)	5 (1982)	
Egyptian Vulture *Neophron percnopterus*	1002 (1984)	1189 (1982)	554 (1987)
Eurasian Griffon *Gyps fulvus*	3900 (1947)	141 (1982)	3 (1987)
Short-toed Eagle *Circaetus gallicus*	12136 (1984)	3063 (1982)	1202 (1987)
Western Marsh Harrier *Circus aeruginosus*	35 (1981)	19 (1982)	45 (1987)
Hen (Northern) Harrier *Circus cyaneus*	5 (1981)		
Pallid Harrier *Circus macrourus*	11 (1981)	7 (1982)	67 (1987)
Montagu's Harrier *Circus pygargus*		17 (1987)	
Northern Goshawk *Accipiter gentilis*			
Eurasian Sparrowhawk *Accipiter nisus*	51 (1984)	116 (1982)	2135 (1987)
Levant Sparrowhawk *Accipiter brevipes*	41 (1981)	141 (1982)	7 (1987)
Common Buzzard *Buteo buteo*	856 (1984)	80887 (1982)	98339 (1987)
Long-legged Buzzard *Buteo rufinus*		53 (1982)	131 (1987)
Lesser Spotted Eagle *Aquila pomarina*	40000 (1981)	10000 (1981)	
Greater Spotted Eagle *Aquila clanga*	86 (1981)	74 (1981)	20 (1987)
Steppe Eagle *Aquila nipalensis*	64880 (1981)	15778 (1982)	76586 (1987)
Imperial Eagle *Aquila heliaca*	556 (1981)	36 (1982)	70 (1987)
Golden Eagle *Aquila chrysaetos*	5 (1981)		
Booted Eagle *Hieraaetus pennatus*	1104 (1984)	457 (1982)	1123 (1987)
Bonelli's Eagle *Hieraaetus fasciatus*	7 (1984)	3 (1982)	2 (1987)
Lesser Kestrel *Falco naumanni*	26 (1981)		8 (1987)
Common Kestrel *Falco tinnunculus*	24 (1981)	57 (1982)	183 (1987)
Red-footed Falcon *Falco vespertinus*	462 (1981)		
Eurasian Hobby *Falco subbuteo*	31 (1981)		69 (1987)
Eleonora's Falcon *Falco eleonorae*	27 (1981)	6 (1982)	2 (1985)
Saker Falcon *Falco cherrug*			3 (1985)
Peregrine Falcon *Falco peregrinus*			16 (1985)
Maxima annual counts of raptors	133982 (1981)	132242 (1982)	246478 (1987)

From Bilsma (1983), Welch & Welch (1988), Wimpfheimer *et al.* (1983).

[a] A count is available for Bab el Mandeb in spring but covered only a few days (Welch & Welch 1998).

than those in Africa and South America. At least 33 raptor species migrate along portions of this flyway through eastern Asia, with the bulk of the flight dominated by Oriental Honey Buzzards *Pernis ptilorhynchus*, Grey-faced Buzzard Eagles *Butastur indicus*, Chinese Sparrowhawks *Accipiter soloensis* and Japanese Sparrowhawks *Accipiter gularis*. Eurasian (Japanese) Buzzards *Buteo b. japonicus* use the northern half of the flyway, and Black Bazas *Aviceda leuphotes* the southern half. Major concentrations of birds are seen in autumn at Chumphon in Thailand (>170 000), at Taiping (>30 000) and the Selangor Plain (121 000) in Peninsular Malaysia, and at the Straits of Malacca, including Singapore, suggesting that most of them winter in the Indonesian Islands (DeCandido *et al.* 2004). More than 90 000 individuals of several species have been counted crossing from Bali to Lombok, presumably to winter in the islands to the east (Germi & Waluyo 2006).

5. **The East Asian Oceanic Flyway**: More than 225 000 raptors travel along this 5000 km largely overwater flyway that stretches from coastal eastern Siberia and Kamchatka to Japan, the Philippines and into eastern Indonesia (**Figure 7.1**). At least 26 species migrate at least part way along the main flyway, which extends from southern Japan through the Ryukyu Islands and Taiwan to the Philippines and beyond. The bulk of the flight is dominated by Grey-faced Buzzard Eagles *Butastur indicus* (30 000 at Kenting on the southern tip of Taiwan) and Chinese Sparrowhawks *Accipiter soloensis* (up to 201 000 at Kenting). The long sea-crossings of up to 300 km may restrict the variety of species that use this flyway (Lin & Severinghaus 1998 updated). However, some of the overwater crossings fall within the tradewind zone, so it is possible that raptors can take advantage of the weak thermals that develop there, although they are not known for sure to do so.

Some species use only certain parts of the major flyways, while other species use other parts, depending on the locations of their breeding and wintering areas. Along each flyway, however, the numbers of birds seen in autumn tend to increase southwards towards the tropics, as more and more individuals join the migration, outnumbering those that stop. Numbers reach a peak in the northern tropics and, as the streams continue southwards, numbers then gradually decline, as birds progressively stop migrating and settle in their wintering areas. Nevertheless, substantial numbers of birds continue to the southern parts of the major flyways in South America or Africa. In spring, the reverse geographical trend in numbers occurs as the birds return northwards. Observation sites in temperate latitudes tend to produce seasonal counts up to tens of thousands of individuals, whereas those at lower latitudes can produce hundreds of thousands or millions, with the record from Veracruz in Mexico (about 22°N) where more than 6.6 million raptors and other soaring birds were counted in autumn 2001 (website http://www.hawkcount.org; for earlier counts see Ruelas Inunza *et al.* 2000).

Other major flyways may remain to be discovered. In particular, only minor routes have yet been described around or through the Himalayas, yet India forms an important wintering area for raptors and other soaring birds. Similarly, Lake Baikal in Siberia is the largest body of freshwater in the world, yet to my knowledge only minor concentration points (<10 000 birds) on its edges have yet been described. Moreover, little is known of the routes taken by raptors through South America or Africa, although the radio-tracking of individuals suggests a continuation of narrow corridor routes in some species (see Fuller *et al.* 1998 for Swainson's Hawks *Buteo swainsoni* in South America (see **Figure 8.6**), Meyburg *et al.* 2002 for Lesser Spotted Eagles *Aquila pomarina*, Berthold *et al.* 2002, 2004 for White Storks *Ciconia ciconia* in Africa). In addition to the major flyways, many minor ones can be discerned, such as the Strait of Messina between Italy and

Sicily which is crossed by thousands of raptors in autumn and spring en route between Europe and North Africa (Corso 2001).

Many species perform 'loop migrations', taking mainly different routes on their outward and return journey, as mentioned earlier for the Amur Falcon *Falco amurensis*. There seems to be at least two loops for migration between Europe and Africa. In one, the southward movement occurs through Gibraltar and the northward one through Sicily (an anticlockwise loop). In the other, a southward route occurs across Arabia, down the east side of the Red Sea and crossing to Africa at Bab el Mandeb, and the northward route occurs up the west side of the Red Sea to cross from Africa at the Gulf of Suez (a clockwise loop). Both loops are used by many raptors, as well as by passerines and other kinds of birds. They have been demonstrated by counts on the different flyways, by ringing recoveries and by the satellite tracking of radio-marked birds (see Meyburg *et al.* 2003 for Steppe Eagle *Aquila nipalensis*). Similarly, in North America, many Peregrines *Falco peregrinus* migrate southward down the east coast, crossing the Caribbean Islands to South America. But on their return in spring, most Peregrines travel up through the centre of the continent, again like many other species (Chapter 8).

The numbers of birds counted at established watch sites vary greatly from day to day in the migration season, and from year to year. Even in the commoner species, counts may vary by more than two-fold from one year to the next. Much of this variation may be due to fluctuations in observer effort and to variations in weather, which influence the volume of migration on particular days, its altitude and route. Lateral displacement of the centre of the migration stream by only a few kilometres can mean that most birds are missed by observers based at a fixed site. Occasional high flights can also reduce the proportions of passing birds that are readily visible to ground observers. Compared with such effects, the proportion of the year-to-year variation in counts that is due to annual variation in population sizes is probably small, but such counts have nevertheless proved useful in revealing long-term trends in populations. As the birds are usually drawn from a wide area, any long-term trends they might show are likely to be widespread, overriding purely local changes. The counts from Hawk Mountain in Pennsylvania, and from Falsterbo in Sweden, which span several decades, have been used to assess long-term population trends (Roos 1978, Bednarz *et al.* 1990, Kjellén & Roos 2000), as have the shorter runs of counts from other sites elsewhere (for Israel see Shirihai *et al.* 2000). In some species, annual counts also indicated changes in the ratio of juveniles to adults during the 1950s–1970s, as reproductive rates were lowered by the effects of DDE (a metabolite of the insecticide DDT, known to cause reproductive failure through shell-thinning) (Dunne & Sutton 1986, Bednarz *et al.* 1990, Bildstein 1998). The numbers of raptors counted at many concentration sites in the past 50 years were probably small fractions of the numbers present several hundred years ago, before their populations were so greatly reduced by human activities.

USE OF THERMALS AND OTHER UPDRAFTS

Birds that travel over land by soaring–gliding flight mostly gain lift from rising air, whether in the form of thermals or other updrafts (Chapter 8). On many days, updrafts are formed when wind strikes a slope or cliff and is deflected upwards. Long mountain ridges thus provide excellent flyways for soaring migrants in those (relatively rare) places where the ridge lies roughly north–south, in a direction appropriate for migration. One such site is the Kittatiny Ridge in eastern North America, which provides a long unbroken source of lift on which hawks can glide with hardly a flap for up to 300 km, beginning in New York State and continuing past Hawk Mountain in Pennsylvania (Kerlinger 1989). The ridge deflects the prevailing west–north autumn winds upwards, at rates often exceeding 3–4 m per second, and sometimes to heights exceeding 300–400 m above ground. The hawks are mostly moving northwest–southeast when they encounter the ridge running northeast–southwest, at right angles to their path. For a time, they ride the upcurrents generated by the ridge, even though this activity takes them off their main direction. It gets them to lower latitudes, but eventually they must leave the ridge to continue their journey towards the southeast. Other long ridges used by migrating raptors occur in the Rockies of western North America, the Andes in South America, and in various mountains in the Middle East, including the Rift Valley which extends southwards into Africa. With knowledge of the needs of migrant raptors, topographic maps can be examined for likely-looking sites, parts of the world being largely unexplored in this respect.

Thermals are localised columns of rising air created mainly through the uneven heating of the ground by the sun (**Box 7.1**). These columns rise to high elevation, until they have cooled to the temperature of the surrounding air, where they often produce a cumulus cloud, marking their position. They usually begin in the morning once the ground has heated sufficiently, but gather strength during the day. They climb gradually faster and higher, often reaching more than 1000 m at noon, and then wane in the evening as the ground cools. They typically rise at 1–4 m per second (Kerlinger 1989). Birds progress on migration by circling in one thermal to gain height, and then gliding with loss of height to the next thermal where they rise again, repeating the process along the route (see **Figure 3.5**). This enables birds to travel across country at around 30–50 km per hour, depending on the rate and extent of rise within thermals and the distance covered between thermals (in turn dependent on the 'glide coefficient', which is the ratio between the horizontal distance covered by the bird and its altitude loss over that distance). Small species with light wing-loading ascend more rapidly so spend less time in each thermal; but in travelling between thermals they glide less rapidly than large species and so lose more height per unit distance; they can therefore glide less far before having to climb again (Chapter 3; **Figure 7.2, Table 7.5**). Yet small species can get underway earlier in the morning and continue later into the evening than large species with heavier wing-loading that are restricted to a shorter period each day when the thermals are strongest. The flight times of the smaller species migrating through Israel typically extend over 8–10 hours each day (beginning around 9 a.m.) and the larger ones over 6–7 hours (beginning around 10 a.m.). However, all species tend to make more rapid progress in the middle part of the day, when climbs are fastest and highest, and glides are longest (**Figure 7.3**). Particular species can travel across country twice as fast around noon than in the morning or evening. When a bird glides, it may partly fold its wings which reduces drag and thereby increases speed; but the reduced wing and tail area also provides less lift, so the bird sinks more rapidly.

Box 7.1 More about thermals

Thermal formation ultimately depends on heat from the sun, which is greater at mid–lower latitudes and passes through the atmosphere to heat the ground. The input of solar energy depends on whether the surface absorbs or reflects heat, and because the ground surface varies, heating is uneven. Undulating surfaces are more conducive to thermal formation than flat ones; land is more conducive than water, and bare earth is more conducive than snow and ice. Rock and dry sand heat more rapidly than damp soil. The bottom layer of air is heated locally by contact with the warm ground. This heated air then expands, reaching lower density than surrounding air, so rises as a thermal. Replacement air is sucked in at the base of the thermal, which in turn warms and rises, causing surrounding air to sink. As the ground heats through the day, thermals grow higher and faster. Condensation starts at a height which is determined by the temperature and water vapour content of the air, and gives rise to a cumulus cloud, which marks the position of the thermal. As the air temperature rises during the day, the cloud base rises too; and as heating declines during the evening, thermals gradually die out, until the next day. Particular thermals may last anything from about 20 minutes to several hours.

In light winds, thermals tilt in the direction of the wind, which causes passive drift in the birds using them. If necessary, this drift can be corrected in the next glide. Thermals sometimes form into lines (thermal streets) along the wind direction, and may even coalesce into roll vortices, giving continuous lines of lift. These are marked as 'cloud streets' which form above the condensation level. Soaring birds exploit cloud streets by flying along them above the line of thermals, but they are useful only when they run more or less along the path of migration, as at Panama. Strong winds can prevent thermal formation altogether, and even with winds in their migration direction, soaring birds are often grounded in such conditions.

Another condition in which rising air currents develop is when two air masses with different frontal characteristics converge, giving rise to a narrow zone of lift along the convergence. They include the sea breeze fronts that meet a heated land surface, so that coastlines are commonly used by gulls and other birds for soaring.

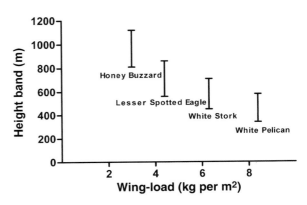

Figure 7.2 The relationships between wing load and average height band in thermals for four species of soaring migrants. Average climbing time in thermals (not shown) increased with increase in body weight. From Leshem & Yom-Tov (1996b).

Table 7.5 Average altitude (± SD) and daily progress velocity (± SD) of four species of soaring migrants over Israel, together with estimates of the daily flight time and distances travelled

Species	Number	Start height in thermal (m)	End height in thermal (m)	Height band in thermal (m)	Velocity (km/hour)	Mean migration time (hours/day)	Mean daily flight distance over Israel	Estimated mean duration of migration days[a]
Great White Pelican *Pelecanus onocrotalus*	467	344 ± 175	562 ± 186	218	29.2 ± 9.1	7.5	249	21
White Stork *Ciconia ciconia*	1059	463 ± 209	713 ± 221	250	38.7 ± 9.6	9.0	348	23
Lesser Spotted Eagle *Aquila pomarina*	78	567 ± 201	871 ± 184	304	50.9 ± 6.7	7.5	381	21
Honey Buzzard *Pernis apivorus*	215	836 ± 211	1123 ± 225	287	45.2 ± 9.0	10.0	446	23

Measurements made from a motorised glider and light aircraft, as well as by radar. Species listed in decreasing order of body weight. From Leshem & Yom-Tov (1996a).
[a] Estimated time taken to travel between the centre of the breeding range and the centre of the wintering range, assuming that migration occurred on every day over the daily distance shown. These estimates are lower than those uncovered by radio-tracking of the same species, mainly because migration did not occur on every day (Chapter 9).

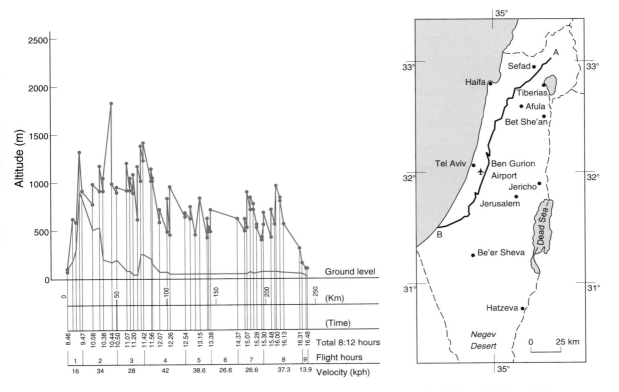

Figure 7.3 Route and altitude in relation to ground level for a flock of Great White Pelicans *Pelecanus onocrotalus* followed by glider through Israel on 12 October 1987. Each dot represents a point where the first birds started to climb or descend in a thermal. Time of day and average velocity are given along the bottom, and the route from take-off to landing that day is shown in the map on the right. From Leshem & Yom-Tov (1996b).

In Steppe Eagles *Aquila nipalensis* studied in Israel, cross-country speed was related to the climb rate in thermal circling. Over the whole diurnal cycle, the mean climb rate in thermals was 1.9m per second, but rates up to 5.0m per second were reached around noon. Mean gliding air speed between thermals was 56km per hour which, allowing for climb times, gave a mean cross-country speed of 45km per hour. The upper limit of migration was about 1600m above ground, but was mostly below 1000m (Spaar & Bruderer 1996). Smaller Honey Buzzards *Pernis apivorus* and Steppe Buzzards *Buteo buteo vulpinus* achieved lower cross-country speeds, at 37 and 35km per hour. However, in a 12-hour day, Steppe Eagles soared for only 6 hours and covered 270km, whereas the two smaller species migrated for 10 hours and covered 360km. Hence, although larger birds can travel faster and further between thermals, they do not necessarily cover more kilometres per day (Spaar & Bruderer 1996). In cooler climates, where thermals occur during a smaller part of each day, migration times are shorter, and larger raptors, such as Golden Eagles *Aquila chrysaetos*, travel for at most a few hours per day, chiefly between 12.00 and 14.00 hours.

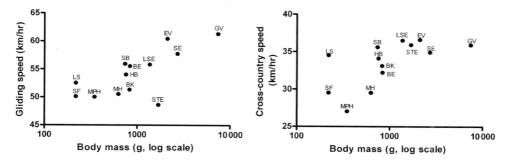

Figure 7.4   Gliding speed (left) and cross-country speed (right) relative to the air in soaring–gliding flight in raptors of different body mass. BE, Booted Eagle *Hieraaetus pennatus*; BK, Black Kite *Milvus migrans*; EV, Egyptian Vulture *Neophron percnopterus*; GV, Griffon Vulture; HB, European Honey Buzzard *Pernis apivorus*; LS, Levant Sparrowhawk *Accipiter brevipes*; LSE, Lesser Spotted Eagle *Aquila pomarina*; MPH, Montagu's/Pallid Harrier *Circus pygargus*; MH, Marsh Harrier *Circus aeruginosus*; SB, Steppe Buzzard *Buteo buteo vulpinus*; SE, Steppe Eagle *Aquila nipalensis*; SF, Small falcon (ca. 220 g), StE, Short-toed Eagle *Circaetus gallicus*. Cross-country speed = 2.67 × log body mass + 1.37; 95% confidence interval of the slope: 1.38–11.23; $r_{11} = 0.63$, $P < 0.05$. Note that with a following wind, all speeds would be higher. From Spaar (1997).

As expected, weather conditions also affect progress. On one favourable occasion, White Storks *Ciconia ciconia* over Israel were able to climb to 1550 m and then glide for 36 km before needing to climb again. Their average flight speed was 57 km per hour, about 47% faster than average. The crossing of short stretches of sea, as at the Bosphorus or Gibraltar, ideally needs to be accomplished in one long descending glide, in order to avoid laborious flapping flight. Early in the day, when conditions are not ideal, birds sometimes start a crossing, and having reached a few kilometres from shore, turn back to try again later.

Comparisons between the various species of soaring raptors in Israel revealed that: (1) the average climbing rate in thermal circling was independent of body size, at 1.5–2.1 m per second, although smaller species had a smaller turning radius, so in theory could benefit more from the faster currents in the centre of the column; (2) in inter-thermal gliding, air speed was positively related to body mass, and gliding angle negatively related to body mass – heavier species glided faster, and at shallower angles, losing less height per unit distance; (3) overall cross-country speed relative to the air was positively related to the species body mass, the larger species travelling faster but for fewer hours each day (**Figure 7.4, Table 7.5**; Spaar 1997). The better gliding performance of larger raptors fits theoretical considerations about gliding flight (Chapter 3). The same principles apply to sailplanes, and to increase gliding speed, pilots sometimes add water as ballast. The faster a bird glides, the more lift it gains from the airflow below its wings, over and above any obtained from localised rising air currents.

In some topographic situations, given appropriate wind and temperature conditions, updrafts called leewaves are sometimes formed on the downwind sides of mountains, reaching much greater altitudes than thermals. Because of their restricted distribution, leewaves are unlikely to be of widespread importance in soaring bird migration, but they could be used by raptors in some situations, as when crossing the Strait of Gibraltar (Evans & Lathbury 1973) or moving through Panama (Smith 1985a). Migrants using leewaves would normally be invisible to ground-based observers.

When migrating over land or water, where no thermals are available, some raptors can make use of a following wind, using dynamic soaring. The birds glide downwind, picking up speed, but gradually losing height, then turn into the wind to gain lift, thus regaining the altitude lost in the glide, before turning again into the next glide to continue their journey. They continue this looping flight for long distances, but are constrained to travel more or less downwind. Progress can be quicker than in thermal soaring, because the birds spend less time in gaining altitude. They can also fly in moderate-angled crosswinds which provide the lift, but if conditions turn against them, they have to switch to energy-demanding flapping flight if they are not to be blown off course. Using these flight modes, birds are less restricted to the middle parts of the day, and flights may begin at dawn, as at Kenting on the southern tip of Taiwan. The main participants at this site are Grey-faced Buzzard Eagles *Butastur indicus* and Chinese Sparrowhawks *Accipiter soloensis*, but many other species are seen in smaller numbers. The Chinese Sparrowhawks, detected over the sea by radar, occurred in long straggling flocks up to 21 km long, travelling at an average of 20 km per hour.

While most soaring bird migration occurs by day, for the reasons given earlier, some species have been seen flying over the sea or other hostile terrain at night. For example, Western Honey Buzzards *Pernis apivorus* have been recorded by radar migrating over Malta at night (Elkins 1988), and have appeared on dark nights in autumn at the lighthouse on Heligoland Island off Germany (Gätke 1895). In addition, Levant Sparrowhawks *Accipiter brevipes* have been found to enter tree roosts at night in Israel after descending from a daytime desert flight (Yosef 2003). Other raptor species that have occasionally been seen flying at night include Northern Harrier *Circus cyaneus*, Osprey *Pandion haliaetus* and Peregrine *Falco peregrinus*, and other overwater migrants, such as Chinese Sparrowhawk *Accipiter soloensis*, Grey-faced Buzzard Eagle *Butastur indicus* and Amur Falcon *Falco amurensis* (DeCandido *et al.* 2006).

Weather affects soaring landbirds in some of the same ways as other birds. Wind strength and direction influence the number of soaring birds that migrate on particular days, their travel routes, flight altitudes and speeds. Low dense cloud and rain suppress migration altogether. To provide ideal migration conditions, however, different wind directions and temperatures are needed at different sites, depending on local topography, and on the importance of thermals as opposed to updrafts from slopes. For compared to other birds, soaring species are more dependent on the interaction between wind and topography than on wind alone. In addition, thermals cannot form in strong winds which, in the absence of other updrafts, bring soaring bird migration to a standstill, however favourable the wind direction. Hence, on some days that are ideal for most migrating birds, soaring species may be grounded (**Box 7.1**).

EXTENSION OF MIGRATION AS A CONSEQUENCE OF SOARING

In travelling between two points, soaring migrants typically cover much longer distances than flapping migrants. The route extension is made up of: (1) distance added as a result of circumventing large water bodies and using longer overland routes; (2) distance added due to circling and climbing in thermals, as well as moving between them, which is not necessarily along the shortest and straightest route. These various distances were calculated for four species in Israel using data collected by following birds with a motorised glider (Leshem & Yom-Tov 1996b). The four species were White Pelican *Pelecanus onocrotalus*, White Stork *Ciconia ciconia*, Lesser Spotted Eagle *Aquila pomarina* and Honey Buzzard *Pernis apivorus*. Between the centres of their breeding and wintering areas, these four species had extended their migration distances by 48–91% compared with the straight line distances. The increased distance caused by circumventing sea areas was estimated at 22–34% in different species, while the increase resulting from use of thermals accounted for an additional 23–57%. Presumably, the energy saved by use of soaring–gliding flight more than compensated for the energy consumed by covering the extra distances.

Avoidance of long water crossings results in some curious detours, as exemplified by Short-toed (Snake) Eagles *Circaetus gallicus* nesting in central Italy (Agostini *et al.* 2002). Instead of migrating south in autumn, and making the oversea flight from Sicily to North Africa, these birds start by migrating northwest up the Italian peninsula, then move westward and southwestward around the Mediterranean to cross the 14 km of sea at Gibraltar. In order to reach 37°N in North Africa they travel a roundabout route of 2000 km rather than take the shorter 700 km direct flight from their breeding areas through southern Italy and over the sea to North Africa. This gives a forceful demonstration, in this species, of the amount of extra travel undertaken to avoid a long sea-crossing.

SOCIAL FACTORS

Seven of the 35 soaring raptor species that migrate through Israel form long drawn-out flocks, as do White Storks *Ciconia ciconia*, Black Storks *C. nigra*, Great White Pelicans *Pelecanus onocrotalus* and Common Cranes *Grus grus*. One presumed advantage of soaring birds migrating in such concentrations is that it makes finding thermals easier, thus conserving energy. By watching the birds ahead that are already circling upward, a bird can head for them without wasting time and energy in thermal location. The longer is the migration to be performed, the greater is the amount of time and energy saved by this behaviour. Observations made by radar and by use of a glider in Israel revealed that, on peak migration days, the lines formed by flocks extended up to 200 km, so that most individuals had before them a continuous route marked out by their predecessors (Leshem & Bahat 1999).

It is difficult with raptors to tell whether the birds migrate in flocks simply because they share the same narrow migration route, and the same thermals and updrafts within it, or whether they are attracted to one another for other reasons.

Not surprisingly, the biggest flocks are seen in relatively numerous species which migrate within a short time period, such as the Western Honey Buzzard *Pernis apivorus* and Levant Sparrowhawk *Accipiter brevipes* in the Old World, or the Broad-winged Hawk *Buteo platypterus* and Swainson's Hawk *B. swainsoni* in the New World, all of which can be seen in concentrations exceeding 1000 individuals. Once they leave a thermal, the birds seem to behave as individuals, and head off for the next thermal, whose position is marked by the presence of other circling birds. In this sense, the birds clearly benefit from travelling on the same days as others, and from watching one another, but in some species, such as the Mississippi Kite *Ictinia mississipiensis* and Lesser Kestrel *Falco naumanni*, the social bonds seem stronger. These latter species are mostly insectivores and normally feed in flocks.

Pelicans tend to migrate in their own flocks and seldom intermix with raptors. They use a somewhat different system to locate thermals. After leaving one thermal, they split into dozens of secondary flocks, flying in V-formation on a broad front that can extend for a kilometre or more. This allows the pelicans to randomly sample air currents, so when part of a flock locates a thermal and begins circling upward, the other pelicans immediately glide towards them, and in this manner the flocks proceed with maximal efficiency. This behaviour is especially important for pelicans because they migrate in individual flocks over a period of about two months, and are not concentrated within a short interval, as are storks and most raptors.

COUNTS IN ISRAEL

Israel lies near the east end of the Mediterranean Sea, at the junction of the Eurasian and African land masses. In consequence, most of the species that breed in Eurasia and winter in Africa pass through Israel, some in very large numbers (**Figure 7.5**, **Table 7.3**). The Rift Valley runs roughly north–south through the length of Israel and Jordan, creating optimal conditions for rising updrafts and thermals. It is narrow in this region, and the fault escarpment reaches altitudes of hundreds of metres. This combination of steep cliffs cut by narrow gorges creating updrafts, and high average temperatures creating thermals in the valley, provide an ideal route. In addition, Israel's central mountain ridge stretches almost parallel to the Mediterranean coast, and fairly close to it, providing further good conditions for migrating raptors.

Three main migration tracks for soaring birds run through Israel (**Figure 7.5**). Two extend mainly along the western and eastern sides of the central mountain chain (the latter along the Jordan Rift Valley) and the third cuts across the southern end of the country in a northeast–southwest direction between Jordan and Sinai through Eilat (Leshem & Yom-Tov 1998). The birds using these routes cross to and from Africa via the Gulf of Suez, mostly at its northern or southern ends. Further east, as mentioned earlier, another flyway runs across the Arabian peninsula, crossing to Africa at the narrow Bab el Mandeb Straits at the southern end of the Red Sea (the so-called Caspian–Arabian route, Alon *et al.* 2004). By these various routes, the birds use the shortest water crossings and avoid the wider parts of the Red Sea. The actual positions of the migration streams shift east or west to some extent during the course of each day, as well as from one day to the next,

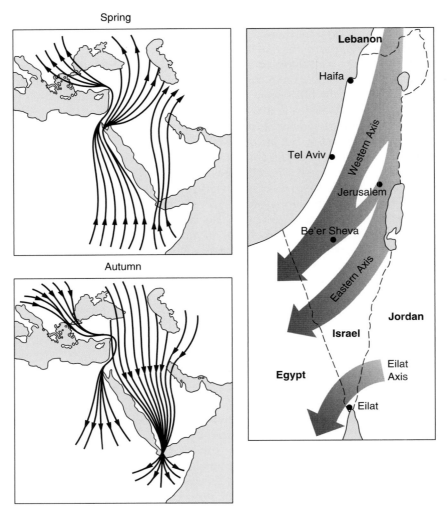

Figure 7.5   Main migration routes for soaring birds through the Middle East, autumn and spring (left), and through Israel alone (right); the western route – over the western slopes of Israel's central mountain spine; the eastern route – mainly along the Jordan (Rift) Valley, continuing south during most of the day, but crossing to join the western route during part of the day; the Eilat mountain route – crosses southern Israel in the region of the Eilat Mountains, running northeast–southwest through Jordan and Sinai. From Leshem & Bahat (1999).

and from the early to the later part of each season, as wind and other conditions change. In autumn, birds enter northern Israel on a single narrow route, with 87% of all soaring birds encountered within a 20-km-wide strip, some 11–31 km in from the Mediterranean coast. For reasons not fully understood, some species fly along routes to the east or west of others. Surveillance radar photographs show long lines of thousands of soaring migrants progressing south-southwest parallel to the coast, moving inland with the sea breeze during the course of the day (Leshem & Yom-Tov 1998).

Over several years from the early 1980s, attempts were made to count the numbers of soaring birds passing through Israel on each day of the migration season in autumn (August–November) and spring (March–May) (Leshem & Yom-Tov 1996a, Shirihai *et al.* 2000, Alon *et al.* 2004). Because many birds take somewhat different routes at the two seasons, different sites were used for counting in autumn and spring, by groups of observers spaced 2–3 km apart across the migration route. Although not every passing raptor is likely to have been counted, the volume of the migration, as assessed each day by these ground observers, was correlated with the volume of migration as assessed on the same days by radar (in autumn, $r^2 = 0.66$, $P < 0.001$). This correlation indicated the general reliability of ground counts for large soaring birds that in this region move mainly within visual range.

The resulting counts were some of the most thorough and complete ever made of soaring bird migration, involving 35 species (**Table 7.3**). On average, about half a million raptors (mainly Lesser Spotted Eagles *Aquila pomarina*, Western Honey Buzzards *Pernis apivorus* and Levant Sparrowhawks *Accipiter brevipes*) passed through northern Israel each autumn, as did 250 000 White Storks *Ciconia ciconia* and 70 000 Great White Pelicans *Pelecanus onocrotalus*. In the spring in southern Israel, about a million raptors (mainly Western Honey Buzzards *Pernis apivorus*, Steppe Buzzards *Buteo b. vulpinus*, Steppe Eagles *Aquila nipalensis* and Black Kites *Milvus migrans*) passed northeast, plus about 450 000 White Storks. Only Steppe Eagles, which use breeding areas further to the east than the other species, used the southern (Eilat) route in large numbers in both autumn and spring. The whole Eurasian populations of Lesser Spotted Eagles, Levant Sparrowhawks and Great White Pelicans are thought to pass through Israel each year. For the two raptors, this represents the entire global population.

The totals of each species seen in autumn or spring varied greatly from year to year, depending partly on wind conditions (which sometimes pushed birds off their normal route), altitude of flight (which sometimes put birds beyond the visual range of ground-based observers), and probably also on the numbers and experience of observers (**Table 7.3**). Overall, however, the counts indicated that around half as many soaring raptors pass through Israel each autumn as in spring. The seasonal difference arose largely because the majority of Black Kites, Steppe Buzzards and Steppe Eagles missed Israel in autumn and took more easterly routes, as shown by counts elsewhere (**Table 7.4**). The counts for Western Honey Buzzards alone in spring 1985 exceeded 851 000 individuals, with up to 145 000 seen on a single day (at Eilat on 6 May 1985).

The timing and duration of passage varied between species (**Table 7.6**), but within each species both autumn and spring peak dates were remarkably consistent from year to year. The confidence intervals of the mean dates of appearance for raptors in different autumns ranged between 1.5 and 5.5 days, depending on species, whereas for White Stork *Ciconia ciconia* and Great White Pelican *Pelecanus onocrotalus* they were 4.2 and 13.8 days respectively, reflecting the fewer years of data and the longer and more variable migration periods of these species. The equivalent confidence intervals for spring migration were 2.1–5.5 days for raptors and 6.1 days for White Storks.

Table 7.6 Timing of passage of migrants through Israel, spring and autumn. Only species with the largest totals at each season are shown

Species	Autumn migration			Spring migration			Mid-point between spring and autumn dates
	Mean date	90% pass between (days)	Total duration (days)	Mean date	90% pass between (days)	Total duration (days)	
Honey Buzzard Pernis apivorus	7 Sep	30 Aug–14 Sep (16)	60	5 May	2–12 May (11)	36	6 July
Levant Sparrowhawk Accipiter brevipes	20 Sep	14–26 Sep (13)	47	24 Apr	18–26 Apr (9)	71	7 July
Lesser Spotted Eagle Aquila pomarina	29 Sep	21 Sep–5 Oct (15)	57	–	–	–	–
Steppe Eagle Aquila nipalensis	–	13 Oct–1 Nov (19)	–	10 Mar	23 Feb–27 Mar (33)	73	–
Short-toed (Snake) Eagle Circaetus gallicus	23 Sep	15 Sep–9 Oct (25)	52	–	–	–	–
Egyptian Vulture Neophron percnopterus	19 Sep	6–29 Sep (24)	50	1 Apr	8 Mar–3 May (57)	101	25 June
(Western) Marsh Harrier Circus aeruginosus	23 Sep	7 Sep–6 Oct (30)	57	11 Apr	10 Mar–10 May (62)	100	2 July
Booted Eagle Hieraaetus pennatus	22 Sep	8 Sep–1 Oct (24)	56	10 Apr	25 Mar–2 May (39)	87	2 July
Black Kite Milvus migrans	14 Sep	31 Aug–4 Oct (35)	53	30 Mar	21 Mar–10 Apr (21)	91	21 Jun
Red-footed Falcon Falco vespertinus	2 Oct	25 Sep–13 Oct (19)	49	–	–	–	–
Steppe Buzzard Buteo buteo	–	–	–	3 Apr	22 Mar–15 Apr (25)	91	–
White Stork Ciconia ciconia	29 Aug	16 Aug–12 Sep (28)	92	2 Apr	14 Mar–25 Apr (43)	82	14 Jun
Great White Pelican Pelecanus onocrotalus	14 Oct	18 Sep–7 Nov (51)	126	–	–	–	–

From Leshem & Yom-Tov (1996a).

Comparing the different species of raptors, three general trends emerged: (1) species that eat warm-blooded prey (birds and mammals) generally passed earlier in spring and later in autumn than species that eat cold-blooded prey, with the insectivores being last to move north and first to move south (Chapter 14); (2) species that travel in obvious flocks (Lesser Spotted Eagle *Aquila pomarina*, Western Honey Buzzard *Pernis apivorus*, Levant Sparrowhawk *Accipiter brevipes*, Red-footed Falcon *Falco vespertinus*) passed within a shorter period each year than non-flocking species; (3) species that breed over large areas of Eurasia had the longest passage periods, presumably because birds from different localities had started at different dates and travelled different distances; and (4) whereas for some species the spring passage period was shorter than the autumn one (as in many other birds, Chapter 14), in most species the spring passage period was longest. Longer spring passage was most marked in large species in which individuals do not begin nesting until they are more than two years of age. Their populations therefore contained a larger proportion of immatures, which usually migrated later in spring and over a longer period than the adults, spending a much shorter period in the breeding areas (for Steppe Buzzard *Buteo b. vulpinus*, see Gorney & Yom-Tov 1994; for Lesser Spotted Eagles *Aquila pomarina* see Meyburg *et al.* 2001). The same held for the White Stork *Ciconia ciconia* in spring when adults began passing over in February, while immatures appeared in April–May. Year-to-year consistency in passage periods seems usual in the migrations of other obligate migrants, but counting difficulties make them harder to assess. Typically, however, whatever the length of the total migration period, the bulk of the birds may pass on a small number of days within it.

Food is important to the timing of raptor migration, not so much on the route itself, but in breeding areas (Chapter 14). At the end of summer the first animals to disappear with the onset of cold weather at high latitudes are large insects, followed by reptiles and amphibians, while fish retreat to deeper water. By then, many small birds have begun to migrate, and mammals begin to disappear, some hibernating and others spending increasing periods in sheltered sites where they are unavailable to raptors. During spring, the situation is reversed, with mammals appearing first and large insects last. The Steppe Eagle *Aquila nipalensis*, which eats chiefly mammals, is the first to migrate north, passing through Israel mainly in early March, while the Western Honey Buzzard *Pernis apivorus*, which eats insects, migrates last, in May. The Steppe Eagle spends six months in its breeding areas, passing south through Israel mainly in mid-October to mid-November, while the Western Honey Buzzard spends only three months in its breeding areas, passing through Israel in early September. The eagle also has a longer breeding cycle than the Honey Buzzard, with longer incubation, nestling and post-fledging periods.

NUMBERS ENTERING AFRICA

The number of raptors, storks and pelicans counted at other watch sites in Europe and the Middle East are summarised in **Tables 7.2, 7.3** and **7.4**. The numbers and species composition vary greatly from site to site, depending on location with respect to breeding and wintering areas, and the importance of the site as a concentration point. Adding the maximum counts from Gibraltar, Messina (Italy), Israel (or Suez) and Bab el Mandeb for certain soaring species gives a minimum rounded estimate of the numbers entering Africa from Eurasia in autumn, as follows: Short-toed (Snake) Eagle *Circaetus gallicus* 19000, Lesser Spotted Eagle *Aquila pomarina* 142000, Steppe Eagle *Aquila nipalensis* 75000 (spring count), Booted Eagle *Hieraaetus pennatus* 16500, Levant Sparrowhawk *Accipiter brevipes* 60000, and Steppe Buzzard (spring count) 466000. These species alone total 778000 birds, with White Stork *Ciconia ciconia* at 530000, Black Stork *Ciconia nigra* at 17000, and White Pelican *Pelecanus onocrotalus* at 70000. Only for these species were the migration seasons reasonably well covered at each site, and no other major routes are known. In addition, Western Honey Buzzards *Pernis apivorus* alone probably total more than one million, because maximum counts at three major sites sum to 996000, and many others are known to cross the Mediterranean Sea on a broader front, especially in autumn (Chapter 15). Meaningful counts of falcons and others, which migrate by flapping, or by a mixture of flapping and gliding, cannot be made, as they migrate on a broad front, and can cross the Mediterranean Sea anywhere. Hence, their total numbers are as yet unknown. One of the main gaps in knowledge concerns the extent of spring passage at Bab el Mandeb at the southern end of the Arabian peninsula. These birds are likely to take a more eastern route than those that pass through Eilat, so will not have been counted elsewhere on their journey. It seems likely, however, that the total numbers of raptors and other soaring birds entering Africa from Eurasia each year could greatly exceed two million.

MIGRATION AT OTHER SITES

In other parts of the world, migrating raptors face very different conditions from those encountered in the hot, cloudless climates of the Middle East. In general, the atmosphere contains more energy in tropical regions than at higher latitudes, which enables raptors to migrate in some conditions through the tropics that would ground them elsewhere. For example, at Talamanca in Costa Rica, thermals can form over shallow water in the early morning, which at that time is warmer than the nearby land, enabling soaring raptors to set off earlier in the day. There may also be sufficient updrafts even during rain, enabling some birds to continue migrating in the wet. In Panama, corridors of rising air produce long clouds, and for much of the time raptors fly through the cloud base, taking advantage of the 'thermal cloud streets' that enable them to glide for tens of kilometres at a time. They also use the updrafts created by winds hitting the central mountain spine. Counts at Panama are inevitably underestimates, partly because the birds are often hidden by cloud, and also because they often fly higher than visual range. Some fly over the central mountains at 4–6km above sea level, while others may over-fly the mid-afternoon storms at even higher altitudes (Smith 1985a). Unpublished counts near Panama City (Ancon Hill) in 2004–2005 produced a total of around 2.5 million birds (K. Bildstein).

The Veracruz totals represent the largest migration stream of soaring birds seen between North and South America (and the largest known stream of soaring birds from anywhere in the world). The overall autumn total in 2002 amounted to nearly 6.6 million raptors, including nearly 2.7 million Turkey Vultures *Cathartes aura*, 2.4 million Broad-winged Hawks *Buteo platypterus* and 1.1 million Swainson's Hawks

B. swainsoni (the latter species exceeding 1.2 million in 2003 and 2005). Up to 25 other raptor species have been seen at this site (**Table 7.1**), and several other soaring bird species, including more than 28 000 American White Pelicans *Pelecanus erythrorhynchos*). In addition, however, other raptors are known to migrate down the western side of the Americas at similar latitude to Veracruz, and further east, yet others also migrate down the Florida peninsula across the Caribbean Islands to South America. No good estimates are available for these other routes, but the Ospreys *Pandion haliaetus* counted on passage through southeast Cuba (La Gran Pedra) in autumn would represent about 90% of the known eastern North American population. It seems that the total raptor migration between North and Central–South America could well exceed seven million birds, which makes the known Eurasian–African total of two million seem erroneously small.

ENERGY RESERVES

The energy cost of soaring flight has been calculated at only 1.5–2.0 × BMR (basal metabolic rate), a cost spent mainly in maintaining the wings in gliding position. Per unit time, this is only about 5–25% of the energy consumed in flapping flight, making gliding by far the cheapest mode of flight (Chapter 3). One would expect, therefore, that soaring birds would lay down less migratory fuel as other same-size birds making similar overland journeys, or that they would travel further on a given amount. Comparing different species of raptors and owls, larger species tend to have significantly higher fat levels (and females higher than males), findings that hold summer and winter, and possibly also at migration times (Overskaug *et al.* 1997).

One detailed study concerns Steppe Buzzards *Buteo buteo vulpinus* caught on spring migration through Israel, just after crossing the Sahara (Gorney & Yom-Tov 1994). Fat levels were generally low at this stage in the journey, but significantly higher in adults (mean 4.5% of body mass) than in immatures (3.8%); neither group could have completed the rest of their migration to northern Europe (1400–6000 km) without eating, and the fact that the birds were caught in baited traps showed that they were prepared to feed. The mean weights of the birds caught in Eilat were around 25% lower than those of birds caught in the Cape Province of South Africa before spring departure (adults 579 g vs. 762 g, immatures 529 g vs. 709 g). Even if the entire weight difference was due to fat, at expected rates of usage, the birds could not have completed the 7800 km journey from the Cape to Israel without feeding. These data indicated that, although substantial body reserves were apparently accumulated by Steppe Buzzards at the start of their spring migration, representing more than 30% increase in body weight, further feeding en route was necessary to reach Israel, and still further feeding to reach the northern Eurasian breeding areas. This species performs one of the longest migrations of any bird of prey, at 9200–14 200 km each way.

Of 37 Broad-winged Hawks *Buteo platypterus* caught on their winter territories in Panama in March–April before setting off on migration, 62% were 'moderately to very fat' (Smith *et al.* 1986). This compared with 35% of 46 birds caught in mid-migration in autumn. Fasting for at least part of the journey was suggested by the absence of food in the guts of Broad-winged Hawks found dead in Panama and by the absence of pellets or faeces at large roosts of migrants (Smith 1980, Smith *et al.* 1986). If these birds had fed in the mornings, however, such food could have passed through by the evening, so it remains uncertain whether these soaring raptors accumulate large fat reserves before departure, and perhaps at certain sites en route, but fast during much of the journey. Turning to other species, Western Honey Buzzards *Pernis apivorus* and Oriental Honey Buzzards *P. ptilorhynchus* make no obvious attempt to feed over much of their route, although they do come down to drink. Museum labels on skins of Swainson's Hawks *Buteo swainsoni* and Mississippi Kites *Ictinia mississippiensis* collected at migration times often carried the note 'very fat' (W. S. Clark), and Bald Eagles *Haliaetus leucocephalus* have been tracked on migration for up to 12 days and were not seen to feed (Harmata 2002). Long-distance fasting has also been proposed for other soaring raptors, and recent satellite tracking of various species shows many examples of long, uninterrupted travel steps by soaring migrants (Berthold *et al.* 1992b, Meyburg *et al.* 1995a, Kjellén *et al.* 1997, Håke *et al.* 2003; Chapter 8). They could only do so on the basis of stored fuel reserves. Even though they make much of their flight on energy-saving gliding, therefore, it is clear that some raptors accumulate migratory fat like other birds.

In theory, such migrants could forage in the early mornings when conditions are unsuitable for soaring, and suffer no reduction in overall migration speed, but whether they do so is another question. Many species migrate partly through desert or other terrain offering little food and often birds travel in such large numbers over such narrow routes that most would have little chance of a meal. Of the soaring species that pass through Israel, only three make regular feeding stops in autumn (White Stork *Ciconia ciconia*, Great White Pelican *Pelecanus onocrotalus*, Booted Eagle *Hieraaetus pennatus*), but others can be caught in baited traps in spring or are occasionally seen with swollen crops. Other raptors that eat regularly on migration include the bird-eating falcons and accipiters which migrate at the same time as their prey, and also those insectivores which can take advantage of migrating insects (for Plumbeous Kites *Ictinia plumbea* eating dragonflies, see Smith 1980).

Some smaller raptor species are said to accumulate large fat reserves, notably the Amur Falcons *Falco amurensis* that cross the sea between India and Africa (Moreau 1972) and the Mississippi Kites *Ictinia mississippiensis* that migrate between North and South America (Bent 1938). On the other hand, American Kestrels *Falco sparverius* were found with relatively low fat reserves in early autumn, females slightly but significantly more than males (7.0% vs. 5.3% of body weight), resulting from a doubling of usual fat levels immediately before migration (Gessaman 1979). Fat levels of 3–12% were reported for Cooper's Hawks *Accipiter cooperi* and Sharp-shinned Hawks *A. striatus* in New Mexico in autumn and spring, with females containing significantly more fat than males, and adults more than juveniles (DeLong & Hoffman 2004). Other observers have noted large variations in the visible fat reserves of raptors caught on migration (Geller & Temple 1983, Clark 1985), and many immature Red-tailed Hawks *Buteo jamaicensis* from northern populations were thin during their autumn passage through Wisconsin (Geller & Temple 1983). Other emaciated raptors have been reported by ringers, notably among Northern Goshawks *Accipiter gentilis* caught during irruptions (Mueller *et al.* 1977).

In many studies of migrating raptors, the birds were caught in baited traps. As lightweight birds are perhaps more likely to enter such traps than heavy ones, the weights of trapped birds may not be typical of the population as a whole. In two species studied at Eilat in spring, Gorney *et al.* (1999) compared body weights between individuals caught in baited traps and those caught without bait in mist nets. In the Levant Sparrowhawk *Accipiter brevipes* the trapped birds were significantly lighter than the netted ones, whereas in the Eurasian Sparrowhawk *Accipiter nisus* no difference was found between the two groups. In at least one of these species, therefore, individuals caught in baited traps were not representative. However, the fact that both species entered baited traps showed that they were prepared to feed, at least during this part of their journey.

Turning to non-raptorial species, White Storks *Ciconia ciconia* seem to feed every morning when passing through suitable areas, but would not be expected to do so in desert. Some other species clearly accumulate substantial reserves. For example, Sandhill Cranes *Grus canadensis* increased in body weight by about 34% (males) and 30% (females) in spring while at the Platte River in Nebraska and Last Mountain Lake in Saskatchewan before migrating to their northern breeding areas (Krapu *et al.* 1985). Carcass analyses revealed that most of this weight increase was due to fat deposition. This was not surprising considering that cranes use soaring–gliding flight much less than the other species discussed in this chapter, especially at the high latitudes where most of them breed. They frequently travel by flapping flight, or by a mixture of flapping and gliding, and often migrate at night as well as by day.

Pelicans killed in Israel at migration times were found to contain only 3.4% fat, compared with 5.4% in wintering birds, which does not suggest extensive pre-migratory fattening. On the basis of a soaring flight cost of $1.5 \times$ BMR, Great White Pelicans *Pelecanus onocrotalus* were calculated to need only about 480 g of fat (about 5.5% of body mass) to migrate the 2500 km between Israel and their wintering areas in the Sudd swamps of Sudan, with a travel time of 7–8 days, but as they probably travelled south along the Nile Valley, they may have been able to feed en route (although no regular stopping sites are known) (Schmueli *et al.* 2000). Clearly, more information is needed on the migratory fuelling of soaring migrants.

SUMMARY

Because soaring birds migrate entirely (or almost entirely) by day, often at heights low enough for the birds to be identified by ground-based observers, and because they pass certain points in concentrated streams, some aspects of their migrations have been studied in detail.

On migration, soaring raptors, storks and pelicans depend mainly on updrafts where crosswinds are deflected upwards from cliffs or slopes, or on thermals where columns of rising air provide lift, enabling the birds to glide to the next thermal losing height, and then rise again. Such species are constrained to make as much of their journey as possible overland, making detours to avoid or minimise water crossings.

On a world scale, five major flyways for soaring birds are currently recognised, along which enormous numbers pass each year: one converges through Panama, the second across the Straits of Gibraltar, the third along the Great Rift Valley in the Middle East, the fourth through Southeast Asia and western Indonesia, and the fifth through the islands of eastern Asia to the Philippines and eastern Indonesia. Particularly large numbers of raptors have been counted at Eilat in Israel (more than one million in spring), at Corpus Criste in Texas (840 000 in autumn), at Panama (2.5 million in autumn and spring) and at Veracruz in Mexico (>6 million in autumn).

At various sites in Israel and elsewhere, different species pass in the same sequence each year, both autumn and spring, and passage dates are remarkably consistent from year to year. In general, species that pass northward earliest in spring pass southward latest in autumn. In some species, autumn passage is spread over a longer period than spring passage, and in others the reverse. Spring passage tends to be longest in large species in which the non-breeding immatures migrate north much later than the adults. Otherwise the timing and duration of passage depend on the extent of the breeding range (longer passage with wider latitudinal spread) and diet (bird-eating and mammal-eating species spending longer in breeding areas than reptile-eaters and insect-eaters). Some species take slightly or markedly different routes in autumn and spring, depending largely on wind conditions.

Many raptor species are known to feed on migration, especially bird-eating falcons and accipiters which migrate at the same time as their prey, and insectivores which also encounter food en route. Eagles and buteos that migrate long distances feed more episodically, and probably make large parts of their journeys without eating. At least some raptors are known to accumulate migratory fat (with up to 30% increase in body weight in Steppe Buzzards *Buteo b. vulpinus*). Because of their dependence on topography, the wind and other conditions that favour heavy passage differ to some extent from one site to another. However, low cloud and rain are largely inimical to migration by soaring birds, as well as by other birds.

Wandering albatross *Diomedia exulans*, one of the fastest of all migrating birds

Chapter 8
Speed and duration of journeys

There must come a point at which the new facts that have been collected are felt to be both too raw and too numerous, and it is at this point that the need for coordinating principles begins to be felt. (Charles Elton 1930.)

In most birds, migration consists of periods of flight, interspersed with periods of feeding, when body reserves are replenished. Both activities need to be taken into account in assessing the speed and duration of the overall journey, which might be important for several reasons. First, travel through unfamiliar areas carries costs and risks which could be minimised if migration were completed as quickly as possible. Second, rapid migration can lead to early arrival, ahead of most other individuals, which in turn facilitates access to the best habitat. Third, the time taken for migration could in some species restrict the time available for other activities, such as breeding and moult. It could thus limit the total distance that can be covered on migration, and hence constrain the geographical ranges of certain species.

The times spent on migration by different bird populations are enormously variable, depending partly on features of the birds themselves (such as body size, wing shape and flight speed), but largely on the distances travelled and the conditions encountered en route. At one extreme, some birds can complete their migration in less than one day (such as a radio-tagged Bald Eagle *Haliaetus leucocephalus* that flew 435 km between its wintering site in Michigan and its nesting place in Ontario, Grubb *et al.* 1994). At the other extreme, some landbirds take more than three months to reach their distant winter quarters, and a similar period to return, so that more than half of every year is spent on migration. Long journey times are also shown by some marine species, including shearwaters and petrels which have a fixed base only during the breeding season, and are effectively on migration for the rest of the year, pausing to feed wherever food is available en route. Whatever the advantages in migrating as quickly as possible, external conditions provide severe constraints, notably the rate at which food can be obtained and converted into body reserves to fuel the flights, and also the weather at the time, which can speed or slow the journeys. Rain or snow, cold or ice, or unfavourable winds can delay migration for days or weeks at a time.

A broad idea of the overall speed of migration in some species has come from noting the dates when birds leave one area and arrive in another. Such information could in theory be collected for many bird populations, but only for a few is it likely to yield accurate estimates. One problem concerns the lack of certainty that the same population of birds is involved throughout. Another is that individuals from the same population often vary greatly in their dates of departure and arrival, making mean dates hard to estimate. Nevertheless, such records have provided useful indications of the duration of flights by large conspicuous birds, such as cranes, geese and swans, which use only a limited number of traditional stopping sites. Confidence in such records increases if some of the birds are marked in a way that they can be recognised individually. While mostly based on observation, some useful estimates of this type have been based on museum skins, from the dates that migrants were collected in different areas (e.g. Byrkjedal & Thompson 1998).

Other information on the overall speed of migration has come from birds ringed and recovered on passage. Such recoveries carry the risk of error because it is seldom known precisely when a ringed bird leaves one place or arrives in another. The temptation is therefore to use only the fastest records, as an indication of maximum possible migration speeds. Not only do such records then come from extreme individuals, they mostly also cover only a small part of a migratory journey.

More representative data have come from calculating the mean geographical positions of ring recoveries obtained from a defined breeding population at successive dates through a journey. Thus, if recoveries of birds already on migration were centred in mid-October at latitude 30°N, say, and those in mid-November due south at latitude 20°N, then an average of one month would have been needed to cover the distance spanned by 10° of latitude. Similar estimates can be made from the dates that particular populations pass through different watch sites or trapping stations along a route. But many species break their autumn journeys for weeks at a time if conditions are favourable, and for much longer than is needed merely to replenish their body reserves. Some species, notably shorebirds, may moult during a several-week break in their autumn migrations (Chapter 11). Moreover, young birds often migrate more slowly than adults, and because juveniles form a larger proportion of any migratory population in autumn than in spring, they may also have more effect on the population's average migration speed in autumn.

The most reliable data on migration speeds come from individual radio-marked birds tracked by plane or satellite over their whole migration. With daily records of position, it is theoretically possible to separate the flight periods from stopover periods during the entire journey (but in practice short breaks of a few hours are easily missed). As yet, such data are available only for birds big enough to carry a radio-transmitter or geolocation tag, and their interpretation depends on the (probably justified) assumption that migration behaviour is unaffected by the attachment (Chapter 2).

One problem with all four types of data is that they provide no measure of the duration of the initial fuelling period before birds leave their breeding or wintering areas. Strictly speaking, this period is part of the migration, yet can be determined only by separate study. To judge from recorded rates of weight gain, small passerines take from a few days to three weeks to accumulate the required body reserves, while long-distance shorebirds and waterfowl can take up to two months or more, depending on the length of the subsequent flight.

THEORETICAL BASIS

In addition to empirical studies, attempts have been made to estimate theoretical migration speeds of particular species from knowledge of their likely flight speeds, and rates of fuel deposition and use (**Boxes 8.1** and **8.2, Figure 8.1**; Hedenström & Alerstam 1998, Alerstam 2003). Flight speeds can be taken from radar measurements (ideally corrected to their still-air values, Bruderer & Boldt 2001) or from theoretical estimates (Pennycuick 1969, 1975). Rates of fuel deposition can be measured from the repeat weighing of individual birds before migration or at stopover sites. Such measurements are usually taken over periods of days, so include sleeping and other non-feeding times. Rates of fuel use during flight can be measured only with difficulty, and are more easily obtained as theoretical estimates, based on body mass and other features of the bird (see Appendix 5.1). Rates of fuel (energy) gain and loss are best expressed in some common currency, such as multiples of BMR (basal metabolic rate, the rate of energy use by a resting, inactive bird). Daily rates of energy gain (above the $2 \times$ BMR typically required for everyday life) are usually around $1 \times$ BMR, but in good feeding conditions they can reach $2.5 \times$ BMR (Alerstam & Lindström 1990). For birds that can feed by night as well as by day, such as shorebirds, rates of fuel deposition can be much higher, occasionally reaching up to $10 \times$ BMR (Kvist & Lindström 2003). In these conditions, birds show extremely rapid accumulation of body reserves.

On such a basis, for a small bird with an energy deposition rate of $1 \times$ BMR (above the $2 \times$ BMR required for normal daily activity), and travelling by flapping flight, the predicted average migration speed is 200 km per day (Hedenström & Alerstam 1998). This includes both flight and fuelling periods. At the higher energy deposition rate of $2.5 \times$ BMR, the predicted speed rises to 300–400 km per day. In larger birds, theoretical migration speeds are lower, at 70–100 km per day for an energy deposition rate of $1 \times$ BMR, and at 150–200 km per day for an energy deposition rate of $2.5 \times$ BMR.

Box 8.1   Calculation of theoretical migration speed. From Hedenström & Alerstam (1998)

The speed of migration (V_{migr}) is determined by the rate at which fuel is accumulated (P_{dep}), the flight speed (V) between consecutive stopover sites (which depends on wind conditions and flight mode, etc.), and the flight power (P_{flight}), the rate at which energy is consumed), according to the following general relationship:

$$V_{migr} = \frac{VP_{dep}}{P_{dep} + P_{flight}} \qquad (1)$$

This equation relates to still-air migration speed, and wind conditions could greatly influence the actual speed. A graphical illustration of this equation is given in **Figure 8.1**. For soaring flight, the above equation can be used, but replacing the flapping flight speed (V) with the cross-country soaring–gliding speed (V_{cc}).

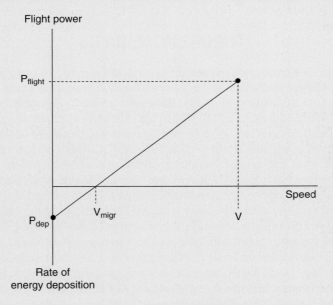

Figure 8.1   Graphical solution of the theoretical overall migration speed (V_{migr}) on the basis of the power required for flight (P_{flight}), rate of energy deposition at stopovers (P_{dep}) and flight speed (V). Migration speed is found where a straight line connecting the point on the downward-extended ordinate indicating P_{dep} to the point in the power–speed plane (V, P_{flight}) intersects the speed axis. The migration speed can be calculated according to the equation in **Box 8.1**. From Hedenström & Alerstam (1998).

Power consumption in gliding flight is generally assumed to be a constant multiple of the basal metabolic rate (BMR), and consequently migration speed will be directly proportional to the cross-country performance ($V_{mig} \propto V_{cc}$) in soaring migration. The cross-country speed in typical thermal soaring is given as:

$$V_{cc} = \frac{VV_c}{V_c - V_z} \tag{2}$$

Where V is the gliding speed between thermals, V_z is the sink component (negative downwards) and V_c is the climb rate in thermals (Pennycuick 1972, 1975, 1989).

Box 8.2 Calculation of theoretical energy and time costs of migration. From Alerstam *et al.* (2003)

Energy costs

Generally, if migration is subdivided into periods of movement, interspersed with periods of stopover, the total energy consumption during migration can be written as:

$$E = \frac{PD}{V}\left(1 + \frac{x}{P_{dep}}\right) \tag{1}$$

where P is the power of locomotion (rate of energy consumption), D is the migration distance, V is the flight speed, P_{dep} is the rate of energy deposition at stopovers, and x is the field metabolic rate at stopovers. Equation (1) can be used to compare the total investment in migration among, for example, birds of different size and using different modes of flight. The ratio $x:P_{dep}$ determines the ratio between energy consumed during stopovers and cost of flight, which in a typical passerine bird may be about 2:1 or larger. From Equation (1), it is also evident that minimising the ratio $P:V$, a measure closely related to cost of transport, will minimise the energy cost of migration.

Time costs

Generally, the time required for migration can be written as:

$$T_{migr} = \frac{D}{V}\left(1 + \frac{P}{P_{dep}}\right) \tag{2}$$

where P, P_{dep} and V are defined as for Equation (1). The relationship between stopover and transportation time is P/P_{dep}, which was estimated to be about 7:1 or larger for small birds (Hedenström & Alerstam 1998). Given limited time available for migration, there is a maximum return distance (D_{max}) that a bird can achieve. The time needed for migration is reduced by low locomotion cost (P) and high travel speed (V) and fuelling rate (P_{dep}). Hence, adaptations favouring low energy cost of transport, high rates of fuel deposition, and optimal scheduling of life history events are expected in long-distance migrants.

On theoretical grounds, large birds that travel by flapping flight would be expected to migrate more slowly than small ones because the rate of energy consumption needed for flight increases more strongly with body mass than does the rate of fuel deposition (Lindström 1991, Hedenström & Alerstam 1998). On this basis, large birds would require longer than small ones to acquire the fuel necessary to fly the same distance. But large birds also fly faster than smaller ones, and can therefore partly compensate for their lower fuelling rates and lower maximum fuel loads. The theoretical maximum speed of migration (including flights and stopovers) was estimated by Lindström (1990) as proportional to $M^{-0.14}$, where M is body mass. However, migration speed depends not only on bird body mass, but also on wing and body morphology. For this reason, the Arctic Tern *Sterna paradisaea*, with its long and slender wings, is expected to match or even surpass the migration speed of the Willow Warbler *Phylloscopus trochilus* with its broader and shorter wing shape, in spite of the fact that the tern weighs more than 10 times as much as the warbler (**Figure 8.2**). Moreover, soaring flight by some large birds can give more rapid overall progress than flapping flight because of its smaller energy needs, and consequent savings on refuelling times.

Even allowing for body mass and flight mode, theoretical calculations are likely to provide only very approximate estimates of maximum possible migration speeds, but they point to the enormous influence that feeding rates could have. Within any wild bird population, rates of fuel deposition are highly variable (Chapter 5), and the weather has an additional major influence, speeding or

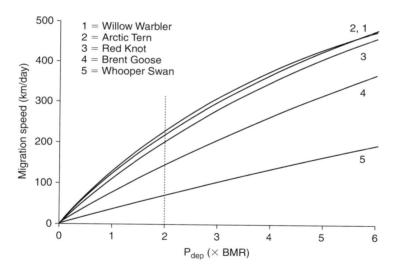

Figure 8.2 Expected overall speed of migration by flapping flight in relation to fuel deposition rate P_{dep} for some selected species on the basis of flight mechanical calculations (still-air conditions). As a reference level of basal metabolic rate (BMR) in birds of different body weight, the allometric equation according to Lasiewski & Dawson (1967) was used. In most cases fuel deposition rates (P_{dep}) of migrating birds are limited to levels up to $2 \times$ BMR, but some greatly exceed this rate (see text). From Alerstam (2003).

slowing a journey (Chapter 4). Especially in spring, the timing of snow melt, or the appearance of specific food types, can greatly influence migratory progress. Nevertheless, most of the higher migration speeds recorded in field conditions are similar to the speeds that have been calculated theoretically from know-ledge of flight speed, fuel deposition during stopovers and fuel use during flight (Hedenström & Alerstam 1998).

From the same type of information, one can also calculate the expected propor-tion of total migration time that is spent on fuel intake and deposition, as opposed to flight. Migration speed as a fraction of flight speed is given by the ratio of energy deposition rate to the sum of this rate and the rate of energy consump-tion for flight (called the flight power, **Box 8.1**). For example, for a bird with an energy deposition rate corresponding to $1 \times BMR$ and an energy consumption rate for flight corresponding to about $7 \times BMR$, the maximum speed of migration will amount to at most one-eighth of the flight speed. In other words, for every hour of migratory flight, the bird would require 7 hours of feeding (and sleeping) in order to accumulate the necessary fuel reserves. This ratio is in fact typical of that found from ring recoveries for many species of small passerines, in which the energy requirements of flight are relatively small (Hedenström & Alerstam 1998).

In the calculation of daily fuelling rates from weight data, allowance is auto-matically made for sleeping and other activities during fuelling periods. Ideally, however, birds need to be studied over several weeks to provide a reasonable estimate of their flight-to-stopover ratios, and in those species that accumulate large fuel reserves before their initial departure, this fuelling period should be included in the estimate of overall migration time. When measured over short periods of up to a few days, migration speed can easily be overestimated because, when captured, the bird has usually already stored some energy for the flight.

In contrast to birds that migrate by flapping flight, soaring–gliding birds would be expected to spend less time on accumulating body reserves, because much of the energy for their flight is external in origin, derived from air currents. Power consumption in gliding flight is generally assumed to be a constant multiple of the BMR, and is unrelated to flight speed. Consequently, the migration speed of soaring–gliding species should be directly proportional to their cross-country speed. Because cross-country speed increases with body mass (Chapter 3), migra-tion speed in soaring species might also be expected to increase with body mass, the opposite to the situation in birds that migrate by flapping flight (Hedenström & Alerstam 1998). However, overall migration speed is also influenced by the period each day that birds can migrate, and over land where birds depend on thermals, this period is usually longer for smaller species which can make use of weaker thermals than larger ones (see later). This constraint on flying time may be more limiting than body reserves for overall migration speed.

EMPIRICAL ESTIMATES

Migration speeds from individual ring recoveries

Migration speeds have been calculated for various species (mainly passerines and shorebirds) that were ringed on autumn passage in Sweden and Finland, and

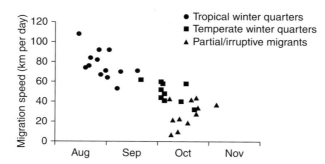

Figure 8.3 Migration speeds recorded for passerines ringed in Finland in relation to the seasonal timing of migration as recorded at Ottenby in Sweden. The data clearly indicate the reduction of migration speed with advance in mean departure date, and the differences between different categories of migrants. From Alerstam & Lindström (1990), based on data in Hildén & Saurola (1982).

recovered up to 50 days later at various points on their migration routes (Hildén & Saurola 1982, Ellegren 1993, further analysed by Alerstam & Lindström 1990, Alerstam 2003). From these and other records (**Figure 8.3**), the following generalisations emerged:

1. For birds studied over at least 10 days, average migration speeds in the range 21–263 km per day were found for different species. In general, shorebirds migrated more rapidly (median 79 km/day, $N = 13$ species) than passerines (medians 27–75 km per day in different groups, see below) (Alerstam & Lindström 1990). This difference may be because shorebirds fly faster than passerines, are more selective of favourable winds, and migrate at higher altitudes where winds are stronger. In moving between successive staging areas, shorebirds typically fly 500–1000 km in 1–2 days, interrupted by several-day stops, whereas passerines typically migrate over land almost every day at a more even speed, with shorter flights and shorter stops. Maximum speeds up to 200–300 km per day were recorded in passerines, and up to 400–1000 km per day in shorebirds. These maximum speeds were consistent with theoretical expectations.

2. Among passerines, average migration speeds varied with autumn starting date, early-departing species travelling more rapidly than late-departing ones (**Figure 8.3**). Thus, speeds were considerably higher for long-distance species with tropical winter quarters that left early (median speed 75 km per day, $n = 13$) than for short-distance temperate zone migrants that left later (median speed 53 km per day, $n = 19$), or for irruptive and partial migrants that left latest of all (median 27 km per day, $n = 11$). This trend could result from seasonal declines in food supplies or daylengths, leading to reduced rates of fuel deposition from late summer into autumn (Alerstam & Lindström 1990, Ellegren 1993, Kvist & Lindström 2000). In addition, fast travel may be more important for early, long-distance species bound for the tropics than for later, short-distance species migrating only within Europe (Alerstam & Lindström 1990). Not only do the earlier migrants have further to go, but they probably gain

from crossing the Sahara in August–September, before the dry season takes hold in the Sahel zone to the south.

3. In association with the above relationships, migration speeds were correlated with the length of journey (**Figure 8.3**). Juvenile passerines travelling 1000 km on migration covered about 80 km per day, on average, whereas those travelling 5000–6000 km covered about 140 km per day (Alerstam 2003). Long-distance migrants seem to migrate in longer steps, and are more selective of good weather conditions. In addition, long-distance migrants are generally nocturnal flyers, using time for flying that they would otherwise spend at roost, without reducing the time available for foraging (this would not apply to shorebirds which can feed day and night given suitable tidal conditions).

4. Among passerines, nocturnal migrants made faster progress, on average, than diurnal migrants (72 km per day vs. 53 km per day; samples sizes 17 and 6 species). In addition, migration flight lengths were longer in nocturnal (177 km) than in diurnal migrants (111 km), even though nocturnal migrants did not always use the entire night for flying (Cochran *et al.* 1967, Ellegren 1993).

5. Despite their faster migration speeds, Eurasian–Afrotropical migrants, with mean migration distances of 6000–10 000 km, still travelled for much longer autumn periods (median 88 days, $n = 13$) than regular temperate zone migrants (median 42 days, $n = 10$, distances 1700–3000 km), or than partial migrants (median 32 days, $n = 11$, distances 200–300 km).

6. Some extremely rapid migrations were recorded in some long-distance passerine migrants. For example, a Willow Warbler *Phylloscopus trochilus* travelled 8000 km from Finland to Congo within 56 days, a mean speed of 145 km per day (Hildén & Saurola 1982), while another Willow Warbler travelled from Finland to South Africa in 47 days, a mean speed of 218 km per day (Hedenström & Pettersson 1987). A Marsh Warbler *Acrocephalus paludicola* travelled 1400 km within Africa in five days, an average of 280 km per day (Cramp 1992); and three Barn Swallows *Hirundo rustica* travelled average distances of 250, 350 and 433 km per day over journeys of 8500, 12 000 and 3000 km respectively (Turner 2006). However, all these records refer to exceptional individuals, the majority of their species progressing much more slowly.

7. The decline in migration speed with advance in departure date did not apply within species. On the contrary, late migrants tended to progress more rapidly than early ones of the same species, at least in the initial part of the autumn journey (for various passerines see Ellegren 1993; for juvenile Reed Warbler *Acrocephalus scirpaceous* and Sedge Warbler *A. schoenobaenus* see Bensch & Nielsen 1999). This trend would give birds that were delayed in their departure from breeding areas an opportunity to 'catch up' during migration (Fransson 1995). Blue Tits *Parus caeruleus* migrating in autumn along the southern Baltic coast travelled an average of about 28 km per day during 15–24 September, increasing significantly to about 38 km per day during 15–24 October (Nowakowski & Chruściel 2004).

8. In some species, adults travelled faster than juveniles. For example, adult Dunlins *Calidris alpina* took an average of 4.5 days to travel 660 km, while juveniles took 13 days over the same journey (Hildén & Saurola 1982). A similar age-related difference held in various *Sylvia* warblers migrating in autumn from northern Europe (Fransson 1995). By the time they start migrating, juvenile

birds are to all appearances 'full-grown', and in this respect are almost as well equipped for migration as older birds.

Average migration speeds from population-based ring recoveries

Average estimates are more useful than those obtained from individual flights, because they derive from many individuals and not just a few extreme ones. Ringing data indicate that Barn Swallows *Hirundo rustica* complete the 10000 km journey between northern Europe and southern Africa at an average speed of at least 150 km per day (as against individual ringing recoveries which indicate exceptional speeds up to 340 km per day) (Alerstam & Lindström 1990). Other estimates relate to Spotted Flycatchers *Muscicapa striata* between southern Africa and Fennoscandia (spring) at 140 km per day (Fransson 1986), and to Willow Warblers *Phylloscopus trochilus* between Fennoscandia and Africa (autumn) at 85 km per day, both over distances exceeding 10000 km (Hedenström & Pettersson 1987).

Birds do not necessarily maintain the same rate of progress over their whole migration. To judge from ring recoveries, some species speed up during the course of their journey, while others slow down. For Willow Warblers *Phylloscopus trochilus* travelling from northern Europe to Africa, mean speeds were estimated at 41 km per day over distances of 400–1000 km ($N = 37$), 54 km per day over distances of 1000–2000 km ($N = 30$), 59 km per day over distances of 2000–3000 km ($N = 40$) and 85 km per day over distances greater than 3000 km ($N = 22$) (Hedenström & Pettersson 1987). Similarly, for birds recovered in the first three weeks after ringing in autumn, mean migration speeds increased with distance in Bluethroats *Luscinia svecica*, Sedge Warblers *Acrocephalus schoenobaenus*, Reed Warblers *A. scirpaceous* and Lesser Whitethroats *Sylvia curruca* from Sweden, and in Garden Warblers *S. borin* and Blackcaps *S. atricapilla* from Britain (Ellegren 1990a, 1993, Fransson 1995, Bensch & Nielson 1999). Migration speed increased more rapidly through the journey in long-distance migrants wintering in Africa than in short-distance migrants wintering within Europe (Alerstam 2003). The reasons are unknown, but weather-induced delays may be fewer as birds reach lower latitudes.

On the other hand, other *Sylvia* warblers ringed in northern Europe and Britain seemed to slow down during their autumn passage through Europe. Speeds exceeding 100 km per day were recorded mainly in the first 10 days after ringing and, after about 20 days from ringing, average speeds had dropped markedly (Fransson 1995). Much depends on the nature of the migration, whether by long flights and long stopovers, or short flights and short stopovers. Whatever the migration mode, however, speeds calculated over the early parts of a journey (which often provide most ring recoveries) may not be typical of the whole route, for birds may already have accumulated some migratory fuel when they were caught and ringed. In addition, for obvious reasons, journeys over seas or deserts are almost always undertaken more rapidly than equivalent journeys through more favourable terrain.

In a process so dependent on food and weather, migration speeds would be expected to vary greatly from year to year over the same route. Annual variations have been little studied, but over the period 1963–1998 a four-fold variation in average migration speeds was recorded among Blue Tits *Parus caeruleus* travelling between various bird-ringing stations along the southern Baltic coast

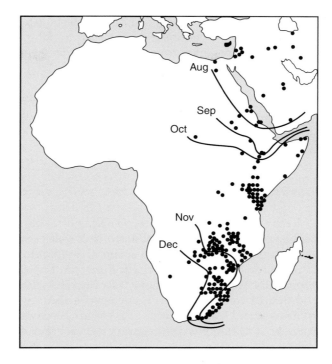

Figure 8.4 The course of migration of European Marsh Warblers *Acrocephalus palustris* to South African wintering areas. The whole southward journey takes at least five months, and lines show progress, as revealed by ring recoveries (Dowsett-Lemaire & Dowsett 1987). The return journey is faster, taking about three months. Map reproduced by permission of NISC, publishers of *Ostrich*.

(Nowakowski & Chruściel 2004). In the slowest year, the birds averaged less than 10 km per day, compared with 38 km per day in the fastest year. In most years, the average speed was in the range 25–35 km per day, making the Blue Tit relatively slow among European passerines. This may be associated with its habit of migrating in short flights just above the tree-tops. The annual variations in speed paralleled those in the Great Tit *P. major*, which, however, migrated generally faster. Interestingly, Coal Tits *P. ater* studied in the same way migrated twice as fast in invasion years (40–80 km per day) as in other years (30–40 km per day), possibly because the migratory drive was stronger in invasion years (Rute 1976), a difference also noted in Siskins *Carduelis spinus* (Payevski 1971).

Most of the estimates of migration speed obtained from ring recoveries refer to autumn, when juveniles predominate, and the few records from spring suggest that faster progress is made then, at least by passerines and shorebirds. Thus, Barn Swallows *Hirundo rustica* from Britain take an average of 10 weeks to reach South Africa in autumn, travelling at 150 km per day, and 5–6 weeks to return in spring, travelling at 300 km per day (Mead 1970). Similarly, Marsh Warblers *Acrocephalus paludicola* migrating between Europe and South Africa were estimated to take 20 weeks over the autumn journey, and 12 weeks over the spring journey (**Figure 8.4**).

Table 8.1   Speed of migration (km per day) of five *Sylvia* species from northern Europe and Great Britain, estimated according to differences in median trapping dates and median dates of recoveries in the Mediterranean area

Species	Great Britain		Northern Europe	
	Autumn	Spring	Autumn	Spring
Barred Warbler *Sylvia nisoria*	–	–	92	–
Lesser Whitethroat *Sylvia curruca*	56	97	80	98
Greater Whitethroat *Sylvia communis*	47	186	85	129
Garden Warbler *Sylvia borin*	39	232	116	163
Blackcap *Sylvia atricapilla*	52	162	85	162

From Fransson (1995).

On the basis of median recovery and trapping dates, four species of *Sylvia* warblers also showed much faster migration speeds in spring than in autumn for journeys between the Mediterranean region and the north European breeding areas (**Table 8.1**, Fransson 1995), as did Spotted Flycatchers *Muscicapa striata* through the same region (Fransson 1986). In contrast, Pied Flycatchers *Ficedula hypoleuca* travelled between northern Europe and tropical Africa at average speeds of 120–170 km per day in autumn but only at 100 km per day in spring (Lundberg & Alatalo 1992).

For many diurnal birds, fuelling rates may be influenced not only by the food supplies available en route, but also by the prevailing daylengths which determine the maximum daily feeding times, and hence the maximum possible migration speed (Kvist & Lindström 2000). In many bird species, spring migration occurs closer than autumn migration to the summer solstice, and hence over longer days (Chapter 14; Bauchinger & Klaassen 2005). The migratory speeds of three species of *Sylvia* warblers were, on average, 47% faster in spring than in autumn, calculated from ring recoveries of adult birds migrating between the Mediterranean and northern Europe. For these species, the spring journey was 34% faster in *S. communis*, 47% faster in *S. borin* and 59% faster in *S. atricapilla* (Fransson 1995). In parallel, the amount of daylight over the same migratory distance was 26% longer in spring than in autumn, calculated by comparing daylength at the mid point (52°N) of the migratory journey, and of the spring and autumn migration periods (15 May and 18 September respectively). A fourth species, *S. curruca*, showed no significant difference in migration speed between spring and autumn. If daylength has an influence (through its effect on fuelling), migration speed should change during the course of a journey, as the bird moves through latitudes with different daylengths. To some extent, the same would be expected in soaring birds which are limited each day by the period when thermals occur (both daylength and temperature dependent). Birds that migrate between hemispheres change daylength regime part way through the journey. Nevertheless, there will still be some specific departure date that is optimal for maximising the overall daylength to which the bird is exposed, and hence for minimising the duration of migration (Alerstam 2003).

Although we can estimate or measure the maximum migration speeds of birds, we should not assume that all birds are under pressure to migrate as fast as they are able, at least in autumn. Comparisons between different populations of some species indicate that other factors are involved, and that the time available for migration may also have an influence. For example, Common Kestrels *Falco tinnunculus* from northern Sweden travel twice as far as those from the south, but in the same time (Wallin *et al.* 1987). This difference held even though the entire journey of the short-distance birds lay within the route of the long-distance ones. In spring, the two populations migrated at the same speed, so the northern birds took longer to reach their breeding areas. Similarly, White Storks *Ciconia ciconia* from central Europe using the southwest route through Iberia take about three months to reach their winter quarters in West Africa. Those taking the southeast route through the Middle East to East Africa also take about three months to cover twice the distance (**Figure 8.5,** Bairlein 2001).

Analysing ringing data for four species of *Sylvia* warblers, Fransson (1995) discovered that British birds travel through Europe at slower average speeds (43–62 km per day) than Scandinavian birds (66–93 km per day). He attributed this to the fact that British birds have shorter total migration distance (and hence may be under less pressure) and a slightly different migration schedule from Scandinavian ones. These various findings suggest that some populations are not under pressure to migrate in autumn at the maximum speed of which they are capable. Perhaps at this time of year they have a certain period in which to migrate, according mainly to other events in the annual cycle, and can adjust their travelling speed accordingly. The benefits of migrating slowly are that birds can take advantage of rich feeding areas they encounter en route, yet do not need to accumulate the massive fuel reserves required by long flights, thereby avoiding the associated predation risks (Chapter 5).

Radio-tracked birds

When radio-tagged migrating birds are followed in a small aeroplane, considerable detail on day-to-day behaviour can be obtained, in addition to measures of speed. In one of the earliest studies, 10 radio-marked *Hylocichla* thrushes (of three species) were followed on their nocturnal migrations over favourable terrain in North America (Cochran *et al.* 1967). Some birds migrated for 4–8 hours per night, over distances of about 180–540 km, while others flew for less than an hour each night (especially when interrupted by bad weather). In another study, seven individual Peregrines *Falco peregrinus* were followed from a small plane while they were migrating across the USA. In each 24-hour period, these birds typically spent 17 hours on a perch, 6 hours in migratory flight, and about 1 hour in hunting. In especially favourable conditions, migration increased to 9 hours per day, and when held up by weather, perching increased to 23 hours per day. On average, these falcons migrated on six days out of seven, generally from mid-morning to late afternoon (W. Cochran, records compiled by White *et al.* 2002).

In another study, 15 radio-marked adult Bald Eagles *Haliaetus leucocephalus* were tracked from a vehicle for an average distance of 2019 km in spring (Harmata 2002). These birds migrated individually (like those just mentioned), appeared not to feed during the journey, and did not migrate on days of overcast or high wind. All their flights occurred at some time within a 9-hour period each day (10.30–17.30 hours), covering an average of 180 km (range 33–435 km), at an average flight speed of 50 km per hour (range 20–144 km per hour). The birds flew at altitudes of 30–4572 m above ground, but mostly at 1500–3050 m, so for most of the time they were beyond the range of human vision.

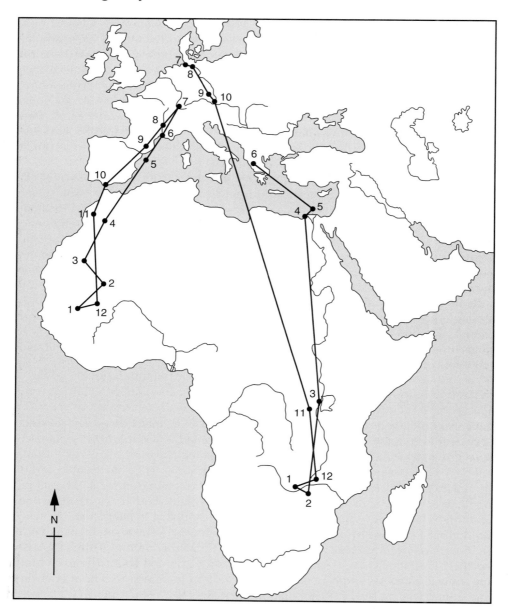

Figure 8.5 Course of first-year migration in White Storks *Ciconia ciconia* from southwestern Germany (westerly route) and northwestern Germany (easterly route), showing the migratory divide. Dots show the average monthly location of ringed birds. The lines join the dots of successive months, but do not necessarily reflect the exact routes followed. First-year storks are on the move for much of the year, and most do not reach the breeding areas until it is too late to nest that year. From Bairlein (2001).

Other data from radio-marked birds refer to individuals mostly tracked from satellites, in which records of locations came at longer intervals, and without the behavioural details. Such studies have involved only large species, complementing the data from ringing which refer mainly to small species. At face value, the two types of data provide only weak support of the theoretical prediction that smaller species migrate more rapidly than larger ones. However, both data sets exclude the initial fattening period (which can extend to several weeks in some large species). Also, most records refer to autumn, when many birds seem not to migrate at their maximum possible rate, but to linger in areas of plentiful food which they encounter en route. Even in spring, when birds are assumed to migrate as rapidly as possible to reach their breeding areas, the migration speeds of passerines are generally no faster than those of larger waterfowl and others (including raptors and cranes which travel by soaring flight). The data are more consistent with the view that spring speeds are largely influenced with the spread of warmth to higher latitudes (as reflected, for example, in the northward movement of particular isotherms, Chapter 14). On this view, species would move to progressively higher latitudes as conditions allowed, with no obvious differences between small and large species, whatever their theoretical capabilities.

Radio-tracking studies have also revealed great variation in migration speeds among adults from the same population making essentially the same journey, some individuals taking 2–3 times longer than others, or occasionally up to 12 times as long. Much of this variation was associated with the weather encountered en route, birds departing at different dates encountering different conditions, but some may also have been associated with the varying capabilities of the birds themselves. The inclusion of juveniles in the comparison would in some species have increased the variation in migration speed even more. Some studies have involved species that migrate by flapping flight, while others have involved species that migrate mainly by soaring–gliding flight. The two groups are discussed separately below.

Birds that migrate by flapping flight

The average speeds of four waterfowl species tracked over land in spring averaged 55–118 km per day (**Table 8.2**). Individual variation was sometimes great, as illustrated by Bewick's Swans *Cygnus columbianus bewickii* migrating between northern Russia and the Netherlands. In spring two birds tracked along part of the route gave mean migration speeds of only 28 and 39 km per day. In autumn, five birds followed over the entire 3200 km journey took 41–78 days, giving average speeds of 44–72 km per day. However, observational records of departure and arrival dates indicated that some flocks under favourable winds completed the whole 3200-km journey in about eight days (400 km per day). Hence, the individuals from this population varied between eight and 78 days in the time taken to complete their autumn journey, giving average speeds of 44–400 km per day, a nine-fold variation depending on the amount of stopover, which in turn depended largely on the weather. In addition, pairs with young took longer over the journey than adults alone, stopping more frequently and arriving later in the wintering areas: probably juveniles were less able than adults to cover such long distances non-stop (Beekman *et al.* 2002).

Table 8.2 Details from the tracking of radio-tagged waterfowl and other birds on their migrations

Species	Migration route	Autumn				Spring				Days on longest stop	Source
		N	Distance (km)	Time (days)	Speed (km/day)	N	Distance (km)	Time (days)	Speed (km/days)		
Houbara Bustard *Chlamydotis undulata*	W. Kazakhstan–Iraq/Iran/Afghanistan	8	1990	40	50					51	Combreau et al. 1999
Houbara Bustard *Chlamydotis undulata*	Abu Dhabi–Turkimenistan					1	2328	49	48	27	Osborne et al. 1997
Whooper Swan *Cygnus cygnus*	Japan–Siberia					8	2940	53.4	55.1	34	Kainai et al. 1997
Whooper Swan *Cygnus cygnus*[a]	Iceland–Scotland	4	1092	4.3	254	2	1299	5.5	236		Pennycuick et al. 1996
Whistling (Tundra) Swan *Cygnus columbianus*	Japan–Siberia					1	3083	21	147	30	Higuchi et al. 1991
Bewick's Swan *Cygnus columbianus bewickii*	Siberia–Baltic region	5	2023	34	61	2	1871	32	58		Beekman et al. 2002
Brent Goose *Branta bernicla*	Baltic region–Taimyr					6	5004	42	118	16	Green et al. 2002

Species	Route										Reference
Brent Goose *Branta bernicla*[a]	Denmark–Svalbard/Greenland					10	2762	3.4	812		Clausen et al. 2003
Barnacle Goose *Branta leucopsis*[a]	Svalbard–Scotland	4	2750	5–15						12	Butler et al. 2003
White-fronted Goose *Anser albifrons*[a]	Ireland–W. Greenland				275	12	3000	25	120	22	Fox et al. 2003
Bar-headed Goose *Anser indicus*	India–Tibet					1	788	11	72		Javed et al. 2000
Barrow's Goldeneye *Bucephala islandica*[b]	Within Quebec Province	12	986	18.6	53	13	64.8	5.9	11		Robert et al. 2002

All figures are mean values except where otherwise indicated. Details are only from birds that were followed over one or more complete journey (i.e. part-journeys excluded), and were calculated in the same way for all studies. In some cases, therefore, the figures may differ from those given in the original publications. All the species shown migrate by self-powered flapping flight. N: number of birds tracked.

[a] Mainly over water.

[b] Autumn movement: moult migration, so not comparable to spring migration. Wintering area includes moulting area.

Brent Geese *Branta bernicla* migrating mainly over land between the Wadden Sea and the Taimyr Peninsula covered the mean flight distance of 5004 km in an average of 42 days (range 34–52 days), with a mean speed of 118 km per day (range 97–148) (Green *et al.* 2002). The slowest birds thus took half as long again as the fastest. Seven males on their longest apparent non-stop flight covered an average of 1056 km (range 768–1331 km). Much greater individual variation was recorded among Barnacle Geese *Branta leucopsis* migrating largely over water from Svalbard via Bear Island to Scotland in autumn over a total distance of 2500–3000 km. At one extreme, some birds completed this journey in as little as 2–3 days – more than 1000 km per day – during which their stops were of such short duration that they gave little chance to replenish reserves (Butler *et al.* 1998). At the other extreme, four birds held up en route by unfavourable winds took 9–36 days over the journey, giving mean migration speeds of 306–76 km per day. Overall journey time thus varied at least 12-fold between individuals. Possibly most birds left Svalbard with enough fuel to complete the journey to Scotland if they could do it within 2–3 days. But if they were held up by weather, their reserves would have become depleted, requiring additional feeding.

Other relevant data have been obtained from Houbara Bustards *Chlamydotis undulata*, which also migrate using flapping flight (**Table 8.2**). Of eight individuals trapped in Kazakhstan, two flew directly to their wintering areas with no stopover longer than one day (Combreau *et al.* 1999). They covered distances of 1600 and 1970 km in 20 and 13 days respectively, giving mean speeds of 24 and 151 km per day (6.3-fold). Six other birds made a longer stop, lasting more than a week, and one bird stopped twice for 51 and 16 days. With such wide variations, the duration of autumn migration varied from 13 to 73 days, depending largely on the time spent on stopovers, mean travel speeds ranging between 24 and 151 km per day. These birds also showed a roughly two-month spread in departure dates from breeding areas (15 July–18 September) and of arrival dates in wintering areas (14 September–14 November).

Birds that migrate by soaring flight

Although many seabirds travel by soaring flight, few estimates have been made of their migration speeds. With an average flight cost estimated at only 3 × BMR, and a rate of energy accumulation of 2 × BMR, and assuming that they were on the wing for 50% of the time, albatrosses are expected on theoretical grounds to migrate at average speeds in the range 440–880 km per day (Hedenström 1993). This exceeds by a broad margin the average migration speeds of birds using flapping flight, discussed above (Alerstam 2003). Tracking studies have confirmed that albatrosses achieve overall average speeds between 220 and 950 km per day on trips over distances of 3000–25 000 km (Jouventin & Weimerskirch 1990, Prince *et al.* 1992, Weimerskirch *et al.* 1993, Croxall *et al.* 2005). Some Grey-headed Albatrosses *Thalassarche chrystostoma* that flew around the earth included some extraordinary examples of flight performance (Croxall *et al.* 2005). Typical journeys from South Georgia to the southwest Indian Ocean took 6.2 days at 950 km per day; the second leg to the southwest Pacific lasted 13.2 days at 950 km per day, and the last leg back to South Georgia 10.3 days at 750 km per day. Without stopping, a complete circumnavigation of the Southern Ocean could, in theory, be completed in 30 days; which provides a context for the exceptional performance of one bird that made this journey in just 46 days. Based on average rather than occasional speeds, albatrosses that travel by gliding flight are evidently among the fastest of all long-distance animal travellers.

Estimates of migration speeds were also obtained from a smaller species, the Sooty Shearwater *Puffinus griseus*, by use of geolocation tracking tags (Shaffer *et al.* 2006). On their figure-eight migrations around the Pacific, 19 tagged birds travelled an average of 64 037 km in 198 days, giving a mean speed of 323 km per day. On parts of the journey, speeds rose to an average of 910 km per day, not very different from the highest recorded from albatrosses.

Turning to species that migrate by soaring–gliding migration over land, many species of raptors have been fitted with radio-tags and monitored by satellite on their journeys (**Table 8.3**, **Box 8.3**, **Figures 8.6** and **8.7**). Most of these birds made long journeys, some from one continent to another. Their migration was limited to the period each day during which thermal soaring was possible – at most about 9 hours (Spaar & Bruderer 1996). Their need for thermals also led many such birds to take long indirect routes to avoid long sea-crossings. Their journeys were thereby lengthened by up to 50% over the shortest (great circle) routes.

The shortest migration recorded was undertaken by the Bald Eagle *Haliaetus leucocephalus* mentioned earlier, which spent less than one day over its 435 km journey (Grubb *et al.* 1994). The longest migration recorded was undertaken by Swainson's Hawks *Buteo swainsoni* which travelled an average of 13 500 km (maximum 15 000 km) between breeding areas in western North America and wintering areas in southern South America (**Box 8.3**; Fuller *et al.* 1998). Overall, excluding the Bald Eagle mentioned above, mean migration speeds of raptors varied between 12 and 294 km per day in autumn, and between 23 and 196 km per day in spring (**Table 8.3**). Not unexpectedly, migration was more rapid over unavoidable seas and deserts, but otherwise daily rates were much affected by weather, being greatly reduced by strong winds, dense cloud and rain.

As expected, the time spent on migration varied with the length of the journey, a trend apparent both within and between species, but with considerable individual variation (**Figure 8.8**). Within species, this was most apparent among Peregrines *Falco peregrinus* and Ospreys *Pandion haliaetus*, large numbers of which have been tracked from different parts of the breeding range. Like some passerines that travel by flapping flight, soaring species tended to migrate faster on longer journeys, but nevertheless no significant correlation emerged between migration speed and distance travelled (**Figure 8.8**). This was presumably because soaring species required much less food to fuel the flights, and did not need to stop for long feeding periods during the journey (although some individuals clearly did). Similarly, among 24 juvenile Steller's Sea Eagles *Haliaeetus pelagicus* tagged at nests in various parts of the breeding range, mean speed per day increased with length of journey (as measured by degrees of latitude travelled) (McGrady *et al.* 2003). Theory predicts that, in soaring birds, migration speed should increase with body mass (see above), and although this relationship was found for the cross-country speeds of raptors studied in Israel (**Figure 7.4**), no such relationship was apparent for the whole migrations of species in **Table 8.3**. This may be because, while larger species can glide more rapidly than small ones, small ones can utilise weaker thermals, and so can migrate over a longer part of each day (Chapter 3).

Table 8.3 Details from the tracking of radio-tagged raptors and other soaring birds on migration

Species	Migration route	Autumn			
		N	Distance (km)	Time (days)	Speed (km/day)
Egyptian Vulture *Neophron percnopterus*	France–Mauritania	2J	3572	29	123
Osprey *Pandion haliaetus*[a]	Sweden–Africa	13[a]	6742	45	150
Osprey *Pandion haliaetus*	East coast USA	35A	5134	31	166
	Midwest USA	30A	5872	26	226
	West coast USA	2A	3824	13	294
Osprey *Pandion haliaetus*	Florida–S. America	7A	4105	25	164
	Within Florida	3A	145	1	145
White-tailed Eagle *Haliaeetus albicilla*	Japan–Kamchatka	1A	2244	57	39
Steller's Sea Eagle *Haliaeetus pelagicus*	Japan–Russia				
Steller's Sea Eagle *Haliaeetus pelagicus*	Siberia–Japan	24J			23
Bald Eagle *Haliaeetus leucocephalus*	Colorado–Saskatchewan				
Bald Eagle *Haliaeetus leucocephalus*	Arizona–NW Territories				
	Michigan–Ontario				
Bald Eagle *Haliaeetus leucocephalus*[c]	Montana–various	9J	553	14	40
Bald Eagle *Haliaeetus leucocephalus*[d]	California–W.Canada	5J	1048	7	150
Bald Eagle *H.leucocephalus*	Labrador–Eastern USA	5J	720	62	12
Short-toed Eagle *Circaetus gallicus*	France–Niger	1A	4685	20	234
Lesser Spotted Eagle *Aquila pomarina*	Germany–Zambia	1A	8986	53	170
Lesser Spotted Eagle *Aquila pomarina*	Namibia–Hungary				
	Namibia–Ukraine				
Lesser Spotted Eagle *Aquila pomarina*	Slovakia–Zambia	3×1A	8629	50	174
Greater Spotted Eagle *Aquila clanga*	Yemen–West Siberia				
Steppe Eagle *Aquila nipalensis*	Mongolia–Tibet	1A	1800	20	90
Golden Eagle *Aquila chrysaetos*	Quebec–northeastern USA	4A	2015	31	65
Wahlberg's Eagle *Aquila wahlbergi*	Namibia–Nigeria	1A	3148	35	90
European Honey Buzzard *Pernis apivorus*	Sweden–West Africa	6A	6747	42	161
		3J	6596	64	104
Oriental Honey Buzzard *Pernis ptilorhyncus*	Japan–southeast Asia	2A	10636	60	177
		1J	7912	77	103
Swainson's Hawk *Buteo swainsoni*	North America to South America	34A	13504	72	188

	Spring			Other details				Source
N	Distance (km)	Time (days)	Speed (km/day)	Days on longest stops	Days on breeding area	Days on wintering area	Total days on migration	
				14				Meyburg et al. (2004a)
				0–44				Kjellén et al. (2001)
				20				Martell et al. (2001)
						54		Martell et al. (2004)
2A	5430	67	81	10	165	7676	124	Ueta et al. (1998)
4A	1501	23	65					Ueta et al. (2000)
4J	1697	45	38					
				28				McGrady et al. (2003)
4A	2019	15	135					Harmata (2002)
1I	3032	37	82					Grubb et al. (1994)
1I	435	1	435					
								McClelland et al. (1996)
				<1				Hunt et al. (1992)
5J	720	32	23	25	195	76	94	Laing et al. (2005)
								Meyburg et al. (1998)
1A	8986	55	163		133	121	111	Meyburg et al. (2000)
1I	10084	121	83					Meyburg et al. (2001)
1I	8252	70	118	14				
2×1A	9075	49	185		157	120	88	Meyburg et al. (2004b)
1A	5526	77	72					Meyburg et al. (1995a)
1A	2200	16.5	133	4	226	102	37	Ellis et al. (2001)
3A	2236	26	86		196	112	57	Brodeur et al. (1996)
1A	4024	50	80	5	172	108	85	Meyburg et al. (1995a)
								Håke et al. (2003)
2A	10794	83	130	37	120	102	143	Higuchi et al. (2005)
	11952	80	149				152	Fuller et al. (1998)

(Continued)

Table 8.3 (*Continued*)

Species	Migration route	Autumn			
		N	Distance (km)	Time (days)	Speed (km/day)
Broad-winged Hawk *Buteo platypterus*	North America to South America	4A	7000	70	100
Prairie Falcon *Falco mexicanus*	Within western N. America	26A	1035	11.3	92
Peregrine Falcon *Falco peregrinus*	Northern North America & Greenland to South America	61A	8624	50	172
Peregrine Falcon *Falco peregrinus*	Mexico–North America	6A	5059	40	126
Peregrine Falcon *Falco peregrinus*	N. Russia–SW Europe	2A	3841	20.5	191
White Pelican *Pelecanus onocrotalus*	Romania–Israel	1I	2400	18	133
White Stork *Ciconia ciconia*	Europe–E. Africa[b]	75	4600	19	243
White Stork *Ciconia ciconia*	Africa–Germany	1A×6yrs	9579	149	64
White Stork *Ciconia ciconia*	Russia–China	7I	3256	117	28
White-naped Crane *Grus vipio*	Japan–China				
White-naped Crane *Grus vipio*	Central–east Russia	5A	2512	42	60
White-naped Crane *Grus vipio*	Russia–China	8A	2563	32	80
	Russia–Japan	3A	2544	53	48
Red-crowned Crane *Grus japonensis*	Russia–China	2A	2242	30	75
	Russia–Korea	7A	874	6	146
Hooded Crane *Grus monacha*	Japan–Siberia	2A			
Siberian Crane *Grus leucogeranus*	Siberia–China	5A	5313	51	104
Sandhill Crane *Grus canadensis*	Saskatchewan–Texas	1A	3378	103	33
Sandhill Crane *Grus canadensis*	Saskatchewan–Mexico	1	3998	80	50
Demoiselle Crane *Grus virgo*	Mongolia–India	3A	2500[e]	7	350[e]
	Kazakhstan–India	1A	2000[e]	7	290[e]

Details are only from birds that were followed over one or more complete journey (i.e. part-journeys excluded) and were calculated tions. N: number of birds tracked, A: adult, I: immature, J: juvenile (first-year).

[a] No difference between adults and juveniles.

[b] To first major staging area from which some birds later migrated on to southern Africa.

[c] Excludes one extreme bird, all juveniles. The author's calculated faster ratio of travel, presumably over only parts of the journey.

[d] Followed by plane, northward migration post-fledging, all juveniles.

[e] Approximate only.

	Spring			Other details				Source
N	Distance (km)	Time (days)	Speed (km/day)	Days on longest stops	Days on breeding area	Days on wintering area	Total days on migration	
	11 952							
	11 952							
1A	7 868	74	105					Haines et al. (2003)
								Steenhof et al. (2005)
	8 247	42	196				92	Fuller et al. (1998)
13	5 059	30	169		87	208	70	McGrady et al. (2002)
								Ganusevich et al. (2004)
1A	2 240	22	102		168	157	40	Schmueli et al. (2000)
								Berthold et al. (2001)
	11 105	69	161	106	42	217		Berthold et al. (2004)
								Higuchi et al. (2000)
2A	1 925	22.5	86					Higuchi et al. (1992)
2J	2 435	29.5	82.5	20				
				30				Fujita et al. (2004)
				30				Higuchi et al. (2004)
				25				Higuchi et al. (1998)
				4				
2A	3 484	35	99.5	21				Higuchi et al. (1992)
				28				Kanai et al. (2002)
1A	3 378	42	80	18			145	Hjertaas et al. (2001)
								Hjertaas et al. (2001)
								Kanai et al. (2000)

in the same way for all studies. In some cases, therefore, the figures may differ slightly from those given in the original publica-

followed by car and plane.

Box 8.3   Migrations of Peregrine Falcons *Falco peregrinus* and
Swainson's Hawks *Buteo swainsoni* from North America. From Fuller
et al. (1998)

1. **Peregrine Falcon.** Sixty-one adult female Peregrines were tracked
on migration between nest-sites in northern North America or Greenland
and wintering sites in southern North America, Central and South
America (**Figure 8.6**). Breeding sites spanned about 35° of latitude
and wintering sites about 80° of latitude (from 40°N in the mid-Atlantic
United States coast to 40°S in central Argentina). On average, on their
southward journey these birds covered 8624 km and on their northward
journey 8247 km (taking slightly different routes at the two seasons).

Figure 8.6 Migration routes of Peregrine Falcons *Falco peregrinus*
(above) and Swainson's Hawks *Buteo swainsoni* (overleaf) from North
American breeding areas, as shown by the tracks of satellite-tracked
radio-marked adults. See also **Box 8.3**. From Fuller *et al.* (1998).

Their southward journey took an average of 50 days at around 172 km per day, and their northward journey took an average of 42 days at around 198 km per day. In general, these birds migrated on a broad front, but in autumn tended to concentrate along coastal routes, and many crossed the Gulf of Mexico and Caribbean. In spring they took more inland routes, on the western side of the Gulf, from which they headed towards various northern destinations. Birds migrating north through a single locality (Padre Island on the Texas coast) diverged for destinations ranging from Alaska to west Greenland. These findings confirmed those obtained over a longer period from ringing, i.e. that many individual Peregrines which migrate southward down the east coast of North America in autumn migrate northward up the Gulf Coast in spring, performing a 'loop migration' like many shorebirds.

2. **Swainson's Hawk**. In contrast to Peregrines, Swainson's Hawks migrate by soaring flight and converge on a relatively narrow route from North to South America through Panama, thus avoiding sea-crossings and showing no great divergence of routes between spring and autumn

(**Figure 8.6**). Of 34 birds tracked from various localities in the breeding range in western North America, all wintered in a relatively small area in South America lying at 30°–40°S and 61°–64°W. On average, these birds travelled 13 504 km on their southward journey, at 188 km per day, and 11 952 km on their northward journey, at 150 km per day. On southward migration, the birds became concentrated on the Gulf Coast of Central Mexico, and remained in a relatively narrow stream through Panama and down the eastern flanks of the Andes.

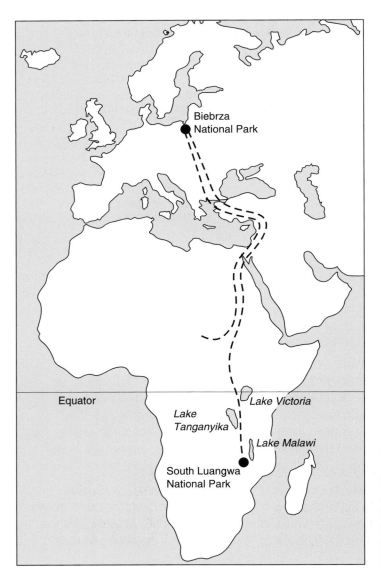

Figure 8.7 Migration routes of a mated pair of Greater Spotted Eagles *Aquila clanga* from a nest-site in Poland to wintering sites in Africa. Note that the two mates spent the non-breeding period in widely separated areas. From Meyburg *et al* (1998).

To save cost and battery power, most raptors on migration were satellite-checked every few days, but Meyburg *et al.* (1998) monitored a Short-toed Eagle *Circaetus gallicus* every night on its autumn journey between France and Niger in West Africa. They were thus able to record every roost location, and the distances travelled each day throughout the journey. These daily distances varied from 17 to 467 km (mean 234 km), and the whole 4685 km journey took 20 days (**Figure 8.9**). Other raptors for which sufficient data were obtained have occasionally moved more than 400 km per day, with 746 km recorded from an Osprey *Pandion haliaetus* through Europe (Kjellén *et al.* 2001), and 537 km for a Lesser Spotted Eagle *Aquila pomarina* through Africa (Meyburg *et al.* 2001), as well as the 435 km for the Bald Eagle *Haliaetus leucocephalus* that migrated within a day (Grubb *et al.* 1994).

Not surprisingly, those species tracked between Europe and Africa travelled more rapidly over the Sahara (which probably offered excellent soaring conditions but no food) than over other parts of the journey (for Short-toed Eagle *Circaetus gallicus* see Meyburg *et al.* 1998, for Osprey *Pandion haliaetus* see Kjellén *et al.* 2001, for Honey Buzzard *Pernis apivorus* see Håke *et al.* 2003). Elsewhere some species broke their autumn journeys for up to several weeks at a time, apparently when they encountered good feeding areas. Once in winter quarters, individuals of most species remained in one place throughout their stay, and those individuals studied in successive years returned to the same breeding and wintering localities each year (e.g. Meyburg *et al.* 1998). However, an adult female Wahlberg's Eagle *Aquila wahlbergi* tracked from its breeding area in Namibia to a non-breeding area centred mainly in Nigeria wandered over a large area there of 60 000 km² (Meyburg *et al.* 1995b). Similarly, some Lesser Spotted Eagles *A. pomarina* wandered over a large area in southern Africa, frequently making extensive moves during the course of their stay (Meyburg *et al.* 1995c, 2004b). This species is

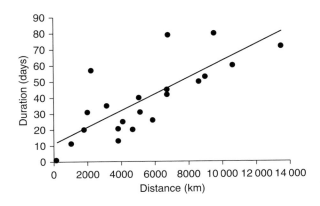

Figure 8.8 Relationships between duration and distance in the autumn migrations of different raptor species, based on the mean values from different studies listed in **Table 8.3**. Regression relationship: Duration (days) = 11.8 + 0.00493 × distance (km), $r = 0.75$, $P < 0.001$. No significant relationship emerged between migration speed and distance: speed (km per day) = 116 + 0.0057 × distance (km), $r = 0.31$, $P = 0.162$.

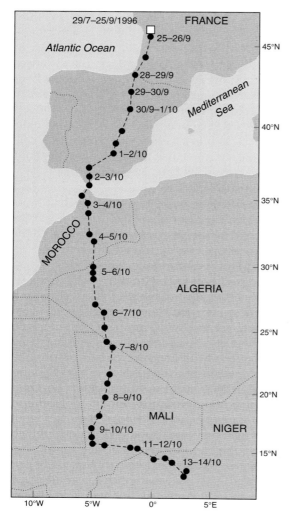

Figure 8.9 Migration of a satellite-tracked radio-tagged Short-toed Eagle *Circaetus gallicus* from France to Niger, showing the daily distances flown and the nightly stopping places. From Meyburg *et al.* (1998).

known from previous observations to move around over large areas, concentrating temporarily in areas of abundant food (Newton 1979). In effect, the individuals tracked were on migration for most of the time between leaving the nesting area in one year and returning there the next, as they followed the rain-belts in southern Africa in search of termites and other prey (Meyburg *et al.* 2004b).

Among various raptors, no consistent difference was apparent between the duration of autumn and spring journeys, either from the mean values calculated for different populations or using the values for those individuals of each species tracked on both outward and return journeys (**Figure 8.10**, **Table 8.3**). In some populations (or individuals) the autumn journey took longer, in others the spring journey. The biggest divergence was recorded in a White-tailed Eagle *Haliaetus albicilla* which took markedly different routes at the two seasons, so that the autumn journey lasted 57 days (39 km per day) and the spring journey 67 days

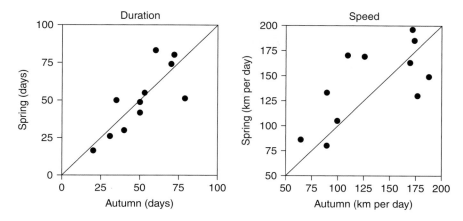

Figure 8.10 Relationship between the durations and speeds of the autumn and spring migrations of different raptor populations, based on the mean values from different studies listed in **Table 8.3** (but excluding the White-tailed Eagle which took markedly different routes in autumn and spring). The line on each graph shows the relationship expected if the two migrations were of equal duration and speed. Overall, no significant difference was apparent in the duration or speed between seasons (on a paired t-test for duration, $t = 0.08$, $P = 0.94$; for speed, $t = -0.93$, $P = 0.37$).

(81 km per day). However, even including this bird, the regression line relating the duration of autumn and spring journeys was not significantly different from one, which implies equality in the mean duration of both journeys.

A striking feature in those species with reasonable samples was again the large individual variation in the duration of migration. Among Ospreys *Pandion haliaetus* that breed in Sweden and winter in West Africa, the total length of journey varied between 5813 and 7268 km (longest journey 1.25 times longer than the shortest), but this journey took between 14 and 55 days to complete (greatest value 3.9 times longer than shortest) in a sample of 13 individuals (Kjellén *et al.* 2001). Mean speeds varied between 108 and 431 km per day. Among Honey Buzzards *Pernis apivorus* over the same route, the journey varied between 6299 and 7091 km (longest journey 1.13 times longer than the shortest), and took between 34 and 70 days to complete (greatest value 2.06 times longer than the shortest) in a sample of nine individuals (Håke *et al.* 2003). Mean speeds varied between 93 and 209 km per day. In other species, equivalent journeys took up to three times longer in some individuals than others (**Table 8.3**).

Over the same journey, no difference was apparent in the autumn migration speeds of juvenile and adult Ospreys. In Honey Buzzards the average speeds of adults and juveniles on travelling days were similar, about 170 km per day in Europe, 270 km per day across the Sahara and 125 km per day in Africa south of the Sahara. However, as the adults had fewer stopover days en route, they maintained higher overall speeds and completed migration in a shorter time (42 days) than the juveniles (64 days) (Håke *et al.* 2003). The spring journeys of several eagle species were slower in juveniles than in adults, mainly for the same reason (**Table 8.3**). Although Peregrines and Ospreys often progress by flapping flight when soaring is not possible, their mean migration speeds were within the range of values recorded for other soaring species in **Table 8.3**. In general, within regional populations, those adults that spent the longest periods on migration spent the shortest periods in their wintering areas, while no effect was apparent on the amount of time spent in breeding areas.

Turning to non-raptorial species, White Storks *Ciconia ciconia* covered the first part of their autumn migration from Europe to East Africa by almost daily travelling. They took an average of 18.9 days over this part of their journey, which included only 1.4 non-travelling days (93% of all days were travelling days), giving an overall speed of 243 km per day (Berthold *et al.* 2001b). No differences were apparent between adults and juveniles. The storks did not eat excessively before their departure, but seemed to forage opportunistically in the mornings and evenings en route, regaining any net loss in body mass at their intermediate destination in northern Africa (mostly Sudan). Some individual White Storks showed great variation in the timing, length and duration of their migratory journeys from year to year, as well as in the duration and location of stopovers. One adult female was followed on six outward and six return journeys from its breeding place in Germany to its wintering place in Africa, which in two years was within a few degrees of the equator and in four years in South Africa. This bird took 121–169 days over the southward journey (mean 149 days) and 45–115 days over the return journey (mean 69 days) (**Table 8.4**). Its starting dates from the nesting area in different years varied by up to 15 days, and from the wintering area by up to 25 days. The longest stopover (in autumn) lasted 75 days, apparently in an area with abundant food (Berthold *et al.* 2004). Mean speeds in different autumns varied from 56 to 92 km per day (average 74 km per day) and in different springs from 97 to 165 km per day (average 147 km per day). The autumn journeys thus took longer, but chiefly because of the extended breaks in northern Africa which were not repeated in spring. Another population of White Storks in eastern Asia migrated through mainly favourable habitat (green river valleys), and radio-tagged individuals travelled much more slowly than the European–African birds, averaging only 28 km per day, and less than 10 km on most days, providing a clear demonstration of the effect of terrain on migratory behaviour (Higuchi *et al.* 2000).

Among various species of cranes, mean speeds in the range 33–146 km per day were recorded, somewhat slower than the raptors and storks, but cranes often migrate by flapping flight and stop for longer periods to refuel (**Table 8.3**). Demoiselle Cranes *Grus virgo* provided an exception, however, as the individuals tracked migrated over mainly high desert and the Himalayas, with few potential feeding places, and completed the main part of their journey within a week, averaging 350 km per day (Kanai *et al.* 2000). This flight may have been preceded by a long fuelling period.

PROPORTION OF MIGRATION SPENT IN FLIGHT

Knowledge of the duration of migration, obtained from ring recoveries or radio-tracking, together with knowledge of the usual flight speed, enable us to calculate how much of the total journey time of different species is spent on the wing (**Table 8.5**). For example, Chaffinches *Fringilla coelebs* usually leave northern

Table 8.4 Six outward and six return migrations of the same radio-tagged adult female White Stork *Ciconia ciconia* tracked by satellite from its nesting place in Germany (57°N) to its wintering localities in different parts of Africa

	Southward outward journey						Northward return journey					Overall	
Year	Start date	End date	Days taken	Distance (km)	Speed (km/day)	Winter latitude	Start date	End date	Days taken	Distance (km)	Speed (km/day)	Days on longest stopover	Days in wintering area
1994–95	29 Aug	28 Jan	152	8438	55.5	3°S	17 Mar	25 May	69	7428	107.7	60	48
1997–98	25 Aug	24 Dec	121	7598	62.8	4°N	10 Mar	25 Apr	45	6382	141.8	51	76
1998–99	23 Aug	16 Jan	146	13381	91.7	34°S	26 Feb	1 May	64	11147	174.2	87	41
2001–02	25 Aug	17 Jan	145	11958	82.5	34°S	21 Feb	18 Apr	56	10970	196.0	75	35
2002–03	25 Aug	31 Jan	159	13285	83.6	34°S	2 Mar	28 May	115	11156	97.0	58	30
2003–04	15 Aug	31 Jan	169	11969	70.8	30°S	20 Feb	24 Apr	63	10390	164.9	73	20

From Berthold *et al.* (2004).

Table 8.5 Proportion of migration spent in flight, based on ringing and radio-tracking of individual birds

Species (season)	Distance (km)	Time on journey (days)	Mean (km/day)	Flight speed (km/h)	% time in flight	Ratio flight to stopover	Source
Passerines							
Chaffinch *Fringilla coelebs* (a)	3000	35	86	30.0	11.9	1:7	Dolnik & Blyumental (1967)
Willow Warbler *Phylloscopus trochilus* (a)	10500	121	85	30.0	12.1	1:7	Hedenström & Petterson (1987)
Spotted Flycatcher *Muscicapa striata* (a)	10500	75	140	48.0	12.2	1:7	Fransson (1986)
Barn Swallow *Hirundo rustica* (a)	10000	70	150	40.7	14.6	1:6	Mead (1970)
Barn Swallow *Hirundo rustica* (s)	10000	42	238	40.7	24.4	1:3	Mead (1983)
Large non-passerines							
Common Wood Pigeon *Columba palumbus* (a)	1000	15	67	63.4	4.4	1:22	Gatter (2000)
Greater White-fronted Goose *Anser albifrons* (s)	3000	25	120	(70.0)	7.2	1:13	Fox et al. (2003)
Brent Goose *Branta bernicla* (s)[a]	5004	42	118	70.0	7.1	1:13	Green et al. (2002)
Barnacle Goose *Branta leucopsis* (a)[b]	2750	10	275	(70.0)	16.4	1:5	Butler et al. (2003)
Bewick's Swan *C. c. bewickii* (a)	2023	34	61	64.0	3.9	1:25	Beekman et al. (2002)
Bewick's Swan *C. c. bewickii* (s)	1871	32	58	64.0	3.8	1:24	Beekman et al. (2002)
Whistling Swan *C. c. columbianus* (s)[b]	3083	21	147	(64.0)	9.6	1:9	Higuchi et al. (1991)
Large soaring species							
European Honey Buzzard *Pernis apivorus* (a)	6747	42	161	39.4	17.0	1:5	Håke et al. (2003)
Oriental Honey Buzzard *Pernis ptilorhyncus* (a)	10636	60	177	(39.4)	18.7	1:4	Higuchi et al. (2005)
White-tailed Eagle *Haliaeetus albicilla* (a)	2244	57	39	47.2	3.5	1:28	Ueta et al. (1998)
White-tailed Eagle *Haliaeetus albicilla* (s)	5430	67	81	47.2	7.2	1:13	Ueta et al. (1998)
Bald Eagle *Haliaeetus leucocephalus* (s)	2019	15	135	50.0	11.2	1:8	Harmata (2002)
Egyptian Vulture *Neophron percnopterus* (a)	3572	29	123	36.4	14.1	1:6	Meyburg et al. (2004a)
Short-toed Eagle *Circaetus gallicus* (a)	4683	20	234	39.5	24.7	1:3	Meyburg et al. (1998)

Species							Reference
Lesser Spotted Eagle Aquila pomarina (a)	8700	50	174	34.4	21.0	1.4	Meyburg & Meyburg (1998)
Lesser Spotted Eagle Aquila pomarina (s)	8896	55	163	34.4	19.6	1:4	Meyburg et al. (1998)
Lesser Spotted Eagle Aquila pomarina (a)	8896	53	170	34.4	20.3	1:4	Meyburg et al. (1998)
Lesser Spotted Eagle Aquila pomarina (s)	9065	49	125	34.4	22.4	1:3	Meyburg et al. (2004b)
Golden Eagle Aquila chrysaetos (a)	2015	31	65	46.8	5.8	1:16	Brodeur et al. (1996)
Golden Eagle Aquila chrysaetos (s)	2236	26	86	46.8	7.7	1:12	
Steppe Eagle Aquila nipalensis (a)	1800	20	96	35.9	10.4	1:9	Ellis et al. (2001)
Steppe Eagle Aquila nipalensis (s)	2200	16.5	133	35.9	15.5	1:5	Ellis et al. (2001)
Osprey Pandion haliaetus (a)	5134	31	166	49.3	14.0	1:6	Martell et al. (2001)
Osprey Pandion haliaetus (s)	5872	26	226	49.3	19.1	1:4	
Osprey Pandion haliaetus (a)	3824	13	294	49.3	24.9	1:3	
Osprey Pandion haliaetus (a)	4105	25	164	49.3	13.9	1:6	
Osprey Pandion haliaetus (a)	6742	45	150	49.3	12.7	1:7	
Peregrine Falcon Falco peregrinus (a)	8624	50	172	43.6	16.5	1:5	Fuller et al. (1998)
Peregrine Falcon Falco peregrinus (s)	8247	42	196	43.6	18.8	1:4	Fuller et al. (1998)
Peregrine Falcon Falco peregrinus (a)	5059	40	126	43.6	12.1	1:7	McGrady et al. (2002)
Peregrine Falcon Falco peregrinus (s)	5059	30	169	43.6	16.1	1:5	McGrady et al. (2002)
Peregrine Falcon Falco peregrinus (a)	3841	21	191	43.6	17.5	1:5	Ganusevich et al. (2004)
White Stork Ciconia ciconia (a)	9579	149	63	48.6	5.5	1:17	Berthold et al. (2004)
White Stork Ciconia ciconia (s)	11105	69	161	48.6	13.8	1:6	Berthold et al. (2004)
White Stork Ciconia ciconia (a)	4600	79	242	48.6	5.0	1:19	Berthold et al. (2001b)
White Stork Ciconia ciconia (a)	3256	117	28	48.6	2.4	1:41	Higuchi et al. (2000)
Great White Pelican Pelecanus onocrotalus (a)	2400	18	133	57.6	9.7	1:9	Schmueli et al. (2000)

Passerines and Woodpigeon from ringing data, all others from radio-tracking (adults only).

a: autumn, s: spring.

Flight speeds mainly from Bruderer & Boldt (2001). Bracketed speeds are from closely related species.

[a]Including the time needed for pre-migratory deposition would lower the mean flight speed to 62 km per day, the fraction of time spent in active flight to 4%, and the flight to stopover ratio to 1:24.

[b]Mainly overwater journeys.

Finland in the first half of September and reach their wintering areas in France and Iberia around mid-October, the complete journey taking five weeks. The birds thus cover 3000 km at an average of 86 km per day. As they could cover 86 km in 3 hours, and the whole distance (if they could fly non-stop) in 105 hours, they must have spent up to 88% of their total five-week journey stationary, giving a ratio of flight-to-stopover of about 1:7. This agrees with the theoretical estimate given earlier for birds of this size.

Because the energy requirement per hour of flight is generally greater in large birds, and their refuelling rates are lower, they take longer to accumulate the fuel necessary for a standard journey than do small birds. In Brent Geese *Branta bernicla* migrating mainly over land between the Wadden Sea and the Taimyr Peninsula, the mean flight distance of 5004 km is equivalent to about three days and nights (72 hours) on the wing, assuming a mean ground-speed of 70 km per hour (Green *et al.* 2002). Because it took six geese on average 42 days to cover this distance, the mean rate of progress was 118 km per day (range 97–148), and only 7% of the time spent on spring migration consisted of active flying (but excluded initial fattening periods). This gave a flight-to-stopover ratio of 1:13. Long-time (foraging) stopovers made up 79% of the total migration time, and short-time stops another 14%. Including the time needed for pre-migratory fuel deposition in the Wadden Sea area would lower the mean flight speed to 62 km per day and change the flight-to-fuelling ratio to 1:25.

To take a more extreme example, consider five Bewick's Swans *Cygnus columbianus bewickii* radio-tracked on their autumn migrations between Siberia and the Baltic region (Beekman *et al.* 2002). These birds travelled the 2023 km journey in 34 days, on average. With a mean flight speed of 64 km per hour, an estimated 32 hours was spent on the wing, giving a flight-to-stopover ratio of 1:25. In spring, the equivalent figures from two birds were also 1:25. These estimates excluded the initial fattening period, but they broadly agree with theoretical predictions, and with the longer established observational finding that these Bewick's Swans take around two months, on average, on both their autumn and spring journeys. They also give some idea of the small proportion of the overall journey time spent by large birds in flapping flight when migrating over land, the rest of the time being spent refuelling and resting at stopover sites. We should not be misled by the fact that some swans can cross 1000 km of sea in less than two days (see later), because this makes no allowance for the period of prior fuel deposition. In practice, of course, birds often stay longer at stopovers than is needed solely to accumulate fuel, and in addition, their energy costs in flight are greatly influenced by wind conditions. On the spring journey, some birds may also be accumulating extra body reserves for breeding (Chapter 5).

Similar estimates of flight-to-stopover ratios for other species are given in **Table 8.5**, and generally support the notion that, among birds that travel by flapping flight, small species spend a greater proportion of the total journey time in flight than larger species. The largest species, such as geese and swans, show a flight-to-stopover ratio of 1:13 to 1:25 on overland flights. These estimates ignore long sea-crossings by geese and swans discussed later, because in estimates of their flight times, no feeding periods are included.

The situation in soaring birds is different. Their lower internal energy needs reduce the time they must spend feeding to accumulate large body reserves. In almost all the species for which relevant data are available, the flight-to-stopover ratios in adult birds varied between 1:3 and 1:7, reflecting the much reduced time spent on stopovers, but a few studies gave much longer stopover periods (**Table 8.5**). Considerable variation was apparent within species, with mean values of 1:4–1:7 from three studies of Peregrines, and of 1:3–1:7 from five studies of Ospreys. These stopover times are generally much lower than equivalent figures from similar-sized birds that migrate by flapping flight, and some are lower even than those from small passerines. They re-affirm the great advantage of soaring flight for large birds: it needs much less feeding time, so gives faster overall migration speeds. All the values quoted were from adult birds; and the non-breeding immatures of some species, which were under less pressure to return to breeding areas, showed much more leisured spring journeys (**Table 8.5**).

APPARENT NON-STOP SEA-CROSSINGS

Apparent non-stop sea-crossings provide further striking examples of the flight capabilities of birds. They enable some journeys to be made extremely rapidly, providing the journey time is taken from take-off to landing, and no allowance is made for the pre-flight fattening period. Most data on the time taken for sea-crossings derive from comparing the departure dates of birds from coastal staging sites near their breeding areas with their arrival dates at their destinations, a method applicable only to conspicuous species (see above). On this basis, the Bar-tailed Godwits *Limosa lapponica baueri* that migrate 10 400 km from Alaska to New Zealand were estimated to cover this distance in about 175 hours (7.3 days) of uninterrupted flight, giving a mean speed of 1512 km per 24-hour day (or 63 km per hour). This is the longest and fastest overwater flight known from any land-bird (Gill *et al.* 2005). Similarly, Great Knots *Calidris tenuirostris* were estimated to fly 5420 km between northwest Australia and Chongmen Island at the mouth of the Yangtze River in China in about four days, or 1355 km per day (equivalent to 56.5 km per hour) (Battley *et al.* 2001). Both these species lost around half their body weights over the journey.

Autumn migrations of Brent Geese *Branta bernicla* over the Pacific Ocean from Izembek Lagoon, Alaska, to Baja California, Mexico, were studied in the same way (Dau 1992). Observers at both sites documented departures and arrivals in three different years, estimating times en route to be 60, 60 and 95 hours respectively, which implied average speeds of 1992, 1992 and 1344 km per 24-hour day (or of 83, 83 and 56 km per hour) over the 5000 km journey. Variability in mean route times between years was associated mainly with differences in wind speed and directions, and possibly also with times spent resting on the sea. This migration is also energetically costly, with birds of both sexes losing an estimated one-third of their total pre-departure body weights during the 2–4 day journey. On a somewhat shorter journey, Canada Geese *Branta canadensis minima* were estimated to take 48 hours to fly 2800 km over the sea from the Alaska peninsula to the Klamath Basin in Oregon, averaging 1392 km per 24 hours, or 58 km per hour (Gill *et al.* 1997). One can never be certain in such studies that the birds did not rest on the sea for a time, but their journey times were so short that any such rests cannot have lasted long.

Other information on sea-crossings has come from the tracking of radio-tagged birds. Ten Whooper Swans *Cygnus cygnus* tracked by satellite took 12.7–101 hours to cross the sea between Iceland and the British Isles, a journey of 962–1700 km in different individuals (Pennycuick *et al.* 1999). These figures relate to the sea-crossing alone, and exclude times spent over land, as well as the initial fattening period. The great variations in journey times were due mainly to variations between individuals in the amount of time spent sitting on the sea. Two birds were blown far off course and waited on the sea until conditions improved, when they flew back on course. Two others made the journey with no obvious stops, one reaching 1856 m above sea level, the maximum height recorded.

PENGUINS

Because penguins migrate entirely by swimming, their migration speeds are not strictly comparable with those of other birds. Magellanic Penguins *Spheniscus magellanicus* were tracked northward up the coast of Argentina. Some reached more than 1800 km north of the nesting colony but, by following the coastline, they had travelled at least 2700 km. Average travelling speeds of 10 birds were in the range 1.1–1.9 km per hour, higher initially and slower later in the journey when the birds encountered better feeding areas. Maximum speeds recorded over short periods were up to 6.1 km per hour (Putz *et al.* 2000). Higher mean speeds were recorded for Adélie Penguins *Pygoscelis adeliae* on migration, at 1.8–3.4 km per hour (Davis *et al.* 1996). These mean rates translate to average daily distances of 26–46 km and 43–82 km in the two species, which are not dissimilar to the average daily distances recorded for many short-distance passerine migrants (including stopovers). Even higher travelling speeds have been reported for various penguins in the breeding season when they had chicks to feed, with King Penguins *Aptenodytes patagonicus* recorded at up to 10 km per hour on return journeys (Putz *et al.* 1998). Another mode of locomotion, related to swimming, is also employed by penguins. Using their short stiff wings, they can paddle on their bellies across snow fields. On well-used 'penguin highways' the birds achieve speeds between running and swimming.

MIGRATION WITHIN THE ANNUAL CYCLE

In birds followed on both outward and return journeys, it was possible to calculate how much of each year, on average, they spent on migration. The figures for passerines are based on individual ring recoveries, and for many species may be biased in favour of the faster migrating individuals, and so of limited value in this respect. However, for a few species, data on average migration times from ring recoveries or radio-tracking are available. They suggest that Barn Swallows *Hirundo rustica* migrating between northern Europe and southern Africa can spend around 16 weeks per year on migration (or 31% of the entire year). Among waterfowl, radio-tracked Bewick's Swans *Cygnus c. bewickii* spent about 33% of the year on migration, while storks and cranes gave figures up to 59% of each year in different populations, and raptors up to 42% of each year in different populations (**Table 8.3**). As expected, the migration period depended on the journey, with species making the longest migrations taking the longest periods. The duration of a breeding cycle (from egg-laying to independence of young) and of moult also increase with body size in birds, leaving less time to complete a long return migration within an annual cycle. Regardless of the number of feeding stops, this may constrain the length of total migratory journey that can be undertaken by large species, and hence their geographical ranges. It would be hard to imagine some large species being able to perform even longer migrations without compromising reproduction (Hedenström & Alerstam 1998). The problem is perhaps especially acute in some species of geese and swans which seem unable to begin migratory fattening until vegetation growth begins in spring, providing the necessary increase in food supply. Such species also make some of the slowest journeys (including feeding periods), and it is perhaps not surprising that no arctic-nesting goose or swan species migrates further than about 6000 km, although journeys twice this length are performed by many smaller birds, including soaring species. The implication is that, in some large bird species that travel by flapping flight, fuelling times may make it impossible to breed, moult and migrate over longer distances within one year. Fuelling times may thereby limit geographical ranges (Hedenström & Alerstam 1998).

CONCLUDING REMARKS

Most species of birds must stop migrating in order to feed, the two activities being mutually exclusive. However, a minority of species, such as swallows and swifts that depend on aerial prey items, can pick up food while on migration, at least when flying over vegetation. Although they do not necessarily make the entire journey without pausing to refuel, an ability to catch and eat food on the wing could enable them to migrate with smaller body reserves and make faster progress than other birds of similar size. Swifts are normally on the wing day and night, except when they have nests to tend, so in this respect migration is probably little different from normal daily life. Some raptors and shrikes may also be able to get food at places offering little for other species, if only by catching and eating their fellow travellers.

Calculations of maximum possible theoretical migration speeds are informative, even though not all birds necessarily attempt to migrate as quickly as possible. Instead of being 'time minimisers', to use the jargon, they may be 'load minimisers', migrating in short stages requiring no great fuel deposition, thus reducing their overall energy needs and predation risk (Lindström & Alerstam 1990; Chapter 5). In addition, few birds can feed at the maximum possible rate recorded for their species, poor food supplies or competition limiting their daily intakes. They may also at times be delayed by adverse weather long beyond the date at which they are physiologically prepared to depart. All these types of constraints tend to slow the overall migration period, and increase the variability within populations.

In addition, at least in autumn, it may be disadvantageous for some species to pass rapidly through areas of abundant food, if they cannot be sure of finding similar plentiful supplies further along the migration route. They would be better to stay and feed until the food was exhausted. This is a particular consideration in irruptive species, which depend on unpredictable food supplies, such as tree-seed crops (Chapter 18). Their migration timing therefore varies greatly from year to year, according to the seed crops encountered, and in most years migration speed is by no means close to the maximum possible (recall the slow rates of progress shown by irruptive species in **Figure 8.3**). For these various reasons, therefore, we can expect that most birds do not normally migrate at the theoretical speeds of which they are capable. It is perhaps in ideal conditions in spring, when birds are migrating to reach their breeding areas, that the maximum rates are most likely to be approached, but even then most birds cannot migrate long ahead of the appearance of suitable food supplies.

If a bird had no other events in its annual cycle, it would seem advantageous for daylength reasons to undertake both migrations as close to the summer solstice as possible, thereby gaining longer days for feeding and fuel deposition. This may be one reason why, in some large bird species, the immatures in their non-breeding years migrate from wintering to breeding areas much later than the breeding adults, and return much earlier, spending only a few weeks per year in breeding areas.

SUMMARY

Depending on the distance travelled, a one-way migratory journey can last for several hours up to several months, including resting, fuelling and flight times. Some species that migrate long distances between northern and southern hemispheres spend around half of each year on migration.

The ratio of flight to stationary time during the migration periods of small passerines is typically around 1:7, matching theoretical predictions, and in larger species that travel by flapping flight the ratio is around 1:14 to 1:30, as large species have relatively higher flight costs than small ones, and take longer to accumulate the necessary body reserves. However, these ratios vary greatly between individuals in the same population, probably depending largely on their fuel deposition rates and on weather during the journey. Individuals from the same breeding area have been found to vary up to several-fold (in extreme cases up to 12-fold) in the time they take to complete roughly the same journey. Nevertheless, maximum observed migration speeds are fairly close to maximum theoretical estimates.

Although they need to spend less time feeding, small passerines do not always achieve higher migration speeds than larger species that also travel by flapping flight: typically 50–150 km per day versus 50–250 km per day. Among large birds, species that travel by soaring–gliding flight achieve faster average speeds (200–400 km per day), mainly because they do not need to accumulate and maintain such large fuel reserves. Some pelagic seabirds travel even faster, with mean speeds up to 950 km per day recorded from albatrosses.

In birds that fly by flapping flight, as well as in those that migrate by soaring–gliding, migration speed generally increases with length of journey. Birds with twice as far to go take longer over the journey, but not twice as long.

In favourable winds, wild geese have been found to fly 5000 km over water within 60 hours, giving mean speeds of around 2000 km per day, or around 80 km per hour. Bar-tailed Godwits *Limosa lapponica* travelling from Alaska to New Zealand covered more than 10 400 km non-stop in seven days, around 1500 km per day, or 60 km per hour.

In some species that travel by flapping flight, fuelling times associated with large body size may make it impossible to breed, moult and migrate long distances within one year, and could thereby limit migration distances (and hence geographical ranges).

Common Starlings *Sturnus vulgaris* migrating at night

Chapter 9
Finding the way

on their journeys, the migrants not only travel vast distances overland, but also cross pathless seas and oceans. The question is – how do they find their way? How are they guided? Here we are face to face with one of the greatest mysteries to be found in the animal kingdom. (William Eagle Clark 1912.)

Many birds are capable of migrating, year after year, from their customary nesting sites to exactly the same winter quarters, sometimes using the same stopover sites on successive journeys. Young birds migrating alone can also find their way unaided by experienced adults to the usual wintering areas for their species, and back to their natal areas the following spring. This holds true even though breeding and wintering areas may lay half a world apart on different continents. Some pelagic seabirds cover enormous distances during their foraging flights and during their migrations over what appears to us as featureless ocean, yet they can return repeatedly to the tiny islands where they breed. How do birds achieve these remarkable feats of orientation and navigation over such huge distances?

It is not just a question of finding the way. Birds must know where in their journeys they need to do particular things, such as change direction or accumulate extra fuel reserves in preparation for a long non-stop flight. The fact that they can respond appropriately at specific places on their route again implies that they possess some geographical sense – an ability to detect and respond in an appropriate manner to conditions at particular locations.

To migrate effectively, birds need a sense of where they are, or need to be, a sense of direction, an ability to navigate from one place to another, and a sense of time, both seasonal and diurnal (essential for navigation by some celestial cues, see later). In short, they need the equivalents of a map, compass, calendar and clock, together with a good memory, all packed into a brain that in some birds is no bigger than a pea. In this chapter, I can provide only a brief review of this vast subject area, concentrating on ecological aspects (for more extensive reviews of particular aspects, see Berthold 1993, Wiltschko & Wiltschko 1995, 2003, Åkesson 2003). Despite much research, many unanswered questions remain.

The ability of birds to navigate depends ultimately on their sensory abilities. The eyesight of normally diurnal birds at night is probably not much better than ours: good enough to allow them to fly through the open airspace, and to recognise major topographic features, such as coastlines and mountains, and potential habitat below. Collisions with obstacles such as radio-towers occur mainly on dark misty nights, when vision is restricted. In addition, at least some bird species are able to perceive ultraviolet light and the plane of polarised light.

That birds and other animals can detect and respond to the earth's magnetic field is well established, but hard for us to appreciate because we have no obvious magnetic sense ourselves. There are two competing hypotheses for the primary process underlying the avian magnetic sense. One involves the magnetic material magnetite (Fe_3O_4) which is found in many organisms, including pigeons in which a magnetite structure is located between the brain and the skull (Walcott et al. 1979). In theory, this structure could function in orientation through the forces exerted by the earth's magnetic field (Kirschvink & Gould 1981). The other proposed mechanism is based on a magnetically sensitive chemical reaction. It involves electron transfer between donor and acceptor molecules, and operates only under short-wavelength light (blue–green end of the spectrum). Being light dependent, it is based in the eyes, presumably enabling the bird to visualise the magnetic field in some way. This 'radical pair' model of magneto-reception has received some support in recent experiments (Ritz et al. 2004), but like the magnetite hypothesis, is still to be properly tested. Neither proposed mechanism may make it though to established fact.

Other senses that some bird species might use in navigation include smell and hearing. Most birds seem to have a poor sense of smell, but in some species, notably petrels, the olfaction sense is well developed, although nothing is known of the range of odours that they can detect (Wenzel 1991). Birds also have a generally well-developed sense of hearing, allowing individual migrants to detect the calls of other birds at night. They also seem able to detect changes in barometric pressure (see later), and can also perceive wind direction and speed during flight, perhaps by reference to the ground below. This latter ability enables birds to select optimal flight altitudes and to correct for wind drift when on migration (Chapter 4).

Birds may possess other sensory capacities that are as yet unknown to us, but which could also play a role in route finding.

COMPASS ORIENTATION AND BI-COORDINATE NAVIGATION

Some of the earliest studies in the field involved trapping wild birds on autumn migration, transporting and releasing them in a distant location, and using the resulting ring recoveries to find where they went (**Figure 9.1**). Such simple but large-scale experiments revealed a fundamental difference in behaviour between experienced adult and naïve young birds. Whereas the older birds were able to reach their traditional winter areas, even from sites outside their normal migration route, young birds on their first autumn migration proved unable to do so. Instead, they continued on their usual migratory heading for about the same distance they would normally travel from the capture site. The inference was that inexperienced young migrants are guided on migration by innate information expressed as a direction and distance from the starting point, distance being controlled by the duration of migratory activity (Chapters 12 and 20). The innate information would be equivalent to an instruction like: 'travel for six weeks towards the southwest', or, in cases of non-straight routes: 'travel for six weeks toward the southwest and then for five weeks toward the south-southeast' (Wiltschko & Wiltschko 2003). This system is known variously as clock-and-compass, bearing-and-distance or vector migration. On the other hand, experienced migrants that have travelled the route before can make use of their experience, as well as innate information, on subsequent migrations. They reveal an ability to navigate, that is to head towards a specific point on the earth's surface from some distant location. Homing to a known site is more complicated than the ability to head only in particular compass directions because it involves true bi-coordinate navigation, requiring a map sense (also known as goal orientation).

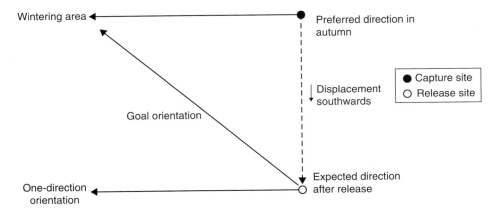

Figure 9.1 Displacement experiment, showing the difference in response to displacement between adult birds (which homed to a familiar wintering area) and juvenile birds which flew in an appropriate direction, but without correcting for displacement off course.

Important experiments

One of the biggest displacement experiments ever undertaken involved Common Starlings *Sturnus vulgaris* which in autumn migrate west-southwest from northeast Europe through the Netherlands to winter in northern France and southern Britain. Over a period of years, Perdeck (1958, 1967) caught more than 19 000 individuals within the Netherlands, released about 7500 on site to act as controls, and transported 11 500 others by air 500 km south-southeast of the capture site for release in Switzerland. The subsequent ring recoveries from translocated juveniles were on a line west-southwest of the release site and extended for a similar distance as usual (into southern France and northern Iberia). This indicated that the translocated birds had kept their inherent directional preference and normal migration distance, but not corrected for their displacement (**Figure 9.2**). They therefore migrated parallel to the normal route, and wintered south of the regular wintering area for their population.

The adults, in contrast, which had already experienced the normal wintering area, corrected for the experimental displacement, and headed northwest towards the normal wintering area with which they were familiar. They had evidently 'realised' they were off course at the release site, and took a bearing different from usual in order to correct for this. The age-related (or experience-related) difference in behaviour persisted whether the birds were released in separate juvenile and adult groups or in mixed-age groups. Recoveries from subsequent years showed that, while juveniles tended to return to the new wintering areas reached after their displacement, both age groups continued to return to their original breeding areas (with which they were familiar). The conclusions were that: (1) autumn-migrating adult Common Starlings *Sturnus vulgaris* used true goal orientation (homing) to reach their wintering areas, whereas the juveniles used one-directional orientation, moving on a fixed bearing for a fixed distance; and (2) juveniles were able to fix both their breeding and their winter quarters in their first year and return there in subsequent years by goal orientation.

In another experiment, Starlings *Sturnus vulgaris* were transported in the migration season from the Netherlands to Barcelona in eastern Spain (Perdeck 1967), an area much favoured by wintering Starlings. Despite this, the young birds moved on west for several hundred kilometres across the width of Spain. The implications were again that the length of the migratory journey was controlled by internal influence, and that migration continued in the normal direction from the release site while ever the drive to migrate persisted.

Similar displacement experiments were done with Chaffinches *Fringilla coelebs*, White Storks *Ciconia ciconia*, Sparrowhawks *Accipiter nisus*, Hooded Crows *Corvus c. cornix*, and various ducks. All gave similar results, with a difference in response between juveniles and adults (Drost 1938, Schüz 1938, 1949, 1950, Rüppell 1944, Rüppell & Schüz 1948, Perdeck 1958, Bellrose 1958, Matthews 1968, Wolff 1970, see also Chapter 12). They were consistent with the view that young birds inherited directional preferences which enabled them to head off towards wintering areas appropriate to their population. The fact that adults, with experience of a specific wintering area, could correct for displacement does not detract from this conclusion. It adds to what ring recoveries have often shown: namely that birds with experience of a location can re-find it. They must therefore have

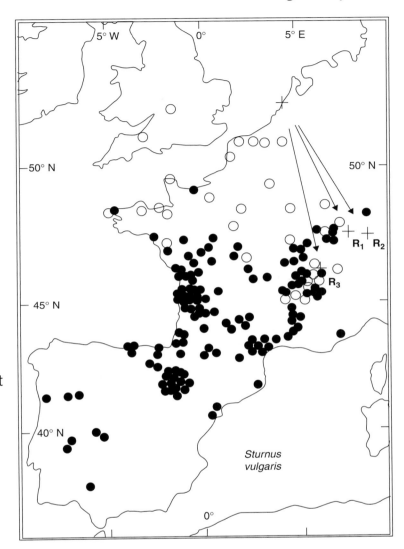

Figure 9.2 Difference in recovery patterns of adult and juvenile Common Starlings *Sturnus vulgaris* displaced about 500 km south-southeast (Netherlands to Switzerland) off their normal west-southwest migration route through the Netherlands. Birds were caught on migration at site F, transported and released at Basel (R), Zürich (R$_2$) and Genf (R$_3$). Filled circles show recoveries of juveniles and open circles of adults during the ensuing autumn–winter. From Perdeck (1958).

not only a directional sense, but also a map sense: an ability to determine their spatial position with respect to home (or at least their direction towards home) based solely on information from their current location. Among other things, such map-based navigation would enable birds to correct for wind-induced drift off their normal migration route.

The fact that all these birds, whether caught on breeding, wintering or migration areas, travelled approximately the same distance as they normally would, has been attributed to the involvement of an internal clock, which ended migratory behaviour after an appropriate period (Chapter 11). Transportation of other birds to a suitable wintering locality did not switch off the behaviour immediately, only after the lapse of an appropriate time period. This was true even of young Pied Flycatchers *Ficedula hypoleuca* and Garden Warblers *Sylvia borin* which were taken from Denmark to Kenya (south of the normal wintering area) (Rabøl 1993). The same behaviour would not necessarily be expected in adult birds, which had experienced the area previously (Chapter 12).

Further evidence for inherent directional preferences came from other early field experiments, in which young birds of various species were held in their natal areas until all other individuals of their species had left. After their release, the young were found from subsequent sightings and ring recoveries to have migrated in the direction normal for their population. For example, 144 White Storks *Ciconia ciconia* were taken from nests in the Baltic region and reared in western Germany. Released after all local migrant storks had passed, these birds showed a strong tendency to head south-southeast, the appropriate direction for the population from which they were drawn, but quite distinct from the south-west direction taken by storks from the release area (Schüz 1949).

Similarly, 1071 young Blue-winged Teal *Anas discors* were caught on migration through Illinois, and held until November, when the species had virtually left the USA for South America. Half of the delayed young birds recovered in the same winter were within 40 km of the release point, but the other 54 were almost exclusively in the southeast sector normal for their species (Bellrose 1958). These experiments eliminated the possibility that the released birds may have been influenced by other individuals of their species, because other individuals had already left the release area. More stringent tests of innate directional preferences involved the release of young birds in areas where their species does not breed. Thus, 21 White Storks were transported from the Baltic region for release in England (Schüz 1938), 953 Herring Gulls *Larus argentatus* were transported from the Friesian Islands for release in inland Germany (Drost 1955, 1958), 192 White Storks were transported from Algeria for release in Switzerland (Bloesch 1956, 1960), and 377 Blue-winged Teal ducklings were transported from Minnesota for release in Missouri (Vaught 1964). Again ring recoveries from all these birds indicated migration in the direction normal for the population from which the birds were drawn.

More recent displacement experiments have used smaller numbers of birds, and tested their directional preferences in orientation cages. The results were inconsistent, as some young birds showed no change in directional preference after displacement (and thus, like the experiments above, supported the vector hypothesis, Hamilton 1962a, Mouritsen & Larsen 1998), while others appeared to change their preferred direction so as to correct for displacement, taking a direction that would lead them to some place on the normal migration route (Rabøl 1969a, 1994, Åkesson *et al*. 2001, 2005, Thorup & Rabøl 2001). It is possible that not all the experimental birds were in a migratory state when tested, or that the distance, direction, speed or manner of displacement influenced the results. The most impressive directional corrections were shown by long-distance migrants that were displaced slowly, predominantly east–west, whereas the Starlings and others mentioned above were transported rapidly over shorter distances, and mainly north–south. Both adults and juveniles were involved in the long-distance displacements, and both corrected appropriately. Similar findings were obtained when naturally drifted migrants (including juveniles) were tested in orientation cages, or tracked by radar (Evans 1968, Chapter 10), and when birds were subjected to simulated displacement in a planetarium (Rabøl 1992). The difference between adults and juveniles found in the earlier experiments was not therefore borne out in all subsequent ones, especially with substantial east–west displacements.

Non-experimental evidence for inherited directional preferences

The existence of innate migratory directions is further supported by those species in which juveniles migrate to wintering areas independently of adults. A striking example is provided by the European Cuckoo *Cuculus canorus*, in which the last young leave the breeding areas up to a month after the last adults have left, yet they still find their own way to their African wintering areas, despite their various foster species wintering in a range of different areas. The same is true for other cuckoos, including the Shining Bronze Cuckoo *Chrysococcyx lucidus*, which migrates from New Zealand 4000 km over open water to winter on small Pacific Islands.

Yet other evidence for an innate directional preference comes from the ring recoveries of migrants, which in many species lie on a straight line route from the breeding area towards a staging or wintering area (for Pied Flycatcher *Ficedula hypoleuca* see Mouritsen 2001, for Linnet *Carduelis cannabina* see Newton 1972, for Honey Buzzard *Pernis apivorus* see Thorup *et al.* 2003). Ring recoveries tend to fan out to some extent with increasing distance, but this would be expected assuming a normal distribution of individual directional preferences. Similarly, radio-tracking has shown that, in some species, each step of the migration follows a rather straight track, with only slight and temporary deviations from the overall direction (for various thrushes see Cochran *et al.* 1967, for Whooping Crane *Grus americana* see Kuyt 1992, for Osprey *Pandion haliaetus* and Honey Buzzard *Pernis apivorus* see Thorup *et al.* 2003). Other species, however, show much greater variation in directions between individuals, and less step-to-step consistency within a journey (Chapter 18). Final confirmation of the genetic determination of migratory directions came from the cross-breeding experiments reported in Chapter 20, in which juveniles showed directional preferences intermediate between those of their parents.

Because some adult birds migrate between specific breeding and wintering territories, separated by hundreds or thousands of kilometres, they cannot rely solely on simple clock-and-compass navigation, which is far too imprecise. The navigation strategy of experienced adult birds must include a precise map component, based on either local or global cues. Both juveniles and adults may have the capacity for bi-coordinate navigation, but juveniles may use it only when they have learnt the coordinates of particular breeding and wintering areas, or when they have drifted far off course. The precise mechanism of goal-oriented navigation remains unknown, although various hypotheses have been proposed (for discussion see Wallraff 1991, Berthold 1993).

Return of displaced adults to breeding sites

Further evidence for bi-coordinate navigation in birds, and an associated map sense, comes from experiments in which breeding adults were taken from their nests, marked and released elsewhere. A watch at the nests could then reveal whether and when the marked owners returned. Experiments on more than 50 species, from songbirds to seabirds, have shown some remarkably high rates of return over distances varying from a few kilometres to more than 5000 km, depending on the species (Matthews 1968, Wiltschko 1992, Akesson 2003). Some non-migratory species, such as Eurasian Tree Sparrow *Passer montanus*, returned only in small numbers and only from short distances of up to a few kilometres, so they could have got back using their local knowledge (Creutz 1949). Other species returned after transportation over hundreds or thousands of kilometres, sometimes from areas they could never have visited, so they must have used other means (Matthews 1968). Among landbirds, high return rates over fairly long distances were recorded in various hirundines. Of 16 adult Purple Martins *Progne subis* released at distances of 2.5–350 km, all returned to their nests, but at greatly varying speeds, the most rapid involving an overnight flight of 350 km in 8.6 hours (Southern 1959). One Barn Swallow *Hirundo rustica* returned after release at 1725 km away (Southern 1959).

Among seabirds, some of the longest displacements involved Laysan Albatrosses *Diomedia immutabilis* (Kenyon & Rice 1958). Fourteen of 18 adults returned to the breeding island in mid-Pacific within 30 days, after having been transported between 2116 and 6629 km to six locations lying in different directions from the breeding place (**Figure 9.3**). These birds showed an extraordinary ability to return quickly over long distances, suggesting that they did not spend much time searching but knew from their release site in what direction to head in order to get home. The fastest bird covered 5148 km in 10 days after release off the Washington coast, giving a mean speed of 515 km per day. Similarly, Manx

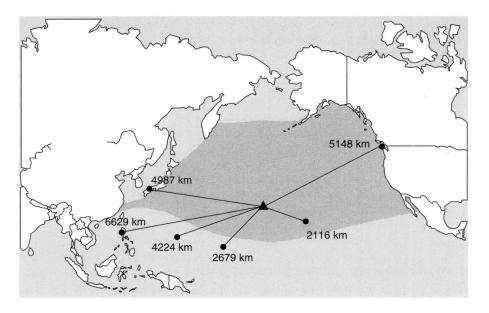

Figure 9.3   Homing of Laysan Albatrosses *Diomedia immutabilis* breeding at Midway Atoll and displaced to different sites around the Pacific Ocean. Eighteen albatrosses were transported to six different sites between 2116 and 6629 km away, and 14 of them returned to the breeding island within 30 days. Grey shading indicates regular range. From Kenyon & Rice (1958).

Shearwaters *Puffinus puffinus* displaced from Wales to eastern North America managed the return distance of more than 4800 km in only 12 days, an average of 400 km per day (Matthews 1968).

Return of displaced birds to wintering sites

Turning to winter, some of the largest displacement experiments involved ducks (McIlhenny 1934, 1940). In the United States, Mallard *Anas platyrhynchos*, Northern Pintail *A. acuta* and Green-winged Teal *A. carolinensis* use as main flyways the Atlantic and Pacific coasts and the Mississippi valley. In one experiment, 440 ducks were taken from their winter quarters in the Mississippi valley, ringed and released on the coasts. Of 90 subsequently shot, 79 had returned to the Mississippi flyway. No distinction was made in this experiment between adults and juveniles or between recoveries in the same or subsequent winters. However, in another experiment in which 895 drake Mallard were transported from the Mississippi flyway about 1800 km to the Pacific flyway, recoveries in the same winter of both juveniles and adults were mostly in the area of release. Recoveries in subsequent winters showed that birds transferred as juveniles mostly maintained their westward displacement, whereas birds transferred as adults mostly returned to the original site (Bellrose 1958). Evidently, both age groups could return to areas previously visited, but the adults had greater attachment to their original area, having spent more time there than juveniles.

Displacement of wintering White-crowned Sparrows *Zonotrichia leucophrys* from San José in California gave some remarkable returns in later years (R. Mewaldt 1964; **Figure 9.4**). Of 411 birds displaced by aircraft 2900 km east-southeast to Baton Rouge in Louisiana, 26 were recaught the following winter in the same small garden at San José where they were originally captured. They formed 21% of the 123 which would normally have been expected to survive and return to the traps if no displacement had been made. In the next winter, of 660 birds transported 3860 km eastward to Laurel, Maryland, 15 were recaught at the capture site in the following season. Of special interest were six of 22 birds displaced to Laurel after they had already returned from Baton Rouge. All birds are presumed to have returned to their breeding areas in northwestern North America in the interim, and one was found during the spring migration period in an intermediate locality; again adults returned in greater proportion than juveniles. The numbers were small, but the annual mortality in this species was around 50%, and low site fidelity as well as poor navigation may have depressed the recorded return rates.

The rate of return next year by young sparrows to the capture site varied according to date of displacement. Those displaced before mid-January were less likely to return, possibly because site attachment had not occurred before then. In addition, however, birds displaced early in winter must live for a longer period than those displaced in late winter before their return can be observed next year (Ralph & Mewaldt 1975, 1976).

In another study, 30 adult and 69 juvenile Sanderlings *Calidris alba* were transplanted from Bodega Bay in California 200 km south (Myers *et al.* 1988). Overall, 60% of adults but only 4% of juveniles later returned and settled in Bodega Bay,

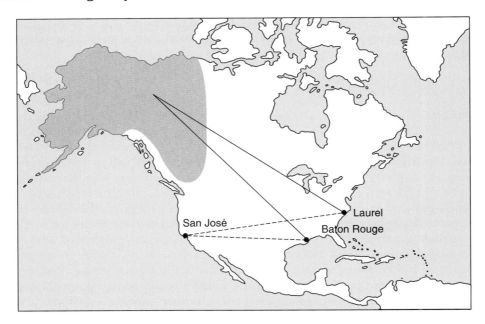

Figure 9.4   Long-distance displacements of Golden-crowned Sparrows *Zonotrichia atricapilla* and White-crowned Sparrows *Z. leucophrys gambelii* from a wintering site at San Josè in California to Baton Rouge (Louisiana) and Laurel (Maryland). Shaded area depicts the breeding range of the populations under study. The birds were presumed to have returned to their breeding areas after displacement before they returned and were recaptured in the original wintering area. A recovery of a bird displaced to Laurel was reported from the spring migration period to the northwest. After R. Mewaldt (1964).

either in the same or a subsequent winter (**Table 9.1**). The age difference was most apparent among birds moved in autumn (from which no juveniles returned) ($\chi^2 = 40.9$, $P < 0.001$), but was non-existent among birds moved in winter. By that time the juveniles had been in Bodega Bay long enough to have become attached to the site, and returned in the same proportion as adults. The adults were already familiar with the site in autumn, from having spent their previous winters there. Clearly, the time spent at a site influenced the tendency to return there after displacement.

Discussion of displacement experiments

Overall, it seems that motivated birds can return to any site that they have previously visited, providing it lies within their capabilities; however, migrants usually return successfully from longer distances, in greater proportions and at higher speed than non-migratory birds (Matthews 1968, Wiltschko 1992, Åkesson 2003). Species thus vary greatly in their homing abilities, according to their normal movement behaviour. Moreover, even within species, the return of displaced

Table 9.1 Effects of age and season on the proportions of Sanderlings *Calidris alba* that returned to a wintering site (Bodega Bay) in the same or subsequent winters after being transplanted to another wintering site 200 km to the south

	Returned and settled in Bodega Bay after transportation	Did not return to Bodega Bay after transportation
Autumn (October–November)		
Adults	13	5
Juveniles	0	45
Winter (January)		
Adults	4	8
Juveniles	3	21

Difference between age groups in autumn: $\chi^2 = 40.9$, $P < 0.001$; in winter $\chi^2 = 2.2$, $P < 0.13$.
From Myers *et al.* (1988).

birds from long distances is more or less restricted to migratory stock, as shown from comparison of results from migratory and sedentary populations of the same species, such as White-crowned Sparrows *Zonotrichia leucophrys* (Mewaldt 1964) or Herring Gulls *Larus argentatus* (Matthews 1968). Experiments with completely sedentary species, such as House Sparrows *Passer domesticus*, gave no returns from further than about 12 km. In general, the proportions of individuals that returned declined with increase in distance of displacement and, allowing for distance, adults returned in greater proportions than juveniles.

In most homing experiments, even after allowing for expected mortality, the proportion of displaced birds that returned successfully to the capture site was often low. Some birds may have had no fixed home at the time, so had no incentive to return to the capture site; or if they returned, they may have stayed too short a time to be recaught and identified. Moreover, the navigational skills of individuals are likely to have varied because of differences in innate ability and experience. Such variation is well established in homing pigeons, only a small proportion of which are capable of achieving really long-distance returns at high speed (Matthews 1968). Nevertheless, the ability of birds to return to a known site, either after migration or artificial displacement, indicates that individuals are able to find their way back to the place where they were trapped, provided that they are motivated to do so and are tested within the limits of their abilities. Whether migratory or non-migratory, many birds evidently have a map sense, which they can use outside the migration seasons as well as within. The same is true for many other kinds of animals subject to displacement experiments.

CUES USED IN DIRECTION FINDING

The most obvious way in which birds and other animals could find their way around on a day-to-day basis is by use of landmarks or other consistent features of their home areas. This explains how some migrant birds manage to return to exactly the same nesting places year after year. But such features are useful only in familiar areas, and when moving over longer distances into unknown terrain, a reliable geographical reference system is needed for navigation. At least two types of factors can act as compasses – celestial and geomagnetic – and both are used by birds as directional aids (for reviews see Emlen 1975, Able 1980, Wiltschko & Wiltschko 1995, 2003, Berthold 1996, Åkesson 2003). In migratory birds, compasses based on the sun (and various sunset cues), stars and magnetic information have been studied in detail, but a prior requirement for using any compass is that the bird should 'know' beforehand – either by inheritance or experience – what direction it has to take. Also, effective use of any of these compasses requires a period of learning, and frequent revision as the bird continually changes location while on migration.

One feature of celestial cues, such as the sun and stars, is that they appear to change in position through each 24-hour cycle, as the earth spins on its axis. In the northern hemisphere, the sun lies in the south and moves during the day from east to west, and at night the stars rotate anticlockwise around the geographical north. In the southern hemisphere the sun lies in the north and moves from east to west, while the stars rotate clockwise around the geographical south. In using the sun and related factors in direction finding, therefore, birds in both hemispheres must allow for time of day. But the same is not necessarily true for star patterns if they are used solely to indicate geographical north or south, determined by the centre of rotation of the night sky. Time-keeping depends on the 'internal clock', kept to time by the regular day–night cycle.

The sun-azimuth compass

The height of the sun's arc in the sky varies with latitude and season, but it is always symmetrical with respect to true north or south. The use of the sun as a compass by birds has been known for more than 50 years. Under a sunny sky, Starlings *Sturnus vulgaris* kept in circular wire cages during the migration period oriented in the same direction as free-living birds. They varied the angle they took to the sun according to the time of day. If the sky became overcast, their directional preference disappeared. When their view of the sun's direction was changed using mirrors, the birds oriented at the same angle to the apparent sun as they would to the real sun (**Figure 9.5**). These experiments confirmed that Starlings made use of a sun compass which gave accurate information only if regulated by an internal clock, allowing adjustment of directional preference as the sun moved through the sky (the 'time-compensated sun compass'). Additional experiments in which Starlings were kept under artificial cycles of day and night of the same duration as natural time, but out of phase, revealed clearly that, in order to orientate, the birds used both the sun's position on the azimuth (direction from the observer) and the time of day. Given a simulated stationary sun, a caged migrant orientated at different angles to it according to the time of day.

The use of a sun-azimuth compass has now been confirmed experimentally in several species, including penguins walking over ice fields, and may be commonly used by diurnal migrants. Starlings and homing pigeons are able to employ the sun compass anywhere on the globe: under polar conditions when the sun does not set, and under equatorial conditions when it reaches its zenith. If birds of these species are transported to the southern hemisphere, however, they

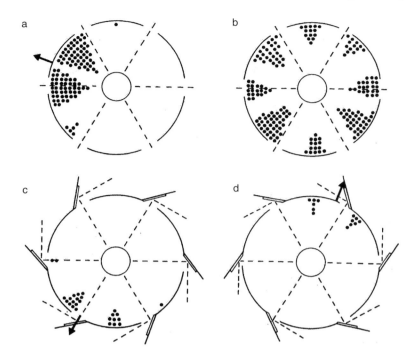

Figure 9.5   Orientation of spontaneous diurnal migratory activity in a caged Starling *Sturnus vulgaris* under various conditions of sun exposure. The bird was tested in a pavilion with six windows during the spring migration season: (a) Behaviour under clear skies; (b) Behaviour under total overcast, when the sun was not visible; (c) Behaviour when the image of the sun was deflected 90° counter-clockwise by means of mirrors; (d) Behaviour when the sun was deflected 90° clockwise by means of mirrors. Each dot represents 10 seconds of fluttering activity. Dotted lines show incidence of light from the sky; arrows denote mean direction of activity. Redrawn from Kramer (1951).

orient themselves incorrectly, interpreting the sun as if they were in the northern hemisphere (i.e. indicating south rather than north). Regular trans-equatorial migrants must presumably be able to make the necessary adjustment, but how they do so is still unclear (Schmidt-Koenig *et al.* 1991). If birds adjusted for time of day on the basis of their experience in the previous few days, their response would be appropriate for the latitude and time of year (which also influence sun position).

Skylight polarisation patterns

The ability of birds to detect sky polarisation patterns, which change with respect to the sun's position (being particularly striking around the time of sunset), has been demonstrated by experiment (Able 1993). At least seven species of normally nocturnal migrants have been shown to respond to manipulations of polarised light from the sky. The birds were tested outdoors in otherwise normal conditions in cages covered by sheet polaroids (Able 1982b, 1989, Moore & Phillips 1988, Phillips & Moore 1992, Helbig & Wiltschko 1989). In each case, the birds changed orientation as predicted by alterations in the alignment of the polaroids. However, the visual stimulus created by this procedure is quite unnatural, and birds sometimes oriented differently under artificial polarised light than under naturally polarised skylight (Helbig & Wiltschko 1989). Nevertheless, the experimental birds were clearly responding to polarised light as an orientation cue, rather than to other sunset features, as discussed in a later section.

The star compass

Providing enough of the night sky is visible, nocturnal migrants proved able to use the stars as a guide. When tested in orientation cages, they orientated correctly on clear starry nights, but became inactive or disoriented under overcast skies. They also became confused if star patterns were varied experimentally in a planetarium (Sauer & Sauer 1955, 1960, Hamilton 1962c, Emlen 1967a, 1967b). When Indigo Buntings *Passerina cyanea* were tested under a natural starry sky during autumn migration, they preferred southerly directions. They maintained this southerly preference under an artificial star pattern imitating the natural sky in a planetarium. But when the artificial star pattern was changed by 180°, the birds changed their directional preference to the north. Under a static night sky, no obvious migratory restlessness occurred. The development of a star compass evidently involved learning, with celestial rotation as a directional reference, and captive Indigo Buntings without early experience of the night sky failed to orientate correctly in a planetarium (Emlen 1967b, 1975). But detecting the rotation of the night sky probably takes considerable time – it could not be determined at a glance. Not surprisingly, therefore, the birds did not depend on the axis of rotation *per se*; rather, they quickly learnt star patterns that indicated where the axis lay, and thereafter relied on those patterns. As with the sun compass, however, the ability and tendency to acquire this knowledge was apparently innate. Similar results were later obtained with Garden Warblers *Sylvia borin* (Wiltschko *et al.* 1987).

Some birds developed this ability under unnatural star patterns, as shown for example by the selective blocking out of constellations in the artificial sky. The key factor was the rotation of the night sky about the Pole Star, and even an extremely simplified and reduced star pattern would suffice, so long as it rotated about a single conspicuous star. Thus, nestling Indigo Buntings raised under an artificial sky with the star Betelgeuse (in the constellation Orion) as the point of rotation treated Betelgeuse as the Pole Star when subsequently tested. The same was found in other migratory species (Emlen 1975).

The use of a star compass has now been demonstrated experimentally in at least six different bird species, but may be widespread in nocturnal migrants. If the birds use the rotating star pattern only to define the position of the poles, then no correction for time of day is necessary. They may, however, gain further information from star patterns. As birds proceed on their journeys, lasting up to several weeks, stars that were once visible disappear below the horizon behind them, while others appear above the horizon in front, another indication that birds are unlikely to rely throughout on particular star patterns. Moreover, many bird species, such as the Lesser Whitethroat *Sylvia curruca*, change direction during their migration from Europe to Africa, heading first southeast into the Middle East, and then south-southwest into Africa. When shown northerly skies in a planetarium during the autumn migration season, individuals of this species headed southeast, but when shown skies characteristic of more southerly latitudes, they headed south-west (Sauer 1957). This finding indicated that birds could respond appropriately to skies encountered at different points in their journey. Moreover, Pied Flycatchers *Ficedula hypoleuca* and Redstarts *Phoenicurus phoenicurus* compensated by changing direction after a simulated east–west displacement in a planetarium, indicating again an appropriate response to sky patterns (Rabøl 1994).

Integrated use of celestial cues

Nocturnal migrants often set off around dusk, when celestial cues related to the sun (such as sunset position, horizon glow, and skylight polarisation pattern) are clearly visible, and when at the same time the star pattern is gradually emerging. They could therefore make use of all these celestial cues within a relatively short period. The fact that migratory birds keep flying in the same direction over the transition from day to night or night to day (e.g. Myres 1964) implies that they can switch between sun and stars for navigation, or that they rely on some other cue, such as the earth's magnetic field, to maintain their course. In addition, some arctic species which prefer to migrate by night at lower latitudes necessarily migrate in daylight at high latitudes in summer. Moonlight can hinder the use of star patterns and produce the same disturbing effects as cloud (Sauer 1957). The moon itself seems to play no obvious role in bird orientation.

Time shifts

In using celestial cues, long east–west migrations present greater navigational problems than north–south flights because they involve time shifts, as the birds pass through successive time zones. If long-distance migrants using celestial cues to navigate did not allow for time shifts during the course of their east–west journeys, they would make ever greater directional errors, and would thereby veer progressively further off course. The problem created by time shift is greatest at the highest latitudes, where the longitude lines are closest together, requiring more rapid adjustment. Hence, a second presumed function of an internal clock is to measure the changes in timing of sunrise and sunset, as the bird flies long distances west or east. High-latitude east–west or west–east flights are not uncommon, being performed every year, for example, by the many waterfowl and seabirds that migrate along the northern coasts of Eurasia and North America to reach the Atlantic or Pacific Oceans on either side (Chapter 1).

Apart from measuring changes in the time of sunrise and sunset through an internal clock during westward or eastward travel, there is no global cue through which birds are definitely known to detect their longitudinal position. This is in contrast to latitude, which birds could determine at any time, either through the height of the celestial centre of rotation above the horizon or from the angle of magnetic inclination, both of which birds can apparently sense (Emlen 1975, Wiltschko & Wiltschko 1972). But whatever their sensory capacities, birds may be better able to fix their direction when stationary on the ground than when in migratory flight (Lack 1960a), and major changes in direction normally follow stopovers in the journeys. Perhaps birds need time to learn the movement patterns of the sun or stars or local magnetic conditions in a new area in order to be able to fix their position precisely.

The magnetic compass

The second major system of bird orientation makes use of the earth's magnetic field, which has both horizontal and vertical components. Imagine the earth as a hugely powerful magnet, whose north magnetic pole is situated fairly close to the geographic North Pole, and whose south magnetic pole is similarly close to the geographic South Pole.[1] Running through the atmosphere between the two magnetic poles are invisible longitudinal lines of magnetic force, which circle the globe rather like the segments of an orange (**Figure 9.6**). At the equator, the magnetic force lines run horizontal to the earth's surface, but toward higher latitudes they dip more and more strongly into the earth, until they tip vertically downward at the magnetic poles. The vertical component of the field thus varies in strength within each hemisphere according to latitude. Hence, for any creature that can measure the inclination of the force lines, the earth's magnetic field can give a cue to latitude and direction (toward the equator or pole) within each hemisphere. However, it cannot give a reliable cue to longitude, so it cannot provide

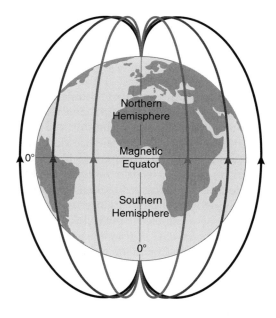

Figure 9.6   Diagram depicting force lines in the earth's magnetic field.

a firm basis for bi-coordinate navigation.[2] Unlike celestial cues, the magnetic field can give consistent information in all weather conditions, both day and night, and unlike the sun compass, it needs no correction for time of day.

As revealed by radar studies, nocturnal migrants can occasionally orientate correctly even under completely overcast conditions, as can caged birds with no view of the sky. Caged birds lost this ability when isolated from both the sky and the earth's magnetic field behind metal-reinforced walls. When the magnetic field experienced by caged Robins *Erithacus rubecula* was rotated using a powerful electromagnetic coil so that, for example, magnetic north was shifted to the east while the field's total intensity and inclination, as well as other potential directional cues, were kept unchanged, the birds altered their orientation accordingly (Wiltschko & Wiltschko 1968). This crucial experiment showed conclusively that birds can respond appropriately to the earth's magnetic field, even though they may often use celestial cues.

Following from their early work, Wiltschko & Wiltschko (1972) found that Robins accustomed to the normal magnetic field in Frankfurt (0.46 gauss) initially gave random bearings in fields 26% lower (0.35 gauss) or 48% higher (0.68 gauss). However, if the birds were kept in the altered field for a few days, they could then orient in fields as low as 0.16 gauss or as high as 0.81 gauss. A similar sensitivity to altered field strengths was found in Whitethroats *Sylvia communis* and Garden Warblers *S. borin* (Wiltschko & Merkel 1971, Wiltschko & Gwinner 1974). It was suggested that the birds' ability to adapt to a range of field strengths would permit them to adjust to a range of magnetic intensities that they would encounter at different latitudes during migration.

The role of the geomagnetic field as a reference for migratory direction was further shown in experiments with several species of young passerines (Wiltschko & Wiltschko 1995, 1999), and more recently in a non-passerine (Gudmundsson & Sandberg 2000). Passerines were hand-raised without access to celestial cues and, when tested during autumn migration, they headed in their population-specific migratory direction with the magnetic field as the only cue (Wiltschko & Wiltschko 2003). These findings indicated that the geomagnetic field alone was sufficient for establishing the migratory course, at least in temperate latitudes. Birds tested at higher latitudes, where the angle of magnetic inclination was steeper, needed to have observed celestial rotation to adopt the correct heading

---

[2]*Two magnetic parameters that vary in different directions across the earth's surface in a largely predictable way are the angle of inclination and the magnetic field strength, so in theory these parameters could be used for bi-coordinate navigation, at least in some regions. However, in other regions these two parameters coincide, so they do not form a grid system. Furthermore, in areas with geomagnetic anomalies, where the magnetic parameters can be largely distorted, and in areas close to the geomagnetic poles where the field lines are vertical, such a bi-coordinate map might be difficult to use. Further difficulty arises from the fact that geomagnetic parameters at a given location change through time, drifting significantly during the life of a long-lived bird. Even the positions of the two magnetic poles change somewhat over the centuries with respect to true geographical north and south. On a much longer timescale, the polarity of the geomagnetic field has changed several times in geological history (with north and south reversing), so polarity has not formed a consistent source of reference. The magnetic field is also prone to natural disturbances, such as solar flares.*

in relation to magnetic cues alone (Weindler *et al.* 1996). Overall, the use of a magnetic compass has now been demonstrated experimentally in about 20 bird species, and its use may be widespread, but mainly in association with other information.

Trans-equatorial migrants using magnetic cues alone would face the problem that the magnetic field is horizontal for some distance north and south of the equator, and hence ambiguous with respect to direction. In the equatorial region, therefore, birds may have to use celestial cues to maintain an appropriate heading. Between the equatorial and polar regions, however, the earth's magnetic field represents a reliable and omnipresent reference for orientation, both in terms of directional and latitudinal information. No wonder, then, that various kinds of migratory animals appear to make use of it. Although the method of its sensory perception is still unclear, birds are sensitive to both inclination and intensity, but apparently not to polarity (Wiltschko & Wiltschko 1972).

Response to specific areas

As mentioned above, Lesser Whitethroats *Sylvia curruca* showed different directional preferences under planetarium skies representing different points on their migration route (Sauer 1957). Other experiments using an artificial magnetic field implied that some birds use magnetic information to indicate regions where they must stop migrating, change direction or accumulate large fat reserves before crossing a barrier. Adult Tasmanian Silvereyes *Zosterops lateralis* were tested near the mid-point of their south–north migration in southeast Australia. Birds exposed in captivity to artificially-generated magnetic field values of inclination and intensity normally experienced near the start of their migration, oriented north-northeast. In contrast, birds exposed to magnetic field values that they would experience near the end of their migration ceased to show any significant directional preference; in this respect, they acted as though they had arrived in wintering areas (Fischer *et al.* 2003). These findings thus pointed to the involvement of the geomagnetic field in the migration of adult Silvereyes. In contrast, no effects of changing the artificial magnetic field were noted on inexperienced young birds caught prior to their first migration.

Young migratory European Pied Flycatchers *Ficedula hypoleuca* from western Europe showed a distinct change in compass heading when exposed in captivity to values of the magnetic field normally encountered in southern Europe, where the normal migratory route shifts from southwest to south. The altered magnetic field was followed by the shift in orientation only when applied at the appropriate time during the migratory period (Beck & Wiltschko 1988). In Pied Flycatchers, therefore, the magnetic conditions of the location where the change is to occur and the time programme evidently interact to produce an appropriate response at an appropriate latitude. As mentioned above, magnetic inclination gives a good indication of latitude, and could thus trigger change in direction or fattening (Wiltschko & Wiltschko 1995).

Some juvenile Thrush Nightingales *Luscinia luscinia*, caught in autumn in Sweden and exposed there to the geomagnetic conditions they would normally experience in northern Egypt, accumulated high fat levels appropriate to the subsequent desert crossing. They contrasted with control birds, exposed to local geomagnetic conditions, which accumulated much smaller fuel loads typical for south Sweden (Kullberg *et al.* 2003). However, birds trapped late in the onset period of autumn migration accumulated a high fuel load irrespective of magnetic treatment. It seemed that the relative importance of endogenous and environmental factors in individual birds was affected by time of season, as well as by geographical area (see also Chapter 12).

These various experiments indicate that inexperienced birds on their first migration can detect and make use of the geomagnetic field, at least to indicate when major changes are needed during their journeys. The implication is that such birds have an inborn response to external geographic cues (especially geomagnetic cues) that are characteristic of certain latitudes or regions, and that they can use particular conditions to trigger a halt to migration, or a change in direction or fattening regime (Beck & Wiltschko 1982, 1988, Fransson *et al.* 2001).

We thus have two potential mechanisms which could act in these ways. The first is an endogenous time programme which switches particular activities on and off at appropriate times in the migration cycle (Chapter 11). The second is a response to particular latitudes or more specific regions, at least partly through regional magnetic or night sky conditions, which can similarly trigger appropriate changes in migratory behaviour. How much these separate mechanisms act independently or in conjunction with one another is an open question, but different species would not necessarily be expected to respond to particular experiments in the same way.

The concentrations of ring recoveries of some bird populations in areas north of the Sahara Desert in autumn indicate that individuals from these populations favour these areas to fatten before their desert crossing (Chapter 5). If birds were to use magnetic information as a cue to where to fatten, they must be able to distinguish the small differences in magnetic field between such areas. A long-distance migrant of the New World, the Bobolink *Dolichonyx oryzivorus*, has been shown to detect changes of 200 nT (nanotesla) electrophysiologically (Semm & Beason 1990), and studies on White-crowned Sparrows *Zonotrichia leucophrys gambelii* indicate that the magneto-receptors are extremely sensitive to small changes (less than 3°) in the angle of geomagnetic inclination (Åkesson *et al.* 2001). Differences in the total intensity and inclination of the field between the centres of different major stopover areas for trans-Saharan migrants taking the eastern Mediterranean route into Africa (Libya–Egypt, Cyprus, and Israel–Jordan–Syria–Turkey) are within these limits, indicating that geomagnetic cues could be relevant for populations migrating to these population-specific fuelling areas (Fransson *et al.* 2005). Studies on other kinds of animals have reported sensitivities to the magnetic field as small as 120–50 nT, so once again birds seem not unusual in their magnetic sensitivity. If young birds are responding in the way that ringing and experimental findings suggest, they are not just using a clock-and-compass system on their first migration, but are also benefiting from an inborn response to magnetic cues, which leads them to converge on specific regions.

Overview of orientation cues

To summarise so far, experiments indicate that birds can use a number of different compasses for orientation during long-distance migrations, based on information from the sun and related pattern of skylight polarisation, from star patterns and from the earth's magnetic field. These different cues would normally give the same directional message. The use of the sun compass requires a time-compensation mechanism, through which the bird can allow for changes in the sun's position during the day, while the rotation centre of the night sky indicated by the stars gives the direction towards the geographical poles (Emlen 1975). Geomagnetic compass courses are given by the angle of inclination (varying from horizontal at the equator to vertical at the poles). The reliability of these compasses might vary between regions, seasons and local conditions, so that for instance the sun compass cannot be used if the sky is totally overcast, the star compass might not be visible during the round-the-clock daylight in high-latitude summers, and a magnetic compass based on the angle of inclination is unusable around the geomagnetic poles and the geomagnetic equator (Akesson *et al.* 2001). All the known compasses used by migratory birds seem to be modifiable by experience during early development, or later. Bi-coordinate navigation could be provided by any two non-parallel gradients, and in theory the different gradients could be provided by different types of cues, for example one coordinate being based on a celestial cue and another on a magnetic cue. Latitude can be fixed by both celestial and magnetic cues, but longitude seems much more difficult to determine. Displacement experiments with White-crowned Sparrows *Zonotrichia leucophrys* in arctic North America indicated that a combination of geomagnetic and celestial information might be used to define longitude, but the precise mechanism remains unclear (Åkesson *et al.* 2005).

Cue conflicts

Which of several potential compasses is most important to a bird at a given time has been investigated in so-called cue-conflict experiments, which involve presenting a bird with two (or more) orientation cues at once, manipulating one of them while leaving others unchanged, and monitoring the response of the bird (Able 1993). A typical experiment might involve placing an orientation cage surrounded by electric coils outdoors under a clear night sky. In this situation, the bird would have access to two known orientation cues: the stars and the magnetic field. The coils can be used to shift the direction of magnetic north so that magnetic compass directions differ from star-based ones. If, as compared to control birds tested in an unaltered magnetic field, the birds experiencing the cue-conflict changed direction in line with the magnetic field shift, one would conclude that in this situation magnetic information took precedence over stellar information.

Just such an experiment was conducted on three species of *Sylvia* warblers and European Robins *Erithaca rubecula* captured on migration (Wiltschko & Wiltschko 1975a, 1975b). When the directions of stellar and magnetic north were at variance, the birds seemed to orient preferentially with respect to magnetic cues. The warblers changed direction during the first test in the conflict situation, but the Robins did not shift until tested in cue-conflict for several consecutive nights (a finding later replicated on the same species elsewhere, Bingman 1987).

Cue-conflict experiments have proved especially useful in elucidating the cues used around sunset, a time when many birds set off on migratory flights. Orientation based on visual cues between the time of sunset and the appearance of the first stars could be based on the sun itself (e.g. the azimuth of sunset) or on patterns of polarised skylight which are particularly prominent at this time. Both could provide the same directional information, and both have the potential to indicate true compass directions. However, in cue-conflict experiments polarised light appears to be the predominant stimulus when placed in conflict with the sun's position or magnetic directions. When Yellow-rumped Warblers *Dendroica coronata* were exposed to rotated polarised light patterns for three nights, their subsequent orientation (apparently based on the sun, as indicated by mirror-shift tests) continued to exhibit the shifts induced by the polaroids (Phillips & Moore 1992). Further, when European Robins *Erithaca rubecula* and Blackcaps *Sylvia atricapilla* were tested under depolarisers in a situation lacking directional magnetic information, they were disoriented even though sunset position was clearly visible to them (Helbig 1991a). Such findings point to polarised light as being a crucial visual orientation cue at dusk.

The findings from several cue-conflict experiments are listed in **Table 9.2**. Few have compared the same combination of cues, and in only two species (Savannah Sparrow *Passerculus sandwichensis* and Robin *Erithaca rubecula*) have all known cues been examined. In addition, different techniques and equipment were used in the experiments, further complicating comparisons. However, insofar as it is possible to generalise from the results, in short-term decision-making, magnetic cues took precedence over stellar cues, visual information at sunset overrode both of those stimuli, and polarised light (rather than sunset) was the relevant orientation cue used at that time.

Some studies produced results contrary to these general patterns, and possibly for good reasons. First, species may differ in the weight they give to different orientation cues, according to the types of journey they undertake. Second, individuals may also weigh cues differently, depending on their early experience. Third, while some birds are able to detect and respond rapidly to changes in some types of cue, they may take longer to assess and adjust to changes in other cues (as in the Robins mentioned above). In most of the experiments in **Table 9.2**, birds were tested for only one night or less, perhaps long enough for them to adjust to changes in some types of cues but not others. Experiments have shown the importance of recalibrating compass systems when birds are exposed to conflicting information. In nature, birds would not normally find themselves in situations where their various compass mechanisms suddenly gave disparate signals. Usually they could safely pool information from as many sources as possible in making orientation decisions.

In the wild, birds may regularly calibrate one compass cue against another. Thus, while the rotation of the earth relative to the sky provides a stable reference for defining geographic north and south, changing geomagnetic declination renders the earth's magnetic field less reliable as a geographic reference. Accordingly, young birds were found able to use celestial information to calibrate a migratory orientation response to the earth's magnetic field (Weindler *et al.* 1996, Bingham *et al.* 2003). The combined experience of the night sky and the natural geomagnetic field seemed crucial for songbirds at high latitude to find the appropriate migration direction to a population-specific wintering area (Weindler *et al.* 1996). Experienced adult Savannah Sparrows *Passerculus sandwichensis* also used celestial cues to recalibrate their migratory orientation to an experimentally

Table 9.2 Cue-conflict experiments performed on migratory birds

Species	Cues available	Cue(s) varied	Apparent primary cue	Source
Magnetic versus stellar cues				
Indigo Bunting *Passerina cyanea*	Magnetic field, planetarium stars	Stars	Stars	Emlen (1967a, 1967b)
European Robin *Erithacus rubecula*	Magnetic field, stars	Magnetic field	Magnetic field	Wiltschko & Wiltschko (1975b), Bingman (1987)
Sylvia species	Magnetic field, stars	Magnetic field	Magnetic field	Wiltschko & Wiltschko (1975a)
Bobolink *Dolichonyx oryzivorus*	Magnetic field, planetarium stars	Stars	Magnetic field	Beason (1987)
Magnetic versus sunset cues				
Savannah Sparrow *Passerculus sandwichensis*	Sunset, magnetic field	Sunset by mirrors; sunset, magnetic field	Sunset	Moore (1982, 1985)
Savannah Sparrow *Passerculus sandwichensis*	Sunset, magnetic field, stars	Sunset by mirrors	Sunset	Moore (1985)
White-throated Sparrow *Zonotrichia albicollis*	Sunset, magnetic field	Sunset by clock shift	Sunset	Able & Cherry (1986)
Dunnock *Prunella modularis*	Sunset, magnetic field	Magnetic field	Magnetic field	Bingman & Wiltschko (1988)
European Robin *Erithacus rubecula*	Sunset, magnetic field	Sunset by mirrors	Sunset	Sandberg (1991)

Species				Reference
Northern Wheatear *Oenanthe oenanthe*	Sunset, magnetic field	Magnetic field	Sunset	Sandberg et al. (1991)
Yellow-faced Honeyeater *Lichenostomus chrysops*	Sun, magnetic field	Magnetic field	Magnetic field	Munro & Wiltschko (1993)
Polarised skylight versus sun and magnetic cues				
White-throated Sparrow *Zonotrichia albicollis*	Sunset, magnetic field	Polarised light by polaroids	Polarised light	Able (1982b, 1989)
White-throated Sparrow *Zonotrichia albicollis*	Polarised light, magnetic field	Polarised light by polaroids	Polarised light	Able (1989)
Yellow-rumped Warbler *Dendroica coronata*	Sunset, magnetic field	Polarised light by polaroids	Polarised light	Phillips & Moore (1992), Moore & Phillips (1988)
Northern Waterthrush *Seiurus noveboracensis* Kentucky Warbler *Oporornis formosus*	Sunrise, magnetic field	Polarised light by polaroids	Polarised light	Moore (1986)
Blackcap *Sylvia atricapilla*	Sunset, magnetic field	Polarised light by polaroids	Polarised light	Helbig & Wiltschko (1989)
Blackcap *Sylvia atricapilla* European Robin *Erithacus rubecula*	Sunset	Depolarised skylight	Polarised light	Helbig (1990, 1991a)

From Able (1993).

shifted magnetic field (Able & Able 1995). Subsequent experiments revealed that Savannah Sparrows used polarised light cues from the region of sky near the horizon to calibrate the magnetic compass at both sunrise and sunset (Muheim *et al.* 2006). Once a bird has established geographic north and south with respect to the local geomagnetic field, magnetic cues could come to assume a greater role in navigation. In another experiment, *Catharus* thrushes caught on migration were exposed to a deflected magnetic field during twilight, and then released and radio-tracked on their subsequent night flights (Cochran *et al.* 2004, Cochran & Wikelski 2007). Their tracks indicated that the thrushes recalibrated their magnetic compass in relation to twilight cues, and then relied on their (miscalibrated) magnetic compass for their nocturnal flight, apparently ignoring stellar cues. The experimental birds changed to normal orientation again on succeeding nights, apparently having recalibrated their magnetic compass (correctly) back to north. Daily recalibration of the magnetic compass could explain how birds cope with changes in magnetic declination during the route, as well as various local magnetic anomalies; it could also explain how birds operating with a magnetic inclination compass can cross the equator without becoming disoriented. The birds' view of the twilight sky near the horizon may also be decisive for the calibration rank between magnetic and celestial cues (Muheim *et al.* 2006).

In practice, birds probably use information from several compass mechanisms, with emphasis on whatever cues are most reliable in the conditions prevailing, switching from one type of cue to another during the course of a journey, depending on location, weather and light values (Wiltschko *et al.* 1998, Muheim *et al.* 2003). The present variety of orientation mechanisms found among birds may reflect some traits that birds have evolved from reptilian ancestors (for example, magnetic and sun-based orientation), whereas other traits (such as stellar orientation) are as yet known only from birds, and may be specific to them.

Other potential cues

From time to time, birds may use yet other directional cues, including auditory and olfactory ones. For example, infrasound emanating from wind against mountain ridges or from waves breaking on shorelines can travel long distances, and could give strong directional clues to birds able to detect them (Keeton 1980, Wallraff 2003). At present, however, olfactory orientation is a debated hypothesis for birds, but it is well tested and accepted for some other animals, at least over short distances. Although they may not normally use this method, homing pigeons have been shown to be capable of olfactory navigation, again over relatively short distances, apparently on the basis of odours transported through the atmosphere by wind (Papi 1989). It is unknown, however, whether large-scale smell-gradients are sufficiently well developed over the globe to provide a useful long-distance navigation mechanism for birds. As humans, we are poorly equipped to appreciate any such gradients that might occur, but even with our impoverished senses we can detect aromatic differences (due largely to the changing vegetation) as we journey from Mediterranean through temperate to boreal regions.

Birds often behave in a way that suggests they can predict impending weather, for example by not setting off on migration when a storm is brewing. One way in which they might do this is by monitoring change in barometric pressure, and experiments have suggested that pigeons have this ability. Such a pressure sense could assist migrants in maintaining their flight altitude and in moving from areas of high to low pressure or vice versa. By this means, in arid areas, birds could move towards areas where rain is falling, a frequent behaviour which has so far defied explanation (Chapter 16). However, air pressure also changes with altitude, and homing pigeons can apparently detect barometric pressure changes equivalent to a startling 10-m altitude or less (Keeton 1980).

Other, as yet unappreciated, environmental cues could provide goal-orienting birds with an analogue of map information. Various possible means of navigation have been suggested, including inertial navigation, and different uses of the magnetic field, sun and stars, and use of the Coriolis force. But these proposed mechanisms have found no support so far, and some would require levels of sensory perception well beyond those known in birds (for review see Berthold 1993). Whatever the basis, however, a compass alone cannot reveal to a bird its current position, nor in which direction it should fly.

Observational evidence on orientation

Observations made directly or with radar have shown that birds prefer to migrate under clear skies, when sun or stars are visible (Chapter 4). Nevertheless, those that do fly in overcast sometimes seem no less well orientated in that their flight relative to the ground and air mass appears 'as straight, level and fast as comparable birds flying on clear nights' (Able 1982a). Such behaviour might be expected if birds were migrating on magnetic cues. However, at other times nocturnal migrants seem to become disorientated when they enter low cloud, mist or rain and when neither sky nor ground is visible to them. When over land they normally settle, but when over the sea they mill around in various directions gradually drifting downwind (as revealed by radar). However, a brief break in the cloud to expose the sky is enough to enable them to get back on course. Such observations suggest that birds depend on celestial navigation and are not normally guided by cues from the earth's magnetic field alone. The additional fact that migrating birds caught in mist or rain are attracted to lighthouses and other illuminated structures further implies that nocturnal migrants respond to visual cues.

The perception of a horizon may also be important to migrating birds. Some radio-marked Whooper Swans *Cygnus cygnus* travelling between Iceland and the British Isles tended to keep going when over the sea, providing that the altitude above the horizon of either the sun or moon was higher than −4° (giving sufficient light to see), and also that the visibility was greater than about 2 km; otherwise these birds tended to sit on the water and wait until conditions improved (Pennycuick *et al.* 1999). When out of sight of land, they may have needed a visible horizon to navigate by, as expected if they used celestial cues.

That flying birds, both day and night, are influenced by the landscape below is well accepted, partly from the observation that migrants often follow coasts or other leading lines. Birds might maintain direction in the same way as humans, by projecting a compass direction onto a series of landmarks, and heading successively towards them. This would enable birds to compensate for drift by crosswinds, and so remain on course, at least over land. It has been suggested that birds crossing extensive tracts of water could compensate for the drifting effects of wind by using the wave pattern below them to maintain a constant heading, but drift is generally more marked in birds migrating over water than over land (Chapter 4). Use of visual cues also provides continuity between night and day and allows the recognition of familiar areas.

Pelagic birds

Pelagic seabirds travel over vast stretches of featureless ocean, yet they can return unerringly to their tiny nesting islands after long migrations or foraging trips. The use of satellite tracking has enabled the movements of individual birds to be followed. Albatrosses leave their breeding islands on flights that take them over distances of several thousand kilometres, foraging in seas all around the nesting colony (Weimerskirch *et al.* 1993, 1994). After wandering in all directions over days or weeks, they can then return on a straight line to the breeding island. Penguins are similarly efficient oceanic navigators, even though they remain on or below the water surface while at sea. Foraging King Penguins *Aptenodytes patagonicus* extend to the polar front, with individuals foraging up to 1500 km from their nesting places, and with total journey lengths up to 4000 km (Jouventin *et al.* 1994). Again, the inbound part of such trips usually has a straight course. Emperor Penguins *A. forsteri* can also walk in a straight line over featureless sea-ice for distances exceeding 100 km to their nesting areas (Ancel *et al.* 1992). These various findings imply that marine birds, on returning to their distant nesting places, know the direction they should go from any point in the surrounding ocean.

While we have no reason to suppose that pelagic birds navigate differently from landbirds, using celestial and geomagnetic cues, other mechanisms have been suggested based on olfactory cues (smell-gradients) or on 'route-based navigation'. Among petrels, in particular, the brain structure suggests a well-developed olfaction sense, and undoubtedly these birds use their keen sense of smell to find food sources and nest-sites; but whether they use olfaction for long-distance orientation remains an open question. For route-based navigation, an animal must encode its home location with respect to its current position. When moving, it must continuously update the home-pointing vector by subconsciously processing information collected en route about its changes of direction and location. Such an updating process, called path integration, works independently of the presence of landmarks, and could be useful in featureless seascapes. Although known in insects, the process has not been tested in birds. However, seabirds passively displaced from their nesting colonies can get back again fairly efficiently, as shown in experiments with Manx Shearwaters *Puffinus puffinus*, Laysan Albatrosses *Diomedea immutabilis* and others (Kenyon & Rice 1958, Matthews 1968), and more recently using satellite-based radio-tracking of Cory's Shearwaters *Calonectris diomedea* (Dall'Antonia *et al.* 1995). It seemed that these birds did not depend on path integration, but on some other kind of site-dependent mechanism.

Cory's Shearwaters carrying magnets on both wings and heads (to disrupt perception of the earth's magnetic field) were released 160 km and 900 km from the breeding colony (Massa *et al.* 1991). They homed with the same success as control birds released without magnets. The same held for nine Wandering Albatrosses *Diomedia exulans* that had magnets fixed to their heads, along with back-transmitters to reveal their movements (Bonadonna *et al.* 2005). On their normal foraging flights over several thousand kilometres, these birds showed no impairment in their ability to return to specific nest-sites compared with control birds, equipped only with transmitters. The two groups showed no differences in trip duration or length or in directness of the homeward flight, implying that these birds did not require magnetic cues to navigate back to their nesting colonies. The same held for White-chinned Petrels *Procellaria aequinoctialis* that were caught and displaced 300–360 km from the nesting colony (Benhamou *et al.* 2003). From these and other experiments, geomagnetic information does not seem to be crucial in such species, if alternative cues are available.

SOCIAL FACTORS

Migrating in flocks may improve direction finding, allowing for the emergence of either skilled or experienced leaders or of collective decisions leading to optimal 'mean directions'. This view is supported by field observations showing that the navigational accuracy in migrating flocks of birds often exceeds that expected from the navigational abilities of single individuals (Thorup & Rabøl 2001, Simons 2004), or that the directional scatter decreased with increase in group size (Rabøl & Noer 1973). It is also supported by an experimental comparison of the orientation of single homing pigeons compared with small flocks, the latter showing less scatter in directions and homing times (Tamm 1980). Moreover, in many bird species, young migrate from their natal areas at the same time as experienced adults and, in some species, notably cranes, swans and geese, the young travel in family parties with their parents. This raises the possibility that, in some species, young birds could learn migration routes from experienced individuals, and that such knowledge could be passed down the generations by cultural transmission. Individuals migrating in flocks often call continually during the journey, which may help to maintain flock cohesion, especially at night or in mist.

Research on some species has shown that young birds can sometimes be influenced in migratory behaviour by the example of others. For instance, 754 young White Storks *Ciconia ciconia* were transported from the Baltic region (where storks normally migrate southeast) and released at migration time in western Germany (from where storks normally migrate southwest) (Schüz 1950). In contrast to the experiment described earlier, in which the release of White Storks was made after the local birds had left, these releases were made while the local birds were still present. The resulting ring recoveries from the released birds showed a strong tendency toward the southwest, approximating the direction taken by the local storks, and contrary to the southeast direction prevalent in the homeland. Similar results were achieved in North America with 131 Canada Geese *Branta canadensis* and 213 ducks of various species (Williams & Kalmbach 1943). Inherent directional preferences were apparently overridden by social influences. This is in any case evident in the males of some duck species, which pair with females in wintering areas, and then accompany their partner to her natal area to breed (Chapter 17).

In addition, migration can even be induced in non-migratory stock by the example of other migratory individuals of the same species. Eggs obtained from English non-migratory Mallards *Anas platyrhynchos* were hatched in Finland and the Baltic region (Valikangas 1933, Pützig 1938). The resulting 116 young were allowed freedom with the local migratory Mallard. They gave 19 recoveries up to 2300 km away, all within the normal winter range of the host populations. In contrast, other young released in autumn after the local young had left remained during winter near the release points. Once again, innate behaviour was apparently overruled by social influences.

Different conclusions could be drawn from experiments on gulls. In some cross-fostering experiments, migratory Lesser Black-backed Gulls *Larus fuscus* were raised by resident Herring Gulls *L. argentatus*, and vice versa (Harris 1970). The young of both species became imprinted on their foster species. In autumn, however, the young Lesser Black-backed Gulls migrated as normal, even though they had associated with Herring Gulls until then, and later returned to pair with them. In contrast, the cross-fostered Herring Gulls were mostly sedentary as was usual for their species in this region, but some moved further south than expected (Harris 1970). The implication was that, in these species, in contrast to the waterfowl, genetic influences overrode social ones.

The knowledge that young geese, swans and cranes appear to find their specific wintering localities by accompanying their parents has been used to re-establish populations and migration routes in regions from which former populations had been eliminated (Ellis *et al.* 2003). Captive-bred young were taught to follow ultralight or microlight aircraft (in the same way that they would naturally follow their parents), and then at the appropriate time, they were led on a long 'migration' to an appropriate wintering area. For example, young Whooping Cranes *Grus americana* were taught to fly behind an ultralight plane, and released on former breeding areas in central Wisconsin. The plane was then used to guide these young from their release area 1800 km south to former wintering grounds in Florida that had not been used for many years. The cranes over-wintered successfully there. In the spring they returned unaided to their release area, and in autumn migrated again to their new wintering site. Apparently one journey was enough to fix a migration route upon these young birds. Further releases were made subsequently and, at the time of writing, the new population consists of 64 individuals, some of which have nested. In another experiment, Sandhill Cranes *Grus canadensis* did not necessarily fly the exact route in spring that they had used on their first autumn training flight. In general, the birds used the most direct route rather than repeat the circuitous path necessitated by motorised craft needing to refuel and avoid obstacles (Ellis *et al.* 2003). This implied that particular landscape details were not of primary importance in the directed migration of these birds.

From 15 similar experiments conducted during 1990–2001, which also included Canada Geese *Branta canadensis* and Trumpeter Swans *Cygnus buccinator*, most of the birds reached a new wintering area chosen for them and returned on their own to their starting area next spring (Sladen *et al.* 2002, Ellis *et al.* 2003). When Canada Geese were trained to follow an ultralight aircraft for southward migrations of 680 km or 1320 km, 81% returned in the next spring to their locality of training. Again, the birds evidently learnt the route from their first guided journey. These birds probably had an innate migratory direction, like other birds that have been studied, but social factors influenced the actual route and stopover sites used on migration and, in some species, even whether migration occurred.

While many bird species migrate in flocks, others migrate singly, so could gain little or nothing from other individuals. This applies to the cuckoos mentioned earlier, in which the young also depart 3–4 weeks later than adults, and find their own way to wintering areas, presumably relying entirely on inherent directional preferences and time programmes.

LOXODROMES AND ORTHODROMES

An unresolved question concerns the type of route that long-distance migrants take on their journeys. The most straightforward procedure would be to set off in the appropriate direction and maintain the same heading throughout the journey on a rhumbline (loxodrome) route. This type of journey has simple navigational needs, and if it ran directly north–south it would also be the shortest route between two points and would not involve a time-shift. If the journey had an easterly or westerly component, so that it involved crossing lines of longitude (as most routes do), a constant heading would still be the simplest but not the shortest route. The great circle (or orthodrome) route covers the shortest distance between two longitudinally separated points, but requires continual change in direction during the journey (**Figure 9.7**). Such a route is thus more demanding in its navigational needs. Great circle routes can be accomplished by aeroplanes, with sophisticated navigation equipment, but whether by birds remains an open question. Moreover, on any journey that involves longitudinal displacement, whether on loxodrome or orthodrome routes, the bird is also subject to time-shifts. Both the distance and the time-shift problems are greatest at high latitudes where the longitude lines are closest together.

The tracking of individual birds on their journeys has so far done little to resolve the question of whether long-distance migrants take constant direction (rhumbline) or great circle routes. Brent Geese *Branta bernicla* migrating between the Wadden Sea and the Taimyr Peninsula took a rhumbline route, with constant compass direction (**Figure 9.7**). However, it was not clear whether they took this route because they are not capable of navigating the shorter (overwater) route or in order to stay near the coast with its feeding areas. The shortest (great circle) route was about 4300 km, compared to the rhumbline of about 4700 km. Although the birds that were tracked kept closer to a rhumbline than a great circle, they also made continual minor deviations, bringing their average flight distance to 5000 km, at least 700 km (16%) further than the shortest possible route (Green *et al.* 2002). Similarly, Brent Geese travelling from Iceland to the Queen Elizabeth Islands in northeast Canada tended to migrate along fairly straight rhumbline routes to their breeding areas, as did Red Knots *Calidris canuta*. Again these routes took the birds mostly over land, where they could come to ground in inclement weather (Gudmundsson *et al.* 1991). Only when not influenced by topographic features, important feeding sites or weather patterns, would migrants be expected to follow either straightforward orthodrome or loxodrome routes, and this ideal may be quite rare. Much radio-tracking has involved soaring birds, most of which migrate along dog-leg routes to avoid long sea-crossings. In these and other birds, large-scale topography seems more important than distance in shaping migration routes.

Evidence that any birds take great circle routes (other than north–south) is as yet rather slender. Use of radar on the coast of northern Siberia revealed the occurrence of an east-northeast, post-breeding migration, indicating direct flights between Siberia and North America, 1800–3000 km across the pack-ice of the Arctic Ocean (Alerstam & Gudmundsson 1999). If the migrants gradually

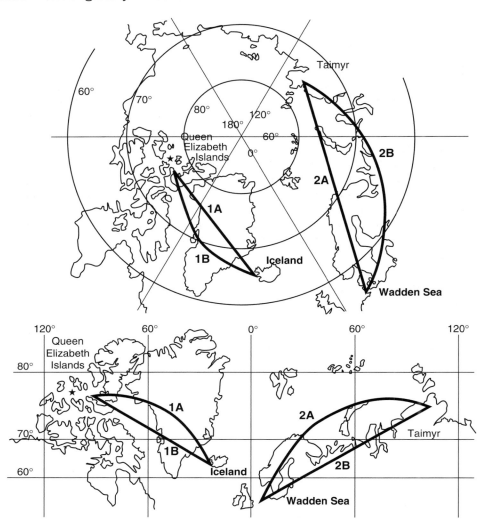

Figure 9.7   Great circle (orthodrome) and rhumbline (loxodrome) routes between points of departure and destination for migratory flights by certain high-arctic shorebirds and Brent Geese *Branta bernicla*, drawn on an azimuthal stereographic map projection (above) and on a Mercator map projection (below). Between Iceland and the Queen Elizabeth Islands, great circle (1A) distance and courses are 2535 km and 328°/265° (initial/final course). Rhumbline (1B) distance and course are 2665 km and 300° throughout. Between the Wadden Sea and Taimyr Peninsula, great circle (2A) distance and courses are 4234 km and 23°/110° (initial/final course). Rhumbline (2B) distance and course are 4634 km and 59° throughout. Spring flight routes by the high-arctic migrants are in agreement with rhumbline but not with great circle routes. The position of the magnetic North Pole is indicated by a star. For further details see **Box 9.1**. From Alerstam (1990b).

Box 9.1   The significance of map projections

Projecting the rounded surface of the earth onto a flat, two-dimensional map causes inevitable distortions which are important to bear in mind when charting and analysing migration routes, or when plotting the geographical ranges of birds.

The familiar Mercator projection (a cylindrical projection) is useful because it represents routes with constant geographical courses (rhumblines or loxodromes) as straight lines (**Figure 9.7**). This type of projection has been traditionally used for nautical charts. However, because the longitude lines are drawn parallel to one another rather than converging towards the poles, the scale varies with latitude, so that polar distances and areas are exaggerated in comparison with equatorial ones.

The gnomonic projection (a central azimuth projection) is useful because it depicts great circle routes as straight lines (**Figure 9.7**). Different tangent points must be selected on the earth's surface for gnomonic maps covering different parts of the globe. Central polar projections give a satisfactory correspondence with true global geography around the poles, and at the same time show great circles as straight lines. Gnomonic maps are not equidistant, and do not reflect the true area proportions.

In considering bird geographical ranges, maps of equal area projection are desirable, because on such a projection all areas are drawn to precisely the same scale, so that they appear on a flat map in the same relative proportions as they occur on the earth's spherical surface.

changed their orientation to the right during these flights, they would travel towards Alaska and neighbouring parts of Canada along the shortest possible great circle route to South America. The commonest species involved were the Pectoral Sandpiper *Calidris melanotos* and Grey Phalarope *Phalaropus fulicaria*, which winter on and near South American coastlines, respectively. This is one piece of evidence indicating that some long-distance migrants might travel along approximate great circle routes, but it will remain inconclusive until birds have been followed along more of the route. Ring recoveries from these or other candidate species are also insufficient to confirm travel by great circle routes.

Different compass systems allow different possibilities. The sun compass allows birds to identify the azimuth of the sun during the day in association with local time measured by their internal clock (Schmidt-Koenig *et al.* 1991). As long as the birds do not compensate for the change in local time when travelling across longitudes, a sun compass would direct birds along migration routes that are similar to great circle routes at high latitudes (Alerstam & Pettersson 1991, Åkesson *et al.* 2001). In contrast, if birds were to compensate for the longitudinal shift in time and reset their internal clocks regularly as they crossed longitudes, they would follow a constant geographic rhumbline route when using the sun compass (Alerstam & Pettersson 1991). Ground-based radio-tracking suggested that a *Catharus* thrush oriented at a constant angle in relation to the sunset azimuth during six successive nocturnal flights over a total distance of 1500 km, again suggesting a rhumbline route (Cochran *et al.* 2004).

The star compass provides birds with geographic north–south information based on the rotation centre of the starry sky (see above). A migration route following a star compass will therefore lead the birds along a constant geographic rhumbline route. On geomagnetic information, birds could use the inclination angle of the field lines to gain information about latitude as well as the direction towards the magnetic pole or equator. A migratory route based on geomagnetic orientation alone will lead birds along a constant magnetic course, a magnetic loxodrome (Alerstam & Gudmundsson 1999), or possibly along so-called magneto-clinic routes, assuming the birds follow an apparent angle of inclination, a constant angle between the direction of the field lines and the heading of the bird. Hence, among the known mechanisms, only a sun compass with no compensation for changes in local time could lead birds along a track similar to a great circle route. In addition, however, birds might achieve an approximate great circle route by using one or more appropriately positioned stopover sites, flying straight from one to another, but making a directional change at each one. Many landbirds take roundabout routes in order to avoid long water crossings or high mountains, or to make use of refuelling sites that are off the most direct route. The journey is thus divided into successive legs with different main orientations.

Despite the advantage of a great circle route, a straight rhumbline route based on a constant compass heading appears more likely in many birds, and is consistent with the routes frequently recorded by ringing and radio-tracking. It also fits the experimental evidence (based mainly on passerines) of a genetically fixed directional preference that steers inexperienced juveniles towards their wintering areas (although they may change directions at specific points on their journeys). Clearly, more research is needed on the precise routes taken by long-distance migrants before their navigation systems can be more thoroughly assessed.

Problems at high latitudes

Migration routes in the Arctic and Antarctic are of special interest. For it is at high latitudes that great circle routes bring the greatest proportional reductions in distance. It is also near the poles where the longitude lines are closer together that migrants become exposed to the most rapid time-shifts, and where birds are faced with difficulties in using any sort of recognised compass. Use of a sun compass brings problems of time compensation during rapid longitudinal (east–west) displacement, but is still usable. Use of a star compass is not possible in the polar summer because stars are not visible for months on end. And a magnetic compass is unreliable in a wide region around the north and south magnetic poles owing to the rapid changes in declination of the geomagnetic field in those regions, as well as the very steep angles of inclination. After considering the possibilities, Alerstam & Gudmundsson (1999) concluded that shorebirds tracked by radar off northern Siberia were using sun compass orientation as they travelled along apparent orthodrome-like routes. Orientation experiments at high latitudes revealed that birds made use of celestial cues (Ottosson *et al.* 1990, Åkesson *et al.* 2001), but could also detect magnetic information, despite the high angle of inclination (Muheim *et al.* 2003). White-crowned Sparrows *Zonotrichia leucophrys* under simulated overcast could select a magnetic compass course where the inclination angle was less than 3° from the vertical, implying a very accurate receptor. Interestingly, this species does not normally live at such high latitudes.

CONCLUDING REMARKS

In conclusion, the navigational tools available to migrating birds include: (a) a celestial compass based on sun, skylight polarisation and star patterns; (b) a magnetic compass based on the earth's magnetic field; (c) an internal clock, recording diurnal (circadian) and longer-term time changes; and (d) an inherited mean migratory direction and time programme, which together ensure that the bird flies in an appropriate direction for an appropriate time. Some, if not all, birds also have a map sense used for homing to a previously experienced place. Moreover, the fact that birds can re-find places they have already visited implies a good spatial memory.

Equipped with these navigation aids, a bird could use at least four different route-finding strategies:

1. In guiding or 'follow the leader', some birds might complete their migration by following others which know the way, thereby learning the route. Providing the leaders were experienced migrants, travel routes could be passed from old to young by cultural transmission. This strategy is used by swans, geese and cranes, in which young migrate with their parents (see above). Any birds using this method would benefit from a back-up mechanism (such as clock-and-compass) in case they were left to migrate on their own.

2. In clock-and-compass (vector) navigation, birds aim to head in a constant migratory direction (which may change one or more times during a journey) for an innately determined amount of time controlled by the internal clock. By this mechanism birds could reach previously unknown but appropriate wintering areas. Theoretically, birds of all ages could use this orientation strategy, which has been demonstrated experimentally in young passerines and others. On this mechanism alone, birds are unable to determine their position and are therefore unable to correct for wind drift, directional mistakes, over-flight, or experimental displacement.

3. In bi-coordinate navigation, birds can sense at least two global coordinates forming a reliable grid through which they can determine their geographical position. Bi-coordinate navigation could provide continual positional feedback, enabling birds to correct for drift or directional mistakes. Theoretically, birds of all ages could use this strategy, but experimental evidence from several species suggests that it is used primarily by experienced birds returning to a known area.

4. In piloting, a migration route is retraced by using a sequence of learnt landmarks. This method would require birds to build a landmark-based map during a previous migratory journey which is retraced during each subsequent migration. Such landmarks could be visual, auditory, magnetic or olfactory. This is not a method for inexperienced migrants on a first journey, but could help birds returning to a known area.

Ideally, birds should also have emergency strategies for when things go wrong – for example, when over the sea, they might reverse direction and retrace the route or, if the wind is too strong to fly back, continue on the same heading or downwind until land appears. Migrants commonly experience wind drift during flight, and often compensate for this displacement at the time or during a later flight, showing that some course recording and correction is at work, even if they do not 'know' the coordinates of their migratory goal (Thorup & Rabøl 2001; for possible mechanism see Matthews 1968). The fact that juvenile (as well as adult) birds can correct for drift may seem at odds with some of the displacement experiments discussed above, in which adults corrected for displacement and headed towards their former wintering areas, whereas juveniles continued on their usual direction and made no correction. However, drifted juveniles experience the drift and can see the ground below them, whereas artificially displaced birds travel in a vehicle or airplane and are unable to see the outside world. The method of lining up successive landscape features and flying from one to another is one obvious way of maintaining a consistent course in a crosswind.

Perhaps the main message to emerge from this review is that migratory birds have a number of orientation and navigation mechanisms available to them, and are not restricted to just one. They also have an inherent ability to learn to make use of the more important navigational cues, re-assessing them and if necessary cross-checking them at points along the route. Some of the pressing questions in bird navigation currently centre on the role of the external signals from map-related factors in young birds (to supplement the clock-and-compass mechanism), and on the extent to which migrants can travel on great circle as opposed to rhumbline routes.

SUMMARY

Birds have at least two orientation systems: inherited clock-and-compass orientation enables naïve young birds to find appropriate wintering areas, and bi-coordinate navigation enables experienced birds to home to a known area. Both groups use celestial (sun, skylight polarisation and stars) or magnetic information (inclination and intensity of force lines) as cues to compass direction. The sun compass must be time-compensated to allow for time of day, so an important component of this mechanism is the bird's internal time sense. Birds also use known landmarks to recognise previous travel routes and home areas, and perhaps also to help maintain a straight course. Use of all known compasses employed by migratory birds can be modified by experience.

At the very least, inexperienced juvenile birds have an innate ability to fly on a straight compass course for an appropriate time to reach their wintering areas; they can change directional preference between autumn and spring, and even at different stages in the same migratory journey. This is apparently achieved by having an inherent seasonal directional preference and an internal calendar clock (endogenous rhythm) to switch migratory behaviour on and off at appropriate times. Birds also know the positions of areas previously visited, and can return to their natal areas, or to their previous breeding or wintering sites at a later date, or if experimentally displaced. In addition to a simple 'clock-and-compass' system, therefore, birds also have a map sense which enables them to find familiar places again, often by direct flight. Exactly how they achieve this remains a mystery, despite our knowledge of the directional cues involved.

Many landbirds that cross deserts or oceans accumulate the necessary fuel reserves just before embarking on the crossing, which may occur part way through their journey. Others change direction at particular points in their journeys. They evidently 'know' that they have reached an appropriate place to accumulate large body reserves or change their heading. One possible mechanism involves the inherent time responses: after a given period after the start of migration, the bird changes its behaviour. Another involves response to the specific conditions found at particular sites, notably magnetic conditions.

Birds can also correct for drift by crosswinds, a facility which is often more apparent in adults than in juveniles, but is clearly present in both. In-flight compensation for drift could occur through reference to the ground below, but in strong winds birds are sometimes blown far off course, yet they are later able to re-orient and get back on course. This applies to juveniles making their first migration as well as to experienced adults, and implies the existence of some course recording and correction mechanism.

In some bird species, young migrate singly, and later in the year than adults, so could not learn specific migration routes from more experienced individuals. In other species, especially cranes, geese and swans, the young are influenced by the behaviour of their parents from whom they learn migration routes and stopping places. In such species, experiments have shown that social influences can sometimes override inherent migratory and directional tendencies.

Pallas's Warbler *Phylloscopus proregulus*, an increasing vagrant to northwest Europe from Asia

Chapter 10
Vagrancy

The erratic wanderings of migratory birds, resulting in their appearance in countries far removed from their accustomed haunts, and off the routes followed to reach them, are in many cases to be attributed to their failure, from some cause or other, to inherit unimpaired this all-important faculty of unconscious orientation. (William Eagle Clarke 1912.)

Every bird-watcher delights in seeing birds that normally live far away but periodically turn up as rarities. For some, seeing such species becomes an obsession, and bird vagrancy supports a time-consuming passion, which rewards the participants with gradually lengthening personal bird lists. The term vagrant is applied to any exotic visitor of a species which does not normally breed, overwinter or pass through the region concerned. Outside the bird-watching community, little attention has been paid to vagrancy, partly because records are collected unsystematically (depending largely on the numbers and distribution of skilled observers), and partly because vagrants have usually been assumed to be abnormal in some way, or chance victims of unusual weather. Some may be escaped cage birds, or free-living birds that have hitched a ride on a ship. But many vagrants belong to species that cannot be kept in captivity, and could not survive on a ship for days on end. Moreover, their occurrence often follows specific seasonal patterns, as they appear in particular weather conditions, in the same places and at about the same dates year after year, different species in mainly different places. Almost all vagrants are from migratory populations, and often in their first year of life.

Vagrancy may result from: (1) normal dispersal, but over unusually long distances; (2) population growth or expansion, with vagrants being noticed because populations are larger or breeding closer to the area concerned; (3) drift, in which migrants are blown off course by winds; (4) migration overshoots, in which individuals migrate further than usual, appearing well beyond their normal breeding or wintering range; (5) deviant directional tendencies in which individuals migrate at the right time but in the wrong direction, including (6) mirror-image migration, in which birds migrate at a mirror angle to their normal direction; and (7) reversed-direction migration, in which birds migrate in the opposite direction to usual for that time of year. Some of these mechanisms could involve natural phenomena, while others could imply some inherent defect in the normal migration control mechanism.

Many vagrants are encountered on isolated islands off the fringes of continents, some of which are visited by a remarkable diversity of birds. Such islands include Heligoland off Germany ('a spot which, from the ornithological point of view, is without rival in the world', Gätke 1895), the northern Isles of Scotland (notably Fair Isle, Clarke 1912, Dymond *et al.* 1989), the Scilly Isles off southwest England (Gantlet 1991), the Nova Scotia Islands (notably Sable and Seal Islands) off eastern Canada (McLaren 1981), the South Farallon Islands off California (De Sante & Ainley 1980) and the Aleutian Islands off Alaska (Gibson 1981, Obmascik 2004). On the most westerly of the Aleutian Islands (notably Attu), occasional westerly winds bring migrants from Siberia, and give North American birders abundant opportunities to view Eurasian birds on American soil, sometimes dozens at a time. One feature of all these specially situated places, however, is that they receive a great variety of birds not just from neighbouring land areas, as expected, but from up to several thousands of kilometres away. They have often provided the first and sometimes only records of certain species for their particular regions. Of all the species added to the British List during the twentieth century, about one in every five was first seen on Fair Isle, which occupies less than 0.01% of the total land area of the British Isles.

During the nineteenth and early twentieth centuries, virtually all recorded vagrants were shot and preserved as skins, some of which are still available for examination. The older records may well have been biased in favour of large and conspicuous species. Subsequently, binoculars, field guides and mist nets became available, many of the extreme rarities were trapped, examined and photographed in the hand, and nowadays field photographs and written descriptions by skilled observers have often proved sufficient to identify rare species unequivocally. The recent development of 'digiscoping' (digital photography through a telescope) has further helped, as has the increasing ability to record calls and produce sonograms for subsequent inspection. The use of the web and other means to advertise fresh sightings has also increased the numbers of observers able to see and photograph individual rarities. Little wonder that the annual totals of recorded vagrants have increased over the years, and that some species once regarded as rare vagrants are now regarded as regular passage migrants (**Figure 10.1, Table 10.1**).

But what are the chances of any rare birds that turn up now being detected by bird-watchers? This question was addressed by Fraser (1997) using data from some of the best-watched sites around Britain. At such localities, about 40% of all passerines and waders were seen on one day only, around 15% on two days, and a decreasing proportion over longer periods (**Figure 10.2**). From these data, it was calculated that, even on a site as well watched as Fair Isle, with bird-watchers continually present at several per square kilometre through the migration seasons, at least 11% of all vagrants that settled on the island were probably missed. This figure rose to more than 40% for less favoured, but still well-watched coastal sites.

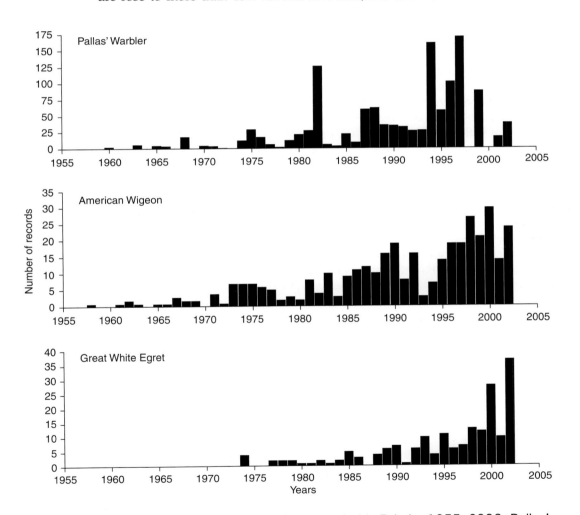

Figure 10.1   Numbers of some vagrant species recorded in Britain, 1955–2002: Pallas's Warbler *Phylloscopus proregulus* from Fraser *et al.* (1999) and subsequent reports; American Wigeon *Anas americana* from Fraser & Rogers (2005); Great White Egret *Egretta alba* from Rogers and the Rarities Committee (2003).

Table 10.1 Vagrants that greatly increased or decreased in their occurrence in Britain during the twentieth century in association with changes in population sizes or distributions

Increasing populations and increasing frequency of occurrence	Squacco Heron *Ardeola ralloides*, Cattle Egret *Bubulcus ibis*, Black Stork *Ciconia nigra*, Snow Goose *Chen caerulescens*, Red-breasted Goose *Branta ruficollis*, Ring-necked Duck *Aythya collaris*, Black Kite *Milvus migrans*, Common Crane *Grus grus*, Terek Sandpiper *Tringa cinerea*, Forster's Tern *Sterna forsteri*, Ring-billed Gull *Larus delawarensis*, Red-rumped Swallow *Hirundo daurica*, Greenish Warbler *Phylloscopus trochiloides*, Melodious Warbler *Hippolais polyglotta*, Eurasian Penduline Tit *Remiz pendulinus*, European Serin *Serinus serinus*, Common Rosefinch *Carpodacus erythrinus*, Rustic Bunting *Emberiza rustica*, Thrush Nightingale *Luscinia luscinia*
Decreasing populations and decreasing frequency of occurrence	American Bittern *Botaurus lentiginosus*, Little Crake *Porzana parva*, Little Bustard *Tetrax tetrax*, Pallas's Sandgrouse *Syrrhaptes paradoxus*

Although islands and coastal localities are most likely to attract incoming migrants, in most inland areas, where bird-watching is more desultory and sporadic, the chance of detecting any rare passerines that turn up is presumably extremely low.

In addition, more than 40% of all rarities recorded in Britain in recent decades were seen at weekends, significantly more than the 28% expected if coverage was uniform through the week (Fraser 1997). Weekend bias was hardly noticeable at the best-watched sites, such as Fair Isle and Scilly Isles, and became much greater at other less-watched sites. It did not affect all birds in the same way, being much more marked among ducks than among passerines and waders. This was probably because finding rare ducks frequently involves sifting through large flocks of commoner species for the elusive rarity, a practice most commensurate with leisured weekend birding. As a general conclusion, we can assume that, even in a country with as many bird-watchers as Britain, only tiny (and unknown) proportions of vagrants that turn up away from migration hotspots are likely to be detected, with some species being more often missed than others.

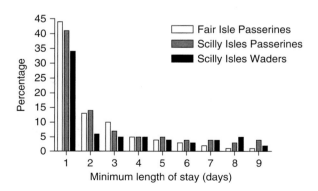

Figure 10.2   Apparent length of stay of individual vagrants at two localities. Based on data from passerines on Fair Isle and from passerines and shorebirds on the Scilly Isles, 1958–1993. From Fraser (1997).

Another point is obvious everywhere: the longer a place has been studied, the greater the number of species recorded there. New species are continually added to local lists. Even in an area as well watched as Britain, more than 70 newly recorded species were added in the 25 years after 1980, an average of nearly three per year (Pitches & Cleeves 2005). The rate of addition would be expected to slow over time, as the cumulative total begins to reach a plateau, but the terminal stages of this process clearly take a long time – more than a century. And all the time, bird distributions are changing, bringing additional species within range. In the first 17 years that the bird ringing station had operated at Eilat in Israel 139 354 individuals of 268 species had been trapped. The cumulative species total increased at a progressively slowing rate over the years, but even in the later years some 2–3 new species were added annually (Yosef & Tryjanowski 2002a).

For obvious reasons, vagrancy is much harder to study in pelagic seabirds than in landbirds, because it depends largely on birds being sighted from land or washed up dead on shorelines (to which they may have been transported long distances by currents). Away from their nesting places, pelagic birds sometimes fly over land, either deliberately or accidentally during storms. This habit may explain some occurrences in the 'wrong' ocean. Thus, did the Mottled Petrel *Pterodroma inexpectata* from the Pacific reach New York State by crossing the Isthmus of Panama to get into the Atlantic Ocean, or did the Great Shearwater *Puffinus gravis* that reached California cross from the Atlantic in the opposite direction, or simply start its northward migration from the wrong sector of the Southern Ocean? Other examples of long-distance vagrancy in seabirds include two Short-tailed Shearwaters *Puffinus tenuirostris* and two Laysan Albatrosses *Diomedia immutabilis* seen in the Indian Ocean rather than the Pacific, and Jouanin's Petrel *Bulweria fallax* seen in the Pacific (near the Hawaiian Islands) rather than the Indian Ocean (Bourne 1967, Warham 1996). As oceanic bird distributions are still poorly known, however, some records may relate to individuals from small undetected nesting colonies well away from the known range, rather than genuine vagrants.

PROBABILITY OF ARRIVAL

At the time of writing, more than half of the 560 or so species on the British list are classed as vagrants. Some such species occur in numbers every year, while others are seen less often, in extreme cases perhaps only once in several decades. The same has been noted in every well-watched region of the world, leaving no doubt that many species continually reach new areas. Even a land mass as isolated from the rest of the world as New Zealand receives a continual supply of avian vagrants, about 29% of all species recorded there being classed as 'irregular visitors' (Bell 1991). Small oceanic islands similarly receive a steady trickle of exotic birds from distant lands. Almost all these birds turn up in autumn or spring, during the normal migration seasons, and derive from known migratory populations.

The importance of distance from a source area in influencing the frequency of arrivals by non-native species is shown by three types of observation. First, as expected, islands close to continents (such as Heligoland off Germany) receive many more vagrant birds than remote islands (such as Ascension in mid-Atlantic). Second, on continents, the numbers of vagrant species which appear in particular localities each year, and their proportion of the overall species numbers, decline

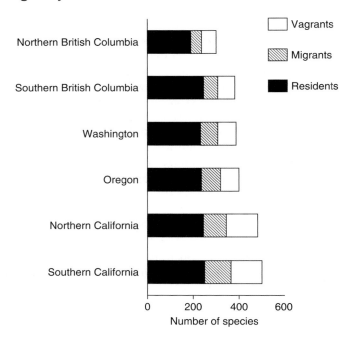

Figure 10.3 Relative contributions of resident, migrant and vagrant species to the regional bird lists of western North America. The percentage of vagrants in the total avifauna of southern California is 27%, northern California 29%, Oregon 20%, Washington 21%, southern British Columbia 20%, and northern British Columbia 9% ($r = -0.84$, testing percentage against latitude, $P < 0.05$). Redrawn from Stevens (1992).

with increasing latitude, as shown for western North America in **Figure 10.3**. This matches the latitudinal decline in overall species numbers. Third, the chance of any one species appearing in a given area declines, on average, with increasing distance from its regular range, as shown for various warblers in California in **Figure 10.4**.

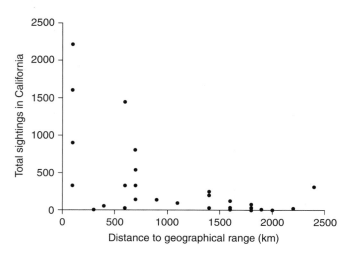

Figure 10.4 Total numbers of sightings of different species of eastern North American warblers in California in relation to distance between California and the western edge of the breeding range ($r = -0.55$, $P < 0.01$). Redrawn from Stevens (1992).

These birds migrate between eastern North America and Central or South America, but species that breed nearest to California are those most often seen there. Other groups of birds, which are more or less mobile than warblers, show different levels of vagrancy, but again distance from regular range has a major influence, as does prevailing wind direction (De Sante 1983a). Another factor of likely importance to occurrence is body size, if only because larger birds can fast for longer than small ones, and are thus more likely to survive a difficult journey. Relative to their population sizes, large vagrants (such as ducks) seem to turn up more often than smaller ones (such as warblers), but I know of no formal analysis of this aspect. As explained later, however, there are notable exceptions to all these general trends, and almost certainly, the majority of long-distance overwater vagrants end up as fish-food.

LONG-DISTANCE DISPERSAL

Ring recoveries of dispersed birds tend to occur in all directions from the natal area, but fall off exponentially with increasing distance (Chapter 17). On this basis, while most birds disperse relatively short distances, occasional individuals are likely to disperse much further than others, beyond the regular range. Such dispersing individuals would be expected to occur in any direction from the regular range, and not just in the zone between the breeding and wintering range. They would also be expected to occur mainly in the post-breeding period, before autumn migration. Records fitting such a pattern have been obtained for the Scissor-tailed Flycatcher *Muscivora forficatus* which breeds in the southern prairie region of North America and winters on southern Caribbean islands. Individuals of this distinctive species have been seen in almost every North American State and in Canada from British Columbia to Nova Scotia, mainly in late summer after breeding but also in spring. These birds evidently move long distances in almost any direction from their breeding areas. In the British Isles, dispersal vagrants are most likely to derive from species whose nearest breeding areas lie in western continental Europe, such as Tawny Pipit *Anthus campestris*, Melodious Warbler *Hippolais polyglotta*, Woodchat Shrike *Lanius senator* and Ortolan Bunting *Emberiza hortulana*. Individuals of these species occur mainly in late summer, but also at other times (see **Figure 10.5** for Ortolan Bunting).

Some waterbirds that breed early in the year in subtropical and Mediterranean latitudes disperse mainly northwards after breeding, sometimes for hundreds of kilometres, as many southern wetlands dry out. Several heron and ibis species participate in these movements, including Great White Herons *Ardea alba*, Squacco Herons *Ardeola ralloides* and Black-crowned Night Herons *Nyctocorax nycticorax*, as also do Brown Pelicans *Pelecanus occidentalis*, Little Blue Herons *Egretta caerulea*, Snowy Egrets *Egretta thula*, Black-necked Stilts *Himantopus mexicanus* and Bald Eagles *Haliaeetus leucocephalus* in North America (Townsend 1931). This northward dispersal seems more marked in North America than in Europe, possibly because of the greater areas of wetland habitat remaining in southern parts of the continent, and their larger populations of waterbirds. Some seabirds also disperse northwards after breeding, resulting for example in the late summer appearance of occasional Blue-footed Boobies *Sula nebouxii* and Magnificent Frigate Birds *Fregata magnificens* at the Salton Sea in southern California, along

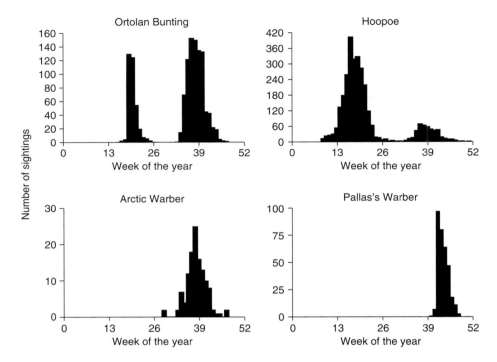

Figure 10.5   Seasonal appearance of four species of irregular visitors to Britain, as determined from the number of records in different weeks of the year. All four are more or less restricted to migration seasons, and the two warblers only in autumn. From records published in *British Birds*.

with more frequent summer–autumn dispersive visitors, such as Brown Pelicans *Pelecanus occidentalis*, Laughing Gulls *Larus atricilla*, Wood Storks *Mycteria americana*, White Ibises *Eudocimus albus*, Roseate Spoonbills *Ajaia ajaia* and others.

EFFECT OF POPULATION SIZE

Population size is likely to influence the number of individuals that appear as vagrants outside their regular range. For example, several species which have increased in eastern North America in recent decades have appeared with increasing frequency as vagrants in western Europe. Examples include the Red-eyed Vireo *Vireo olivaceus* and Ring-billed Gull *Larus delawarensis*.

Comparing species, the size of the source population accounted for nearly 60% of the variance in the occurrence of vagrant warblers and vireos in California (DeBenedictis 1971), and for about 30% of the variance in occurrence of various birds in the South Farallon Islands (De Sante 1983a). Similarly, changes in the numbers of vagrant warblers in California over the years have been attributed to known changes in regional population trends (Patten & Marantz 1996). In particular, Bay-breasted Warblers *Dendoica castanea* and Cape May Warblers *Dendoica tigrina* appeared in greatest numbers in California in years when their breeding populations further north were swollen in association with outbreaks of

Spruce Budworm *Choristoneura fumiferana* which forms their food (Robertson 1980). In addition, the relationship between annual variations of vagrant numbers and population size was studied in the Yellow-headed Blackbird *Xanthocephalus xanthocephalus* which appears on the Eastern Seaboard of North America every autumn from breeding areas far to the west. The numbers in different autumns in Massachusetts were correlated with annual reproductive output at the northern and eastern periphery of the breeding range (Veit 1997). The greater the estimated post-breeding population, the greater the numbers of vagrants reported. Such annual variations had previously been attributed to weather and other conditions at the time. But if the tendency of individuals to disperse in given directions and distances is constant from year to year, more individuals would be expected to turn up at distant localities in years when overall numbers are high.

Range changes are also likely to influence the frequency with which vagrants occur in particular areas. As several species spread westward across Eurasia during the twentieth century, they appeared with increasing frequency in Britain. Examples included the Red-flanked Bluetail *Tarsiger cyanurus*, River Warbler *Locustella fluviatilis*, Greenish Warbler *Phylloscopus trochiloides*, Penduline Tit *Remiz pendulinus*, Common Rosefinch *Carpodacus erythrinus*, Rustic Bunting *Emberiza rustica* and Little Bunting *E. pusilla*. Likewise, the first records of the Shiny Cowbird *Molothus bonariensis* in Florida followed the numerical increase and expansion of the species northwards through the West Indies (Post *et al.* 1993).

It is of course hard to separate the effects of increased population size over periods of decades from increased numbers of observers, but records have increased much more rapidly for some species than for others, while records of other species have fallen. Examples of the latter in Britain include Little Bustard *Tetrax tetrax* and Pallas's Sandgrouse *Syrrhaptes paradoxus*, whose numbers are known to have declined in their regular range further east (**Table 10.1**). Moreover, sustained observations for more than a hundred years at such well-watched places as Fair Isle surely indicate that some of the increases are real and not just a result of greater observer coverage and awareness.

DRIFT BY WIND

The influence of wind on the occurrence of birds outside their regular range is evident to any bird-watcher. Drift by crosswinds is one of the commonest ways in which migrants turn up as vagrants in places off their normal routes, and often several species from the same region appear simultaneously. Almost every autumn, on days with easterly winds, migrants that would normally pass down the western seaboard of continental Europe turn up on the east side of Britain (Lack 1960a , Evans 1968), and in periods with prolonged easterly winds, migrants that breed as far away as Siberia may appear. In contrast, at times of prolonged westerly winds, vagrants from North America turn up.

In clear weather, drift may not even be noticed, but if the birds meet mist or rain, 'falls' of migrants occur, leading to concentrations in particular localities, especially sea-coasts. In one incident on 11 October 1982, an estimated 15000 Goldcrests *Regulus regulus*, 4000 Robins *Erithacus rubecula* and other birds settled on the Isle of May off eastern Scotland (Zonfrillo 1983). In another incident in

September 1965, an estimated 15000 Common Redstarts *Phoenicurus phoenicurus*, 8000 Northern Wheatears *Oenanthe oenanthe*, 4000 Pied Flycatchers *Ficedula hypoleuca*, 3000 Garden Warblers *Sylvia borin* and many others suddenly appeared on one 3.2 km stretch of coast in eastern England (Davis 1966). Over a somewhat wider area, at least half a million birds of 78 species were estimated. Many of these birds seemed exhausted, and others were washed up dead on beaches. When the cloud broke and the wind veered, the majority of survivors disappeared, presumably having re-oriented and continued their journeys. Some individuals of these species normally pass over southeast England in migrating from Scandinavia towards Iberia, but in easterly winds the numbers are greatly swollen, and in rain large numbers alight and wait for conditions to improve. This type of weather-induced displacement and fall is likely to affect adults as well as immatures, but not necessarily in similar proportions.

The same occurs under easterly winds in spring, when species that do not breed in Britain, such as Red-spotted Bluethroats *Luscinia svecica* and Icterine Warblers *Hippolais icterina*, but which are heading for Scandinavia, may be deflected across the North Sea. Again, large numbers of birds may sometimes be involved, including rarer species with more easterly distributions, such as Red-throated Pipit *Anthus cervinus* and Rustic Bunting *Emberiza rustica*. At least one species, the Red-footed Falcon *Falco vespertinus*, is much commoner as a drift migrant in Britain in spring than in autumn, probably because it is a loop migrant whose spring migration route through Europe lies far to the west of the southward autumn route.

Arrivals of vagrants from more distant areas are likewise correlated with particular weather systems (Elkins 2005). For example, the appearance of Siberian birds in western Europe is associated with the occurrence of prolonged fine anticyclonic weather with easterly winds over the western half of Eurasia in September. The numbers of Siberian species seen in Britain thus vary greatly from year to year, with unusually large numbers in the autumns of 1975, 1982, 1992 and 2003 (Baker 1977, Howey & Bell 1985, Elkins 2005). In contrast, the appearance of North American birds in western Europe follows prolonged westerly gales at mid-latitudes. Westerly jet stream winds over the Atlantic can reach speeds of 250 km per hour, potentially bringing migrants over 5000 km from Canada to Britain in less than 24 hours. Almost all the North American vagrants that have appeared naturally in Europe are long-distance migrants which carry substantial body fat, as are the European species that have appeared in North America.

Again, the numbers of transatlantic migrants recorded in Britain vary greatly from year to year, depending on weather conditions, with unusually large numbers recorded in 1976, 1982, 1985 and 1995. They include passerines, shorebirds and waterfowl, as well as gulls and terns, with more than 60 species in total having been recorded (**Table 10.2**). The most frequent passerines include Gray-cheeked Thrush *Catharus minimus*, Blackpoll Warbler *Dendroica striata*, Red-eyed Vireo *Vireo olivaceus* and Rose-breasted Grosbeak *Pheucticus ludovicianus*, and the most frequent shorebirds are the Pectoral Sandpiper *Calidris melanotus* and White-rumped Sandpiper *C. fuscicollis*. Most of these species have a strong west–east component in the initial part of their autumn migration. Some of the North American wader species that occasionally cross the Atlantic (such as Pectoral Sandpiper *Calidris melanotos* and Baird's Sandpiper *Calidris bairdii*) breed in the

Table 10.2 The main seasons and proposed means of occurrence of some vagrants to the British Isles. Some species occur in more than one category

Category	Species
Transatlantic autumn vagrants	Green-winged Teal *Anas crecca carolinensis*, Ring-necked Duck *Aythya collaris*, Sora Rail *Porzana carolina*, American Golden Plover *Pluvialis dominica*, Semi-palmated Sandpiper *Calidris pusilla*, Least Sandpiper *Calidris minutilla*, White-rumped Sandpiper *Calidris fuscicollis*, Baird's Sandpiper *Calidris bairdii*, Pectoral Sandpiper *Calidris melanotos*, Lesser Yellowlegs *Tringa flavipes*, Solitary Sandpiper *Tringa solitaria*, Spotted Sandpiper *Actitis macularia*, Wilson's Phalarope *Phalaropus tricolor*, American Robin *Turdus migratorius*, Red-eyed Vireo *Vireo olivaceus*, Black-and-white Warbler *Mniotilta varia*, Northern Parula *Parula americana*, Blackpoll Warbler *Dendroica striata*, Northern Waterthrush *Seiurus noveboracensis*, Rose-breasted Grosbeak *Pheucticus ludovicianus*, Bobolink *Dolichonyx oryzivorus*, Baltimore Oriole *Icterus galbula*
Possible autumn mirror-image migrants	Greenish Warbler *P. tochiloides* (breeds Finland-Siberia), Arctic Warbler *Phylloscopus borealis* (breeds northern Scandinavia eastwards), Richard's Pipit *Anthus novaeseelandiae*, Pechora Pipit *Anthus gustavi*, Rustic Bunting *Emberiza rustica* (breeds Scandinavia eastwards), Little Bunting *Emberiza pusilla* (breeds Scandinavia eastwards), Yellow-breasted Bunting *E. aureola* (breeds Finland eastwards)
Possible reverse-direction autumn migrants	Grey-tailed Tattler *Heteroscelus brevipes* (breeds northeast Siberia), Pacific Golden Plover *Pluvialis fulva* (breeds eastern Siberia and western Alaska), Red-necked Stint *Calidris ruficollis* (breeds Siberia), Rose-coloured Starling *Sturnus roseus*[a] (breeds Turkey-southern Asia), White's Thrush *Zoothera dauma* (breeds Siberia), Barred Warbler *Sylvia nisoria* (breeds mid-Europe eastward), Pallas' Leaf Warbler *Phylloscopus proregulus* (breeds Siberia), Yellow-browed Warbler *Phylloscopus inornatus* (breeds Siberia), Dusky Warbler *Phylloscopus fuscatus* (breeds Siberia), Lanceolated Warbler *Locustella lanceolata* (breeds Siberia), River Warbler *Locustella fluviatilis* (breeds central Europe into Siberia), Paddyfield Warbler *Acrocephalus agricola* (breeds central Asia), Tennessee Warbler *Vermivora peregrina* (breeds eastern North America), Red-breasted Flycatcher *Ficedula parva* (breeds southeast Europe eastward), Yellow-browed Bunting *Emberiza chrysophrys* (breeds Norway eastwards)
Possible transatlantic spring overshoots	American Coot *Fulica americana*, Cape May Warbler *Dendroica tigrina*, Brown-headed Cowbird *Molothrus ater*, Cedar Waxwing *Bombycilla cedrorum*, American Robin *Turdus migratorius*, Yellow-headed Blackbird *Xanthocephalus xanthocephalus*
Possible long-distance Eurasian spring overshoots	Collared Pratincole *Glareola pratincola*, Great Spotted Cuckoo *Clamator glandarius*, Marmora's Warbler *Sylvia sarda*, Black-eared Wheatear *Oenanthe hispanica*, Black-headed Bunting *Emberiza melanocephala*
Other long-distance autumn vagrants from Siberia	Olive-backed Pipit *Anthus hodgsoni*, Siberian Stonechat *Saxicola toquata maura*, Siberian Thrush *Zoothera sibirica*, Black-throated Thrush *Turdus ruficollis atrogularis*, Pallas's Grasshopper Warbler *Locustella certhiola*, Lanceolated Warbler *Locustella lanceolata*, Yellow-browed Warbler *P. inornatus*, Raddes Warbler *Phylloscopus schwarzi*

[a]Autumn records only; also occurs in spring.

northwestern sector of North America and first make an easterly flight across that continent, so it is easy to see how they might cross the Atlantic by continuing in the same direction, rather than veering south or southwest for South America. Although most North American vagrants to Europe turn up in autumn, others appear in spring, raising the question whether they may have overwintered in the Old World (Nisbet 1959a).

Far more landbirds seem to cross the Atlantic at mid-latitudes from west to east than from east to west. There are probably two reasons for this. First, the prevailing winds at these latitudes blow from west to east, making it easier for North American birds to be carried to Europe than for European birds to reach North America. Second, many thousands of landbird migrants fly south or south-southeast over the western Atlantic each autumn, as they travel from northeastern North America to Caribbean Islands or to northern South America. Some reach more than 2000 km out from the North American coastline, as shown by ship-born radar, and are thus vulnerable to westerly winds and poor visibility (McClintock *et al.* 1978). However, no equivalent overwater movement occurs off the western seaboard of Europe, because most birds take the overland route through France and Spain to West Africa. Even the Canary Islands, which lie only 50 km off the West African coastline, receive extremely few individuals from the many millions of migrants that pass southwest each autumn through northwest Africa.

Widespread mist over the sea can also cause birds to fly in the wrong direction (Alerstam 1990b). Once within the mist, migrants can become disoriented, and fly in various directions, as shown by radar. If the mist is sufficiently widespread, birds could end up tens or even hundreds of kilometres from their normal routes, coming down and settling on the first land they encounter, to be recorded as vagrants.

It is not just landbirds that are susceptible to the effects of wind over water. Strong onshore gales offer the best prospects for bringing normally pelagic seabirds within view of coastal headlands. Counts exceeding 10 000 birds per hour have been achieved at some sites, including rarities. In addition, small seabirds are also frequently blown far off course, sometimes for hundreds of kilometres, while some normally pelagic species, such as Sabine's Gulls *Larus sabini* and Grey Phalaropes *Phalaropus fulicarius*, sometimes appear at inland lakes (Elkins 2005). Other small seabird species are occasionally found as 'wrecks' of dead and exhausted birds over land, the Little Auk *Alle alle* being a frequent victim in Europe (Murphy 1936, Bailey & Davenport 1972, Underwood & Stowe 1984; review Newton 1998a). Juveniles often predominate, and many appear in poor condition, but it is usually uncertain whether poor condition predisposed birds to the effects of strong winds, or whether the winds caused the poor condition through preventing feeding. Not all such wrecks relate to birds on migration, however, because some involve birds near breeding or wintering areas.

Because drift affects mainly birds that are on their normal migration, it usually involves lateral (west or east) displacements, but displacements to the north or south of the usual range occur in extreme conditions. The most striking examples involve seabirds carried by hurricanes hundreds of kilometres beyond their usual range. Such conditions push coastal birds northwards on both the east and west sides of North America, and sometimes far inland. In the Atlantic, such cyclonic storms follow a circular route. They usually begin in equatorial waters off the African coast, travel westward across the Atlantic into the Caribbean or Gulf of Mexico, then swing northward through eastern North America, weakening all the time, to emerge eastward into the Atlantic in the region of Newfoundland. Because winds circle clockwise around a hurricane, they could carry seabirds from either west (on the northern edge) or east (on the southern edge) onto land. As the winds spiral inwards, some seabirds may become caught in the calm of the eye, tend to stay there, and move north with it. Others move ahead of the storm, notably Sooty Terns *Sterna fuscata*, which on some Caribbean Islands are known as hurricane birds for the warning they provide. Most of the records of tropical seabirds off the coasts of the northern States and Canada occur soon after hurricanes, which would have carried them north. They include White-tailed Tropicbirds *Phaethon lepturus*, Sooty Terns *Sterna fuscata*, Black Skimmers *Rynchops nigra*, Magnificent Frigate Birds *Fregata magnificens*, and various petrels. Hurricanes may thus be important agents of vagrancy in seabirds (Murphy 1936, Enticott 1999).

It is uncertain how much landbirds are affected by hurricanes, which occur during the autumn migration season, but records abound of offshore islands and ships being inundated with passerine migrants at such times. During one incident on 27 August 1926 in the Gulf of Mexico: 'Birds which seemed to be migrating landbirds, chiefly swallows, filled the air about the vessel and were so thickly strewn on deck that they could be scooped up by the armful' (Murphy 1936). Similarly, on 25 September 1987, hurricane 'Emily' deposited 10 000 Bobolinks *Dolichonyx oryzivorus* and thousands of Connecticut Warblers *Oporornis agilis* on Bermuda (Case & Gerrish 1988). While hurricanes may kill many migrating passerines and carry others off course, they are not obviously responsible for the bulk of transatlantic landbird vagrancy (Nisbet 1963).

Age and wind drift

In general, juvenile birds are more often drifted off course by crosswinds than adults are, and many of the 'big days' at coastal observatories occur when the wind has a strong easterly or westerly component. In many bird species, apparently as a result of greater drift, disproportionately more juveniles turn up at coastal sites than at inland ones, as often noted among passerines in both Europe and in North America (Murray 1966, Ralph 1971, Smith & Schneider 1978). In addition, about 95% of all the raptors seen in autumn at Cape May on the New Jersey coast are birds of the year, compared with 50% of those seen at Hawk Mountain in inland Pennsylvania (Allen *et al.* 1996).

The greater effect of crosswinds on juveniles than adults is also evident from radio-marked birds tracked on autumn migration from satellites. Among Ospreys *Pandion haliaetus* and European Honey Buzzards *Pernis apivorus* migrating between Europe and Africa, juveniles were drifted off course to a much greater extent than adults, and followed a much more zig-zag day-to-day route to winter quarters (Håke *et al.* 2003, Thorup *et al.* 2003). This suggested that adults were better able to detect and compensate for wind drift than juveniles. Perhaps the previous experience of the adults and their familiarity with landmarks below helped them to maintain a straighter course. Adult birds may also be better at selecting for migration days with favourable winds, avoiding days with crosswinds. In addition, however, crosswind compensation may be of less value to juveniles than adults, because, being on their first migration, juveniles have no specific prior winter site as their destination. Third, some of the juveniles may be such poor navigators that they get lost on their first migration, and do not survive to be represented in the adult population. These age-dependent effects of wind may help to explain some of the contradictory findings on wind drift reported from radar studies (Chapter 4).

Correction for drift

Drifted birds that have settled usually leave their drift areas when the wind slackens or veers to a more favourable direction. Four types of evidence indicate that birds can detect their displacement and attempt to get back on course. First, radar observations have shown re-oriented movements after drift has occurred (Myres 1964, Evans 1966b, Able 1977). Second, some radio-tagged birds tracked as they were displaced by wind subsequently re-directed themselves to their normal route (for Bald Eagle *Haliaetus leucocephala* see Harmata 2002; for Whooper Swan *Cygnus cygnus* see Pennycuick *et al.* 1996), and some birds ringed as drifted migrants were later recovered in localities closer too, or within, their regular range, including migration areas (Evans 1968; for extreme examples see **Table 10.3**). Third, known drifted birds trapped and tested in orientation cages did not head mainly in their normal migration direction, but in whichever direction would have got them back on their normal route (Evans 1968, Able 1977). Such birds tested on the ground showed the same appropriate directional preferences as birds tracked by radar (Evans 1968, Able 1977). Fourth, birds caught on migration and transported 450 km westward off route compensated for their displacement when tested in orientation cages (Rabøl 1969a, 1994). This was true of juveniles, as well as of adults. Hence, although other evidence suggests that juveniles inherit only a general compass direction to guide them on their first journey (Chapter 9), they nevertheless seem able to recognise and correct for any drift to which they are subjected. Unless they get far from land in continuing adverse winds, therefore, migrants are unlikely to be drifted long distances off course before they re-orientate.

   When correcting for effects of wind drift, birds take different directions in different regions, according to local geography. Off eastern North America, autumn migrants drifted over the sea generally turned at dawn and headed northwest, which in general gave the shortest route to the predominantly southwest–northeast running coastline (Baird & Nisbet 1960, Able 1977). Migrants drifted from continental Europe to the east coast of England re-orientated southeast, which would have got them back on their normal southwest route through western Europe (Evans 1968), while those blown beyond the west coast of Britain over the Atlantic usually back-tracked until they reached land (Myres 1964). Similarly, vagrants to the Farallon Islands off California often approach the islands from the west in the mornings, suggesting that they were back-tracking after having found themselves over the sea at dawn (Robertson 1980). Much of this re-orientation by night migrants occurs in daytime, and the birds may often be able to see the coast from the altitude at which they normally fly.

Table 10.3 Recoveries of vagrant birds ringed in Britain that suggest re-orientation towards the regular range

Species	Usual migration route	Subsequent recovery of vagrant
Ring-necked Duck *Aythya collaris*	Breeds across Canada and the northern USA and winters in southern States and Central America	An adult male ringed at Slimbridge, Gloucestershire, 1 March 1977, was shot at Isertoq, southeast Greenland, in May of the same year, suggesting that it was heading to the regular breeding range
Booted Warbler *Hippolais caligata*	Breeds central Palaearctic, and winters in India; nearest population to Britain breeds in Russia	A juvenile ringed at Spurn, Yorkshire, on 16 September 1993 was recaught at Wetheren, Belgium, on 5 October of the same year, suggesting that it had re-orientated eastwards
Penduline Tit *Remiz pendulinus*	Resident in middle and southern latitudes of continental Europe; nearest population to Britain breeds in Denmark	An adult female ringed at Icklesham, Sussex, in October 1988 was recaught at Kuismaren, Sweden, in May 1989, suggesting that it had re-oriented towards the usual breeding range
Rustic Bunting *Emberiza rustica*	Breeds from Sweden across northern Eurasia and winters in Southeast Asia; nearest population to Britain breeds at 60°N in Sweden	A female ringed on Fair Isle, present there 12–19 June 1963, was recovered on the Greek Island of Chios four months later, in October of the same year, suggesting that it had re-oriented eastwards
Red-breasted Flycatcher *Ficedula parva*	Breeds from eastern Europe across Eurasia, and winters in India and Southeast Asia	An adult male ringed on Shetland on 6 September 1997 was trapped again in Norway 13 days later, suggesting that it had re-oriented eastwards
Barred Warbler *Sylvia nisoria*	Breeds from central Europe east to western Siberia and winters in East Africa	A bird ringed in September 1978 on Fair Isle was recovered in February 1979 in Yugoslavia indicating that it had re-orientated eastward. Similarly, another ringed in Sweden on 26 August 1976 was recovered in Syria on 16 September 1976

From Wernham *et al.* (2002).

OVERSHOOTING

In overshooting, birds follow their usual migration direction but travel too far, turning up at places beyond their regular range. In Europe, the phenomenon is most evident during spring migration, in stable anticyclonic conditions when temperatures are higher than usual, and winds are favourable (Elkins 2005). In these conditions, birds may occur hundreds of kilometres beyond their normal breeding range, as seen in Britain by the periodic appearance in spring of species that breed

Table 10.4   The main seasons and proposed means of occurrence of some vagrants in western North America

Spring overshoots from Eurasia to the North Pacific islands (Aleutians, Pribilofs and St Lawrence Islan)	Garganey *Anas querquedula*, Common Pochard *Aythya ferina*, Common Scops Owl *Otus scops*, Oriental Cuckoo *Cuculus saturatus*, Common Cuckoo *Cuculus canorus*, Fork-tailed Swift *Apus pacificus*, Brown Shrike *Lanius cristatus*, Eurasian Bullfinch *Pyrrhula pyrrhula*, Arctic Warbler *Phylloscopus borealis*
Spring overshoots to Point Barrow (northern Alaskan coast)	From Eurasia: Wood Sandpiper *Tringa glareola*, Grey-tailed Tattler *Tringa brevipes*, Little Stint *Calidris minuta*, Curlew Sandpiper *Calidris ferruginea*, Fieldfare *Turdus pilaris*, Dusky Thrush *Turdus naumanni*, Eyebrowed Thrush *Turdus obscurus*, Arctic Warbler *Phylloscopus borealis*, White Wagtail *Motacilla alba*, Pallas's Reed Bunting *Emberiza pallasi*;
	From North America: Wilson's Phalarope *Phalaropus tricolor*, Scarlet Tanager *Piranga olivacea*, Eastern Kingbird *Tyrannus tyrannus*, Bobolink *Dolichonyx oryzivorus*
Possible autumn mirror-image migrants from Eurasia to western North America	Tufted Duck *Aythya fuligula*, Eurasian Wigeon *Anas penelope*, Baikal Teal *A. formosa*, Slaty-backed Gull *Larus schistosagus*, Dotterel *Charadrius morinellus*, Spotted Redshank *Tringa erythropus*, Eurasian Jacksnipe *Lymnocryptes minimus*, Curlew Sandpiper *Calidris ferruginea*, Ruff *Philomachus pugnax*, Northern Wheatear *Oenanthe oenanthe*, Red-throated Pipit *Anthus cervinus*, Dusky Warbler *Phylloscopus fuscatus*, White Wagtail *Motacilla alba lugens* (breeds Japan-Kamchatka), Olive-backed Pipit *Anthus hodgsoni*, Brambling *Fringilla montifringilla*, Rustic Bunting *Emberiza rustica*
Possible autumn mirror-image migrants from eastern to western North America	Brown Thrasher *Toxostoma rufum*, Dickcissel *Spiza americana*, Blackpoll Warbler *Dendroica striata* and 28 other species of eastern warblers
Possible autumn long-distance reverse migrants from Eurasia to North Pacific Islands	Wood Warbler *Phylloscopus sibilatrix* to Shemya (Aleutian Islands)

Mainly from Robertson (1980).

much further south, including Alpine Swift *Apus melba*, Woodchat Shrike *Lanius senator*, Eurasian Hoopoe *Upupa epops*, Rock Thrush *Monticola saxatilis* and others (**Table 10.2**).

Much longer overshoots occasionally occur, as shown by the appearance in spring of Cave Swallows *Hirundo fulva* from normal Caribbean breeding areas north as far as Nova Scotia (McLaren 1981), and the appearance of various Eurasian and North American species at Point Barrow on the northern coast of Alaska, hundreds of kilometres beyond their normal breeding range (**Table 10.4**). Such occurrences in birds migrating mainly overland have usually been dismissed as 'judgement errors' induced by unusually strong winds. Alternatively, while migrating in the proper direction, they may simply fail to 'switch off' their migration drive at the appropriate time, and continue travelling.

Whenever overshoot birds are closely examined, they are almost always found to be in their first year of life, and are usually males. Spring overshoots also often occur relatively early in the year, arriving in their overshoot areas at about the time they would have been expected in their normal breeding areas, another indication that they were normal spring migrants that travelled too far.

The appearance of occasional individuals well beyond their usual winter range is also frequent, as may been seen from the records of normally more northerly wintering birds in southern Europe or the southern States of America. Northern rarities in California include Snowy Owl *Nyctea scandiaca*, Common Redpoll *Carduelis flammea* and Snow Bunting *Plectrophenax nivalis* among landbirds, and Ivory Gull *Pagophila eburnea*, Ross Gull *Rhodostethia rosea* and Thick-billed Murre (Brunnich's Guillemot) *Uria lomvia* among seabirds.

The likelihood of a species occurring in a particular area as an overshoot depends primarily on the orientation of its migration route. For example, if a Woodchat Shrike *Lanius senator* heading from West Africa to Spain overshoots its breeding area, and continues on its normal heading, it is likely to end up in the British Isles. On the other hand, a Masked Shrike *Lanius nubicus* heading from East Africa towards eastern Europe is more likely to end up in Scandinavia, where indeed there have been several spring records.

Other natural movements leading to vagrancy

Other more familiar types of movements are analogous to overshoots, except that the movement occurs in response to external conditions outside the normal migration season. At the start of exceptionally severe winter weather, birds are sometimes driven to lower latitudes than usual, with occasional individuals reaching much further than the rest, to be classed as vagrants. Such movements are common among various waterfowl and waders that are affected by freezing waters and soils, or ground-feeding thrushes and seed-eaters whose food is covered by snow too deep to penetrate (Chapter 16). Similarly, during hard winters in mountain areas, birds may be driven to lower altitudes than usual, appearing as vagrants outside their regular winter range. European examples include the Alpine Accentor *Prunella collaris* and Wallcreeper *Tichodroma muraria*, and North American examples include Clark's Nutcracker *Nucifraga columbiana* and various rosy finches *Leucosticte*. Irruptive species also occasionally appear well beyond their usual invasion areas.

The occurrence of severe droughts in the desert and semi-desert regions of southern Asia sometimes results in the appearance of arid land species in western Europe. Examples include the Pallas's Sandgrouse *Syrrhaptes paradoxus* and Rose-coloured Starling *Sturnus roseus*, and also Ruddy Shelduck *Tadorna ferruginea* and Glossy Ibis *Plegadis falcinellus*, which often appear in the same summers as one another, as in 1886, 1892 and 1994 (Cottridge & Vinicombe 1996). Irruptions of Pallas's Sandgrouse were noted in 12 years between 1859 and 1909, the largest in 1863 and 1888, but the species has appeared much less often and in only small numbers subsequently (Chapter 16).

The distributions of many marine birds are closely tied to particular current systems, whose distribution periodically changes from the normal, as during El Niño events. When this happens, the seabirds associated with those currents also change distribution, disappearing from their usual areas of occurrence and appearing elsewhere, hundreds of kilometres distant. This results in mass distributional changes, together with the appearance of vagrants in more distant areas than usual.

Yet another mechanism that occasionally brings vagrants to western Europe occurs particularly in waterfowl when an individual of one species attaches itself to a flock of another, and ends up in the wrong breeding or wintering area. For example, a Ferruginous Duck *Aythya nyroca*, which would normally winter in the eastern Mediterranean region, may attach itself to a flock of Common Pochards *A. ferina* in late summer and head off with them towards the west. Similarly, in western Europe, Falcated Teal *Anas falcata* occasionally turn up with Eurasian Wigeon *A. penelope*, Lesser White-fronted Geese *Anser erythropus* with Greater White-fronts *A. albifrons*, and Red-breasted Geese *Branta ruficollis* with Brent Geese *B. bernicla*. These are all closely related species which either breed in the same area as one another, or cross routes on migration, giving the potential for intermixing to occur. Similarly, Common Cranes *Grus grus* and Sandhill Cranes *Grus canadensis* both breed in parts of eastern Siberia, and Common Cranes (which normally winter in the Old World) have occasionally been seen among Sandhill Cranes in North American wintering areas.

DEVIANT DIRECTIONAL TENDENCIES

Some bird species, leaving a particular point on the migration route, show much greater spread in their migration directions than others, depending partly on the locations and distributions of their respective wintering areas. This is obvious both from observations of birds on migration, and from subsequent ring recoveries (Busse 2001). However, in some species, it is also apparent that juveniles on their first autumn journey show much greater spread in departure directions than adults. Such an age difference has been reported from Swedish or Finnish ringing recoveries for Spotted Flycatchers *Muscicapa striata* (angular spread 8–18° for adults, 13–23° for juveniles), Willow Warblers *Phylloscopus trochilus* (9° for adults, 21° for juveniles), Honey Buzzards *Pernis apivorus* (8° for adults, 12° for juveniles) and others (Alerstam 1990b, Fransson & Stolt 2005). This difference in autumn directional spread between adults and juveniles is also obvious among birds tested in orientation cages to find their directional preferences (De Sante 1973, Able 1977, Moore 1994). If juveniles were to continue migrating on their more extreme directions, many would end up well outside the usual migration range, as is again evident from observations and ring recoveries. However, it is not clear whether the juveniles that take the wrong direction suffer greater mortality, and so are less represented in the adult samples, or whether they learn from their mistakes, and perform better in later life. In any case, one plausible hypothesis is that some autumn vagrancy is simply a consequence of the wide range of directional preferences normally existing within some bird populations and evident chiefly in juveniles. Directional errors have been classed as 'disorientation' (inability to follow any consistent direction) or 'misorientation' (ability to follow a consistent direction, but not the correct one) (De Sante 1983a).

Because the directional tendencies of individuals are known to be partly under genetic control (Chapters 12 and 20), extreme individuals could be classed as mutants. We can expect some directional mutants to arise in every generation, but in general mutation rates are too low to account for the frequency with which some long-distance vagrants occur at particular sites. Thus, annual mutation alone within each year's crop of young could not be expected to account for the relatively large numbers of Richard's Pipits *Anthus novaeseelandiae*, Yellow-browed Warblers *Phylloscopus inornatus* and Pallas's Warblers *P. proregulus* which occur each autumn in western Europe, or of Blackpoll Warblers *Dendroica striata* and other species in California. Some other explanation is therefore required, either a cumulative effect of favourable mutations, increasing through the generations, or frequent non-genetically controlled mistakes in directional preferences. While such 'misoriented' individuals could in theory turn up in any direction from the usual range, in practice they tend to concentrate in specific directions, showing at least two distinct types of pattern, as described below.

MIRROR-IMAGE MIGRATION

At a broad geographical scale, juveniles of many species appear annually in localities that indicate they have followed a direction that is a mirror-image of the correct direction. Mirror-image migration supposedly occurs when a departing bird takes a correct bearing with respect to the north–south axis, but chooses the wrong east–west (right–left) side of that axis. Instead of migrating southwest, say, the bird migrates southeast, with distance, timing and other aspects of the migration remaining unchanged. This apparent situation was studied in some northern North American warblers, about 30 species of which normally migrate far to the east of California, but have been recorded there in autumn, with Blackpoll Warblers *Dendroica striata* and Palm Warblers *D. palmarum* together forming more than half the records. Simple drift on westerly winds could not account for this pattern, for species would then be expected to occur in numbers proportional to their overall abundance, and over a much wider span of latitude than that over which they occur. Instead, such records are more confined geographically, and more consistent with the hypothesis of mirror-image navigation (De Sante 1973, 1983a, 1983b, Patten & Marantz 1996). Moreover, vagrant Blackpoll Warblers that were tested in orientation cages showed both the correct migratory direction and the mirror image of that direction. Other apparent mirror-image migrants to California from eastern North America include the Brown Thrasher *Toxostoma rufum* and Dickcissel *Spiza americana* (**Table 10.4**).

The same hypothesis has been proposed to explain the occurrence of various western North American species, including *Dendroica* warblers, in eastern North America in spring, which would imply a mirror-image error in the direction taken from wintering areas (McLaren 1981). Some 38 records of western *Dendroica* warblers lay on a direct mirror-image route at around 45° from wintering areas in western Mexico northwest towards Nova Scotia (McLaren 1981). The same mechanism could explain the spring occurrence of southwestern species, such as Green-tailed Towhee *Pipilo chlorurus*, Rock Wren *Salpinctes obsoletus* and Band-tailed Pigeon *Columba fasciata* on Nova Scotian Islands.

Other possible examples of mirror-image misorientation include: (1) the occurrence in western Europe in autumn of Eurasian species that normally migrate broadly southeast or east then south from their breeding areas (such as Arctic

Warbler *Phylloscopus borealis*, Greenish Warbler *P. trochiloides*, Richard's Pipit *Anthus novaeseelandiae* and Pechora Pipit *A. gustavi* (**Table 10.2**); (2) the occurrence of Citrine Wagtail *Motacilla citreola* that normally migrates from eastern Siberia south-southwest in autumn but occasionally reaches eastern Australia, suggesting mirror-image flights to the south-southeast; (3) vagrants of numerous species that breed in far eastern Siberia and normally migrate south-southwest to Southeast Asia, but occasionally follow a mirror-image route to western Alaska and down the North American coastline. On this last route, Eurasian Wigeon *Anas penelope* and Tufted Duck *Aythya fuligula* are regular, while many others are occasional (**Table 10.4**). Re-examination of off-route migration records elsewhere may yield further examples of possible mirror-image misorientation.

REVERSED-DIRECTION MIGRATION

Another type of navigational error that could result in a one-step change in migration direction is reversed-direction migration, in which some young birds leave the breeding range in the direction opposite to usual; that is, they take their spring direction in autumn, or vice versa (Nisbet 1962, Rabøl 1969b, 1976). This could be regarded either as an error in timing or in navigation (correct bearing, but wrong side of both north–south and east–west axes). Among European birds, the Barred Warbler *Sylvia nisoria* and Red-breasted Flycatcher *Muscicapa parva* appear each autumn across northern Europe from Iceland to Finland. They breed far to the southeast and east-southeast and migrate more or less southeast and east-southeast to East Africa and India respectively. Their autumn occurrence in northern Europe thus involves a displacement almost directly opposite to their normal direction for autumn migration. Similarly in North America, the Yellow-breasted Chat *Icteria virens* breeds over much of the USA north to about 42°N. In autumn, it migrates to the southwest, but is seen in surprising numbers along the Eastern Seaboard from New York City to Nova Scotia and Newfoundland, up to 1500 km northeast of its breeding range. Again, this has been attributed to a reversal of the normal migration direction (Nisbet 1962).

One of the most striking supposed examples of reverse-direction migration is provided by the Pallas's (Lemon-rumped) Warbler *Phylloscopus proregulus* which breeds in eastern Asia but is now regularly recorded in late autumn in northwest Europe. Again, these birds could be making a simple mistake of 180° in their migratory flight; the great circle route projected from the normal wintering area through the breeding area would reach northwest Europe (Rabøl 1969; **Figure 10.6**). In fact, almost any species which normally migrates in autumn from western or central Siberia to Southeast Asia could end up in northwest Europe if its migration was reversed (Cottridge & Vinicombe 1996). The area where such species would be expected on reverse migration has been termed the 'vagrancy shadow'. The presence of these species in northwest Europe cannot be attributed entirely to wind drift, because such long journeys are likely to be spread over several weeks, during which the birds could correct for their displacement if they perceived themselves as being off course.

The proposed reality of reverse or mirror-image misorientation stems from observations that, in certain situations, specific large-angle misorientations seem

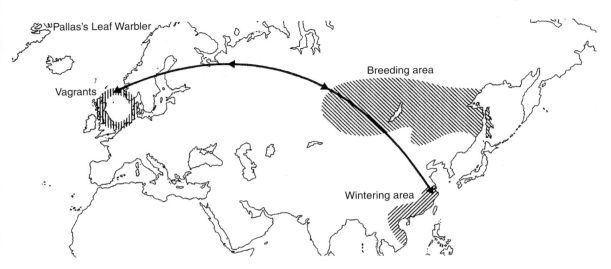

Figure 10.6   Hypothetical reversed 'great circle' migration route of Pallas's Warbler *Phylloscopus proregulus*. From Rabøl (1969b).

more common than small or intermediate deviations from the normal migration course. Moreover, misorientations could best explain why species that breed in far away areas turn up more frequently than other closely related species that breed commonly only a short distance away. For example, Pallas's Warblers *Phylloscopus proregulus* from Siberia are much more frequent in Britain than are Bonelli's Warblers *Phylloscopus bonelli* which breed as close as France.

If reversed-direction migration were a real phenomenon, rather than an untested hypothesis, it would limit the variety of species expected in any given area to those whose reversed routes passed through that area. The vagrancy shadows of other species would fall elsewhere. Take as examples the Red-breasted Flycatcher *Ficedula parva* and Collared Flycatcher *F. collaris*, which have similar breeding distributions in Europe (although the former also extends further east). In autumn, most Red-breasted Flycatchers head southeast for southern Asia, whereas Collared Flycatchers head south for wintering areas in East Africa. The much more frequent autumn occurrence of Red-breasted Flycatchers in northwest Europe (more than 2000 records in Britain up to 1995) compared with Collared Flycatchers (19 records up to 1995) has been attributed to reversed-direction migration of the former (**Figure 10.7**, Cottridge & Vinicombe 1996). This difference in occurrence between species is greater than expected from the respective sizes of their geographical ranges (and hence population sizes). And because almost all recorded individuals were juveniles in their first autumn, we can exclude persistence after spring overshooting as a cause of their presence.

Many apparent examples of long-distance reversed-direction migration are based on the assumption that migration routes normally follow a great circle. Such routes represent the shortest distance between two points on the earth's surface, but it is by no means certain that birds follow such routes, and to my knowledge they have not yet been convincingly demonstrated for any long-distance migrant over most or all of a migration route (Chapter 9).

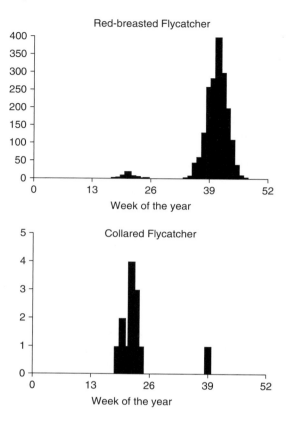

Figure 10.7 Seasonal appearance of two species of flycatchers in Britain, as determined from the number of records in different weeks of the year. One species occurs mainly in autumn, the other mainly in spring. Note the far greater number of records of the Red-breasted Flycatcher *Muscicapa parva*. From Fraser & Rogers (2004) and previous reports.

Another problem with the reversed migration hypothesis is that the vagrancy shadow produced by 180° reversal of orientation for Siberian migrants (which normally migrate to Southeast Asia) falls in Britain and other parts of northwest Europe where there are lots of observers (Gilroy & Lees 2003). On almost all other compass points to which a Siberian vagrant might head, observer coverage is much less thorough. However, observer bias could not account for the concentration of eastern North American warblers in mid-coastal California, for areas to the south and north are also well watched, or for the occurrence of western American *Dendroica* warblers on the northeast coast, including the Nova Scotia Islands (McLaren 1981). Many other regular autumn migrants from Siberia occur occasionally in western Europe, even though they are not within their migration shadow. Examples include the Pied Wheatear *Oenanthe plechanka*, River Warbler *Locustella fluviatilis*, Eastern Olivaceous Warbler *Hippolais pallida elaeica* and Isabelline Shrike *Lanius isabellinus*. However, reverse-direction migration need not be the only means by which Siberian vagrants reach northwest Europe in autumn.

Among Eurasian birds, species migrating in an east–west direction are more likely to reverse their migration than are those migrating in a north–south or northeast–southwest direction (Thorup 2004). This difference may be related to the availability of orientation cues on different migratory axes, or to birds using mainly different cues on east–west and north–south axes. Furthermore, the very long distances covered by some displaced birds (up to twice the normal length of journey) throw doubt on the effectiveness of the birds' internal clock in terminating migration (Thorup 2004). It may be the same faulty mechanism that leads them to take a spring direction in autumn.

These considerations also raise the question whether factors other than west-blowing winds could underlie the occurrence of some North American species in Europe. Both spring overshooting and autumn reversed-direction migration have been suggested. Most of the eastern North American coastline runs south-west–northeast. If one draws a line on a globe up the Eastern Seaboard, say from South Carolina to Newfoundland, and then projects that line across the Atlantic, it reaches the British Isles (**Figure 10.8**). Hence, a bird migrating northeastwards up the Eastern Seaboard in spring and overshooting into the North Atlantic could reach northwest Europe, providing it had sufficient body fuel and favourable winds. Correspondingly, nearly all the North American passerine vagrants that appear in Britain in spring breed well to the northeast in Canada. This phenomenon could thus account for the numerous spring records in Britain of North American sparrows, together with the Brown-headed Cowbird *Molothrus ater*, Cape May Warbler *Dendroica tigrina* and Cedar Waxwing *Bombicilla cedrorum*, all appearing as long-distance spring overshoots.

As for autumn vagrants from North America, some could well be reversed-direction migrants. Such birds would normally migrate from northeastern North America southwestward to Central America, but on reverse migration they could again end up in northwest Europe. Examples of autumn species that might have arrived in this way in Britain include Tennessee Warbler *Vermivora peregrina*, Hooded Warbler *Wilsonia citrina* and Sandhill Crane *Grus canadensis* (Cottridge & Vinicombe 1996). Reversed migration might also account for the autumn appearance at Point Barrow on the northern coast of Alaska of North American species from further south (**Table 10.4**), for the appearance in British Columbia in September 1889 of a Gray Kingbird *Tyrannus dominicensis* from the southeastern US and Caribbean region, and for the occasional autumn–winter appearance of Dusky-capped Flycatchers *Myiarchus tuberculifer*, Greater Pewees *Contopus pertinax*, Streak-backed Orioles *Icterus pustulatus* and Varied Buntings *Passerina versicolor* in California from regions to the southeast (Robertson 1980).

Reversed migration may also occur among birds that have reached their normal wintering areas, but then simply continue in that direction when they migrate again in spring (equivalent to extending their autumn journey). This may account for the appearance of southern South American birds in North America in autumn. For example, the Fork-tailed Flycatcher *Tyrannus savana* breeds in southern South America and winters in northern South America, but in spring instead of returning southwestward, some individuals travel northeast through the eastern States, appearing as far north as the Nova Scotian islands (Kaufman 1977). Similarly, on the west coast the Tropical Kingbird *T. melancholicus* (nominate form), which normally spends the non-breeding season in Central America,

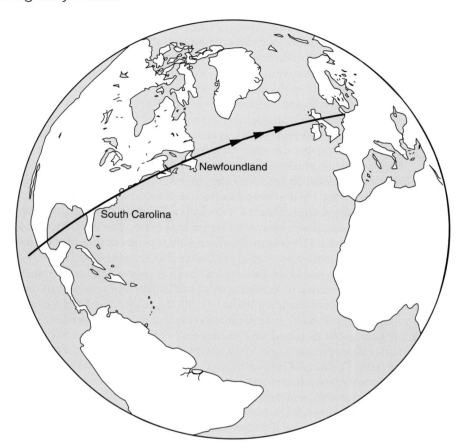

Figure 10.8   Two mechanisms through which eastern North American vagrants might appear in northwest Europe. If a bird flying north in spring up the North American seaboard overshoots into the North Atlantic, it is likely to arrive in northwest Europe if it continues flying along the same great circle route. The same is true for any bird reverse migrating along the same route in autumn. From Cottridge & Vinicombe (1996).

has been found at various localities north to Vancouver, with at least one record in Alaska. At this time of year, the Central American wintering populations should be moving southeast towards their breeding areas, rather than northwest on an apparent reversed migration.

NEW ROUTES AND RANGE EXPANSION

The fact that regular migrants are much more likely to turn up in unexpected places than are non-migrants emphasises the connection between migration and vagrancy, as does the concentration of vagrant records in the normal migration seasons. Although migrants travel longer distances than non-migrants, and are often blown off course, they are not necessarily more likely to stay and establish themselves in new areas. When migrants arrive in a new area, they are usually in a migratory state, programmed to re-orientate and continue their journeys. Ring recoveries confirm that many soon move on again (**Table 10.3**).

Not all vagrants return to their usual range, however, and some have established territories, and occasionally paired up with local species and bred. Examples from Britain include: a male Pied-billed Grebe *Podilymbus podiceps* which paired with a Little Grebe *Tachybaptus ruficollis* and produced three hybrid young; a female Blue-winged Teal *Anas discors* which paired with a Northern Shoveler *Anas clypeata* and raised three hybrid young; a male and female American Black Duck *Anas rubripes* each of which paired with a Mallard *Anas platyrhynchos* and produced young; and a female Lesser Crested Tern *Sterna bengalensis* which paired with a Sandwich Tern *Sterna sandwichensis* and produced young in at least three years. In addition, individual Black-browed Albatrosses *Diomedea melanophris* from the Southern Ocean set up territories and displayed (without success) in Northern Gannet *Morus bassanus* colonies over periods of 3, 23 and 35 years respectively (at Bass Rock, southeast Scotland, in 1967–1969, Hermaness, Shetland, in 1972–1995; and Faeroes in 1860–1894; the first two cases may have involved the same bird in different colonies, and may also have been the bird seen in Iceland in 1966).

In addition, some vagrants, having once wintered in an area, return there year after year, as in normal homing behaviour, with examples from Britain involving a King Eider *Somateria spectabilis* (1986–1991), Whistling Swan *Cygnus columbianus* (1985–1986) and Wallcreeper *Tichodroma muraria* (1976–1977), and in spring–summer even an Ancient Murrelet *Synthliboramphus antiquus* which normally occurs in the North Pacific, but was seen off Lundy Island in three successive summers (Cottridge & Vinicombe 1996). The record holder, however, was a Ring-billed Gull *Larus delawarensis*, which turned up in 13 successive winters in Stromness, Orkney, Scotland (Fraser & Rogers 2002). These birds were not ringed, but their individual identification was assumed because they occupied the same places in successive years. Other records of winter site fidelity by vagrants have come from California, involving Tufted Duck *Aythya fuligula* (with one drake returning to the same locality for five years), Black-headed Gull *Larus ridibundus*, Little Gull *Larus minutus* and Brown Thrasher *Toxostoma rufum* (Robertson 1980). On the other hand, the two records of a North American Belted Kingfisher *Ceryle alcyon* in Britain were 71 years apart (1908 and 1979), yet still came from the same inland locality (Cottridge & Vinicombe 1996).

Other recoveries imply fidelity to off-route stopover sites by vagrants. An adult male Bluethroat *Luscinia svecica* was ringed on 17 May 1958 in Devon, southwest England, more than 300 km west of the usual migration route, and retrapped at the same site on 5 May 1963. Another first-winter male Bluethroat ringed at the same site on 21 September 1966 was retrapped there on 14 September 1968. This suggests that, once having taken a particular migration route, at least some individuals repeat it in successive years, even though it involves a detour (see Chapter 8 for other instances in birds that were radio-tracked in successive years). Such recoveries confirm that at least some out-of-range individuals survive and return at later dates.

Some of the Siberian vagrants that appear in western Europe have managed to survive into winter. Records are frequent for Richard's Pipits *Anthus richardi*,

Pallas's Leaf Warblers *Phylloscopus proregulus* and Yellow-browed (Inornate) Warblers *Phylloscopus inornatus*, but even rarer species have occasionally been seen then, including Olive-backed Pipit *Anthus hodgsoni*, Dusky Warbler *P. fuscatus*, Little Bunting *Emberiza pusilla*, Rustic Bunting *E. rustica* and others. Moreover, Richard's Pipits now winter regularly in Spain (Cramp 1988), and Yellow-browed (Inornate) Warblers and Pallas's Reed Buntings *Emberiza pallasi* are seen there occasionally, as are Richard's Pipits and Common Rosefinches *Carpodacus erythrinus* in Morocco, and Red-breasted Flycatchers *Ficedula parva* in the Canary Islands. It is, of course, almost impossible to distinguish a true vagrant (a bird well off the normal migration route) from a regular migrant on a route used by only a small number of individuals.

Genetically controlled deviations from the normal migration direction occasionally lead to the establishment of new migration routes and wintering areas, as illustrated by Blackcaps *Sylvia atricapilla* that now migrate from central Europe to winter in Britain (Chapter 20). Is it possible, then, that many of the vagrants which turn up in particular localities in increasing numbers are in the process of establishing new wintering populations, and that birds with appropriate deviant directions are gradually increasing in the population? As in the Blackcap, this is likely to occur only if: (1) the new wintering area permits good overwinter survival, and (2) the deviant individuals more often pair with one another than with individuals with 'normal' directional tendencies. This could happen, for example, if birds from the new wintering area returned to the breeding area at a somewhat different date than those from the regular wintering area (for example, see Bearhop *et al.* 2005). Although such assortative mating is not essential for any genetically controlled new behaviour to be passed on, it would greatly facilitate its establishment and spread through the population.

Not long ago, bird-watchers considered the Pallas's Leaf Warbler *Phylloscopus proregulus* to be one of the rarest vagrants reaching northwest Europe. Breeding in Siberia, it provided only three records in Britain prior to 1958, yet after 1980, the annual total in several years exceeded 100 individuals (**Figure 10.1**; Cottridge & Vinicombe 1996). The Yellow-browed (Inornate) Warbler *Phylloscopus inornatus* from Siberia increased even more, culminating in a record number of 615 in 1985. It is now the commonest Siberian vagrant to western Europe. Possibly both species have established new wintering areas, for which Britain lies on the route. They seem to arrive slightly earlier in the north of Britain than in the southwest, suggesting some onward movement, perhaps to Iberia or western Africa. Occasional spring records may reflect a return passage.

In North America, many parulid warblers have extended their breeding ranges far to the northwest since the end of the last glacial period. Most mirror-image vagrants from such populations probably end up far offshore over the Pacific Ocean and die. There are, however, several records of banded birds being recovered at the same west coast location in 2–3 consecutive winters, and the 'Myrtle' race of the Yellow-rumped Warbler *Dendroica coronata* now has a regular west coast winter range separated from the main winter range to the south and east, at locations consistent with mirror-image migration (Diamond 1982). Hence, this process, if under genetic control, might also be producing new migration routes, wintering ranges, and migratory divides (Chapter 20).

Such developments may also be occurring in Lesser Black-backed Gulls *Larus fuscus* which are wintering in increasing numbers on the east coast of North

Table 10.5 Transatlantic ring recoveries of various bird species some of which breed on both sides, so would not otherwise have been recognised as vagrants. The list gives vagrants only, and regular transatlantic migrants are listed in the footnote

	Movement from Europe to North America	Movement from North America to Europe
Little Blue Heron *Egretta caerulea*	0	1
Black-crowned Night Heron *Nycticorax nycticorax*	0	1
Caspian Tern *Sterna caspia*	0	1
Common Tern *Sterna hirundo*	0	4
Sandwich Tern *Sterna sandvicensis*	0	1
Roseate Tern *Sterna dougallii*	1	0
Gull-billed Tern *Sterna nilotica*	3	0
Ring-billed Gull *Larus delawarensis*	0	5
Herring Gull *Larus argentatus*	0	1
Northern Gannet *Morus bassanus*	0	3
Leach's Storm Petrel *Oceanodroma leucorhoa*	0	3
Northern Fulmar *Fulmarus glacialis*	0	3
Green-winged Teal *Anas crecca*	0	3
Northern Pintail *Anas acuta*	0	4
Mallard *Anas platyrhynchos*	0	1
American Black Duck *Anas rubripes*	0	2
Ring-necked Duck *Aythya collaris*	0	1
American Coot *Fulica americana*	0	1
Peregrine Falcon *Falco peregrinus*	0	1
Semi-palmated Plover *Charadrius semipalmatus*	0	1
Northern Lapwing *Vanellus vanellus*	1	0
Upland Sandpiper *Bartramia longicauda*	0	1
Canada Goose *Branta Canadensis*	0	1
Blue-winged Teal *Anas discors*	0	13
American Wigeon *Anas americana*	0	5

The above individuals were classed as vagrants and were recovered at various sites in eastern North America and in western Europe south to the Azores. Regular transatlantic migrants include Manx Shearwater *Puffinus puffinus*, Great Shearwater *Puffinus gravis*, Black-legged Kittiwake *Rissa tridactyla*, Arctic Tern *Sterna paradisaea*, Sooty Tern *Sterna fuscata*, Brent Goose *Branta bernicla*, Ruddy Turnstone *Arenaria interpres*, Red Knot *Calidris canutus*, Purple Sandpiper *Calidris maritima*, Common Ringed Plover *Charadrius hiaticula*, Northern Wheatear *Oenanthe oenanthe*, and several other species that move between Greenland and Europe.
Mainly from Dennis (1981, 1987, 1990).

America, among Eurasian Wigeon *Anas penelope* and others on the west coast of North America, and among Ring-billed Gulls *Larus delawarensis* now wintering annually in small numbers in western Europe (with at least five ring recoveries, **Table 10.5**). However, with small or inconspicuous species in little known areas, one can seldom be certain that they have not always been there, although they may have increased in recent years. The presence of an unknown population of Swinhoe's Petrel *Oceanodroma monorhis* in western Europe may account for the appearance of individuals in northeast England. Several have now been caught and identified from their DNA, including one individual in four successive years, but the nearest known nesting colonies occur on the other side of the Eurasian landmass, off Korea (Pitches & Cleeves 2005).

Not all migration-related range expansions result from genetically influenced directional changes. While most birds drifted off course may either die or re-orientate and get back on course, drift may occasionally lead to the colonisation of new areas, and so have lasting biogeographical consequences. Many of the species now found on oceanic islands are likely to have descended from ancestors that were long ago blown off course by winds. Such a process led to the colonisation of the New World by Old World Cattle Egrets *Bubulcus ibis* (Maddock & Geering 1994), and to the establishment of many species introduced to New Zealand on various oceanic islands to the south, involving water crossings of up to 1200 km (Williams 1953). Many other birds probably reach remote areas and breed there for a time, but without establishing a lasting population, such as the Fieldfare *Turdus pilaris* in Greenland (Salomonsen 1951). Imagine the odds against two Spotted Sandpipers *Actitis macularia* from North America reaching the same Scottish locality and nesting thousands of kilometres from their normal range. Yet it happened on at least one occasion (Wilson 1976).

An increase in vagrancy is sometimes followed by colonisation. In several bird species that started to breed in Britain during the twentieth century, breeding was usually preceded by increasing populations in continental Europe and increasing numbers of vagrants to Britain (O'Connor 1986). The Little Egret *Egretta garzetta* provides a striking example.

HUMAN-ASSISTED VAGRANCY

Not all apparent transoceanic vagrants have travelled unaided, for some types of birds hitch rides on ships. One documented incident involved the ship 'Mauritania' as it began its seven-day transatlantic journey from New York to Southampton (England) on 7 October 1962 (Durand 1963). On the morning of the second day, when the ship was 400–500 km out from New York and under heavily overcast skies, more than 130 birds of at least 34 species appeared on deck, mainly passerines but also woodpeckers. The crew provided fruit and other types of food, which some of the birds ate. As the days passed, these birds gradually disappeared, and others were added, but by 12 October nine birds were still present as the ship passed the Fastnet Lighthouse off southwest Ireland. At this point a Yellow-shafted Flicker *Colaptes auratus* flew ashore. When the ship docked at Southampton on 15 October, four birds were still present, including two White-throated Sparrows *Zonotrichia albicollis*, one Song Sparrow *Melospiza melodia* and a Slate-coloured Junco *Junco hyemalis*. The next day, a White-throated Sparrow was seen in a nearby park. The author mentioned similar incidents from other journeys.

Most records of birds alighting on ships refer to passerines which stay for a few days or so, but occasionally for longer if they are fed. At least one species, the House Crow *Corvus splendens* from southern Asia, has apparently moved around the world entirely on ships, and has established itself in a number of widely separated port towns. Some other ship-assisted visitors are surprising, such as the Double-crested Cormorant *Phalacrocorax auritus* which travelled undetected in the hold of a cargo ship from Newfoundland to Scotland (Cottridge & Vinicombe 1996). Incidents such as these raise the possibility that almost any migratory passerine could cross the Atlantic on a ship or by using different ships as stepping stones. But this does not mean that no passerine vagrants have crossed the Atlantic naturally, and it is inconceivable that shorebirds and other types of transoceanic vagrants have travelled in this way. In any case they could not take the types of food provided on ships.

The cage-bird trade

An estimated 2–5 million birds of many different species are trapped worldwide each year for the cage-bird trade, large numbers of which are transported for sale elsewhere. Some apparent vagrants might therefore be escapees that are reported at varying dates and distances after their escape. It is often possible to tell if a bird has been kept in captivity for a long time, but this is much less easy for a bird that escaped soon after capture, or had lived in the wild long enough to moult before being noticed. Again, however, such processes are unlikely to account for the appearance of the same species at the same offshore islands in numbers at about the same dates every year. Most cage birds are passerines or parrots, and other species, such as waders, are kept in captivity only by a very small number of specialists. In addition, in some parts of the world, including North America, almost no native birds are caged or exported, except small numbers for scientific purposes. Yet vagrancy is just as frequent a phenomenon in the New World as in the Old.

EVIDENCE FROM RINGING

Some vagrants wore rings which confirmed their place of origin, with more than 30 records of ringed North American vagrants in Europe (**Tables 10.5** and **10.6**), plus at least one record of a successful return to the breeding range, namely a Ring-billed Gull *Larus delawarensis* ringed in Norway and recovered in Canada (Cottridge & Vinicombe 1996). Such records mostly involve seabirds and waterfowl, which typically have high recovery rates. These records show conclusively, for anyone who may have doubted it, that vagrancy is a real and natural phenomenon, which extends to species commonly kept in captivity (such as waterfowl).

CONCLUDING REMARKS

If bird vagrancy were an entirely random process, birds could in theory turn up anywhere at anytime, but with a probability that declined with increasing distance from the regular range. However, random processes could not explain why almost all vagrants derive from migratory species, which turn up at migration seasons, often in specific localities year after year and in association with specific weather patterns. Nor can it explain why several species from the same region

Table 10.6 Some interesting transatlantic ring recoveries of vagrants (excluding pelagic species and regular transatlantic migrants; see **Table 10.5** footnote)

Species	Ringing site	Recovery site
Black-crowned Night Heron *Nycticorax nycticorax*	New York, 11 June 1988	Azores, 16 October 1988
Little Blue Heron *Egretta caerulea*	New Jersey, 26 June 1964	Azores, November 1964
Peregrine Falcon *Falco peregrinus*	New Brunswick, 18 July 1986	England, 24 December 1986
American Coot *Fulica americana*	Ontario, 30 August 1971	Azores, 25 October 1971
Upland Sandpiper *Bartramia longicauda*	Michigan, 5 July 1988	Spain, 2 December 1988
Common Tern *Sterna hirundo*	Massachusetts, 3 July 1956	Azores, October 1964
Common Tern *Sterna hirundo*	Massachusetts, 12 July 1986	France, 26 October 1986
Caspian Tern *Sterna caspia*	Michigan, 14 July 1927	England, August 1939
Wood Duck *Aix sponsa*	North Carolina, 11 August 1984	Azores, August 1985
Northern Pintail *Anas acuta*	Labrador, 19 August 1948	England, 15 September 1948
	Labrador, 7 September 1951	England, 25 September 1951
	Prince Edward Island, 18 August 1969	Ireland, 29 January 1974
	Nova Scotia, 2 August 1982	France, 6 January 1985
Blue-winged Teal *Anas discors*	New Brunswick, 26 July 1971	Denmark, 25 August 1972
	New Brunswick, 26 July 1971	England, 10 October 1971
	New Brunswick, 1 August 1977	France, 30 September 1977
	New Brunswick, 29 July 1977	Spain, 23 October 1977
	Nova Scotia, 1 August 1969	Spain, 16 September 1969
	Nova Scotia, 14 August 1978	Portugal, 29 October 1976
	Prince Edward Island, 28 August 1973	Spain, 6 January 1974
	Prince Edward Island, 12 September 1971	Spain, 31 October 1971
	Prince Edward Island, 1 September 1970	Morocco, 10 October 1970
	Ontario, 20 September 1971	Azores, 22 November 1971
	Newfoundland, 11 September 1983	Ireland, 15 January 1984
	Maine, 25 August 1981	Azores, 23 September 1981
	Quebec, 6 August 1982	Scotland, September 1982

Species		
American Wigeon *Anas americana*	New Brunswick, 5 August 1966	Scotland, 7 October 1966
	New Brunswick, 29 August 1968	Ireland, 12 October 1968
	Prince Edward Island, 30 August 1977	Ireland, 8 October 1977
	New Brunswick, 13 August 1986	Scotland, 21 September 1986[a]
	New Brunswick, 8 August 1982	France, 9 December 1982
Green-winged Teal *Anas crecca*	New Brunswick, 22 August 1970	England, 2 January 1971
	Prince Edward Island, 16 August 1970	Iceland, 17 April 1979
	Newfoundland, 22 September 1985	Ireland, 31 October 1985
American Black Duck *Anas rubripes*	New Brunswick, 27 July 1976	France, 12 October 1976
	New Brunswick, 5 September 1970	Germany, 10 February 1988
Mallard *Anas platyrhynchos*	Michigan, 19 August 1975	Iceland, 3 January 1978
Ring-necked Duck *Aythya collaris*	New Brunswick, 7 September 1967	Wales, 26 December 1967
Ring-billed Gull *Larus delawarensis*	Michigan, 14 June 1950	Spain, 18 January 1951
	Ontario, 10 June 1945	Azores, 4 November 1945
	Ontario, 27 June 1964	Spain, 20 January 1965
	New York, 21 June 1980	Ireland, 28 December 1981
	Ontario, 8 July 1980	Spain, 21 January 1981
Lapwing *Vanellus vanellus*	Cumbria, England, May 1926	Newfoundland, December 1927
Semi-palmated Plover *Charadrius semipalmatus*	Quebec, 24 July 1972	Azores, 23 September 1972

[a]Later shot County Wexford, Ireland, 30 November 1986.

often turn up together, nor why the vast majority of vagrants are inexperienced youngsters. At the least, vagrancy is a vivid tribute to the dispersive powers of some migratory birds.

The preponderance of juveniles among vagrants would be expected for several reasons, such as their inexperience, and their greater dependence on compass orientation as opposed to the bi-coordinate navigation (=homing behaviour) used by adults. Also, inherent errors in orientation and migratory timing would be expected more frequently in juveniles, because at that age deviants from the norm would not have been exposed to natural selection. For the most part, only successful individuals, behaving normally, are likely to reach adulthood.

While the arrival of birds in unexpected places has led to speculation on how they might have got there, only limited testing of ideas has occurred. The roles of different proposed mechanisms of vagrancy clearly vary between the species occurring at any one site, and some species may be affected through more than one mechanism. Off-route displacement by inclement weather is usually little more than a temporary phenomenon for which birds are normally able to correct at a later date. Other types of vagrancy probably result from deliberate behaviour on the part of the birds, but acting on the basis of flawed instructions – some genetic or other change in the orientation or time-keeping mechanism which leads them to migrate in an atypical direction, or fail to switch off their migratory behaviour at an appropriate time. In theory, more than one kind of genetically-based flaw could occur simultaneously in the same population, with some individuals showing behaviour consistent with mirror-image migration (say) and others reversed-direction migration. Although some observers have tested directional preferences of vagrants in orientation cages, and the location may eliminate some possible explanations of their presence, it is seldom possible to say which particular theoretical mechanism might have brought most vagrants to a given site.

Several authors have suggested that vagrancy is more common in nocturnal than in diurnal migrants. However, many more bird species migrate by night than by day, and nocturnal migration is particularly common among long-distance migrants. It is therefore hard to tell without more detailed analysis whether vagrancy is associated with nocturnal migration as such, or with the longer and more difficult journeys that nocturnal migrants often make. What is clear, however, is that vagrancy cannot be explained in terms of only one phenomenon, even though some of the proposed mechanisms are no more than working hypotheses.

SUMMARY

Vagrants are birds that appear from time to time far removed from their usual haunts, in localities where they do not normally breed, winter or occur on passage. They are seen most numerously on islands near the edges of continents, or in other coastal localities. Their appearance is to some extent predictable, and particular species may turn up at more or less the same places and in the same months in different years. Almost all vagrants derive from long-distance migratory populations, and occur away from their regular range at their normal migration seasons. Their appearance can often be linked to natural processes, such as post-fledging dispersal (in any direction) and off-course drift by wind.

For some species, vagrancy can be linked to apparently recurring 'flaws' in the inherent timing and directional control mechanisms. One apparent error in timing produces longer-than-usual migrations, in which birds fly in the usual direction but much further than usual. This leads to long-distance overshooting, and the appearance of individuals far beyond their regular breeding or wintering areas. Spring overshoots are often associated with prolonged periods of favourable tailwinds.

In many species, juveniles migrating for the first time show a greater spread of directions than adults, so that juveniles more frequently end up outside the usual route for their population. In some species, however, distinct directional errors seem to occur, including: (a) mirror-image migration in which birds take the same angle as normal, but on the wrong side of the north–south axis from either the breeding or the wintering area; and (b) reversed-direction migration, in which birds take the spring direction in autumn, or the autumn direction in spring. Again, both types of movement can be assisted by winds, with birds travelling for many hundreds of kilometres in a wrong (but consistent) direction. Possible bias in observer coverage throws doubt on some apparent examples of mirror-image and reversed-direction migration, and neither mechanism can be considered as proven or disproven. There can be no doubt, however, that vagrancy is a natural phenomenon resulting from several different types of causal factors.

Part Two
The Timing and Control of Migration

Ringed Plover *Charadrius hiaticula* incubating

Chapter 11
Annual cycles

For everything there is a season, and a time for every purpose. (Ecclesiastes 3: 1–2.)

In our present context, migration cannot be considered in isolation from breeding and moult – the other major events in the annual calendar of birds. Comparing species, these three activities vary in their timing, in the sequence in which they occur, and in the extent to which they overlap with one another. These variations are in turn linked with features of the species themselves, and the circumstances in which they live. Most bird species have only one breeding period each year, during which they raise one or more broods, but some (mainly tropical) species have two separate breeding periods. Many species also have only one moulting period each year, but some have two or more; and most species migrate twice each year (to and from their breeding areas), while others move three or more times or only once (from one breeding area to another). This chapter is concerned with how the variations in annual cycles between bird species are regulated by both external and internal influences. In this context, migratory birds are of special interest because, unlike most other animals and plants, they are exposed within each calendar year to conditions in more than one part of the world.

In matters of timing, whether of migration, breeding or moult, it is helpful to distinguish between ultimate and proximate causal factors (Baker 1938, Lack 1954). The ultimate factors include those aspects of environment, such as seasonal fluctuations of food supply, that influence the timing of various events through their effects on the survival and reproductive success of individual birds. They thereby influence the optimal period for the bird to undertake particular activities and, through the action of natural selection, favour individuals that organise their annual cycles in the most effective manner. The proximate factors are those, such as daylength, that birds can use as reliable cues to begin their preparation for migration, breeding and moult at appropriate dates each year.

VARIATIONS IN ANNUAL CYCLES

Most parts of the world are seasonal, in terms of daylength, warmth or precipitation, and hence also in terms of biological productivity. In particular regions, food supplies are therefore more plentiful at certain times of year than at others. At high latitudes the favourable season is relatively short, lasting only about one fourth of the whole year. Nutritionally, the most demanding event in the annual calendar of birds is reproduction, which normally overlaps the season of most abundant food supplies (Lack 1954). The other events fit around this. Migration is timed so that birds can be present on their breeding areas at least for long enough to breed successfully, taking advantage of the favourable season, but are absent for the unfavourable season when their survival chances would be much lower there. The first individuals normally arrive in their breeding areas as soon as conditions become suitable. In some species, especially at high latitudes, this is only a few days before nest-building and egg-laying begin, but in others up to several weeks might elapse between arrival and egg-laying (Chapter 14). Post-breeding migration is timed so that birds leave their breeding areas before their chances of survival there would become precarious.

Birds are further constrained in migration timing by the need for an integrated annual cycle, minimising the overlap with breeding and moult. All three processes need additional food intake, so for nutritional reasons these events are best separated as much as possible. In addition, while breeding, birds must remain in a fixed locality, at least until their young are well grown and mobile, which precludes simultaneous migration. Likewise, moulting birds might have missing and part-grown flight feathers, which could hamper migration (Hedenström 2004). Hence, for both extrinsic and intrinsic reasons, migration is normally confined to particular parts of the annual cycle, and in most species is separated from breeding and moult (for exceptions see below). Before they reach breeding age, immature birds of some species have more freedom in the times that they moult and migrate, and often do so at somewhat different times of year from adults, which are constrained by breeding.

The eight most common sequences of annual cycle events described among migratory birds are shown in **Figure 11.1**, and others could be added to this list. They reveal the various ways in which birds arrange the major events in their annual cycles to suit the circumstances in which they live and to take maximum advantage of breeding areas offering only a short favourable season. Comparing species, moult is much more variable in timing than is breeding or migration, probably because its

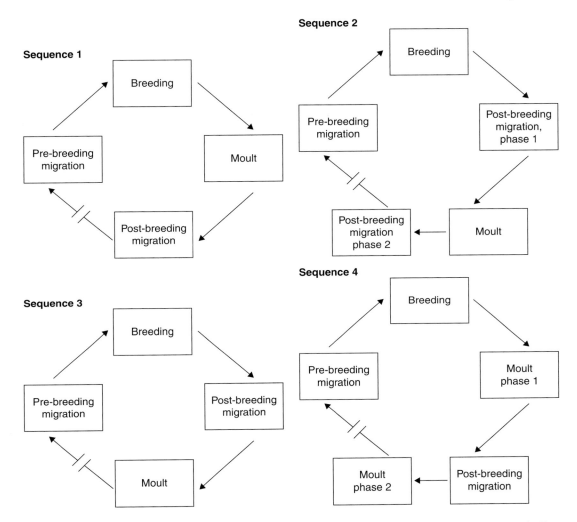

Figure 11.1   Some variations in the annual cycles of birds. The break in the arrows indicates a quiescent period, when the bird is not breeding, moulting or migrating. Sequence 1: Examples: Chaffinch *Fringilla coelebs*, Common Redpoll *Carduelis flammea*, Thrush Nightingale *Luscinia luscinia*, Fieldfare *Turdus pilaris*, Jack Snipe *Lymnocryptes minimus*. Sequence 2: Examples: Lazuli Bunting *Passerina amoena*, Great Reed Warbler *Acrocephalus arundinaceus*, Northern Lapwing *Vanellus vanellus*, Green Sandpiper *Tringa ochropus*, various ducks. Sequence 3: Examples: Scarlet Rosefinch *Carpodacus erythrinus*, Wood Warbler *Phylloscopus sibilatrix*, Garden Warbler *Sylvia borin*, Least Flycatcher *Empidonax minimus*, Orchard Oriole *Icterus spurius*, Barn Swallow *Hirundo rustica*, Common Swift *Apus apus*, Whimbrel *Numenius phaeopus*, and some shearwaters, terns and skuas that breed and winter in opposite hemispheres. Sequence 4: Examples: Yellow-bellied Flycatcher *Empidonax flaviventris*, Bonelli's Warbler *Phylloscopus bonelli*, Purple Martin *Progne subis*, Alpine Swift *Apus melba*, Scops Owl *Otus scops*, Bee-eater *Merops apiaster*, Red-necked Nightjar *Caprimulgus ruficollis*, Turtle Dove *Streptopelia turtur*, Osprey *Pandion haliaetus*, Honey Buzzard *Pernis apivorus*, Collared Pratincole *Glareola pratincola*, Marsh Sandpiper *Tringa stagnatilis*.

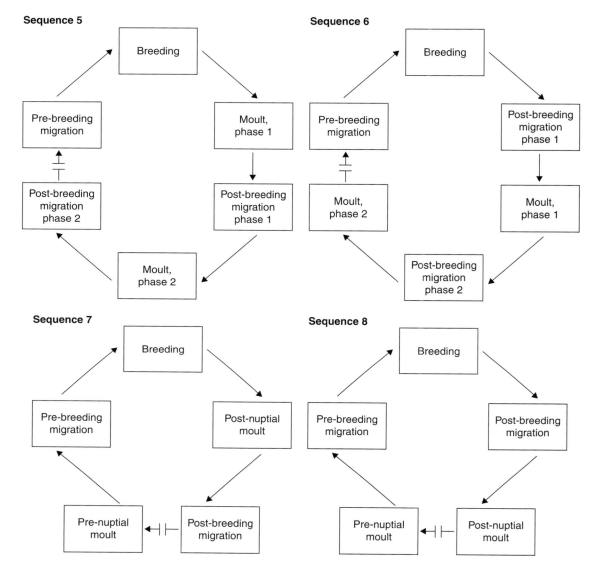

Figure 11.1   (continued) Sequence 5: Examples: Kentish Plover *Charadrius alexandrinus*, Spotted Redshank *Tringa erythropus*, some individuals of Curlew Sandpiper *Calidris ferruginea*, Red-necked Stint *Calidris ruficollis* and Solitary Sandpiper *Tringa solitaria*. Sequence 6: Examples: Wilson's Phalarope *Phalaropus tricolor*, Spotted Sandpiper *Actitis macularia*, and other populations of shorebirds. Sequence 7: Examples with two complete moults per year: Willow Warbler *Phylloscopus trochilus*, Bobolink *Dolichonyx oryzivorus*, Sharp-tailed Sparrow *Ammospiza caudacuta*; examples with one complete and one partial moult per year: Rose-breasted Grosbeak *Pheucticus ludovicianus*, Melodious Warbler *Hippolais polyglotta*. Sequence 8: Examples: Lanceolated Warbler *Locustella lanceolata* and some other *Locustella* warblers, Great Knot *Calidris tenuirostris*, Curlew Sandpiper *Calidris ferruginea*, Western Sandpiper *Calidris mauri*.

timing is less crucial than that of other events. Moult is necessary in birds because feathers wear and deteriorate, becoming less effective for flight and insulation. In general, residents and short-distance migrants moult in summer after breeding (residents more slowly, sequence 1), while long-distance migrants moult in late summer in the breeding area (as in sequence 1), in autumn at a migratory staging area (sequence 2) or in winter (sequence 3), depending on population (Pienkowski *et al.* 1976, Bensch *et al* 1991, Jenni & Winkler 1994). In many other migratory species, the moult is split, occurring partly in one area and partly in another, separated by migration. It can be split between breeding area and wintering area (sequence 4), between breeding area and a staging area (sequence 5), or between a staging and wintering area (sequence 6). The autumn staging area where moult occurs may lay hundreds or thousands of kilometres from the breeding area. The moult is normally arrested during the migration itself, so that the bird can fly with a full set of flight feathers, some new and others old. The bird resumes the second phase of moult wherever it left off in the first phase. In patterns 5 and 6, a split moult is associated with a split migration, which is halted while moult occurs. In other (mostly large) species, split moults are associated with breeding (as moult stops temporarily during chick feeding), or with periods of winter food shortage (for examination through modelling of moult scheduling, see Holmgren & Hedenström 1995).

While some migratory species have a single split moult, replacing their feathers once, but in two bouts, others have two separate moults, replacing the same feathers twice in one year. One moult occurs either before or after autumn migration (the so-called post-nuptial or pre-basic moult), and the other before or during spring migration (the pre-nuptial or pre-alternate moult) (sequences 7 and 8). In a small proportion of species that moult twice each year, such as the Willow Warbler *Phylloscopus trochilus* and Bobolink *Dolichonyx oryzivorus*, both moults are complete, involving the replacement of both wing and body feathers. But in most twice-yearly moulting species, the autumn moult is complete and the spring moult is partial, involving the replacement of the body feathers only (and sometimes a few tertial, secondary or tail feathers). In some species with two moults per year, both plumages are the same, but in other species at pre-nuptial moult the males or both sexes don a special breeding plumage.[1] Spring body moult occurs in many species of passerines, shorebirds and others, and usually overlaps with migration. In many species of diving and dabbling ducks, the second (pre-nuptial) moult (mainly body feathers) follows a few weeks after the first (post-nuptial) moult (complete),

---

[1]*In many scolopacid waders, the pre-nuptial moult is a double moult, involving the production over part of the body of two generations of feathers in quick succession. The first such plumage is distinguishable from the basic winter plumage but still cryptic, and the second 'supplementary' plumage comprises the showy breeding garb (Stresemann & Stresemann 1966, Jukema & Piersma 2000). In many species, these body moults begin in winter quarters and continue through migration.*

Other bird species acquire a special breeding dress by abrasion, when the dull tips of the body feathers wear off to expose the richer colours below. This is how the male Eurasian Linnet Carduelis cannabina *acquires its red breast and forehead, the male Brambling* Fringilla montifringilla *its black head, and the Snow Bunting* Plectrophenax nivalis *its striking black-and-white breeding plumage. In some species, the bill and other unfeathered areas also change colour in the breeding season, under the action of gonadal hormones. Thus, in both the male Brambling and the male House Sparrow* Passer domesticus, *the beak changes from pale to black for the breeding season.*

beginning in the sequence body – wing – body. In consequence, drakes are in dull 'eclipse' plumage for only a few weeks each year, and in bright breeding plumage for most of the year (Cramp & Simmons 1977, Bluhm 1988). In association with this, many species of ducks form pairs while in winter quarters.

The above generalisations apply to small or medium-sized birds in which the moult occurs as a distinct event in the annual cycle (**Table 11.1**). In some larger birds, breeding and moult may take so long that they cannot both be fitted within the annual cycle without overlapping (Stresemann & Stresemann 1966). In most raptors, for example, moult begins during incubation (earlier in females than males) and overlaps with most of the breeding cycle, although it may be arrested during chick-rearing. Small or medium-sized raptor species which migrate short distances can normally finish their moult before the post-breeding migration, but long-distance migrants typically arrest moult during migration, and continue after reaching winter quarters. In some of the largest flying birds, such as vultures, condors and albatrosses, each moult cycle lasts more than a year, but again may be arrested during difficult periods, such as chick-rearing. Otherwise such birds appear to moult more or less continuously, and may have two or more moult waves in the flight feathers at once (so-called serial moult, **Box 11.1**).

Box 11.1 Serial moult

In the first post-juvenile moult, a moult wave starting from the innermost primary spreads outwards feather by feather, but stops before it has reached the outermost. In the following year, a new wave starts from the innermost primary, but at the same time the first wave resumes from where it broke off. The bird then has two separate but simultaneous moult-waves in the primaries of each wing. This is the basis of the serial (stepwise) moult, found in various seabirds (including albatrosses, terns and cormorants), most large accipitrid raptors and large owls ('staffelmauser' of Stresemann & Stresemann 1966). In some species, 1–3 waves can be found in each wing at the same time, often with some asymmetry between wings. A serial moult enables a bird to change a number of primaries at a time, but with little impairment of flight performance, because growing feathers are interspersed among complete ones, rather than occurring adjacent to one another, creating a large gap. Each growing feather takes several weeks to reach full length. The secondaries moult from three (or more for long-winged species) different loci, and single feathers may last through the moult periods of one, two or three years. In White-tailed Eagles *Haliaetus albicilla* in Germany, moult occurs during April–October each year, but immatures can take 2–3 years to replace all their primary wing feathers, while adults take 3–4 years (Struwe-Juhl & Schmidt 2003). Black-browed Albatrosses *Diomedea melanophris* take four years to fully replace their large wing feathers for the first time. Within species, serial moults tend to be very variable in duration because the shedding of successive feathers occurs at longer intervals, or is suspended altogether, at times of food stress, such as chick-rearing and (in some species) migration. In addition, occasional anomalies occur in the normal sequence of feather replacement.

Table 11.1   Variation in moult schedules of various warblers, shorebirds and raptors, in relation to migration

Warblers

Moult in the breeding area before autumn migration	Dartford Warbler *Sylvia undata*, Subalpine Warbler *S. cantillans*, Lesser Whitethroat *S. curruca*, Whitethroat *S. communis*, Blackcap *S. atricapilla*, Eurasian Chiffchaff *Phylloscopus collybita*
Moult in the wintering area after autumn migration, following some body moult in breeding area	Greenish Warbler *Phylloscopus trochiloides*, Arctic Warbler *P. borealis,* Wood Warbler *P. sibilatrix*. Icterine Warbler *Hippolais icterina*, Sedge Warbler *Acrocephalus schoenobaenus*, Eurasian Reed Warbler *A. scirpaceus*
Split moult, partly in the breeding area before migration and partly in the wintering area after migration	Grasshopper Warbler *Locustella naevia*, Orphean Warbler *S. hortensis*, Bonelli's Warbler *Phylloscopus bonelli*
Two moults: the post-nuptial moult in the breeding area after breeding, and the prenuptial moult in the wintering area	Willow Warbler *Phylloscopus trochilis*
Two moults: post-nuptial and pre-nuptial, both mainly or entirely in the wintering area	Lanceolated Warbler *Locustella lanceolata* and some other *Locustella* species

Shorebirds[a]

(1) Moult in the breeding area before autumn migration	Killdeer *Charadrius vociferous*, Jack Snipe *Lymnocryptes minimus*, some populations of Purple Sandpiper *Calidris maritima* and Dunlin *Calidris alpina*
(2) Moult in the temperate wintering area after autumn migration	Some populations of Golden Plover *Pluvialis apricaria* and Red Knot *Calidris canutus*
(3) Moult in the tropical wintering area after autumn migration	Lesser Sand Plover *Charadrius mongolus*, Great Knot *Calidris tenuirostris*, Sanderling *Calidris alba*, Curlew Sandpiper *C. ferruginea*, Little Stint *C. minuta*, Western Sandpiper *Calidris maura*
(4) Moult at staging areas during autumn migration	Green Sandpiper *Tringa ochropus*, Northern Lapwing *Vanellus vanellus*, some populations of Semipalmated Sandpiper *Calidris pusilla* and Dunlin *Calidris alpina*
(5) Split moult, partly in or near the breeding area and partly in the wintering area	Collared Pratincole *Glareola pratincola*, Lesser Golden Plover *Pluvialis dominica*, Black-tailed Godwit *Limosa limosa*, Marsh Sandpiper *Tringa stagnatilis*
(6) Split moult, partly in or near the breeding area and partly at a staging area on autumn migration	Kentish Plover *Charadrius alexandrinus*, Spotted Redshank *Tringa erythropus*, some Red-necked Stints *Calidris ruficollis*, some Ruff *Philomachus pugnax* and some Common Snipe *Gallinago gallinago*
(7) Split moult, partly on a staging area during autumn migration and partly in the wintering area after migration	Wilson's Phalarope *Phalaropus tricolor*, Spotted Sandpiper *Actitis macularia*, some populations of Red Knot *Calidris canuta* and Dunlin *Calidris alpina*

(Continued)

Table 11.1   *(Continued)*

Raptors

Moult in the breeding area before autumn migration	Eurasian Sparrowhawk *Accipiter nisus*, Common Kestrel *Falco tinnunculus*, White-tailed Kite *Elanus leucurus*, Sharp-shinned Hawk *Accipiter striatus*, Common Black Hawk *Buteogallus anthracinus*, Red-shouldered Hawk *Buteo lineatus*, and other short-distance migrants
Moult partly in the breeding area before autumn migration, and partly in the wintering area after migration	Eurasian Hobby *Falco subbuteo*, Honey Buzzard *Pernis apivorus*, Mississippi Kite *Ictinia mississippiensis*, Swainson's Hawk *Buteo swainsoni*, Osprey *Pandion haliaetus*, Swallow-tailed Kite *Elanoides forficatus*, and other long-distance migrants

[a] In addition to a complete post-nuptial moult, almost all shorebirds have a partial pre-nuptial body moult, which starts in winter quarters, and in some species continues into spring migration.

From Cramp (1992), Cramp & Simmons (1983), Jenni & Winkler (1994), Kjellén (1994b).

It is thus not only the sequence of events in the annual cycle that varies among species, but also their duration. The breeding cycle in different bird species can last from a few weeks to more than a year, as can a complete cycle of feather replacement (the latter often with breaks). In some short-distance migrants, both autumn and spring journeys can each take less than one day, but in some long-distance migrants, the autumn journey can take up to five months and the spring journey up to three, so that individuals are on migration for more than half the year (again with breaks). Some pelagic bird species also have extremely long migration periods, as they are 'stationary' only in the breeding season and effectively on the move for the rest of the year. Moreover, even during the 'stationary' periods, individuals of some pelagic species regularly commute hundreds of kilometres between their nesting and feeding areas (Chapter 1).

Geographical variation within species

Even within species, geographical variations in annual cycles are apparent. With increasing latitude, the migrations of many species lengthen, and take up more of the year, while the periods devoted to breeding and moult decline in association with the decreasing length of the favourable season. Again, however, moult shows the greatest variation in its position in the annual cycle. In some species, populations at lower latitudes moult in breeding areas, whereas those from higher latitudes, where the favourable season is shorter, postpone moult for winter quarters. For example, Barn Swallows *Hirundo rustica* in the most southern breeding populations, which are resident or short-distance migrants, moult during June–August after breeding; whereas those in more northern populations begin moulting in September–October, after they have reached their distant wintering areas. In between, varying proportions of individuals show a split moult, starting in breeding areas, arresting during migration, and resuming in winter quarters (Cramp 1988). Likewise, most European populations of Ringed Plovers *Charadrius hiaticula* moult rapidly in their breeding areas in August–September, before migrating no further than southern Europe, whereas arctic-nesting ones leave their nesting areas after breeding, and postpone their moult until November–March after reaching their southern African wintering areas (Stresemann & Stresemann 1966). Other geographical variants in the timing and duration of moult occur in other shorebirds species, mainly in association with the latitude of breeding and wintering areas (see Cramp & Simmons 1983, especially for Knot *Calidris canutus*, Dunlin *C. alpina* and Little Stint *C. minuta*). In addition, in species with arrested moults, the position of the split varies to some extent between populations, according to regional variations in the length of the favourable season (Mead & Watmough 1976, Swann & Baillie 1979).

Variations within populations

In species in which only one sex looks after the young, or in which one sex gives up before the other, the two sexes also moult and migrate at different times. Such differences are especially marked in ducks and in some species of shorebirds (Chapter 15). In some shorebird populations, moreover, same-sex individuals from the same breeding population vary greatly in the relative amounts of moult undertaken in breeding, stopover and wintering areas. Some individuals suspend wing moult, while others go straight through. In some Dunlin *Calidris alpina* populations, some birds break migration for several weeks, when they moult rapidly, growing several feathers simultaneously, whereas others migrate by brief stopovers and shorter flights, moulting simultaneously but growing fewer feathers at once (Johnson & Minton 1980, Holmgren *et al.* 1993). Most Redshanks *Tringa totanus* that migrate along the coasts of western Europe and West Africa in autumn by relatively short flights seem not to arrest moult while on migration (Pienkowski *et al.* 1976). In some bird species, therefore, breaks in moult may be facultative, and dependent on individual circumstances.

An important feature of moult is that it can be arrested in a way that a successful breeding attempt cannot. In terms of control, it is as though the demands of breeding (and to some extent migration) can override those of moult, and slow or stop it when necessary (Wingfield & Farner 1979, Hahn *et al.* 1992). In some desert species that have flexible breeding seasons (see later), immediately following unexpected rainfall, breeding starts and moult stops at whatever stage it has reached (Payne 1972).

Split migrations

Split migrations are common in shorebirds and passerines, some of which moult over several weeks at a stopover site en route to winter quarters. Shorebird examples include some populations of Dunlin *Calidris alpina*, Curlew Sandpiper *C. ferruginea* and Purple Sandpiper *C. maritima*. Passerine examples include the Lazuli Bunting *Passerina amoena*, Western Tanager *Piranga ludoviciana* and Bullock's Oriole *Icterus bullocki* of western North America, the Great Reed Warbler *Acrocephalus arundinacous*, Garden Warbler *Sylvia borin*, Marsh Warbler *Acrocephalus palustris* and Thrush Nightingale *Luscinia luscinia* of western Eurasia–Africa, and the Yellow-breasted Bunting *Emberiza aureola* of eastern Eurasia (Young 1991, Stresemann & Stresemann 1966). In some western North American species, juveniles moult in breeding areas before migration, and adults at a staging site or in their wintering areas (Rohwer & Manning 1990, Butler *et al.* 2002). Some Common Starling *Sturnus vulgaris* populations in northern Europe show the opposite pattern, with adults remaining to moult in their breeding areas and juveniles migrating and moulting on a staging area (Chapter 15). Split migrations are also common in waterfowl, many of which first migrate to special sites where they moult, becoming flightless as they replace their large wing feathers (see 'Moult migrations', Chapter 16). Only after completing wing moult do the birds move on to wintering areas.

Relationship between breeding and moult

In general, in the northern hemisphere, overlap between breeding and moult increases with latitude (and associated shortening of the favourable season). This is evident in comparisons between closely related species, and between different populations of the same species. For example, breeding and moult overlap extensively in the arctic-nesting Ivory Gull *Pagophila nivalis* and Glaucous Gull *Larus hyperboreus*, but much less in more southern temperate-zone gull species, such as Black-headed Gull *L. ridibundus* and Lesser Black-backed Gull *Larus fuscus*. In the White-crowned Sparrow *Zonotrichia leucophrys*, overlap between breeding and moult increases with latitude (and altitude), while moult duration decreases from about 83 days in California to about 47 days in Washington State; that is, by 2.6 days per degree of latitude over a latitudinal span of about 14 degrees (Mewaldt & King 1978). Similar geographical trends in the overlap and duration of moult have been documented in many other species, including Common Starling *Sturnus vulgaris* (Lundberg & Eriksson 1984), Chaffinch *Fringilla coelebs* (Dolnik & Blyumental 1967), Willow Warbler *Phylloscopus trochilus* (Underhill *et al.* 1992), Blackcap *Sylvia atricapilla* (Berthold 1995), Dunlin *Calidris alpina*, and other shorebirds (Cramp & Simmons 1983).

Birds nesting in the high arctic have a very short season in which to raise young. They often have more synchronised and shorter breeding and moulting periods than closely related forms further south, and show more overlap between the different activities. For example, Snow Buntings *Plectrophenax nivalis* breed, moult and accumulate migratory fat within a period of only 10–12 weeks from mid-June, during which time food is very plentiful. They have one of the shortest nestling periods recorded among passerine birds (about 10 days), and while still feeding their young, the adults start moulting, shedding their flight feathers in such quick succession that for some days they can hardly fly. The whole moult is completed in about 4–5 weeks (Ginn & Melville 1983, Cramp & Perrins 1994). At the same time, they begin to put on weight for migration, leaving the bleakest breeding areas before the end of August. Many other arctic birds can only breed within this short period, postponing their moult until they have reached a more southern staging area or winter quarters. Moreover, if the spring is late, some species cannot breed at all that year, and usually pass earlier than usual to the next stage in their annual cycle, whether moult or migration. Other adaptations to a short arctic season found in some species include pair formation in winter quarters or en route, nest-building and egg-laying immediately after arrival, and accumulation of body reserves for egg production along with migratory fat deposition (as in geese and swans, Chapter 5).

Relationship between moult and migration

Whatever the annual sequence, moult is scheduled in most small and medium-sized bird species in such a way that no migrant has to fly with large gaps in its wings (Stresemann & Stresemann 1966, Payne 1972). The extent to which moult is completed before autumn departure may depend directly on the time available between the end of breeding and the start of migration (Berthold & Querner 1982a). While most passerines migrate only with fully grown flight feathers, birds that are under pressure to leave before conditions deteriorate sometimes start autumn migration before completing their moult (for Greater Whitethroat *Sylvia communis* see Hall & Fransson 2001, for Rose-breasted Grosbeak *Pheucticus ludovicianus* see Cannell *et al.* 1983). This applies particularly to late-moulting individuals (Chapter 12). Among many bird species, the juveniles replace only their body plumage, and not their flight and tail feathers, before autumn migration. Again, juveniles often start migration before their body moult has finished, especially those moulting late in the season (for examples see Ginn & Melville 1983).

Exceptions to general patterns

Species which feed on the wing, such as terns and hirundines, commonly replace their flight feathers while on migration, at least where food is abundant (for Black Tern *Chlidonias niger* see Zenatello *et al.* 2002, van der Winden 2002). In some tern species, the adults also feed their recently fledged young while on migration, giving overlap between these different activities. Some species of hirundines moult their feathers while breeding or while migrating through favourable terrain, but not when they have a long flight over sea or desert (Elkins & Etheridge 1977, Cramp 1988, Jenni & Winkler 1994). For example, Rough-winged Swallows *Stelgidopteryx serripennis* in eastern North America moult while migrating south over land, but pause for about two months on the north side of the Gulf of Mexico, crossing when they have finished wing moult (Yuri & Rohwer 1997). In addition, some auk species swim to their wintering areas, enabling the young to set off before they can fly, accompanied and fed by a parent en route. This behaviour facilitates overlap between breeding and migration, but is possible only because the young develop their insulating body plumage and can swim from an early age, migrating through continuously suitable habitat. Because the adults stay with their young, and can escape their main predators (large gulls) by diving, they also moult on migration, shedding and replacing all their flight feathers simultaneously. Some species of auks, such as the Common Guillemot (or Murre) *Uria aalge*, thus engage in breeding activity, moult and migration at the same time. Other exceptions to the main patterns can be found in the ornithological literature, while others are likely to emerge through future work.

Concluding comments on annual cycles

The main message to emerge from studies of the annual cycles of birds concerns the flexibility of such cycles and the enormous variation between species in the sequence of different events through the year, their duration and extent of overlap. This flexibility is manifest mostly by differences between species, but also between geographical populations of the same species, between sexes, and to

Table 11.2   Likely selection pressures and trade-offs influencing the timing of various events in the annual cycles of birds

	Advantage	Disadvantage
Arrive early in breeding area	Obtain good nesting territory and have a longer breeding season	Increased risk of adverse weather after arrival, loss of body condition or starvation
Arrive late in breeding area	Decreased risk of adverse weather, and hence loss of body condition and starvation after arrival	Relegation to poor territory, late start to breeding, and shorter breeding season
Breed as soon as possible after arrival	Often better able to exploit the peak food supply. Fit in additional (or repeat) breeding attempts	Increased risk of adverse weather during first nesting attempt, with higher costs to parents, especially laying females
Curtail period of parental care	May allow adults time for further breeding attempt or earlier moult or migration	May jeopardise survival of young
Finish breeding early in favourable season	Moult at optimal time, with good quality feathers	Fewer nesting attempts within season, or shorter period of parental care
Finish breeding late in favourable season.	More nesting attempts within season, or longer period of parental care.	Moult overlapped with other activities or compressed to less suitable time, producing poor-quality feathers
Moult in breeding area	Take advantage of summer food supply. Migrate with new feathers	May occupy time that could be used for extended parental care, or for raising an extra brood, or delay departure from breeding areas
Moult in wintering area	Allows longer breeding season in summering area; more time available for moult than in breeding areas, so requires less extra food per day	Adults must migrate with old worn feathers. Food may be scarcer than in summer, and winter movements could be hampered
Overlap between breeding and moult	Saves time in the favourable season	Increases the daily food needs, and could reduce fitness and feather quality of parents and young
Depart earlier in autumn	Avoid unseasonably cold weather; reach stopover sites before food depleted, and obtain best territories in winter quarters	Fail to make maximum use of the potential breeding–moulting season
Depart later in autumn	Make maximum use of potential breeding–moulting season	Risk of cold weather and reduced food supplies, which may prevent departure or cause starvation. May be relegated to poor territory in winter quarters
Migrate in short flights with minimal fat deposition	Saves time spent on fattening before starting migration, reduces predation risk, and saves fuel transportation costs	Lengthens journey time. Suitable refuelling places may be far apart
Migrate in long flights with high fat deposition	Shortens journey time, and allows crossing of extensive unfavourable areas	Long periods of fattening, with high predation risk, and high fuel transportation costs

Partly after Mead (1983).

a lesser extent even between individuals in the same population. Interruptions can occur in moult or migration to permit another activity to occur, and a small number of species, with short breeding cycles, can make long movements between successive broods within a single breeding season (Chapter 16). In general, large bird species take longer to breed and moult than small ones, and show greater overlap between these activities. The timing of migration clearly cannot have evolved independently of breeding and moult, but only in concert with them. Conflicting pressures operate in the timing of any event (**Table 11.2**), giving different optimal solutions in different circumstances. If they had nothing else to do, the best time for most birds to migrate would be in early summer, when food reaches its peak abundance. The fact that adult birds do not migrate then, but breed instead, is evidence that breeding takes precedence.

INTERNAL TIME KEEPING

While the ultimate (extrinsic) factor controlling the annual cycles of birds is the seasonality of the environment, the primary intrinsic (proximate) factor is apparently an endogenous rhythm within the bird. This self-sustaining rhythm tends to ensure that the major processes of migration, breeding and moult occur in the correct sequence each year, and at roughly the right times. The evidence for the existence of an internal rhythm has come largely from studies on captive birds kept for up to several years under rather specific constant daylengths (Gwinner 1968, 1971, 1972, 1981, 1986, Berthold *et al.* 1974b, Berthold & Terrill 1991). Such birds have no clue from the outside world as to what the date might be. Yet they usually moult and reach breeding and migratory condition in the correct sequence, and at roughly appropriate intervals, with corresponding cycles in body weights, gonad sizes and hormone levels. This finding is taken to imply the existence of some underlying 'endogenous' controlling system. However, in conditions of constant daylength, the cycles do not stick strictly to a year, but tend to drift, getting either shorter (rarely) or longer, hence the term circannual cycles, which typically last 9–13 months.

The existence of internal circannual rhythms, underlying the natural yearly cycles, and persisting for at least two cycles, has now been shown experimentally in more than 20 different bird species, including resident and migratory, temperate and tropical, passerines and non-passerines, as well as in other animals and plants (**Table 11.3**; Gwinner 1981, 1986, 1996b, Berthold & Terrill 1991). They evidently underlie the control of seasonal activities in a great variety of organisms. Among birds, such rhythms are expressed in gonad development, moult or 'migratory' fat deposition and restlessness. Garden Warblers *Sylvia borin* and Blackcaps *Sylvia atricapillus* kept under constant photoperiodic conditions (10 light:14 dark) have shown up to 10 successive moult cycles in eight calendar years, suggesting that, in these species, a circannual rhythm keeps running throughout the entire lifetime, requiring no obvious environmental stimulus. In such experiments, different birds housed in the same room got out of phase with one another, providing further evidence that the rhythms are endogenously controlled and expressed independently of environmental cues. Also, the fact that such cycles are shown by birds that have been hand-reared and kept under

Table 11.3 Evidence for endogenous control of various events in the annual cycles of birds kept under constant photoperiod regimes

Species	Light regime	Maximum duration (months)	Pre-nuptial moult	Spring fattening	Spring restlessness	Gonadal cycle	Post-nuptial moult	Autumn fattening	Autumn restlessness	Weight cycle	Source
Robin *Erithacus rubecula*	8L:16D	30				−	−			+	Merkel (1963)
	12L:12D	21								−	Merkel (1963)
	18L:6D	16					+			+	Merkel (1963)
Stonechat *Saxicola torquata*	12L:12D	144				+					Gwinner & Helm 2003
Dartford Warbler *Sylvia undata*	10L:14D	19					+		+	+	Berthold (1974a, 1974b)
Subalpine Warbler *Sylvia cantillans*	10L:14D	19	+	+			+		+	+	Berthold (1974a, 1974b)
Sardinian Warbler *Sylvia melanocephala*	10L:14D	19					+		+	+	Berthold (1974a, 1974b)
Marmora's Warbler *Sylvia sarda*	10L:14D	19					+		+	+	Berthold (1974a, 1974b)

Species	Light regime	n									Reference
Whitethroat *Sylvia communis*	9L:15D	24	+	+		+	+				Merkel (1963)
	12L:12D	24					+				Merkel (1963)
	19L:5D	24	+								Merkel (1963)
Garden Warbler *Sylvia borin*	10L:14D	34	+	+	+	+		+	+		Berthold et al. (1971), Berthold (1978a)
	12L:12D	32	+	+	+	+	+	+			Berthold et al. (1971, 1972)
	16L:8D	36	+	+	+	+		+			Berthold et al. (1971, 1972)
Blackcap *Sylvia atricapilla*	10L:14D	96				+				+	Berthold (1977, 1978a)
Chiffchaff *Phylloscopus colibyta*	12L:12D	27	−	+		+			−	−	Gwinner (1972, 1976)
Willow Warbler *Phylloscopus trochilus*	12L:12D	28	+	+	+	+		+	+	+	Gwinner (1968, 1971)
Collared Flycatcher *Ficedula albicollis*	13L:11D	31	+	+		+					Gwinner & Schwabl-Benzinger (1982)

(Continued)

Table 11.3 (Continued)

Species	Light regime	Maximum duration (months)	Pre-nuptial moult	Spring fattening	Spring restlessness	Gonadal cycle	Post-nuptial moult	Autumn fattening	Autumn restlessness	Weight cycle	Source
European Pied Flycatcher *Ficedula hypoleuca*	12L:12D	30					+		−	+	Gwinner & Schwabl-Benzinger (1982)
Starling *Sturnus vulgaris*	11L:11D	43				+	+				Gwinner (1981)
	12L:12D	43				+	+				Gwinner (1981)
	13L:13D	43				+	+				Gwinner (1981)
Dark-eyed Junco *Junco hyemalis*	9L:15D	18		+	+	−	−	+	+		Weise (1962), Wolfson (1959)
	12L:12D	12					−				Wolfson (1959)
White-throated Sparrow *Zonotrichia albicollis*	9L:15D	18	+	−	−	−					Weise (1962)

Species								Reference
White-crowned Sparrow *Zonotrichia leucophrys*	8L:16D	13	+			−	+	+	King (1968)
	12L:12D	47			+	−	−		Farner *et al.* (1980), Moore *et al.* (1982)
	20L:4D	13	+		−	+	+	+	King (1968)
Fox Sparrow *Passerella iliaca*	9L:15D	18	+	+	−	−			Weise (1962)
Dickcissel *Spiza americana*	12L:12D	21	+	+		+	+	+	Zimmerman (1966)
Red-billed Quelea *Quelea quelea*	12L:12D	29			+				Lofts (1964)
Chaffinch *Fringilla coelebs*	12L:12D	42	+			+		+	Dolnik & Gavrilov (1980)
	20L:4D	42	+			+		+	Dolnik & Gavrilov (1980)

+ = examined and response recorded, − = examined and no response recorded.

In addition to the above studies, male Dark-eyed Juncos *J. hyemalis* kept up to three years on constant dim light showed in that time up to three cycles of gonad growth and regression, spring migratory fattening, moult and spring and autumn restlessness (Holberton & Able 1992). The circannual cycles of individual birds drifted out of phase with one another, and averaged longer than a calendar year.

constant conditions from hatching further implies that they are innate, and do not arise from early experience (Gwinner 1981, 1996b, Berthold 1984b). The control centre of such cycles is unknown, but it almost certainly differs from that of the better-researched circadian (diurnal)[2] rhythm, control of which lies within the medial basal hypothalamus (Ball & Balthazart 2003).

Circannual cycles may be reflected in gonad condition alone, in moult alone, in migratory condition alone, or in any combination of these activities. The cycles can thus be viewed as consisting of separate but integrated components involving the different activities (Wingfield 2005). Which of these components are expressed in captive birds depends largely on the constant photoperiod to which the birds are exposed, and perhaps also on the time of year (=internal physiological state of the birds) when the experiment starts. In Starlings *Sturnus vulgaris*, for example, the range of constant photoperiods under which gonad size and moult are expressed is very narrow, about 11.5–12.5 hours (Schwab 1971, Gwinner 1996b). Outside this range, no proper gonad growth occurs (Dawson 2007). Evidently, photoperiods act as permissive factors, setting limits to the expression of the different components of circannual rhythms; each component may occur only if the experimental photoperiod is close to the range of natural daylengths to which the bird is normally exposed for that component.

In many species, long light periods tend to suppress gonad growth altogether, even over periods of several years (Sansum & King 1976, Wingfield & Silverin 2002), but they seldom suppress moult and fattening (**Table 11.3**). On specific experimental photoperiods, different activities can thus be eliminated from the annual cycle, separated or overlapped with other activities. For example, prenuptial moult and fat deposition, which normally occur in sequence, can be made to occur simultaneously, while moult or migratory fattening can be made to overlap with any stage of the gonad cycle, or to occur in the absence of a gonad cycle. Such findings are relatively rare, and restricted to particular photoperiods, but they again suggest that the different components of the annual cycle are to some extent independent of one another, and that the chosen photo-regime can affect the phase relations among them (King 1972). The different components are normally kept in phase with one another by strong controlling mechanisms, including the natural photoperiod.

It is also clear that different species react differently to the same constant daylength regime, presumably reflecting their adaptations to different daylength regimes in natural conditions (**Table 11.3**). For example, under a regime of 9L:15D, Dark-eyed Juncos *Junco hyemalis* and Fox Sparrows *Passerella iliaca* showed fat deposition and migratory restlessness (delayed by about two months on the usual spring timing), whereas White-throated Sparrows *Zonotrichia albicollis* showed no such response, but underwent two pre-nuptial moults at an appropriate interval during the 18-month experimental period (Weise 1962).

Spontaneous endogenous rhythms are most apparent in long-distance migrants, which are normally exposed to varying photoperiodic regimes on migration, and in which the need for some form of endogenous control is greatest (as appreciated

---

[2]*Kept under constant light, birds show inherent 'circadian' cycles in which periods of activity alternate with periods of rest, as in the natural light–dark cycle. These cycles approximate to 24 hours but, like circannual cycles, often tend to lengthen with time.*

long ago by Rowan 1926). They are also apparent in some resident species of equatorial regions, where daylengths are constant year-round. For example, equatorial Stonechats *Saxicola torquata axillaris* kept caged in constant 11.8L:11.2D conditions in Germany went through up to 12 reproductive-moult cycles in a 10-year period (Gwinner 1996b). However, in temperate zone residents and short-distance migrants caged in constant conditions, the cycles tend to continue for less long, and are more variable among individuals; they seldom proceed for more than one year, and the different events tend to become increasingly out of phase with one another. They are thus less rigid and persistent, as found in some populations of Blackcaps *Sylvia atricapilla* (Berthold *et al.* 1972) and in European (as opposed to African) Stonechats *Saxicola torquata* (Gwinner 1996b).

The importance of daylength

Under natural conditions, the endogenous cycles of many birds are kept in phase by seasonal daylength changes,[3] and in experimental conditions particular events can be advanced or retarded by appropriate use of an electric light (Farner & Follett 1966, Lofts & Murton 1968, Wolfson 1970). For example, if migratory birds of some species are exposed in late winter to photoperiods longer than natural days, their gonads begin to grow earlier than usual, and they show migratory and reproductive behaviour prematurely (Rowan 1925, 1926, Wolfson 1953, Lofts *et al.* 1963, King 1972).

The importance of daylength as a time-keeper (*Zeitgeber*) derives from its reliability. Its seasonal changes are consistent between years, making it the most obvious environmental feature that, at most latitudes, gives a reliable cue to date. The synchronisation of the internal annual cycle to photoperiod has been shown most convincingly in experiments in which birds were exposed to seasonal photoperiodic cycles with periods deviating from 12 months (e.g. six month cycles). As a rule, the birds' biological rhythms then conformed to the altered photoperiodic regime (Gwinner 1986, 1990b). For example, when the normal annual

---

[3]*Exceptions can be seen in cases like the Sooty Tern* Sterna fuscata *on Ascension Island, where the period of the cycle is about 9.6 months. This can be interpreted as a free-running endogenous rhythm, in which breeding and moult require a certain time, in an equatorial environment in which ecological conditions for breeding are nearly uniform throughout the year (Ashmole 1963). Selection has favoured maximum reproductive rate, and the cycle has lost (or not evolved) any coupling to environmental synchronisers (Lofts & Murton 1968). The same may apply to various species in equatorial South America where, in a semi-arid area with erratic rainfall at 3°30′N, eight out of ten bird species bred year-round, with the cycles of different individuals out of phase with one another (Miller 1954). In the Rufous-collared Sparrow* Zonotrichia capensis, *individual cycles averaged six months in duration (four months for breeding and two months for moult), and two complete cycles were manifest each year, 'uncoerced by small variations in photoperiod' and 'only incompletely controlled by the seasonal occurrence of rainfall' (Miller 1959). Outside equatorial regions, the same species has a single breeding period each year, which shortens with increasing latitude, as in other birds. Breeding–moult cycles shorter than one year have been described in other species elsewhere (for the Babbler* Stachyris erythroptera *and Little Spider-hunter* Arachnothera longirostra *see Fogden 1972b, for the Bat Hawk* Machaeramphus alcinus *see Hartley & Hustler 1993). As more studies are done in equatorial regions, other resident bird species may be found to show more than one cycle per year.*

cycle of daylength was shortened to six months without altering its amplitude, Garden Warblers *Sylvia borin* went through four instead of two moult periods within one calendar year, two instead of one gonad cycle, and four instead of two periods of migratory restlessness (Berthold 1996). The same occurred in Sardinian Warblers *Sylvia melanocephalus*, in which the usual one annual moult occurred twice within one calendar year (six months apart). It also occurred in Stonechats *Saxicola torquata*, which underwent two gonad and moult cycles in one calendar year (Gwinner & Helm 2003). More remarkably, Dark-eyed Juncos *Junco hyemalis*, which were exposed to four periods of short (9-hour) days and five periods of long (20-hour) days in one year, showed in this time five periods of gonadal activity, five of fat deposition and two of moult (Wolfson 1954). By shortening the photoperiodic cycle to two months, Starlings *Sturnus vulgaris* showed up to six gonadal cycles in one calendar year, but on this extreme regime the testes did not fluctuate over the full range and moult was disturbed and incomplete (Gwinner 1996b).

A second line of evidence for the role of the daylength as a time-keeper involves birds from the northern hemisphere that became established in the southern hemisphere by human action. Such birds normally adjusted to the local daylength regime within a year or so of their release. Their annual cycles remained essentially unaltered, except that they were six months out of phase with those in their original home (Aschoff 1955). This finding has been duplicated in many phase-shifting experiments in various captive birds (Gwinner 1986). Similar change occurred in most southern hemisphere birds transported to the northern hemisphere, although some exceptional species (such as Northern Rosella *Platycercus venustus* and Gouldian Finch *Poephila gouldiae*) bred in the same calendar months in both hemispheres, their breeding being unexpectedly resistant to change (Baker & Ransom 1938).

Despite the year-to-year consistency of daylength change, at particular dates the weather, food supplies or other conditions vary from year to year. It is therefore advantageous for wild birds to respond, not only to daylength but also to other secondary factors, and modify their activities to suit conditions at the time. Plant growth and invertebrate activity begin much later in cold springs than in warm ones, and any bird that did not respond appropriately to this variation could be seriously disadvantaged. Hence, secondary environmental factors may fine-tune the timing of various events to prevailing conditions: for example, enabling the bird to arrive on its breeding area and start nesting at what is an ecologically suitable time that year (Lack 1954, Wingfield *et al.* 1993, Wingfield & Jacobs 1999). In general, the effect of daylength (interacting with an endogenous rhythm) could be said to initiate the preparatory processes that precede each event. Then, as development proceeds, other modifying factors, such as food supply, mate and nest-site availability, come to play a greater role (for examples of food effects see Newton 1998a, for mate and social influences see Ashmole 1963, Lewis & Orcutt 1971, for nest-sites see Village 1990 and H. Gwinner *et al.* 2002).

Daylength has another useful property that is almost certainly used by birds, namely that its annual cycle varies in a consistent manner with latitude. This provides a means by which migrants could identify their latitudinal position with respect to breeding or wintering areas, enabling individuals from the same population to time their migrations in an appropriate manner from different latitudes within the breeding or wintering range. More generally, it provides a basis for selection to encode the seasonal behaviour of populations to latitudinal conditions. This facility is particularly evident in migratory populations (Chapter 12).

The endogenous rhythm in migrants

Migratory birds that breed at high latitudes and winter in regions close to and beyond the equator face two particular problems in the control of their annual cycles. First, birds that winter near the equator can spend half of each year under constant daylengths. Hence, in contrast to other species, such birds cannot rely on changes in daylength for timing those seasonal activities that occur in their wintering areas, notably the spring departure for breeding areas. Moreover, in many of these regions, other environmental factors also vary little or unpredictably through the year, and so are unreliable as seasonal cues. Second, unlike birds that remain at mid to high latitudes year round, long-distance migrants that cross the equator are exposed to relatively long days in both summer and 'winter' (= austral summer). They breed during the long days of the boreal summer, but not in those of the austral summer, even though conditions in their austral 'wintering' areas may be similar to those that stimulate reproduction in their northern breeding areas. These facts raise questions in such long-distance migrants: what prevents reproduction in wintering areas, yet stimulates pre-nuptial moult, migratory fattening and departure at appropriate dates? Again, the evidence suggests the intervention of an endogenous time-keeper.

In resident birds of higher latitudes, present in the same locality year-round, control of the annual cycle by an endogenous rhythm, kept to time by daylength changes, would seem relatively straightforward, because at any one locality, daylength varies in a consistent fashion from year to year, providing a reliable indication of date. In fact, in this situation an endogenous rhythm seems unnecessary. In long-distance migrants, however, the situation is more complicated because within the space of a few days birds can pass rapidly from one daylength regime to another. The most obvious way to cope with such problems is to restrict the period of response to daylength to only part of the year, using that period for 'clock-setting', and then allowing the internal rhythm to run for a period, regardless of external daylength. The existence of a 'refractory period', when birds do not respond by gonad growth to otherwise stimulatory daylengths, is consistent with such a mechanism, as is the finding above, that long-distance migrants (exposed to the most rapid and complex changes in daylength regime) rely more heavily on a self-sustaining internal rhythm than do short-distance migrants and residents (Gwinner 1972). The critical photoperiod for the ending of photorefractoriness, which prevents the gonads of migrants from developing in the austral summer, is related to the photoperiodic conditions of the 'wintering' areas (Gwinner & Helm 2003).

Not surprisingly, endogenous factors seem to have more influence on the timing of migration in species that migrate to the tropics and beyond, and winter far from their breeding areas, than in species that migrate short distances within the northern continents (Gwinner 1972, Hagan *et al.* 1991). Thus, the long-distance migratory Willow Warblers *Phylloscopus trochilus* that winter in the tropics, displayed in experimental conditions a firm endogenous control of migratory restlessness, with changes in body weight and moult persisting for more than two years in a constant 12-hour photoperiod (L12:D12). In contrast, closely related but short-distance migratory Chiffchaffs *P. collybita*, that winter in temperate and Mediterranean zones, lost any endogenous control of these activities within a year, so that body weight became almost constant, and migratory restlessness and moult became irregular or ceased (Gwinner 1971, 1972). Kept under natural daylengths, Willow Warblers moulted earlier and more rapidly than Chiffchaffs; they prepared for migration earlier, and showed more fattening and restlessness. Most of these differences persisted when birds were kept under constant daylengths, indicating some degree of endogenous control. The same was found for different races of Stonechats *Saxicola torquata* kept in the same, constant conditions (Gwinner & Helm 2003).

It seems, then, that in long-distance migrants that winter in the tropics and beyond, endogenous time programmes may operate for the entire life of the individual, and regulate the timing of all seasonal activities, whether reproduction, moult, fat deposition or migratory restlessness. Although the cycles do not require environmental cues to operate, but are intrinsic to the individual, they are normally synchronised by changes in daylength that keep the endogenous components in step with the seasons. Evidence of genetic influence on annual cycles derives partly from the fact that, on natural daylengths, first-generation hybrids have shown patterns of migratory timing and fat deposition that are intermediate between those of their parents (for waterfowl see Murton & Westwood 1977, for *Sylvia* warblers see Berthold 1984, 1990, for *Phoenicurus* redstarts see Berthold 1990, for Stonechats *Saxicola torquata* see Helm & Gwinner 1999). Other evidence of genetic influence comes from heritability studies based on parent–offspring comparisons (for breeding dates see van Noordwijk *et al.* 1981; for moult see Larsson 1996; for migration dates see Møller 2001, Pulido *et al.* 2001, Pulido & Berthold 2003, Chapter 20). Still further evidence comes from the changes that have occurred in wild birds in response to particular selection pressures (for laying dates of Great Tit *Parus major* see Visser *et al.* 1998, for spring migration dates of Cliff Swallow *Petrochelidon pyrrhonota* see Brown & Brown 2000).

Effects of delays

A major advantage of intrinsic control of the annual cycle is that events normally follow one another in an appropriate sequence. Nevertheless, birds whose spring migration is delayed also delay the start of breeding, birds that are breeding late delay the start of moult, and birds that are moulting late normally delay the start of migration (Chapter 12). However, there is a limit to the delaying process. Birds that arrive too late on their breeding areas may abandon nesting altogether that year (as in geese, Newton 1977). Birds that breed late may start to moult while they are still feeding young (as in Bullfinch *Pyrrhula pyrrhula*, Newton 1966), or birds that moult late may start migrating before they finish feather growth (as in Greater Whitethroat *Sylvia communis*, Hall & Fransson 2001). Similarly, Common Swifts *Apus apus* normally moult in their winter quarters, but when held up in Finnish breeding areas by bad weather, they started moulting in October at the normal time (Kolunen & Peiponen 1991). This again implies the influence of an underlying endogenous rhythm, which allows some delay but limits each event to an appropriate season (its 'time window'). Moreover, as is evident from both wild and captive birds, it is not necessary for a bird to complete one event in the annual cycle before it can start the next. Birds that do not complete a breeding cycle may nevertheless moult and reach migration condition at an appropriate time (although somewhat earlier than usual). The relationships between successive events in the annual cycle are explored further in Chapter 12.

Geographical variation in photoperiodic responses

With increasing latitude, the annual warm season becomes progressively shorter and the cold season longer. Correspondingly, most migratory bird species arrive later in spring, often breed over a shorter period, and leave earlier in autumn at higher latitudes than at lower ones (Chapter 14). Yet in the northern hemisphere, birds from the north of the range do not start breeding when they pass through the southern parts, at a time when individuals that breed in those areas have already started nesting. Nor do southern birds continue migrating northwards with others of their species once they have reached their own particular breeding areas. The implication is that birds nesting at different latitudes have different inherent responses to daylengths, thereby adjusting their annual activity cycles to the latitude at which they breed. Again, this view is amply confirmed by experiments involving the manipulation of photoperiod (for Dark-eyed Junco *Junco hyemalis oreganus* see Wolfson 1942, for rosy finches *Leucosticte* see King & Wales 1965, for Chaffinch *Fringilla coelebs* see Dolnik 1963, for Willow Warbler *Phylloscopus trochilus* see Gwinner 1972, for waterfowl see Murton & Westwood 1977, for *Ficedula* flycatchers see Gwinner 1990, for Great Tit *Parus major* see Silverin *et al.* 1993, for Stonechat *Saxicola torquata* see Helm 2003).

In general, populations breeding at higher latitudes require longer daylengths before gonad maturation occurs; but their gonads then remain active for a shorter period than in birds that breed at lower latitudes. Birds could adjust their response to different latitudes, under the action of natural selection, in at least three different ways. First, they could alter their rate of response to the same daylength stimulus. Second, they could alter the length of a latent (refractory) period, before a response became possible. Third, they could alter their photosensitive threshold, so that they needed a shorter or longer daylength (or different number of days with daylength above a minimum value) to trigger a response (Wolfson 1959, Marshall 1960, Murton & Westwood 1977). Which of these mechanisms birds use as they extend their breeding ranges into lower or higher latitudes is still uncertain. In addition, different species living together in the same area show differences in photosensitivity that correspond to differences in the timing of events in their respective annual cycles (for different species of pigeons see Lofts *et al.* 1967).

Even birds from mountain and lowland 100 km apart in the same region may show different circannual rhythms when kept in identical conditions in captivity (Widmer 1999). Nestling Garden Warblers *Sylvia borin* were collected from both the Upper Rhine Valley (200 m above sea level) and from the Central Swiss Alps (1,500 m), hand-raised and kept under standard conditions for nearly a year.

Under each of two different photoperiod treatments, lowland birds began spring migratory activity significantly earlier than montane birds. From this and other evidence, it was concluded that lowland and montane birds differ in their endogenously controlled migration programmes.

Similar findings emerged from comparison of the 'reproductive windows' (interval between testes growth and regression) of tropical and temperate Stonechat *Saxicola torquata* races kept under identical constant photoperiod conditions in captivity (Gwinner 1991, Helm *et al.* 2005). The tropical birds, which are normally single-brooded, had shorter reproductive windows and longer moults than the temperate zone birds, which normally raise 2–3 broods and then moult fairly quickly. First-generation hybrids between the two races were intermediate in their moults, again indicating genetic control of this process (reproductive windows were not checked). Under natural daylengths, juveniles of Siberian Stonechats *Saxicola torquata maura* (long-distance migrants) moulted faster, and at an earlier age, than those of European *S. t. rubicola* (short-distance partial migrants) and African Stonechats *S. t. axillaries* (non-migratory), and heritability values for moult duration in Siberian birds were high (based on full sibling comparisons) (Helm 2003). It need scarcely be added that the evolution of regional responses of this type depend on 'site-fidelity', in which individuals and their offspring remain in, or return to, the same general areas to breed each year (Chapter 17). Only then can populations evolve specific regionally appropriate responses.

Equatorial birds

Other questions hang over the role of external factors in regulating the annual cycles of birds that live year-round near the equator and experience more or less constant daylengths throughout their lives. Some such species live in relatively aseasonal habitats, such as rainforest, whereas others live in open country, where seasonal rainfall causes seasonal fluctuations in food supplies. In these latter species, the regular wet–dry seasons and associated food changes may act as external time-keepers. Nevertheless, some equatorial and desert birds show endogenous circannual rhythms when kept in constant conditions in captivity (Gwinner & Helm 2003), and can also respond to photoperiodic changes (Lofts 1964, Gwinner & Scheuerlein 1999, Bentley *et al.* 2000). In Panama at 9°N, the Spotted Antbird *Hylophylax naevoides* experiences daylengths that vary only between a minimum of 12 hours in December and a maximum of 13 hours in June. Yet in experimental conditions, individuals responded by song to an increase in photoperiod of only 17 minutes, and by gonad growth to an increase of 28 minutes (Hau *et al.* 1998).

Daytime light intensity, rather than duration, could act as a time-keeper for some equatorial birds. It changes greatly with cloud cover, and hence reflects the cycle of dry and wet seasons in relation to which many tropical birds breed. The experimental exposure of African Stonechats *Saxicola torquata* to a constant 12.25-hour photoperiod, but with cyclic changes in daytime light intensity, caused their gonadal and moult rhythms to become synchronised with the light intensity cycle. Control birds exposed to the same photoperiod, but to a constant high light intensity, were not synchronised, and showed variable responses. These results suggest a role for daytime light intensity as a circannual timing mechanism, and also provide a possible explanation for the strong responsiveness of African

Stonechats to photoperiodic change: light intensity and daylength may act syner-gistically on one and the same mechanism (Gwinner & Helm 2003).

FLEXIBLE CYCLES

Most birds breed regularly at the same times each year, but others breed at vary-ing times depending on the food resource, with knock-on effects on the timing of moult and movements. Given sufficient food, three main patterns have been recorded among such opportunists. The first involves a substantial extension of the normal breeding season, as found in Galapagos finches and others (Gibbs & Grant 1987). The second involves a main breeding season in spring and an additional one in autumn, separated by moult, gonad regression and re-growth, as noted in the Tricolored Blackbird *Agelaius tricolor*, Pinyon Jay *Gymnorhinus cyanocepha-lus* and others (Payne 1969, Ligon 1971). The third involves breeding in different months in different years, whenever food is sufficiently plentiful (often depending on irregular rainfall), as found in the Black-and-White Manakin *Manacus manacus*, Zebra Finch *Taenopygia guttata* and others (Snow 1962, Zann *et al*. 1995).

In the central Australian desert, rainfall is irregular, and at particular local-ities can fall at different times each year.[4] It seems that some desert birds remain ready to breed for much of the year, but nest only when stimulated by fresh rain-fall or the resulting surge in food supplies which can occur up to several weeks later (e.g. Serventy 1971, Zann *et al*. 1995, Leitner *et al*. 2003). They include not only landbirds, such as the Zebra Finch, which can breed for more than 10 con-secutive months if conditions remain suitable (Zann *et al*. 1995), but also various waterbirds which move rapidly into areas where rain has fallen. They breed in the resulting shallow lakes, and move on again when the lakes dry out (Frith 1959, Burbidge & Fuller 1982; Chapter 16). Examples include the Freckled Duck *Stictonetta naevosa*, Grey Teal *Anas gibberifrons*, Pink-eared Duck *Malacorhynchus membranaceus* and Banded Stilt *Cladorhynchus leucocephalus*. Such birds have a long 'reproductive window', but nest only at times within this period when other conditions are suitable. In contrast to breeding, moult in these desert birds occurs on a more fixed schedule, but if rain falls unexpectedly, moult may or may not be suspended as breeding begins (for Galapagos Finches see Snow 1966, for Budgerigar *Melopsittacus undulatus* see Wyndham 1981, for Zebra Finch *Taenopygia guttata* see Zann *et al*. 1995).

A similar opportunist strategy is adopted by Common Crossbills *Loxia curvi-rostra* in boreal regions. These birds depend on conifers, but the seeds of different species become available at different times of year. Seed abundance also varies greatly from year to year, so that crossbills face both temporal and spatial unpre-dictability in food supply (Chapter 18). They can breed over much of the year,

---

[4]*In deserts elsewhere in the world, rainfall is more regular in timing, even though it may be slight to non-existent in some years. The local birds therefore tend to breed at the same times in different years, though not necessarily every year. In some arid parts of the southwestern USA, the breeding seasons of some species, such as the Roadrunner* Geococcyx californianus, *tend to be bimodal, with little nesting in the hottest period from mid-June to late July (Ohmart 1973).*

but most nesting occurs some time between September and May, depending on local seed supplies (Newton 1972, Berthold & Gwinner 1978, Benkman 1987).[5] Yet despite this variable breeding season, and associated movement patterns, the single annual moult occurs consistently in July–September, sometimes overlapping with breeding (Newton 1972, Hahn 1998). However, despite seasonal overlap at the level of the population, individuals seldom migrate and moult at the same time. Experiments have shown that Common Crossbills and some other opportunist breeders can respond to photoperiod (Hahn 1998), even tropical ones which under natural conditions experience little or no daylength variation (for Red-billed Quelea *Quelea quelea* see Lofts 1964).

To judge from these various findings, aseasonal opportunistic breeders are not fundamentally different from regular breeders. Both groups depend partly on an endogenous rhythm, and both can respond to photoperiod in captivity, but opportunist breeders respond more strongly to prevailing ecological cues (notably food supplies) to fine-tune the timing of their nesting and movements (Wingfield 1980, Hahn *et al.* 1997). Some opportunist breeders can delay or arrest moult or movement in order to breed if food happens to be plentiful at the time these other activities would normally occur. Many northern birds, too, show great flexibility in the timing and extent of their movements, which vary from year to year in relation to food availability (so-called irruptive migrants, Chapters 18 and 19).

CONCLUDING REMARKS

In comparing the different annual cycles of birds, one of the most striking findings is the great variation in the sequence and duration of breeding, moult and migration among species, or even among different populations of the same species. This variation shows clearly how the cycles are constrained by features of the birds themselves and adapted to the areas and circumstances in which different populations live. In each population, the cycle is adjusted primarily to the seasonal changes in climate and food to which that population is exposed, migration timing having evolved in concert with the other events that make up the yearly cycle.

In the majority of bird species, breeding, moult and migration occupy short enough periods to be fitted into a calendar year with little or no overlap. In larger species, with longer breeding and moulting periods, overlap between these events is greater, and in the few species in which moult lasts longer than a year, it overlaps with both breeding and migration. Such extensive overlap occurs in some large seabirds, such as albatrosses, and large raptors, such as vultures and

---

[5]*The testes of male Common Crossbills are active from late autumn to the following summer (Berthold & Gwinner 1978). This allows the birds to breed in late summer or autumn, soon after they have found new areas of ripening cones, or in the winter and following spring. Juveniles may be fertile only a few weeks or months after fledging, and can start breeding before moulting out of juvenile plumage (Berthold & Gwinner 1978, Jardine 1994, Hahn* et al. *1997). Parallels to this situation are found among some nomadic species in the deserts of Australia (for example Zebra Finches* Taeniopygia guttata *move around with partially developed gonads, so they can breed soon after rain has begun, bringing an abundance of grass seed (Immelmann 1963).*

eagles. Even in these species, however, moult can be slowed or arrested tempo-rarily during chick-feeding, migration or other difficult periods.

Some parts of the annual cycle are apparently controlled intrinsically, namely the broad time windows of the different processes and the sequences in which they occur. The assumption is that each species has developed the timing and sequence of events that best fit the conditions in which it lives, and has evolved to respond appropriately to the daylength regime to which it is exposed. In this way, the various events occur year after year at appropriate dates. Nevertheless, cycles can be advanced, retarded, lengthened or shortened by changing the experimental photoperiodic regime to which the bird is exposed. This is regarded as response to a time-keeper.

The main uncertainty concerns the relative importance of endogenous control in different species, and the extent to which an internal rhythm can run automat-ically on constant daylengths. One major source of variation between species is the time that autonomous cycles continue in captive birds in the absence of pho-toperiodic change. In many residents or short-distance migrants from mid–high latitudes, this may be less than one year, but in populations that are resident in the tropics or migrate there for part of the non-breeding period, the cycles may persist much longer – in some populations year after year, throughout a bird's life. In resident temperate zone birds, living year-round at the same latitude, strong endogenous control would seem unnecessary, because they are exposed to the same daylength regime year after year, which could therefore control all pro-cesses directly, without the need for an internal rhythm. It is uncertain whether photoperiod acts in any species without an endogenous rhythm (i.e. entirely as a driver rather than as a permisser and synchroniser), but it could only do so if birds could distinguish shortening from lengthening daylengths, otherwise they could give the same responses in autumn and spring, rather than different responses. But in long-distance migrants, which can winter on the equator with little or no daylength change, it is hard to see how individuals could set off for their breeding areas every year at an appropriate date without an internal clock to trigger the return, or at least prevent it from happening until after a certain time has elapsed. Other environmental factors, such as rainfall or food supply, might theoretically act as external cues for migration, but in practice are far too variable from year to year to act as reliable cues to date.

It is also hard to see how northern hemisphere migrants that winter in the southern hemisphere could avoid breeding during the austral summer with-out some internal clock mechanism to prevent gonad development until a more appropriate date. The same applies more generally in the existence of seasonality in response periods, when long days at one time of year might stimulate breed-ing, moult or migration, but not at another (so-called photo-refractoriness), and also influence how long these various processes take (Chapter 12). The adaptive value of a temporary suppression of response (refractoriness) is to prevent breed-ing (or other processes) at seasons when stimulatory daylengths occur, but when it is disadvantageous for other reasons to attempt that activity.

The combination of an endogenous rhythm as a template for seasonal activ-ity, together with daylength as a synchroniser, provides many birds with a basis for seasonal timing that operates well under variable seasonal conditions and movement patterns. This dual system gives reliability, precision and flexibility.

Endogenous rhythms help to buffer short-term environmental influences, and ensure that different events occur within an appropriate time-period, regardless of unusual external conditions. But dependence exclusively on internal control would carry the risk that any slight deviation in the endogenous rhythm would uncouple the bird's activities from the external seasons. Similarly, dependence entirely on photoperiod (or other external cues) would make a bird vulnerable to exceptional events, such as an unavoidably longer than usual stay in wintering or migration areas with a different daylength regime. Only the combination of endogenous rhythm and photoperiodic response provides the bird with a sound basis for seasonal timing, enabling preparation for each event in good time, and promoting the slowing or speeding of successive processes depending on whether they are early or late with respect to prevailing daylengths and other environmental conditions.

SUMMARY

Depending largely on the environments in which they live, bird populations show great variation in the timing and sequence of the major events in their annual cycles. In most species, for nutritional and other reasons, migration, breeding and moult occur at different times of year, but in the same sequence every year. In some species, wing moult overlaps extensively with breeding but, in general, neither process overlaps extensively with migration, although exceptions occur. In many migratory species, the autumn migration and moult are split into two or more stages, each occurring in a different area.

The main ultimate factor governing the annual cycles of birds is assumed to be the seasonality of the environment, and its effect on food supplies. The main proximate factor is apparently an endogenous rhythm within the bird, which is entrained by seasonal changes in daylength. This self-sustaining rhythm is revealed in captive birds kept for years on constant photoperiods. Such birds may depart from the usual annual periodicity, but typically show repeated periods of gonad activity, moult or migratory restlessness and fattening, depending on the photoperiod on which they are kept.

In general, endogenous cycles continue for longer under constant conditions in tropical species and in long-distance migrants than in temperate zone residents or short-distance migrants. Species that breed and winter in opposite hemispheres, and experience long days in both summer and 'winter', apparently rely on this internal rhythm to suppress reproduction in winter quarters, and those that winter there or in equatorial latitudes (where daylengths are constant year-round) apparently rely on the internal rhythm to prevent homeward migration until an appropriate date in spring.

At particular latitudes (away from the equator), daylength alters in a consistent manner from year to year, and therefore gives a reliable indication of date. Evidence that birds respond to daylengths (through adjustment of the internal rhythm) comes from several findings. First, gonad growth, moult or migratory activity can be advanced or delayed by experimental exposure to longer or shorter photoperiods. Second, birds can be made to undertake more than one 'annual' cycle in a calendar year by alternating exposure to long and short photoperiods.

And third, birds adjust their annual cycles when transported to the opposite hemisphere, becoming about six months out of phase with conspecifics in the original home area (an observation that can be duplicated by experiment). Birds also respond to secondary stimuli, such as temperature, food supply or social conditions, to fine-tune the timing of various events to conditions at the time, and can delay or delete particular events, as the need arises.

Different species have different inherent cycles, and respond differently to the same daylength regime, so that their annual cycles are adjusted to their particular needs. Captive-bred hybrids behave intermediately, implying genetic control of the endogenous response mechanism. Similarly within species, birds breeding or wintering at different latitudes respond to the local daylength regime, in such a way as to reach breeding, moulting or migratory condition at times appropriate to the latitudes concerned.

White-crowned Sparrow *Zonotrichia leucophrys*, commonly used in experimental work on migration

Chapter 12
Control mechanisms

Yea, the stork in the heavens knoweth her appointed time; and the Turtle (Dove), and the Crane, and the Swallow observe the time of their coming. (Jeremiah (viii.7).)

This chapter is concerned with the control of migration – with the factors that stimulate migration at appropriate times of year and influence the preferred directions. It is concerned with the external factors to which birds respond, such as daylength and food supply, and also with the internal regulating mechanisms. It thus attempts to integrate the findings from both field and laboratory studies, and develops some aspects of this subject area already touched upon in Chapter 11.

The behavioural changes necessary for migration involve an urge to depart given suitable weather, and a tendency to fly in one particular direction rather than others. In addition, many normally diurnal birds also become active and migrate at night. The symptoms of this 'migratory state' are easily noticed in captive birds which at appropriate times of year develop 'migratory restlessness', when they hop and flutter round their cages and show long periods of wing-whirring (fluttering the wings rapidly while perched). Some species, such as White-crowned Sparrow *Zonotrichia leucophrys*, also spend long periods pointing the bill skywards, and fluttering upwards, which in a netting-topped cage provides a broad view of the night sky (Ramenofsky *et al.* 2003).

Because most birds cannot forage while flying, the chief physiological change necessary for migration involves the accumulation of fat and other body reserves to sustain the bird and fuel its flight (Chapter 5). The symptoms of this state include increases in the food intake and weight of the bird, and the appearance of a yellow colour (due to fat) beneath the skin. Fat also accumulates at other sites around the body, mainly in the tracheal pit (at the base of the neck) and among the viscera (Berthold 1996). Once adequate fat has accumulated, migration (or migratory restlessness) usually follows but, in the wild, adverse weather might delay the date of departure (Chapter 4).

OBLIGATE AND FACULTATIVE MODES

In considering the proximate control of migration, a useful distinction can be drawn between obligate migration (formerly called instinct or calendar migration) and facultative migration (formerly called weather migration). In obligate migration, all main aspects are viewed as under firm internal (genetic) control, mediated by daylength changes, which gives a high degree of annual consistency in the timing, directions and distances of movements (Chapter 20). For the most part, each individual behaves in the same way year after year, migrating at similar dates and for similar distances. Obligate migrants often leave their breeding areas well before food supplies collapse, and while they still have ample opportunity to accumulate body reserves for the journey. They tend to migrate long distances, often to the tropics or beyond.

In contrast, facultative migration is viewed as a direct response to prevailing conditions, especially food supplies, and the same individual may migrate in some years but not in others (Chapter 20). Within a population, the proportions of individuals that leave the breeding range, the dates they leave and the distances they travel, can vary greatly from year to year, as can the rate of progress on migration, all depending on conditions at the time (e.g. Svärdson 1957, Terrill 1990, Moore et al. 2003). In consequence, facultative migrants have been seen on migration at almost any date in the non-breeding season (at least into January in the northern hemisphere), and their winter distributions can vary greatly from year to year (Chapters 18 and 19). Although in such facultative migrants, the timing and distance of autumn movements may vary with individual circumstances, other aspects must presumably be under firmer genetic control, notably the directional preferences and the tendency to return at appropriate dates in spring. Compared with obligate migrants, facultative migrants tend to migrate shorter distances, although many exceptions occur. The two types of migrants thus have different distribution patterns in midwinter. Whereas obligate migrants are concentrated in a distinct wintering area, usually at long distance from the breeding area, facultative migrants are typically found over the whole migration route from breeding to wintering areas, usually tailing off with increasing distance from breeding area, but with marked annual variations.

In general, it seems that obligate migration occurs in populations whose food supplies in breeding areas are predictably absent in winter, whereas facultative migration occurs in populations whose food supplies in breeding areas vary greatly from one winter to another, according to weather or other variables. The distinction between obligate and facultative migrants is important because it reflects the degree to which individual behaviour is sensitive to prevailing external conditions, and hence varies from year to year. However, obligate and facultative migrants are best regarded, not as distinct categories, but as opposite ends of a continuum, with predominantly internal control (= rigidity) at one end and predominantly external control (= flexibility) at the other.

Another reason for not drawing a sharp distinction between the two categories is that many birds seem to change from obligate to facultative mode during the course of their journeys, as the endogenous drive to migrate wanes with time and distance, and the stimulus to continue becomes more directly dependent on local conditions (Helms 1963, Terrill & Ohmart 1984, Gwinner *et al.* 1985a, Terrill 1990). Theoretically, the initial obligate phase of any journey might take the migrant across regions where the probability of overwinter survival is practically zero: where any individuals that attempted to winter there in the past were eliminated by natural selection. As migration continues into more benign areas, and survival probability increases, the bird switches to a facultative mode, in which it benefits by responding to local conditions, stopping where food is abundant. The obligate phase would therefore be expected to be undertaken much more rapidly, on average, than the facultative phase, which involves longer and more variable stops. Such a two-phase migration, with obligate and facultative stages, would also ensure that, in any particular year, the bird migrated no further than necessary. In some species only the tail end of the migration may be facultative, in others the entire journey. Most irruptive migrants are near the latter end of the spectrum.

Arctic-nesting geese provide circumstantial evidence for a two-phase migration, in which the first part is obligatory and the second part facultative. Geese need to leave the arctic every year before survival there becomes impossible, and they tend to depart en masse on about the same dates every year. But once they reach suitable wintering areas, their movements become much more variable in timing and extent, depending on local food availability. They appear to change from a primarily endogenous migratory phase (obligate migration) to a stage when the stimulus for further migration is primarily environmental. In effect, as birds travel south in autumn, the drive to continue becomes increasingly dependent on food and other local conditions (Terrill 1990). The same holds for many other migrants, notably irruptive seed-eaters (Svärdson 1957, Newton 1972, Köenig & Knops 2001), but also American Tree Sparrows *Spizella arborea* (Niles *et al.* 1969), Chipping Sparrows *Spizella passerina* (Pulliam & Parker 1979), Snow Buntings *Plectrophenax nivalis* (Haila *et al.* 1986), Blackcaps *Sylvia atricapilla* (Klein *et al.* 1973), Yellow-rumped Warblers *Dendroica coronata* (Terrill & Crawford 1988), and many others (Chapters 18 and 19).

Further support for the idea that migration is often two-phase comes from studies of captive birds. In White-throated Sparrows *Zonotrichia albicollis*, Helms (1963) identified two subdivisions of migratory behaviour in both autumn and spring. The first phase (which he called the motivational subdivision) was characterised by intense and continuous night-time activity, while the second phase (the 'adaptational subdivision') was less intense, with numerous interruptions and greater variability. Helms (1963) aligned these two phases with the behaviour of free-living birds during spring migration, as they switched from an intense, highly directed phase to a more casual 'wandering phase', in which they searched for suitable habitat and took advantage of local opportunities. In addition, observations of the directional preferences of caged migrants revealed an increasing variance in headings towards the end of the migratory period (Wiltschko & Wiltschko 2003).

Experiments on captive birds have also confirmed that individuals can develop migratory restlessness in response to food deprivation in winter, well outside the normal migration period (Biebach 1985, Gwinner *et al.* 1985a). For example, Garden Warblers *Sylvia borin* showed spontaneous nocturnal restlessness during the autumn migration period (September–December), but not in January when they would normally have settled in winter quarters. However, they again became active at night in winter if deprived of food (Gwinner *et al.* 1985a). The fact that deprived birds could put on fat may seem surprising, but they seemed to do so through changes in physiology and behaviour, feeding for much longer each day than normal. After late January, migratory activity could not be reactivated when birds were subjected to restricted food treatment.

Several researchers, working with different species, have been able to initiate or enhance nocturnal restlessness in autumn by restricting food or lowering temperature (reviews Farner 1955, Helms 1963). It is not known whether the birds responded to temperature directly or through their energy needs and body condition.

Role of dominance in facultative migrants

Because birds compete for food, and vary in dominance or feeding efficiency, some individuals could survive in conditions where others would die unless they moved out. In facultative migrants, the subordinate sex and age groups typically migrate in greater proportions, at earlier dates, or extend further from the breeding areas, than the dominants (Chapter 15). Thus, in many bird species, adult females are more migratory than adult males, juveniles more than adults, and late-hatched young more than early-hatched ones (Chapter 15; Michener & Michener 1935, Gauthreaux 1982a, Smith & Nilsson 1987). Such differences have led to the notion that competition (or its effect on body condition) is involved as a proximate mechanism stimulating migration in those individuals least able to survive in local conditions (Gauthreaux 1982a).

Depending on circumstances, the same individual might migrate in one year but remain in the breeding area another (for Song Sparrow *Melospiza melodia* see Nice 1937, for Mockingbird *Mimus polyglottos* see Brackhill 1956, for Common Starling *Sturnus vulgaris* see Kessel 1953). The commonest pattern is for a bird to migrate in its first year and not thereafter, but other sequences occur (e.g. for Blackbird *Turdus merula* see Schwabl 1983), presumably because foraging conditions in the breeding area vary from winter to winter. The sequence of behavioural events stimulating migration in facultative migrants could thus be hypothesised as: social status → competition → failure to obtain a winter territory or sufficient food → difficulty in maintaining body condition → departure. On this mechanism, the overall proportion of birds stimulated to migrate would depend on feeding conditions that year, with more birds from the dominant age

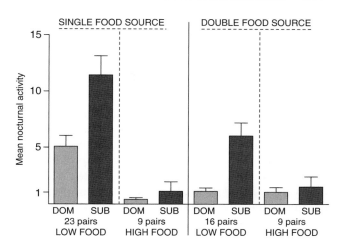

Figure 12.1 Effect of social environment, food availability and distribution on migratory restlessness of Dark-eyed Juncos *Junco hyemalis*. Low food: 8 g of food per day for each twosome; high food: 14 g of food per day for each twosome. Single source: food placed in a single container in the centre of each pair's cage; double source: food divided between two containers. SUB: subordinate partners of twosomes; DOM: dominant partners. From Terrill (1990).

groups departing in poor food years than in good ones (Kalela 1954). That the effects of food supply could be mediated by social dominance was shown in captive Dark-eyed Juncos *Junco hyemalis*, in which subordinate individuals that suffered the greatest deprivation were most likely to accumulate fat and show migratory restlessness (**Figure 12.1**, Terrill 1987).

MIGRATION TIMING, DISTANCES AND DIRECTIONS

Much has been learned about the proximate control of migration from studies on captive birds. Under natural daylengths, caged birds from obligate migratory populations develop fat reserves and migratory restlessness at appropriate dates in autumn and spring, at about the same times as their wild counterparts. Evidently, the same factors that stimulate departure in wild birds trigger restlessness in captive ones kept on natural daylengths (Gwinner 1972, Berthold 1996).

The role of daylength in promoting these processes (in association with an endogenous rhythm) is shown by findings discussed in detail later, namely that: (1) captive birds experimentally exposed to photoperiods longer than natural days in spring develop migratory condition earlier than their wild counterparts; and (2) birds exposed to photoperiods shorter than natural days in late summer or autumn also develop migratory condition earlier than their wild counterparts.

Time and distance programmes

In general, the longer the distance between breeding and wintering areas, the greater the duration and intensity of migratory restlessness shown by caged birds (**Box 12.1**). Different *Sylvia* warblers migrate average distances varying from a few hundred to nearly 6000 km, and show corresponding average periods

Box 12.1 Migratory restlessness

When caged birds reach a migratory state, they usually show periods of wing-whirring, together with increased perch-hopping, body-turning and other activities. Cages are too small to allow flight and, while birds that are caged straight from the wild flutter around and try to escape, they eventually adjust to cage life, and show wing-whirring instead – a cage-adapted behaviour, viewed as 'migration in sitting position' (Berthold 1996). Migratory restlessness (or *Zugunruhe*) has now been recorded in caged birds from more than 100 different species. It can be quantified from the alteration of the normal pattern of diurnal locomotor activity, especially the increase of activity peaks at species-specific diurnal migration times. The behaviour can be recorded automatically in various ways. By subtracting activity patterns obtained in non-migratory periods from those recorded during migratory periods, migratory restlessness can be quantified, basing the comparison on daytime or night-time activity, as appropriate.

The fact that this behaviour in caged birds is equivalent to migration in wild ones is indicated by three types of observation. First, it is typically most developed in birds from migratory populations and much less developed in birds from non-migratory ones. Second, in more than 25 species and populations studied in detail, migratory restlessness is broadly related to the specific migratory seasons of the population concerned. Third, at least in some species, the duration of the autumn period of migratory restlessness corresponds to the length of the autumn migratory journey (**Figure 12.2**). Fourth, in nocturnal migrants the behaviour occurs at night and in diurnal migrants in the daytime.

However, not all nocturnal activity by normally diurnally active birds can be viewed as migratory restlessness. Some species in captivity continue to show restlessness beyond the normal migration seasons, especially in spring when it may continue in non-nesting birds until the late summer moult. Others show evening restlessness when they would normally fly to a roost site, and yet others show restlessness at times of natal dispersal in late summer. For these latter activities, however, birds show none of the marked directional preferences which are usual only at migration times (although few researchers have yet tested both migratory restlessness and directional preferences in the same birds). Moreover, the relationship between duration of autumn *Zugunruhe* in captive birds and length of migration in wild birds held in some but not all species studied (Gwinner 1986). These findings led some authors to question the relationship between *Zugunruhe* and migration at certain times of year and in certain resident species, suggesting that it may represent a state of readiness for various activities, and not only migration. At the least, it seems that *Zugunruhe* is a character which is strongly, but not consistently, related to migration.

of restlessness varying from less than 20 to more than 1000 hours (**Figure 12.2**). The Marsh Warbler *Acrocephalus palustris,* which breaks its autumn journey between Europe and southern Africa for up to several weeks in equatorial Africa, shows a prolonged and two-phase pattern of restlessness in autumn in captivity;

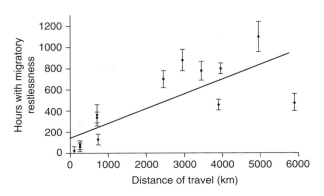

Figure 12.2 Relationship between the usual distance migrated and the number of nights that captive juveniles showed migratory restlessness for different populations of *Sylvia* warblers. The longer the journey undertaken by wild birds, the more nights of migratory restlessness were shown by captive birds from the same population. From Berthold (1973).

but in spring when it returns in a shorter single period, it shows a shorter single period of restlessness in captivity (Berthold & Leisler 1981, Berthold 1993). In addition, the amount of fat accumulated by captive birds at migration times is related to the types of journeys they make in the wild. Birds that migrate by long flights, as from Europe to sub-Saharan Africa, typically accumulate more fat in captivity than do birds that migrate short distances within Europe (Chapter 20; Berthold 1973, 1984a). This contrast is evident in comparisons between related species, such as Willow Warbler *Phylloscopus trochilus* and Chiffchaff *P. collybita* (Gwinner 1972), and between different populations of the same species which migrate different distances, as in the Blackcap *Sylvia atricapilla* (**Figure 20.2**; Berthold & Querner 1981).

Even more remarkably, sex differences in the timing and duration of migratory restlessness emerged in captive White-throated Sparrows *Zonotrichia albicollis*, Dark-eyed Juncos *Junco hyemalis* and Blackcaps *Sylvia atricapilla*, even though males and females were exposed to identical conditions (Helms 1963, Ketterson & Nolan 1986, Holberton 1993, Terrill & Berthold 1989). These findings match those from wild birds, and suggested an inherent difference between the sexes, presumably arising from different selection pressures acting on each sex in the wild. Taken together, these various findings indicate that the timing and duration of migratory behaviour are adaptive, and partly under endogenous control. Genetic influence is supported by hybridisation experiments, heritability studies, and changes in the migration timing of wild populations in response to known selection events (Chapter 20).

Directional preferences

The timing and duration of migratory restlessness, and patterns of fattening, are not the only features under endogenous control, as the same applies to directions. Birds taking different directions in the wild show the same directional preferences when tested in captivity. Some migrants in spring retrace their path from the previous autumn, but others take different routes at the two seasons (so-called loop migrants, Chapter 22). When tested for directional preferences in orientation cages, hand-reared Garden Warblers *Sylvia borin* kept in constant (12L:12D) conditions changed their mean heading from southwest to southeast part way through their autumn migration period. This corresponded with a change they would normally make part way through their journey between central Europe and Africa (Gwinner & Wiltschko 1978, 1980). They made no such change in spring, when they return by a more direct northerly route, requiring no change in direction during the journey. The whole pattern was in line with the loop migration routes between Europe and Africa revealed by ring recoveries. Other spontaneous shifts in directional preferences were also recorded during the migration seasons of captive Blackcaps *Sylvia atricapilla* (Helbig *et al.* 1989), Pied Flycatchers *Ficedula hypoleuca* (Beck & Wiltschko 1988) and Yellow-faced Honeyeaters *Lichenostomus chrysops* (Munro & Wiltschko 1993), and may well be widespread among birds.

The most obvious difference between seasons, namely the direction of travel, is apparently controlled by daylength changes and their effects on the physiological state of the bird. By appropriate manipulation of photoperiod, Emlen (1969) brought two groups of captive Indigo Buntings *Passerina cyanea* into spring and autumn migratory condition at the same time as one another. He then tested the directional preferences of both groups under identical planetarium skies. Birds in autumn condition oriented southward, those in spring condition northward. In some earlier experiments, Dark-eyed Juncos *Junco hyemalis* and American Crows *Corvus brachyrhychos*, which had been exposed to long photoperiods in midwinter, moved northward when released. However, castrates of these species migrated southeast after release, as did non-photostimulated control birds (Rowan 1925, 1932). This finding suggested that the effects of daylength could be mediated by the differing levels of gonadal hormones present in autumn and spring. A later study showed that the orientation of captive White-throated Sparrows *Zonotrichia albicollis* could be reversed by altering the temporal pattern of administration of the hormones prolactin and corticosterone. Birds injected with prolactin 4 hours after they had been injected with corticosterone oriented southward, whereas birds given prolactin 12 hours after corticosterone oriented northward (Martin & Meier 1973).

Integration of time–distance and direction programmes

The combination of inherent time–distance programmes and directional preferences provides a viable mechanism to explain how juvenile birds migrating on their own can reach wintering areas unknown to them but specific to their population. Naïve autumn migrants do not therefore need to experience the particular conditions of their wintering areas before they stop migrating. After an appropriate time, caged birds from both European and North American breeding areas lost their autumn restlessness and fat reserves, even though they had moved no further than the confines of their cages. Moreover, when captive juveniles were experimentally transported to their species-specific wintering areas, or even beyond their normal wintering range, the migratory activity they showed in cages persisted as long as that of individuals kept in the breeding area (for experiments on young Garden Warblers *Sylvia borin*, Lesser Whitethroats *S. curruca* and Pied Flycatchers *Ficedula hypoleuca* see Gwinner 1971, Rabøl 1993). These birds also showed appropriate directional preferences. Similarly, juvenile Starlings *Sturnus vulgaris* and other species that were trapped on migration, flown by

aeroplane and released immediately in the usual wintering areas for their population, or at some other locality off the normal route, resumed migration. Ring
recoveries revealed that these transported birds moved in the same direction and
covered about the same additional distance that they would have travelled had
they not been displaced (Chapter 9; Perdeck 1964, 1967). These experiments again
suggested control of timing and direction by an endogenous programme, rather
than by location.

In a different type of experiment, juvenile Blue-winged Teal *Anas discors* were
caught in autumn, held in captivity for some days, and released at the same place.
Many then migrated after release, in the same direction but over shorter distances
than normal (Bellrose 1958). This would also be expected if they were migrating
to a time programme. While all these findings suggest that non-breeding destinations are broadly set by inherent time–distance and directional programming,
this does not mean that birds are unable to move on if they encounter unsuitable
conditions, or if they are homing to a specific site experienced in a previous year.
Nor does it mean that all migratory species behave in this way, irruptive and other
facultative migrants being more flexible in timing and distance (Chapter 18).

Influence of local conditions

As indicated above, many birds change direction or accumulate extra fuel reserves
part way through their journey, in preparation for a sea- or desert-crossing.
Such changes in direction and fattening patterns observed at different stages of
a journey may not depend solely on an inherent time programme. Experiments
have indicated that birds can also respond to the star patterns or magnetic conditions found at particular regions en route (Chapter 9). In one experiment, Thrush
Nightingales *Luscinia luscinia* were captured in southern Sweden at the start of
their first autumn migration (Fransson *et al.* 2001, Kullberg *et al.* 2003). During
the next 10 days, some individuals were exposed to the earth's local magnetic
field (controls) and others to an artificial magnetic field typical in inclination and
strength to that of northern Egypt, from where the birds are thought to depart
for their Sahara crossing (experimentals). Fat deposition was significantly accelerated in the experimental birds, suggesting an effect of magnetic conditions on
fattening. Whether this represented a specific response to the magnetic conditions
of that particular area, or an unspecific response to a change in magnetic conditions, remains to be clarified. In another experiment, captive Pied Flycatchers
Ficedula hypoleuca changed their directional preference at an appropriate date
only when magnetic conditions (inclination and strength) were switched to those
of the region where a directional change normally occurs, but not when magnetic
conditions were held constant. This result suggested an involvement of both the
endogenous timing mechanism and the 'expected' external cues; both had to be
appropriate before a direction change occurred (Wiltschko & Wiltschko 2003).
The implication is that birds have a built-in (genetically controlled) response to
conditions in particular areas normally encountered on migration, which triggers
an appropriate change in direction or fattening regime. If this is so, a similar dual
mechanism could operate to signal arrival in winter quarters, but I know of no
experiments that have tested this possibility.

Diurnal patterns

Another facet of migration that seems to be inherently programmed is the diurnal pattern of activity. It is not simply that captive diurnally migrating species show migratory restlessness by day, and nocturnally migrating ones by night. The hour-to-hour patterns of restlessness seen in captive birds, whether day or night, match fairly precisely the hour-to-hour patterns of migration seen in the wild (for White-throated Sparrow *Zonotrichia albicollis* see MacMillan *et al.* 1970, for Reed Bunting *Emberiza schoeniclus* see Berthold 1978, for Goldfinch *Carduelis carduelis* see Glück 1978). This finding has been taken as another indication of endogenous control over the migratory process.

Throughout the year, diurnal migrants in captivity show a peak of activity in the morning and a smaller one in late afternoon, but at migration times these peaks are much more marked (e.g. Munro 2003). In addition, nocturnal migrants in captivity tend to become inactive in the afternoons before nights of migratory activity. They cease feeding about 2 hours before dusk, which in the wild would give time for the gut to empty before take-off (Ramenovsky *et al.* 2003). They go to roost in the normal way, and become active some time later. Low light intensity is necessary for a good display of nocturnal activity (Gwinner 1967).

Role of experience

On current thinking, then, the urge to migrate in autumn and spring, at least in obligate migrants, is genetically controlled. It is reflected in an autonomous rhythm of physiology and behaviour, which is kept on schedule by daylength changes (Gwinner 1972, 1986, Berthold 1996). This inherent system controls both fattening and migratory restlessness, as well as the general direction and time-course of migration. Most of the experimental work which gave rise to this paradigm concerned naïve juvenile birds that had no previous knowledge of the wintering range of their population. The situation differs somewhat in experienced birds migrating to a known site, as first shown in the experiments with Starlings *Sturnus vulgaris* and others described in Chapter 9. Adult birds displaced off their normal route in autumn could change direction and find their previous wintering areas, whereas juveniles continued in the usual direction after displacement, making no correction and moving parallel to the normal route. Having no previous experience of the wintering area, juveniles evidently used one-directional orientation, maintaining their displacement. However, on return migration to their breeding areas, both juveniles and adults proved able to return to the region of their birth (Chapter 9; Perdeck 1958). Previous experience of the breeding area enabled them to compensate for their displacement, and find their way home.

These early findings on the role of experience were extended by additional work on Dark-eyed Juncos *Junco hyemalis*, in which adults usually return to the same breeding and wintering sites in successive years. Individuals caught in July on their breeding areas and transported to a locality within the wintering range showed the usual autumn restlessness and fattening (Ketterson & Nolan 1986). On the other hand, birds caught in winter and held at the capture site over summer showed no migratory restlessness and fattening during the next autumn.

This held for birds kept on local 'wintering area' photoperiods, and for birds kept on wintering areas, but exposed to longer photoperiods typical of their breeding areas. This suggested that the locations of these birds on their previous wintering sites suppressed their autumn migratory behaviour. Both groups showed migratory behaviour in spring.

A further experiment involved a similar procedure, except that Juncos were kept over summer in their wintering areas, and then released in September, the normal time of autumn migration. Of 129 birds released, 47 were subsequently re-sighted in the release area, often within the same home range where they had been caught in the previous winter. Hence, these birds, which were already in wintering sites known to them, did not migrate that autumn. However, they apparently did migrate in the following spring (as none were found in wintering areas), and in the next autumn (when some re-appeared on their winter ranges).

The holding of Juncos on a previous (known) wintering site was apparently sufficient to suppress the usual autumn migration (Ketterson & Nolan 1986). It was uncertain what happened to the birds that were not re-sighted, but if they had dispersed only a short distance from the release sites, they would not have been found. These experiments thus provided further indication that the endogenous template of migration could be altered by experience. Similar results were obtained with migrating Dunnocks *Prunella vulgaris* in Europe (Schwabl *et al.* 1991b).

The same may hold for the spring return to familiar breeding areas, whether by juveniles or by adults, with migratory behaviour ceasing only when the birds have reached their specific goal. Dark-eyed Juncos that were held captive in wintering areas for up to two months beyond the normal spring departure date disappeared (presumably migrated) upon release, even though their gonads were by then in full breeding condition (Wolfson 1942, 1945). Likewise, among various warblers and finches that were caught in spring and held captive at a locality en route, migratory restlessness and appropriate directional preferences continued for up to several weeks longer than normal (Merkel 1956, Shumakov *et al.* 2001). Bramblings *Fringilla montifringilla* caught on spring migration, but south of their breeding range, also showed northeast directional preferences until late August, long after their spring migration would normally have finished.

These various findings could be interpreted as continuing attempts by these birds to reach their breeding areas, again indicating the importance of locality in suppressing the migratory behaviour of experienced birds returning in spring to a familiar area. This is different from the simple vector migration shown by inexperienced juveniles on their first autumn migration, as described above. It may account for the fact that some other birds caught from the wild and tested in captivity continued spring restlessness well into summer, or autumn restlessness well into winter, much beyond the normal migration seasons. They could have been birds held far away from their previously experienced breeding or wintering sites.

That spring migration can be suppressed by recognition of a familiar area was shown in an experiment with Indigo Buntings *Passerina cyanea* (Sniegowski *et al.* 1988, Ketterson & Nolan 1990). Males were caught on their nesting territories, held over winter, and released there in spring at a date when migration was just beginning in conspecifics wintering far to the south. Controls were transported and released 1000 km to the south. Seven out of 20 buntings released in spring on their nesting territories remained, while eight out of 20 released to the south returned to their nesting localities. The remaining birds in each group were unaccounted for. Nevertheless, these results indicated that, when migrants were exposed before spring migration to their previous nesting place, they did not migrate.

During reintroduction projects at various localities in western Europe, White Storks *Ciconia ciconia* were reared and kept in aviaries for at least one winter, which prevented them from migrating as they normally would. After they had been released, they remained in the same areas year-round, breeding and wintering there, and supported by supplementary food in winter. However, the free-living offspring of these birds migrated as normal for their population, yielding ring recoveries along the usual southwestern migration route (Fiedler 2003). These findings provided further indication that experience based on learning and memory can modify the inherent migratory behaviour of individuals. The findings from these different types of experiment on the role of experience on migration behaviour are summarised in **Table 12.1**.

Some remarkable feats of memory became evident in some radio-tracked birds which took the same routes in successive years, even to the extent of repeating apparent mistakes. For example, a Lesser Spotted Eagle *Aquila pomarina* in successive autumns took the same diversion off its route and back again, which added an apparently unnecessary 500 km and 2–3 days to its migration (Meyburg *et al.*

Table 12.1   Results of some trap and retention experiments (see text)

Experiment	Result	Interpretation	Species	Source
Held over first winter in breeding area	Did not subsequently migrate	Fixation to former wintering area greater than motivation to migrate in autumn	White Stork *Ciconia ciconia*	Fiedler (2003)
Held in wintering area over summer and then released	Did not migrate but stayed in wintering area until spring	Fixation to former wintering area greater than motivation to migrate in autumn	Dark-eyed Junco *Junco hyemalis*	Ketterson & Nolan (1986)
Held in breeding area over winter and then released	Did not migrate but stayed in breeding area until autumn	Fixation to former breeding area greater than motivation to migrate in spring	Indigo Bunting *Passerina cyanea*	Sniegowski et al. (1988), Ketterson & Nolan (1990)
Held on spring migration route	Showed prolonged migratory restlessness	Failure to reach breeding area resulted in migratory restlessness beyond the normal period	Dark-eyed Junco *Junco hyemalis*, Chaffinch *Fringilla coelebs*, Brambling *Fringilla montifringilla*	Merkel (1956), Shumakov et al. (2001)
Held on spring migration route and released about two months late	Left site, presumably on migration	Failure to reach breeding area resulted in prolongation of migratory behaviour	Dark-eyed Junco *Junco hyemalis*, Brambling *Fringilla montifringilla*	Wolfson (1942, 1945), Shumakov et al. (2001)

2002). This bird seemed to have remembered and repeated the same detour from one year to the next.

Migratory fattening and restlessness

Within the migration season, migratory flights may be influenced by prevailing weather and food supply (Chapters 4 and 27). The role of feeding conditions in affecting autumn fattening rates, body condition and departure dates has been established in the field for a range of species from Sedge Warbler *Acrocephalus schoenobaenus* (Bibby & Green 1981) to Greylag Goose *Anser anser* (van Eerden *et al.* 1991). In general, individuals that fatten most rapidly leave first, whether in autumn or spring, or from starting or stopover sites. Poor feeding conditions contribute to the variation between individuals within years, and can delay the progress of whole populations in some years (Chapter 27).

When they reach an appropriate weight, migrants normally leave immediately if weather permits. This would be expected because, once acquired, fat stores are dangerous and expensive to maintain, conferring increased vulnerability to predation (Chapter 5). Research has confirmed that the motivation to proceed with migration is related to fat stores, as is the strength of directional preference (Dolnik & Blyumental 1967, Bairlein 1885a, Yong & Moore 1993, Berthold 1996, Sandberg *et al.* 2002). In one study, four songbird species were caught on stopover at an oasis in Algeria, and kept in cages fitted with activity recorders (Bairlein 1985a, 1992). Generally, lean birds were active only during daylight, feeding as normal. In contrast, fat birds remained inactive by day without feeding, but became active at night when they would normally have migrated. Birds with moderate fat reserves fed during the day and were also active at night. Nocturnal migratory activity was thus clearly associated with fat levels. In another study, European Robins *Erithacus rubecula* were trapped in autumn an hour or two before sunset, placed in cages that provided a view of the sky, and their behaviour recorded (Bulyuk & Mukhin 1999). Nocturnal restlessness appeared in 18% of the birds tested, mainly in the fattest ones, which also became restless earlier with respect to sunset than leaner birds. In both autumn and spring, birds trapped and tested during waves of passage tended to start nocturnal restlessness in greater proportion, and earlier with respect to sunset, than conspecifics caught during migratory pauses.

At the start of migration, the precise relationship between extra fat deposition and migration varies between species, and between autumn and spring in the same species, in apparent adaptation to the journeys they have to make. Some populations accumulate large fat levels before they start migrating, others only at a later stage in the journey, when they have to make a long flight over sea or desert without feeding (Chapter 6). For example, captive Pied Flycatchers *Ficedula hypoleuca* in Germany began increasing body mass in autumn soon after showing migratory restlessness, whereas Collared Flycatchers *F. albicollis* began much later, long after the first appearance of restlessness, and then fattened rapidly. This difference between species was linked to their different migration patterns, as revealed by ring recoveries (Gwinner 1996b). Similarly, in captive Bobolinks *Dolichonyx oryzivorus*, migratory restlessness began soon after the start of fat deposition, but did not reach maximum intensity until much later, after the peak of fat deposition (Gifford & Odum 1965). This fits the fact that, like many other species, Bobolinks start their migration with short flights, which lengthen later in the journey.

During the course of a journey, birds alternate periods of migratory flight (when body reserves are expended) with periods of stopover (when reserves are replenished through feeding). In general, periods of flight and refuelling must be integrated in such a way that leanness leads to feeding and fatness to migration. At departure and stopover sites, migrants have often been caught and weighed, and whereas lean birds are often recaught on subsequent days, as they gain weight, heavy (fat) ones are seldom recaught (Bibby & Green 1981, Yong & Moore 1993). Such findings imply that the fat levels of a bird influence its tendency to resume migration (Chapter 27; see also Dolnik & Blyumental 1967, Berthold 1996, Sandberg *et al.* 2002).

Captive birds have provided additional information. Spotted Flycatchers *Muscicapa striata* and Garden Warblers *Sylvia borin* that were fed and then kept temporarily without food showed migratory restlessness (Biebach 1985, Gwinner *et al.* 1985a, Totzke *et al.* 2000). When they were again provided with ad lib food, their migratory restlessness initially ceased, but re-appeared later as their body mass and fat levels continued to rise. The implication was again that the fat content of a bird influenced its inclination to depart. In another type of experiment, Red-eyed Vireos *Vireo olivaceus* were trapped as they were about to cross the Gulf of Mexico in autumn. Each was weighed, and tested for directional preference. It was then fitted with a small chemiluminescent light stick, released on its own 1–2 hours after sunset, and followed with binoculars until lost from view (Sandberg & Moore 1996). Before their release, significantly more fat vireos (81%) than lean ones (61%) showed migratory activity in cage tests. After release, all the fat birds flew out of sight in the migration direction. Some 38% of the lean birds stayed at their current location, and most of the others took a direction diametrically opposed to the migration direction. Similar results on the relationship between fat levels, inclination to migrate and directional preferences were obtained in release experiments using Robins *Erithaca rubecula* and Pied Flycatchers *Ficedula hypoleuca* (Sandberg *et al.* 1991), and in cage tests on several North American warblers (Able 1977), Chaffinches *Fringilla coelebs* (Bäckman *et al.* 1997), Snow Buntings *Plectrophenax nivalis* (Sandberg *et al.* 1998) and Swainson's Thrushes *Catharus ustulatus* (Sandberg *et al.* 2002). In general, the greater the level of stored fat, the more likely was the bird to show in test cages a preferred direction. Many lean birds preferred the opposite direction, but temporarily reversed migration is not uncommon in the wild when birds encounter a sea-coast or other barrier. It has been interpreted as an attempt to find more profitable feeding areas away from the crowded coast (Chapter 4; Åkesson *et al.* 1996).

All these various findings confirm that the fat levels of a bird on migration influence both its tendency to depart on each leg of the journey, and the strength of its directional preference. However, most of the above studies were conducted at localities where the birds were about to cross the open sea or desert. The behavioural difference between lean and fat birds might be much less marked in birds migrating overland where feeding sites are frequent (for Red-eyed Vireo *Vireo olivaceous* see Sandberg & Moore 1996a). Birds might then simply vary the length of their flights, according to their available fuel levels.

If fat is extracted from plant or animal oils, or from the carcasses of other birds, and injected into the subcutaneous deposits of living birds, it is apparently utilised in the same way as the bird's own fat (Dolnik & Blyumental 1967). During the migration period, this procedure increased the activity of lean birds to resemble that of fat ones, but at other times of year and in sedentary species, it had no such effects. Again, the feeding and flight activity of migrants was related to the amount of fuel they contained. This conclusion was in line with findings from seven passerines species studied at Rybachi on the southern Baltic coast, where the intensity of migration on particular days (as shown by the numbers of birds trapped) was strongly correlated with the fat levels of the birds themselves. Nevertheless, the fattening patterns of birds do not entirely conform to the simple model: arrive lean, fatten rapidly to a threshold, and then depart as soon as possible. The proximate factors influencing fattening rates and stopover times in each species include date in season, available food supplies, weather, prior body condition, age, sex and social factors (Chapters 4 and 27).

Moreover, fattening and restlessness are not inseparably linked in birds, because captives of various passerine species kept lean by extreme food deprivation still showed nocturnal restlessness (King & Farner 1965, Lofts et al. 1963, Gwinner 1968, Berthold 1977). Moreover, restlessness is not an infallible reflection of migratory condition, for it sometimes occurs in other circumstances, and it is often a matter of interpretation what is considered true migratory restlessness (e.g. Berthold 1988; **Box 12.1**). Nevertheless, fattening at migration seasons is normally absent in birds from non-migratory populations in either natural or experimental conditions (e.g. Wolfson 1945), although some nocturnal restlessness has occasionally been recorded from non-migratory populations (Smith et al. 1969, Helm & Gwinner 2006).

AUTUMN MIGRATION

Migrants normally leave their breeding areas when conditions deteriorate, but before their continued survival there would become precarious. In association with the earlier onset of winter at high latitudes, many species withdraw from high-latitude parts of their breeding range first, and from lower latitude parts later. The assumption is that, in order to prepare for autumn migration, populations have evolved responses to different daylength regimes, appropriate to the latitude at which they breed. From any one area, departure dates also differ widely between species, depending largely on their type of food and when it becomes scarce (Chapter 14). As elsewhere in this book, I use the term autumn migration for the post-breeding return to wintering areas, even though in some species this migration occurs in the latter half of summer.

In obligate migrants, in which all individuals leave the breeding range each autumn, the dates of migration are in general fairly consistent from year to year. This is apparent not only in the dates that birds leave their breeding areas, but also in the dates they pass particular places on their migration routes and arrive in wintering areas. For example, among raptors migrating through Israel, the timing and duration of passage varied greatly between species, but within species the autumn passage dates were remarkably similar between years (as were spring dates). Over nine years, the confidence intervals of the mean autumn dates ranged between 1.5 and 3.4 days (versus 2.1–5.5 days for spring dates), depending on species (Leshem & Yom-Tov 1996a).

In many species of obligate migrants (in contrast to facultative migrants), adults leave the breeding areas before juveniles. This is especially marked in species which leave immediately after breeding, postponing their moult until later in the year, including many long-distance passerines, shorebirds and others (Chapter 11). Sex differences in timing are most evident in the adults of those species in which males and females have different roles in parental care, and one sex is free to leave before the other (Chapter 15). No such sex difference is apparent in the juveniles of these species. However, the net result of such age and sex differences in autumn migration timing is that they add to the spread in migration dates within populations.

In some species, marked individuals showed great year-to-year consistency in their arrival dates in staging or wintering areas (for Bewick's Swans *Cygnus columbianus* see Rees 1989, for Snow Geese *Chen caerulescens* see Maisonneuve & Bédard 1992). Individuals which arrived early with respect to other individuals in one year also arrived early with respect to other individuals the next year. Rees (1989) attributed this individual consistency to variation in their response thresholds to daylength, but other explanations were possible, including variation in their abilities to feed and fatten for the journey. In addition, individual swans that arrived early in autumn also departed late in spring. In summary then, obligate migrants show relative consistency in autumn migration dates from year to year, both as populations and in some species also as individuals, and some show age-related and sex-related differences in mean migration dates.

In facultative migrants the situation differs. Such populations generally moult after breeding and, in contrast to obligate migrants, individuals do not necessarily depart immediately after finishing moult. If conditions are favourable, they may linger longer in the breeding or stopover areas, leaving only when food supplies dwindle or are shut off by snow and ice. Mean autumn migration dates may therefore vary greatly from year to year, depending on local food supplies, as may rates of travel. This situation is exemplified by most short-distance migrants, especially irruptive seed-eaters (Chapter 18), and by waterfowl and others affected by frost (e.g. Hario *et al.* 1993). For example, the peak date for passage of Siskins *Carduelis spinus* through Falsterbo Bird Observatory in south Sweden during 1949–1988 varied from 15 August (in 1988) to 17 November (in 1958), the last date the station was manned that year (Roos 1991). At another site, Siskins passed in largest numbers, and at the earliest dates, in years when birch seeds (the main autumn food) were scarce (Svärdson 1957; Chapter 18). Likewise, the date on which the last Whooper Swans *Cygnus cygnus* left Lake Chuna on the Kola Peninsula each year during 1931–1999 varied between about 20 September and 9 November, depending on when the lake froze over (Gilyazov & Sparks 2002). These observations illustrate the point that some migrants depart only when deteriorating conditions encourage them to leave. Together with variable conditions on migration routes, this facultative response gives wide variation in the dates that birds arrive in their various wintering areas, the more distant of which may be reached only in occasional years. Where an age difference is apparent, juveniles of facultative migrants generally leave before adults (Chapter 15).

Split migrations

Although autumn migration is usually considered as a single event, consisting of alternating periods of flight and fattening, some species break their autumn journeys for periods of several weeks, much longer than is needed for refuelling. Some species moult during this break in migration. This behaviour is shown by many Eurasian migrants to East Africa, such as the Marsh Warbler *Acrocephalus palustris* and Garden Warbler *Sylvia borin*, which remain in the northern tropics for several weeks, and only later move on to the southern tropics (Jones 1995). In this way, they get the best from both regions, remaining in the northern tropics until conditions deteriorate, and arriving in the southern tropics at the optimal time, after fresh rains have promoted vegetation growth. In captivity, as mentioned already, such species show two separated peaks of migratory restlessness (Berthold 1993). Equivalent behaviour is shown by some small insectivores that migrate between North and South America (Terrill & Ohmart 1984). Interrupted migrations are also evident in various irruptive and other facultative migrants that break their journeys to exploit food supplies they encounter en route, and travel much further from their breeding areas in some years than in others (Svärdson 1957, Newton 1972; Chapter 18). One consequence of split migration is that the outward journey (including the break) takes more than four months in some species, whereas the return journey in spring can take only 1–2 months.

Relationship between breeding, moult and autumn migration

Many birds lay a repeat clutch if the first is lost, or raise more than one brood in a year. This means that adult pairs get increasingly out of synchrony with one another, and that some breed much later into the summer than others in the same area. In species that moult in their breeding areas, individuals that finish breeding late also moult and migrate later, but may begin moult earlier with respect to a nesting cycle, well before their young have fledged, and may also moult more rapidly than earlier birds (for Dunlin *Calidris alpina* see Johnson & Minton 1980, for White-crowned Sparrow *Zonotrichia leucophrys* see Morton & Morton 1990, Morton 2002). By these means, individuals breeding until late in the season reduce the delay in their migration. In some passerines, the delay in moult caused by late breeding is greater in females than in males, producing differences in departure dates between the sexes, including members of the same pair (Ginn & Melville 1983, for Cetti's Warbler *Cettia cetti* see Bibby & Thomas 1984, for Willow Warbler *Phylloscopus trochilus* see Norman 1990).

Similar things happen in the juveniles. In multi-brooded populations, young are produced over periods of several weeks. Compared to early-hatched young, later ones start to moult at a later date but at an earlier age, and may also replace their feathers more rapidly or less completely (for Chaffinch *Fringilla coelebs* see Dolnik & Gavrilov 1980; for various warblers see Berthold, 1988, Mukhin 2002, for White-crowned Sparrow *Zonotrichia leucophrys* see Morton 2002, for Stonechat *Saxicola torquata* see Helm 2003). Among Great Tits *Parus major* in northern Europe, first brood young start moulting in July, and second brood young in mid-August, the mean age at the start of moult decreasing over this period from 56 to 42 days, and the mean duration of moult from 67 to 59 days (Bojarinova *et al.* 1999).

Because of these differences, late-hatched young also migrate at an earlier age than early-hatched ones, with a mean difference of four weeks between their departure dates (Bojarinova *et al.* 2002). Within populations, a shortening of moult duration is achieved by shedding successive feathers at shorter intervals, so that more are growing at one time; with no obvious change in the growth rates of the feathers themselves (e.g. Newton 1967, Pienkowski *et al.* 1976).

In addition, while early young normally prepare for migration after completing moult, late birds may begin to fatten well before the end of moult (for Garden Warbler *Sylvia borin* and Blackcap *S. atricapilla* see Berthold 1975, for Reed Warbler *Acrocephalus scirpaceus* see Mukhin 2002). Among Mountain White-crowned Sparrows *Zonotrichia leucophrys oriantha* in California, autumn fattening usually took 8–9 days, as found by the repeated trapping and weighing of individuals (Morton 2002). In a few early individuals, moult had been finished for up to two weeks before fattening began but, in some late birds, fattening began up to two weeks before moult ended (**Figure 12.3**).

In various species, late individuals may also accumulate greater fat reserves before departure from breeding areas or stopover sites than early ones. Among Reed Warblers *Acrocephalus scirpaceus* migrating through southern France, mean stopover duration increased from 6.1 to 11.1 days, and rate of fat deposition from 0.29 to 0.40 g per day over the period late July to late October (Balança & Schaub 2005). The potential flight range of Reed Warblers migrating later in the season was therefore substantially longer than that of early migrants, enabling them to make longer flights and perhaps achieve a higher overall migration speed. Increased fattening rates or greater fat reserves in late season have been recorded in many other species, including Bluethroat *Luscinia svecica* (Ellegren 1991), Greater Whitethroat *Sylvia communis* and Lesser Whitethroat *S. curruca* (Ellegren & Fransson 1992), Blackcap *Sylvia atricapilla* (Izhaki & Maitav 1998a, 1998b), and Temminck's Stint *Calidris temminckii* (Hedenström 2004). Similarly, in several passerine species studied at Ottenby in Sweden, body mass and fat level were found to increase during the course of the several week autumn migration season (Åkesson *et al.* 1995, Fransson 1998, Danhardt & Lindström 2001). Evidence from ring recoveries revealed that birds which migrated late in the season also travelled more rapidly than earlier ones from the same population (for various species see Hildén & Saurola 1982, for *Sylvia* warblers see Fransson 1995). Speed of migration may therefore be another aspect of the annual cycle that responds to date, faster migration late in the season being achieved by faster and greater refuelling rates.

The time-saving resulting from quicker development of late young can be substantial. For example, in German Blackcaps, the earliest young leave the nest in late May and the latest in August. Their hatching dates are spread over about 72 days, but owing to accelerated development (especially of moult), the late young develop migratory activity only 18 days later than the early ones. Late young are ready to migrate in September, but without accelerated development they could not depart until mid-November, a dangerously late date (Berthold 1988). One consequence of the relationship between hatching date and development rate is that, in years of late breeding, an entire population can moult, on average, later and more rapidly than usual in preparation for migration (for Lapland Bunting *Calcarius lapponica* see Fox *et al.* 1987). Conversely, if birds fail in their breeding,

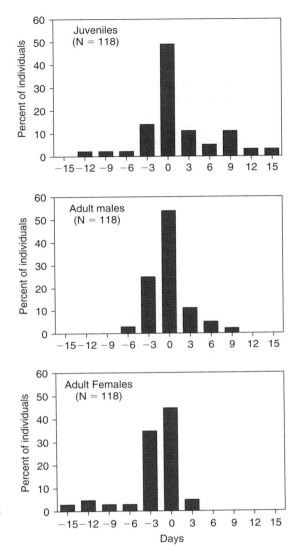

Figure 12.3 Relationship between onset of pre-migratory fattening and end of moult (0) in Mountain White-crowned Sparrows *Zonotrichia leucophrys oriantha*. Minus values indicate fattening before the end of moult and plus values after the end of moult. From Morton (2002).

they usually start moulting or migrating earlier than usual and, in poor breeding years, many individuals can begin their post-breeding migrations weeks ahead of normal.

In experimental conditions, various passerine migrants started post-juvenile moult and autumn migration activity at a younger age if held under short photoperiods (Berthold *et al.* 1970, 1988, Gwinner 1972, Kroodsma & Pickert 1980, Jenni & Winkler 1994, Berthold 1996, Noskov *et al.* 1999, Coppack *et al.* 2001, Gwinner & Helm 2003). They also moulted and fattened more rapidly, or showed more overlap between moulting and fattening (for Bobolink *Dolichonyx oryzivorus* see Gifford & Odum 1965, for White-crowned Sparrow *Zonotrichia leucophrys* see Moore *et al.* 1982, for Bluethroat *Luscinia svecica* see Lindström *et al.* 1994, for Lesser Whitethroat *Sylvia curruca* see Hall & Fransson 2001). These findings indicated that it was the shortening days of autumn which accelerated the development of late-hatched young in the wild. In all these species, the ending of moult probably does not itself promote the start of fattening because the two events do not invariably coincide, and as fattening rates increase with reducing daylength (= advancing date), this can be regarded as another time-saving adaptation in late birds. Such a response to declining daylengths is, of course, not possible in birds breeding at such high latitudes that they are still subject to 24-hour days at the time they leave.

There are costs to more rapid development. Feathers grown during a period of rapid moult are sometimes of poorer quality, which presumably reduces a bird's chance of survival (Dawson *et al.* 2000). In general, under food restrictions, birds seem to reduce the quality of feathers produced rather than their individual growth rates (Payne 1972). Hence, the greater acceleration and overlap of different processes in late birds probably involves a trade-off. It saves time but only at the costs of additional daily energy need and reduced feather quality. Early birds can enjoy greater separation between the different processes, whereas in late birds time-saving becomes paramount. The greater fat levels accumulated by late migrants may also make them more vulnerable to predation.

While studied mainly in passerines, links between breeding, moult and autumn migration timing are also apparent in other species. For example, among arctic-nesting geese and swans, sub-adults and failed breeders moult and migrate earlier than breeders with young (for Canada Geese *Branta canadensis* see MacInnes 1966, for Snow Geese *Chen caerulescens* see Maisonneuve & Bédard 1992, for Tundra Swans *Cygnus columbianus* see Rees 1980, for Whooper Swans *Cygnus cygnus* see Black & Rees 1984); and in years of widespread breeding failure, the mean migration date of the whole population may be earlier than usual by a week or more (Maisoneuve & Bédard 1992), as also noted in shorebirds (see above) and ducks (Blomqvest *et al.* 2002).

In conclusion, response to daylength not only influences the timing of autumn migration; in late-breeding adults, and in juveniles hatched late in the season, it can also speed up the moult, and increase the overlap between breeding and moult (adults only), or between moult and fat deposition, and increase the rate and extent of fat deposition, thereby reducing the delay in migration caused by late breeding. The acceleration of development by short daylengths, known as the calendar effect (Berthold 1993), is of functional significance in causing late birds to moult and leave breeding areas before conditions deteriorate (Berthold *et al.* 1970, Berthold 1988). In species which depart immediately after breeding, findings with respect to moult do not hold, because moult is postponed until after arrival in a staging area or winter quarters.

SPRING MIGRATION

Birds normally leave their wintering areas so as to reach their nesting areas in time to breed at the most favourable season. Many migrants winter so far from their breeding areas that they could not judge conditions there from their position in wintering areas. They can only leave their wintering areas at a time that natural selection has decreed is appropriate, using an internal timer combined with local conditions as a cue. Among species wintering in the temperate zone, the main known environmental stimulus for spring migration (superimposed on an endogenous rhythm) is increasing daylength which promotes extra feeding, fattening and migratory restlessness at appropriate dates for the population concerned (Rowan 1925, Wolfson 1952, Lofts *et al.* 1963, King 1972). In species that undergo a spring moult, this is also initiated by increasing daylengths, as is gonad growth, each of these processes occurring in appropriate overlapping sequence through the season. The role of daylength has been shown repeatedly in experiments on captive birds, in which longer-than-natural photoperiods advance all spring-occurring processes, whether gonad growth, pre-nuptial moult or migration.

Migratory birds can also respond to daylengths encountered on the journey. Helm & Gwinner (2006) exposed captive Stonechats *Saxicola torquata* at spring migration time to two different photoperiodic regimes, one typically experienced during migration through the temperate region (fast change) and the other typical of lower latitudes (slow change). These small short-term differences in daylengths had longer-term effects on the birds. Slow-change migrants continued migratory activity longer than fast-change migrants, delayed the growth and subsequent regression of their testes, moulted, and developed autumn migratory activity later in the year. These changes were appropriate to those needed if the birds had started migration at widely different latitudes.

In many wild species, then, gonad growth and sperm formation begin before the birds leave their wintering areas, and continue during migration (for Eurasian migrants see Rowan & Batrawi 1939, Lofts 1962, Marshall 1952b, for North American migrants see Blanchard 1941, Wolfson 1942). Some species have been seen to copulate while on migration, and females have been found with sperm in their reproductive tracts (Quay 1989, Moore & McDonald 1993). This seems to be usual in wild geese, for example, in some populations of which copulation and egg-formation begin even before birds have left their winter quarters (McLandress & Raveling 1983). Egg formation in these geese takes around 12 days, and laying can occur within a few days after arrival in nesting areas. In yet other species, most gonad growth occurs after the birds have arrived in breeding areas, and copulation is seen only after establishment of a territory. Much depends on the interval between arrival and egg-laying, which can span days or weeks, depending on the species and the ecological circumstances in which it lives. In general, it is in late-arriving species (relative to the latitude) that gonad development is most advanced on arrival, and in which the interval between arrival and egg-laying is shortest (Berthold 1996).

In species that undergo a spring 'pre-nuptial' body moult into breeding plumage, the process also starts in winter quarters and proceeds through migration. However, its timing varies greatly between individuals in the same population. Among shorebirds seen at spring stopover sites, some individuals are in predominantly winter plumage, while others at the same sites at the same time are in predominantly breeding plumage. Individual nutritional status may have a big influence on the timing and duration of this moult (Chapter 20).

Daylength is not the only factor to which birds respond in preparing for spring migration. The effects of daylength may be modified by both temperatures and food supplies. For example, captive White-crowned Sparrows *Zonotrichia leucophrys* exposed to air temperatures increasing from 5°C to 26°C advanced the time of migratory restlessness compared with control birds (Eyster 1954, Lewis & Farner 1973), but it is hard to tell whether temperature acted directly on the birds, or through its effect on their energy needs and the food available for fuelling. In *Catharus* thrushes studied during spring migration using doubly-labelled water, energy expenditure increased by nearly 20%, reflecting greater thermoregulatory costs, as ambient temperature dropped from 20°C to 10°C (Wikelski *et al.* 2003). On nights below 21°C, passage birds stayed at their stopover sites rather than migrating (Cochran & Wikelski 2005). Research on both wild and captive birds of many species has shown the importance of food supplies in affecting fuelling rates (Chapter 27).

Consistency and spread in spring departure dates

Migrants tend to leave particular wintering areas over a limited period of days or weeks, but in roughly the same period each year, some species before others (e.g. King & Mewaldt 1981). This holds in both obligate and facultative migrants. The year-to-year consistency in migratory timing within species is evident in studies on captive passerines. In White-crowned Sparrows *Zonotrichia leucophrys gambelii*, for example, the standard error was 1.0 day for the mean date of onset of pre-migratory fattening during eight different springs in males kept outdoors in winter quarters (King & Farner 1965, King 1972). Among wild birds, however, subject to differing conditions, the situation is far more variable.

As in autumn, preparation for migration in spring may be delayed in some individuals by poor food supply and body condition (Chapter 27), or by lateness of preceding events in the annual cycle, such as winter moult (Dugger 1997, Saino *et al.* 2004). Hence, in many populations, departure dates from particular wintering areas are often spread over several weeks (for White Stork *Ciconia ciconia* see Berthold *et al.* 2002; for Peregrine *Falco peregrinus* see McGrady *et al.* 2002). The same is true for the dates of passage through particular localities en route. Mean migration dates may also differ greatly between years. For example, the mean passage dates (= capture dates) of male Blackcaps *Sylvia atricapilla* through Israel over seven springs varied between 6 April and 3 May, and in late years the spread in passage dates was also less, as was the difference in mean dates between the sexes (Izhaki & Maitav 1998a). In general, as found in many localities, species that migrate earliest in the spring show bigger variation in mean passage dates between years, and also a bigger range of individual migration dates within years than species that migrate latest in spring (e.g. Francis & Cooke 1986, Hagan *et al.* 1991).

As in autumn, ringed adults of some species have shown some consistency in migration dates from year to year, as measured either by their departure dates from wintering areas (for Tundra Swans *Cygnus columbianus* see Rees 1989), or their arrival dates in breeding areas (for Barn Swallows *Hirundo rustica* see Møller 2001), or both (for Black-browed Albatrosses *Thalassarche melanophrys* see Phillips *et al.* 2005). Individuals that are early relative to other individuals in one year tend also to be early relative to other individuals in other years. As in autumn, such consistency could be due to inherent differences in migration timing between individuals, or to other differences between them, perhaps in their abilities to feed and fatten. Differences in timing between age and sex groups also contribute to the spread of spring migration dates within populations, as explained in Chapters 15 and 27.

In long-distance migrants, a wide spread in departure dates from wintering areas does not necessarily result in a similar wide spread in arrival dates on breeding areas. Birds encounter different conditions on their journeys, and early birds are more likely to be delayed by poor weather or food supply, enabling later ones to catch up with them. In Red Knots *Calidris canutus*, which migrate in spring from Tierra del Fuego to arctic Canada, departures from successive stopover sites become progressively more concentrated so that from Delaware Bay, the last staging site before the breeding areas, almost the entire population leaves within a 3-day period, usually 28–30 May (Baker *et al.* 2004). In this species, then, wide variation in departure dates from wintering areas is reduced along the migration route, so that most birds arrive in breeding areas within the same few-day period.

A 'calendar effect' may operate in spring, as well as in autumn, affecting the timing and extent of fattening, and the duration of stopovers. For example, in Semi-palmed Sandpipers *Calidris pusilla* passing through South Carolina, stopover durations decreased, and fat levels increased, as spring progressed (Lyons & Haig 1995). This would have had the effect of accelerating the migration of 'late' birds within the population, but whether it was an inherent response to daylength (time of year) or a response to increased food supply in late spring is an open question. Similarly, White-throated Sparrows *Zonotrichia albicollis* departing from their Florida wintering areas late in the migration season accumulated more fat than those that left earlier (Johnston 1966).

In other species, the opposite may hold, as the latest birds to leave wintering areas get even later with respect to earlier ones as they progress along the migration route – the domino effect of Piersma (1987). This situation is likely to hold where food supplies at stopover sites become progressively depleted through the season, so that later birds take longer to fatten (Chapter 27). One consequence would be that the arrival dates of birds in their breeding areas would show greater spread than their preceding departure dates from wintering areas.

Different populations of a species wintering in the same area

Different species wintering in the same area, and hence subject to the same daylength regime, may start their migrations at different dates, weeks or sometimes months apart, depending on the distance they have to travel and the dates their breeding areas become fit for occupation (for shorebirds see Piersma *et al.* 1990a). Such differences are also found in different races (or populations) of the same species wintering in the same area, as shown for White-crowned Sparrows *Zonotrichia leucophrys* in California (Blanchard 1941), and for Yellow Wagtails *Motacilla flava* in tropical Africa (Curry-Lindahl 1958, 1963, Wood 1992). Although exposed to the same winter conditions, the members of the various races differ in the dates at which their gonads develop, and at which they accumulate fat and depart for breeding areas (Blanchard 1941, Curry-Lindahl 1963, Fry *et al.* 1972). In general, in these northern hemisphere species, races that breed furthest south are first to leave, and those that breed furthest north are last. Thus, in the

Yellow Wagtail, the first to leave in spring is the southern race *M. f. feldegg*, then *M. f. lutea*, followed by *M. f. flava* and *M. f. flavissima*, and finally *M. f. thunbergi*. These various races arrive in their breeding areas in the same sequence, spanning the period March–June, from south to north. Inherent differences in endogenous rhythms could explain such population differences, as could inherent differences in the threshold daylengths required to trigger departure. Either mechanism could account for how birds of different races can leave their shared wintering area in appropriate sequence and reach their respective breeding areas (at different latitudes) at appropriate dates. However, the Yellow Wagtails are the more remarkable because several races winter together on the equator, where daylength is constant year round. In these birds, endogenous control seems essential, with different races responding differently, according to where they breed, and setting their 'internal clocks' before they reach the equator, so as to leave at appropriate dates some weeks or months later.

The idea of endogenous influence is supported by the observation that the onset of spring migratory restlessness in caged warblers in Europe coincides with the spring departure of conspecifics from their equatorial wintering grounds (e.g. Gwinner 1968). Captive migrants spontaneously resume spring migratory activity after a winter rest, even when kept under constant daylengths. Moreover, different species or different races of the same species kept under identical captive conditions reach breeding, moulting or migratory condition at different dates appropriate to the latitude at which they breed (for *Phylloscopus* warblers see Gwinner 1972). It seems that the seasonal timing of winter events in equatorial or trans-equatorial migrants is mainly accomplished by the involvement of an endogenous timing mechanism, as discussed in Chapter 11. Among populations wintering in the temperate zone, endogenous control would seem less important than in equatorial regions, because in the same area in the temperate zone, different wintering populations could respond to different threshold daylengths, so that they left in appropriate sequence. However, an endogenous influence, or an ability to separate increasing from decreasing days, would still be necessary in the temperate zone to prevent such birds from returning northward under the same daylengths in autumn that stimulate northward migration in spring.

Return migration from variable wintering areas

In many bird species, the migrants from particular breeding areas can winter over a wide span of latitude and daylength regimes. For example, Siskins *Carduelis spinus* breeding in the northern boreal forest of western Europe may winter anywhere between mid-Sweden and Morocco, a latitudinal span of about 30 degrees, and the same individuals may winter at widely separated places in different years (Chapters 17 and 18). The general pattern in such species is that return migration begins earliest from the most distant (most southern) parts of the wintering range, and latest from the most northern parts. This sequential withdrawal from lower to higher latitudes can be spread over many weeks. For example, in the Red-breasted Nuthatch *Sitta canadensis* in North America, the last birds withdrew from the southern parts of the wintering range, in northern Florida (30°N) as early as 21 February, from North Carolina (35°N) by 24 March, from West Virginia (38°N) by 28 April, and from Pennsylvania (41°N) by 15 May (Harrap & Quinn 1996).

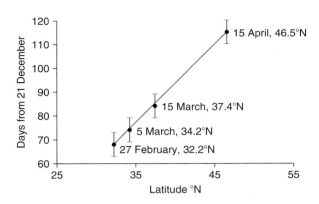

Figure 12.4   The median dates (±5 days) of onset of spring pre-migratory fattening in White-crowned Sparrows *Zonotrichia leucophrys gambelii* wintering at different latitudes in western North America. Regression analysis indicates a delay of 3.3 days, on average, for every additional degree N of latitude. From King & Mewaldt (1981).

Progressive withdrawal in spring from a latitudinal span of 11 degrees was thus spread over a period of about 12 weeks, or 7.6 days later per degree northward. In wintering White-crowned Sparrows *Zonotrichia leucophrys* in western North America, the date of onset of pre-migratory fattening varied linearly with latitude (and hence with solstical daylength), averaging 3.3 days later for each degree of latitude northward. The start of withdrawal was thus spread over seven weeks from the 14 degrees of latitude involved (**Figure 12.4, Table 12.2**; King & Mewaldt 1981). Mean rates of fattening were the same in all areas, regardless of latitude.

In many bird species, wintering entirely within the northern hemisphere, the timing of spring fattening and departure is therefore latitude-specific and, at any one wintering latitude, fairly consistent between years. Regardless of any endogenous influence, therefore, individuals must react appropriately to whichever daylength regime they find themselves under at the time. This is consistent with the experimental finding that longer photoperiods advance migratory fattening and restlessness in captive birds (because before the spring equinox, daylength at particular dates is longer at lower than higher latitudes).

Table 12.2   Pre-migratory fattening in White-crowned Sparrows *Zonotrichia leucophrys* wintering at different latitudes in relation to aspects of photoperiod. Although birds fattened at widely different dates at different latitudes, and had experienced different total amounts of daylight since the winter solstice, the mean daylength between the solstice and start of fattening was about the same in birds from all latitudes

Mean wintering latitude (°N)	Days of start of pre-migratory fattening[a]	Cumulative daylight hours from winter solstice[a]	Mean daylength (hours) from winter solstice to start of migratory fattening[a]
32.3	68	707	10.39
34.2	74	766	10.35
37.4	84	866	10.31
46.5	115	1200	10.43

[a]From 21 December (winter solstice), only one possible date which birds might use as an anchor point.
From King & Mewalt (1981).

The response to daylength could enable individuals from the same population (or the same individual in different years) to leave different wintering areas in order of southernmost to northernmost, in accordance with the distances they have to travel. Experimental support has been provided in the Pied Flycatcher *Ficedula hypoleuca*, in which captive males exposed to winter daylengths typical of southern Europe began pre-nuptial moult, migratory activity and gonadal maturation about one month earlier than control birds held under the normal African photoperiods (Coppack & Both 2002, Coppack *et al.* 2003). Because spring daylengths are longer in southern Europe than in tropical Africa, these flycatchers presumably perceived the date as being later than it actually was, and prepared earlier for migration. In addition, captive Garden Warblers *Sylvia borin* kept in winter in photoperiodic conditions found at latitude 20°S (that is, longer but decreasing daylengths) initiated gonad growth and migratory activity significantly earlier than conspecifics held under a shorter and more constant equatorial photoperiod (Gwinner 1987). Such a built-in response to photoperiod not only regulates the normal date of migration, but could enable birds to react appropriately to location changes between winters.

In some long-distance migrants, part of the population winters north of the equator and another part south of the equator. Even the same individuals may winter north or south of the equator in different years. This has been shown, for example, in radio-tagged White Storks *Ciconia ciconia*, in which one individual wintered at localities between 10°N and 30°S over four different years, each time returning successfully at an appropriate date to its nesting site in central Europe (Berthold *et al.* 2002). One female, tracked on return migration in six different years, set off, on average, around 24 February in four years when she was at 29–34°S, but about 18 days later, on average on 14 March, when she was at 3°S–4°N. She took an average of 75 days over the 11 000 km longer journey, arriving on 24 April–28 May in different years, and an average of 57 days over the 7000 km journey, arriving on 25 April–25 May in different years (**Table 8.4**). There was no significant difference in arrival dates, according to length of journey. She bred at the same site in all six years, and the spread in her autumn departure dates from the breeding locality was only 14 days (15–29 August) (Berthold *et al.* 2004).

While daylength response may reduce the delay in dates of arrival in breeding areas among birds wintering furthest away, it may not eliminate it altogether. Among eight radio-tagged Willow Ptarmigan *Lagopus lagopus*, autumn migration distances and spring arrival dates were correlated, the furthest migrating individuals arriving back latest in spring (Gruys 1993). More strikingly, Pied Avocets *Recurvirostra avosetta* breeding in the same colonies in Germany are known to winter either in mid-European latitudes or about 1000 km to the southwest in Mediterranean latitudes (Hötker 2002). Individuals wintering in mid-latitudes arrived significantly earlier at their breeding sites than those wintering further south. Their arrival dates also varied more from year to year in relation to local spring temperatures than did those from further south. As in many other birds, earlier arrival was associated with better breeding success, which implied a cost in more distant wintering. This raises the question of why any avocets winter so far south of their breeding areas when they do not survive any better there (Hötker 2002). This could be related to competition and dominance relationships on wintering areas, which lead some individuals to migrate further than others (Pienkowski & Evans 1984). A similar but slight reproductive advantage, with earlier egg-laying and larger clutches, was evident in King Eiders *Somateria spectabilis* that wintered closest to the breeding site, compared with those that wintered in a different region further away (Mehl *et al.* 2004).

Relationship between the internal rhythm and prevailing daylength

All northern migrants wintering north of the equator set off on their spring journey in conditions of increasing daylengths, whereas those wintering south of the equator set off in conditions of decreasing daylengths. Such flexibility in wintering area within species again implies the importance of an internal clock in influencing the timing of spring departure, for outside the breeding season individuals may be exposed to markedly different daylength regimes from year to year. If these birds use daylength change to calibrate their internal clocks, they must do so in their breeding areas, at a latitude which they occupy consistently from year to year. Because birds can measure daylength, the longest day (at the summer solstice) could provide a useful baseline against which any internal clock could be re-set, but so far as I am aware, this possibility has not been tested. Despite the use of an internal clock as the primary timer, birds must presumably also respond to prevailing daylengths in order to adjust the timing and speed of migration to their particular wintering latitude, as explained above.

RELATIONSHIP BETWEEN SPRING ARRIVAL, BREEDING AND AUTUMN DEPARTURE

In long-distance migrants that raise only a single brood each year and then depart from their breeding areas, the timing of arrival in breeding areas can influence the timing of all subsequent events up to post-breeding departure. This is because the breeding cycle, consisting of egg-laying, incubation and chick growth, is of fairly consistent duration, both from bird to bird and from year to year. Thus, in springs when birds arrive early in their breeding areas, they can usually breed earlier and depart earlier in late summer, the whole sequence of events being affected by spring weather (Nisbet 1957, Sokolov & Payevsky 1998, Sokolov *et al.* 1999). Moreover, in years of widespread breeding failure in arctic-nesting shorebirds and others, the post-breeding migration occurs much earlier than usual, as mentioned above.

Consider some specific examples. The Common Swift *Apus apus* migrates soon after breeding, postponing moult for winter quarters. Each pair leaves Britain within a few days after raising its brood. In this species, the date of departure is clearly determined primarily by the completion of breeding, which is in turn influenced by the date in May when breeding starts (Lack 1956). Since the latter is influenced by temperature, the mean date of departure of Swifts in August depends on the weather in the preceding May. Another well-known example is provided by the Spotted Redshank *Tringa erythropus*, which migrates between tropical Africa and northern Europe (Hildén 1979).

More recently, year-to-year correlations between spring arrival or laying dates and autumn departure dates have also been noted in the Bluethroat *Luscinia svecica*,

Willow Warbler *Phylloscopus trochilus*, Arctic Warbler *P. borealis*, Pied Flycatcher *Ficedula hypoleuca* and Little Bunting *Emberiza pusilla*, among others (Ellegren 1990a, Sokolov *et al.* 1999). The implication is that, in such single-brooded obligate migrants, autumn departure dates depend not so much on environmental conditions at the time, but on the completion of previous events in the annual cycle, whether the end of breeding or moult in adults or the end of growth or moult in juveniles, as the case may be. In line with this, the timing of peak autumn migration in different years at Rybachi on the Baltic coast was related to the preceding April temperature in 15 species examined, emerging as statistically significant in five such species (Sokolov *et al.* 1999). In none of these species was the timing of autumn migration related to local temperature at the time. Such close coupling between the timings of spring and autumn migration would not be expected in birds that can raise more than one brood each year, nor in some facultative migrants in which individuals may linger in breeding areas well beyond the end of moult in autumns when the weather is warm or food is plentiful.

The relationship between arrival and departure dates is also apparent in different populations of the same single-brooded species in different breeding areas. For example, those Ospreys *Pandion haliaetus* nesting in Florida that are migratory leave their breeding areas earlier in autumn, and return earlier in spring than do Ospreys from more northern breeding areas, but both populations spend about the same number of days in winter quarters (Martell *et al.* 2004). With only one brood raised, the breeding cycle also takes the same time in both regions. The wintering areas of both populations overlap, but southern breeding birds undertake both outward and return journeys several weeks earlier than the northern ones, neither spending more time than necessary in breeding areas.

DEFERRED RETURN TO BREEDING AREAS

In some long-lived bird species, individual migrants leave their natal areas towards the end of their first summer, and do not return in the next spring, but only in a later one, when they are two or more years old (up to four years in Bristle-thighed Curlews *Numenius tahitiensis*, Marks & Redmond 1996). These young birds either remain in their 'wintering' areas year-round over one or more years, or they may return only part way towards the breeding areas, or they may visit the breeding areas for only a small part of the breeding season, migrating later in spring and earlier in autumn than older birds. They perform both migrations in less hurry than the breeding adults, and in more favourable conditions. Such patterns are shown by various raptors, seabirds, shorebirds and others in which individuals do not breed until they are several years old (Chapter 15).

Most of the first-year shorebirds that stay in 'wintering areas' show no sign of pre-migratory fat deposition or spring moult into breeding plumage, but remain light in weight and in well-worn winter plumage until the next 'post-breeding' moult in late summer into new winter plumage. In other individuals, 'pre-breeding' moult and fattening are much delayed, sometimes into July, too late for the birds to breed that year (McNeil *et al.* 1994). Lack of both weight gain and 'pre-breeding' moult was apparent among juvenile Curlew Sandpipers *Calidris ferruginea* in South Africa, among Turnstones *Arenaria interpres* in Scotland, and among Western

Sandpipers *Calidris mauri* in Panama, while adults wintering in the same places began moult, accumulated body fat and left in spring in the usual manner (Elliott *et al.* 1976, Metcalfe & Furness 1984, O'Hara *et al.* 2002). Although the birds that stay year-round in 'wintering areas' do not always undergo the 'pre-breeding' moult into summer plumage, they undergo the late summer 'post-breeding' moult up to several weeks earlier than do adults returning from breeding areas. Turnstones *Arenaria interpres* over-summering in England moulted seven weeks earlier than adults returning from their arctic nesting grounds (Branson *et al.* 1979), and Western Sandpipers *Calidris mauri* over-summering in Panama moulted 3–4 weeks earlier than returning adults (O'Hara *et al.* 2002). They provide an example of birds moulting at a more favourable time of year when not constrained by breeding to a less favourable time later in the year.

The failure of immatures to migrate, or their tendency to return later and travel shorter distances than adults, might result from inability to obtain enough food and accumulate fat at the same rate as more experienced adults (Chapter 15).

Alternatively, it might result from an inherent, endogenous response that matures with age, leading birds of some species not to return to nesting areas in their earlier years or to visit nesting areas only to prospect for suitable sites but not to breed. Resolving these questions must await more research, but whatever the answer, endogenous factors are presumably involved in initiating moult and movements at appropriate dates, at least in birds that winter in equatorial or opposite-hemisphere regions. A proximate factor of possible importance is the reproductive state of the individual at the time of spring migration. This migration may be triggered only when the bird achieves reproductive capability, with high levels of gonadal hormones, at two or more years of age (similar to the attainment of hormone-dependent breeding plumage in some species). In first-year gulls and others, the testes develop only partially in the first year of life, and not until late in the breeding season (for review, see Lofts & Murton 1968). Moreover, in passerines, experimental castration has revealed the dependence of spring moult and migratory fattening on gonadal hormones, as explained in the next section. After their first nesting attempt, most individuals of species showing deferred maturity evidently return to breeding areas every year, as judged by the annual recurrence of marked individuals at their usual nest-sites (Chapter 17).

HORMONAL ISSUES

Relatively little is known about the hormonal control of bird migration, but the endocrine system presumably mediates all physiological aspects, including: (1) fuel deposition and use, coupled with associated changes in body composition; (2) increased haematocrit production for enhanced oxygen transport during long flights; and (3) modification of different aspects of behaviour, notably the change in diurnal rhythm (for example to promote night flights in otherwise day-time species) (Wingfield *et al.* 1990). Both autumn and spring migrations have the common requirements of fuel deposition and rapid mobilisation, and other associated changes in body composition (such as breast muscle enlargement, Chapter 5). However, spring migration usually coincides with gonadal growth and other preparations for breeding, whereas autumn migration follows gonadal regression or the ending of moult. Gonadal hormones are therefore present in the blood-stream at much higher levels in spring than in autumn (Wingfield *et al.* 1990, Wingfield & Silverin 2002).

In spring, gonadal hormones seem to influence hyperphagia, fat deposition and the haematocrit content of the blood. Intact and castrated males of various finches showed similar amounts of fattening and migratory restlessness, but castrated birds began about a week later (Lofts & Marshall 1961, Morton & Mewaldt 1962, King & Farner 1965). This held for birds castrated in late winter, after daylength had begun to increase. Birds castrated before the winter solstice, still experiencing shortening days, showed no pre-nuptial moult in spring, no fattening and much reduced migratory restlessness (for White-throated Sparrow *Zonotricia albicollis* see Weise 1967, for White-crowned Sparrow *Z. leucophrys* see Stetson & Erickson 1971). However, implants of testosterone in gonadectomised White-crowned Sparrows (no testosterone previously detectable) stimulated fattening and restlessness to varying degrees in spring, at the same time as in control birds (Mattocks 1976, Schwabl *et al.* 1988). It thus seemed that the hormone testosterone was involved in the preparation for spring migration.

In contrast, gonadal hormones cannot be involved in triggering autumn migration because the gonads have regressed by then. In fact, free-living male Pied Flycatchers *Ficedula hypoleuca* dosed with testosterone did not start migration after breeding, but remained in the breeding areas as long as testosterone levels remained high (Silverin 2003). The same was true of female Song Sparrows *Melospiza melodia* treated with oestradiol (Runfeldt & Wingfield 1985). In the latter species, untreated males that were paired with oestradiol-implanted females continued to show territorial behaviour and elevated testosterone levels well into autumn. Behavioural interactions between partners ensured that both remained in breeding condition for up to three months longer than control pairs in which neither partner was hormonally treated. While gonad removal in either sex suppresses hyperphagia and fattening in spring, it has no such effect in autumn (see also Rowan 1932, Lofts & Marshall 1960).

During migration, thyroid hormones seem to play no more than their normal role in regulating metabolism (changes in which may be substantial during migration). They seem less important in migration than in moult, but may interact with growth hormone to influence fat deposition and muscle growth. The hormone prolactin (from the anterior lobe of the hypophysis) is often considered fundamental to the migratory condition of birds, especially migratory fattening (Dobrynina 1990). The injection of prolactin can induce or accelerate fat deposition and, when combined with adrenal hormones, it can also stimulate migratory restlessness, but its actions are not fully understood. As reported earlier, it may also be involved in influencing the directional switch between autumn and spring.

The role of the adrenal hormone corticosterone in migratory restlessness was shown in a study of Garden Warblers *Sylvia borin* (Schwabl *et al.* 1991a). These nocturnal migrants had high levels of circulating corticosterone during the night, and lower levels during the day. When corticosterone secretion was artificially interrupted, nocturnal restlessness ceased. Other links between high corticosterone, fattening and migration have been shown in recent field studies (Holberton 1999, Piersma *et al.* 2000, Lohmus *et al.* 2003, Long & Holberton 2004). In Bar-tailed Godwits *Limosa lapponica* at a stopover site, corticosterone levels were higher in birds that had just arrived than in refuelling birds, but increased again as birds fattened (Landys-Ciannelli *et al.* 2002). Adrenocorticosteroids may therefore help to regulate hyperphagia, fattening and restlessness at appropriate times. Corticosterone has also been found at high levels in birds sampled during sudden weather movements (Wingfield 2003).

Almost certainly, other hormones are involved in migration but their role is even less clear, partly because of the difficulty of separating cause and effect in correlations (Ramenofsky 1990, Wingfield *et al.* 1990). In the response to daylength, the pineal–melatonin system seems to be involved (Brandstätter 2003), and in nocturnal migrants melatonin profiles change during the migration season (Fusani & Gwinner 2005). Melatonin is a hormone that modulates day–night rhythms. In Blackcaps *Sylvia atricapilla* from migratory populations, night levels of melatonin were lower during the migration season, when birds showed nocturnal activity, than at other times of year when they were active only by day. In contrast, Blackcaps from non-migratory populations showed no seasonal reduction in melatonin levels. In other experiments on migratory Blackcaps, long migratory flights and long refuelling stopovers were simulated by depriving birds of food for two days, and subsequently re-providing food. In both autumn and spring, nocturnal activity was suppressed and melatonin increased in the night following food reintroduction, the response depending on the amount of body fat. These studies revealed a relationship between melatonin and migratory restlessness, influenced by fat levels and food availability.

CONCLUDING REMARKS

In several respects, migration differs markedly between the two seasons. In autumn, birds leave their breeding areas under decreasing daylengths as conditions are deteriorating. In moving toward lower latitudes, they generally meet more clement conditions (except where they have to cross a large area of sea or desert). In spring, by contrast, most birds leave their same-hemisphere wintering areas under increasing daylengths, when conditions are improving or at least benign. In moving towards higher latitudes, they generally encounter progressively worsening conditions; they are often held up by bad weather, and may even have to backtrack for part of the route. It is normally some time after arrival in nesting areas that conditions improve enough to allow them to breed. In fact, environmental conditions, and the appearance of food supplies at particular latitudes, may have much more influence on the progress of migration in spring than in autumn (Chapter 14).

In autumn, populations are large, and juveniles are making their first migration; but in spring, populations are smaller and all participants have experienced at least one previous journey. This could influence the time taken and the extent of deviations from the primary direction, young birds being more susceptible to wind drift than older ones (Chapter 10). But while in some species, the autumn migration season is more extended than the spring one, in others the reverse holds (especially in those species with a large non-breeding contingent, see above).

The birds themselves are also in different physiological states at the two seasons. In autumn, the migrants travel with regressed gonads, and in spring with growing gonads, which leads to seasonal differences in hormonal states, which may in turn influence aspects of the migration, including direction (see above). In some species, patterns of fat deposition, flight lengths and speeds of travel also differ in a consistent manner between the two seasons, associated with different travel routes and stopping sites. For these various reasons, therefore, the spring journey is not simply the reverse of the autumn one.

Differences in the spread of migration dates between autumn and spring are evident at many localities on migration routes (Enquist & Pettersson 1986, Lavée *et al.* 1991, Morris *et al.* 1994, Leshem & Yom-Tov 1996a). They could be due to: (1) greater spread in the start dates at one season than the other; (2) more variation in mean migration speeds at one season than the other; or (3) a combination of both these influences. Even captive birds behave differently at the two seasons: for example, caged White-crowned Sparrows *Zonotrichia leucophrys* showed more intense nocturnal restlessness in spring, but over a shorter period of days than in autumn (Ramenofsky *et al.* 2003). At least some of the differences in migratory behaviour and fattening patterns observed in many species between autumn and spring are evidently under endogenous influence, and presumably have an adaptive basis.

While the main ultimate effect of climatic seasonality on migration timing is through its effect on food supplies, some researchers have suggested a more direct influence, with migratory seasons evolving to coincide with those times that, on average, are favourable for travel, and to avoid those times that, on average, are unfavourable for travel. This view remains speculative, but the correlation between migration timing and ecological conditions is so strong, especially in spring, that average weather conditions are unlikely to have any more than minimal effect on the broad timing of migration seasons. If birds had greater freedom in migratory timing, they would presumably avoid the hurricane seasons, which span the main autumn migration period in several parts of the world. Weather does, however, influence the actual dates of flights within the migration season (Chapter 4), and possibly also the routes taken, which often differ between autumn and spring, in line with seasonal differences in prevailing winds or feeding conditions (Chapter 22). In addition, the immature non-breeders in some species have greater freedom in migratory timing than breeding adults, and tend to migrate later in spring and earlier in autumn, when conditions are better.

SUMMARY

In obligate migration, all main aspects are viewed as under firm internal (genetic) control, mediated by daylength changes, giving a high degree of consistency in timing, directions and distances of movements from year to year. For the most part, each individual behaves in the same way year after year. In contrast, facultative migration is viewed as a direct response to prevailing autumn conditions, especially food supplies. Within a population, the proportions of individuals that leave the breeding range, the dates they leave and the distances they travel, can vary greatly from year to year, as can the rate of progress on migration, all depending on conditions at the time. The same individual may migrate in some years, but not in others.

In general, obligate migration occurs in populations whose food supplies in breeding areas are predictably absent in winter, whereas facultative migration occurs in populations whose food supplies in breeding areas vary greatly from one winter to another, according to weather or other variables. Obligate and facultative modes can be regarded as opposite ends of a continuum, with predominantly internal control at one end and predominantly external control at the other. In addition, many migrants may change from obligate to facultative mode during the course of their journeys.

An endogenous programme influences the time-course (and hence distance) of autumn migration among obligate migrants. Evidence for endogenous control has come from findings that: (a) the timing and duration of migratory restlessness in captive birds resembles the temporal pattern of migration in free-living birds; (b) populations which migrate over different distances show corresponding differences in the amount and duration of migratory activity in cages (and hybrids show intermediate patterns); and (c) experimental interruption of migration or migratory restlessness is not subsequently compensated for. These findings apply primarily to juveniles, and do not necessarily hold in experienced birds returning to a known area.

In many species, departure dates from breeding areas on autumn migration are spread over periods of up to several weeks, because they are influenced by variations in the dates that individuals finish breeding and (in some species) moulting. Adult passerines breeding until late in the season also start moulting later, but earlier with respect to the stage of a nesting cycle, and replace their feathers more rapidly than earlier birds, thus minimising the delay in autumn migration. Similarly, young produced late in the season moult at an earlier age, and more rapidly, than earlier hatched young, again reducing the delay in migration. They also begin fattening before the end of moult and may migrate more rapidly. As confirmed experimentally, this acceleration in development in late birds is triggered by the shortening daylengths of late summer and autumn.

Preparation for spring migration appears to be influenced primarily by increasing daylengths in association with an endogenous rhythm, the latter being particularly important in populations wintering in equatorial and opposite hemisphere regions. The spread in departure dates between individuals wintering in the same area may be attributable partly to variations in completion of previous events in the annual cycle, and in their feeding and fattening rates. Adverse weather can cause further delays.

In some single-brooded populations, the dates of arrival or egg-laying in spring influence the dates of autumn departure because, with a breeding cycle of roughly constant length, early arrival (or egg-laying) allows early departure. Post-breeding departure dates in such populations thus depend more on preceding spring weather than on prevailing weather.

In some bird species with deferred maturity, individuals remain in wintering areas and do not return to breeding areas until they are two or more years old; other individuals may return part way towards breeding areas, or may visit breeding areas only for a short time each year, leaving wintering areas later and returning earlier than breeding adults. Mechanisms that control the occurrence and timing of migration in such species during the early years of life await study.

Sex hormones may help to promote preparation for spring migration but delay autumn migration. They may also influence direction. Corticosterone is involved in migratory fattening and restlessness, and melatonin in nocturnal activity.

Part Three
Large-Scale Movement Patterns

Northern Lapwings *Vanellus vanellus* on migration

Chapter 13
Geographical patterns

The migratory habit enables a species to enjoy the summers of northern latitudes while avoiding the severity of the winters. (Frederick C. Lincoln 1935.)

Migration is the most spectacular of bird movements. It can be defined as a large-scale return journey, which occurs each year between regular breeding and wintering (or non-breeding) areas. Involving seasonal shifts of millions of individuals, it produces a massive twice-yearly re-distribution of birds over the earth's surface. High-latitude regions receive birds mainly in the breeding season, while lower latitude regions support wintering birds from higher latitudes, as well as year-round residents. Migration thus increases the numbers of species that occur in particular regions, even though some are present for only part of the year.

Throughout the world, migration is most apparent wherever the contrast between summer and winter (or wet season and dry season) conditions is great. Migration thus allows individual birds to exploit different areas at different times of year, whether to benefit from seasonal flushes of food or to avoid seasonal shortages. In fact, some migratory birds occupy habitats over winter that they could not use for breeding, and then occupy breeding areas that would not support them in winter. This applies to all arctic nesting shorebirds which spend the winter on coastal mudflats where, due to tidal flooding, nesting would be impossible, and then migrate north to breed on the arctic tundra which is frozen and snow-covered for the rest of the year. Thus some bird species exist over much or all of their range only by exploiting widely separated habitats at different seasons.

Although most marked at high latitudes, migration also occurs in the tropics, especially in the savannahs and grasslands exposed to regular wet and dry seasons. In the northern tropics, for example, many species move south for the non-breeding season, some crossing the equator, while in the other half of the year many species of the southern hemisphere move north. In contrast, birds confined to lowland equatorial rainforest are probably the least migratory, especially the small insectivores of the understorey where conditions remain relatively stable and suitable year-round. This year-round consistency in the rainforest environment removes any advantage in moving, and many individuals may remain within the same few hectares throughout their adult lives. In the same forests, however, some nectar-eaters and fruit-eaters move within small latitudinal or altitudinal bands in response to flowering and fruiting patterns, while other birds from higher latitudes move in for their 'winter' (Levey & Stiles 1992).

Most migratory bird species thus form parts of different communities at different seasons, and may interact with different species in their breeding and non-breeding homes. Each migratory species must therefore be compatible with the various species it encounters through the year, as well as with different climatic and dietary regimes. Overall, more than half the world's 10 000 or so bird species are likely to perform migratory movements in at least part of their range, causing a general shift in the centre of gravity of populations towards the north for the boreal summer and towards the south for the austral summer.

LATITUDINAL TRENDS

Worldwide, the numbers of breeding landbird species found per unit area generally decreases progressively from equatorial to polar regions (but is also generally low in barren deserts). At the same time, the proportions of species that are migratory increase from equatorial to polar regions, as the contrast between summer and winter conditions widens (Newton & Dale 1996a, 1996b). Migration thus tends to steepen the latitudinal gradient in species numbers between summer and winter. Progressing northward up the western seaboard of North Africa and Europe, for example, the proportion of breeding bird species which move out totally to winter further south increases from 29% of species at 30°N (North Africa) to 83% of species at 80°N (Svalbard), a mean increase of 1.3% of breeding species for every degree of latitude (**Figure 13.1**).

The relationship between proportion of summer visitors and latitude holds, it has been suggested, because at high latitudes the numbers of resident birds are held at low level by winter severity. The flush of food in summer is greater than the small

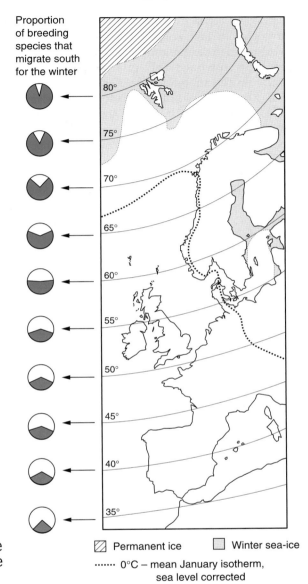

Figure 13.1   Map of western Europe showing the proportions of breeding bird species at different latitudes that migrate south for the winter. From Newton & Dale (1996).

number of resident species can exploit, leaving a surplus available for summer migrants. The latter therefore increase in proportion with latitude, as the severity of the winters increases, and the numbers of year-round resident species decline. At lower latitudes, a high proportion of breeding species can remain year-round, leaving fewer openings for summer visitors (Herrera 1978, O'Connor 1985, Morse 1989).

A similar relationship between migration and latitude holds in the largely different avifauna of eastern North America, where the proportion of migrants among breeding species also increases with distance northwards from 12% at 25°N to 87% at 80°N, a mean increase of 1.4% per degree of latitude (**Figure 13.2**).

The difference between the two regions (**Figure 13.3**) reflects the climatic shift between east and west sides of the Atlantic: over most of the latitudinal range, at any given latitude winters are colder in eastern North America than in western Europe. Correspondingly, at any given latitude, a greater proportion (an average of about 17% more) of breeding species leaves eastern North America for the winter than leaves western Europe. The slopes of the two linear regression lines calculated from the data in **Figure 13.3** do not differ significantly, but the intercepts do ($F_{1,19}=27.5$, $P<0.001$), reflecting this climatic difference.

On both continents, this northward trend in migration is easily understood in terms of winter conditions. In Europe, over much of the twentieth century, mean January temperatures exceeded 10°C only in southern Spain and North Africa; they lay within the range 0–5°C in much of western Europe, but fell below freezing and as low as −15°C in most of Fennoscandia, and to −20°C in Novaya

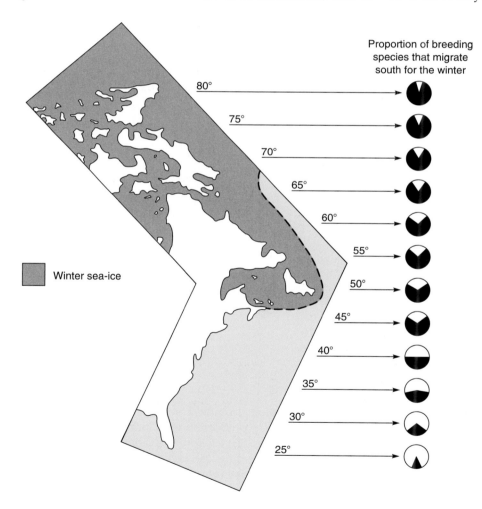

Figure 13.2   Map of eastern North America showing the proportions of breeding species at different latitudes which migrate south for the winter. From Newton & Dale (1996b).

Zemlya in the far north. Minimum winter daylengths were around 11 hours at 35°N in southern Europe but decreased to zero at the Arctic Circle. The season of plant growth lasted 6–9 months at 35–50°N, but shrank to less than three months in Svalbard, a mean decline in growing season of about one month for every 11° of latitude. In continental western Europe, most fresh waters north of 55°N froze in winter, although they mostly remained open in Britain and Ireland. Much of the Baltic and Barents Seas also iced over during the course of the winter, closing these areas for seabirds. In North America, similar latitudinal trends occurred, but were more marked because the continent spans a wider range of latitude than Europe. Throughout much of these areas, temperatures are now rising as part of global warming, and bird migration patterns are changing accordingly (Chapter 21), but they still relate to gradients in prevailing conditions.

The few species that remain to winter in the far north include the Common Raven *Corvus corax*, Rock Ptarmigan *Lagopus mutus*, Gyrfalcon *Falco rusticolus* and Snowy Owl *Nyctea scandiaca* among landbirds, and the Northern Fulmar *Fulmarus glacialis*, Ivory Gull *Pagophila eburnea* and Glaucous Gull *Larus hyperboreus* among seabirds. The most northerly seabirds depend in winter on the open water provided by polynyas, and some of the gulls also scavenge the remains of seals killed by Polar Bears *Ursus maritimus*. Some individuals of these species may move south to some extent in the weeks of complete darkness.

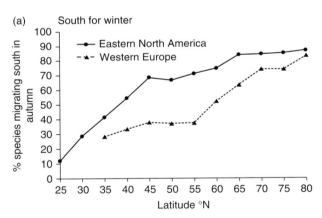

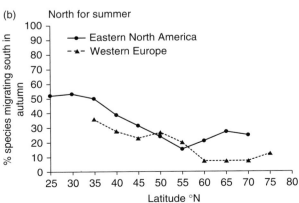

Figure 13.3 Proportions of breeding bird species (y-axis) at different latitudes (x-axis) in western Europe and eastern North America that migrate south for the winter (a) or north for the summer (b). For southward migration, on regression analysis for western Europe: $y = 41.49 - 1.03 + 0.02x^2$, $r^2 = 0.97$; for eastern North America: $y = -75.05 + 4.33x - 0.3x^2$, $r^2 = 0.98$. For northward migration, on regression analysis for western Europe: $y = 55.65 - 0.66x$, $r^2 = 0.81$; for eastern North America, $y = 123.72 - 3.22x + 0.03x^2$, $r^2 = 0.87$. For three of these relationships, a quadratic equation gave a significantly better fit than a linear one. From Newton & Dale (1996b).

Similar relationships between migration and latitude have been shown among the birds of particular habitats from grassland to forests (Wilson 1976, Herrera 1978), and among the birds of particular families, such as flycatchers in South America (Chesser 1998). They presumably hold worldwide in both hemispheres.

The decline in proportion of migrants with decreasing latitude, established above for Europe and North America, continues southwards towards the equator. Thus, by 8° latitude in Panama, only five (0.6%) out of 807 breeding species are wholly summer migrants (Ridgely & Gwynne 1989). This is consistent with the regression line between proportion of migrants and latitude derived from the data for North America in **Figure 13.3**. The five migratory species found at Panama are all insectivores which leave for the winter dry season and head further south, namely Swallow-tailed Kite *Elanoides forficatus*, Plumbeous Kite *Ictinia plumbea*, Common Nighthawk *Chordeiles minor*, Piratic Flycatcher *Legatus leucophaius*, and Yellow-eyed (Yellow-green) Vireo *Vireo flavoviridis*.

The converse of these relationships for western Europe and eastern North America is shown in **Figure 13.3** as the proportions of the birds wintering at different latitudes that move north for the summer. As expected, this proportion is greatest in the south, in western Europe affecting 36% of species wintering at 35°N, and declining northward to 8% of species wintering at 70°N (mostly seabirds) and none at 80°N. In eastern North America, the equivalent figures are 52% at 25°N decreasing to none at 70°N. Again, the slopes of the regression lines do not differ between continents but the intercepts do ($F_{1,16} = 9.9$, $P < 0.01$). Throughout the latitudinal range, the proportion of wintering species that leaves northward for the summer averages around 10% greater in eastern North America than in western Europe, again reflecting the climatic difference between the two regions. The precise regression relationship varies somewhat between birds of different habitats. In addition, 23% of all species breeding in western Europe, and 24% of those in eastern North America, leave these areas completely in autumn for the tropics, returning in spring (Newton & Dale 1996b).

The proportions of all bird species that are migratory are correlated not only with latitude, but also with various climatic factors that vary with latitude, such as the temperatures of the hottest or coldest months or the temperature difference between the hottest and coldest months (**Appendix 13.1**). These various measures are of course interrelated, but what really matters is the degree of climatic difference between summer and winter. It is this difference that, for many birds, governs the difference in food supply between summer and winter at particular latitudes, and hence the difference in environmental carrying capacity between the two seasons.

The seasonal difference in carrying capacity may also vary from west to east, according to changes in climate (as between west and east sides of the Atlantic). From west to east across Europe, summer climates become warmer and drier, and winter climates become colder. In consequence, progressing eastward through Europe into Asia, increasing proportions of the local breeding species become migratory. This is especially obvious in comparing populations of coastal areas that live under mild oceanic climates with those further inland that live under more extreme continental climates. For example, Common Starlings *Sturnus vulgaris* live year-round on the Shetland Islands at 60°N, while at the same latitude in Russia (and for 10–15° south of it) they are wholly migratory.

At some mid-latitude areas, similar numbers of species may be present in summer and winter, but species composition changes somewhat between seasons, as some species from lower latitudes are present only in summer and other species from higher latitudes only in winter (Gauthreaux 1982a, Newton & Dale 1996a, 1996b). In southern England, for example, insectivorous swallows and warblers arrive from the tropics for the summer, whereas fruit-eating thrushes from further north arrive for the winter. Seasonal changes in bird communities in particular regions are thus tied to seasonal changes in the types of food available. This emphasises the point that migrants often exploit seasonal abundances in both their breeding and non-breeding areas. It is a strategy that, for obvious reasons, is much more developed in birds than in most other animals.

Although seasonality in the movements of birds is evident worldwide, in warmer regions, rainfall becomes more important than temperature. In tropical regions, away from the equator, overall rainfall declines and becomes increasingly unpredictable and localised. So superimposed on their north–south pattern, many bird species tend to concentrate wherever rain has fallen, and food is most available at the time, varying in distribution from year to year. In the most arid regions, some species are truly nomadic, with no regular directional movements. This is apparent in parts of Africa and South America, but particularly in Australia (Chapter 16). In addition, at any given latitude, islands tend to have lower percentages of migrants than comparable mainland localities, presumably because the surrounding seas buffer islands from the extremes of seasonal climatic fluctuation.

MIGRATION AND DIET

Superimposed on the overall latitudinal trend is another related to diet. Broadly speaking, those species that are resident year-round in a particular region exploit food supplies that are available there all year, whereas those that leave after breeding exploit foods that disappear then. In the northern coniferous forests, for example, residents include mainly species that feed directly from trees, on bark-dwelling arthropods (tits, tree-creepers), fruits and seeds (some corvids, finches, tits), buds or other dormant vegetation (grouse), or that eat mammals and other birds (some corvids, raptors and owls). Almost the entire resident landbird fauna at high northern latitudes falls into one or other of these dietary categories. In contrast, species that depart for the winter include those which eat active leaf-dwelling or aerial insects (warblers, hirundines) or which eat foods that become inaccessible under snow or ice (ground-feeding finches and thrushes, some raptors, waterfowl and waders). Towards the equator, as winters become less severe, the range of bird dietary types that remain for the winter increases, as a wider range of food types remains available year-round.

To illustrate in more detail the link between migration and seasonal changes in food availability, consider two examples. Most European songbirds feed either on: (1) insects or other invertebrates, (2) seeds and fruits, or (3) a mixture of both categories (Newton 1995a). Of these food types, seeds and fruits are clearly much more available in winter at high latitudes than are insects. **Figure 13.4** shows the proportions of bird species in these different diet categories that are migratory at different latitudes. Within each group, the proportion of migrants increases with latitude, following the general trend in birds as a whole. But at each latitude from about 35°N, a larger proportion of insect-eaters than of seed-eaters leaves, while species with mixed insect–seed diets are intermediate. Furthermore, the insect-eaters generally move longer distances, many wintering in the tropics and some south of the equator (**Figure 13.4**). The result is that insect-eaters are concentrated at more southern latitudes in winter than are seed-eaters. The same holds for New World migrants, in which most small insectivorous warblers winter in Central America (30°N–10°S) and most seed-eaters further north (mostly 40°–15°N) (Keast 1995). Among these general relationships, specific exceptions occur, such as the seed-eating European Turtle Dove *Streptopelia turtur* which winters in Africa.

As a second example, consider the various European raptors, which also differ in the extents to which their foods remain available at high latitudes in winter (Newton 1998b). Species can be divided according to whether they feed primarily on warm-blooded prey (birds and mammals which remain active and available in winter at high latitudes) or on cold-blooded prey (reptiles, amphibia and insects, which become inactive and unavailable in winter). Within each raptor group, the proportion of migratory species again increases with latitude, but at any one latitude a larger proportion of the species that eat cold-blooded prey than of those that eat warm-blooded prey leaves for the winter, while species with mixed diets are intermediate (**Table 13.1**). The cold-blooded feeders also migrate furthest. The reasons for this difference are fairly obvious, in that species which eat cold-blooded prey and breed at high latitudes must winter in the tropics or in southern hemisphere temperate areas if they are to have access to the same types of prey year round. Of the 22 species of west Palaearctic raptors that eat mainly warm-blooded prey, most winter entirely within the Palaearctic and only one species (Booted

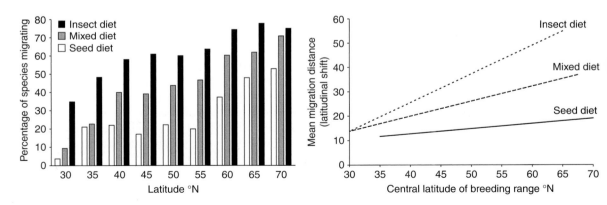

Figure 13.4   Migration in relation to diet in west Palaearctic songbirds. Left: Proportion of species breeding at different latitudes which migrate south for the winter. Right: Distances moved by migrants as measured by the difference between the central latitudes of the breeding and wintering ranges. Lines calculated by regression analyses. From Newton (1995b).

Table 13.1 Wintering areas of west Palaearctic raptors in relation to diet. Figures show numbers of species in each category

Wintering area	Main prey types		
	Warm-blooded	Mixed	Cold-blooded
North of Sahara	16	0	0
North and south of Sahara	5	8	2
South of Sahara	1	4	7

Significance of variation between categories (examined by Monte-Carlo randomisation test): $\chi^2_4 = 35.9$, $P < 0.001$.
Of 22 species that eat mainly warm-blooded prey, only one species (Booted Eagle *Hieraaetus pennatus*) winters entirely in Africa. In contrast, seven of nine species that eat mainly cold-blooded prey winter entirely in Africa. They include six insectivores, four of which winter entirely (and two largely) south of the equator, where the seasons are reversed. The 12 species with mixed diets show intermediate patterns.
From Newton (1998a).

Eagle *Hieraaetus pennatus*) winters entirely in Africa (**Table 13.1**). Of the nine species that eat mainly cold-blooded prey, some winter partly in the Palaearctic and partly in Africa, but most winter entirely in Africa. Moreover, all six insectivorous species winter south of the Sahara, four of them entirely (and two largely) south of the equator, where the seasons are reversed. So most insectivorous species live in almost perpetual summer, in conditions in which they have easy access to their insect food supply year-round. The 12 species with mixed diets show intermediate patterns. Such patterns again underline the link between migration and the seasonal changes in specific food sources (Newton 1979).

That movements are related to diet is also apparent in tropical regions, even though many such movements are relatively short. In tropical forests, nectar-eaters and fruit-eaters move around more than other species, probably because flowers and fruit are much more seasonal and patchy in occurrence than are the insects and other small creatures eaten by other birds (Morton 1980, Levey & Stiles 1992). Because flowers and fruit are generally more abundant in the canopy and forest edge, it is these parts of forest environments that show the greatest seasonal variation in their bird populations, the understorey species maintaining the most sedentary lifestyles year-round (see above). Species of open habitats, such as savannah and grassland, tend to fluctuate strongly through the year because such habitats tend to be more seasonal than forest, notably in rainfall, which affects the food supplies of a wide range of species. In these grassy habitats, fires also attract many kinds of birds in the dry season to feed on the insects and other small animals disturbed and killed by the flames. Again, whether a species is resident or migratory depends mainly on its diet, and whether its particular foods remain locally available year-round. Its phylogenetic status seems largely irrelevant (Chapter 20).

LATITUDINAL SHIFTS

The overall effect of bird migration is to alter the latitudinal distribution of birds between summer and winter, so that species numbers in the northern hemisphere are at their greatest in the northern summer and in the southern hemisphere in

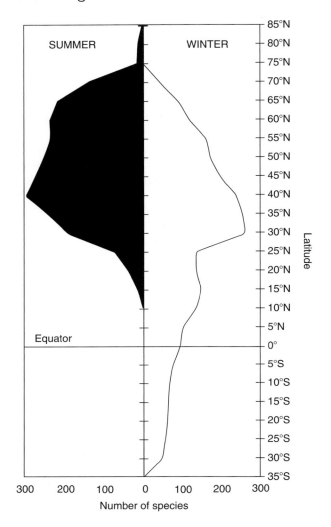

Figure 13.5 Latitudinal shift between summer and winter distributions of bird species that breed in Europe. Includes wintering areas in Europe, Asia and Africa. From Newton (1995b).

the austral summer (northern winter). Take the west European migrants as an example. Some species move relatively short distances within Europe, but others move longer distances to Africa or southern Asia. But the net result, each autumn and spring, is a huge latitudinal shift in avifaunal distribution (**Figure 13.5**). In summer, the whole European assemblage of breeding birds is (by definition) concentrated north of 25°N, but in winter the same assemblage extends southwards as far as the southern tip of Africa (35°S). Forty-eight species of Palearctic birds reach the southern Cape of South Africa (Harrison *et al.* 1997), and some seabirds extend into the seas beyond. When they are in their wintering areas, the migrants add to the local species, increasing the overall species numbers, especially in the tropics (Chapters 23 and 24).

Many arctic-nesting species pass the northern winter in the southern hemisphere, including some populations of many shorebird species, three skuas *Stercorarius*, Arctic Tern *Sterna paradisaea* and other terns, Sabine's Gull *Larus sabini*, Peregrine

Falcon *Falco peregrinus*, Northern Wheatear *Oenanthe oenanthe*, White Wagtail *Motacilla alba*, Petchora Pipit *Anthus gustavi* and Red-throated Pipit *A. cervinus*. They gain the advantage of summer conditions year-round.

Trends within species

The above analyses (**Figures 13.1–13.5**) were based on the presence or absence of species at particular latitudes in winter. They were therefore based only on complete migrants, while for purposes of analysis partial migrants (in which only a proportion of individuals leave for the winter) were counted as year-round residents. However, in many species that breed over wide areas, a greater proportion of individuals migrate from higher than from lower latitudes. Thus, some such species in the northern hemisphere are completely migratory in the north of their breeding range and completely sedentary in the south, while in intervening areas some individuals leave and others stay (partial migration). European examples include Eurasian Blackbird *Turdus merula* and Peregrine Falcon *Falco peregrinus*, and North American examples include American Robin *Turdus migratorius* and Red-tailed Hawk *Buteo jamaicensis*. In general, therefore, the extent to which any population migrates for the winter broadly corresponds to the degree of seasonal reduction in food supplies. Taking account of partial as well as complete migrants, the latitudinal trends discussed above would be even more marked.

ALTITUDINAL MIGRATION

By moving a few hundred metres down the sides of a mountain, birds can achieve as much climatic benefit as by moving several hundred kilometres to lower latitudes, but without the extra winter daylength. Mirroring the latitudinal trend, with rising altitude, increasing proportions of breeding species move out for the winter, but in contrast to latitudinal migration, altitudinal movements can be in any directions that reach lower ground. They occur on mountain ranges worldwide, and can involve a large proportion of local montane species. Examples include the Citril Finch *Serinus citrinella* and Water Pipit *Anthus spinoletta* in Europe, and the Rosy Finch *Leucosticte arctoa* and White-tailed Ptarmigan *Lagopus leucurus* in western North America. In many mountain ranges, nectar-eaters also move upward through spring into late summer in response to the progressively later flowering found at higher elevations. Seasonal altitudinal movements occur even on relatively low mountains, such as the Great Dividing Range in southeast Australia, where several montane species appear in lowland towns and farms in winter (see later).

Some species make both altitudinal and latitudinal movements. In western North America, the two rosy finches *Leucosticte tephrocotis dawsoni* and *L. australis* are altitudinal migrants (in any direction), two others (*L. t. wallowa* and *L. atrata*) are altitudinal migrants with small latitudinal shifts, and two others (*L. t. littoralis* and *L. t. tephrocotis*) are mainly latitudinal migrants (King & Wales 1964).

Owing to altitudinal migrations, the valleys and foothills of montane regions often hold a far greater variety of birds in winter than comparable areas in flatter terrain. They include in winter not only the year-round lowland residents, but the immigrants from higher altitudes and from higher latitudes. Some montane species appear to be obligate migrants, the entire population leaving for the non-breeding season, whereas others appear to be partial or facultative migrants, with year-to-year variations in the proportions of individuals that move to lower elevations.

COMPARISONS BETWEEN HEMISPHERES

Migration of landbirds from their breeding areas is a much more obvious phenomenon in the northern hemisphere than in the southern. This is partly because land covers three times the area in the northern hemisphere as in the southern hemisphere, and the difference is most marked at high latitudes (we can ignore Antarctica because it holds no landbirds) (**Figure 13.6**). In North America, Greenland and Eurasia, some landbird habitat extends north of 80°N, but in the southern hemisphere, South America reaches only to 55°S, Africa to 35°S, Australia to 43°S, and New Zealand to 47°S. The net result is that latitudes 30–80°N hold 15 times more land than do latitudes 30–80°S, and it is at these latitudes that winters are coldest, and migration is most developed (for discussion of area effects in South America see Chesser 1994, and in Australia see Chan 2001). The greater latitudinal spread of land in the northern hemisphere results not only in more marked migration, but also in generally longer journeys than are undertaken by southern hemisphere breeders, which are closer to the equator. These factors are likely to explain the relative proximity of breeding and wintering ranges typical of many southern hemisphere migrants (Chesser 1994, Jahn *et al.* 2004).

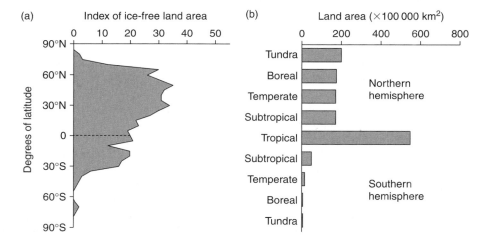

Figure 13.6   The greater land areas in the northern than in the southern hemisphere, shown by land area (a) and habitats (b). The habitats refer to vegetation zones before human impact, only parts of which still remain. From Rosenzweig (1992).

Another factor is temperature, which has a steeper downward gradient north of the equator than south of it. For example, in the New World the mean midwinter (January) temperature at the Tropic of Cancer is about 13°C, while at 50°N it is −15°C, a 28°C difference. In contrast, the mean midwinter (July) temperature at the Tropic of Capricorn is 16°C, while at 50°S it is 0°C, a 16°C difference. Similar hemisphere differences are apparent in much of the Old World too. This difference in steepness of temperature gradient between hemispheres may explain why greater proportions of species leave from temperate latitudes in the northern hemisphere than in the southern. For example, about 29% of species leave completely for the winter from Morocco, but only 6% of species leave from equivalent latitudes around the Cape in South Africa (Newton & Dale 1996, Harrison *et al.* 1997). It is presumably largely for both these land-related and temperature-related reasons that landbird migration is much more marked in the northern than in the southern hemisphere.

While many landbird species that breed in the northern hemisphere migrate south of the tropics, no landbird species that breeds in the southern hemisphere moves north of the tropics. This difference in long-distance migration might also be attributed to the difference in available land areas between the two hemispheres. Birds migrating south from the northern continents encounter progressively smaller habitable land areas, which could force some individuals to extend far to the south. In general, the numbers of migrants from the northern continents decline with increasing distance southwards into the southern continents, as more and more species settle for the winter. In contrast, birds migrating northward from the southern parts of the southern continents encounter widening land areas, so may need to move less far north before they find sufficient wintering habitat. In Africa, no breeding landbird species migrates north of the Sahara, in South America none (apart from stragglers) penetrate north of Panama, and in Australasia none migrate beyond Lombok in Indonesia (Wallace's Line) in winter. It is not that birds which breed in the southern hemisphere do not move around, but rather that most migration is short distance and partial (involving only part of a population), altitudinal (and also short-distance) or nomadic (as birds concentrate locally in line with sporadic rainfall patterns). Relatively few species make regular long-distance moves. It remains to be discovered to what extent the two sets of migrants, from northern and southern regions, occupy the same niches in the tropics but at different times of year.

Also contributing to the reduced numbers of long-distance migrants among the breeding birds of the southern hemisphere is the virtual lack of tundra-nesting shorebirds, which form a large proportion of the long-distance migrants from the northern hemisphere (although one species, the Red-chested Dotterel *Charadrius modestus*, occurs on the tundra of the Falkland Islands). Most species of tundra-nesting shorebirds seen on southern hemisphere shorelines year-round are migrants from the northern hemisphere which do not return north in their first year of life (Chapter 15).

In extent of seasonal migration, pelagic seabirds provide a telling contrast with landbirds. The reduced land areas in the southern hemisphere mean that the sea areas are correspondingly larger there than in the northern hemisphere. Linked with these greater sea areas and large numbers of scattered island breeding sites, pelagic seabirds are much more numerous in the southern hemisphere than in the northern, both in terms of species and of individuals. Correspondingly, a greater proportion of southern than of northern hemisphere breeding seabird species make long migrations. Five (11%) of 47 species that breed north of the tropics extend to south of the tropics in the northern winter, whereas 14 (23%) of 61 species that breed south of the tropics extend to north of the tropics in the austral winter (calculated from maps in Harrison 1983). The implication is again that the sheer numbers of birds, in relation to the habitat available, influence the distances moved and area occupied outside the breeding season.

Populations in both hemispheres

Very few migratory bird species have separate breeding populations north and south of the tropics. Examples include the Little Tern *Sterna albifrons* and Whiskered Tern *Chlidonias hybridus* in Asia–Australia, the Black Stork *Ciconia nigra* and Booted Eagle *Hieraaetus pennatus* in Europe–Africa, and the Turkey Vulture *Cathartes aura* and Black Vulture *Coragyps atratus* in North–South America.[1] In all these species, the northern birds winter in the south when the southern birds are breeding but, among the terns, the southern birds also winter in the north when the northern birds are breeding. It is as though there is a single population of terns occupying the same range but with part of the population breeding at one end of the migratory terminal and another part at the other end. In each area, terns from each population can be distinguished according to whether they are in breeding or non-breeding plumage.

RELATIONSHIP BETWEEN BREEDING AND WINTERING AREAS

Patterns in distribution

By definition, resident bird populations occupy the same geographical range year-round, while migratory species occupy partly or wholly different ranges at different times of year. The variations are in the degree of separation of breeding and wintering ranges, from coincident, through overlapping, to completely

---

[1]*Three species of Palaearctic migrants have recently been recorded breeding in their South African winter quarters, namely White Stork* Ciconia ciconia, *Whiskered Tern* Chlidonias hybridus *and Black-necked Grebe* Podiceps nigricollis, *while the Common Sandpiper* Actitis hypoleucos *has bred in various parts of East Africa. Other species, possibly derived from migrants, are now resident or migratory within southern Africa, namely Great-crested Grebe* Podiceps cristatus, *Black Stork* Ciconia nigra, *Booted Eagle* Hieraaetus pennatus, *Pied Avocet* Avosetta avosetta, *Black-winged Stilt* Himantopus himantopus, *European Bee-eater* Merops apiaster, *Alpine Swift* Apus melba *and Common Stonechat* Saxicola torquata. *Two others have subspeciated, namely the African Bittern* Botaurus stellaris *and Mountain Buzzard* Buteo b. oreophilus. *So perhaps migrants have been colonising southern Africa for a long time. The Barn Owl* Tyto alba *is also resident, but is almost cosmopolitan and did not necessarily derive from European migrants.*

1. Resident year-round throughout the entire latitudinal span of the range.

2. Present in summer only in the northern part of the range, and year-round in the southern part.

3. Present year-round only in the northern part of the range, and in winter only in the southern part.

4. Present in summer only in the northern part of the range, year-round at intermediate latitudes, and in winter only in the southern part.

5. Summer range immediately to the north of the winter range, with little or no overlap.

6. Summer range separated from winter range by a latitudinal gap in which the species occurs only on passage.

Figure 13.7   Main migration patterns found in northern hemisphere birds, based on the degree of separation between breeding and wintering ranges. See also **Table 13.2**.

separate (**Figure 13.7**; **Table 13.2**). In most species, breeding and wintering ranges are coincident or overlapping, while smaller numbers show a latitudinal gap between the two, differing in extent between species (**Figure 13.8**).

In both Old and New Worlds the greatest separation of breeding and wintering ranges is found, as expected, in species that breed only at high latitudes in one hemisphere and winter only at high latitudes in the opposite hemisphere. Among landbirds, the Swainson's Hawk *Buteo swainsoni* is one of the most extreme examples, as it breeds between 25° and 65°N in North America and winters between 24° and 40°S in South America, giving a 49° latitudinal gap between the breeding and wintering ranges (apart from small numbers that winter in some southern States). Among seabirds, the Arctic Tern *Sterna paradisaea* is probably

Table 13.2 Migration patterns of birds in the northern hemisphere, arranged roughly in order of increasing segregation of breeding and wintering ranges

	Total no. in each category[a]
1. Present year-round throughout their whole latitudinal range	195
Old World examples: Black Grouse *Tetrao tetrix*, Eurasian Green Woodpecker *Picus viridis*, Black-billed Magpie *Pica pica*, House Sparrow *Passer domesticus*	
New World examples: Ruffed Grouse *Banasa umbellus*, Northern Bobwhite *Colinus virginianus*, Common Raven *Corvus corax*, Carolina Wren *Thryothorus ludovicianus*	64
2. Present only during the summer breeding season in the north of their range, year-round in the south	22
Old World examples: Common Wood-pigeon *Columba palumbus*, Eurasian Skylark *Alauda arvensis*, European Serin *Serinus serinus*, European Robin *Erithacus rubecula*	
New World examples: Red-tailed Hawk *Buteo jamaicensis*, Common Moorhen *Gallinula chloropus*, Blue Jay *Cyanocitta cristata*, Common Grackle *Quiscalus quiscula*	47
3. Present year-round in the north of the range, only during winter in the south	21
Old World examples: Eurasian Blackbird *Turdus merula*, Siberian Tit *Parus cinctus*, Willow Tit *Parus montanus*, Pine Grosbeak *Pinicola enucleator*	
New World examples: Evening Grosbeak *Hesperiphona vespertina*, House Finch *Carpodacus mexicanus*	23
4. Present only during the summer breeding season in the north of their range, year-round at intermediate latitudes, and only during winter in the south	111
Old World examples: Eurasian Woodcock *Scolopax rusticola*, Rook *Corvus frugilegus*, Redwing *Turdus iliacus*, Common Starling *Sturnus vulgaris*	
New World examples: Canada Goose *Branta canadensis*, Short-eared Owl *Asio flammeus*, Cooper's Hawk *Accipiter cooperii*, Song Sparrow *Melospiza melodia*	52

5. Summer breeding range immediately to the north of the wintering range	Old World examples: Ruddy Turnstone *Arenaria interpres*, Great-spotted Cuckoo *Clamator glandarius*, Bluethroat *Luscinia svecica*, Ring Ouzel *Turdus torquatus*		22
	New World examples: Red-breasted Merganser *Mergus serrator*, House Wren *Troglodytes aedon*, Yellow-throated Warbler *Dendroica dominica*, Vesper Sparrow *Pooecetes gramineus*		34
6. Summer breeding range separated geographically from the wintering range by a gap in which the species occurs only on passage. Some of these species cross large stretches of inhospitable sea or desert in which they cannot feed during migration	Old World examples: Arctic Tern *Sterna paradisaea*, Eurasian Dotterel *Eudromias morinellus*, Sanderling *Calidris alba*, Red-footed Falcon *Falco vespertinus*, Lesser Grey Shrike *Lanius minor*, Icterine Warbler *Hippolais icterina*		117
	New World examples: Whistling Swan *Cygnus c. columbianus*, Brent Goose *Branta bernicla*, Swainson's Hawk *Buteo swainsoni*, Long-tailed Jaeger *Stercorarius longicaudus*, Iceland Gull *Larus glaucoides*, Baird's Sandpiper *Calidris bairdii*, Bristle-thighed Curlew *Numenius tahitensis*, Snow Bunting *Plectrophenax nivalis*		144

See also **Figure 13.7.**

[a]The proportions in the six categories differ significantly between western Europe–Africa and eastern North America–South America ($\chi^2_5 = 85.7$, $P < 0.001$), reflecting mainly the smaller proportions of class 1 species, and the greater proportions of classes 5 and 6 species, in North America. The three species of southern hemisphere seabirds that spend the northern summer (austral winter) off Europe, and the eight that spend the northern summer off eastern North America, are excluded from analysis.

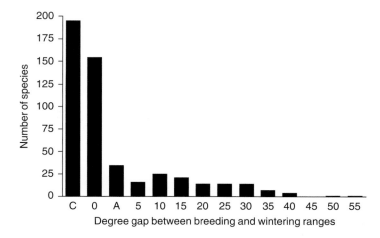

Figure 13.8 Frequency distribution of latitudinal gaps between breeding and wintering ranges, calculated for west Palaearctic breeding birds. C, coincident; O, overlapping; A, adjacent.

the most extreme example, breeding between about 50° and 80°N and wintering between about 40° and 70°S, giving a 90° latitudinal gap between the breeding and wintering ranges.

Comparison of sizes of breeding and wintering areas

Whether overlapping or separate, those migrant species that have the largest breeding ranges also tend to have the largest wintering ranges, and vice versa (Newton 1995a). This point is illustrated in **Figure 13.9** for 57 species of landbirds that breed entirely within Eurasia and winter entirely within Africa, so that their breeding and wintering ranges are completely separated. As another reflection of the same phenomenon, European landbirds and freshwater birds which breed over the widest span of latitude also winter over the widest span of latitude, and vice versa (**Figure 13.10**). These correlations may have their basis in the ecology of the species themselves, in that those species that have the widest climatic and habitat tolerances may be able to spread over the largest areas, summer and winter. Alternatively, the correlations may depend on the abundance of the species concerned, in that those that have the largest populations (for whatever reason) spread over the largest areas, summer and winter (Newton 1995a). These two explanations are not mutually exclusive, and in practice are difficult to separate.

The correlations between sizes (or latitudinal spans) of breeding and wintering areas hold only as general tendencies, however, and some species do not fit the overall patterns. Moreover, because of the geographical scale involved, measures of range size can only be crude, and take no account of areas within the range that lack suitable habitat. They also take no account of the fact that the bulk of the population may occupy only part of the winter range at one time, either shifting south during the course of the northern winter or occurring at any one time only in those parts where rainfall or other factors have created suitable conditions (Chapter 16).

Despite the general correlation between the sizes of breeding and wintering areas, in about 69% of Eurasian–African migrants the breeding range is noticeably

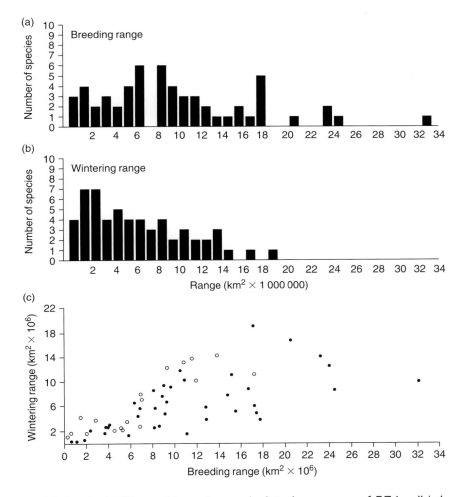

Figure 13.9 (a, b) Sizes of breeding and wintering ranges of 57 landbird species which breed entirely in Eurasia and winter entirely in Africa. (c) Relationship between sizes of breeding and wintering ranges of the same 57 species. Warblers are shown separately as open circles. From Newton (1995a).

larger than the wintering range. This is significantly greater than the expected 50% ($\chi^2 = 4.0$, $P < 0.05$). The most extreme example is the Lesser Grey Shrike *Lanius minor*, whose breeding range in Eurasia covers an area at least seven times greater than its known wintering range in southwest Africa, which is centred on the Kalahari basin (Herremans 1998). In contrast, in only 31% of species is the wintering range larger than the breeding range. The most extreme examples include the Olive-tree Warbler *Hippolais olivetorum* and Subalpine Warbler *Sylvia cantillans*, whose known wintering ranges cover more than twice the area of their respective breeding ranges. However, these low-density species are little known in Africa, and their effective wintering ranges may have been overestimated by the inclusion of records of vagrants or of occurrences in occasional years only.

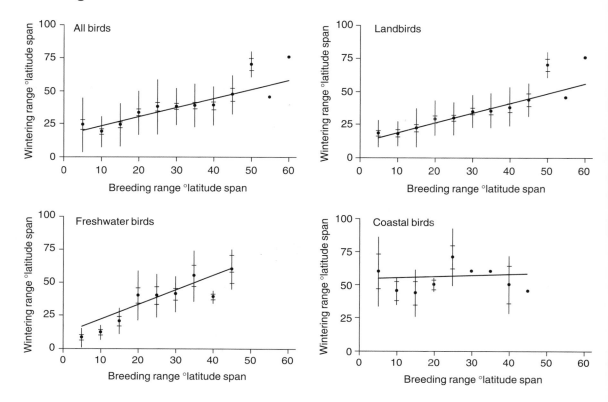

Figure 13.10   Relationship between the latitudinal spans of breeding and wintering ranges of west Palaearctic breeding birds. Some species winter entirely in the west Palaearctic and others partly or entirely in Africa. Excludes seabirds. Spots show mean values, and lines show one standard error and one standard deviation on either side of the mean. The lack of relationship in coastal birds can be attributed to the fact that in winter they switch from an areal distribution in inland areas to a linear distribution along coastlines, often with a very wide latitudinal spread. From Newton & Dale (1997).

For the 57 Eurasian–Afrotropical landbird migrants as a whole, however, wintering ranges are, on average, about one-third smaller than breeding ranges, and in some species only parts of the wintering range may be occupied at any one time (**Figure 13.9**). These findings imply that most species live at greater densities in their African wintering areas than in their Eurasian breeding areas, but whether this reflects differences in available land area, or in the per unit area capacities of the two regions to support the birds at the times they are present remains unknown. It may be that individual birds need more space in Eurasia when they are feeding young than in Africa when they have only themselves to feed. Moreau (1972) estimated that, owing to warmer weather, the individual daily energy needs of passerines in Africa were about 60% of their breeding season needs. Whatever the reason for the differences between sizes of breeding and wintering ranges, a similar phenomenon occurs in the New World, where migrants from large parts of North America concentrate each winter in a relatively small area in the northern Neotropics. On average, then, the geographical ranges of most species are smaller and more overlapping in winter than in summer. The sizes of breeding and wintering ranges of different species are broadly correlated nevertheless.

Among wetland birds (waterfowl and waders) that winter in inland areas, the sizes of breeding and wintering ranges are also correlated, but in contrast to landbirds, wintering ranges are generally larger, and cover a greater latitudinal span (**Figure 13.10**). This may be because, as wetlands become scarcer southwards (from tundra to savannah), freshwater birds have to spread over greater areas than landbirds in winter in order to find enough suitable habitat. Freshwater birds may also make longer movements within a winter than landbirds, in response to rainfall patterns (Newton & Dale 1997).

It is hard to get an appropriate measure of the availability of shallow (and often temporary) wetlands in Africa, compared to Eurasia. The total annual renewable water resource, calculated as the 'average annual flow of rivers and ground water generated from endogenous precipitation', was given by the World Resources Institute (1994) as $4184 \, km^3$ for Africa, compared with $17219 \, km^3$ for Eurasia. This translates to $0.14 \, km^3$ per km^2 of land area in Africa and $0.32 \, km^3$ per km^2 of land area in Eurasia, more than a two-fold difference between these land masses. Moreover, greater evaporation in Africa would greatly increase this difference in terms of surface water, and the big variations in rainfall, both from year to year and from place to place, would further contribute to the sporadic nature of much wetland habitat in Africa.

Most shorebird species switch from an areal distribution on the tundra in summer to a linear distribution on coastlines in winter (Newton & Dale 1997). They therefore breed only in a narrow span of latitude, mostly between 70° and 80°N, but in winter extend southwards over 116° of latitude between about 60°N and 56°S, reaching the southern tip of Africa (35°S), Australasia (47°S) or South America (56°S). Some species, such as Ruddy Turnstone *Arenaria interpres*, can be found in winter on many a rocky coast within this wide latitudinal span, while others, such as Red Knot *Calidris canutus*, may be restricted to the relatively few sites where suitable conditions occur, but are found there in great numbers. Because of the seasonal switch in habitat, it is difficult to compare the sizes of their breeding and wintering ranges but, as a group, they show no relationship between the latitudinal extents of breeding and wintering ranges (**Figure 13.10**). Overall, then, the correlation between summer and winter range sizes is most apparent in landbirds.

MIGRATION WITHIN THE SOUTHERN CONTINENTS

On all the southern continents, the north–south migrations of the local breeding birds (austral migrants) more or less coincide with the north–south movements of the intercontinental migrants from the northern hemisphere, the movements of both groups being driven by the same seasonal changes in climate and food supplies. However, away from the equatorial rainforests, as mentioned above, bird movements are linked not so much to temperature, but to the corresponding wet–dry seasons, and the predictability or otherwise of rainfall. In addition, most parts of the southern continents escape cold winters, and span a wide enough range of latitude to accommodate the native bird species year-round. Apart from the seasonal influx of birds from the northern continents, therefore, the southern continents have self-contained migration systems. The migrations of many species are relatively short-distance and partial, so are often hard to detect without special study. Nevertheless, some species cross the equator on seasonal journeys exceeding 1000 km. Differences in the ecological circumstances prevailing between the three southern continents impose broad-scale variations on their overall migration patterns.

Africa

The rainfall and vegetation zones in Africa mirror one another on both sides of the equator, progressing from rainforest in the wettest equatorial regions, through deciduous woodland to increasingly dry savannahs and grasslands (Chapter 24). In consequence, many birds can find equivalent habitat on both sides of the equator, and because the wet seasons are reversed between north and south sides, they can, by migrating between the northern and southern tropics, benefit from the wet seasons in both. Some such species breed in the northern tropics and spend their non-breeding season in the southern tropics. Others breed in the southern tropics and spend their non-breeding season in the northern tropics. Yet others have separate breeding populations both north and south of the equator, each crossing to the other side on migration (**Figure 13.11**).

Overall, more than 500 African breeding species are known to perform migrations within the continent (Curry-Lindahl 1981). Most move entirely within the northern tropics, or entirely within the southern tropical and temperate zones. In each case, the general trend is for species to move towards wetter (lower latitude) areas for the dry season (**Figure 13.11**, for raptors see **Figure 13.12**). Those that cross the equator to equivalent habitats on the other side are relatively few in number. In addition, in the mountainous areas of the east and south, many species make seasonal altitudinal movements. For example, in Natal in southern Africa, no less than 76 species have been described as 'altitudinal migrants' (Johnson & MacLean 1994).

In Africa, as in Europe, the proportions of species in each region that are migratory can be predicted fairly accurately from the average temperature of the coldest month (Hockey 2000). In regions where this temperature exceeds 20°C, less than 10% of species are migratory. The likelihood of any one species being migratory again depends on its diet, with frugivores being mainly sedentary and insectivores migratory. This applies particularly to those that eat aerial insects (swallows, swifts, nightjars), large active insects (halcyonid kingfishers and rollers) or larvae of flying insects (cuckoos). Over the entire latitudinal range from northern Europe to southern Africa, a strong linear relationship exists between the proportion (P) of birds that are resident and the mean temperature (T) of the coldest month ($P = 1.92T + 53.66$, $r^2 = 0.96$, df = 19, $P < 0.001$, Hockey 2000). It extends the relationship discussed above for European birds alone.

South America

The mirror-image symmetry of vegetation zones north and south of the equator, which is so marked in Africa, is less prominent in South America, where altitude

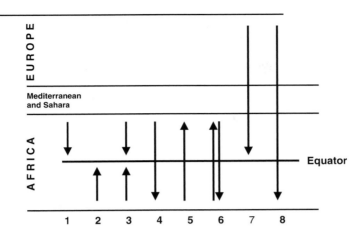

Figure 13.11   Diagrammatic representation of the main migration patterns within Africa, and between Africa and Europe. Examples include partial migrants. ———► breeding area.
1. Within the northern tropics, breeding in the northern wet season. Examples: Grasshopper Buzzard Eagle *Butastur rufipennis*, White-throated Bee-eater *Merops albicollis*, African Collared Dove *Streptopelia roseogrisea*, Red-shouldered Cuckoo Shrike *Campephaga phoenicea*.
2. Within the southern tropical–temperate zone, breeding in the southern wet season. Examples: Fiscal Flycatcher *Sigelus silens*, Greater Striped Swallow *Hirundo cucullata*, Black Cuckoo Shrike *Campephaga flava*, Square-tailed Nightjar *Caprimulgus fossii*. 3. Within both the northern and the southern tropics, migrating nearer to the equator for the dry season. Examples: Spotted Ground Thrush *Zoothera guttata*, African Striped Cuckoo *Oxylophus levaillantia*, Woodland Kingfisher *Halcyon senegalensis*. 4. Transequatorial migration, breeding in the northern wet season. Examples: Abdim's Stork *Ciconia abdimii*, Lesser Crested Tern *Sterna bengalensis*, Plain Nightjar *Caprimulgus inornatus*, Dusky Lark *Pinarocorys nigricans*. 5. Transequatorial migration, breeding in the southern wet season. Examples: Openbill Stork (African Openbill) *Anastomus lamelligerus*, Standard-winged Nightjar *Macrodipteryx longipennis*, Pennant-winged Nightjar *M. vexillarius*. 6. Transequatorial migration with two populations, one breeding in the northern wet season and the other in the southern wet season. Examples: Black Kite *Milvus migrans*, Wahlberg's Eagle *Aquila wahlbergi*, Jacobin Cuckoo *Oxylophus jacobinus*. 7. Eurasian-northern tropics, breeding in the Eurasian warm season. Examples: Melodious Warbler *Hippolais polyglotta*, European Pied Flycatcher *Ficedula hypoleuca*, Woodchat Shrike *Lanius senator*, European Turtle Dove *Streptopelia turtur*. 8. Eurasian-southern tropics, breeding in the Eurasian warm season. Examples: Icterine Warbler *Hippolais icterina*, Barn Swallow *Hirundo rustica*, Common Swift *Apus apus*, Red-footed Falcon *Falco vespertinus*.

effects on climate and vegetation are greater. Correspondingly, smaller proportions of South American breeding species are known to undertake trans-equatorial migrations (although this may be partly due to inadequate information). As in Africa, most movements occur within the northern tropics, or within the southern tropical–temperate zones, with most species moving to lower latitude (wetter) areas for the non-breeding season (Levey & Styles 1992, Joseph 1997, Jahn *et al.* 2004). For some species, this can still involve movements of over 1000 km each

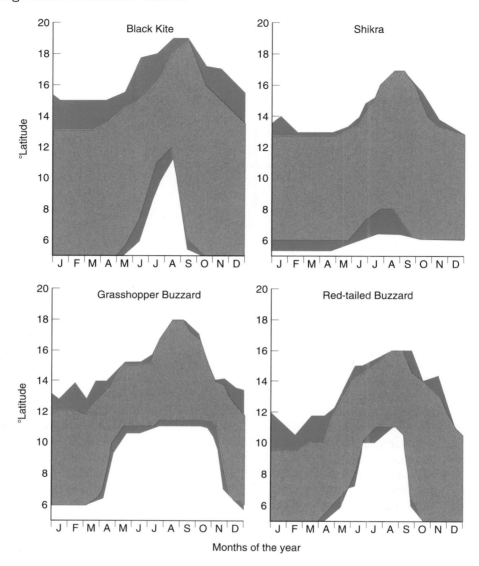

Figure 13.12   Seasonal changes in the distribution of four migratory raptors in West Africa. Grey shading – frequent sightings; blue shading – infrequent sightings. In this region, migratory raptors stay in the southern woodlands during the local dry season, while food is plentiful and hunting conditions are good. When it rains heavily and the grass grows rapidly to 1.3 m high, the migrants move north to the short grass areas, where the rains are later and lighter and produce less growth of vegetation. The birds thus manage to remain in a fairly favourable environment all year and, while in the north, they breed, taking advantage of a short seasonal surplus of food which is not fully utilised by the sparse resident population. Species differ in the extent of their migrations, and in the periods spent at different latitudes, depending on their particular needs, but the general northward passage occurs at the start of the rains in 'spring' and the southward passage at the end of the rains in 'autumn' (From Thiollay 1978).

way. As expected, many species in the Andes, where winters can be harsh, move to lower elevations for the non-breeding season.

Recent estimates of the numbers of intra-continental migrants in South America range from 220 to 237 species, as compared with 338 species of Nearctic–Neotropical migrants (Rappole 1995), but the South American figure could well be an underestimate, considering the richness of the avifauna, and the relatively poor state of knowledge. One-third of all known migratory species belong to the Tyrannidae (New World Flycatchers), reflecting the overall preponderance of this family across the continent (Chesser 1994, Jahn *et al.* 2004).

Australasia

Overall, about 40% of Australian landbird species are known to migrate in at least part of their range. Partial migration occurs in about 44% of 155 non-passerine species and 32% of 317 passerines examined (Chan 2001). Similar findings emerged from an analysis of bird count data in which 37% (145/393) of species were detected as making movements (Griffioen & Clarke 2002).

East of the Great Dividing Range, rainfall is generally adequate, and seasonal temperature changes largely influence seasonal productivity. Movement is generally northward toward the subtropics and tropics for the winter. Some species in the southern parts of their range are completely migratory, but many species are partial migrants that occupy much the same geographical range year-round, with seasonal south–north shifts in the centre of gravity of their populations. These shifts are mostly over distances of less than 1000 km. Eastern examples include the Scarlet Honeyeater *Myzomela sanguinolenta* and Yellow-faced Honeyeater *Lichenostomus chrysops* (Clarke *et al.* 1999, Munro 2003). Nevertheless, some species perform longer movements. They include the many waterbirds that migrate from Australia to New Guinea in the non-breeding season, including egrets, ibises, pelicans and ducks. For three species of egrets that come from southeastern Australia, this entails a journey of more than 3000 km (Geering & French 1998). Further south, five species leave Tasmania completely for the winter, spending their non-breeding season in Australia or beyond, namely the Swift Parrot *Lathamus discolor*, Orange-breasted Parrot *Neophema chrysogaster*, Pallid Cuckoo *Cuculus pallidus*, Shining Bronze Cuckoo *Chrysococcyx minutillus* and Satin Flycatcher *Myiagra cyanoleuca* (Chan 2001, Dingle 2004). Other eastern species are altitudinal migrants on the Great Dividing Range. They include such conspicuous species as Yellow-tailed Black Cockatoo *Calyptorhynchus funereus*, Golden Whistler *Pachycephala pectoralis*, Regent Bowerbird *Sericulus chrysocephalus* and others. Whereas these and others move to lower ground for the winter, others, such as the Eastern Spinebill *Acanthorhynchus tenuirostris*, move uphill to exploit *Banksia collina* pollen at higher elevations in winter.

Except for the monsoonal north and small areas in the southwest and south-centre that are Mediterranean in climate, rainfall over most of Australia west of the Dividing Range is sparse and erratic, making it the driest continent overall. Because bird breeding in these circumstances is frequently tied to erratic rainfall, migration can be complex and variable, but many species show an underlying north–south pattern (Nix 1976; Chapter 16). Comparing different parts of Australia, the proportions of birds that are migratory decline with increase in the amount and evenness of the annual rainfall. Where the annual total exceeds 125 cm, and is well distributed through the year, 70–85% of honeyeater species (Meliphagidae) are year-round residents, but in the central desert, where annual rainfall is 20–28 cm and erratic, fewer than 50% are residents, and many seem to perform nomadic movements in response to rainfall patterns (Keast 1968b; Chapter 16).

New Zealand now holds relatively few native landbird species, and only three of these spend the winter elsewhere. They include the Long-tailed Cuckoo *Eudynamis taitensis* which winters in a wide arc of islands from New Guinea across the Marshall and Caroline Islands to the Marquesas in the east; the Shining Bronze Cuckoo *Chrysococcyx lucidus* which migrates to the Bismarcks and Solomons; and the Double-banded Plover *Charadrius bicinctus* which migrates to southeast Australia for the winter. The last two species also breed within Australia.

CONCLUDING REMARKS

The overriding role of food supplies in influencing the movement patterns of birds is evident from: (1) the link between migration and climatic seasonality, with increasing proportions of migrants in increasingly seasonal environments; (2) the relationship in different species between migration and diet; (3) within species, the regional variations in the proportions of birds that leave for the non-breeding season; (4) the locations of wintering areas; and (5) the precise seasonal timing of movements, such that birds are absent from breeding areas at a time when their particular foods are scarce there (Chapter 14). Some authors have stressed the role of climate and competition in influencing migration patterns, but both these factors are likely to act primarily through the food supply.

Migratory habits are so closely related to diet, and to the seasonal fluctuations in the food types involved, that it is hard to tell whether any other factors have any influence beyond those that act through food (such as climate and competition). The main problem is that diet is related to almost all aspects of the morphology, ecology and behaviour of birds, as well as to migration, so any factor could seem to be important to migration through association. For example, in a recent analysis of British breeding birds, Siriwardena & Wernham (2002) found no consistent association between migration and body size, but that territorial species tended to be more sedentary than colonial and semi-colonial ones, that hole-nesters tended to be less migratory than open-nesters, and that localised species tended to be more migratory than widespread ones. However, their analysis was based on a wide range of species from passerines to gulls, and many of their findings depended on the inclusion or exclusion of particular groups. They also found that migrants in general had higher survival and lower reproductive rates than residents, a finding expected on other grounds (Chapter 20). Their most robust finding, however, was a relationship between migration and diet, with insectivores and piscivores being more migratory than omnivores and herbivores.

In species that migrate between the northern and the southern hemispheres, the question arises why the same individuals do not breed twice in one year, in both summer and winter quarters. One reason in many species is that individuals moult while in winter quarters, a process that takes several weeks or months and could not be undertaken at the same time as breeding (the two processes being mutually exclusive in most birds, Chapter 11). Another reason is that many migratory species do not remain for long in the same area in winter, but periodically move to other areas in response to changes in food supplies (Chapter 24; Jones 1995). This exploitation of temporary abundances is one way in which migrants in the southern hemisphere could avoid competing with the local birds which, breeding at that time, are tied to fixed nesting areas. Neither explanation applies to all transequatorial migratory species, however, and there are still some that are sedentary while they are in both breeding and wintering areas, and would seem able to breed in both, six months apart, but do not.

A related question is why many birds, having had a winter break, do not breed more than once at different localities on their migration routes. Migrants that travel in spring northward through Europe and North America, and have short breeding cycles, would seem able to breed in the southern parts of these continents, before moving on to breed again further north. They could then raise more young per year than by exploiting only the short favourable season in the north. The fact that a few species are known to do this (see Chapter 16) makes it even harder to explain why most do not. However, many candidate populations also have competing conspecifics nesting at lower latitudes, where they can raise two or more broods in a season.

Long-distance migration is found not only in birds, but in various mammals and insects, as well as in various marine fish, turtles and invertebrates. Although food supplies and reproduction underlie the movements of all these animals, in some of them the pattern is different from birds. For instance, various whales migrate to high latitudes in summer, where they feed hard and accumulate body fat, before returning to winter in the tropics, where their young are born. However, the fact that they accumulate most of the food that permits breeding at high latitudes may make them less different from birds in this respect than appears at first sight. Various fish migrate only twice in their lives, from the breeding grounds as young, returning several years later as adults, when they spawn and die, or return repeatedly to spawn in successive years. Various butterflies breed at different points on the migration route, with successive generations making different parts of each journey. All this contrasts with the situation in birds, most of which migrate regularly back and forth every year between their higher latitude breeding areas and lower latitude wintering areas.

Appreciation of such large-scale, long-distance movements highlights the conservation problems that migratory animals present. Depending on a chain of areas, spanning a wide range of latitude, such species are vulnerable to human impact, not only in breeding and wintering areas, but also at crucial stopover sites en route. Nature Reserves established in one region can be expected to protect some species for no more than a few weeks each year, before they move on. It underlines the need for international collaboration in the conservation of migrants, a concept that has already led to concerted action for waterfowl and shorebirds in some parts of the world.

SUMMARY

Migration is most pronounced in environments in which food supplies vary greatly through the year. It enables birds to exploit seasonal abundances and to avoid seasonal shortages. Broadly speaking, birds move so as to keep themselves in favourable habitat for as much of the year as possible, allowing for the fact that their requirements may differ between the breeding and non-breeding seasons. The proportions of breeding species that leave for the winter increase with latitude, as seasonal changes in food availability become more marked. At any particular latitude, migration is also related to diet; only species whose food remains available year-round remain in their breeding areas year-round.

Migrations cause huge seasonal latitudinal and, to some extent, longitudinal changes in the distributions of birds. From low to high latitudes, a progressively greater proportion of breeding species leaves for the winter, and few species remain in winter north of the tree line. In mountain regions, many bird species migrate from high to lower ground for the winter, and back again in spring. Most latitudinal migrants occupy larger geographical ranges in summer than in winter when populations become more concentrated, but across species, the sizes of breeding and wintering ranges are correlated. The main exceptions include many shorebirds which breed over wide continental areas, switching in the non-breeding season to a linear distribution along coastlines.

Among landbirds, migration is much more marked in the northern than in the southern hemisphere. This is associated mainly with the far greater land areas in the northern hemisphere, especially between latitudes 30 and 80°N, where migration is most developed. Many birds from the northern continents migrate into the southern continents for the non-breeding season, but none from the southern continents migrate into the northern continents. In the southern continents, most migration is short-distance and partial, altitudinal or nomadic. Among pelagic birds, in contrast, migration is more marked in southern hemisphere species. This is associated with the greater sea areas, and more abundant nesting islands in the southern hemisphere, which support much greater numbers there. Some northern hemisphere seabirds migrate beyond the southern tropics, but more southern ones migrate beyond the northern tropics.

Appendix 13.1 Relationships between species numbers (or % migrants) and various environmental variables in western Europe

	Latitude	Land area	Elevation range	Mean annual temperature	Mean January temperature	Mean July temperature	Seasonal temperature range	Annual precipitation range
Number of breeding species	−51* / −90***	+78*** / −95***	+25	+62** / −88**	+75** / −94***	+64** / −99**	−59*	+79*** / −94***
Number of wintering species	−93*** / −97***	+68**	+29	+94***	+90***	+88***	−51*	+77***
% breeding species that leave for winter	+93*** / +97***	−61**	−28	−95***	−87***	−85***	+50*	−75**
% wintering species that leave for summer	−85***	+32	+22	+80*** / +83**	+66**	+65**	−36	+37

Relationships (Pearson correlations) between the various environmental variables listed above

	Latitude	Land area	Elevation range	Mean annual temperature	Mean January temperature	Mean July temperature	Seasonal temperature range	Annual precipitation range
Latitude		−0.76*	−0.60	−0.99***	−0.94***	−0.97***	0.63*	−0.80**
Land area			0.66*	0.80**	0.82**	0.77**	−0.65*	0.81**
Elevation range				0.61	0.53	0.64*	−0.16	0.66**
Mean annual temperature					0.97***	0.97***	−0.71*	0.83**
Mean January temperature						0.96***	−0.82**	0.84**
Mean July temperature							−0.61	0.83**
Seasonal temperature range								−0.5

Upper figures show percentage variation accounted for by linear regression. Lower figures show the percentage variation accounted for by quadratic regression, and are given only where the quadratic term was statistically significant. All relationships were essentially monotonic over the observed range, and + or − values indicate signs of linear and (where appropriate) quadratic coefficients.
*$P < 0.05$, **$P < 0.01$, ***$P < 0.001$.

Canada Geese *Branta canadensis* en route to breeding areas

Chapter 14
Seasonal occupation of breeding areas

Every bird student has noted the feverish impatience with which certain species push northwards in spring, sometimes advancing so rapidly upon the heels of winter as to perish in great numbers when overtaken by late storms. (Frederick C. Lincoln 1935.)

With the approach of spring, as days lengthen and temperatures rise, migratory birds begin to move from lower to higher latitudes in order to re-occupy their breeding areas. One species after another spreads in wave-like manner towards higher latitudes, progressively re-settling areas vacated since the previous autumn. When individuals reach their home areas, they establish territories and acquire mates in preparation for breeding. After breeding is completed, they withdraw to lower latitudes where they spend the non-breeding period. These large-scale distributional changes enable birds to exploit the surge in fresh food supplies produced each spring and summer at high latitudes, yet avoid the shortages of winter.

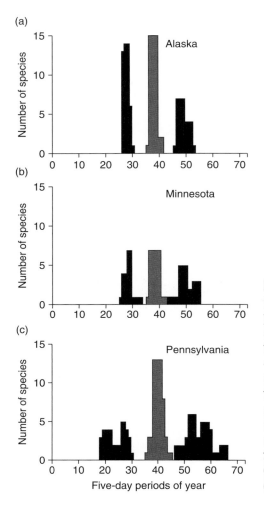

Figure 14.1   Number of species with median migration dates in each five-day period through the year, based on: (a) 18 species that pass each spring and autumn through Alaska at 64° 50′N; (b) 18 species that pass each spring and autumn through Minnesota at 44° 55′N; and (c) 33 species that pass each spring and autumn through Pennsylvania at 40° 40′N, together with their bisectrix dates (midway between median dates of spring and autumn passage). Migration dates shown in black, bisectrix dates in grey. Note the vastly greater spread in migration dates than in bisectrix dates, and the greater spread in autumn than in spring dates. (a) from Preston (1966), (b) from Winker *et al.* (1992a), (c) from Benson & Winker (2001).

In particular regions, seasonal changes in the food supplies of birds are driven by seasonal changes in daylength and weather – mainly by temperature at higher latitudes and by rainfall at lower latitudes. These seasonal changes ultimately influence how much of each year migrants can remain in their breeding areas without jeopardising their survival prospects. With increasing latitude, spring begins later and autumn begins earlier, shortening the growing season for plants, the activity season for insects, and the potential breeding season for birds.

Away from the equator, the annual temperature cycle typically lags about one month behind the daylength cycle. In the northern hemisphere, while the longest day falls on 21 June, the warmest day in any particular locality falls, on average, around 21 July. Similarly, while the shortest day occurs on 21 December, the coldest day falls, on average, around 21 January (Preston 1966). The peak dates of spring and autumn migration for various bird species at different localities in North America are shown in **Figure 14.1**. At each locality, the mid date between spring and autumn migration dates averages around 17 July, the warmest time

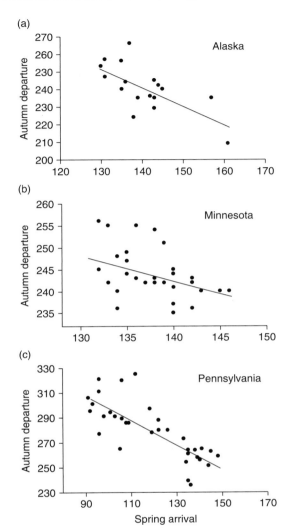

Figure 14.2   Relationship between median spring and autumn migration dates for various species in (a) Alaska (64° 50′N, $N = 18$); (b) Minnesota (44° 55′N, $N = 18$); and (c) Pennsylvania (40° 40′N, $N = 33$). In general, species that arrived early tended also to depart late, and vice versa. Regression relationships: Alaska, $b = 1.09$, $r = 0.69$, $P < 0.002$; Minnesota, $b = 0.61$, $r = 0.43$, $P < 0.01$; Pennsylvania, $b = 0.97$, $r = 0.78$, $P < 0.001$. Details from Preston (1966), Winker *et al.* (1992a), Benson & Winker (2001).

of year. The same is true for other regions. Evidently, the migrations of most species hinge around the annual temperature cycle rather than around the daylength cycle. It is the temperature cycle that, at high latitudes, has most influence on vegetation growth and bird food supplies.

Within any one breeding area, the timings of spring arrival and autumn departure differ between species according to when their particular food supplies re-appear in spring and collapse in autumn. In general, regardless of latitude, those species that arrive relatively early in their breeding areas depart relatively late, but exceptions occur (**Figure 14.2**). Moreover, the peak migration dates for different species at particular localities are spread over a shorter period in spring than in autumn (**Figure 14.1**), and within most (but not all) species spring arrival occurs over a shorter period than autumn departure (Chapter 12).

LATITUDINAL TREND IN THE TIMING OF SPRING

The temperature of 10°C is often taken to indicate the start of spring conditions, suitable for rapid plant growth and insect activity. The spread of the 10°C isotherm from south to north through Europe is shown in **Figure 14.3**, based on the average figures over many years. This particular isotherm takes more than three months to spread northeastwards through the whole continent, beginning in the southwest in February–March and not reaching the northernmost areas until July. It also takes almost as long to spread from the lowest to the highest parts of mountain areas, such as the Alps. It gives a good indication of the timing of spring conditions, suitable for bird breeding, in different parts of the continent.

The problem of using particular temperature values, like this, as a signal of spring's arrival is that they take no account of previous temperatures and daylengths which may also have influenced plant and insect development. The alternative is therefore to use some biological measure of the timing of spring, and for this purpose data have been assembled on the date of first apple flowering (**Figure 14.4**). This measure takes more than two months to progress from the southwest to the northern limits of the apple in southern Scandinavia. Whatever measures are taken, however, they indicate that conditions become suitable for bird breeding in the northernmost parts of Europe at least three months later than in the south. The spread is even greater in North America, which covers a greater span of latitude than Europe.

Species differences in spring migration dates

The rates at which most bird species move towards higher latitudes in spring seems to be associated with the dates that their particular foods become available at successive latitudes. Migrating birds need food not only for daily maintenance, but also to fuel successive stages of their journeys (Chapter 5). There would be no advantage in migrating birds getting far ahead of their food supplies, for they would then only lose body reserves, and might even have to turn back (as has sometimes been recorded, Chapter 4). For each species, therefore, the timing of spring arrival at particular latitudes generally coincides with the re-appearance of appropriate food supplies, and numbers of birds appear in breeding areas as soon as their local survival becomes likely.

Because some types of food become available at lower temperatures than others, different species arrive at particular localities in a fairly consistent sequence from year to year. Thus, species which depend on spring thaw to release their food supplies (such as some waterfowl and waders) arrive earlier than those that depend on aerial insects (such as midges), and earlier still than those that depend on larval insects from later-developing leaves (such as some warblers). Nectar-feeders depend on the opening of flowers, and the spring migration of the Ruby-throated Hummingbird *Archilochus colubris* through much of North America is nearly synchronous with peak flowering of Jewelweed *Impatiens biflora*, which is an important source of nectar at that time of year (Bertin 1982).

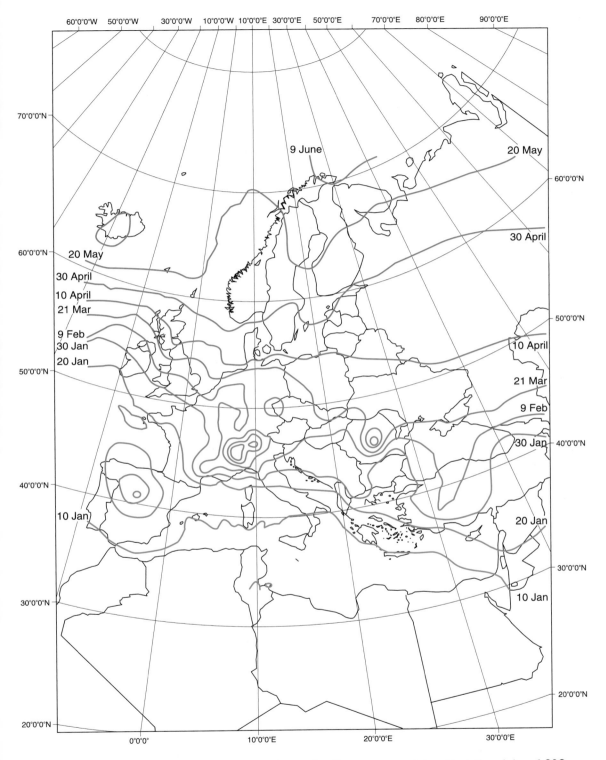

Figure 14.3 The advance of spring, as shown by the average dates that the rising 10°C isotherm reached different parts of Europe during 1971–2000. From T. Sparks, unpublished.

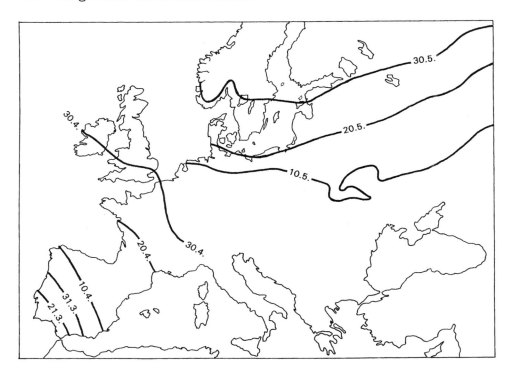

Figure 14.4   The advance of spring, as shown by the mean peak dates of apple flowering in different parts of Europe. From Gatter (2000).

Annual variation in spring migration dates

Because some food supplies depend on weather, many bird species travel more slowly and arrive in their breeding areas later in cold springs than in warm ones. This annual variation is especially apparent in insectivores, such as the Barn Swallow *Hirundo rustica* and various warblers, whose food is clearly temperature dependent, or in species that depend on snow and ice melt, such as waterfowl and shorebirds (**Figure 14.5**; Cooke *et al.* 1995; Huin & Sparks 1998). At particular localities, temperatures in early spring show more annual variation than temperatures in late spring, and correspondingly, early-arriving species typically show significantly more year-to-year variation in arrival dates than do late arriving ones (Gilyazov & Sparks 2002, Tryjanowski *et al.* 2002; **Figure 14.6**). For example, at localities in southern England, first arrival dates of the early-arriving Sand Martin *Riparia riparia* varied by more than 29 days between years, but in the late-arriving Common Swift *Apus apus*, they varied by about 20 days (Loxton *et al.* 1998). Disruptions in the normal arrival sequence occur in years when warm and cold periods alternate through the spring, and in some years a spell of bad weather during the arrival period can produce a twin-peaked arrival pattern (see Elkins 2005 for patterns in Sand Martin *Riparia riparia*).

In some species, the year-to-year relationship between spring arrival and temperature has been studied at more than one locality. In Slovakia, first arrivals of

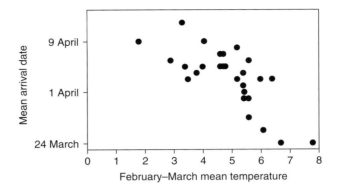

Figure 14.5 Arrival dates of Barn Swallows *Hirundo rustica* in Britain in different years in relation to mean February–March temperatures. From Sparks *et al.* (1999).

Barn Swallows *Hirundo rustica* advanced about 2.1 days per 1°C temperature rise over February–March, in England about 1.7 days per 1°C, and in Finland about 1.2 days per 1°C (Sparks 1999).[1] This suggests a more rapid response at higher latitudes, where arrival dates are much later, and occur over a shorter period. In yet other species, an effect of local food supply on arrival has been detected over and

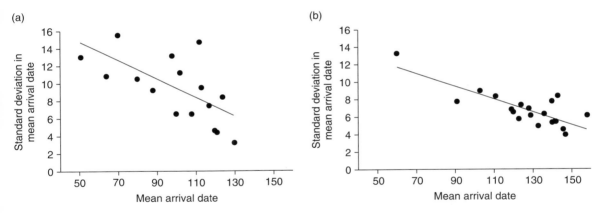

Figure 14.6 Annual variations (as reflected in standard deviations, SD) in the first arrival dates of migratory species in relation to their mean first arrival dates (MFAD) over many years. (a) An area in western Poland, 1913–1996 (Tryjanowski *et al.* 2002); (b) an area in northern Russia, 1931–1999 (Gilyazov & Sparks 2002). In both areas, earlier arriving species had more variable first arrival dates. They were mainly short-distance migrants that wintered within Europe, while the later arriving species were mainly long-distance migrants wintering in Africa south of the Sahara. Regression equations: western Poland, $SD = 19.6 - 0.103x$, where x is the mean first arrival date, $r = 0.65$, $P < 0.01$; northern Russia, $SD = 11.7 - 0.0724x$, $r = 0.80$, $P < 0.001$.

---

[1] *Most studies of arrival dates have used the first date each spring that an individual of each species was seen. There are obvious problems with this in that the first record may represent only a single individual, and the chance of seeing such a bird depends on the area monitored, and the numbers of observers involved. The first dates from wide areas, such as counties, tend to be earlier (cont./)*

above any temperature effect. The Brambling *Fringilla montifringilla*, for example, arrives on its northern breeding areas earlier in years of good spruce crops than in other years (Mikkonen 1981). As the cones open, the seeds provide an early food supply, enabling the birds to survive in their breeding areas until their main summer food (caterpillars) becomes available. In one area over a number of years, the size of the spruce crop, together with air temperature and snow cover, explained 89% of the annual variation in Brambling arrival dates.

It is seldom certain to what extent early arrival in particular years is due to earlier departure of the species from wintering areas, or to faster progress en route but, because the same weather patterns influence large parts of the migration route, migrants come under the influence of warmer or colder conditions long before they reach their breeding areas. Several studies have found a relationship between arrival dates in breeding areas and temperatures back along the migration route, including the wintering areas (Hüppop & Winkel 2006, Sokolov 2006). At least among short-distance migrants wintering within Europe or North America, birds may leave their wintering areas earlier in warm than in cold years, as found for 17 species wintering in Spain (Sokolov 2006). Both long-distance and short-distance migrants arrived at the Courish Spit in the southern Baltic earlier in years when spring temperatures were higher, and apple flowering was earlier (Sokolov 2006).

Those species that advance northward in short flights of perhaps 50–200 km at a time may be able to keep in more precise step with their food supplies than species which arrive in their breeding areas after a long flight from a locality several hundreds of kilometres away. This difference in flight lengths may account for why the arrival dates of short-distance migrants generally show good year-to-year correlations with spring temperatures in the breeding locality, while the arrival dates of long-distance migrants show much poorer correlations or none at all (Tryjanowski *et al.* 2002). In any case, earlier arrival in breeding areas normally leads to earlier breeding, as documented for 10 out of 15 species studied at the Courish Spit. In addition, however, the later that birds arrived, the shorter was the period between arrival and breeding (Sokolov 2006; see also Dalhaug *et al.* 2001, Tryjanowski *et al.* 2004, Hupp *et al.* 2006).

RECOLONISATION PATTERNS

In the first half of the twentieth century, much attention was given, with the help of observer networks, to recording the northward advance of various bird species in spring (**Figure 14.7**). In general, earlier migrants took longer over

(/cont.) *than those from point localities, such as bird observatories. In species which undergo marked changes in numbers over the study period, first dates tend to be earlier in years of greatest abundance, possibly because of the statistical effect on the chance of observation (Loxton & Sparks 1999, Sparks 1999). The same points apply to last observations in autumn. Because by definition first or last birds are atypical, a more representative picture is obtained by using median or mean dates, such as those based on captures at bird observatories, but this greatly reduces the numbers of localities from which such records are available.*

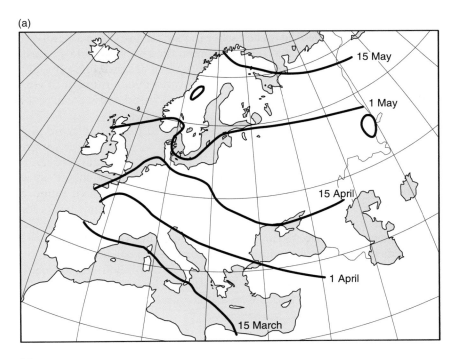

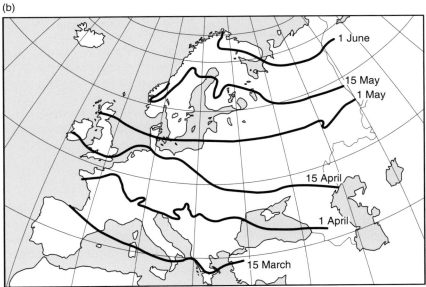

Figure 14.7 The northward advance of various migratory bird species in spring. (a) Common Redstart *Phoenicurus phoenicurus* through Europe (Southern 1939); (b) Barn Swallow *Hirundo rustica* through Europe (from Southern 1938);

(c)

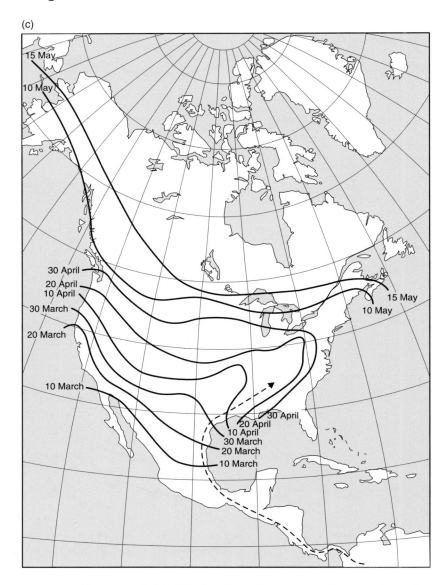

Figure 14.7 (Continued) (c) Cliff Swallow *Hirundo pyrrhonota* through Central America into North America (Lincoln 1935a);

the journey, and spread north at a slower rate per day. Five African–European migrants, namely the Barn Swallow *Hirundo rustica*, Willow Warbler *Phylloscopus trochilus*, Common Redstart *Phoenicurus phoenicurus*, Wood Warbler *Phylloscopus sibilatrix* and Red-backed Shrike *Lanius collurio*, arrived in southern Europe on the progressively later dates of 13 February, 5 March, 15 March, 1 April and 1 April respectively. They spread north through western Europe at average speeds of about 40, 46, 66, 70 and 88 km per day respectively, and took 109, 88, 61, 45 and 45 days to get from the southern to the northern parts of their breeding ranges

(d)

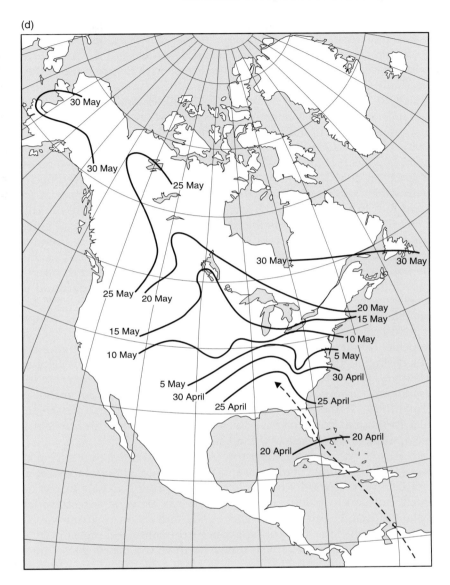

Figure 14.7   (Continued) (d) Blackpoll Warbler *Dendroica striata* through the Caribbean Region into North America (Lincoln 1935a).

(Southern 1938–41). Their northward progression generally kept step with particular isotherms (different isotherms for different species), but like the isotherms, their migrations tended to accelerate with distance northwards. Of these species, the Barn Swallow bred over the widest latitudinal range, and its period of spread over this range (109 days) more or less fitted expectations from the different measures of the northward advance of spring conditions, mentioned above.

More recent estimates of migratory progression, based on first arrival dates at different localities within Britain, gave mean arrival dates in the south of the country of 22, 26 and 19 April and 14 May, and mean rates of progress of 42, 43, 52 and 145 km per day for Common Cuckoo *Cuculus canorus*, Common Nightingale *Luscinia megarhynchos*, Barn Swallow *Hirundo rustica* and Spotted Flycatcher *Muscicapa striata* respectively. Yet again, the last species to arrive made the fastest progress (Huin & Sparks 2000). Remember that these rates of spread refer to the dates that birds first appeared in successive localities, and not to the movement speeds of individuals, which can be faster, as revealed by ringing (Chapter 8).

Similar trends were found in North America (Lincoln 1935a). In some species, such as Canada Goose *Branta canadensis*, the birds push north 'on the heels of winter' and keep step with the 35°F (3°C) isotherm, as advancing warmth melts the ice on lakes and rivers and creates bare ground for feeding (Lincoln 1935a). In other species, such as the Blackpoll Warbler *Dendroica striata*, the northward movement occurs much later in spring and much more rapidly, often with increasing rapidity towards the northernmost breeding areas (**Figure 14.7**). Mean rates of advance in different North American species were found to vary from about 30 km per day in the earliest migrants to 300 km per day in the latest, towards the ends of their journeys. This again reflects the facts that the rate of spring warming gets progressively more rapid with advancing date and increasing latitude.

These relationships with isotherms presumably hold because temperature gives a good indication of the date at which each area normally becomes suitable for the species in question to settle. The fact that the migratory advance is slower in colder springs is further indication of the importance of the development of a food supply. It does not necessarily mean that the species migrate in direct response to temperature, or that individual migrants travel north at the speed of 'isothermal lines'. Rather the average rate of northward spread keeps step with particular isothermal lines, different lines in different species, in accordance with the development of their particular food supplies.

The advance of species from lower toward higher latitudes each spring is therefore linked with warming conditions, and the ecological changes they set in train. For example, the northward advance of geese in spring is closely linked not only to ice melt, but also to plant phenology. The birds arrive at successive latitudes just as plant growth begins (usually above 3°C), and follow a 'green wave' northwards (Owen 1980). After this spring flush, food quality and digestibility decline as the plants age, but on their well-timed northward journey, geese are able to benefit from successive latitudinal peaks in plant growth and digestibility. From each site, most birds leave as the food passes its peak, and arrive at the next site as food reaches its peak. The geese therefore move according to the phenology of their food plants (van der Graaf *et al.* 2006, Hübner 2006, Hupp *et al.* 2006). Unlike geese, which feed mainly on land, swans obtain most of their food from water, so also track the melt line, as it advances northward. Naturally, in many species the speed of migration varies from year to year with the temperatures encountered en route. This was shown, for example, for 15 species studied over a 40-year period in North America (Marra *et al.* 2005). By comparing the passage dates of these species between a trapping site in coastal Louisiana and other sites 2500 km to the north, birds were found to cover this distance most rapidly in the warmest springs. Species that nest at high latitudes are in some years detained some distance short of their breeding areas, as they wait for a thaw to set in and expose their food.

If one watches passage migration at low-latitude localities, species are found to differ not just in their mean passage dates but also in the spread of their passage dates, with some taking much longer to move through than others. Typically, species with short passage periods are those that breed over a narrow span of latitude (with relatively small breeding ranges), whereas those with long passage periods breed over a wider span of latitude, giving a wider range of dates at which different localities become fit for occupation (for shorebirds see Nisbet 1957, for raptors see Leshem & Yom-Tov 1996a).

Patterns within species

With increasing latitude, as the annual warm season becomes shorter, many species spend progressively shorter periods in breeding areas, arriving later in spring and leaving earlier in autumn. The first individuals to arrive at successive localities in the breeding range are normally those that nest there, these settlers being followed by others destined for even higher latitudes. By the time when individuals arrive at the highest latitude breeding areas, perhaps in late May or June, other individuals of their species at lower latitudes have already started nesting, and may even have young.

This low-to-high latitude settling pattern is particularly apparent in species in which different races breed at different latitudes. On passage migration, the different races move through in sequence according to the latitude at which they breed. For example, among Yellow Warblers *Dendroica petechia* migrating north through Arizona, the first birds to arrive in March and early April belong to the local breeding race *D. p. sonorana*. Races breeding further north do not arrive until late April, and the Alaskan race *D. p. rubiginosa* until May–June (Phillips 1951). Similar differences occur among populations of White-crowned Sparrows *Zonotrichia leucophrys* (Blanchard 1941), Swainson's Thrushes *Catharus ustulatus* (Ramos 1983), Yellow Wagtails *Motacilla flava* (Curry-Lindahl 1963, Moreau 1972), Blackcaps *Sylvia atricapilla* (Klein *et al.* 1973) and many others. Such temporal differences extend back along the migration route, and where several populations winter in the same region, those that nest at lower latitudes depart first, and those that nest at the highest latitudes depart last (Chapter 12).

Duration of residence

Because of the shortness of the favourable season at high latitudes, populations that breed there remain for relatively short periods. This can be illustrated by the migration records from the Alaska Bird Observatory situated at 64° 50′N, where the average frost-free period each year spans only 105 days (Benson & Winker 2001). The six species of passerines that migrate there from within North America were present on their breeding areas for an average of 119.8 days (standard error (SE) 3.4 days), or 33% of the year. The 12 species of passerines that migrate there from Central or South America were present for an average of only 90.6 days (SE 4.4 days), or less than 25% of the year. These various estimates were based on the intervals between the median spring and autumn migration dates, as assessed from the numbers of birds caught each day at the bird observatory. Extreme values were provided by the American Robin *Turdus migratorius* at 129 days (35% of the

Table 14.1   The number of days spent by five migratory species on their breeding areas at different latitudes in North America, calculated from the median dates of spring and autumn migration

Species	Alaska 64°50′N	Minnesota 44°45′N
Grey-cheeked Thrush *Catharus minimus*	98	124
Swainson's Thrush *Catharus ustulatus*	95	116
Northern Waterthrush *Seiurus noveboracensis*	86	113
Wilson's Warbler *Wilsonia pusilla*	98	105
Blackpoll Warbler *Dendroica striata*	94	121

In addition to the above records, estimates were given for the Alder Flycatcher *Empidonax alnorum* of 48 days in Alaska and 73 days in S. Ontario at 40°42′N, and for the Yellow Warbler *Dendroica petechia* of 84 days in Alaska and 104 days in S. Manitoba at 50°1′N.

From Winker *et al.* (1992a), Benson & Winker (2001).

year) and the Alder Flycatcher *Empidonax alnorum* at only 48 days (13% of the year). In the flycatcher, the adults left immediately after breeding and the young a fortnight later, both groups postponing their moult until later in the year. Comparable estimates of residence periods for the same species at lower latitudes were longer than in Alaska (**Table 14.1**). In Minnesota, at 44° 55′N, 18 long-distance migrants remained for 105 days or less (compared to 91 in Alaska). Among five species that bred in both areas, the Minnesota birds stayed, on average, 22 days longer in their breeding areas than their Alaskan equivalents. The shortest periods in Minnesota of 95 days or less were shown by the Yellow-bellied Flycatcher *Empidonax flaviventris* and Least Flycatcher *E. minimus*, both of which left immediately after breeding (Winker *et al.* 1992a). This pattern of longer residence in lower latitude areas is widespread in birds, but not universal (see later).

RE-OCCUPATION OF LOCAL BREEDING AREAS

In spring, individual birds are assumed to be under pressure to return to breeding areas early, in order to gain precedence in competition for territories or nest-sites, and to gain time, increasing their chance of raising young (Chapter 27). But there are limits to earliness because, unless they have substantial body reserves, birds cannot survive in breeding areas until food becomes sufficiently plentiful there. Typically, among the early-arriving species, individuals first concentrate in particular places where food is available, moving only later to their nesting places as conditions permit. For example, hirundines often first appear in spring over wetlands, where midges first become abundant, and only later spread to their nesting sites, which may be scattered through the surrounding landscape. Montane species often appear first in valleys, moving up as the higher ground becomes snow-free and habitable. These 'pre-breeding areas' could be important in enabling birds to build up body condition in preparation for breeding. Later arriving species seem to settle directly on their territories, especially those returning to their territories of the previous year. Nocturnal migrants, absent in the evening, are often

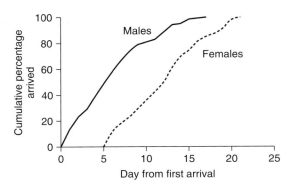

Figure 14.8   The pattern of arrival by Prairie Warblers *Dendroica discolor* in a breeding area at Bloomington, Indiana. The diagram is based on the combined data from 1958 to 1965, but corrected for annual variation in arrival periods by counting the first day of arrival each year as Day 1. In different years, the first arrival date varied between 11 and 22 April for males, and between 21 and 28 April for females. The mean interval separating the first male and female in each year was 5.1 days (extremes 1 and 9 days). As may be seen, males in general arrived before females. From Nolan (1978).

found on their territories at dawn, when they start singing and chasing intruders (e.g. Nolan 1978). They can switch instantly from migration to reproductive mode.

In many species, re-occupation of local breeding areas each spring follows the pattern depicted in **Figure 14.8**. In the population as a whole, arrival may be spread over three or more weeks, but most individuals arrive around the middle of the arrival period, within a few days of one another, unless disrupted by poor weather. In most species, males tend to arrive before females, and older individuals in better body condition are commonly seen to arrive and pair up before younger or poorer condition ones (Francis & Cooke 1986, Hill 1988, Lundberg & Alatalo 1992, Møller 1994). This may be because older, better condition birds depart earlier from the wintering areas (as noted, for example, in the Great Reed Warbler *Acrocephalus arundinaceus*, Nisbet & Medway 1972, and American Redstart *Setophaga ruticilla*, Marra *et al.* 1998). Alternatively, it may be because they spend the winter nearer to their breeding areas (as in many species, Ketterson & Nolan 1983, Chapter 4), or migrate faster (as in many species, Hildén & Saurola 1982, Ueta & Higushi 2002, Chapter 8). Whatever the mechanism, birds in good condition reach the breeding areas before others, and can presumably also better survive the costs associated with early arrival, including a poorer food supply than prevails later in spring (Møller 1994, Kokko 1999).

Where birds pair up soon after arrival in their breeding areas, early-arriving females would be expected to pair with early-arriving males, and vice versa. This pattern seems to hold not just in birds pairing for the first time, but also in established pairs re-uniting on their nesting areas after spending the winter apart. In several species studied in detail, members of a pair arrived closer in time than expected by chance, even though (as shown by colour-ringing or radio-tracking) partners may have migrated independently of one another, and wintered hundreds of kilometres apart (for Black-tailed Godwit *Limosa limosa* see Gunnarsson *et al.* 2004, for White Stork *Ciconia ciconia* see Tryjanowski 2005). In many species that have been studied, males are surplus to females, and it is the latest arriving males that often end up without a breeding partner.

How might competition for territories and mates influence the arrival patterns of migrants? In mathematical models of this situation, increasing the number of competitors for territories can generate cascading pressure for early arrival, which advances arrival dates even further ahead of optimal breeding dates. If the habitat is saturated, so that latecomers risk not obtaining any territory, or if the worst territories are of much lower quality than the rest, competition may lead to most breeders arriving within a short interval, followed by a much later non-breeding contingent (as seen in some birds of prey and others). The penalties for later arrival are not necessarily greatest for the earliest birds, but for those that have the most to lose if they drop a few places in the arrival sequence (Kokko 1999).

The fact that some populations seem to arrive on breeding areas up to several weeks before they start nesting has been attributed to competition for territories, which provides strong selection pressure for early arrival but, as emphasised above, birds can only respond to that selection if the breeding areas offer sufficient food at that time. An extreme example is provided by the Snow Bunting *Plectrophenax nivalis* in arctic Greenland, where males arrive in early April, 6–8 weeks before nesting and 2–4 weeks ahead of females (Salomonsen 1967). During this lengthy pre-nesting period, males commonly experience severe storms and temperatures down to −30°C, and considerable mortality can occur. Yet year after year, the males continue to arrive at this early date, apparently in order to compete for territories in the limited high-quality nesting habitat.

The importance of an early return is also evident in some colonial cliff-nesting seabirds, in which pairs compete for limited space on cliff ledges. Apparently in order to secure their sites, birds return weeks or months before egg-laying, and in some species return dates became progressively earlier as populations grew and competition intensified. In some species, such as the Northern Fulmar *Fulmarus glacialis*, birds are now present on their nest-sites almost year-round in Britain, mates taking turns to guard the site, foraging between times. Eggs are laid in May. Similarly, over most of their breeding range, Common Guillemots (Murres) *Uria aalge* return to their breeding colonies in late winter or early spring, some two months before the first eggs are laid. Return dates to cliff colonies on the Shetland Isles, off Scotland, became earlier by 25 weeks from March to October during a 10–12 year period. This change coincided with a period of continued population growth, and was attributed to intensified competition for nest-sites (Harris *et al.* 2006). Autumn returns persisted for about 10 years, after which return dates gradually reverted to late winter, as the population declined. Over the whole period, the correlation between mean annual return date and population size was highly significant ($r = -0.695$, $P < 0.001$, $n = 29$ years). This link between arrival dates and potential competition levels did not rule out an influence of food supply, which, through a period of change, could itself facilitate earlier arrival or overwintering, and at the same time promote population growth.

Not all species follow this pattern, probably because, in the conditions prevailing, birds cannot survive in breeding areas until shortly before nesting can begin. In such species, the gap between arrival and nesting is much shorter, and late-arriving individuals can start breeding activities within days. There are occasional extreme examples, as illustrated by a male Wood Warbler *Phylloscopus sibilatrix* that was caught on spring migration and ringed on the Isle of Man on 8 May. The following morning, less than 24 hours after ringing, the same bird had established a territory 205 km further north, in Scotland, and by that evening it had attracted a mate and started nest-building (Morton 1986).

The basic problem inherent in the timing of spring migration is that individuals arriving early in breeding areas stand to benefit from greater production of young that year, but they may die if conditions at the time of arrival are bad. Birds that nest at high latitudes often risk facing an untimely cold snap or snowfall that cuts off the food supply, and the earlier in spring they arrive, the more likely this is to happen, killing a large proportion of the early-arriving individuals (Chapters 20 and 28). On the other hand, within the limits of the possible season, the earlier birds begin to nest, the greater the number of young they are likely to produce. Not only do first arrivals tend to acquire better territories, nest-sites or mates than later arrivals, they also lay earlier in the season, and often achieve better nest success. Examples of species showing relationships between arrival date, territory quality, laying date and reproductive success are given in **Table 14.2**, and include both passerines and non-passerines (see also Sergio *et al.* 2007). In some of these species, the latest birds to arrive failed to get either a territory or a mate.

The Barn Swallow *Hirundo rustica* illustrates the conflicting pressures on migration timing, as it benefits greatly from an early start to breeding, but the costs of early breeding are represented in cold seasons by mortality among early-arriving males (Møller 1994, 2001). In these inclement seasons, birds suffer from snowstorms, when their increased energy demand is coupled with an absence of insect prey. Catastrophic losses early in the season can exert a measurable selection pressure on migration dates if losses hit early-arriving individuals more severely than others, as evident among migrant Cliff Swallows *Hirundo pyrrhonata* in Nebraska (Chapter 12; Brown & Brown 1998, 2000). Arrival date has a hereditary component in both these swallow species, leading in cold springs to selection for later arrival. If early arrival confers the competitive advantage of prior occupancy, but increases the risk of mortality, arrival date can be viewed as a trade-off between opposing pressures, but sometimes the 'best' males may be able to survive when others could not (Drent *et al.* 2003). Other features of the individual, besides inheritance, also influence arrival dates. In particular, many individuals may be unable to get sufficient food on migration to travel at the optimal rate and keep up with improving conditions. As a result they arrive later than they otherwise would, and suffer the reproductive consequences. Other individuals, having arrived in breeding areas, may be unable to acquire enough food for egg production at the optimal date. In some species, such as Styan's Grasshopper Warbler *Locustella pleskei* and Bar-tailed Godwit *Limosa lapponica*, individuals that arrived in the first part of the season were most successful, but not the very first birds, which did less well (Takaki *et al.* 2001, Drent *et al.* 2003). Whether the poor performance of the very earliest arrivals is general in these species, or a feature of the particular study years, remains to be seen.

Territory settlement

In any particular area, it is commonly found after detailed study that territories vary in quality: that is, in the fitness benefits they confer on their occupants. The quality of nesting habitat has often been assessed from measurements of cover,

Table 14.2 Relationship in various bird species between spring arrival date in breeding areas and subsequent breeding performance. In general, an early arrival relative to other individuals resulted in better performance

Species	Mate acquisition	Territory quality	Laying date	Clutch size	Young per nest	Source
Northern Wheatear *Oenanthe oenanthe*	+	+	+	+	+	Currie et al. (2000)
Black Redstart *Phoenicurus ochruros*	+	+				Andersson (1995), Landmann & Kollinsky (1995)
American Redstart *Setophaga ruticilla*	+	+			+	Lozano et al. (1996), Norris et al. (2004), Moore et al. (2005)
American Redstart *Setophaga ruticilla*		+	+	+		Smith & Moore (2005)
Great Reed Warbler *Acrocephalus arundinaceus*	+	+	+		+	Hasselquist (1998)
Savi's Warbler *Locustella luscinioides*		+			+	Aebischer et al. (1996)
Styan's Grasshopper Warbler *Locustella pleskei*					+	Takaki et al. (2001)
European Pied Flycatcher *Ficedula hypoleuca*		+	+		+	Lundberg & Alatalo (1992), Moore et al. (2005)
Painted Bunting *Passerina ciris*		+				Lanyon & Thompson (1986)
Barn Swallow *Hirundo rustica*	+				+	Møller (1994), Saino et al. (2004)
Black-tailed Godwit *Limosa limosa*		+			+	Gill et al. (2001), Gunnarsson et al. (2005)
Pied Avocet *Recurvirostra avosetta*					+	Hötker (2002)
Black Kite *Milous migrans*		+	+		+	Sergio et al. (2007)
Barnacle Goose *Branta leucopsis*			+	+		Dalhaug et al. (1996, 2001)
White Stork *Ciconia ciconia*			+		+	Tryjanowski et al. (2004)

+ = effect observed.
For other examples see Sergio et al. 2007

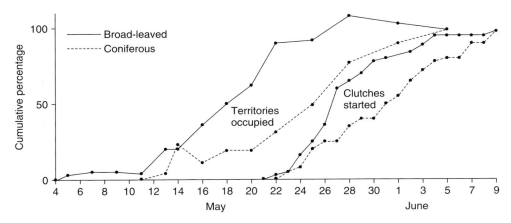

Figure 14.9   Settling and egg-laying dates of Pied Flycatchers *Ficedula hypoleuca* in broad-leaved and coniferous forests in central Sweden. In the preferred broad-leaved areas, food supply increased earlier in spring, the birds settled earlier and at greater density, and annual breeding success (in terms of young per female) was higher than in coniferous areas. Redrawn from Lundberg *et al.* (1981).

disturbance, local food supplies or proximity to good feeding areas (whichever was most relevant in the population concerned), or more directly from the feeding rates, survival or reproductive rates of previous occupants. On the basis of such assessments, places classed as best are generally occupied first each spring. As these places become occupied, later arrivals are relegated to poorer places. For example, among migrant Willow Ptarmigan *Lagopus lagopus* in Alaska, the first birds to arrive settled near a stream, while later settlers were forced by territorial competition onto adjacent hillsides, where food was shown to be poorer (Moss 1972). In a previous year, when numbers were lower, only the stream area was occupied. Similarly, Pied Flycatchers *Ficedula hypoleuca* arriving in central Sweden in spring settled in deciduous areas in preference to coniferous (**Figure 14.9**). In deciduous areas, food supply increased earlier in spring, the birds settled at greater density, laid earlier and raised larger broods than in the conifers. Also, the males (but not the females) that occupied deciduous areas were larger than those in coniferous areas, a difference attributed to the effect on settling patterns of fights and other interactions (Lundberg *et al.* 1981).

Patterns of spring settlement, with the best areas occupied first, have been described in many other bird species, including Painted Bunting *Passerina ciris* (Lanyon & Thompson 1986), Great Reed Warbler *Acrocephalus arundinaceus* (Bensch & Hasselquist 1991), Collared Flycatcher *Ficedula albicollis* (Wiggins *et al.* 1994), Savi's Warbler *Locustella luscinioides* (Aebischer *et al.* 1996) and Northern Wheatear *Oenanthe oenanthe* (Currie *et al.* 2000). In several such species, the same sequence of territory settlement held year after year in the same area, even though the occupants changed, and even though some early settlers were displaced by former owners which returned later. In some species, however, the sequential pattern was disrupted by site-fidelity, as some returning birds settled in poor areas where they had previously bred, despite the presence of vacancies in better habitat (Lanyon & Thompson 1986). Site-fidelity sometimes led individuals to re-occupy for several further years habitats that had deteriorated (Hildén 1965, Wiens & Rotenbury 1986). The use of such areas declined over time, however, as existing occupants died or left, and were not replaced.

In not all species do individuals establish a nesting territory as soon as they arrive in breeding areas. Instead they feed in flocks for some days or weeks before they eventually take up a nest-site (e.g. Pied Avocets *Recurvirostra avosetta*, Hötker 2002). Acquisition of the best nesting places in such species cannot therefore be due directly to early arrival, but birds that arrive earliest may also be the ones most able to compete for the best nesting places.

Manipulation of arrival date

The better breeding of early-arriving individuals, recorded in many species, could have been because early birds were of higher quality and would have bred well at any time in the season, or because they acquired the best territories, or because date itself was important, perhaps in relation to food supplies for the young. In an attempt to test whether individual quality or date was most important, Cristol (1995) delayed the effective arrival date of female Red-winged Blackbirds *Agelaius phoeniceus*. The females arrived when their Indiana breeding marshes were still ice-covered, and almost two months before nests were built. Newly arrived females were caught, and some were released immediately (the controls), while others were removed to an aviary and released later, but before the control females had started nest-building. The subsequent breeding performance of both groups was compared. In this polygynous breeder, the late-released birds were subdominant to the control females mated to the same male, suffered from less male help later in the cycle, and raised fewer young. The 20 delayed females nested, on average, one week later than the 30 controls, a date difference expected to reduce breeding output. In this species, therefore, early arrival conferred a reproductive advantage via an enhancement of social dominance over late-arriving birds, and when the early arrivals were removed from social contact, they lost this prerogative.

These findings from a polygynous species may not apply to monogamous ones. Another experimental study involved manipulation of clutch sizes and hatching dates of Collared Flycatchers *Ficedula albicollis* on Gotland Island, Sweden (Wiggins *et al.* 1994). Findings suggested that both bird/territory quality and environmental changes during the season contributed to the seasonal decline in reproductive success.

The role of territory establishment in the regulation of breeding density

The first arrivals in breeding areas can settle in the preferred areas partly because they encounter little or no resistance from other individuals. Typically, they establish large territories which they might later contract to some extent under pressure from later settlers; but as more birds arrive, newcomers find it increasingly difficult to find a place, and eventually local density reaches a plateau (**Figure 14.10**). The area is then occupied to capacity and no further birds can settle that

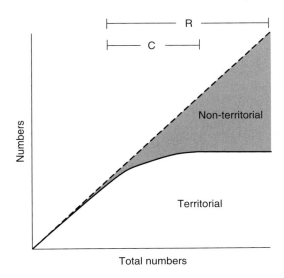

Figure 14.10   Model showing how the numbers of territorial and non-territorial birds might change according to changes in the total numbers of potential settlers. C indicates the range of total numbers over which territories could be compressed under pressure from further contenders, and R indicates the range of total numbers over which replacements of dead (or removed) territorial birds could be expected. From Newton (1998b).

At low numbers, all individuals can establish a territory, and territorial behaviour serves merely to space out birds within the habitat. At higher numbers (zone C) an increasing number of birds can establish territories, but an increasing proportion is excluded from doing so, providing a mechanism for the regulation of local density. Territorial behaviour can thus limit density from the start of zone C, even though higher densities can be reached (to the end of zone C) under pressure from further rise in the number of contenders. Beyond C, no further increase in territorial numbers occurs, despite further rise in the number of contenders. At this level, for every additional territorial bird that settles, one must die or leave, perhaps joining a surplus of non-territorial, non-breeders.

Over short periods of years, any one population would normally be expected to fluctuate within only part of the density range shown on the horizontal axis.

year unless they displace others already there. Both adult numbers and subsequent overall breeding output are thereby limited. This pattern of spring settlement has been observed in a wide range of bird species (for reviews, see Klomp 1972, Davies 1978, Newton 1998b). It provides a clear behavioural mechanism that could not only limit densities in a given year, but also regulate densities over a number of years, with increasing proportions of potential settlers excluded as their overall numbers rise. However, the level at which density stabilises can vary from year to year in line with prevailing food supply and other features of the habitat itself (Newton 1998b).

Although this model was developed for territorial birds (Brown 1969), in which competition for space is most apparent, it could apply equally to colonial or flocking birds, summer or winter, given the competitive interactions that occur over food or other resources (Newton 1998b). In fact, any competitive interaction can regulate density, providing it results in an uneven sharing of resources among competing individuals, leading some to remain and others to leave or die. Local densities are thereby brought in line with local resource levels, and the proportion of birds excluded could increase as total numbers rise.

POST-BREEDING RETREAT

Regional variation in the onset of winter conditions can again be tracked by following particular isotherms as they move from higher to lower latitudes, bringing colder conditions southwards across the northern continents (**Figure 14.11**). In autumn, the 10°C isotherm takes nearly three months to spread southward through Europe. Although not studied in detail, the autumn withdrawal of migratory birds from the northern continents occurs, as expected, from the top down, as northern populations generally leave before southern ones (for exceptions see below).

As mentioned earlier, species tend to leave particular localities in autumn in reverse order to that in which they arrived in spring: that is, the latest to arrive are first to leave (**Figure 14.2**). This sequence is again broadly linked with the decline in their respective food supplies, associated with falling temperatures. The earliest species to leave particular localities also show less year-to-year variation in their departure dates than do the last to leave (Gilyazov & Sparks 2002). Insufficient data are available to plot the withdrawal from much of a continent of any species, and see how closely the process keeps step with moving isotherms. But in particular species the whole process can again take several weeks, or up to three months in species such as the Barn Swallow *Hirundo rustica* which breed over a wide span of latitude. Low-latitude populations may still be on their breeding areas when individuals from higher latitude breeding areas pass through again, en route to winter quarters (examples include Yellow Wagtail *Motacilla flava* and others in Europe and Wilson's Warbler *Wilsonia pusilla* and others in North America; Curry-Lindahl 1963, Kelly 2006).

There are exceptions to this general trend. They include some single-brooded species, which spend about the same amount of time on their nesting areas at all breeding latitudes, but can both arrive and depart earlier from southern than from more northern parts of the breeding range. For example, migratory Ospreys *Pandion haliaetus* nesting in Florida arrive in breeding areas about one month earlier than those in New York and New Jersey, and also depart for their wintering areas about one month earlier (Martell *et al.* 2001, 2004). It is as though the annual cycle of high-latitude breeders is shifted later relative to lower latitude breeders. Other North American species in which southern populations migrate earlier than northern ones in autumn include Orange-crowned Warbler *Vermivora celata* and Common Yellowthroat *Geothlypis trichas* (Kelly 2006).

Early departure from breeding areas also occurs in some species that breed in drought-stricken parts of western North America (Rohwer & Manning 1990, Butler *et al.* 2002), while their conspecifics from further north leave later. The same holds in arid parts of southeast Europe, as exemplified by the Eurasian Reed Warbler *Acrocephalus scirpaceus* (Akriotis 1998). Trans-Saharan migrants generally benefit from early departure because they can then reach the Sahel zone at

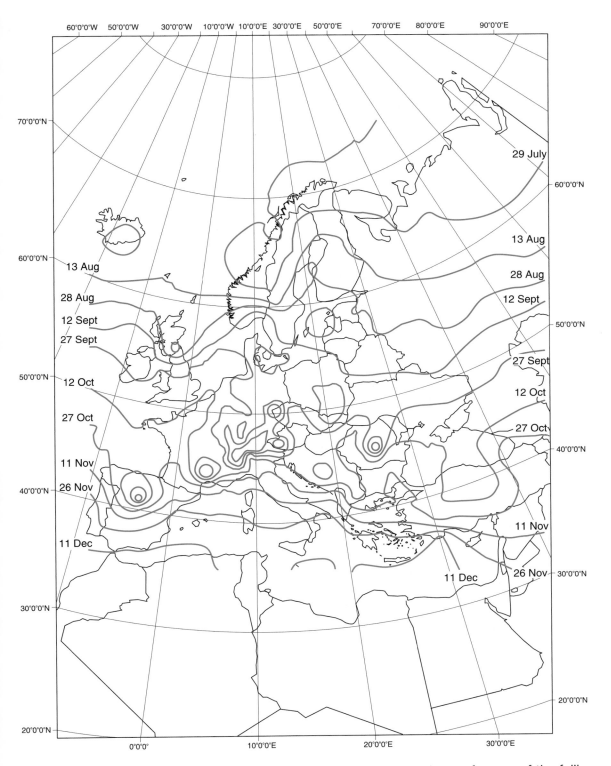

Figure 14.11 The advance of autumn, as shown by the average dates of return of the falling 10°C isotherm to various parts of Europe during 1971–2000. From T. Sparks, unpublished.

its greenest, and while feeding conditions are optimal. By the time most migrants arrive in the Sahel, from September on, the region has already begun to dry.

Patterns of autumn departure seem to differ between obligate (complete) migrants, in which all individuals leave every year, and facultative (partial) migrants in which the proportions of individuals that leave, and their dates of departure, vary with local food supplies at the time (Chapter 12). In obligate (often long-distance) migrants, individuals seem to leave their breeding areas as soon as they can after breeding or after breeding and moulting, as the case may be (Lack 1960b, Preston 1966). In many such species, departure occurs before the collapse of local food supplies. If birds do not have time to raise another brood, they probably benefit by leaving as soon as possible. Unlike the situation in spring when food is scarce but increasing, in late summer food is plentiful but often declining, and weather is still fine. If the birds have finished breeding, and therefore have no reason to wait longer, they have every advantage in migrating while conditions are still good, and before food supplies on staging areas have been depleted. They might also establish themselves on wintering areas at an early date, perhaps benefiting in competition for the best habitat. This holds especially in species which have territories in winter quarters and is probably the main selective pressure for rapid departure and progress on the autumn journey. Hence, while spring arrival usually coincides fairly precisely with the re-appearance of new food supplies, autumn departure of obligate migrants may precede the autumn collapse of food supplies by up to several weeks (Schneider & Harrington 1981).

Facultative migrants show less year-to-year consistency in autumn departure dates, and many individuals remain on breeding areas as long as food lasts, leaving up to several weeks later in some years than in others, and long after the end of moult (Chapter 12). This holds for seed-eating finches, which, in years of good tree-seed crops, stay all winter or leave much later than in years of poor crops (Chapter 18). They can stay after snowfall, providing that seeds remain on the trees, whereas other finch species, which pick seeds from low herbaceous plants or the ground, must leave by the first heavy snowfall. Likewise, many northern waterbirds leave as their wetland habitats begin to freeze over, which occurs much earlier in some years than in others. Moreover, because shallow waters freeze before deep ones, the shallow-water dabbling ducks usually migrate before deep-water diving ducks. In these facultative migrants, local food supplies seem to have a major influence on the rate of southward migration, with birds lingering en route while food lasts, and travelling markedly less far in some years than in others.

Competition for winter habitat

In species that compete for territories or feeding areas in winter, studies have shown an advantage of arriving early relative to other individuals. For example, wintering American Kestrels *Falco sparverius* arriving in Florida from the north occupied habitats in decreasing order of their quality in terms of food supply. Early-arriving birds (mostly females and juveniles) acquired most of the good places, while later arrivals took the less good places (Smallwood 1988). Hence, as in breeding areas, the best places were occupied first, and by the earliest birds.

It is not only territorial species that benefit from an earlier arrival in wintering areas. On an estuary in southern England, Oystercatchers *Haematopus ostralegus* accumulated first on the two most preferred mussel beds where feeding rates were highest (Goss-Custard *et al.* 1982, 1984). But as more birds arrived and interactions increased, birds progressively occupied the less favoured mussel beds. Later arriving adults displaced immatures already present on the favoured beds. Moreover, as mussel stocks declined on the favoured beds, more birds left, but in reverse order of their dominance rank, to feed on less preferred beds or in other habitats. In this species, therefore, the effects of arrival date were modified by dominance relations among age groups, but within age groups the benefits of early arrival relative to other individuals were still apparent. Further evidence of first arrivals taking the best habitats, and thereby having greater chance of surviving the non-breeding season, is available for Ruddy Turnstone *Arenaria interpres* (Whitfield, in Wernham *et al.* 2002).

CONCLUDING REMARKS

The benefits of early arrival in breeding and wintering areas could lead to the whole breeding cycle becoming earlier and earlier, if it were not environmentally constrained in some way. The brake on this process occurs in spring, when birds cannot occupy breeding areas and start nesting until conditions become suitable. In addition, multi-brooded species gain from remaining in breeding areas later than otherwise, if they could thereby raise another brood, even though this may delay to some extent their departure to wintering areas. Again the birds face a trade-off between the advantages of extra reproduction and the disadvantages of a late migration, when food supplies are likely to be depleted, and when better habitat in wintering areas is likely to be already occupied. The advantages of early arrival in wintering areas are perhaps most obvious in populations in which individuals occupy the same areas throughout a winter. Yet many migratory bird species occupy different areas in different winters, or move around within a winter, and it is not yet clear whether in these circumstances early arrival could carry any advantage.

SUMMARY

Migratory bird species have been found to advance northward over the northern continents in spring at average rates of 30–300 km per day, in step with improving conditions, as reflected in the northward advance of particular isotherms. In general, early-migrating species progress more slowly than later ones. Barn Swallows *Hirundo rustica*, which breed over a wide span of latitude, take more than three months each year to re-colonise all parts of their breeding range. The rate at which a continent is colonised each spring, from low to high latitudes, is often much slower than the migration speeds of individual birds, as birds must often wait for snow-melt or other conditions to improve.

Many birds benefit from arriving relatively early in their breeding areas, as they are able to obtain better territories and mates than later arriving ones, and often show better breeding success. However, if they arrive too early, they risk food shortage, and may sometimes die from late cold snaps. These conflicting selection pressures are likely to influence the average spring migration (and arrival) dates of different species. Nevertheless, species migrate at different dates, depending partly on when their particular foods become sufficiently available, and in general earlier in warm springs than in cold ones. The arrival dates of many species in particular localities are spread over a shorter period than their autumn departure dates. Some species spend only 2–3 months each year on their high-latitude breeding areas.

Most migratory birds vacate their breeding areas on the northern continents in autumn from north to south, in reverse order to that in which they re-colonised in spring. However, not all species follow this pattern and, in some single-brooded migrants, southern-nesting populations (having completed their breeding earlier) move out first.

Male and female Chaffinch *Fringilla coelebs*

Chapter 15
Sex and age differences in migration

For many years past, I have observed that, towards Christmas, vast flocks of Chaffinches have appeared in the fields. . . . when I came to observe them all narrowly, I was amazed to find that they seemed to me to be almost all hens. (Gilbert White 1789.)

In many bird populations, sex and age groups differ in aspects of their migrations. Such differences include the proportions of each sex and age group undertaking migration, the timing of outward and return journeys, and the distances travelled. The latter produce geographical gradients in the sex and age ratios of species in the non-breeding season, with one group predominating nearest to the breeding areas and the others furthest away. One or more such sex and age differences have been recorded in a wide range of bird species, from at least nine different Orders, including passerines, shorebirds, ducks, raptors, herons, various seabirds and others. In some species the differences are great, with little or no overlap between sexes or age groups, but in other species the differences are manifest chiefly in mean values, with extensive overlap between sex or age groups. Establishing statistical significance then depends largely on sample sizes or consistency in patterns between years. In this chapter, I have cited as examples only those studies in which statistical significance has been established, but similar trends occur in many other species.

Sex differences in migration have been known for a long time. The Swedish tax-onomist Linnaeus in 1758 named the Chaffinch *Fringilla coelebs* (meaning 'bachelor finch') because it was chiefly the males that stayed to winter in Sweden where he lived, while females moved to lower latitudes. Subsequent work, based on ring-ing returns, showed that the males which did migrate moved, on average, less far than females (Payevsky 1998). The same was true for the closely related Brambling *Fringilla montifringilla*, and for many other passerine species (**Table 15.1**).

In a review of so-called differential migration, Cristol *et al.* (1999) listed 146 bird species in which sex or age differences were known or suspected to occur, against 16 for which no evidence was found (though often on small samples). Our knowledge of differential migration is of course restricted to species in which the sex and age classes can be readily distinguished by plumage or size differences. In some species, including many passerines, shorebirds and seabirds, it is practi-cally impossible to separate the sexes in this way, so without special techniques, any sex differences that might occur are effectively undetectable (for use of DNA methods see Catry *et al.* 2004, Remisiewicz & Wennerberg 2006).

Sex and age differences in migration are associated with at least three aspects of species biology, based upon: (1) the different roles of the sexes in breeding, which can be linked to their migration timing; (2) the timing of other events in the annual cycle, especially moult, which can again influence the timing of depart-ure from breeding areas; and (3) body size and dominance which often also differ between sex and age groups, and can be linked to migratory timing and distances. Some of these features thus relate to only one aspect of migration, while others relate to more than one. Moreover, they are not mutually exclusive, and more than one may apply to the same species. These different aspects are explored below, followed by discussion of the movements of immature non-breeders, and then of sex- and age-related differences in local distribution patterns.

MIGRATORY TIMING AND BREEDING ROLES

Arrival in breeding areas

In those bird species in which pair formation occurs in the wintering or migra-tion areas, as in some waterfowl, the sexes arrive together in their breeding areas. However, in most bird species pair formation occurs in the breeding areas, with males arriving first to establish territories in which they then attract females (for White-crowned Sparrow *Zonotrichia leucophrys* see **Figure 15.1**). The sex differ-ence in mean arrival dates can vary from a few days to a few weeks, depending on species, year and area. Among 18 species of North American warblers studied over many years at Prince Edward Point in Ontario, the sex difference in mean arrival dates was greater in species that arrived early in the season than in those that arrived later, and within species, the sex difference in arrival dates was great-est in years when males arrived earliest (Francis & Cooke 1986). These findings were typical of those from many other species studied elsewhere (Cramp 1988, Morgan & Shirihai 1997).

Because in most species males compete for territories (or mates), they may gain more advantage than females in staying on or near their nesting territories

Table 15.1 Sex and age differences in the timing and distance of migration in various bird species. Migration timing is in some species inferred from passage dates, rather than from departure and arrival dates. A – adult, J – juvenile, M – male, F – female

	Autumn departure (earliest–latest)	Distance travelled (nearest–furthest)	Spring arrival (earliest–latest)	Source
Passerines				
Chaffinch *Fringilla coelebs*	JF, JM, AF, AM	JM, AM, JF, AF	AM, AF, JM, JF	Schifferli (1965), Payevski (1998)
Brambling *Fringilla montifringilla*		J, AM, AF		Jenni (1982), Jenni & Neuschulz (1985), Payevski (1998)
Yellow-headed Blackbird *Xanthocephalus xanthocephalus*		M, F	M, F	Twedt & Crawford (1995)
Red-winged Blackbird *Agelaius phoeniceus*		M, F, J	M, F	Dolbeer (1982), James et al. (1984)
Evening Grosbeak *Hesperiphona vespertina*		M, F	M, F	Prescott (1991)
Eurasian Siskin *Carduelis spinus*	J, A		A, J	Payevski (1994)
American Goldfinch *Carduelis tristis*		M, F		Prescott & Middleton (1990)
European Goldfinch *Carduelis carduelis*		M, F		Newton (1972)
European Greenfinch *Carduelis chloris*		M, F		Main (2000)
House Finch *Carpodacus mexicanus*		M, F		Belthoff & Gauthreaux (1991)
Reed Bunting *Emberiza schoeniclus*	MF	MF	MF	Villarán & Pascual-Para (2003)
Snow Bunting *Plectrophenax nivalis*		AM, JM, AF, JF	M, F	Meltofte (1985), Smith et al. (1993)
White-throated Sparrow *Zonotrichia albicollis*		M, F		Jenkins & Cristol (2002)
White-crowned Sparrow *Zonotrichia leucophrys*	JM, AM		AM, JM, AF, JF	King et al. (1965), Morton (2002)
Dark-eyed Junco *Junco hyemalis*			AM, JM, AF, JF	Ketterson & Nolan (1983)
Savannah Sparrow *Passerculus sandwichensis*	F, M	JM, AM, JF, AF	M, F	Bedard & LaPoint (1984)

(Continued)

Table 15.1 Continued

	Autumn departure (earliest–latest)	Distance travelled (nearest–furthest)	Spring arrival (earliest–latest)	Source
Eurasian Skylark Alauda arvensis	FM		M, F	Spaepen & Cauteren (1968)
Pied (White) Wagtail Motacilla alba		A, J		Wernham et al. (2002)
Yellow Wagtail Motacilla flava		M, F	M,F	Wood (1992)
American Redstart Setophaga ruticilla			M, F	Lozano et al. (1996)
Red-eyed Vireo Vireo olivaceus	A, J			Woodrey & Chandler (1997)
European (Common) Starling Sturnus vulgaris	J, A		A, J	Feare, in Wernham et al. (2002)
European Robin Erithacus rubecula		M, F		Adriaensen & Dhondt (1990)
Whinchat Saxicola rubetra			AM, JM, AF, JF	Spina et al. (1994)
Bluethroat Luscinia svecica	JM, JF, AM, AF			Bermejo & de la Puente (2004)
Black Redstart Phoenicurus ochruros	J, A			Cramp (1988)
Isabelline Wheatear Oenanthe isabellina	J, AF, AM		M, F	Cramp (1988)
Song Thrush Turdus philomelos		A, J		Wernham et al. (2002)
Mistle Thrush Turdus viscivorus		A, J		Wernham et al. (2002)
Eurasian Blackbird Turdus merula		M, F		Wernham et al. (2002)
Fieldfare Turdus pilaris	A, J	J, A	A, J	Milwright (1994)
Hermit Thrush Catharus guttatus	A, J	M, F	M, F	Stouffer & Dwyer (2003)
European Pied Flycatcher Ficedula hypoleuca			AM, JM, AF, JF	Spina et al. (1994)
Blue Tit Parus caeruleus		AM, AF, JM, JF	A, J	Smith & Nilsson (1987)
Magnolia Warbler Dendroica magnolia	J, A			Woodrey & Chandler (1997)
Greater Whitethroat Sylvia communis	M, F		AM, JM, AF, JF	Cramp (1992), Spina et al. (1994)
Marsh Warbler Acrocephalus palustris	A, J			Cramp (1992)
Eurasian Reed Warbler Acrocephalus scirpaceus		A, J		Insley & Boswell (1978)

Species				Reference
Great Reed Warbler *Acrocephalus arundinaceus*	A, J		M, F	Nisbet & Medway (1972), Cramp (1992)
Sedge Warbler *Acrocephalus schoenobaenus*	A, J			Insley & Boswell (1978)
Spectacled Warbler *Sylvia conspicillata*	J, AF, AM			Cramp (1992)
Barred Warbler *Sylvia nisoria*	A, J			Cramp (1992)
Greenish Warbler *Phylloscopus trochiloides*	A			Cramp (1992)
Willow Warbler *Phylloscopus trochilus*	J, A		M, F	Cramp (1992)
Eurasian Chiffchaff *Phylloscopus collybita*	J, A	M, F	M, F	Catry et al. (2005)
Goldcrest *Regulus regulus*	J, A			Cramp (1992)
Spotted Flycatcher *Muscicapa striata*	A, J			Hyytiä & Vikberg (1973)
Sand Martin *Riparia riparia*	J, A		A, J	Wernham et al. (2002)
Red-backed Shrike *Lanius collurio*	A, J		AM, AF, JA, JF	Jakober & Stauber (1983)
Woodchat Shrike *Lanius senator*			AM, JM, AF, JF	Spina et al. (1994)
Western Flycatcher *Empidonax difficilis*	A, J			Johnson (1973)
Least Flycatcher *Empidonax minimus*	A, J		M, F	Hussell et al. (1967), Ely (1970)
Traill's (Willow) Flycatcher *Empidonax traillii*	A, J			Ely (1970)
Owls				
Snowy Owl *Nyctea scandiaca*		AF, AM, JF, JM		Kerlinger & Lein (1986), Parmalee (1992)
Northern Hawk Owl *Surnia ulula*		M, F, A, J		Byrkjedal & Langhelle (1986), Duncan & Duncan (1998)
Game birds				
Blue Grouse *Dendragapus obscurus*	M, F	F, M	M, F	Cade & Hoffman (1993)
Greater Prairie Chicken *Tympanuchus cupido*	F, M	M, F	M, F	Schroeder & Braun (1993)

(Continued)

Table 15.1 Continued

	Autumn departure (earliest–latest)	Distance travelled (nearest–furthest)	Spring arrival (earliest–latest)	Source
Willow Ptarmigan *Lagopus lagopus*		M, F	M, F	Weeden (1964)
White-tailed Ptarmigan *Lagopus leucurus*		M, F		Hoffman & Braun (1977)
Raptors				
Eurasian Kestrel *Falco tinnunculus*	J, AM, AF	AM, AF, JM, JF	AM, AF, JM, JF	Wallin et al. (1987), Village (1990), Kjellén (1992, 1994a)
American Kestrel *Falco sparverius*	JF, JM, AF, AM	AM, AF, JM, JF		Arnold (1991), Mueller et al. (2000), Hoffman et al. (2002)
Merlin *Falco columbarius*	AF, J, AM			Mueller et al. (2003)
Eurasian Hobby *Falco subbuteo*	A, J			Kjellén (1992)
Peregrine Falcon *Falco peregrinus*	AM, JM, AF, JF	F, M		Hunt et al. (1975), Restani & Maddox (2000)
Osprey *Pandion haliaetus*	AF, AM, J	AM, AF		Hoffman et al. (2002)
Common Buzzard *Buteo buteo*	A, J	A, J	A, J	Kjellén (1992, 1994a), Yosef et al. (2002)
Rough-legged Buzzard *Buteo lagopus*	AF, AM, J	F, M		Kjellén (1992, 1994), Gauthreaux (1985)
Red-tailed Hawk *Buteo jamaicensis*	J, A			Mueller et al. (2000), Hoffman et al. (2002)
Red-shouldered Hawk *Buteo lineatus*	J, A			Mueller et al. (2000)
Eurasian Sparrowhawk *Accipiter nisus*	JF, JM, AF, AM	F, M	AM, AF, JF, JM	Belopolskij (1971), Saurola (1981), Kjellén (1992)
Sharp-shinned Hawk *Accipiter striatus*	JF, JM, AF, AM	AM, AF, JM, JF		Clark (1985), Mueller et al. (2003), Hoffman et al. (2002)
Cooper's Hawk *Accipiter cooperii*	JF, JM, AF, AM	AM, AF, JM, JF		Hoffman et al. (2002)
Northern Goshawk *Accipiter gentilis*	JF, JM, AM, AF	A, J		Kjellén (1992, 1994a)

Species				References
Northern Harrier *Circus cyaneus*	J, AF, AM	F, M	AM, AF, JF, JM	Hamerström (1969), Kjellén (1992), Wernham et al. (2002), Mueller et al. (2000)
Western Marsh Harrier *Circus aeruginosus*	J, AF, AM	F, M		Kjellén (1992), Wernham et al. (2002)
Red Kite *Milvus milvus*	J, A	A, J		Kjellén (1992, 1994a)
White-tailed Eagle *Haliaeetus albicilla*	J, A	A, J	A, J	Gätke (1895), Kjellén (1994a)
European Honey Buzzard *Pernis apivorus*	A, J			Kjellén (1992)
Waterfowl				
Eurasian Wigeon *Anas penelope*		M, F, A, J		Campredon (1983)
American Wigeon *Anas americana*		M, F	M, F	Hepp & Hair (1984), Thompson & Baldassarre (1992)
Green-winged Teal *Anas crecca*		M, F		Hepp & Hair (1984)
Mallard *Anas platyrhynchos*	M, F	M, F	M, F	Pattenden & Boag (1989)
Northern Pintail *Anas acuta*		M, F		Thompson & Baldasarre (1992)
Northern Shoveler *Anas clypeata*		M, F	M, F	Thompson & Baldasarre (1992)
Canvasback *Aythya valisineria*		M, F	M, F	Alexander (1983)
Tufted Duck *Aythya fuligula*	AM, AF, J	M, F	M, F	Wernham et al. (2002)
Common Pochard *Aythya ferina*		M, F		Carbone & Owen (1995)
Ring-necked Duck *Aythya collaris*		M, F		Alexander (1983)
Common Goldeneye *Bucephala clangula*		M, F		Nilsson (1969)
Shorebirds				
Spotted Sandpiper *Tringa macularia*		F, M	F, M	Oring et al. (1997)
Western Sandpiper *Calidris mauri*	M, F	M, F	M, F	Harrington & Haase (1994), Bishop et al. (2004)

(Continued)

Table 15.1 Continued

	Autumn departure (earliest–latest)	Distance travelled (nearest–furthest)	Spring arrival (earliest–latest)	Source
Semi-palmated Sandpiper *Calidris pusilla*			AM, AF, J	Lyons & Haig (1995)
Curlew Sandpiper *Calidris ferruginea*	AM, AF, J			Wernham et al. (2002)
Common Greenshank *Tringa nebularia*	AF, AM, J			Wernham et al. (2002)
Common Redshank *Tringa totanus*		A, J		Wernham et al. (2002)
Northern Lapwing *Vanellus vanellus*		A, J		Wernham et al. (2002)
Grey Plover *Pluvialis squatarola*		M, F		Cramp & Simmons (1983)
Eurasian Oystercatcher *Haematopus ostralegus*		A, J		Wernham et al. (2002)
Ruddy Turnstone *Arenaria interpres*	AF, AM, J			Whitfield, in Wernham et al. (2002)
Ruff *Philomachus pugnax*			M, F	Gill et al. (1995)
Seabirds and others				
Herring Gull *Larus argentatus*		A, J		Moore (1976), Kilpi & Saurola (1984)
Sooty Tern *Sterna fuscata*		A, J		Robertson (1969)
Northern Gannet *Morus bassanus*			A, J	Wernham et al. (2002)
Great Cormorant *Phalacrocorax carbo*	J, A	AM, AF, J	AM, AF, J	Bregnballe et al. (1997)
Black-browed Albatross *Thalassarche melanophrys*	F, M	M, F	M, F	Phillips et al. (2005)

For migration dates of other North American species see Benson & Winker (2001) and Carlisle et al. (2005), and for sex-related differences in migration distances of Nearctic species wintering in the tropics see Komar et al. (2005).

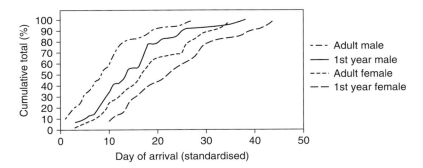

Figure 15.1   Spring arrival schedules of White-crowned Sparrows *Zonotrichia leucophrys* at Toiga Meadow Pass, California. The records for four years are standardised so that the date the first bird was captured each year was called Day 1 of arrival for that year. Older males ($N = 82$) arrived on average 5.4 days earlier than first-year males ($N = 62$, $t = 4.20$, $P < 0.001$), and older females ($N = 43$) arrived, on average, 5.0 days earlier than first-year females ($N = 51$, $t = 2.51$, $P = 0.014$). Overall, however, the arrival dates of different sex and age groups overlapped considerably. From Morton (2002).

in winter or returning there earlier in spring (the so-called arrival time or territorial defense hypothesis, Myers 1981). By remaining nearer the breeding areas, the argument goes, males should be better able to monitor and respond to variations in weather and thus return there as soon in any year as conditions permit, and before the females which winter further away (Alerstam & Högstedt 1982). Because the females are not involved in territory establishment, they are under less pressure, and can arrive later, closer to the time when they can begin nesting. This does not mean that females do not also compete among themselves – for example, for the most desirable territory-holding males. The idea is that there are both costs and benefits associated with arrival time, and the males gain most from early arrival and the females from somewhat later arrival (Morbey & Ydenberg 2001). Adults that arrive after their former territory has already been occupied by another individual of the same sex are sometimes able to oust the newcomer, the outcome depending partly on how long the new bird has had to establish itself (Newton 1998b). In general, however, birds that arrive late relative to others of their sex have less choice and are relegated to less favoured territories, and might thereby produce fewer young (Chapter 14).

That this sex difference in arrival time has something to do with reproduction is consistent with the finding that in many species in which males migrate earlier than females in spring, when reproduction is about to begin, the two sexes migrate at the same dates as one another in autumn when reproduction has finished (for Bluethroat *Luscinia luscinia* see Ellegren 1990a, for Blackcap *Sylvia atricapilla* see Izhaki & Maitav 1998b, for Red-backed Shrike *Lanius collurio* see Tryjanowski & Yosef 2002). It is supported also by the fact that in species in which females rather than males establish territories, it is the females that arrive first. For example, in some shorebird species, females are bigger than males, and take on some aspects of breeding that in most bird species are undertaken by males. In these species, the females compete with one another for territories and for the later arriving males, as observed in Eurasian Dotterel *Eudromias morinellus*, the three phalarope species *Phalaropus* spp., Spotted Sandpiper *Tringa macularia* and others (Myers 1981, Oring & Lank 1982). In addition, in five European songbird species, the degree of difference between the spring passage dates of males and females was found to correlate with the levels of extra-pair paternity in these species (Coppack *et al.* 2006). It seemed that males arrived relatively more in advance of females in species in which sexual selection through female choice was likely to be intense.

Another possible explanation of the arrival difference is that it results from competition for food at stopping sites, in which the larger sex is less affected by aggressive interactions, so can refuel and complete the journey more rapidly (see later). Because food is scarcer in spring than in autumn, competition is more intense in spring, producing a bigger difference between the sexes then. Yet other explanations of the sex difference in arrival dates have been proposed, but are mainly applicable to animals other than birds (Morbey & Ydenberg 2001).

Departure from breeding areas

In many species of birds, failed breeders that are freed from parental duties leave their nesting areas earlier than successful breeders, which can depart only when their young are grown. In consequence, in years of widespread breeding failure, as sometimes occur in arctic-nesting shorebirds, the bulk of the post-breeding migration occurs noticeably earlier than usual. It illustrates the dependence of post-breeding migration time on previous events in the annual cycle. More significantly, however, in some bird species, only one sex looks after the young, which frees the other sex to leave the breeding areas at earlier dates. This is evident, for example, in most duck species in which the males play no part in parental care, and leave their breeding places up to several weeks before the females and young (Cramp & Simmons 1977). Typically, the males assemble at special moulting sites, where they pass the flightless period (when they replace all their flight feathers), before moving on to the wintering areas (Chapter 16). Many females stay with their young in the breeding areas, where they moult their flight feathers, and only later in the year move directly to wintering areas, arriving later than the males. Other females move to the moulting sites, but later than the males.

Some shorebird species show a similar pattern to ducks in that the females remain with the young until they are full-grown, allowing the males to leave the breeding areas at earlier dates (as in Curlew Sandpiper *Calidris ferruginea* and Ruff *Philomachus pugnax*). In other shorebirds, however, the males raise the young to this stage, thus allowing the females to depart at earlier dates (as in Spotted Redshank *Tringa erythropus*, Wood Sandpiper *Tringa glareola* and Grey Plover *Pluvialis squatarola*). In yet other shorebirds, both partners help to the same stage with parental care, and the two sexes migrate at about the same time (as in Lapwing *Vanellus vanellus* and Black-tailed Godwit *Limosa limosa*). In all these shorebird species, the young tend to depart after the adults, requiring longer to prepare themselves. Hence, the general sequence of post-breeding migration in shorebirds begins with failed breeders, followed by the successful breeders of the non-parenting partner, then the parenting partner (or both partners), and finally the young. Departure dates in a population can thereby spread over several weeks, as can the subsequent arrival dates in moulting or wintering areas (Cramp & Simmons 1983).

Among many small raptor species, the adults migrate after completing moult. Females start moult around the time of egg-laying, while the males, who provide the food, delay the start of moult until around the time of hatch. This in turn enables the females to finish moulting and depart on migration earlier, leaving the remaining parental care to the male (Kjellén 1992). In the Honey Buzzard *Pernis apivorus*, in which the sexes share breeding duties equally, the two sexes depart at the same time. Other bird species which show uniparental care show marked sexual divergence in autumn migration dates. The fact that this pattern is repeated independently in different taxonomic groups further emphasises the link between migration and breeding system. Moreover, amongst species which show bi-parental care, some show only minor sex differences in autumn departure dates, while others show no such differences (e.g. Murray 1966 for various passerines).

In addition to the sex difference in arrival and departure times, young adults nesting for the first time typically arrive on their spring breeding areas some days later than older and more experienced birds of the same sex (**Figure 15.1**). Such age-related differences have been noted in a wide range of species, from passerines to shorebirds and seabirds (see later). Two explanations have been offered, which are not mutually exclusive. The first is that, because young adults cannot compete effectively with older ones for nesting territories, they are better to arrive later in the season, when most old birds have already settled, thereby saving the energy that would otherwise be wasted on futile battles. On this basis, later arrival is under endogenous (genetic) control. Secondly, because of their inexperience and subordinate status with respect to older birds, young adults cannot feed as efficiently before departure from wintering areas, or at stopover sites en route, so are delayed on migration (Chapter 27). On this basis, later arrival is under external influence, and results partly from prevailing conditions. In single-brooded species, a late start to breeding would be expected to result in a late finish, and hence in a later departure from breeding areas by young adults compared with earlier nesting older ones. In multi-brooded species, young adults might attempt fewer broods and thus be ready to migrate from breeding areas no later than older ones.

Age differences in autumn migration dates have been recorded in a wide range of bird species, from passerines to raptors and shorebirds (**Table 15.1, Figure 15.2**). In some species of long-distance migrants, the difference in departure dates between age groups is substantial. Among passerines, the adult–juvenile difference seems to depend on whether a wing moult occurs in breeding areas before the start of migration, or whether migration occurs immediately after breeding, with moult either started and then suspended, or delayed altogether until after arrival in winter quarters (Chapter 11). Most species that moult completely before autumn migration are short-distance or partial migrants, while most of those that arrest or delay moult are long-distance migrants that winter in the tropics.[1] In the former,

---

[1]*While some long-distance passerines moult none of their feathers before leaving their breeding areas in late summer, others moult some of their body feathers, or body feathers and tertials, while others also moult some of their flight feathers, before moult is arrested for migration (Jenni & Winkler 1994). Moult is then resumed either at a staging area part way through migration, or at the ultimate destination in winter quarters. There is considerable variation in these patterns within species, depending on breeding area and individual breeding dates (Chapter 11).*

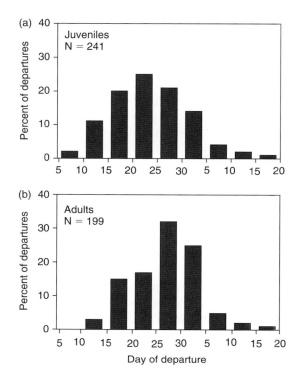

Figure 15.2   Migration departure schedules of (a) juvenile (*N*=241) and (b) adult (*N*=199) White-crowned Sparrows *Zonotrichia leucophrys* based on the pooled data for seven years from Toiga Meadow Pass, California. No sex difference in departure dates was detected, but juveniles left 3.2 days earlier, on average, than males. The range of departure dates was 45 days for juveniles, and 37 days for adults, with most of the variation traceable to inter-year variation in breeding dates. Departure was delayed about one day for every two days that nesting had been delayed by environmental conditions, such as persisting snow cover earlier in the summer. Over the seven-year period, mean departure date varied by about 14 days in juveniles and eight days in adults. From Morton (2002).

juveniles migrate first, presumably because they replace only their body feathers, a process which takes less time than the adults take to replace their entire plumage, including their flight feathers. But in species that suspend or postpone moult, the adults can leave their nesting areas soon after their young are independent, although the young themselves take another week or more before they are ready to undertake their first migration. This dichotomy holds in all the passerine species listed in **Table 15.1**, whether Eurasian or North American. In addition, of 18 species of passerines studied in Alaska, 11 moulted in their breeding areas, and the juveniles left before the adults. One species (Alder Flycatcher *Empidonax alnorum*) postponed moult until later in the year, and the adults left around 13 days before the juveniles. In the remaining six species, which also moulted later in the year, no significant difference in departure dates occurred between adults and juveniles (Benson & Winker 2001). Likewise in Idaho, in nine passerine species that migrated immediately after breeding, adults left significantly earlier than

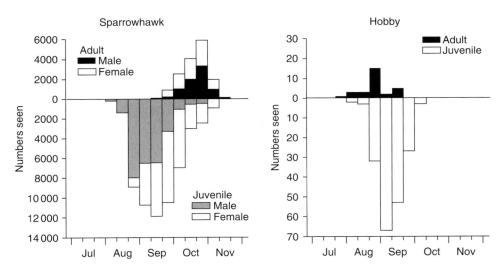

Figure 15.3 Numbers of Eurasian Sparrowhawks *Accipiter nisus* and Eurasian Hobbies *Falco subbuteo* seen migrating through Falsterbo, Sweden, at different dates in autumn. In the Sparrowhawk, which is a partial migrant, juveniles moved before adults and females before males, on average. Similar patterns were shown by other partial migrants, including Northern Goshawk *Accipiter gentilis* and Red Kite *Milvus milvus*. In the Eurasian Hobby, which is a complete migrant, adults moved before juveniles, on average, but sexes could not be distinguished. Similar patterns were shown by other complete migrants, including European Honey Buzzard *Pernis apivorus*, Montagu's Harrier *Circus pygargus* and Osprey *Pandion haliaetus*. From Kjellén (1992).

juveniles, while in 27 species that moulted before migrating, juveniles left before adults (except for the Pine Siskin *Carduelis pinus*, Carlisle *et al.* 2005).

Similar patterns occur among European passerines migrating to Africa. The Common Cuckoo *Cuculus canorus* is an extreme example, for the adults leave before their last young, reared by other species, have even left the nest, giving a difference in mean departure dates between the two age groups of about one month (Wyllie 1981). Both age classes moult in winter quarters.

Similar departure patterns are seen in raptors (**Table 15.1, Figure 15.3**). In short-distance and partial migrants that finish moult before migrating, such as the Eurasian Sparrowhawk *Accipiter nisus*, juveniles leave the breeding areas earlier than adults. The juveniles do not moult at all in their first autumn of life, but retain their juvenile plumage (acquired in the nest) for another year. But in most long-distance migrants that suspend moult and migrate immediately after breeding, such as the Osprey *Pandion haliaetus*, adults tend to leave before juveniles, departing as soon as the young are independent but still gaining the experience necessary to undertake migration. In the Black Kite *Milvus migrans*, the age difference is substantial, with adults leaving 3–4 weeks before juveniles (Schifferli 1967).

After nesting in the arctic, adult shorebirds usually leave the breeding areas well before the juveniles, and moult at a migratory staging site or in winter quarters. This temporal difference can be increased by the juveniles' slower rate of progress, resulting from longer and more frequent stopovers, use of poorer staging sites and less direct migration routes (Saurola 1981, Evans & Davidson 1990, Baccetti *et al.* 1999). In some populations in which the adults stop and moult at a staging site, the two age groups arrive in wintering areas at about the same date, as observed in Dunlin *Calidris alpina* in southern Europe (Baccetti *et al.* 1999). Hence, in all these species from different taxonomic groups, whether adults or juveniles depart first on autumn migration is linked to whether or not they migrate immediately after breeding, and where and when the moult occurs. Species whose summer food supply collapses soon after breeding have to leave before moulting, whereas those whose food supply lasts long enough beyond breeding can moult before migrating. The pressures are somewhat different on juveniles and adults because, while the juveniles have a shorter body moult (or in some species no moult), the adults have a longer complete moult, including flight feathers.

Although in geese, swans and cranes the young of the year migrate with their parents, the older immatures tend to migrate earlier than families in autumn (as do failed breeders) and later in spring (for Canada Goose *Branta canadensis* see MacInnes 1966, for Snow Goose *Chen caerulescens* see Maissoneuvre & Bedard 1992, for Tundra Swan *Cygnus columbianus* see Black & Rees 1989).

MIGRATORY DISTANCE, BODY SIZE AND DOMINANCE

Migration and body size

The sex and age groups of a population often differ in body size. This may affect their ability to withstand cold and food shortage, their dominance status and many other features that could in turn influence movement patterns. In most bird species, males are bigger than females, but in raptors, owls, some shorebirds and others, females are bigger than males. The differences in body size between the sexes are usually slight, with some overlap between them, but in some species (notably most raptors and some shorebirds) the differences are substantial, with no overlap between sexes. In most bird species, juveniles are also slightly smaller than older birds of the same sex.

In many species, as mentioned already, the different sex and age groups migrate different distances from one another, but with some overlap, giving rise to gradients in the sex and age ratios from high to low latitudes across the wintering range. Such patterns have been repeatedly shown from recoveries of birds ringed at the same breeding or staging locations (**Figure 15.4**).[2] In most such species, females migrate further, on average, and winter at lower latitudes than males, and in some the sex difference is substantial. For example, in the Ruffs *Philomachus pugnax* migrating through western Europe, most of the males remain within Europe, whereas most females winter in sub-Saharan Africa, where females (which are smaller) can outnumber males by more than 10 to 1 (Gill *et al.* 1995). In some

---

[2]*For various finches and other seed-eaters see King* et al. *(1965), Ketterson & Nolan (1979), Prescott & Middleton (1990); for icterids and starlings see Dolbeer (1982); for Wood Duck* Aix sponsa *see Hepp & Hines (1991); for Canvasback* Aythya valisineria *see Nichols & Haramis (1980); for American Woodcock* Scolopax minor *see Diefenbach et al. (1990).*

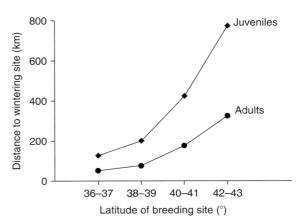

Figure 15.4 Mean distances between ringing and recovery sites for adult (*N*=961) and juvenile (*N*=84) Common Starlings *Sturnus vulgaris*. The birds were ringed in their breeding areas in eastern North America and recovered in winter (January–February) further south. Birds from further north moved longer distances, but juveniles moved further than adults. Modified from Dolbeer (1982).

albatross species, the sex differences in migration are also great. For example, in Wandering Albatrosses *Diomedia exulans* from the Crozet Islands, males were found to spend their non-breeding periods mainly in Subantarctic–Antarctic waters in latitudes south of the nesting colony, whereas females visited subtropical–tropical waters in latitudes north of the colony (Weimerskirch & Wilson 2000).

From ring recoveries, differential migration by age classes was found in 10 species of British seabirds, with immatures moving further, on average, than adults (Gannet *Morus bassanus*, Great Cormorant *Phalacrocorax carbo*, Shag *P. aristotelis*, Black-headed Gull *Larus ridibundus*, Lesser Black-backed Gull *L. fuscus*, Herring Gull *L. argentatus*, Great Black-backed Gull *L. marinus*, Guillemot *Uria aalge*, Razorbill *Alca torda* and Puffin *Fratercula arctica*, G. Siriwardena & C. Wernham, in Wernham *et al.* 2002). The magnitude of these age-related differences varied between species, with median distances of 141 km and 84 km in immature and adult Black-headed Gulls, ranging up to 1380 km and 639 km in immature and adult Gannets. Guillemots were unusual in that juveniles were recovered significantly further north and east than adults. Similar differences have emerged in other studies of seabirds (e.g. Great Cormorant, Bregnballe *et al.* 1997, three large gull species, Kilpi & Saurola 1984), and in several species of terns the wintering areas of different age groups were almost completely segregated (for Common Terns *Sterna hirundo* in the West Atlantic flyway see Hays *et al.* 1997, for Roseate Terns *S. dougallii* in the eastern Asian–Australasian flyway see Minton 2003).

Other studies have examined sex and age ratios among birds from different wintering localities, either by observation or by examination of trapped samples or museum skins. They include a wide range of species,[3] some of which winter in temperate areas and others in tropical areas (Komar *et al.* 2005). Such studies have

---

[3]For *Dark-eyed Junco* Junco hyemalis *see Ketterson & Nolan (1976); for Evening Grosbeak* Hesperiphona vespertina *see Prescott (1991); for Yellow Wagtail* Motacilla flava *see Wood (1992); for Snow Bunting* Plectrophenax nivalis *see Smith* et al. *(1993); for White-throated Sparrow* Zonotrichia albicollis *see Jenkins & Cristol (2002); for Robin* Erithacus rubecula *see Catry* et al. *(2004); for various passerines see Komar* et al. *(2005), for American Kestrel* Falco sparverius *see Arnold (1991); for shorebirds see Myers (1981); for diving ducks see Alexander (1983), Nichols & Haramis (1980), Carbone & Owen (1995).*

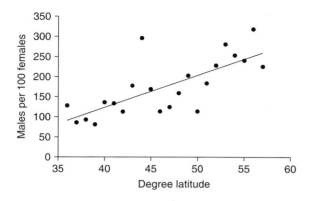

Figure 15.5   Sex ratios among wintering Pochard *Aythya ferina* in relation to latitude in western Europe. Redrawn from Carbone & Owen (1995).

also revealed geographical gradients in winter sex ratios, with a predominance of males nearest the breeding areas and of females furthest away. They are common in diving ducks (**Figure 15.5**; Bellrose *et al.* 1961, Nichols & Haramis 1980, Alexander 1983, Carbone & Owen 1995), but less pronounced in dabbling ducks, many of which form pairs in early winter (Owen & Dix 1986, Hepp & Hines 1991). The association between the sexes, and the protection given to paired females by their mates, may help to reduce any segregation through male dominance that might otherwise occur in these species (see below).

Geographical clines in age ratios of a different kind were found in Black-throated Blue Warblers *Dendroica caerulescens* on their breeding areas in eastern North America (Graves 1997). The proportion of yearling males increased as overall abundance decreased towards the margins of the breeding range. This was attributed to older males being sufficiently numerous to occupy almost all the habitat in the central (supposedly better) parts of the range, but not in the marginal parts, which therefore offered more opportunities for yearlings. The concentration of young breeders in poor habitat has been shown in many species on a local scale (see later), but not previously on a wide geographical scale, in the breeding season.

The supposed greater ability of large individuals to withstand cold has been proposed as an explanation of why, in so many bird species, males tend to winter at higher latitudes (nearer their breeding areas) than females, and why within the sexes, adults tend to winter at higher latitudes than juveniles (the so-called winter cold hypothesis, Ketterson & Nolan 1976, 1983). The idea is that, because large birds have a lesser surface area-to-volume ratio than smaller birds of similar shape, they experience a lower rate of heat loss from the body, giving a better energy return on foraging. Also, if energy stores are proportional to body mass, larger birds should have greater body reserves relative to their daily energy needs, enabling them to fast for longer periods (Ketterson & Nolan 1983). While this explanation of differential migration distances cannot be disproved, the laboratory evidence for differential energy management within species is at best equivocal (Cristol *et al.* 1999). In addition, the differences in body size between the sex and age groups of a population are often very small, possibly insufficient to have the effects attributed to them. Differential migration distances could also result from other pressures, such as dominance relationships as explained below, or from the need for males to obtain territories in order to breed (with males wintering nearer to nesting areas so as to return quickly as soon as conditions permit in spring, as mentioned above).

Moreover, in some species, sex differences in body size are associated with differences in bill size, which could influence feeding habits, leading the sexes to prefer different areas. Differences in bill lengths are especially marked in some shorebird species, such as the Ruff *Philomachus pugnax* mentioned above, in which the sexes show marked differences in their migration distances. It is an open question how much this distributional difference results from bill size, rather than from body size or other gender-based differences. For all we know, there may be latitudinal differences in the depth distribution of prey, which favour longer migration of one sex than the other (for Western Sandpiper *Calidris mauri* see Nebel *et al.* 2002). In situations like this, in which the sexes differ in several respects (migration distance, body size, bill size and feeding habits), it is practically impossible to say with confidence which aspects are causes and which are consequences. Moreover, where the sexes of a species occur in largely different regions, they could come under different selection pressures, leading to greater divergence in body size and other features than might otherwise occur.

Migration and dominance relationships

One of the main effects of body size differences within a population is that they influence the competitive relationships among individuals, with dominants gaining food at the expense of subordinates, which may therefore have to move elsewhere. In certain species, more individuals stay in their breeding areas in winters when food is plentiful than in winters when food is scarce, in mild winters than in cold ones, or in good habitats than in poor ones. Such findings imply that, in such species, competition for food and other resources can have an immediate influence on the proportions of birds that stay or leave each year, and the distances they travel (Chapters 12 and 18). The benefits of migration extend further up or down the social hierarchy, according to prevailing conditions.

Such competition can in turn influence the sex and age groups involved (the so-called dominance hypothesis, Gauthreaux 1978a, 1982a). For example, among Blue Tits *Parus caeruleus* in a breeding area in Sweden, the proportion of migrants increased from adult males (virtually none), through adult females, juvenile males and juvenile females (>40%). In addition, more late-hatched juvenile males than early-hatched juvenile males migrated (Smith & Nilsson 1987). These proportions were correlated with the dominance relations within the population, with adult males the most dominant and late-hatched juvenile females the least. Such migratory patterns are frequent among passerines, raptors, gulls and other seabirds (Gauthreaux 1982a, Dolbeer 1991, Kjellén 1994a, Catry *et al.* 2004), but the dominance relationships which supposedly produce them are often assumed (as a correlate of body size), rather than measured directly.

Where only part of the population leaves the breeding areas, the commonest pattern is for juveniles to migrate in greater proportion, to leave earlier and return later, and to winter further from the breeding areas than adults. Moreover, in several species, late-fledged young migrate further than earlier ones, and may return later to breeding areas the following spring (Jakober & Stauber 1983,

Bairlein 2001). The implication that the same individuals may move further in their first than subsequent years has been confirmed in some species by ring recoveries (Newton 1972, Schwabl 1983). During an unusually severe winter in southern France, 260 Little Egrets *Egretta garzetta* were found dead. Their carcasses revealed that young birds were affected first, then adults, with adult males succumbing last as the cold persisted. This provided independent indication that adult males could survive the usual cold periods better than adult females, and better still than juveniles. It fitted the facts that, in this species, adult males normally migrate in lower proportion than adult females, and adults in lower proportion than juveniles (Pineau 2000).

The role of competition in influencing migration distances emerged in a study of Grey Plovers *Pluvialis squatarola* on the Tees Estuary in northeast England (Townshend 1985). Some newly arrived juveniles established territories only to be soon displaced by larger juveniles, or after a few weeks by later arriving adults. Two of the displaced juveniles (which had been colour-marked) were subsequently seen on another estuary 900 km to the south. Because migration is costly in terms of energy needs and mortality risks, birds can be expected to minimise the distances moved, and settle in the first suitable site they reach within the wintering range (Greenberg 1980, Gauthreaux 1982a, Pienkowski & Evans 1985). Pressure from other birds pushes them further along the route, so that subordinate individuals are likely to move furthest.

In subsequent years, individuals may be assumed to gain other advantages by returning to a site with which they are familiar, and at which they overwintered safely as inexperienced juveniles, even if this site was not as near to the breeding areas as possible. On this scenario, therefore, winter distributions are partly a consequence of events in previous years, and if a population declined over several years, the site-fidelity of adults would ensure a lag in the withdrawal of birds from the furthest parts of a wintering range. This model of 'within-species' winter distribution is similar to the 'despotic model' used to explain the distribution of birds among different habitats, whereby birds occupy the best areas in preference and, as these become filled, other individuals spread to less good areas (Brown 1969, Newton 1998b).

If body size-dominance relationships are important in differential migration, species with a greater degree of sexual size dimorphism should show a greater sex difference in migration distances than species that show little sexual size dimorphism. The winter distributions of two highly dimorphic icterid species (the Common Grackle *Quiscalus quiscula* and Red-winged Blackbird *Agelaius phoeniceus*) fit this prediction, whereas the sexes of the monomorphic Common Starling *Sturnus vulgaris* show no difference in winter distributions (as shown by ring recoveries of birds from the same breeding areas, Dolbeer 1982). However, female Brown-headed Cowbirds *Molothus ater* migrated the same average distance as males, even though they are considerably smaller. In addition, in some raptors in which females are bigger than males, males migrate further (e.g. Peregrine Falcon *Falco peregrinus*, Restani & Mattox 2000; Northern Goshawk *Accipiter gentilis*, Mueller *et al.* 1997; Eurasian Sparrowhawk *A. nisus*, Belopolsky 1971, Payevsky 1990; Hen (Northern) Harrier *Circus cyaneus*, B. Etheridge, in Wernham *et al.* 2002; Rough-legged Buzzard *Buteo lagopus*, Kjellén 1994a; Snowy Owl *Nyctea scandiaca*, Kerlinger & Lein 1986; Northern Hawk Owl *Surnia ulula*, Byrkjedal & Langhelle 1986).

Among the Peregrines that breed in Greenland, the sex difference in migratory distance is extreme. Nearly 400 birds ringed in nesting areas have given 125 recoveries abroad. All females were found between the Gulf of Mexico (28°N) and the northernmost parts of South America (2°S), whereas all males were recovered in South America between 2°S and 26°S. On average, males were found 4000 km further south than females, and their migrations often exceeded 25 000 km annually (Lyngs 2003). Despite their longer migrations, males arrive back on nesting places no later than females. The findings on raptors are thus consistent with both the dominance and winter cold hypotheses, but not with the territorial defence hypothesis (that males winter near breeding areas in order to get back quickly in spring). Such differences are not apparent in all raptors, however, and in Common Kestrels *Falco tinnunculus*, females migrate in greater proportion, further and earlier than males (Wallin *et al.* 1987, Village 1990, Kjellén 1992, 1994). In various diurnal raptors, the juveniles move furthest, the sequence (nearest–furthest) being adult females, adult males, juvenile females, and juvenile males (**Table 15.1**).

As well as latitudinal gradients in sex and age ratios, some species show altitudinal gradients. For example, among Snow Buntings *Plectrophenax nivalis* wintering in Britain, the proportion of males decreased from north to south within Britain, and also from mountain to coastal sites, reflecting the tendency of males to winter nearest the breeding areas, and for females to occur in climatically milder places (Smith *et al.* 1993). In Dark-eyed Juncos of the race *Junco hyemalis carolinensis*, females moved further downslope than males from their ridge-top breeding areas, a tendency more marked in severe winters than mild ones (Rabenold & Rabenold 1985). Likewise, among Willow Ptarmigan *Lagopus lagopus* and Rock Ptarmigan *L. mutus* on Alaskan mountains, most males remained in winter on the alpine tundra where they breed, while most females moved downslope into the forest zone (Weeden 1964). These findings were again consistent with the hypothesis that social dominance in competition for food or feeding places affects winter distribution patterns. However, not all montane birds follow these patterns: the Blue Grouse *Dendrogapus obscurus* is unusual in that both sexes often move upslope for the winter, vacating fairly open breeding areas for dense forest, where they eat conifer needles (Cade & Hoffman 1993). However, the males moved furthest, so that, like the other montane species just mentioned, they wintered at higher elevations than the females.

Among the species examined by Cristol *et al.* (1999), females migrated further than males in at least 77% of 53 species, and young migrated further than adults in at least 38% of 53 species. The individuals migrating further were usually members of the class whose body size was smaller (71% of 69 size comparisons between population classes), socially subordinate (82% of 44 comparisons), and later arriving in breeding areas (74% of 58 comparisons). However, there was a great deal of deviation from these patterns. In most species, the smaller sex migrated further, but in others less far. There can be no doubt, therefore, that the patterns described above are widespread among birds, but not universal.

Little evidence of sex differences in migration distances was found in the American Woodcock *Scolopax minor*, Sanderling *Calidris alba* and Grey Phalarope *Phalaropus fulicarius*, despite females being larger than males (Myers 1981, Diefenbach *et al.* 1990). Nor were such differences found in Eurasian Siskins *Carduelis spinus* and Savannah Sparrows *Passerculus sandwichensis* in which the sexes are about the same size (Payevsky 1998, Rising 1988). On the other hand, among Indigo Buntings *Passerina cyanea* migrating from North to Central America, the usual sequence was reversed, as females predominated in the northern part of the wintering range and males in the southern part (Komar *et al.* 2005).

Nor is it the case that juveniles invariably migrate further than adults. Among Chaffinches *Fringilla coelebs*, Bramblings *F. montifringilla* and Eurasian Siskins *Carduelis spinus* that were ringed on autumn migration on the southern Baltic coast, adults were recovered at significantly greater distances than juveniles (Payevsky 1998). The same was true in the males of various finches in North America, including the Dark-eyed Junco *Junco hyemalis*, White-crowned Sparrow *Zonotrichia leucophrys* and American Goldfinch *Carduelis tristis* (King *et al.* 1965, Ketterson & Nolan 1982, 1983, Morton 1984, Prescott & Middleton 1990). In Dark-eyed Juncos, the most abundant classes from highest to lowest latitudes in winter were juvenile males, adult males, juvenile females and adult females (Ketterson & Nolan 1982, 1983).

In experiments, Rogers *et al.* (1989) found no evidence that social dominance was responsible for variance in migration distances in juncos. In one experiment, juncos caught in winter in Michigan were each matched in captivity against another junco of the same sex and age class caught in Indiana (further from the breeding range). Michigan birds were dominant in only half the experimental dyads (21 out of 41 dyads), not in the majority as expected on the dominance model. In a second experiment, young males wintering in Michigan were tested against adult males from Indiana. In 19 out of 25 dyads, the more southern-wintering old males were dominant, which is also counter to the prediction of the dominance hypothesis. The authors concluded, therefore, that social dominance did not play an important role in influencing the wintering latitude of Dark-eyed Juncos (see Chapter 12 for other experiments on juncos).

Among waders, the Red Knot *Calidris canutus* provides another example of an unexpected age difference in migration. Most juveniles of the race *rogersi* spend their first year in eastern and southeastern Australia, and only make the additional journey to more distant non-breeding areas in New Zealand at the beginning of their second year. Thus, the proportion of first-year birds in eastern Australia is particularly high (up to 70% in years of good breeding success), while in New Zealand only a small proportion of first-year birds is ever recorded (Minton 2003). These various species thus seem not to fit either the dominance or the territorialism hypotheses, and the only explanation offered is that juveniles are physiologically less capable of migrating long distances than adults (Prescott & Middleton 1990). They cannot therefore move on to places where competition might be less.

No age-related differences in wintering latitude were apparent from ring recoveries of the Evening Grosbeak *Hesperiphona vespertina* (Prescott 1991), Cedar Waxwing *Bombicilla cedrorum* (Brugger *et al.* 1994), Barn Swallow *Hirundo rustica* (C. Mead in Wernham *et al.* 2002), Osprey *Pandion haliaetus* (Poole & Agler 1987), Mallard *Anas platyrhynchos* (Nichols & Hines 1987), American Black Duck *Anas rubripes* (Diefenbach *et al.* 1988) and American Woodcock *Scolopax minor* (Diefenbach *et al.* 1990). In fact, age-related differences in migration distances and winter distributions do not appear to be nearly as frequent as gender-based differences. Whether this is because age-related differences are smaller, and therefore harder to detect, or because they are genuinely less frequent, remains to be seen.

While dominance relationships appear to be general in bird populations, they are likely to result in differential survival and migration chiefly in competitive situations, where food or some other resource is limiting. In fights over food, dominants usually win over subordinates, and in the same area dominants may survive the winter in greater proportions (Kikkawa 1980). Among Mallards *Anas platyrhynchos* in much of northern Europe, males predominate in wintering populations but, in areas where birds are artificially fed, the sex ratio is more nearly equal (Nilsson 1976). Among non-migratory Song Sparrows *Melospiza melodia*, the survival rate of subordinates in the presence of dominants was increased when food supplies were experimentally supplemented (Smith *et al.* 1980). Among European Robins *Erithacus rubecula*, apparent overwinter survival was greater among individuals that remained near their breeding areas than among those from the same locality that migrated to distant wintering areas (Adriaensen & Dhondt 1990). The assumptions of the dominance hypothesis thus gain some support from field studies.

MIGRATION AND DEFERRED BREEDING

In some long-lived bird species which do not breed until they are several years old, differences in wintering areas are apparent between several successive age groups, as individuals move progressively less far from the breeding areas as they age, or spend shorter periods in the more distant parts of the wintering range (for Herring Gull *Larus argentatus* see Coulson & Butterfield 1985, for Western Gull *L. occidentalis* see Spear 1988). This pattern is shown for the Herring Gull in **Figure 15.6**, based on month-by-month analysis of ring returns from breeding colonies around the Great Lakes in North America (Moore 1976). In November, when the weather was mild, all age classes were recovered near the breeding areas. But as winter progressed, the age classes became increasingly separated, with the youngest birds moving furthest south. Then, as conditions improved again from March, all age classes moved northwards towards the breeding places, where most of the population spent the summer. Nevertheless, some first- and second-year birds remained well south of the nesting areas during the breeding season.

The difference between age groups in date of return to breeding areas is greatest in those (mostly large) species that do not breed until they are several years of age, and in which the immatures are therefore under less pressure to return early to the breeding areas in spring. In some such species, the young migrate much later than the adults, and take longer over the journey. In the various eagle species that pass through Israel each spring, the age groups migrate in order of oldest first to youngest last, but there is considerable overlap between them (Shirihai *et al.* 2001). The youngest age groups of some eagles and vultures return to breeding areas up to several weeks later than adults, and too late to attempt breeding that year. Radio-tracking revealed that immature Lesser Spotted Eagles *Aquila pomarina* and Steppe Eagles *A. nipalensis* migrate more slowly and later in spring, and arrive in breeding areas up to 10 weeks later than adults (Meyburg *et al.* 1995c, 2001, Meyburg & Meyburg 1999). As they make no attempt to nest, the immatures suffer no obvious penalty by arriving late. The same held for Steller's Eagles *Haliaetus pelagicus*, some juveniles of which migrated a shorter distance

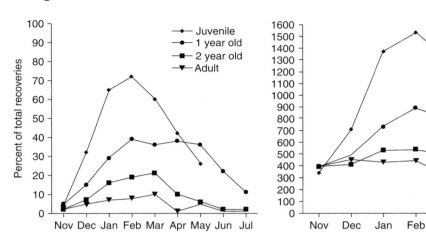

Figure 15.6 The seasonal distribution of Herring Gulls *Larus argentatus* breeding in the Great Lakes region of North America. Left: Monthly proportion of ring recoveries from each age class south of the Great Lakes region. Right: Mean monthly distance of ring recoveries from each age class. Based on 6949 recoveries of juveniles, 1900 of one-year-olds, 879 of two-year-olds, and 2956 of older birds (adults) obtained during 1929–1971 from birds ringed at colonies mainly within 44–46°N. Year classes separated at 31 May each year, the approximate date of peak hatch. From Moore (1976).

and spent the summer in areas well south of their natal areas (Ueta *et al.* 2000, McGrady *et al.* 2003).

In three species studied in Japan by satellite-tracking, namely Steller's Eagle *Haliaeetus pelagicus*, Black-faced Spoonbill *Platalea minor* and White-naped Crane *Grus vipio*, the immatures on spring migration stayed, on average, about twice as long on stopover sites, and took about twice as long over the whole journey as adults. In this study, the two age groups showed no difference in the mean distances between stopover sites or in the overall distance of migration (Ueta & Higuchi 2002). The longer stopovers of immatures may have reflected a lack of urgency in reaching breeding areas, as well as a lower feeding efficiency and rate of fuel accumulation.

SPECIES SUMMERING IN 'WINTERING' AREAS

In some long-distance migrants, non-breeding individuals can be seen in 'winter quarters' in every month of the year. These are mostly young individuals that do not breed until they are two or more years of age. They stay in their wintering areas beyond their first winter, returning to the breeding area in a later year (deferred return). In this way, the age groups are separated geographically in the breeding season, and the young birds, which would in any case not reproduce in their first year, avoid the costs and risks of an unnecessary return journey. Alternatively, the immatures may migrate only part way from their wintering areas, passing the summer at sites en route to the breeding area, sometimes getting nearer year by year (graded return). Once sexual maturity is attained, these birds make a full return migration, and usually settle to breed close to where they were hatched (Chapter 17). From then on, they normally show the usual twice-yearly migrations.

Over-summering in 'winter' quarters is regular in at least 15 families of birds, being best known among raptors, shorebirds and seabirds. It is not always certain that the same individuals stay throughout, and the continued presence of birds in winter quarters may be due to overlap in the departure and arrival dates of different individuals, with some returning before others have left. However, some species have now yielded enough ring recoveries, or sightings of colour-marked birds, to indicate that some individuals remain in 'wintering' areas year-round, and that the younger age groups are not present in breeding areas in spring–summer.

Among raptors, the Osprey *Pandion haliaetus* has been studied in detail. Almost all ring recoveries from the winter range during the northern breeding season refer to first-year birds, younger than the usual age of first breeding, while no first-year recoveries come from the breeding range. This is true in both Old and New World populations, and implies that such birds stay in wintering areas all their first year, rather than returning north for brief periods varying in timing between individuals (Henny & van Velzen 1972, Saurola 1994), a finding also supported by recent radio-tracking studies (Chapter 8). The same is true for Egyptian Vultures *Neophron percnopterus*, judging from the virtual absence of the distinctive first-year individuals among the birds migrating north through Israel in spring (Shirihai *et al.* 2001), and in Oriental Honey Buzzards *Pernis ptilorhynchus* judging from a single radio-tracked individual (Higuchi *et al.* 2005).

Among practically all migrant raptor populations that winter in Africa, a proportion of individuals remain there through the northern summer (the local wet season) (Thiollay 1989). They are found even in small species such as the Common Kestrel *Falco tinnunculus*, but are relatively more numerous in larger species, such as the Short-toed Eagle *Circaetus gallicus*, in which individuals do not breed until they are two or more years old. There is also a northward shift within Africa of these 'summering' birds, with many more individuals in the Sahel and Sudan zones in the summer wet season than in the winter dry season (tantamount to a part-way return, coincident with a northward migration of intra-African migrants). Similarly, most Black-crowned Night Herons *Nycticorax nycticorax* migrate from Europe to winter in the Sudan–Sahel zones of Africa, where a high proportion of juveniles remain during their first 2–3 years (Brosselin 1974). The same holds for White Storks *Ciconia ciconia* and others.

First-year seabirds that perform long migrations often also remain in 'winter quarters' for the summer or at least in sea areas separate from those used by the breeding adults. As they get older, the non-breeding immatures migrate progressively nearer to breeding areas (as found, for example in Fulmars *Fulmarus glacialis* and Northern Gannets *Morus bassanus*). As the years pass, increasing proportions begin to return to nesting colonies during summer, 'prospecting' for nest-sites, but they usually arrive later in spring and depart earlier in summer than the adults. For example, at a colony of Common Terns *Sterna hirundo* nesting in Germany, fledglings were marked with transponders, allowing the automatic registration of their subsequent return. Most individuals returned to the colony

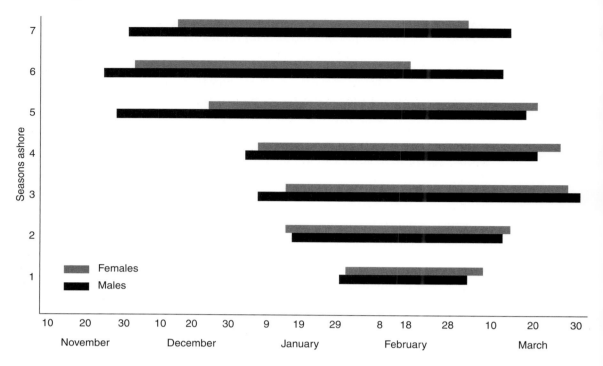

Figure 15.7 Attendance patterns of Wandering Albatrosses *Diomedia exulans* at a nesting colony in relation to experience, South Georgia Island. Each horizontal bar spans the median arrival to median departure dates of males and females. In successive seasons after first arrival, the date of arrival became earlier ($F_{6,371} = 43.0$, $P < 0.001$) and the number of days spent ashore increased ($F_{3,371} = 63.3$, $P < 0.001$). Redrawn from Pickering (1989).

for the first time as two-year-olds, and remained as 'prospectors' for a year or more before starting to breed. Prospectors arrived later than breeders (too late to breed), and first-time breeders arrived, on average, 17 days later than experienced breeders (laying, on average, 19 days later). The arrival dates of older Common Terns got progressively earlier with increasing age and experience of the colony but, on average, males arrived earlier than females (Dittman & Becker 2003).

At a Wandering Albatross *Diomedia exulans* colony, the variability was even greater, as recruitment to the breeding population took 2–8 years after first return to the natal colony (Pickering 1989). In successive seasons, from first return to pairing, the date of arrival became earlier, and the number of days spent ashore and interacting with other birds increased (**Figure 15.7**). Birds that paired arrived earlier and spent more time ashore than birds with similar experience which did not pair that year. In the season following pairing, birds returned at the same time as breeders, but most did not produce eggs. They left in mid-season before other non-breeders and bred the following year. In this and other seabird species, the successive arrivals of established breeders, new breeders and then prospectors can be spread over periods of weeks or months each year, as can their subsequent departure (for Manx Shearwater *Puffinus puffinus* see Perrins *et al.* 1973, for Razorbill *Alca torda* see Lloyd & Perrins 1977, for Common Guillemot (Murre) *Uria aalge* see Halley *et al.* 1995, for Common Tern see Ludwigs & Becker 2002).

In all marine terns from high latitudes, most juveniles stay in their 'wintering' areas for 1–2 years before returning part way or all the way back to breeding areas. In the Arctic Tern *Sterna paradisaea*, the majority of immatures remain in the southern hemisphere in the northern summer, extending from equatorial cold water areas south to the pack-ice. Some appear to circumnavigate Antarctica, possibly taking up to two years before moving north towards the nesting colonies where they breed at 3–4 years old (P. Monaghan, in Wernham *et al.* 2002). Most young skuas (jaegers) also remain in the southern hemisphere, although some young Great Skuas *Stercorarius skua* and Arctic Skuas *S. parasiticus* return not just to the breeding areas, but spread to even higher latitudes in the northern summer (B. Furness, in Wernham *et al.* 2002).

Similar patterns of (a) pre-breeding birds over-summering in 'wintering' areas, (b) returning part way towards breeding areas, or (c) returning to breeding areas for a shorter time than nesting adults, occur in some shorebirds, especially among populations that migrate the longest distances and winter in the southern hemisphere (**Table 15.2**). Some colour-dyed Short-billed Dowitchers *Limnodromus griseus* were absent from their wintering site in Apalachee Bay in Florida for less than a month in the period late June to early July (Loftin 1962). If these birds reached their breeding areas, they could have spent only a few days there before returning.

Most waders seen during summer in their 'wintering' areas are in their first year of life, but some are older. In the Bristle-thighed Curlew *Numenius tahitiensis* that breeds in Alaska and winters on central Pacific islands, individuals do not leave their wintering areas until they are three or even four years old (Marks & Redmond 1996). In some other large shorebird species, immatures leave the wintering areas in increasing proportion, and progressively earlier each year as they grow older, as shown in individually marked Oystercatchers *Haematopus ostralegus*, Grey Plovers *Pluvialis squatarola* and Pied Avocets *Recurvirostra avosetta* (Goss-Custard *et al.* 1982, Evans & Davidson 1990, Hötker 2002). Not all migrating immature shorebirds travel as far as breeding areas, but little is known of their fattening patterns. In one wintering population of Oystercatchers, individuals began to return to breeding areas from about the fourth summer on, but males did not begin to breed until ages 5–8 years and females until 3–6 years (Goss-Custard *et al.* 1982).

Comparing different shorebird species, a higher tendency to over-summer in winter quarters occurs in those populations with the longest migratory flights that winter furthest from their breeding range (Summers *et al.* 1995, Hockey *et al.* 1998). The same is true for some gulls, cormorants and others (Dolbeer 1991). Perhaps the longer journeys are more arduous and hazardous, tipping the balance in favour of staying rather than returning to the breeding range. Such latitudinal trends are apparent within species. In the Western Sandpiper *Calidris maura*, most juveniles wintering in California and western Mexico gain fat and migrate north in their first spring. In contrast, nearly all juvenile Western Sandpipers wintering in Panama do not gain fat or migrate north. Instead they spend their first summer in predominantly winter plumage in non-breeding areas (Nebel *et al.* 2002, O'Hara *et al.* 2005). The difference between these sites was assumed to reflect a latitudinal trend within Western Sandpipers, with the birds showing an increasing propensity to over-summer the further from the breeding areas they

Table 15.2   Holarctic nesting shorebird species that are commonly found in the southern hemisphere during the northern breeding season (austral winter)

	South Africa	Australia	South America
Hudsonian Godwit *Limosa haemastica*			+
Bar-tailed Godwit *Limosa lapponica*	+	+	+
Whimbrel *Numenius phaeopus*	+	+	+
Eurasian Curlew *Numenius arquata*	+		
Far Eastern Curlew *Numenius madagascariensis*		+	
Willet *Catoptrophorus semipalmatus*			+
Common Greenshank *Tringa nebularia*	+	+	
Greater Yellowlegs *Tringa melanoleuca*			+
Lesser Yellowlegs *Tringa flavipes*			+
Marsh Sandpiper *Tringa stagnatilis*	+		
Wood Sandpiper *Tringa glareola*	+		
Terek Sandpiper *Tringa cinerea*	+	+	
Common Sandpiper *Tringa hypoleucos*		+	
Grey-tailed Tattler *Heteroscelus brevipes*		+	
Ruddy Turnstone *Arenaria interpres*	+	+	+
Asian Dowitcher *Limnodromus semipalmatus*		+	
Great Knot *Calidris tenuirostris*		+	
Red Knot *Calidris canutus*		+	
Sanderling *Calidris alba*	+		+
Semi-palmated Sandpiper *Calidris pusilla*			+
Little Stint *Calidris minuta*	+		
Rufous-necked Stint *Calidris ruficollis*		+	
Sharp-tailed Sandpiper *Calidris acuminata*		+	
Curlew Sandpiper *Calidris ferruginea*	+	+	
Ruff *Philomachus pugnax*	+		
Red-necked Phalarope *Phalaropus lobatus*		+	
Grey Plover *Pluvialis squatarola*	+	+	+
Greater Sand Plover *Charadrius leschenaultii*		+	

spent the winter. A similar trend occurs among different wintering populations of Ringed Plovers *Charadrius hiaticula* and others (Evans & Davidson 1990). Over-summering young birds can certainly achieve high survival rates. For example, among Bristle-thighed Curlews *Numenius tahitiensis* that migrate from Alaska to the Hawaiian Islands, the average annual survival of birds after their first arrival in Hawaii exceeded 91% for the first year, 93% for the second year and 97% for the third year, compared with 85% in migrating adults (Marks & Redmond 1996). Clearly, the younger age classes suffered no survival penalty from staying in 'wintering' areas year-round, and may have benefited, but on the samples available, such slight differences did not achieve statistical significance.

The main question about deferred return to breeding areas is whether it occurs because young birds are unable to accumulate the fat reserves necessary for the return journey (owing to lower feeding rates or heavier parasite loads, McNeil *et al.* 1994), or whether deferred return is an inherent feature, evolved because young birds have no chance of breeding and benefit from avoiding the risks of a long return journey to breeding areas. These explanations are not mutually exclusive, and each may apply to different populations. Many studies have shown that young birds are less efficient foragers than older, more experienced ones, and often suffer from their subordinate status in competitive interactions. Young birds may, therefore, have difficulty in accumulating migratory fat, at least in time to breed (Hockey *et al.* 1998). The fact that over-summering in wintering areas seems commoner in populations that winter furthest from their breeding areas (on the southern continents) supports the evolutionary explanation; but the fact that some birds migrate only part way to their breeding areas (still travelling up to several thousand kilometres) supports the food-limitation hypothesis. First-year birds that do not return to breeding areas normally show little or no gonad development, and no spring moult or pre-migratory fat deposition, but remain in drab winter plumage (Chapter 12).

LOCAL DISTRIBUTION PATTERNS

In addition to geographical gradients in the age and sex ratios of wintering migrants, extreme sex and age ratios can be seen at smaller scales, and have again mostly been attributed to dominance relationships (e.g. Marra *et al.* 1993, Catry *et al.* 2004). Differences between the sexes in habitat during the non-breeding season have been described in a wide range of species, from passerines to raptors and ducks (Nisbet & Medway 1972, Nichols & Haramis 1980, Ardia & Bildstein 1997). Among passerines, for example, such differences were studied in the Great Reed Warbler *Acrocephalus orientalis* (Nisbet & Medway 1972), American Redstart *Setophaga ruticilla* (Sherry & Holmes 1989), Hooded Warbler *Wilsonia citrina* (Lynch *et al.* 1985), Black-throated Blue Warbler *Dendroica caerulescens* (Wunderle 1995), Red-backed Shrike *Lanius collurio* (Herremans 1997), Cape May Warbler *Dendroica tigrina* (Latta & Faaborg 2002), Eurasian Robin *Erithacus rubecula* (Catry *et al.* 2004), and several other species (Lopez & Greenberg 1990). Habitat differences might arise from male dominance forcing females into different places (through current competition) or from innate differences between the sexes in habitat selection (evolved through competition in the past).

An attempt to distinguish between these explanations was made with Hooded Warblers *Wilsonia citrina* in Mexico. Morton *et al.* (1987) removed male warblers from their winter territories in primary forest, to find whether they would be replaced by females from the adjacent scrub. In fact, the males were not replaced, and females remained in scrub, despite the presence of vacancies in better habitat nearby. The authors concluded that, in this species, the habitat difference between males and females resulted from inherent preference, rather than current male dominance. This conclusion was later supported by experimental work in which the habitat preferences of each sex were tested in isolation (Morton 1990). In the American Redstart *Setophaga ruticilla*, in contrast, removed males were replaced mainly by females, implying that in this species males had directly excluded females from mutually acceptable wintering habitat (Marra *et al.* 1993).

In species that show sex differences in habitat, it is the larger sex (usually male) that occupies the most food-rich habitats, is usually in best body condition and shows the strongest site fidelity (Wunderle 1995, Latta & Faaborg 2002, Catry *et al.* 2004). These findings suggest that the difference results from competition, whether here and now (leading to a dominance-enforced difference) or in the past (leading to an evolved difference). However, another explanation is possible in some species, namely that, being of different sizes, the sexes are adapted to different habitats. Differences in size may have evolved through sexual selection, but then, as an evolutionary by-product, influence the habitat preferences of males and females. This is apparent in those shorebirds in which one sex has a longer bill than the other, and can probe for food to greater depth. It is then not just a case of subordinates being 'driven' from certain areas, but of their preferring other areas because they can feed more efficiently there. Whatever the origin of the habitat difference, it means that the sexes could have different food supplies, reducing competition between them, but affecting local or overall sex ratios.

At the local level, even among flocking species, good competitors tend to accumulate in the richest feeding sites, and gain the highest rates of food intake, while weak competitors end up in poor sites, with lower rates of intake. This is not necessarily because poor competitors are driven from the richer sites; it may be because, if they attempted to feed there, they would suffer more interference and have lower rates of intake, and for this reason they prefer to feed elsewhere (for Herring Gull *Larus argentatus* see Monaghan 1980, for White-winged Crossbill *Loxia leucoptera* see Benkman 1997). In some species, dominance relationships seem to influence local movements, dominants tending to remain at local food sources and subordinates moving on. To bird-ringers operating at a single site, winter populations in some species can seem to consist of residents and transients. The two categories differ in sex and age ratios and sometimes also in average weights and fat levels (for Siskin *Carduelis spinus* see Senar *et al.* 1990, 1992).

In addition to foraging areas, birds compete over roost sites, which vary in the amount of protection and shelter they provide. Among species that roost communally, the distribution of individuals within a roost can be markedly non-random with respect to age, sex and body condition. Statistically significant segregation of age and sex groups within winter roosts has been noted in many bird species, including Brambling *Fringilla montifringilla* (Jenni 1993), Red-winged Blackbird *Agelaius phoeniceus* (Weatherhead & Hoysak 1984), Starling *Sturnus vulgaris* (Feare *et al.* 1995) and Dunlin *Calidris alpina* (Ruiz *et al.* 1989). In each species, adults predominated in the safer and more sheltered sites and the juveniles in the more exposed sites. Such local observations give some inkling of how larger-scale distributional differences between sex and age groups might arise, leading to gradients in sex and age ratios across a wintering range.

OTHER DIFFERENCES BETWEEN SEX AND AGE GROUPS

Males arrive earlier than females in breeding areas either because they winter closer to the breeding areas, begin their return migration earlier, travel faster, or show any combination of these features. In many passerines and shorebirds, in which the sexes winter side by side in the same area, it has been observed that the males fatten and leave first (Rogers & Odum 1966, Nisbet & Medway 1972, Cramp 1988, Bishop *et al.* 2004, Catry *et al.* 2005), and on stopover sites that males often fatten more rapidly and stay for shorter periods than females, thus making more rapid progress (Chapter 27). These differences may also be due to the greater dominance of males, widening the sex differences in migration timing as birds travel towards their destinations. Greater fat levels in males may allow them to cross wider stretches of sea or other hostile habitat, as suspected in the Goldcrest *Regulus regulus* (J. H. Marchant, in Wernham *et al.* 2002).

Age groups, in particular, may differ in other aspects of migration behaviour. Juveniles often take longer to accumulate migratory fuel than adults, migrate with lower fuel loads, stay longer at stopover sites, and take longer to complete their journeys (Chapter 27). Such differences have been attributed partly to the lower experience and social status of juveniles, compared with adults. When they leave their breeding areas, juveniles also show more spread in departure directions than adults (Chapter 10), and the two age groups sometimes migrate along partly different routes, with juvenile passerines and others more concentrated in coastal areas (e.g. Murray 1966, Ralph 1971). This could result from the greater drift of juveniles by wind and their greater concentration near the edges of a migration route (Woodrey 2000).

A remarkable difference in migration routes between juvenile and adult European Honey Buzzards *Pernis apivorus* was found by following radio-tagged individuals from Sweden to West Africa (Håke *et al.* 2003). All tagged adults crossed the Mediterranean Sea at Gibraltar and continued across the Sahara Desert to their wintering areas. Analysing three main steps of the migration: (1) from the breeding site to the southern Mediterranean region; (2) across the Sahara; and (3) from the southern Sahara to the wintering sites, the adults changed direction significantly between these steps, and took a large detour via Gibraltar. In contrast, the juveniles travelled in more southerly directions, crossed the Mediterranean Sea at various places, and ended up in the same wintering range as adults. Average speeds maintained on travelling days were similar in the two age groups, about 170 km per day in Europe, 270 km per day across the Sahara and 125 km per day in Africa south of the Sahara. However, as the adults had fewer stopover days en route, they maintained higher overall speeds and completed migration in a shorter time (mean 42 days) than the juveniles (mean 64 days). Although the juveniles set off on more direct courses towards the wintering areas, they did not cover shorter distances than the adults, as they tended to show a larger directional scatter between shorter flight segments, as expected if they were more severely drifted by wind.

Age differences in migration routes and wintering areas may not be unusual in raptors. At watch sites that get most of their birds as a result of wind drift, the majority of birds seen in autumn are juveniles (95% at Cape May Point in New Jersey), much greater than at other sites (50% juveniles at Hawk Mountain in Pennsylvania) (Chapter 10; Allen *et al.* 1996). In many raptor species, juveniles have somewhat longer flight and tail feathers than adults, an unusual situation which affects wing shape, and may adapt the two age groups to different flight modes. Still, however, juveniles are more likely to be blown off course than are adults (Kerlinger 1989, Hoffman & Darrow 1992).

GENETIC CONTROL OF SEX AND AGE DIFFERENCES

As indicated above, dominance and other factors could influence movements through direct effects on individuals at the time of migration, or through evolution, by providing a consistent selection pressure for the subordinate sex to migrate furthest, at a different time, or to occupy a different type of habitat, so that such features become innate and under genetic influence. Hence, dominance could drive differential migration at either a mechanistic (proximate) or evolutionary (ultimate) level, or both. The one could be a precursor to the other. Research on captive birds has provided evidence for genetic control of sex differences in migration behaviour in at least three species, namely the Dark-eyed Junco *Junco hyemalis*, White-throated Sparrow *Zonotrichia albicollis* and Blackcap *Sylvia atricapilla* (Chapter 20). This matched the situation in the wild, and suggested an inherent difference between the sexes, possibly arising from different selection pressures on each sex.

CONCLUDING REMARKS

Overall, birds show a range of different migration patterns, which are associated with mating and parental systems, annual cycle features and body size dominance relationships, all of which may modify the timing and distance of migration within the constraints imposed by environmental conditions. Where species in the same family or genus differ in mating and parental systems, they also show corresponding differences in migration patterns. The fact that similar patterns recur in unrelated families emphasises the role of mating and parental systems in influencing migration. Differences in average migration dates between the sexes vary from a matter of days in some species to weeks in others, depending in autumn on the relative roles of the two sexes in parental care and in spring on the time between arrival and nesting.

Apart from the relationships with patterns of parental care and moult, no unambiguous evidence favours one explanatory hypothesis on migratory timing over others, and almost certainly different explanations apply in different species. The body size, dominance and arrival time hypotheses are not mutually exclusive, and each is hard to test independently of the others. This is because in many bird species, adult males are the largest individuals, are socially dominant and return to breeding areas first. In fact, dominance and body size are so closely correlated in birds that separating their importance in differential migration is practically impossible (Cristol *et al.* 1999). Unless birds arrive in breeding areas already paired, almost all the studies known to me show sex differences in the mean arrival dates of males and females, but many of the same species show no obvious differences in autumn migration dates.

On the dominance hypothesis, as usually interpreted, dominance acts here and now to influence bird behaviour, which thus depends on conditions at the time. In many species, dominance relationships supposedly lead subordinates to migrate furthest: females further than males and juveniles further than adults. Such effects are most evident in short-distance, partial and irruptive migrants (Chapters 12 and 18), in which travel distances also vary from year to year according to prevailing food supplies. In species in which genetic factors have most influence on migration, and in which all individuals migrate every year, dominance effects on migration are much less apparent. In such birds, evolved migration behaviour is likely to depend more on past conditions, which might include a range of factors, and not just dominance relationships. For example, the travel itself may have costs which increase with the length of journey, thus leading to selection for shorter movements; and if these travel costs are greater in young than older birds, selection could in turn result in young birds performing shorter journeys than older ones (as in Dark-eyed Juncos *Junco hyemalis* and some other seed-eaters, see above). With genetic factors (rather than prevailing conditions) having the major influence on migration distances, each sex/age class would be expected to winter each year in whichever regions offered the best prospects for survival and future reproduction, rather than where they were pushed by competition. It is unlikely that the optimal wintering area, resulting from the balance of these various selective forces, would coincide exactly for each sex/age class of a species, considering their differences in morphology, behaviour and reproductive roles. Understanding differential migration then becomes an 'optimality' problem to be solved separately for each sex and age class, with competition as only one of several factors involved. There is clearly scope here for further research.

On the face of it, it may seem strange that dominance relationships can have different effects at different seasons. For example, among partial and short-distance migrants, juveniles leave the breeding areas before adults and often migrate further, whereas in practically all migrants studied in spring, adults leave their wintering areas and arrive on breeding areas before juveniles. At each season, however, it is the dominant adults (especially males) that are able to pursue the optimal strategy and other individuals are predisposed by competition to behave otherwise. In autumn, the optimal strategy among partial migrants is to stay in breeding areas as long as possible, preferably all winter, for resident individuals generally survive better than those that migrate, and also retain the best territories for breeding (Chapter 20). In spring, the optimal strategy is to be among the first birds to return to breeding areas, for those individuals can then get the best of the remaining territories.

In addition to the aspects of migration discussed in this chapter, sex and age differences are also apparent in site-fidelity – in the tendency of individuals to return to the same breeding and wintering sites from year to year. In general, males tend to show greater site-fidelity than females, and in both sexes site-fidelity tends to increase with age (Chapter 17). These and other types of bird movements would benefit from more research on the relationships between the dominance status, body condition and behaviour of individuals.

SUMMARY

In many bird species, sex and age differences occur in the proportions of individuals that migrate, in the timing of outward and return movements, and in the distances travelled, leading to geographical gradients in sex and age ratios in the non-breeding season. Among passerines, short-distance partial migrants migrate after moulting in breeding areas. Typically, females migrate in greater proportion, depart earlier, travel further from their nesting areas, and return later than males; and juveniles migrate in greater proportion, depart earlier, travel greater distances, and return later than adults. Many species of obligate long-distance migrants leave after breeding, and suspend or delay moult until after arrival in a staging area or winter quarters. Typically, in such species, males leave before females, and adults before juveniles. Sex differences can be interpreted in terms of the different roles of the sexes in breeding, and both sex and age differences partly in terms of dominance and competition within populations, duration and extent of moult, and perhaps also by other (as yet unknown) factors. Many exceptions to these general patterns are associated with the particular life history characteristics and circumstances of the populations concerned. Thus, in species in which only one partner looks after the young, the other leaves breeding areas earlier (males in ducks, and in some shorebirds, females in others). Typically, in these species, adults leave breeding areas earlier than young of the year.

In most bird species, males arrive in breeding areas, on average, earlier than females, the difference varying in different species from a few days to a few weeks. The reverse sequence holds in species in which females rather than males compete for territories and mates. Within each sex, young birds arrive later than older ones, on average.

Sex and age differences in migration distances may result from current competition or from past conditions and their effects on genetic control mechanisms. In species in which individuals do not breed until they are several years of age, the immatures typically spend one or more year in their wintering areas, or migrate only part way towards the breeding areas. They also migrate later in spring than older birds, and depart earlier in autumn, their timing getting more like that of adults with increasing age. Winter distributional differences between sex and age groups are apparent: (1) on a geographical scale across a wintering range; (2) on a local scale, among particular habitats or localities; and (3) on an even smaller scale, within particular feeding or roosting flocks.

Shelduck *Tadorna tadorna* on moult migration

Chapter 16
Variations on a migratory theme

Bird migration as a world phenomenon has existed as long as birds have flown ... it is the specific circumstances that are so fascinating. (R. E. Moreau 1972.)

Variants occur around the basic outward–return movement between fixed breeding and wintering ranges, each in a limited number of species. They include: (1) moult migration; (2) long movements within the breeding season; (3) long movements within the non-breeding season, including irruptions and weather movements; and (4) nomadism. These variants can all be related to the particular ecological circumstances faced by the species concerned. They are all described in this chapter, but irruptions are discussed in greater depth in Chapters 18 and 19, because of the further light they throw on the role of food supplies in bird movements.

MOULT MIGRATIONS

In some waterfowl species, many individuals perform a 'moult migration' each summer, when they travel long distances from their nesting places to assemble in large numbers at traditional sites that offer food and safety. Here they pass the flightless period, replacing all their large wing feathers simultaneously within the space of a few weeks. Body moult begins before flight feather moult and continues afterwards. For habitat-related reasons, moulting sites typically hold far fewer local breeders than the summer food supply will support, leaving a big surplus for the incomers. Among geese, moult migrations are undertaken entirely by non-breeders and failed breeders, whose moulting areas nearly always lie at higher latitudes than the nesting places, from a few tens to more than 1000 km away (**Figure 16.1**). For example, some Canada Geese *Branta canadensis* from mid and west continental North America undertake a moult migration of at least 1000 km to the Thelon River area of the Northwest Territories (Sterling & Dzubin 1967), while those breeding in northwest Wisconsin scatter northward for up to 1300 km into northern Manitoba (Zicus 1981). By leaving breeding areas, such birds avoid competing with families which stay there, and by moving to higher latitudes where plant growth begins later than in the breeding areas, the birds gain from better protein-rich food, as well as from longer days. After moult, the birds migrate to their wintering areas, perhaps stopping for a time in breeding areas en route south.

Similar moult migrations occur in some duck species, but are more variable in direction than those of geese. The majority of Common Shelducks *Tadorna tadorna* from Europe gather each summer on the vast tidal mudflats of the Grosser Knechtsand in the German Wadden Sea, where they feed on the abundant mudsnail *Hydrobia ulva*. The birds converge on this site from all directions, travelling up to several hundred kilometres from their breeding areas, their numbers peaking at more than 200 000 individuals. Yearlings and young adults arrive first, followed by failed breeders, then successful adults, which leave their well-grown young behind (Patterson 1982). Smaller concentrations assemble at various other sites within the breeding range. After moult the birds drift back to their breeding areas over a period of weeks or months or move on to wintering areas, depending on whether their particular breeding areas are habitable in winter.

In various diving and dabbling ducks, the males, together with non-breeding or failed females, also perform moult migrations, leaving the successful females to hatch and tend their broods. Once their young are grown, some of the adult females perform a moult migration, being last to arrive in moulting areas, while others remain in their breeding areas and moult their flight feathers there.

Some moult migrations are not trivial. In some duck species, individuals have been found by ringing to travel distances of more than 3000 km from breeding to moulting sites and some of the biggest concentrations hold tens of thousands or hundreds of thousands of birds. In general, sea-ducks perform longer journeys than freshwater ducks. Usually birds fly quickly and directly from breeding to moulting sites, and even sea-ducks which normally stick to the coast, sometimes fly long distances overland to take the shorter routes (Salomonsen 1968). A particularly big arrival of birds on moult migration was noted at Mud Lake, Idaho, where during one night (5 August) no less than 52 000 dabbling ducks of six species suddenly appeared (Oring 1964). Examples of some of the more impressive moult migrations include the following:

King Eiders *Somateria spectabilis* from most of eastern arctic Canada travel to coastal areas in mid-western Greenland to moult. By that time the participants have spent as little as three weeks on their breeding areas, and they travel eastward up to 2500 km, forming a concentration of more than 100 000 birds. After moult, the birds migrate to wintering areas off southwest Greenland where they are joined by the females and young (Salomonsen 1968). Birds from the western part of the

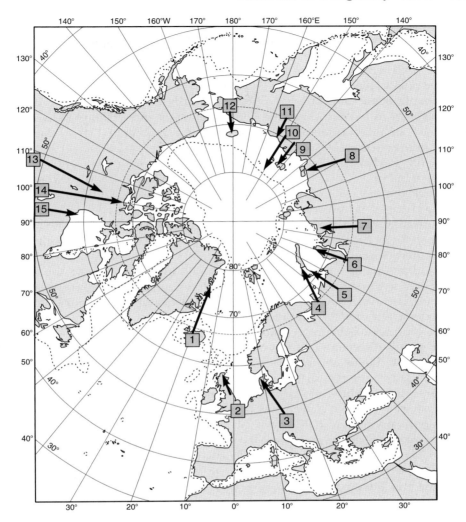

Figure 16.1 Moult migrations of geese. The arrows show direction and distance from origin to destination. Greylag Goose *Anser anser* (3), Bean Goose *A. fabalis* (5–9, 11), Pink-footed Goose *A. brachyrhynchus* (1), Greater White-fronted Goose *A. albifrons* (5–9, 11), Canada Goose *Branta canadensis* (2, 13–15) and Brent Goose *B. bernicla* (4, 9, 10, 12). From Salomonsen (1968).

North American range migrate westward to concentrate off eastern Siberia and in Disco Bay, reaching up to 200 000 individuals.

Long-tailed Ducks *Clangula clangula* in eastern Siberia perform an extensive migration across the Arctic Ocean to Wrangel Island, north of the breeding range, where they gather in tens of thousands. Another large concentration occurs near Thule in northwest Greenland, again attracting birds mainly from further south (Salomonsen 1968).

Goosanders *Mergus merganser* from much of northwest Europe migrate from their breeding areas to North Cape in northernmost Norway, forming a concentration of about 35 000 birds (mostly males), from which they later move southward to their wintering areas. Drakes are virtually absent from breeding areas from June to October (B. Little & J. H. Marchant, in Wernham *et al.* 2002).

Common Scoters *Melanitta nigra* from the northeast migrate each summer to the west coast of Jutland, Denmark, where they reach more than 150 000 individuals. After moulting, some birds remain there, while others move on elsewhere, but not necessarily as soon as moult has finished (Salomonsen 1968).

Common Pochards *Aythya ferina* assemble each summer on various waters in the mid-latitudes of Europe, including the IJsselmeer in the Netherlands (50 000 individuals) and Ismaninger Reservoir in Bavaria (20 000 individuals). Collectively, they far exceed the European population, and contain many birds from further east. Initially males form nearly 100% of both populations but, as the season wears on, females gradually increase to nearly 50% by late September. Many Tufted Ducks *A. fuligula* and other species use the same sites, with similar seasonal changes in sex ratios (van der Wal & Zomerdijk 1979).

It takes individual ducks 3–5 weeks to regrow their flight feathers, depending on species, shelducks and geese 4–6 weeks, and swans about six weeks. However, some individuals arrive in moulting areas weeks before starting their wing moult, or leave long after finishing it, providing that food remains available. Typically, flocks arrive and leave over a several-week period, so that sites can be occupied by moulting birds for a total of 2–4 months. For example, Barrow's Goldeneyes *Bucephala islandica* in eastern Canada travel north from their nesting areas around 1000 km to arctic waters, and spend 3–4 months there, longer than they spend in nesting areas, and much longer than the 31-day flightless period associated with wing-moult (Robert *et al.* 2002).

While many species of waterfowl throughout the world gather on safe sites to moult, it is mainly species in boreal or arctic latitudes that make the longest migrations, often to sites outside the breeding range. While some birds migrate north of their breeding areas to moult, others move south of their breeding areas to moult at staging or wintering sites. For some species, therefore, the moult migration could be regarded as an extra stage of the northward spring migration or as the first stage of the southward migration, involving the same directional preferences. But for other species, the moult migration takes them in a quite different direction, sometimes at right angles to the normal migration axis. Moreover, because sites within the breeding range can attract birds from all directions, there is clearly no directional consistency between individuals. In addition, birds can periodically change their moulting sites in response to changes in water depths, or as established sites are destroyed, and new ones created, by human action. Given these facts, it seems unlikely that birds from different parts of the breeding range inherit different directional preferences for moult migration, especially considering that individual waterfowl can sometimes breed in areas hundreds of kilometres from where they were raised, or in widely separated places in different years (Chapter 17). Rather they could learn the locations of favoured sites from other individuals, this knowledge being passed on by tradition. However, not all birds in a particular region participate in a moult migration, some males remaining to moult in their breeding areas, along with many females and young.

Other wetland birds also gather in large concentrations to moult, and in some species lengthy migrations are involved. This is true of both European and American Coots (*Fulica atra* and *F. americana*) and of various grebes, all of which become flightless for a time. In western North America, most Eared (Black-necked) Grebes *Podiceps nigricollis* from prairie breeding areas migrate to moult on the Great Salt Lake in Utah or on Mono Lake in California, which together hold more than two million birds in late summer, supported by the masses of brine shrimps. From there, after moulting, the birds move on to winter on the sea off California (Chapter 5; Storer & Jehl 1985, Jehl 1997). In Europe, up to 186 000 Black-necked Grebes have been counted on Burdur Göla in Turkey (Hagemeir & Blair 1997), and smaller numbers elsewhere. Also, some 10 000–40 000 post-breeding adult Great-Crested Grebes *Podiceps cristatus* assemble to moult each year on the Dutch IJsselmeer (Piersma 1987). This latter site holds many thousands of waterbirds during the moult period, including the Common Pochards *Aythya ferina* mentioned above. Because of moult migration, many waterbirds occupy at least three main areas each year – for breeding, moulting and wintering; and from time to time some also make long movements within their wintering range (see later).

In all these various waterbirds, moult migration thus differs from typical migration in several respects. First, the flight direction is often different from autumn or spring migration, and may vary by up to 360° among birds converging on a single site from different parts of the breeding range. Second, only certain sectors of the population participate; goose pairs with young stay behind, as do some female dabbling and diving ducks with broods. Third, the localised and unusually high population densities found in the moulting areas are quite unique in some species, whereas in breeding and wintering areas the birds scatter over much wider areas. As in some other migrations, however, individuals follow a rigid schedule, with the timing of moult migration varying only slightly from year to year, according to the timing and success of breeding. Some species travel such long distances that they must presumably accumulate body reserves to fuel the journey, although I know of no published information on this.

The adults of many shorebird species also travel considerable distances to specific places to moult. These sites are usually at latitudes to the south of the breeding areas, so they act as staging areas on southward migration, or as wintering sites. Often among the birds from a single population, some individuals may remain after moult throughout the winter, while other individuals move on further south (as recorded, for example, in the Green Sandpiper *Tringa ochropus*, Bar-tailed Godwit *Limosa lapponica* and Sanderling *Calidris alba* in Britain, Wernham et al. 2002). Moulting flocks of shorebirds can consist of a few tens, hundreds or thousands of individuals, but at some sites much larger numbers occur. For example, several hundred thousand Wilson's Phalaropes *Phalaropus tricolor* gather at saline lakes in western North America, where they replace only the first 3–4 primaries along with the tail and nearly all the body plumage. The moult migration involves only post-breeding adults, with females arriving by mid-June and males a fortnight later (Jehl 1996). Individuals remain for up to six weeks, then double their weight before departing on an apparent non-stop flight of 4800 km to South America. Although most shorebirds do not become flightless during moult (except for the Bristle-thighed Curlew *Numenius tahitiensis* on Hawaii), the same factors as in waterfowl are likely to influence their choice of site: abundant food over a long period and relative security from disturbance and predation.

What could be construed as moult migrations have also been reported in divers, flamingos, cranes, terns and auks, all of which may gather to moult at places away from their nesting or wintering areas (Jehl 1990, Cherubini *et al.* 1996). For example, in the Little Auk *Alle alle*, non-breeders leave Greenland colonies in late July or August, in advance of breeding adults, and move northward to moult in rich feeding areas along the edge of the pack-ice. The main difference between this movement and that shown by many waterfowl is that the moulting locality varies from year to year, according to ice conditions (Jehl 1990).

Some passerines also perform what has been called a moult migration, but in effect they moult at stopover sites on autumn migration. Many Eurasian–Afrotropical passerine migrants break their autumn journeys in the northern tropics for several weeks and moult, later continuing their migrations to localities further south in the northern tropics or across the equator to the southern tropics (Chapter 24). In western North America, adult Western Tanagers *Piranga ludoviciana* and other passerines migrate to the southwest of the continent after breeding, moult there during the late summer rains, and then continue their journey southward to winter quarters. In contrast, juvenile Western Tanagers move to montane habitats near their natal areas to moult their body feathers before migrating (Butler *et al.* 2002). In northeast Europe, some Common Starlings *Sturnus vulgaris* migrate immediately after breeding either before their moult begins or in the early stages. Ring recoveries suggest a movement of 300–1200 km. The second phase of migration occurs in autumn after moult has finished. The two phases of movement are indistinguishable in direction, altitude of flight and temporal pattern, and both include some nocturnal flights (Kosarev 1999). Juveniles participate mainly in the first period, moulting in their migration areas, and adults mainly in the second period, after moulting in their breeding areas. The post-breeding migration of the whole population therefore occurs as two well separated waves (**Figure 16.2**).

These various moult migrations could be viewed either as an extra post-breeding movement in the annual cycle of the species involved, which may otherwise be resident or migratory, or in some species as a part of the autumn

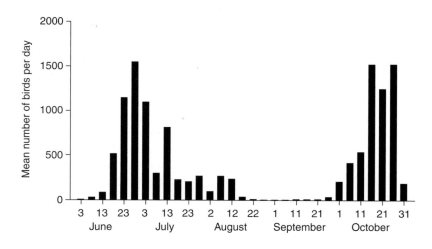

Figure 16.2 Autumn migration of Starlings *Sturnus vulgaris* through Ottenby in south Sweden. The first peak consists mainly of juveniles before moult and the second peak mainly of adults after moult. From Svärdson (1953).

migration, involving a several-week stop in a favourable area en route to winter quarters.

MOVEMENTS WITHIN THE BREEDING SEASON

When they are breeding, most bird species tend to remain resident in an area throughout the season, so that the whole breeding range is occupied in a consistent manner from spring arrival to autumn departure. There are, however, exceptions to this general pattern, in that individuals of some multi-brooded bird species nest in more than one locality each year. Typically, they raise one brood in one locality, then migrate up to several hundred kilometres, and raise another brood elsewhere. So-called itinerant breeding was first suspected in the European Quail *Coturnix coturnix*, in which females arriving in Italy in June–July, apparently to breed, often showed regressing brood patches from an earlier breeding attempt, and were frequently accompanied by young no more than two months old. These young must have hatched from clutches begun in March, a date at which breeding would have been possible only in North Africa. Reports of a general exodus of Quail from Tunisia in spring after breeding support this view, and single birds ringed there in May and early June were recovered in Italy and Albania respectively, 2–3 months later (Moreau 1951, Cramp & Simmons 1980). Late clutches in northern Europe during August–September could result partly from an influx in mid-summer of birds that have previously bred in the Mediterranean region. Apparently the males leave southern Europe once the females have laid, and move on to establish new territories further north, where females join them later. Many females do not reach the most northern areas, so males predominate there (evidence summarised in Aebischer & Potts 1994). European Quail can mature and breed at three months old, so that young produced in the southern parts of the breeding range could breed the same summer in the more northern parts, along with the adults on their second or third attempt. So far, however, they have not been proved to do so.

Itinerant breeding is also shown by the Red-billed Quelea *Quelea quelea*, which feeds on grass seeds on the African savannahs. As the rain belts spread across tropical Africa, they stimulate the growth and seeding of grasses, providing a sufficient temporary food supply to enable Queleas to breed. After raising their young, the birds move en masse, stopping again in an area where rain has recently fallen, and raise another brood (Ward 1971, Jaeger *et al.* 1986). Throughout the dry season, these birds subsist on dry seeds picked off the ground but, when the rains break, this seed suddenly germinates, thus removing the food supply, and causing the birds to move on (called the early rains migration). The birds then fly over the approaching rain front to areas where rain fell about two months earlier and new grass seed has already formed (the 'breeding migration'). This new seed, and insects associated with the growing vegetation, enable the birds to breed again. In theory, the birds could follow the rains in successive breeding attempts, each time moving some hundreds of kilometres. Within this framework, the pattern is variable from year to year, depending on regional variations in rainfall and grass seed production. Conditions suitable for rearing young do not last long in any one place, and despite a short breeding cycle of only five weeks, queleas cannot raise two broods in the same place. The adults abandon their young at about three weeks old, with enough body fat to ensure their survival, and continue on the breeding migration. Only by remaining within the slowly shifting zone of seeding grasses are individual queleas able to raise more than one brood per year.

Evidence for this process in Red-billed Quelea is largely circumstantial: (1) many females begin developing the yolks for a second clutch while still feeding the first brood, yet they do not breed again in the vicinity; and (2) in those regions where colonies contain birds likely to be breeding for the second time, adults arrive in badly worn plumage, consistent with a previous breeding attempt. However, at three colonies in Ethiopia birds were sprayed from the air with fluorescent particles, and 2–3 months later some were re-sighted at other colonies 500 and 700 km to the north (Jaeger *et al.* 1986). In West and southern Africa, there is time in wet years for up to three broods to be raised in different places along a 'breeding migration', and in East Africa up to five, but it is unknown whether these maxima are ever reached.

Some other birds in semi-arid regions of the world, which breed at a particular stage in the dry–wet seasonal cycle, may also raise successive broods in places far apart. Multiple breeding along a migration route, following a rain belt, has been suspected in the Eared Dove *Zenaida auriculata* in northeastern Brazil (Bucher 1982), and substantial shifts in colony sites during a single season have been suspected in the White-crowned Pigeon *Columba leucocephala* on Hispaniola (Arendt *et al.* 1979), in the Tricolored Blackbird *Agelaius tricolor* in California (Hamilton 1998), and in the Spanish Sparrow *Passer hispaniolensis* in Kazakhstan and elsewhere (Summers-Smith 1988, Cramp & Perrins 1994).

Individuals of some species seem to move around continually during a breeding season. For example, in a semi-arid area of southeast Australia, membership of three permanent nesting colonies of Zebra Finches *Taeniopygia guttata* changed continually, due to the frequent arrival and departure of birds from distant colonies (Zann & Runciman 1994). Some 66% of adults stayed for no more than one month and many that stayed longer also left for extended periods. Another Australian species, the Regent Honeyeater *Xanthomyza phrygia*, has been found to move more than 20 km between successive broods in the same season (maximum recorded distance 260 km), as pairs tracked the occurrence of rich food sources, which appeared sequentially through the season (Geering & French 1998).

Other European birds known sometimes to move long distances between successive breeding attempts include some cardueline finches, such as Common Redpoll *Carduelis flammea* and Siskin *Carduelis spinus* that depend on different types of tree seeds (Chapter 18). Other carduelines have been found to move shorter distances (up to a few tens of kilometres) during a breeding season (Newton 2001). Also, over shorter distances, some Sand Martins *Riparia riparia* have been found to change colonies between successive broods in the same season (Mead 1979b). In North America, the raising of two or more broods, each in a different region, has been suspected in the Phainopepla *Phainopepla nitens*, Dickcissel *Spiza americana* and Sedge Wren *Cistothorus platensis* (Walsberg 1978, Fretwell 1980, Bedell 1996). In addition, females of some shorebirds also move on to breed elsewhere after laying a clutch of eggs, leaving incubation and chick care to their mates. Examples include Eurasian Dotterel *Eudromias morinellus* (in which some males also move long distances during a breeding season, Whitfield, in Wernham 2002), and probably also the North American Snowy (Kentish) Plover *Charadrius alexandrinus*.

Itinerant breeding enables species that could raise only one brood in a given locality to become multi-brooded. Pre-requisites are a short breeding cycle with early independence of young. Species with long breeding cycles, with prolonged parental care, would not be expected to become itinerant breeders in seasonal environments, because they would not have time to raise any more than one brood in the time available. The above examples seem to fall into two categories: one involving local movements of up to a few tens of kilometres, and the other longer movements of up to several hundred kilometres between different points on an established migration route. Both types could occur in a much wider range of species than recorded, but could be detected only if newly arrived adults had brood patches, recently fledged young or other signs of recent reproduction, and then proceeded with a new nesting attempt. It is clearly rare or non-existent in the vast majority of well-studied European and North American migratory birds.

It might be thought that many species migrating north through Eurasia or North America in spring could raise more broods by stopping at different latitudes en route than by breeding only for a short season in the north, where only one brood can be fitted into the time available. The prior occupation of more southern habitat by conspecifics may be the main factor preventing this. Moreover, the southern habitat usually remains suitable after the first broods for subsequent ones (Chapter 13). This is not true for known itinerant multi-brooded species, such as queleas *Quelea* spp., which settle in areas that have only just become suitable, and hence where there is no established prior population, and which are forced to move on if they are to breed again in the same season because local conditions soon deteriorate. A remarkable feature of some of these examples is that individuals move so far between successive breeding attempts that the movement can be fairly described as a migration, inserted within a breeding season. In queleas, the 'early rains migration' is preceded by fat deposition, the amount varying between three populations, according to the distances they have to travel (Ward & Jones 1977).

The pre-laying exodus of petrels

A different kind of 'within-season' movement is undertaken by some pelagic seabirds. Once they have returned to breeding areas, re-occupied nest-sites and re-established pair bonds, many procellariiform species leave their breeding areas for periods of days or weeks to feed up for egg production and incubation. In the process, they may travel to foraging areas hundreds or thousands of kilometres from the nesting colonies. For example, Manx Shearwaters *Puffinus puffinus* from Skokholm Island off Wales fly southwest to the Bay of Biscay off Spain during this two-week period, a major foraging area some 700 km from the colony (Perrins & Brooke 1976), while White-chinned Petrels *Procellaria aequinoctialis* nesting on South Georgia fly 2000 km northwest to the Patagonian Shelf off central Argentina, which is also a major wintering area for the same birds (Phillips *et al.* 2006). We must assume that any body reserves brought to the colony by birds on first arrival have largely gone by the time of egg-laying, making it necessary to accumulate fresh reserves then, and that it is more economic to fly long distances to rich feeding areas than to attempt to accumulate reserves in the vicinity of the colonies at that time. Both sexes need reserves to get back to the colony; the female also to produce the single large egg, and the male to undertake the first major incubation stint.

During the rest of the breeding cycle, birds tend to feed closer to the colony. This is shown mainly by the duration of their return foraging trips and in some species by satellite-based radio-tracking (for White-chinned Petrel *Procellaria aequinoctialis* see Phillips *et al.* 2006).

In some procellariiform species, both sexes participate in the pre-laying exodus, travelling independently of one another, but females are often away for longer than males. In other species, the journeys are made mainly or only by the females. The period away varies from a few days to several weeks, depending on the species and the distances involved (Warham 1990). The females apparently form the egg while away and usually lay within a day or two after their return, whether the male is back or not. This confirms that fertilisation occurred in the pre-exodus period, up to several weeks earlier, the sperm being kept in special utero-vaginal storage glands. The birds from different nesting sites may head for different feeding areas, which can lie in various directions from the colony. Those feeding areas that have been identified lie in a customary migration or wintering area. Almost certainly, then, these rich feeding areas are learnt from previous wanderings. Nevertheless, it is amazing that birds may return from spring migration, remain a few weeks in their nesting places, and then migrate back to staging or wintering areas, before returning to lay an egg and proceed with the nesting attempt. Although widely known among petrels, a prolonged pre-laying exodus may occur chiefly among individuals nesting in areas remote from rich feeding grounds. Only later in the summer may food become abundant locally.

MOVEMENTS WITHIN THE NON-BREEDING SEASON

Whereas the individuals of some species migrate only between a single fixed breeding area and a single fixed wintering area, individuals of other species make substantial movements during the course of a non-breeding season, often extending ever further from their breeding areas as the season progresses. In some species such movements are regular, somewhat analogous to different stages in a single migration, but the birds stay in the same place for up to several weeks between each move. The separate moves occur at about the same dates each year, and each move is preceded by fat deposition. Many species that breed in Eurasia and winter in Africa perform such two-or-more-stage migrations, pausing for a time in the Sahel zone (where some of them moult, as described above), and then move on within the northern tropics or beyond the equator to the southern tropics (Jones 1995; Chapter 24). This latter movement brings the birds into the austral summer at a time when food is at its most plentiful.

Other species move between three areas each year that are not on the same direct route. As found by radio-tracking, Prairie Falcons *Falco mexicanus* leave their breeding areas in southwest Idaho between late June and mid-July (Steenhof *et al.* 2005). They migrate northeast across the continental divide to spend the rest of the summer in Montana, Alberta, Saskatchewan and the Dakotas. They stay there for 1–4 months, and then in October move southwards around 1000 km to the southern

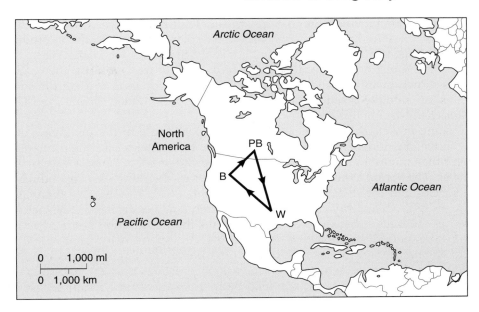

Figure 16.3   Three-part migration of Prairie Falcons *Falco mexicanus*, as revealed by the satellite-tracking of radio-tagged birds nesting in southern Idaho. B, breeding area; PB, post-breeding area; W, wintering area. After breeding, the birds move northeast to spend the summer in cooler areas at higher latitude or elevation where ground squirrels remain available, and then after 1–4 months, they move southward to winter mainly in northern Texas, returning in spring directly to their breeding areas. Based on Steenhoff *et al.* (2005).

Great Plains, mainly in Texas. After five months there, they return on a direct course for their breeding areas (**Figure 16.3**). This three-step migration enables the falcons to exploit seasonally abundant food supplies in each region. Their absence from breeding areas at the hottest time of year coincides with the period when their main prey, ground squirrels, are underground and unavailable. But in the cooler areas occupied in late summer, ground squirrels remain active until winter. In their wintering areas, Prairie Falcons feed largely on migratory birds, especially wintering Horned Larks *Eremophila alpestris*. In addition, Ferruginous Hawks *Buteo regalis*, which also eat ground squirrels, perform a similar clockwise movement, first eastward and northward to higher latitude or higher altitude post-breeding areas, and then southward to wintering areas (Schueck *et al.* 1998).

Several species of hummingbirds also move through three or more different regions each year (Grinnel & Miller 1944, Stiles 1973). For example, Anna's Hummingbird *Calypte anna* breeds in spring in the coastal chaparral of southern California, summers in the high mountains of California, and winters in the deserts of Arizona and Mexico, thereby ensuring that its need for nectar is met year-round. Movements following flowering patterns have been described also in other nectar-eating birds, including the sunbirds of Africa and Asia and the honeyeaters of Australia.

Facultative movements in relation to food supply

In many species, movements in the non-breeding season are optional (facultative). They can take place at almost any time in the non-breeding season if feeding becomes difficult. In some northern species, the proportions of birds that leave the breeding range, and the distances they travel, vary greatly from year to year (Chapter 18). Most individuals stay in the north in years when food is plentiful there, wintering in, or just south of, their breeding areas, but moving further south in years when food is scarce. Such annual variation in migration is most pronounced in finches and other birds that depend on fluctuating tree-seed crops (such as Common Redpolls *Carduelis flammea* and Bohemian Waxwings *Bombycilla garrulus*), and in raptors and owls that depend on fluctuating (cyclic) prey species (such as rodent-eating Rough-legged Buzzards *Buteo lagopus* and Snowy Owls *Nyctea scandiaca*) (Chapters 18 and 19, **Figure 16.4**). Their so-called invasions or irruptions, in which every few years they appear in large numbers well outside their usual range, follow periodic widespread crop failures (finches) or crashes in prey populations (raptors). Irruptions therefore occur in response to annual, as well as to seasonal, reductions in food supplies. In some such years the birds can, in effect, be on passage for much of each winter, as they move from one area of temporary abundance to another, and reach the most distant parts of their wintering range only in the late winters of extreme years.

Although irruptive migrants provide extreme examples, year-to-year distributional changes dependent on flexible migration patterns occur in a wide range of birds, at least in the non-breeding season. Many bird species can migrate at any date in the non-breeding season, if stimulated to do so by reduced food supplies, and usually they move further along the regular migration route (Chapter 18). One species that shows this behaviour is the Yellow-rumped Warbler *Dendroica coronata* of the New World (Terrill & Ohmart 1984). In autumn the densities of this warbler at a number of sites in Arizona and northern Mexico were correlated

(a) Snowy Owl

(b) Common Redpoll

Figure 16.4   North American winter ranges of (a) Snowy Owl *Nyctea scandiaca* and (b) Common Redpoll *Carduelis flammea* that winter mainly at high latitudes (shaded area), but in years of food shortage extend far to the south (dotted line).

with local food supplies. Subsequent changes in the abundance of both insects and warblers at particular sites were influenced by weather, with declines in both occurring during cold spells. Declines in warbler numbers at Arizona sites corresponded with increases at more southerly Mexican sites, suggesting movements, and the magnitude of changes were correlated with insect availability. These particular population shifts occurred in January. However, records of Yellow-rumped Warblers killed overnight at television towers showed that these birds were able to migrate at any date in winter, usually in association with cold snaps, the numbers varying greatly from year to year. Birds tested for directional preferences after dark showed southerly orientation (Terrill & Ohmart 1984).

Similar findings have emerged for many other species that migrate within the temperate zone (e.g. Pulliam & Parker 1979, Niles *et al.* 1969), as well as in many that reach the tropics (Chapter 24). For example, the numbers of Yellow Wagtails *Motacilla flava* at a roost in West Africa declined progressively from 16 000 in November to 2000–3000 in March, in line with a progressive decline in local food supply. A southward shift was revealed by ring recoveries (Wood 1978). The facultative nature of the movement was implied in the fact that the decline was substantially more marked in non-territorial birds that fed in flocks, than in dominant adult males that remained on feeding territories near the roost. Experiments on several species have shown that captive birds can develop migratory restlessness in response to food deprivation in winter, well beyond the normal migration period (Chapter 12).

In Africa many species move southwards in stages, pausing for weeks or months before proceeding (Chapter 24). In some species these movements are apparently 'obligate' and consistent in timing from year to year (reflected in the periods of migratory restlessness in captive warblers), but in others they are facultative, varying in timing and extent with prevailing conditions (Chapter 24; Lack 1983, Herremans 1993, 1998). In all such species, further movement in the usual direction of migration may give the birds the best chance of finding food.

Facultative movements in relation to weather

Facultative movements are also shown by species that obtain their food from water or the ground, and which must therefore move in response to freezing temperatures, snow or drought. Such escape (or weather) movements occur as soon as waters freeze or snow covers the ground, cutting off food supplies. The dates of such movements therefore vary between years, depending on the weather, and in mild winters need not occur at all. Even in the arctic, some waterfowl and gulls remain in autumn to feed in remaining patches of open water until long after most other migrants have left. They are forced out as the last open water areas freeze in early winter, and must usually migrate south over habitats that have been ice- or snow-covered for some weeks.

Hard weather movements are often extremely obvious. Whenever cold weather strikes, thousands of birds can be counted as they stream past particular observation points, as large regions are evacuated within a matter of hours. Counts made at such times in Britain include the 20 000 Eurasian Skylarks *Alauda arvensis* that passed westward over the Axe Estuary in Devon on 28 December 1964, or the 8300 counted in 2 hours as they passed south over a site in northeast Scotland on

24 January 1976; the 4500 Northern Lapwings *Vanellus vanellus* that passed south-west over Tring in 50 minutes on 9 December 1967, and the 10 200 that passed in 2.5 hours over Portsmouth, Hampshire on 30 January 1972 (from Wernham *et al.* 2002). Over much of the world, ducks usually travel at night, so are seldom seen on migration, but their weather movements are marked by sudden massive overnight increases in the numbers on particular wetlands, following the onset of hard weather back along the migration route.

Many hard weather movers usually return soon after conditions improve again, sometimes less than a week later. One probable reason for their return over several hundred kilometres is the avoidance of competition for food, which is likely to be more intense in the overcrowded hard weather refuges than in the areas previously left. It is not that birds are necessarily driven back by competition, but this could be the ultimate factor involved. The proximate stimulus may be a favourable change in temperature or wind direction, for it would be advantageous for the birds to leave before they were weakened by food shortage. In Europe, winds from the south and west bring warmer temperatures and suitable tailwinds. It is almost as though such birds shuttle back and forth along part of their migration route, on average getting further from their breeding areas as winter advances. It seems that some facultative migrants have some sense of where they ought to be – that is, as near to the breeding areas as conditions allow – and can migrate effectively in either direction during the course of a winter.

Weather movements were found by radar to occur almost every day and night in November–February between Britain and continental Europe (Lack 1963). The Northern Lapwing *Vanellus vanellus* and Common Starling *Sturnus vulgaris* were the most frequent participants, but many other species were involved too, including finches, thrushes, larks, plovers, grebes and other waterfowl (Elkins 1988, Evans & Davidson 1990, Ridgill & Fox 1990). All these birds are partial migrants which leave in winter more or less in the same direction as they would normally migrate. Lapwings fleeing from hard winters may reach Spain, where they are known as *avefria* ('birds of the cold'). In the rare years when the cold weather extends to the usual hard-weather refuge areas (such as southwest Ireland), enormous mortality may occur among the huge numbers of migrants concentrated there (Clarke 1912). Needless to say, hard weather movements are much more pronounced in severe winters than in mild ones, and at mid-latitudes they have become less frequent in recent decades as winters have become milder.

Some wetland species perform the equivalent of hard weather movements in summer, often to higher latitudes, in their attempts to escape drought. Such movements have most often been recorded among various herons and ducks (in which they merge with moult migrations). In addition, at least one non-wetland species in Europe performs extensive weather movements in summer. The Common Swift *Apus apus* feeds on small high-flying insects which are available only in fine weather. As found by radar studies, these birds escape cold rainstorms by flying into the wind ahead of the rain then clockwise round the depression, and returning behind it (Lack 1956). In the process, they can travel up to 2000 km. It is mainly the non-breeders that participate in these movements, but also some breeders, forming flocks of up to 50 000 birds. Young Common Swifts can survive without food for periods of a week or more by becoming torpid, lowering their body temperature and metabolic rate, an adaptation lacking in most other birds.

Facultative movements in relation to disturbance

Some birds seem to respond by movements to predation and disturbance (Chapter 27). For example, the banning of hunting in two coastal areas of Denmark resulted in waterfowl staying in larger numbers than in previous years, delaying their onward migration south (Madsen 1995). Over a five-year period, these reserves became important staging areas, where most quarry species stayed up to several months longer each winter than in earlier years. In contrast, no changes in waterfowl numbers were noted in other areas of Denmark still open to wildfowling, so the accumulation of birds in the reserves was attributed to individuals remaining there, rather than moving further down the migration route. Birds presumably respond in the same way to natural predators which, like human hunters, could kill some individuals and continually disturb others, preventing them from getting enough food (Chapter 27).

The important point to emerge from the above sections is that migration is not always a simple two-season movement between fixed breeding and wintering areas. Some species move between three distinct areas: a breeding area, a late summer or autumn area, and a wintering area. Others move only when necessary, and no further than necessary, wintering much nearer to their breeding areas in some years than in others, and thereby saving on the costs of migration. Yet others seem to spend much of their lives on the move, following the same route each year, pausing for a few weeks here and a few weeks there, before moving on. Their mobile lifestyle enables them to exploit short-lived food supplies at different places at different times, as they occur (Newton 1979, Lack 1990, Pearson & Lack 1992, Jones 1995). It is a lifestyle that birds more than most other kinds of animals can adopt to the full. These various examples of movements within the breeding or non-breeding seasons provide further circumstantial evidence for the underlying role of food supplies in governing the migration patterns of birds.

NOMADISM

Even greater flexibility in movement patterns is shown by some species that exploit sporadic habitats or food sources. Such species often appear to be truly nomadic, as they show little or no year-to-year consistency in their movement patterns, but shift from one area to another, residing for a time in whichever parts of their range food is plentiful at the time. The areas successively occupied may lie in various directions from one another. No one area is necessarily used every year, and some areas may be used only at intervals of several years, but for months or years at a time, if conditions are suitable. The birds themselves may in theory move at any time and in any direction between successive breeding sites, and over shorter or longer distances.

The essence of nomadism, then, is its inconsistency, involving 'unpredictable, unseasonal and irregular movements across landscapes and regions' (Dean 2004). Of course, many bird species wander locally over distances of a few kilometres in search of winter food, but so-called nomadic species can travel hundreds of kilometres from one breeding area to another or from one non-breeding area to another. It is partly a matter of scale. However, much of what we know about nomadic movements is based on inference, to a large extent unsupported by ring recoveries. This is a consequence of the mobility of the birds themselves, and of the fact that most of them live in environments that support only low densities of people.

It is in desert regions where nomadism is best developed, being governed by sporadic rainfall, which influences plant growth and invertebrate densities, as well as the availability of surface water for waterfowl. Nomadism seems to occur in response to the variability, rather than the severity, of the desert environment. In the most unpredictable conditions, both the movements and breeding of birds become increasingly aseasonal. Following rain, landbird densities in particular localities can increase more than a hundred-fold within a few days, and water-birds that have not been seen for years can re-appear in huge numbers on new flood waters (for examples see Dean 2004). In such environments, the survival of many bird species depends primarily on their mobility.

In many species, in both deserts and elsewhere, nomadic movements are super-imposed on regular north–south migrations so that the population is concentrated at different latitudes at different times of year, but always patchily distributed, wherever food occurs. This is obvious in northern regions in some seed-eating species such as Redpoll *Carduelis flammea* and Siskin *Cardulis spinus*, and in some rodent-eaters such as Long-eared Owl *Asio otus* and Hawk Owl *Surnia ulula* (Chapters 18 and 19). It is also obvious in some Australian species, such as the seed-eating Emu *Dromaius hollandiae* (Davies 1976, 1984), Budgerigar *Melopsittacus undulatus* (Wyndham 1982) and Cockatiel *Nymphicus hollandicus* (Rowley 1974), or the blossom-eating Black Honeyeater *Certhionyx niger* (Ford 1978). On the other hand, no regular latitudinal shifts are apparent in other Australian species, such as the seed-eating Flock Bronzewing *Phaps histrionica* and the rodent-eating Letter-winged Kite *Elanus scriptus*, which seem to be entirely nomadic (Keast 1959). The latter tends to concentrate at local outbreaks of Long-haired Rats *Rattus villossis-simus*, which follow the vegetation growth resulting from rain. In all, some 36 spe-cies of landbirds move about the interior of Australia, either on a combination of regular migration and nomadism, or on nomadism alone.

Yet other bird species behave nomadically only in the non-breeding season, and return to the same localities to nest each year. Examples from among Eurasian–African migrants include the White Stork *Ciconia ciconia*, Black Kite *Milvus migrans*, Steppe Eagle *Aquila nipalensis*, Lesser Spotted Eagle *A. pomarina*, Lesser Kestrel *Falco naumanni*, Red-footed Falcon *Falco vespertinus*, Amur Falcon *F. amurensis* and Black-winged Pratincole *Glareola nordimanni*. Typically, these species specialise in their non-breeding areas on sporadic food supplies, such as locusts and grasshoppers, emerging termites, large nesting colonies of Red-billed Quelea *Quelea quelea* or Wattled Starlings *Creatophora cinerea*, or small animals disturbed by grass fires. Some of these prey species are available at the same localities for a short time every year, but others only after exceptional rainfall.

According to Dean (2004), some 233 arid-land bird species worldwide could be classed as primarily nomadic. They form varying proportions of desert avi-faunas in different regions (**Figure 16.5**). In deserts in the northern hemisphere, regular seasonal migrants tend to outnumber year-round nomadic species, espe-cially in the colder regions (Dean 2004); but in the warm deserts of the southern

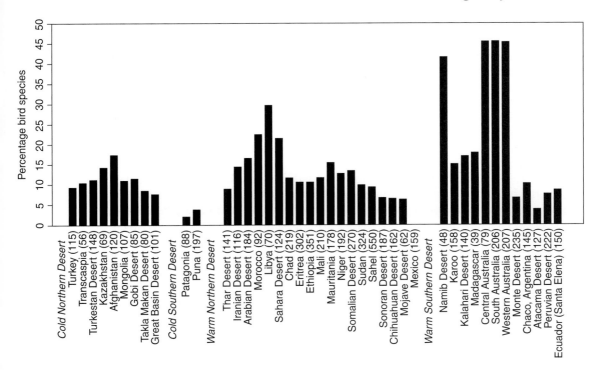

Figure 16.5 Percentage of breeding species in different arid regions that have been classed as nomadic. Total numbers of breeding species in parentheses. Nomadic species reach significantly greater numbers (and proportions) in the warm deserts of the southern hemisphere than in similar deserts in the northern hemisphere (χ^2 with Yates' correction = 12.17, P = 0.005), and also significantly greater numbers (and proportions) in all southern hemisphere arid regions regardless of winter temperatures (χ^2 with Yates' correction = 17.76, P < 0.001). The deserts of Australia hold the highest proportion of nomadic species (compared with all other southern hemisphere deserts, χ^2 with Yates' correction = 89.09, P < 0.001). From table 2.3 in Dean (2004).

hemisphere, nomadic species tend to outnumber regular migrants. This difference between hemispheres is attributed mainly to the greater fluctuation and unpredictability of rainfall in southern regions (associated with El Niño–La Niña events). In Australia, where conditions are most variable, up to 46% of desert bird species can be classed as nomadic, a greater proportion than in deserts in any other region (Dean 2004).

The nomadic birds of the African and Asian deserts are similar, being dominated by sandgrouse (Pteroclidae), larks (Alaudidae) and sparrows, weavers and finches (Passeridae). The Australian deserts are dominated by honeyeaters (Melophagidae), parrots (Psittacidae) and crows (Corvidae), and the New World deserts by finches (Fringillidae). Overall, nomadism is found in about half the bird families that breed in arid and semi-arid environments. It seems more related to diet than to phylogeny, but occurs disproportionately in some families, such as sandgrouse, all of which are nomadic to greater or lesser extent, as are many larks.

In general, nomadism is associated with ecosystems in which the underlying productivity, and the densities of resident bird species, are very low, but in which periodic pulses of high productivity occur, with resources sufficiently abundant to support major influxes of birds from elsewhere. Most nomadic species are specialists, feeding on only a small range of food items, which appear occasionally in extreme abundance. Thus, throughout the world many nomadic birds are seed-eaters, and in deserts they mostly specialise on the masses of grass seeds produced after rain has fallen (equivalent to tree seeds in the northern forests). Others are insectivores which tend to specialise on 'plague' insects, such as noctuid moth larvae, grasshoppers or locusts (equivalent to various defoliating caterpillars in northern forests). Yet others are predators that feed on the other birds that move in, or on the rodents that increase in response to vegetation growth and seeding.

The impact of rainfall on predators can be illustrated by findings at Red-billed Quelea *Quelea quelea* colonies in the savannah grasslands of Africa. As explained above, these small birds move into areas after rain has fallen in order to feed on the resulting grass seed. They form enormous colonies, containing many millions of pairs, and covering many hectares of scrub, from which the birds fan out over the surrounding land to forage. These colonies attract concentrations of predators. At one large quelea colony in Kruger National Park, Pienaar (1969) counted 200–240 Wahlberg's Eagles *Aquila wahlbergi*, 800–1160 Steppe Eagles *A. nipalensis*, 300 Marabou Storks *Leptophilus crumenifenus* and many other species. At least 60% of quelea nests were torn open by these birds. In the same region at another large quelea colony, Kemp (2001) counted more than 1000 eagles, the majority being Lesser Spotted Eagles *A. pomarina*, with smaller numbers of Tawny Eagles *Aquila rapax*, Wahlberg's Eagles and Steppe Eagles, and at least 100 Marabou Storks. Most of these predators were non-breeding migrants from other parts of Africa or Eurasia. Other concentrations of raptors have been found at the smaller colonies of Wattled Starlings *Creatophora cinerea*, which normally cover a few hectares or less. These birds move in after rain to exploit the resulting mass hatches of locusts or noctuid and other lepidopterous larvae.

Other concentrations of different raptor species occur in association with local rodent outbreaks, again following rainfall. For example, Malherbe (1963) recorded concentrations of several diurnal and nocturnal birds of prey in a semi-arid area in Northwest Province, South Africa, associated with an outbreak of Multi-mammate Mice *Praomys natalensis*. In such conditions, several local species bred and produced larger than usual broods. During another mouse outbreak in Zimbabwe, one pair of Barn Owls *Tyto alba* over a 12-month period nested continuously from March to December, laying 32 eggs, almost all of which produced fledglings (Wilson 1970). Normally they would raise only one brood, containing less than six young. Few other bird species live on such an extreme boom-and-bust economy.

All these various nomadic species may appear in enormous numbers when conditions are suitable, but then disappear for up to several years before they come again. They all move in relation to sporadic changes in food supplies, but without this knowledge, their movements would appear random and unpredictable. The importance of food supplies is further shown by the fact that some typical nomadic species, such as the Wattled Starling, are regular breeders in parts of their range where food is more consistently available, usually through human action (Broekhuysen *et al.* 1963). Similarly, in Southern Australia, the White-fronted Chat

Epthianura albifrons is nomadic in the drier parts of its range, but sedentary near the coast (Dean 2004).

In the Australian central desert, rainfall is more irregular in amount, timing and distribution than almost anywhere else on earth (Dingle 2004). Persistent heavy rain can produce widespread flooding, giving rise to large areas of wetland habitat, while subsequent droughts can eliminate these wetlands for years on end (Roshier & Reid 2002). To survive and prosper, birds must therefore move at irregular intervals, to wherever suitable wetlands exist. The many thousands of temporary wetlands can vary in size from a few square metres to thousands of square kilometres. Some are terminal water bodies filled by major drainage systems (e.g. Lake Eyre), others are lakes filled by local drainage, or overflow areas from swollen rivers. These floodlands last at most a few months before they dry out. They are re-created to varying extents, up to several times per 20-year period but at irregular intervals. For example, in eastern Australia, total wetland varies from almost nothing in extreme drought years to around 30 000 km² in wet years (**Figure 16.6**).

During their short lives, these wetlands can produce great quantities of aquatic invertebrates and fish. Most Australian waterfowl are highly mobile, and it is not uncommon, as ring recoveries have shown, for individual ducks to move more than 2000 km from one area to another (Roshier & Reid 2002). The most extreme examples include the Grey Teal *Anas gracilis*, Freckled Duck *Stictonetta naevosa* and Pink-eared Duck *Malacorhynchus membranaceus* (Frith 1967). But irregular nomadic movements also occur among Australian Shelduck *Tadorna tadornoides*, Chestnut Teal *Anas castanea*, Australasian Shoveler *Anas rhynchotis*, Hardhead *Aythya australis*, Hoary-headed Grebe *Poliocephalus poliocephalus*, Australian Pelican *Pelecanus conspicillatus*, Black-tailed Native Hen *Gallinula ventralis*, Red-kneed Dotterel *Erythrogonys cinctus*, Gull-billed Tern *Sterna nilotica* and others, each requiring rather specific conditions. In Australia, as in deserts elsewhere, stilts and avocets also benefit from the flooding of normally dry hollows, feeding on the masses of crustaceans that result. Both movement and breeding patterns have arisen to take advantage of such temporary bonanzas (Chapter 11).

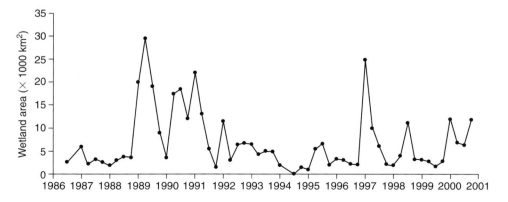

Figure 16.6   Temporal variation in the total water area in the major inland drainage basins of eastern Australia, 1986–2001. From Roshier & Reid (2002).

Lake Eyrie has been called the sump of Australia: a dry salt plain at times of drought, and a rich shallow lake extending over thousands of square kilometres at times of flood, perhaps once every 20 years. When floods begin, thousands of waterbirds appear from nowhere and immediately begin to breed. One colony of pelicans reached 50 000 pairs, the largest on record. Some birds must travel 1500 km to reach this area from the coast. It remains a mystery how such birds find suitable localities within vast desert areas: whether they search at random over appropriate range or whether they respond to climatic and other clues that indicate the most rewarding directions to fly. Older birds might remember the site, but this still leaves the question of how they know the time is right, as there is nothing regular about the rains that bring the flood. In theory, birds could find suitable areas by travelling with the wind, which blows towards the low pressure areas where rain is falling. This would not always work, however, because flood-waters in inland Australia can sometimes take weeks to reach the lower catchments, long after the weather that created them has passed. Olfaction has been suggested as an alternative mechanism, but is untested in this context (Roshier *et al.* 2006). Few desert species have been studied in detail however, and it is possible that further research might reveal more regularity in their movements than casual observations suggest.

Typically, many species of the central Australian desert build up in numbers as a result of good breeding in occasional wet years, then move outwards to the more humid peripheral districts in the following dry years (Nix 1976). Many birds probably die before they can return, but some species spend most years in coastal localities and periodically move inland to breed in vast numbers when rain creates suitable conditions. For example, Banded Stilts *Cladorhynchus leucocephalus* live for years as non-breeders on scattered briny coastal lagoons. But within days of rain falling inland, they concentrate in tens of thousands on newly formed shallow lakes, feeding on the freshly hatched swarms of brine shrimps (Burbidge & Fuller 1982, Robinson & Minton 1989). They breed while conditions last, making repeated nesting attempts, while the young form huge crèches. Water evaporates rapidly, however, and the land soon resorts to its normal parched state. The birds then return to the coast, if necessary leaving the last thousands of eggs and young to die. Years may pass before they can breed again, and not necessarily in the same sites. In recent years, known breeding events have occurred at Lake Eyrie in South Australia in 2000, in the Coorong in South Australia in 2005, and at Lake Corangamite in Victoria in 2006, the latter for the first time in recorded history.

Poor conditions over most of the range, coupled with good conditions in other parts, can sometimes lead to enormous concentrations of birds. For example, in the austral summer of early 2004, an estimated 2.9 million Oriental Pratincoles *Glareola maldivarium* gathered along 235 km of coastal grassland (Eighty-mile Beach, southwest of Broome) in northwest Australia, feeding on the abundant grasshoppers (Sitters *et al.* 2004). These birds breed widely over Southeast Asia and migrate mainly to Australia for their non-breeding season. In most years, the birds are thinly scattered over a wide area. This makes them hard to find, and the total flyway population had been previously estimated at only 15 000 individuals. But the circumstances of early 2004 led to a huge localised concentration, and a greatly revised overall population estimate.

Irruptive movements from deserts to neighbouring areas

Normally, desert species are able to remain within their regular range, in the wettest years penetrating deep into the driest areas, but at times of widespread drought, they may move out into neighbouring less arid habitats. This is strikingly illustrated in Australia, as mentioned above, by the periodic outward movement of many interior desert species to the less arid edges of the continent, and in Eurasia by the westward movements of Rose-coloured Starlings *Sturnus roseus*, Pallas's Sandgrouse *Syrrhaptes paradoxus* and others from the steppes into western Europe.

In their regular range, Rose-coloured Starlings may settle in thousands in suitable localities, breed and then move on. They feed their young mainly on grasshoppers and locusts, and seldom occur in abundance in the same localities in successive years. Their irruptions outside the usual range occur in spring and early summer, apparently in years when birds return from winter quarters in the Indian subcontinent to find their food supply has failed over wide areas (Schenk 1934). Big irruptions to western Europe occurred in 1853, 1907–1909, 1925, 1932 and 1948, but subsequently they became less frequent and extended less far, only stragglers reaching western Europe (Cramp & Perrins 1994). However, a major movement into Europe occurred in summer 2002, with birds reaching as far as Iceland and Britain. In this year, many thousands of pairs bred outside the usual range, with an estimated 14 000 pairs in Romania, 2000 pairs in Bulgaria and 10 000 non-breeding birds in Hungary (Fraser & Rogers 2004). Nesting at various places in Bulgaria coincided with an invasion of locusts (Davies & Sharrock 2000). Irregular occurrences have long been the norm in Europe. In 1875, 6000–7000 pairs settled to breed in holes in the ramparts of the castle of Villafranca di Verona in northern Italy, then for the next 33 years not a single Rose-coloured Starling was recorded in that country (Voous 1960). Ring recoveries are scarce, but one bird ringed during an invasion in Hungary in June was recovered the next April in a wintering area some 4800 km to the east in Pakistan (Ali & Ripley 1978).

The irruptions of Pallas's Sandgrouse *Syrrhaptes paradoxus*, which occurred from the steppes of Turkestan and Kazakhstan into western Europe in the spring–summers of 1859, 1863, 1872, 1876, 1888, 1891, 1899 and 1908 (with small numbers in some of the intervening years and subsequently) were attributed to seed shortage (mainly of the chenopod *Agriophyllum globicum*) resulting from prolonged drought. Again, after each of these irruptions there were occasional breeding records, some from as far west as Britain. Irruptions into eastern Asia occurred in different years, indicating that regional factors affected different segments of the population. No substantial emigrations are known to have occurred subsequently, possibly because populations have declined following degradation of nesting habitat.

CONCLUDING REMARKS

It will be clear from the foregoing that not all bird movements are restricted to spring and autumn; not all long-distance movements occur on a north–south axis, or even on a consistent axis; and not all movements occur at the same times from year to year. The irregularity of some of these movements implies that they occur as a direct response to habitat and food conditions at the time, and are in no sense 'anticipatory' as is much regular seasonal migration. For the most part, movement patterns have been inferred from local numerical changes, and in very few species have they been supported by abundant ring recoveries. This is not surprising in species which live mainly in areas where human population density is extremely low, and whose movements are so variable from year to year. More extensive use of satellite-based radio-tracking is one obvious means of gaining more understanding of the nomadic movements of individual birds. In the mean-time, many questions remain, notably: how do nomadic species know when and where to go? From how far away can they detect suitable conditions? How far do individuals move? And to what extent are their lifetime movements directional? As mentioned above, many birds classed as nomadic still have a north–south component in their movements, in line with latitudinal–seasonal trends in tem-perature and rainfall. Once again we see the different kinds of bird movements grading into one another, whether migration and nomadism or migration and dispersal. In any one population, however, one or two kinds usually prevail.

SUMMARY

Some waterfowl perform 'moult migrations' after breeding, when they move in large numbers to traditional sites, offering abundant food and safety. They stay at least long enough to moult their flight feathers, shedding them all at once, and remaining flightless for several weeks until replacement feathers are grown. Moulting sites can lie in any direction from breeding areas. In many species of geese, however, non-breeders migrate to moult mainly at sites to the north of their breeding areas, starting their southwards migration to wintering areas soon after completing wing moult. In addition, some shorebirds and passerines moult at stopover sites on their southward migration, retaining their powers of flight throughout.

Some bird species move within seasons, in summer raising successive broods in different places, or in winter moving further from their breeding range as the winter progresses. After re-occupying their nesting colonies, many petrel species fly long distances in a 'pre-laying exodus' to obtain the food necessary to produce an egg (females) or undertake the first incubation stint (males). In many birds, winter movements occur in response to food shortage caused either by deple-tion or snow and ice, or in response to frequent disturbance. Irruptive species, dependent on annually varying food supplies, migrate much further in some years than in others, concentrating wherever they encounter sufficient food.

Nomadism occurs mainly in some desert species which live under the influ-ence of sporadic rainfall. Typically, they show little or no annual consistency in their movement patterns, but each year concentrate wherever food (and in some species water) is available at the time. Some species of boreal and tundra regions are also nomadic to some extent, in association with sporadic tree-seed crops or rodent peaks. Many nomadic species are also seasonal north–south migrants, moving much further in some years than in others, depending on food supplies, and each year concentrating where food is available.

Ruddy Turnstone *Arenaria interpres* with identifying colour rings

Chapter 17
Site-fidelity and dispersal

There is probably no single aspect of the entire subject of bird migration that challenges our admiration for birds so much as the unerring certainty with which they cover thousands of miles of land and water to come to rest in exactly the same spot where they spent the previous summer or winter. (Frederick C. Lincoln 1935.)

In many bird species, as may be judged from ring recoveries, individuals tend to breed in the same general region where they were raised; and once they have bred, they often use the same territories or even the same nest-sites in successive years. This holds more for one sex than the other, but is true of both territorial and colonial species. Moreover, individuals of some species also winter in the same localities in successive years. Although such site-fidelity is apparent in both resident and migratory populations, in migrants it implies the existence of precise navigational skills and homing behaviour, through which individuals seek out, year after year, the same breeding and wintering localities hundreds or thousands of kilometres apart. Individuals of some migratory species also use the same refuelling sites each year on their journeys, while some waterfowl use the same moulting sites.

The terms site-fidelity and dispersal denote opposite sides of the same coin, as they are both concerned with the distances that separate different places of residence of the same individual. Birds that occur in the same places in successive breeding seasons are said to show site-fidelity, while those that move from one place to another are said to disperse, but because distances vary along a continuum, there is no clear division between the two. Another term in frequent use is philopatry, which signifies continued residence in the place of birth, or return there after a period of absence.

In studies of the site-fidelity and dispersal of individuals, therefore, it is useful to distinguish: (1) natal dispersal, measured by the linear distances between natal and first breeding sites; (2) breeding dispersal, measured by the distances between the breeding sites of successive years; and (3) non-breeding dispersal, measured by the distances between the wintering sites of successive years. In almost all bird species that have been studied, individuals move much greater distances between natal site and breeding site than between the breeding sites of different years (e.g. Newton 1986, 2002, Paradis *et al.* 1998). Dispersal distances are of interest in their own right, as they relate to the ecology of the species concerned. They also have wider consequences, as they influence gene flow and the genetic structure of populations, the persistence and dynamics of local populations, and the potential for colonisation and range expansion (Newton 2003). This chapter is concerned mainly with the dispersal patterns of different species, and the environmental factors that underlie them.

Site-fidelity and dispersal have been studied primarily with the help of ringing. Most studies have been made by observers working in defined areas, ringing nestlings or adults in one year and noting where they are found in a later year in the same area. Such records are invariably biased in favour of short-distance moves because, being confined to the study area, they are not balanced by the longer moves of other individuals which may have settled outside the area and gone undetected. Moreover, some study areas are so small compared with the natal dispersal distances of the birds themselves that only a tiny proportion of ringed chicks is found breeding there in later years, the majority of survivors having settled elsewhere (see Weatherhead & Forbes 1994 for review of passerine studies). In general, for each species, the larger the study area (up to a point), the greater the proportion of locally raised young later found breeding within its boundaries (see Sokolov 1997 for European Pied Flycatcher *Ficedula hypoleuca*). In areas where practically every chick was ringed, other (unringed) individuals breeding within the area are taken as immigrants hatched elsewhere. In most study areas, immigrants greatly outnumber the locally raised birds. Some researchers, aware of the effects of size of study area on the proportions of locally raised young recovered as breeders, have attempted to devise ways of correcting for it (Barrowclough 1978, van Noordwijk 1983, Baker *et al.* 1995, Köenig *et al.* 2000), or have used other means of studying dispersal, such as radio-tracking (Walls & Kenward 1998).

Less biased information on dispersal distances comes from recoveries of ringed birds reported by members of the public. Even if all the birds are ringed in a particular locality, the recoveries are not confined to that locality, so can give a much more representative picture of dispersal distances. Although the natal sites are known precisely for birds ringed as chicks, the assumption usually has to be made that individuals of reproductive age recovered in the breeding season were in fact nesting, or had the potential to nest, at the localities where they were reported. Collectively, such records reflect the settling patterns of individuals with respect to the ringing site, regardless of their movements in the period between ringing and recovery, which remain unknown. In some large, long-lived species, care must also be taken to separate the immatures, which may summer in areas partly different from the breeding adults of their population (Chapter 15).

Benefits and costs of site-fidelity

There are obvious benefits to a bird in nesting near where it was raised, and in returning to the same areas each year, providing conditions permit. One is that the individual can benefit from local knowledge. This holds on both breeding and wintering areas, and at any sites the bird might stop on migration. Familiarity and prior ownership might also give a bird an advantage in competitive interactions with other individuals, making it better able to defend its feeding and breeding sites against potential takers. Moreover, through long-term residence over many generations, populations tend to become adapted through natural selection to the conditions prevailing in their particular region. This is evident, for example, in the consistent inherent patterns of size and colour variation found within species across their geographical ranges. So by remaining in (or returning to) the same general breeding area, individuals occupy regions to which both they and their likely breeding partners are best adapted. The benefits of local experience and of local adaptation, acting at the level of the individual, could thus be the main selective forces underlying site-fidelity in birds, where this is feasible.

Another potential advantage of site-fidelity stems from the social cohesion that it facilitates. Many birds pair with the same partner each year, even though they may live separately outside the breeding season. Partners can re-unite if they share a common breeding site (like most birds) or a common breeding and wintering site (like some sea-ducks in which partners separate after egg-laying and re-unite on wintering areas, Robertson & Cooke 1999). In at least some species with long-term pair bonds, breeding success improves as pairs remain together, but declines for a time following a change of mate (Black 1996). For these species then, site-fidelity enables partners that migrate independently to re-mate and gain any resulting reproductive benefits. But whether mate fidelity is a selective force behind the evolution of site-fidelity or an incidental behavioural consequence of site-fidelity is an open question.

Dispersal also has advantages. One is that birds can leave areas where conditions are poor or overcrowded to find somewhere better, in the process exploring around their home area. Some birds occupy successional habitats, which get less suitable over time, while others exploit patchy or ephemeral habitats or food sources, which are available in different places in different years. By moving from one good area to another, as appropriate, individuals may thereby enhance their survival and reproductive prospects. Another advantage of dispersal is that it could reduce inbreeding, which can lower the production and viability of offspring (Greenwood *et al.* 1978, Keller *et al.* 1994, Brown & Brown 1998, Daniels

& Walters 2000). Despite the various intensive studies that have been made of colour-ringed bird populations, extremely few brother–sister or parent–offspring matings have been recorded in most wild species. The chances of inbreeding are reduced further if, as in many bird species, one sex tends to settle further from its natal site to breed than the other (Greenwood & Harvey 1982), or if birds are more likely to disperse if close relatives are present in the same group or vicinity (Pärt 1996, Wheelwright & Mauck 1998, Cockburn et al. 2003). This last pattern is commonly found among species that live in groups of mainly closely related individuals. Conversely, individuals that mate with genetically very different individuals may suffer reduced fitness owing to the breakup of co-adapted gene complexes. Theoretically, an optimal balance between the contrasting risks of inbreeding and outbreeding would allow sufficient genetic mixing without disrupting local adaptations. Dispersal distances influence where this balance is drawn. Again, whether such genetic considerations have influenced the dispersal patterns of birds or are merely consequences of dispersal patterns resulting from other influences remains an open question.

There are clearly both benefits and costs to site-fidelity and dispersal, which are influenced by the ecological needs of the bird and by the environmental conditions prevailing. In some circumstances, it pays to remain in the same general area, in others to move elsewhere. Species vary their behaviour accordingly, and some closely related species which exploit different types of environment or food can differ greatly in their dispersal distances (see later). And whatever factors influence individual movements, they clearly have both ecological and genetic consequences on populations.

NATAL DISPERSAL

In many bird species, when the ring recoveries of young birds in a subsequent breeding season are plotted in relation to the hatch site, the numbers of recoveries tend to be greatest in the vicinity of the hatch site, and decline progressively with increasing distance. Typically, the recoveries come from all sectors of the compass, indicating no directional preference at the population level. Such patterns are shown for several species in **Figures 17.1** and **17.2**; they reflect the settling patterns of individuals with respect to natal site, regardless of their movements in the interim. In each species, the density of recoveries declines in approximately exponential manner in concentric circles out from the natal site. Such dartboard settling patterns do not result merely from an inability of some migrants to re-find their home areas for they also occur in sedentary species. Factors such as landscape structure, habitat and nest-site availability, and prior occupancy by other individuals, can all influence how close to its natal site a bird is likely to settle.

While this type of settlement pattern has been found in almost all species of birds that have been studied, whether resident or migrant, some species disperse over much longer distances than others (e.g. Newton 1979, Paradis et al. 1998, Wernham et al. 2002). In general, larger species tend to breed further from their natal sites than do small ones, as might be expected, but this relationship is rather loose because of the other factors that influence dispersal distances.

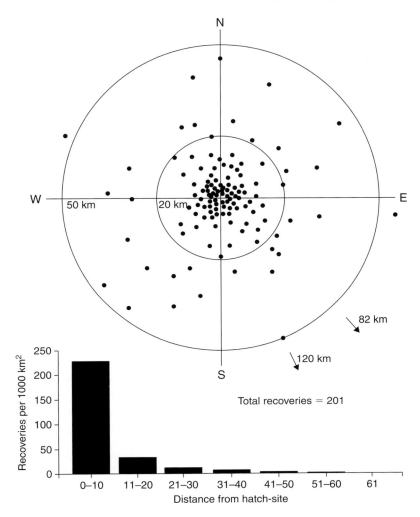

Figure 17.1 Locations of Sparrowhawks *Accipiter nisus* ringed as chicks and recovered in a later breeding season, shown in relation to natal site (centre). Recoveries came from all sectors of the compass and declined in density with increasing distance from the hatching site. From Newton (1979).

Among north European raptors, for example, a circle drawn at 50 km radius around the birthplace would include 98% of the breeding season recoveries of Eurasian Sparrowhawks *Accipiter nisus*, 95% of Common Buzzards *Buteo buteo*, 89% of Common Kestrels *Falco tinnunculus*, 75% of Merlins *Falco columbarius*, 71% of Northern Goshawks *Accipiter gentilis* and 43% of Ospreys *Pandion haliaetus* (Newton 1979).

Within species, the settling pattern may differ somewhat from year to year or from region to region, depending on circumstances, including the density of the population and the patchiness of habitat in the region concerned (for European Pied Flycatcher *Ficedula hypoleuca*, see Sokolov 1997; for Long-tailed Tit *Aegithalos caudatus*, see Russell 1999; for Blue Tit *Parus caeruleus* and Great Tit *P. major*, see Matthysen *et al.* 2001). Some small forest species are reluctant to cross open land. Blackwell & Dowdeswell (1951) described how a playing field just 130 m wide presented a barrier to local movements of Blue Tits so that their ringed study

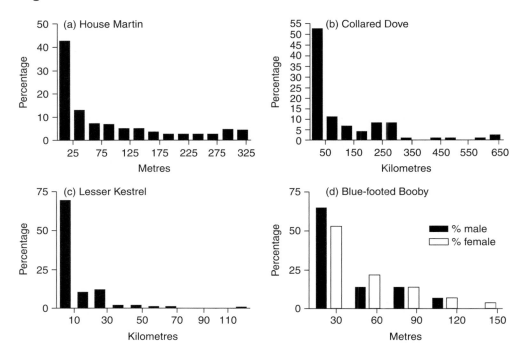

Figure 17.2   Natal dispersal patterns of several species ringed as chicks and recovered in a later breeding season. All species show a decline in numbers with increasing distance from the natal site, but the form of the relationship differs between species, and distances are greater in some species than in others. (a) Northern House Martin *Delichon urbica* (Rheinwald 1975); (b) Eurasian Collared Dove *Streptopelia decaocto* (Hengeveld 1993); (c) Lesser Kestrel *Falco naumanni* (Negro *et al.* 1997); and (d) Blue-footed Booby *Sula nebouxii* within a colony (Osorio-Beristain & Drummond 1993).

population, which was otherwise stable in composition through the winter, was completely divided in two. Big differences in the natal dispersal differences of Ospreys *Pandion haliaetus* between eastern North America (median distance about 10 km) and northern Europe (median distance about 100 km) were attributed by Poole (1989) to differences between regions in the availability of nest-sites. In eastern North America, many people put up platforms for nesting Ospreys, giving an unusually high density of nest-sites, enabling birds to settle near their natal sites and take advantage of rich food supplies in coastal areas. This does not happen in Europe, however, where nest-sites are few and far between and where food is less plentiful, so Ospreys must disperse over longer distances. Another indication of the influence of landscape structure on dispersal is that, within species, natal philopatry is much more marked, with higher return rates, in isolated populations (such as those on islands) than in their counterparts occupying similar areas in continuous habitat (Weatherhead & Forbes 1994).

    In general, natal dispersal distances tend to be greater in migratory species than in closely related resident ones, and the same is true for migratory and resident populations of the same species. For example, in Blackbirds *Turdus merula*, natal dispersal distances increased from Denmark through Norway and Sweden to Finland, in line with increasing migratoriness of breeding populations (as measured by proportion migrating and distances travelled) (Main 2002). In partial migrants, such as the Song Sparrow *Melospiza melodia*, migratory individuals showed greater natal dispersal distances than resident ones (Nice 1933). This was probably because resident ones were resident partly because of their dominance, and were already settled on territories by the time the migrants returned.

However short the usual dispersal distances, they are no handicap to range expansion through continuous habitat. In the House Sparrow *Passer domesticus*, the median natal dispersal distance is less than 1 km (with 97% of individuals at <20 km), yet in many parts of the world this species has expanded its range into new areas at rates of 15–80 km per year (Summers-Smith, in Wernham *et al.* 2002).

Seabirds and other colonial species

In colonial species, as expected, natal dispersal is influenced by the distribution of colonies. Nevertheless, as confirmed by ring recoveries, the settlement pattern is essentially the same as in other birds, with most individuals breeding in their natal or neighbouring colonies, and fewer individuals moving to colonies further away. This skewed settling pattern holds for colonial landbirds, such as Sand Martin *Riparia riparia* (Mead & Harrison 1979) and Lesser Kestrel *Falco naumanni* (Negro *et al.* 1997), and for a wide range of colonial seabirds, including gulls and terns, auks, shags and sulids, petrels and others. To give one example, only 36% of surviving Black-legged Kittiwakes *Rissa tridactyla* (mostly males) bred in their natal colony, a further 43% in other colonies up to 100 km away, and the remainder moved to colonies up to 900 km away (Coulson & de Mévergnies 1992). This pattern was found from ring recoveries of birds old enough to be breeding, and from resightings of individually colour-marked birds.[1]

In some seabirds, many individuals settle to breed within the same part of a colony where they were raised. This finding is much more frequent than expected if individuals settled at random within their natal colony. It also occurs in a wide range of species, including penguins (Williams 1995), albatrosses (Fisher 1971), shearwaters (Richdale 1963), auks (Gaston *et al.* 1994, Halley *et al.* 1995), skuas (Klomp & Furness 1992), shags (Aebischer 1995), and others. Extreme examples are provided by some tropical boobies. In a large colony of Blue-footed Boobies *Sula nebouxii*, the median distance between natal site and subsequent breeding site was less than 30 m (**Figure 17.2**; Osorio-Beristain & Drummond 1993), and in a colony of Nazca Boobies *Sula granti* on Galapagos the median dispersal distances of males and females were 26 m and 105 m respectively. As adults, both sexes retained the same sites year after year (Huyvaert & Anderson 2004).

---

[1]*For further specific examples for gulls and terns see Mills (1973), Spear et al. (1998), Duncan & Monaghan (1977), Coulson & Mévergnies (1992), Austin (1949), Spendelow et al. (1995); for auks see Harris (1984), Swann & Ramsay (1983), Gaston et al. (1994), Halley et al. (1995); for shags and sulids see Aebischer (1995), Huyvaert & Anderson (2004); for petrels see Richdale (1963), Fisher (1971), Brooke (1978), Thibault (1994).*

Such findings on the site-fidelity of seabirds are more remarkable than the bland figures suggest. Take the Short-tailed Shearwater *Puffinus tenuirostris*, for example, 23 million of which breed annually in burrows and headlands around southeastern Australia, migrating to the northern Pacific for the non-breeding season (Skira 1991). On one tiny island in Bass Straight, a population of a few hundred birds has been monitored for more than 50 years. Over 40% of young hatched on this island later returned there, usually breeding for their first time in their seventh year, and a constant 45% of the breeding population consisted of locally-hatched recruits (Serventy & Curry 1984). Such precision in the selection of a breeding location is extraordinary, considering the wide-ranging migration, the average seven-year period between fledging and first breeding, and the fact that less than a kilometre from the study site was a much larger island holding several hundred thousand nesting shearwaters. Moreover, most Short-tailed Shearwaters returned not just to the island, but to the same small part of the colony where they were hatched. Much the same could be said of most other pelagic seabirds that have been studied.

In any colony, then, breeders typically include some individuals raised within the colony and others that have moved in from elsewhere. The proportion of immigrants recorded in nesting colonies has varied greatly between species and with circumstances at the time. During the establishment and growth phases of a colony, immigration is high, so that most of the occupants have been raised elsewhere, whereas during a decline phase the reverse may be true (for seabirds, see Porter & Coulson 1987, Phillips *et al.* 1999; for geese, see Larsson *et al.* 1988, Johnson 1995).

Those gulls and terns that nest on the ground in exposed and often unstable substrates, such as sandbanks, often have to move their breeding places, as sites become washed away or flooded, accessible to mammalian predators, or infested with parasites. Hence, whole colonies can sometimes disband and re-form elsewhere, affecting the dispersal distances of both first-time and established breeders. Among southern African species, entire colony shifts are frequent in King Gulls *Larus hartlaubii*, Great Crested Terns *Sterna bergii*, Roseate Terns *Sterna dougallii* and Cape Cormorants *Phalacrocorax capensis* (Crawford *et al.* 1994). These species contrast with others in the same region which occupy more stable substrates and show strong colony persistence, including Jackass Penguins *Spheniscus demersus*, Cape Gannets *Morus capensis*, Bank Cormorants *Phalacrocorax neglectus*, Great Cormorants *P. carbo* and Great White Pelicans *Pelecanus onocrotalus*. In some cliff-nesting seabirds, generation after generation has used the same sites for centuries.

In many seabird species, individuals are known to visit colonies for one or more years before they attempt to breed, supposedly acquiring experience and local knowledge on which to assess the relative merits of potential nesting sites. The number of colonies visited by individuals during this pre-breeding phase seems to vary between species. In the Great Skuas *Stercorarius skua* on Foula, virtually all individuals seemed to visit only their natal colony in their pre-breeding years (Furness 1987, Klomp & Furness 1992), but in European Storm Petrels *Hydrobates pelagicus* and Atlantic Puffins *Fratercula arctica*, individuals regularly visited more than one colony, sometimes hundreds of kilometres apart, before settling to breed (Mainwood 1976, Fowler *et al.* 1982, Harris 1984). The most extreme record was for a Puffin ringed on the Treshnish Isles (Scotland) in late June and caught while visiting the Westman Islands (Iceland) 1087 km away 21 days later (M. P. Harris, in Wernham *et al.* 2002). By the time they start to visit colonies, some seabird species have spent up to several years of their early life in distant seas, apparently without coming to land, yet they still can find their natal areas.

Sex differences in natal dispersal

In some bird species, both sexes show similar dispersal distances, but in many others one sex moves further than the other, at least as a general tendency. The commonest pattern is for young females to disperse further between hatch site and breeding site than males. This pattern has been found in a wide range of species, including many passerines, owls and raptors, gallinaceous birds, shorebirds and colonial seabirds (Greenwood 1980, Clarke *et al.* 1997). In all such species, therefore, more males than females in local populations have been raised locally.

Most northern waterfowl show sex-biased dispersal, but with males moving furthest, as documented in swans, geese, shelducks, and in various diving and dabbling ducks (Mihelsons *et al.* 1986, Rohwer & Anderson 1988, Clarke *et al.* 1997, Nilsson & Persson 2001). In many such species, in contrast to most other birds, pairing occurs in wintering areas, and the male then accompanies the female to her natal area. Because birds from different breeding areas may share the same wintering places, the males of some species have settled to breed up to several hundreds of kilometres from their natal sites (Salomonsen 1955, Rockwell & Cooke 1977, Cooke *et al.* 1995). Moreover, the males of some migratory duck species have a different mate each year, so they often change their breeding sites substantially from one year to the next. Unlike dabbling ducks, some sea-ducks, geese and swans normally keep the same mate for several years and change their breeding sites much less often (Savard 1985, Anderson *et al.* 1992). In some sea-ducks, the partners separate after egg-laying and re-unite on wintering areas, as seen in marked pairs of Common Eiders *Somateria mollissima*, Barrow's Goldeneyes *Bucephala islandica* and Harlequin Ducks *Histrionicus histrionicus* (Robertson & Cooke 1999).

Other species in which males disperse further than females between natal and breeding sites include some shorebirds that show 'sex-role reversal', with the female defending the territory or mate, and the male doing the incubation and chick care. Examples include the Spotted Sandpiper *Actitis macularia* and various phalaropes *Phalaropus* spp. (Oring & Lank 1982, Colwell *et al.* 1988). Like the ducks, they have a social system based on mate defence, but unlike many ducks, they form into pairs in the breeding area. Males also disperse further than females in some lekking species, such as Great Bustard *Otis tarda* (Alonso & Alonso 1992).

In some group-living birds (cooperative breeders), the young remain with their parents for up to several years before they disperse, mostly over short distances (for Florida Scrub Jay *Aphelocoma coerulescens*, see Woolfenden & Fitzpatrick 1978; for Acorn Woodpecker *Melanerpes formicivorus*, see Köenig & Mumme 1987; for Arabian Babbler *Turdoides squamiceps*, see Zahavi 1989; for Siberian Jay *Perisoreus infaustus*, see Ekman *et al.* 1994). Delayed dispersal has developed, it is supposed, in situations where all suitable habitat is occupied by territorial groups, leaving nowhere for unattached birds to live (for experimental evidence in Seychelles Warbler *Acrocephalus sechellensis* see Komdeur *et al.* 1995). The adults then gain by allowing their young to remain in the territory, with access to its resources, until an opening becomes available elsewhere. The young pay for their long-term accommodation by

helping with territorial defence and (in some but not all species) by feeding subsequent broods. While they forgo reproduction themselves, they may gain some
'inclusive fitness' if they help to raise younger siblings. To become breeders, young
males sometimes inherit the territory from their father, or take over another territory
nearby, but young females almost always move to another territory, as does any
breeding female whose son takes over the home territory. The dispersal of young
birds from the natal territory is thus delayed by up to several years, and inbreeding is largely prevented by females moving more often or further than males. In
general, however, natal dispersal distances in such species tend to be short.

Competition and natal dispersal

Although a tendency to disperse from the natal site seems to be inherent in
birds, several types of evidence point to competition in influencing the distances
moved. First, in some studies the young moved further from their natal sites in
years of high than low population density (e.g. European Greenfinch *Carduelis
chloris*, Boddy & Sellars 1983; Great Tit *Parus major*, Greenwood *et al.* 1979,
O'Connor 1981; Marsh Tit *P. palustris*, Nilsson 1989; Bearded Tit *Panurus biarmicus*, Cramp & Perrins 1993, Spotted Sandpiper *Tringa macularia*, Oring & Lank
1982). Alternatively, the young moved further in successive years as a local population grew (e.g. Eurasian Sparrowhawk *Accipiter nisus*, Wyllie & Newton 1991;
Lesser Kestrel *Falco naumanni*, Negro *et al.* 1997).

Second, the young in other studies moved further from their natal sites in years
of low than high food supply (e.g. Coal Tit *Parus ater*, Sellers 1984; Northern
Goshawk *Accipiter gentilis*, Kenward *et al.* 1993a, 1993b, Byholm *et al.* 2003; Great
Horned Owl *Bubo virginianus*, Houston 1978; Common Kestrel *Falco tinnunculus*, Adriaensen *et al.* 1998; Barn Owl *Tyto alba*, Schönfeld 1974, Tengmalm's Owl
Aegolius funereus, Löfgren *et al.* 1986, Sonerud *et al.* 1988, Saurola 2002; Tawny
Owl *Strix aluco* and Ural Owl *S. uralensis*, Saurola 2002). The last five species
depend on cyclically fluctuating vole populations, and in most the length of their
movements fluctuated with the vole cycle in roughly three-year periodicity, with
the longest natal dispersal in the low vole years (this affected young hatched in
the peak years, which dispersed during the subsequent crashes). In addition, the
experimental provision of supplementary food to non-migratory Song Sparrows
Melospiza melodia greatly reduced emigration from the study area (Arcese 1989).

Third, in yet other studies, young fledged late in the season dispersed further, on average, than young fledged earlier. In the autumn, when Eurasian
Tree Sparrows *Passer montanus* occupied nest boxes as roost sites, twice as many
birds as boxes were present (Pinowski 1965). The available boxes were taken by
adults and by young hatched early in the season. By the time late-hatched young
attempted to obtain boxes, most were already taken, so that the majority of late-
hatched young had to move away, subsequently breeding further from their
natal sites. Similar increases in dispersal distances with fledging date were noted
in many other species.[2] Moreover, in an experimental study of Great Tits *Parus*

_____

[2]*Examples include Blue Tit* Parus caeruleus *(Dhondt & Hublé 1968), Great Tit* P. major *(van Balen
& Hage 1989), Marsh Tit* P. palustris *(Nilsson 1989), Northern House Martin* Delichon urbica
(Rheinwald & Gutscher 1969), European Pied Flycatcher Ficedula hypoleuca *(Sokolov 1997),*

major, when most (90%) of the first brood young had been removed, and competition was thereby reduced, young from late broods settled much nearer to their natal sites than in other years (Kluijver 1971).

These various findings, which derive from both resident and migratory populations, imply that dispersal distances may be density-dependent, and influenced by levels of competition, whether for territories, food or nest (roost) sites. None of this is surprising, but it has taken detailed long-term studies to confirm it. In sedentary populations, competition effects may be manifest mainly in the late summer and autumn, as post-fledging dispersal occurs, but in long-distance migrants that migrate soon after breeding, it may occur mainly after the return migration as birds compete for nesting territories in their home areas. The intensity of competition is dependent on the densities of contenders relative to available nesting sites or territories and, as a rule, adults are dominant over first-year birds, but other factors, such as date of hatching or date of settling on territory, also have an influence on who settles where. It is presumably in part for reasons of dominance that young move further than adults, on average, and late-hatched young further than early-hatched ones.

Other factors known to relate to dispersal behaviour in young birds include individual features, such as: (1) inherent 'personality traits', such as exploratory behaviour (e.g. Great Tit *Parus major*, Dingemanse *et al.* 2003); and (2) hatching order, body size, weight or dominance within broods (Greenwood *et al.* 1979, Nilsson 1989, Pärt 1990, Strickland 1991, Altwegg *et al.* 2000, Forero *et al.* 2002; Byholm *et al.* 2003). These aspects act in addition to the various external influences, such as landscape structure and habitat patchiness, territory or food availability and population density, discussed above. However, not all studied species have shown such relationships.

BREEDING DISPERSAL

One of the earliest findings to emerge from bird-ringing was that, once individuals had bred in an area, they tended to stay there, or to return there to breed year after year. Individuals were found to occupy the same territories in successive years, or to move to other territories nearby. In many species of seabirds and raptors, individuals often returned to exactly the same nest-sites.

Site-fidelity of adult birds has been shown mostly in specific study areas, by counting the proportion of marked individuals present in one year that returned the next. On this basis, any birds that moved outside the area would be missed, and the smaller the area the lower the expected return rate, as explained above. Nevertheless, in some species the proportion returning to particular areas was so high that, allowing for known mortality rates, few if any individuals could have moved elsewhere (**Figures 17.3** and **17.4**, **Table 17.1**). To judge from general ring recoveries, the settling pattern in breeding dispersal shows a similar dartboard pattern to natal dispersal, but over shorter distances (e.g. **Table 17.2**; Jackson

---

Eurasian Sparrowhawk Accipiter nisus *(Newton & Rothery 2000), Northern Goshawk* Accipiter gentilis *(Byholm* et al. *2003), Common Kestrel* Falco tinnunculus *(Village 1990), and Western Gull* Larus occidentalis *(Spear* et al. *1998).*

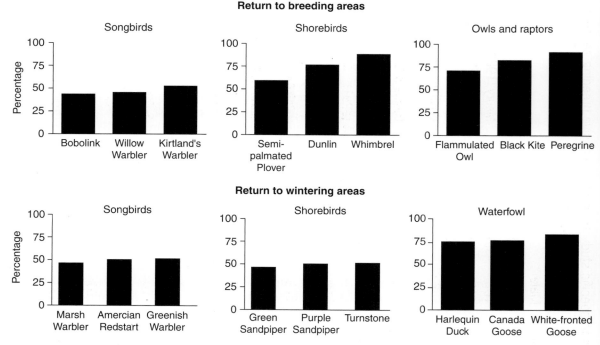

Figure 17.3   Proportions of adults in migratory bird populations that returned to the same breeding sites (upper) or wintering sites (lower) in successive years. Allowing for mortality, the species depicted reveal some of the most extreme examples of site-fidelity in birds, in which all (or almost all) surviving individuals may be inferred to have returned to the same site in successive years. Some of the species shown bred in one continent and wintered in another. Details from **Tables 17.1** and **17.6**.

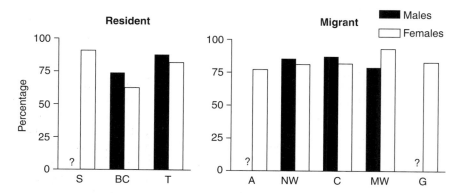

Figure 17.4   Proportions of adult Peregrine Falcons *Falco peregrinus* in resident and migratory populations that were present at nests in the same study areas in successive years. A, Alaska; BC, British Columbia; C, Colorado; G, Greenland; MW, Midwest United States; NW, Northwest Territories; S, Scotland; T, Tasmania. Return rates were high, with no obvious difference between resident and migratory populations. From Newton (2003), in which the original references may be found.

Table 17.1 Annual return rates of migrant juvenile and adult birds to specific study areas in successive breeding seasons

Species	Location (size of study area)	Natal area Unsexed young[a] N	Natal area Unsexed young[a] %	Breeding area Male N	Breeding area Male %	Breeding area Female N	Breeding area Female %	Source
Tree Pipit *Anthus trivialis*	Switzerland	114	12.5	102	51	74	32	Meury (1989)
House Wren *Troglodytes aedon*	Illinois (128 ha)	6299	2.8	643	38	1468	23	Drilling & Thompson (1988)
Whinchat *Saxicola rubetra*	Germany/Switzerland (<100 ha)		7.9	88	74	59	57	Bastian (1992)
Redwing *Turdus iliacus*	Norway	727	0.7	63	25	74	16	Bjerke & Espmark (1988)
Great Reed Warbler *Acrocephalus arundinaceus*	Sweden (200 ha)	477	9	59	54	66	57	Bensch & Hasselquist (1991)
Marsh Warbler *Acrocephalus palustris*	S. England (5.4 ha)			41	44	–	–	Kelsey (1989)
Willow Warbler *Phylloscopus trochilus*	Finland (25 ha)			122	41	97	17	Tiainen (1983)
Willow Warbler *Phylloscopus trochilus*	S. England (47 ha)	232	5.0	104	36	57	23	Lawn (1982)
Willow Warbler *Phylloscopus trochilus*	S. England (8.8 ha)			176	30	133	17	Pratt & Peach (1991)
Wood Warbler *Phylloscopus sibilatrix*	Mid England (35 ha)			29	28			Norman (1994)
Garden Warbler *Sylvia borin*	Finland (27 ha)	201	0	80	25	84	6	Solonen (1979)
European Pied Flycatcher *Ficedula hypoleuca*	Germany (20 km²)	2406	2.5	160	28	340	27	Creutz (1955)
European Pied Flycatcher *Ficedula hypoleuca*	Germany	229	0.4	37	35	40	38	Trettau (1952)
European Pied Flycatcher *Ficedula hypoleuca*	Germany	710	10.4	121	45	120	30	Curio (1958)

(Continued)

Table 17.1 (Continued)

Species	Location (size of study area)	Natal area Unsexed young[a] N	%	Breeding area Male N	%	Female N	%	Source
European Pied Flycatcher *Ficedula hypoleuca*	England (24 ha)	4086	3.9	374	24	646	23	Campbell (1959)
European Pied Flycatcher *Ficedula hypoleuca*	Finland	2842	1.8	378	36	576	14	von Haartman (1960)
European Pied Flycatcher *Ficedula hypoleuca*	Russia (50 km²)	1840	9.1	256	28	327	15	Sokolov (2001)
Prairie Warbler *Dendroica discolor*	Indiana (50 ha)	272	4	55	60	105	19	Nolan (1978)
Eastern Phoebe *Sayornis phoebe*	Indiana (>200 km²)			198	40	237	40	Beheler et al. (2003)
American Redstart *Setophaga ruticilla*	New Brunswick (>1.5 km²)	172	5	398	36	119	21	Lemon et al. (1996)
American Redstart *Setophaga ruticilla*	New Hampshire (34 ha)			134	50	48	19	Holmes & Sherry (1992)
Kirtland's Warbler *Dendroica kirtlandii*	Michigan (47 ha)	296	2.7	51	53	110	31	Berger & Radabaugh (1968)
Black-throated Blue Warbler *Dendroica caerulescens*	New Hampshire (55 ha)			49	39	50	36	Holmes & Sherry (1992)
Prothonotary Warbler *Protonotaria citrea*	Michigan (36 ha)			18	50	59	20	Walkinshaw (1953)
Hooded Warbler *Wilsonia citrina*	Pennsylvania (1 km²)			174	52	195	43	Howlett & Stutchbury (2003)
Eastern Kingbird *Tyrannus tyrannus*	New York (>20 km²)			142	69	138	54	Murphy (1996)

Species	Location							Reference
Willow Flycatcher *Empidonax traillii*	Oregon (5 km long)	1190	7.8	534	52	679	51	Sedgewick (2004)
Barn Swallow *Hirundo rustica*	New York (>30 km²)	331	2.0	216	42[b]			Schields (1984)
Yellow-breasted Chat *Icteria virens*	Indiana (18 ha)			18	11	29	0	Thompson & Nolan (1973)
Bobolink *Dolichonyx oryzivorus*	New York State (43 ha)			85	44	86	25	Gavin & Bollinger (1988)
Red-backed Shrike *Lanius collurio*	Czechoslovakia (15 km²)	799	2.8	134	28	144	20	Šimek (2001)
Loggerhead Shrike *Lanius ludovicianus*	Manitoba	3716	0.9	140	16[b]			Collister et al. (1997)
	Alberta (36 ha²)	249	1.2	96	32[b]			Collister et al. (1997)
Chaffinch *Fringilla coelebs*	Finland (68 ha)			104	42	69	35	Mikkonen (1983)
Common Shelduck *Tadorna tadorna*	Scotland (12 km²)	422	18.7	186	91	183	88	Patterson (1982)
Common Tern *Sterna hirundo*	Germany (300 km²)	575	38.0	509	91[b]			Becker et al. (2001)
Northern Lapwing *Vanellus vanellus*	England (877 ha)	801	28.1	171	74	129	70	Thompson et al. (1994)
European Golden Plover *Pluvialis apricaria*	Scotland (100 ha)	100	26	77	78	35	77	Parr (1980)
Semi-palmated Plover *Charadrius semipalmatus*	Manitoba (384 km²)	445	1.6	127	59	126	41	Flynn et al. (1999)
Snowy (Kentish) Plover *Charadrius alexandrinus*	Utah (ca. 50 km²)			224	40	278	26	Paton & Edwards (1996)
Semi-palmated Sandpiper *Calidris pusilla*	Manitoba (200 ha)	802	4.0	415	61	401	56	Sandercock & Gratto-Trevor (1997)
Temminck's Stint *Calidris temminckii*	Finland (12 ha)	170	21	112	79	61	70	Hildén (1978)
Burrowing Owl *Speotyto cunicularia*	Manitoba	538	3.5	87	40	78	24	De Smet (1997)

(Continued)

Table 17.1 (Continued)

Species	Location (size of study area)	Natal area		Breeding area				Source
		Unsexed young[a]		Male		Female		
		N	%	N	%	N	%	
Flammulated Owl *Otus flammeolus*	Colorado (452 ha)			21	71	16	81	Reynolds & Linkhart (1987)
Black Kite *Milvus migrans*	Spain (1000 km²)			142	83	143	90	Forero et al. (1999)
Osprey *Pandion haliaetus*	Eastern US (>1000 km²)			460	88[b]			Poole (1989)
Lesser Kestrel *Falco naumanni*	Spain (<1000 km²)	997	34.1	262	71[b]			Hiraldo et al. (1996)

The data are drawn from studies in which attempts were made to identify all individuals present. Records for different years are pooled. N = number marked, % = percentage recovered. Note the lower return rates for juveniles returning to their natal areas than for adults returning to their former breeding areas. This is due partly to higher mortality rates of juveniles, which allow fewer to return, and partly to their longer dispersal distances, which lead fewer to settle in the study area, compared with returning adults. Among established breeders, in most species return rates were higher for males than for females.

[a] In most bird species nestlings cannot be sexed on morphology, but the adults can. If the nestling sex ratio is assumed to be equal, it is clear from individuals subsequently trapped in the study area that males show much greater survival or site-fidelity than females. For example, in three studies of the European Pied Flycatcher *Ficedula hypoleuca*, return rates of male and female nestlings were 12.4% and 8.4%, 4.4% and 3.3%, and 2.5% and 1.1% respectively (Curio 1958, Campbell 1959, von Haartman 1960).

[b] Both sexes included.

Table 17.2   Dispersal distances of House Wrens *Troglodytes aedon* in Ohio

Distance away	% Recoveries in later years of birds ringed as:		
	Breeding males (N = 278)	Breeding females (N = 279)	Nestlings (N = 181)
In same nest box	31	26	2
Up to 1000 feet (0.3 km)	53	44	13
1000–2000 feet (0.3–0.6 km)	6	11	19
2000–4000 feet (0.6–1.2 km)	7	9	25
4000–7000 feet (1.2–2.1 km)	3	5	19
7000–11 000 feet (2.1–2.5 km)	–	3	7
2–5 miles (3–7.5 km)	–	1	4
5–10 miles (7.5–15 km)	–	>1	9
Over 10 miles (>15 km)	–	–	2

From Kendeigh (1941).

1994, Paradis *et al.* 1998, Winkler *et al.* 2004). Typically, within species, mean adult dispersal distances are about one half of natal distances, but in some species only one sixth as great (**Table 17.3**, **Figure 17.5**). The majority of adults nested close to where they bred the previous year.

Such patterns are again evident in a wide range of species, including passerines, raptors, game birds, shorebirds, waterfowl and colonial seabirds, both resident and migratory. For example, in the Northern House Martin *Delichon urbica* studied in some German villages, the median dispersal distance between natal and first breeding sites was found to be 75 m, whereas for breeding adults moving between nesting sites of different years, the median distance was 35 m. Some 7% of adults returned to the same nests in successive years (Rheinwald & Gutscher 1969). Within species, differences between age groups may arise largely because, each year when nesting begins, older birds are first to establish territories, or are better able to compete for them, so that the later settling youngsters then have to search over wider areas to find vacancies. The alternative, that young birds show poorer navigational skills, seems unlikely because the age differences in dispersal distances occur in resident, as well as in migratory populations.

For some species, more detailed studies have revealed patterns in the year-to-year territory fidelity and territory changes of breeders. Five main patterns have emerged: (a) sex differences in site-fidelity within species, with males in most species more likely than females to stay on the same territory from year to year,[3] possibly related to sex differences in territory acquisition and defence; (b) a tendency for greater site-fidelity in later life,[4] possibly related to increasing benefits through life of site-familiarity; (c) a greater tendency to change territories after

[3]Examples: *Greenwood (1980), Gavin & Bollinger (1988), Payne & Payne (1993), Jackson (1994), Murphy (1996).*
[4]Examples: *Newton (1993), Harvey et al. (1984), Thompson et al. (1994), Aebischer (1995), Lemon et al. (1996), Morton (1997), Bried & Jouventin (1998), Forero et al. (1999), Winkler et al. (2004).*

Table 17.3 Dispersal distances (km) calculated from recoveries of birds ringed and found dead in Britain, 1909–1994

Species	Natal dispersal (ND)				Breeding dispersal (BD)				Ratio
	N	AM	GM[a]	SD	N	AM	GM[a]	SD	ND:BD
Mute Swan Cygnus olor	49	34.3	16.772	35.9	497	18.0	2.719	48.0	6.17
Canada Goose Branta canadensis	173	7.0	0.969	10.6	365	8.9	1.503	10.8	1.50
Mallard Anas platyrhynchos	666	19.9	6.058	21.6	328	18.6	5.192	21.6	1.17
Common Black-headed Gull Larus ridibundus[b]	1478	47.0	10.527	69.2	110	44.5	7.968	72.5	1.32
Lesser Black-backed Gull Larus fuscus[b]	1882	28.2	2.384	40.7	190	38.2	13.805	37.4	0.17
Common Wood Pigeon Columba palumbus	718	10.7	2.277	19.3	233	10.9	1.805	24.1	5.93
Sand Martin Riparia riparia[b]	70	20.9	6.650	22.8	144	7.7	1.221	13.4	5.45
Barn Swallow Hirundo rustica	395	14.1	3.194	28.4	76	4.8	0.564	9.4	5.66
Northern House Martin Delichon urbica	72	10.4	3.185	12.2	191	4.2	0.688	8.3	4.63
Dunnock Prunella modularis	237	2.1	0.380	7.2	190	1.4	0.191	8.31	1.99
European Robin Erithacus rubecula	409	6.0	0.571	20.2	147	8.0	0.359	35.9	1.59
Eurasian Blackbird Turdus merula	2189	3.3	0.264	20.3	1806	3.2	0.224	20.6	1.18
Song Thrush Turdus philomelos	779	7.0	0.591	21.6	397	4.0	0.253	21.8	2.34
Mistle Thrush Turdus viscivorus	92	8.3	1.490	17.4	89	2.3	0.384	5.8	3.88
Eurasian Reed Warbler Acrocephalus scirpaceus	77	47.0	5.215	68.6	53	32.4	2.935	61.6	1.78
Greater Whitethroat Sylvia communis	89	14.4	2.815	19.0	51	11.1	1.145	19.0	2.46
Blackcap Sylvia atricapilla	74	41.2	17.539	37.9	64	27.5	8.027	32.0	2.19

Species									
Willow Warbler *Phylloscopus trochilus*	79	20.8	2.172	46.3	58	16.9	0.816	39.6	2.66
European Pied Flycatcher *Ficedula hypoleuca*	1551	20.6	14.272	16.5	238	20.6	11.668	17.7	1.22
Blue Tit *Parus caeruleus*	703	5.3	0.796	15.2	201	2.3	0.232	10.2	3.43
Great Tit *Parus major*	560	5.3	0.797	17.9	173	2.5	0.246	12.3	3.24
Eurasian Jackdaw *Corvus monedula*[b]	51	8.6	2.127	11.6	51	6.0	0.721	12.8	2.95
Rook *Corvus frugilegus*[b]	84	8.5	1.964	13.0	96	3.1	0.650	4.7	3.02
Common Starling *Sturnus vulgaris*	401	9.5	1.100	28.1	1672	3.4	0.273	19.1	4.03
House Sparrow *Passer domesticus*	531	1.7	0.206	6.9	526	1.9	0.147	22.4	1.40
Chaffinch *Fringilla coelebs*	64	3.6	0.787	5.6	120	2.8	0.302	9.9	2.46
European Greenfinch *Carduelis chloris*	99	4.2	0.954	6.4	283	7.5	0.732	22.1	1.30
European Goldfinch *Carduelis carduelis*	85	11.1	1.663	18.2	63	10.6	0.835	20.8	1.99
Eurasian Linnet *Carduelis cannabina*	147	4.4	0.694	8.8	110	3.5	0.393	8.3	1.77
Eurasian Bullfinch *Pyrrhula pyrrhula*	195	4.6	0.852	9.8	194	2.5	0.382	5.2	2.23
Reed Bunting *Emberiza schoeniclus*	58	5.4	0.952	13.1	79	3.8	0.468	9.3	2.03

N = number of recoveries. AM, arithmetic mean distance; GM, geometric mean distance; SD, standard deviation of distances. All figures refer only to species for which at least 50 recoveries were available for each age group analysed. From Paradis et al. (1998).

In these analyses, ringing and recovery occurred in the same area (Britain) for all species, so findings should have been comparable between them. This large-scale study covered a large number of ringing and recovery sites, thereby reducing to a minimum any systematic regional bias in reporting probability. All the birds were ringed in one breeding season and found dead at breeding age in a later one, excluding live recaptures by ringers which would have biased the records towards short moves. Records for the two sexes were pooled, even though they may have differed in dispersal distances. An assumption was that birds found dead in the breeding season were near their breeding sites.

[a]Geometric means calculated as the arithmetic mean of the log$_e$ distances, back transformed.
[b]Colonial.

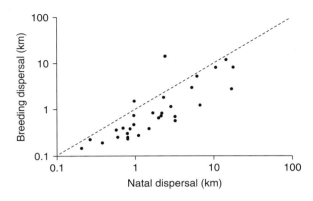

Figure 17.5   Relationship between the geometric means of the natal and breeding dispersal distances of various bird species, based on recoveries from the British ringing scheme. From Paradis *et al.* (1998).

a breeding failure than after a success,[5] possibly related to site quality, (d) a tendency to move to better territories (where previous breeding success was high) through early life,[6] possibly related to increase in status and competitive ability; and (e) a strong tendency for a change of territory to be associated with a change of mate.[7] Sex differences in the frequency of territory changes are accompanied by differences in the distances moved, and in most studied species, females moved more often and further than males. These various patterns have been confirmed in a wide range of studied species (but not all). In addition, some birds changed territory immediately after their spring arrival, in response to local food depletion, ectoparasite abundance, absence of mate, or other factors likely to reduce breeding success (Feare 1976, Korpimäki 1993, Brown & Brown 1992), and again females generally moved further than males.

Sex differences in breeding site-fidelity have been noted in more than half the species so far studied, from passerines to raptors, shorebirds and seabirds (for other examples see **Tables 17.1** and **17.2**). In some species they are pronounced. For example, in the European Pied Flycatcher *Ficedula hypoleuca*, 93% of surviving males returned each year to their previous nesting locality, while only 39% of surviving females did so, the rest moving elsewhere (von Haartman 1949). The difference in behaviour between the sexes of this species was also shown by experiment (Berndt & Sternberg 1968). In one year, when most of the ringed females in 40 ha of woodland were laying or incubating, all nest boxes were removed, including the occupied ones. The males remained in the area and continued singing, but only one day after box removal, not a single female could be found. Of the 146 females involved, 37 were found later in the same season nesting in other woods nearby, having moved 1.4–18.3 km, some to areas where they had bred in previous years. This was about the same range of distances as female flycatchers would normally have moved between years.

---

[5]*Examples: Newton & Marquiss (1982), Shields (1984), Beletsky & Orians (1991), Hepp & Kennamer (1992), Payne & Payne (1993), Reed & Oring (1993), Jackson (1994), Haas (1998), Serrano* et al. *(2001), Blums* et al. *(2002), Sedgewick (2004), Winkler* et al. *(2004).*
[6]*Examples: Hildén (1979), Matthysen (1990), Montalvo & Potti (1992).*
[7]*Examples: Johnston & Ryder (1987), Bradley* et al. *(1990), Payne & Payne (1993), Newton (2001), Howlett & Stutchbury (2003).*

Table 17.4 Within-season nest site-fidelity and mate-fidelity in the Eastern Phoebe *Sayornis phoebe* in Indiana

	Males	Females
Site-fidelity		
Same nest-site in same territory	127	135
Different nest-site in same territory	22	28
Different nest-site in different territory	1	2
Mate-fidelity		
Same mate throughout season	124	115
Different mate after loss of first mate	14	6
Different mate after 'divorce'	7	3

From Beheler *et al.* (2003). Note that sample sizes differ between categories.

Dispersal within a breeding season

Many birds raise more than one brood per season, or attempt a second nest if the first fails, normally close by in the same territory. Some even use the same nest-sites for repeat broods, especially some cavity nesters where sites are scarce. In most species, the same partners stay together for successive nests in a season but in other species some individuals change territories and partners, especially after nest failure, the female often moving further than the male (e.g. Yellow-breasted Chat *Icteria virens*, Thompson & Nolan 1973; House Sparrow *Passer domesticus*, Sappington 1977; Common Starling *Sturnus vulgaris*, Feare & Burham 1978; Barn Swallow *Hirundo rustica*, Shields 1984; Eastern Phoebe *Sayornis phoebe*, Beheler *et al.* 2003; **Table 17.4**).

A more extreme situation is shown by species in which some individuals raise successive broods in widely separated localities within the same breeding season (Chapter 16). Examples include some cardueline finches which feed on sporadic food supplies, settling temporarily wherever suitable seeds are plentiful, and moving elsewhere for the next nesting attempt that year (for Eurasian Bullfinch *Pyrrhula pyrrhula*, see Newton 2000; for Common Redpoll *Carduelis flammea* and Eurasian Siskin *C. spinus*, see Chapter 18). Other species change location part way through a breeding season, in line with changes in habitat suitability (including crop growth on farmland). For example, Eurasian Penduline Tits *Remiz pendulinus* studied in central Europe may raise first and second broods in different patches of riverine and lakeside scrub, as conditions change through the season, with movements generally greater than 0.5 km, and some extending up to 208 km (Franz 1988). Yet other examples include the few species (such as European Quail *Coturnix coturnix*) that habitually stop and breed at two or more points on a migration route (Chapter 16). In some polyandrous shorebirds, such as Eurasian Dotterel *Eudromias morinellus* and Kentish (Snowy) Plover *Charadrius alexandrinus*, the females lay one clutch for the male to incubate, and then leave in search of another male to raise a second clutch. In the process, as shown by ringing, the females can travel hundreds of kilometres between nesting attempts, with movements up to 400 km and 1140 km recorded for the two species (Mead & Clark 1988, Stenzel *et al.* 1994).

LONG-DISTANCE DISPERSERS

The precise year-to-year homing behaviour, shown by the species in **Figure 17.3**, is not universal in birds. It may be favoured only where habitats remain fairly stable from year to year, and where returning birds can expect to survive and reproduce. It would not be expected in species that depend on unpredictable habitats or food supplies, which are available in different areas in different years. This is the case, for example, in some tundra-nesting species affected by variable patterns of spring snow conditions (Tomkovitch & Soloviev 1994), in some waders and waterbirds affected by fluctuating water levels (Frith 1967, Johnson & Grier 1988, Nager *et al.* 1996, Robinson & Oring 1997), in desert species affected by irregular rainfall (Davies 1988, Zann & Runciman 1994, Dean 2004), in some boreal finches that exploit sporadic tree-seed crops (Svärdson 1957, Newton 1972), or in some predatory birds that exploit locally abundant rodents (Newton 1979, 2002, Saurola 2002). The local population densities of such species often fluctuate greatly from year to year, in line with fluctuating habitat or food supplies (Chapters 16, 18 and 19). The speed with which local numbers increase in response to improving conditions has led to the view that such species are nomadic, with individuals concentrating in different areas in different years, wherever conditions are good at the time. For some such species, ring recoveries have now confirmed that some individuals do indeed breed at hundreds or thousands of kilometres from their natal sites and in widely separated areas in different years (Chapters 18 and 19).

Among various duck species nesting on the North American prairies, the extent and depth of wetlands varies greatly from year to year, according to previous rain and snowfall. Diving duck species, such as Redhead *Aythya americana*, Canvasback *A. valisineria* and Lesser Scaup *A. affinis*, that occupy the deepest and most stable wetlands, show the greatest degree of site-fidelity; while dabbling species, such as Northern Pintail *Anas acuta* and Blue-winged Teal *A. discors*, which use the shallowest and most ephemeral wetlands, tend to settle wherever conditions are suitable at the time (Chapter 19; Johnson & Grier 1988). The opportunist settling behaviour of dabbling ducks is reflected in their long natal and breeding dispersal distances and by their low return rates to particular areas, as measured by ringing (**Table 17.5**). In years of extreme drought on the prairies, many Northern Pintail and other ducks continue their spring migration northward and settle to breed on the tundra. This means that many Pintails that were raised on the prairies move much longer distances in drought years to breed further north, as shown by many ringing recoveries (see **Figure 19.7**; Smith 1970). Dabbling ducks, and to a lesser extent some species of freshwater diving ducks, contrast with some sea-ducks, which breed in more stable habitats and typically show much higher natal and breeding site-fidelity (Dow & Fredga 1983, Savard & Eadie 1989, Cooke *et al.* 2000).

Similarly, among shorebirds, species that nest in habitats that normally remain stable from year to year, such as Eurasian Oystercatcher *Haematopus ostralegus*, Dunlin *Calidris alpina* and Common Sandpiper *Tringa hypoleucos*, show strong philopatry and site-fidelity, while species that nest in patchy and ephemeral habitats, or are affected by variable patterns of snow-melt, such as Curlew Sandpiper *Calidris ferruginea*, Sanderling *Calidris alba* and Little Stint *C. minuta*, show much less natal philopatry and less adult site-fidelity (Evans & Pienkowski 1984, Tomkovich & Soloviev 1994). In such species, the numbers nesting in any one locality may fluctuate greatly from year to year, depending on how many birds move to or from

Table 17.5 Sex and age differences in the percentage return rates of waterfowl to particular study areas in successive breeding seasons

	Natal dispersal				Breeding dispersal			
	Males		Females		Males		Females	
	N	%	N	%	N	%	N	%
Northern Shoveller *Anas clypeata*	134	1	116	3	19	11	20	15
Gadwall *Anas strepera*	42	2	28	7	236	9	54	41
American Wigeon *Anas americana*	6	0	3	33	11	9	21	38
Canvasback *Aythya valisineria*	206	1	101	27	52	10	75	76
Lesser Scaup *Aythya affinis*	91	4	76	49	351	9	58	66

Species arranged according to habitat preference, shallow (ephemeral) to deep (permanent) waters. Only studies with both sex and age groups represented are shown.

From Johnson & Grier (1988) in which the original references may be found. For other waterfowl studies, see Rohwer & Anderson (1988).

other nesting areas. Species that show low site-fidelity also show low mate-fidelity, because pairing occurs annually on or near nesting areas and former partners do not normally move together to the same places. In the American Avocet *Recurvirostra americana*, individuals were found to move up to 325 km between breeding sites in different years (Robinson & Oring 1997). Many other waterside birds, which require specific conditions for nesting, change their breeding places between years, in association with changes in water levels and vegetation, or the flooding and exposure of sandbanks.

In conclusion, high site-fidelity and short dispersal distances are characteristic of bird populations whose habitats or food supplies stay reasonably stable in distribution from year to year, whereas low site-fidelity and long-distance dispersal are shown by populations whose habitats or food supplies vary in distribution from year to year. The adaptive value of moving when conditions deteriorate is obvious, because this is likely to improve individual survival and breeding success in the conditions prevailing. Because of the moves, numbers in particular localities fluctuate greatly from year to year, in line with changes in habitat or food conditions, and some localities which support high densities of breeding birds in one year may be wholly devoid of birds the next. Differences in dispersal patterns between species have led some authors to divide birds into two types: those with limited natal dispersal, and even more limited breeding dispersal (strong site-fidelity), and those with wide-ranging natal and breeding dispersal (little or no site-fidelity). In practice, however, variation between these extremes is continuous, and the same species may lie at different points on the continuum in different parts of its range, depending on the stability of local habitats and food supplies (Chapters 18 and 19).

NON-BREEDING DISPERSAL

In migrating to wintering areas, young birds differ from adults in that they have no prior experience of where they are heading. The young of some birds, such as swans, geese and cranes, accompany their parents on their first autumn migration, so in this way the young could learn the locations of specific wintering sites. This behaviour could help to maintain the integrity of particular breeding populations, most of which also keep to distinct wintering areas, separated from those of other populations of their species (Chapter 23). In most bird species, however, the young do not accompany their parents, and may migrate earlier or later in the season than adults. Such young therefore migrate to wintering areas unknown to them, and often unaided by experienced birds (Chapter 9). The same is not true of breeding sites, because in most species both young and adults are returning to localities they already know.

In some bird species, once individuals have wintered in an area, they return from year to year with the same consistency as they return to their breeding localities. Most detailed studies of winter site-fidelity, like breeding site-fidelity, have been conducted by individual researchers working in the same small areas year after year. So again, while such records are useful in confirming site-fidelity, most give little idea of the proportion of surviving birds that settle elsewhere. However, in some species, the proportion of individuals that returned to the same area in successive winters was close to the expected annual (or over-summer) survival, implying that almost all surviving individuals were site-faithful. Thus, annual return rates to particular wintering areas of 37–52% have been recorded from various small songbirds, and of 77–95% for various longer-lived shorebirds (**Table 17.6, Figure 17.3**). More generally, return to the same specific wintering localities has been recorded by casual records for a wide range of passerines, raptors, gulls, waders, waterfowl and others, including many intercontinental migrants (Schwartz 1964, Moreau 1972, Holmes & Sherry 1972, Pearson 1972, Ely *et al.* 1977, Raveling 1979, Finlayson 1980, Cuadrado 1992, Rappole 1995, Rimmer & Darmstadt 1996, Sauvage *et al.* 1998, Salewski *et al.* 2000).

Many species in winter base themselves on communal roosts, from which they range out into surrounding areas to feed. For example, Starlings *Sturnus vulgaris* and Barn Swallows *Hirundo rustica* can range more than 50 km from their winter roosts, so records of individuals several tens of kilometres apart in different winters may still represent 'site-fidelity' in a way that breeding season records at such distances would not. A survey of Barn Swallows wintering in South Africa showed that nearly all returned to the same place, or to within 100 km of it, from year to year (Oatley 2000). Many shorebirds are faithful to their roost sites in different years, and the median distance moved by ringed Dunlin *Calidris alpina* trapped in different winters (14 km) was well within their daily travel range in large estuaries (Rehfisch *et al.* 1996). Hence, the existence of communal winter roosts serving extensive feeding areas gives more flexibility to the notion of site-fidelity than do nesting territories in summer.

Site-fidelity in winter seems to show the same pattern as in the breeding season, with ring recoveries in subsequent winters being centred on the first-recorded site, and declining sharply with increasing distance. Such dartboard patterns have been documented in many species,[8] but again the distance scale varied greatly between

---

[8]*Examples include the Common Starling* Sturnus vulgaris *(Spaans 1977), European Greenfinch* Carduelis chloris *(Boddy & Sellers 1983), Bullfinch* Pyrrhula pyrrhula *(I. Newton, in Wernham* et al. 2002*), Harlequin Duck* Histrionicus histrionicus *(Iverson et al. 2003), Ruddy Turnstone* Arenaria interpres *(Clapham 1979), Eurasian Curlew* Numenius arquata *(Bainbridge & Minton 1978), Sanderling* Calidris alba *(Myers* et al. 1988*), Grey Plover* Pluvialis squatarola, *Dunlin* Calidris alpina *and Common Redshank* Tringa totanus *(Rehfisch et al. 1996, Burton 2000), among others.*

Table 17.6 Annual return rates of migrant birds to specific study areas in successive winters (all except Blue Tits *Parus caeruleus* refer to migrants)

Species	Location (Size of study area)	Unsexed N	%	Source
Siberian Blue Robin *Luscinia cyane*	Malaysia (15 ha)	156	46	Wells (1990)
Great Reed Warbler *Acrocephalus arundinaceus*	Malaysia, area 1 (10 ha)	62	50	Nisbet & Medway (1972)
	Malaysia, area 2 (2 ha)	83	37	Nisbet & Medway (1972)
Marsh Warbler *Acrocephalus palustris*	Zambia (7.6 ha)	17	47	Kelsey (1989)
Melodius Warbler *Hippolais polyglotta*	Ivory Coast	69	6	Salewski et al. (2000)
Greenish Warbler *Phylloscopus trochiloides*	India (200 ha)	25	52	Price (1981)
Willow Warbler *Phylloscopus trochilus*	Ivory Coast	110	0	Salewski et al. (2000)
European Pied Flycatcher *Ficedula hypoleuca*	Ivory Coast	94	23	Salewski et al. (2000)
Prairie Warbler *Dendroica discolor*	Puerto Rico (8 ha)	25	40	Staicer (1992)
Black-throated Blue Warbler *Dendroica caerulescens*	Jamaica	57	46	Holmes & Sherry (1992)
Cape May Warbler *Dendroica tigrina*	Puerto Rico (8 ha)	8	38	Staicer (1992)
Northern Parula Warbler *Parula americana*	Puerto Rico (8 ha)	65	49	Staicer (1992)
American Redstart *Setophaga ruticilla*	Jamaica	111	51	Holmes & Sherry (1992)
Blue Tit *Parus caeruleus*				
males	England (one garden)	69	48	Burgess (1982)
females	England (one garden)	67	19	Burgess (1982)
White-crowned Sparrow *Zonotrichia leucophrys*	California (<1 ha)	?	51	Mewaldt, L.B. (1964)
adults	California (8 ha)	306	30	Mewaldt, L.B. (1964)
immatures	California (8 ha)	868	24	Mewaldt, L.B. (1964)
Golden-crowned Sparrow *Zonotrichia atricapilla*	California (8 ha)	2456	18	Mewaldt, L.B. (1964)
Eurasian Oystercatcher *Haematopus ostralegus*				
5-year-olds	England	734	89	Goss-Custard et al. (1982)
2- to 4-year-olds	England	475	83	Goss-Custard et al. (1982)
Grey Plover *Pluvialis squatarola*	England (<100 km²)	71	80	Evans (1981)
Pacific Golden Plover *Pluvialis fulva*				
territorial (1–2 years)	Hawaiian Islands (<10 km²)	34	90	Johnson et al. (2001)
territorial (older)	Hawaiian Islands (<10 km²)	78	80	Johnson et al. (2001)
non-territorial (1–2 years)	Hawaiian Islands (<10 km²)	16	82	Johnson et al. (2001)
non-territorial (older)	Hawaiian Islands (<10 km²)	35	67	Johnson et al. (2001)
Eurasian Curlew *Numenius arquata*	England (<100 km²)	119	82	Evans (1981)
Common Redshank *Tringa totanus*	Wales (4 km)	61	86	Burton (2000)

(Continued)

Table 17.6 (Continued)

Species	Location (Size of study area)	Unsexed		Source
		N	%	
Green Sandpiper Tringa ochropus	England (306 km²)	115	84	Smith et al. (1992)
Purple Sandpiper Calidris maritima	Helgoland (150 ha)	117	85 (adult)	Dierschke (1998)
Purple Sandpiper Calidris maritima	Helgoland (150 ha)	30	63 (first-year)	Dierschke (1998)
Purple Sandpiper Calidris maritima	England (13 km)	61	66	Burton & Evans (1997)
Sanderling Calidris alba	England (<100 km²)	93	91	Evans (1981)
Ruddy Turnstone Arenaria interpres	England (13 km)	71	86	Burton & Evans (1997)
Ruddy Turnstone Arenaria interpres	Scotland (6 km)	42	95	Metcalfe & Furness (1985)
Bewick's (Tundra) Swan Cygnus columbianus	England (<100 ha)	690	67	Rees (1987)
Canada Goose Branta canadensis	Minnesota (<100 ha)	271	78	Raveling (1979)
Barnacle Goose Branta leucopsis	Scotland (<10 km²)	540	76	Percival (1991)
Greater White-fronted Goose Anser albifrons	Scotland (<100 km²)	531	85[a]	Wilson et al. (1991)
Barnacle Goose Branta leucopsis	Netherlands (many areas)	576	90	Ebbinge et al. (1991)
Snow Goose Chen caerulescens	Texas-Louisiana (<100 km²)	77	86	Prevett & MacInnes (1980)
Harlequin Duck Histrionicus histrionicus, males	British Columbia	82	77	Robertson & Cooke (1999)
Harlequin Duck Histrionicus histrionicus, females	British Columbia	66	62	Robertson & Cooke (1999)
Bufflehead Bucephala albeola, males	Maryland	91	26	Limpert (1980)
Bufflehead Bucephala albeola, females	Maryland	37	11	Limpert (1980)
Great Black-backed Gull Larus marinus	Northeast England	111	73[b]	Coulson et al. (1984)

The data are drawn from detailed studies in which attempts were made to identify all individuals present. Records for different years are pooled. N = number marked, % = percentage known to return to the next winter in the same area. Sizes of study areas are given in hectares or as length of coastline.
Other examples for Neotropical migrants in Rappole et al. (1983), for waterfowl in Robertson & Cooke (1999).

[a] This is the percentage of re-sightings (not total birds) that were in the same site, so is not comparable to most of the other data presented in which return rates include mortality as well as movement elsewhere.

[b] Includes 43/56 males and 38/55 females, all adults. Other figures of 9/17 (53%) for 2- to 14-year-old birds and 1/12 (9%) for 1-year-old birds.

them. In one study, shorebird species were ranked in order of decreasing winter site-fidelity (= increasing dispersal) from Purple Sandpiper *Calidris maritima* (most site-faithful), through Ruddy Turnstone *Arenaria interpres*, Curlew *Numenius arquata*, Ringed Plover *Charadrius hiaticula*, Oystercatcher *Haematopus ostralegus*, Redshank *Tringa totanus*, Dunlin *Calidris alpina*, Bar-tailed Godwit *Limosa lapponica* to Red Knot *Calidris canutus* (least site-faithful), the differences being again linked with the degree of year-to-year stability in their habitats and food supplies. Red Knots regularly moved from estuary to estuary during the course of a winter, but many visited the same several estuaries every winter. Except for the last three species, more than 94% of recovered individuals were found in the same section (within 10 km) of a large estuary within and between winters (Rehfisch *et al.* 2003). In the same study, juvenile Dunlins were found to move longer distances among roosts than did adults.

In some passerines which are territorial in the non-breeding season, site attachment may be so strong that individuals remain throughout each winter in the same 50 m radius, and return there year after year (see later). Other species, while apparently solitary during the non-breeding period, do not appear to defend territories, perhaps because they are more mobile, depending on patchily distributed food supplies (Salewski *et al.* 2002). Yet others join roaming flocks in the non-breeding season, foraging over wide areas.

The fact that birds have greater freedom to move around in the non-breeding than in the breeding season gives problems in measuring site-fidelity, especially in migrants. In some species that show site-fidelity, such as the Blackcap *Sylvia atricapilla*, the numbers of birds at particular sites have varied greatly from year to year, and from one part of the winter to another, depending on the size of the local fruit crops that provide their food. This fluctuation was apparently due largely to the behaviour of young birds which settled in their first year mainly in sites where fruit was plentiful, and partly to individuals (young or old) varying their length of stay from year to year, moving on each year when the local crops were eaten (Cuadrado *et al.* 1995).

Site-fidelity with variable lengths of stay has been noted in some other species that exploit annually variable food supplies, such as Rosy Finch *Leucosticte arctoa* (Swenson *et al.* 1988) and Eurasian Siskin *Carduelis spinus* (Senar *et al.* 1992). Typically at any one locality, some individuals seem merely to pass through, others stay for days or weeks and yet others for months, with the proportions varying from year to year in line with food supplies. It fits with the notion that many northern seed-eaters take broadly the same migration route each year, but pass particular latitudes much earlier in some years than in others, and reach the limits of their potential wintering range only in occasional years of widespread food shortage (Chapter 18). Species can therefore vary markedly in distribution from one winter to the next, in association with regional variations in food supplies (for large-scale patterns, see Rey 1995). This is true not only of the finches that depend on sporadic tree-seed crops, but also of some ground-feeding finches whose food may be covered by snow, of some raptors and owls which depend on temporarily and locally high rodent populations, of shorebirds of soft substrates whose prey varies in spatial distribution from year to year, and of some waterfowl whose food supplies are affected by variations in water levels and ice cover. Moreover, in any one year, many species show age-related differences in wintering range (Chapter 15;

Ketterson & Nolan 1976, Gauthreaux 1982a, Dolbeer 1991). The inference that indi-viduals of such species must use widely separated areas in different years has been increasingly supported by ring recoveries (Chapters 18 and 19).

The chance of recording ringed individuals at particular sites increases with their length of stay, and many more individuals could use a site than are present at one time. Because duration of stay is often linked with measured food sup-plies and social status, greater return rates are recorded in good than in poor food years, in adults than in first-year birds (e.g. 76% vs. 48% in Great Cormorant *Phalacrocorax carbo*, Yésou 1995), or in territorial than in non-territorial individuals (Cuadrado 1995, Johnson *et al.* 2001). The potential for observational bias gives uncertainty in how much the recorded variations in return rates between years and localities, or between sex and age groups, are due to differential site-fidelity, to differential survival, or merely to different durations of stay. All three factors could contribute in some degree to recorded return rates, adding to the effect of size of study areas. The question is whether birds move around in winter over dis-tances of metres or kilometres, or over tens, hundreds or thousands of kilometres.

Studies of insectivorous and frugivorous passerines (mainly Syliviidae and Parulidae) in the tropics have reported high winter recurrence, suggesting that at least most surviving individuals have returned to their territories in successive years (Nisbet & Medway 1972, Pearson 1972, Price 1981, Kelsey 1989, Holmes & Sherry 1992, Salewski *et al.* 2000, Wunderle & Latta 2000, Latta & Faaborg 2001, 2002). On the other hand, studies of wintering passerines (mostly Sylviidae and Turdidae) in the Mediterranean region have generally reported lower return rates, indicating that many birds could have changed winter quarters between years, or within years (Herrera 1978, Herrera & Rodriguez 1979, Finlayson 1980, Cuadrado 1992, 1995, Catry *et al.* 2003). As the Mediterranean species depend more heavily on fruit in winter, it would not be surprising if they were less terri-torial and site-faithful than their tropical wintering equivalents which take more insects, and hence have a more consistent food supply. Moreover, some of the studies involved only territorial individuals (e.g. Nisbet & Medway 1972, Kelsey 1989), while others involved a mixture of territorial and non-territorial ones (e.g. Constant & Ebert 1995, Cuadrado 1995). Non-territorial ones would be expected to show less site-fidelity.

Swans, geese and sea-ducks generally show high return rates to wintering sites, while freshwater diving ducks show lower return rates, and dabbling ducks still lower rates (Robertson & Cooke 1999). The differences again relate partly to the stability and permanence of the various habitats that the different species occupy, and to changing water levels and ice cover. Wide dispersal between one winter and the next has been well documented from ring recoveries of several species, including Common Pochard *Aythya ferina* (see **Figure 19.8**) and Tufted Duck *A. fuligula* (M. Kershaw and R. Hearn, in Wernham *et al.* 2002).

Sex-related differences

Few data are available to compare winter site-fidelity between the sexes. Sex-differences were apparent in the return rates of Harlequin Ducks *Histrionicus histrionicus* to a wintering site in British Columbia (77% males vs. 62% females, Robertson & Cooke 1999), and of Buffleheads *Bucephala albeola* to a site in

Maryland (26% males vs. 11% females, Limpert 1980), but at least part of this sex difference may have been due to differential survival rather than differential site-fidelity. In contrast, ring recoveries gave no indication of sex differences in the winter site-fidelity of Black Ducks *Anas rubripes*, Canvasbacks *Aythya valisineria* and American Woodcocks *Scolopax minor* in eastern North America (Nichols & Haramis 1980, Diefenbach *et al.* 1990), or of Whooper Swans *Cygnus cygnus* and Bewick's Swans *Cygnus columbianus* in Britain (Scott 1980, Black & Rees 1984). Such sex differences would not be expected in adult geese and swans because they remain in pairs year-round. So while sex differences in winter site-fidelity may occur in some species, they are apparently absent in others.

Age-related differences

Outside the breeding season, when many birds move away from their nesting areas, the adults often move less far, or stay away for less long, than the young (Chapter 15). This is true whether the species performs a fixed-direction migration or a multi-directional dispersive migration. Some long-lived species, in which individuals do not breed until they are several years old, show a progressive change to shorter distance moves or to shorter periods away from the breeding areas, with increasing age (Chapter 15). Such patterns have been noted in a wide range of bird species, including Grey Heron *Ardea cinerea* (Olssen 1958), Black Kite *Milvus migrans* (Schifferli 1967), Eurasian Oystercatcher *Haematopus ostralegus* (Goss-Custard *et al.* 1982), Herring Gull *Larus argentatus* (Coulson & Butterfield 1985), Common Guillemot *Uria aalge* (Birkhead 1974), Great Cormorant *Phalacrocorax carbo* (Coulson 1961) and Northern Fulmar *Fulmarus glacialis* (Macdonald 1977). This means that some individual birds occupy different areas in successive winters, as they age, or that they spend progressively less time on their wintering areas as they age.

Studies on other species have indicated increasing winter site-fidelity with increasing age and social status, greater in males than in females (for Great Black-backed Gull *Larus marinus* see Coulson *et al.* 1984, for Great Cormorant *Phalacrocorax carbo* see Yésou 1995, for Mallard *Anas platyrhynchos* see Nichols & Hines 1987, for swans see **Table 17.7**), and matching the findings from breeding areas (see above). However, it is usually uncertain how much the higher return

Table 17.7 Percentage return to wintering sites by different age/social classes of swans

	Pairs without young		Pairs with young		Single adults		Yearlings		Source
	N	%	N	%	N	%	N	%	
Tundra Swan *Cygnus columbianus bewickii*	162	78	116	67	43	58	20	55	Scott (1980)
Whooper Swan *Cygnus cygnus*	24	88	7	86	63	70	5	60	Black & Rees (1984)

Note: Tundra Swan data from Welney Refuge, southeast England; Whooper Swan data from Caerlavorock Refuge (12 km²), southwest Scotland. Some returns occurred after more than one year, and in no group were sex differences detected.

rates of adult birds, compared to young ones, are due to greater site-fidelity (and movements elsewhere) and how much to greater survival from the previous year.

Comparison of breeding and non-breeding site-fidelity

Many species show greater fidelity to breeding than to wintering sites. This reflects not only spatial variation in wintering habitat from year to year, but also the fact that individuals of many short-distance migrants migrate in their first but not in subsequent years, or migrate different distances in different years (Chapters 12, 16 and 18). However, it is also apparent in some long-distance migrants, such as Willow Warbler *Phylloscopus trochilus* and White Stork *Ciconia ciconia* (Salewski *et al.* 2002, Berthold *et al.* 2002). White Storks show extreme fidelity to their breeding places, normally returning year after year to the same nests, but they are much less faithful to particular wintering places. During studies based on satellite-tracking, four White Storks were tracked from Europe to their African winter quarters several times, one bird on nine successive journeys (Berthold *et al.* 2002). These birds occupied different areas in Africa from year to year, depending on the food supply. One bird wintered in Tanzania in one year, further north in the Sudan in the second year, but then went as far as South Africa in the third year and Botswana in the fourth. Hence, this one individual wintered in different years in places that extended from the northern tropics to the southern temperate zone.

Other species show the opposite tendency, with less fidelity to their breeding areas than to their wintering areas. In the American Redstart *Setophaga ruticilla* and Black-throated Blue Warbler *Dendroica caerulescens*, lower site-fidelity in summer reflected a breeding habitat that varied strongly in suitability from place to place, and from year to year (Holmes & Sherry 1992). The same holds for some arctic-nesting shorebirds, such as Curlew Sandpiper *Calidris ferruginea* and Sanderling *Calidris alba* whose breeding sites may change from year to year according to snow-melt patterns (Tomkovich & Soloviev 1994), but whose wintering sites are more consistent and are used by the same individuals year on year (Elliott *et al.* 1977, Evans 1981). To set against these patterns, other species show extreme site-fidelity in both breeding and wintering areas, while others show little or no site-fidelity in either breeding or wintering areas, depending on the degree of predictability in their habitats and food supplies (Chapters 18 and 19).

FIDELITY TO STOPOVER SITES

Most migrants pause for refuelling up to several times during their journeys. Individuals of some species have been identified in successive years at the same staging sites which they visit for at most a few days or weeks at a time before moving on. Moreover, because some species take different routes on their outward and return migrations, individuals may use different stopping sites at the two seasons. This leads to the remarkable implication that some migratory birds remember the specific locations of several sites scattered over two continents, which they visit successively each year on a circuit that is repeated annually throughout their lives. To some extent, the landscape itself and its associated habitat areas are likely to impose patterns of recurrence, regardless of any inherent tendencies in the birds themselves. Species with an infinite number of potential stopping places on their migration routes are perhaps less likely to show strong fidelity to particular sites than species that have only a small number of possible sites.

Most passerines migrate in a broad front over more or less continuously suitable habitat, offering a wide choice of potential stopping places. In addition, the turnover of birds at stopover sites is high, as individuals normally remain only for short periods, often less than one day, so the chances of recording particular individuals are low. Not surprisingly, therefore, most studies in which passerine migrants were trapped year after year at the same stopping sites have provided little or no evidence of individual year-to-year site-fidelity; others gave small, but highly variable, proportions of retraps, exceeding 10% only in extreme cases (Moreau 1969, Nisbet 1969, Winker *et al.* 1991, Winker *et al.* 1992b, Cantos & Tellería 1994, Merom *et al.* 2000, Dowsett-Lemaire & Dowsett 1987). Some of the low return rates were perhaps no greater than expected by chance, if birds had paused at random in suitable habitat encountered en route, but this possibility cannot be tested statistically. Recurrence might be higher at habitat patches situated on the edges of seas or deserts that offer the last chance to feed before a crossing, or the first chance to feed after the crossing.

Another problem with existing data on stopover fidelity in passerines is that they derive from sites where large numbers of birds were caught each year, and the chance of catching any one individual, even if present, was low. Moreover, in some studies migrants could not always be separated from summering or wintering birds of the same species, so that the samples did not necessarily consist entirely of birds on migration. However, one study in which attempts were made to address these problems, and in which trapping was done in a consistent manner from year to year, was undertaken at a reed-bed site in south Portugal over five years (Catry *et al.* 2004b). In the five species and populations known to include only passage migrants at this site, the proportions of individuals retrapped there in a subsequent year varied between 0 and 1.2% of those ringed over a five-year period (**Table 17.8**). Two other populations that showed higher recovery rates could have included summer visitors (Reed Warbler *Acrocephalus scirpaceus*) and winter visitors (Bluethroat *Luscinia luscinia*). After attempts to correct for low recapture probabilities and annual mortality, maximum possible return rates in the five migrant populations were estimated at 0–12.9% of surviving birds (**Table 17.8**). Hence, fidelity to this stopover site (considered as good habitat) was generally low.

A similar study on Pied Flycatchers *Ficedula hypoleuca* was made in central Spain (away from breeding areas), where mist nets were operated in a 1-ha garden throughout the migration seasons of 1983 and 1984. To judge from fattening rates, this garden provided good habitat, but of 122 Pied Flycatchers caught in 1983, only one was retrapped there the following year (Veiga 1986). The maximum possible proportion of surviving birds estimated to have returned to the site was 10.6%, a figure within the range found for other passerines. These rates were much lower than those reported from breeding and wintering sites. However, in Eurasian Reed Warblers *Acrocephalus scirpaceus*, for birds caught two

Table 17.8   Overall between-year re-capture rates for different species and populations (short-winged and long-winged) caught in the autumn migration season mainly at Santa André, southwest Portugal

	Total ringed	Number (%) re-captured	Estimated maximum return rate[a]
Migrants known to be on passage			
Sedge Warbler *Acrocephalus schoenobaenus*	498	1 (0.2)	5.4
Common Grasshopper Warbler *Locustella naevia*	432	0 (0)	0
Willow Warbler *Phylloscopus trochilus*	3365	3 (0.1)	4.9
European Pied Flycatcher *Ficedula hypoleuca*	122	1 (0.8)	10.6
Migrants likely to be on passage			
Bluethroat (long-winged) *Luscinia svecica*	118	1 (0.8)	9.1
Eurasian Reed Warbler (long-winged) *Acrocephalus scirpaceus*	1244	15 (1.2)	12.9
Probable mixture of passing migrants and regionally breeding or wintering birds			
Bluethroat (short-winged) *Luscinia svecica*	200	14 (7.0)	61.8
Eurasian Reed Warbler (short-winged) *Acrocephalus scirpaceus*	2793	126 (4.5)	26.2

[a]Taken as the between-year re-capture rate divided by the within-year re-capture rate, and by the estimated survival rate (from other studies) to give maximum likely estimates of the return rates of surviving birds (after Catry *et al.* 2004b).
From Catry *et al.* (2004). Data for Pied Flycatcher are from central Spain (Veiga 1986).

or more times, return rates to a stopover site in Israel were 27/123 or 22%, which was not very different from the return rates of summer breeders in the same area (210/773 or 27%) (Merom *et al.* 2000). Among Chaffinches *Fringilla coelebs* migrating over the Courland Spit in the southeast Baltic, some individuals were recaught up to seven years after having been ringed there (Payevsky 1971).

For other migratory birds, notably waterfowl and shorebirds, suitable feeding areas are often localised and far apart, and it is in such species that the highest recurrence rates have been recorded. For example, some shorebirds and geese have only one or two main stopover sites on their migration routes, so inevitably almost the entire population may stop at these sites on each journey. Not surprisingly, then, many ringed birds of these species are known to have visited the same staging sites year after year on migration.[9] The minimum annual return rate of adult Semi-palmated Plovers *Charadrius semipalmatus* to 10–15 ha of beach at Manomet in eastern North America, where the birds stayed for about three weeks each autumn, averaged 71% (Smith & Houghton 1984). This proportion was similar to the likely annual survival rate, implying that most surviving individuals returned. Another large estimate of 65% was obtained for Sanderlings *Calidris alba* at a stopover site in southwest Iceland (Gudmundsson & Lindström

[9]*For shorebirds, see Pienkowski (1976), Evans & Townsend (1988), Harrington et al. (1988), Smith & Houghton (1984), Gudmundsson & Lindström (1992), Pfister et al. (1998), Wernham et al. (2002).*

1992). Similarly, among Greenland White-fronted Geese *Anser albifrons* staging in two areas of Iceland, and seen in more than one season, 89% were re-sighted within 4 km of the capture site from spring to the following autumn, 88% from autumn to the following spring, 96% from one spring to the next, and 100% from one autumn to the next (Fox *et al.* 2002). In general, same-season site-fidelity (97%) was significantly greater from year to year than different-season site-fidelity (87%, Fishers exact test, $n = 177$, $P < 0.05$). But the overall rates were again so high as to imply that all (or almost all) surviving birds used the same sites in successive years. Lower estimates for geese include 40.7% (or 48.5% when adjusted for mortality) in Brent Geese *Branta bernicla* in the Netherlands (Ebbinge 1992), 52.4% as a mortality-adjusted estimate for Barnacle Geese *B. leucopsis* in Norway (Gullestrad *et al.* 1984), and 72% as a non-adjusted estimate for Canada Geese *B. c. minima* in Alaska (Gill *et al.* 1997). Among birds in general, however, this degree of stopover site-fidelity may well be exceptional, and despite following the same route, many migrants may stop at different places in different years, either by choice or by force of circumstance.

High fidelity to stopover sites might also be expected in soaring raptors, storks, cranes and pelicans, in which entire populations funnel each spring and autumn through narrow bottlenecks (Chapter 7). Such narrow front migration is likely to limit the number of stopover sites available, thus raising the probability that individuals use the same places on successive journeys. Many individuals may use these sites for roosting, but not necessarily for feeding.

In summary, passerines that encounter many possible stopping places on their journeys apparently show less fidelity to specific stopover sites than do some waterfowl and shorebirds for which potential stopping places are few and far between. Among many passerines, fidelity to stopover sites is much less marked than fidelity to breeding or wintering sites. Patterns of stopover site-fidelity seem to reflect the effects of landscape, habitat and food supplies, and not only features of the species themselves.

POST-FLEDGING DISPERSAL

Once they become independent of their parents, the young of many bird species disperse from their natal sites, for a time moving in various directions and progressively further with increasing age (**Table 17.9**). As birds hatched in particular study areas gradually leave, others hatched elsewhere move in. This moving and mixing happens every year at this time regardless of local conditions. Yet post-fledging dispersal is the least conspicuous of all bird movements, for it occurs at a time when many birds are at their most silent and often moulting. They accumulate no special body reserves, and although they appear in a wider range of habitats and places than when nesting, they undergo no large-scale distributional change. Evidently, the tendency to disperse is inherent (and presumably endogenously controlled), but the actual distances that individuals move may be influenced by environmental conditions, as explained above. After a period of post-fledging dispersal, individuals of many sedentary species appear to stay in those areas thereafter, whereas individuals of migratory species depart for their winter quarters.

Table 17.9 Percentage of European Pied Flycatchers *Ficedula hypoleuca* caught at different distances from the natal site at successive ages

Age of birds (days)	Number caught	Distance between natal site and capture site (km)						
		0–1	1.1–2.0	2.1–3.0	3.1–4.0	4.1–5.0	5.1–10.0	>10
20–30	123	74.8	9.8	6.5	4.9	4.0	0	0
31–40	177	49.2	20.3	11.9	10.2	6.8	1.1	0.5
41–50	76	36.8	25.0	11.8	9.2	2.6	9.2	5.3

From Sokolov (2000b).

On their return from winter quarters the next spring, the migrants settle near their natal areas in the 'dartboard' pattern described earlier, sometimes in the localities to which they dispersed in the post-fledging period. This was shown, for example, in European Robins *Erithacus rubecula* studied near Lake Ladoga in Russia (Zimin 2001, 2002). In this area of 7 × 1.5 km, most local young were replaced by immigrants during the period of post-fledging dispersal. The immigrants stayed on their territories until they had completed moult, and then left on migration. Survivors from these summer immigrants returned the following spring to nest in the study plot, often in the very same territories where they had moulted the previous autumn. In its effect on settling patterns, their post-fledging dispersal was tantamount to natal dispersal. However, individuals of some resident species may have a second period of dispersal in spring.

In many migratory birds, from songbirds to raptors, centrifugal post-fledging dispersal precedes their first southward migration (for Barn Swallow *Hirundo rustica* see Ormerod 1991, for Pied Wagtail *Motacilla alba yarrelli* see Dougall 1992, for Sand Martin *Riparia riparia* see Mead & Harrison 1979, for Willow Warbler *Phylloscopus trochilus* see Norman & Norman 1985, for Common Kestrel *Falco tinnunculus* see Snow 1968, Village 1990). The young from early broods have a longer period between fledging and departing on migration than young from later broods, and tend to wander over wider areas from the nest. In some species, a sex difference is also apparent, with females dispersing at an earlier age and further from their natal sites than males in the Willow Warbler *Phylloscopus trochilus* (Norman 1994) and males earlier and further than females in the Sparrowhawk *Accipiter nisus* (Newton 1986).

For a few weeks in late summer, the pattern of ring recoveries in migrants is similar to that of natal dispersal, with no directional preference (except in a few species in the northern hemisphere which have shown a northerly bias in post-fledging dispersal directions; e.g. Lesser Kestrel *Falco naumanni*, Olea 2001; certain herons, Chapter 1). Then, as the young age, recoveries come from longer distances and are more directionally orientated, as the initial centrifugal dispersal changes to long-distance directed migration. The change from random to directed movements is also evident among birds caught from the wild and tested in orientation cages (Shumakov 2001). In those passerine species that moult in their breeding areas, the transition from dispersal to migration occurs around the end of moult, and hence earlier in young from earlier than from later broods

(Boddy 1983, Dougall 1992). But in some other long-distance migrants which do not moult in their breeding areas, directional movement begins soon after the young become independent of parental care, as shown, for example, by the general southwesterly movement of newly independent European Pied Flycatchers *Ficedula hypoleuca*, Lesser Whitethroats *Sylvia curruca* and others revealed by ringing on the Courish Spit (Sokolov 2000b).

In some species, such as the European Robin *Erithacus rubecula* and Eurasian Reed Warbler *Acrocephalus scirpaceus*, dispersal can occur at night, which again implies that this is a movement that birds are programmed and adapted to make (Rezvyi & Savinich 2001, Mukhin & Bulyuk 2001). Dispersing Reed Warblers are attracted by nocturnal song playback, and some were caught in mist nets placed in reedbeds near the tape recorder. The majority of birds were caught in the last third of the night. That such birds were dispersing rather than migrating was indicated by their age (mostly 39–52 days), their state of moult, lack of migratory fat and lack of consistent directional preferences when tested in orientation cages (Bulyuk *et al.* 2000, Mukhin 2004). Because Robins and Reed Warblers migrate at night, it is perhaps not surprising that they also disperse then; but it raises the question how many other species do so.

In the Greater Flamingo *Phoenicopterus ruber*, dispersal distances of young birds in their first autumn were correlated with body condition at fledging, with birds in poor condition staying near the natal colony or making short moves, and birds in good condition making the longest moves, involving sea-crossings (Barbraud *et al.* 2003). However, flamingos occur in restricted and isolated habitats separated by long distances, and in other birds occupying more continuous habitats dispersal may be less dependent on body condition. Like many other birds, however, flamingos often returned to their first wintering sites in subsequent years, if they migrated at all then.

DISPERSIVE MIGRATION

Ring recoveries resulting from post-fledging dispersal usually drop off with increasing distance from the origin, as in natal dispersal, but they may extend further in some years, regions or habitats than in others, and further in one sex (usually females) than the other. In many resident species, recoveries come from greater average distances in successive months from fledging into winter, and then towards spring they come from localities progressively closer to the origin. The out-and-back movement patterns that are implied are apparent in a wide range of 'resident' species in Britain, including Great Tit *Parus major*, Blue Tit *P. caeruleus*, Song Thrush *Turdus philomelos*, Greenfinch *Carduelis chloris*, Chaffinch *Fringilla coelebs*, Common Murre *Uria aalge*, Herring Gull *Larus argentatus*, Great Black-backed Gull *Larus marinus*, European Shag *Phalacrocorax aristotelis*, Great Cormorant *P. carbo*, Grey Heron *Ardea cinerea*, Common Buzzard *Buteo buteo* and Mute Swan *Cygnus olor* (Wernham *et al.* 2002).

Such patterns could result from birds moving first progressively outward from their natal sites and then, as the next breeding season approaches, moving back towards their natal sites. Or they could result from birds that disperse furthest being more likely to die (not implausible if they are subordinate to others), and becoming less represented in the samples from late winter and spring. However, an outward–return movement has been confirmed independently in many species from repeated captures or sightings of live individuals, and also from radio-tracking studies (e.g. Spruce Grouse *Dendragapus canadensis*, Herzog & Keppie 1980, Schroeder 1985; Blue Grouse *D. obscurus*, Cade & Hoffman 1993; Greater Prairie Chicken *Tympanuchus cupido*, Schroeder & Braun 1993; Red Kite *Milvus milvus*, Evans *et al.* 1999; Great Bustard *Otis tarda*, Morales *et al.* 2000; Common Buzzard *Buteo buteo*, R. Kenward & S. Walls, in Wernham *et al.* 2002). In all these species, individuals moved outward from their natal or nesting areas after the breeding season, settled elsewhere over winter, and then moved back in time for the next breeding season. In other parts of their range, some of these species perform directed migrations.

Because this dispersal is a back-and-forth seasonal movement, consistent in timing from year to year, it parallels migration. But at the population level it differs in that: (1) it is not directional (unless direction is imposed by landscape or coastline); (2) the movements are also relatively short – say mostly in the range of a few kilometres or tens of kilometres (at least in passerines and game birds); (3) recoveries drop off rapidly with increasing distance rather than concentrating in a specific distant wintering area; and (4) all or most individuals remain year-round within the breeding range of the species (although some seabirds may disperse far from their colonies to remote feeding areas). The term dispersive migration seems appropriate for this type of movement, emphasising that it is seasonal, in various directions, and involves outward and return stages. While not obviously associated with seasonal changes in food supplies, one function of the movement is presumably to seek out better feeding sites for the winter, or to exploit places that are suitable for wintering but not for breeding. It is obviously not related to latitudinal trends in food supplies.

In Chapter 1, I alluded to the fact that the different types of bird movements intergrade, and that no clear distinction separates dispersal from migration. Nowhere is this more apparent than in dispersive migration. In some populations of dispersive migrants, some individuals shuttle back and forth between regular breeding and wintering areas, and show fidelity to both sites (as in Spruce Grouse *Dendragapus canadensis*, Herzog & Keppie 1980). As in many other birds, males migrate in smaller proportion than females, leave somewhat later and return earlier to their breeding areas next spring. In these respects, the return movements resemble the partial migrations of other birds. There are also sex differences in the distances moved, and in some lekking species in which males play no part in parental care, males travel further, on average, than females, as in Blue Grouse (Cade & Hoffman 1993) and Great Bustard (Morales *et al.* 2000). Significant differences in the distances moved are also apparent between age classes in some species. For example, adult and yearling female Black Grouse *Tetrao tetrix* that were radio-tagged in their wintering sites moved median distances of 2.6 km and 9.2 km to breeding sites, and maximum distances of 29.6 and 33.2 km, respectively. Females from the same lek moved in different directions from one another, with no overall directional preference (Marjakangas & Kiviniemi 2005).

In Common Buzzards *Buteo buteo* and other large raptors, ring recoveries suggest that birds disperse in their first year, and then move back towards their natal area as they approach breeding age (R. Kenward & S. Walls, in Wernham *et al.* 2002, Haller 1982). However, the radio-tracking of 124 young Common Buzzards showed that the situation was more complicated than this (Walls & Kenward 1998). In one study, the young started by making long excursive flights when they were about 70 days old, after their feathers had hardened, and then explored up to 25 km from their nests without dispersing (i.e. continually returning to their natal nests). However, about half the young also dispersed right away from natal areas by their first October. These birds moved longer distances than the remaining birds which left later in winter and the following spring. After leaving their natal areas, birds often changed home range more than once before settling, but none moved after their third summer. In the spring, they made temporary visits back to their natal areas, earlier in each successive year until they bred, either there or, more frequently, where they had settled in winter (Walls & Kenward 1998). Patterns of outward–return movement each year, of females moving further than males, and spending longer each year in the natal (future breeding) area with increasing age have also been recorded among Red Kites *Milvus milvus* (Evans *et al.* 1999), and some aspects in Spanish Imperial Eagles *Aquila adalberti* (Ferrer 1993). Again, these patterns of behaviour recorded from dispersive migrants during their first few years of life resemble those recorded from regular migrants wintering at lower latitudes (Chapter 15).

Dispersive migration on a much grander scale occurs among some seabirds which move in various directions away from their nesting colonies after breeding, but often remain within the same sea-water zones as they use in the nesting season (**Figure 17.6**). This has been indicated by ring recoveries from many species, and more convincingly by use of geolocation loggers attached to birds at their breeding colonies. This latter method has been used on adult Wandering Albatrosses *Diomedia exulans* nesting on the Crozet Islands in the southern Indian Ocean (Weimerskirch & Wilson 2000). After the young have fledged, the adults of this species leave the foraging areas frequented while breeding and head for sea areas 1500–8500 km away, elsewhere in the southern Indian Ocean or in the southwest Pacific. Here they spend the next year, returning to their nesting colonies for the following year, as they breed only every second year, being replaced in the nesting colonies during their away-years by other individuals. The ocean areas they occupy range from subtropical–tropical waters (females) to sub-Antarctic–Antarctic waters (males), and individuals probably visit the same areas in each sabbatical (Weimerskirch & Wilson 2000).

Similarly, 22 adult Grey-headed Albatrosses *Thalassarche chrysostoma*, studied by use of geolocation loggers, were followed throughout the approximately 18-month interval between successive breeding attempts on South Georgia (Croxall *et al.* 2005). During their sabbaticals, these birds showed three distinct dispersal strategies: (1) some stayed in the southwest Atlantic in an enlarged version of the breeding home range; (2) others made return movements (mainly eastward and back from the colony) to a specific region of the southwest Indian Ocean; and (3) yet others made one or two eastward heading journeys around the earth, foraging in various regions en route (the fastest in just 46 days) (**Figure 17.7**). Females more often remained in a restricted range, while most males performed at least one round-the-earth trip. The timing of journeys was generally well synchronised between individuals, and of consistent duration, and most made the same journeys in different years. Again, these birds on their dispersive migrations remained throughout within the same latitudinal band, and their round-the-earth journeys followed the prevailing winds.

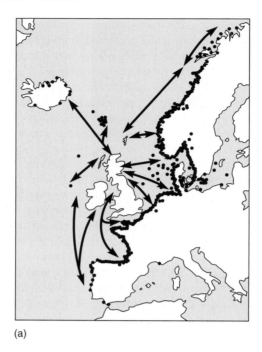

(a)

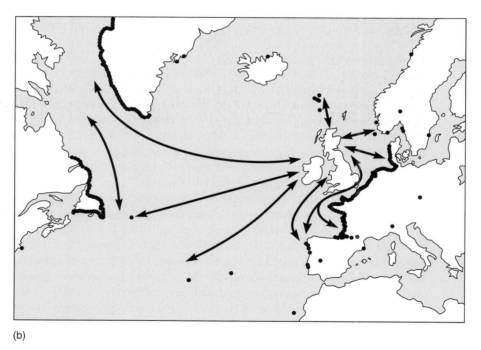

(b)

Figure 17.6 Dispersive migration of two species of seabirds, based mainly on recoveries of birds ringed as nestlings in Britain (Wernham *et al.* 2002). (a) Common Guillemot *Uria aalge*. (b) Black-legged Kittiwake *Rissa tridactyla*. Modified from Flegg (2004).

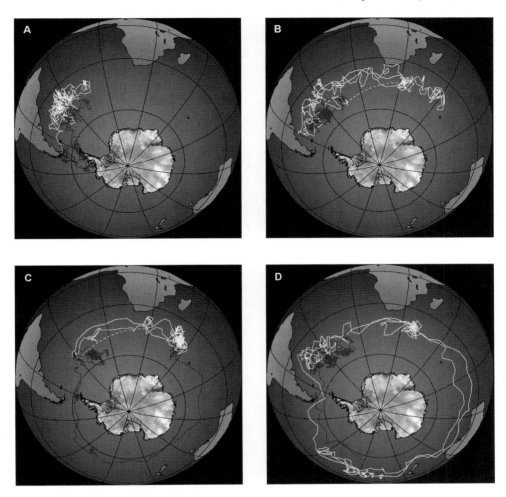

Figure 17.7   Representative migration routes of four Grey-headed Albatrosses (A–D) *Thalassarche chrystostoma* in the 18 months between successive breeding attempts. White – winter; dark – summer; dotted lines link locations obtained before and after the equinox. From Croxall *et al.* (2005). Reprinted with permission from AAAS.

Similar features involving broad-scale site-fidelity and consistency in timing of movements were also noted in Black-browed Albatrosses *Thalassarche melano-phrys* from South Georgia, which spent their sabbaticals off southwest Africa or Australia (Phillips *et al.* 2005). The migrations of all three albatross species could be said to be food-related, as birds moved from near breeding areas where competition was intense to other areas with abundant food away from colonies, but they achieved this by predominantly longitudinal rather than latitudinal shift. Moreover, individuals from the same colony did not all depart in the same direction as one another, and some of these directions differed from those taken by individuals from other colonies. For these reasons, it is hard to imagine that the directions taken by individual albatrosses have an innate basis, and more likely that cultural transmission and personal experience leads birds to focus on particular feeding areas (rather like moult migration in ducks, Chapter 16). Their movements all fall within the category of dispersive migration, as defined here.

SITE ATTACHMENT

Attachment of young birds to natal sites

The young of migratory birds must presumably learn the location of their natal area before they leave it, for only then could they return there after a long migration. This assumption has been confirmed experimentally in various species. When eggs or young were transferred from one area to another, tens or hundreds of kilometres away, the resulting adults were usually found breeding near their foster home rather than near their original home. The age of the birds at transference emerged as crucial, because attachment to a locality occurred at a specific stage of development: in passerines in the post-fledging period, between leaving the nest and leaving the area (Löhrl 1959, Berndt & Winkel 1979, Sokolov *et al.* 1994).

In one experiment, young Collared Flycatchers *Ficedula albicollis* were hand-reared at one site, and released at different ages at another, situated about 90 km to the south and previously unoccupied by this species (Löhrl 1959). Individuals released before or early in post-juvenile moult returned next spring to the release site, whereas those released late in moult or after moult did not return there. A short period of freedom during the first fortnight or so after leaving the nest appeared sufficient to fix the locality on these birds. Further research showed that fixation to a potential breeding locality occurred between 45 and 55 days of age in both Collared Flycatchers and European Pied Flycatchers *Ficedula hypoleuca* (Berndt & Winkel 1979). The young of these species are raised in cavities, so would have no opportunity to view the outside world until they left the nest.

In another experiment, young Chaffinches *Fringilla coelebs* released in a new area at less than 30 days old returned to the area in a subsequent year, while those older than 40 days returned after migration mainly to their natal place (Sokolov 1997). Evidently, site imprinting in Chaffinches occurred between ages 30 and 40 days, again mainly in the post-fledging period, before the young dispersed. However, if Chaffinches were hand-reared without sight of the outside world and then released at 50 days of age, some still returned to the release site in a later year. Hence, if birds were denied the chance to learn the site at the usual age, they could do so later. If they learnt at the usual age, subsequent experience did not alter the preference established then. Age at site attachment for various other small passerines lay in the range 30–55 days, depending on species and rate of development (Sokolov 2000b).

Attachment to a locality does not mean that we can expect all surviving young to breed there subsequently. They may return there initially in spring but, on finding breeding sites already occupied, they may have to move elsewhere. Alternatively, they may not return with exact precision to the natal area, but to somewhere nearby, perhaps to a place they located during their post-fledging dispersal before they left on migration (see above).

Other experiments have involved pelagic seabirds, in which individuals range over large areas of ocean before they return to their natal colonies up to several years later. Owing to the building of a military airbase, 3124 Laysan Albatross *Diomedea immutabilis* fledglings were transferred from Midway Atoll in the Pacific to other colonies up to 400 km away (Fisher 1971). Most of the survivors returned to Midway Island years later, as they reached breeding age. They had been raised in open nests, and by the time they were moved, they had already become attached to their natal area, and were unaffected by the move. If young were relocated long before they reached fledging age, some of the transported chicks later returned to the release site rather than to the natal site. The sensitive stage, when site attachment occurred in this species, fell in the last fourth of the nestling period.

Even in burrow-nesting Short-tailed Shearwaters *Puffinus tenuirostris*, the young developed an attachment to the natal colony long before they could fly (Serventy *et al.* 1989). Attachment occurred in the nestling period, as revealed by translocation experiments in which eggs and chicks of different ages were moved between different islands (Serventy *et al.* 1989). It presumably occurred at night when chicks sat outside their burrows, for only then could they see the outside world. After several years at sea, with annual migrations covering 30 000 km, returning birds distinguished their own colony from other colonies only 1 km and 3 km away.

The tendency of young birds to breed near where they were raised has contributed to the success of conservation programmes in which certain bird species have been reintroduced to areas from which they were earlier eliminated. Even where young raised in the release area moved away after becoming independent, most of the survivors returned there to breed, leading to the establishment of a new local population. This has occurred in a wide range of species, including waterfowl, raptors and seabirds (Newton 1979, Kress & Nettleship 1988, Cade 2000). Again, the implication was that the young became fixated to the locality during their early life.

Attachment of young birds to wintering sites

The attachment of young migrants to specific wintering localities seems to occur soon after their arrival there. Of 111 wintering Palm Warblers *Dendroica palmarum* that were caught soon after arriving on the Bahamas, 34 were released where they were caught, and the rest were transported and released at sites 9.7 and 22.5 km away. The proportion recaught at the capture site was about the same in all three groups, showing that displaced birds had returned to the capture site. Moreover, birds translocated at early dates in their stay (28 November–2 December) showed no difference in return rates from birds moved at later dates, suggesting that in this species site fixation had occurred fairly soon after arrival there (Stewart & Connor 1980).

Similarly, juvenile and adult Dunlins *Calidris alpina* that had arrived at a wintering site mainly in September–October were displaced 133 km from the site at different dates in November–December (Baccetti *et al.* 1999). From the proportions that returned from birds displaced at different dates, the juveniles seemed to have become attached to the site during November, within 1–2 months after arriving there. By December, return rates were the same as those of adults, which showed no seasonal pattern, presumably because they had known the site from previous years (for similar findings on Sanderling *Calidris alba* see **Table 9.1**; Myers *et al.* 1988).

OTHER ASPECTS OF DISPERSAL

Links between breeding and wintering sites

To what extent do birds from particular breeding areas migrate to the same wintering areas as one another, as opposed to different wintering areas? Their 'migratory connectivity' could be considered strong or weak, depending on the extent to which individuals from a given breeding area distribute themselves among different wintering areas, mixing with other individuals from different breeding areas. The answer to this question has implications for the ecology and genetic structure of populations, as well as for their conservation.

Among geese and swans, birds from separate (often isolated) breeding areas migrate along traditional routes to separate (often isolated) wintering areas. For example, Barnacle Geese *Branta leucopsis* from Greenland, Spitzbergen and Novya Zemlya migrate each autumn to separate parts of western Europe, with very little exchange of ringed individuals between the three populations (see **Figure 23.3**). Hence, even though pair formation occurs in winter in this species, breeding populations remain almost genetically isolated. Some populations of other goose species are subspecifically distinct, implying that this isolation is of long standing, as in the Greenland and European races of the Greater White-fronted Goose *Anser albifrons*, both of which winter in Europe but in different parts. Because juvenile geese migrate with their parents, they winter in the same areas, and find their breeding partners there. In most species of geese, swans and cranes, which use narrow migration routes and a limited number of staging and wintering sites, many individuals clearly remain in close proximity to one another for much of the year. In a sense, they migrate as a community, but do not necessarily stay together all the time. In a much less extreme situation, populations of many bird species remain partially segregated year-round, through parallel, chain and leap-frog migration patterns (Chapter 23), so that birds from different breeding areas clearly do not mix at random in winter.

In other species, however, ringing and satellite-based radio-tracking have shown that different birds from the same breeding locality can migrate to widely separated wintering places, and conversely that birds from a single wintering locality can migrate to widely separated breeding places. Individuals may return year after year to their own breeding and wintering sites, but have different sets of neighbours at the two seasons. For example, Peregrine Falcons *Falco peregrinus* caught wintering on a 50-km stretch of coast in eastern Mexico were tracked to breeding areas that lay across much of North America and western Greenland, with a west–east spread of more than 5000 km (see **Figure 23.5**; McGrady *et al.* 2002). Similarly, Peregrines from any one breeding area wintered at sites scattered over a wide range of latitudes from southern North America to northern South America, mixing with Peregrines from other breeding areas (Chapter 8). Again, individuals tracked in more than one year showed fidelity to their own breeding and wintering sites. In some species, even breeding partners tracked by satellite wintered in areas separated by more than 1000 km, as shown in Ospreys *Pandion haliaetus*, Greater Spotted Eagles *Aquila clanga* and others (Chapter 8; Kjellén *et al.* 1997, Meyburg *et al.* 1998). In contrast to the geese discussed above, such species could be said to show weak migratory connectivity. Species that breed or winter in different areas in different years, according to conditions, are also likely to show weak connectivity.

Dispersal and genetic structure

Dispersal patterns depend largely on the ecology of the species themselves, particularly on the level of stability in their habitats and food supplies. But such patterns in turn have a major influence on the genetic structure of populations and their propensity to subspeciate: when there is little movement of individuals among populations, considerable genetic substructuring can arise. Several studies have drawn attention to the link between dispersal and subspeciation (Rensch 1933, Belliure *et al.* 2000, Newton 2003). This subject is outside the scope of this book, except to point out that species with large geographical ranges in which individuals disperse over long distances between different breeding areas typically have few, if any, subspecies, whereas species in which individuals disperse over short distances have many subspecies (Newton 2003). This is evident even within families in which species vary in their dispersive behaviour, including finches, owls, raptors and waterfowl. More recent studies of dispersal patterns of Great Tits *Parus major* have shown the potential for genetic substructuring at a more local scale, over distances of kilometres or tens of kilometres, related to spatial variation in habitat quality (Garant *et al.* 2005, Postma & van Noordwijk 2005). How long this local substructuring lasts remains to be seen.

SUMMARY

For individual birds, dispersal can be measured by the distances between natal sites and subsequent breeding sites (natal dispersal), between the breeding sites of different years (breeding dispersal), and between the wintering sites of different years (non-breeding dispersal), regardless of any movements made in the interim. In all these types of dispersal, as well as in post-fledging dispersal, directions appear random (and centrifugal at the population level). Dispersal enables individuals to leave areas of overcrowding or poor food supply to explore and find somewhere better; it also reduces the chance of pairing with a close relative. At the population level, dispersal movements influence gene flow and the genetic structure of populations, facilitating or suppressing the development of locally adapted populations and subspecies. Dispersal movements also influence patterns of abundance and distribution across the range, enable depleted local populations to recover, vacated areas to be re-colonised or new areas to be occupied, leading to range expansion.

In many resident and migratory bird species, individuals tend to settle and breed in the neighbourhood where they were raised, and the numbers of dispersed individuals declines with increasing distance from the natal site. However, individuals of large bird species generally move longer distances than those of small species. In addition, individuals of species that depend on ephemeral habitats or food sources tend to disperse further, on average, than do individuals of species with annually consistent habitats and food sources. Within species, dispersal distances also vary with population density and other factors that promote competition, and differ according to gender and other features of the individual. In many bird species, females move generally further between natal and breeding sites than do males, but in waterfowl and some others, males move furthest.

In general, adults move over much shorter distances between their breeding attempts of successive years than young of their species move between natal and first breeding sites. Where conditions remain fairly stable from one breeding season to the next, adults of many species use the same nesting territories in successive years, although some individuals move to nearby territories, females more often than males, and especially after a breeding failure. Where local conditions fluctuate from year to year, adults can move long distances between their nesting sites of successive years. Long natal and breeding dispersal distances (of up to hundreds of kilometres) are frequent, for example, in some seed-eaters that depend on sporadic tree-seed crops, and in some ducks that depend largely on ephemeral waters. All such species tend to concentrate in different areas in different years, wherever conditions are good at the time.

In wintering areas, some birds return to the same localities from year to year, while others are more mobile, within and between winters. The broad behavioural spectrum, from winter residency to almost continual and variable movement, appears to reflect a gradient in consistency in food supplies, ranging from stable and predictable to unstable and unpredictable.

In many resident bird species, post-fledging dispersal turns into dispersive migration, in that individuals move back toward their natal site as the breeding season approaches. In such out-and-back movements, directions appear random, but one sex tends to disperse further than the other and juveniles further than older birds. Dispersive migration thus has some features of dispersal (directions highly variable but generally short distances) and some features of migration (return movement, differing between the sexes, often repeated year after year).

Young birds become fixated on their natal area in the late nestling or post-fledging period, depending on species, and on the wintering area within a few weeks after their first arrival there.

Eurasian Siskins *Carduelis spinus*, irruptive migrants from boreal regions

Chapter 18

Irruptive migrations: boreal seed-eaters

In the course of this year, about the fruit season, there appeared, in the orchards chiefly, some remarkable birds which had never before been seen in England, somewhat larger than larks, which ate the kernel of the fruit and nothing else, whereby the trees were fruitless to the loss of many. The beaks of these birds were crossed, so that by this means they opened the fruit as if with pincers or a knife. (The first documented record of a Crossbill invasion in England, Matthew Paris, 1251.)

One of the most striking features of bird migration is its regularity. Most populations of birds migrate at about the same dates, in the same directions, and for similar distances each year, with many individuals returning year after year to the same breeding and wintering localities. However, consistency in movement patterns is advantageous only in predictable environments, where birds can be sure of finding suitable conditions year after year in the same breeding or wintering areas. Some bird species exploit habitats or food supplies that are highly variable in distribution and abundance from year to year. The so-called irruptive migrants show great flexibility in their movement patterns, leaving their breeding areas in varying proportions and at variable dates from year to year, and concentrating wherever resources are plentiful at the time. The movements of irruptive migrants appear largely facultative in nature, occurring in direct response to prevailing conditions (Chapter 12). The term eruption is used for mass emigration from an area, and irruption for mass immigration.

Typical irruptive migrants of northern regions include: (1) boreal finches and others that depend on fluctuating tree-seed and fruit crops; (2) owls and others that depend on cyclically fluctuating rodent populations; and (3) waterbirds that depend on ephemeral wetlands created by irregular rainfall. This chapter is concerned with the seed-eaters, and the next chapter with the other two groups. They all provide striking evidence for the influence of prevailing food supplies on bird movements. But the movements of these species can be understood only in terms of their underlying ecology, which is therefore described here in greater detail than for other migrants.

Among the specialist seed- and fruit-eaters, most individuals stay in the north in years when food is plentiful there, wintering within, or just south of, their breeding areas, but moving further south in years when food is scarce. Their so-called invasions or irruptions, in which they appear in large numbers well beyond their usual range, follow periodic widespread crop failures. Irruptive migrations therefore occur in response to annual, as well as to seasonal, reductions in food supplies. The effect of food shortage is often accentuated because the birds themselves tend to be numerous at such times, as a result of good breeding and survival in previous years when food was plentiful (Lack 1954, Keith 1963, Berndt & Henss 1967, Köenig & Knops 2001). The greater the imbalance between the birds and their food, the greater the proportion of individuals that leaves, presumably as a result of competition (for Bohemian Waxwing *Bombycilla garrulus*, see Siivonen 1941, Cornwallis 1961, Cornwallis & Townsend 1968; for Great Tit *Parus major* see Perrins 1966, Ulfstrand 1962; for Purple Finch *Carpodacus purpureus* see Köenig & Knops 2001). Tyrväinen (1975) noted that Fieldfares *Turdus pilaris* left an area in southern Finland when their main food source (Rowan *Sorbus aucuparia* berries) had been reduced to an average of about two fruits per inflorescence. The date at which this occurred depended on both the initial crop size and the number of consumers.

Most irruptive seed-eaters are hard to study because they breed mainly in high-latitude regions, where human population density is low, and where the chance of obtaining ring recoveries is extremely small. Moreover, because of their eruptive behaviour, many such species are seldom in the same area long enough for detailed study. For these reasons, an understanding of their movement patterns must be pieced together from scraps of information collected over a long period, and scattered widely through the ornithological literature, although ring recoveries are slowly adding new information.

SEED CROPS

Every naturalist knows that tree-fruit crops vary greatly in size from year to year (**Figure 18.1**). In some years trees and shrubs are laden with seeds and fruits of

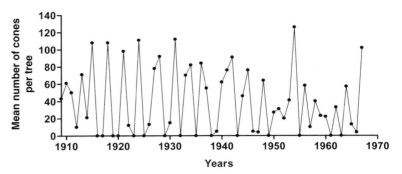

Figure 18.1 The mean number of Norway Spruce *Picea abies* cones per tree each year in southern Sweden, 1909–1967. From Hagner (1965) and Götmark (1982).

various kinds, but in other years bear almost none. Fruiting depends partly on the natural rhythm of the trees themselves and partly on the weather. Trees of most species require more than one year to accumulate the nutrient reserves necessary to produce a fruit crop. In addition, for a good crop, the weather must also be fine and warm in the preceding autumn when the fruit buds form, and again in the spring when the flowers set. Otherwise the crop is delayed for another year. In any one area most of the trees of a species fruit in phase with one another, and often those of different species also fruit in phase, partly because they come under the same weather. The result is an enormous profusion of tree fruits in some years, and practically none in others: good crops almost never occur in consecutive years, and are almost always followed by poor crops (**Figure 18.1**; Köenig & Knops 1998, 2000). Nevertheless, annual seed crops of some tree species fluctuate less than others, with Scots Pine *Pinus sylvestris* producing smaller, but more consistent crops than Norway Spruce *Picea abies*, for example, and Alder *Alnus glutinosa* more consistent crops than Birch *Betula pendens* and *B. pubescens* (for annual cropping patterns of different tree species, see Svärdson 1957, Hagner 1965, Perrins 1966, Götmark 1982, Knox 1992, Thies 1996, Köenig & Knops 1998, 2000).

The trees in widely separated areas may be on different fruiting regimes, partly because of regional variations in weather, so that good crops in some areas may coincide with poor crops in others. Nevertheless, good crops may occur in many more areas in some years than in others, so that the total continental seed production also varies greatly from year to year. An analysis of the fruiting patterns of various boreal conifer species at many localities in North America and Eurasia revealed high synchrony in seed production in localities 500–1000 km apart, depending on tree species. The synchrony declined at greater distances, and by 5000 km no correlation was apparent in the fruiting patterns of particular tree species (Köenig & Knops 1998). These figures give some idea of the range of distances that must separate the successive breeding and wintering areas of some boreal finches if the same individuals are to have access to good tree-seed crops every year of their lives. Besides total production, the timing of ripening and release of seeds is also important to seed-eating birds. Some tree species release most of their seeds within a short period in autumn, others more slowly over winter, and yet others in spring, or even more slowly over two or more years. This is true of some Pine *Pinus* and Larch *Larix* species which can thereby provide some food for seed-eaters even in non-cropping years. For most finches, the seeds are most available while they remain on the tree; once they fall to the ground they become rapidly removed by other animals or covered by snow. Among the European irruptive finches that depend on tree seeds, only the Brambling *Fringilla montifringilla* feeds for preference on the ground, though others do so occasionally; among the North American irruptive species, all seem to prefer the trees.

THE IRRUPTIVE SEED-EATERS AND FRUIT-EATERS

Species known for their irruptive migrations are listed in **Table 18.1**, along with their main food plants. Most species eat mainly different foods in summer and winter and, like most birds, migrate twice each year, away from the breeding range in autumn and back in spring. Some such species eat mainly seeds year-round, but different types of seeds at different seasons (e.g. Common Redpoll *Carduelis flammea*, Eurasian Siskin *C. spinus*). Others eat mainly insects in summer and seeds (or fleshy fruits) in winter (e.g. Bohemian Waxwing *Bombycilla garrulus*, Brambling *Fringilla montifringilla*, Evening Grosbeak *Hesperiphona vespertina*). The latter tend to concentrate in summer in areas with insect outbreaks (e.g. the Brambling in areas with high densities of the moth *Epirrita autumnata*, and the Evening Grosbeak and others in areas with Spruce Budworm *Choristoneura fumiferana*, Morris *et al.* 1958, Enemar *et al.* 1984, Lindström *et al.* 2005). In contrast to all other species, however, crossbills and nutcrackers eat conifer seeds year-round. They apparently make only one major movement each year, from regions where last year's crops were good to regions where the current year's crops are good (see later).

TWICE-YEARLY MIGRANTS

Twice-yearly migrants generally move towards more southerly latitudes in autumn and to more northerly ones in spring, but they may concentrate to breed and winter in different areas in different years, wherever their food is plentiful at the time. Their local numbers fluctuate greatly from year to year according to local food supplies, and in winter may range between total absence in years when appropriate tree seeds are lacking, to thousands of birds per square kilometre in years when such seeds are plentiful. The bird species involved seem to move each autumn only until they find areas rich in food, then settle there (Svärdson 1957, Newton 1972, Jenni 1987). In consequence, the distance travelled by the bulk of the migrants varies from year to year, according to where the crops are good, and only when the migrants are exceptionally numerous, or their food is generally scarce, do they reach the furthest parts of their wintering range, as an 'invasion' (**Figure 18.2**). The advantages of stopping in the first suitable feeding areas are presumably that birds travel no further than necessary, and do not pass over good feeding areas which might be the only ones available that year.

Other species also exploit the same foods as the irruptive species, but do not depend so heavily on them, so are less affected by the fluctuations in fruiting. Similarly, some of the species listed in **Table 18.1** may be irruptive in some parts of their range but not in others, depending on the breadth of the diet and the level of fluctuation in the entire food supply. Populations that have access to a wide range of dietary items are less likely to experience a shortage of all types in the same year.

Table 18.1 Established year-to-year correlations between bird abundance and food supply in seed-eating and fruit-eating birds

	Preferred winter food[a]	Summer densities	Winter densities	Autumn emigration	References
Great Spotted Woodpecker *Dendrocopos major* (P)	Spruce, Pine and other seeds	•	•	•	Pynnönen (1939), Formosov, (1960), Eriksson (1971)
Bohemian Waxwing *Bombycilla garrulus* (H)	Rowan and other berries			•	Siivonen (1941), Tyrväinen (1975), Bock & Lepthien (1976)
Fieldfare *Turdus pilaris* (P)	Rowan and other berries			•	Tyrväinen (1975)
Coal Tit *Parus ater* (P)[b]	Spruce seeds, insects			•	Formosov (1965)
Black-capped Chickadee *Parus atricapillus* (N)	Conifer seeds, insects			•	Bock & Lepthien (1976)
Great Tit *Parus major* (P)	Beech seeds		•	•	Ulfstrand (1962), Perrins (1966), Berndt & Henss (1967)
Blue Tit *Parus caeruleus* (P)	Beech seeds		•	•	Ulfstrand (1962), Perrins (1966)
Wood Nuthatch *Sitta europaea* (P)	Spruce seeds			•	Berndt & Dancker (1960), Enoksson & Nilsson (1983)
Red-breasted Nuthatch *Sitta canadensis* (N)	Pine and other conifer seeds			•	Bock & Lepthien (1976), Widlechner & Dragula (1984), Davis & Morrison (1987)
Brambling *Fringilla montifringilla* (P)	Beech seeds	•	•	•	Silvola (1967), Enemar et al. (1984), Nilsson (1984), Lindström (1987), Lindström et al. (2004), Hogstad (2000), Jenni & Neuschulz (1985), Jenni (1987), Eriksson (1970d), Mikkonen (1983)

(continued)

Table 18.1 (Continued)

	Preferred winter food[a]	Summer densities	Winter densities	Autumn emigration	References
Eurasian Siskin *Carduelis spinus* (P)	Birch, Alder and conifer seeds	•	•	•	Svärdson (1957), Haapanen (1966), Hogstad (1967), Eriksson (1970c), Petty et al. (1995), Förschler et al. (2006)
Pine Siskin *Carduelis pinus* (N)	Conifer, Birch and Alder seeds			•	Bock & Lepthien (1976), Widlechner & Dragula (1984)
Common Redpoll *Carduelis flammea* (H)	Birch and Alder seeds	•		•	Evans (1966a), Eriksson (1970b), Enemar et al. (1984)
Arctic (Hoary) Redpoll *Carduelis hornemanni*		•		•	Bock & Lepthien (1976), Nyström & Nyström (1991)
Eurasian Bullfinch *Pyrrhula pyrrhula* (P)	Various tree seeds and berries			•	Svärdson (1957)
Pine Grosbeak *Pinicola enucleator* (H)			•	•	Grenquist (1947), Bock & Lepthien (1976)
Evening Grosbeak *Hesperiphona vespertina* (N)	Maple and other tree seeds	•		•	Parks & Parks (1965), Bock & Lepthien (1976)
Purple Finch *Carpodacus purpureus* (N)	Various tree seeds			•	Bock & Lepthien (1976), Köenig & Knops (2001)
Common (Red) Crossbill *Loxia curvirostra* (H)	Spruce and other conifer seeds	•		•	Reinikainen (1937), Formosov (1960), Newton (1972), Bock & Lepthien (1976), Benkman (1987), Petty et al. (1995), Förschler et al. (2006)
Two-barred (White-winged) Crossbill *Loxia leucoptera* (P)	Larch and other conifer seeds	•	•	•	Newton (1972), Larson & Tombre (1989), Bock & Lepthien (1976)
Parrot Crossbill *Loxia pytyopsittacus* (P)	Scots Pine seeds	•	•	•	Newton (1972)
Eurasian Jay *Garrulus glandarius* (P)	Oak seeds			•	Cramp & Perrins (1994)

Species	Food[a]	Reference
Thick-billed Nutcracker *Nucifraga c. macrorhynchos* (P)	Hazel and Swiss Stone Pine seeds	• Schütz & Tischler (1941), Mattes & Jenni (1984)
Thin-billed Nutcracker *Nucifraga c. caryocatactes* (P)	Siberian Stone Pine seeds, Brush Pine seeds	• Formosov (1933)
Clark's Nutcracker *Nucifraga columbiana*	Whitebark Pine and other conifer seeds	• Lanner (1996)

H, Holarctic; N, Nearctic; P, Palaearctic.

[a] Scientific names of trees: Alder *Alnus*, Beech *Fagus sylvatica*, birch *Betula*, Hazel *Corylus avellana*, larch *Larix*, maple *Acer*, oak *Quercus*, Rowan *Sorbus aucuparia*, Scots Pine *Pinus sylvestris*, Siberian Stone Pine *Pinus sibirica*, spruce *Picea*, Swiss Stone (Arolla) Pine *Pinus cembra*, Whitebark Pine *Pinus albicaulis*. Where several species in the same genus are involved, only the generic name is given.

[b] In this species, mass emigration has been more frequently linked with high numbers (which may cause food shortage) or with high spring temperatures which promote high breeding success (Markovets & Sokolov 2002).

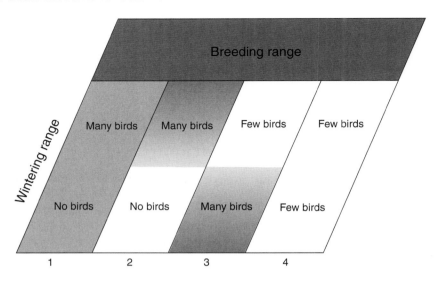

Figure 18.2   Model of migration system in irruptive finches. The birds leave the breeding range across boreal Europe each autumn on a broad front, travelling roughly southwest, but only until they meet areas with abundant seed crops. The sporadic cropping pattern of trees means that each year birds from some sectors of the breeding range must travel further than others. The next year the pattern of tree cropping and migration may differ, so that birds from particular sectors of the breeding range travel different distances in different years. From Jenni & Neuschulz (1985).

Scenario 1 – good crops throughout the wintering range: all birds accommodated in northern areas, nearest the breeding areas, and crops further south remain unused; survival good.
Scenario 2 – good crops in the north of the wintering range, but not in the south: all birds accommodated in northern areas; survival good.
Scenario 3 – good crops in the south of the wintering range but not in the north: birds accommodated in the southern areas; survival moderate.
Scenario 4 – crop failure throughout the wintering range: birds spread widely, but at low densities; survival poor.

On this model, it is in the northern parts of the wintering range that correlations between seed crops and bird densities in different years would be expected to be most marked; further south, good crops can sometimes occur with few birds, or vice versa.

   Most tree species, such as Birch *Betula* and Rowan *Sorbus aucuparia*, retain their seeds into the winter, so the finches that eat them take the seeds directly from the trees, regardless of snow. Other trees, such as Oak *Quercus robur* and Beech *Fagus sylvatica*, shed their seeds in autumn so that the birds that eat them must take them from the ground. Such birds are therefore also affected by snowfall, which can render abundant seed crops unavailable. Thus, the movements and winter distributions of Bramblings *Fringilla montifringilla* in different years are a function of both Beech crops and snowfall (Jenni & Neuschulz 1985, Jenni 1987).

Examples of species that behave as irruptive migrants in only parts of their range include Eurasian Bullfinch *Pyrrhula pyrrhula*, Wood Nuthatch *Sitta europaea* and various titmice *Parus* spp. In other parts they are either residents or regular migrants.

Breeding densities

Among irruptive seed-eaters, local breeding densities can vary from nil or few in poor food years to dozens of pairs per square kilometre in good food years (for references see **Table 18.1**). For example, local densities of Common Redpolls *Carduelis flammea* fluctuated about 39-fold over a 20-year period, with the highest densities coinciding with exceptionally good Dwarf Birch *Betula nana* crops, the low bushes and their seed-catkins protected under snow since the previous year (Enemar *et al.* 1984). Densities of Eurasian Siskins *Carduelis spinus* fluctuated 6- to 50-fold in four different areas in parallel with the spruce cone crop (Haapanen 1966, Hogstad 1967, Shaw 1990, Förschler *et al.* 2006); and densities of Bramblings *Fringilla montifringilla* fluctuated by 5- to 26-fold in three different areas in parallel with the abundance of moth larvae *Epirrita autumnata* (**Box 18.1**; Silvola

Box 18.1   Bramblings and caterpillars

In the Brambling *Fringilla montifringilla*, unlike most other seed-eaters, high breeding densities in northern Europe seem particularly linked to outbreaks of a single insect, the autumnal moth *Epirrita autumnata*. The caterpillars of this small geometrid moth are found mainly on birch trees, and in the alpine and arctic regions of northernmost Europe outbreaks occur on average once per decade (variation 5–15 years, Ruchomaki *et al.* 2000, Selås *et al.* 2001), when large areas of forest may be defoliated. Each outbreak lasts about 3–4 years and in the intervening years larval densities are hardly measurable. In addition, the outbreaks seem well synchronised over large areas of northern Europe, hundreds of kilometres across. In such years, Bramblings nest at much higher densities than usual (Silvola 1967, Hogstad 2000, Ytreberg 1972, Enemar *et al.* 1984, Lindström 1987, Lindström *et al.* 2005). In a 19-year study, breeding densities of Bramblings varied by three-fold (density index 45.1 to 126.2), in association with densities of *Epirrita* larvae which varied by more than four orders of magnitude (between 0.1 and 235.1 larvae per 1000 short shoots). Bramblings produced more young per pair in the outbreak years, and 49% of the annual variation in adult to juvenile ratios in late summer was explained by annual variation in *Epirrita* densities. However, Bramblings still produced some young in the lowest *Epirrita* years (Lindström *et al.* 2005).

Another predominantly seed-eating species that concentrates in areas of abundant caterpillars in the breeding season is the Evening Grosbeak *Hesperiphona vespertinus* in North America. It is especially attracted to outbreak areas of the Spruce Budworm *Choristoneura fumiferana* (Parks & Parks 1965).

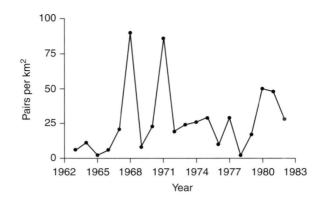

Figure 18.3   Densities of Common Redpolls *Carduelis flammea* in 9 km² of birch scrub in Swedish Lapland, 1963–1982. Over this period, the fluctuations in numbers were mostly moderate, but with exceptionally large numbers in 1968 and 1971, which could only have resulted from massive immigration. In these two years, the birch crop was unusually good, the seeds having remained on the trees from the previous summer when they were formed. In these years, Redpolls overwintered in the area, started breeding earlier than usual, and produced larger clutches, some raising more than one brood. In 1978, few Redpolls occurred in the area, despite good numbers in the preceding and following years. From Enemar *et al.* (1984).

1967, Lindström 1987, Hogstad 2000). In all these species, local densities sometimes increased so much from one year to the next that the increase could not be explained by good reproduction and survival from the previous year, but must have involved mass immigration (**Figures 18.3** and **18.4**). This in turn implies that some birds changed their breeding areas from one year to the next, or that young moved in from distant as well as local natal areas.

Breeding dispersal

Among populations of regular migrants, if the birds occupying a particular study area are trapped and ringed in the breeding season, large proportions of the same individuals are usually found breeding in the same area next year. Return rates are usually within the range 30–60% for passerines, and 60–90% for non-passerines (Chapter 17). Allowing for expected mortality, such high figures imply that most surviving individuals return to breed in the same limited area year after year. In some regular migrants, the same also holds for wintering areas (Chapter 17).

The situation differs in irruptive migrants, whose return rates are normally much lower (**Table 18.2**). For example, among Bramblings *Fringilla montifring-illa* trapped in the breeding season in various areas, individuals were seldom or never caught in the same locality in a later year, so that each year's occupants were different from those the year before (Mikkonen 1983, Lindström 1987, Lindström *et al.* 2005, Hogstad 2000). In one such study, only seven (0.6%) of 1238 adults were retrapped in the same area in a later year, and none of 1806 juveniles, despite an annual trapping programme over many years (Lindström *et al.* 2005). In another area where Bramblings were studied, the closely related non-irruptive Chaffinches *Fringilla coelebs* showed much more stable densities and greater

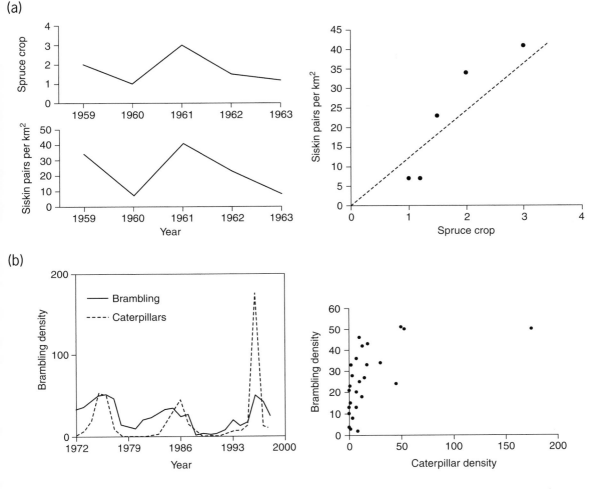

Figure 18.4 (a) Numbers of breeding Eurasian Siskins *Carduelis spinus* in relation to the spruce crop in Finland. Siskins shown in pairs per km² and cone crops classified in four categories. From Haapanen (1966). Similar data were also given for a Norwegian area by Hogstad (1967). (b) Numbers of breeding Bramblings *Fringilla montifringilla* in relation to the abundance of geometrid Autumnal Moth *Epirrita autumnata* caterpillars in northern Norway. Bramblings measured as number of singing males, and *Epirrita* as the number of caterpillars per 100 net sweeps in birch trees during 1972–1998. Redrawn from Hogstad (2000). Similar data were given for other areas by Silvola (1967) and Lindström (1987).

site-fidelity from year to year, with most surviving individuals returning in successive years (Mikkonen 1983). In all the irruptive species listed in **Table 18.2**, return rates were extremely low, compared with what would be expected from their annual survival rates. The implication is that large proportions of individuals changed their nesting locations from year to year.

The small numbers of relevant ring recoveries available for irruptive species confirm that some individuals have indeed occurred in widely separated localities in

Table 18.2 Annual return of individual birds to the same area – irruptive migrants

	Number ringed	Number (%) recaught in a later year in the same place	Location (years)	Reference
Breeding areas				
Brambling *Fringilla montifringilla*	1238	7 (0.6)	Sweden (19)	Lindström et al. (2005)
Eurasian Siskin *Carduelis spinus*	391	30 (7.7)	Scotland (6)	Shaw (1990)
Pine Siskin *Carduelis pinus*	1322	4 (0.003)	Oklahoma (?)	Baumgartner & Baumgartner (1992)
Common Redpoll *Carduelis flammea*	?	? (<1)	Alaska (?)	Troy (1983)
Cedar Waxwing *Bombycilla cedrorum*	54	2 (3.7)	Ohio (6)	Putnam (1949)
Wintering areas				
Evening Grosbeak *Hesperiphona vespertina*	2637 >1700	0 (0) 48 (0.003)	New York (18) Pennsylvania (14)	Yunick (1983) Speirs, in Newton (1972)
Pine Siskin *Carduelis pinus*	3810 4045 1322	0 (0) 0 (0) 4 (0.3)	New York (18) New York (2) Oklahoma (4)	Yunick (1983) Yunick (1997) Baumgartner & Baumgartner (1992)
Common Redpoll *Carduelis flammea*	7946 1800 5200	0 (0) 0 (0) 16 (0.3)	New York (18) New Hampshire (?) Alaska (?)	Yunick (1983) Troy (1983) Troy (1983)
Purple Finch *Carpodacus purpureus*	2822 1015	13 (0.5) 51 (5.0)	New York (18) North Carolina (5)	Yunick (1983) Blake (1967)
Brambling *Fringilla montifringilla*	2330	16 (0.5)	England (7)	Browne & Mead (2003)

different breeding seasons. Three Bramblings were found respectively at places 280 km, 420 km and 580 km apart; one Northern Bullfinch *Pyrrhula pyrrhula* at places 424 km apart (natal dispersal); two Common Redpolls *Carduelis flammea* at places 280 km and 550 km apart; one Lesser Redpoll *C. cabaret* at places 300 km apart (natal dispersal); and one Eurasian Siskin *C. spinus* at places 120 km apart. On the other hand, some adult Siskins in south Scotland were found in the same breeding area in successive years, despite large fluctuations in cone crops and breeding numbers from year to year (Shaw 1990). In North America, one Pine Siskin *Carduelis pinus* was found at the same place in different breeding seasons, while two others were found at places 346 km and 1138 km apart (Brewer *et al.* 2000). Of nine recoveries of Evening Grosbeaks *Hesperiphona vespertina* in different breeding seasons, two were at the same place, three were within 100 km and four had moved 322–946 km (Brewer *et al.* 2000). Clearly, long-distance moves from one breeding area to another are not unusual in irruptive finches.

Autumn emigration

While some individuals remain within the breeding range year-round, but change breeding localities, others winter at lower latitudes, at least in certain years. Typically, as mentioned above, the numbers of birds undertaking migration, and the distances involved, vary greatly from year to year. Huge annual variations in the numbers of migrants are evident at migration watch sites, such as Falsterbo and Ottenby in Sweden, in which the numbers of some irruptive species can vary from nil in some years to many thousands of individuals in other years (**Figures 18.5** and **18.6**). During 1950–1960, the Eurasian Jay *Garrulus glandarius* count in seven years at Falsterbo was nil, but in 1955 it reached more than 10 000. Over the same period, Eurasian Siskins *Carduelis spinus* varied from 4220 to 35 904, Common Redpolls *C. flammea* from 0 to 489, Eurasian Bullfinches *Pyrrhula pyrrhula* from 0 to 360, Great

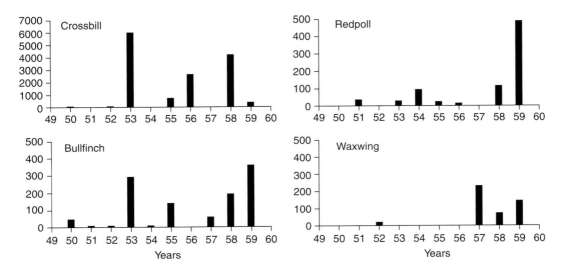

Figure 18.5 Annual fluctuations in the numbers of four irruptive species counted while migrating over Falsterbo in southern Sweden, 1950–1960. From Ulfstrand *et al.* (1974).

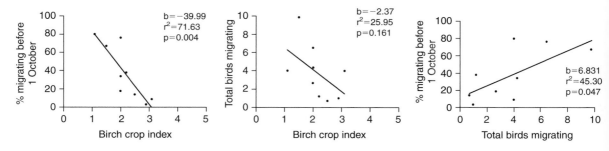

Figure 18.6 Numbers and timing of Eurasian Siskins *Carduelis spinus* migrating through Ottenby each year in relation to the size of the Birch *Betula* seed crop further north. Drawn from data in Svärdson (1957).

Tits *Parus major* from 2 to 7438, Blue Tits *P. caeruleus* from 7 to 3595, and Coal Tits *P. ater* from 0 to 18785 (Ulfstrand *et al.* 1974). These annual fluctuations were far greater than those recorded in more regular long-distance migrants (counts of which typically varied by up to 3- to 4-fold, occasionally up to 10-fold, between years, Edelstam 1972, Ulfstrand *et al.* 1974, Roos 1991). Similar differences between irruptive and regular migrants have been recorded at other watch sites (e.g. Dorka 1966, Gatter 2000).

The amount of autumn emigration has been related to food supplies in almost all seed-eating and fruit-eating species discussed here, being greatest in years when high densities coincide with poor seed crops (**Table 18.1**, **Figures 18.6** and **18.7**). Different species that depend heavily on the same seed or fruit crops as one another

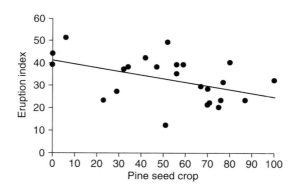

Figure 18.7 Irruptions of Red-breasted Nuthatches *Sitta canadensis* in relation to pine cone production in western North America. The pine seed index was calculated as the proportion of sites surveyed each year that had greater than average seed crops. The irruption index was calculated from the Christmas Bird Counts, as the proportion of sites south of 50°N at which greater than expected numbers of Red-breasted Nuthatches were recorded. Expected numbers for each year were calculated from a regression of the annual totals on year for the whole 21-year count period, 1968–1988. Based on data in Köenig & Knops (2001), in which the relationship was given as significant (P<0.05) on a Spearman rank correlation test.

tend to irrupt in the same years. They include the Blue Tit *Parus caeruleus* and Great Tit *P. major* which both feed heavily on Beech *Fagus sylvatica* mast, and the Common Crossbill *Loxia curvirostra* and Great Spotted Woodpecker *Dendrocopus major* which both feed heavily on spruce seeds. Where different tree species fruit in phase with one another, the numbers of participating species is increased further.

Over much of the boreal region of North America, conifer and other tree crops tend to fluctuate biennially, and in the alternate years with poor crops several species that depend on them migrate to lower latitudes. The migrations of irruptive species are therefore much more regular in North America than in Europe and much more synchronised between species (Bock & Lepthien 1976, Kennard 1976, Larson & Bock 1986, Köenig & Knops 2001; **Figure 18.8**). At least eight species of boreal seed-eating birds tend to irrupt together, namely Common Redpoll *Carduelis flammea*, Pine Siskin *C. pinus*, Purple Finch *Carpodacus purpureus*, Evening Grosbeak *Hesperiphona vespertina*, Red-breasted Nuthatch *Sitta canadensis* and Black-capped Chickadee *Parus atricapillus*, as well as the Red Crossbill *Loxia curvirostra* and White-winged (Two-barred) Crossbill *L. leucoptera* discussed later. These species vary in the proportions of conifer and broad-leaved tree seeds in their diets (**Table 18.1**; Bock & Lepthien 1976, Kennard 1976, Köenig & Knops 2001). Over periods of years, different species of trees in the same area can drift in and out of synchrony with one another, affecting the movements of the birds that specialise on them. During the period 1921–1950, the biennial pattern and synchrony between the various North American seed-eaters was less marked than before or after this period (Larson & Bock 1986). In some years, irruptions were evident across North America, but in other years only in western or eastern regions (Köenig & Knops 2001). In addition, crop failures and irruptions from montane areas south of the boreal region were not well synchronised with those within the boreal region (Bock & Lepthien 1976). No consistent relationships were apparent between irruptions and environmental factors other than seed crops, such as warm summers or cold winters (e.g. Köenig & Knops 2001).

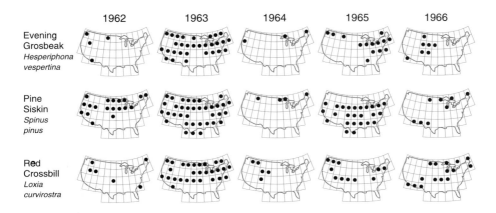

Figure 18.8   Year-to-year variation in the winter ranges of three species of seed-eating birds that breed in northern North America. Dots show large wintering populations in local regions. In years of low food supply in the north, all species tend to winter far to the south. From Bock & Lepthien (1976).

The role of food shortage in stimulating autumn emigration is shown in another way, namely that some species which take food from garden feeders have become less irruptive over the past 40 years, as increasing amounts of food have been provided. Winter feeding was held responsible for turning a previously migratory population of Great Tits *Parus major* in the city of Oulu in northern Finland into a resident one (Orell & Ojanen 1979), and reduced movements by various tit species, attributed to the same cause, have been recorded in other parts of western Europe over the same period (van Balen & Hage 1989, Wernham *et al.* 2002). No big invasions of Blue Tits and Great Tits have occurred in Britain since 1977, and no big invasions of Great Spotted Woodpeckers *Dendrocopus major* since 1974. All these species feed heavily from garden feeders.

Migration timing

In at least some irruptive migrants, the timing of autumn migration appears much more variable than among regular migrants, being markedly earlier in the big years. For example, the peak autumn days for Siskins *Carduelis spinus* at Falsterbo during 1949–1988 varied from mid-August to mid-November in different years (Roos 1991), and elsewhere heavy southward movements have been seen as late as December–January. Over a nine-year period, Svärdson (1957) attempted to relate the numbers of Siskins passing through Ottenby Bird Observatory to the size of the birch crop further north (**Figure 18.6**). The birds tended to pass in largest numbers, and at the earliest dates, in years when the birch crop was poor. Similar relationships between numbers and migration dates have been noted in other species elsewhere (Berndt & Henss 1967, Gatter 2000), and the corresponding tendency for birds to arrive earlier than usual in their wintering areas in invasion years has been observed in various irruptive species. Annual variations in autumn migration dates are clearly much greater in irruptive than in other migrants, which travel at approximately the same dates each year (Chapter 12).

Most irruptive seed-eaters begin to take their main winter food in late summer or autumn before their migration begins, so the size of the crop can have a direct influence. Other species start migrating before the winter seed crop is ready, so numbers *per se* (and the resulting competition for whatever food is eaten then) may have a bigger influence on migration. Such species tend to migrate at more consistent dates from year to year. They include the Brambling *Fringilla montifringilla*, Wood Nuthatch *Sitta europaea* and various tits, all of which eat Beech *Fagus sylvatica* mast in winter, but begin to leave before the seeds are ready (Berndt & Dancker 1960, Eriksson 1971).

Migration directions

In most European bird species, individuals from the western parts of the breeding range tend to use migration routes and wintering areas that lie to the west of birds from the eastern parts of the breeding range (Chapter 23). It is as though the birds from different sectors of the range migrate back and forth along roughly parallel routes. However, this tendency is less marked in irruptive than in regular migrants because irruptive migrants typically show a greater east–west component in their movements, and also greater directional spread, both from year to year and between individuals in the same year. In contrast to most migrants, moreover, irruptive species may end their journey at widely different places on the route each year. Thus, in any given year, birds from one part of the breeding range may stay in the north if food permits, while those from another part may extend far to the south, though not necessarily vacating the north completely (**Figure 18.2**). The following year, the pattern may differ. It is these behaviours that contribute to individuals turning up in widely separated areas in different winters. Most are at different points on the same migration axis (for Common Redpoll *Carduelis flammea* see Eriksson 1970b, for Siskin *Carduelis spinus* see Eriksson 1970c, for Brambling *Fringilla montifringilla* see Jenni & Neuschulz 1985), but some may end up in one winter far to the west or east of their position in a previous winter. Even within a winter, the birds may wander in various directions, and end up markedly to the east or west of their initial route (for Bohemian Waxwing *Bombycilla garrulus* see Cornwallis & Townsend 1968).

The greater directional spread of irruptive, compared to regular migrants, is evident from observations at migration watch sites, and also from subsequent ring recoveries of birds trapped on migration. Typically, the recoveries of irruptive migrants ringed in autumn and recovered in the following months show two or three times the angular spread as those of regular migrants. For example, from birds ringed at the Courish Spit in the southeast Baltic, winter recoveries of various non-irruptive species lay within an arc of 100° from the ringing site, but in irruptive species the spread was much greater: in the Eurasian Jay *Garrulus glandarius* 198°, in the Bohemian Waxwing *Bombycilla garrulus* 237°, in the Brambling *Fringilla montifringilla* 188°, and in the Eurasian Siskin *Carduelis spinus* 298° (Payevsky 1998). Similar differences between irruptive and non-irruptive migrants have been noted elsewhere, the recoveries of some non-irruptive species being mainly confined to a narrow corridor from the ringing site (for Eurasian Linnet *Carduelis cannabina* see Verheijen 1955, for Goldfinch *C. carduelis* see Newton 1972, for Chaffinch *Fringilla coelebs* see Bairlein 2001, for Pied Flycatcher *Ficedula hypoleuca* see Mouritsen 2001).

It is not yet clear to what extent the wide directional spread in irruptive migrants is apparent within years, or results from birds taking mainly different main directions in different years. From Common Redpolls *Carduelis flammea* ringed in Fennoscandia, most ring recoveries in 1965 came from directions east-southeast (mostly in Russia), whereas in 1972 and 1986 they came from directions to the southwest (mostly in western Europe) (**Figure 18.9**, Zink & Bairlein 1995, Lensink *et al.* 1986, Thies 1991). This difference could be explained if the birds were tracking seed crops, which took them in different directions from their breeding areas in different years, but this explanation remains untested. During the 50-year period 1951–2000, Common Redpolls appeared in large numbers in the mid-latitudes of Europe in at least 14 years, but concentrated mainly in the western, central or eastern parts of this region in different years. One of the biggest movements (in 1986) seems to have started mainly in northern Russia, and spread west-southwest across the continent; some 5.5 million birds are estimated to have entered the Netherlands then (Lensink *et al.* 1986), but very few extended further west to Britain.

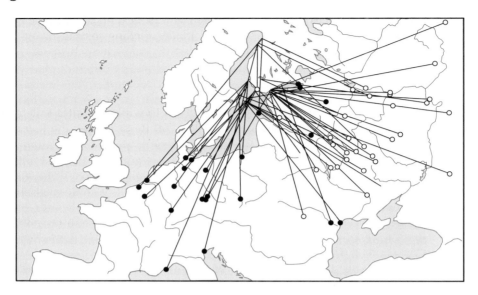

Figure 18.9   Winter recoveries of Common Redpolls *Carduelis flammea* ringed in Finland. Circles – winter 1965/66, dots – winter 1972/73. After Zink & Bairlein (1995).

Another feature of irruptive migrants is that they often make long movements within a winter, staying in one area as long as food lasts, and moving further down the migration route when food runs out (for Pine Grosbeak *Pinicola enucleator* see Grenquist 1947; for Common Redpoll *Carduelis flammea*, Bohemian Waxwing *Bombicilla garrullus* and Fieldfare *Turdus pilaris*, see Haila *et al.* 1986). Even in big invasion years, Bohemian Waxwings *Bombycilla garrulus* seldom reach middle Europe before January (Glutz von Blotzheim 1966). In effect, the birds move progressively further from their breeding areas during a winter, stripping seed and fruit crops as they go. Because most species take their food from the trees, they are little affected by weather, but Bramblings *Fringilla montifringilla*, which feed on the ground, may be forced to leave areas of good seed crops when snow cover renders their food unavailable. Their winter distribution in particular years thus depends on both seed crops and snowfall (Jenni & Neuschulz 1985).

Winter densities

Correlations between numbers of birds and size of fruit crops in wintering localities have been documented in several species of irruptive migrants, as shown in **Table 18.1**, but mostly in northern parts of the wintering range. Further south in the wintering range, such correlations would be expected to be weak or non-existent, because the birds reach the more distant localities only in exceptional years. In some years, seed crops are available, but no birds from further north turn up to use them, having settled in good feeding areas already encountered. Conversely, irruptive migrants may turn up in remote areas in some years only to find their favoured foods are lacking (as shown, for example, by the large numbers of Jays *Garrulus glandarius* which irrupted into western Europe in 1983, a year almost devoid of acorns, John & Roskell

1985). Other species (such as crossbills and nutcrackers) often migrate beyond the geographical range of their main food plants, feeding instead on whatever other seeds they can find. Mortality is almost certainly heavy in such years.

Changes in wintering areas: evidence from ringing

Among the irruptive species studied, return to wintering sites in later years was even rarer than return to breeding sites (**Table 18.2**). Despite some very large numbers ringed (often in small gardens), return rates of seed-eaters were mostly nil or less than 1% (but with Purple Finch *Cardopdacus purpureus* at 5% in one area). Evidently, extremely few individuals of such species returned to the same localities in subsequent years. They showed little or no winter site-fidelity, compared with regular migrants (Chapter 17).

Among Eurasian species, although some ringed individuals were found in the same area in successive autumns or winters, indicating fidelity to the same migration or wintering sites (for Brambling *Fringilla montifringilla*, Eurasian Siskin *Carduelis spinus*, Common Redpoll *Carduelis flammea* and Eurasian Bullfinch *Pyrrhula pyrrhula*, see Cramp & Perrins 1994), other individuals were found in widely separated sites in different years. Most captures of the same individuals in successive winters lay at localities on the same migration axis, as mentioned above (including some more than 3000 km apart), but others included substantial east–west displacements. Extreme examples included a Bohemian Waxwing *Bombicilla garrulus* found in Poland one winter and in Siberia in a later winter (6000 km apart), a Pine Siskin *Carduelis pinus* found in Quebec in one winter and in California in a later winter (3950 km apart), and a Common Redpoll *Carduelis flammea* found in Belgium in one winter and in China in a later one (8350 km apart) (**Table 18.3**). Another Common Redpoll was recorded in North America in one winter and Eurasia in a later winter, having been ringed in Michigan and recovered near Okhotsk in Siberia, some 10200 km to the northwest (Troy 1983). All these birds are likely to have returned to the breeding range in the interim (though not necessarily to the same locality), and taken a markedly different migration direction in the second year. Further records of this type are given in **Box 18.2**.

North American ringing records also reveal the transcontinental nature of boreal finch movements. Although most species breed across the entire continent in boreal forest, populations breeding in Alaska and other parts of the northwest tend to migrate east-southeast to winter in the eastern States. Birds from eastern Canada migrate south-southwest to also winter in the eastern States, the same region as the western birds (Brewer *et al.* 2000). Many of the records of individual Common Redpolls *Carduelis flammea* and Pine Siskins *Carduelis pinus* found on opposite sides of the continent in different winters (Troy 1983, Kaufman 1984) may therefore represent northwestern breeders that remain in the northwest in one year and migrate far along their east-southeast route in another year. However, some individuals have been recovered well south of the boreal forest on different sides of the continent in different winters, indicating that they had taken different directions from the breeding range in different years. For example, a Cedar Waxwing *Bombycilla cedrorum* ringed in California one April was recovered in Alabama 3000 km to the east two years later (for other interesting ringing records from North American species see **Box 18.2**).

Table 18.3   Examples of irruptive species in widely separated localities in different winters (December–March)

	Ringed	Recovered	Distance (km)
Bohemian Waxwing *Bombycilla garrulus*	Sweden	Siberia	3060
	Sweden	Siberia	4070
	Poland	Siberia	4500
	Ukraine	Siberia	6000
	British Columbia	South Dakota	1360
	Sweden	Russia	2980
	Sweden	Russia	2280
	Sweden	Russia	2910
Cedar Waxwing *Bombycilla cedrorum*	California	Alabama	3000
Brambling *Fringilla montifringilla*	Belgium	Turkey	3000
	Britain	Greece	2500
Eurasian Siskin *Carduelis spinus*	Belgium	Lebanon	3000
	Sweden	Iran	3000
Pine Siskin *Carduelis pinus*	Ontario	California	3537
	Quebec	California	3950
	New York	British Columbia	3470
	Tennessee	British Columbia	3780
	Pennsylvania	Washington	2800
Common Redpoll *Carduelis flammea*	Sweden	Russia	1800
	Hungary	Siberia	3300
	Belgium	China	8350[a]
	Alaska	New Brunswick	5200
	Quebec	Alaska	4850
	New Jersey	Alberta	3250
	Saskatchewan	Vermont	2550
	New Jersey	Alberta	3250
	New Jersey	Manitoba	2100
	Alaska	Saskatchewan	2730
	Michigan	East Siberia	10200
Evening Grosbeak *Hesperiphona vespertina*	Maryland	Alberta	3400
	Virginia	Newfoundland	2200
	Quebec	Georgia	1750
Eurasian Bullfinch *Pyrrhula pyrrhula*	Finland	Siberia	1900
	Finland	Siberia	2350

These recoveries, which are selected as extreme examples from among many, refer mainly to birds that seemed to be on a different migration axis in different winters, being recovered in winter far to the east or west of where they were ringed in a previous winter. Other examples in **Box 18.2**.

[a] This movement is matched by at least three others almost as long, from Norway to eastern China, Finland to eastern China and eastern China to Sweden respectively.

Sources: Rydzewski (1939), Cornwallis & Townsend (1968), Newton (1972), Eriksson (1970b), Zink (1973–85), Troy (1983), Baumgartner & Baumgartner (1992), Cramp & Perrins (1994), Zink & Bairlein (1995), Glutz von Blotzheim *et al.* (1997), Yunick (1997), Brewer *et al.* (2000); plus records from Olaf Runde from the Norwegian Ringing Scheme and Thord Fransson from the Swedish Ringing Scheme.

Yunick (1997) gives additional information on the Pine Siskin, while Brewer *et al.* (2000) list 12 Redpolls that were trapped in different winters at places 1345–4836 km apart in North America.

Box 18.2   Other examples of individuals of irruptive species found in widely separated areas in different winters (December–February)

Bramblings *Fringilla montifringilla* ringed in Britain or Belgium in one winter were recovered as far east as Turkey and the Balkans in a later one, and Bramblings ringed in Switzerland in one winter were recovered at various localities from Ireland to Greece and Georgia in a later one (Jenni & Neuschulz 1985). Similarly, many Bramblings that were ringed in western Europe (Belgium–Germany–Switzerland) at 5°–15°E in one winter (November–February) were retrapped at a similar latitude at 60–70°E in a later winter (Zink & Bairlein 1995). These various records at similar latitude involved up to 65° of longitudinal displacement between one winter and another.

Of 17 long-distance recoveries of Eurasian Bullfinches *Pyrrhula pyrrhula* ringed in Finland in winter, nine were at 550–1000 km and eight at more than 1000 km away in a later winter. Fifteen were between south-southwest (Hungary and Poland) and east (Russia) of the ringing site (including two at 1920 km and 2350 km east in Siberia), one was north-northeast and another north-northwest. Another three ringed in Sweden were found at 475–1491 km, between northeast to southeast in a later winter (Cramp & Perrins 1994). These records involved longitudinal displacements of up to 45°.

Many Eurasian Siskins *Carduelis spinus* were ringed in western Europe (Belgium, France or Iberia) in one winter and recovered as far east as Turkey and the Balkans in a later winter, at places up to 2500 km apart; one ringed in Belgium in April was recaught about 3000 km to the southeast in Lebanon in the next November and another ringed in Sweden in October 1980 was recovered 3000 km to the southeast in Iran in January 1982 (Glutz von Blotzheim *et al.* 1997). These records involved longitudinal displacements of up to 35°.

Four Pine Siskins *Carduelis pinus* in North America were caught at localities 2055–3780 km apart in different winters, involving 15–44° of longitudinal displacement (Brewer *et al.* 2000).

Three Redpolls *Carduelis flammea* ringed in Fennoscandia are known to have occurred in different winters at places 1300, 1500 and 1800 km apart (Eriksson 1970b), and another caught in Hungary in February 1978 was recovered 3300 km to the east–northeast in Sverdlovsk in west Siberia in March 1979 (Glutz von Blotzheim *et al.* 1997). Even more remarkable are three movements between western Europe and China, all exceeding 8000 km as shown in **Table 18.3**. In North America, ten Common Redpolls *Carduelis flammea* were caught at localities 1345–3251 km apart in different winters, involving 8–43° of longitudinal displacement (Brewer *et al.* 2000).

Five Evening Grosbeaks *Hesperiphona vespertina* were caught at localities 925–3402 km apart in different winters, involving 2–42° of longitudinal displacement.

Many Redwings *Turdus iliacus* were ringed in Europe west of 10°E in one winter, and recovered east of 55°E in a later winter, at places more than 3000 km apart (Zink 1973–85). Of Redwings ringed in winter in Britain, dozens have been recovered in subsequent winters as far east as Italy, Greece and Turkey, and some as far east as Israel and Iran, at localities up to 5000 km and up to 50° of longitude apart (Milwright 2002). Even birds from the same brood have been recovered in widely separated localities in the same winter.

Many Fieldfares *Turdus pilaris* were found as far apart as Ireland and Italy, England and Turkey, or as Switzerland and Georgia, in different winters, at localities 2000–3000 km and up to 35° of longitude apart (Zink 1973–85, Milwright, in Wernham *et al.* 2002). Again, birds from the same brood have been recovered in widely separated areas in the same winter.

A Bohemian Waxwing *Bombycilla garrulus* ringed in Switzerland in one winter was found 1300 km to the east in Romania in a later winter. Another Waxwing ringed in Poland one February was recovered in the next winter 4500 km further east in Siberia (Rydzewski 1939), a third was ringed in the Ukraine in one winter and recovered 6000 km to the east in Siberia the next, and a fourth was ringed in England one November and recovered about 3200 km to the east in October three years later. Many others were recorded west of 20°E in one winter and 45–65°E in another, at places more than 2000 km apart (Zink 1973–85). These recoveries involved some in which longitudinal displacements exceeded 50°. Other Bohemian Waxwings have been found at the same place at intervals of one, two or three years, in successive irruptions.

The extent to which irruptive finches wander for food is well illustrated by the North American Evening Grosbeak *Hesperiphona vespertinus*, which breeds in conifer forests and moves south or southeast in autumn. This species feeds mainly on large, hard tree fruits, but also visits garden feeding trays, a habit which makes it easy to catch. Over 14 winters, 17 000 individuals were ringed at a site in Pennsylvania. Of these, only 48 (0.003%) were recovered in the same place in subsequent winters, yet 451 others were scattered among 17 American States and four Canadian Provinces. Another 348 birds that had been ringed elsewhere were caught at this same locality, and these had come from 14 different States and four Provinces (D. H. Speirs, in Newton 1972). These recoveries show how widely individual grosbeaks range, and how weak is their tendency to return to the same place in later years.

Common Redpolls *Carduelis flammea* in North America showed another interesting pattern. Most birds were ringed at garden feeders, and Troy (1983) examined the distribution of all recoveries to 1978, arguing that because invasions of more southern parts of the range occurred every second year, this should be reflected in the recoveries (**Table 18.4**). In the northern parts of the winter range (north of the Canadian border), recoveries fell off steadily from one to five years after banding, as expected from mortality. However, in the southern parts (south of the Canadian border), recoveries peaked two years after banding, with a minor peak four years after, as expected from the biennial migration pattern. Some individuals were caught at sites more than 2000 km apart in different years; from birds ringed in winter in the eastern States, two were recaught in the breeding season in Alaska, and two others in a later winter in Alaska and Okhotsk (Siberia) respectively, the latter giving the straight-line distance of 10 200 km mentioned above. Most recoveries were of birds caught at different sites in different winters, but some were found at the same sites in successive winters (mainly in the north) or after gaps of two winters (mainly in the south). Similarly, among Black-capped Chickadees *Parus atricapillus* ringed in various winters, only 3–4% were retrapped one year later, but around 20% two years later, again reflecting the biennial pattern in movements. Such biennial site-fidelity has also been noted in Purple Finches *Carpodacus purpureus* in North Carolina (Blake 1967). Irruptive species also show greater turnover at particular sites within a winter than do non-irruptive ones (for Brambling *Fringilla montifringilla* see Browne & Mead 2003, Jenni & Neuschultz 1987; for Siskin *Carduelis spinus* see Senar *et al.* 1992), and some individuals may be continually on the move through the winter.

Table 18.4 Number of Common Redpolls *Carduelis flammea* ringed in winter in North America that were recovered in successive years after ringing

	Years until recovery				
	1	2	3	4	5
Northern region	34	26	4	2	1
Southern region	6	28	0	5	2

North of the Canadian border, recoveries declined steadily with time, as expected from mortality. South of the Canadian border, recoveries peaked every second year as expected from the biennial pattern in migration. From Troy (1983).

Sex and age differences

As in many other migrants, females of irruptive species often move earlier, in greater proportion or further than males, while juveniles of both sexes often move earlier, in greater proportion and further than adults. Although not all these tendencies emerged in every species examined, they suggest that competitive social interactions over food supplies influence the migration dates and distances of individuals, causing some to move earlier and further than others (Chapter 15). Females outnumbered or moved further than males in Northern Bullfinches *Pyrrhula pyrrhula* (Gatter 1976, Møller 1978, Riddington & Ward 1998), Bramblings *Fringilla montifringilla* (Payevsky 1971, Cramp & Perrins 1994) and Evening Grosbeaks *Hesperiphona vespertina* (Prescott 1991). In some other species, almost all the immigrants in some irruptions were juveniles, as found in Eurasian Jays *Garrulus glandarius*, Great Tits *Parus major*, Blue Tits *P. caeruleus*, Coal Tits *P. ater* and Great-spotted Woodpeckers *Dendrocopos major*, as well as in crossbills and nutcrackers as discussed later (Lack 1954, Newton 1972, Hildén 1974, Van Gasteren *et al.* 1992, Markovets & Sokolov 2002).

Breeding in migration and wintering areas

It is not only the autumn distances that vary from year to year in irruptive seed-eaters. The same holds for the spring distances, with many individuals settling in different areas in different years, depending on food supplies. The Common Redpoll *Carduelis flammea* provides a striking example, for this species curtails its migration by up to several hundred kilometres to breed in southern Fennoscandia in some of the years when the spruce crop there is good. Once the seeds have fallen, the birds in some years move north with their young to their usual birch-scrub breeding areas, where they raise another brood. Such movements have not been proved by ringing, but have been inferred from the simultaneous changes in the populations of the two regions and, in particular, from the late arrival in these years of birds in the birch *Betula nana* and *B. tortuosa* areas with their free-flying young. Such events have been documented in at least seven different years (Peiponen 1967, Hildén 1969, Antikainen *et al.* 1980, Götmark 1982). In years with little or no spruce seed, the birds bred only in the birch, which shortens the overall breeding season.

A similar split migration may sometimes occur in Eurasian Siskins *Carduelis spinus*, in which adults and recently fledged juveniles were in several years seen migrating northeast in May–July over the Courland Spit in the southern Baltic (Payevsky 1994). Some of the adults were clearly paired at the time, and many trapped females had a well-developed brood patch, signifying a recent nesting attempt. The first females with brood patches usually appeared at this site towards the end of April. In the years 1984–1987, some 23–91% of adult females trapped in late April–July had brood patches, as did 35–86% of yearling females (total females caught = 1230). One juvenile caught in June 1959 had been ringed 25 days earlier, 760 km to the southwest, in Germany. While it could not be proved that the adults among these Siskins went on to breed elsewhere in the same year, they clearly had enough time to do so (for other movements of seed-eating birds within a breeding season see Chapter 16).

Some other irruptive species have been recorded breeding well to the south of their usual range in certain years, giving further evidence of variable spring settling patterns (for Brambling see Otterlind 1954; for Northern Bullfinch *Pyrrhula pyrrhula* see Svärdson 1957; for Lesser Redpoll *Carduelis f. cabaret* see Newton 1972; for Bohemian Waxwing *Bombycilla garrulus* see Cornwallis 1961). They have also bred in their wintering areas for one or more years following a major influx, as noted in Lesser Redpoll (Newton 1972), Mealy (Common) Redpoll *Carduelis flammea* (Thom 1986), Northern Bullfinch (Svärdson 1957), and Coal Tit *Parus ater* (Gantlet 1991), in addition to the crossbills and nutcrackers discussed later.

Conclusions on twice-yearly migrants

In conclusion, the most striking feature of irruptive seed-eaters, compared with regular migrants, is the flexibility of their movements. In both breeding and wintering areas, irruptive migrants typically show huge numerical fluctuations from year to year, and much less site-fidelity than regular migrants, as individuals may breed or winter in widely separated areas in different years. In most species, the ringing records from different years mostly lie at different points on the same migration axis (northeast–southwest in much of Europe, northwest–southeast in much of North America), but some individuals show marked east or west displacement off this axis in different years. Typical irruptive migrants also show much more variation in their autumn migration dates and directions from year to year than do regular migrants, and make more frequent long movements within a winter. All these features are linked with the fluctuating and sporadic nature of the seed and fruit crops that provide their food. They contrast with the marked directional tendencies and strong homing and site-fidelity shown by birds that feed on more predictable food supplies, and that benefit from returning to the same familiar areas each year (Chapter 17).

ONCE-YEARLY MIGRANTS

Crossbills

Perhaps the most famous of irruptive migrants in the northern hemisphere are the crossbills, which feed year-round almost entirely on conifer seeds obtained directly from the cones (Newton 1972, 2006b). Three main boreal species are recognised: the Common (Red) Crossbill *Loxia curvirostra* and the Two-barred (White-winged) Crossbill *L. leucoptera* which occur in suitable coniferous habitat across North America and Eurasia, and the Parrot Crossbill *L. pytyopsittacus* which occurs in pine forests of northern Europe. Broadly speaking, these species differ in body and bill size, and specialise on different types of conifers, the Two-barred mainly on soft-coned species, the Common Crossbill mainly on medium-coned species, and the Parrot Crossbill on the very hard, thick-scaled cones of Scots Pine *Pinus sylvestris*. In addition, the Common Crossbill also varies in body and bill size from region to region across its extensive range, in association with the particular species of conifers that grow there (**Box 18.3**).

Box 18.3 The different types of Common (Red) Crossbills

In Europe, large-billed races of Common (Red) Crossbills *Loxia curvirostra* occur in the southern parts of the range, in the isolated mountain pine forests of: (1) southern Spain and the Balearic Islands (*L. c. balearica*); (2) southern Italy, Sicily and Northwest Africa (*L. c. poliogyna*); (3) the southern Balkans, Greece and Cyprus (*L. c. guillemardi*); (4) Corsica (*L. c. corsicana*); (5) the Crimea (*L. c. mariae*); and (6) Scotland, the latter regarded by some as a distinct species *Loxia scotica*. The first two of these large-billed subspecies (1 and 2) feed mainly on seeds of Aleppo Pine *Pinus halapensis*, the next two (3 and 4) mainly on Black Pine *P. nigra*, the Crimean Crossbill (5) mainly on Black Pine and Scots Pine *P. sylvestris*, and the Scottish Crossbill (6) mainly on Scots Pine.

In addition to these recognised types, Common Crossbills of the nominate subspecies, but with slightly different bill-types, have been represented in different invasions to western Europe, perhaps reflecting different boreal source areas (Davis 1964, Herremans 1988). The different types of crossbills also have call-notes which sound slightly different to the practised ear. In northwestern Europe alone, Robb (2000) described six different vocal types of Common Crossbills, as well as three other recognised 'species' (Two-barred, Parrot and Scottish). He showed that these calls were distinctive and constant, and suggested that they represent as yet undescribed 'cryptic species' (for other details, see Summers *et al.* 2002, Summers & Piertney 2003). Clearly, there may be more geographical variation in crossbills than at first meets the eye. Moreover, birds of different bill types tend to concentrate in localities with different conifer species, apparently reflecting different food plant preferences. In northern Scotland after an invasion, spatial segregation was evident even over distances as small as a few tens of kilometres, with large-billed crossbills mainly in pine areas and small-billed ones mainly in spruce and larch areas (Marquiss & Rae 1994, 2002, Summers *et al.* 1996).

Among Red Crossbills in North America, birds of eight different types are currently recognised, each type occupying a mainly different area from the others, and adapted in body size, bill size and palate structure to different conifer species (Benkman 1993, Groth 1988, 1991, 1993a, 1993b). The biggest type is almost twice the size of the smallest, and has a much larger bill. Usually all the birds in a flock are of the same type with their own distinct calls, voice again being a useful key to identification. Each type of crossbill apparently depends on at least one key conifer species, defined as one which normally produces cones somewhere in its wide range every year, holds its seeds in the cones at least until late winter, and for one reason or another does not get totally eaten by other seed-eaters (Benkman 1993). Each bill type appears adapted primarily to one species of conifer, even though all types are able to use several other conifers, albeit with lower rates of seed intake (as measured on both wild and captive birds).

The different types of crossbills may represent different species or at least different nomadic populations that are almost, if not completely, reproductively isolated from other types of crossbill (for occasional cases of interbreeding, see Benkman 1993, Adkisson 1996). Paired birds are almost always of the same morphological or call type, and juveniles develop flight calls like their parents (Groth 1993b). In some places after irruptions, birds of two or more call types may occasionally breed in the same place simultaneously, staying with their own kind and interbreeding rarely, if at all (Griscom 1937, Groth 1988, Knox 1992), or birds of different call types may use the same localities in different years from one another (Griscom 1937, Knox 1992). Genetic differences between the types are small and inconsistent with their calls, allozyme analyses suggesting that they diverged relatively recently, within the last 100 000 years (Groth 1993a). In addition, little genetic difference was evident between the Red Crossbill and White-winged Crossbill which, from study of their mitochrondrial DNA, were estimated to have diverged only about one million years ago (Groth 1991, 1993a, Questiau et al. 1999, Piertney et al. 2001).

White-winged Crossbills *Loxia leucoptera*, which have finer bills than most types of Red Crossbill, appear to show less geographical variation, but the North American race *L. l. leucoptera* has a smaller and narrower bill than the Eurasian *L. l. bifasciata*, and an even larger-billed, isolated pine-feeding form of the White-winged Crossbill (*L. l. megaphaga*) occurs on the island of Hispaniola, where it specialises on the seeds of the West Indian Pine *Pinus occidentalis*. Such geographical variations in crossbills imply that, despite their widespread wanderings, they are sufficiently faithful to particular regions to have become adapted to the conifers growing there.

Where several conifer species occur in the same region, crossbills switch from one species to another through the season, according to the different patterns of cone ripening and seed fall (Benkman 1987). Both Common and Two-barred Crossbills can breed in any month of the year, depending on the species of conifers available. However, most breeding occurs in late summer–autumn as cones mature, or in late winter–spring as cones open to release their seeds (Newton 1972, Benkman 1987, 1990).

Annual cycle

Over much of Europe, Common Crossbills depend mainly on Norway Spruce *Picea abies* which has more fluctuating and sporadic seed crops than pine. Each year, new cones begin to form in May, providing food for Crossbills from about June into the winter, and particularly from January on, when the cones begin to open and shed their seeds. By late May, when most cones and seeds have fallen, they are lost to Crossbills, which then switch to alternative foods, and eventually to the new Norway Spruce crop (Marquiss & Rae 1994, 2002).

In most years, Common Crossbills remain within the boreal forest, each year concentrating in areas where spruce cones are plentiful. In particular areas, their breeding numbers tend to fluctuate on a roughly 2–4 year cycle, in line with the local cone crops (**Figure 18.10**; Reinikainen 1937, Formosov 1960, Thies 1996, Förschler et al. 2006). The birds have one major period of movement within the summer of each year, as they leave areas where the previous year's spruce crop was good but coming to an end, and concentrate in areas where the current

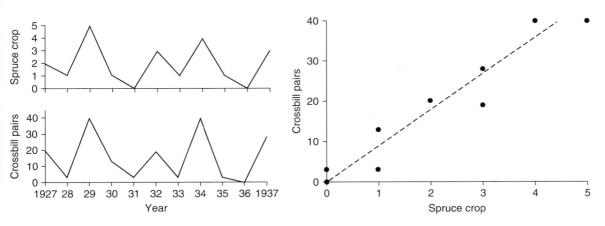

Figure 18.10   Relationship between the population density of the Common Crossbill *Loxia curvirostra* and the cone crop of Norway Spruce *Picea abies* in Finland. Crossbills in number of pairs per 120 km transect; spruce crop ranked in five categories. From Reinikainen (1937).

year's crop is good but forming. Between these times, some birds concentrate temporarily in Scots Pine areas. It is only at this period of transition, in summer, that Common Crossbills in Europe have been found with substantial fat reserves, presumably deposited as 'migratory fat' (Marquiss & Rae 2002, Newton 1972, 2006b). However, lesser movements occur at other times of year in relation to local changes in food availability, as seeds are shed or consumed, or as cones open and close.

In a good cone year for Norway Spruce, Crossbills may begin nesting in autumn, within weeks after arriving in a new area, and may continue for as long as the food holds out. A pair may start a second nest while still feeding young from the first, and young birds may begin breeding while they themselves are still in juvenile plumage (Berthold & Gwinner 1978, Jardine 1994, Hahn *et al.* 1997). The main breeding period, however, is in January–April, when the cones begin to open, making seeds more readily available. As seeds and cones fall through May, breeding in Norway Spruce areas comes to an end and the birds move on.

Irruptions

In some years, apparently when high populations coincide with poor Norway Spruce crops over wide areas, Common Crossbills leave the boreal forest, and move southwestward through Europe, appearing in many places that lack suitable conifer habitat. For example, between 1880 and 2000, Common Crossbills irrupted into Britain on at least 40 occasions, at intervals of 1–9 years (Newton 1972, 2006b). Each time, the birds appeared sometime during late May–October, mainly in June–July. The fact that trapped migrants differed slightly in bill dimensions from one irruption to another (Davis 1964, Herremans 1988) suggests that not all movements originated from the same region (or that not all bill size categories were stimulated to move in the same years). Most of these irruptions included small numbers of other crossbill species (*L. pytyopsittacus* or *L. leucoptera*) (Newton 1972).

Irruptive movements occur at about the same time as normal annual movements, but are more directional (towards the southwest), and cover much longer distances (with extremes at more than 5000 km). These differences could be explained by the birds achieving a higher migratory state in irruption years. A similar change occurs in normal migrants as they switch in late summer from random dispersal movements to directional long-distance migration (for Barn Swallow *Hirundo rustica* see Ormerod 1991; for Willow Warbler *Phylloscopus trochilus* see Norman & Norman 1985).

Once on the move, Crossbills spread over a wide area, some reaching the extreme southwest of Europe. Thousands of irrupting Crossbills have now been ringed at trapping sites in central Europe, giving hundreds of recoveries (Newton 1972, 2006b, Payevsky 1971, Weber 1972, Schloss 1984). Almost all recoveries within the current cone year (June–May) were to the south and west (mostly southwest) of the ringing site, while many of those in later years were far to the northeast, in the boreal zone of northern Russia, almost all in a region west of the Urals between 57° and 62°N and 40° and 60°E (**Figure 18.11**; Newton 2006b). Overall, at least 29 adults caught on migration during different invasions through central Europe have now been recovered back in the boreal forest to the northeast, 1–3 years later. None was found in northern boreal forest within the same spruce year as it was ringed in central Europe (Newton 2006b). If the outward movement was stimulated by food shortage, birds would gain no advantage in returning to their area of origin before the next year's crop was ready. In addition to ring recoveries, observational evidence for eastward or northeastward return movements in the summers of non-invasion years is available from several localities (Newton 2006b). The return movement is much less conspicuous than the outward one, perhaps because it involves smaller numbers and occurs in more than one year. The main axis of migration is clearly northeast–southwest, as in most other seed-eaters, but unlike them, Crossbills remained for a year or more before returning.[1]

If irrupting Crossbills had remained in the regular range, they would almost certainly have starved, but by moving out, a proportion survived to return in a later year, and may even have bred in the interim. Hence, the adaptive value of periodic mass emigration is presumably the same as in regular annual migration, namely the avoidance of food shortage. Those irrupting birds that find areas of seeding conifers often remain to breed. Mostly they move away after one breeding season, but some irruptions have resulted in the longer-term colonisation of new areas (for example, conifer plantations in Britain).

---

[1]*Gatter (1993) has suggested, on the basis of the flight directions of Common Crossbills* Loxia curvirostra *seen over an observation point in Germany, that the birds return from whence they came within a few months of arriving in an invasion year, in the autumn and winter. Ring recoveries give little support for this view, however, because almost all of the recoveries of birds ringed on irruptions (mostly in Germany) were to the south or west of ringing sites until May of the next year, while birds recovered in later years were more evenly distributed between southwest and northeast. In fact, all recoveries in the boreal forest of northern Russia came more than one year after ringing in central Europe (see above).*

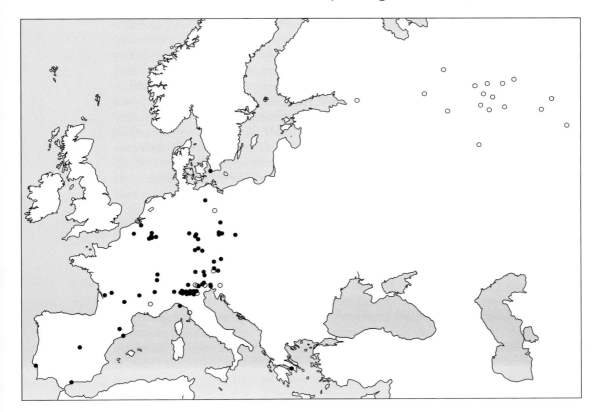

Figure 18.11 Recoveries of Common Crossbills *Loxia curvirostra* ringed at migration time in June–October of invasion years in Germany, and recovered in the following year (filled circles) and later years (open circles) respectively. All the 83 birds recovered in the following year (to 31 May) were in the invasion areas of western Europe, whereas 22 out of 44 birds recovered in later years were far to the northeast, in the boreal zone of northern Russia (significance of difference in distribution, $\chi^2 = 34.2$, $P < 0.001$). Only movements more than 50 km are included. These recoveries indicate that some irrupting Crossbills return to their region of origin in a later year. Map compiled from the recoveries listed in Schloss (1984), which include recoveries from the invasions of 1930, 1935, 1953, 1956, 1959, 1962, 1963, 1972, 1977 and 1979, although two-thirds of all recoveries in Russia probably stem from the invasion of 1963. Schloss (1984) lists nine other recoveries in northern Russia of birds ringed in Germany in non-invasion years or outside the migration season, some of which are in **Figure 18.12**. In addition to the records shown, another Common Crossbill ringed at Falsterbo in Sweden on 12 August 1963 was recovered in a later year at Gayny in northern Russia at 60°18′N, 54°18′E (Roos 1984), and one ringed in Northamptonshire in England on 2 June 1991 was recovered 1958 km east-northeast on 14 April 1998 in Pskov in northern Russia (58°16′N, 28°54′E) (Clark *et al.* 2000). For other recoveries of birds ringed in Switzerland, see Newton (1972).

Change of breeding localities

Ring recoveries also lend support to the view that individual Common Crossbills have bred in widely separated areas in different years, taking the main breeding season in Norway Spruce areas as January–April inclusive. Movements of Common Crossbills that were ringed in the January–April of one year and recovered in the same period in a later year are shown in **Figure 18.12**. They include 10 recoveries of birds ringed as adults (representing breeding dispersal) and three of birds ringed as juveniles and recovered as adults (representing natal dispersal). These various birds were reported up to four years after ringing; their successive capture sites were between 28 and 3170 km apart, and included eight birds that had moved more than 2000 km. The main interest in these records lies in the great distances that separated the breeding areas of different years, which are thoroughly in line with the long and irruptive movements of the species. None of these birds was reported as nesting at the time of ringing or recovery, but they were all

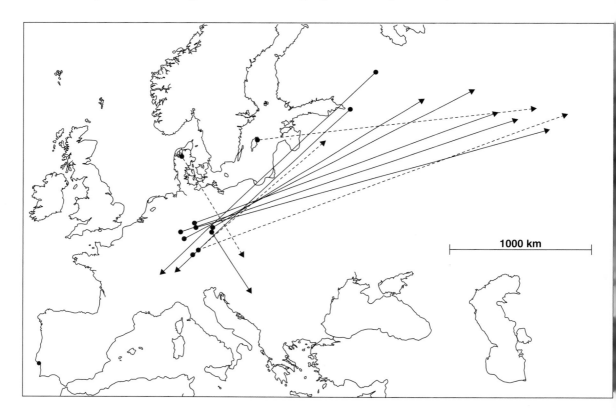

Figure 18.12   Ringing and recovery sites of Common Crossbills *Loxia curvirostra* that were both ringed and recovered in different breeding seasons (taken as January–April in areas of Norway Spruce *Picea abies*). Continuous lines – ringed as adults (representing breeding dispersal); dashed lines – ringed as juveniles (representing natal dispersal). Compiled mainly from information in Schloss (1984); also from Danish Ringing Report (1931–1934), Swedish Ringing Report (1965), Swiss Ring recoveries provided by Dr L. Jenni, and Russian ring recoveries provided by Dr K. Litvin. Details listed in Newton (2006b).

within the usual breeding season from within potential breeding range. Moreover, no records have emerged of individuals in the same spruce locality in successive spruce years, which is not surprising as good crops seldom occur in the same locality two years running. This contrasts with the situation in Pyrenean Mountain Pine *Pinus uncinata* forests, where seed crops are relatively stable from year to year, and where Common Crossbills are more sedentary, with many individuals trapped in the same place in successive years (Senar *et al.* 1993). These findings from a pine area again imply that the frequent movements in spruce areas are driven by the huge annual fluctuations in regional crop sizes. Similarly, colour-marked Scottish Crossbills *L. c. scotica* were re-sighted in the same pine areas in successive years, or had moved over distances of at most a few tens of kilometres (Marquiss & Rae 2002). The same presumably holds in other pine areas in Europe, most of which hold local races of large-billed crossbills (**Box 18.3**).

The return of crossbills after an outward movement is analogous to the 'homing' of other migrants, but unlike most other species, crossbills sometimes breed before returning. This means that, if they are to reach their ancestral home, young raised in invasion areas have to make their first migration in a direction opposite to the first migration of their parents and of all other individuals raised on the regular range. Several ring recoveries of birds from Germany to northern Russia, and another from south Sweden to northern Russia, are consistent with such a movement. All these birds were hatched in breeding seasons following invasions. Perhaps the juveniles simply travelled with returning adults (as reported by Weber 1971–72). An alternative possibility is that the directional preferences of Crossbills differ according to geographical location, with birds leaving northern Russia moving mainly southwest, and those leaving central Europe moving mainly northeast. In one respect, this would be no different from other migrants, which head in one direction from their breeding areas and in the opposite direction from their non-breeding areas. But it differs in that Common Crossbills would need to take different directions at the same time of year, and both times after breeding, rather than at markedly different seasons, as in twice-yearly migrants. To judge from recent east–west displacement experiments on other bird species (Chapter 9), this might be expected, but it would be interesting to test at migration time the directional preferences of young Crossbills raised in the two regions.

Other crossbills

In Eurasia, where the Two-barred Crossbill *Loxia leucoptera* feeds primarily on larch, it may not suffer such extreme fluctuations in food supply as the spruce-feeding Common Crossbill *L. curvirostra*. This is because larch cones remain on the tree with some of their seeds for 2–3 years, so poor crops can be enhanced by carry-over from previous years. This may explain why this species is much less irruptive in the western Palaearctic than the Common Crossbill, and appears in only very small numbers south of the boreal zone, mostly along with Common Crossbills. Irruptions into Scandinavia occur during cone failures in larch further east in Russia. They tend to occur from late summer, as new cones ripen, but at longer intervals than the 2–4 years recorded in North America (Bock & Lepthien 1976, Larson & Tombre 1989). Two-barred Crossbills *L. leucoptera* have come to Britain mainly in the same years as Red (Common) Crossbills, though much less frequently (Newton 1972), with larger numbers than usual recorded in 1889, 1956, 1979, 1985–1987 and 1990 (Cramp & Perrins 1994). Following the irruption of Two-barred Crossbills in 1990, successful nesting of this species was recorded near Berlin in 1991, some 1700 km southwest of the usual breeding range (Fischer *et al.* 1992), and after the invasion of 1956 nesting was recorded in Sweden, 600 km east of the usual breeding range (Markgren & Lundberg 1959).

Small numbers of Parrot Crossbills *L. pytyopsittacus* also appear in Britain, mainly in the same years as Common Crossbills. Relatively large invasions occurred in 1962, 1982 and 1990 (Catley & Hursthouse 1985, Rogers 1992). Unlike the other European crossbills, Parrot Crossbills move mainly in the autumn, in keeping with the different phenology of their food plant, as the seeds in new pine cones are not normally developed much before September. One individual that was ringed as a nestling in Norway was recovered in the breeding season three years later 340 km to the east, in Sweden (Norwegian Ringing Report, 1966–67). Like other crossbills, Parrot Crossbills may nest in peripheral invasion areas for a year or two, before disappearing, as recorded occasionally in Denmark, Britain and the Netherlands.

North American crossbills

North America has a far greater variety of conifer species than Europe, and many seem to crop synchronously over wide areas. It also has a wide range of Red (Common) Crossbills, at least eight different types having been described, each occurring in mainly different regions from the others and adapted to the conifer species that grow there (**Box 18.3**). Conifer species differ in the dates their cones mature, and in the dates the cones fall or shed their seeds. Hence, while spruce-feeding, Common Crossbills in Europe move mainly in May–July, Red and White-winged Crossbills in North America make their main movements anytime from June to November, depending on the conifer, with more minor movements at other times in response to changes in seed availability (Benkman 1987, 1992). In eastern North America, White-winged Crossbills often arrive in areas of Tamarack *Larix laricina* and White Spruce *Picea glauca* in late May–July, as the new cones are forming, and breed on the strength of the new crop. They stay only until the seed has fallen, usually by November, switching then to Black Spruce *P. mariana*, whose seeds can remain available in the cones until the following summer. But if the Black Spruce crop is poor, White-winged Crossbills then irrupt into regions south of the usual range. The timing of this movement, in October–November, coincides with the timing of seed fall from Tamarack and White Spruce. In the same region, Red Crossbills (with their larger bills) can exploit all the conifers eaten by White-winged Crossbills, but can also tackle pines efficiently, so they have a greater range of conifers available to them. However, if pine cones are scarce, Red Crossbills also emigrate, mostly from November onwards (Benkman 1987). Thus, the southward irruptive movements of crossbills in North America occur from October–November onwards, compared with May–July onwards in Europe, but in both regions the timing of the movements matches the phenology of the food plants, as the birds respond to a dwindling food supply.

Some individual ringed Red Crossbills in North America have been recorded at the same site two years later, but others were recovered more than 2000 km away

(Adkisson 1996). Of particular interest are two birds, recorded in the same months of different years. One bird was recorded in May of 1991 and 1992 at localities 1409 km apart, and another was recorded in August 1990 and 1993 at localities 521 km apart. Without knowledge of local breeding seasons, it is hard to interpret these movements, but they would not be expected in a regular migrant moving annually between fixed breeding and wintering areas.

Nutcrackers

Other irruptive seed-eaters include the Spotted Nutcracker *Nucifraga caryocatactes* of Eurasia and the Clark's Nutcracker *N. columbiana* of western North America. These species feed primarily on the fruits of various large-seeded pines, which they cache in the ground in late summer to eat during the ensuing winter into the next breeding season. Their overwinter survival and breeding success vary from year to year in line with the size of seed crops (Lanner 1996). Like crossbills, they make one major annual movement, in July–September (as the new cones form), with occasional irruptions outside the usual range. Having settled in appropriate areas, their strategy of living off food stores ideally requires continued residence until the next crop is ready. Like Crossbills, Nutcrackers breed early in the year, and can rear their young entirely (or almost entirely) on seeds.

The Eurasian *N. caryocatactes* has several subspecies, each occupying a different part of the range and depending on different conifer species. The most widespread type is the slender-billed *N. c. macrorhynchos*, which occurs over most of Siberia, and depends on the seeds of the Siberian Stone Pine *Pinus sibirica* and, further east, the Korean Stone Pine *P. koraiensis*. In northwestern Europe, where large-seeded conifers are lacking, the thick-billed *N. c. caryocatactes* lives in mixed deciduous–coniferous areas, but depends in winter mainly on Hazel *Corylus avellana* nuts, which it caches in late summer in the same way that other subspecies cache pine seeds. Further south, in the Alps, it also uses the hard-shelled Arolla Pine *Pinus cembra* seeds.

All these nutcrackers adjust to annual variation in local cone crops by redistributing themselves each year within the regular range (mainly in July–September) and by exploiting alternative small-seeded conifers when their main food is scarce. However, in years when major food sources are scarce over wide areas, the birds leave their regular range in large numbers, and live as best they can off any suitable plant or animal material they can find.

The Siberian *N. c. macrorhynchos* performs the most spectacular irruptions, reaching in some years as far as western Europe, on journeys of several thousand kilometres. Thirty-one irruptions into Europe were recorded in the 250 years from 1750, at intervals of 1–33 (mean 8) years (Mayaud 1947, Cramp & Perrins 1994). In the twentieth century alone, the figures were 13 irruptions, at intervals of 2–16 years (mean 7.7). Only in the largest irruption of 1968, which started as far east as Lake Baikal, did large numbers of birds reach Britain, a journey exceeding 7000 km (Hollyer 1970), with smaller numbers occurring in at least five other years. Again, not all invasions necessarily originated from the same region. According to Formosov (1933), the invasion of 1911 came from southeast Siberia, that of 1931 from northwest Siberia, and that of 1933 from central and northern Siberia.

In such invasions, surviving birds normally stay until the next year, like cross-bills, but in 1968 many nutcrackers that reached western Europe reversed their migration and returned homeward in the same autumn, but probably not as far as their region of origin. Recoveries of 14 migrants ringed in Finland were all between east and south-southeast of the ringing sites, including three at 2200–3300 km east. Another bird ringed on the southern Baltic coast was recovered 2500 km east two weeks later (Hildén 1969, Cramp & Perrins 1994). Other irruptive birds trapped on the island of Gotland in the Baltic were recovered in a later year at 70–85°E in west Siberia, some 4000 km to the east. A partial return eastward movement within weeks of the westward late-summer exodus has been recorded at Lake Ladoga in Russia (Noskov *et al.* 2005).

Like crossbills, nutcrackers can remain in invasion areas for up to a year, breeding if food supplies permit, and returning to the regular range when the next year's crop is ready. Small longer-term populations have been established after invasions in areas where suitable food trees had been planted where none occurred before (Lanner 1996). Following the invasion of 1968, colonies of Slender-billed Nutcrackers *N. c. macrorhynchos* from Siberia became established in Finland and Belgium in areas where Stone Pines *Pinus sibirica* and *P. cembra* had been planted (Cramp & Perrins 1994); and after the invasion of 1977, similar colonies were established in Sweden (Elmberg & Mo 1984).

Those Thick-billed Nutcrackers *N. c. caryocatactes* that live in northern Europe move shorter distances to the south, while those in the mountains of central Europe move mainly to adjacent lower ground in years of extreme food shortage. Big movements of central and northern European birds do not usually coincide with one another, nor with the bigger movements into Europe of Siberian birds (Mattes & Jenni 1984).

In North America, Clark's Nutcracker *Nucifraga columbiana* breeds in the western mountains, but not in the northern boreal forest which lacks large-seeded pines. Like the Eurasian species, it depends on different types of conifers in different parts of its range, moving around each summer and autumn to areas where new crops are good (Vander Wall *et al.* 1981, Lanner 1996). In years of widespread crop failure that follow good crops, the birds extend mainly to lower ground nearby, in the deserts, plains and coastal areas, but some have been recorded far to the east, more than halfway across the continent (Fisher & Myres 1980). Invasions into the lowlands of California and other southwestern States occurred in 1898, 1919, 1935, 1950, 1955 and 1961, at intervals of 5–21 years (mean 12.6 years) (Davis & Williams 1957, 1964). Another big one occurred in 1996, the largest for many years (National Audubon Field Notes). At the northern end of the range, in Alberta, invasions of low ground during 1904–1976 occurred in 1919, 1960, 1965, 1972 and 1976; that is, in mainly different years from those in the southwestern States (Fisher & Myres 1980). Like some other irruptive migrants, Clark's Nutcrackers sometimes breed in invasion areas before returning to their regular range and, like crossbills, they thereby move in opposite directions in successive years (Vander Wall *et al.* 1981, Tomback 1988).

In North America, other corvids also harvest and store pine seeds, namely the Pinyon Jay *Gymnorhinus cyanocephalus*, Steller's Jay *Cyanocitta stelleri* and Scrub Jay *Aphelocoma californica*, but none of these species is so heavily dependent on them, or so specialised as the nutcrackers, and none of them performs obvious irruptive migrations.

OVERVIEW OF SEED-EATERS

Coping with a boom-and-bust economy

By changing breeding and wintering areas between years, irruptive seed-eaters lessen the effects of the massive food shortages they would experience if they occupied the same areas every year. Nevertheless, they may still be exposed to a hugely fluctuating food supply, as reflected in their reproductive rates. For example, in an area of northern Sweden, Bramblings *Fringilla montifringilla* bred every year over a 19-year period, but in greatly varying numbers, depending on food supply (Lindström *et al.* 2005). Post-breeding juvenile-to-adult ratios varied more than 10-fold over this period, from 3.54 in good food years to 0.33 in poor ones. The Redpolls *Carduelis flammea* and Siskins *Carduelis spinus* mentioned earlier can breed for more than twice as long in good spruce years than in other years (giving time for 2–3 broods instead of 1–2), and could thereby double their production of young (Peiponen 1967, Shaw 1990). In good seed years, both species begin nesting as early as March, when conifer cones open. After these seeds have fallen, the birds raise another one or two broods on the fresh seeds of herbaceous plants which form in May–July, depending on area, or (in the case of Redpolls on the tundra) on the seeds of Dwarf Birch *Betula nana* left from the previous year. In poor spruce years, only the later broods are reared. Among Common Crossbills *Loxia curvirostra*, the annual variations may be even greater, for in mixed conifer areas in which different tree species release their seeds in widely different months, individual Crossbills could in theory breed for more than nine months each year, raising brood after brood (Newton 1972). But in areas containing only Norway Spruce *Picea abies*, individuals breed for no more than half this time or, in poor cone years, not at all.

The fact that juveniles often predominate among irruptive species caught on migration has been taken as evidence that irruptions follow good breeding seasons (Lack 1954). Care is needed, however, because in such facultative migrants juveniles often leave the breeding areas in greater proportion and earlier than adults, and move greater distances. Nevertheless, when most of the migrants in particular years are adults, this probably gives a reliable indication of a poor breeding season. This situation was recorded, for example, among Common Crossbills in 1963, when young formed only 6%, 8%, 31% and 37% of the birds caught at four localities in western Europe (Newton 1972). This compares with up to 88% recorded in other irruption years. Similarly, study of Nutcracker *Nucifraga caryocatactes* skins in museums suggested that the movements of 1864, 1911 and 1968 consisted entirely of adults, and about half the adult females collected in the 1968 irruption had never laid eggs, evidence that they had not bred that year. In contrast, from the irruptions of 1885, 1913 and 1954 only first-year birds were collected in western Europe, while in other irruptions both age-groups were represented (Cramp & Perrins 1994).

Regularity in irruptions

Because the cropping patterns of many northern tree species are more regular than random, they would be expected to impart some regularity to large-scale emigration.

Some evidence points to a cyclic rhythm in the irruptions of seed-eating species, notably the tendency to a biennial pattern in several species in North America. In addition, a seven-year rhythm emerged in the influxes of Two-barred Crossbills *Loxia leucoptera* in autumn to Finland during 1960–1988 (Larson & Tombre 1989), a 5–6 year rhythm in the irruptions of Pine Grosbeaks *Pinicola enucleator* to southeastern Canada during 1889–1936 (Speirs 1939), and a 4–5 year rhythm in the movements of Eurasian Jays *Garrulus glandarius* in the Swabian Alps of central Europe during 1954–1973 (Gatter 1974). Such regularity reflects patterns in tree-seed production, but may not necessarily persist over longer periods, and in many areas irruptions seem irregular.

However, invasions of irruptive seed-eaters have been studied chiefly in the reception areas, where the degree of regularity probably varies with distance from the breeding range. While every irruption might reach the nearest areas, only the largest would reach the furthest areas, as illustrated above for nutcrackers and others. Sometimes, invasions come not just in one year but in two or more successive years, hence the term echo flights. But these might involve birds from different parts of the breeding range, thus obscuring any regularity there might be in more local breeding populations. For example, most invasions of the Pine Grosbeak *Pinicola enucleator* into Germany came from the north, and the birds belonged to the European subspecies, but in 1892, a larger subspecies invaded from Siberia (Grote 1937). Similarly, Redpoll invasions in central Europe in 1985 and 1986 were mainly long-billed *holboellii* types and hence probably came from within the range of the Larch *Larix dahaurica* in Siberia (Two-barred and Red Crossbills also came from this area in 1985). Moreover, different Common Crossbill invasions to western Europe have involved birds with different bill sizes, as mentioned above, which might also indicate different areas of origin (Davis 1964, Herremans 1988).

Directional preferences

In western Europe, the usual migration direction of most irruptive species is southwest, a direction in which birds could usually be expected to find suitable habitat and food supplies, especially in montane conifer forests. Birds from further east, however, tend to have a stronger westward component in their movements, which brings them to western Europe, either within the boreal forest itself, or south of it into the temperate region. They thereby largely avoid the dry steppe and desert lands of Asia in which their survival chances would presumably be low. This westerly movement from Siberia is especially marked in Nutcrackers, Common Crossbills, Two-barred Crossbills, Siberian Nuthatches and Pine Grosbeaks, which leads the invasions of some of these species to be more obvious in Fennoscandia than further south.

Such patterns led Svärdson (1957) to suggest that some boreal birds performed regular pendulum movements, moving first west then east across the boreal forest of Eurasia, a view for which there is only limited evidence beyond the eastward return after occasional westward irruptions. However, some ringed Redpolls *Carduelis flammea* showed easterly autumn movements within the boreal zone. Of two ringed in northern Norway in August 1977, one was reported 3091 km due east near Novosibirsk two months later on 20 October 1977, and another was reported

2753 km east-southeast near Chaklovo on 12 March 1978. In the 1965 irruption, some birds ringed in Finland reached the edge of the Altai in central Asia, over distances up to 3573 km.

In North America, too, there is a strong east–west component in the movements of some irruptive species. Evening Grosbeaks *Hesperiphona vespertina*, Common Redpolls *Carduelis flammea* and Pine Siskins *C. pinus* from the northwest of the continent tend to move east–southeast in autumn, as mentioned earlier, while Purple Finches *Carpodacus purpureus* move first east-southeast to the Great Lakes region, then veer south towards Texas and Louisiana (Houston & Houston 1998, Brewer *et al.* 2000). Similarly, observations suggested a large movement of White-winged Crossbills *Loxia leucoptera* from Alaska and Yukon across the boreal forest to Ontario and Quebec in May–June 1984, and a return movement in May–June of the following year (Benkman 1987). In addition, the small-billed Red Crossbill *L. c. sitchensis* may regularly make movements between the Pacific Northwest, including Alaska, and the Great Lakes region (Benkman 1987). By these movements, all species avoid the desert and prairie areas of the mid-west, and remain largely in forested areas that provide their food.

Other seed-eaters

Although variations in autumn movements among seed-eating birds are most marked in those species that depend on the seeds of trees, they are also apparent to some extent in those that eat the seeds of herbaceous plants. In arid regions, the local production of such seeds can vary greatly from year to year, according to rainfall patterns, and in other regions their availability on the ground is affected by patterns of agriculture or snowfall. Many seed-eaters concentrate at different points on their migration routes in different years, depending on seed supply (Pulliam & Parker 1979, Dunning & Brown 1982), and ringed individuals have occurred at widely separated points on that route in different winters (for Goldfinch *Carduelis carduelis* and Eurasian Linnet *Carduelis cannabinna* see Newton 1972, for Dark-eyed Junco *Junco hyemalis* see Ketterson & Nolan 1982), or have moved further along the route during the course of winter (Haila *et al.* 1986, Terrill & Ohmart 1984). This is irruptive behaviour in less extreme form, with more directional consistency. Moreover, occasional eruptions are recorded from species that are normally sedentary, if they reach high numbers relative to available food supplies, as noted in Bearded Tits *Panurus biarmicus* and others (Axell 1966, Bjorkman & Tyrberg 1982).

CONCLUDING REMARKS

Observations and ring recoveries give some idea of the huge areas over which individual boreal seed-eaters may roam to find food supplies necessary for their survival and reproduction. With their ability to move rapidly over long distances, these species can exploit a specialist niche in a way that other, more sedentary, animals could not. But because of the sporadic nature of their food supplies, only parts of their vast geographical ranges may be occupied at one time. Individual movements of hundreds or thousands of kilometres have been recorded between the natal and breeding sites in different years, and between the breeding sites of successive years. Again, however, because the ring recoveries are supplied by members of the public, they carry the assumption that birds found in the breeding season were in fact nesting at the locality concerned. The movements of irruptive species clearly contrast with those of more regular migrants which exploit more stable food supplies and, in association, show much greater site-fidelity and smaller dispersal distances.

The advantage of strong, endogenous control of migration, as shown by regular obligate migrants, is that it can permit anticipatory behaviour, allowing birds to prepare for an event such as migration, before it becomes essential for survival, and facilitating fat deposition before food becomes scarce. But such a fixed control system is likely to be beneficial only in predictable circumstances, in which food supplies change in a consistent manner, and at about the same dates, from year to year. It is not suited to populations that have to cope with a large degree of spatial and temporal unpredictability in their food supplies. It is these aspects of food supply which probably result in irruptive migrants showing greater variations in autumn timing, directions and distances, selection having imposed less precision on these aspects than in regular migrants. Both regular and irregular systems are adaptive, but to different types of food supplies, the one to consistent and predictable, and the other to inconsistent and unpredictable (**Table 18.5**). Nevertheless, regular (obligate) and irruptive (facultative) migrants are best regarded, not as distinct categories, but as opposite ends of a continuum, with predominantly endogenous control (=rigidity) at one end and predominantly external control (=flexibility) at the other.

Many irruptive migrants vary geographically in their behaviour, being more strongly irruptive in some regions than in others. It is easy to appreciate how regional variation in the obligate/facultative balance might evolve within species, as food supplies across the breeding range change from the more predictable to the less predictable, partly in association with the diversity of food types available, and with the degree of their year-to-year fluctuation. This is little different in principle from the transition from resident to migratory found in many

Table 18.5   Comparison between typical regular and typical irruptive migration

	Regular (obligate) migrants	Irruptive (facultative) migrants
Habitat–food	Predictable	Unpredictable
Breeding areas	Fixed	Variable
Wintering areas	Fixed	Variable
Site-fidelity	High	Low
Migration		
Proportion migrating	Constant	Variable
Timing	Consistent	Variable
Distance	Consistent	Variable
Direction	Consistent	Variable

bird species from low to high latitudes. It accounts for why some species, such as the Great Tit *Parus major* or Eurasian Bullfinch *Pyrrhula pyrrhula*, are essentially resident in some parts of their geographical range, irruptive in other parts, and perhaps regular migrants in yet other parts (Cramp & Simmons 1980, Cramp & Perrins 1993). Each mode of behaviour is adapted to the nature of the food supply in the region concerned, in accordance with their different food plants.

SUMMARY

The seed crops of some northern tree species vary greatly in size from year to year, both in particular regions and over the two northern land masses. However, seed crops in different regions fluctuate independently of one another, so that good crops in some regions coincide with poor crops in others. Each year, the birds that depend on tree seeds tend to settle at greatest densities in areas with the largest crops and, in line with the variable fruiting patterns, some individuals breed and winter in widely separated areas in different years. In different breeding seasons, as in different winters, ringed individuals of several species have been found at localities hundreds or thousands of kilometres apart. In years of widespread food shortage (or high numbers relative to food supplies), extending over many thousands or millions of square kilometres, large numbers of birds migrate to lower latitudes, as an irruptive migration. The timing of autumn migration, and the distances moved, vary greatly from year to year; in years of widespread crop failure birds often depart earlier and travel much further than usual. In North America, irruptions of several species occur about every second year, but in Europe irruptions are much less regular and less synchronised between species.

Crossbills depend year-round on conifer seeds. In much of Europe, Common Crossbills *Loxia curvirostra* make one major movement each year, in summer, between the shedding of one seed crop in certain areas and the formation of the next crop in other areas. In years of widespread crop failure, the movements become more directional, and birds leave the regular range in large numbers, occurring as irrupting flocks well outside their usual range, only to return in a later year. The same holds for Spotted Nutcrackers *Nucifraga caryocatactes* in the boreal zone of Eurasia, while Clark's Nutcrackers *N. columbiana* in North America move from the western mountains to the neighbouring lowlands in years of crop failure.

Compared with regular (obligate) migrants, irruptive (facultative) migrants show much greater year-to-year variations in the proportions of individuals that migrate, and greater individual and year-to-year variations in the autumn timing, directions and distances of movements. The control systems are flexible in irruptive migrants, enabling individuals to respond to feeding conditions at the time. Regular and irruptive migrants probably represent opposite extremes of a continuum of migratory behaviour found among birds, from narrow and consistent at one end to broad and flexible at the other. Both systems are adaptive, the one to conditions in which resource levels are predictable temporally and regionally, and the other to conditions in which resource levels vary unpredictably. Depending on the predictability and stability of its food supply, the same species may behave as a resident or regular migrant in one part of its range, and as an irruptive migrant in another.

Snowy Owl *Nyctea scandiacus*, a well-known irruptive migrant

Chapter 19

Irruptive migrations: owls, raptors and waterfowl

Its distribution is irregular, it being abundant at one season and almost totally unknown the next. (E. W. Nelson 1887, writing of Snowy Owls.)

This chapter is concerned with two other groups of irruptive birds, namely those owls and other predators that specialise on voles and other cyclically fluctuating prey species, and waterfowl that depend on ephemeral wetlands. Again, the large-scale movements of these species can best be understood in light of their underlying ecology, which I therefore describe in some detail. Ducks are not usually regarded as irruptive migrants but, particularly in arid regions, they show many of the typical features: a tendency to concentrate temporarily wherever conditions are suitable, and in widely separated areas in different years. In winter in many regions they are also greatly influenced by prevailing temperatures, as individuals move to avoid areas where surface waters are frozen, appearing in numbers in the furthest parts of their wintering ranges only in exceptional years.

OWLS AND OTHER PREDATORS

Owls and raptors that depend on cyclically fluctuating prey species often suffer food shortages in years when their prey numbers crash. In some northern species this leads to massive southward emigration. Two main prey systems underlie their behaviour: (1) an approximately 3–5 year cycle of small (microtine) rodents in the northern tundras, open parts of boreal forests and temperate grasslands;[1] and (2) an approximately 10-year cycle of Snowshoe Hares *Lepus americanus* in the boreal forests of North America (Elton 1942, Lack 1954, Keith 1963, Stenseth 1999). The numbers of certain grouse species also fluctuate cyclically, in some regions in parallel with the rodent cycle and in others with the longer hare cycle (Hörnfeldt 1978, Keith & Rusch 1988, Newton 1998b). Any predators that special-ise on such prey species, as well as more generalist predators that also eat other things, are affected by the fluctuations in these prey.

Rodents and rodent-eaters

Populations of microtine rodents do not reach a peak simultaneously over their whole range, but the cycles may be synchronised over tens, hundreds or many thousands of square kilometres, out of phase with those of more distant areas. However, peak populations may occur simultaneously over many more areas in some years than in others, giving a measure of synchrony, for example, to lem-ming cycles over large parts of northern Canada, but with regional exceptions (Chitty 1950). In addition, vole cycles tend to lengthen northwards from about three years between peaks in temperate and southern boreal regions to 4–5 years in northern boreal regions (**Figure 19.1**). The amplitude of the cycles also increases northwards, from barely discernable cycles in some southern temper-ate regions, to marked fluctuations further north, where peak densities typically exceed the troughs by more than 100-fold (Hanski *et al.* 1991). On the tundra, the periodicity of lemming cycles is in some places even longer (5–7 years between peaks on Wrangel Island, Menyushina 1997), and the amplitude even greater, with peaks sometimes exceeding troughs by more than 1000-fold (Shelford 1945). In most places, the increase phase of the cycle usually takes 2–3 years and the crash phase occurs within one year. Importantly, the crash phase often occurs during spring and summer, which can cause widespread breeding failure among rodent predators (e.g. Lockie 1955, Maher 1970). Specialist rodent-eating birds

---

[1]*Most of the rodent species eaten by specialist predators live in open areas (including openings in forest) and feed mainly on plant leaves, but other rodent species eat mainly seeds. To some extent, the populations of the seed-eating rodents may fluctuate in relation to tree-seed crops, giving them a connection with the seed-eating finches described in Chapter 18. Different species of open-country microtine rodents are involved, depending on their occurrence in particular regions,* Microtus agrestis *and* M. arvalis *being widely eaten in Europe, and* M. pennsylvanicus *in North America.*

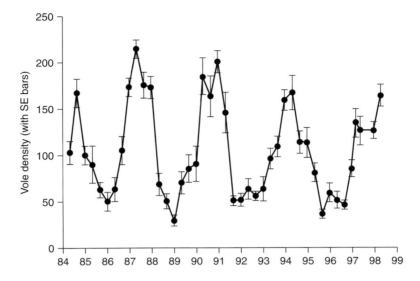

Figure 19.1   Index of Field Vole *Microtus agrestis* densities in spring, summer and autumn in Kielder Forest, northern England, over 15 years showing the regular peaks in numbers. Note that in most years vole densities increased from spring to summer (the owl breeding season), but in some years they decreased from spring to summer. From Petty (1999).

include various owls, diurnal raptors and skuas (jaegers). In general, these birds would have to shift their breeding areas by at least several hundred kilometres every few years if individuals were to breed under adequate food conditions every year, and avoid the lows.

Owls and other predators show two main types of response to fluctuations in their rodent food supply. One type is shown in resident species, which tend to stay in the same territories year-round and from year to year. While preferring rodents, they also eat other things. They can therefore remain in the same area through low rodent years, switching to alternative prey, but their survival may be lower, and their productivity much lower than in good rodent years (Newton 2002). The Tawny Owl *Strix aluco*, Ural Owl *Strix uralensis* and Barn Owl *Tyto alba* are in this category, responding to prey numbers chiefly in terms of the number of young raised (Southern 1970, Saurola 1989, Petty 1992, Taylor 1994). This type of response, shown by resident rodent feeders, produces a lag between prey and predator numbers, so that high predator breeding densities follow 1–2 years after good food supplies and low densities follow poor supplies. Prey and predator densities go up and down in parallel, but with the predator behind the prey (Newton 2002). The fluctuations in prey are reflected in the movement patterns of the predator, whose dispersal distances tend to be longer in poor rodent years (see later).

The second type of response is shown by 'prey-specialist' nomadic species, which concentrate to breed in different areas in different years, depending on where their food is plentiful at the time (**Figures 19.2** and **19.3**). Typically, individuals might

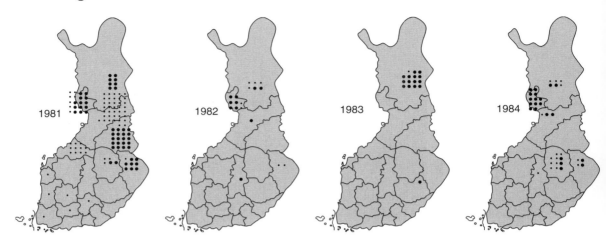

Figure 19.2   Annual variations in the breeding distribution of Great Grey Owls *Strix nebulosa* in Finland. Small dots – territorial pairs; large dots – nests. From Solonen (1986).

have 1–2 years in the same area in each 3–5 year vole cycle, before moving on when prey decline. Response to change in food supply is almost immediate (with no obvious lag), and the increases in numbers from one year to the next are often far greater than could be explained by high survival and reproduction from the previous year, so must also involve immigration. As in irruptive seed-eaters, such observations lead to the inference that year-to-year changes in local breeding densities are due primarily to movements – immigration or emigration – depending on food conditions at the time.

The Short-eared Owl *Asio flammeus*, Long-eared Owl *A. otus*, Northern Hawk Owl *Surnia ulula* and, to some extent, Snowy Owl *Nyctea scandiaca* and Great Grey Owl *Strix nebulosa* are in this category, as are the Common Kestrel *Falco tinnunculus*, Hen (Northern) Harrier *Circus cyaneus* and Rough-legged Buzzard *Buteo lagopus* in some regions. Their local breeding densities can vary from nil in low rodent years to several tens of pairs per 100 km² in intermediate (increasing) or high rodent years. In a 47 km² area of western Finland, for example, over an 11-year period, numbers of Short-eared Owls varied between 0 and 49 pairs, numbers of Long-eared Owls between 0 and 19 pairs, and Common Kestrels between 2 and 46 pairs, all in accordance with spring densities of *Microtus* voles (**Figure 19.4**, **Table 19.1**; Korpimäki & Norrdahl 1991). All these raptors were summer visitors to the area concerned, and settled according to vole densities at the time.

Other variations in local breeding densities recorded over periods of years for owls and raptors that exploit cyclic prey species are referred to in **Table 19.1**. Their fluctuations contrast with findings from other owls and raptors that depend on a wider range of prey species and show much more stable breeding densities from year to year (Newton 1979, 2003). The main points to emerge are that year-to-year fluctuations in breeding densities are typically very much greater in irruptive than in regular migrants, that the year-to-year fluctuations parallel food supplies at the time, and that (by inference) the primary proximate cause of the

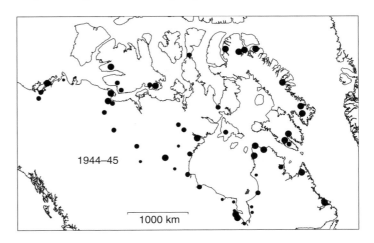

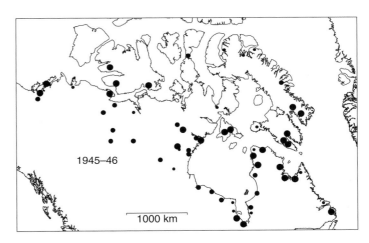

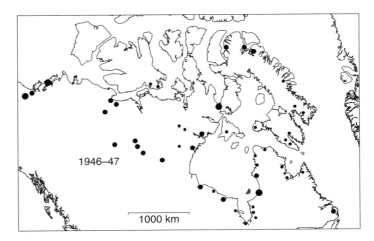

Figure 19.3 Year-to-year changes in the densities of Snowy Owls *Nyctea scandiaca* in different parts of northern Canada, as judged from questionnaire surveys of trappers and other local residents. Each dot marks the centre of an area covered by an individual trapper. Large dots – marked increases from previous year; small dots – marked decreases from previous year. Changes in owl numbers generally matched those of lemming numbers, and changes were synchronised over much of northern Canada, with a few regional exceptions. From Chitty (1950).

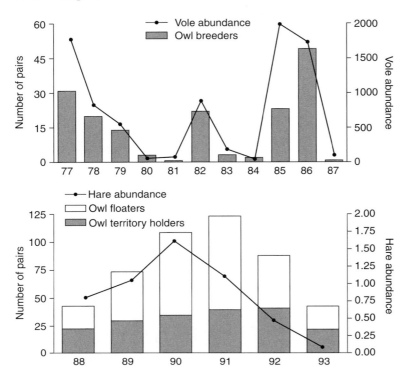

Figure 19.4   Annual fluctuations in the breeding densities of the (upper) Short-eared Owl *Asio flammeus* in relation to Field Vole *Microtus arvalis* densities in an area of western Finland, and (lower) the Great Horned Owl *Bubo virginianus* in relation to Snowshoe Hare *Lepus americanus* densities in an area of the Yukon, Canada. From Korpimäki & Norrdahl (1991), Rohner (1995).

fluctuations must be movement. Some species, such as Gyrfalcon *Falco rusticolus* and Ferruginous Hawk *Buteo regalis*, fluctuate in breeding densities from year to year in some regions, but it is not known whether individuals change their breeding places over long distances from year to year.

Some species that exploit cyclically fluctuating vole populations move around mainly within the breeding range, as exemplified by the Tengmalm's Owl *Aegolius funereus* and Northern Hawk Owl *Surnia ulula* in forest. In other species, parts of the population migrate to lower latitudes in winter, thereby avoiding the worst effects of snow cover, and return to the breeding range each spring, settling in areas where voles are numerous at the time. This pattern is exemplified by the Short-eared Owl *Asio flammeus*, Long-eared Owl *A. otus*, Common Kestrel *Falco tinnunculus* and Northern Harrier *Circus cyaneus* over much of their breeding ranges (Hamerström 1969, Korpimäki & Norrdahl 1991). All these species hunt on the wing, a relatively high-energy method compared to the sit-and-wait methods of some other rodent-eaters (Sonerud 1984). Their expensive hunting methods may be why they tend to leave areas with prolonged winter snow cover. Such species may therefore re-distribute themselves in relation to vole densities at least twice each year; first in autumn when they search for and settle in good vole areas for the winter, and again in spring when they either stay where they are and breed if vole densities permit, or move on in search of another good area. In some regions, snowfall can modify this pattern because under deep snow voles can become unavailable to most northern owl species (the Great Grey Owl *Strix nebulosa* is exceptional in being able to penetrate up to 40 cm of snow). In their

Table 19.1 Established year-to-year correlations between bird abundance, emigration and food supply in rodent-eating birds

	Summer	Winter	Autumn emigration	References
Diurnal				
Hen (Northern) Harrier *Circus cyaneus*	•			Hamerström (1969), Hagen (1969)
Rough-legged (Hawk) Buzzard *Buteo lagopus*	•		•	Schüz (1945), Hagen (1969), Court et al. (1988), Potapov (1997)
Eurasian (Common) Kestrel *Falco tinnunculus*	•	•		Cavé (1968), Rockenbauch (1968), Hagen (1969), Korpimäki & Norrdahl (1991), Village (1990)
Black-shouldered Kite *Elanus axillaris*	•			Malherbe (1963), Mendelsohn (1983)
Nocturnal				
Short-eared Owl *Asio flammeus*	•		•	Village (1987), Korpimäki & Norrdahl (1991)
Long-eared Owl *Asio otus*	•		•	Village (1981), Korpimäki & Norrdahl (1991)
Great Grey Owl *Strix nebulosa*	•		•	Hildén & Helo (1981), Nero et al. (1984), Duncan (1992, 1997), Bull & Duncan (1993)
Snowy Owl *Nyctea scandiaca*	•		•	Shelford (1945), Chitty (1950), Parmelee (1992), Newton 2002
Northern Hawk Owl *Surnia ulula*	•		•	Korpimäki (1994)
Tengmalm's (Boreal) Owl *Aegolius funereus*	•		–	Korpimäki & Norrdahl (1989)

movement patterns, the specialist rodent-eaters thus parallel the boreal seed-eaters discussed in the Chapter 18.

On the northern tundras, some species of skuas also eat microtine rodents. These birds spend most of their lives at sea, but return to the tundra to breed. They seem to return to the same breeding areas each year, but nest only if rodents are plentiful, remaining as non-breeders in other years. This pattern has been found in Pomarine Skuas *Stercorarius pomarinus* in Alaska (Pitelka *et al.* 1955, Maher 1970) and in Long-tailed Skuas *S. longicaudus* in northern Europe (Andersson 1976). In the latter, a similar number of pairs returned to the study area each spring, but the percentage that bred varied from 0% to 100%, according to rodent abundance, giving nest densities over nine years of 0–60 per 100 km². In years of low rodent numbers the birds took a range of different foods, including birds' eggs, but in years of high rodent numbers they ate almost nothing else. At a number of localities across the Eurasian tundra, the numbers of nesting Pomarine and Long-tailed Skuas found in one year were correlated with local lemming numbers at the time, as were the local numbers of Snowy Owls *Nyctea scandiaca* and Rough-legged Buzzards *Buteo lagopus*, although the latter two also avoided one another (Wiklund *et al.* 1998).

Evidence of movements from ringing and radio-tracking

The implication from local fluctuations in the breeding densities of many species – that individual adults may nest in widely separated localities in different years – is supported by ring recoveries, although the proportions of birds that move can vary from year to year, depending on food conditions. As in the seed-eaters, evidence for movements comes partly from the high turnover among birds caught each year in the same localities. In all the species listed in **Table 19.1** that were studied in this respect, return rates were extremely low, compared to what would be expected from their annual survival rates. Among Common Kestrels *Falco tinnunculus*, of 146 individual breeders trapped and ringed in a 63 km² area in Finland over an 11-year period, only 13% of males and 3% of females were found back in the same area in a later year (Korpimäki & Norrdahl 1991). The implication is that a large proportion of breeders changed their nesting localities from year to year.

Other evidence for widescale movements stems from adults that were found in widely separated areas in different breeding seasons. Most information of this type relates to the Tengmalm's (Boreal) Owl *Aegolius funereus*, which nests readily in boxes and has been studied at many localities in Europe (**Figure 19.5**). In this species, the males in some regions are mainly resident and the females more dispersive. Both sexes tend to stay in the same localities if vole densities remain high, moving no more than about 5 km between the nest boxes used in successive years. But if vole densities crash, females move much longer distances, with many having shifted 100–600 km between breeding sites in different years (**Figure 19.5**). In contrast, fewer long movements were recorded from males, with only two at more than 100 km. The greater residency of males was attributed to their need to guard cavity nest-sites which are scarce in their conifer nesting habitat, while their smaller size makes them better able than females to catch small birds, and hence to survive (without breeding) through low vole conditions (Lundberg

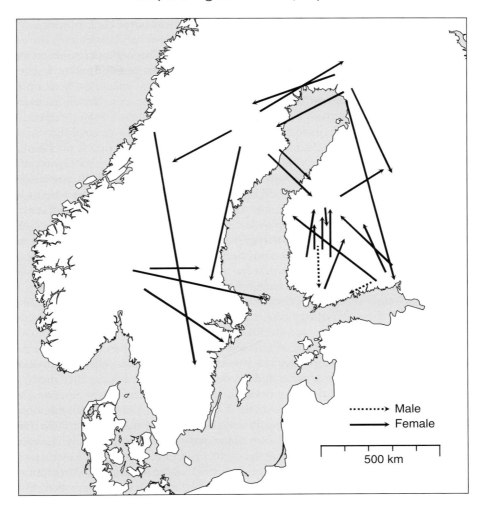

Figure 19.5 Ringing and recovery sites of adult Tengmalm's Owls *Aegolius funereus* that were identified in different breeding seasons. Continuous lines – females; dashed lines – males. Only movements greater than 100 km are shown. From Newton (2003), compiled from information in Löfgren *et al.* (1986), Korpimäki *et al.* (1987) and Sonerud *et al.* (1988).

1979, Korpimäki *et al.* 1987). In the following year, when voles become plentiful again, male owls still have their previous nest-sites, and are again able to attract females which move in to exploit the abundant prey. In the northern boreal zone, however, where small birds are scarce in winter, both sexes tend to vacate areas with low vole numbers.

Far fewer records are available for other nomadic owl species, because the chances of recording marked individuals at places far apart are low. However, in a study of Short-eared Owls *Asio flammeus* in south Scotland, 21 breeders were tagged in 1976. Vole numbers then crashed, and only one of the tagged birds remained to breed in the area in 1977. Two others were reported in spring 1977 in nesting habitat 420 km and 500 km to the northwest, and the latter, at least, was proved to breed there (Village 1987). Of seven breeders tagged in 1977, when vole numbers began to increase, three bred in the area in 1978. Hence, as in Tengmalm's Owl *Aegolius funereus*, individuals seemed more likely to remain to breed in successive years when voles were increasing than when they were declining.

Recoveries of Northern Hawk Owls *Surnia ulula* and Great Grey Owls *Strix nebulosa* ringed as breeding adults include examples of both males and females residing in an area from one nesting season to the next when microtine abundance remained high, and of both sexes leaving when microtine populations declined (Sonerud 1997, Duncan 1992). Adult radio-marked Great Grey Owls in Manitoba and northern Minnesota dispersed 41–684 km (mean 329 km, SD 185 km, N = 27) between breeding sites in response to prey population crashes. Eleven marked birds that did not disperse died (Duncan 1992, 1997). At least one male returned southward to its original home range in the following summer, but did not re-nest then, and two others were found breeding on their original range three years later. At least one female returned, and was found breeding on her original range two years later, and another, after breeding in the north, had set off south in the direction of her original range when her radio stopped working. These observations raise the possibility that these owls are not truly nomadic (in the sense that they may occupy a different area each year), but may shuttle back and forth between two or more regular breeding areas up to several hundred kilometres apart, but more work is needed to check this possibility.

More revealing information is available for four adult female Snowy Owls *Nyctea scandiaca* which were radio-tagged while nesting near Point Barrow in Alaska, and tracked by satellite over the next 1–2 years (Fuller *et al.* 2003). These birds mostly stayed in the arctic but dispersed widely in different directions from Point Barrow, reaching west as far as 147°E and east as far as 116°W, a geographical spread encompassing nearly one-third of the species' Holarctic breeding range. Two birds that bred at Point Barrow in 1999 were present during the next breeding season in northern Siberia (147°E and 157°E respectively), up to 1928 km west of Point Barrow, and then in the following breeding season, they were on Victoria Island (116°W) and Banks Island (122°W) respectively, in northern Canada (having passed eastward through Point Barrow) (**Table 19.2**, **Figure 19.6**). The two birds that bred at Point Barrow in 2000 were present on Victoria and Banks Islands in the breeding season of 2001. The successive summering areas of these four birds were thus separated by distances of 628–1928 km (**Table 19.2**). From the dates they were present, some could have bred successfully, while others were unlikely to have done so, having arrived too late or left too early. None returned to the same breeding or wintering site used in a previous year, but three passed through Point Barrow in 2001. In winter these Snowy Owls made long-distance moves at various dates; one bird remained continually on the move, venturing as far south as 59°, but spending no more than a fortnight at any one place, while another remained for 2.5 months in one place. Only two ventured south of the breeding range, one briefly but the other staying for more than two months at 60°N in southern Alaska. Some spent time in winter on sea-ice far from land, presumably hunting seabirds.

Table 19.2   Locations of four adult female Snowy Owls *Nyctea scandiaca* in successive breeding seasons. All were radio-tagged at Point Barrow, Alaska in 1999 (numbers 54 and 57) or 2000 (numbers 80 and 81), and tracked by the Argos satellite system

Owl number	Breeding seasons		
	1999	2000	2001
54	Alaska[a] (71°N, 156°W)	Siberia[c] (70°N, 157°E)	Victoria Island[c] (73°N, 108°W)
57	Alaska[a] (71°N, 156°W)	Siberia[b] (71°N, 147°E)	Banks Island[b] (73°N, 121°W)
80		Alaska[a] (71°N, 156°W)	Victoria Island[c] (73°N, 115°W)
81		Alaska[a] (71°N, 156°W)	Banks island[b] (73°N, 122°W)

[a]Known to have bred; [b]probably bred; [c]unlikely to have bred successfully.

Great circle distances between places where successful breeding was known or probable were: 628 km between Point Barrow and Banks Island (bird 81); 1548 km between Siberia and Banks island (bird 57); and 1928 km between Point Barrow and Siberia (bird 57).

From Fuller *et al.* (2003).

The only other ringing-based records of breeding dispersal known to me involve three adult Long-eared Owls *Asio otus* in North America, which were found at localities more than 450 km apart in different breeding seasons (Marks *et al.* 1994), and two adult Great Grey Owls *Strix nebulosa* in northern Europe which were

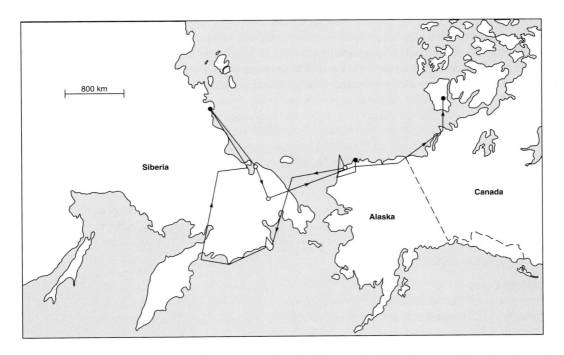

Figure 19.6   Movements of a radio-tagged female Snowy Owl *Nyctia scandiaca* tracked by satellite between the summers of 1999 and 2001. Filled circles show known likely breeding sites in three consecutive years (see **Table 19.2**), and open circles show other sites where the bird spent more than four weeks at a time. Constructed from information in Fuller *et al.* (2003).

found at localities 300 km and 430 km apart in different breeding seasons (Hildén & Solonen 1987). Regarding natal dispersal, movements exceeding 1000 km have been documented for Short-eared Owls *Asio flammeus* (up to about 4000 km), Long-eared Owls (up to about 2300 km) and Hawk Owls *Surnia ulula* (up to 2700 km), as well as for Rough-legged Buzzards *Buteo lagopus* (up to 2700 km) (Saurola 1983, 1997, 2002, Cramp & Simmons 1980, 1985). Some natal dispersal distances for young owls of various species ringed in Finland are summarised in **Table 19.3**. They indicate much longer median and maximum distances from irruptive than from 'sedentary' species. Among diurnal raptors, Galushin (1974) summarised data from Russian ring recoveries showing that irruptive (mostly vole-eating) species had much longer dispersal distances than non-irruptive ones. The mean natal dispersal distance of the Rough-legged Buzzard was given as 1955 km (but with no mention of sample size).

These various irruptive owls and raptors thus contrast greatly with more sedentary populations, which exploit more stable food supplies. In such species, adults usually remain in their territories year after year, with only small proportions moving to other territories, usually nearby (for Tawny Owl *Strix aluco*, Ural Owl *S. uralensis* and Barn Owl *Tyto alba* see Saurola 1989, 2002, Petty 1992, Taylor 1994). One consequence of such site-fidelity is strong mate-fidelity, as partners remain together year after year, so long as neither dies or changes territory. Natal dispersal distances are mostly less that 30 km, with few exceeding 100 km: for Tawny Owl see Petty 1992, Saurola 2002; for Ural Owl see Saurola 1987, 2002; for Barn Owl see Taylor 1994; also **Table 19.3**. In both nomadic and resident species, movements are generally longer in poor food years than in good ones (Taylor 1994, Saurola 2002). Typically, it is the young produced in peak prey years that move furthest, because it is they that experience the crash that follows the peak; they are also more numerous than young produced in other years,

Table 19.3 Average natal dispersal distances (km) of owls ringed as nestlings in Finland and found dead in a subsequent breeding season

Species	Number recovered	Median distances (km)	Maximum distance (km)
Sedentary species			
Eurasian Eagle Owl *Bubo bubo*	563 (20)	52 (47)	416 (114)
Tawny Owl *Strix aluco*	1126 (1288)	22 (17)	386 (270)
Ural Owl *Strix uralensis*	538 (1036)	28 (22)	339 (205)
Pygmy Owl *Glaucidium passerinum*	10 (131)	8 (15)	183 (288)
Nomadic species			
Northern Hawk Owl *Surnia ulula*[a]	2	490	869
Great Grey Owl *Strix nebulosa*	16 (6)	227 (124)	912 (315)
Long-eared Owl *Asio otus*	48	287	1759
Short-eared Owl *Asio flammeus*	16	822	3453
Tengmalm's Owl *Aegolius funereus*	96 (541)	71 (78)	874 (588)

Figures in parentheses refer to additional birds re-captured alive at nest-sites.
From Saurola (2002).
[a]The two birds moved 111 km and 869 km.

which could further add to their difficulties. For both reasons, birds presumably have to search further, on average, in poor food years before they find suitable areas.

Geographical variation in movement patterns within species

The two responses to change in food supply (delayed and simultaneous) found among rodent-eaters are not completely distinct, and different species of owls and raptors may be better described as forming a gradient in response, from the most sedentary at one end to the most mobile at the other. Moreover, the same species may show regional variation in movement behaviour depending on food supply, and the extent to which alternative prey are available when favoured prey are scarce. Examples of regional variation among diurnal raptors include Common Kestrel *Falco tinnunculus* and Hen (Northern) Harrier *Circus cyaneus*, in which the proportion of rodents in the diet differs from region to region. The more varied the diet, the less the chance of all prey types being scarce at the same time, the more stable are their local breeding densities and the greater the site-fidelity shown by individuals. Examples among owls include the Tengmalm's Owl *Aegolius funereus*, which has been described as a resident generalist predator of small mammals and birds in central Europe, as partially nomadic (with males mainly resident and females moving around) in south and west Finland, and as a highly nomadic microtine specialist in northern Fennoscandia, in areas with pronounced vole cycles and fewer alternative prey (Korpimäki 1986).

Interestingly, the return rates of adults to former nesting areas can vary regionally within species, according to the degree of year-to-year stability in food supply. In comparison with the return rates mentioned above for Kestrels in Finland of 13% males and 3% females, Village (1990) recorded rates of 29% and 18% for males and females in Scotland, and of 43% and 36% in southern England, commenting that the more sedentary nature of the English population was 'due to the greater stability of the food supply both within and between years'. In a study in the Netherlands, as many as 70% of adult Kestrels remained from year to year when vole numbers were on the increase, and as few as 10% when vole numbers crashed (Cavé 1968). Similarly, the Long-eared Owl *Asio otus* shows greater year-to-year site-fidelity in the Netherlands than in Finland (Wijnandts 1984, Korpimäki 1992), as does the Great Grey Owl *Strix nebulosa* in some parts of North America compared with others (Collister 1997, Duncan 1997), while the Barn Owl *Tyto alba* is highly sedentary in Britain (Taylor 1994) but more dispersive in parts of continental Europe (where movements up to 2000 km have been recorded) and in parts of North America (Bairlein 1985b, Marti 1999).

In addition to the owl species mentioned earlier, mean distances between birthplace and breeding place were greater for the Common Kestrel *Falco tinnunculus* (277 ± 57 km) and Common Buzzard *Buteo buteo* (295 ± 105 km) in northern and eastern Europe than in western and central Europe (146 ± 39 and 60 ± 11 km) (Galushin 1974). Prey remains indicated that microtine rodents form a much greater proportion of the diets of these raptors in northern and eastern Europe than in western and central Europe. In their regional variation in movement behaviour, owls and raptors show parallels with some of the seed-eaters discussed in the previous chapter.

Irruptive migrations

Like the irruptive finches, some rodent-eating species respond to periodic crashes in their main food supply by winter emigration, appearing south of their breeding range in much larger numbers than usual. Irruptions of Snowy Owls *Nyctea scandiaca* from the tundra to the boreal and temperate regions of eastern North America have been documented at least since 1880. Throughout the next 120 years, irruptions occurred every 3–5 years, at a mean interval of 3.9 (SE ± 0.13) years (Newton 2002, **Table 19.4**). Moreover, in periods when information on lemmings was available from potential breeding areas to the north, mass movements of owls coincided with crashes in lemming numbers (Shelford 1945, Chitty 1950). In western North America, irruptions were not well synchronised with those in the east, presumably reflecting asynchrony in lemming cycles between breeding regions (but this has not been proved). The irruptions were also less regular and less pronounced in the west than in the east, with some birds appearing on the northern prairies every winter, and some of the same marked individuals appearing on the same territories in different (not necessarily consecutive) winters (Kerlinger *et al.* 1985).

In eastern North America, two other vole-feeders, the Rough-legged Buzzard *Buteo lagopus* and Northern Shrike *Lanius excubitor*, have irrupted at similar 3–5 year intervals, mostly (but not always) in the same years as Snowy Owls (Davis 1937, 1949, Speirs 1939, Shelford 1945, Lack 1954). Perfect synchrony between the

Table 19.4   Irruptions of Snowy Owls *Nyctea scandiaca* into eastern United States, 1881–2000

1881–1920	1921–1960	1961–2000	1881–1920	1921–1960	1961–2000
–	1921	–	1901	1941	1981
1882	–	–	–	–	–
–	–	–	–	–	–
–	–	1964	–	–	–
–	–	–	1905	1945	–
1886	1926	–	–	–	1986
–	–	1967	–	–	–
–	–	–	–	–	–
1889	–	–	1909	1949	–
–	1930	–	–	–	–
–	–	1971	–	–	1991
1892	–	–	1912	–	–
–	–	–	–	1953	–
–	1934	1974	–	–	–
–	–	–	–	–	–
1896	–	–	–	–	1996
–	1937	–	1917	1957	–
–	–	1978	–	–	–
–	–	–	–	–	–
–	–	–	–	1960	2000

From various sources, summarised in Newton (2002).

three species would perhaps not be expected, because their breeding ranges only partly overlap. The buzzard and the shrike breed mainly in the transition zone between forest and tundra, while the owl breeds on the open tundra, but part of the owl population winters in the transition zone. Nevertheless, with most invasions of each species coinciding with those of the other species, the level of synchrony is striking. In the shrike, the cyclic pattern in invasions was most marked before 1950, after which the fluctuations became more irregular and eventually hardly apparent (Davis & Morrison 1987).

In North America, Northern Hawk Owl *Surnia ulula* irruptions tend to occur at 3–5 year intervals (modified by snow cover, and in different years in different regions) (Duncan & Duncan 1998). Great Grey Owl *Strix nebulosa* irruptions are also occasionally recorded south and east of the usual breeding range, with big flights noted in eastern regions in 1978, 1983, 1991, 1995 and 2004 (Nero *et al.* 1984, Davis & Morrison 1987, Bull & Duncan 1993, Jones 2005), while Saw-whet Owl *Aegolius acadicus* migrations are also much more marked in some years than in others (National Audubon Field Notes). In Europe, the Great Grey, Tengmalm's Long-eared and Short-eared Owls seem to migrate on regular 3–4 year patterns (Harvey & Riddiford 1996, Schmidt & Vauk 1981, Hildén & Helo 1981, Lõhmus 1999), as expected from the 3–4 year crashes in microtone rodent populations. As in seed-eaters, movement often begins earlier than usual in irruption years, and mortality among the participants is often heavy, as illustrated by the 500 Great Grey Owls found dead in Ontario in 2004–2005 (Jones 2005).

Occurrence in different areas in different winters

As far as I am aware, few ring recoveries from different winters are yet available for any species of irruptive owls, which have been trapped only in relatively small numbers. However, the satellite-tagged Snowy Owls *Nyctea scandiaca* mentioned above were present in widely separated localities in different winters, and often moved long distances within a winter (Fuller *et al.* 2003). Another Snowy Owl was ringed near Edmonton in January 1955 and recovered 330 km to the southeast in Saskatchewan in January 1957 (Oeming 1957). There is also an intriguing record of a Long-eared Owl *Asio otus* ringed in California in April and recovered in Ontario in October of the same year (Marks *et al.* 1994). Presumably it returned to the breeding range in the interim.

Whatever the species, irruptive migrants do not necessarily remain in the same localities throughout a winter. Trapping has revealed considerable turnover in the individuals present at particular sites or the occurrence of the same individuals at widely separated sites in the same winter (for Great Grey Owl *Strix nebulosa*, see Nero *et al.* 1984; for Snowy Owl *Nyctea scandiaca*, see Kerlinger 1985, Smith 1997). The implication is that, in the non-breeding period, individuals move around, perhaps in continual search for good hunting areas. Local abundances of microtines can attract high densities of irruptive owls, and in these conditions some species form communal winter roosts, as recorded often in Long-eared *Asio otus* and Short-eared Owls *A. flammeus* (Cramp 1985) and also in Great Grey Owls (Nero *et al.* 1984). Recurrence at the same site in different winters has been recorded, but involved only a small proportion of the total caught (for Snowy Owl, see Smith 1997). In some years, the migrants may travel long distances without encountering any areas with abundant food, and in such cases their mortality rates are likely to be much higher than usual.

Nesting outside the regular range

Various species of owls have also been recorded nesting well outside their regular range in years of abundant food, often following invasions (Duncan & Duncan 1998). For example, several hundred pairs of Snowy Owls *Nyctea scandiaca* bred on the tundra of Swedish Lapland in 1978, where they had been rare to non-existent in many previous years (Andersson 1980). Snowy Owls bred in Finnish Lapland in 1974, 1987 and 1988, but before these dates none had been seen for several decades (Saurola 1997). Similarly, Northern Hawk Owls *Surnia ulula* bred in an area in Norway in the peak years of only four out of seven observed vole cycles (Sonerud 1997). This lack of response may arise because in many years the entire owl population can be accommodated in certain parts of the range with abundant prey, without needing to search out other parts. In Fennoscandia, the numbers of Snowy and Hawk Owls at any time seems to be influenced not only by the occurrence of a rodent peak, but by the arrival of large numbers of immigrants from further east (Sonerud 1997). In more central parts of the range, the owls may exploit a much greater proportion of the rodent peaks. When voles were plentiful, Hawk Owls have also bred in invasion areas well south of their usual breeding range in North America (Duncan & Duncan 1998). Among diurnal raptors, numbers of Rough-legged Buzzards *Buteo lagopus*, Common Kestrels *Falco tinnunculus* and Pallid Harriers *Circus macrourus* were also found breeding outside the regular range in localities where voles were plentiful (Galushin 1974).

Boom-and-bust

Specialist rodent-eaters are known for raising large families (up to eight or more young per brood in some species) in years when prey are plentiful, but few or none in years when prey are scarce, or they may crash from abundance to scarcity during a summer (Cramp 1985, Newton 2002). Juveniles formed 85% of 80 Northern Hawk Owls *Surnia ulula* obtained on irruption in northern Europe in 1950, 100% of 52 obtained in 1976, and 88% of 150 museum skins collected over several years (Cramp 1985). Not all owl invasions follow good breeding years, however, and after a known poor year, only four out of 126 Great Grey Owls *Strix nebulosa* trapped in Manitoba in 1995 were juveniles (Nero & Copland 1997), and a big invasion into Ontario in 2004 included virtually no juveniles (Jones 2005). In Snowy Owls *Nyctea scandiaca*, juveniles predominated in invasion years, but in other years when few owls appeared, the majority were adults and many were underweight (Smith 1997). Such observations reveal that, while irruptive owls move away from areas poor in food, they do not altogether avoid the adverse effects of fluctuating supply. The birds are likely also to suffer poorer survival in low food years, but I know of no relevant studies apart from accounts of unusual numbers found dead.

Hares and hare-eaters

Other striking examples of the link between widespread winter emigration and food supply are provided by the Northern Goshawk *Accipiter gentilis* and the Great Horned Owl *Bubo virginianus*, which feed on Snowshoe Hares *Lepus americanus*. The fluctuations in Snowshoe Hare numbers across North America have been well recorded for more than 200 years, initially from the number of hare pelts provided each year to the fur market, and latterly from detailed studies (Elton 1942, Keith 1963, Krebs *et al.* 2001). However, unlike the situation in rodents, the hare cycle appears synchronised over much of boreal North America, with populations across the continent peaking in the same years, at roughly 10-year intervals (Keith & Rusch 1988, Krebs *et al.* 2001). In particular localities, peak densities can exceed troughs by more than 100-fold (Adamcik *et al.* 1978). In addition to the Northern Goshawk and the Great Horned Owl, other predators of Snowshoe Hares include Northern Hawk Owls which eat young hares, as well as rodents.

The Great Horned Owl seems to show much greater numerical fluctuation in the north of its range, where it depends primarily on Snowshoe Hares, than further south where it has a wider range of prey, but I know of no detailed studies in the southern parts. Because Snowshoe Hares seem to fluctuate in synchrony over their whole range, there would be little advantage in Great Horned Owls in northern regions breeding in widely separated localities in different years. Any that left one area because of a shortage of hares would be unlikely to find many more elsewhere (thus contrasting with the situation in rodent-eaters). If they moved southwards, and attempted to breed outside the range of the Snowshoe Hare, they would come up against other Great Horned Owls, already breeding there. Nevertheless, the hare-eaters do perform massive southward irruptions, abandoning their northern breeding areas at least for the winter, and returning there in summer when alternative prey are available.

Because Goshawks fly by day and are more readily seen, their invasions have been better documented than those of owls. They occur for 1–3 years at a time, but at about 10-year intervals (**Table 19.5**), coinciding with known lows in the Snowshoe Hare cycle (Keith 1963, Mueller *et al.* 1977, Keith & Rusch 1988). It is usual for juveniles to predominate among migrants in most years when relatively few birds leave the breeding range, but for adults to predominate in invasion years, which follow poor breeding seasons. Thus, among more than 12 000 Goshawks trapped over a number of years at Cedar Grove in Wisconsin, juveniles formed more than 80% of the total in most years, but less than 50% in the invasion years of 1962–1963 and 1972–1973 (Mueller *et al.* 1977).

Migrating Great Horned Owls fly by night, and because they move into more southern regions already populated by resident Great Horned Owls, their irruptions have been less well documented. However, all those that I could find recorded coincided with Goshawk invasions, and hence with low hare numbers, again providing indications that food shortage stimulated large-scale emigration in this mainly resident species (**Table 19.5**). In a study at Kluan, Yukon Territory, annual emigration rates of radio-marked Great Horned Owls increased over a period of years from 0 to 33% for territory holders ($N = 2$–54 in different years) and from 0 to 40% for non-territorial floaters ($N = 2-18$ in different years), as hares declined (Rohner 1996). To judge from ring recoveries, migration of Great Horned Owls from Saskatchewan is mainly to the southeast, but over greater distances in years with low hare numbers, with some individuals travelling more than 1000 km from their breeding sites (Houston & Francis 1995, Houston 1999).

Table 19.5 Irruptions of Northern Goshawks *Accipiter gentilis* and Great Horned Owls *Bubo virginianus* in North America, 1881–2000

1881–1920		1921–1960		1961–2000	
Great Horned Owl	Goshawk	Great Horned Owl	Goshawk	Great Horned Owl	Goshawk
–	–	–	–	–	–
–	–	–	–	1962	1962
–	–	–	–	1963	1963
–	–	–	–	–	–
–	–	1925	–	–	–
–	1886	1926	1926	–	–
1887	1887	1927	1927	–	–
–	–	–	1928	–	–
–	–	–	–	–	–
–	–	–	–	–	–
–	–	–	–	–	–
–	–	–	–	1972	1972
–	–	–	–	1973	1973
–	–	–	–	–	1974
–	–	–	1935	–	–
–	1896	1936	1936	–	–
1897	1897	1937	–	–	–
–	–	1938	–	–	–
–	–	–	–	–	–
–	–	–	–	–	–
–	–	–	–	1981	1981
–	–	–	–	1982	1982
–	–	–	–	–	1983
–	–	–	1944	–	–
–	–	–	–	–	–
–	1906	–	–	–	–
1907	1907	–	–	–	–
–	–	–	–	–	–
–	–	–	–	–	–
–	–	–	–	–	–
–	–	–	–	–	–
–	–	–	–	1992	1992
–	–	–	–	–	1993
–	–	–	1954	–	1994
–	–	–	–	–	–
1916	1916	–	–	–	–
1917	1917	–	–	–	–
1918	–	–	–	–	–
–	–	–	–	–	–
–	–	–	–	–	–
–	–	–	–	–	–

From various sources, summarised in Newton (2002).

WATERFOWL AND OTHERS

Our understanding of the distribution and movement patterns of the more obvious irruptive migrants has been pieced together over many years from local observations and ring recoveries. For North American waterfowl, however, detailed continent-wide information on breeding and wintering distributions has been obtained over many years by aerial survey. This has enabled their wide-scale distribution patterns to be examined in relation to prevailing wetland conditions, in both breeding and wintering areas. From the results of aerial surveys conducted annually in May over 27 years, Johnson & Grier (1988) examined duck numbers and wetland conditions across a large part of western North America. Much of this region is arid and subject to large annual fluctuations in precipitation. Johnson & Grier (1988) envisaged three possible patterns of spring settlement in this vast region, where wetlands varied greatly in numbers and extent from year to year, the shallower ones disappearing altogether in dry years.

1. Homing, with adults returning to breeding areas used in the previous year, and yearlings returning to near their natal areas. This pattern was expected in predictable environments where habitat conditions remained fairly stable from year to year. On this system, the large-scale distribution of particular species would be expected to be fairly consistent from year to year.
2. Opportunistic settling, in which birds occupy the first suitable sites encountered on their migration routes, providing that such sites are not already taken by other birds. This pattern was expected where habitat conditions in particular localities were unpredictable from year to year. Such opportunistic settling could minimise migration costs because it ensures that individuals migrate no further than necessary, but it could result in large-scale changes in the distribution of populations from year to year, depending on the distribution of suitable wetlands. It could also result in some long-distance shifts of individuals between their natal and breeding sites, and between their breeding sites in different years. Local abundance levels would be influenced not just by local conditions, but also by conditions elsewhere in the range.
3. Flexible settling, in which birds home to the area used in the previous year, but move on if conditions there are not suitable. This pattern could be viewed as a compromise between the first two (homing and opportunistic settling), and could affect yearlings more than adults. If birds simply maintained the same migration direction when they moved on, this would result in a general onward displacement of breeding birds when conditions in the usual breeding areas were unsuitable (although birds might not breed as well in these more distant areas as in their usual areas). If yearlings that had not bred previously were more willing to move on than were site-faithful adults, yearlings would be largely responsible for the year-to-year changes in distribution, again expected to be large.

Breeding distributions

Evidence for some level of opportunistic settling emerged for all 10 species of ducks examined, in that their numbers in particular breeding areas fluctuated

from year to year in relation to annual pond numbers. The most obvious opportunists, showing the biggest distributional changes from year to year, inhab-
ited shallow and ephemeral wetlands (that dried out during droughts), notably
Northern Pintail *Anas acuta*, Blue-winged Teal *A. discors*, Mallard *A. platyrhynchos*
and Northern Shoveller *A. clypeata*. At the other extreme, showing the biggest distributional stability from year to year, were the diving ducks, such as Redhead
Aythya americana, Canvasback *A. valisineria* and Lesser Scaup *A. affinis*. Their numbers showed generally poorer correlations with local pond numbers than the dabbling species, because they occupied only the deepest (most permanent) ponds
and ignored the shallow temporary ones.

The distributional patterns that emerged for some species year after year were
consistent with breeding habitat being filled as it was encountered on spring
migration. Four species spent the winter mainly in the southern States and in
spring migrated roughly south–north. They included the Gadwall *Anas strepera*
and Green-winged Teal *A. crecca* (dabble-feeders), and the Canvasback *A. valisineria*
and Lesser Scaup *A. affinis* (dive-feeders). In all these species, the best correlations
between bird numbers and pond numbers were found in the southern parts of
the breeding range, and the poorest correlations in the northern parts, supposedly
reached mainly by birds unable to find accommodation further south. Three
other species wintered in large numbers in the southwestern States and in spring
migrated mainly north and northeast, namely the American Wigeon *A. americana*, Northern Shoveller *A. clypeata* and Northern Pintail *A. acuta*. The numbers
of these species showed high correlations with pond numbers in southwestern
count areas, but the correlations again decreased northwards. It seemed, therefore, that large segments of the populations of these seven species could have
responded to wetland conditions as they encountered them on return from their
wintering areas, moving progressively further in the same direction as habitats
became filled. Three other species (Mallard *A. platyrhynchos*, Blue-winged Teal
A. discors and Redhead *A. americana*) appeared to respond more directly to conditions in the central parts of their breeding ranges, as this was where the year-to-
year correlations between population numbers and pond numbers were highest.
Hence, the habitat which these three species filled first was not that which lay
closest to their wintering areas.

Flexible settling, indicated by over-flight of the usual breeding areas in dry
years, was shown to some extent by all species. It was most marked in dabbling
ducks, notably Northern Pintail, Mallard, Gadwall, Blue-winged Teal and Green-
winged Teal, but was also evident in the diving Redhead and Canvasback. All
these species were displaced to the north and northwest in prairie drought years,
occurring then in much larger numbers than usual in boreal and tundra regions.
The American Wigeon, by contrast, was displaced to the east and northeast during
drought years. These results confirmed earlier findings that, in years of drought
on the prairies, more Pintails than usual overflew the prairie breeding areas to
settle further north in the boreal forest and tundra of Canada, Alaska and eastern
Siberia, a movement confirmed both by counts and ring recoveries (**Figure 19.7**;
Smith 1970, Henny 1973). The magnitude of Pintail migration into eastern Asia,
for example, thus depended on water conditions some 4500 km away. Breeding
success was evidently poorer in the north, as shown from the ratio of young to
adults shot in the subsequent winter. This rate was lowest following springs when

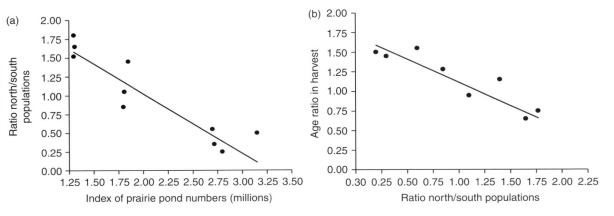

Figure 19.7   Distribution and breeding success of Northern Pintails *Anas acuta* in relation to wetland conditions on the Canadian prairies. (a) Relationship between numbers of ponds and lakes on the prairies in May and the proportion of the total Pintail population found breeding north of the prairies, 1958–1968 ($r = 0.91$, $P < 0.001$) (b) Relationship between the proportion of the total Pintail population in northern areas and overall breeding success ($r = -0.92$, $P < 0.001$). Redrawn from Smith (1970).

the greatest proportions of Pintail occurred in northern breeding areas. Some similar findings were obtained for Mallard (Pospahala *et al.* 1974).

In general, these various patterns were supported by other ring recoveries, with diving ducks showing strong site-fidelity, and the dabbling ducks variable but generally much lower site-fidelity (**Table 17.5**; Johnson & Grier 1988). The Blue-winged Teal was extreme in various local studies, in that very few adults or young were found to return to the same nesting areas in successive years. The northward displacement of several species in drought years from the prairies to the tundra was also confirmed by ringing; it represented a substantial shift in the breeding sites of adults in successive years and between natal and breeding sites for yearlings.

Winter distributions

The winter distributions of North American ducks also vary from year to year, depending on prevailing conditions. Many species move further south in cold winters than in mild ones or further when low rainfall reduces the numbers of wetlands available. This pattern in Mallard *Anas platyrhynchos* was confirmed by winter ring recoveries which were centred further north during winters when December–January temperatures were high than during winters when they were low (Nichols *et al.* 1983). Birds also tended to concentrate in areas that experienced greater autumn–winter rainfall, and hence good wetland conditions. Similarly, in the Wood Duck *Aix sponsa*, both adults and young migrated further in years of low summer rainfall and the young also migrated further south when autumn temperatures were lower than normal (Hepp & Hines 1991). Both rainfall and temperature affect the availability of wetland habitat and food, competition for which may well stimulate further migration, with young ducks more affected than older ones.

Eurasian ducks

Although best studied in North America, ringing has confirmed similar distributional patterns among ducks in Eurasia. In particular: (1) some species of ducks take up to 4–5 months over their post-breeding migration, the bulk of the population moving more rapidly or further in some years than in others; (2) individuals may winter in widely separated areas in different years; and (3) spring settling patterns may also vary from year to year. As in North America, all these features have been linked with patterns in water levels, freezing and thawing, all of which influence the distribution of habitat and associated food supplies. Thus, in Teal *Anas crecca*, spring settling patterns depended on conditions when the birds arrived: ring recoveries were concentrated to the south and west parts of the breeding range in wet, cool years, and to the north and east in dry, warm years (when wetlands were dryer than average) (M. A. Ogilvie, in Wernham *et al.* 2002). Depending on shallow waters, Teal are clearly sensitive to local conditions, whether drought, flood or freeze, which involves them at times either in

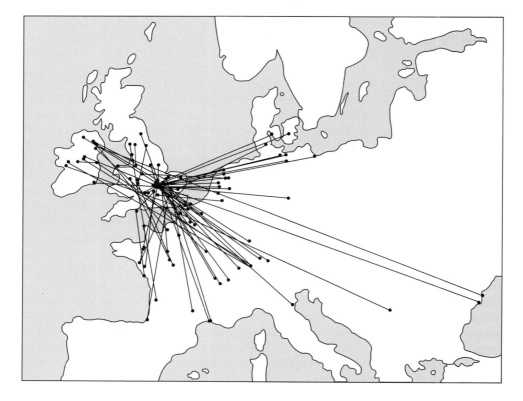

Figure 19.8   Recovery locations of Pochards *Aythya ferina* ringed in Britain in winter and found elsewhere in a later winter, showing that individuals can be found in widely separated areas in different winters. Only movements greater than 20 km are shown. There were 128 movements over 20 km, and another 29 under 20 km. From R. Hearn, in Wernham *et al.* (2002). Reproduced with permission of the British Trust for Ornithology.

continual movement or alternatively in residency for weeks on end, all manifest on a continental scale. Individual Teal may be found in different winters in localities hundreds or thousands of kilometres (up to 3500 km) apart. The same holds for some diving ducks, with individual Common Pochards *Aythya ferina*, Tufted Ducks *A. fuligula* and others found up to several hundred kilometres apart in different winters (**Figure 19.8**; Wernham *et al.* 2002). In Common Pochard breeding dispersal distances of 200–600 km have been found, and in female Tufted Ducks up to 2500 km, but only a small proportion of females made such long moves (Blums *et al.* 2002).

In Europe, during hard winters many waterfowl appear in greater numbers than usual in milder southern and western parts of the continent. Ring recovery distances of several ducks (Eurasian Wigeon *Anas penelope*, Teal *A. crecca*, Northern Pintail *A. acuta*, Common Pochard *A. ferina* and Tufted Duck *A. fuligula*) were greater in cold winters than in mild ones, and greater in cold spells than in equivalent mild periods (Ridgill & Fox 1990). Other species showed no significant differences. Such hard weather (or 'escape') movements were associated with greater mortality than usual, but this may have resulted from greater vulnerability to hunting.

The number of waterfowl wintering near the northern limits of the wintering range depends on ice cover, which normally reaches its greatest extent in February–March. Many more birds stay in mild winters, with most open water, than in more severe ones (see Hario *et al.* 1993 for southwest Finland). The same is true for sea-ducks. In many regions, recent warmer winter temperatures have resulted in waterfowl wintering further north than previously (Chapter 21). Like other facultative migrants, food has a big influence on the winter distributions of ducks. Such species contrast with others which migrate well before food becomes scarce, at consistent times each year, as is typical of obligate migrants.

CONCLUDING REMARKS

All the species discussed above – whether owls, raptors or waterfowl – show variable migration distances, moving towards milder climes in autumn, often in several stages, separated by weeks or months of residence. Some such species also migrate variable distances in spring, partly because many individuals settle in different parts of the breeding range each year, depending on conditions encountered en route. This flexible settling behaviour differs strikingly from that of other migrants, in which individuals occupy fixed breeding and wintering areas from year to year, but it is adaptive for the species concerned, which face unpredictable variations in habitat or feeding conditions from year to year, in either breeding or non-breeding seasons, or both. In contrast to obligate migrants, whose behaviour is apparently under stricter genetic limits (Chapter 12), and hence more consistent from year to year, facultative migrants are more responsive to prevailing conditions. Again, the two types are not wholly distinct, but lie at different points on a continuum of behaviour from fixed to flexible, in adaptation to a spectrum of resource conditions, and the same species may behave differently in different parts of its range. A further extreme in behaviour is evident in the deserts of Australia, where some rodent-eaters, along with waterfowl and other birds, are more appropriately classed as nomadic in response to sporadic rainfall, as described in Chapter 16.

SUMMARY

In northern regions, microtine rodent densities normally fluctuate on an approximate 3–5 year cycle, but the peaks occur in different years in different regions. The raptors that depend on cyclic rodents may breed or winter in widely separated areas in different years, wherever prey are plentiful at the time. Some owl species that exploit sporadic rodent supplies move around mainly within the breeding range (e.g. Tengmalm's Owl *Aegolius funereus*, Northern Hawk Owl *Surnia ulula*), but in other owl and raptor species, part of the population migrates to lower latitudes for the winter, thereby avoiding the worst effects of snow cover. These birds return to the breeding range each spring, settling wherever voles are plentiful at the time (e.g. Short-eared Owl *Asio flammeus*, Long-eared Owl *A. otus*, Common Kestrel *Falco tinnunculus*, Northern Harrier *Circus cyaneus*). In years of widespread food shortage, some rodent predators leave their breeding range in large numbers, appearing in more southern areas as irruptions. In eastern North America, irruptions of Snowy Owls *Nyctea scandiaca*, which have been well documented since 1880, have occurred every 3–5 years, at a mean interval of 3.9 years. Similar 3- to 5-year periodicity has been noted in the movements of some other rodent-eating owl and raptor species in parts of North America and Europe.

In North America, Snowshoe Hares *Lepus americanus* fluctuate greatly in numbers, with roughly 10-year periodicity, but the cycle is more or less synchronised over the whole boreal region. Irruptions of Northern Goshawks *Accipiter gentilis* and Great Horned Owls *Bubo virginianus*, which prey upon Snowshoe Hares, have occurred for 1–3 years at a time, coinciding with known lows in the hare cycle.

The movements of waterfowl are influenced by the fluctuating availability of suitable wetland and associated food supplies. According to the types of habitat they occupy, species show varying degrees of site-fidelity and, in some species, distribution patterns change markedly from year to year. This is especially true of species that occupy shallow, ephemeral waters, which disappear altogether in dry years. Many individuals breed or winter in widely separated areas in different years, and migrate further in some years than in others. These various species provide further striking evidence for the importance of prevailing conditions in influencing movement patterns.

Part Four
Evolution of Movement Patterns

Blackcap *Sylvia atricapilla*, used in experiments on the heritability of migration behaviour

Chapter 20
Evolutionary aspects

Bird migration, in its more highly developed forms, is both too regular in its performance and too provident in its anticipation of events to be conceivable as being created anew each year by the mere pressure of external forces. ... Migration must, then, be a recurrent manifestation of a mode of behaviour which has become inherent in the nature of various species. This implies that the behaviour serves useful ends which give it a survival value; that it became implanted in the inheritance of the species by some originating cause; that it recurs annually in the life of the individual in response to immediate stimuli; and that its exhibition brings into play still other factors which determine the actual path and goal of the movement performed. (A. Landsborough Thomson 1936.)

Migration might be expected to occur wherever individuals benefit more, in terms of survival or reproduction, if they move seasonally between different areas than if they remain in the same area year-round (Lack 1954). The usual reason why breeding areas become unsuitable during part of the year is lack of food. Such food shortages occur for many birds because plant growth stops for part of the year, and many kinds of invertebrates die or hibernate or become inaccessible under snow and ice. At high latitudes, daylengths also shorten in winter to such an extent that many diurnal birds would have too little time to get enough food, even if it were available. Hence, the purpose of the autumn exodus from high latitudes is fairly obvious.

The reason why birds leave their wintering areas to return in spring is less obvious, because many wintering areas seem able to support the birds during the rest of the year. But if no birds migrated to higher latitudes in spring, these latitudes would remain almost empty of many species, and a large seasonal surplus of food would go largely unexploited. Under these circumstances, any individuals that moved to higher latitudes, with increasing food and long days, might raise more young than if they stayed at lower latitudes and competed with the birds resident there. So whereas the advantage of autumn migration can be seen as improved winter survival, dependent on better food supplies in winter quarters, the main advantage of spring migration can be seen as improved breeding success, dependent on better food supplies in summer quarters. Compared to survival, reproduction also has more stringent requirements in terms of specific food needs and predation avoidance.

In effect, migration reduces the seasonal fluctuations in food supplies to which a breeding population could otherwise be exposed. Species that breed in one hemisphere and 'winter' at an equivalent latitude in the opposite hemisphere, where the seasons are reversed, would seem to get the best of both worlds. Further, some habitats offer excellent conditions for survival but cannot be used for reproduction – tidal mudflats used by wintering shorebirds providing an example. Thus, as well as lessening the exposure of individuals to seasonal variation in resource levels, migration facilitates the exploitation of different and widely separated habitats for survival and reproduction (and sometimes also moulting, Chapter 16).

While seasonal change in food supply is clearly important for bird migration, this does not rule out an influence of other factors, such as reduced predation (Fretwell 1980, Pienkowski 1984), parasitism (Piersma 1997), or competition (for food or other resources, Cox 1968, von Haartman 1968). All these pressures may decline with increasing latitude because of the general latitudinal decline in the total numbers of animal species, whether these animals act as predators, parasites or competitors. At one time or another, all these factors have been proposed as contributing to the evolution of migration, but on scant evidence (for more details, see above references). Moreover, it does not necessarily follow that any species suffers fewer losses through having only one predator (or parasite) species to contend with, rather than many. Much depends on the kinds of predators (or parasites).

Whatever the main selective forces, therefore, the migratory habit ensures in the long term that species in seasonal environments adopt and maintain movement patterns that allow individuals to survive and breed better than if they remained in the same area year-round. The main cost of migration is seen as the increased mortality associated with the journey itself. Most recorded mortality incidents relate to storms and unseasonable weather, which can kill thousands of birds at a time, sometimes causing widespread reductions in breeding numbers (Chapter 28). Other risks include exposure to a greater range of pathogens, predators and competitors, as mentioned above, all of which could have additional fitness costs. For the migratory habit to persist, therefore, despite the risks involved in long journeys, the net fitness benefits to individuals of moving both ways must outweigh the costs. Conversely, year-round residence presumably persists in populations where the net benefits of staying in one area outweigh the costs of seasonal movement.

At the population level, one consequence of migration, it may be assumed, is greater overall numbers. Because the breeding area (and sometimes also the wintering area) could not support that number of birds year-round, the overall population is larger as a result of seasonal movement. For many species, the geographical range is also larger, because birds can breed in areas where they could not remain year-round, nesting at higher latitudes only by virtue of migration. Species that are entirely migratory, with separate breeding and wintering ranges, exist today only through their seasonal movements. Other animals, that are less mobile than birds, cope with seasonal shortages in other ways, notably by hibernation – a metabolic shutdown involving torpor. Hibernation occurs in at most a handful of bird species, the best known participant being the Common Poorwill *Phalaenoptilus nuttallii*, though hummingbirds and others show shorter periods of torpidity (Chapter 1).

ADAPTATIONS FOR MIGRATION

Assuming that migration evolved as an adaptation to life in seasonal environments, it may have been an important aspect of bird behaviour for a long time; it may even have occurred in the earliest species. Nevertheless, the ice ages evidently played a major role in its recent development in high-latitude regions. At the height of the last glaciation, much of the northern hemisphere above about 50°N was covered with ice, and generally devoid of terrestrial life. Colonisation of these areas by plants and animals followed the retreat of the ice, which began about 10000–14000 years ago and continues to this day. This means that some of the longest and most impressive of current bird migrations must have developed within this period, as birds spread gradually from lower to higher latitudes to occupy the newly available habitats. At the start of this process, when birds were confined to lower vegetated latitudes, they may have been resident or shorter distance migrants. Moreover, the last 2.5 million years have seen more than 20 successive glaciations of varying severity, so over this whole period, bird migration systems must have been in continual flux, and many populations are likely to have passed through alternating sedentary and migratory phases, each lasting for many thousands of years. The ability to migrate was perhaps ever-present in these populations, never disappearing but becoming suppressed or re-activated according to prevailing conditions and the changing pressures of natural selection (Berthold 1999).

It would be misleading to divide bird species neatly into migratory and sedentary. Both types of behaviour can be found in a single species or even in a single population. In the northern hemisphere, many widely distributed bird species are completely migratory in the north of their breeding range, completely resident in the south, and partially migratory in between, with the proportion of birds leaving any particular locality corresponding to the degree of seasonal reduction in food supplies (Chapter 13). Where populations are entirely migratory or entirely resident, their behaviour is usually considered as obligatory, and under firm genetic control. In contrast, partial migration can apparently arise in two different ways. In one way 'obligatory migratory' and 'obligatory resident' individuals occur intermixed in the same area. In the other way, the behaviour of each individual is optional (facultative), enabling it to behave differently in different circumstances, staying in the breeding area in years when conditions are favourable there, and leaving in other years (Chapter 12). This latter system can result in big year-to-year changes in the ratio of migrating to resident individuals, in particular breeding populations, with the same individuals behaving differently in different years, depending on conditions.

Although migration requires adaptations in morphology, physiology and behaviour, it is not hard to see how it might have evolved, because transitional or intermediate stages still occur. For instance, some bird species do not migrate at all, others travel only short distances, and yet others long distances. The full range of variation can be found among different populations of the same species, or even among individuals in the same population, as mentioned above. Further, the main adaptations needed for long-distance migration, such as seasonal fat reserves, timing mechanisms and orientation skills, are all found in less developed form in non-migratory birds, as well as in other animals. These features are all necessary for effective migration to develop, but may have arisen independently of migration, in different contexts.

The main features that set migration apart from other long-distance movements are that it involves a two-way journey in more or less fixed directions. An ability to perform a return movement is necessary before any migration pattern (as defined above) can evolve. Again, however, such an ability is already present in non-migratory birds, as well as in other animals, but operates over smaller distances. All birds are able to return repeatedly to their nests, as well as to particular feeding or roosting sites, and the same is true for other animals. In addition, many non-migratory birds move locally away from their nesting areas after each breeding season and return for the next. Even juveniles, after wandering widely in the non-breeding season, normally return to settle near their natal sites to breed, and other non-migratory birds revisit the same wintering sites in successive years (Chapter 17). In resident populations, such individual movements are short distance and localised, but they provide a basis from which the longer return movements of migration might evolve, step by incremental step, each extension being beneficial in its own right.

The ability of even resident birds to return to their home areas from at least a few kilometres away has been shown repeatedly in displacement experiments in which individuals were trapped, transported to a different location and released (Chapter 9; Matthews 1968, Wiltschko & Wiltschko 1999). Many of the displaced individuals quickly re-appeared at their capture sites. There is, however, a difference between the needs of short-distance and long-distance moves. In theory, short-distance moves could be performed entirely on the basis of learned landscape features, as a bird gets to know its local area. But to return from long distances requires something more, an ability to orientate by different globally available signposts, as provided by celestial and magnetic cues. Such an ability is almost certainly inherent to some degree in all birds, as it is present in lower animals, including the reptiles from which birds evolved. If birds differ in any way from other animals, it is in their use of stars to orientate, a mechanism documented in birds (Chapter 9), but not so far in other animals.

Similarly, it is not hard to imagine how, given a sense of location, directional preferences could evolve from random dispersal movements. Any birds living in seasonal environments that move long distances after breeding are more likely to meet favourable conditions in some directions than in others. Moreover, at high latitudes, individuals with an inherent tendency to move towards lower latitudes after breeding are likely to survive better than any that move in the opposite direction, so that over generations directional preferences could become fixed by natural selection. It is not just the directions, but also the distances, and hence the specific wintering areas, that could be fixed in this way, the birds from each population wintering wherever they can reach and survive best, taking account of the mortality costs of getting there and back. Suppose that the birds from a certain breeding area have heritable tendencies to fly particular directions and distances at migration time and back again in spring, but that these directions and distances differ from bird to bird. Some birds will then reach suitable areas and survive to breed again, others will reach less suitable areas and survive in small numbers, and yet others will reach unsuitable areas and die. Thus those individuals with the most beneficial migratory behaviour will perpetuate themselves, and in this way the migratory habits of a population could become fixed. Only in populations (like some nomadic ones), which on balance are as likely to find food in one direction as in any other, is no directional preference likely to become fixed by natural selection.

All birds have timing mechanisms which ensure that breeding and moult occur at appropriate and consistent times of year and, in migrants, such mechanisms also promote movements at appropriate dates. They ensure that birds arrive in their breeding areas each spring in time to take advantage of the favourable season, and leave after breeding in late summer or autumn before deteriorating conditions reduce their survival chances. These timing mechanisms are basically endogenous (within the bird) but are entrained by seasonal changes in daylength and modified by other environmental conditions (Chapter 11). Migrants also need behavioural adaptations for responding appropriately to weather conditions, both before take-off and during the journey.

Many species travelling long distances over unfavourable terrain lay down substantial body reserves for the flight. Again the accumulation of body fat is common in all kinds of animals in preparation for periods of privation, when food is scarce or unavailable, and many birds lay down body reserves in preparation for breeding or in winter in association with severe weather and long nights. In accumulating migratory fat, therefore, birds are merely modifying a pre-existing facility for a different purpose, rather than developing a completely new adaptation. It goes with long periods of fasting and endurance performance.

Because of the time it takes, long-distance migration often involves modification of other parts of the annual cycle, particularly the timing of moult (Chapter 11). Most small passerines, resident and migrant, moult in summer after breeding, but some long-distance migrants set off soon after breeding, and postpone their moult until after they have reached their tropical wintering areas. Among the Eurasian warblers, as indicated by phylogenetic analyses, summer moult is the ancestral pattern, while winter moult has evolved independently 7–10 times within this group (Svensson & Hedenström 1999). As these birds colonised northern breeding areas in post-glacial times, thereby lengthening their migrations, summer moult gave way to winter moult. Other patterns, such as split moults and twice-yearly moults, also seem to have evolved from the ancestral state of summer moult, in adaptation to various migration patterns. Many of the evolutionary transitions from resident to migratory or vice versa, as well as changes in the extent and pattern of migration, apparently occur without phylogenetic constraint. Many bird genera bear striking witness of this, as they include a wide spectrum of residents, short-distance migrants and long-distance migrants among closely related species. And in some species, as mentioned already, the same range of variation occurs between populations occupying different parts of the breeding range.

Migration also involves morphological changes, as birds become more adapted for long-distance flight. Compared with residents, closely related migrants usually have longer and more pointed wings, and somewhat shorter tails and smaller bodies (**Figure 20.1**; Leisler & Winkler 2003). Pointed wings and short tail are most efficient during level flight because they reduce drag, whereas more rounded wings and longer tail give greater manoeuvrability and more rapid lift at take-off (Kerlinger 1989, Rayner 1990). Wing shape is thus a compromise between conflicting selection pressures, the balance being drawn differently in resident and migratory populations. Change is achieved mainly by altering the relative lengths of different feathers, but in some types of birds bone lengths (femur, ulna and carpo-metatarsus) also correlate with migratory distance, as does the size of the sternum and coracoid bones, giving greater surfaces for flight muscle attachment (Calmaestra & Moreno 2000).

In several groups of birds of widely different body shapes, correlations have thus emerged within groups between morphology and migration distance

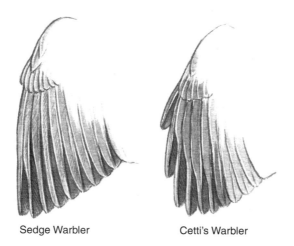

Sedge Warbler   Cetti's Warbler

Figure 20.1   The wing formulae of two Old World wetland warblers clearly fall into two groups – rounded wings of the comparatively sedentary (or totally resident) Cetti's Warbler *Cettia cetti* and pointed wings of the highly migratory Sedge Warbler *Acrocephalus schoenobaenus*. Note the relative lengths of the outer flight feathers, which give greater 'aspect ratios' in the migrants. From Svensson (1975), in Mead (1983). See also Winkler & Leisler (1992).

(Winkler & Leisler 1992, Marchette *et al.* 1995, Mönkkönnen 1995, Lockwood *et al.* 1998, Leisler & Winkler 2003), and in *Calidris* sandpipers also between wing morphology and relative fuel load (itself correlated with migration distance) (Burns 2003). Yet again, such differences occur even between sedentary and migratory populations of the same species (Alerstam 1990a, Fiedler 2005). For example, among different populations of Blackcaps *Sylvia atricapilla*: (1) wing length, aspect ratio and wing pointedness increase; (2) wind-load decreases; (3) slots on the wingtips become relatively shorter; (4) the alula becomes shorter in relation to wing length; and (5) the tail becomes shorter in relation to wing length, with increasing migratory distance. These changes are significantly greater than expected from the simple trend of increasing body mass from southern to northern populations (Fiedler 2005). Moreover, it is not only external features which are modified in migratory birds, but also internal ones, including the brain. Migratory birds have a more highly developed hippocampus, and a more effective spatial memory than non-migratory ones, another difference evident in comparisons between resident and migratory populations of the same species (Cristol *et al.* 2003).

For the most part, then, migratory birds do not possess any fundamentally different adaptations from residents, whether orientation mechanisms, physiological, morphological or other features. Migration simply involves the further development or modification of features already present in non-migratory populations. These features are necessary before effective migration can evolve, but the key novel adaptation is the evolution of narrow directional preferences to unknown areas from the situation of no such preferences shown at the population level in dispersal and nomadism. Directional preference away from the breeding area is a heritable migratory trait in its own right, distinct from other traits (Bell 2000). Like other inherited traits, each is amenable to the action of natural selection, either independently or in association with other traits. Moreover, all can be modified in an incremental manner, in which each small appropriate change brings fitness benefits. This is a firm basis from which the different movement patterns of birds, and their associated adaptations, can be moulded.

If existing bird movements were arranged in order of increasing specialisation, the sequence might run: multi-directional one-way dispersal, nomadism, multi-directional return dispersal (including altitudinal movements), facultative return migration over restricted directions, and obligate return migration over restricted directions. Dispersal movements are performed by all bird species, usually by the young after they become independent of parental care, and at the population level they can occur in any direction from the starting point (Chapter 17). Nomadism can be regarded as a form of dispersal performed repeatedly through life and usually over longer distances, as birds apparently move in any direction from one area of temporary suitability to another (Chapter 16). Dispersive migration is also multi-directional, but involves a return journey (Chapter 17). True return migrations across latitudes involve movements in restricted directions, but whereas facultative movements are stimulated largely by conditions at the time, and hence vary from year to year, obligate movements occur consistently every year, usually between fixed breeding and wintering areas, and seem more firmly under endogenous control. It is possible that this is the sequence in which the most impressive and fixed bird migrations have evolved (Terrill 1990).

Listing the different types of bird movements in this way reinforces the point that migration itself (or any other type of movement) is not a trait in its own right, but an attribute made up of several components, each of which can be independently modified by selection to give the variety of movement patterns we see today. These variable traits are all pre-requisites for the evolution of effective migration, but their origins are independent of migration. One important goal for comparative studies of migration should therefore be to identify the genes responsible for each component of migration, and establish their distribution across different bird families.

Although migration is probably an ancient phenomenon, developed in many kinds of animals long before birds arose, this need not mean that all kinds of animals use the same control systems. Similarly, orientation and navigation mechanisms are probably also ancient, but not all animals necessarily rely on the same environmental cues. Ideas on the development of migration within birds have been put forward by various authors (e.g. Lack 1954, Merkel 1966, Baker 1978, Gauthreaux 1982a, Terrill 1990, Bell 2000, Rappole & Jones 2002), and most contain elements that can be incorporated into a coherent modern theory, consistent with recent research findings. Some aspects of the origin and inheritance of migration remain speculative, but others that were uncertain only a few decades ago have since found support in controlled breeding experiments (see later).

ADAPTIVE TIMING

Some food shortages are predictable because they occur at about the same time every year, in response to the changing seasons. This enables birds, through evolutionary mechanisms, to anticipate periods of shortage, and to get out while they are still able to accumulate fat reserves for the journey. Similarly, birds can leave their wintering areas at a time that allows them to reach their breeding areas at an appropriate date in spring, when breeding again becomes possible. It is in response to these predictable seasonal changes in food supplies that most migration is presumed to have evolved. Many obligate migrants react to proximate environmental cues, such as daylength, in conjunction with an endogenous rhythm, in such a way that they can prepare for migration ahead of time (Chapters 11 and 12).

On current understanding, this is the situation in complete migrants, in which the whole population leaves the breeding area each autumn in anticipation of a predictable loss of winter food supplies there. In facultative migrants, however, the proportion of birds that leaves the breeding range varies according to food and other conditions at the time. In such populations, food shortage would then act as both an ultimate and as a proximate factor in influencing whether any particular individual migrates; food conditions might also influence the date of departure, the length of journey, and the date at which birds move on from place to place along the route (Chapter 18). Such an immediate response allows for the fact that some food shortages are unpredictable in that they occur as a result of unseasonable extreme weather, fruit-crop failures and other irregular events. Because these shortages cannot be anticipated, birds can only respond to conditions at the time, moving out as food gets scarce, and not always with the benefits of prior fat deposition. The obligate–facultative continuum allows birds to cope with increasing degrees of uncertainty and unpredictability in the conditions encountered.

PARTIAL MIGRATION

In many bird populations, as mentioned already, some individuals stay behind for the winter while others from the same area leave.[1] This situation is found in breeding areas that are able to support some individuals in winter, but not as many as in the breeding season. Such partial migration can be divided into two types: obligate and facultative partial migration.

In obligate partial migration, certain individuals in a population migrate some distance every year regardless of prevailing environmental conditions, while other individuals remain resident every year. Particular individuals behave in the same way throughout their lives. This situation is presumed to occur where migratory behaviour is genetically controlled, and where there is no overwhelming advantage in staying or migrating. In some years, perhaps, the migrants do best and in others the residents, so that in the long term both types persist in the same breeding area (Berthold & Querner 1982b, Biebach 1983). They can maintain their distinctiveness only if they mate assortatively, with most migrants pairing with other migrants and most residents with other residents. This situation could occur, for example, if residents paired up in spring before the migrants arrive (a not uncommon event: for example, see Bearhop *et al.* 2005). If the two types interbred freely, they would lose their distinctiveness. Such a dual strategy is often viewed as the crucial intermediate stage in the transition from full resident to full migrant, or vice versa (see later).

In this system, the advantage of each type of behaviour depends partly on what other individuals do. If more and more birds became resident, say, the time would come when far more birds stayed than the winter food supply would support. Competition would then increase, leading to greater mortality in the resident sector, eventually tipping the balance in favour of migration. In this way, changes in the relative advantages of the two types of behaviour could ensure that the ratio of residents to migrants in any one population was kept roughly in line with local conditions (Lundberg 1987). The relative advantages of resident versus migrant may, of course, change over the years if environmental conditions change, such as winter temperatures or winter food supplies in breeding areas (see Chapter 21 for examples). They may also differ between areas and, in many partial migrants, a greater proportion of individuals migrates from higher than from lower latitudes. Obligate partial migration could thus be viewed as an 'evolutionarily stable strategy' (ESS) because, with the appropriate ratio of migrants and residents in the local breeding population, in the long term the pay-offs for both genotypes, in terms of lifetime reproductive rates, are balanced, and in the conditions prevailing no one type can completely replace the other (Berthold 1984a, Kaitala *et al.* 1993).

---

[1]*The term partial migration is used here for the individual variation in behaviour within a population, not for the variation between different populations of the same species.*

In facultative partial migration the same individuals might migrate in some years and not in others, depending on conditions at the time. Because birds compete, and vary in dominance or feeding efficiency, some individuals are able to survive in conditions where others would perish unless they moved out (the 'behavioural dominance hypothesis' of partial migration, Kalela 1954, Gauthreaux 1978a, 1982a; Chapter 12). In many partially migrant populations, a higher proportion of juveniles than of adults, and of females than of males, migrate (Chapter 15). This fits the dominance order in such populations, so that when food is scarce, juveniles fare less well than adults, and females less well than males. The subordinate individuals therefore migrate in greater proportion, or for greater distances, than the dominants (Chapter 15). In this situation, facultative partial migration is a 'conditional response' to food supply, dependent partly on social status or on differing priorities related to age or sex. Individuals leave only if it becomes difficult for them to survive locally, and the same individual can vary its behaviour according to circumstance and age (Newton 1979, Smith & Nilsson 1987, Adriaensen & Dhondt 1990, Schwabl & Silverin 1990).

In this facultative system, partial migration need not necessarily involve equal pay-offs for resident and migratory behaviour. As a group, the migrants may survive or reproduce less well than the residents, but better than they themselves would have done if they had stayed behind. The migrants are simply making the best of a bad job. The argument is that each bird responds to conditions at the time, so as to maximise its own survival chances. And as these conditions include the competitive pressure from other individuals, migration becomes to some extent density dependent: the more birds there are relative to resources, the greater the proportion that leaves for the winter. Dominance relationships are held to account for the facts that, in many partial migrants, larger proportions of individuals leave their breeding areas in years of high population or of poor food supplies than in other years, and that sex and age differences are often apparent in the proportions of individuals that leave, the dates they leave and the distances they travel, giving rise to geographical gradients in the sex and age ratios of populations in the non-breeding season (Chapter 15).

The European Robin *Erithacus rubecula* is a facultative partial migrant in Belgium (Adriaensen & Dhondt 1990). In one study, most males that nested in parks and gardens were resident year-round, whereas most males that nested in woodland were migratory, as were all females from both habitats. Resident males survived, on average, about three times better than migratory individuals (50% versus 17%), and even during the extremely cold winter of 1984–1985, residents still survived best. In addition, resident males were much more likely than migratory ones to obtain a mate (74% versus 44%). On the basis of both survival and mating success, the expected reproductive success of resident males was 2–4 times higher than that of migratory males. So why did migration persist? In ecological jargon, partial migration was a 'conditional strategy with unequal pay-offs': if individuals could find a territory locally in which they had a good chance of surviving the winter, they could stay; if not, they had to migrate, for only then did they stand any chance at all of surviving the winter. On this basis, socially dominant individuals were more likely to become resident, and subordinate ones to migrate. This could explain why females migrated more than males, why migratory and resident males tended to use different breeding habitats, and why young of early broods were less migratory than those of later broods. A similar system seems to operate in other species. For example, Great Crested Grebes *Podiceps cristatus* are partial migrants at Lough Neagh in Northern Ireland. By the time those that winter on the coast return, resident ones have begun breeding in the best areas, pushing the later arrivals into poorer nesting habitat (K. Perry, in Wernham *et al.* 2002). The same probably happens in many partial migrants. The two types of partial migration, involving an obligate genetic dimorphism in migratory tendency or a facultative migration dependent on individual circumstances, are not necessarily mutually exclusive. They may represent extremes of a continuum, or both could operate to varying degrees in the same population (Lundberg 1988). In addition, migration might be conceived as a split journey in which the first part was obligatory and under genetic control, and the second part facultative and under environmental control (Terrill 1990). This would account for the fact that, in some migratory species, the whole population departs in autumn, but the distance that most individuals travel varies greatly from year to year depending on the conditions (especially food supplies) encountered en route (Chapter 18).

Attempts have been made to assess the advantages or otherwise of migration by cost–benefit analyses, comparing the survival, reproductive success or energy budgets of birds that behave in different ways. Such comparisons may sometimes be useful, but often they are misleading, as explained in **Appendix 20.1**. In any case, different outcomes would be expected in obligate and facultative systems, further information on which is given in Chapter 12.

THE GENETICAL CONTROL OF MIGRATION: EXPERIMENTAL EVIDENCE

Migration has often been seen as evolving from a resident ancestral mode, involving a single mutation or a stepwise development from dispersal (Terrill 1990). It could have evolved many times over and in many different kinds of animals. More recently, the study of obligate partial migrants has given a different perspective, in which the existence of a range of genotypes (from migratory to resident) in the same population could form the basis from which complete migratory or complete resident behaviour could evolve.

The main questions in the genetic control of migration are how does a bird come to leave and return to breeding areas at particular dates rather than others, to fly in particular directions rather than others, and to stop migrating once it has reached a suitable wintering area it might never have seen before? In addition, what is the difference between obligate migration in which the bird leaves every year at about the same date regardless of conditions, and facultative migration in which the bird may or may not migrate, or may leave at widely different dates in different autumns, and travel varying distances, depending on conditions?

For many years, ideas on the role of natural selection in shaping bird migration patterns were based on little more than surmise. One indication of genetic control was the year-to-year consistency in all aspects of the migratory behaviour of particular populations, including the timing, directions and distances travelled. Where young and old birds migrate together, the young could possibly learn from older birds the seasonal timing, migration routes and wintering areas. But in many species, the two age groups migrate independently of one another, sometimes weeks apart, giving little or no opportunity for cultural transmission of information. In experiments on some species, young birds were held in their natal area until all other individuals of their species had left, or transported outside their usual range. The young were then released, and found from subsequent sightings and ring recoveries to have migrated in the direction normal for their population (Chapter 9). The only plausible explanation was that migratory directions were inherent.

Other early evidence revealed a genetic difference between resident and migratory individuals in the same breeding population of Common Eiders *Somateria mollissima* (Milne & Robertson 1965). The egg albumen proteins showed a genetic polymorphism, in which the ratios between two alleles differed between resident and migratory individuals. Field observations revealed that residents and migrants did not normally interbreed, because individuals of both types paired in winter, so were already mated, according to wintering area, when they met on their common breeding area.

In recent years, the role of genetic factors in the control of migration has been shown experimentally, mainly by Peter Berthold and his colleagues, who found that Blackcaps *Sylvia atricapilla* and other songbirds could be bred on a large scale in aviaries (Berthold & Helbig 1992, Berthold 1995, 1999). Different populations of Blackcaps varied from completely migratory to completely sedentary in different parts of the range. The various migratory populations also differed in their migration dates and in the directions and distances travelled. Most of these traits could be measured in captive birds by assessing the timings and amounts of migratory restlessness (or *Zugunruhe*), a specific behaviour involving fluttering and wing-whirring which appears in caged birds mainly at migration seasons (see **Box 12.1, Figure 20.2**). Directional preferences of individuals could be assessed by placing them in orientation cages, and checking where they most frequently headed. Both restlessness and headings could be recorded automatically, as described in Chapter 2.

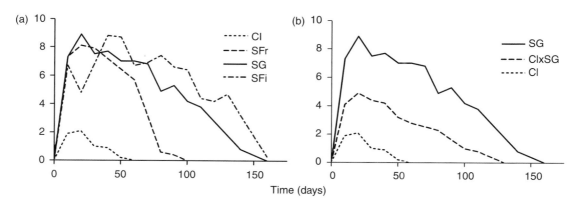

Figure 20.2   Patterns of migratory activity (*Zugunruhe*) of Blackcaps *Sylvia atricapilla*. (a) Activity of birds of four different populations. (b) Activity of birds of two of these populations and their hybrids. SFi – birds from southern Finland, SG – southern Germany, SFr – southern France, CI – Canary Islands. From Berthold & Querner (1981).

The main finding was that when birds from different populations were cross-bred, the offspring showed migration features that were intermediate between those of their parents. Moreover, if individuals with particular traits were selected and bred together, these traits became enhanced in future generations. This held for the timing and duration of migratory restlessness, and for directional preferences. The overriding implication was that all these aspects of migration were under genetic control, and could thus be changed in the wild by natural selection, depending on the survival and reproductive benefits of different behaviours in the conditions prevailing. The details of particular experiments are summarised below.

Migratory inclination

Experiments in which individuals showing high levels of migratory restlessness were paired together showed that the trait persisted or increased in the offspring. Similarly, when individuals showing low levels of restlessness were paired together, the trait persisted or decreased in the offspring. These findings confirmed, for both Blackcaps *Sylvia atricapilla* and European Robins *Erithacus rubecula*, that the amount of migratory restlessness was to a large extent genetically pre-programmed (Berthold & Querner 1981, Biebach 1983). Moreover, cross-breeding of migrant with non-migrant strains (using Blackcaps from south Germany and the Cape Verde Islands respectively) resulted in the partial transmission of migratory activity to 33% of the first-generation offspring, again indicating that the urge to migrate was inherited (Berthold *et al.* 1990b). Starting with a partially migratory population, high or low levels of migratory restlessness (assumed to reflect migratoriness or sedentariness) could be selected to phenotypic uniformity within as few as 3–6 generations. More precisely, after three generations of selection, sedentary behaviour was fixed in the migratory line, whereas selection over 5–6 generations was needed to produce an almost exclusively migratory stock. These experiments thus revealed not only high heritability of migratory behaviour, but also how rapidly migratory behaviour could change in response to strong selection pressure. In nature, selection is never likely to be that strong, so substantial changes would be expected to take longer, perhaps decades or more depending on generation times.

In another study, Blackcaps *Sylvia atricapilla* from widely separated populations (southern Finland, southern Germany, southern France and the Canary Islands off North Africa) were hand-raised and tested under identical conditions. The proportion of migrants (showing a given level of migratory restlessness) among birds from the four areas differed markedly: from 100% of birds from Finland and Germany, to 80% of those from southern France, and to only 23% of those from the Canary Islands. Furthermore, the number of nights of migratory restlessness shown by migratory individuals differed between populations, declining north to south (**Figure 20.2**; Berthold & Querner 1981). Appropriate differences in physiology (moult and fat accumulation) and morphology (wing shape) were also found between the different populations (Berthold & Querner 1982a). Cross-breeding of individuals from two of these populations (southern Germany and Canary Islands) gave offspring of intermediate characteristics, again implying that migratory features were under genetic control (**Figure 20.2**). Intermediate migratory behaviour was also found in hybrids between migratory

European Quail *Coturnix c. coturnix* and non-migratory Japanese Quail *C. c. japonica* (Derégnaucourt *et al.* 2005).

Timing and distance

In some bird species, the number of nights (or days) that individuals show migratory restlessness correlates with the actual distance travelled by their population (see **Figure 12.2, Box 12.1**). This led to the view that migration was controlled in part by endogenous time programmes, entrained by the seasonal change in daylength (Chapter 11). Again, the timing of migratory activity, whether early or late in the season, or long or short in duration, could be influenced by selective breeding in captivity. The cross-breeding experiment with Blackcaps, mentioned above, confirmed that migratory activity is a population-specific quantitatively inherited characteristic. Restlessness began earlier and continued longer in one parent population than in the other (Berthold & Querner 1981), while the first-generation hybrids showed intermediate patterns of behaviour (**Figure 20.2**). In another experiment, Blackcaps from a migratory population (southern Germany) were selected for later onset of migratory activity (Pulido *et al.* 2001). After only two generations of artificial selection, the mean onset of migratory activity was delayed by more than one week.

Experiments involving the cross-breeding of different species also proved instructive. In southern Germany, the Black Redstart *Phoenicurus ochruros* is a short-distance migrant, travelling about 1000 km to Mediterranean winter quarters, whereas the Common Redstart *P. phoenicurus* is a long-distance migrant, travelling up to 7000 km to tropical Africa. The young of both species hatch at about the same time in May–June, but the Common Redstart moults earlier and more rapidly in preparation for its longer journey, beginning in August. In captivity, this is reflected in the early appearance and long duration of nocturnal migratory restlessness. In contrast, the Black Redstart's period of restlessness is later (October–November) and briefer. The Common Redstart also becomes much heavier through fat deposition, and retains its fat for much longer than the Black Redstart. The two species were therefore ideal for cross-breeding experiments to elucidate the genetic control of these processes. In the event, hybrids between the two proved to be intermediate in all respects between their parent species, in the timing and duration of migratory restlessness, the timing of moult and the degree of fat deposition (Berthold 1999). Yet again, the strong implication was that all these aspects were under genetic control.

In addition to the Blackcap and two Redstarts, in at least two other obligate partial migrants, the occurrence of migratory and non-migratory individuals has been shown to be genetically determined, namely the central European populations of the European Robin *Erithacus rubecula* (Biebach 1983) and Eurasian Blackbird *Turdus merula* (**Table 20.1**; Schwabl 1983). Indications of genetic control have also been found in particular populations of the Song Sparrow *Melospiza melodia* (Nice 1937, re-analysed by Berthold 1984b), Silvereye *Zosterops lateralis* (Chan 1994), Great-crested Grebe *Podiceps cristatus* (Adriaensen *et al.* 1993) and others.

How quickly changes from migrancy to residency could occur in the wild would depend not only on the initial level of genetic variation in the behaviour of the population and on the strength of the selection pressure, but also on generation

Table 20.1 Partial migration among Blackbirds *Turdus merula* in different parts of Fennoscandia

Breeding area	Percentage of breeding population that is migratory	Median migration distance (km) (90% within)
Denmark	16	533 (30–1143)
Norway	61	894 (587–1442)
Sweden	76	1113 (196–1754)
Finland	89	1738 (791–2288)

Only in the least migratory population (Denmark) was an age difference found, with adults moving in greater proportion than juveniles. In the three most migratory populations, in Norway, Sweden and Finland, resident individuals occur only in the most southern areas, all from more northern regions being migratory.
From Main (2002).

times (or longevities) and on whether matings were selective or random among genotypes. If matings were assortative, with individuals of similar migration habits pairing together, the change to new migration habits could occur more rapidly than if matings were random between genotypes.

Measures of migratory restlessness, along with breeding experiments, revealed that in Blackcaps, non-migrants, short-distance migrants and long-distance migrants were part of the same continuum of variation in a single trait, as reflected in restlessness. The binary response (migrate or not migrate) is caused by a threshold which divides individuals into those above and below. Thus all birds without measurable migration in the wild have activity levels in captivity at the low end of a continuous distribution, below the limit of expression or detection (Pulido *et al.* 1996). Similarly, the division between short-distance and long-distance migrants is caused by variation in the numbers of nights on which they show appropriate behaviour. These findings have profound implications for the evolution of migration, because they suggest that ecologically significant transitions between resident or migratory, and between short- and long-distance migration, could come about by selection on a single trait. Of course, other selection would be needed to change the migratory direction and fattening pattern, as discussed later.

Most probably, then, migratory traits are threshold characters, which are determined by multiple genetic loci. This mode of inheritance, based on quantitative genetics, supersedes the idea of a single locus determination of a genetic dimorphism (migrate or not), since the offspring of given pairs are normally intermediate in their behaviour (Berthold *et al.* 1990b, 1996). Consistency tests on captive birds showed that individuals with a low level of restlessness did not change to a high level of restlessness in subsequent years, although a proportion of individuals that initially showed a high level subsequently showed a lower level (equivalent to wild birds becoming less migratory in later life) (Berthold 1993). In the Blackcap *Sylvia atricapilla* and Dark-eyed Junco *Junco hyemalis*, migratory behaviour (= high restlessness) was also expressed in a greater proportion of females than males, raising the possibility of sex-linked inheritance (Berthold 1986, Terrill & Berthold 1989, Holberton 1993). In this situation, any sex difference in migratory tendency would not result entirely from dominance relationships at the time, but also from genetic influence.

These various findings support the view that inborn, temporal patterns of migratory activity serve as time programmes that, together with inherited migration directions, enable inexperienced, first-time migrants to 'automatically' cover the distance between their breeding areas and their specific wintering areas. They embody the so-called vector system of migration, and help to explain why young birds, with no previous experience of a wintering area, can find their own way to an appropriate locality, without help from experienced adults (Chapter 9).

Migratory directions

The finding that captive migrants, during periods of migratory restlessness, flutter in their cages with an appropriate heading, suggested that migratory directions were also under genetic control. This could be easily tested on Blackcaps because in Europe this species shows a migratory divide, with those from the western part of the continent migrating southwest in autumn and those from the eastern part southeast. Moreover, the eastern populations migrate around the eastern edge of the Mediterranean to East Africa, so they must change their course (by about 60° clockwise) from southeast to south-southwest about halfway through their autumn journey. Western populations migrate southwest and winter mainly in the western Mediterranean region; they therefore have a shorter journey and do not obviously change direction en route.

With birds from two localities lying west and east of the migratory divide, hand-raised and tested under identical conditions, this behavioural difference did indeed seem to have a genetic basis (**Figure 20.3**). In orientation cages, Blackcaps from western Germany kept a constant southwest course throughout the season, as expected from their relatively direct route to the Iberian Peninsula and northwest Africa, whereas those from eastern Austria started with a southeast course in September–October, and changed within a 10-day period to a south-southwest course in November (Helbig *et al.* 1989). It seemed that young Blackcaps not only inherited from their parents a general starting direction, but a fairly detailed time–direction programme appropriate to their dog-leg migration route. Moreover, when Blackcaps from both sides of this divide were cross-bred, the first-generation hybrids moved directly southward, behaving in a phenotypically intermediate fashion between both parent groups, with no difference in the amount of directional spread (Helbig 1991b). These findings thus confirmed the genetic basis of migratory directions and, through cross-breeding, revealed a phenotypically intermediate mode of inheritance. The implication is that, if birds from southwest and southeast migrating populations interbred where they meet, their offspring would inherit intermediate, and probably inappropriate, directional preferences. Any selection against the hybrids could help to maintain the differences between populations, perhaps leading eventually to taxonomically distinguishable subspecies (Chapter 22).

Evidence for an intra-seasonal change in direction, genetically determined and endogenously programmed, was also found in Garden Warblers *Sylvia borin* (Gwinner & Wiltschko 1978) and Pied Flycatchers *Ficedula hypoleuca* (Beck & Wiltschko 1988) in Europe, and in Yellow-faced Honeyeaters *Lichenostomus chrysops* in Australia (Munro & Wiltschko 1993).

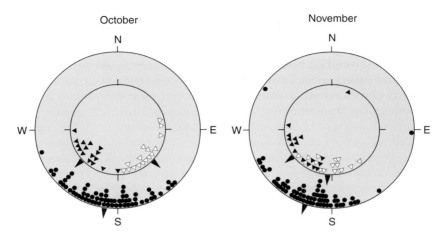

Figure 20.3   Orientation directions of Blackcaps *Sylvia atricapilla* from two populations breeding east and west of a migratory divide in Europe, and from hybrids between these two populations. Within the inner circle, triangles indicate the parental population from southwest Germany (filled) and eastern Austria (open), while in the outer circle, dots show first-generation offspring of mixed pairs bred in aviaries. Arrowheads indicate group mean directions. The hybrids took mean directions intermediate between the two parental forms. Note the seasonal shift in direction shown in November by the southeastward migrating parents and the intermediate mean orientation of the hybrid offspring. From Helbig (1991b, 1996).

It is not sufficient that birds inherit a directional tendency that allows them to reach winter quarters. They must also be able to reverse this direction in spring in order to return to their breeding areas. Such reversal in direction has been demonstrated in tests of hand-raised Indigo Buntings *Passerina cyanea*,, Blackcaps and Garden Warblers (Chapter 12; Emlen 1969, Helbig *et al.* 1989, Gwinner & Wiltschko 1978, 1980). In autumn, Garden Warblers orientated southwest, changing some days later toward south or south-southeast, but in spring they headed towards the north, fitting with the loop migration recorded in wild Garden Warblers (Chapters 12 and 22).

Less direct indications for genetic control of migratory directions in various other species stem from the facts that: (1) inexperienced migrants tested in orientation cages showed regular species-specific and population-specific migratory directions; (2) displaced first-time migrants (White Storks *Ciconia ciconia*, Schüz *et al.* 1971, Starlings *Sturnus vulgaris*, Perdeck 1958) showed migratory directions parallel to those of their parental populations (Chapter 9); (3) partial and irruptive migrants also migrate regularly in specific directions (Chapter 18); (4) many young birds (notably cuckoos) migrate from Eurasia and tropical Africa independently of their parents (Chapters 9 and 15); and (5) inexperienced warblers and flycatchers show population-specific seasonal changes in their directional preferences in captivity matching the course of migration in wild birds. All these findings indicate that such migrants possess pre-programmed migratory directions or directional sequences, although they may, of course, deviate from their genetically prescribed course in response to local topography, weather and other conditions (Chapter 4). In some populations, ring recoveries of migrating birds lie in such a narrow and straight line between breeding and wintering areas that a simple clock-and-compass system might be sufficient to account for this pattern, providing that the birds are also able to correct for directional errors and wind displacements to stay on course (Mouritsen & Larsen 1998, Thorup & Rabøl 2001). Some radio-marked birds have also shown remarkably straight migration tracks (for Whooping Crane *Grus americana* see Kuyt 1992, for Common Eider *Somateria mollisima* see Desholm 2003, for Honey Buzzard *Pernis apivorus* see Håke *et al.* 2003).

Morphological features

The cross-breeding experiments with exclusively migratory Blackcaps *Sylvia atricapilla* from southern Germany (with long wings and high body weights) and poorly migratory conspecifics from the Canary Islands (with short wings and low body weights) demonstrated the genetic influence on both characteristics. Wing lengths, as well as body weights, were intermediately expressed in the hybrids (Berthold & Querner 1982a). Similarly, in a later study, hybrids between migratory and non-migratory Blackcap populations, from Moscow and Madeira, showed intermediate values between parental populations in wing length, wing shape and wing area, while in other variables they resembled either parent population (Fiedler 2005).

Natural variability

Another interesting finding confirmed in captive birds concerns the level of individual variation in migratory restlessness and directions apparent within some populations. Among Blackcaps *Sylvia atricapilla* that breed in the Mediterranean region, some are resident, while others travel mainly short distances up to 1300 km. Correspondingly, migratory restlessness among captive individuals from this region varied by a factor of 100. Migratory directions in the Mediterranean population range from west through south to east, spanning an angle of about 180°, as shown by ring recoveries. In contrast, Blackcaps that breed in Germany are wholly migratory, travelling distances from about 700 to 4500 km to reach various winter quarters which extend from southern France to the Ivory Coast. Their migration directions span about 90° to judge from ring recoveries (Berthold 1996). The amount of migratory restlessness displayed by captive individuals from this region varies by a factor of about six (between about 150 and 900 hours), and occurs over generally longer periods than in the Mediterranean birds. Individuals from resident populations of other species have also shown low levels of migratory restlessness in spring and autumn, including tropical Stonechats *Saxicola torquata* (Helm & Gwinner 2005). The important point is that not all individuals in the same population are equal in their migratory behaviour, and that substantial variation exists on which natural selection can act if necessary.

A natural change in the migration of Blackcaps

A new migration route of central European Blackcaps *Sylvia atricapilla* to winter quarters in Britain and Ireland (as opposed to the western Mediterranean area) was discovered from ring recoveries (Zink 1962, Berthold 1995). Prior to 1960, Blackcaps were rare during winter in Britain and Ireland, but from the late 1980s thousands of individuals wintered regularly (Lack 1986). For nearly 40 years, ringing provided no indication that these birds were raised in Britain or Scandinavia. However, many Blackcaps ringed during the breeding season in Belgium, southern Germany and western Austria have been recovered in a west to northwest direction, some of them in Britain and Ireland. In all these continental areas, some southwest migrants also breed.

It therefore appeared that, within a period of 30 years, a portion of the southwest migrating population in central Europe had shifted their migration toward the west and northwest and that the majority of British-wintering Blackcaps originated from this region. To see whether this new migration had a genetic basis (as opposed to being due to increased wind drift or other prevailing conditions), some wintering Blackcaps were caught in southwest England, bred in captivity in Germany, and then tested for directional preferences in standard conditions. Both the adults and their offspring showed a west-northwest preference, indicating that the new migration route represented an evolutionary change that had apparently arisen within recent decades (**Figure 20.4**; Berthold *et al.* 1992a, Helbig 1994, 1996).

Such a change may have started from one or more individuals with an unusual directional tendency that found themselves in a new area where they could survive the winter. These pioneers may have been individuals from the extreme end of a range of directional preferences already existing within the central European source population, or mutant individuals with a new directional preference not previously represented. Alternatively, as suggested by Busse (1992), they may

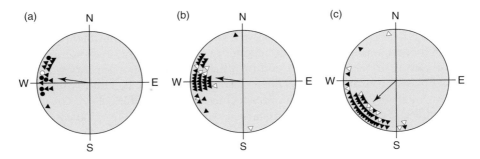

Figure 20.4   Autumn orientation of Blackcaps *Sylvia atricapilla* caught in winter in Britain (a), their captive-bred first-generation offspring (b), and a control group from southwest Germany (c). Each symbol shows the direction of the mean vector of one individual during 15–20 tests. The findings indicate that British-wintering Blackcaps have different inherent directional preferences from Blackcaps that breed in the same region of central Europe but winter to the southwest in Iberia. Modified from Helbig (1996).

have been individuals from a normal southeastward migrating population with an inherent fault in their migration control system, which led to their migrating in the reverse direction to normal for that time of year (that is, taking the spring direction in autumn and vice versa, see Chapter 10). But whatever the origin of the pioneers, it would not have been enough merely for them to have been blown off course, because the new route could not have been inherited by their offspring. It must have been genetically influenced. Once started, the selection pressures that may have favoured wintering of continental Blackcaps in Britain rather than in the Mediterranean region could have included: (1) factors acting in the new wintering area, such as progressively milder winters, or improved food supply provided at garden feeders, and at winter fruit bushes planted in recent decades; (2) factors related to the location of the new wintering areas, which involve a shorter migration distance by up to 1500 km and possibly an earlier return to the breeding areas; or (3) factors acting in the former Mediterranean wintering areas such as drought-induced declining food supplies and increasing competition. Other ringed Blackcaps from central Europe have been found in autumn further north in western Europe than the British birds, but have not established wintering populations, possibly because at these higher latitudes the winters are too cold for them.

Captive British-wintering Blackcaps show migratory restlessness at a date that would enable them to arrive in their central European breeding areas by 1 April, whereas Mediterranean wintering birds would not arrive until about 17 April. In addition, the mean testis size at the date of arrival was about 25% greater in the first group than in the second. An early return could enable British-wintering Blackcaps to pair preferentially with one another (assortative mating based on differential arrival times), and perhaps raise an extra brood each year, speeding the evolution of the new habit. A field study revealed that, on average, birds from British wintering areas did indeed arrive in central European breeding areas earlier than those from Spanish wintering areas, and were 2.5 times more likely to mate with one another than at random (Bearhop et al. 2005). The British-wintering birds also settled in the better territories, produced larger clutches and broods. In view of these findings, the British-wintering birds may eventually out-compete and replace the Spanish-wintering ones. It is also surprising that British-breeding Blackcaps have apparently not yet become partially resident, with some individuals remaining all year in their breeding areas. So far, only one recovery of a confirmed British-bred Blackcap has been obtained in Britain in winter, but this may well mark the start of a new habit. Meanwhile, most British Blackcaps continue to winter in Iberia and North Africa.

Heritability and other studies

Further evidence for the genetic control of various migratory traits has come from field studies of resemblance among genetically related individuals, such as siblings or parents and their offspring (heritability). It can be assumed that part of any resemblance between closely related individuals is due to their shared genes. More specifically, variation in the behaviour of individuals can be ascribed partly to genetic differences between them, partly to differences in the environments to which they are exposed, and partly to an interaction between genetic and environmental influences. In wild birds, environmental variation might be so great as to mask the effects of genetic variation. Alternatively, different individuals may behave in the same way because they live in the same environment, rather than because they share the same genes. But in captive birds, kept in standard conditions, environmental variation can be kept to a minimum, enabling the effects of genetic variation to be more clearly revealed.

Of the 16 heritability estimates in **Table 20.2**, five derive from field studies and 11 from captive ones. Only one of the five field estimates was significantly different from zero (implying an inherited component in the behaviour studied), whereas 10 out of 11 laboratory studies showed statistically significant inheritance. The lack of significance in most studies of wild birds may have been due partly to greater variation in the environmental conditions to which the individual birds were exposed, as mentioned above. It may also have been due to low sample sizes, to reduced precision of measurements taken in the field, and to greater age variation of some wild samples (assuming age effects on traits). Overall, however, these various heritability estimates provided further evidence for genetic influence on most of the traits examined.

Additional evidence for genetic influence has come from field studies of the year-to-year consistency (repeatability) of behaviour in the same individuals. This was especially true for individual arrival dates in successive springs which were much more consistent within individuals than expected by chance (for White-eyed Vireo *Vireo griseus* the correlation in individual arrival dates in consecutive years was 0.6, Hopp *et al.* 1999; for Barn Swallows *Hirundo rustica* repeatability was 0.51, Møller 2001; for Dusky Warblers *Phylloscopus fuscatus* repeatability was 0.34, Forstmeier 2002; and for Garden Warblers *Sylvia borin* repeatability was 0.40, Pulido & Berthold 2003). In contrast, European Pied Flycatchers *Ficedula hypoleuca* showed low consistency in spring arrival dates (among-year repeatability = 0.03, Potti 1998). The only study of a non-passerine, the Bewick's Swan *Cygnus columbianus bewickii*, showed some consistency in autumn arrival dates (ANOVA, $F = 3.16, P < 0.001$) and in spring departure dates (ANOVA, $F = 1.72, P = 0.005$) (Rees 1989). This year-to-year consistency in the migration dates of individuals has been taken as evidence for genetic influence, but other explanations are possible: for example, some individuals might feed more effectively and accumulate migratory fat before others. The effect of environmental influence on arrival dates could be reduced somewhat if arrival dates were expressed relative to those of other individuals each year (Nolan 1978), rather than using simple calendar dates, as in most of the studies concerned. This might strengthen the case for genetic influence.

Other field studies provided a different type of evidence. For example, Cliff Swallows *Petrochelidon pyrrhonata* in central North America were exposed in 1996 to a severe cold snap in the middle of the spring arrival period, when insects were inactive, killing many of the newly arrived birds (Brown & Brown 1998). Other swallows arrived after the event. Observations in later years showed that the generation hatched after the selection event contained a greater proportion of later arriving birds than earlier generations, so that the mean arrival date of the whole population was shifted later. The cold snap seemed to have changed the genetic composition of the population, in favour of later arriving genotypes. Although in some years early arrival gave better nest success, this advantage was apparently offset by the greater risk of mortality in occasional cold years, the resulting arrival dates being a compromise between these opposing selection pressures (Chapter 14).

Table 20.2 Heritability estimates for migratory traits. From Pulido & Berthold 2003

Trait	Species	Environment[a]	Heritability[b]	Statistical significance	Method[c]	Reference
Onset of autumn migratory activity	Sylvia atricapilla	1	0.34–0.45	+	FSC, POC	Pulido et al. (2001)
	Sylvia atricapilla	1	0.67*	+	FSC, POC	Pulido & Widmer (in prep.)
Termination of autumn migratory activity	Sylvia atricapilla	1	0.16–0.44	+	FSC, POC	Pulido & Berthold (2003)
Arrival at wintering site	Cygnus bewickii	w	0.19	–	POC_{INC}	Rees (1989)
Departure from wintering site	Cygnus bewickii	w	0.10	–	POC_{INC}	Rees (1989)
Onset of spring migratory activity	Sylvia borin	1	0.67*	+	FSC, POC	Widmer (1999)
Arrival at breeding site	Ficedula hypoleuca	w	<0.03	–	FSC, POC	Potti (1998)
	Hirundo rustica	w	0.54	+	POC	Møller (2001)
	Erithacus rubecula	1	0.52	+	POC	Biebach (1983)
Amount of autumn migratory activity	Sylvia borin	1	0.44*	+	FSC	Widmer & Pulido (unpub.)
	Sylvia atricapilla	1	0.36–0.47	+	FSC, POC	Berthold & Pulido (1994), Pulido (2000)
Intensity of autumn migratory activity	Sylvia atricapilla	1	0.38–0.53	+	FSC, POC	Pulido (2000)
Migratory status (migrant/resident)	Melospiza melodia	w	0.58– >1.0	–	POC_{INC}	Nice (1937), Pulido (unpub.)
Incidence of autumn migratory activity	Sylvia atricapilla	1	0.43– >1.0	+	SEL	Berthold et al. (1990a), Pulido & Berthold (unpub.)
	Sylvia atricapilla	1	026–0.37	+	POC_{INC}	Pulido et al. (1996), Pulido & Berthold (unpub.)
Response of onset of migration to hatching	Sylvia atricapilla	1	0.07–0.16	–	POC	Pulido (unpub.)

[a]Whether heritabilities were derived from laboratory (1) or wild (w) populations. The range of heritabilities given is the range of estimates obtained by different methods.

[b]Asterisk indicates that estimates give the average heritability for two populations.

[c]Methods for estimating heritabilities are as follows: full-sib correlations (FSC), parent–offspring co-variance (POC), parent–offspring co-variance using the liability model (POC_{INC}), selection experiments (SEL).

Only one of five field-based estimates was significantly different from 0 (implying an inherited component in the behaviour studied), whereas 10 of 11 laboratory-based studies showed statistically significant inheritance. The lack of significance in most wild studies may be partly due to low sample sizes, partly to reduced precision in measurement of traits in the field, partly to greater variation in the environmental conditions experienced by wild populations, and partly to greater age-variation of some wild samples (assuming age-effects on traits).

From Pulido & Berthold (2003).

Likewise, bad weather in central Europe in the autumn of 1974 eliminated many of the late-departing migrants among hirundine populations, as millions of birds died (Chapter 28). The effects of this selective mortality were apparent for several subsequent years, when migration finished earlier in autumn than in the years before this event. This change was most evident in Barn Swallows *Hirundo rustica*, but also in House Martins *Delichon urbica* and Sand Martins *Riparia riparia* (Gatter 2000).

CONCLUDING REMARKS

Most of the experimental research discussed above was concerned with the modification and further development of migratory behaviour, and not with its inception. Most research also involved only one species, although some findings were confirmed on other species. It implies that, given sufficient underlying genetic variation and strong enough selection pressure, some wild birds could change their migratory habits within only a few generations in response to environmental change (for more examples see Chapter 21). Moreover, the intermediate behaviour of hybrids suggests control by several genes, and not just one (for any behaviour controlled by a single gene would be expected to show an on–off pattern, with no scope for intermediates). This does not mean that all populations could adapt rapidly to environmental change. Insufficient genetic variation in a trait can prevent adaptation occurring altogether, whatever the selection pressure. Once the available genetic variation within the population had run out, time would then be needed for mutation to create yet more variation on which selection could act. Whether a population will respond to a new selection regime also depends on whether the trait under selection co-varies with other traits. If selection to change one trait beneficially brings concurrent detrimental changes in other traits, the resulting behaviour is likely to be a compromise. In terms of the length of journey, there are also limits to how much of each year a bird could spend on migration, and still have time to reproduce and moult, and how much body fuel it could carry for non-stop flights. These aspects act as constraints on evolutionary change, at least in the short term.

Artificial selection for higher and lower levels of migratory activity among captive Blackcaps resulted in changes in subsequent generations not only in the frequency of individuals showing high levels of activity, but also in the amount of migratory activity they showed. The correlation between these two traits was very strong, suggesting that both traits were controlled by the same genes: in other words, genetically, both incidence and amount were aspects of the one trait. This finding has important implications for the evolution of migration: in particular, that obligate migrants and non-migrants could be present in all bird populations, although the frequency of one or other may be very low in populations classed as completely sedentary or completely migratory. As a consequence, all bird populations could be considered as partially migratory, differing only in degree of migratory behaviour.

High correlations between migratory traits (date of onset, incidence, and duration of migratory activity) indicate that such traits are expressed as a syndrome: that is, different traits do not occur in isolation, but as a suite of connected features. This means that if selection changes one trait, others will change at the same time. If, for instance, the survival of sedentary individuals increases due to milder winters in the breeding areas, the frequency of non-migrants in the population will rise as a direct response to selection. At the same time, migratory individuals in that population would be expected gradually to delay their autumn departure from the breeding area and to shorten their migration distance as correlated responses (Pulido & Berthold 1998). This fits the facts that partial migrants often travel only short distances and, compared with other migrants, usually leave late and return early to their breeding areas. It also implies that the difference between facultative and obligate migration might be a question of degree, rather than type, involving genetical influence over a moveable threshold. Migratory syndromes may also include other features, such as morphology and physiology but, as far as I am aware, these aspects have not yet been investigated in this context (for further discussion of migratory syndromes see Dingle 2005).

On the basis mainly of the experimental results described above, Berthold (1999) proposed what he called a 'comprehensive theory' of migration control. The theory centres on the concept of obligate partial migration, in which individuals vary in their inherent migratory activity. This situation is clearly widespread (and possibly universal) in birds, enabling complete migratory or complete resident behaviour to be achieved or lost, depending on selection. That occasional migrants exist in basically sedentary populations has long been evident, as exemplified in recent analyses of British ring recoveries (Wernham *et al.* 2002). No clear-cut difference separated resident and migratory species, but different species showed a continuum of variation in the proportion of individuals that had made long directed moves. This accounts for many familiar aspects of bird movements: for why occasional individuals of normally migratory species (such as Barn Swallow *Hirundo rustica*) can be seen in breeding areas in winter, long after other individuals of their species have left; and why, even in the most sedentary of species, occasional ringed individuals are recovered in autumn or winter at long distances, all in a restricted direction south of their natal areas. In the Reed Bunting *Emberiza schoeniclus* and Stock Dove *Columba oenas*, for example, these long-distance birds provided less than 1% of all recoveries of British-ringed birds.

What is clear is that, once obligate partial migration has evolved, the whole range of behaviour from strictly resident to strictly migratory can develop in short-lived birds in much less than a human lifetime. If the propensity for resident or migratory behaviour exists in a population as a gradient rather than as a dichotomy (as research implies), selection to change from resident to migratory (or vice versa) could begin anywhere on the behavioural gradient, and is not necessarily dependent on the prior presence of a few migratory individuals in an otherwise resident population (or vice versa).

This theory of in-built flexibility contrasts with an earlier suggestion that migration might have evolved independently several times in birds by convergent evolution. If this were so, it could be controlled by different mechanisms in different types of birds. Because most of the relevant work so far has involved passerines, this earlier theory of multiple control mechanisms cannot yet be considered as having been invalidated, however unlikely it may seem. Moreover, all the known examples of variable or changing migratory behaviour, whether from wild or captive birds, refer to species that live in seasonal environments, where migration is the norm. Whether birds that have spent their entire evolutionary history in the relatively stable environment of lowland tropical rainforest could develop latitudinal migration so rapidly is much less certain, and provides an obvious opening for further research. Examples of Neotropical families or subfamilies which contain no known migratory species include the Piprinae, Pipromorphinae, Dendrocolaptinae, Formicariidae, Rhinocryptidae, Conophagidae and Thamnophilidae.

Another gap in our understanding concerns the extent to which differences in migratory characteristics (such as distance and direction) are reflected in the overall genetic differentiation between populations. The few studies undertaken so far suggest that differences in migration behaviour between populations occupying contiguous breeding ranges generally do not correlate with strong overall genetic differentiation, as reflected in microsatellite or mitochondrial DNA (for Blackcap *Sylvia atricapilla*, see Helbig 1994, Pérez-Tris *et al.* 2004; for Willow Warbler *Phylloscopus trochilus* see Bensch *et al.* 1999; for Prairie Warbler *Dendroica discolor* see Buerckle 1999; for Great Bustard *Otis tarda* see Pitra *et al.* 2000). Rather, changes in their migratory behaviour seem to result from selection on relatively few loci (Helbig 2003). This conclusion agrees with the findings that few genes may be involved in the expression of migratory traits (Helbig 1994, 1996), and that strong correlations exist between these traits (Pulido & Berthold 2003). It also agrees with the finding that some evolutionary changes in migratory behaviour can happen rapidly (within a few generations), and that such adaptations are population-specific rather than species-specific.

This discussion leads to the question of how many different aspects of migratory behaviour (or traits) there might be on which selection could act (see also Chapter 1). We have one apparent gradient in behaviour, as reflected perhaps in *Zugunruhe*, which determines the timing and distance of migration (from residents to short-distance and long-distance migrants). Associated with this is another gradient in patterns and levels of fuel deposition in preparation for different types of journeys. We have a third apparent gradient between mainly endogenous control (regular long-distance migrants) to mainly external control (facultative irruptive migrants). We have a fourth apparent gradient in breadth of directional preference (from no preference to strong and narrow preference in a particular direction). Fifth, we have an apparent binary response on whether the movement is one-way or return (but even this might be regarded as a graded response from no return, through partial return to full return, Chapter 15). These are the behavioural gradients on which selection can act. The discovery of correlations between different aspects of migration may tempt us to reduce these different aspects of variation to as small a number as possible. But given the enormous diversity of movement patterns found among birds, pooling aspects of behaviour in this way (from correlations in a small number of similar species) brings the risk of masking other kinds of variation. Each additional aspect of modification allows an extra 'degree of freedom', and hence finer adjustment of behaviour to environments and species needs. Acknowledgement of all aspects of variation may be necessary to account for the wide range of movement patterns found among modern birds.

SUMMARY

Migration is a product of natural selection, leading species in seasonal environments to adopt movement patterns that enable individuals to survive and breed most effectively. Autumn migration gives improved winter survival through providing access to greater food availability in winter quarters. Spring migration gives improved breeding success, through greater seasonal food availability in summer quarters. The roles (if any) of predation, parasitism and competition in the evolution of migration are less obvious. Migratory birds seem to possess no major adaptations that resident birds lack; they differ only in the degree of development and modification of particular features, whether concerned with orientation, physiology or morphology. They include an internal clock and other timing mechanisms, the ability to deposit extra fat for the journeys, and to orientate and navigate over long distances, west–east as well as north–south.

Some bird populations are obligate migrants and others are obligate residents. Yet others are partial migrants in which some individuals stay year-round in their breeding areas while others migrate elsewhere. Partial migration can apparently arise in two ways, either with a mixture of obligate migrants and obligate residents in the same population, or with the entire (or part of the) population consisting of facultative migrants in which migration is optional, with individuals varying their behaviour according to conditions at the time. However, because migratory behaviour shows continuous, rather than dichotomous variation, both obligate and facultative phenotypes may also exist in the same population.

Breeding experiments on captive birds have shown that various aspects of migratory behaviour, such as the amount, timing and duration of migratory activity, along with directional preferences, are under genetic control, and can be altered by selection to give substantial changes within a few generations. Such changes may be facilitated by genetic correlations between some traits, so that selection for one trait can alter others at the same time. In particular, the occurrence of migratory behaviour is probably controlled by the same genes that control migration distance, operating via the amount of migratory activity. Directions are evidently controlled independently, and individual variation in directions is less in long-distance than short-distance migrants.

Since 1960, a natural change has occurred in the migratory behaviour of Blackcaps *Sylvia atricapilla*: some central European birds have changed their direction of migration, and now travel west-northwest to winter in southern Britain (the local British breeding birds continuing to migrate southwest to Iberia). Breeding experiments have confirmed that this new migratory behaviour is inherited. Further evidence for the genetic control of migration has come from studies of the year-to-year consistency (repeatability) in the migratory behaviour of individuals, and from the resemblance in behaviour between genetically related individuals (heritability), both in captive and in wild birds. Most migratory traits that have been examined show moderate to high heritability.

APPENDIX 20.1   PITFALLS IN MEASURES OF THE COSTS AND BENEFITS OF MIGRATION

Cost–benefit analyses of migration are best conducted on individuals from within the same breeding population, some of which remain year-round while others migrate. For each type, measures are needed of: (1) mortality over winter, which in a genetically controlled system may be expected to be lowest in the migrants wintering in a milder area; (2) mortality over the migration periods, which is expected to be greatest in the migrants because of the risks involved; and (3) reproductive success, which is expected to be similar in both groups or greater in the residents, which might acquire the best territories and start nesting earlier than the migrants. If the long-term net outcome of these three processes favours residents, in a genetically controlled system the population should change towards becoming totally resident, but if the long-term outcome favours migrants, the population should change towards becoming totally migratory. Genetically controlled partial migration would be expected to persist if both types, while perhaps varying in success from year to year, had equal lifetime reproductive success in the long term.

A different situation would be expected in facultative partial migrants, in which the same individuals might migrate in some circumstances and not in others. In these populations, as described above, dominants are most likely to remain in breeding areas, and subordinates to leave. This is because the dominants could prevail in competition for winter territories, food or other resources. Hence, any difference in mortality or reproductive success between the two groups could result from dominance status or 'quality' effects, and not from any genetic differences in migratory behaviour, so would not affect the movement patterns of the population in the longer term. The crucial question here in any one year concerns the optimal strategy for the individual, which would be to stay in some circumstances and to leave in others.

These various considerations indicate that, in cost–benefit analyses of partial migration based on measurements of mortality and breeding success, different outcomes would be expected in obligate and facultative extremes. In a facultative system in which migration was driven by dominance, say, the residents would usually be expected to perform best in all respects, as the migrants are in effect refugees making the best of their misfortune, as in the Robins *Erithacus rubecula* discussed earlier.

Some studies on the relative merits of migratory versus resident behaviour have involved comparisons of survival and reproductive rates between resident and migratory populations of the same species. In general, such studies are of little value in this respect, mainly because populations (breeding in different areas) differ in survival and reproduction for reasons other than migration (for such comparisons see Harrington *et al.* 1988, Nichols & Johnson 1990, Hestbeck *et al.* 1992, Mönkkönen 1992). Within species, migratory habits tend to increase with latitude (Chapter 13). In single-brooded species, brood sizes also tend to increase with latitude, so if populations are to remain stable in the long term, average annual mortality must also increase with latitude (for Blue Tit *Parus caeruleus*, see Snow 1956), in parallel with the migratory habit. In multi-brooded species, in contrast, the number of broods raised per year (and hence the total number of young raised per year) may decline with increasing latitude, in line with decline in the duration of the favourable season. So in these species, if populations are to remain stable, annual mortality must also decline with latitude. The salient point is that latitudinal gradients in reproductive and mortality rates would be expected in many widespread species, regardless of whether they are resident or migratory, or of the length of their migratory journey. In any population that remains numerically stable (with no long-term upward or downward trend), whether migrant or resident, average mortality rate must balance average reproductive rate, whatever these rates happen to be.

Nor can one reliably assess the costs and benefits of migration by comparing the breeding and mortality rates of resident and migratory species nesting in the same area. This is because, over a wide range of latitude, most resident species (because of their diets) can try to breed over a longer period each year than can closely related migrant species in the same area (Chapter 14). Residents would therefore be expected to produce more broods per year and, in stable populations, have correspondingly higher annual mortality rates than closely related migratory species breeding in the same area, an expectation that more or less fits the facts (von Haartman 1968, Greenberg 1980, O'Connor 1986, Mönkkönen 1992). Differences in reproduction and mortality between the two groups are not consequences of their migratory or resident behaviour, but of the different seasonal patterns of their particular food supplies, which influence both their migratory behaviour and their reproductive and mortality rates.

Nor can the costs and benefits of migration be assessed from comparisons of the annual energy budgets. Because migrants normally overwinter in warmer climates than occur in winter in their breeding areas, they save maintenance energy by migrating (Greenberg 1986). Even when the energy costs of the return journey are added in, the overall energy budget for the whole non-breeding season could often be lower than it would have been if the birds had overwintered in their breeding areas. Again, however, this fact has not necessarily had any influence on the evolution of migration. For it is not the overall energy budget that matters, but the ease of getting the daily needs in the conditions prevailing. There may be no cost in acquiring a large daily intake, if food is sufficiently plentiful within the area concerned. Whatever the overall requirement, it is an inability to get enough on a day-to-day basis that leads to mortality or reduced reproduction, and exerts the selection pressure for or against migration. In general, therefore, cost–benefit analyses of migration should be treated with caution and, in facultative migrants, individuals that leave breeding areas may reasonably be expected for the same reason to show lower survival and reproduction, on average, than residents.

Pied Flycatcher *Ficedula hypoleuca*, a species adversely affected in some regions by climate warming

Chapter 21
Recent changes in bird migrations

The far-reaching works of man in altering the natural conditions of the earth's surface can so change the environment necessary for the well being of the birds as to bring about changes in their yearly travels. (Frederick C. Lincoln 1935.)

We saw in the previous chapter how the migratory habits of birds can be rapidly altered under the influence of selection, continually shaped and re-shaped in response to changing conditions. Change in the migratory behaviour of wild birds has attracted attention in recent years as a result of growing interest in the effects of climate change. If weather has become warmer, as it has over much of the world, one might expect birds to have responded accordingly, with migratory species wintering at higher latitudes than previously, or arriving earlier and departing later from their breeding areas. This gives special interest in the measurement of changes in migratory behaviour, derived from long-term observations or ring recoveries.

A second aspect of interest concerns the basis of any changes observed. Many have assumed that recorded changes in migratory behaviour are likely to be genetically controlled, providing examples of 'evolution in action'. Genetically-based changes are not hard to imagine. Consider a partially migratory species in which some individuals have a genetic propensity to migrate and others not. During a series of severe winters, the migrants would survive better than the residents. As a result, the offspring of the migratory genotypes would comprise an ever greater proportion of each succeeding generation, gradually changing the genetic composition and average migratory habits of the population. Conversely, during a series of mild winters, the resident genotypes, able to occupy the best territories and start nesting early, could come rapidly to outbreed and outnumber the migrants. However, many of the changes observed in bird migratory behaviour need entail no genetic change, for in every aspect of migration there is scope for individual flexibility, through which individuals can adjust their migratory behaviour to some extent according to prevailing conditions (facultative variation). For example, the same birds might arrive on their breeding areas earlier in warm springs than in cold ones, or they might migrate further in cold winters than in mild ones, in response to differing food supplies. Hence, as climate changes from year to year, or over longer periods, birds have considerable scope for adjusting their behaviour to match these changes, without the need for any modification in the genetic control mechanisms.

Many of the changes in migratory behaviour witnessed in recent decades, which parallel changes in climate or food supplies, could therefore be facultative in nature, lying within the pre-existing range of response behaviour, and requiring no genetic change. Without special study, on a case-by-case basis, there is no way of telling to what extent any of the observed changes in migratory behaviour have resulted from the action of natural selection on the gene pool of the population. For the most part, we can only measure the extent of changes, and their association with environmental change. The following sections illustrate the different types of change observed in bird populations in recent decades, with various examples given in **Table 21.1**.

MIGRATORY TO SEDENTARY

Over a wide range of latitudes, many bird populations have become more sedentary over recent decades. Prior to 1940, the Lesser Black-backed Gull *Larus fuscus* was almost entirely migratory in Britain, only a few individuals remaining year-round. But nowadays large numbers of all age groups stay for the winter, feeding mainly on refuse dumps which have increased the winter food supply (Hickling 1984). A similar change has occurred among Herring Gulls *Larus argentatus* in Denmark (Petersen 1984). Another example is the Eurasian Blackbird *Turdus merula*, in which the British and mid-European populations have become progressively more sedentary during the last two centuries, as winters have mellowed (Berthold 1993, Main 2000). In both Europe and North America, many seed-eaters are now wintering further north in their breeding range, in association with the provision of suitable food at garden feeders (**Table 21.1**). Among many other short-distance and medium-distance migrants, increasing numbers of individuals now winter in areas where they once were wholly migratory, these species developing into typical partial migrants. Some such changes could be genetic in nature, others facultative. Their net effect is to expand the winter avifauna of many high-latitude regions.

Table 21.1 Examples of recent changes in migration patterns. Excludes changes in timing of migration described in the text.

A. Shortening of autumn migration to winter nearer breeding areas

Cape Gannet *Morus capensis*	Early ring recoveries of South African ringed birds from West Africa, but no recent recoveries, perhaps because fisheries discards in the Benguela upwelling region now allow them to stay south, nearer their breeding areas	Oatley (1988)
Great Cormorant *Phalacrocorax carbo*	Shortening of migrations in the Baltic Sea region	Schmidt (1989)
Bewick's Swan *Cygnus columbianus bewickii*	Rare in Poland until 1960s. Now regular migrant and winter visitor	Tomiałojc (1990)
Greylag Goose *Anser anser*	Wintering grounds changed from Spain to the Netherlands, a previous region of stopover. Similar shortening of migration route in eastern Germany	Nilsson & Persson (1993), Rutschke (1990)
Bean Goose *Anser fabalis*	In the nineteenth century large numbers wintered in Britain, now mainly in Germany–Netherlands	
Greater White-fronted Goose *Anser albifrons*	Gradual shortening of migration of north European and west Siberian birds to give greater proportions wintering in continental areas and smaller proportions in Britain	Stroud et al., in Wernham et al. (2002)
Snow Goose *Chen caerulescens*	Historically wintered primarily in Louisiana, Texas and Mexico. Many now winter further north in rice-growing areas of Iowa, Nebraska, Missouri and Arkansas	Cooke et al. (1995)
Canada Goose *Branta canadensis*	Formerly wintered mainly in the southern tier of States, now winters mainly in the middle tier of States	Hestbeck et al. (1991)
Red-breasted Goose *Branta ruficollis* Red-crested Pochard *Netta rufina*	Used to winter in Egypt (illustrated in tombs), but now confined to more northern latitudes Increasing proportion of population wintering in central, as opposed to southwest, Europe	Houlihan (1986), Kear (1990) Keller (2000)
White-tailed Sea Eagle *Haliaeetus albicilla* Sharp-shinned Hawk *Accipiter striatus*	Swedish birds used to winter in southeast Europe but, after winter food supplied, now stay in Sweden Now remaining in northeastern breeding areas in North America, and declining at migration sites further south, associated with increased garden bird feeding and more northern wintering of prey species	Helander (1985) Viverette et al. (1996)

(continued)

Table 21.1 (Continued)

Species	Description	Reference
Various waterfowl	Number of species wintering in Lithuania increased from 17 in the 1940s to 42 in the 1990s, and total numbers increased from several thousands in the 1930s to 150000 in the 1990s. This was due largely to reduction in migration distance (shown by ringing), with smaller proportions reaching western Europe. It was especially obvious in Mallard *Anas platyrhynchos* and Mute Swan *Cygnus olor*	Švažas (2001)
Merlin *Falco columbarius*	Northern boundary of winter range extended northward into the Canadian prairies	James et al. (1987)
Great White Egret *Egretta alba*	Increasing proportion wintering in mid-latitudes of western Europe	Marion et al. (2000)
Black-crowned Night Heron *Nycticorax nycticorax*	Increasing proportion wintering in southern Europe and North Africa as opposed to sub-Saharan Africa	Pineau (2000)
Squacco Heron *Ardeola ralloides*	Increasing proportion wintering in southern Europe and North Africa, as opposed to sub-Saharan Africa	Hafner (2000)
Common Crane *Grus grus*	West European population formerly wintered in Spain–Morocco, now winters mainly in France–Spain and eastern Germany. Following construction of the dam at Lac du Der (Champagne, northern France) cranes started to use this site as a major stopover, and also as a wintering area, thus shortening their migration by about 1500km each way	Alonso et al. (1991) and others
White Stork *Ciconia ciconia*	Formerly wintered entirely in Africa, thousands are now wintering regularly in the Mediterranean region, notably Spain, Bulgaria and Israel. Progressive reduction in migration distance of birds breeding in eastern Europe during the 1950s–1980s as shown by winter ring recoveries in Africa	Berthold (1996), Fiedler (2003)
Lesser Black-backed Gull *Larus fuscus*	Winters further north in Europe. Was a summer visitor to northern Europe but over last 40 years changed, so that large numbers now winter	Hickling (1984)
Eurasian Chiffchaff *Phylloscopus collybita*	Increasing proportion winter in Britain	Lack (1986)
Great Reed Warbler *Acrocephalus arundinaceus*	Formerly wintered entirely in Africa, south of the Sahara. Some now wintering in Spain	de la Puente et al. (1997)
Bluethroat *Luscinia svecica*	Increasing proportion wintering in Spain as opposed to sub-Saharan Africa	Bermejo & de la Puente (2004)

Species	Description	Reference
Yellowhammer *Emberiza citrinella*	Increasing proportion winter in Finland	Väisänen & Hildén (1993)
Eurasian Bullfinch *Pyrrhula pyrrhula*	Increasing proportion winter in Finland	Väisänen & Hildén (1993)
European Greenfinch *Carduelis chloris*	Increasing proportion winter in Finland and Sweden	Väisänen & Hildén (1993)
Evening Grosbeak *Hesperiphona vespertina*	Winters much further north than formerly in response to garden bird feeding	Root (1989)
Hooded Crow *Corvus corone cornix*	Much reduced numbers wintering in eastern England now than in the nineteenth century, attributed to greater proportion of European birds wintering further north and east, confirmed by ringing	O'Donoghue, in Wernham et al. (2002)

B. Shortening of spring migration to breed nearer wintering areas

Species	Description	Reference
Whooper Swan *Cygnus cygnus*	Increasing numbers now breed in Poland, about 800 km southwest of main breeding range of that population	Tomiałojc & Stawarczyk (2003)
Barnacle Geese *Branta leucopsis*	Breeding population established on Gotland and other sites around the Baltic and in the Netherlands, thus shortening migration to Novya Zemlya by 1300 km, or more. Birds from a different population have also started to breed in Iceland, thus shortening migration to Greenland by more than 500 km	Larsson et al. (1988), Forslund & Larsson (1991)

C. Change in direction to establish new wintering area

Species	Description	Reference
Greater White-fronted Goose *Anser albifrons*	Change in wintering sites from central Europe to the Netherlands and western Germany associated with a directional change of 15° to give a new route along the Baltic coast	Švažas et al. (2001)
Black Stork *Ciconia nigra*	Re-colonised west Europe in 1970s–1980s probably from eastern populations. These birds started migrating southwest through Gibraltar, whereas eastern European populations migrate southeast through the Bosphorus	G. Neve, in Sutherland (1998)
White Stork *Ciconia ciconia*	The descendents of birds that started breeding in South Africa in 1933 have been shown by satellite-tracking to migrate northward for about 3000 km in the non-breeding season	Underhill (2001)
Little Egret *Egretta garzetta*	Southern European birds migrate south, often across the Sahara. Many now move northwest to winter in northern France and southern Britain. Most of these birds are now resident in these areas	Marion et al. (2000)

(continued)

Table 21.1 (Continued)

Blackcap *Sylvia atricapilla*	Some central European birds switched from migrating southwest to winter in western Mediterranean area to migrating north or northwest to winter in Britain	Berthold *et al.* (1992a)
Shore (Horned) Lark *Eremophila alpestris*	Colonised northern Europe in the mid-nineteenth century (1847) and then some thought to have started migrating southwestwards to winter around North Sea	Gätke (1895), Glutz von Blotzheim & Bauer (1985)

D. Change from migratory to resident population

Great Crested Grebe *Podiceps cristatus*	Increasing proportion in the Netherlands has become resident	Adriaensen *et al.* (1993)
Mute Swan *Cygnus olor*	From the first record in 1955, up to 2000 birds now winter near their breeding areas in Lithuania	Švažas *et al.* (2001)
Canada Goose *Branta canadensis*	Birds from migratory North American populations introduced to Britain have become resident, apart from newly developed moult migration	Wernham *et al.* (2002)
Red Kite *Milvus milvus*	Increasing proportion in Scandinavia has become resident	Kjellén (1992)
Eurasian Blackbird *Turdus merula*	Formerly wholly or mainly migratory in mid-latitudes of western Europe, now mainly resident	Berthold (1993, 1999), Main (2000)
Eurasian Chiffchaff *Phylloscopus collybita*	First recorded in 1846, the numbers wintering in Britain have increased greatly since 1940. They include some local birds and some migrants from Europe which may indicate a change in direction	Green, in Wernham *et al.* (2002)
Great Tit *Parus major*	Once-migratory population in the Finnish city of Oula near the Arctic Circle is now resident, following food provision by householders	Orell & Ojanen (1979)
Common Starling *Sturnus vulgaris*	Central European populations have become partially resident, with increasing numbers staying in towns in winter	Merkel & Merkel (1983), Berthold (1993)

E. Change from resident to migratory population

Canada Goose *Branta canadensis*	Birds introduced to Scandinavia winter in increasing numbers on north Polish coast	Polish ringing scheme
Cattle Egret *Bubulcus ibis*	Migration system has arisen in North America, and between Australia and New Zealand (2000 km)	Root (1989), Maddock & Geering (1994)
Common Starling *Sturnus vulgaris*	Birds introduced into North America from British resident population have become migratory over much of continent	Kessel (1953), Dolbeer (1982)

House Finch *Carpodacus mexicanus*	Resident population introduced to eastern North America became migratory in less than 30 years, with different migration directions (south–southwest) from west to east across the breeding range	Able & Belthoff (1998)
European Serin *Serinus serinus*	As spread north in Europe, switched from being resident to migratory	Berthold (1999)

F. Establishment of breeding population in former wintering range

Leach's Storm Petrel *Oceanodroma leucorhoa*	Now breeds on islands off South Africa	Harrison et al. (1997)
Manx Shearwater *Puffinus puffinus*	Now breeds on west side of Atlantic, in former migration area	Storey & Lien (1985)
White Stork *Ciconia ciconia*	Started breeding in South Africa in the 1933. Now regular. Winters in the tropics of Zaire and Rwanda	Harrison et al. (1997)
European Bee-eater *Merops apiaster*	Thousands now breed in South Africa	Harrison et al. (1997)

G. Lengthening of autumn migration

Great White Pelican *Pelecanus onocrotalus*	Shifted wintering sites following development in the Nile delta in Egypt to further south	Crivelli et al. (1991)
Pink-footed Goose *Anser brachyrhynchus*	Population has shifted in Britain to wintering in large numbers in southeast England, associated with decline in food supply further north	Gill et al. (1997)
Red-breasted Goose *Branta ruficollis*	Shifted wintering grounds from Azerbaijan to Romania and Bulgaria following habitat changes	Vangelewe & Stassin (1991)
Chaffinch *Fringilla coelebs*	Vast predominance of females in flocks in southern England in late eighteenth century changed to more equal ratio in late twentieth century	Wernham et al. (2002)

Based partly on Sutherland (1998).

SEDENTARY TO MIGRATORY

Examples of changes from sedentary to migratory behaviour are somewhat less common, and are generally associated with an extension of breeding range into higher latitudes. For example, the European Serin *Serinus serinus* was once restricted to the south of Europe where it is resident, but in the early twentieth century it spread north, where it became migratory. In more recent years, with milder winters, this migratory population has become partially resident (Berthold 1999). Likewise, since the nineteenth century, many bird species have spread north in Fennoscandia, including the Northern Lapwing *Vanellus vanellus*, Starling *Sturnus vulgaris*, Eurasian Blackbird *Turdus merula* and Dunnock *Prunella modularis*. In their newly colonised breeding areas they are essentially migratory, whereas further south they are partial migrants or residents (Schüz *et al*. 1971).

Common Starlings *Sturnus vulgaris* introduced to North America at the end of the nineteenth century supposedly came from resident British stock. They were initially sedentary in the eastern USA but, in connection with range expansion, different proportions of migrants appeared in different regions (Kessel 1953, Dolbeer 1982). Similarly, as Cattle Egrets *Bubulcus ibis* spread north through North America, they remained resident in Florida and other southern parts, but became migratory further north (Root 1989, Maddock & Geering 1994). During the nineteenth century, Snowy Egrets *Egretta thula* were eliminated by human persecution from the northern parts of their breeding range in North America where they were migratory, and survived only in the southern parts where they were resident. As the remnant resident population has recovered in recent decades, the birds have spread northwards, where they have become migratory in the newly colonised areas (Rappole 2005). Severe range contractions are likely to have eliminated the migratory or resident sectors from other species in which both types of behaviour were once represented, one or other type of behaviour being regained if the remnant population expands to re-occupy its former range.

It is not only the migrations themselves that can change, but also the sex and age differences within populations. The House Finch *Carpodacus mexicanus* was introduced from California, where it is resident, to eastern North America. Within less than 30 years in its new home, as numbers grew, greater proportions of individuals migrated; they also migrated progressively further, and females further than males (Belthoff & Gauthreaux 1991, Able & Belthoff 1998). Moreover, the birds in western parts of this eastern distribution now show more north–south migration directions than the more eastern birds which migrate northeast–southwest (Brewer *et al*. 2000). These changes are almost certainly based partly on genetic changes, evolved under the action of natural selection.

SHORTENING OF MIGRATION ROUTES

So called short-stopping has occurred in many species as more food has become available at higher latitudes in the wintering range, either through human activities or climate change. Several populations of Canada Geese *Branta canadensis* in North America have responded in this way to agricultural changes or to the creation of refuges where food was provided (e.g. Terborgh 1989, Hestbeck *et al*. 1991), as did

Greylag Geese *Anser anser* and Common Cranes *Grus grus* in Europe (Rutschke 1990, Alonso *et al.* 1991). Other species of waterfowl have shortened their migrations, apparently in response to warmer winters, as open water has become available nearer the breeding areas. This is manifest by increased numbers wintering in northern and eastern parts of Europe, and declining numbers of the same species wintering in the south and west. Yet other waterfowl have shortened their migrations in apparent response to reduced disturbance and predation, as sanctuaries have been established in areas previously open to hunting (Chapter 27).

Shortened migrations in many other species are reflected in the changing distributions of ring recoveries. Among 30 species of short-distance or partial migrants breeding in Germany, a tendency towards wintering at higher latitudes was found in 10 species, and at lower latitudes in three species, although ringing recoveries are affected by changes in human land use and hunting, as well as in climate (Fiedler *et al.* 2004). More and more European migrants that formerly wintered entirely in tropical and southern Africa are now also wintering in small but increasing numbers in the Mediterranean region. Examples include the Yellow Wagtail *Motacilla flava*, House Martin *Delichon urbica*, Osprey *Pandion haliaetus*, Lesser Kestrel *Falco naumanni* and White Stork *Ciconia ciconia* (**Table 21.1**; Berthold 2001).

In some regions irruptive migrations have become less frequent than formerly, presumably because the birds have become less numerous, or more often remain in their breeding areas year-round or migrate less far. Comparing the nineteenth with the twentieth century, the Pine Grosbeak *Pinicola enucleator* became a much less frequent visitor to the middle latitudes of Europe. No noticeable invasions of Scandinavian Great Tits *Parus major* and Blue Tits *Parus caeruleus* to Britain have occurred since 1977, and no big invasions of Great Spotted Woodpeckers *Dendrocopos major* since 1974. In Germany, invasions of Blue Tits, Waxwings *Bombicilla garrulus* and Redpolls *Carduelis flammea* have also become less frequent (Fiedler 2003). On the other hand, Two-barred Crossbills *Loxia leucoptera* have appeared in Fennoscandia in increasing numbers and frequency, possibly associated with the increased planting of larch *Larix* species outside their natural range. Likewise, in eastern North America, Evening Grosbeaks *Hesperiphona vespertina* have become less numerous, and their invasions less frequent, than previously. This may be associated with reduced outbreaks of Spruce Budworm *Choristoneura fumiferana*, a favoured summer food, and with increased winter bird feeding by householders.

Some Whooper Swans *Cygnus cygnus* and Barnacle Geese *Branta leucopsis* have shortened their migrations in another way, by establishing nesting populations hundreds of kilometres south of their historic breeding range, nearer to wintering areas (**Table 21.1**). This type of change is more difficult to explain, but may in some cases be due to suitable habitat being created by human activities, to expansion into former range from which the species was eliminated by humans in the past, or simply to unprecedented population increase leading to an expansion of breeding range into lower latitudes.

Such changes often occurred within the lifetimes of individual birds, some of which substantially changed their breeding or wintering areas from one year to the next, so they presumably represented facultative responses to prevailing conditions, rather than genetically driven changes (although genetic change may follow).

Other types of change have also occurred. For example, like many other birds that do not start to breed until they are two or more years old, young White Storks *Ciconia ciconia* remain in 'winter quarters' through their first summer, or migrate only part way towards breeding areas (Chapter 15). In recent decades, second-summer birds, whose predecessors used to remain in Africa, have returned in increasing numbers to southern Europe to pass the summer. The mean distance of recoveries of second-summer birds from their natal sites in north Germany was 2517 km in 1923–1975 ($N=120$), reducing to 720 km in 1978–1996 (Fiedler 2001).

LENGTHENING OF MIGRATION ROUTES

In species that have expanded their breeding areas to higher latitudes yet have retained the same wintering areas, some extension of migration routes is inevitable. Eurasian examples include: (1) Black-winged Stilt *Himantopus himantopus*, which is expanding its breeding range northward (France, Ukraine, Russia) but still winters south of 40°N latitude; (2) European Bee-eater *Merops apiaster*, which has expanded northwards in almost all central European countries, yet still winters entirely in Africa south of the Sahara; (3) Citrine Wagtail *Motacilla citreola*, which is expanding its breeding range from Asia westward into Europe, but still winters in India and Southeast Asia (Fiedler 2003). The intra-European routes have increased by up to 1000 km. These examples represent the sort of changes that must have occurred in many species after each glaciation, when ice receded, and plants and animals spread from lower to higher latitudes (Chapter 22).

In the Red-breasted Goose *Branta ruficollis*, most individuals now winter in Romania–Bulgaria, some 300–600 km further from their breeding areas than in the 1950s, as former wintering sites in Azerbaijan have been altered by land use changes (Sutherland & Crockford 1993). In even earlier times, the species was found in winter yet further from its breeding areas, being depicted in the art of ancient Egypt (Houlihan 1986). So over recorded history, this species has both shortened and lengthened its migration routes. Such changes in the length of migrations could initially involve only facultative responses to local conditions but, as migrations lengthen over time, some genetic change seems likely, as they would require changes to the regulatory mechanisms.

In some other species, greater proportions of ring recoveries are now being obtained from the distant parts of migration routes than formerly, but it is hard to tell whether this is due to altered migration behaviour, or to changed recovery chances along the routes (Fiedler *et al.* 2004). In particular, over recent decades hunting has declined much more in the northern and mid-latitudes of Europe than further south. This could affect the migratory behaviour of hunted species, or the distribution of their ring recoveries.

CHANGES IN MIGRATORY DIRECTIONS

In addition to the Blackcap *Sylvia atricapilla* mentioned in the previous chapter, changes in the direction of migration, leading to the adoption of new wintering areas, were recorded in several species in the last century (**Table 21.1**). For example, Little Egrets *Egretta garzetta* breeding in southern France migrated southward, some crossing the Sahara to winter in the Afrotropics. But from the 1970s, increasing numbers began to migrate northwest to winter in northern France, southern Britain and Ireland (Marion *et al.* 2000). Some later became resident in these areas, and from the 1990s started to breed there. Similarly, Lesser Black-backed Gulls *Larus fuscus* from Europe have begun increasingly to winter on the coasts of eastern North America, with records from Nova Scotia to Florida, a change which requires a much stronger westerly component in the directional preferences. Almost certainly, such marked directional changes have involved genetic changes, as confirmed for the Blackcap by breeding and direction-testing in captivity (Chapter 20).

A different type of change is shown by those northern hemisphere species introduced to the southern hemisphere, which have reversed the direction of their spring and autumn journeys, as appropriate, so that they continue to winter in lower rather than in higher latitudes. This is true, for example, of the European Goldfinch *Carduelis carduelis* and others introduced from Europe to New Zealand in the nineteenth century, and also for the White Stork *Ciconia ciconia*, which is thought to have colonised South Africa naturally in the 1930s, and now migrates north to winter in equatorial Africa (Underhill 2001).

MIGRATION TIMING

Studies of long-term trends in arrival times of birds are mostly based on dates of first sightings, as it is these dates that are most frequently recorded, in some European localities for periods exceeding 300 years (Lehikoinen *et al.* 2004). The problem with first arrival dates is that many refer to only single individuals, which may not be typical. They also depend on the number of observers, which in many regions has increased in recent decades, raising the likelihood of extreme early birds being detected. The median or mean arrival dates of populations of individuals in their breeding areas are more representative, but have been recorded much less often, and chiefly in recent decades. They depend on detailed studies in particular areas, where the arrival of each individual (or occupant of each territory) was recorded as it occurred. Bird observatories provide a third source of migration dates, where observations or trapping were maintained throughout the migration seasons each year, enabling median or mean passage dates (and standard deviations) to be calculated. Another approach has been to combine the records from different bird observatories in the same region and calculate mean values (see **Table 21.2** for examples). Whereas arrival (or departure) dates refer to birds from a single population breeding at a particular locality, passage dates usually refer to birds from a much wider area, counted at a point on their migration. Some studies have compared first and median or mean passage dates from the same site over a period of years, and found the various dates to be correlated (Hüppop & Hüppop 2003, Jenni & Kéri 2003, Vahatalo *et al.* 2004, Sparks *et al.* 2005). In years that were early, the total arrival period became more prolonged. But whatever the recording process, long-term counts from all these sources reveal similar general long-term trends in migration timing.

Table 21.2   Mean first arrival dates recorded at seven British bird observatories during 1970–1996, summarised by decade

	1970–79	1980–89	1990–96	Significance of trend
Chiffchaff *Phylloscopus colybita*	21 March	17 March	11 March	*P*<0.001
Swallow *Hirundo rustica*	6 April	3 April	31 March	*P* <0.01
Willow Warbler*Phylloscopus trochilus*	7 April	2 April	30 March	*P*<0.001
Blackcap *Sylvia atricapilla*	15 April	10 April	3 April	*P*<0.001
Swift Apus apus	1 May	28 April	26 April	*P*<0.01
Garden Warbler *Sylvia borin*	4 May	28 April	20 April	*P*=0.06

Changes through time were examined by regression of annual mean date on year. From Sparks *et al.* (1999).

Under the presumed influence of long-term climate warming, many birds now arrive in their breeding areas earlier in spring and depart later in autumn than in the past, spending from a few days to a few weeks longer per year in their summer quarters. Such changes have become apparent in a wide range of species at many localities in both Eurasia and North America.[1] However, not all species have shown such changes. Some of the exceptions may result from observational deficiencies, for they refer to species that have undergone marked changes in status over the period concerned: a decline in numbers makes it increasingly difficult to detect the earliest arrivals and latest departures, where first and last dates were the measures recorded. Other exceptions are inexplicable, and may be due to changes in conditions that are independent of temperature, such as some types of food supplies.

Spring dates

From 983 Eurasian bird populations in which first arrival dates in breeding areas were monitored over periods of years, 59% showing no significant change, but 39% had become significantly earlier, while only 2% had become significantly later (Lehikoinen *et al.* 2004). Both short-distance and long-distance migrants showed the same trends. From 222 populations for which mean passage dates could be calculated over periods of years, 69% showed no change, while 26% had become significantly earlier, and only 5% had become significantly later. The average change of first arrival date over all species and sites was −0.37 days per year, while the equivalent figure for mean passage dates was −0.10 days per year, both figures being statistically significant. It is not obvious why the two figures differed, but in general the mean migration dates were based on larger, more standardised data-sets, and were less susceptible to increase in observer numbers.

---

[1]*For examples, see Moritz (1993), Loxton & Sparks (1999), Vogel & Moritz (1995), Sparks (1999), Sparks & Mason (2001), Fiedler (2001), Inouye et al. (2000), Jenkins & Watson (2000), Sokolov (2001, 2006), Sokolov et al. (1998, 1999), Bairlein & Winkel (2001), Zalakevicius & Zalakeviciute (2001), Hüppop & Hüppop (2003), Tryjanowski et al. (2002), Bradley et al. (1999), Root et al. (2003), Lehikoinen et al. (2004), Vahatalo et al. (2004), Mills (2005), Stervander et al. (2005).*

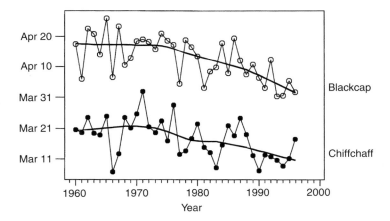

Figure 21.1   Mean first arrival dates of Common Chiffchaff *Phylloscopus collybita* and Blackcap *Sylvia atricapilla* at eight coastal bird observatories in Britain during 1960–1996. In both species the response to temperature is in the order of 2–3 days earlier per 1°C rise in mean spring temperature. From Sparks (2000).

Despite the long-term trends, arrival and migration dates still fluctuated from year to year in line with local temperatures (**Figures 14.5, 21.1** and **21.2**).[2] Typically, most birds arrived about 2.5–3.3 days earlier for every 1°C increase in spring temperature (based on 203 regression analyses for different Eurasian bird populations, Lehikoinen *et al.* 2004). The smaller number of studies available from North America revealed some similar trends (Bradley *et al.* 1999, Inouye *et al.* 2000, Butler 2003, Marra *et al.* 2005, Mills 2005, Murphy-Klaassen *et al.* 2005), although in some eastern parts of the continent, long-term temperature change has been less marked than in western Europe. In general, earlier arrival of migrants in spring leads to earlier breeding, as described as a recent trend in a wide range of species (Crick & Sparks 1999, Sokolov 2006), and earlier breeding often gives rise to better success at the population level (Chapter 14; Thingstand 1997, Sokolov 1999, 2000, Bairlein & Winkel 2001).

Despite strong correlations between arrival dates and temperature on the breeding area, much of the variance in arrival dates remains unaccounted for. Arrival dates are also influenced by weather further down the migration route

[2]*Most researchers have used annual temperatures from localities on the migration route or breeding area, while some have used the winter–spring index of the North Atlantic Oscillation (NAO), a large-scale climate phenomenon influencing weather in this region (e.g. Hüppop & Hüppop 2003, Vahatalo et al. 2004, Stervander et al. 2005, Sokolov 2006, Zalekevicius et al. 2006). It is calculated as a difference in normalised values of atmospheric pressure between the Azores and Iceland for each month. Positive values indicate warmer and wetter winter–spring weather (followed by earlier spring migration) in northwest Europe and the opposite weather conditions and later arrival dates than usual in southern Europe.*

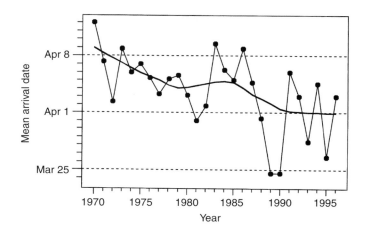

Figure 21.2   Mean first arrival dates of Barn Swallows *Hirundo rustica* averaged for seven bird observatories in Britain, 1970–1997, during which time mean arrival dates advanced by about one week in association with gradual spring warming. From Sparks *et al.* (1999).

or in wintering areas (Sokolov 2006), as well as by other aspects of weather, such as wind, and by different factors such as food supply, as yet largely unstudied in this context. Moreover, poor weather on one part of the route can hold up migration, even though conditions may be favourable further along the route. The variations in response between species, recorded in every relevant study, could well have been partly related to diet, and the dates that their different foods become available, but further investigation of species variation is needed. Moreover, not all studies have shown a single sustained long-term trend in migration dates. For example, at the Rybachy Bird Observatory on the Courish Spit in the southeastern Baltic, warmings in the 1930s and 1940s, and then in the 1960s and 1980s, were associated with significantly earlier spring migration in many species of passerines, while colder periods during the 1950s and 1970s were associated with later passage (Sokolov *et al.* 1998).

In comparing the changes that have occurred in the spring migration dates of different species, several fairly general patterns have emerged:

- Greater changes have occurred in the migration dates of early-migrating species than of later migrating ones (**Figure 21.3**). This is associated with weather (including temperature) being more variable earlier than later in the spring (for passage dates see Sokolov *et al.* 1998, for arrival dates at breeding sites see Slagsvold 1976, Loxton & Sparks 1999).
- Greater changes have occurred in the arrival dates of short-distance migrants than of long-distance migrants – presumably because short-distance migrants generally arrive earlier in spring (same point as above), and have more flexibility in their migration timings (Tryjanowski *et al.* 2002, Butler 2003).

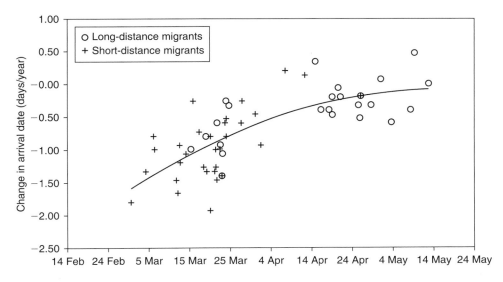

Figure 21.3   Rate of change in first arrival dates (days per year) of 56 bird species in Lithuania (Zuvintas Strict Nature Reserve) in 1966–1995, in relation to their average arrival dates. Regression (day 1 = 1 January): change = $-0.0003 \times day^2 + 0.071 \times day - 4.933$, $r^2 = 55.2\%$. Negative values indicate earlier arrival. From Lehikoinen *et al.* (2004, based on Zalakevicius & Zaleveciute (2001).

- Greater changes have occurred in the arrival dates of small bird species than of large ones. This is possibly because the smaller species are more sensitive to annual temperature differences and their effects on food supplies (although their shorter generation times would also favour more rapid genetic change than is possible in large, longer-lived species).
- Year-to-year arrival dates of short-distance migrants generally showed a correlation with spring temperatures in the breeding locality, but such correlations were less obvious in long-distance migrants (Tryjanowski *et al.* 2002). Moreover, where it has been explored, weather back down the route often showed a better relationship with arrival dates than did weather at the arrival location itself (e.g. arrival dates of Barn Swallows *Hirundo rustica* in Britain were better related to weather in France–Spain than to weather in Britain, Huin & Sparks 1998).
- Spring weather has not changed everywhere in the same way. Correspondingly, the degree of change in bird arrival dates in breeding areas has varied across Europe, with arrival dates in most areas getting earlier as springs have warmed, but in some areas getting later as springs have cooled. In the Mediterranean region, springs are now cooler than in the past, which may be another factor slowing the return of long-distance migrants from tropical Africa to the mid-and higher latitudes of Europe.
- Most species still arrive in breeding areas earlier in warm springs than in cool ones.

Three explanations may account for the fact that more short-distance migrants than long-distance migrants now arrive earlier in spring, and in closer correlation to temperatures on breeding areas. First, the stronger endogenous control of migration in long-distance migrants might inhibit a rapid reaction to a changing environment (Gwinner 1986, Berthold 1996). Short-distance migrants are typically more flexible (facultative) in their response, and more able to alter their behaviour in relation to prevailing conditions (Chapter 12). Second, the closer a species winters to its breeding areas, the more closely correlated are the day-to-day weather changes in the two areas, enabling short-distance migrants to react more rapidly and appropriately. Third, as mentioned above, weather is more variable early in the spring, when most short-distance migrants arrive in their breeding areas, than later in the spring, when most long-distance migrants arrive.

Because, in most bird species, males arrive in breeding areas before females (Chapter 15), studies of migration timing based on first arrival dates concern males only. But the two sexes may not necessarily respond in the same way to climate change. A long-term study of arrival dates of male and female Barn Swallows *Hirundo rustica* in Denmark revealed that only males responded to an amelioration of weather conditions during migration, while females did not (Møller 2004). Therefore, the sex difference in arrival dates increased as a consequence of climate change, giving no change in breeding date, because females arrived no earlier than they did 30 years previously.

Earlier arrival in breeding areas in response to warmer weather there could be brought about by birds: (a) increasing the speed of spring migration, (b) leaving wintering areas earlier but migrating at the same speed, (c) wintering closer to the breeding area and migrating at the same speed, or (d) a combination of these various possibilities. More rapid progress in warm than cold springs has been recorded in many migrants from the dates they pass through successive observation sites in different years. Only facultative responses could account for the year-to-year variation in arrival dates seen in many migrants, but this need not exclude the possibility of genetic change in response to longer-term environmental trends, such as climate warming (for evidence of genetic variation in timing of spring migration, see Chapter 20).

Autumn dates

Changes in autumn migration dates over recent decades have been generally less, and more variable, than changes in spring dates (Gatter, 2000, Bairlein & Winkel 2001, Fiedler 2001, Sparks & Mason 2001, Jenni & Kéri 2003, Lehikoinen *et al.* 2004, Sokolov 2006), apart from a study in southern Canada in which changes in autumn dates were more marked and more frequent (Mills 2005). Two patterns have emerged, involving either earlier or later departure over the years. In some single-brooded populations, earlier arrival is followed by earlier breeding and moult, so that birds are ready to depart earlier. In such populations, the timing of successive events through the summer, from arrival, egg-laying, hatching, fledging, moult and autumn migration, are correlated with spring temperatures, and show little or no relationship with the prevailing autumn temperature (Chapter 12). An earlier spring arrival pulls the whole cycle forward to give an earlier autumn departure (Ellegren 1990b, Sokolov *et al.* 1998, Sokolov 2000a, 2001, Bojarinova *et al.* 2002). At Rybachi on the southern Baltic coast, warming in the 1960s and 1980s led to significantly earlier mean dates in spring passage, breeding and autumn passage. Conversely, colder springs during the 1970s caused a shift towards later spring passage, breeding and autumn passage (Sokolov *et al.* 1999). These trends occurred in both short-distance and long-distance migrants. Most of the migrants through Rybachy came from breeding areas that gave time for only one brood. Similar relationships were found for single-brooded long-distance migrants passing through the Swiss Alps in autumn (Jenni & Kéri 2003). The long-distance migrants may have benefited in autumn from an earlier crossing of the Sahara before its seasonal dry period. In contrast, shorter-distance migrants passing over the Alps and wintering north of the Sahara mostly showed a later autumn passage. These are chiefly passerine species that can raise more than one brood per year, so could better take advantage of a longer season by remaining longer in their breeding areas. Further south and west in Europe, where individuals can make up to two or three breeding attempts in the same season, departure dates of passerines have tended to get later as local temperatures have risen (Bairlein & Winkel 2001, Sparks & Mason 2001), but it is not known whether this has been associated with a lengthening of the breeding season.

BREEDING IN WINTERING RANGE

Occasional individuals of some species have nested in their wintering range (through suspending their normal spring migration), leading in some cases to the establishment of new nesting populations. Examples include the North American Barn Swallows *Hirundo rustica* found nesting in Argentina, within the usual wintering range (Martinez 1983), and Eurasian White Storks *Ciconia ciconia*, Bee-eaters *Merops apiaster* and others now nesting in South Africa (Snow 1978, Harrison *et al.* 1997). Among seabirds, Leach's Storm Petrels *Oceanodroma leucorhoa*, which breed widely in the North Atlantic and Pacific, have been found breeding on an island off South Africa, and in potential nesting burrows on the Chatham Islands off New Zealand, within the wintering range of northern hemisphere birds (Imber & Lovegrove 1982, Whittington *et al.* 1999). This phenomenon of 'migration suspension' may have been going on for millions of years, considering the numbers of bird species that have conspecifics, or closely allied forms, breeding in equivalent habitat in the opposite hemisphere (Snow 1978, Newton 2003). It may result from occasional individuals becoming 'time-trapped' in winter quarters, switching under the influence of southern hemisphere daylengths to an annual cycle appropriate to the southern hemisphere (for an example in experimental conditions see Gwinner & Helm 2003).

DISCUSSION

Most of the studies cited in this chapter were concerned with particular species or suites of similar species, so it is hard to tell what proportion of an avifauna has changed in migratory habits in recent decades (apart from migratory timing which has been studied in many species). Over the past 50 years, climate changes have been more marked in some regions than in others, and in particular regions studies reporting changes in migratory behaviour were more likely to be published than those finding no change. However, among the bird species that breed in Britain, 73 provided enough ring recoveries from a sufficiently long period to look for changes in the lengths and directions of migrations. In total, 51 (70%) of these species showed no significant change in these respects during the twentieth century, but in 15 species movements had become shorter, in five they had become longer, while in two the movements had changed in other ways. The 22 species that showed changes were significantly more than the four expected on a significance level of 5%. They included passerines, raptors, waders, waterfowl and seabirds (G. Siriwardena & C. Wernham, in Wernham *et al.* 2002). Similarly, of 30 species that breed in Germany and provide enough ring recoveries, eight species showed decreasing mean recovery distances with time, while five species showed increasing mean recovery distances (Fiedler *et al.* 2004). Again the numbers that showed change were significantly greater than the two expected on a significance level of 5%. Such studies confirm that changes in the migration behaviour of birds have been common over the last several decades.

These various observations, along with the breeding experiments discussed in Chapter 20, all serve to confirm that migration is a dynamic phenomenon, subject to continual change in response to prevailing conditions. Some aspects, such as an abrupt change in the direction of migration, imply rapid evolutionary change, but other aspects could represent either genetic or facultative responses to changing conditions. In any case, in any marked long-term change, both are likely to be involved, the birds responding initially by facultative means, and eventually by genetic change, as natural selection comes to bear. Facultative responses are relatively limited (though variable in extent between species) and, if environmental conditions continue to change in the same direction, such responses become inadequate to deal with the new conditions. Only genetic change may then enable the population to respond appropriately to conditions beyond the previous range.

Genetic responses

Although all main aspects of migratory behaviour have been shown to have heritable components, mainly through artificial selection and cross-breeding in captivity (Chapter 20), genetic change is not easy to demonstrate in wild birds. The most convincing way is to test wild birds in standard conditions in captivity, but this requires compliant species and suitable facilities. The assumption is that, if individuals taken from the wild in different years or from different regions express behavioural differences when held under identical controlled conditions, these differences are likely to have a genetic basis. This conclusion is strengthened if the trend is maintained in captive-bred offspring from these birds, unaffected by parental effects or experience in the wild, because only genetic effects are maintained through the generations. Such a test has been made on samples of Blackcaps *Sylvia atricapilla* randomly collected as nestlings from south Germany and hand-raised each year over a 13-year period (Berthold 1998, Pulido & Berthold 2004). In successive samples of birds, the amount of autumn migratory activity was found to decline towards a later onset and reduced intensity (less activity per night). This was precisely the result expected if the population had responded genetically to ameliorating environmental conditions, so at least in this species later departure and shorter migration may partly represent a genetic response resulting from natural selection.

In any population the rate of evolutionary change is limited by: (1) the amount of genetic variation within the population at the time; (2) the strength and consistency of the selection pressure; and (3) the extent to which selection on one trait causes parallel changes in others, which could be beneficial or detrimental. Genetic variance is often reduced in populations that have suffered recent numerical declines in which much of the variance was lost (genetic bottlenecks). Such variance can be increased again by immigration and gene flow from another population, or in the longer term by mutation and other means. Immigration and gene flow can also have deleterious effects if they break up locally adapted gene complexes, and render the local population less well adapted to local conditions.

Single selection events, such as spring storms, can cause rapid genetic change in the arrival dates of populations, as explained in Chapter 20, but counter-selection pressures could rapidly reverse the situation, and change arrival dates back to their original state. Selection pressures must act consistently in the same direction over several generations if they are to have any more than temporary effects on the genetic composition of a population. Most selection probably acts to stabilise the gene pools of populations rather than to change them. Moreover, some migratory traits (notably incidence, intensity and timing) are part of a syndrome of co-adapted traits (Pulido & Berthold 2003), so selection on one trait is likely to have strong simultaneous effects on the others. If this is disadvantageous in the new conditions, it may take many generations of selection to dissociate the beneficial traits from the detrimental ones before evolutionary change can occur. Evolutionary change may thus be rapid or slow, depending on the circumstances.

An important aspect of global warming is that temperatures have increased more in some regions than others, and more at some times of year than others. While the timing of spring migration could be influenced by weather conditions along the whole migration route, the timing of egg-laying depends on conditions in the breeding area. Any discrepancy between conditions en route and in the breeding area can worsen any mismatch between breeding and food supply. Moreover, in the breeding areas themselves, birds may respond more or less rapidly than their food organisms to climatic changes, so that birds cease to arrive and breed at the optimal time. An apparent example is provided by Pied Flycatchers *Ficedula hypoleuca* nesting in the Netherlands, where climate change has advanced the food supply on which breeding depends, but spring migration has not advanced sufficiently to allow the birds to make best use of this food supply, as they did in the past (Both & Visser 2001). The birds thereby suffered reduced breeding success and, in areas with the biggest mismatch, population levels declined by about 90% over a 20-year period (Both et al. 2006). Such mismatches can only be rectified in the longer term by changes in the genetic controlling mechanism, so as to trigger spring migration at an earlier date with respect to conditions in winter quarters. The longer the migratory journey, the less likely is weather in the breeding and wintering areas to be correlated. Long-distance migrants could have no indication from their wintering areas of how spring is developing in their breeding areas. Their departure dates from wintering areas are triggered by a photoperiodically timed endogenous rhythm, evolved through natural selection, which ensures that they arrive in breeding areas at an appropriate date (with minor variation according to prevailing conditions) (Chapter 12). Only by further evolution acting on this endogenous control mechanism is the trigger date for departure likely to be changed. In this situation, the selection pressure to migrate earlier is applied in the breeding area, but the action to accomplish an earlier arrival occurs weeks earlier in the wintering area, hundreds or thousands of kilometres away (Visser *et al.* 2004). Changing this control mechanism may be a relatively slow process, perhaps explaining why the arrival dates of long-distance migrants are less well correlated with temperatures in breeding areas than are the arrival dates of short-distance migrants, wintering nearer to breeding areas. Another mismatch was found in the American Robins *Turdus migratorius* that breed at high elevations in the Rocky Mountains of Colorado and whose spring arrival dates advanced by two weeks over a 20-year period. At the same time, winter snowfall increased and took longer to melt, giving a mismatch between arrival dates and the exposure of bare ground-feeding areas (Inouye *et al.* 2000).

These examples raise a general point that the photoperiodic responses of many birds, through which their annual cycles are timed, may become less reliable predictors of seasonal change in food supplies as climate change alters the phenology of their food supplies. This is not a new problem, as it is faced by all birds which expand their breeding ranges into different regions, but it may take time for them to adjust genetically to new situations, during which time they could perform less well than usual (though not necessarily with effects on population levels).

Facultative responses

The most convincing evidence for the existence of facultative responses comes from marked individuals in the wild which behave differently in different years, according to prevailing conditions. Plenty of evidence has emerged that individuals adjust their spring arrival dates to current conditions, or migrate in some autumns or habitats but not others, or migrate at different dates and for different distances in different years (Chapter 12). All these aspects could be under genetic influence, but the range of individual flexibility of behaviour in some populations is very wide (so-called facultative migrants, Chapter 12).

One consequence of facultative migration is that the proportion of migrants in a population could change over time without the need for any genetic change. In particular, if the population increased or food declined, increased competition could cause a greater proportion of birds to leave. Ring recoveries over the period 1960–1990 revealed that an increasing proportion of Greenfinches *Carduelis chloris* in southeast England performed seasonal movements. This change coincided with a period of population growth, so competition may also have intensified over this period (Main 1996). Similarly, from Eurasian Linnets *Carduelis cannabina* breeding in Britain, much greater proportions of ring recoveries came from the far end of the migration route in Iberia in recent years than earlier (Wernham *et al.* 2002). This coincided with a big reduction in winter food supplies over much of the route (caused by loss of weed seeds through herbicide use), which could also have caused increased competition within the population, stimulating more birds to migrate longer distances in autumn. These changes could have resulted from either facultative or genetic responses to the altered conditions.

It is difficult to separate totally genetic from facultative changes, considering that the limits to flexibility may themselves by genetically controlled, as in the comparison between 'obligate' and 'facultative' migrants (Chapter 20). But whatever the basis of the observed changes, it is presumably in these various ways that birds continually adjusted their migration patterns to the massive climate changes of the past, and through which we can expect them to respond in future. It is hard to imagine how any new pattern of migration could arise, except by evolution from an existing pattern, or by the facultative modification of an existing pattern. The next step is to identify the genes that control migration behaviour, and the constraints to further adaptive change. The potentially rapid evolutionary change in movement patterns may be a key factor buffering migratory species against extinction during periods of climatic or other change, including the present.

SUMMARY

Records indicate that many bird species have changed some aspects of their migratory behaviour during the last century or more, in response to changed conditions, with (1) earlier arrival in spring, (2) earlier or later departure in autumn, (3) shortening or lengthening of migratory routes, (4) directional changes, and (5) reduced or enhanced migratoriness, reflected in changes in ratios of resident to migratory individuals in particular breeding areas, and in the occurrence of wintering birds in regions previously lacking them. Almost all these changes are associated with changes in food availability, or with climatic conditions that are likely to affect food supplies, such as milder winters. Most examples of shifts to increasing migratoriness involve species that have extended their breeding ranges into higher latitude areas where overwintering is not possible or risky.

Some of the observed changes in migratory behaviour could represent an immediate (facultative) response to prevailing conditions, and others may have resulted from genetic changes brought about through the action of natural selection. Despite the difficulties of detecting genetic changes, evidence for a few species has indicated genetic changes in migration timing and, for at least one species, in migratory intensity and direction. In practice, most changes in migratory behaviour are likely to start as facultative responses which become genetically entrenched as selection comes to act in a consistent manner from year to year.

Northern Wheatear *Oenanthe oenanthe* whose migration routes are thought to reflect post-glacial colonisation patterns

Chapter 22
Biogeographical legacies

Migration is too complex to be explained except historically. (A. Landsborough Thomson 1936.)

It is comparatively easy to trace the probable steps in the evolution of the migrations of some species, and some routes have developed so recently that they still plainly show their origin. (Frederick C. Lincoln 1935.)

Some bird species have changed their migratory habits in recent decades, but others have not, although it seems to the human onlooker that they would benefit from doing so. Yet other species undertake such long journeys over sea or desert that one wonders how their migrations could have arisen. This chapter examines some of the more puzzling migration patterns of birds in order to understand how they might have evolved, and why they continue to persist. Some extraordinary journeys can be explained most plausibly in terms of past conditions which may have influenced their development.

Five aspects are considered here: (1) indirect migration routes and seemingly unnecessarily distant wintering areas; (2) long migrations over seas or other hostile areas; (3) loop migrations; (4) migratory divides; and (5) the development of migration, from lower to higher latitudes, as well as from higher to lower latitudes.

INDIRECT ROUTES TO DISTANT WINTERING AREAS

All else being equal, energy costs would normally favour birds taking the shortest, most direct routes between their breeding and wintering areas, which would be as close together as possible. Yet many birds take long roundabout routes between one area and the other, or use breeding and wintering areas that are separated by extraordinarily long distances, while other apparently suitable areas much closer remain unused. The problem is to understand why. Some roundabout routes might be explained on grounds of safety, if the shortest route is more risky (say with unfavourable winds or a lengthy sea-crossing), or on grounds of competition avoidance (if closer areas are already occupied by other populations of the same or similar species). However, these explanations seem unable to account for all such patterns. Almost certainly, the explanation of some long routes rests in past colonisation patterns.

A striking example is provided by the Northern Wheatear *Oenanthe oenanthe*, which from Eurasia has colonised Greenland and northeast Canada in the west and Alaska in the east, yet birds from all these breeding areas continue to migrate to Africa (**Figure 22.1**). The Greenland–Canadian birds cross the Atlantic, one of the longest sea-crossings undertaken by a passerine, while the Alaskan birds cross the Bering Sea and travel via Siberia and then the Middle East, covering a distance of 15 000 km twice each year. If Wheatears from the two ends of the breeding range were to migrate instead to South America or to Southeast Asia respectively, where no Wheatears currently winter, they could halve their migration distances. Perhaps Africa is the only continent with suitable wintering habitat for Wheatears, but a more likely explanation is that the outlying populations have failed to evolve a new route. The problem with a switch from Africa to South America is that there are now no suitable intermediate routes, so no opportunity for gradual change. Thus Northern Wheatears from northeast Canada, for example, would need a big change in direction to end up in South America rather than in Africa and any progressive change by intermediate directions (from east-southeast to south) would take them to the South Atlantic. Similarly, any Northern Wheatears from Alaska would need a big directional change to reach South America, while small stepwise changes would land them in the Pacific.

It is not just the Northern Wheatear that behaves in this way. Several species that have colonised Alaska from the east end of Eurasia (such as the Willow Warbler *Phylloscopus trochilus* and Yellow Wagtail *Motacilla flava*) continue to winter in the Old World, as do other species (such as Common Ringed Plover *Charadrius hiaticula* and Red Knot (race *Calidris canutus islandica*) that have colonised Greenland and northeastern Canada from Eurasia. Likewise, several other species (such as Grey-cheeked Thrush *Catharus minimus* and Pectoral Sandpiper

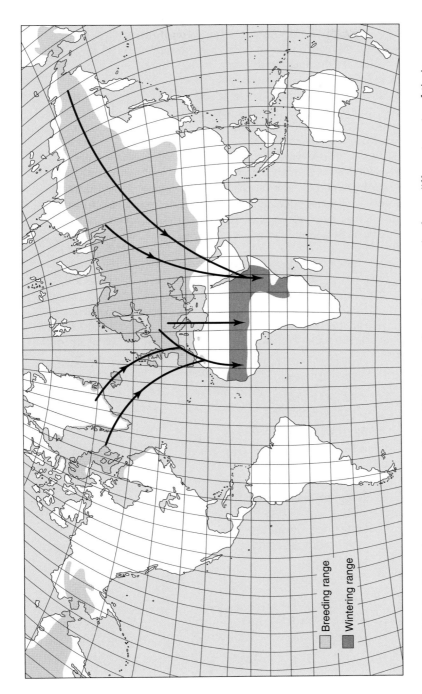

Figure 22.1 Migration routes of Northern Wheatears *Oenanthe oenanthe* from different parts of their breeding range (light shading) to their wintering range (dark shading) in Africa.

Calidris melanotos) that have colonised eastern Siberia from North America continue to winter in the New World (**Table 22.1**). It is as though the migration routes of all these various species retrace their ancestral routes of spread. As a species expands its breeding range, it simply adds step after step onto its already existing migration route. On this view, the route we see now could be interpreted largely as a consequence of post-glacial colonisation history, as has long been appreciated (Thomson 1926, Cox 1968).

On the basis of such findings, Böhning-Gaese *et al.* (1998) suggested that existing long-distance migration patterns prevent species from spreading from Eurasia to North America, or vice versa, thus explaining why many long-distance migrants occur on one land mass or the other but not on both. However, several species breed right across both North America and Eurasia, and winter in both the New and Old Worlds respectively. In this group, birds from North America winter in central and South America, while their conspecifics from Eurasia winter mainly in Africa and Southeast Asia. Examples include many shorebirds and waterfowl, Peregrine Falcon *Falco peregrinus* and Merlin *F. columbarius*, and several passerines, such as Barn Swallow *Hirundo rustica* and Sand (Bank) Martin

Table 22.1   Species that breed in both Eurasia and North America, but winter entirely in the Old World or entirely in the New World

Recent colonists from Siberia to Alaska that winter in the Old World	Recent colonists from Alaska to Siberia that winter in the New World	Recent colonists from Europe to Greenland and northeast Canada that winter in the Old World
Northern Wheatear *Oenanthe oenanthe*	Gray-cheeked Thrush *Catharus minimus*	Northern Wheatear *Oenanthe oenanthe*
Bluethroat *Luscinia svecica*	Pectoral Sandpiper *Calidris melanotos*	Common Ringed Plover *Charadrius hiaticula*
Arctic Warbler *Phylloscopus borealis*	Baird's Sandpiper *Calidris bairdii*	Red Knot *Calidris canutus islandica*
Yellow Wagtail *Motacilla flava*	Buff-breasted Sandpiper *Tryngites subruficollis*	Ruddy Turnstone *Arenaria interpres*
White Wagtail *Motacilla alba*	Long-billed Dowitcher *Limnodromus scolopaceus*	Brent Goose *Branta bernicla hrota*
Red-throated Pipit *Anthus cervinus*	Snow Goose *Chen caerulescens*	
Siberian Tit *Parus cinctus*	Sandhill Crane *Grus canadensis*	
Rustic Bunting *Emberiza rustica*		
Rufous-necked Stint *Calidris ruficollis*		
Bar-tailed Godwit *Limosa lapponica*		

Riparia riparia. At some time in the past, all these species presumably colonised one northern land mass from the other. They show that not all species have been constrained in range expansion by the difficulty of evolving new migration routes. It may be largely a matter of time, with the longest-established species having evolved new routes to new wintering areas, and the most recent colonists still retaining ancestral routes to the old wintering areas.

Similar patterns are apparent within the northern land masses. Some bird species breed across the whole of northern Eurasia, for example, yet winter entirely in Africa or entirely in Southeast Asia (**Table 22.2**). Birds breeding at the most distant end of Eurasia thus travel across the entire west–east span of this land mass, as well as to lower latitudes, to reach their wintering areas. The most likely explanation is that these species have spread to breed across Eurasia in post-glacial times from one end to the other, yet retained their ancestral wintering grounds. Some species, such as Greenish Warbler *Phylloscopus trochiloides* and Common Rosefinch *Carpodacus erythrinus*, are known to have spread into Europe since the nineteenth century from breeding areas further east, and all have retained their migration to Southeast Asia. Yet again, their long migration routes can be regarded as a retracement of ancestral routes of spread, another legacy of biogeographical history (Landsborough Thompson 1926, Cox 1968). Such routes would presumably not persist unless they served their purpose in present conditions, but a difficulty in evolving (by gradual change) new routes to totally new wintering areas may be the main reason for their continuing existence.

Most of the species that breed across Eurasia, from one end to the other, are resident species, while relatively few are long-distance migrants. The latter tend to have more restricted distributions, breeding in one half or the other. This finding is consistent with the view that expansion of breeding range is constrained in long-distance migrants by the difficulty of evolving appropriate new migration routes (Bensch 1999), although this situation may change through time. An analysis of the migrations of arctic-nesting shorebirds indicated that migration routes were constrained not by distance as such, but by distance across seas and other unfavourable areas, possibly because of the complex adaptations required for barrier-crossing and extensive detour migrations (Henningsson & Alerstam 2005).

Some other migratory species that breed across Eurasia have split wintering grounds, with western populations moving southwest into Africa and eastern populations into Southeast Asia. These species may have survived the glaciations in more than one refuge, one lying near or within Africa and the other near or within Southeast Asia. For example, the Great Reed Warbler *Acrocephalus arundinaceus* winters in both regions, so may have had two refugia (or groups of refugia), from which it subsequently spread across the Palaearctic. But it is also possible that such species survived the last glaciation at one end of the land mass, spread across the whole land mass following ice melt and vegetation growth, and then developed a secondary wintering area at the other end of the land mass. Other examples of similar patterns are given in the later section on migratory divides.

Such patterns are also evident on a smaller scale, for example within Europe. Thus, all Red-backed Shrikes *Lanius collurio* migrating from Europe to Africa cross at the eastern side of the Mediterranean, including those from Spain that start their autumn journey by flying northeast then east. Some other summer visitors that

Table 22.2 Different migration patterns of passerine species that breed across much of Eurasia and winter in the tropics

(1) Western and eastern populations winter entirely in Africa	(2) Western and eastern breeding populations winter entirely in Southeast Asia [a]	(3) Western breeding populations winter in Africa and eastern populations in Southeast Asia, with a migratory divide
Greater Whitethroat *Sylvia communis*	Scarlet Rosefinch *Carpodacus erythrinus*	Tawny Pipit *Anthus campestris*
Garden Warbler *Sylvia borin*	Lanceolated Warbler *Locustella lanceolata*	Tree Pipit *Anthus trivialis*
Willow Warbler *Phylloscopus trochilus*	Arctic Warbler *Phylloscopus borealis*	Red-throated Pipit *Anthus cervinus*
Spotted Flycatcher *Muscicapa striata*	Greenish Warbler *Phylloscopus trochiloides*	Yellow Wagtail *Motacilla flava*
Common Redstart *Phoenicurus phoenicurus*	Red-breasted Flycatcher *Ficedula parva*	Bluethroat *Luscinia svecica*
Northern Wheatear *Oenanthe oenanthe*	Taiga Flycatcher *Ficedula albicilla*	Black Redstart *Phoenicurus ochruros*
Rock Thrush *Monticola saxatilis*	Pechora Pipit *Anthus gustavi*	Desert Wheatear *Oenanthe deserti*
Ortolan Bunting *Emberiza hortulana*	Siberian Accentor *Prunella montanella*	Blue Rock Thrush *Monticola solitarius*
	Red-flanked Bluetail *Tarsiger cyanurus*	
	Rustic Bunting *Emberiza rustica*	
	Little Bunting *Emberiza pusilla*	
	Yellow-breasted Bunting *Emberiza aureola*	
	Grasshopper Warbler *Locustella naevia*	
	Great Reed Warbler *Acrocephalus arundinaceus*	
	Lesser Whitethroat *Sylvia curruca*	
	Common Chiffchaff *Phylloscopus collybita*	
	Eurasian Golden Oriole *Oriolus oriolus*	
	Sand Martin *Riparia riparia*	
	Barn Swallow *Hirundo rustica*	
	Red-rumped Swallow *Hirundo daurica*	
	Northern House Martin *Delichon urbica*	

[a]Including India

Note: In addition to the species listed above, Richard's Pipit *Anthus novaeseelandiae* breeds in the eastern Palaearctic, Africa and Australasia, but migrates in small numbers through western Europe, presumably en route to Africa.

breed in western Europe, such as Marsh Warbler *Acrocephalus palustris* and Lesser Whitethroat *Sylvia curruca*, also migrate eastwards before turning south into Africa. Why do they not fly directly southwest through Spain into Africa like many other species do? Perhaps there is an adaptive explanation for this roundabout migration route, but another possibility is that these species survived the last glaciation only in southeast Europe–eastern Africa, and in post-glacial times spread to occupy for breeding first the eastern then western side of the European continent. Their current migration routes could again trace the ancestral routes of colonisation, with the western birds having failed to evolve a new and more direct route through Iberia into western Africa. Such a change in west European birds would have needed a marked change in migratory direction, together with a longer trans-desert flight on the west side of Africa than is necessary on the east side.

The opposite situation is shown by the European Pied Flycatcher *Ficedula hypoleuca* in which practically the whole Palaearctic population from as far as 93°E (north of Mongolia) apparently passes through Iberia in autumn en route to West African wintering areas. It contrasts with the closely related Collared Flycatcher *F. albicollis* in which the whole population travels through Italy and southeast Europe into African wintering areas that lie to the east and south of those used by the European Pied Flycatcher. One of the most curious migrations is shown by the Aquatic Warbler *Acrocephalus paludicola* which breeds in eastern Europe but migrates first west to the North Sea coasts, including Britain, before turning south to Africa. It is possible that this is the optimal route to its African wintering areas, but another possibility is that this species colonised eastern Europe from the west, then died out as a breeder in the west but retained its original migration route.

Many other examples of curious routes to distant wintering areas can be found among Eurasian birds, and in many different families. But such long and indirect migration routes are not restricted to the Old World. In North America, some species breed across the northern parts of the continent but concentrate to winter at lower latitudes entirely in the west or entirely in the east. About 33 species of the eastern forests have spread westward, north of the prairies, in the boreal forests, yet migrate through the east of the continent en route to Neotropical wintering areas. The Bobolink *Dolichonyx oryzivorus* has spread westward in the last 200 years. This species winters in southern South America, and the western birds start their migration by flying east, instead of taking a more direct route (Lincoln 1935a). It is easy to imagine how a new route could evolve, step by step in whatever direction range expansion occurred, with each successive extension adding to the pre-existing route.

The underlying assumption here is that, in the genetic control of migration, progressive minor extensions or modifications to an existing programme are easier to achieve than are abrupt big-step changes. Such 'evolutionary inertia' may account for some of the variants we see in existing migration routes, some species having achieved the change to markedly different routes, while others have achieved only extensions to former routes.

For many years, the view that some migration routes retraced ancestral routes of spread was no more than plausible speculation, but recent DNA analyses have added support. Parts of the DNA in organisms are thought to have no effects on the fitness of the individual but to change at fairly constant rates over time, owing to mutations which alter the sequence of individual nucleotides along the length of the molecule. By comparing this sequence in equivalent pieces of DNA (usually mitochondrial DNA) from individuals from different parts of the breeding range, it is sometimes possible to separate populations, and even to work out the order in which different populations arose, and hence the historical pattern of range expansion. For example, Swainson's Thrush *Catharus ustulatus* is separated into two genetically distinct populations, one of which occupies the west coast region of North America north into Canada, and the other the rest of the continental range from Alaska to Newfoundland. The two populations meet and interbreed around the British Columbian–Alaskan border. To judge from molecular dating, the two populations diverged during the last glaciation, when they were probably confined to separate southwestern and southeastern refuges. They now have migration routes that could retrace their likely routes of spread from these two areas (Ruegg & Smith 2002). Ring recoveries and genetic data from this species show nearly complete segregation of migratory routes and wintering areas. Western coastal populations migrate along the Pacific coast to Central America and Mexico, whereas continental populations migrate eastward and then southward to Panama and South America. The circuitous spring migration of the eastern continental population of this species, northward then westward, mirrors the expansion route of the boreal forest (where the species breeds), following the last glacial retreat. Other North American species with similar distributions and indirect migration routes include the Hermit Thrush *Catharus guttatus* and Gray-cheeked Thrush *C. minimus* (Brewer *et al.* 2000). Western and eastern types of other North American birds also differ in mitochondrial DNA (e.g. Yellow Warbler *Dendroica petechia*, Milot *et al.* 2000; Fox Sparrow *Passerella iliaca* Zink 1994), as do some European birds (e.g. Great Reed Warbler *Acrocephalus arundinaceus*, Bensch & Hasselquist 1999), suggesting that they diverged in separate western and eastern refuges. Some species may still breed in these refuges, as well as in more northern areas colonised since then, but other species may have abandoned their glacial refuges as breeding areas, now using them only as wintering or passage areas.

Similarly, two populations of Willow Warblers *Phylloscopus trochilus* breed in Sweden, the subspecies *trochilus* in the south and the subspecies *acredula* in the north. The two populations meet at about 62°N. They differ morphologically, and show different migration routes to different wintering areas; *trochilus* moving southwest into West Africa and *acredula* southeast into eastern and southern Africa (Bensch *et al.* 1999, 2002, Chamberlain *et al.* 2000). They can also be distinguished by their DNA or by the isotope ratios in their feathers, reflecting their different wintering areas where moult occurs (**Figure 22.2**). Their contact zone, at about 62°N and spanning >350 km, is paralleled in other species that breed in Scandinavia, both birds and other animals. Such examples are most plausibly explained by the postglacial colonisation of Scandinavia along two routes, one from the southwest and the other from the east, over the northern end of the Baltic.

Further comments on the role of glacial changes

Recent estimates suggest that more than 20 glaciations occurred in the last 2.5 million years, together occupying about 90% of this period (Newton 2003). In each one, ice sheets spread from the poles to much lower latitudes than today, obliterating vegetation or pushing it southward, along with its animal inhabitants. During the peak of the glaciations, most European bird species were confined

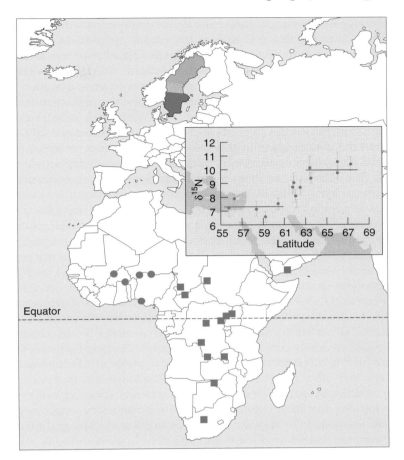

Figure 22.2 Recoveries of Willow Warblers *Phylloscopus trochilus* ringed in Sweden. Those from birds ringed in southern Sweden are shown as filled circles and those from northern Sweden as squares. The breeding distribution in Sweden shows the approximate location of the hybrid zone (shaded grey) between two subspecies (*P. t. acredula* in the north and *P. t. trochilus* in the south), where there is a migratory divide. The diagram shows the ^{15}N signature in birds from southern Sweden, the hybrid zone, and northern Sweden. These data reflect the location of the winter moult in Africa, and hence the differing wintering areas of the two populations. The hybrid zone marks an area where ice during the last glaciation was unusually thick, so that land to the north and south cleared much earlier, allowing colonisation by plants and animals. When the ice went completely, organisms from the south and north could meet, producing hybrid zones, and in the case of the Willow Warbler, a migratory divide. Based on Chamberlain *et al.* (2000).

to one or more of the three southern peninsulas of Iberia, Italy and the Balkans, although the two latter were joined across the top of the Adriatic Sea. Other glacial refuges existed in North Africa and on Mediterranean Islands, as well as to the east of the Mediterranean. The islands were larger than today, and in some cases joined together, owing to lower sea levels. During each interglacial, the ice retreated, allowing vegetation to spread back again, providing new habitat for birds and other animals. Both plants and animals spread to higher latitudes from their glacial refuges. During each period of spread, migration systems would have had to adjust. Hence, many existing migration routes, assumed to follow ancestral routes of spread, are likely to pass through the breeding areas occupied in glacial times. Depending on species and area, some such refuges may still be used for breeding and wintering, but others may not. Some species are thought to have survived the last glaciation in only one refuge system, and others in more than one. This is a plausible hypothesis for why in some species the whole migration filters through one region, whereas in other species migrations separate into different streams towards different regions.

In species occupying more than one glacial refuge, the separate populations sometimes differentiated, giving separate species or subspecies, which subsequently spread out from their refuges to form overlap or hybrid zones where they met (for review see Newton 2003). Many such 'suture' zones run roughly north–south through central Europe, marking the overlap between such taxa as Nightingale *Luscinia megarhynchos* and Thrush Nightingale *L. luscinia*, and Carrion Crow *Corvus corone* and Hooded Crow *C. c. cornix*. The precise position of the suture zone varies from species to species, possibly reflecting differential rates of spread from separate glacial refuges. Other suture zones occur through western North America, marking the overlap zone between such species as Lazuli Bunting *Passerina amoena* and Indigo Bunting *P. cyanea*, and Bullock's Oriole *I. g. bullockii* and Baltimore Oriole *I. g. galbula*. These and other populations, on either side of these suture zones, whether classed as species or subspecies, apparently retained their ancestral migration routes, giving one plausible explanation for the existence of 'migratory divides', on either side of which the population follows a different route to different winter quarters (see later). Such divides are also apparent in taxonomically undifferentiated populations, such as the White Stork *Ciconia ciconia* and Blackcap *Sylvia atricapilla* in Europe (although western and eastern populations of such species may be separable through DNA analyses). This is the case in the Great Bustard *Otis tarda*, in which Spanish birds are separable on both mitochondrial and nuclear DNA from Hungarian ones (Pitra *et al.* 2000) but, with bustards having disappeared from the intervening terrain, it is not clear where the divide may once have been.

Knowledge of glacial history thus provides a plausible hypothesis of why some species, which now breed across a continental land mass, migrate to one end of that land mass while others migrate to the other end or to both ends. During evolution, selection would be expected to shorten and straighten migration routes, in order to reduce the energy costs, but in some species this process seems yet to have occurred to only a limited extent.

Based on present-day examples, many birds may have crossed the ice sheets during the glacial periods in order to breed on the restricted vegetated areas that persisted north of the ice. This idea is supported by the patterns of subspeciation seen in waders and other high-arctic breeding species today, in many of which different subspecies are centred on areas known to have remained ice-free in summer throughout the glacial periods (Ploeger 1968, Newton 2003). In this case, birds could have nested at some locations in the far north throughout the Pleistocene glaciations, as well as in the interglacials. During the glaciations, however, they would have to have crossed hundreds or thousands of kilometres of ice between their breeding and wintering areas, longer distances than are covered today by those species that cross the Greenland ice sheet.

While many populations of birds owe their present genetic structuring to fragmentation of their populations into 'refuges' during the climatic extremes of the glacial or interglacial periods, the subsequent maintenance of this structure is due largely to behavioural philopatry, in which individual birds return from migration year after year to breed near their natal areas. The Dunlin *Calidris alpina* provides a good example. This species now breeds widely in northern Eurasia and North America, but five main genetic lineages have been recognised, each centred on a different breeding area (Wenink *et al.* 1993, 1996). All these lineages appear to have differentiated in isolation in different refugia within the latter half of the Pleistocene, and to have maintained their differences as a result of strong philopatry; the birds from each area have expanded to colonise new breeding areas and, to some extent, mixed with other populations, but the ancient genetic subdivisions are still apparent today in the mitochondrial DNA. Birds sampled on migration can therefore be allocated to one of these five lineages, and hence to a particular part of the breeding range.

Abrupt changes in migration routes

Are there any known ways in which a sudden change in migration route could evolve? Clearly, if a single step mutation took a bird to a new wintering area where it could survive, this genetic change could be passed on to subsequent generations, leading to the establishment of a new wintering area, as in the central European Blackcaps *Sylvia atricapilla* discussed in Chapter 20. This is one way in which a big directional change could occur. However, most such mutations are likely to take birds in inappropriate directions or across hostile terrain that they would be ill-equipped to cross, leading to their elimination. Even if they survive, they will be few in number and most likely, for lack of choice, will breed subsequently with birds with normal directional preferences, and the resulting hybrids will move in an inappropriate direction. This type of selection would also hinder the merging of populations with different directional tendencies (see later).

Other suspected mechanisms through which new routes and wintering areas could evolve in one step far from the original include 'mirror-image migration' and 'reversed direction migration', as discussed in Chapter 10, the latter operating mainly on a west–east axis. Again, the directional error would need to be under genetic control, and favoured by selection for it to be passed on to future generations, and the process could be facilitated if the aberrant individuals mated with one another, rather than with birds with 'normal' directional preferences.

EVOLUTION OF BARRIER CROSSING

It is easy to envisage how migration might evolve in species that have suitable areas for breeding and wintering adjacent to one another, giving a continuum of habitat between the two. This would then enable a gradual incremental lengthening of the journey. But it is less easy to understand the origin of those migrations in which the breeding and wintering areas are separated by hundreds or thousands of kilometres.

How can birds from one part of the world come to winter in another part so far away when the intervening area is unsuitable?

One possible way is through the loss of intermediate habitat or populations. In many species whose breeding and wintering areas are now contiguous, the birds from the most northerly areas often migrate furthest, over-flying the birds in intervening areas (leapfrog migration, see later). But if the intermediate populations for some reason died out, this would leave the outer areas, one for breeding and the other for wintering, with a gap between the two. The intervening population may die out through the gradual loss of intervening habitat, so that birds initially migrating in continuous habitat found themselves crossing first short then increasingly long stretches of unsuitable terrain.

The main difficulty is in explaining how landbirds could evolve in one step the ability to cross large stretches of sea or desert, which for most species offer no opportunities for refuelling. If a species were to evolve such a migration today, it would entail not a gradual change in migration physiology but an abrupt change, with markedly greater fat deposition and flight lengths. One obvious possibility is that barriers which are now wide, such as the Sahara Desert, were once narrow, so that a progressively longer migration could have evolved gradually as the barrier widened. Migrants would then have been able to lengthen their flights bit by bit, to match the need. Climate changes, with their effects on vegetation, could thus lead to the splitting of a once-contiguous vegetation type, in the way that the Sahara Desert now separates scrub to the north and south, or the boreal forest now separates open tundra from open steppe. This would enable birds to evolve gradually longer migrations as their breeding and wintering areas became progressively more separated. Alternatively, although a barrier may always have been wide, it may once have held more potential stopping places than are present now, perhaps more oases in the desert or more islands in the sea. If such stopping places disappeared gradually, this would again have allowed the progressive evolution of longer flights. This is the second way in which long flights over hostile areas might have evolved. It is happening today as stopover sites are lost, either through climatic change (as in the expansion of the Sahara Desert), or by more direct human action in destroying already fragmented habitats.

The difficulty of evolving a migration over the Sahara Desert (as it now is) may explain why most species known to have spread westward in Eurasia in historical times still return to traditional winter quarters in Southeast Asia, even though apparently suitable habitats exist in Africa to the south. It may also explain why so few forest birds (as opposed to scrub birds) cross the Sahara, because throughout the Tertiary Period, the west Eurasian forests were separated from the African forests by an even wider belt of arid land (now mainly the Sahara Desert).

The evolution of some migration routes, notably onto remote islands in the Pacific, is particularly hard to understand. Yet the entire population of the Bristle-thighed Curlew *Numenius tahitiensis*, which nests in western Alaska, now winters on islands in the central Pacific, along with large numbers of Pacific Golden Plovers *Pluvialis fulva*, Wandering Tattlers *Tringa incana*, and other species. These islands are volcanic, having arisen directly from the sea-bed, so were never connected to a continental land mass. Some are no more than a few million years old. Still other species, such as the Bar-tailed Godwit *Limosa lapponica*, fly from Alaska almost straight south across the Pacific to winter quarters in New Zealand

(Chapter 6). Such migration routes offer only a few scattered volcanic islands or coral atolls where birds could rest and feed. While in glacial times, when sea levels were lower, the numbers and sizes of islands were larger than today, the hazards of the long oversea migration must always have been great, posing problems of how it could have evolved.

One obvious mechanism involves progressive corner cutting. Imagine a shorebird that once migrated from Alaska via the Aleutian Islands in the north to winter on the coasts of Southeast Asia and Australasia. By cutting across the Pacific initially over short distances in the north, and then over progressively longer distances further south, it could gradually shorten its total journey, but by making progressively longer overseas flights. The gradual elongation of overwater flights could also explain how migration to Pacific Islands evolved, if the islands were once stepping stones on a longer journey to Southeast Asia and Australasia, which was then foreshortened. These different hypothetical stages are shown by different species today, which make overwater flights of different lengths between their breeding and wintering areas. The hypothetical starting situation is still shown by stints and others that migrate from Alaska first west, then southward down the coasts of eastern Asia. Representing an intermediate stage, other species cross the northern Pacific to join the Asian coastline at some point, before proceeding southward along that coastline. At the extreme are the Bar-tailed Godwits *Limosa lapponica*, mentioned above, which now migrate, apparently mostly non-stop in autumn, from Alaska across the Pacific to New Zealand (Chapter 6).

Migration over other sea areas, such as the Gulf of Mexico, could have evolved in the same way, from an initial movement around the western Gulf coast to a gradual lengthening of the overwater flight, but a consequent reduction in the total journey length. So too could the long overwater route between northeastern North America and South America now taken in autumn by certain shorebirds and songbirds. This pattern could have started as a mainly coast-hugging eastern North American flight, crossing the Caribbean Islands, and then shifted progressively eastward over the Atlantic, to give a progressively more direct (and hence shorter) overall route, but again requiring a progressively longer, non-stop overwater flight (**Figure 22.3**). The advantage of a direct overwater flight is not only that it cuts the distance, but also that it reduces the risks of daytime predation from falcons and other predatory birds. This and other long overwater flights are undertaken mainly in autumn when winds are favourable, but in spring when winds are unfavourable the birds take a longer and safer overland route.

Corner-cutting might also explain the current migration pattern of Icelandic birds. This island has been colonised during the last 10 000 years by landbirds entirely from Europe. The predominant migration direction in these species within Europe is southwest, yet in departing from Iceland the same species fly southeast. Only in this way could they reach the nearest suitable wintering area in the British Isles. If these species migrated back along their most likely colonisation route, they would migrate first east and then southwest, but a progressive change from east to southeast would bring them more swiftly to their current wintering areas.

All these various mechanisms for the evolution of long-distance barrier crossing can be inferred from existing patterns of variation. They would allow gradual development of a long-distance migration system over seas or deserts, without the need for a sudden step-change in one or more aspects of migration behaviour.

This makes it easier to understand how such long and difficult migrations might have evolved. It does not of course prove that they did evolve in this way.

Topographic influences

There are circumstances where a long roundabout migration route is of advantage, an end result in its own right, rather than a historical legacy or stage on the way to a more direct route. By taking a detour, birds may avoid crossing an area of unsuitable habitat, even though such a non-stop flight may seem within their capabilities. The detour may offer several benefits, such as: (1) reduced risk from adverse weather or predators; (2) reduced energy costs, despite the longer journey (if, for example, winds are more favourable); or (3) suitable habitat in which the birds can stop and feed each day, thereby saving on fuel transportation costs. From examination of a number of common detour routes taken by birds, Alerstam (2001) concluded that the latter benefit – the saving on fuel transportation

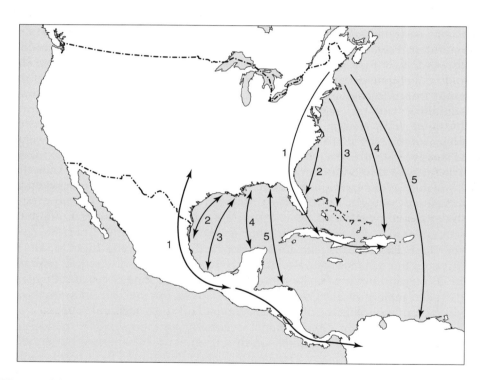

Figure 22.3   Development of long oversea migrations by progressive corner cutting. By this process, the sea-crossing is gradually lengthened, but the total journey is gradually shortened. Different stages in this process (1–5) are shown today in different species, and may be presumed to represent the steps through which the longest overwater journeys developed. Note that the long oversea flights off eastern North America are made mainly in autumn, most birds taking a longer overland route in spring when winds over the sea are less favourable (Chapter 4).

costs – was probably paramount in the evolution of some detour routes (and also loop migrations discussed below).

In Eurasia, the major obstacles to migration, such as the various mountain ranges, the Asian deserts and the Mediterranean Sea, run west–east, at right angles to the mainly north–south migration routes. While some birds migrate over these obstacles, others migrate to one side or the other to avoid them. This gives another reason, in addition to the glacial legacy hypothesis, for a separation of autumn migration directions in Europe and elsewhere between southwest and southeast. An examination of the departure directions of central European passerines revealed that 45 species start their migration in a southwest direction, 10 in a southeast direction, while three species show funnel-shaped migration toward Italy and the central Mediterranean. In addition, 13 species show a spread of directions between southwest and southeast, and 16 others show a migratory divide that to a greater or lesser extent separates migrants moving southwest from those moving southeast (**Figure 22.4**, Bairlein 1985b).

A migratory divide is largely imposed on many North American birds by the Rocky Mountains which run roughly north–south down the western side of the continent. In many species, birds breeding to the west of this range migrate south down the western side of the Americas, while those from the eastern two-thirds of the continent migrate mostly southeast, into the southeastern States, Caribbean Islands and South America. Otherwise, southwest–southeast splits in autumn migratory movements are less obvious within North America than in Eurasia, although they occur in many northern forest species which tend to avoid the central prairies (Chapter 18). Nevertheless, because most of South America lies east of North America, most intercontinental migration occurs on a northwest–southeast axis, at least for part of the route.

In general, barriers may have at least three main consequences for bird migration: (1) they may stop onward movement altogether; (2) they may lead to genetically controlled or optional detours, where at one or both migration seasons the crossing of the barrier is circumvented, thus lessening the risks, and providing opportunities

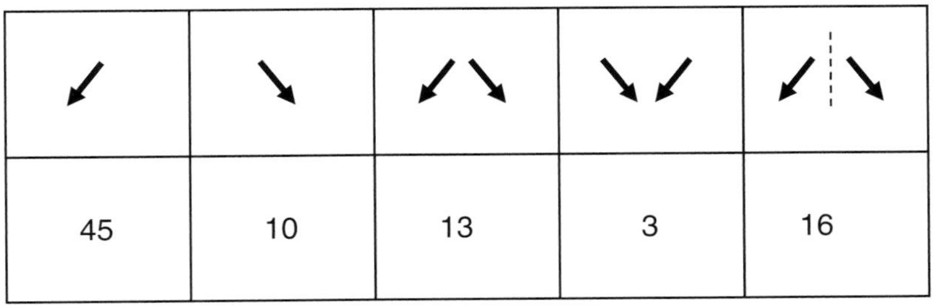

Figure 22.4 Schematic representation of initial departure directions (arrows) of central European passerine species on their way to the Mediterranean, and the number of species concerned. From left to right: southwest migration (45 species); southeast migration (10 species); both directions (13 species); between southwest and southeast = funnel migration into the central Mediterranean Region (3 species); clear migratory divide (16 species). From Bairlein (1985d).

for feeding en route; or (3) they may lead to modification of the fattening pro-gramme, ensuring that birds accumulate sufficient fuel to get across the barrier without any en route feeding. The two latter consequences depend on modification of the endogenous control programme, especially with respect to orientation and fuel deposition.

LOOP MIGRATIONS

This term is applied to the many populations that take markedly different routes on their outward and return journeys. Loop migrations are widespread among birds, having been described for passerines, shorebirds, raptors, seabirds and others from many different parts of the world (**Figures 22.5** and **22.6**). Some European species travel into Africa via Iberia, but return by a more central route mainly through Italy (e.g. Garden Warbler *Sylvia borin*, European Pied Flycatcher *Ficedula hypo-leuca*), the whole migration following an anticlockwise loop. Yet other species travel south over the eastern Mediterranean, and back further east mainly via Arabia on another anticlockwise loop (e.g. Red-backed Shrike *Lanius collurio*). In loop migra-tions, most individuals in a population take different routes in spring and autumn, but not necessarily all of them. They cause big seasonal differences in the numbers of birds seen at particular stopover sites, as some sites that are favoured in autumn are almost deserted in spring, and vice versa. In association with the different

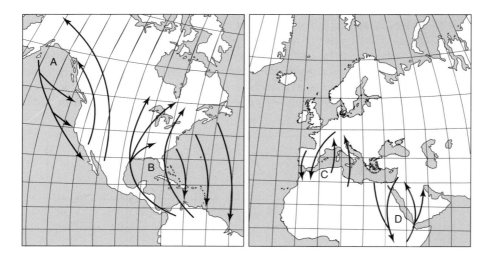

Figure 22.5   Some well-documented loop migrations in different regions. A. Western North America, shown by various waterfowl, shorebirds and others, notably Brent Geese *Branta bernicla* and Western Sandpiper *Calidris mauri*. B. Eastern North America, shown by various passerines, shorebirds and others, notably Blackpoll Warbler *Dendroica striata*. C. Western Europe–North Africa, shown by various passerines and others, notably Pied Flycatcher *Ficedula hypoleuca* (see also **Figure 22.7**). D. Middle East–North Africa, shown by various passerines and others, notably Red-backed Shrike *Lanius collurio*.

routes, patterns of migratory fuelling may also differ between autumn and spring, depending on the numbers and spacing of stopping sites (Chapter 5).

Loop migrations can usually be explained in terms of the conditions encountered en route. In eastern North America, as well as in Europe, prevailing winds north of 35–40°N blow from west to east, whereas south of 30–35°N they blow from east to west. Many species in eastern North America make their southward journey to the east of their northward journey. This is most strikingly apparent in those species that fly over the Atlantic between northeastern North America and northeastern South America in autumn, but return in spring over the land route further west. Prevailing winds favour their autumn journey over the Atlantic, but would be inimical there in spring (Chapter 4). In other species, feeding conditions seem to be important. For example, in western North America, several hummingbird species migrate up the Pacific coast in spring, taking advantage of the spring flowers on low ground, and down the Rocky Mountains in autumn when flowers are more plentiful on high ground (Phillips 1975). Male Hammond's Flycatchers *Empidonax hammondii* move

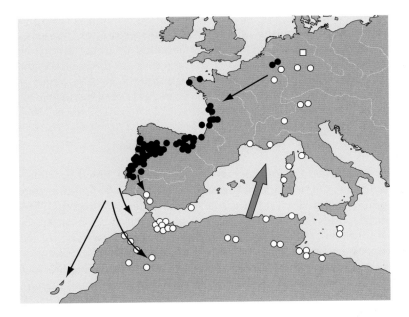

Figure 22.6 'Loop migration' of European Pied Flycatchers *Ficedula hypoleuca* ringed near Brunswick in northern Germany (open square). Black dots: recoveries during autumn migration (*n* = 71); circles: recoveries during spring migration (*n* = 40); arrows: main migration directions in autumn and spring. Autumn migration follows a fairly narrow westerly route, whereas spring migration follows a much broader and more easterly route. The recoveries suggest that birds stop mainly in southwest Europe on their autumn journey to west tropical Africa and mainly in North Africa on their return spring journey. From Bairlein (2001), based on Winkel & Frantzen (1991).

north up the Pacific coast of North America, while females, which migrate later in spring when conditions have improved, take a more direct inland route (Johnson 1965). Another impressive example of loop migration dependent on ground conditions in central Asia was described in Chapter 6 (see **Figure 6.4**).

Evidence that both the outward and return routes are under genetic influence comes from the directional preferences shown by birds in captivity, which are not simply reversed between autumn and spring. For instance, captive Garden Warblers *Sylvia borin* from central Europe, tested in orientation cages throughout their periods of migratory restlessness, initially showed a southwest heading in autumn, but after some days they switched to southeast (Chapter 20). This change fits with their migration southwest via Iberia and then southeast into tropical Africa. In spring, by contrast, captive birds headed directly north throughout, which fits with their normal return through Italy (Gwinner & Wiltschko 1978, 1980). In many populations that perform loop migrations, one route might be considered as the ancestral route, perhaps following the path of colonisation, and the other the derived route, evolved to shorten the distance or to avoid adverse conditions. Their importance is in showing that selection pressures on the outward journey can act independently of those on the return journey. The one route only partly predetermines the other. Loop migrations require no special explanation, then, but they do require more complex directional inheritance than direct back and forth journeys by the same route. This does not, of course, exclude the possibility that some individuals respond facultatively to conditions at the time. Habitat conditions may influence the numbers seen on the ground, and in many areas birds break their journeys in greater numbers, or for longer periods, in years when food is abundant than in other years, as is obvious to any experienced bird-watcher.

MIGRATORY DIVIDES

As already mentioned, many species that have wide breeding ranges show a 'migratory divide'. On one side of the divide birds head in one direction to winter quarters, while on the other side they head in a different direction. Often (but not always), such divides lead birds to circumvent geographical obstacles such as seas, deserts or mountain ranges. They have been described from ringing recoveries in many European species that fly to Africa via the shortest sea-crossings at the western and eastern ends of the Mediterranean, including the White Stork *Ciconia ciconia*, Black Stork *C. nigra*, various soaring raptors, and some small warblers, such as the Blackcap *Sylvia atricapilla* and Greater Whitethroat *Sylvia communis* (**Figures 8.5** and **22.2**). Hence, while one explanation of migratory divides is that they represent the meeting point of populations that spread in post-glacial times from different glacial refuges but have retained their ancestral migration routes, another explanation is that they evolved more recently to circumvent barriers. The two explanations are not mutually exclusive, for while divides may persist as post-glacial legacies, they may now serve a secondary role in directing birds around unfavourable areas. However, the taxonomic or genetic differentiation in many species on either side of the divide implies long-term independent evolution, and hence their refuge-based origin. Once again, their current migration routes follow their likely past colonisation routes. In some species, moreover, DNA analyses have dated the time of the separation of west and east populations roughly to the last glacial period when they were refuge-based (as for Swainson's Thrush *Catharus ustulatus* discussed above).

In North America, the explanation of migratory divides is less equivocal because no major ecological barrier lies west–east across the main flyways north of the Gulf of Mexico. Yet many birds were confined to southwest and southeast parts of the continent during glacial times, and still have distinct southwestern and southeastern wintering areas. Compared with the species mentioned earlier, the Tundra Swan *Cygnus columbianus* is an extreme example, breeding across much of the North American tundra, but wintering on either the west or east coasts. The migratory divide lies far to the west, at Point Hope in Alaska, and the bulk of the population (including that part directly north of the Californian wintering area) winters on the east coast (**Figure 22.7**).

Whatever the origin of migratory divides, an interesting question is how they persist, given that directional preferences are inherited, and the populations on either side can (at least in theory) interbreed where they meet. The persistence of the divide may be due to the fact that migratory directions and distances inherited by hybrids are intermediate between those of their parents (Chapter 20). Such hybrids

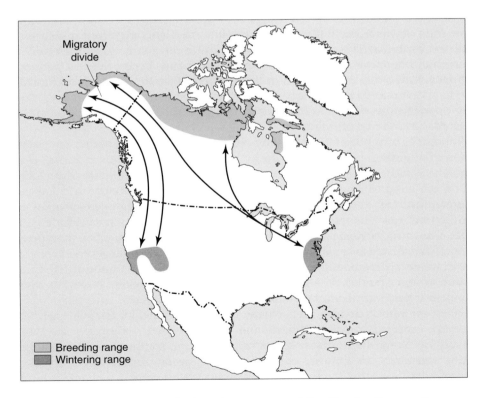

Figure 22.7 Breeding and wintering ranges of the Tundra Swans *Cygnus columbianus* in North America, showing the migratory divide in Alaska. Birds from the southwest of the breeding range in Alaska winter on the western side of the continent, but birds from most of the breeding range winter on the eastern side of the continent.

arising in the contact zone may then be less viable than the parental forms, contributing to the maintenance of the divide, and the genetic integrity of both populations. In this case, migratory divides might contribute to the speciation process through reducing interbreeding between the populations on either side of the divide.

One of the best documented examples of non-hybridisation concerns the Greenish Warbler *Phylloscopus trochiloides*, in which populations spreading northwards in post-glacial times around the west and east sides of the Tibetan Plateau met on the north side of the Plateau, in southern Siberia (Irwin *et al.* 2001). During their isolation, the two forms had not only developed minor genetic differences (recognisable in the mitochondrial DNA), but also different songs. Where they meet on the north side, birds from the two populations do not recognise one another's song, and so remain distinct without any significant interbreeding. In effect, they behave where they meet as different species, but are linked by a ring of interbreeding populations encircling the south side of the Plateau. The two forms are assumed to migrate southward down the west and east sides of the Plateau to reach their wintering areas. The Tibetan Plateau is evidently a major barrier to migration. Of 97 long-distance migrants breeding in Siberia, most (85%) use only one route round Tibet (42 through Kazakhstan, 40 through eastern China). Of the 15 species that use both routes, seven are known to have migratory divides between western and eastern subspecies (Irwin & Irwin 2005). In four additional cases, migratory divides occur between western and eastern sister species. These findings suggest that two very different migratory programmes seldom persist in the same gene pool, and that migration may play a role in speciation in this region. Moreover, although all described divides in migratory birds relate to breeding areas, equivalent divides must presumably operate in some wintering areas wherever adjacent populations separate to reach their individual breeding areas.

Sometimes bird populations nesting relatively close to one another have totally different migration patterns, marked by a divide. For example, Harlequin Ducks *Histrionicus histrionicus* nesting on the east side of Labrador migrate south to winter on the coast of Maine, whereas those from northern Quebec and Labrador migrate northeast to winter on the coast of Greenland (being one of the few bird populations that winters at higher latitudes than its breeding area (Brodeur *et al.* 2002). The authors suggested that these two population segments may represent previously isolated populations, one originating from the Pleistocene glacial refuge in western Greenland and the other from south of the Laurentide ice sheet in eastern North America. It is not known whether other species share this divide, or whether it is confined to Harlequin Ducks.

Other migratory divides are evident in the far north of Eurasia and North America, especially among seabirds and waders which, on leaving the tundra, initially fly along the northern coasts before turning south down the Atlantic or Pacific coastlines. In Eurasia, many species show an abrupt migratory divide at about 100°E on the Taimyr Peninsula, with post-breeding movement west of that site being mainly west-southwest and east of that site mainly east-southeast, more or less parallel to the coast (Alerstam & Gudmundsson 1999). One plausible explanation for the divide being situated in the Taimyr Peninsula is that this region lies midway across the Eurasian land mass. Hence, in whatever way this divide evolved, it may persist largely as a result of distance-dependent flight costs

(Alerstam & Gudmundsson 1999). However, not all tundra species show a divide at this point, and some take more southerly routes across the Eurasian land mass to their winter quarters. Any shorebirds migrating from Taimyr to India, say, would have to cross at least 5000 km of the Eurasian land mass before reaching another marine coastal site. Other migratory divides occur in Greenland, with birds from west and southern parts of the island wintering in North America, and those from the northeast wintering in Europe (for Snow Buntings *Plectrophenax nivalis* and others see Lyngs 2003).

Even in the absence of geographical barriers, or twin glacial refuges, it is not hard to see how migratory divides persist. Imagine a species with a wide breeding range that has two wintering areas near either end, as in **Figure 22.8**. Imagine now that migration costs increase with length of journey, in proportion to the lengths of lines in **Figure 22.8a**. If birds with higher migration costs return in smaller proportion to the breeding range, generation after generation, then segregation of breeding populations with a clear migratory divide will develop (**Figure 22.8b**). If competition on wintering areas is intense, this could lead to a gap in the breeding range of the species, because individuals migrating the longest distances to fill that gap could be disadvantaged during competition on the wintering range, and continually be eliminated by selection.

What could change this situation, leading to breakdown of the divide, is if the conditions in one wintering area were so good that birds from all parts of the breeding range benefited from migrating there. A migratory divide could shift by populations changing their inherited migration direction under the action of natural selection, but a more likely mechanism would be through the spread of the more successful wintering population into the breeding range of the other, gradually replacing it (**Figure 22.8c**).

Another type of divide exists in populations in which some individuals take one direction to a wintering area, while others from the same place take a different direction to another wintering area. An example is provided by the central European Blackcaps *Sylvia atricapilla*, some of which began to migrate westward to winter in Britain in the mid-twentieth century (Chapter 20). These birds differ from those that continue to migrate southwest to Iberia. Although birds from both types currently breed in the same general area, they show assortative mating, pairing with birds of their own directional preference much more than expected by chance, owing to a difference in arrival times (Bearhop *et al.* 2005). The divide may be temporary, and in time one migratory genotype may gradually replace the other.

In conclusion, the various migratory routes and patterns of distribution described above could all represent a carry-over from conditions operating at an earlier stage in the earth's history. They would be unlikely to persist, however, if they were markedly disadvantageous in present conditions. Nonetheless, the reason that some birds stick to their present migration routes may stem from the difficulty of making a single large-step change in migratory habits that would otherwise be beneficial: for example, a big change in direction that would greatly reduce the length of the migratory journey. For most of the patterns described above, alternative explanations have been proposed. They again involve untested assumptions and, although sometimes couched in mathematical terms, they are still little more than guesses. However, the idea that glacial conditions and post-glacial colonisation

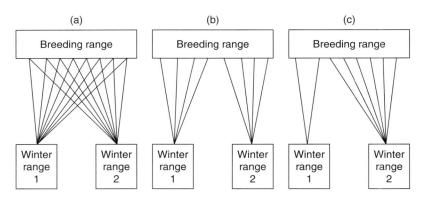

Figure 22.8   Model depicting the evolution and maintenance of a migratory divide without the necessary involvement of glacial refuge areas. (a) Hypothetical starting scenario, with a continuous breeding range, and two separate wintering areas, each visited by birds from all parts of the breeding range. Migration costs proportional to lengths of lines depicting routes. (b) Migratory divide resulting from differential survival consequent upon differential migration costs (or from migration costs plus asymmetrical competition favouring use of nearest breeding and wintering sites). In this model the carrying capacity of the two wintering areas is assumed to be equal. The divide then appears midway through the breeding range, so that the populations on each side are of about equal size. (c) Situation resulting from differential survival consequent upon differential migration costs, together with wintering areas of markedly different qualities or carrying capacities. Since the carrying capacities of the two wintering areas are unequal, the migratory divide would be expected to shift from the centre of the breeding range towards whatever side had the wintering area of smallest capacity. Other modifications to the model could be envisaged by changing the relative distances of the wintering sites from the breeding range, and hence the migration costs. Modified from Lundberg & Alerstam (1986).

patterns could have influenced migratory behaviour is increasingly supported by palaeontological and DNA studies.

MIGRATION DEVELOPMENT TOWARDS HIGHER OR LOWER LATITUDES

In contemplating the development of migration, two scenarios have been proposed: (1) that the breeding range is the original year-round home of a population, and the wintering range is the secondary home, visited to enhance survival through the most difficult season (e.g. Lack 1954, Cohen 1967); or (2) that the present wintering range is the original year-round home, and the breeding range is the secondary home visited to enhance reproductive prospects (Cox 1968, 1985, Levey & Stiles 1992, Rappole & Tipton 1995, Safriel 1995, Poulin & Lefebvre 1996, Bell 2000, Böhning-Gaese & Oberrath 2002, Rappole & Jones 2002). In theory, current migration patterns could have evolved in some species from one of these starting points, and in other species from the opposite starting point. However, as most of the migration that we now observe developed as birds spread from low to high latitudes at the end of the last glaciation, range expansion and migration almost certainly developed hand in hand. This not a question of the origin of migration, but of how past migratory patterns were progressively modified to produce those of today.

The process was evident recently in the northward expansion of the European Serin *Serinus serinus* (Chapter 21). This species is resident in southern Europe, but as it has extended its breeding range northwards, it has become increasingly migratory in the newly colonised areas. The same holds for other species elsewhere, including the Cattle Egret *Bubulcus ibis* in North America. Such recent colonists reveal that the occupation of higher latitudes can occur over a matter of decades, rather than millennia, if suitable habitat already exists. The colonisation of new breeding areas is achieved by normal dispersal processes, and migration develops secondarily in response to conditions in the newly occupied breeding areas, wherever individuals survive better, on average, by moving out for the winter than by staying in their breeding areas year-round. According to Berthold (1999), most bird populations are likely to contain a mixture of migratory and non-migratory individuals, or else a range of individuals with different thresholds for the initiation of migratory behaviour. This inherent variation within a population provides the raw material on which natural selection can act to produce a predominantly migratory population from a predominantly resident one, or vice versa (Chapter 20).

We can suppose that the post-glacial colonisation of northern regions involved a general northward movement of bird breeding ranges, associated with the northward advance of suitable vegetation, following ice melt. Evidence of this process derives from the post-glacial range expansions of bird-dispersed plants following ice melt, both in Europe and North America. Bird species that were non-migratory could have remained so, or become migratory in the more northern of the areas colonised. Species that were migratory during the glaciation (wintering at lower latitudes than their breeding refuges) could have moved both their breeding and wintering areas northward, or they could have retained the same wintering areas, and simply lengthened their migrations as their breeding range moved northward. The latter scenario could have held in species that, throughout the glaciation until now, have continued to winter in tropical or opposite-hemisphere regions.

Along similar lines, Rappole & Jones (2002) proposed the following steps in the evolution of tropical–temperate migration, although they were concerned with range spreads over several million years, and not merely those that followed the last glaciation. On their view:

1. Long-distance migrants from tropical to temperate regions originated as predominantly sedentary, tropical-breeding residents.
2. These birds tended to expand their breeding ranges into all adjacent and suitable habitat by normal dispersal processes.
3. Increased food availability, along with reduced competition and predation, in seasonal temperate habitats during summer allowed higher reproductive rates in temperate breeding populations relative to tropical populations of the same species (a difference still evident today).

4. These fitness benefits favoured rapid (in an evolutionary sense) invasion of temperate habitats by those tropical species able to exploit them.

5. Spreading out from the tropics into the more seasonal environments of higher latitudes, migratory genotypes, which already existed within the population, were favoured over resident ones, and under natural selection the expanding populations changed from predominantly resident to predominantly migratory, returning to lower latitudes for the winter. The latitude at which migration was favoured over residency is likely to have varied between species, depending on their particular food needs, but giving in each species a geographical gradient from wholly resident at the lowest latitudes, through partially migratory to wholly migratory at the highest latitudes.

Some authors have proposed the development of small-scale seasonal dispersal movements (tracking locally available resources) as a precursor to the spread to higher latitudes (e.g. Levey & Stiles 1992, Rappole 1995, Safriel 1995). However, individuals of every bird species disperse to some extent, and there is no reason why range expansion should not also have occurred in the most sedentary of resident species that developed migration secondarily as they spread into more seasonal environments (Bell 2000).

Evidence from species relationships

On the above view, many long-distance migrants, especially among passerines, have evolved from tropical or subtropical ancestors. One line of evidence cited by Rappole & Jones (2002) in support of this view is based on species relationships. Of the 338 species of migratory landbirds that breed in temperate and boreal North America, 162 (48%) have conspecifics and 78% have congeners that breed in the Neotropics. For the Asian migration system, the numbers are similar: 106 (31%) of 338 species of migrants have conspecific populations that breed in the tropics, while 262 species (78%) have congeners. For the Palaearctic–African system, 30[1] (16%) of 186 species of migrants have conspecific populations that breed in the tropics, while 139 (75%) have tropical breeding congeners. These numbers, they argued, imply that most of the long-distance migrants that breed in the Nearctic and Palaearctic regions are tropical in origin, although several (e.g. *Phylloscopus*, *Dendroica*, *Vermivora*, *Empidonax*, *Vireo* and *Myiarchus*) have undergone radiations after colonising the northern continents, and subsequent to the evolution of a migratory lifestyle (Rappole 1995). Almost certainly, much of the evolution of these groups pre-dates the last glaciation, and may date back to a much earlier period. There is, of course, no reason why colonisation of the tropics by the descendants of extra-tropical species should not have contributed to this pattern of shared species and genera, so without supplementary information (see later), it is hard to assess the value of this evidence alone. In any case, the evolution of species should not be confused with

---

[1]*Correcting Rappole & Jones (2002), who gave 42 such species, but this number included some breeding in North Africa, north of the Sahara. The total figures given by Rappole & Jones for North America and Asia are also greater than those given by other researchers, and cited in Chapters 24 and 25.*

the different process of the evolution (or modification) of current migration patterns within species. A species may have evolved in a region far removed from that in which it became migratory.

Evidence from habitat preferences

A second line of evidence proposed in support of a tropical origin for many long-distance intercontinental migrants comes from their choice of habitat. The overall species numbers in the Palaearctic–Asian and Nearctic–Neotropical systems are similar (about 338 species in both), and much greater than in the Palaearctic–Afrotropical system (186 species, excluding seabirds). The main deficiency in the Palaearctic–Afrotropical system is in forest birds. In total, about 48 species of long-distance migrants breed in west Palaearctic forest, while about 112 such species breed in Nearctic forests and 107 in east Asian forests (Moreau 1972, Rappole 1995). However, while nearly all the Asian and Nearctic species that breed in forests also winter in some type of forest, only five Palaearctic migrants winter in African tropical forests, mainly round the edges, namely European Honey Buzzard *Pernis apivorus*, Eurasian Golden Oriole *Oriolus oriolus*, European Pied Flycatcher *Ficedula hypoleuca*, Collared Flycatcher *F. albicollis* and Wood Warbler *Phylloscopus sibilatrix* (Mönkkönen *et al.* 1992). The remaining species (apart from aquatic and coastal birds) winter in savannah and scrub (for comparison between Palaearctic–Afrotropical and Nearctic–Neotropical systems, controlling for phylogeny, see Bohning-Gaese & Oberrath 2003).

These differences between migration systems throw further light on their evolution. Throughout the Tertiary Period, including the glacial cycles, the east Palaearctic and Nearctic forest species have had continuity of forest from their breeding to their wintering areas, while the west Palaearctic forest species have been separated from the Afrotropical ones by a broad belt of arid scrub and desert. These intervening habitats could have acted in the past as a filter, screening out any species not adapted to those habitats and letting through those that were: that is, mainly forest species from the American and Asian tropics and scrub species from Africa. Any forest-dependent species from tropical Africa would throughout their history have confronted, in any attempted northward range expansion, thousands of kilometres of arid scrub or desert (during the glacial periods, the Sahara was much better vegetated than today, but apparently by scrub rather than high forest). The same filter could have also prevented European forest species from invading Africa, but the other facts on species numbers and relationships presented above, together with genetic evidence from some species, strongly suggest movement mainly from the tropics outwards (or at least from lower to higher latitudes). This would in any case be expected as the commonest pattern, because many more bird species occur at low than at high latitudes, and because the bout of re-colonisation following each glaciation would have been mainly from low to high latitudes.

Nevertheless, the poverty of the west Palaearctic forest avifauna cannot be attributed entirely to the effects of the Sahara as a barrier to spread. The forests in Europe were much more thoroughly obliterated by the glaciations than were the forests of North America and eastern Asia. This process is known from palaeontological evidence to have eliminated large numbers of the plants and animals found in west Palaearctic forests in pre-glacial times (many of which still survive today in Asian forests (at least at the generic level). Among birds, extinctions are likely to have included migratory and resident species. We thus have two explanations for the small numbers of forest species in Europe compared with other regions, but it is clear that not many forest species could have colonised Europe from Africa in recent post-glacial times (say the last 15 000 years). The opposite is true for scrub species, however, which may account for the predominance of scrub species among current Eurasian–Afrotropical migrants.

Evidence from ecological niches

Most species seem to occupy ecological niches in their wintering areas that are similar to those they use in their breeding areas. They occupy places with similar climate at the time they are there, habitats of similar structure, and take similar types of foods in similar ways. The actual plant species that comprise the habitat, or the plant or animal species that comprise the diet, may differ between breed-ing and wintering areas, but superficially they look much the same. However, some bird species make switches in climate types or habitats between seasons, such as inland marsh to open ocean, or forest to scrub, or they make switches in the types of foods eaten, such as insects to seeds. Extreme examples include the sandpipers that switch from tundra to seashore and from insects to worms and molluscs, or the skuas that switch from tundra to the open sea and from lem-mings to fish. Hypotheses regarding the evolution of migration would gener-ally predict 'niche-following' as primitive, and 'niche-switching' as derived, and thus as representing a further step in the evolution of migratory from resident populations.

Examination of 21 species of New World migrants (nine *Dendroica*, four *Spizella*, three *Vermivora* and five *Vireo* species) revealed that the majority occupied a par-ticular climatic regime year-round, seeking areas of similar climate in summer and winter, while others make a clear switch in climatic regime (and presumably also in other aspects of their ecology) between seasons (Nakazawa *et al.* 2004). In the 'niche-switchers', the winter habitat seemed ancestral and the summer habi-tat derived. Niche-switching is unavoidable for many of the small insectivores that breed in boreal forest, because no coniferous habitat is available at the low tropical latitudes where they winter, and in which many probably evolved. Niche constancy in terms of climate had earlier been noted in Swainson's Flycatcher *Myiarchus swainsoni* (Joseph & Stockwell 2000).

A further study involving the directionality of evolution of gross habitat prefer-ences within the *Passerina* buntings again indicated greater evolutionary changes in the niches of breeding populations than of wintering populations (Martínez-Meyer *et al.* 2004). These results were consistent with the hypotheses of: (1) niche conserva-tism (in winter at least) across a recently speciated lineage; and (2) the derived state of the breeding (rather than wintering) niches of each species – a further indication that, in range expansion, many birds have moved to new breeding areas rather than to new wintering areas.

Uncertainty surrounds species that occupy habitats in the non-breeding season in which they could not breed, such as shorebirds which nest on the arctic tundra and winter on lower latitude intertidal mud, or skuas which nest on the tundra and winter in lower latitude sea areas. During glacial periods, the breeding and wintering areas of such species may have been less separated than today, but if these birds always used the same habitats as they do now, they must always have migrated to lower latitudes for the winter when arctic shorelines and seas freeze over. This might seem like a clear case of birds developing migration from breeding to wintering areas. However, if such species originally nested in other open habitats (such as saltmarsh) near seas that did not freeze in winter, they could have fitted the same pattern as proposed for other species, spreading to higher latitudes and developing migrations as they reached areas where wintering was not possible. Short-distance change in habitats between breeding and non-breeding seasons is known now in many shorebirds that winter in intertidal areas and nest on nearby shingle (Ringed Plover *Charadrius hiaticula*, Oyster-catcher *Haematopus ostralegus*) or saltmarsh (Redshank *Tringa totanus*, Dunlin *Calidris alpina*). In these respects, such populations are little different from the many pelagic seabirds which feed from the sea and nest on nearby coasts and islands. With many shorebirds, migrations probably developed so far back in time that it is impossible to tell from current patterns how they evolved.

Evidence from DNA studies

The above circumstantial evidence for the development of migration from breeding areas as birds spread from lower to higher latitudes has gained additional support in recent decades from DNA analyses. Many bird species now live in areas that remained favourable during the glaciations, as well as in higher latitude areas that could have been colonised only since glacial retreat. The majority of such species show more mitochondrial DNA variation in the refugial than in the more recently colonised areas. They are assumed to have colonised from the edge of their refugial ranges, and are like other founder populations which, because of the small number of initial colonists, show less genetic variation than the source populations. Examples include the Greenfinch *Carduelis chloris* in Europe (Merilä *et al.* 1997), and the Yellow Warbler *Dendroica petechia* and MacGillivray's Warbler *Oporornis tolmiei* in Central and North America (Klein & Brown 1994, Milá *et al.* 2000). Reductions in genetic variance with increasing latitude, assumed to result from post-glacial leading edge spread, have also been described in plants, insects and mammals from both continents (Hewitt 2000).

Several other widespread species examined from different parts of their deglaciated North American ranges have shown unexpectedly low levels of mitochondrial DNA differentiation, despite, in some species, high levels of phenotypic differentiation (for Red-winged Blackbird *Agelaius phoeniceus* see Ball *et al.* 1988; for Song Sparrow *Melospiza melodia* see Zink & Dittman 1993a; for Swamp Sparrow *Melospiza georgiana* see Greenberg *et al.* 1998). Yet again, the implications are that large deglaciated areas were colonised by individuals from limited areas, and that such species underwent morphological differentiation only after they had spread to their current ranges, in the last 10 000 years or so. This is too short a period to expect much geographical variation in mitochondrial DNA. They contrast with some species mentioned earlier, which are likely to have spread from two or more different refuges, as they show two or more distinct types, based on mitochondrial DNA (e.g. Great Reed Warbler *Acrocephalus arundinaceus* in

Europe, Bensch & Hasselquist 1999; and Yellow Warbler *Dendroica petechia* in North America, Milot *et al.* 2000).

Interesting evidence for genetic isolation between northern migratory and southern resident populations of the same species has emerged for the Prairie Warbler *Dendroica discolor* in North America (Buerckle 1999). Within the continent, the species is resident year-round only in mangrove habitat around the Florida coastline, but migratory Prairie Warblers breed further north over a large part of the continent. Some of these migrants also winter in Florida, where they could come into contact with the resident population. The two populations are morphologically indistinguishable, but they can be separated genetically, having split from one another probably in the late Pleistocene. This situation shows clearly how a southern resident spreading northward to breed can become genetically isolated from its parent population, even though the two populations overlap in wintering range. The isolation develops presumably because the two populations breed in different regions, and at different times of year.

In some species, estimates have been made from mitchondrial DNA of both the dates and routes of spread. Thus, based on diversity patterns in mitchondrial DNA (four sections), Common Chaffinches *Fringilla coelebs* appear to have spread from North Africa into southern Europe within the last 370 000 years or so, and spread further north through Europe only during the past 15 000–3000 years, behind the retreating ice sheets (Marshall & Baker 1999). Only the more northern populations, breeding in the most seasonal environments, have become migratory. Similarly, studies of several North American bird species, including the Song Sparrow *Melospiza melodia* and various parids, have revealed basal haplotypes in Newfoundland, a known glacial refuge area, suggesting that these species may have spread from there to other parts of the continent in post-glacial times (Gill *et al.* 1993, Zink 1997). In these populations, spread was not only northward, but mainly westward and to some extent southward.

A similar approach was applied to two pairs of closely related plover species: (1) the non-migratory Australasian *Charadrius ruficapillus* and the similar largely migratory *C. alexandrinus* which occurs across Australia, Eurasia and the Americas; (2) the non-migrating *C. australis* of Australian deserts and *C. veredus* which migrates between breeding areas in eastern Asia and non-breeding areas in Indonesia and Australia (Joseph *et al.* 1999). With these distributions, one possibility is that the strongly migratory *alexandrinus* and *veredus* evolved from Australian ancestors through northward shifts of their breeding distributions from Australia to continental Asia. Alternatively, there may have been southward shifts in breeding distribution from eastern Asia to Australia, with the non-migratory Australian species *ruficapillus* and *australis* having arisen from migratory ancestors. In fact, DNA analyses suggest that, in both these species pairs, the more northern migratory species evolved from the non-migratory ones.

This latter type of study is as relevant to the evolution of species as to the evolution or modification of migration, for we know that changes from resident to migrant or vice versa can occur rapidly within species (Chapter 20). While the path of evolution in the plovers may well have been as DNA suggests, any of the four species could probably have become migratory if they had spread into a seasonal environment that encouraged it. In practice, many species spreading from low to high latitudes have to develop migration if they are to avoid the cold, high latitude winters. This holds in comparisons between species, as well as between different populations of the same species.

Another fact adduced in support of a tropical origin of many long-distance migrants is that long-distance tropical landbird migrants are phylogenetically more different between North America and Eurasia than are the short-distance migrants and residents of these two land masses (Böhning-Gaese *et al.* 1998). The long-distance migrants are likely therefore to have evolved further south, in tropical regions, where greater distances separate the two land masses, ensuring greater isolation (in evolutionary time) between their respective avifaunas. Moreover, most of the long-distance passerine migrants belong to families that are essentially tropical in distribution, such as the Parulidae and Tyrannidae in the New World and the Silviidae and Muscicapidae in the Old World. This has fuelled the view that these mainly insectivorous species have evolved from tropical families, representatives of which spread north to breed in northern latitudes (Rappole & Warner 1980).

It is in any case impossible to separate in the northern hemisphere the situation in which a tropical species spreads gradually north and develops southward migration secondarily, from that in which a species develops from within its existing range a northern migration to a new breeding area. In practice, as mentioned above, both the northward spread in breeding range and the associated southward migration were likely to develop hand in hand once the species had reached areas unfavourable for wintering.

Southern hemisphere migration systems have been less well studied, but again some of the same conclusions seem to hold, at least in South America where the question has been addressed. In this region, around 70% of southern temperate zone birds, especially insectivores, have conspecific or closely related species in the Amazonian tropics (Joseph 1997, Jahn *et al.* 2004).

In summary, it can be hypothesised that the majority of current bird migrations arose in the same way, with dispersing individuals spreading from lower to higher latitudes, where they could exploit the abundant seasonal resources for breeding, but returning to lower latitudes for the non-breeding season. The types of species involved depended on the diversity of source populations, the types of intervening habitats, and the types of seasonally available breeding habitats. This view of migration is centred of the notion that 'migratory genes' persist even within essentially resident populations, allowing them to continually change in migratory habits, as well as in distribution, in response to ever-changing eco-geographical conditions. It is concerned mainly with the modification of pre-existing movement patterns, and not with the evolution of the species themselves, which could, of course, have arisen at low or high latitudes, and spread from there, developing resident or migratory behaviour as the need arose.

Colonisation of wintering from breeding areas

The above discussion argues in favour of current migration systems developing as species spread from lower to higher latitudes, becoming more migratory as they reached increasingly seasonal environments. This is the situation that would have affected all those species that colonised high latitude areas after the last glaciation, and is evident today in species that are still spreading north. However, we can imagine that the same process in reverse occurred repeatedly through the Pleistocene, as glacial conditions advanced from high to lower latitudes. In those conditions, previous resident populations would need to become migratory in order to survive the colder winters, before being finally obliterated from their higher latitude breeding areas entirely by the advancing cold, and surviving only at the lowest latitudes. As we are now in an interglacial, the existing migration system is likely to have been established the other way round, as explained above, but there are also recent examples of resident species introduced to higher latitudes by human action then developing migration to lower latitudes secondarily, and thus acquiring new wintering areas. The most recent involves the House Finch *Carpodacus mexicanus*, introduced from California (where it is sedentary) to the New York area. As the population expanded, the migratory habit developed, and now a large proportion of the once sedentary population migrates to winter further south (Able & Belthoff 1998). The same happened earlier in the Common Starling *Sturnus vulgaris*, after birds supposedly from the resident British stock were introduced to the New York area. Now that the species has spread over much of North America, the most northerly breeding populations are migratory or partially migratory, and only the most southern ones are sedentary. Providing that 'genes' for migration persist in populations through both resident and migratory phases in their history, the one type can develop from the other whenever distribution or conditions change. There is probably no time when conditions are not changing somewhere in the range.

The role of partial migration

Partial migration, it may be recalled, is the tendency for only some individuals in a population to migrate. Where individuals leave a high-latitude population each year to winter at a lower latitude, they encounter on their return to breeding areas other individuals that have stayed there year-round. The two sectors of the population can then interbreed (though not necessarily as often as expected by chance), so that resident and migratory sectors are not reproductively isolated. In such species, migration is typically stimulated by competition, including dominance interactions in relation to prevailing food supplies, so that in any one year the subordinate members of a population are most likely to migrate (Chapter 12).

   In contrast, where individuals leave a low-latitude population to breed at higher latitude, while other conspecifics remain behind to breed at low latitudes, the two sectors of the population become reproductively isolated because they breed in different areas usually at partly different times. This separation could lead to divergence of the two populations, as in the Prairie Warblers mentioned above, first to subspecific then perhaps to specific level. Because individuals from one population can then reproduce only if they migrate to breed at the higher latitudes to which they have become adapted, migration can at the same time become genetically (rather than behaviourally) controlled – obligate rather than facultative. The two processes leading to genetic control of migration and genetic divergence are, therefore, likely to occur hand in hand. This type of partial migration was called population partial migration by Jahn *et al.* (2004), in order to draw the distinction between 'individual level' and 'population level' processes, the one in which migration is condition dependent (from breeding to wintering areas), involving only a proportion of individuals moving each year from high to low latitudes, and the other in which it is genetically dependent (from wintering to breeding areas), involving all high latitude individuals wintering at low latitude. Possibly both result at different stages in the same process of spread from low to high latitudes. Initially, as birds spread out of an ancestral range, natal philopatry leads them to develop a measure of reproductive isolation in their new breeding area, which accentuates as they spread further, and breed at a progressively different time, and in a different daylength regime.

DEVELOPMENT OF MIGRATION PATTERNS

As a breeding population expands from lower to higher latitudes, and birds at the expanding front become increasingly migratory, more and more individuals are likely to concentrate for the winter in the original lower latitude range. Theoretically, the resulting competition for the limited resources there could lead to one of four situations, depending on circumstances:

1. The total numbers wintering in the original range could be limited, preventing any further expansion of breeding range to higher latitudes, so that the amount of wintering habitat limits the extent of the breeding range.
2. If the winter immigrants from high latitudes are competitively superior to the original population, when the two come together on their joint wintering area, the original resident population could itself develop migration to areas beyond its current range, being replaced in its breeding area each winter by the immigrants. This scenario depends on the presence of suitable wintering habitat beyond the original range, and would give rise to a system of 'chain migration' in which breeding populations are seasonally replaced by wintering populations of the same species. In this way, populations from successive latitudes maintain the same south–north sequence in wintering areas as they do in breeding areas, but the whole series is displaced towards the equator in winter (for examples see Chapter 23).
3. If the winter immigrants were inferior in competition with the original breeding population, their numbers could be held at a low level (as in 1 above), or they could develop an even longer migration to winter in previously unoccupied areas beyond the original breeding range. This possibility again depends on suitable wintering habitat being available, and would give rise to a system of 'leapfrog migration' in which the migrant population moves through the area occupied by the resident population to winter beyond it. In this way, populations of a species breeding at different latitudes reverse their sequence in winter, with those breeding furthest north wintering furthest south (or vice versa in the southern hemisphere) (for examples see Chapter 23).
4. Alternatively, if the immigrants were competitively superior, they could eliminate the original population completely. The species would then persist as a single population breeding at high latitude and wintering at low latitude, and the original breeding range would be vacant during that part of the year when

the wintering population was away on its higher latitude breeding range. One way in which the migrant population could achieve competitive superiority is through the greater fecundity expected at higher latitudes. Even if the two populations then competed on equal terms on their shared wintering range, with no difference between them in average survival rates, the more fecund population (with greater average reproductive rate) would in time outbreed the other, either holding it at much lower level or eliminating it altogether in their shared range.

Given these possibilities, the transition from resident to long-distance migrant emerges as a three-stage process, beginning with the origin (maintenance or re-expression) of the migratory habit, followed by the establishment of a fully migratory population, and in some cases ending with the disappearance of ancestral resident population. On the latter process, the general pattern that emerges involves a 'rolling forward' of the breeding range of a species as new breeding populations are established at higher latitudes and old ones are lost at the trailing edge. This process could have been greatly facilitated by the development of productive high-latitude habitats in the wake of glacial retreat, but may also continue into periods of climatic stability (Bell 2000). On this mechanism, improvements for breeding in one part of the range can, through winter competition, lead to retreat or extinction in another part. The implication is that, were it not for the seasonal influx of migrants from higher latitudes, the same species might be resident over a wider range of latitudes. The species can winter at lower latitudes, but year-round residency is reduced or prevented there by competition for winter food supplies with conspecifics that breed at higher latitude.

This consideration may help to explain why many migratory birds do not breed in their wintering areas even though conditions seem suitable there year-round. Take the Osprey *Pandion haliaetus*, for example, which does not breed in the tropics and subtropics of South America and Africa, even though the bulk of the northern continental populations winter there, and the immatures from these populations remain there year-round. Ospreys can obviously survive through the year at low latitudes, and the most likely reason they do not remain to nest there is that they can achieve higher individual reproductive output by migrating to nest at higher latitudes, returning for the intervening winters. These various considerations imply that intraspecific competition for resources in low-latitude wintering areas may be a major driving force behind patterns of latitudinal distribution in migratory birds.

Shorebirds contain many high-latitude migrants. Global populations of such species may well be limited by the availability of coastal food supplies in the non-breeding season, and as a consequence are entirely accommodated in the breeding season in the high-latitude habitats where they achieve the greatest breeding output, taking migration losses into account. Many of these high-latitude species could probably breed at lower latitudes and be more widespread at that season if highly productive tundra habitat were not available (Myers 1980). On this view, intra-specific competition on the wintering grounds could have a major role in influencing the overall population sizes, distributions and migrations of birds (for further discussion see Chapter 23).

Many seabird species, once restricted to arctic-nesting areas, spread southward during the twentieth century, greatly increasing their numbers (Newton 2003). These range expansions were in a direction opposite to that expected under global warming, but could be explained by birds being originally confined to areas where they could achieve the highest fecundity and, as these areas became filled, spreading southward during a period of overall population growth. Examples of such species include Northern Fulmar *Fulmarus glacialis*, Gannet *Morus bassanus*, various large gull species, and Common Eider *Somateria mollisima*. The increase and spread of most of these species has been linked to greatly increased food supplies resulting largely from the development of the fishing industry, and the resulting waste produced from the processing of fish at sea. Populations breeding over a wide span of latitude share a common wintering area, in which competition would have been reduced as food supplies increased. These various events may illustrate how the breeding range of a migrant may be limited by competition for restricted food supplies in winter quarters. If this food supply increases and the overall population grows, there may be competition for nesting sites in the existing range, which may promote an expansion in breeding range, if necessary into lower latitude areas where breeding success may be lower than in the high-latitude range.

CONCLUDING REMARKS

It is clearly unwise to assume that, because some birds have altered their migrations in recent decades, all birds can do so. Most of the observed patterns of alteration involve gradual change, in which each incremental step is itself of selective advantage. Patterns that seem resistant to change are those that would involve a marked step-change in some feature, such as direction or fuel deposition, to be viable, while the intermediate steps would be lethal. The birds may thus be locked into some patterns simply because the step-changes in genetic control needed to break free of them are unlikely to arise by a single mutation. Some extreme journeys of today, involving long sea- or desert-crossings, can most plausibly be explained by gradual development, followed by eventual loss of the intermediate stages. Some of the strangest migration routes are likely to have developed in early post-glacial times, when many species were expanding into high-latitude regions recently freed of ice, but these routes might not persist today if the birds could make the step-changes in genetic control necessary to exploit more efficient options. The persistence of long, roundabout migration routes provides a cautionary reminder against the notion that birds invariably behave optimally (but see Alerstam 2001). Apparent legacies of the past are clearly evident in the migration routes of many birds worldwide. Selection cannot act on a clean slate, but must start on a pre-existing gene pool, adapted to different (historical) conditions. Much potential inherent variation may already have been lost.

Some other aspects of the spatial genetic structure of bird populations have been attributed to post-glacial colonisation and migration patterns. For example, many migratory species have more uniform population structure (often manifest in fewer subspecies) than closely related resident species (Belliure *et al.* 2000, Newton 2003). This may be partly because migrants intermix more, and are exposed to selection pressures in several different areas, rather than just one, but also because many migrant species have recently colonised huge areas from one or two glacial refuges. They have therefore not had the time, or consistency in local selection pressures, needed to impart phylogeographic structure to their populations (for examples from warblers, gulls and others see Helbig 2003). Second, since most post-glacial colonisation is north–south rather than east–west, phylogeographic differentiation tends to be less on the main migration axis than off it (for examples in the Yellow Warbler *Dendroica petechia* and North American sparrows see Helbig 2003). Third, highly migratory species have not necessarily been derived from ancestors that were themselves highly migratory, but frequently come from ancestors that were less migratory or sedentary (again, examples from DNA analyses are found among *Sylvia* and *Phylloscopus* warblers, Helbig 2003). In the evolution of migratory behaviour, therefore, the ancestry of a species (its 'phylogeny') seems of far less importance than the ecological circumstances in which it lives. On this view, many species can be treated as statistically independent units in analyses of migratory behaviour, because phylogeny does not to any measurable extent seem to constrain the evolution of migratory adaptations (Helbig 2003). The same probably holds for many other adaptations related to breeding latitude, such as duration and timing of moult, number of broods per season, and so on. In contrast, many other features of birds, such as body form or egg colour, are clearly influenced by ancestry, so that all members of a taxonomic family share the same features, and cannot be considered as statistically independent in these respects.

Many of the ideas proposed in this chapter, although based on a firm observational foundation, are necessarily speculative, but they are essential to a full consideration of the evolutionary and ecological background to current migration patterns. They show how complex migration patterns might have arisen step by step, each of which brings benefits and enhances fitness in its own right. They cannot be tested quantitatively in the field, particularly those that depend on past changes in the nature and distribution of habitats. The results are therefore less satisfying and in a sense 'less scientific' than are the quantitative or experimental studies that are possible on other aspects of migration. Yet they help us to understand how some otherwise puzzling aspects of current bird migration systems might have arisen and, if we disregard them merely because they cannot be formally tested, we risk ignoring some of the most important evolutionary aspects of the subject. Only 30 years ago, almost all aspects of the evolution and genetical control of migration were speculative, and increasingly they have been confirmed or modified by experiment or by genetic studies. Already DNA analyses of certain species have helped to define post-glacial colonisation routes, and confirm that they match some current migration routes. Perhaps, therefore, in the coming years, other as yet intractable aspects of the subject will yield to ingenuity or new methodology.

The underlying message from examination of migration routes is that small step-changes involving successive extensions, or minor modifications to existing routes, each of which brings fitness benefits in its own right, more readily occur than complex and abrupt changes. This conclusion exemplifies a more general principle of evolution, that genetic change must have a starting point, developing from the raw material already available, and within the constraints that this might provide. The idea of constraint for evolutionary change in migratory behaviour stands in some contrast to the great flexibility of those aspects of migration discussed in Chapter 20.

SUMMARY

Some long and indirect migration routes can most plausibly be attributed to birds following ancestral routes of post-glacial range expansion. Some species continue to follow these routes, even though, for birds breeding at the end of the colonisation route, apparently suitable alternative wintering areas are available much closer. Their failure to switch to new routes may be due to the difficulty of making in one step the necessary huge changes in directional preference and fattening regime. In effect, the birds are locked into an existing genetically controlled system, in which small incremental modifications are possible, but not the big single-step change necessary to bring a fitness benefit.

From current migration patterns, possible intermediate stages can be envisaged that might have occurred in the evolution of disjunct migration patterns, in which breeding and wintering areas are well separated, even though such migration might now involve long sea or desert crossings. The same is true for chain and leapfrog migration patterns, both of which can be explained historically. Similarly, migratory divides may mark the meeting points of two populations that recolonised northern areas after the last glaciation, with each population spreading from a different refuge and then following on migration its ancestral colonisation route. However, such patterns are likely to persist only if they serve adequately in present conditions. Populations that were isolated during glacial or other climatic extremes often have distinct DNA (and sometimes also distinct taxonomic status), as well as separate breeding and wintering areas.

Loop migration, in which birds take different routes in spring and autumn, probably occurs in response to seasonal differences in conditions, notably weather (especially wind) and food supply.

In many species, migration is likely to have developed hand in hand with range expansion, as birds spread by normal dispersal into higher latitude areas where they could breed, but from which it was necessary for them to return to lower latitudes to winter. This was probably the prevailing pattern as birds re-colonised high-latitude areas following each glacial retreat. It can be seen in process of development today in some species that are spreading northward within Europe or North America, developing migration secondarily as they reach more seasonal environments. Conversely, other species, introduced to high latitudes by human action, have changed from resident to migratory, utilising lower latitudes in winter. The same probably happened at the start of each glaciation, and as advancing cold changed resident to migratory populations, before eventually obliterating the higher latitude ones altogether.

Whereas roughly one-third of the Nearctic/Neotropical and eastern Palaearctic/Indomalayan migrants winter in tropical forest, almost none of the European/African migrants do so, wintering instead in scrub. This is attributable to the fact that, throughout the glacial periods, forest occurred continuously from tropical to boreal latitudes in the first two regions, but the European–African forests were broken by a broad band of scrub-desert, limiting the movements of forest species, but favouring the movements of scrub-dwelling ones.

When breeding populations expand and become more migratory as they spread from lower to higher latitudes, competition is likely to occur in common wintering areas. This competition could result in: (1) a constraint to further growth in numbers and expansion of breeding range; (2) development of chain migration; (3) development of leapfrog migration; or (4) elimination of breeders from the original range, depending on circumstances, particularly the competitive relationships between breeders from different latitudes and the availability of suitable wintering habitat beyond the original breeding range.

Bar-tailed Godwits *Limosa lapponica*, known for long-distance migrations with few refuelling stops

Chapter 23
Distribution patterns

With increasing knowledge as to the winter distribution of species, and more especially subspecies of migratory birds, it becomes possible to determine more precisely the relationship between the summer-quarters and the winter-quarters of particular forms: this is a line of enquiry that is usefully supplemented by records of individual marked birds, and one that may throw interesting light upon the evolutionary origin of migrations.' (A. Landsborough Thomson 1936.)

One of the central tenets of ecology which has stood the test of time is known as Gause's (1934) Principle. It states that no two closely related species, which have identical requirements and are limited by resources, can persist together in the same area. Being different species, they are by definition unidentical, so inevitably one would be better adapted to exploit their mutual resources, and would eventually out-compete and eliminate the other. To coexist, therefore, closely related species must normally differ either in geographical range, habitat or foraging behaviour, or in more than one of these respects (Lack 1971, Newton 1998b). The relative importance of these segregation mechanisms in migratory birds may differ not only between species, but also between their breeding and non-breeding ranges. Two similar species may separate by geographical range at one season, perhaps, but occupy the same range and separate by habitat or feeding habits at another.

The notion that competition could have influenced the migration patterns of closely related bird species stems from the following observations: (1) some closely related species (and subspecies) segregate geographically in non-breeding as well as in breeding areas; (2) more species segregate geographically in winter, when food is scarcest, than in summer when food is more plentiful; and (3) species that coexist in the same geographical region (called sympatry) typically differ markedly in habitat or feeding behaviour, whereas species that occur in different regions (allopatry) often have similar habitat and feeding behaviour (for details and discussion see Lack 1944, 1971, Lack & Lack 1972, Cox 1968, Greenberg 1986, Newton 1998b). Allopatry arises, it is supposed, because individuals of each species survive or breed better in areas that lack closely competing species than in areas that have them (Salomonsen 1955, Cox 1968, Greenberg 1986, Gauthreaux 1982a). Hence, some common patterns of distribution that occur among migratory birds, or among migrants and residents, have often been attributed to the need to avoid competition, but the evidence is at best circumstantial, and other explanations often cannot be excluded. Nevertheless, all closely related species studied so far show some form of geographical or ecological segregation, except in unusual circumstances, such as temporary superabundance of food (Lack 1971).

Some closely related sympatric species occupy such different habitats that they are spatially segregated almost as effectively as if they occupied different geographical ranges. It is not just a matter of spatial scale, however: habitat differences can result in a patchwork distribution pattern, with areas of each species intermixed in the same region, whereas range differences result in the different species occupying totally (or almost totally) separate regions. It is these patterns of large-scale geographical segregation among migrants that are explored in this chapter.

NON-BREEDING DISTRIBUTION IN CLOSELY RELATED SPECIES

Examples can be found of closely related migratory bird species that: (a) occupy completely (or largely) different regions in both breeding and non-breeding seasons (e.g. European Pied Flycatcher *Ficedula hypoleuca* and Collared Flycatcher *F. albicollis* in Eurasia–Africa); (b) occupy the same region in the breeding season but different regions in the non-breeding season (e.g. Meadow Pipit *Anthus pratensis* and Tree Pipit *A. trivialis* in Eurasia–Africa); (c) occupy different regions in the breeding season but the same region in the non-breeding season (e.g. Tawny Pipit *A. campestris* and Red-throated Pipit *A. cervinus* in Eurasia–Africa); or (d) occupy the same regions in both breeding and non-breeding seasons (e.g. Redwing *Turdus iliacus* and Fieldfare *T. pilaris* over much of their ranges in Eurasia, or Sedge Warbler *Acrocephalus schoenobaenus* and Reed Warbler *A. scirpaceus* over most of their ranges in Eurasia–Africa). Where species occupy the same range, they usually differ markedly in habitat or feeding habits, as indicated above.

Allopatric (or 'alloheimic') winter distributions are found in several groups of closely related migratory species. For example, among the Old World *Hippolais* warblers, the Melodious Warbler *H. polyglotta*, Olivaceous Warbler *H. pallida*, Upcher's Warbler *H. languida* and Icterine Warbler *H. icterina* occupy similar habitats in mainly different parts of Africa, while the Olive-tree Warbler *H. olivetorum* overlaps only with the Icterine, but is much larger (so presumably takes different foods). In the New World, closely related species of several songbird genera occur together in breeding areas, yet show range segregation in the non-breeding season, including species of *Wilsonia*, *Vermivora*, *Dendroica* and *Oporornis* warblers, *Piranga* tanagers, *Catharus* thrushes, and *Empidonax*, *Tyrannus* and *Myiarchus* flycatchers. Despite the many examples, it is impossible to prove that their segregation into geographically distinct wintering areas is a result of competition, but this remains the most likely explanation (Greenberg 1986). Also, in those that show geographical segregation, it is sometimes only the central parts of the range that are separated, with extensive overlap elsewhere. There are thus varying degrees of geographical segregation.

Some closely related species, which overlap in breeding range and diet, separate geographically in winter, with one staying and the other migrating. One possible explanation is that food supplies, which are sufficient for both species in summer, can support only one in winter, so to avoid competition the other leaves. An example is provided by the Herring Gull *Larus argentatus* and Lesser Black-backed Gull *Larus fuscus*, which in Europe share much of the same breeding range, but the latter winters, on average, much further to the south. The fact that the Herring Gull stays behind may be linked with its feeding more on land-based food supplies than the Lesser Black-backed Gull, which feeds more on fish at sea. Another example is provided by the Common Stonechat *Saxicola torquata* (resident over much of Europe) and the Whinchat *S. rubetra* (long-distance migrant between Europe and Africa).

Longitudinal patterns

Many migratory species show consistent east–west patterns in distribution year-round. Most long-distance migrants with wide west–east breeding ranges in Europe also extend in winter from west to east across Africa. Those species that breed only in western Europe (e.g. Melodious Warbler *Hippolais polyglotta*) winter mainly in the western half of Africa; and those species that breed only in eastern Europe or Asia winter mainly in the eastern half of Africa (e.g. Masked Shrike *Lanius nubicus*). There are exceptions, however, such as Lesser Whitethroat *Sylvia curruca* and Red-backed Shrike *Lanius collurio*, in which the birds from western Europe migrate via eastern Europe to eastern Africa, where they join more eastern breeders.

Similarly, in the New World, species that breed only in western North America winter only on the western side of Central America or beyond (e.g. Hammond's Flycatcher *Empidonax hammondii*, Townsend's Solitaire *Myadestes townsendi*, Grey Vireo *Vireo vicinior*), while others that breed only in eastern North America winter only in the eastern side of Central America and the Caribbean Islands or beyond (e.g. Yellow-throated Vireo *Vireo flavifrons*, Northern Parula *Parula americana*, Prairie Warbler *Dendroica discolor*). As in Eurasia, however, some species with a wide longitudinal spread in breeding distribution concentrate in winter in either eastern or western parts of a potential wintering range (e.g. Rusty Blackbird *Euphagus carolinus*, which breeds from coast to coast across boreal North America but winters entirely in southeastern parts of the continent). These species form exceptions to the more or less parallel migrations shown by most species with transcontinental breeding ranges.

Closely related migrant species tend to segregate not only from one another in their breeding or wintering areas, but also from similar species that are resident year-round in those areas. Among Eurasian–Afrotropical migrants, chats, wheatears and shrikes generally occur in African habitats devoid of resident species, and segregate from one another, while various silviid warblers overlap broadly with residents, but forage in different parts of the habitat (Moreau 1972). In the New World, such patterns are commonplace, and ecological segregation is often evident from detailed study. For example, 20 different warbler species occur in Jamaica, two being present year-round and the other 18 in winter only (Lack & Lack 1972). These various species show marked ecological differences from each other: one climbs branches and twigs, one feeds from hanging dead leaves and twigs, one lives in thick herbage, and one fly-catches. Of the five ground-feeders, four are separated by the type of terrain, while two that inhabit the same terrain are separated by feeding method – one probes and the other picks. Of the 11 leaf-gleaners, the two resident species are separated largely by habitat, the rest mainly by feeding method, the parts of the tree in which they forage, or the type of leaf from which they take insects. Hence, even where a relatively large number of similar species occur in the same geographical area, each differs ecologically from all the others in a way that would be expected to lessen competition for food between them.

Latitudinal patterns

Many pairs of closely related species show so-called chain migration, in which the different species maintain the same latitudinal sequence in the non-breeding season as in the breeding season but at lower latitudes: the species that breeds furthest north also winters furthest north, and the species that breeds furthest south also winters furthest south (Chapter 22). An example is provided by the Lapland Longspur *Calcarius lapponicus* and Chestnut-collared Longspur *C. ornatus* in North America; both are migratory, but the latter both breeds and winters south of the former. In some cases the northern species replaces the southern one as it vacates its breeding range in autumn to move even further south.

In other pairs of closely allied species, the member that breeds furthest north in the northern hemisphere winters furthest south, to give 'leapfrog migration', reversing their latitudinal sequence between summer and winter (Chapter 22). Some species pairs among leapfrog migrants differ markedly in body size, and in each case the smaller species breeds furthest north, in the coldest areas, and on average migrates to winter furthest south in the warmest area. Together they follow 'Bergman's Rule', in that higher latitude forms are larger, but in such leapfrog migrants the pattern holds only for their winter and not their summer distributions. Examples include the finches *Carpodacus rosea* and *C. erythrina*, the knots *Calidris tenuirostris* and *C. canutus*, the godwits *Limosa limosa* and *L. lapponica*, and the curlews *Numenius arquata* and *N. phaeopus*, but with some geographical overlap between the species in each pair (Salomonsen 1955).

NON-BREEDING DISTRIBUTION IN DIFFERENT POPULATIONS OF THE SAME SPECIES

Because of their greater similarity to one another, individuals from different populations of the same species are likely to overlap more in habitat and food preferences than are individuals from different species, and hence compete more strongly. Not surprisingly, therefore, some of the same patterns of geographical segregation that occur among closely allied species also occur among different populations of the same species, and individuals from different parts of the breeding range often migrate to mainly different parts of the wintering range. Such patterns are not always easy to detect because their demonstration usually requires abundant ring recoveries from different parts of the breeding and wintering ranges.

Longitudinal patterns

Typically, birds that breed furthest west in the breeding range tend to winter furthest west in the non-breeding range, and those that breed furthest east also winter furthest east (**Figure 23.1**). Allowing for the uneven distribution of land masses, this holds in both the Old and New Worlds, reflecting the more or less parallel migrations of populations. Such longitudinal patterns have been revealed by ring recoveries from most groups of birds (Lincoln 1935b, Moreau 1972, Newton 1972, Holmes & Sherry 1992, Hoffman *et al.* 2002), and in some groups they have also been broadly confirmed by isotope or DNA analyses, as described in Chapter 2 (for Dunlin *Calidris alpina* see Wenink *et al.* 1996, Wennerberg 2001; for Swainson's Thrush *Catharus ustulatus* see Ruegg & Smith 2002; for Wilson's Warbler *Wilsonia pusilla* see Kimura *et al.* 2002).

Based mainly on findings from waterfowl, Lincoln (1935b) proposed that migratory birds in North America followed four main flyways, west to east: the Pacific, Central, Mississippi and Atlantic Flyways. This division soon came to form the political basis for waterfowl management and hunting throughout the continent. For the most part, it reflects biological reality, because of the more or less parallel migrations. The concept was further developed by Bellrose (1968) who used radar surveillance and ringing records of migratory waterfowl to describe the presence of migration corridors within the flyways. He defined a corridor as a narrow strip of airspace used by waterfowl as they migrated between their breeding and wintering grounds. In some places the corridors were little more than 16 km wide, in others more than 200 km. Such more-or-less parallel migrations need not necessarily involve competition, however, for they could be explained by the birds from different segments of the breeding range taking the shortest routes to their wintering areas at lower latitudes. Other factors being equal, any birds that deviate from the parallel pattern, and cross the routes of other populations, are likely to lengthen their journeys, and thus have greater migration costs.

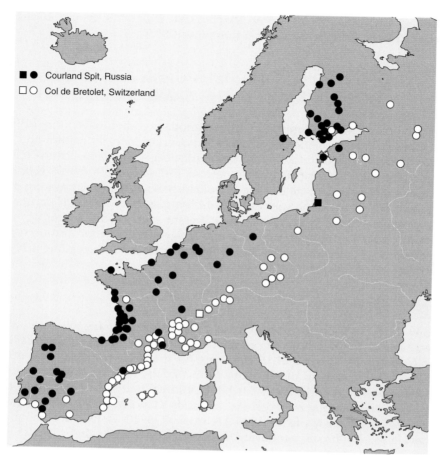

Figure 23.1   Examples of parallel migration among west Palaearctic breeding birds, from ring recoveries. (a) Chaffinches *Fringilla coelebs* ringed at the Courland Spit, Russia (filled square, filled dots) and at the Col de Bretolet (open square, open dots). From Bairlein (1998).

The migration corridors of some species are narrow over almost the whole route, as in some cranes, geese and shorebirds that travel between traditional widely spaced stopover and wintering sites. For example, Whooping Cranes *Grus americana* from a small breeding area in northern Canada migrate along a narrow corridor direct to their only wintering site at Aransas, on the Texas coast, then back along the same route (**Figure 23.2**). Different populations of geese remain separated year-round as they move from different breeding localities, along different flyways to their own traditional wintering areas. Examples include Barnacle Geese *Branta leucopsis* and Pink-footed Geese *Anser brachyrhynchus* in both of which birds from three separate breeding areas have distinct migration and wintering areas (**Figure 23.3**). Breeding and wintering localities may increase in number as a population expands, but localities that lie off the migration routes are seldom, if ever, visited by such species. Similarly, many shorebirds tend to

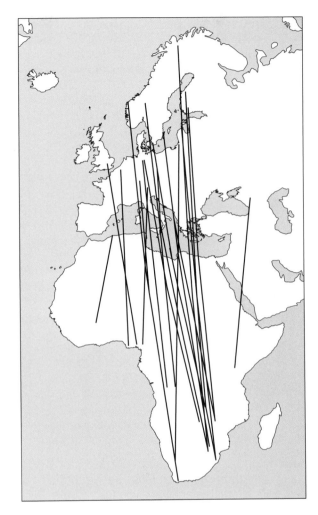

Figure 23.1   (Continued) (b) Northern House Martins *Delichon urbica* from Europe to Africa. From Hill (1997).

travel along coastlines which provide potential stopping sites, if needed. This situation of narrow, separated routes is no different in principle from the parallel but broad-front migration of most widespread species, except that the different breeding or wintering areas are localised and well separated.

Whatever their history, the different populations (or subspecies) of some widespread species show almost total separation year-round, whether in breeding, migration or wintering areas. An example is the Red Knot *Calidris canutus*, which breeds at high arctic latitudes throughout the world, and performs long and complex migrations to reach wintering areas that extend from northern temperate to southern temperate regions. The population associated with each breeding area migrates to a discrete wintering area (**Figure 23.4**). In other shorebird species, different populations have separate breeding and wintering areas, but come together

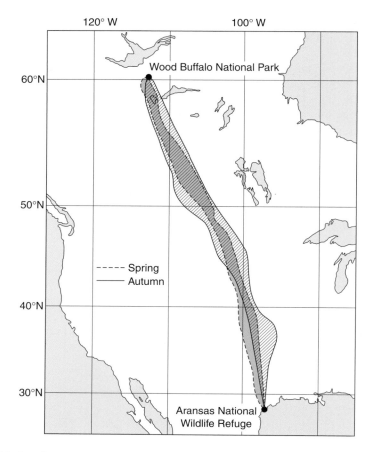

Figure 23.2 Autumn and spring migration corridors of Whooping Cranes *Grus americana*, as defined by radio-tracking, between their sole breeding area in Canada and their sole wintering area at Aransas, on the Texas coast. The map is a Mercator projection. Based on Kuyt (1992) and Alerstam (1996).

at migration times, often mingling on their shared staging sites. For example, in autumn and spring, Ruddy Turnstones *Arenaria interpres* from two populations occur in Britain, one from northeast Canada and Greenland and the other from the Baltic region. The former stay all winter, while the latter pass through to winter in West Africa.

Many other examples of partially allopatric winter distributions are found among waders in Britain, with birds from breeding areas to the northwest (Iceland, Greenland and Canada) wintering mainly in Ireland and western Britain, and birds from breeding areas to the northeast (Eurasia) wintering mainly in eastern Britain, the Common Snipe *Gallinago gallinago* providing a good example. Similarly, the Black-tailed Godwits *Limosa limosa* that breed in Iceland winter in the British Isles, while those that breed in continental Europe winter in Europe and western Africa. Although not strictly parallel migrations, they still result in birds maintaining the

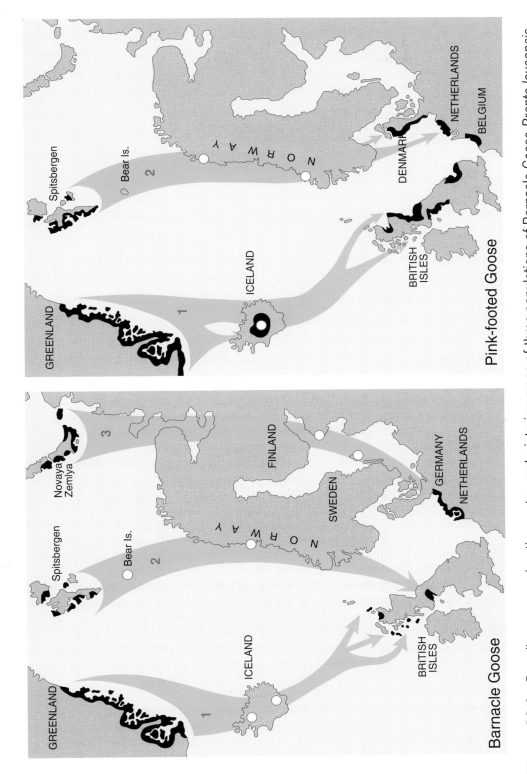

Barnacle Goose

Pink-footed Goose

Figure 23.3 Breeding areas, migration routes and wintering areas of three populations of Barnacle Geese *Branta leucopsis* and two populations of Pink-footed Geese *Anser brachyrhynchus*. Black shading depicts the separate breeding and wintering areas, grey the migration routes, and white spots the main stopover sites. From Madsen *et al.* (1999).

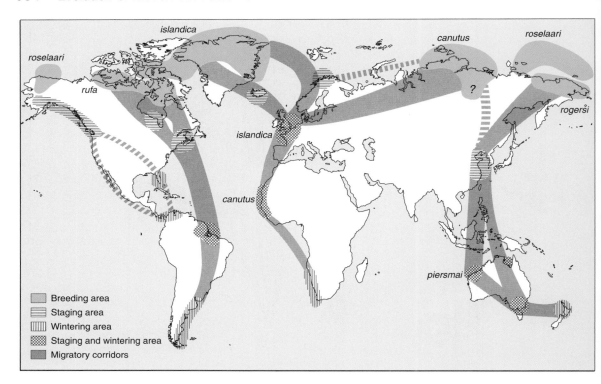

Figure 23.4 Year-round separation of different subspecies of Red Knots *Calidris canutus*. These birds breed at high arctic latitudes around the world, different subspecies in different regions. Vertical stripes – wintering areas; horizontal stripes – stopover areas used only during southward and northward migration; cross-hatched – areas used for both stopover and wintering; shaded corridors – proven migration routes; broken shaded corridors – possible migration routes on which more information is needed. The population *islandica* breeds in eastern Canada–Greenland, and migrates through Iceland to winter in western Europe. The Siberian *canutus* stages in the Wadden Sea, across the English Channel from *islandica* birds, but migrates on to winter on the western coasts of Africa. The subspecies *rogersi* breeds in eastern Siberia and winters in southeast Australia and New Zealand. Other birds (considered *piersmai*) breed on the New Siberian islands and winter in northwest Australia. The race *roselaari* breeds in far eastern Siberia and northern Alaska and probably overwinters in Florida–Texas, and possibly the northern coast of South America. The race *rufa* breeds in the central Canadian arctic and winters on the Atlantic coast of Argentina. No Red Knots are known to winter around the Indian Ocean, except for vagrants. This example, along with many other shorebirds, geese and others, indicates how populations maintain their distinctive distributions year-round (Piersma & Davidson 1992).

same relative west–east sequence in both breeding and wintering areas (Wernham *et al.* 2002).

In parallel migration, then, allowing for geography, species maintain the same west–east distribution with respect to one another year-round. In contrast to this pattern is fan (or funnel) migration, in which birds from a small part of the breeding

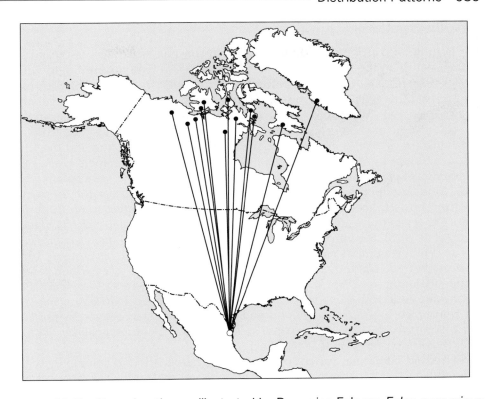

Figure 23.5   Fan migration as illustrated by Peregrine Falcons *Falco peregrinus* trapped and attached with radio-transmitters during winter on the Gulf Coast of Mexico (north of Tampico) and subsequently tracked to breeding areas across the Arctic. Lines connect the winter trapping site with the subsequent breeding sites but do not necessarily depict the routes flown. From details in McGrady *et al.* (2002).

range can spread out to occur across a wide part of the wintering range or vice versa (**Figure 23.5**). This results in the individuals in particular wintering localities being drawn from a wide longitudinal spread in the breeding range or, conversely, individuals in particular breeding localities being drawn from a wide longitudinal spread in the wintering range. There is no clear dividing line between parallel and fan migration, but rather different species reveal a continuum of variation between these extremes, depending partly on geography. In other words, species show different degrees of year-round longitudinal segregation. Extreme mixing would be expected to be rare among most kinds of birds, because it would entail some individuals migrating much further than they need. However, many raptors and other soaring species become funnelled on migration through points with narrow land bridges (such as Panama) or short sea-crossings (such as Gibraltar) (Chapter 7). Their various migration routes are shaped like an hourglass, constrained in the middle.

Latitudinal patterns

The latitudinal distribution of populations in the non-breeding season is more complicated than the longitudinal. At least three main patterns are found within species (**Figure 23.6**). In the first pattern, of chain migration, more northern (usually larger) birds may replace others as they move south, so that the same areas are occupied year-round, but by one population in summer and another in winter. An example of chain migration within a species is provided by the Common Redshank *Tringa totanus* in which Icelandic birds move to Scotland and the North Sea area, while local breeding birds move further south (Summers *et al.* 1988). Another example is provided by the Purple Sandpiper *Calidris maritima*, in which birds from four populations move to the nearest ice-free coast for the winter, replacing birds from two other populations that move further south (Summers 1994). Other examples are provided by the Chiffchaff *Phylloscopus collybita* in Iberia (Catry *et al.* 2005), the European Goldfinch *Carduelis carduelis*, Linnet *Carduelis cannabina*, White Wagtail *Motacilla alba*, Eurasian Sparrowhawk *Accipiter nisus*, Tufted Duck *Aythya fuligula* and Common (Mew) Gull *Larus canus* over wider parts of Europe (Salomonsen 1955, Kilpi & Saurola 1985, Siriwardena & Wernham 2002), and by the American Coot *Fulica americana*, Sharp-shinned Hawk *Accipiter striatus*, Hermit Thrush *Catharus guttatus* and American Redstart *Setophago ruticilla* in North America, the two latter patterns established by isotope analyses (Phillips 1951, Ryder 1963, Smith *et al.* 2003, Norris *et al.* 2006).

The second latitudinal pattern of leapfrog migration is found commonly within various species of passerines, raptors, waders, waterfowl, gulls and other seabirds (Salomonsen 1955, Moreau 1972, Alerstam & Högstedt 1980, Kilpi & Saurola 1985, Bourne 1986, Wallin *et al.* 1987, Boland 1990, Wood 1992, Siriwardena & Wernham

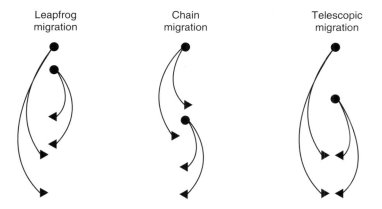

Figure 23.6   Three common patterns of latitudinal replacements in the migrations of various bird species. Partly from Salomonsen (1955). Chain migration in which populations occur in winter in the same latitudinal sequence as on their breeding areas. Leapfrog migration in which populations occur in winter in reverse latitudinal sequence as on their breeding areas. Telescopic migration, in which populations from different breeding areas occur together in the same wintering area.

2002). Well-known examples are provided by the Fox Sparrow *Passerella iliaca* west of the Rockies in North America (Swarth 1920, Bell 1997) and by the Common Ringed Plover *Charadrius hiaticula* in Europe (**Figure 23.7**; Salomonsen 1955). In the last species, the British population is mainly resident, while the southern Scandinavian birds winter in southwest Europe and the northernmost Scandinavian–Siberian birds winter largely in Africa. Icelandic birds also migrate to Africa, thereby leap-frogging the European ones. Among seabirds, Common Terns *Sterna hirundo* from Britain winter around the western bulge of Africa mainly north of the equator, while those from Scandinavia winter mainly south of the equator on the western coast (Wernham *et al.* 2002). Cory's Shearwaters *Calonectris diomedea* from the Cape Verde Islands winter mainly in latitudes to 24°S, whereas those from more northern colonies winter south to 48°S (Bourne 1986). In all these examples, leapfrog patterns were established by ringing, but in recent years such patterns have also emerged through isotope analyses (for Wilson's Warbler *Wilsonia pusilla* see Clegg *et al.* 2003).

Leapfrog migrations could have arisen in post-glacial times through competition, if populations spreading progressively further north to breed had to migrate ever further south to avoid competing with other populations and find unoccupied wintering areas (Chapter 22). Without such competition, the most northern breeding populations would be expected to shorten their journeys. In some species that show leapfrog migration, the northern breeding birds are smaller than the southern ones (see later). As in closely related species pairs, they follow

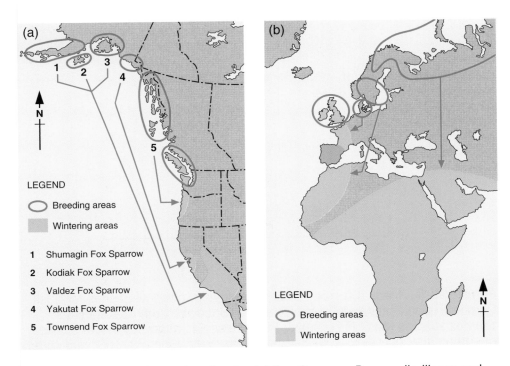

Figure 23.7   Leapfrog migration in: (a) Fox Sparrow *Passerella iliaca*; and (b) Common Ringed Plover *Charadrius hiaticula*. From Swarth (1920) and Salomonsen (1955).

Bergman's Rule that high-latitude forms are larger, but the trend holds for wintering rather than for breeding distributions.

In the third latitudinal pattern, populations that breed over a wide span of latitude become telescoped in the non-breeding season into a narrower span, so that birds from different breeding areas intermix in winter. Populations that are allopatric in summer thus become sympatric (or 'synheimic') in winter. This pattern is illustrated by Common Grackles *Quiscalus quiscula* and Common Starlings *Sturnus vulgaris* in eastern North America (Dolbeer 1982; **Figure 23.8**), Rosy Finches *Leucosticte arctoa*, *L. tephrocotis tephrocotis* and *L. a. littoralis* in western North America (King & Wales 1964), several races of Yellow Wagtails *Motacilla flava* in parts of Africa (Salomonsen 1955, Curry-Lindahl 1958, Cramp 1988), and by various races of some Nearctic–Neotropical migrant species in parts of Mexico (Ramos & Warner 1980). Mixing among birds from widely different breeding areas also occurs in winter in various seabirds, including Black-legged Kittiwake *Rissa tridactyla*, Brünnich's Guillemot (Thick-billed Murre) *Uria lomvia*, Northern Fulmar *Fulmarus glacialis*, Sandwich Tern *Sterna sandvicensis* and others (Salomonsen 1955). In many other species, northern migratory races winter within the range of more southern resident ones, as in Richard's Pipit *Anthus novaeseelandiae* (Salomonsen 1955, Cramp 1988).

Telescopic migration also applies to migration from wintering to breeding areas, as populations wintering over a wide span of latitude become concentrated for breeding within a narrower span. Many examples occur among shorebird species which occupy coastlines over much of the world in winter, but withdraw to breed in a relatively narrow span of arctic tundra.

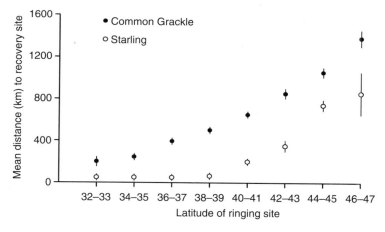

Figure 23.8   Mean distance (±SE) from ringing to recovery sites for adult Common Grackles *Quiscalus quiscula* (N = 855) and Common Starlings *Sturnus vulgaris* (N = 1116), illustrating telescopic migration in two species ringed on their breeding areas in eastern North America (75–100° longitude), and recovered in winter (January–February) further south, in a narrower span of latitude. In both species, birds from more northern breeding areas travelled further, on average, but wintered in the same latitudinal band as birds from more southern breeding areas. From Dolbeer (1982).

In chain migration, the distances may be more or less similar among populations, even though these populations migrate between different areas, but in the other two latitudinal patterns the high-latitude populations migrate furthest, with the greatest population differences occurring in leapfrog migrants. It is impossible to say from the data available how common these different patterns of latitudinal segregation are among migrants, or whether they differ in frequency between regions. These questions could best be answered by comprehensive widescale analyses of ring recoveries.

EVOLUTION OF ALLOHEIMY

Whatever the pattern of geographical segregation among wintering migrants, if such differences are genetically influenced, through migratory behaviour, it is not hard to envisage how they might have come about (**Figure 23.9**). Imagine that populations which breed in separate areas come together on common wintering grounds. If they were limited by food in the wintering area, and individuals of one population were better adapted to that area, they would in time be expected to eliminate individuals of the other population completely, or force them by selection to winter elsewhere. As a second scenario, imagine that, under food shortage in their joint wintering area, they competed on equal terms, with the same proportion of each population surviving the winter. If individuals of one population had, on average, a consistently higher reproductive rate than the other, then in time they would be expected to replace individuals of the other population completely in their shared wintering area, either eliminating them altogether or forcing them to winter elsewhere (Chapter 22). This is one way in which a sedentary population

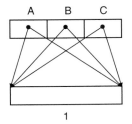 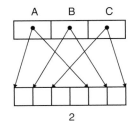 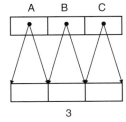

Figure 23.9 Migration of three neighbouring populations (A, B and C) showing: (A) synheimy (complete sharing of wintering areas), (C) alloheimy (complete separation of wintering areas) and (B) an intermediate situation (partial separation of wintering areas). The upper rectangles indicate breeding areas, the lower rectangles wintering areas, and the arrows the scattering of migrants from each breeding area. A gradual change from pattern 1 to 3 could come from selection resulting from competition between individuals from different breeding populations or from selection to reduce migration distances, eventually giving rise to 'parallel migration' (see text). Pattern 1 could be maintained or developed in species that utilise sporadic habitats or food supplies, whose distribution within the winter range varies markedly from year to year (as for some irruptive migrants, Chapters 18 and 19). To judge from ring recoveries, the intermediate situation (2) is the commonest. From Salomonsen (1955).

could become migratory, under pressure of competition from individuals of another population increasingly wintering in the same area (Sutherland & Dolman 1994, Bell 2000). Likewise, imagine that birds in a given breeding area, where their numbers were limited, divided between two wintering areas. If the birds with a genetic predisposition to go to one wintering area survived consistently better than the birds visiting the other area, the first group would eventually replace the second completely. In these ways, competition could act as a selective force behind a genetic change in migratory habits, leading ultimately to year-round geographical segregation of populations.

Whether the development of alloheimy leads to chain or leapfrog patterns could depend on the competitive relationships between individuals from the different populations, as explained in Chapter 22 (Cox 1968, Holmgren & Lundberg 1993,[1] Bell 2000). If individuals in a higher latitude population settling in a wintering area were competitively superior to the individuals of the same species resident there, the immigrants could either eliminate the residents, or force them to become migratory, moving to lower latitudes for the non-breeding season, and thus setting up a system of chain migration (Bell 2000). If, on the other hand, the northern birds were inferior in competition with the more southern residents, the northern ones would benefit from moving to yet lower latitudes, thus becoming leapfrog migrants. Hence, whether one or other system developed could depend on the competitive relationships between individuals from different populations. Such competition might affect the behaviour of all members of the competing inferior population, leading to total segregation in wintering areas, or only some of them, leading to partial segregation in wintering areas.

This raises the question of how synheimy could persist, with populations from different breeding areas wintering together in the same area and taking the same foods. One way would be if they were each limited in numbers on their breeding areas, so that collectively they did not reach the limit of their shared wintering area, and hence did not compete seriously for resources there. Another way would be if they were limited either in breeding or wintering areas by factors other than resources, say by predators or parasites, which held the populations below the level at which serious competition for food occurred. A third way in which synheimy could persist is if the different populations differed in their ecology, reducing any potential competition between them. For example, in many areas resident populations of some species stay near their breeding places year-round, eating whatever foods are available there, while winter visitors of the same species from elsewhere move around from place to place, exploiting temporary abundances of food such as fruit crops or insect outbreaks (for Eurasian species in Africa see Chapter 24). The survival of both residents and migrants in the

---

[1]*Holmgren & Lundberg (1993) formalised patterns of migration depending on whether dominance at a given wintering area was mediated by prior occupancy or by body size, latitudinal gradients in suitability for breeding and wintering and the cost of migration. In their models, chain immigration would result if both breeding and wintering suitability gradients increase towards lower latitudes and dominance is due to body size, or if breeding suitability increases to higher latitudes and wintering suitability to lower latitudes, and dominance is due to prior occupancy.*

same area depends on the mobility of the migrants, and their ability to seek out locally abundant food supplies which are often greater than the local residents can consume themselves in the time available. Resident and migratory populations may also differ in habitat use, as found among Blackcaps *Sylvia atricapilla* wintering in Spain (Pérez-Tris & Tellería 2002). Fourthly, synheimy of different breeding populations may also persist in species whose habitats and food supplies are sporadically distributed, changing in location from year to year. This situation exists for seed-eating boreal finches, vole-eating raptors and owls, and many freshwater ducks (Chapters 18 and 19). It leads individuals to move around, within and between winters, and prevents their developing strong winter site-fidelity, which underpins the evolution of alloheimy. Finally, it is also possible that some of the examples of synheimy in closely allied taxa that we observe now are recent, arising from human land use, or are transitory, as birds change their behaviour and shift gradually from synheimy to alloheimy.

Habitat segregation between resident and immigrant Robins *Erithacus rubecula* was examined in Spain (Tellería & Pérez-Tris 2004). In September, before the migrants arrived, local Robins were found only in forests, but after migrants had arrived, large numbers were also found in shrub habitats. Some migrants replaced residents in the forest, but in general the majority of residents remained in the optimal forest habitat, while the majority of migrants occupied the scrub (the two types being distinguished on morphological criteria). It seemed that local Robins benefited from prior occupancy of the best habitat, forcing the migrants to occupy apparently less suitable sites. This may be how small southern populations survive despite the annual incursion for the winter of large numbers of conspecifics from higher latitudes.

In consideration of these patterns, body size comparisons between northern hemisphere populations are revealing. In some chain migrants, individuals from northern breeding areas tend to be larger, so it is not hard to imagine that, through aggressive encounters, they could displace smaller individuals from more southern areas, which would then benefit from moving even further south. When both populations move back north in spring, the smaller birds occupy the ground vacated by the large ones. In contrast, in some leapfrog migrants, such as Common Ringed Plover *Charadrius hiaticula*, Common Redshank *Tringa totanus*, Common Kestrel *Falco tinnunculus* and Peregrine Falcon *F. peregrinus*, it is the individuals from northern populations that are smallest, so they would tend to lose in aggressive interactions with more southern birds and have to move on further south, giving a leapfrog pattern, eventually fixed by natural selection. However, it remains to be seen whether such differences in body size patterns are consistent across all chain and leapfrog migrants. In any case, it is impossible to tell whether the body size differences between populations evolved before or after the development of their migration patterns (for alternative models of the evolution of leapfrog migration see Cox 1968, Bell 1996, 1997).

Competitive superiority of one wintering population over another could arise in a number of different ways: the individuals in one population may be larger, and hence dominant in aggressive interactions or better able to resist cold; they may gain the benefits of prior occupancy if they are already established when individuals of the second population arrive, or they may differ in structure and behaviour in a way that enables them to more efficiently exploit local resources, or better avoid predators and parasites. Alternatively, they may, in their particular breeding range, be able to achieve a higher mean per capita reproductive rate, and simply 'outbreed' the other population, gradually replacing it in their common wintering range where overall numbers are limited (Chapter 22). This latter situation has been explored in a demographic model for a migratory shorebird, the Oystercatcher *Haematopus ostralegus*, in which different breeding populations shared the same wintering area. Initially, as winter habitat was progressively removed in simulations, all populations decreased in parallel. However, as habitat loss continued, the populations with lower fledgling production began to be disproportionately affected (Goss Custard *et al*. 1995).

These various considerations suggest that intraspecific competition could be a major driving force behind geographical segregation patterns among migratory bird populations, at least for latitudinal segregation. For longitudinal segregation, minimisation of flight costs could also be involved. Various mathematical models have been devised to explore these patterns further. They indicate that such factors as latitudinal gradients in habitat suitability, migration costs, and annual time budgets, could all add to asymmetric competition as potential selective forces behind different patterns of geographical segregation, including chain, leapfrog and overlap patterns (Cox 1968, Greenberg 1980, Lundberg & Alerstam 1986, Holmgren & Lundberg 1993, Bell 1997, 2000). But on present knowledge, we have no way of judging which selective forces are likely to have applied in particular cases.

NON-BREEDING DISTRIBUTIONS IN AGE AND SEX GROUPS OF THE SAME POPULATION

If seasonal food reduction is the main underlying reason for migration, and if individuals compete for food, one would expect that, through social dominance, some individuals would benefit more than others from migrating. In many bird species, as explained in Chapter 15, juveniles migrate in greater proportion and further than adults, while females migrate in greater proportion and further than males. This leads to partial segregation in the winter distributions of different sex and age groups from the same breeding population.

Social interactions are held as the basis for many of the age and sex differences in migration that occur within species, and for the fact that individuals of many species migrate in the early years of their life but not later (Chapter 15). In general, the dominant individuals, taking precedence over resources, are less likely to leave their breeding range than are subordinates, or if the dominants do leave, they migrate less far, or stay away for shorter periods (Gauthreaux 1978a, 1982a). In winters of unusually poor food supply, the benefits of migration may extend further up the social hierarchy to a greater proportion of individuals than usual. This type of competition could influence the migration and distribution patterns of birds directly, and independently of genetic factors, whereas most of the differences between species are presumed to be genetically determined, as a result of competition in the past.

ALTERNATIVE MODELS:
TIME AND ENERGY CONSIDERATIONS

While one idea assumes that wintering areas nearest the breeding areas are best, and that further migration results from competition, an alternative 'time allocation' model emphasises the benefits and costs of migration in influencing the optimal wintering area. Imagine that, within the potential wintering range, sites progressively further from the breeding range are more benign, so that day-to-day survival is higher there, but that migration costs increase with increasing distance from breeding areas (Greenberg 1980). The longer the period in each year that birds spend away from their breeding areas, the greater their survival benefits, compared to sites closer to the breeding area. Greenberg supposed that natural selection would lead to the non-breeding area being located wherever the benefits of improved survival most exceeded the costs of reduced survival imposed by migration to and from it. If migration costs were the same, regardless of date and length of journey, then longer movements should be favoured. The benefits should be especially great in high-latitude species which generally spend least time on breeding areas and most on wintering areas, and therefore gain the benefits of enhanced survival for the longest time each year. This could give another explanation for leapfrog migration in which the most northerly breeding populations winter furthest south (but does not eliminate a role for competition). The key questions centre on the relative costs of migration early and late in the season and of short and long journeys, and on what are the survival benefits in more benign climates.

A pattern of decreasing living costs with decreasing latitude has been found in arctic-nesting Sanderlings *Calidris alba* from measurements of energy consumption (using the doubly-labelled water technique) in free-living birds wintering in New Jersey, Texas, Panama and Peru respectively (Castro *et al.* 1992). The minimal living cost (observed in Panama) was equivalent to twice the basal (resting) metabolic rate ($2 \times$ BMR), while the peak value (observed in New Jersey) was roughly twice this value ($4 \times$ BMR), reflecting differences in winter temperatures, and hence in thermoregulatory costs at the two sites. These findings thus confirmed the large savings in body heat maintenance enjoyed by individuals migrating to the tropics.

Other studies of Bar-tailed Godwits *Limosa lapponica* and Red Knots *Calidris canutus* have measured the energy costs of both wintering and migration. In the Bar-tailed Godwit, the Eurasian–African populations occur in winter in a typical leapfrog pattern (Drent & Piersma 1990). The European population breeds in northern Europe and winters on the coasts of western Europe, especially the Wadden Sea. The Siberian–African population breeds further east and at higher latitude, and winters on West African coasts. The latter birds undertake two or more long-distance flights on their migrations, stopping at some of the same estuaries in Europe where the north European birds winter. They have to cover 8300 km one way, and spend much more time on migration than birds of the European population (2500 km one way). Energy costs per day in wintering areas were calculated at $3 \times$ BMR in Europe and at $2 \times$ BMR in Africa, the difference attributable to ambient temperatures. The energy cost of migration looms large in the annual energy budget of the long-distance migrants of the Siberian–African population (48% of annual expenditure) compared with the European population (22%). The Siberian–African birds experience peak energy demands during pre-migratory fuel deposition before the final leg of the spring migratory journey (the 4000 km separating the breeding area from the Wadden Sea), in the same places where the European birds sustained a high cost throughout their non-breeding season. In the Siberian–African birds, the costs of pre-migratory fattening in spring were so high that they offset the savings in thermoregulation costs during their stay in Africa. The overall annual energy costs did not differ greatly between the two populations, bearing in mind the greater migratory weight of the Siberian birds. We must therefore assume some advantage to wintering nearer the breeding area to explain why any godwits remain at that time in the seemingly inhospitable coastal region of Europe. Competition between the two populations may be the key factor involved, neither the European nor the African wintering areas being able to accommodate both populations for the whole winter.

However, another explanation is relevant to explaining the distribution of the two godwit populations. This comparison between them highlights the time constraints imposed by the lengthy process of accumulating the reserves required for long migratory journeys. The Siberian–African population expends the energy accumulated during 24 hours of spring feeding on the Banc d'Arguin in Mauritania in just over one hour of migratory flight. On the basis of these findings, Drent & Piersma (1990) suggested that:

> 'Africa is available as a wintering site only for populations nesting in the high arctic where the relatively late advent of conditions favourable for breeding provided the leeway necessary for fitting in the three months of preparatory staging. Conversely, the potential of the vast Siberian breeding range can only be realised because of the extensive capacity of the winter quarters in Africa, which in turn can only be reached thanks to the Wadden Sea situated as a stepping stone between the two.'

In contrast, the birds that winter in Europe migrate to the north of the continent, where spring comes earlier than in Siberia. They do not need such large body reserves, and can fatten in a shorter period.

Clearly, the evolutionary interpretation of migration patterns will not be solved by recourse to the role of competition alone, but must include assessment of the energy considerations related to different journey lengths, breeding and wintering areas (Drent & Piersma 1990). The main constraint on the numbers of godwits wintering in West Africa was the low levels of food available there. The West African tropics provided a low-cost/low-yield wintering option, while western Europe provided a high-cost/high-yield contrast. Because godwits can feed at night, as well as by day, they are not constrained by daylengths, only by tidal rhythms that are similar between areas.

The Red Knot *Calidris canutus* breeds in the high arctic, but spends its non-breeding season at specific localities spanning a wide range of latitude, from northern temperate through tropical to southern temperate regions (**Figure 23.4**). Although many individuals winter at British latitudes, the cold makes this location almost prohibitively expensive for them. During November–February, their daily energy needs amount to $4 \times BMR$ (Piersma *et al.* 1991). This is despite the fact that birds adopt all the energy-saving behaviours they can, such as feeding and roosting in dense packs, facing head to wind (to prevent the cold getting under their feathers), and reducing the blood supply to their exposed legs. At lower latitudes, the energy costs of keeping warm are much reduced. Largely because of the temperature difference, the maintenance costs of Red Knots in West Africa were about 40% less than in Europe (although the food supplies were also less good). In fact, in tropical areas Red Knots and other shorebirds may face the opposite problem to keeping warm – that of getting rid of surplus heat – which some species solve by resting on water when they are not feeding.

Interestingly, Red Knots wintering further from their breeding areas do not necessarily incur greater energy costs on migration (Piersma *et al.* 1991). This was shown in a comparison between two subspecies, one (*C. c. canutus*) breeding in Siberia and migrating via western Europe (the Wadden Sea) to West Africa (a round trip of 18800 km), and the other (*C. c. islandica*) breeding in northeast Canada and Greenland, and migrating via Iceland to winter mainly in eastern England and the Wadden Sea (round trip of 9600 km) (**Table 23.1**). Most individuals performed these migrations in two non-stop flights separated by a single stopover. Although *canutus* knots migrated twice as far as *islandica* ones, the costs of migration were estimated to be of similar magnitude, apparently due to differences in tailwind availability en route. Both there and back, *canutus* birds flew parallel to the main weather systems and could always find following winds, sometimes up to 5 km above ground. Such winds could almost double the flight speed of *canutus* birds, giving them much shorter flight times and reduced travel costs.

These studies thus led to two conclusions: (1) for birds wintering further from their breeding areas on the same route, travel costs may be greater, but maintenance

Table 23.1   Travel distances and approximate annual average expenditure on long-distance flights by temperate-wintering Red Knots *Calidris canutus* of the subspecies *islandica* (Ellesmere Island to the Netherlands) and tropically-wintering Red Knots of the subspecies *canutus* (Siberia to West Africa). Cost factors for long-distance flights were calculated from estimated fat losses over the journey

Route	Subspecies	
	islandica	*canutus*
West Africa–Wadden Sea (km)		4600
Wadden Sea–Taimyr (km)		4800
Wadden Sea–Iceland (km)	2100	
Iceland–Ellesmere Island (km)	2700	
Total one-way (km)	4800	9400
Total return (km)	9600	18800
Travel cost (kJ/km)	0.7–0.8[a]	0.3–0.5[a]
Annual flight cost (kJ/year)	6720–7680	5640–9400
Average cost on annual basis (W)	0.22–0.25	0.18–0.30

[a]Estimated from measures of fat loss on migration of 0.8 kJ per km for southeast England to northern Norway; 0.7 kJ per km for northern Norway to Ellesmere Island; 0.3 kJ per km for West Africa to the Netherlands (Wadden Sea) and 0.5 kJ per km for Wadden Sea to Siberia (Taimyr). Observational evidence shows that these distances are normally covered in a single non-stop flight, the whole one-way journey being made up of two flights with a single stopover.
From Piersma *et al.* (1991).

costs in warmer wintering areas may be lower; but (2) for birds taking different routes (like the two races of Red Knots), travel costs are not necessarily greater on the longer migration, because they depend partly on wind conditions. This comparison was concerned only with the overall energy budget, however, and not with the mortality risks from periods of energy crisis, or from other causes, which may have differed greatly between the two populations. It is not the energy budget as such, but the costs in terms of mortality and reproduction that provide the selection pressures affecting the genetic make-up of populations and their particular migratory and wintering areas. Nevertheless, these calculations on the energy needs of migration question the assumptions on which some models of the evolution of migratory habits have been based. However, they do not eliminate the likelihood that, among closely related taxa, differences in winter range evolved partly in response to competition between individuals in different populations.

SUMMARY

Related species with similar ecology often occupy mainly different geographical ranges and, in particular, many species which breed in the same region as one another segregate through migration to occupy different wintering ranges. This raises the possibility that, within the constraints of climate and food supplies, competition between species influences their migrations and wintering ranges. Similarly, within species, the distribution of each population may be related to the distribution of other populations, the east–west pattern in breeding areas often being maintained in wintering areas by parallel migrations. Hence, some longitudinal segregation of populations is more or less maintained year-round. An alternative pattern is fan migration in which birds from a large part of the breeding range may funnel into the narrow width of a small wintering range, or vice versa.

Three main latitudinal patterns have been described: (1) chain migration (northernmost breeding population wintering furthest north and southernmost furthest south); (2) leapfrog migration (northernmost breeding population wintering furthest south); or (3) telescopic migration (birds from a wide latitudinal span of breeding range concentrating in the same narrow latitudinal span of wintering range, or vice versa). Such patterns are apparent between closely related species and between different populations of the same species. Within species, social interactions could partly explain the migration differences between sex and age groups that give rise to latitudinal gradients in sex and age ratios across the wintering range.

These patterns are supposed to have resulted partly from selection pressure to reduce competition between individuals of different populations and species. However, non-breeding distributions have been examined, not just in terms of competition theory, but in terms of energy budgets, including the costs of wintering at different latitudes, and the costs of migrating different distances. They suggest that, while in general the daily energy needs for winter survival decline with decreasing latitude, as expected from the warmer temperatures, the energy needs for migration do not necessarily correlate well with the distance travelled, but also depend on the route taken and the prevailing winds. Nothing in the consideration of energy budgets is at odds with the idea of competition as a factor influencing migratory bird distributions.

Part Five

Migration Systems and Population Limitation

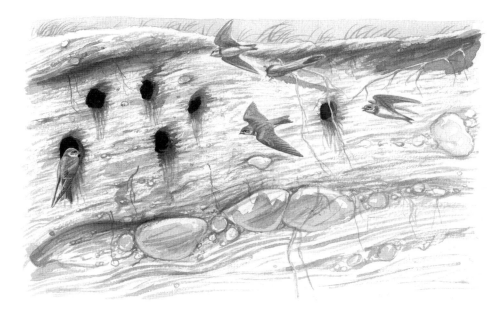

Sand Martins *Riparia riparia*, whose European breeding numbers are strongly affected by conditions in African wintering areas

Chapter 24

The Palaearctic–Afrotropical migration system

If there were no warmer lands to the south, the Palearctic avifauna would be considerably poorer. (David Lack 1954.)

Each autumn, after breeding in Eurasia, many millions of birds, from tiny warblers to large eagles, travel to wintering areas in tropical Africa, and back again next spring. This is perhaps the most impressive migration system in the world, not merely because of the huge numbers of participants, but also because of the long and formidable journeys involved. Birds from the west of Eurasia must cross the Mediterranean Sea and Sahara Desert, and those from further east must cross the deserts of southwest Asia and Arabia. Many of the travellers must also negotiate high mountain ranges, such as the Alps in the west or the Himalayas in the east (**Figure 24.1**). Moreover, once in Africa all these migrants cram into a geographical area less than half the size of their Eurasian breeding grounds, adding to an avifauna already rich in native species (Moreau 1972). Those that winter north of

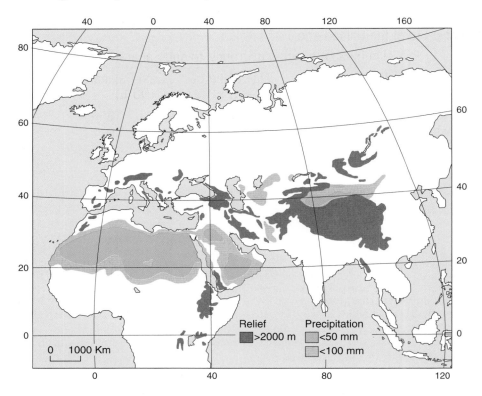

Figure 24.1   Map showing the desert and mountain areas to be crossed or circumvented by Palaearctic migrants that spend the northern winter in Africa south of the Sahara.

the equator do so in a season of progressively deteriorating conditions. On their return, many migrants must fatten for the journey in the Sahel zone, just south of the Sahara, at the driest time, when most types of food are near their lowest level of the year. Many European species that winter in Africa have declined in recent decades, raising questions about whether the causal factors lie in the breeding or wintering areas. The distributions, ecology and movements of the migrants within Africa have therefore attracted considerable interest, as have the factors that influence their population levels.

THE BIRDS INVOLVED

Of the landbirds that breed in the Palaearctic region (comprising most of Eurasia), about 186 species winter wholly or partly in Africa south of the Sahara, and another 29 seabird species winter on the seas and coasts, giving about 215 species altogether (**Table 24.1**). Some 193 of these migrants are drawn from the western Palaearctic, the region lying immediately to the north of Africa, with a smaller number (178) also coming from the mid-Palaearctic between 45°E and 90°E (**Table 24.1**).

Table 24.1 Numbers of landbird species migrating to Africa south of the Sahara from different parts of the Palaearctic

	Numbers of species migrating to sub-Saharan Africa from:			
	West Palaearctic (<45°E)	Mid-Palaearctic (45–90°E)	East Palaearctic[a] (>90°E)	Overall
Warblers	28	23	3	29
Other passerines	39	40	8	42
Raptors	24	21	2	26
Waders	35	42	6	44
Ducks	9	9	0	9
Other non-passerines	22	21	5	24
Herons	12	8	0	12
Seabirds	24	14	1	29
Totals	193	178	25	215

[a]The numbers in this column are probably underestimates because they include only species that breed in the East Palaearctic and winter entirely in Africa (leaving no other possible wintering area) or the few species in which such a long migration has been confirmed by ringing. Future ringing is likely to reveal that some individuals of other species also migrate from the East Palaearctic to Africa. The figures are based primarily on information given in *The Birds of the Western Palaearctic*, and are larger than those given by Moreau (1972) mainly because of increased information resulting from ringing data.

Only about 25 species come from the eastern Palaearctic, beyond 90°E, and some even from across the Bering Sea in Alaska and the Yukon. Some of these birds on their journeys travel as much as 200° west, covering more than 9000 km by the shortest (great circle) route, and then cross the equator and even the tropic of Capricorn. They include such spectacular travellers as the Amur Falcon *Falco amurensis*, Northern Wheatear *Oenanthe oenanthe* and Willow Warbler *Phylloscopus trochilis*.

Many Eurasian species winter partly in Africa and partly in southern Asia. It is assumed that each of these species has a migratory divide, with populations on the west side wintering in Africa. However, only for relatively few of the species is the pos-ition of the divide known, and from how far east the migrants to Africa derive, but in some the divide is surprisingly far to the east. For example, recoveries of Barn Swallows *Hirundo rustica* ringed in Africa have come from a wide range of local-ities from Ireland in the west to Siberia at more than 95°E (Turner 2006). Hence, Barn Swallows from most of the breeding range converge on southern Africa for the northern winter. Similarly, an Imperial Eagle *Aquila heliaca* was radio-tagged and tracked by satellite from western Saudi Arabia to a potential breeding area in northern China beyond 85°E (Meyburg & Meyburg 1998). Again, one would have expected such a bird to have wintered in India, along with others of its kind.

In total, an estimated 5000 million landbirds leave their Palaearctic breeding grounds each year to winter in Africa south of the Sahara, a density equivalent to one migrant per 1 ha of breeding range (Moreau 1972). To these must be added millions of shorebirds and waterfowl,[1] and also seabirds which winter offshore. More recent estimates, made 40 years later and based on radar observations across the Mediterranean region, lay between 3500 million and 4500 million birds

(B. Bruderer, *in litt.*), but these figures would have excluded migrants from east of Europe that enter Africa across Arabia. Radar has also revealed that, although some migration occurs across the whole width of the Mediterranean and Sahara, densities are highest at the western and eastern ends, where sea-crossings are short or avoided (Bruderer & Leichti 1999). Owing to human activities on breeding areas, Moreau (1972) felt that prevailing migrant numbers were only about one-third of those present at their maximum in post-glacial times, perhaps 3000 years ago. The migrants compare in numbers to an estimated 70000–75000 million other birds that live in Africa year-round (Brown *et al.* 1982). In other words, despite their impressive numbers, the migrants form only about 6–7% of the total bird numbers in Africa during the northern winter. However, their relative proportions vary greatly from one region to another and from one type of habitat to another, reaching their highest densities in the northern savannas (see later).

The fact that so many species from the mid-Palaearctic make the long westerly flight to Africa, rather than southwards within Asia, is presumably due partly to the conditions prevailing in southern Asia. The area of land west of 70°E with a genial winter is limited (India and Pakistan cover only 4 million km^2 compared with 21 million km^2 in Africa south of the Sahara), and between 75° and 95°E lies the formidable mountain mass of Tibet, flanked on the south by the even higher Himalayas. The Tibetan plateau averages nearly 5km above sea level, covers 2.5 million km^2, and spans 2000km from north to south, while the Himalayan peaks rise to over 8km and span another 200km from north to south. Nevertheless, birds of about 27 species and subspecies breeding in the mid-Palaearctic do winter in India, mostly 'leaking round the two ends of the Tibetan massif' (Moreau 1972). Many other birds cross the Tibetan Highlands and the Himalayas in autumn when conditions are favourable, but avoid this region in spring when most is still snow-covered and frozen (Bolshakov 2001). Moreover, India attracts for the winter several species from the west Palaearctic that do not go to Africa, including the Scarlet Rosefinch *Carpodacus erythrinus* and Arctic Warbler *Phylloscopus borealis*. From the east Palaearctic, nearly all the migrants move south-southeast into southern Asia, from Burma eastward to Indonesia, on journeys far less strenuous than those undertaken by birds entering tropical Africa from whatever source.

THE AFRICAN WINTERING AREAS

South of the Sahara, Africa is a vast continent, extending over some 21 million km^2 and comprising a wide range of habitats from hot desert to humid forest. Ecological conditions within Africa are determined much more by rainfall than by temperature, so that vegetation communities vary according to the amount of rain and the relative durations of wet and dry seasons. In fact, rainfall is the main factor governing the biological productivity of almost the whole continent.

---

[1]*Separate estimates are available for particular types of birds. Thus, more than 2.5 million Palaearctic ducks winter in Africa, although the numbers vary greatly from year to year (Scott & Rose 1996). In addition, more than two million Palaearctic raptors winter in tropical and south temperate Africa (Chapter 7).*

The alternation of wet and dry seasons is driven by the annual north–south movement of the Inter-Tropical Convergence Zone (ITCZ), a west–east zone of low pressure that forms where the northeast trade winds meet the southeast trade winds, and along which rain falls on a broad front (Jones 1995). During the northern summer, the ITCZ moves from the equator slowly northward, bringing rain to the northern tropics, and then retreats southwards again in autumn (**Figure 24.2**). The wet season thus begins later and ends sooner further north, so that the length of the wet season and the total annual rainfall decline from the equator northwards – in West Africa from more than 2000 mm near the equator to less than 50 mm (failing altogether in some years) near the southern fringe of the desert. In addition, as the mean annual rainfall declines, the more variable is the amount from year to year.

In autumn, the ITCZ crosses the equator southwards, so that from September onwards, southern Africa enters its wet season and the rains continue throughout the austral summer (coincident with the northern winter), ending in the austral autumn (northern spring). Although the rainfall pattern is more complex in southern Africa than north of the equator, in broad terms the rainfall distributions are mirror images of one another, with the wet seasons six months out of step. As in the northern tropics, the rains in southern Africa generally begin later and end sooner with increasing latitude.

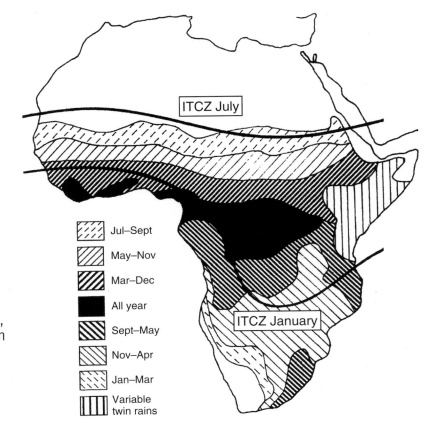

Figure 24.2 Rainfall patterns in Africa. Rainfall determines the vegetation, from rainforest (wettest), through woodland and savannah to desert (driest), giving equivalent vegetation belts both north and south of the equator (see text). ITCZ, Inter-Tropical Convergence Zone. From Jones (1995).

ITCZ July

ITCZ January

Jul–Sept
May–Nov
Mar–Dec
All year
Sept–May
Nov–Apr
Jan–Mar
Variable twin rains

Geographical exceptions to a single rainy season are relatively minor, and include the region between 5°N and 5°S in West Africa, where rain falls year-round and forest flourishes; a region between about 7°N and 7°S in East Africa where two rainy seasons occur, one around May and another minor one around October; and the western Cape and southern coasts where rain falls year-round but more in winter than in summer. Everywhere else the single rainy season is followed by months of unbroken drought, the 'dry season', equivalent in its ecological effects to the northern winter.

Because the total rainfall and duration of the wet season both decline with distance north and south from the equator, the natural vegetation zones also mirror each other in the northern and southern tropics. The region of almost year-round rainfall in equatorial West and Central Africa supports rainforest. Then, with decreasing rainfall towards both north and south, the vegetation becomes progressively shorter and sparser, as forest gives way to broad-leaved deciduous woodland and then to increasingly sparse acacia savannah and eventually desert (**Figure 24.2**). Where the rainforest has been cleared, it has given rise to a 'derived savannah' of remnant secondary forest and broad-leaved trees akin to the adjacent zones to the north and south, known as Guinea savannah in the north and miombo (*Brachystegia*) woodland in the south. Further from the equator occur mixed broad-leaved/*Acacia* woodlands (known as the Sudan zone in the northern tropics), giving way to increasingly arid *Acacia* savannahs (the Sahel in the northern tropics), which reach the margins of the northern (Sahara) and southern (Kalahari/Namib) deserts.

The greatest exception to this broad latitudinal vegetational pattern within the continent is East Africa, where very varied topography and twin rains combine to produce a complex mosaic of vegetation types from rainforest to semi-arid *Acacia* savannah. This complexity makes it harder in this region to discern clear latitudinal trends in rainfall, vegetation zones or bird migration patterns. Most importantly for the migrants, however, the seasons of maximum plant productivity and associated insect life are six months out of step between the northern and southern tropics.

Ecology of wintering areas

To all appearance, far less food is available to wintering migrants north of the equator than south of it. As plant growth occurs in the wet season (northern summer), this is the time of greatest productivity in areas north of the equator, so the majority of migrant species which stay in these regions experience progressively worsening conditions from the time they arrive in Africa to the time they leave. The same is true for wetland birds, for during the rains water collects in every hollow, and rising rivers may flood extensive areas. But as the dry season advances, these productive waters progressively shrink and disappear. Towards the end of the dry season, in the Sudan and Guinea zones, grass fires become widespread. As the grasslands dry at different times in different regions, a 'fire belt' spreads systematically through large parts of the continent each year in more or less regular pattern. Within this belt, the fires are localised at any one time, but over a period of weeks most of the terrain gets burnt, and the belt moves on. Small animals flushed or injured by grassland fires are exploited by a wide range of birds, including raptors, storks, rollers, bee-eaters and others. They include many Palaearctic migrants which are free to follow the fires across the continent, on a roughly similar schedule from year to year. In any one locality, however, the feast is short-lived, and in effect fires destroy many of the remaining food supplies. They affect especially the food available at ground level and, through scorching, also that in the lower trees and bushes. Yet despite the seasonally worsening conditions, many more migrant species somehow maintain themselves in the Sahel zone just south of the Sahara than anywhere further south. Moreover, they are able to fatten there in preparation for the spring desert crossing.

In contrast, those birds that move on south of the equator, where the seasons are reversed, experience progressively improving conditions. The rains are starting as the first migrants arrive in late September and have hardly finished when they leave in March–April. Some 53 landbird species commonly reach the equator, and 31 of these extend to 5–25° south (Pearson & Lack 1992), effectively following the intertropical rain belt and benefiting from humid conditions during most of their stay in Africa. However, these birds must still pass north through the arid Sahel zone on their return migration, getting their last chance to feed and fatten before a long desert-crossing. They do not, therefore, escape the drought conditions entirely, although many fatten in the more mesic Sudan and Guinea zones to the south, where the rains begin in March–April, rapidly increasing local food supplies. In particular, the first rains trigger the swarming of ants and termite alates, a superabundant and energy-rich food source favoured by many species, from passerines to raptors. Many of the migrants that fatten in the Sudan and Guinea zones before setting of on spring migration may then over-fly the Sahel zone (where rain has not yet fallen) along with the Sahara, having little or no chance to feed again until they reach North Africa.

Wetlands

North of the equator, the generally dry environments are alleviated by extensive shallow wetlands which occur patchily from west to east, mainly around the Senegal River (in Senegal), the upper Niger River (in Mali), the Lake Chad basin (in Chad), and the Nile-Sudd swamps (in southern Sudan). These wetlands provide productive feeding areas year-round, even at the height of the dry season, leaving rich grassland and pools as they retreat. The seasonally inundated plains around Lake Rukwa, the Lake Victoria basin and the Wembere swamps in Tanzania also support large numbers of birds, including Palaearctic migrants. Each of these areas extends in the wet season over many thousands of square kilometres.

Important wetlands south of the equator include the Bangweulu swamps and the Kafue floodplain in Zambia and the Okavango Delta swamps in Botswana. Each of these areas again extends in the wet season over many thousands of square kilometres. Together with the northern areas, they comprise some of the largest expanses of freshwater wetland on earth, covering an estimated $657\,000\,km^2$ of West Africa, $208\,000\,km^2$ of East and northern Africa and $172\,000\,km^2$ of southern Africa. However, most of these areas are being rapidly reduced by drainage and other human impact, and now cover only a small part of the ground they occupied in the 1970s.

Many observers have commented on the huge numbers of herons, ducks and waders that concentrate in the Sahel wetlands during the northern winter. In West Africa, the greatest numbers occur in the basins of the Senegal and Niger. These two rivers are subject to a single spectacular flood each year, beginning in July, which transforms huge areas of desiccated earth into rich marshland (Morel 1973). Some of the vegetation brought on by the start of the rains has time to produce seeds before it is flooded, providing a food supply for those ducks and waders that eat seeds. In the swamps, true marsh vegetation develops, producing an additional food supply of seeds and aquatic invertebrates. Fishes reproduce in the wet season, and the flooded areas are rich in small fry. As the flood retreats, it leaves behind a great network of shallow pools and channels in which many fish become trapped. It has been calculated that, in the Senegal Delta of about 146 000 ha, at least 30 000 tons of fish are eaten annually by cormorants, herons, pelicans and other waterbirds (Morel 1973). Although the smaller pools dry up by late December, the larger expanses of flood water last longer and afford suitable habitat for waterbirds throughout the winter.

In the Senegal Delta alone, more than 300 000 Pintail *Anas acuta*, Garganey *Anas querquedula* and Shoveller *A. clypeata* have been counted in good years, along with several hundred thousand Ruffs *Philomachus pugnax*, tens of thousands of Black-tailed Godwits *Limosa limosa*, and vast numbers of *Tringa* sandpipers. However, numbers vary greatly from year to year, depending on wetland conditions, as shown for some ducks in **Table 24.2**. In comparison, even in good years relatively small numbers of African species occur here, mainly from different genera. Among herons, however, the resident and migrant populations of four species occur together, namely Purple Herons *Ardea purpurea*, Night Herons *Nycticorax nycticorax*, Squacco Herons *Ardeola ralloides* and Little Egrets *Egretta garzetta*, so it is impossible to assess the relative proportions of Eurasian and African birds. These wetland areas also support many other birds, both passerines and non-passerines. For example, the Upper Niger floodlands cover 40 000 km^2, and in the dry season support wintering raptors at 12–15 times the densities found in adjacent drier land (Thiollay 1989).

In addition to the inland wetlands, several coastal areas are important. In particular, the Banc D'Anguin off the coast of Mauritania holds very large numbers of shorebirds and waterfowl throughout the northern winter, and is also an

Table 24.2 January counts of the most numerous species of ducks wintering in the Senegal River delta, West Africa, showing the enormous variation between years, 1972–1996

	Highest count (year)	Lowest count (year)
Fulvous Whistling Duck *Dendrocygna bicolor* [a]	12 110 (1983)	70 (1977)
White-faced Whistling Duck *Dendrocygna viduata* [a]	49 990 (1972)	797 (1985)
Northern Pintail *Anas acuta* [b]	210 000 (1989)	204 (1980)
Garganey *Anas querquedula* [b]	136 847 (1972)	910 (1980)
Northern Shoveller *Anas clypeata* [b]	23 202 (1996)	301 (1984)

[a] Intra-African migrants.
[b] Eurasian–Afrotropical migrants.
From Triplet & Yésou (2000).

important staging site for migrant shorebirds in both spring and autumn. Coastal waders face virtually no local competitors in their wintering areas, and have the available food supply largely to themselves.

MOVEMENTS WITHIN AFRICA

The Sahel savannah zone that crosses the continent from west to east is particularly important for the migrants. Not only is it the first area with food that they reach after crossing the Sahara in autumn and the last before crossing in spring, but many species spend the whole or part of the winter there (Moreau 1972, Jones 1999). The migrants arrive between August and November, depending on species, but mainly in September when many African birds (which have spent their breeding season there) are themselves beginning to retreat southward. At this time the rains are finishing, the land is green and food is plentiful, but as the country gradually dries out, feeding conditions deteriorate. Nevertheless, many of the dominant trees have leaves (supporting insects) or produce flowers and fruit in the dry season, which many birds can use. About one-quarter of all migrant species remain in the Sahel and Sudan zones for the whole winter in the local dry season (strategy 1 in **Table 24.3**). Examples include some *Sylvia* warblers, which eat fruit as well as insects from leaves, and some ground-dwelling wheatears and chats which eat small terrestrial insects (Pearson & Lack 1992). They also include granivores, for seeds remain available on the ground throughout the dry season, but disappear rapidly when rain causes their germination. It would make no sense for a wintering seed-eater to move south of the equator in September to the austral summer.

Many other migrant species stay in the Sahel–Sudan zones for only 1–2 months; some moult while they are there, and then move on further south, in late October–November (strategy 2 in **Table 24.3**). Such species are conspicuous in both Sahel and Sudan zones for 4–6 weeks at the end of the rains, and then disappear suddenly with the first arrival of the dry and dust-laden 'Harmattan' wind from the north (Jones 1985, 1995, 1999). In West Africa, most such species undertake a second migration of up to several hundred kilometres to pass the rest of their stay in the less arid Guinea and derived savannahs to the south (their movements being limited to at most a 12° latitudinal span by the Gulf of Guinea coast or the central African rainforest. However, other species, after spending 1–2 months in the Sahel–Sudan zones, perform a much longer second migration to cross the equatorial forest, and spend the rest of their stay in the wet season conditions of the southern African woodlands and savannahs (strategy 3 in **Table 24.3**). Examples include the Sand Martin *Riparia riparia* that feeds on aerial insects, and various warblers that glean small insects from fresh leaves.

Relatively few species seem to fly directly to winter quarters in southern Africa, passing quickly through the northern tropics (strategy 4 in **Table 24.3**). They include the Barn Swallow *Hirundo rustica*, Spotted Flycatcher *Muscicapa striata* and Common Cuckoo *Cuculus canorus*. These species tend to leave Europe relatively early, and pass through the northern tropics while it is still raining, reaching the southern tropics as the rains begin. They thus spend the maximum time possible in wet season conditions. These species also feed mainly on the flush of insects which appear on new leaves, or on the aerial and ground insects brought out by rain.

Table 24.3 Migratory habits of some Palaearctic migrants in Africa. Some species occur in more than one category, usually where migration patterns differ between the west and east sides of Africa

Strategy 1: Wintering entirely in the dry season conditions of the northern Sahel	Greater Whitethroat *Sylvia communis*, Lesser Whitethroat *Sylvia curruca*, Northern Wheatear and Sudan savannahs *Oenanthe oenanthe*, Olivaceous Warbler *Hippolais pallida*, Bonelli's Warbler *Phylloscopus bonelli*, Subalpine Warbler *Sylvia cantillans*, Orphean Warbler *Sylvia hortensis*, Rüppell's Warbler *Sylvia ruppelli*, Short-toed Lark *Calandrella brachydactyla*, Tawny Pipit *Anthus campestris*, Red-throated Pipit *Anthus cervinus*, Bluethroat *Luscinia svecica*, Black-eared Wheatear *Oenanthe hispanica*, Ortolan Bunting *Emberiza hortensis*, Collared Pratincole *Glareola pratincola*, Short-toed Snake-Eagle *Circaetus gallicus*, Egyptian Vulture *Neophron percnopterus*, Black-tailed Godwit *Limosa limosa*
Strategy 2: Wintering in dry season conditions, first in the northern Sahel–Sudan savannahs and then further south in the Guinea and derived savannahs	Barn Swallow *Hirundo rustica* (in West Africa), Tree Pipit *Anthus trivialis*, Woodchat Shrike *Lanius senator*, Common Nightingale *Luscinia megarhynchos*, Common Redstart *Phoenicurus phoenicurus*, Great Reed-Warbler *Acrocephalus arundinaceus*, Melodious Warbler *Hippolais polyglotta*, Wood Warbler *Phylloscopus sibilatrix*, Willow Warbler *P. trochilus* (in West Africa), Garden Warbler *Sylvia borin*
Strategy 3: Wintering in wet season conditions, first in the northern tropics, and then in the southern tropics (transequatorial migrants)	Sand Martin *Riparia riparia*, Tree Pipit *Anthus trivialis*, Great Reed Warbler *Acrocephalus arundinaceus*, Sedge Warbler *A. schoenobaenus*, Willow Warbler *Phylloscopus trochilus* Garden Warbler *Sylvia borin*, Greater Whitethroat *S. communis*, Marsh Warbler *Acrocephalus palustris*, Thrush Nightingale *Luscinia luscinia*
Strategy 4: Wintering entirely in wet season conditions in the southern tropics, after passing rapidly through the northern tropics (trans-equatorial migrants)	Spotted Flycatcher *Muscicapa striata*, Collared Flycatcher *Ficedula albicollis*, Barn Swallow *Hirundo rustica*, Hobby *Falco subbuteo*, Red-footed Falcon *Falco vespertinus*, Amus Falcon *Falco amurensis*, Common Cuckoo *Cuculus canorus*, Icterine Warbler *Hippolais icterina*, Eurasian Golden Oriole *Oriolus oriolus*

Note: During their pause in northeast Africa, many migrants moult, including Great Reed Warbler, Sedge Warbler, Olivaceous Warbler, Greater Whitethroat (eastern race, *icterops*) and Marsh Warbler (Pearson 1973, 1975, 1990, Pearson & Backhouse 1976, 1983, Pearson *et al.* 1988).
Mainly from Moreau (1972) and Jones (1995, 1999).

In East Africa, forest is much more patchy than in the west, and the migrants encounter suitable woodland or savannah habitats from the Sahel southward into South Africa. Movements are longer than in the west, and a greater proportion of insectivorous species crosses the equator to winter in southern Africa (Lack 1990, Pearson 1990); their movements are not broken by any barrier equivalent to the Gulf of Guinea in the west.

Most of the birds that migrate to East and southern Africa derive from Asian breeding areas at 30–80°E, but some European birds move southeast to join their conspecifics from Asia in the Middle East, before passing down the eastern side of Africa. This pattern is evident in the Thrush Nightingale *Luscinia luscinia*, Marsh Warbler *Acrocephalus palustris*, Eurasian River Warbler *Locustella fluviatilis*, Olive-tree Warbler *Hippolais olivetorum*, Barred Warbler *Sylvia nisoria*, Red-backed Shrike *Lanius collurio* and Lesser Grey Shrike *Lanius minor* (Moreau 1972, Pearson 1990). In these species, populations from the whole breeding range spend the northern winter in eastern or southern Africa.

As in West Africa, many species spend the first few weeks after arrival in the *Acacia* savannahs between 12°N and 15°N (Ethiopia–Sudan) in areas where it has recently rained, and some species moult there (**Table 24.3**) (Pearson 1973, Jones, 1995). They then move on in November–December as the dry season sets in, some (mainly ground-feeders) to equatorial eastern Africa (Kenya–Tanzania) where the second wet season is just beginning (Lack 1983), and others (mainly vegetation-gleaners) cross the equator to southern Africa. Some of the latter undergo a second phase of pre-migratory fattening to take them to wintering grounds 2000–4000 km away (Pearson 1990, Pearson & Blackhurst 1976, Lack 1983, Jones 1985, 1999). They thus perform a two-stage migration, with a 2- to 3-month break between stages.

In some of the species involved, onward movements seem to be obligate and endogenously controlled. Such birds leave East Africa at about the same time each year, having accumulated migratory fat before departure. In other species, the onward movements may be facultative, as many more birds remain in northern Africa in wet than in dry years (Lack 1983), and are more variable in their timing of southward departure. Similarly, while some species (such as Red-backed Shrike *Lanius collurio*) are consistent in their arrival times in the Kalahari Basin of Botswana, other species (such as Great Reed *Acrocephalus arundinaceus* and Sedge Warbler *A. schoenobaenus*) show big year-to-year differences in arrival times related to variations in the onset of the local rains (Herremans 1994, 1998).

The centre of gravity of the collective migrant population thus shifts southward within Africa during the course of the northern winter, but over a shorter mean distance in the west than in the east. Some species take up to five months to move between their breeding areas and the farthest parts of their wintering range, travelling in two or more well-separated stages. In contrast, they take less than two months for their return journey, beginning in late March–early April, breaking their travels usually for only a few days at a time, and fattening rapidly (Pearson 1972, Ash 1980, Pearson & Lack 1992, Jones 1995). Throughout their journeys, most species seem to migrate by long flights, up to 1000 km or more, with fuel loads of 30–50% above their lean weights. Like the irruptive migrants of northern regions, some migrants to Africa extend further south in years when food is scarce, their numbers in particular localities fluctuating greatly from month to month and year to year, depending on patterns of rainfall, which affect food supplies (McLachlan & Liversidge 1978, Lack 1983, Liversidge 1989, Herremans 1998). In some years, they hardly appear at all in the southern parts of their wintering range, even though local food supplies may be good, presumably because in those years the whole population is accommodated further north (see Liversidge 1989 for Steppe Eagle *Aquila nipalensis*). Moreover, in some species the distribution of the population in southern Africa varies from year to year within the range, depending on where sporadic rainfall has created suitable conditions (Herremans 1998). Typically, such species move with the rain, seldom staying for long in any one locality, and in drought years not appearing at all (e.g. Lesser Spotted Eagle *Aquila pomarina*, Liversidge 1989). They are ultimately dependent on food supplies that result from fresh rain, whether grasshoppers and locusts, emerging termites or *Quelea* colonies. Although most such species are site-faithful in their Eurasian breeding areas, occupying the same territories throughout their stay year after year, they may be classed as irruptive or nomadic in their African non-breeding areas (Chapter 16).

In these various ways, some migratory species may be on the move for much of the time between leaving their breeding areas in one year and returning there the next, on a long circuitous journey, more or less repeated year after year. In the process, individuals of some species may spend successive periods in different localities, but may return to the same series of localities year after year. As shown by ringing, individuals of some passerine species may pass through the same spots, or return to the same clump of bushes, in successive years (Moreau 1972, Curry-Lindahl 1981). In addition to their nesting territories in Eurasia, individual birds thus remember and seek out two or more territory sites hundreds or thousands of kilometres apart in Africa, which they visit each year in orderly succession as each in turn offers suitable conditions through the changing seasons (Jones 1985). However, the time spent on each of these areas may vary greatly from year to year, depending on conditions at the time.

The strategy of using more than one area during the course of the winter (other than for mere stopovers) was called itinerancy by Moreau (1972). It is clearly a common form of behaviour in many migratory species in Africa, each leg of the journey being separated by up to several weeks from the next, and requiring a separate period of fat deposition.

Both southward and northward migrations through East Africa appear channelled by mountains, and there is marked east–west segregation of species, with some species taking one route and others a different route. Some species show a strong passage mainly through Uganda (e.g. Willow Warbler *Phylloscopus trochilus*, Garden Warbler *Sylvia borin*, Red-backed Shrike *Lanius collurio*) and others through eastern Kenya (Marsh Warbler *Acrocephalus palustris*, Greater Whitethroat *Sylvia communis*, Thrush Nightingale *Luscinia luscinia*). This segregation depends partly on breeding origin and partly on preferred habitat in Africa. Some species seem to pass through Kenya on a front only about 100–200 km wide, the Marsh Warbler *Acrocephalus palustris* being a striking example (Pearson & Lack 1992). On their return northward journey in spring, many birds take a more easterly, coastal route, associated with the more humid conditions there at that season, the whole migration forming a loop. Examples include Red-backed Shrike *Lanius collurio* and Lesser Grey Shrike *Lanius minor*, Great Reed Warbler *Acrocephalus arundinaceus*, Sedge Warbler *A. schoenobaenus* and Willow Warbler *Phylloscopus trochilus*, Garden Warbler *Sylvia borin*, Spotted Flycatcher *Muscicapa striata* and Thrush Nightingale *Luscinia luscinia* (Pearson 1990, Pearson & Lack 1992).

Whatever the particular migration strategies of different species, they more or less conform to a single pattern, evident throughout Africa: north of the equator, birds follow the rainbelt northward from March–April onwards, returning southward in October–November as the rains withdraw; south of the equator birds move further south with the rains from October–November, returning northwards in March–April. The pattern is essentially the same whether birds remain in the northern or southern tropics or whether they cross the equator. It is also the same whether the birds are Eurasian–African migrants or intra-African migrants (Elgood *et al.* 1973, Jones 1985). All are subject to the same ecological pressures. The most obvious difference between the two groups is that the Eurasian migrants move much further north than the African ones reaching the Palaearctic to breed. As the Eurasian migrants return to Africa in autumn, to some extent they replace the African migrants in the Sahel and Sudan zones which then move further south (Jones 1995, 1998). The extent to which the two groups replace each other ecologically awaits more study.

ECOLOGY OF MIGRANTS IN AFRICA

Attention has been directed to how all these migrant birds are accommodated for more than half the year, and more densely than in their breeding areas, in lands already richly endowed with birds of their own. Around 1481 Afrotropical land-bird species (i.e. excluding waterbirds) occur year-round in sub-Saharan Africa, some performing seasonal migrations within the region (Moreau 1972). Of these, 409 are residents of evergreen forests, either lowland or montane, and another 74 are confined to montane non-forest habitats. Because such habitats are barely used by Palaearctic migrants (see later), the two communities hardly come into contact. It is the remaining 998 lowland species, occupying seasonal woodland, savannah and grassland, which interact with the 186 species of Palaearctic immigrants for part of each year (Moreau 1972). The two groups comprise about 84% and 16% respectively of the total bird species that occupy these African habitats in the northern winter.

Distributions

The first point to be made is that the immigrant species do not by any means distribute themselves evenly over the continent (**Figure 24.3**; Newton 1995). As mentioned above, most species occur in the belts of Sahel and Sudan savannah habitats which lie south of the Sahara from west to east across the continent and which, some weeks earlier, were vacated by the intra-African migrants returning southwards. Within these belts, maximum species numbers occur in the east. From the Sudan zone southwards, the numbers of Eurasian migratory species progressively decline so that few reach the southern tip of the continent. It is as though the birds migrate no further in sub-Saharan Africa than is necessary to find suitable conditions. However, it remains conjectural whether this southward decline in species numbers is due to Palaearctic species simply petering out because their populations can be accommodated further north, or whether adverse ecological factors or competition with Afrotropical species curtail their wintering distributions southwards.

In general, in non-forest habitats, regions with high numbers of Palaearctic migrants also tend to support high numbers of Afrotropical species, as in parts of East Africa (Newton 1995a). This may be partly due to the higher topographical diversity in East Africa, and to the wider range of habitats available there,

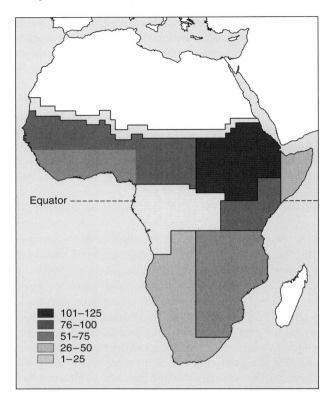

Figure 24.3   Numbers of Palaearctic migrant species found in different parts of sub-Saharan Africa during October–March. From Newton (1995b). In addition, a few migratory species occur around oases in the Sahara in winter, notably Eurasian Chiffchaff *Phylloscopus collybita*, Sardinian Warbler *Sylvia melanocephala*, Olivaceous Warbler *Hippolais pallida*, Northern Shrike *Lanius excubitor* and various desert species (Gaston 1970). These numbers could form at most only a trivial fraction of the total wintering populations of these species. The same is true for the various Palaearctic species that winter in small numbers in some parts of the Arabian peninsula (S. F. Newton 1996).

coupled with the dual wet season. Hence, at this broad distributional scale within vegetation zones, the Eurasian species do not concentrate in regions with the smallest numbers of African species. Similar correlations are also apparent at smaller spatial scales, as found for small insectivores in savannah habitats of southeast Kenya and Nigeria where the numbers of both African and Eurasian birds were correlated with the density of vegetation, and hence with insect food supplies (Lack 1987, Jones *et al.* 1996).

Habitats

For the most part the migrants occupy in Africa habitats that are structurally similar to those they use in their breeding areas, and obtain similar foods in similar ways (Chapter 22). Most of the migrant passerines and raptors live in savannah or woodland. The highest densities and proportions of migrant passerines are found in seasonal savannahs (Lack 1990, Leisler 1992, Jones 1999), many of the passerines being found in scrubby or reedy areas, and in some areas they may outnumber insectivorous resident species during certain times of year (vande Weghe 1979, Aidley & Wilkinson 1987, Rabøl 1987, Kelsey 1989). In contrast, few migratory species from the Palaearctic are seen in rainforest. They include the European Pied Flycatcher *Ficedula hypoleuca*, Collared Flycatcher *F. albicollis*, Spotted Flycatcher *Muscicapa striata*, Wood Warbler *Phylloscopus sibilatrix* and Eurasian Golden Oriole *Oriolus oriolus*, and in smaller numbers, other species such as Common Nightingale *Luscinia megarhynchos*, Blackcap *Sylvia atricapilla* and Willow Warbler *Phylloscopus trochilus*. All these species are found chiefly around edges and openings, and are most numerous in more open woodland, as is the European Honey Buzzard *Pernis apivorus* (Mönkkönen *et al.* 1992, Morel & Morel 1992). As expected, waterfowl and waders occupy coastal or inland wetlands.

Mobility

A major difference between migrants and residents is in frequency of movements. Because they are not tied to nesting areas, the Palaearctic migrants are free to move around, exploiting temporary local abundances of food in a way that most residents cannot. The same is true for intra-African migrants in their non-breeding season. Many migratory species thus pursue an itinerant lifestyle, periodically moving in relation to seasonal changes in food supplies or even in relation to local flushes of food, such as those produced by patchy rainstorms (Sinclair 1978, Liversidge 1989, Leisler 1992, Herremans 1998). In consequence, such migrants may occupy only part of their potential non-breeding range at any one time, avoiding certain areas altogether in some years. In contrast, many resident African species remain on territories all year and, south of the equator, most African species, both residents and intra-African migrants, are still breeding at the time the Palaearctic migrants are present (Moreau 1972). This limits the effectiveness of the African birds in exploiting transient food supplies, and hence their ability to compete for them with the Palaearctic migrants. In affect, the Palaearctic migrants tend to use temporary habitats which cannot be fully exploited year-round by local species: seasonally flooded grasslands, drying pools, recently burnt savannahs, nearly bare fields, and plantations and secondary growth in the forest zone (Thiollay 1989).

Food

Within Africa, the majority of Palaearctic landbird migrants depend on arthropods which are generally more abundant in the wet season, largely because the vegetation achieves most growth then (Morel 1973, Lack 1986b). Even in the dry season the Sahel zone is richer than it appears superficially, as trees and shrubs can be found at every stage of leaf, flower and fruit production. Locally and temporarily, insects can be available in tremendous abundance, as for example around standing water, where for periods at any time of year masses of chironomids coat the emergent vegetation. Periodic swarming of lakeflies (several species of chaoborid and chironomid midges) is well known at Lake Victoria. After larval life in the lake, enormous numbers of flies emerge in synchrony, usually around each new moon, forming huge clouds over the water. They often get blown ashore, forming an important food source for small insectivorous birds lasting for a few days at a time. These insects are exploited by migrants on both their northward and southward passage, enabling rapid fattening, but are not permanently available (Wanink & Goudswaard 2000).

Many of the passerines also eat fruit, particularly of *Salvadora persica*, *Maerua crassifolia* and *Balarites aegyptiaca* (Fry *et al.* 1970, Moreau 1972, Jones 1995, Stoate & Moreby 1995). The fruit crops vary greatly in size from year to year, but may remain available through the dry season, and are particularly favoured during the period of fat deposition for spring migration. Only the Common Quail *Coturnix coturnix* and Turtle Dove *Streptopelia turtur* are entirely granivorous, while the Short-toed Lark *Calandrella cinerea*, Ortolan Bunting *Emberiza hortulana* and Cretzchmar's Bunting *E. caesia* eat both seeds and arthropods during their stay in Africa.

Perhaps the most important element in the insect flush associated with the rains is the appearance everywhere of swarms of winged termites and winged ants. Because of their association with rain, they are in general available mainly to those migrants which extend south of the equator where the seasons are reversed. They are eaten by a wide range of birds which follow the rainbelts, from small warblers to large eagles. They are especially important to insectivorous falcons, and may account for the fact that four of the five migratory species winter mainly south of the equator. Hundreds of Lesser Spotted Eagles *Aquila pomarina* have been found at localities where winged termites are temporarily available (Brooke *et al.* 1972, Meyburg *et al.* 1995c).

On the other hand, the importance of plague locusts to migratory birds has probably been exaggerated (Moreau 1972). These insects erupt after several years of drought when the rains are good. They emerge immediately after rain, their eggs having survived in the ground for many years. Once the larvae have eaten the local vegetation, they swarm and head off downwind to low pressure areas where fresh rain is falling. However, in no part of Africa could locusts ever have provided a reliable source of food, and however superabundant they may be at certain times and places, years pass without outbreaks anywhere. In the intervals, the surviving insects merely take their place among many other solitary grasshoppers, some of which are very abundant. Nowadays, of course, locust outbreaks are often controlled by pesticide applications, which presumably remove many other arthropods too.

Another important food source for some raptors is provided by Quelea *Quelea quelea*, small seed-eating birds that nest in dense colonies, millions strong (Chapter 16). Colonies form after rain has fallen, so in the northern tropics they are available only for a few weeks after the migrants first arrive, but in the southern tropics, they are available for much of the austral summer, in one locality or another. These colonies attract hundreds of migratory raptors and storks which feed on the adults and chicks. Single colonies, occupying only a few hectares of scrub, have attracted more than 100 Steppe Eagles *Aquila nipalensis* or Lesser Spotted Eagles *Aquila pomarina* to localities where these eagles are otherwise seldom seen (Chapter 16). Many raptor species are also attracted to local concentrations of rodents, which occur sporadically, again mainly in response to local rain (Chapter 16).

Within habitats, Eurasian visitors tend to segregate, spatially or ecologically, from closely related African species (Moreau 1972, Thiollay 1989, Lack 1986b, 1987). The latter tend to be in different genera (so may be presumed to have somewhat different ecological requirements), and in arid areas are generally less abundant. In addition, resident waders are almost totally lacking in West Africa, leaving this niche largely free for the Eurasian migrants. In attempts to find how migrants coexist with residents, many researchers have compared the habitats, foraging behaviour and diets of Eurasian migrants with those of closely related African species (e.g. Lack 1986b, 1986, Rabøl 1987, Jones *et al.* 1996, Baumann 2001, Salewski *et al.* 2002). Although these various studies have revealed some ecological differences between migrants and native species, this would be expected in comparisons of any two species, presumably reducing competition between them. Attention has focused on whether the Eurasian migrants have wider niches than the residents, exploiting a wider range of habitats and food sources. This seems true more often than not (Lack 1986b, Rabøl 1993, Leisler 1992, Salewski *et al.* 2002), but depends on the species and area concerned (Salewski & Jones 2006).

Temperatures

In terms of their energy needs, the maintenance costs of individual birds while in Africa must be considerably less than on their Eurasian breeding grounds, as a result of two independent factors. The greater warmth in Africa reduces the intake of food needed to maintain body temperature, and the birds are less active in Africa because they carry none of the stresses associated with breeding, having only themselves to feed. On the basis of information available at the time, Moreau (1972) calculated that the daily maintenance needs of individual birds amount to roughly 60%, on average, of what they are on their breeding grounds. The mean temperature increase from breeding to wintering areas of 50 species of migrants was calculated by Moreau (1972) at over 6°C. Extremes were provided by tundra-nesting species which experienced July temperatures of 5–10°C, and then migrated to tropical Africa, with January temperatures of around 20–30°C.

Social systems

Some Palaearctic species maintain long-term territories during their stay in Africa (for Whinchat *Saxicola rubetra* see Dejaifve 1994; for Pied Flycatcher *Ficedula hypoleuca* see Salewski *et al.* 2000; for *Acrocephalus, Hippolais, Luscinia* see Cramp 1988, 1992), and at least some return to the same territories in successive years (for Marsh Warbler *Acrocephalus palustris* see Kelsey 1989, for Great Spotted Eagle *Aquila clanga* see Meyburg *et al.* 1995). Others set up temporary territories at different successive areas that they occupy, or very short-term territories as they defend fruit and nectar sources. Yet other species are found in flocks or solitarily depending on their food supplies, and seem continually on the move. Others, such as Wood Warbler *Phylloscopus sibilatrix* and Willow Warbler *P. trochilus*, form single-species flocks, or even mixed-species flocks along with African species (Brosset 1984, Demey & Fishpool 1994). In such flocks, aggression is seldom observed, even though many birds forage close to one another.

EFFECTS OF DROUGHTS ON MIGRANT NUMBERS

In much of Africa north of the equator, as emphasised above, rainfall during the northern summer largely determines the state of the vegetation and the extent of wetlands in the following dry season. It thus influences the food supplies of many birds, whether they consume plant or animal matter. But rainfall also varies greatly from year to year, and over much of the region since the late 1960s has in most years been well below former levels, owing to a failure of the rainbelts to extend so far to the north. Hence, drought conditions have been most severe along the northern edge of the Sahel zone, lying immediately south of the Sahara Desert, and have diminished southwards across the savannah zones towards the equator.

In the Sahel zone, rainfall deficits were particularly marked in 1968, 1973, and even more so in 1983 and 1984, and again in 1990. It was in 1969 (after the 1968 drought) that the importance of conditions in African wintering areas was first impressed upon European bird-watchers, when sudden and massive declines were apparent in the numbers of returning summer migrants, especially Greater Whitethroats *Sylvia communis* and Common Redstarts *Phoenicurus phoenicurus*, which are strongly represented in the northern Sahel (Winstanley *et al.* 1974). In Britain, according to the Common Bird Census of the British Trust for Ornithology, Greater Whitethroat numbers dropped by about 70% between 1968 and 1969 (**Figure 24.4**). Sand Martins *Riparia riparia* and Sedge Warblers *Acrocephalus*

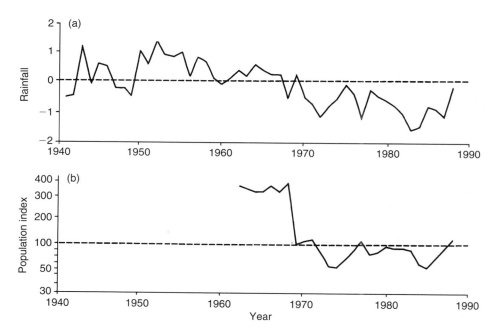

Figure 24.4   (a) Rainfall trends in the Sahel region of Africa expressed as departure from the long-term mean (----), 1940–1988. Data from Grainger (1990). (b) Population trend (log scale) of Greater Whitethroats *Sylvia communis* breeding in Britain, as revealed by the Common Birds Census of the British Trust for Ornithology, 1962–1988. Data from Marchant *et al.* (1990).

schoenobaenus were also badly hit, even though most individuals spend only 4–6 weeks in the Sahel in October–November, before moving further south, and an even shorter period on return passage in March (Cowley 1979, Jones 1985). In Africa, as in Eurasia, both species are seen mainly around water bodies.

With the lapse of further years, it has become apparent that rainfall in the western Sahel has a major impact on the breeding populations of many species that nest in Europe, including some that overwinter in the Sahel and others that pass through. In a Europe-wide analysis of the population trends in breeding bird numbers, many European–African migrants showed sustained declines during 1970–2000, significantly more negative than those of short-distance migrants or residents (Sanderson *et al.* 2006). Analysis of the trends in 30 pairs of closely related species, one wintering in Africa and the other in Europe (e.g. Tree Pipit *Anthus trivialis* and Meadow Pipit *A. pratensis*, Garden Warbler *Sylvia borin* and Blackcap *S. atricapilla*), revealed significantly more negative trends in the former, regardless of breeding habitat. Further examination of species that winter in Africa revealed that those occupying dry open habitats declined significantly more than all those using other habitats. Some idea of the extent of population changes is given in **Table 24.4**, based on data from Britain. Also, the pattern of

Table 24.4   Population trend (%) during the latter part of the twentieth century of some species that breed in Britain and winter in Africa south of the Sahara

Species	Period (years)	Net change (%)
Osprey *Pandion haliaetus*	1973/77–1995/99	+658
Black-tailed Godwit *Limosa limosa*	1973/77–1995/99	−35
Roseate Tern *Sterna dougallii*	1975–2000	−91
Little Tern *Sterna albifrons*	1975–2000	−28
European Turtle Dove *Streptopelia turtur*	1974–1999	−69
Common Cuckoo *Cuculus canorus*	1974–1999	−31
Eurasian Wryneck *Jynx torquilla*	1973/77–1995/99	−64
House Martin *Delichon urbica*	1974–1999	−33
Tree Pipit *Anthus trivialis*	1974–1999	−75
Yellow Wagtail *Motacilla flava*	1974–1999	−36
Common Grasshopper Warbler *Locustella naevia*	1974–1999	−79
Savi's Warbler *Locustella luscinioides*	1973/77–1995/99	−63
Marsh Warbler *Acrocephalus palustris*	1973/77–1995/99	−63
Wood Warbler *Phylloscopus sibilatrix*	1994–2000	−43
Willow Warbler *Phylloscopus trochilus*	1974–1999	−31
Spotted Flycatcher *Muscicapa striata*	1974–1999	−75
Red-backed Shrike *Lanius collurio*	1973/77–1995/99	−90

Other information on trends over the period 1970–2001, using mainly the same source of data, revealed the following percentage changes: Corncrake *Crex crex* (+14, increasing after period of severe decline), European Turtle Dove *Streptopelia turtur* (−77), Common Cuckoo *Cuculus canorus* (−43), Barn Swallow *Hirundo rustica* (+11), Yellow Wagtail *Motacilla flava* (−59), Lesser Whitethroat *Sylvia curruca* (+2), Greater Whitethroat *Sylvia communis* (−18), Spotted Flycatcher *Muscicapa striata* (−82), Red-backed Shrike *Lanius collurio* (−90) (Gregory *et al.* 2004). The slightly different figures for some species between the two studies result from the slightly different time periods and possibly the different methods of analysis (see original papers). The massive increase of the Osprey is attributable to recovery from previous persecution.
From Gregory *et al.* (2002).

decline differed between species, the Greater Whitethroat suffering a catastrophic crash mainly in one year, and others showing more gradual declines over 10 or more years, with some subsequently recovering somewhat and others not. These differences may reflect differences in wintering areas between species, or in the mechanisms involved, and in some species may be confounded by simultaneous changes in breeding areas. For some species drought might be expected to reduce the food supply only in the year concerned, while for others it might also destroy the shrubby habitat which could take years to recover, or not recover if human activity prevented it. Some European–Afrotropical migrants showed no marked net change over the period considered, while a few showed a substantial rise. Most of the latter winter well south of the Sahel zone, however, and some extend south of the equator to the austral summer.

In general, species that winter in the Sahel zone, or pass through in autumn and spring, showed lower population levels in Britain in 1984, 1985 or 1991 than in any previous year since 1962 (when the Common Bird Census started). They included, besides those just mentioned, the Barn Swallow *Hirundo rustica*, Grasshopper Warbler *Locustella naevia*, Eurasian Chiffchaff *Phylloscopus collybita* and Spotted Flycatcher *Muscicapa striata*. The Chiffchaff subsequently recovered but the Willow Warbler *Phylloscopus trochilus* continued to decline (Peach *et al.* 1995), as did the Spotted Flycatcher (**Table 24.5**; Peach *et al.* 1995, 1998). Moreover, annual fluctuations in the numbers of some species followed annual fluctuations in the preceding year's rainfall in the Sahel zone, as shown for the Sedge Warbler *Acrocephalus schoenobaenus* in Britain and the Netherlands (**Figure 24.5**, Peach *et al.* 1991, Foppen *et al.* 1999). Similarly, in British Swallows, breeding numbers were not correlated with rainfall in their South African wintering areas, but with rainfall in the western Sahel, at its driest during spring migration (Robinson *et al.* 2003).

Comparisons of count data from different countries have shown that year-to-year changes in numbers are often correlated over large parts of the European

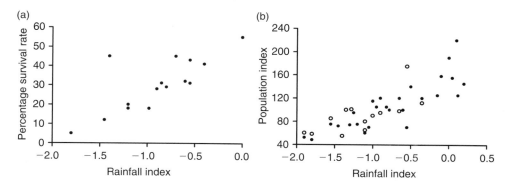

Figure 24.5 Relationships between: (a) annual survival of Sedge Warblers *Acrocephalus schoenobaenus* at various sites in Britain; and (b) annual spring population index in Britain, and preceding annual rainfall in the Sahel zone of Africa, where Sedge Warblers spend part of the winter. Through presumed influence on food supplies, rainfall in West Africa affected the overwinter survival of Sedge Warblers, and hence their subsequent breeding densities in Britain. Redrawn from Peach *et al.* (1991).

Table 24.5 Relationships established between abundance or survival of migrants as measured from studies in Europe, and preceding winter rainfall in African wintering areas

Species (years)	Area in Europe	Area in Africa	Parameter measured		Source
			Abundance	Survival	
Sand Martin *Riparia riparia* (1986–1992)	Hungary	West Africa (Sahel–Sudan)		+	Szép (1995)
Sand Martin *Riparia riparia* (1965–1978)	England	West Africa (Sahel)	+		Cowley (1979)
Sand Martin *Riparia riparia* (1980–1991)	Scotland	West Africa (Sahel)	+	+	Bryant & Jones (1995)
Sand Martin *Riparia riparia* (1966–1973)	Germany	West Africa (Sahel)	+	+	Kuhnen (1975)
Purple Heron *Ardea purpurea* (1963–1978)	Netherlands	West Africa (Sahel–Sudan)		+	Cavé (1983)
Purple Heron *Ardea purpurea* (1961–1979)	Netherlands	West Africa (Sahel–Sudan)	+		Den Held (1981)
Black-Crowned Night Heron *Nycticorax nycticorax* (1967–1979)	France	West Africa (Sahel–Sudan)	+		Den Held (1981)
White Stork *Ciconia ciconia* (1956–1976)	France	West Africa (Sahel)		+	Kanyamibwa et al. (1993)
Barn Swallow *Hirundo rustica* (1970–1988)	Denmark	South Africa (High Veld)[a]		+	Møller (1989)
Sedge Warbler *Acrocephalus schoenobaenus* (1963–1983)	England	West Africa (Sahel–Sudan)	+	+	Peach et al. (1991)
Sedge Warbler *Acrocephalus schoenobaenus* (1973–1985)	Netherlands	West Africa (Sahel–Sudan)	+		Foppen et al. (1999)
Greater Whitethroat *Sylvia communis* (1962–1972)	Britain	West Africa (Sahel–Sudan)	+		Winstanley et al. (1974)
Greater Whitethroat *Sylvia communis* (1946–1977)	Sweden	Lake Chad (Sahel)	+		Hjort & Lindholm (1978)
Willow Warbler *Phylloscopus trochilus* (1986–1993)	England	West Africa (Guinea–Sudan)	+	+	Peach et al. (1995)
Common Nightingale *Luscinia megarhynchos* (1996–2003)	Italy	West Africa (Sahel–Guinea)		+	Boano et al. (2004)
Barn Swallow *Hirundo rustica* (1964–1998)	Britain	West Africa (Sahel)[a]	+		Robinson et al. (2003)

[a] Birds winter in southern Africa, but pass through the Sahel in autumn and spring. The return journey falls at the end of the dry season.

breeding range but, as expected, the counts from some regions show different trends, and some show no evidence for links with Sahel droughts (Svensson 1985, Marchant 1992, Sokolov *et al.* 2001). Europe-wide trends would perhaps not be expected in species whose west European populations winter in West Africa, while east European populations winter in East or southern Africa, and are thereby exposed to different climatic regimes.

An analysis of long-term trapping data from a number of bird observatories scattered between Russia and Belgium (but mainly in eastern Europe) gave a somewhat different picture (Sokolov 2001). In many regions significant increases in numbers of passage passerines of many species occurred in the 1960s and 1980s, and significant declines in the 1970s and 1990s. In not a single species of 36 studied was a significant trend found for the overall period 1959–2000. Because in this area both short- and long-distance migrants adhered to the same broad fluctuations, the author attributed them to events in breeding areas, specifically 'long-term climatic fluctuations in the northern hemisphere'. In general, however, these populations were drawn from eastern, rather than western Europe, and among the tropical migrants, mainly wintered in East or southern Africa, as opposed to West Africa.

It is not just counts that point to the importance of wintering areas in influencing numbers of west European migrants. In key factor analyses, variation in overwinter loss was found to account for most of the variation in total annual loss in seven migrant passerine species monitored in Britain or Denmark (Møller 1989, Baillie & Peach 1992). In addition, annual survival rates of Sedge Warblers *Acrocephalus schoenobaenus* and Greater Whitethroats *Sylvia communis* trapped each summer at specific sites in Britain were also correlated with rainfall in the Sahel zone (**Figure 24.5**; Peach *et al.* 1991, Baillie & Peach 1992), as were survival rates of Nightingales *Luscinia megarhynchos* at sites in Italy (Boano *et al.* 2004) and of Sand Martins *Riparia riparia* trapped at sites in Scotland and Hungary (**Table 24.5**; Bryant & Jones 1995, Szép 1995). Major crashes in Sand Martin numbers (or survival) occurred in 1968/69, 1983/84 and 1990/91, all following years of poor Sahel rainfall. Similarly, annual survival rates of Barn Swallows *Hirundo rustica* breeding in an area of Denmark were correlated with March rainfall in their southern African wintering areas (Møller 1989). Although factors causing long-term trends in bird numbers are not necessarily the same as those causing annual fluctuations, they appear to be the same in these cases. These studies provided further circumstantial support for the proposed causal chain: low rainfall → low winter food supplies → lower overwinter survival → low breeding population.

It is not only insectivorous passerines that have shown such links. The numbers of Purple Herons *Ardea purpurea* nesting in the Netherlands over a 19-year period, and their annual survival rates, were correlated with wetland conditions in their West African wintering areas, as measured by water discharge through the Niger and Senegal Rivers (den Held 1981, Cavé 1983). The wetter the conditions in West Africa, the better was the overwinter survival and the greater the number of herons that returned to Holland to breed each year (**Figure 24.6**). A similar relationship was apparent among Night Herons *Nycticorax nycticorax* counted at colonies in southern France (**Figure 24.6**), and to a lesser extent among Squacco Herons *Ardeola ralloides* in the same localities (den Held 1981). A later study extended these findings to other colonies in western Europe, but not to colonies

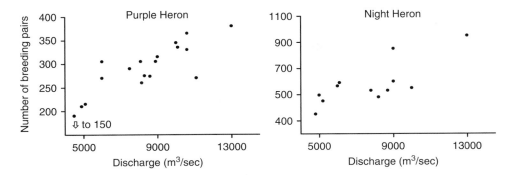

Figure 24.6 Relationship between the numbers of migratory herons nesting at colonies in western Europe and wetland conditions in West Africa during the preceding year. Purple Herons *Ardea purpurea* were counted at one major colony in the Netherlands, and Night Herons *Nycticorax nycticorax* at several colonies in southern France. Wetland conditions measured as the sum of the maximum monthly discharges through the Senegal and Niger rivers. Both relationships were statistically significant (on Kendall's rank correlation test, $P < 0.005$). From den Held (1981).

further east in Europe (whose occupants are likely to have wintered further east in Africa) (Fasola *et al.* 2000). All these species have declined in western Europe since the 1960s, as has the Little Bittern *Ixobrychus minutus* which also winters in drought-prone areas of West Africa (Marion *et al.* 2000). In contrast, the Little Egret *Egretta garzetta*, which winters increasingly within Europe, has expanded over the same period, and its colony sizes showed no relationship with West African conditions (Fasola *et al.* 2000).

Similarly, the numbers of White Storks *Ciconia ciconia* that breed in western Europe have declined greatly since 1960, associated with a general decline in rainfall in their West African wintering areas. Annual variations in survival (measured from ringed birds) and subsequent reproduction (measured from nest studies) were closely linked to annual rainfall in the western Sahel zone, presumably because winter food supplies influenced the condition of adult storks in spring, and not only their survival (Bairlein & Henneberg 2000,[2] Kanyamibwa *et al.* 1993, Barbraud *et al.* 1999). Moreover, following greater rainfall in the western Sahel from the mid-1980s, the west European population began to recover, bringing a 50% increase in breeding numbers between 1984 and 1994. The east European population, which winters mainly in East and southern Africa, having different rainfall regimes, showed a much less marked decline in the earlier period and a less marked increase in the later period (Dallinga & Schoenmakers

---

[2]*The White Stork* Ciconia ciconia *in western Europe has also been affected by loss of habitat caused by drainage of wet pasture, which affected breeding success (Chapter 26). This may explain why populations in central Europe declined more rapidly than those in Spain, even though both populations winter in the same parts of West Africa.*

1989, Schulz 1998). Differences in rainfall patterns may also explain the differences in population dynamics of some other conspecific populations of birds wintering in different parts of Africa, such as western and eastern populations of White Wagtail *Motacilla alba* (Svensson 1985). Overall, however, drought in African wintering areas has been implicated as a major causal factor in the regional population declines of at least 17 Eurasian breeding species over the period 1960–2000.

The importance of winter conditions is further shown by the fact that none of the passerine species mentioned above showed widespread declines in breeding rates over the periods concerned, or fluctuations in breeding rates that matched the subsequent fluctuations in breeding numbers. So far, however, the evidence that conditions on the wintering areas influence the numbers of breeding birds is based almost entirely on correlations between annual rainfall, water levels or river flow on the one hand and year-to-year changes in population or survival on the other. Improvements in some correlative analyses might be made in future if the wintering areas of particular sectors of breeding populations could be defined more precisely, and if weather data were then chosen to match those areas more accurately. Greater understanding of mechanisms would probably emerge if more field studies were conducted in Africa itself, where food supplies could be measured. The clear implication of current findings, however, is that climatic conditions up to several thousand kilometres south of the breeding areas affect the spring population levels of some species of trans-Saharan migrants that breed in Europe.

OTHER POSSIBLE FACTORS IN POPULATION DECLINES

Decreasing rainfall is not the only factor which may have caused declines in the numbers of some European birds that winter in West Africa. The burgeoning human population, and associated overgrazing, burning and woodcutting, together with increasing drainage and water abstraction, have all accentuated the process of desertification, so that even in relatively wet years many areas now hold little habitat for birds (Jones 1985, Thiollay 1989). Through these processes, desert is expanding at about 1% per year, with thousands of extra square kilometres added annually (Le Houerou & Gillet 1986). Indeed, some climatologists believe that, owing to human-induced destruction of vegetation, the climate of the Sahel zone has now flipped to an alternative steady state, represented by more desert-like conditions. Nearer the equator, more than 30% of African forest has been lost since 1970, and the cultivated land area has increased by 21% over the same period (UNEP 2002). Cultivation of rice and the irrigation of other crops has shrunk many wetlands used by migratory birds, reducing the surface area of Lake Chad, for example, from $25\,000\,\text{km}^2$ to $1350\,\text{km}^2$ since 1960 (UNEP 2002).

Over much of the Sahel zone, the human population is now well above sustainable levels, and prevents any recovery of the vegetation and seed stock, which has proved possible locally, even with low rainfall, under total protection. Serious droughts have occurred in the past, as in 1911–1916 and 1940–1946, without any obvious lasting ecological effects. But since then the human and livestock populations have increased by more than 10-fold, leading to considerable impoverishment of the grasslands, disappearance of trees and shrubs, and shrinkage and degradation of wetlands. So even if rainfall returns to its earlier levels, some bird species may not regain their former abundance because of the habitat degradation that has occurred in the interim.

In addition, with the increasing use of pesticides, insects eaten by birds are likely to have become much less available, and many birds have themselves been killed directly by pesticide use (Mendelssohn & Paz 1977, Thiollay 1989, Mullié & Keith 1993). At least four Palaearctic species, which feed extensively on locusts and grasshoppers on their African wintering grounds, have declined in recent decades, including the White Stork *Ciconia ciconia* discussed above, Lesser Kestrel *Falco naumanni*, Montagu's Harrier *Circus pygargus* and Pallid Harrier *Circus macrourus*. However, in all these species, other factors may also have been involved (see below). On the other hand, the felling of rainforest in West Africa may have affected very few Eurasian migratory species adversely, because so few species winter there, and because the forest is replaced by more open secondary habitats ('derived savannah') which many species favour (Morel & Morel 1992). The same is true for miombo woodland in southern Africa (Ulfstrand & Alerstam 1977).

Not all Eurasian–African migrant species are necessarily limited primarily by factors acting on the wintering range, and for any one species the situation may differ from one part of the breeding–wintering range to another. Examples of species whose changes in breeding numbers have been linked with events in breeding areas include the Pied Flycatcher *Ficedula hypoleuca* for year-to-year changes (see **Figure 26.4**; Virolainen 1984), and the Turtle Dove *Streptopelia turtur* for long-term decline (Browne & Aebischer 2001). Other Palaearctic–African migrants whose declines have been linked at least partly to habitat changes in breeding areas include Corncrake *Crex crex*, Yellow Wagtail *Motacilla flava*, Lesser Kestrel *Falco naumanni* and Montagu's Harrier *Circus pygargus* (Norris 1947, Stowe *et al.* 1993, Donazur *et al.* 1993, Arroyo *et al.* 2002, Newton 2004b). In some species, as mentioned above, the relative importance of breeding and wintering conditions in influencing breeding numbers may well change from one period of years to another or from one region to another.

Hunting in the Mediterranean region was said to remove up to 1000 million birds each autumn, mainly migrant songbirds (Magnin 1991, McCulloch *et al.* 1992). This hunting is greatest in Italy, various Mediterranean islands (notably Cyprus and Malta), northern Africa and the Lebanon. The numbers of ringed birds reported each year from these areas are declining, but it is not known to what extent this reflects trends in populations, hunting or reporting rates. Nor is it known what effect this hunting has on breeding numbers. But because hunting occurs mainly in autumn, it could be offset to some extent by reductions in subsequent winter losses, if the total overwinter losses are density dependent (see Newton 1998b for evidence of winter density-dependence in several migratory species). In this case, autumn hunting would not have the large impact on population levels that might be thought from the numbers killed. This seems to be the case in Quail *Coturnix coturnix*, in which the numbers caught each autumn are related more to Sahel rainfall than to hunting pressure (Zuckerbrot *et al.* 1980), with numbers declining in recent decades (Tucker & Heath 1994). The numbers of Corncrakes *Crex crex* taken on the Mediterranean coast of Egypt in the autumns of 1993 and 1994 were estimated at about 9000 and 14 000 respectively, but these numbers formed only 0.5–2.7% of the total European breeding population (Baha el Din *et al.* 1996). There is also considerable hunting of migrants in Africa south of the Sahara. The emphasis seems to be on larger species, with many European Honey Buzzards *Pernis apivorus* killed in Liberia, White Storks *Ciconia ciconia* and Ospreys *Pandion haliaetus* in Nigeria, White Storks and Cranes *Grus grus* in Sudan, terns off West Africa, and many other species in these and other regions (Berthold 1993).

In some species, natural predation takes a considerable toll during migration. Two falcon species specialise on Eurasian–African migrants: Eleonora's Falcon *Falco eleonorae* which nests mainly on rocky islands in the Mediterranean and off the west coast of Morocco, and the Sooty Falcon *F. concolor* which nests along the coasts of the Red Sea and Arabian Gulf, and more sparsely in the deserts of Arabia, Israel, Egypt and Libya. Both species breed late, at the time of the autumn passage, and raise their young on migrants. Three other falcon species – the Peregrine Falcon *F. peregrinus* in the Mediterranean region, the Barbary Falcon *F. pelegrinoides* at scattered localities throughout the Sahara, and the Lanner Falcon *F. biarmicus* in the southern Sahara – are all resident year-round, but nest early, at the time of spring migration, and exploit the northward-bound migrants. Walter (1979) estimated that the world population of 10 000 Eleonora's Falcons *F. eleonorae* caught 1–2 million migrants annually, out of a total migration that Moreau (1972) considered to be in the order of 5000 million birds in the 1960s. Consequently the overall impact of the falcons' predation was assumed negligible (<0.1%). However, three species of shrikes (*Lanius collurio*, *L. minor* and *L. senator*) made up 15–20% of all birds taken (i.e. 200 000–400 000 shrikes per year). There is no reason to believe that the numbers removed by predation have increased during the twentieth century, however, when shrikes were declining over much of their European range. The same applies to the toll taken by the other falcons, which as yet cannot be estimated (Chapter 27).

Recently, additional predation on migratory birds in the Mediterranean region was found unexpectedly to involve the Giant Noctule Bat *Nyctalus lasiopterus* (Popa-Lisseanu *et al.* 2007). This mammal has a wingspan of 46 cm, and was first suspected of eating nocturnal migrants from the appearance of feather pieces in its droppings. More recently, blood samples taken from live bats over the period March–October showed that the isotope signatures of ^{15}N and ^{13}C changed seasonally in precisely the manner predicted if the bats shifted from a diet of insects in summer to birds in autumn. In spring, when migrant numbers were lower, signatures indicated a mixed diet of insects and birds. This bat hunts by aerial pursuit, and the virtual absence of birds from the diet in summer suggests a seasonal specialisation on passing migrants. Interesting though these findings are, the bats themselves are rare and localised, and it is unlikely that they have any appreciable impact on the collective migrant population.

Taking the evidence as a whole, it is hard to escape the conclusion that deteriorating conditions in Africa have been primarily responsible for the marked numerical declines of some Eurasian–African migrant species recorded in recent decades, and that these other potential limiting factors have been of most minor significance in this period. In other species, however, human-induced habitat changes in breeding areas may have caused or contributed to long-term population declines. Such evidence is available for at least seven species, about 4% of the Eurasian landbird species wintering in Africa, and further study may reveal more examples. In many species, the long-term declines may have been caused primarily by progressive land use and habitat changes in Europe, while year-to-year fluctuations about this trend have been associated with climatic and food conditions in Africa (for further discussion see Chapter 26). Moreover, although many Eurasian–Afrotropical migrants have declined during the last 50 years or more, some have increased – the Honey Buzzard *Pernis apivorus* and Hobby *Falco subbuteo* providing striking examples.

THE ASIAN–AUSTRALASIAN MIGRATION SYSTEM

The second major migration system for Palaearctic nesting birds, the Asian–Australasian system, has been less studied, but provides interesting comparisons (Lane & Parish 1991). Approximately 234 species are involved, derived from a breeding area extending eastward from central Asia across to Alaska. These species winter in the southeastern part of the Eurasian land mass, including the Indian subcontinent (about 4 million km^2) and the southeast Asian mainland (about 2 million km^2), but some species extend southwards to Australia and New Zealand and others eastward to some Pacific Islands. The natural vegetation of Southeast Asia, where most of the migrants spend the northern winter, is rainforest or dry deciduous woodland, with some savannah and grassland. But much of this natural habitat has disappeared under human impact, becoming agricultural or arid through deforestation, scrub removal and overgrazing. In general, Southeast Asia is much wetter than Africa but, as in Africa, those migrants that stay north of the equator experience progressively drier conditions during their stay. Vegetation dies back and flood-pools dry. Tropical wintering areas that lie south of the equator, including New Guinea and northern Australia, are at their wettest in the northern winter (austral summer), but temperate southern Australia and New Zealand have their driest weather then.

The further south that birds migrate within Asia, the more mountains they face, the more water gaps they must cross, and the more restricted is the land area that is available to them. Although movement is essentially broad-front, the distribution of land areas tends to channel the birds along particular routes, as confirmed by both observations and ring recoveries. Several island groups provide potential stepping stones, from the Aleutians in the north to the Indonesian and other islands in the south. One major route extends south from Japan and eastern China, through the Philippines; another runs into and through the Malaysian peninsula, while a third runs from south Vietnam southwest across the South China Sea to Borneo (McClure 1974, Medway & Wells 1976, Ellis *et al.* 1994). Some birds also cross the Gulf of Thailand to and from the Malaysian peninsula.

In general, as in Africa, most migrant species stay in the northern part of this wintering area, and species numbers decline with increasing distance southwards to New Zealand. About 161 migratory species filter into Malaysia, but most stop in Indonesia, at about 10°N. Few Eurasian passerine species reach New Guinea or Australia, but many shorebirds reach Australia and New Zealand. Numbers of landbirds also decline eastwards out from the Asian mainland onto Pacific

Islands, with no fewer than 120 species wintering regularly in the Malay archipelago, 84 in the Philippines, 38 on Celebes, 10 in the New Guinea area and three in the Bismarck Archipelago (Mayr 1969). Yet other shorebird species from Alaska winter on various Pacific islands from Hawaii across to Indonesia, and southward to Australia and New Zealand.

The only Asian landbirds that migrate in large numbers to Australia are two species of swifts (Fork-tailed Swift *Apus pacificus* and White-throated Needletail *Hirundapus caudacutus*), while others that reach Australia in smaller numbers include the Oriental Cuckoo *Cuculus saturatus*, the Barn Swallow *Hirundo rustica gutturalis*, Red-rumped Swallow *H. daurica*, Grey Wagtail *Motacilla cinerea*, Yellow Wagtail *M. flava* and the Oriental Reed Warbler *Acrocephalus orientalis* (Dingle 2004). At least one of these, the Oriental Cuckoo, also reaches New Zealand. The scarcity of Palaearctic songbird migrants in Australia during the northern winter contrasts with their abundance in southern Africa.

On the other hand, about 33 species of arctic-nesting shorebirds winter in Australia, totalling some 2–3 million individuals, nearly half the total flyway population (Lane & Parish 1991). The most abundant coastal species include the Bar-tailed Godwit *Limosa lapponica* (165 000), Little Curlew *Numenius minutus* (180 000), Great Knot *Calidris tenuirostris* (270 000), Red Knot *Calidris canutus* (153 000), Red-necked Stint *Calidris ruficollis* (353 000) and Curlew Sandpiper *Calidris ferruginea* (188 000) (Higgins & Davies 1996). These birds mostly arrive in the north of Australia in late August and September. Some species, such as Greater Sand Plover *Charadrius leschenaultii* and Great Knot, mostly remain in the north of Australia through the non-breeding season, while others, such as Red-necked Stint and Curlew Sandpiper, mostly move on to coastal sites further south. Some wader species winter inland, such as the Oriental Pratincole *Glareola maldivarum*, Common Sandpiper *Actitis hypoleucos* and Greenshank *Tringa nebularia*, all of which move to the coast in extreme drought years. The Bar-tailed Godwit and Red Knot reach New Zealand in large numbers (57 000 and 36 000 respectively), as do smaller numbers of the other species (Sagar 1986, Higgins & Davies 1996). Further shorebird species from North America also winter in Australia and New Zealand. Among other waterbirds, the Garganey *Anas querquedula* and White-winged Black Tern *Chlidonias leucopterus* reach Australia, as do several marine tern and skua species. The main threat to these migratory birds, besides coastal drainage, is probably human hunting, with an estimated 1.5 million shorebirds killed each year in Indonesia alone.

It is not hard to imagine why so few Palaearctic species reach Australasia. This land area lies at the end of the migration route, and also well to the east of the Eurasian land mass. Secondly, Australia offers relatively small areas of habitat suitable for most Eurasian landbird migrants; and thirdly, several water crossings are necessary to reach these areas, by far the longest being the 1600–2000 km stretch between Australia and New Zealand.

SUMMARY

Some 186 species of landbirds and 29 seabirds (215 species in all) that breed in Eurasia spend the winter in Africa, south of Sahara. In late summer, they total an estimated 5000 million individuals. These birds derive mainly from the western third of Eurasia, but some also from further east, as far as northwest North America. They all face long and formidable journeys, with long desert-crossings. In the northern winter, they are estimated to form only 6–7% of the total African bird population, but in certain northern savannah habitats they outnumber African species.

Within Africa south of the Sahara, the numbers of migratory species decline southwards, and the greatest diversity occurs in the northeast, coinciding with a region also rich in native African species. Most migrants occur in seasonal woodland, savannah and scrub habitats, which also hold 67% of all African species; however, the migrants have greater mobility, travelling long distances within their stay, and some are nomadic within their wintering range, concentrating temporarily wherever food is abundant (like some intra-African migrants). The Eurasian migrants also differ ecologically from closely allied native species, with at least slight differences in habitats and feeding behaviour.

Many species winter in dry season conditions in the northern tropics which deteriorate during their stay, while smaller numbers move south of the equator to spend the winter in wet-season conditions. Some Eurasian–Afrotropical migrants show single-stage migration to the northern or southern tropics, while others show two-stage migration, either within the northern tropics or between the northern and southern tropics. In consequence, the bulk of the migrant population shifts southward within Africa during the northern winter.

Recent declines in the numbers of some species, detected on European breeding areas, have coincided with reduced rainfall in the northern tropics, especially in the western Sahel zone. In some species, precise relationships have emerged between annual breeding numbers or survival and annual rainfall in African wintering areas. In other species of Eurasian–Afrotropical migrants, long-term declines have been associated with habitat destruction or deterioration in breeding areas. Hunting in the Mediterranean area removes many migrants in autumn, but with no known affects on population trends.

The majority of east Eurasian migrants winter in Southeast Asia (including the Indian subcontinent), with a small number of species (mainly shorebirds) migrating as far as Australia and New Zealand.

Blackpoll Warbler *Dendroica striata*, a Nearctic–Neotropical migrant that migrates over the Western Atlantic

Chapter 25

The Nearctic–Neotropical migration system

Approximately half of all neotropical migrants winter in Mexico and the Antilles. . . . relative to the size of the breeding ground, this is a very restricted area. Populations are consequently highly compressed. (John Terborgh 1989.)

The New World migration system differs from the Eurasian–African one in several respects. First, the presence of a land bridge between North and South America and of sizeable islands in the Caribbean mean that the journey is potentially far less arduous than for the Old World migrants which must cross large areas of hostile desert or sea. The continuum of habitable ground from North through Central to South America means that there is no large geographical break between potential breeding and wintering areas, which for many species are contiguous or overlapping. Moreover, in the New World, the mountain chains run mainly north–south, rather than east–west, and thus present less of an obstacle to the passage of birds. Second, most of the North American migrants winter in Central America and the Caribbean islands, and a relatively small proportion extend in South America south of the equator. Even more than in Africa, therefore, the majority of species crowd into an area smaller than their overall breeding range. Third, most migrants from North America are forest species which winter in forest, rather than in dry savannah or scrub, adding to some of the richest bird communities on earth. For these reasons, the idea that population sizes might be limited primarily on the wintering grounds seems at first sight just as tenable for the New World as for the Old. Yet, as explained later, many of the migratory species that have declined in recent decades are thought to have done so because of changes in the breeding rather than in the wintering areas.

THE BIRDS AND THEIR WINTERING AREAS

Altogether, about 250 landbird species from North America winter at least partly in the Neotropics, rather more than the Palaearctic migrants in Africa. Because these migrants are drawn from a smaller land area, however, their numbers are smaller than in the Palaearctic–Afrotropical system, perhaps around 3000 million (MacArthur 1972). Of the species that reach South America, the broad-winged raptors migrate mainly overland via Panama. Many other species that breed in eastern North America cross the Gulf of Mexico, or the Caribbean Sea (stopping on various islands), and yet other species take a longer overwater route, leaving the eastern North American coast and flying directly across the Atlantic to South America (Chapter 6). In autumn, different species take off at different points along the coast, so that the numbers seen on the coast diminish southward, and relatively few leave from Florida (Terborgh 1989). Hence, although shorter sea-crossings are available, many species in autumn make a substantial sea-crossing, which shortens their overall journey, compared with flying down the eastern seaboard, and crossing via Caribbean Islands. Species from western North America take an overland route to winter mainly in Mexico (although some first cross the northeast Pacific from Alaska to further south on the western seaboard, Chapter 6).

Most of the migrants from eastern North America winter in South America or on Caribbean Islands, while the ones from western North America are accommodated mainly in Central America. Moreover, because most of South America lies in longitudes to the east of North America, the prevailing direction of most intercontinental migration is southeast in autumn and northwest in spring. The latitudinal span that is occupied by Central America and the Caribbean Islands in the New World is occupied by the Sahara Desert in Africa, which accounts for why the New World migrants migrate less far, on average, than the Eurasian–Afrotropical ones.

North America south of the tree-line covers approximately 16 million km^2, and within this area most migratory species have ranges extending over at least several million square kilometres. In winter, these migrants extend over a somewhat larger area, covering about 18.6 million km^2 from Mexico southward to the tip of South America, and including the Caribbean Islands. However, as in sub-Saharan Africa, the migrants are not uniformly distributed over this whole area, but are concentrated in the northern tropics. In fact, the main wintering area for many migrant species extends from Mexico south to Panama, including the Caribbean

Islands, a region covering about 2.7 million km², in which the main natural vegetation is forest. Such a concentration of migrants is known nowhere else on earth. The great disproportion between the breeding and wintering areas of the migrants may result partly from geographical circumstances, notably the progressively narrowing land area in Central America. Not only are the populations of many species highly compressed there, but more species are found together in the same area. For example, in deciduous forest in the eastern USA, some 4–6 warbler species are commonly found together in the same area in summer, but in homogeneous Neotropical forest, 8–10 species are commonly found together, and occasionally up to 15 species, densely packed in communities of extraordinarily high diversity. Some specialists do have extensive distributions in winter, but they are in the minority. They include the Black-and-white Warbler *Mniotilta varia* which feeds nuthatch-like on trunks and branches, and the Northern Waterthrush *Seiurus noveboracencis* and Louisiana Waterthrush *S. motacilla* both of which hunt insects along streams (Terborgh 1989). Unlike Africa, mid- to high-elevation areas (1000–2500 m) are heavily used by migrants in the Neotropics, presumably a consequence of the large proportion of the Neotropical land area that is over 1000 m and well forested.

In the Neotropics, then, the numbers of migrant species and individuals, and their proportion in the local avifauna, decline southwards from Mexico (Slud 1976), but with big variations between habitats (Terborgh 1980). This parallels the north–south decline in migrant numbers that occurs in sub-Saharan Africa. On both continents, it seems that species migrate no further than is needed to find suitable wintering habitat. If the survival chances on migration decrease as the journey lengthens, which may be true at least as a generalisation, this could provide the selection pressure to keep the migration as short as possible (Karr 1976). In addition, the decline in the numbers of migrant species towards the equator is more than offset by an increase in the numbers of resident species, so that competition may be another factor curtailing further southward spread (Keast 1980, Terborgh 1980). On various Caribbean islands, the proportion of migrants in winter bird populations declines with increasing distance from the mainland (Florida), from about 35% on Hispaniola to 10% on Puerto Rico, and 1% on the more distant islands of the Lesser Antilles, Trinidad and Tobago, as well as on the Venezuelan mainland. The effect of island size is difficult to separate from the effect of latitude and distance from the mainland, but comparison of adjacent large and small islands reveals some tendency for migrants to gravitate to larger islands (Terborgh & Faaborg 1980).

In general, Central and South America, with their greater areas of high rainfall, provide greater areas of forest than Africa. Rainforest is again centred on the equator, but is much more extensive, reaching more than 20°N in Central America and more than 10°S in Amazonia, with an additional strip extending to nearly 30°S on the eastern coast. As in Africa, owing to decline in rainfall, habitats become progressively more open north and south of the rainforest, passing through woodland and scrub savannah and grassland (pampas) and in the south to shrubby desert. The driest desert occupies a strip between the western coast and the Andes. The mountains themselves support various types of montane forest and high-elevation grassland. Another area of high rainfall lies on the southwestern coast, supporting mixed evergreen forest.

Like Africa, South America falls under the climatic influence of the Inter-Tropical Convergence Zone, but the effects are less clear-cut because of the more varied topography of South America, and the influence of the Andes range in the west. As in Africa, however, seasonal flooding greatly increases the area of wetland, especially south of the equator. The Amazon Basin receives more than 3 m of rain per year, which provides 57 000 km^2 of seasonally flooded forest. The southern savannahs (llanos) support the greatest area of seasonal swamp on earth, the 60 000 km^2 of flood, drought and fire that forms the Pantanal.

As in Africa, migrant species generally choose winter habitats that resemble their summer ones: species that breed in forest or scrub winter in forest or scrub, and most of those that breed in grassland winter in grassland. According to Hutto (1986), the similarity is even greater, and the majority of western migrants that winter in western Mexico occur in habitats closely resembling their breeding ones, whether desert scrub, riparian, oak or oak–pine woodland or conifer forest. However, some species show striking seasonal differences in feeding behaviour. Among passerines, the Worm-eating warbler *Helmitheras vermivoras* is a fairly typical foliage-gleaning bird during most of the breeding season, but becomes highly specialised in foraging from the leaf litter caught up in tropical forest trees in the non-breeding season (Willis 1980). The Yellow-rumped Warbler *Dendroica coronata* changes from mainly tree-feeding in the breeding season to mainly ground-feeding in the non-breeding season. In addition, among seven migratory species that winter in the Yucatan, four breed in forest and winter in scrub, while three breed in scrub and winter in forest (Askins *et al.* 1990). As elsewhere, many shore-birds breed on the tundra and winter on sea-coasts.

Also as in Africa, those migrants that remain north of the equator experience deteriorating conditions throughout their stay, as the dry season progresses, while the minority that move south of the equator experience improving conditions. The latter again include those species that depend on insects or other food items that reach peak supply in the southern rainy season. They include the Mississippi Kite *Ictinia mississippiensis* and Swainson's Hawk *Buteo swainsoni* which eat grasshoppers, the Common Nighthawk *Chordeiles minor*, Chimney Swift *Chaetura pelagica*, Purple Martin *Progne subis*, Barn Swallow *Hirundo rustica* and Eastern Wood Pewee *Contopus virens*, all of which eat flying insects, and cuckoos that eat caterpillars from young leaves. They also include some shore-birds, which winter on the coasts. Most long-distance raptors are nomadic during the dry season, and are often attracted to large-scale seasonal burns. Others, including Swainson's Hawks, follow migrating swarms of dragonflies, and still others, such as Swallow-tailed Kites (*Elanoides forficatus*), track populations of ephemerally swarming termites (Bildstein 2004).

Most forest species spend longer each year in their tropical wintering areas than in their northern breeding areas. They are clearly as much a part of the tropical bird communities in which they winter as of the temperate zone communities in which they breed, occupying different niches from one another and from resident species (Lack & Lack 1972, Rappole & Warner 1980, Terborgh & Faaborg 1980, Rappole *et al.* 1983, Rappole 1995). Like their Old World counterparts, many Neotropical migrants are insectivores, but whereas in Africa they are mainly in the families Turdidae (thrushes, 18 species) and Sylviidae (warblers, 29 species), in the Neotropics they are mainly in the Parulini (wood warblers, 46 species) and Tyrannidae (tyrant flycatchers, 23 species) (Karr 1980). When the northern migrants leave in spring, their niches are mostly left vacant until the following autumn. Unlike the situation in Africa, few southern hemisphere birds migrate so far north in the austral winter that they could occupy these same niches while the Nearctic–Neotropical migrants are absent.

Many migratory species maintain the same feeding territories throughout their stay and return to them year after year. Examples include the Northern Waterthrush *Seiurus noveboracensis*, Northern Oriole *Icterus galbula*, Summer Tanager *Piranga rubra* and various warblers (Schwartz 1964, Stiles & Skutch 1989, Rappole 1995). Other species, such as the Eastern Kingbird *Tyrannus tyrannus*, defend short-term territories, wherever food is temporarily available. The various hummingbirds provide extreme examples, as individuals defend particular flower patches only for the short periods they produce nectar, and then move on to other more productive patches. In some territorial species, the sexes are found in mainly different habitats (Lynch *et al.* 1985, Wunderle 1992), either because of different preferences, as in Hooded Warbler *Wilsonia citrine* (Morton *et al.* 1987), or because male dominance causes females to occupy largely poorer habitats, as in American Redstart *Setophaga ruticilla* (Marra *et al.* 1993, 1998). The effects of male dominance, in enabling males to occupy the better wintering habitats, may partly account for the great preponderance of males seen in many species of Neotropical migrant passerines in their North American breeding areas (Stewart & Aldrich 1951, Marra & Holmes 1997).

Some wintering forest migrants join group territories containing individuals of several species. Such groups consist predominantly of Neotropical residents, such as ant-wrens and ant-vireos, but for half the year they also contain Nearctic migrants such as Blue-winged Warbler *Vermivora pinus* and Canada Warbler *Wilsonia canadensis* (Greenberg & Gradwohl 1980, Greenberg 1985). Each group normally contains only one pair (or family) of each species. Every morning the birds assemble to forage together. All group members recognise the same territorial boundaries so that in effect the group, rather than its component pairs, holds the territory. This is a social system which is also known from the forests of Southeast Asia (Harrison 1962), but is unknown among the few Palaearctic breeding species that spend the non-breeding season in the Afrotropical forests.

Other forest migrants join single species or mixed species flocks which move around to exploit sporadically available food supplies (Rappole & Warner 1980, Greenberg 1985, Terborgh 1989). Some species, such as Swainson's Thrush *Catharus ustulatus*, join groups of local species which follow columns of army ants to feed on the other insects disturbed (Willis 1966). Like the African migrant insectivores, many also eat fruit in winter (Parrish 2000), and some, such as the Yellow-rumped Warbler *Dendroica coronata*, may make long movements within a winter (Lack & Lack 1972, Terrill & Crawford 1988). The same is true for various thrushes which move slowly southward, consuming fruit crops as they go, until the time comes for them to return to their breeding areas (Karr 1980). One kind of fruit is especially abundant in the Neotropics, namely the aril, in which soft oil-rich tissue surrounds a hard, indigestible seed coat. When the dry inedible pods that contain arillate seeds dehisce in spring to expose the red, yellow or white arils, migratory birds of many kinds eat them, including flycatchers, thrushes, vireos and warblers. Individuals of some species show fidelity to particular sites, even though they may move between several sites during the season (for Indigo Bunting *Passerina cyanea*, see Johnston & Downer 1968). It seems, however, that itinerancy is much less common among Nearctic–Neotropical migrants than among Palaearctic–Afrotropical ones. This difference is associated with the drier, more seasonal habitats occupied by most species in Africa, which in general are hospitable for only a part of the non-breeding period.

As in Africa, many migrants of grasslands and savannahs feed at fires on the animals disturbed or killed. Such fires start up toward the end of the dry season, and in due course range over much of the non-forested parts of the South American continent. A wide range of bird species, including raptors, flycatchers, swallows and many others drawn from among both intra-tropical and Nearctic–Neotropical migrants, accompany the fires as they work their way across country, and feast upon the various animals that are flushed or injured.

POPULATION DECLINES IN MIGRANTS

Tropical migrants form a much larger proportion of the avifauna of deciduous forest in eastern North America than in Europe. In North America, each square kilometre of this forest can support up to 600 pairs of long-distance migrants, which can form 50–75% of the entire bird assemblage (Lynch 1987). Given the immense size of this forest, which in pre-settlement times extended from southern Canada to the Gulf Coast and from the Atlantic to the Great Plains, it could then have accommodated nearly two billion migrants (Terborgh 1989). With the deforestation that followed the arrival of Europeans, this number surely diminished, probably to less than half its original value. At one time all these migrants were presumably accommodated in winter mainly within the tropical forests of Central and South America to which the remainder still migrate.

Deforestation in the Neotropics has been much more recent than in North America, occurring mainly since 1950.[1] So for many years after the destruction of large areas of breeding habitat, the reduced migrant numbers were probably not occupying their more intact wintering habitat to full capacity. But with the more recent and rapidly accelerating deforestation of the Neotropics, declines in migrant populations from this cause become increasingly likely. These statements may hold as a generalisation, but whether particular species are affected probably depends on the extent of forest removal within their particular wintering range, and on whether they can live in the secondary habitats that replace the forest.

In North America, most recent information on trends in bird populations comes from the Breeding Bird Survey (BBS), managed by the United States Fish & Wildlife Service, in cooperation with the Canadian Wildlife Service. Since 1966, this scheme has provided annual data on bird populations, collected mainly by amateur observers, from randomly chosen sites in various habitats across the

---

[1]*In even earlier times, at the height of the Mayan, Aztec and Inca civilisations, deforestation of Central and South America was more extensive than it is now, most of the recent forest being re-growth, less than 1000 years old.*

continent. The scheme has produced a geographically much wider, and statistically more valid, data-set than was available for Europe over the same period.

For 133 species of Nearctic–Neotropical migrants (excluding waterbirds and shorebirds), the data were deemed adequate for analysis of long-term trends. Over the whole 36-year period, 23 species showed significant overall continent-wide increases, while 18 showed significant overall continent-wide declines, the remaining species showing no significant overall trends. However, patterns changed between 1966–1979 and 1980–1991. In the first period, 34 species showed significant continent-wide increases, while 19 showed significant declines (**Table 25.1**; Peterjohn *et al.* 1995, largely confirming Sauer & Droege 1992). However, in the later 1980–1991 period this pattern was reversed, when 15 species showed significant continent-wide increases, while 35 species showed significant decreases. Only the Upland Sandpiper *Bartramia longicauda*, House Wren *Troglodytes aedon*, Solitary Vireo *Vireo solitarius*, Warbling Vireo *Vireo gilvus*, Red-eyed Vireo *Vireo olivaceus* and Blue Grosbeak *Guiraca caerulea* showed significant increases through both periods, while the Chimney Swift *Chaetura pelagica*, Eastern Wood Pewee *Contopus virens*, Lark Sparrow *Chondestes grammacus* and Grasshopper Sparrow *Ammodramus savannarum* showed declines through both periods. Most other species showed both temporal and regional variation in trends. The overall pattern indicated substantial regional and temporal variation in trends, but perhaps with more eastern forest species than expected declining in the 1980s. Many of these findings at national level were confirmed by local counts over shorter or longer periods, either in breeding areas or at migration sites (Terborgh 1989, Askins *et al.* 1990, Peterjohn *et al.* 1995).

From spring and autumn migration counts at Long Point on Lake Eyrie, Ontario, during 1961–1988, only four out of 33 species of Neotropical migrants increased in numbers seen per day, while 29 decreased. These figures differed significantly from those of short-distance temperate migrants, in which 11 out of 23 species increased and 12 decreased (Hussell *et al.* 1992). Despite these net trends, most species fluctuated substantially from year to year. Tropical migrants tended to decrease in the 1960s, increase in the 1970s and decrease in the 1980s, while many intra-continental temperate-zone migrants showed the opposite pattern. Nevertheless, nine (27%) of the 33 tropical migrants showed consistent declines over the 28 years, compared with only three (13%) of the 23 temperate migrants (statistically, a non-significant difference between groups). In general, the same patterns were revealed by both spring and autumn counts, which in turn correlated with BBS indices for Ontario, where many of the migrants were presumed to breed. Counts from some other migration stations showed some similar and

Table 25.1 Population trends, determined from the Breeding Bird Survey, of 133 species of Nearctic–Neotropical migrants during the period 1966–1991

	Western Region (*N* = 91)	Central Region (*N* = 79)	Eastern Region (*N* = 88)	Overall (*N* = 133)
No significant change	61 (67)	56 (71)	60 (68)	92 (69)
Significant net increase	22 (24)	6 (7)	13 (15)	18 (14)
Significant net decrease	8 (9)	17 (22)	15 (17)	23 (17)

Figures show the numbers (%) of species showing different trends in three different regions of North America. From Peterjohn *et al.* (1995).

some partly different patterns, again reflecting regional variations in population trends (Hagan *et al.* 1992).

By use of radar, Gauthreaux (1992) measured the volume of spring migration across northern coastal areas of the Gulf of Mexico, as birds migrated northward past Lake Charles in Louisiana. Although the individual species could not be distinguished by this method, the migration observed consisted almost entirely of Neotropical migrants. The proportion of suitable days (because of favourable weather) on which trans-Gulf flights were observed during 8 April–15 May each year declined from 90%, 95% and 100% in 1965–1967 to 36%, 43% and 53% in 1987–1989. These radar counts thus provided an independent indication of a substantial net decline in the numbers of Neotropical migrants over this 22-year period.

Taken as a whole, then, the trend data from breeding bird counts and surveys, together with migration counts, reveal a recent marked decline in the numbers of some Neotropical migrant species that breed in the forests of eastern North America, but not in others which have shown stable or increasing trends. However, substantial regional variations were apparent. Declines occurred at a time when the total forest area in the eastern USA was on the increase, owing partly to regeneration of abandoned farmland.[2] Other local factors, which seemed to have influenced the trends, included habitat change through forest maturation, and long-term fluctuations in the numbers of Spruce Budworm *Choristoneura fumiferana* and other caterpillars on which many species feed while in their breeding areas (see later).

Role of forest loss and fragmentation

Despite an ongoing debate about the number of species involved, and the significance of regional variation, most studies in North America have attributed declines in tropical migrants to events in breeding areas (Terborgh 1989, Askins *et al.* 1990, Hagen & Johnston 1992, Sherry & Holmes 1992, Martin & Finch 1995, Latta & Baltz 1997). For some eastern forest species, problems are thought to stem from forest loss and fragmentation, the latter probably causing population declines much greater than expected from the areas of habitat lost. The evidence consists partly of an apparent contrast in population trends between birds in small forest fragments and those in extensive forest areas, and partly from current species distributions, which show that many species are now almost absent from small woods. Relevant studies span a wide area from the Atlantic to the Mississippi (Bond 1957, Robbins 1979, 1980, Whitcomb *et al.* 1981, Ambuel & Temple 1983, Blake & Karr 1987, Howe 1984, Lynch & Whigham 1984, Hayden *et al.* 1985, Freemark & Merriam 1986, Robbins *et al.* 1989, Askins *et al.* 1990).

Many studies have entailed assessment of the numbers and kinds of species breeding in forest fragments of different sizes in the same region. The main finding was as expected, namely that total species numbers increased with increasing size of forest. While certain species were found in both small and large forest fragments, others were found entirely or almost entirely in large ones, mainly in the forest

---

[2] *In the USA, the total area of forest reached a low of 200 million ha in 1910–1920, since which time the area has increased to more than 250 million ha, or approximately 60% of the original area. Much of this forest is relatively young, however, and in parts highly fragmented, so is not suitable for all species (Terborgh 1989).*

interior. More importantly, however, species numbers per unit area also increased with increasing size of forest fragment. This was especially true of long-distance migrants that winter in the tropics. In most studies, forest area accounted for most of the variation in species richness and total density of birds, and especially of Neotropical migrants (Askins *et al.* 1990, Lynch & Whigham 1984, Blake & Karr 1987, Robbins *et al.* 1989). Other variables, such as the degree of isolation from other forests, or vegetation structure and heterogeneity, explained relatively little additional variance in species numbers, as judged from the findings of multiple regression analyses. However, vegetation variables influenced the kinds of species that were present, and their local densities (Ambuel & Temple 1983, Lynch & Whigham 1984). As expected, edge species occurred at greatest density in small forests. Similar relationships between species numbers and area have been found in forest and other habitat areas in other parts of the world, but European woods seem to hold few species that could be considered as forest-interior specialists (Newton 1998b).

Predation

Forest fragmentation is held to have promoted increases in the numbers of predators and brood-parasitic Brown-headed Cowbirds *Molothrus ater*. In eastern North America, as in Europe, many small generalist predators, which eat eggs and chicks, reach much higher densities in suburban and farming areas than in more natural areas (Wilcove *et al.* 1986, Small & Hunter 1988, Hoover *et al.* 1995), benefiting from the additional food provided by human activities. Such predators include mammals, notably Grey Squirrels *Sciurus carolinensis*, Raccoons *Procyon lotor* and feral cats, and birds such as American Crows *Corvus brachyrhynchos* and Blue Jays *Cyanocitta cristata*. All these species tend to concentrate their hunting along woodland edges (Gates & Gysel 1978, Whitcomb *et al.* 1981, Paton 1994).

Several studies have shown that the nesting success of songbirds is higher in large forests than in small woods (**Figure 25.1**). Others have shown that in large

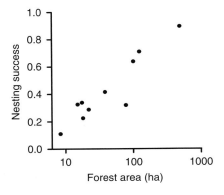

Figure 25.1 Proportion of Wood Thrush *Catharus mustelinus* nests that produced young in forest patches of different size, Pennsylvania ($r^2 = 0.86$, $P < 0.001$). Nest success increased with size of forest fragments, mainly because of differential predation. Both avian and mammalian predators were more abundant in small fragments. From Hoover *et al.* (1995).

forests nest success increased from edge to interior. In total, overall predation rates decreased with increasing area of forest patch in eight out of eight studies, and decreased from centre to edge in 10 of 14 studies involving artificial nests, and in four of seven involving natural nests (Paton 1994). In each case, predators that entered the woods from nearby farmed or suburban areas were held responsible for most of the losses. In one study in southern Wisconsin, for example, the overall nest success of 13 species was only 18% within 100 m of forest edge ($N = 96$), compared with 58% at 100–200 m within forest ($N = 98$) and 70% at more than 200 m into forest ($N = 82$) (Temple & Cary 1988). The entire area of most small woods lay within the most vulnerable zone. It is not hard to imagine, therefore, that as forest becomes more fragmented, overall nest success could decline, eventually to the point where insufficient young were produced to offset the annual adult mortality, leading to regional population declines (Brawn & Robinson 1996). Other (but not all) studies using artificial or natural nests have given similar results elsewhere in North America (**Figure 25.2**; Gates & Gysel 1978, Wilcove 1985, Small & Hunter 1988, Yahner & Scott 1988, Donovan *et al.* 1995, Hoover *et al.* 1995, Robinson *et al.* 1995, Weinburg & Roth 1998, Manolis *et al.* 2002). The same is true for studies with artificial nests in Europe (Andrén & Angelstam 1988, Sandström 1991, Andrén 1992), although findings from artificial nests may not always be typical of those from natural nests.

If nest predation has increased with forest fragmentation, why should it have affected some Neotropical migrants more than other birds? Three reasons have

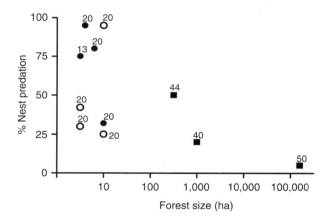

Figure 25.2   Percentage of artificial nests preyed upon in relation to forest size. Closed squares are large forest tracts, open circles are rural fragments and closed circles are suburban fragments. The number beside each point refers to the artificial nests placed in that forest. The artificial nests were wicker baskets of the type sold in pet shops for canaries; they were lined with dry grass and three quail eggs were placed in each. Some 50 or 100 nests were employed in each of 11 tracts of forest, placed at 20 m intervals, alternately on the ground and at heights of 1–2 m in shrubs and saplings. After one week the nests were checked and any that had lost one or more eggs were classed as preyed upon. Redrawn from Wilcove (1985).

been commonly suggested. First, with few exceptions, the Neotropical migrants that breed in the eastern USA construct open nests, either on the ground or in low bushes, in sites that are most vulnerable to predators (Wilcove & Whitcomb 1983). In contrast, none of the residents or short-distance migrants that breed commonly in suburban settings places its nest on the ground, and many use tree-cavities, which provide relatively safe sites. Second, long-distance migrants start breeding later in the year (mostly June) and usually raise only one brood per year, occasionally two (Greenberg 1980, Whitcomb *et al.* 1981). In contrast, many resident species and short-distance migrants start in April or May and routinely attempt second and even third broods. Third, many Neotropical migrants are small and unable to drive off nest predators in the way that some larger resident species can. For these various reasons, then, Neotropical migrants are supposedly disadvantaged with respect to nest predators, which could together cause a disproportionate reduction in their breeding rates, and contribute to the selective disappearance of tropical migrants from fragmented eastern woodlands. Through nest predation alone, Wood Thrushes *Catharus mustelinus* in Pennsylvania produced insufficient young to offset the usual annual mortality (Hoover *et al.* 1995), and through a combination of predation and cowbird parasitism the same was true for Wood Thrushes in Illinois (Robinson 1992). Yet the latter population did not decline, presumably because local numbers were maintained by immigration.

In an analysis of the population trends of 47 North American insectivorous passerine species in relation to various features of their biology, Böhning-Gaese *et al.* (1993) found that migratory status, nest location and nest-type were the best predictors of population trends. Only residents and short-distance migrants increased during the years concerned (1978–1987), whereas long-distance migratory species with low nest locations declined. Also, species with closed nests did better than those with open nests. These are both aspects of nesting biology that influence rates of predation and probably also rates of parasitism, as discussed below.

Parasitism

The Brown-headed Cowbird *Molothrus ater* lays its eggs in the nests of a wide range of passerine host species, thereby reducing their production of young (**Figure 25.3**, Mayfield 1977, Lorenzana & Sealy 1999). Originally confined to the grasslands of mid-continent, following deforestation the species gradually spread to occupy most of the continent up to the boreal forest (Mayfield 1977, Brittingham & Temple 1983). Within this newly occupied terrain, its numbers continued to rise at least to the 1960s, as it benefited from the increased food supplies provided by waste grain in cereal fields, cattle feedlots and garden feeders (**Figure 25.4**). Its overall numbers probably increased more than 10-fold during the twentieth century (Brittingham & Temple 1983), spreading into areas with new host species. Despite this overall national trend, its numbers seem to have declined in the eastern USA since around 1980 (BBS data).

Forest fragmentation increased the amount of edge along which cowbirds have access to the nests of forest species previously immune to attack. Several studies have shown greater parasitism of nests in small than large woods, and near the edges of woods than in the interior (Gates & Gysel 1978, Brittingham & Temple

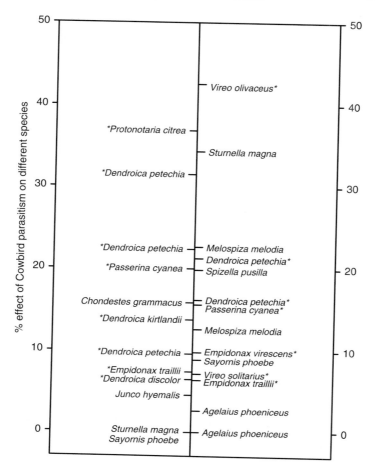

Figure 25.3   Percentage reduction in the breeding success of various North American songbirds caused by the brood-parasitic Brown-headed Cowbird *Molothrus ater*. Effect calculated as % nests parasitised × % difference in success between parasitised and unparasitised nests. *Long-distance Neotropical migrants. Based on data assembled by Payne (1997), in which references to the original studies may be found.

1983, Temple & Cary 1988, Donovan *et al.* 1995, Robinson *et al.* 1995). Many of these new hosts, including Neotropical migrants, have no innate defences, and will not desert or eject strange eggs, but raise the resulting young. Because cowbirds use many host species, moreover, they are not vulnerable to decline in any one, and could in theory parasitise favoured species to extinction, while maintained at high density by other common hosts (May & Robinson 1985). The same features that make Neotropical migrants nesting near the edges of woods more vulnerable to nest predators than residents also make them more vulnerable to parasitism, namely their use of accessible nest-sites and their predominant single-broodedness.

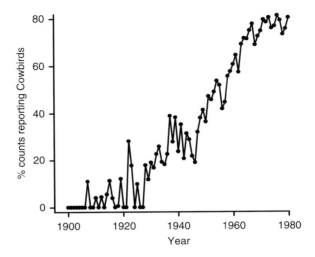

Figure 25.4 Increase in the numbers of Brown-headed Cowbirds *Molothrus ater* in the southeastern USA. Points show the proportion of Christmas Bird Counts on which the species was recorded during 1900–1980. Redrawn from Brittingham & Temple (1983).

During the 1960s, cowbirds laid their eggs in up to 70% of nests of the rare Kirtland's Warbler *Dendroica kirtlandii*, thereby reducing the production of young warblers to less than one per nest (Mayfield 1983, Walkinshaw 1983). This was fewer than needed to maintain the warbler population level, which during 1961–1971 declined by 60%. Then, with the start in 1972 of a programme of cowbird removal, parasitism dropped to 6%, warblers fledged about three young per nest, and their breeding numbers stabilised (Kelly & DeCapita 1982, DeCapita 2000). Later, they increased more than three-fold when the habitat expanded as a result of a forest fire (Rothstein & Robinson 1994) but, to facilitate this increase, and to maintain Kirtland's Warbler in the long term, cowbird control was continued. The historic suppression of wild fires had clearly not helped this species, which prefers young stands.

The effects of cowbird removal have also been examined in various other host species in different areas (**Table 25.2**). Overall, of six experimental removals of cowbirds, five were followed by an increase in the breeding success of the host species, and at least three by an increase in host breeding density. Only in one study, on Hooded Warblers *Wilsonia citrina*, did cowbird removal lead to no obvious increase in either nesting success or breeding numbers, but in this study parasitism rates were low (Stutchbury 1997). Other passerine populations may have declined following greater contact with this generalised brood-parasite. Examples include the Golden-cheeked Warbler *Dendroica chrysoparia* in parts of Texas, the Black-capped Vireo *Vireo atricapillus* in parts of Oklahoma (Grzybowski *et al.* 1986), the Eastern Warbling Vireo *Vireo gilvus* in parts of California and British Columbia (Verner & Ritter 1983, Ward & Smith 2000), the White-crowned Sparrow *Zonotrichia leucophrys* in parts of California (Trail & Baptista 1993), and Bell's Vireo *Vireo bellii* in Missouri (Budnik *et al.* 2000). In addition, some other species seem

Table 25.2 Effects of the experimental removal of parasitic Brown-headed Cowbirds *Molothrus ater* on their songbird host species

Host species	Area	Increased breeding output	Increased breeding numbers	Source
Kirtland's Warbler *Dendroica kirtlandii*[a]	Michigan	+	+	DeCapita (2000)
(Least) Bell's Vireo *Vireo bellii*[a]	S. California	+	+	Griffith & Griffith (2000)
Black-capped Vireo *Vireo atricapillus*[a]	Central Texas	+	+	Hayden *et al.* (2000)
Song Sparrow *Melospiza melodia*	Mandarte Island, BC	+	−	Smith *et al.* (2000)
Southwestern Willow Flycatcher *Empidonax traillii*[a]	California	+	−	Whitfield (2000)
Hooded Warbler *Wilsonia citrina*	Pennsylvania	−	−	Stutchbury (1997)

[a] Cowbird removal became part of the conservation management undertaken for these species (see also De Groot *et al.* 1999, Rothstein & Cook 2000).

no longer to be producing enough young to offset known adult mortality. In a mainly agricultural landscape in central Illinois during 1985–1989, nests of forest-nesting Neotropical migrants contained an average of 3.3 cowbird eggs per nest. End-of-season juvenile to adult ratios averaged 0.1 in Neotropical migrants compared with more than 1.0 for year-round residents and short-distance migrants (Robinson 1992). The Neotropical migrants could not have maintained their numbers under this level of parasitism without large-scale immigration.

It thus seems, then, that many North American songbirds that nest in forest fragments in suburban and agricultural areas are now experiencing the dual effects of increased nest predation and parasitism, particularly near forest edges, and that for various reasons these affects fall most strongly on many Neotropical migrants. So far, attempts to assess these impacts on population levels have been made only in a minority of species, such as the Kirtland's Warbler *Dendroica kirtlandii* and Wood Thrush *Catharus mustelinus*, in which survival and breeding success have been measured. Studies on a greater range of species, and in a greater range of areas, are needed before the generality of these findings can be assessed. Although the predisposing fragmentation of the eastern deciduous forest occurred mainly in the nineteenth century, the associated increase in generalist predators and cowbirds occurred mainly in the latter half of the twentieth century. Decline through excess predation and parasitism would be expected to leave increasing amounts of breeding habitat unoccupied.

Forest fragmentation may also reduce the food available per unit area. This aspect has received much less attention than predation and parasitism, but could be important in some species. It was found among Ovenbirds *Seiurus aurocapillus* nesting in different-sized forest fragments in Ontario (Burke & Nol 1998). Density and pairing success of territorial males increased with area of the woodlot. Within Ovenbird territories, prey biomass was 10–26 times greater in large woods than in small ones, associated with deeper leaf litter. Similar results were also found for Australian Yellow Robins *Eopsaltria australis* whose food supply of ground-dwelling invertebrates was also lower in smaller woods (Zanette *et al.* 2000).

The relationship between forest area and species numbers was found to vary regionally: deficiencies in species numbers in small forests tended to be more marked in agricultural than in more forested landscapes (Freemark & Collins 1992). These regional differences were in turn associated with similar regional variations in songbird nest success. In nine migrant species, predation and parasitism rates were found to increase with reduction in the proportion of forest present. Landscapes with little forest acted as sinks for some species, in that reproduction was insufficient to offset annual mortality, so that breeding numbers could not be maintained in the absence of continual net immigration from more forested areas (notably Wood Thrush *Catharus mustelinus*, Hooded Warbler *Wilsonia citrina*, Ovenbird *Seiurus aurocapillus* and Kentucky Warbler *Oporornis formosus*). There were some exceptions to the trends, as local factors – including those influencing cowbird abundance – modified the overall patterns (compare the studies on Wood Thrushes in Illinois, Indiana, Pennsylvania and Delaware, Hoover *et al.* 1991, Robinson 1992, Roth & Johnson 1993, Weinburg & Roth 1998, Fauth 2000).

Effects of other factors in breeding areas

Other evidence points to events in breeding areas (as opposed to wintering areas) as being a major cause of population fluctuations in some Neotropical migrants. For many years, some migrant warblers have undergone long-term changes in breeding densities in association with changes in caterpillar abundance. Most outbreaks of defoliating caterpillars in eastern North American hardwood forests are sporadic, occurring in one location in one year and somewhere else the next. The probability that any particular stand will experience an outbreak in any given year is low. Every now and then, however, one caterpillar species of boreal regions, the Spruce Budworm *Choristoneura fumiferana*, increases over a period of years to reach plague proportions over wide areas, causing extensive defoliation. Several migrant bird species show strong numerical responses to this insect, including the Bay-breasted Warbler *Dendroica castanea*, Blackpoll Warbler *D. striata*, Tennessee Warbler *Vermivora peregrina*, Cape May Warbler *Dendroica tigrina*, Black-billed Cuckoo *Coccyzus erythropthalmus* and Yellow-billed Cuckoo *C. americanus* (Kendeigh 1947, Morris *et al.* 1958, Crawford & Jennings 1989). In one study, Bay-breasted Warblers *Dendroica castanea* increased in abundance from 2.5 pairs per 10 ha in uninfested stands to 300 pairs per 10 ha during an outbreak, Blackburnian Warblers *D. fusca* increased from 25–30 pairs to 100–125 pairs per 10 ha, and Tennessee Warblers from 0 to 125 pairs per 10 ha (Morris *et al.* 1958). Because in each outbreak the numbers of the birds rose over several years in parallel with the caterpillars, it was hard to tell how much the increase was due to immigration and how much to the high local breeding success, though both were involved.

The role of previous breeding success in the annual population changes of some Neotropical migrants was confirmed in studies in the Hubbard Brooks Experimental Forest in New Hampshire (Holmes *et al.* 1991, 1996, Sherry & Holmes 1992, Rodenhouse & Holmes 1992). Over a period of years, and in different forest plots, warbler breeding densities fluctuated in parallel with caterpillar biomass the previous summer. There was thus a correlation between the production of young in one year and the numbers of breeders returning the next. In other words, breeding success rather than overwinter survival had most influence on the annual changes in breeding numbers, and increases in breeding density were marked by greater recruitment of yearlings (Sherry & Holmes 1988, 1992). These findings were apparent in the American Redstart *Setophaga ruticilla*, Black-throated Blue Warbler *Dendroica caerulescens*, Black-throated Green Warbler *D. virens*, Red-eyed Vireo *Vireo olivaceus* and others, so in these species the year-to-year changes in breeding success and in subsequent breeding numbers were attributable to fluctuations in summer food supplies. Elsewhere, a relationship between the annual recruitment of yearlings to a breeding population and fledgling production the previous year was noted in another Neotropical migrant, the Prairie Warbler *Dendroica discolor* (Nolan 1978), paralleling the Pied Flycatcher *Ficedula hypoleuca* and others in Europe (see **Figure 26.4**; Virolainen 1984, Stenning *et al.* 1988).

The abundance of foliage-gleaning birds in the Hubbard Brook Forest, especially of warblers and vireos, increased in the early 1970s, coincident with a major outbreak of defoliating caterpillars (Holmes 1990). Furthermore, natural and experimentally-induced declines in food (mainly caterpillar) abundance were shown to reduce the frequency of re-nesting and second-brood attempts, as well as hatching and fledging success, and nestling growth rates, while increasing the frequency of nestling starvation. The general pattern seemed to be that food was abundant for birds in this forest only during Lepidoptera irruptions, but such events occurred infrequently, perhaps once every 10–20 years in any one forest-stand. Between these outbreaks, birds probably experienced prolonged periods of food limitation, as caterpillars were scarce for several years in succession. The Neotropical migrants thus showed population trends in this forest that were different from those of residents and short-distance migrants whose breeding numbers fluctuated from year to year mainly in line with winter severity. The Neotropical migrants were able to escape the northern winters, but the residents and short-distance migrants were not.

The importance of these findings in our present context is two-fold. First, they show that events in breeding areas which affect breeding success can influence subsequent breeding densities of migrants, apparently overriding any opposing effects of events in wintering areas. Second, they show that food supply can affect breeding success and breeding numbers in addition to any effects of predation and parasitism. To what extent, then, did recent declines in the numbers of certain migrant species in North America result from recent declines in caterpillar densities? Budworm caterpillars were abundant over wide areas in the late 1940s and again in the 1970s. One would therefore expect a decline in the numbers of some species in the 1950s–1960s and again in the 1980s–1990s. In addition, since the 1950s, insecticides have been used widely in eastern Canada to suppress caterpillar outbreaks. This practice is likely not only to have obliterated the 'good food years', but also to have suppressed all insect populations in the affected areas in the years of spraying. The influence of summer food supply on migrant population trends could thus be substantial and may have contributed to the declines in some species witnessed in the 1980s–1990s.

Another factor thought to operate on breeding areas to reduce migrant numbers is summer drought, again through its influence on insects and other food supplies. Severe declines in the numbers of several Neotropical migrant species in Wisconsin and Michigan during 1986–1988 coincided with drought (Blake *et al.* 1992), as did similar declines in Illinois (Robinson 1992). Because the migrants do not start nesting until June, they are more vulnerable to the effects of drought than are residents and short-distance migrants that start in April or May. Such effects would be expected to be short-lived, however, and followed by population recoveries as the weather conditions improved. In addition, long-term changes in the species composition of particular forest areas would be expected from the successional changes that occur as the forest matures, making the habitat less suitable for some species and more suitable for others (for examples, see Litwin & Smith 1992, Holmes & Sherry 2001).

Further evidence for the importance of events on breeding areas for declines in Nearctic–Neotropical migrants has come from studies in wintering areas. For example, between 1972 and 1990, birds were systematically netted in the Guánica Forest in Puerto Rico (Faaborg & Arendt 1992). Over the period concerned, the numbers of wintering migrants steadily declined. Some species, such as Northern Parula Warbler *Parula americana* and Prairie Warbler *Dendoica discolor*, were common at the start, but virtually absent in later years, while other migrants, such as Black-and-white Warbler *Mniolilta varia*, and various resident species, maintained their numbers. As no obvious change occurred in the forest itself during the study, the declines were attributable to events in breeding areas.

Much of the evidence thus points to changes in the eastern North American breeding areas as being the main factors underlying late twentieth century declines in the numbers of some forest-dwelling Nearctic–Neotropical migrants. Such changes may not have occurred over the entire range or in every wood, but they have clearly occurred in a sufficiently large proportion of woods to put the regional populations of some species into decline. Only time will tell how long the declines will continue.

Factors operating in wintering areas

Over the same period, much tropical forest, where many North American migrants spend the winter, has been destroyed. The highest rates of deforestation have occurred in Central America, the very regions where forest migrants are most concentrated (Myers 1980, Terborgh 1989). On present evidence, it is impossible to say to what degree tropical deforestation has contributed to the declines of forest migrants, but if it continues at the recent rate, it may soon overtake events in breeding areas as the major cause of declines. Not all species are likely to have been affected adversely, however, for some thrive in the secondary habitats that replace the forest.

One species already affected by tropical deforestation, according to Terborgh (1989), is Bachman's Warbler *Vermivora bachmanii*, probably now extinct. At the end of the nineteenth century, this species bred widely in damp, broad-leaved woodland across the southern USA, but by the 1950s it could be reliably seen only at a few places in coastal South Carolina. All wintering records came from Cuba, from the dense evergreen thicket that once covered large parts of the island, and which has since been largely replaced by sugar cane. Almost certainly, this warbler has succumbed primarily through destruction of its wintering habitat, for substantial areas of breeding habitat still remain. More generally, declines in breeding populations of Neotropical migrants have been linked to their winter habitat preferences – a pattern that is consistent with trends in forest cover in the tropics (Robbins *et al.* 1989).

All migrants are affected by conditions in both breeding and wintering areas, and in recent years species have presumably been adjusting to changes in one or other area. However, the habitats occupied by some species, in both breeding and wintering areas, are so vulnerable to human activity that such species seem destined to decline markedly in the coming years. The Cerulean Warbler *Dendroica cerulea* has suffered extensive loss of breeding habitat in the past 200 years (Robbins *et al.* 1992). Its favoured nesting habitat, mature floodplain forest with tall trees, has become scarce over most of its original nesting range, and its apparent sensitivity to fragmentation of remaining tracts gives it an additional disadvantage. In winter the species is restricted to primary humid evergreen forest along a narrow elevation zone at the base of the Andes. This zone is among the most intensively logged and cultivated regions of the Neotropics. From 1966 to 1989, the Cerulean Warbler showed the most precipitous decline of any North American warbler (3.4% per year nationwide).

Surprisingly, few studies have examined annual fluctuations in the numbers of Nearctic–Neotropical species in relation to events in wintering areas. This contrasts with the situation in Palaearctic–Afrotropical migrants. However, the annual survival of Black-and-white Warblers *Mniotilta varia*, studied in a wintering area in Puerto Rico, was related to the local rainfall, with better survival in the wetter years associated with El Niño conditions (Dugger *et al.* 2004). Similar findings emerged in Black-throated Blue Warblers *Dendoica caerulescens* studied in New Hampshire and Yellow Warblers *Dendoica petechia* studied in Manitoba (Sillett *et al.* 2000, Mazerolle *et al.* 2005). In both these areas the wetter years were associated with El Niño conditions, which also affected climate in Central and South American wintering areas, leading to both improved survival and improved breeding success through effects on food supplies.

Factors operating on migration routes

Another factor of concern, which has been mooted as possibly contributing to population declines, is the mortality that occurs on migration (described in detail in Chapter 28). One type of mortality occurs at communication masts, which in certain weather conditions can kill large numbers of nocturnal migrants. Individual towers have been estimated to kill thousands of birds in a season, and collectively such towers have been estimated to kill 4–5 million birds per year in the USA alone (Chapter 28). Most of these casualties involve Nearctic–Neotropical migrants. The second type of mortality occurs during storms at sea, which can kill many thousands of birds at a time, as they migrate between the breeding and wintering areas. These storms affect mainly birds that breed in the eastern half of North America and winter on Caribbean Islands or in South America. They have increased in numbers and severity in recent decades, in association with global climate change (Chapter 28). There is no firm evidence that either of these mortality factors has contributed importantly to population declines of any species, but the possibility cannot be excluded (see Butler 2000).

Non-forest species

Migrants dependent on natural grassland, such as the Sprague's Pipit *Anthus spragueii* and Upland Sandpiper *Bartramia longicauda*, declined greatly following the ploughing of the prairies, but species that can live in agricultural habitats, such as the Indigo Bunting *Passerina cyanea*, greatly expanded following the felling of the eastern deciduous forest. Of more recent changes, some have been attributed to changes in breeding areas and others to changes in wintering areas. Thus the Bobolink *Dolichonyx oryzivorus* has been declining in the eastern USA since the early 1900s. It nests mainly in hayfields, and decline has been attributed to procedural changes that make hayfields less suitable as habitat (Bollinger & Gavin 1992). In contrast, an earlier decline of the Dickcissel *Spiza americana*, which inhabits brushy pastures, was put down to destruction of wintering habitat, resulting from overgrazing by cattle (Fretwell 1980).

Most arctic-nesting shorebirds that winter on Neotropical coastlines have so far been spared the effects of large-scale habitat destruction. As in Eurasian species, their breeding habitats are still largely intact, while their wintering habitats may have expanded in recent centuries as a result of the soil erosion following deforestation, and the resulting expansion of coastal mud. During part of the twentieth century, some shorebird species increased in numbers in response to lessened shooting pressure, which had reduced their numbers in the past. Some raptors, including Swainson's Hawks *Buteo swainsoni*, suffered from pesticide-induced mortality on wintering areas in Argentina. Around 5000 individuals were picked up under communal roosts, following the spraying of their food organisms (grasshoppers) with the organophosphate compound monocrotophos, now banned in this region (Goldstein *et al.* 1996). Earlier declines in Peregrine Falcons *Falco peregrinus* were attributed to the use of organochlorine pesticides on both breeding and wintering areas, and recent recoveries in Peregrine numbers have followed reductions in organochlorine use (Cade *et al.* 1988). It is more difficult to explain recent declines in wetland and coastal species, such as the Least Tern *Sterna antillarum*, which in eastern North America declined by 14% per year during 1978–1988 (Sauer & Droege 1992).

CONCLUDING REMARKS

Although in the last 30–40 years some tropical migrant birds have declined in western Europe and others in eastern North America, the causes seem to have differed. In Europe, declines have mainly involved species that winter in the arid savannas of tropical Africa, which have suffered from the effects of drought and increasing desertification. In several species, annual fluctuations in numbers and adult survival rates were correlated with annual fluctuations in winter rainfall, and by implication winter food supplies. Most species that were sufficiently studied showed no obvious changes in breeding success that could be linked with population changes (examples of exceptions: Corncrake *Crex crex*, Turtle Dove *Streptopelia turtur*, see Chapter 24).

In North America, by contrast, declines have affected many species which breed and winter in forest. In eastern forest regions, declines have been attributed to human activities on the breeding range, particularly forest fragmentation and associated agricultural and suburban developments, which have led not only to loss of forest, but to increases in the densities of nest predators and parasitic cowbirds. Declines in the numbers of some migrants are thought to result from declines in breeding success which is now too low to offset the usual adult mortality, though as yet convincing evidence is available for only a minority of species, such as Kirtland's Warbler *Dendroica kirtlandii* and Wood Thrush *Catharus mustelinus* in some areas. In other species, such as Bachman's Warbler *Vermivora bachmanii*, tropical deforestation seems to have played the major role in population decline, and is likely to affect an increasing range of species in the future. Whereas for the Palaearctic–Afrotropical system a considerable consensus exists on the causes of declines, for the North American system much of the evidence is as yet little more than suggestive, and no one explanation can account for all the facts (for discussion see James & McCulloch 1995).

In consequence, opinions remain divided on the causes of recent declines in Nearctic–Neotropical migrants. Those who favour events on North American breeding areas as major causes emphasise the following:

- Until recently, deforestation was far more extensive in North America than in countries to the south, which should have reduced migrant populations well below the limit set by wintering habitat.
- In North America, declines are much more marked in eastern than in central and western areas, and within eastern areas, are most marked in small woods, but are non-existent or less marked in remaining extensive forest tracts. This points to events on breeding areas as being most important, for if tropical deforestation had been the cause, the argument goes, declines would have been expected across the breeding range and (in some species only) in both small and large woods (Wilcove & Robinson 1990).
- Some migrant species were declining as early as the 1940s, well before deforestation in Central America reached anything like its current levels. Other species have declined in numbers in recent decades even in patches of unchanged wintering habitat (Faaborg & Arendt 1992).
- The declining species include some, such as Red-eyed Vireo *Vireo olivaceus*, Scarlet Tanager *Piranga olivacea* and Wood Pewee *Contopus virens*, which winter south of most other species, in the relatively undisturbed western Amazon basin.
- The numbers of some species are known to respond on a year-to-year basis to conditions in North American breeding areas, particularly food supplies, but also rates of nest predation and parasitism.

On the other hand, those who favour events on Central and South American wintering areas as causing declines point to the following:

- Enormous destruction and degradation of tropical forests has occurred during the last 50 years. Because the migrants are much more concentrated in wintering habitat, the destruction of each hectare of tropical forest could affect far more birds than the destruction of the same area of temperate forest.

- In the breeding areas, by contrast, total forest area has actually increased in recent decades, as abandoned farmland has become tree-covered (Sauer & Droege 1992). This and other factors have tipped the previously increasing cowbird population into decline in eastern districts. Many migrants are now scarce or absent from much apparently suitable nesting habitat, especially in the smaller woods.

- The declines in Neotropical migrants have not been paralleled by similar declines in residents or short-distance migrants that nest in the same woods, but remain year-round in North America.

The whole situation exemplifies the problems often facing ecologists, in understanding the primary cause of a problem when several likely factors change at the same time, when several may act together, and when different factors may affect different species. It also exemplifies the enormous geographical scale on which studies must be undertaken if reliable answers are to emerge. In reality, we are probably dealing with a continuum, with some species or populations declining primarily because of changing conditions in breeding areas (perhaps different conditions in different species), and others because of changing conditions in wintering areas. Also, because most species occupy large geographical areas, summer and winter, pressures and population trends may vary from one part of the range to another, and one period of time to another. It seems, therefore, that further progress is likely to come mainly from detailed studies of particular species, ideally in several different areas, summer and winter. Almost certainly, the division of populations into 'summer-limited' and 'winter-limited' is overly simplistic, because ecological conditions experienced by individuals at one season can influence their performance at another, as carry-over effects (Chaper 26). Bear in mind, too, that the causes of long-term trends may differ from the causes of year-to-year fluctuations about the long-term trends. In consequence, species can show different long-term trends but similar annual fluctuations, or vice versa.

SUMMARY

Most long-distance migrants that breed in North America winter in Central America, where they are much more concentrated in than in their breeding areas, but others migrate further south, occurring to the southern tip of South America. Many both breed and winter in forest and, in some species, individuals tend to occur in the same breeding and wintering territories year after year. Some species have declined in recent decades. They include many forest species that breed mainly in eastern North America. These declines could be due to yet unstudied events in wintering areas, such as deforestation. However, considerable evidence has emerged that the declines may be due to events in North America, particularly forest fragmentation and its associated consequences of increased predation and parasitism of nests.

Many of the relevant predators, as well as the parasitic Brown-headed Cowbird *Molothrus ater*, have increased greatly in recent decades in response to agricultural and suburban developments. They can reduce the breeding success of migrants below sustainable levels, causing populations to decline, and leaving large areas of breeding habitat unused. Declines of Nearctic–Neotropical migrants are most marked in small woods in agricultural landscapes where predation and parasitism rates are greatest. Several life history features appear to make Neotropical migrants more sensitive than resident species to the direct effects of nest losses.

In addition, the abundance levels of some migrant species are also influenced by the abundance levels of caterpillars in their breeding areas, notably Spruce Budworm *Choristoneura fumiferana*, low levels of which during the 1980s may have contributed to declines in the numbers of several migratory species in eastern deciduous forests. Nonetheless, other migratory species have clearly declined in association with destruction of wintering habitat, which is likely to become increasingly important in the coming years. Recent population declines have also been noted in some non-forest migratory species, attributable to various causes, but often unknown.

White Storks *Ciconia ciconia*, familiar Palaearctic–Afrotropical migrants

Chapter 26

Population limitation – breeding and wintering areas

In a migrant species, reproduction and the main mortality may occur in regions several hundred miles apart. This greatly complicates the study of the factors influencing numbers. (David Lack 1954.)

In the limitation of their populations, migratory birds differ from residents in at least one important respect. Their population sizes may be influenced by conditions in more than one part of the world: in areas that are used for breeding, as well as in areas that are visited at other times of year. For many migratory species, the breeding and wintering ranges are widely separated geographically, and might differ greatly in the numbers of birds they can support. Hence, factors operating in the migration or wintering range might limit the numbers that can occur in the breeding range, or vice versa. This is most clearly apparent where changes in the numbers of a species over a period of years are associated with changes in conditions in one area, but not in the other. In this and subsequent chapters, I address some of the issues involved in the population limitation of migrants, using as examples species whose long-term or year-to-year numerical changes can be clearly linked to events in breeding or non-breeding areas. These are increasingly important issues, because many migratory species are declining, and it is important to know where to direct any conservation measures.

Conditions in a particular region can be considered limiting if they slow a population's increase or cause its decline. Such conditions can seldom account entirely for a given population level, however, because among migrants both mortality and reproduction are also influenced by conditions in other regions at other times of year. Moreover, whether in breeding or wintering areas, bird numbers may be limited in a density-dependent or largely density-independent manner. Density-dependent regulation implies competition (perhaps for food or predator avoidance), as a result of which the percentage of birds affected is greater when numbers are high than when they are low. In density-independent limitation, competition is unimportant, and the proportion of birds affected bears no consistent relationship to population density. Probably few situations occur in which there is no competitive (density-dependent) element in reproduction or mortality, although this may often be slight or hard to discern.

In understanding population changes, it is helpful to separate long-term trends from year-to-year fluctuations about the trend, because different factors might be involved (Newton 1998b, 2004a). For example, some species may undergo long-term change in numbers, resulting from progressive habitat change, but may also continue to fluctuate from year to year in response to different factors, such as annual variations in rainfall and associated food supplies. Moreover, within species, population trends and limiting factors may vary across the range, and also at different time periods, so that local research findings cannot necessarily be extrapolated over wider areas or longer periods. Because of the problems of studying birds which routinely occupy two or more different areas each year, it has seldom been possible to follow the same individuals throughout the year, and most of the evidence on population limitation in migrants is indirect.

Spacing and movement patterns

Social organisation is relevant to these issues because it provides the behavioural framework within which the regulation of density and distribution occurs. The spacing of birds in the non-breeding season can vary from solitary and territorial to gregarious and flocking, depending on the species, the habitat, the distribution of food supplies, predation pressures and other factors, as well as on the social status of the individuals concerned (Newton 1998b). In addition, while some migrant birds remain in the same place for the whole non-breeding period from arrival to departure, others occupy two or more sites in succession, at different points on the migration route, or they may remain continually on the move, again depending largely on the spatial and temporal occurrence of food. In the northern continents, such itinerancy is shown by boreal finches and others, while in the drier parts of the southern continents, some migrant species follow rain-belts for the food sources they promote. African examples include the Lesser Spotted Eagle *Aquila pomarina* and White Stork *Ciconia ciconia* (both studied by radio-tracking, Meyburg *et al.* 1995c, Berthold *et al.* 2002), while South American examples include the Swallow-tailed Kite *Elanoides forficatus* and Swainson's Hawk *Buteo swainsoni* (Brown & Amadon 1968, Bildstein 2004). These species illustrate a major difference between the breeding and non-breeding seasons: in the breeding season, mated pairs are based for up to several months in the localities in which they nest, but in the non-breeding season, individuals of at least some species are free to move around, concentrating wherever food is plentiful at the time. In some species, individuals can migrate at almost any date in the non-breeding season, if stimulated to do so by reduced food supplies, when they usually move further along the same migration axis, reaching the most distant parts of their wintering range only in extreme years (Chapter 18). Hence, if migrants are limited by conditions in wintering areas, for some species there may be more than one such area involved, and these areas may not necessarily be the same from year to year.

Individual birds may remain in the same localities throughout the non-breeding period if food supplies remain continually available there. Similarly, birds can normally return to the same sites each year if food supplies are consistent and predictable from year to year. This is true whether the birds live solitarily or gregariously. Solitary species that show both within- and between-year site-fidelity include the European Pied Flycatcher *Ficedula hypoleuca*, Spotted Flycatcher *Muscicapa striata*, Melodious Warbler *Hippolais polyglotta* and Common Nightingale *Luscinia megarhynchos* wintering in Africa (Salewski *et al.* 2000), Pallas' Grasshopper Warbler *Locustella certhiola* and Great Reed Warbler *Acrocephalus arundinaceus* wintering in Malaysia (Nisbet 1967, Nisbet & Medway 1972), and various parulid warblers and others wintering in Central or South America (Rappole 1995). Flocking species that show within-year and between-year site-fidelity include some geese, cranes and shorebirds, among others (Chapter 17).

Some species seem to behave in much the same way in their breeding and wintering areas. For example, Northern Wheatears *Oenanthe oenanthe* are territorial in both, remaining at the same localities throughout their stay, and returning there year after year. But territories are held by pairs in breeding areas and by singles in non-breeding areas. Other species hold pair territories in breeding areas, remaining faithful to these sites throughout the breeding season and from year to year, but are much more flexible and mobile in winter. In West Africa, Eurasian Willow Warblers *Phylloscopus trochilus* often feed in single-species or mixed-species flocks, showing no obvious territorial behaviour. In one area, Willow Warblers were present from November to April, except for 3–4 weeks in February at the height of the dry season, yet at no time did individuals stay for more than a few days (Salewski *et al.* 2002).

Whether species feed in individual territories or gregariously seems to depend on the consistency of the food supply, and whether it is defendable (Newton 1998b). The same species can behave territorially on one type of food and gregariously on another. For example, Steppe Eagles *Aquila nipalensis* in Africa are mostly territorial when feeding on mammals, and gregarious when exploiting local abundances created by emerging termites or Quelea colonies. In addition, dominant individuals of a species may behave territorially, while subordinate ones, unable to get territories, feed solitarily or in flocks, sometimes on different foods (for White Wagtail *Motacilla alba* see Zahavi 1971, for Yellow Wagtail *M. flava* see Wood

1979, for Fieldfare *Turdus pilaris* see Tye 1986, for Sanderling *Calidris alba* see Myers *et al.* 1979, for Common Crane *Grus grus* see Alonso *et al.* 2004).

The occurrence of winter movements, the use of multiple sites within a winter, and often different sites from year to year, all combine to make studies of population limitation difficult in migratory bird populations. It is not surprising, therefore, that such studies have developed only relatively recently, and that the most detailed ones have concentrated on species in which individuals occupy the same breeding and wintering places year after year.

SOME GENERAL PRINCIPLES

If we accept that, for various reasons, the total carrying capacities of the breeding and non-breeding habitats of particular populations need not necessarily correspond, then two scenarios are possible when birds return to their nesting areas each year and compete for territories or nest-sites:

1. Too few birds are left at the end of the non-breeding season to occupy all nesting habitat fully, so that practically all individuals of appropriate age and condition can breed (**Figure 26.1A**). In this case, breeding numbers would be limited by whatever factors operate in the non-breeding areas, and alleviation of these factors (such as the extent or carrying capacity of non-breeding habitats) would

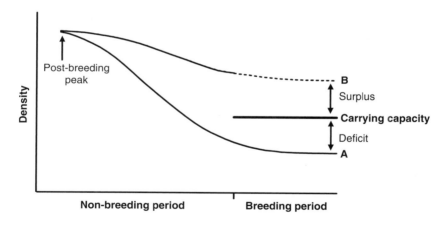

Figure 26.1 Model showing seasonal changes in total bird numbers in relation to the carrying capacity of the breeding area (thick line). In the lower curve (A) numbers left at the end of the non-breeding season are fewer than the nesting habitat could support, and in the upper curve (B) numbers are greater than the nesting habitat could support, leading to a surplus of non-territorial, non-breeders. In (A), breeding numbers are limited by conditions in the area occupied in the non-breeding season, and in (B) by conditions in the area occupied in the breeding season. From Newton (1998b).

be needed before breeding numbers could rise. For convenience, the breeding numbers in such populations could be described as 'winter-limited'.

2. More birds are left at the end of the non-breeding season than the available nesting habitat can support, producing a surplus of non-territorial, non-breeders (**Figure 26.1B**). In this case, alleviation of the factors that influence the extent or carrying capacity of the nesting habitat would be needed before breeding numbers could rise. For convenience, the breeding numbers of such populations could be described as 'summer-limited'.

The same two types of scenario could hold at the end of the breeding season, as birds return to their non-breeding quarters. At that time, the total numbers of adults and young might be insufficient to use the resources of non-breeding areas to the full, leading to good survival through the non-breeding season; alternatively, the total numbers may exceed the carrying capacity of non-breeding areas, leading to intense competition and poor survival.

Because bird reproduction is by definition confined to the breeding range, numbers there might be 'summer-limited' in a different way, namely if breeding success were so poor that subsequent breeding numbers could not reach the level necessary to fill either the available breeding habitat or the available wintering habitat. This situation could arise even with relatively good year-round survival. In other words, while failure to occupy all wintering habitat must be due to events on breeding or migration areas, failure to occupy all breeding habitat could be due to events on either breeding, migration or wintering areas. Understanding the limiting mechanisms in any one population may not be easy, especially if the situation changes from year to year. There is no reason why a species might not be summer-limited in one year or area and winter-limited in another year or area (for examples see Newton 1998b).

Effects of habitat loss on migrants

When areas of habitat are lost or added through human action, bird numbers often change accordingly. To take some recent examples from Britain, through much of the twentieth century Siskins *Carduelis spinus* increased in numbers and expanded in range as new conifer plantations matured and provided additional breeding habitat (Gibbons *et al.* 1993). In contrast, Redshanks *Tringa totanus* declined and contracted as summer nesting habitat was destroyed by land drainage (Norris *et al.* 2004), and Twite *Carduelis flavirostris* declined as areas of winter saltmarsh were lost to reclamation projects (Atkinson 1998). Many other examples have been described in the literature from Europe and North America, including the massive declines in numbers of waterbirds which followed the drainage of marshes. While many such examples may represent causal relationships, some may be due to coincidence between population decline (caused by some other factor) and habitat loss. Clearly, not all bird population changes occur in response to habitat changes, and some species seem to have far more potential habitat than they currently occupy. In the case of migrants, numbers may be limited in wintering areas at a level lower than habitat in breeding areas would support (or vice versa).

In recent years, much thought has been given to predicting the effect of habitat loss (equivalent to food loss) in both resident and migratory bird species. For resident birds, in which breeding and wintering areas are the same, population declines should be roughly in proportion to habitat loss, if habitat were of uniform quality and fully occupied throughout. In other words, if half the habitat (or food supplies) were lost, we would in general expect the population to be roughly halved (but see later). The situation is more complicated for migrant birds because habitat loss may occur in breeding or wintering areas or both (**Figure 26.2**). The actual population change following loss of breeding or wintering habitat would be expected to depend on where the tightest bottleneck occurred; that is, on the relative strengths of density-dependent constraints in the two areas (Sherry & Holmes 1995, Sutherland 1996). Such constraints include those various pressures, such as competition for space and food that can affect an increasing proportion of individuals as their density rises, resulting in an increased per capita mortality or decreased per capita reproduction. In the wintering area, the strength of density-dependence is measured by the per capita rate of increase in mortality that occurs as a result of rising population size (or decreasing area) (slope d). In the breeding area, the strength of density-dependence is measured by the per

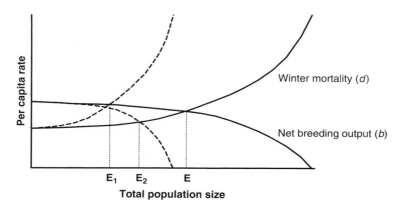

Figure 26.2   Model of relationship between per capita mortality and reproduction (continuous lines), and equilibrium population size (E). Per capita winter mortality increases, and per capita reproduction decreases, with increasing population size. The equilibrium population size is where the two lines intersect. The model shows how loss of habitat (or food supply) results in population decline. If wintering habitat (or food supply) is reduced by 50%, this results in a displacement of the relationship between mortality and total population size in direct proportion to the degree of habitat loss (dashed line). The equilibrium population size is reduced accordingly (E_1). A 50% loss of breeding habitat (or food supply) similarly results in a displacement of the relationship between net breeding output and total population size (dashed line) and a reduced equilibrium population (E_2). In this example, 50% loss of breeding habitat results in a smaller reduction in equilibrium population size than does 50% loss of wintering habitat, because density-dependence is stronger in winter than in summer. Modified from Sutherland (1996).

capita rate of decrease in reproduction with rising population size (or decreasing area) (slope b). If slope $b >$ slope d, loss of breeding habitat would have most impact on overall population size, and if $d > b$, loss of wintering habitat would have most impact. If the two density-dependent relationships were known, the effect of loss of habitat (or food supply) on equilibrium population size could in theory be calculated as $b/(b + d)$ for the breeding area, or as $d/(b + d)$ for the wintering area.

If, in an extreme case, all the density-dependence occurred in winter habitat, say, with no density-dependence in breeding habitat (which at prevailing population levels was present in excess, as in **Figure 26.1A**), then loss of winter habitat would cause a matching reduction in population size. In this situation, loss of breeding habitat would have no effect up to the point where density-dependent decline in breeding success set in. Although as yet there can be few species for which enough information is available to test the model in **Figure 26.2**, or to judge the form of density-dependent relationships over a wide range of densities at both seasons, attempts have been made for the Oystercatcher *Haematopus ostralegus* in Britain (Goss-Custard *et al.* 1995, Sutherland 1996).

The above considerations (model in **Figure 26.2**) lead to a number of conclusions about the effects of habitat (or food) loss on the equilibrium population sizes of migrants: (1) knowledge of the density-dependent response within just the wintering or breeding area cannot be used to predict precisely the effects of habitat or food loss in either, for it is the ratio of density-dependence in the two areas that is important; (2) unless there is no density-dependence acting during one of the seasons, a loss of habitat or food supply in either summer or winter areas could result in population decline; and (3) the consequence of habitat or food loss is greatest for the season in which density-dependence is strongest (winter in **Figure 26.2**). In practice, all migrants are likely to be affected more by changes in one area than the other, although whether breeding or wintering areas are most limiting may change through time. They could also change from year to year in species subject to large annual fluctuations in habitat, food supplies or other conditions.

The above generalisations on the role summer and winter conditions hold most clearly for populations limited by resources – by the available habitats and food supplies. They could also hold for populations limited below the levels that resources would permit by factors such as parasitism, predation and human persecution. In some circumstances, the latter factors can kill an unsustainably large number of individuals each year, sending populations into decline, and leaving a surplus of unused habitat and food. For example, if for some reason the predation pressure on eggs and chicks in the breeding areas increased so much that loss of annual production could not be offset by improved annual survival, the population would decline below the levels that both breeding and wintering habitats would support. Similarly, if shooting pressure on full-grown birds increased in winter quarters, so that the loss could not be offset by improved reproduction or natural survival, the population could again decline below the carrying capacities of both breeding and wintering habitats. In both these examples, decline would continue while ever that situation held (eventually to extinction), the trend being driven primarily in whichever area the effects of adverse factors on individual reproduction or survival were greatest.

The buffer effect and density-dependence

All habitats seem to vary from place to place in quality and attractiveness to the birds they support. One known mechanism through which density dependence in mortality or reproduction could occur during a period of population change involves the 'buffer effect'. This occurs when birds occupy the best habitat areas first and, as they fill these areas to capacity, they spread increasingly to poorer areas as their numbers continue to rise. As survival or reproduction is poorer in the less good areas, the mean per capita performance in the population as a whole declines as overall numbers grow, in a density-dependent manner. Sequential habitat fill of this type is seen: (1) as birds arrive in their breeding areas in spring, or their wintering areas in autumn, when they occupy the best places first, so that later arrivals are relegated to poorer places (Chapter 14); (2) in the annual fluctuations of populations, where numbers remain more stable from year to year in the preferred habitats (or territories) than in the secondary ones (Kluijver & Tinbergen 1953, Zimmerman 1982, Rodenhouse et al. 2003); and (3) in the progressive occupation of habitat areas (or territories) of different quality, as a population grows over a period of years (e.g. Mearns & Newton 1988, Ferrer & Donázur 1996, Lõhmus 2001). All these processes, which result from habitat variation, can help to regulate bird populations (for further examples of each type, see Newton 1998b). Food and other resources are involved in the regulation because they influence the quality and carrying capacity of habitats. In the reverse situation, as populations decline, birds usually withdraw from the poorer habitats first, leaving an increasing proportion of individuals in the good habitats, where survival or reproduction is highest.

The Icelandic population of the Black-tailed Godwit Limosa l. islandica wintering in Britain has risen four-fold since the 1970s, but rates of increase within individual estuaries have varied from zero to six-fold (Gill et al. 2001). In accordance with the buffer effect, rates of increase were greatest on estuaries with low initial numbers, and godwits on these sites were found to have lower prey intake rates and lower survival rates than godwits on longer-occupied sites with stable populations. In this species, therefore, population growth could have resulted in a progressively larger proportion of the population wintering in poorer estuaries, with measured consequences on feeding rates and survival. The buffer effect, acting on a large spatial scale, could thus have been a major density-dependent process acting to constrain population growth in this migratory species.

A similar spread to poor sites during a period of population growth was earlier noted in Grey Plovers Pluvialis squatarola wintering in different parts of Britain (Moser 1988), in Brent Geese Branta bernicla wintering in the Netherlands (Ebbinge 1992), and in Great Cormorants Phalacrocorax carbo wintering in Switzerland (Suter 1995). In the latter, the process was stepwise, and each category of habitat experienced a rapid build-up in numbers, followed by stabilisation, before the next type of habitat was occupied. In each of these studies, the first-filled habitat was assumed to be better, in which case survival would have been poorer in the secondary habitats, leading to progressive decline in mean per capita survival as the population grew, although this was not confirmed (but for effects on reproduction of Brent Geese see Ebbinge 1992). Hence, although the presence of secondary habitat would permit a population to attain a higher level than possible in the primary habitat alone, the poorer performance of individuals in secondary habitat would put ever increasing constraints on further population growth.

EXAMPLES OF SPECIES AFFECTED BY EVENTS IN BREEDING OR WINTERING AREAS

In some species, year-to-year changes in overall population levels have been clearly driven largely by conditions in breeding areas. On the North American prairies, rainfall varies greatly from year to year, and influences the amount of wetland habitat available to nesting waterfowl. In wet periods, populations increase, and in dry periods they decline. So important are these prairie wetlands as nesting habitat that they influence the entire continental wintering populations of several species, including the American Coot *Fulica americana* (**Figure 26.3**). They show how year-to-year conditions in the breeding areas can largely determine year-to-year fluctuations in total numbers.

In contrast, the numbers of several migratory songbird species counted each spring on their European breeding areas have fluctuated according to rainfall (and hence food supplies) in their African wintering areas. Examples include the Sedge Warbler *Acrocephalus schoenobaenus*, Sand Martin *Riparia riparia* and Purple Heron *Ardea purpurea* (den Held 1981, Peach *et al.* 1991, Bryant & Jones 1995, Szép 1995). They show how conditions in wintering areas can largely determine year-to-year fluctuations in total populations, although it is seldom certain whether most mortality occurs during the wintering period or during migration (see Chapter 24).

Other examples of summer-influenced and winter-influenced population changes in migrants, involving more than 67 different populations of 56 species, are given in **Table 26.1**. In most populations, the evidence is entirely circumstantial, based on long-term or year-to-year correlations between changes in breeding numbers and changes in either: (1) conditions in breeding or wintering areas, or in (2) associated

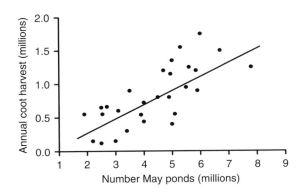

Figure 26.3 Relationship between the numbers of American Coots *Fulica americana* shot each year in the USA (reflecting total population size) and the number of ponds on the prairies in the preceding summer (reflecting the habitat and feeding conditions). Significance of relationship: $b = 0.21$, $r^2 = 0.54$, $P < 0.001$. Redrawn from Alisauskas & Arnold (1994).

Table 26.1 Migratory bird species in which temporal changes in breeding density have been linked with changes in previous winter conditions (A), with changes in summer conditions (B), or with changes in both winter and summer conditions (C)

Species	Location	Long-term upward or downward trend	Annual fluctuations	Source
A. Associated with change in winter conditions/survival				
Twite *Carduelis flavirostris*	Germany–Netherlands	+		Dierschke (2002)
Snow Bunting *Plectrophenax nivalis*	Germany–Netherlands	+		Dierschke (2002)
Shore Lark *Eremophila alpestris*	Germany–Netherlands	+		Dierschke (2002)
Sedge Warbler *Acrocephalus schoenobaenus*	England		+	Peach et al. (1991)
	Netherlands		+	Foppen et al. (1999)
Blackcap *Sylvia atricapilla*	Britain		+	Baillie & Peach (1992)
Greater Whitethroat *Sylvia communis*	Britain	+	+	Winstanley et al. (1974), Baillie & Peach (1992)
Willow Warbler *Phylloscopus trochilus*	Sweden		+	Hjort & Lindholm (1978)
	Britain	+	+	Baillie & Peach (1992), Peach et al. (1995)
Greenish Warbler *P. trochiloides*	India		+	Price & Gross (2005)
Nightingale *Luscinia megarhynchos*	N. Italy		+	Boano et al. (2004)
Loggerhead Shrike *Lanius ludovicianus*	Minnesota	+		Brooks & Temple (1990)
Barn Swallow *Hirundo rustica*	Denmark		+	Møller (1989)
	Britain		+	Robinson et al. (2003)
Sand Martin *Riparia riparia*	Britain		+	Cowley (1979), Bryant & Jones (1995)
	Hungary		+	Szép (1995)
	Germany		+	Kuhnen (1975)
Avocet *Recurvirostra avosetta*	England		+	Hill (1988)
Golden Plover *Pluvialis apricaria*	Scotland	+		Parr (1992), Yalden & Pearce-Higgins (1997)
Roseate Tern *Sterna dougallii*	Britain	+		Ratcliffe & Merne, in Wernham et al. (2002)
Puffin *Fratercula arctica*	Scotland	+	+	Harris & Wanless (1991)
Thick-billed Murre *Uria lomvia*	Eastern Canada		+	Gaston (2003)

Species	Location			Reference
Night Heron *Nycticorax nycticorax*	France, Spain	+		den Held (1981), Fasola et al. (2000)
Purple Heron *Ardea purpurea*	Netherlands, Spain	+		den Held (1981), Cavé (1983), Fasola et al. (2000)
Barnacle Goose *Branta leucopsis*	Svalbard		+	Owen (1984)
Dark-bellied Brent Goose *Branta bernicla*	Netherlands		+	Ebbinge (1991)
Lesser Snow Goose *Chen caerulescens*	Canada		+	Francis et al. (1992)
B. Associated with change in summer conditions/survival/breeding rate				
Chaffinch *Fringilla coelebs*	Russia	+		Sokolov (1999)
Willow Warbler *Phylloscopus trochilus*	Russia	+		Sokolov (1999)
Icterine Warbler *Hippolais icterina*	Russia	+		Sokolov (1999)
Kirtland's Warbler *Dendroica kirtlandii*	Michigan		+	DeCapita (2000)
Prairie Warbler *Dendroica discolor*	Indiana	+		Nolan (1978)
Wilson's Warbler *Wilsonia pusilla*	California	+		Chase et al. (1997)
Bell's Vireo *Vireo bellii*	California		+	Griffith & Griffith (2000)
	Missouri		+	Budnik et al. (2000), Rothstein & Robinson
Back-capped Vireo *Vireo atricapillus*	Texas	+		Hayden et al. (2000)
Thick-billed Murre *Uria lomvia*	Eastern Canada	+		Gaston (2003)
Pied Flycatcher *Ficedula hypoleuca*	Finland	+		Virolainen (1984)
	England	+		Stenning et al. (1988)
	Russia	+		Sokolov (1999)
American Redstart *Setophaga ruticilla*	New Hampshire	+		Sherry & Holmes (1992)
Swainson's Thrush *Catharus ustulatus*	California	+		Johnson & Geupel (1996)
Wood Thrush *Hylocichla mustelinus*	Illinois	+		Robinson (1992)
	Delaware	+		Roth & Johnson (1993)
Corncrake *Crex crex*	Britain		+	Green & Stowe (1993), Green et al. (1997)
American Coot *Fulica americana*	North America	+		Alisauskas & Arnold (1994)
Common Sandpiper *Actitis hypoleucos*	Britain		+	Holland & Yalden (1991)
Dunlin *Calidris alpina*	Fennoscandia	+		Soikkeli (1970), Jönsson (1991)

(Continued)

Table 26.1 (Continued)

Species	Location	Long-term upward or downward trend	Annual fluctuations	Source
Red Knot Calidris canutus	Sweden		+	Blomqvest et al. (2002)
Little Stint Calidris minuta	South Africa		+	Summers & Underhill (1987)
Curlew Sandpiper Calidris ferruginea	Australia		+	Minton (2003)
	South Africa		+	Summers & Underhill (1987)
Red-necked Stint Calidris ruficollis	Sweden		+	Blomqvest et al. (2002)
	Australia		+	Minton (2003)
Lapwing Vanellus vanellus	Britain	+		Peach et al. (1994)
Stone Curlew Burhinus oedicnemus	England	+		Aebischer et al. (2000)
Mallard Anas platyrhynchos	North America		+	Reynolds (1987)
Various duck species	Iceland		+	Gardarsson & Einarsson (1994)
Brent Goose Branta bernicla	Britain		+	Summers & Underhill (1987), Blomqvest et al. (2002)
Greater White-fronted Goose Anser albifrons	NW Europe		+	Blomqvest et al. (2002)
Pink-footed Goose Anser brachyrhynchus	Svalbard		+	Madsen et al. (2002)
Arctic Tern Sterna paradisaea	NW Europe	+		Suddaby & Ratcliffe (1997)
C. Associated with change in both winter conditions (overwinter survival) and summer conditions (previous breeding rate)				
Black-throated Blue Warbler Dendroica caerulescens	Jamaica – New Hampshire		+	Sillett et al. (2000), Rodenhouse et al. (2003)
Northern Pintail Anas acuta	Western North America		+	Raveling & Heitmeyer (1989)
White Stork Ciconia ciconia	France–Germany		+	Dallinga & Schoenmakers (1989), Kanyamibwa et al. (1993), Bairlein (1996b)
Great Skua Stercorarius skua	Scotland	+	+	Klomp & Furness (1992)

breeding or mortality rates. In some populations, however, potential causal relationships were subsequently confirmed by experiments involving manipulation of likely limiting factors (see later).

In some of the bird populations mentioned in **Table 26.1**, breeding numbers increased in years that followed a good breeding season, and decreased in years that followed a poor breeding season (**Figure 26.4**). In these populations, spring–summer conditions in breeding areas evidently had most influence on subsequent year-to-year changes in breeding numbers, and such populations were therefore below the limit imposed by winter habitat (at least in most years). The same was true for other populations whose survival rates changed over the years, according to conditions in breeding areas. In yet other populations, breeding numbers varied from year to year according to previous winter conditions, implying that such populations were close to the limit imposed by winter habitat (or food supplies).

Over many years, the breeding successes of some tundra-nesting geese and waders in Siberia followed the cyclic pattern of lemming populations, with most young produced during lemming peaks (**Figure 26.6**). The relationship apparently resulted from Arctic Foxes *Alopex lagopus* and other predators switching more heavily to lemmings in years when lemming numbers were high, lessening their predation on bird eggs and chicks. Year-to-year fluctuations in the breeding success of geese and waders showed peaks about every three years, as detected from the young to adult ratios (or percentage of young) in flocks of White-fronted Geese *Anser albifrons* and Brent Geese *Branta bernicla* wintering in northwest Europe, in Red Knots *Calidris canutus*, Curlew Sandpipers *Calidris ferruginea* and Sanderlings *Calidris alba* migrating through northwest Europe, and Curlew Sandpipers and Turnstones *Arenaria interpres* wintering in southern Africa (Summers & Underhill 1987, Summers *et al.* 1998, Blomqvest *et al.* 2002). In most of these species, breeding success influenced total population sizes, so once again events on breeding areas had a major influence on year-to-year numerical changes. Because these

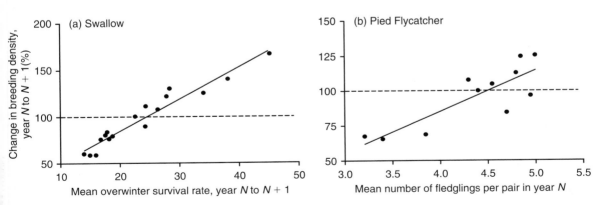

Figure 26.4   Demography and population change. (a) Relationship between annual survival of adult Barn Swallows *Hirundo rustica* and annual change in breeding density. From Møller (1989). (b) Relationship between annual breeding output of Pied Flycatchers *Ficedula hypoleuca* and annual change in breeding density. From Virolainen (1984). Significance of relationships: Barn Swallow, $b = 3.44$, $r^2 = 0.93$, $P < 0.001$; Pied Flycatcher, $b = 0.15$, $r^2 = 0.74$, $P < 0.001$.

various species, which all breed in the same part of Siberia, wintered in widely separated parts of two hemispheres, any effects of wintering conditions must have been small by comparison. Other arctic-nesting shorebirds experience big fluctuations in breeding success, apparently associated with spring weather in breeding areas which affect total population sizes, making any links with lemming cycles much less obvious.

In some migratory species, breeding numbers declined, and it was not immediately obvious whether the causes lay in breeding or wintering areas. However, where fecundity and survival rates were monitored in the same population both during periods of increase (or stability) and of decrease, the comparison provided useful pointers to where the cause of the decline might lie. Thus, where annual reproduction declined while annual survival stayed the same, the problem lay in the breeding areas but, where survival had decreased, the problem lay in either breeding or wintering areas, depending on the time of year the extra deaths occurred. Decline in a European Golden Plover *Pluvialis apricaria* population was associated with a decline in survival rate but no change in reproduction, while decline in a Northern Lapwing *Vanellus vanellus* population was associated with a decline in reproduction but no change in survival (Parr 1992, Peach *et al.* 1994, Yalden & Pearce-Higgins 1997).

In yet other species, long-term population declines were associated with reductions in both breeding and survival. The White Stork *Ciconia ciconia*, for example, has suffered from reduced reproduction on its European breeding grounds, caused by drainage and pesticide-induced food shortages, and also from reduced survival on its West African wintering grounds, caused mainly by drought-induced and pesticide-induced food shortages (mainly locust control) (Dallinga & Schoenmakers 1989, Kanyamibwa *et al.* 1993, Bairlein 1996b). Thus, in any avifauna we can expect to find species whose numbers are changing because of events in breeding or non-breeding areas or both, and routine monitoring of fecundity or survival rates can often point to where the problems lie.

Climatic factors acting in both breeding and wintering areas

Sometimes migratory bird populations are influenced throughout their range by widescale climatic fluctuations, such as the El Niño-Southern Oscillation (ENSO). Climatic shifts of this type can act simultaneously on local weather conditions in widely separated breeding, migration and wintering areas. Changes in some bird populations have been linked to such grand-scale climatic events.[1] For example, in the Black-throated Blue Warbler *Dendroica caerulescens*, reproduction in New Hampshire nesting areas and survival in Jamaican wintering areas were both

---

[1] *El Niño and La Niña represent the extreme phases of the Southern Oscillation, causing year-by-year variations in atmospheric and oceanic conditions in the equatorial Pacific. These conditions in turn have significant consequences for surface temperatures and rainfall worldwide, which in turn modulate the productivity of many terrestrial and aquatic ecosystems. ENSO conditions can be quantified by use of the standardised ENSO index, which is measured as the difference in sea surface temperatures between Darwin (Australia) and Tahiti. High values indicate La Niña conditions and low (negative) values indicate El Niño conditions. This index has proved to be a robust measure for tracking ENSO changes.*

lower during El Niño periods than during the intervening La Niña periods. In this species, therefore, population fluctuations were linked to local conditions in both areas (and probably also on migration routes) that were in turn influenced by the same underlying large-scale climatic cycle, and its effects on food supplies (which were measured in breeding areas) (Sillett *et al.* 2000). Moreover, good breeding seasons were followed by increased recruitment of new individuals into the wintering and breeding populations. Similar associations were found in the annual survival and reproduction of the Yellow Warbler *Dendroica petechia* studied over 10 years in Manitoba breeding areas (Mazerolle *et al.* 2005). These birds wintered in Central America and northeastern South America.

The effects of ENSO on regional weather patterns vary across a continent. Some regions during El Niño conditions receive more-than-average rainfall, while other regions receive less-than-average rainfall. The effects of ENSO on bird food supplies, and on the population dynamics of the birds themselves, would therefore be expected to differ regionally. In contrast to the studies just mentioned, but in association with local weather patterns, Neotropical migrants breeding in the Pacific Northwest had lower (rather than higher) reproduction in El Niño years (Nott *et al.* 2002). Together these various studies demonstrate the role of ENSO in the population dynamics of Nearctic–Neotropical migrants, and highlight the geographical variation in bird responses, depending on the differing effects of ENSO on regional weather.

The same trends in weather do not necessarily have the same effects on both breeding and wintering areas. Among Sand Martins *Riparia riparia* in England, annual survival rates increased with increase in rainfall in the dry African wintering areas, and decreased with increase in rainfall in the wetter breeding areas (Cowley & Siriwardena 2005). Heavy rain in summer also led to greater-than-usual starvation among the young. In the arid wintering areas, relatively high rainfall was assumed to promote greater vegetation growth and insect populations at the time the martins were there, whereas in the English breeding areas rain reduced insect activity and prevented adults from foraging effectively, reducing their body condition. Their subsequent survival in migration or wintering areas was then reduced. Rainfall in both areas thus affected year-to-year population changes, but in opposite directions.

If we can assume that breeding rates were influenced primarily by conditions in breeding areas, and mortality rates primarily by conditions in wintering areas (unless otherwise specified), then 28 of the bird populations in **Table 26.1** were winter-limited, another 35 or more were summer-limited, while four were influenced by both summer and winter conditions. However, because both the extent and carrying capacities of habitats vary from year to year, and from area to area, the same species might be winter-limited in some years or areas, and summer-limited in other years or areas, as in the different Willow Warbler *Phylloscopus trochilus* populations in **Table 26.1**. Each case must be judged on its particular circumstances.

From winter-limited to summer-limited

Over several years, the same population might change from one state to another, as its status with respect to available resources changed. Several species of geese increased during the latter half of the twentieth century in response to reduced hunting pressure in their wintering areas, but then came up against food shortage in the breeding areas, as growing numbers competed for favoured food plants. This increased competition resulted in reduced chick survival in Lesser Snow Geese *Chen caerulescens* in the central Canadian arctic (Francis *et al.* 1992), and in reduced summer survival among adult Pink-footed Geese *Anser brachyrhynchus* on Svalbard (Madsen *et al.* 2002). The major constraint to further population growth thus shifted from the wintering to the breeding areas as populations grew. In some other goose populations, studied in less detail, increasing competition was manifest chiefly in declining proportions of young in wintering flocks (**Figure 26.5**). In Brent Geese *Branta bernicla* wintering in western Europe, total numbers fluctuated from year to year around the long-term upward trend, according to annual variations in breeding success, as discussed above (Summers & Underhill 1987).

The pattern in which breeding density fluctuated from year to year in parallel with the previous year's breeding success was recorded only in short-lived species, in which individuals breed in their first year of life (passerines and dabbling ducks). It would not be expected in longer lived species, in which individuals do not breed until they are two or more years old, and in which annual recruitment rates are naturally low. In such species, breeding success would need to be poor over several years before any effect on breeding numbers became obvious, unless accompanied by a simultaneous increase in mortality or emigration (for Arctic Tern *Sterna paradisaea*, see Suddaby & Ratcliffe 1997).

Divergent patterns

Another indication of the importance of winter conditions in influencing population changes comes when closely related species that breed in the same area show different trends according to where they winter. Of the various waterfowl species that breed in Siberia, those that migrate to western Europe have all increased in numbers in recent decades, following their greater protection from winter hunting. Examples include the western population of the Greater White-fronted Goose *Anser albifrons,* and the Dark-bellied Brent Goose *Branta b. bernicla*. By contrast,

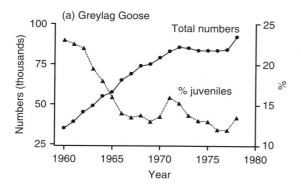
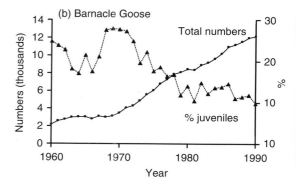

Figure 26.5 Reduced production of young by Greylag Geese *Anser anser* and Barnacle Geese *Branta leucopsis*, as their numbers have grown. The populations concerned breed in Iceland (Greylag) and Svalbard (Barnacle) and winter in Britain, where counts were made. The graphs show five-year moving mean values. Adapted and updated from Owen *et al.* (1986).

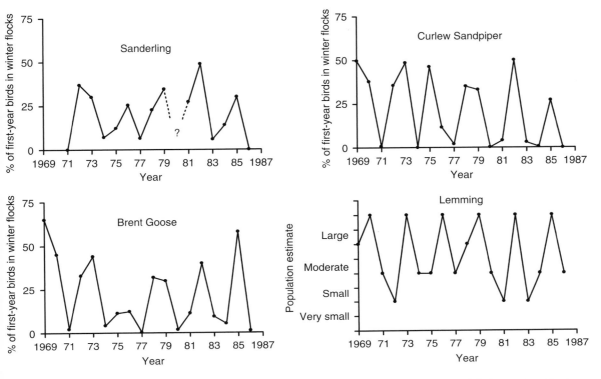

Figure 26.6 Proportion of first-year birds in wintering populations of various bird species that breed on the Taimyr peninsula of Siberia, shown in relation to densities of lemmings on the breeding grounds in different years. High production of young in Sanderlings *Calidris alba*, Curlew Sandpipers *C. ferruginea* and Brent Geese *Branta bernicla* occurred about every three years when lemming numbers were also high, providing alternative food for predators. From Summers & Underhill (1987).

all those populations that winter in Southeast Asia have continued to decline, in association with rising persecution in that region (Syroechkovski & Rogacheva 1994). Examples include the eastern subspecies of the Bean Goose *Anser fabalis serrirostris* and *A. f. middendorffii,* and the Baikal Teal *Anas formosa.* These various waterfowl species share similar nesting habitats, which are largely undisturbed by people, and their divergent population trends are most plausibly attributed to conditions on their different wintering areas (Syroechkovski & Rogacheva 1994).

Another divergent pattern, evident in temperate regions after a hard winter, is that resident species are often found to have declined greatly, whereas summer visitors from tropical wintering areas have not (Dobinson & Richards 1964, Graber & Graber 1979, Cawthorne & Marchant 1980, Holmes & Sherry 2001). This difference provides another indication of the importance of winter conditions, at least for the resident species (Newton 1998b). Some summer visitors may in fact benefit from the scarcity of resident competitors. The larger than usual numbers of European Pied Flycatchers *Ficedula hypoleuca* that bred in Britain in 1917 and 1947 were attributed to the greater availability of nest cavities in those years, associated with the scarcity of resident tits caused by preceding hard winters (Elkins 1988). This early observation also suggested that flycatcher numbers were limited in some areas by shortages of nest-sites, an inference supported by subsequent experiments involving either the removal of competing tit species (Gustafsson 1988), or the provision of additional nest boxes (e.g. von Haartman 1971, Currie & Bamford 1982, Newton 1998b). Many other species of cavity-nesting migrants (excluded from **Table 26.1**) also increased in breeding density and distribution following the provision of artificial nest-sites, as did various hirundines and swifts that use buildings (Erskine 1979, Newton 1994, 1998b, Evans *et al.* 2003b). The implication is that nest-site availabilities (and hence a feature of the breeding area) are important in limiting the overall population levels of such species.

Carry-over effects

Most discussion of population limitation in migrants carries the implicit assumption that conditions in wintering areas have no effects on subsequent breeding success, and that conditions in breeding areas have no effect on subsequent winter survival. These assumptions are not entirely justified. The breeding densities of many bird species are determined partly by winter survival, and winter survival can in turn be related to events in the previous breeding season. But more than this, individual birds can carry over effects from one season to the next, and these residual effects, acting through body condition, can explain some of the variation in breeding success and winter survival. For example, among geese and other waterfowl, foraging conditions at wintering and migration sites have long been known to affect subsequent breeding success through the affect of food supplies on body condition. In most such species, conditions on spring staging areas seem more important than conditions in wintering areas, so further discussion is left for Chapter 27. However, in Whooper Swans *Cygnus cygnus*, the proportions of young in wintering flocks in Sweden each year were correlated with the mean temperature in the preceding winter, implying that winter temperatures influenced subsequent breeding success (presumably though effects on feeding conditions and body reserves) (Nilsson 1979).

Carry-over effects of winter conditions on breeding success have been detected not only in large birds, which carry substantial body reserves to breeding areas, but also in small birds, which carry relatively smaller reserves. Thus, individual Barn Swallows *Hirundo rustica* breeding in Italy were affected by conditions in wintering areas, as estimated from the satellite-recorded 'normalised difference vegetation index' (NDVI) used to assess the annual state of the vegetation (dependent on rainfall) in Africa (Saino *et al.* 2004). The arrival dates and egg-laying dates of the same individuals in consecutive breeding seasons were advanced, more second broods were produced, and the total seasonal production of young was increased, after winters with high NDVI in African winter quarters. Other carry-over effects from winter to summer, affecting morphology, migration dates and breeding, were described in another population of Barn Swallows breeding in Denmark (Møller & Hobson 2004).

One important factor contributing to reproductive success among migratory birds is date of spring arrival and commencement of nesting. Within populations, individuals that arrive and start nesting early in the season do better, in terms of habitat quality, territory acquisition and number of young raised, than do those that

Table 26.2 Some carry-over effects of environmental conditions on subsequent bird performance

	Relationship between winter/spring habitat quality/food supply and:				Source
	Timing of spring migration	Body condition at migration time	Quality of breeding site	Breeding success	
American Redstart *Setophaga ruticilla*	+	+			Marra et al. (1998), Marra & Holmes (2001)
	+			+	Norris et al. (2004)
Black-throated Blue Warbler *Dendroica caerulescens*		+		+	Bearhop et al. (2004)
Nightingale *Luscinia megarhynchos*	+				Gordo et al. (2005)
European Pied Flycatcher *Ficedula hypoleuca*	+		+	+	Lundberg & Alatalo (1992), Cichon et al. (1998)
Barn Swallow *Hirundo rustica*[a]	+			+	Saino et al. (2004)
Sand Martin *Riparia riparia*			+	+	Cowley & Siriwardena (2005)
Common Kestrel *Falco tinnunculus*			+	+	Dijkstra et al. (1990)
Red Knot *Calidris canutus*		+		+	Morrison (2006)
Black-tailed Godwit *Limosa limosa*	+		+	+	Gill et al. (2001), Gunnarsson et al. (2005)
Pied Avocet *Recurvirostra avosetta*	+			+	Hötker (2002)
Pink-footed Goose *Anser brachyrhynchus*		+		+	Madsen (1995)
Greater Snow Goose *Chen caerulescens atlantica*		+		+	Reed et al. (2004)
Barnacle Goose *Branta leucopsis*		+		+	Prop & Black (1998), Prop et al. (2003)
Brent Goose *Branta bernicla*		+		+	Ebbinge & Spaans (1995)
Whooper Swan *Cygnus cygnus*[a]				+	Nilsson (1979)
White Stork *Ciconia ciconia*[a]				+	Bairlein & Henneberg (2000)

[a]Based on comparison of average habitat state between years, rather than habitat of individuals.

arrive and nest late (**Table 26.2**; Chapter 14). Variations in arrival dates sometimes exceed a month, and late-arriving individuals are often in poor condition. All relevant aspects of performance could be influenced by the quality of the individual birds and, without experiments, it is hard to tell to what extent reproductive success follows directly from arrival date, and to what extent from individual quality (though one experimental study confirmed the importance of date as such, Cristol 1995; Chapter 14).

In a few studies, quality of winter habitat and associated food supplies have been found to influence the body condition and spring migration dates of birds. American Redstarts *Setophaga ruticilla* studied in Jamaica competed for territories in moist forest habitats, forcing some individuals into dryer scrub habitat (Marra *et al.* 1998). Birds in the better forest habitat showed better body condition than those in scrub. They also showed lower ^{13}C values in their muscle tissue, a pattern mirrored in the available insect prey. The ^{13}C values in the muscle of newly arrived birds in North American breeding areas provided an indication of the habitat those birds had occupied in winter. Individuals that arrived earliest in breeding areas had lower ^{13}C values than those arriving later, indicating that the early birds had come from the best wintering habitat. In a later study, the first arriving birds were found to have better breeding success than the later ones (Norris *et al.* 2004). This work, involving measurement of carbon isotopes, thus provided another link between the conditions experienced by individuals in winter, their subsequent migration dates and breeding success.

The fact that good winter habitat was limiting for American Redstarts was shown by an experiment in which 28 individuals (largely adult males) were removed from their territories in optimal mangrove habitat (Studds & Marra 2005). They were rapidly replaced by 23 other individuals (females and juvenile males) from poor scrub habitat. Initially, the replacement birds had blood isotope signatures typical of scrub, but two months later they had isotope signatures typical of mangrove. Compared to control birds in scrub, the upgraded birds maintained weight over winter, departed earlier in spring, and returned in greater proportion the next autumn. Insect biomass was greater in mangrove than scrub, suggesting that food availability caused the differences. This experiment gave further indication that prime winter habitat was limited, and that winter conditions influenced the subsequent performance of individuals. Birds able to get a place in good habitat in their first winter may be well positioned to breed productively the next spring, and possibly throughout their lives.

Among Black-tailed Godwits *Limosa limosa* which winter in Britain and breed in Iceland, a similar pattern emerged, as found from study of colour-ringed individuals identified in both breeding and wintering areas (Gill *et al.* 2001, Gunnarsson *et al.* 2005). Those individuals that occupied the best wintering habitat were able to prepare for spring migration earlier, and were therefore first to arrive in breeding areas. They thereby acquired the best nesting habitat, and subsequently had better breeding success, on average, than later arriving individuals that had occupied poorer winter habitat. In this way, as in the American Redstarts, the effects of winter habitat carried through the whole annual cycle.

Of course, things do not always run smoothly. Whatever the date of departure from wintering areas, adverse weather encountered en route may delay arrival in breeding areas, and thus affect reproduction (Johnson & Herter 1990, Richardson

1990). In 1997, many White Storks *Ciconia ciconia* were late in leaving their African wintering areas, which was attributed to poor food supply (Berthold *et al.* 2002). In addition, some individuals (including a radio-tagged bird) were delayed for another week en route, as they hit a severe cold spell. These circumstances led to late arrival in European breeding areas, and depressed breeding success over wide areas, providing another link between conditions in wintering and migration areas and subsequent breeding success.

Similarly, events in breeding areas can in turn affect subsequent survival in migration or wintering areas. An example was mentioned above for the Sand Martin *Riparia riparia*, in which heavy summer rain in breeding areas was associated with poor body condition and reduced subsequent survival of adults (Cowley & Siriwardena 2005). Evidence also emerged for effects of reproduction on the subsequent survival of Barnacle Geese *Branta leucopsis* (Prop 2004). In females, non-breeders (that made no reproductive commitment) survived better than failed breeders, which in turn tended to survive better than successful breeders, with annual rates of 0.95 (SE 0.03), 0.86 (SE 0.02) and 82 (SE 0.04) respectively. In addition, among females that attempted to breed, variance in survival was explained by the date of nest departure (when they could start feeding well again), whether hatching for successful breeders or desertion for failed breeders. The later this date, the lower the subsequent survival probability. This trend was attributed partly to body condition when intense feeding began again, and partly to the seasonal decline in the nutrient content of plant-food, and the limited time available for pre-migratory fattening. In male geese, which played no part in incubation and were free to feed throughout, survival did not differ significantly between birds of different breeding status.

These and other studies based on ringing or isotope analyses have all provided evidence of apparent carry-over effects in individuals from one time of year to another (**Table 26.2**). Evidence for effects of reproduction on subsequent survival emerged when experimentally-increased reproductive effort (achieved through increasing brood sizes) resulted in poor body condition, and in reduced subsequent survival in adult Glaucous-winged Gulls *Larus glaucescens* (Reid 1987), Common Kestrels *Falco tinnunculus* (Dijkstra *et al.* 1990), male Pied Flycatchers *Ficedula hypoleuca* (Askenmo 1979) and female Collared Flycatchers *F. albicollis* (Cichoń *et al.* 1998). Although poor condition in these birds was experimentally induced, it confirmed that events acting at the time of breeding could influence the subsequent performance of individuals. In all these various species, reduced survival was inferred from the lowered return rates of poor condition individuals the next year, but in none was it certain whether the extra mortality occurred on migration or in winter quarters.

Carry-over effects of these various types can be incorporated in the model in **Figure 26.2,** given the appropriate information. Increase in the proportion of individuals experiencing a negative carry-over effect would be expected to magnify a subsequent population decline, while increase in the proportion of individuals experiencing a positive carry-over effect could lessen a population decline. To predict changes in population size resulting from habitat loss, it would therefore help to determine: (1) which factors in which seasons produce strong carry-over effects, and (2) the quality, as well as the amount, of habitat that is lost (Norris 2005).

Carry-over effects are evident not only at the individual level, but also at the population level through well-established density-dependent effects. In particular, in any species good breeding success at the level of the population is likely to result in high post-breeding numbers. In species limited in wintering areas, the ensuing mortality is likely to be density dependent, resulting in high mortality following good breeding years, and lower mortality following poor breeding years. Among migrants, mortality over the non-breeding period has been shown to be density dependent in most populations in which the problem has been examined: namely Sedge Warbler *Acrocephalus schoenobaenus*, Blackcap *Sylvia atricapilla*, Greater Whitethroat *S. communis*, Willow Warbler *Phylloscopus trochilus*, European Pied Flycatcher *Ficedula hypoleuca*, Common Redstart *Phoenicurus phoenicurus*, Barn Swallow *Hirundo rustica*, Redshank *Tringa totanus*, Mallard *Anas platyrhynchos*, Northern Shoveller *Anas clypeata* and Barnacle Goose *Branta leucopsis* (Järvinen 1987, Stenning *et al.* 1988, Mihelsons *et al.* 1985, Kaminski & Gluesing 1987, Owen & Black 1991, Baillie & Peach 1992, Whitfield 2003). In most of these species, overwinter loss was also the key factor governing year-to-year change in breeding numbers (review Newton 1998b), but it included migration as well as wintering periods. Density-dependent mortality in the winter period alone has been documented in the American Redstart *Setophaga ruticilla* (Studds & Marra 2005), and it is probably only a matter of time before other examples come to light.

In general, therefore, it seems that different periods in the annual cycles of migratory birds are inextricably linked, and that events occurring during one period can affect performance in a later period. Such effects can occur at the level of the individual (carry-over effects), and at the level of the population (density-dependent effects). It has become increasingly possible to detect carry-over effects through our ability to 'connect' populations in specific breeding and wintering areas, through ringing or isotope analyses, and as information on global climatic conditions has become more generally available.

CAUSES OF POPULATION DECLINES

Over the past several decades, various species of intercontinental landbird migrants have declined in Europe and North America, but the main causal factors seem to have differed between the two regions (Chapters 24 and 25). The European species winter in the Afrotropics, and most of those that have declined occupy arid scrub habitats prone to drought. Their declines have been widely attributed to events on wintering areas, notably reductions in rainfall and associated food supplies. The North American species winter in the tropics of Central and South America, and most of those that have declined occupy forest, with smaller numbers in scrub and grassland habitats. Their declines have been widely attributed to events in breeding areas, notably forest fragmentation and the associated increases in densities of predators and parasitic Brown-headed Cowbirds *Molothrus ater*. The lines of evidence proposed in favour of these various hypotheses are summarised in Chapters 24 and 25, along with some alternative explanations applicable to at least a minority of species.

Experimental evidence on population limitation in migrants

Most of the evidence obtained on the causes of population changes, whether for Old World or for New World species, is based on correlation and inference. Experiments are almost impossible to perform at regional scales, but sometimes widespread changes in land-use or legislation have been followed by widespread changes in bird populations. Examples include the increases in many waterfowl and shorebird populations that followed the introduction of protective legislation or other conservation measures, or the increases of bird-of-prey populations that followed reductions in the use of organochlorine pesticides (Newton 1979, 1998b, Cade *et al.* 1988). In addition, local population increases have followed experimental manipulations of potential limiting factors or the introduction of specific conservation measures. Examples include increases in breeding success and breeding density that were associated with: (1) removal of predators (for various ducks see Duebbert & Kantrud 1974, Duebbert & Lokemoen 1980; for Sandhill Crane *Grus canadensis* see Littlefield 1995); (2) removal of Brown-headed Cowbirds *Molothrus ater* (for Bell's Vireo *Vireo bellii* see Griffith & Griffith 2000, for Black-capped Vireo *V. atricapillus* see Hayden *et al.* 2000); (3) provision of extra nest-sites for species with special needs (for many species of cavity-nesting and other birds see Newton 1994, 1998b); (4) removal or exclusion of competing species (for effects of tit *Parus* removal on Collared Flycatcher *Ficedula albicollis* densities (via access to nest-sites) see Gustafsson 1988); and (5) change or intervention in destructive agricultural procedures (for Corncrake *Crex crex* see Green *et al.* 1997, Green 1999; for Stone Curlew *Burhinus oedicnemus* see Aebischer *et al.* 2000). Similar measures have led to population increases in some resident bird species too, as has the provision of additional winter food (Newton 1998b). Restrictions in hunting pressure and pesticide use have affected wintering populations, as well as breeding populations, but the planned experiments were restricted to local breeding populations. They confirm that many migratory species were at that time and location limited by conditions in breeding areas.

OTHER ASPECTS

Range size and population limitation

Because migrants occupy different parts of the world at different seasons, their numbers and geographical ranges at one season may be limited by the size of area available to them at the other season. In other words, a species with a limited wintering area, and no scope for expansion, may be unable to achieve the numbers that would enable it to occupy its full potential breeding area, or vice versa. A small wintering range could thereby limit both the breeding population and the breeding range. For example, the Bristle-thighed Curlew *Numenius tahitiensis* breeds in Alaska, but winters on Hawaii and other Pacific Islands. As the wintering habitat on the islands is restricted, the species could never have been numerous there, limiting it to small numbers in its Alaskan breeding range. Moreover, the situation may have worsened in recent centuries because the species becomes flightless during moult, and many of the islands colonised by humans and associated predators have now become unsuitable for it (Marks *et al.* 1990). Another species that winters on islands, and has a small breeding population and range on a continent, is Kirtland's Warbler *Dendroica kirtlandii* (winters Bahamas, breeds Michigan). It is now one of the rarest breeding birds in North America, but even at its peak, it may never have been numerous if it was always restricted to its present wintering range.

Range segregation and sex ratios

Consideration of population limitation in migrants is further complicated by the fact that, in the non-breeding season, the sexes of a species may occupy partly different areas or habitats from one another (Chapter 15). These differences are seldom absolute, but rather statistical tendencies, with the females in many species tending to migrate further from the breeding areas than males, or to occupy less food-rich habitats. Both differences have been widely attributed to male dominance and inter-sex competition, the assumption being that, unless females separated from males, their mortality, and the resulting sex ratio distortion, might be even greater than they are. However, not all species show such differences, and other explanations have been offered of such segregation in particular cases (Chapter 15).

Whatever the underlying cause, the existence of range and habitat segregation means that the two sexes may be exposed to somewhat different pressures, and that the numbers of each sex could become limited at different levels. Distorted sex ratios also occur in resident species, and for various reasons; but the geographical segregation adds another potential causal factor in migrants. Some of the most distorted adult sex ratios recorded among adult birds on their breeding areas were in highly migratory species of ducks and passerines, the numbers of breeding pairs being limited by the scarcest sex (e.g. Stewart & Aldrich 1951, Bellrose *et al.* 1961, Owen & Dix 1986). In some such species, males exceeded females by more than two to one.

CONCLUDING REMARKS

In this chapter, I have been concerned mainly with changes in the numbers of migratory birds, and with the relative importance of events in breeding and wintering areas in influencing these changes. In the longer term, however, population limitation in migrants is probably a dynamic process, involving both areas. If conditions in the wintering range permit increased survival, a species could become more numerous, and may expand its breeding distribution, perhaps into places where the production of young is lower. Similarly, if conditions in the breeding range permit increased production of young, the species could become numerous enough to expand its wintering range, perhaps into areas where survival is lower. In this way, the summer and winter ranges of migrants will tend to expand until reproduction and mortality balance (see also Cox 1985). It is only during a period of change, as in recent decades, that the breeding or wintering range may emerge as providing the stronger limitation. These speculations assume that there are indeed vacant areas into which migrants could expand if their numbers rose. For most species this is probably true; either other habitats within the existing range or other suitable areas outside it.

Migratory birds also depend on encountering suitable conditions at staging points on their routes (Chapter 27). If conditions deteriorate at any one point, a food bottleneck might develop that could begin to limit the population. When conditions are deteriorating everywhere at once, it becomes hard to pinpoint that bottleneck except in the most obvious cases. But the fact that migrants use two or more essential areas each year means that they are inevitably more susceptible to the effects of habitat destruction than are resident birds. Residents suffer only if their particular area is destroyed, but migrants could suffer if any one of several areas important to them were lost. In this sense they have, on the average, more chances than residents of being affected – adversely or otherwise – by human action.

SUMMARY

Changes in the numbers of migratory birds, either long-term or year-to-year, may be caused by changes in conditions in the breeding or wintering areas or both. The strongest driver of numerical change is provided in whichever area the effects of adverse factors on mean survival or fecundity are greatest. Examples are given of some species whose numbers have changed in association with conditions in breeding areas, and of others whose numbers have changed in association with conditions in wintering areas, either year to year or long term. In a few such species, the effects of potential limiting factors have been confirmed locally by experiment.

In some species, habitats occupied in wintering and migration areas, and their associated food supplies, can influence the body condition, migration dates and subsequent breeding success of individual migrants. Similarly, poor weather or stress during breeding can lower the body condition of breeders, and reduce their subsequent survival. In addition, the numbers of young produced in the breeding range could, through density-dependent processes, affect subsequent overall mortality in the non-breeding range. Events in breeding, migration and wintering areas are thus interlinked in their effects on bird numbers. Such effects are apparent at the level of the individual, and at the level of the population.

Some species breed on continents, but winter on islands or other small areas. The limited carrying capacity of such areas could in turn limit the sizes and distributions of breeding populations. In many bird species, probably through male dominance, the sexes occupy partly different winter ranges and winter habitats. These differences may contribute to the surplus of males apparent in many ducks, passerines and others in their breeding areas, with nesting populations limited by the numbers of females.

Bewick's Swans *Cygbus columbianus*, which migrate from western Europe to northern Siberia and are greatly influenced by conditions at crucial staging sites

Chapter 27

Population limitation – conditions on stopover

It must not be supposed that these journeys are free from dangers. Far from it; the perils are many and varied. (William Eagle Clark 1912.)

While the numbers of some migratory bird species are apparently influenced primarily by conditions in breeding or wintering areas (Chapter 26), the numbers of others could be influenced by conditions experienced on migration, mainly at stopover sites, but also during the flights themselves. It would indeed be surprising if bird breeding numbers were unaffected by conditions on migration. The process can in some species occupy several weeks or months each year (including stopovers), outward and return journeys together taking more than six months in extreme cases. Substantial mortality is therefore likely to occur during this period, and annual variations in this mortality could be reflected in annual variations in subsequent breeding numbers. In addition, in order to accumulate the body fat and other reserves necessary to fuel migration, and sustain themselves over the flight periods involved, birds need to obtain more food per day than usual. Although they have various ways to increase their energy consumption, such as change of diet, gut structure or daily foraging routine, their fattening rates are still constrained by intake rates or digestive capacities (Chapter 5). Moreover, because the same stopping sites can be used by large numbers of birds at a time, both solitary and flocking species, densities are often high and competition intense, resulting in severe depletion of food supplies (see later). The problems are magnified at those staging sites where birds must accumulate the extra large reserves necessary to cross an 'ecological barrier', such as a long stretch of water or desert. Hence, obtaining sufficient food in the limited time available could be a major constraint on the timing and success of migration.

The potential for population limitation on staging areas is perhaps especially acute in shorebirds and waterfowl, which in many regions have only a limited number of possible refuelling sites, often at widely spaced intervals along the migration route, but each holding very large numbers of birds. This situation contrasts with many landbird species, which migrate through mainly favourable habitat, and would seem to have feeding opportunities at many places along their route. However, the quality of any stopover site depends not just on the available food supplies, and levels of competition, but also on the security the site offers against predation, disturbance and other threats (for trade-offs between feeding rate and predation danger see Lindström 1990, Cresswell 1994, Ydenberg *et al.* 2002).

Events at stopover sites may thus affect not only the migratory performance of birds, but also their subsequent reproduction or survival, with potential consequences on population levels. In this chapter, therefore, I examine whether conditions at stopover sites can influence: (a) refuelling rates and migration speeds; (b) subsequent reproduction or survival; and (c) eventual population level or trend. Together, these various aspects encapsulate the subject area of 'stopover ecology' (for previous reviews of particular aspects of this subject see Moore & Simons 1992, Moore *et al.* 1995, Drent *et al.* 2003, Jenni & Schaub 2003, Newton 2005, 2006a). Consideration of the effects of storms and other extreme weather events that are sometimes experienced by migrants is left for Chapter 28.

CONCEPTUAL MODELS

Developing the models in the previous chapter, imagine a population in which both reproduction and mortality are density dependent, as shown in **Figure 27.1a**. The place where the lines depicting these relationships intersect marks the equilibrium population level (E), to which numbers tend to return after any perturbation. If mortality during migration is also density dependent, and additive to other mortality, the equilibrium population level (E^1) is reduced as shown. The same holds if either reproduction or mortality is density independent. In **Figure 27.1b**, for example, breeding is shown as density dependent and mortality as density independent. Whether migration mortality is density dependent or density independent, it can reduce subsequent breeding numbers only if it is additive to other mortality, and not compensated by reduction in some other form of mortality before the breeding season. Where mortality in the non-breeding season is

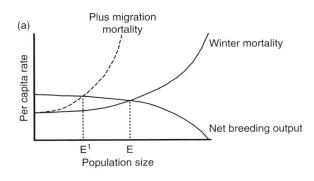

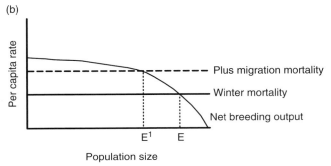

Figure 27.1   Model depicting the relationship between per capita reproduction and per capita mortality in a population. Where the lines cross marks the equilibrium population size (E). In (a), both reproduction and mortality are density dependent, as is additional migration mortality, which reduces the equilibrium population size (E^1). In (b), reproduction is density dependent while mortality is density independent, as is additional migration mortality, which again reduces the equilibrium population size.

density dependent (as found in some migratory species, Newton 1998b), mortality on autumn migration could be offset by reduced winter losses, and have no effects on subsequent breeding numbers. In contrast, mortality on spring migration leaves little or no time for any such compensation to occur before breeding begins.

The above models concern effects at the population level, but events on migration can also influence the subsequent performance of individuals, as carry-over effects. An important element of competition among migrants concerns the timing and speed of the journey, especially in spring. Within populations, those individuals that arrive in breeding areas in the early part of the arrival period usually take territories in the best habitat (identified as such from previous work in the area), begin breeding first, and show the highest nest success (Chapter 14). Later arrivals are relegated to poorer habitat, and may even fail to acquire a territory or a mate, so cannot reproduce that year. Similar pressures may be assumed to affect arrival in wintering areas, if the first arrivals take the best habitats, and thereby have greater chance of surviving the non-breeding season (for American Kestrel *Falco sparverius* see Smallwood 1988, for Ruddy Turnstone *Arenaria interpres* see Whitfield, in Wernham *et al.* 2002). These various considerations give rise to the proposed model of stopover ecology depicted in **Figure 27.2**, which links individual migratory performance to subsequent survival or breeding success. In due course, changes in the performance of individuals may lead to change in population size, the most difficult link to establish in migratory birds.

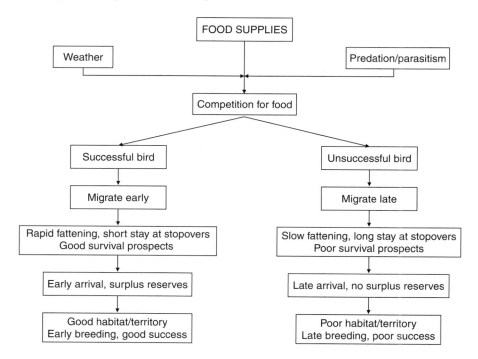

Figure 27.2   Flow diagram showing the hypothetical links between spring food supply, migratory performance, arrival in breeding area and subsequent breeding success. The two columns show the extremes in a spectrum of variation between individuals. From Newton (2005).

FOOD LIMITATION AT STOPOVER SITES

The following findings have been used to support the view that food supplies at stopover sites can limit the migratory performance and numbers of birds (Newton 2006; see also **Appendix 27.1**).

Migrant numbers in relation to food supplies

As at breeding and wintering areas, bird densities at stopover sites are often found to correlate with local food supplies, either from place to place or time to time. At some sites, the numbers of migrants vary between migration seasons, or during the course of a single season, in relation to changing food availability. For example, Rufous Hummingbirds *Selasphorus rufus*, which breed in the northwest of North America, migrate southward in autumn, stopping to feed on nectar from flowers (especially *Castilleja linariaefolia*) that grow in mountain meadows (Russell *et al.* 1992). At each stopover, the birds establish individual territories around flower patches. Competition is usually intense, and not all birds manage to acquire territories. Once a bird has a territory, it takes several days to three weeks to reach an appropriate departure weight and move on, leaving its territory vacant for another individual. During seven years of study, the densities of nectar-providing flowers at a study site varied widely due to natural variation in flowering. Territory sizes and population densities of hummingbirds varied accordingly, both between and within years. During dry years of generally poor flowering, the body weights of incoming birds were low, and stopover durations were long. The peak migration date of hummingbirds roughly coincided with the peak date of local flowering, and both events varied by about a month among years, poor nectar supply delaying migration. These findings thus revealed effects of food supplies at stopover sites on the local densities, behaviour, fattening patterns, staging periods and migration dates of hummingbirds. Instances of hummingbirds starving while on migration were said to be 'not uncommon'.

Relationships between food supplies and migrant numbers have been shown in other species, and in widely different situations (e.g. Martin & Karr 1986, Kelly *et al.* 2002, van Gils *et al.* 2005). In one study in the Arizona mountains, the relative numbers of insectivorous passage migrants in different habitats varied between seasons, in accordance with seasonal changes in insect densities (Hutto 1985). In this area, 54% of bird species showed shifts in habitat use between seasons which matched the changes in insect availability. At other sites, the numbers of birds have increased or decreased over the years in association with changes in food supplies, as noted for example in various species of cranes and waterfowl (Chapter 21). Such relationships confirm that migrants can respond to food supplies en route, concentrating in greatest densities at times and places where food is plentiful. They also indicate that, at some sites, the numbers present at one time could be limited in relation to food supplies.

Depletion of food supplies

Marked declines in food supplies, mainly through depletion, have been measured at stopover sites during the migration season. In some studies, this was done by excluding birds from some places, and comparing the trends in prey populations inside and outside the exclosures. For example, during the shorebird passage (July–September) in Massachusetts, 7–90% declines in different invertebrate species were recorded, mainly due to shorebird predation (Schneider & Harrington 1981). Similarly, following spring migration over the Gulf of Mexico, some passerine migrants were found to depress woodland insect supplies rapidly by up to 67% at coastal stopover sites (Moore & Yong 1991). Depletion by migrants of some other types of foods, such as berry crops, can be observed each year by even the most casual of observers (for case studies see Parrish 2000 and Ottich & Dierschke 2003). There can be no doubt, therefore, that some birds can seriously deplete their food supplies at stopover sites, both in autumn and in spring. Some food supplies, such as the plant leaves eaten by geese in spring, undergo repeated cycles of depletion and re-growth, a system which still limits the numbers of individuals that can feed effectively at a site at one time (Prop 1991).

This type of information is not necessarily useful without corresponding measures of bird performance, such as feeding or fattening rates. This is because, although food stocks might be estimated accurately, we usually have no means of telling what proportion is available to the birds, or how much they could remove before being disadvantaged. Nevertheless, depletion measures are sometimes useful in showing how close birds are to a food limit. If it is found that birds remove almost all their favoured foods from a site, and that no apparent alternatives are available locally, then a big increase in the numbers of migrants able to fatten at that site is clearly not possible. At the other extreme, if food is replenished as it is removed (say by growth, reproduction, or immigration of food organisms), then depletion is probably not an issue.

Some researchers have taken a different approach and measured feeding rates, rather than food itself. For example, depletion of food supplies by Pied Flycatchers *Ficedula hypoleuca* in autumn was inferred from declining insect capture rates the longer a bird spent in a tree, and by increasing feeding rates with time since the last visit to a given tree (Bibby & Green 1980). Competition at stopover sites can come not only from conspecifics, but from other species taking the same foods, the resulting depletion lowering the feeding rates of several species at once (Moore & Yong 1991). Where a food supply is depleted by passage migrants, latecomers, arriving after most of the food has been eaten, could be penalised by reduced rates of feeding and fat accumulation. Bewick's Swans *Cygnus columbianus bewickii* were studied at a staging site in the White Sea, the last stop on spring migration before the Siberian breeding grounds (Nolet & Drent 1998). In this locality, the swans could obtain their main food, tubers of Fennel Pondweed *Potamogeton pectinatus*, by upending to reach the sea-bed, mainly during low tide. In the course of the staging period, the swans tended to forage at progressively lower water levels, indicating that they gradually depleted this food supply, and exploited increasingly deeper parts of the tuber bank as the days went by. This depletion reduced the swans' main foraging period from 6.0 hours per tide on 20 May to 3.3 hours per tide on 28 May. The authors calculated that this must have greatly reduced the refuelling rate during the staging period. Accordingly, swans arriving late stayed longer than those arriving early. It seemed important for the swans to arrive at the stopover site as soon after ice breakup as possible, for a month later, the tubers were greatly depleted, and any remaining began to sprout. The first swans to arrive could also leave the site first, and (in theory) reach the breeding grounds earliest, get the best territories and achieve the highest breeding success. The latest swans seen to leave this site would have arrived in the nesting areas too late to breed that year.

This study hinted at how competition for limited food supplies at a stopover site, used for no more than a few weeks each year, could have helped to regulate the swan population. The White Sea provides the only sizeable stopover site for swans on this part of the spring migration, so with severe food depletion there, the birds would be limited in how much they could respond to any improvements in conditions that might occur in breeding or wintering areas.

Food supplies and fattening rates

The importance of food availability at stopover sites is further shown by the following types of findings, drawn from a range of different studies:

1. Birds were more likely to stay at sites where food was plentiful, and move on rapidly from sites where food was scarce (Bibby & Green 1981, Spina & Bezzi 1990, Ottich & Dierschke 2003).
2. When birds stayed at a site, their refuelling rates (as judged by weight gains) were often correlated with spatial and temporal variation in food supplies

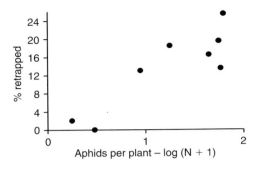 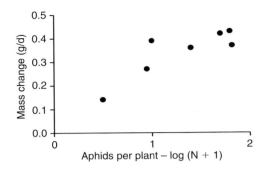

Figure 27.3   Relationship between food supply (aphid density) at a stopover site: (left) tendency to stay; and (right) rate of migratory weight gain in Sedge Warblers *Acrocephalus schoenobaenus*. From Bibby & Green (1981).

(**Figure 27.3**; Bibby *et al.* 1976, Bibby & Green 1981, Cherry 1982, Piersma 1987, Prop & Deerenberg 1991). In one study involving Red Knots *Calidris canutus*, fuelling rates increased with latitude, in parallel with the greater 'harvestable biomass' available at higher latitudes in spring (Piersma *et al.* 2005).

3. Mean stopover durations were inversely correlated with food supplies, as birds put on weight more slowly where food was scarce than where food was abundant (**Figure 27.3**; Piersma 1987, Russell *et al.* 1992).

4. Birds that arrived at particular sites with low body reserves stayed longer than those that arrived at the same sites with larger reserves (Dolnik & Blyumental 1967, Cherry 1982, Bairlein 1985a, Pettersson & Hasselquist 1985, Biebach 1985, Biebach *et al.* 1986, Moore & Kerlinger 1987, Dunn *et al.* 1988, Serie & Sharp 1989, Loria & Moore 1990, Ellegren 1991, van Eerden *et al.* 1991, Kuenzi *et al.* 1991, Morris *et al.* 1994, Yong & Moore 1997).

5. Some populations could not accumulate body fuel for spring migration until food availability increased in some way (for Whimbrel *Numenius phaeopus* see Zwarts 1990).

6. Fuel deposition rates or body weights declined in cold or wet weather expected to reduce feeding rates (Ash 1969, Ormerod 1989, Schaub & Jenni 2001b).

7. The provision of supplementary food to migrants in the field led to increased rates of fuel deposition compared to rates in unfed birds (for Bluethroat *Luscinia svecica* see Lindström & Alerstam 1992, for Greater Whitethroat *Sylvia communis* see Fransson 1998, for Robin *Erithacus rubecula* see Danhardt & Lindström 2001, for Garden Warbler *Sylvia borin* see Bauchinger 2002).

8. Rates of fuel deposition most commonly reported (1–3% of lean mass per day) were much lower than the maximum rates (>7% of lean mass per day) recorded from individuals in the same population. This finding implied that most individuals fed at rates too low to fuel at the maximum possible rate (Lindström 2003).

Whether all this matters depends on how fuelling rates affect migration timing and subsequent performance. Some birds are apparently unable to accumulate sufficient body reserves in time to migrate at an appropriate date. They may then remain in wintering areas (e.g. Dijk *et al.* 1990), or at stopover sites (e.g. Spendelow 1985), or in breeding areas where they may die as winter approaches (e.g. Kolunen & Peiponen 1991). Such findings emphasise the extreme consequences of inadequate food supply to successful migration.

In general, migrants seem to stay at a site for some days if they have a good chance of replenishing their depleted body reserves there, but move on if conditions are unfavourable for replenishment, presumably in search of more suitable habitat elsewhere (Rappole & Warner 1976, Biebach 1985, Kuenzi *et al.* 1991). This inference is supported by feeding experiments in wild birds (for Northern Wheatear *Oenanthe oenanthe* see Dierschke *et al.* 2003), and by laboratory experiments with Spotted Flycatchers *Muscicapa striata* and Garden Warblers *Sylvia borin*, in which migratory restlessness was greatly reduced during periods of weight gain (Biebach 1985, Gwinner *et al.* 1985b). More specifically, birds with high fuel reserves, or with low fuel reserves and insufficient food, showed migratory restlessness, whereas birds with low food reserves and abundant food did not (Chapter 12). In natural conditions, stopover duration depends not only on food levels, but also on weather (as birds wait for conditions favourable for migration), predation and disturbance, the arrival–departure schedules of other individuals, and the location of the stopover site with respect to a sea or other barrier. These other factors often influence the relationships between arrival weight and stopover duration, or rate of weight gain and stopover duration, and from many stopover sites departing birds can show a wide range of fuel loads. In landscapes with many potential stopping places, birds might pause and feed for a fairly constant period each day, and then fly whatever distance their reserves allow before they stop again, the durations of flights rather than the stopovers being the main variable in the journey (Schaub & Jenni 2001a).

Rates of feeding and weight gain are often greatest in lean birds, and then slow as body mass increases towards the departure value, a trend found in captive birds as well as wild ones (Rappole & Warner 1976, Kuenzi *et al.* 1991, Yong & Moore 1993, Moore 1994). This is partly because lean birds take greater risks than fatter ones: they may spend less time scanning for predators, feed for longer each day, more actively, and in more dangerous places (Metcalfe & Furness 1984, Loria & Moore 1990, Moore & Aborn 2000, Ydenberg *et al.* 2002). But the maximum rate at which food can be obtained and processed is still limited by the available feeding time (set by daylength, tidal rhythms or prey activity), and the type of food, and by bird features, such as crop capacity and digestive throughput (Chapter 5). The bird must also do things other than forage, all of which limit the rate of weight gain.

The assumption that long stopovers and associated low rates of weight gain slow the overall progress of migration is supported by observational data. For example, autumn migratory movement by six passerines between traps 12–50 km apart was 1.5–3.0 times faster in fat birds than in lean ones (Blyumental, in King 1972). When weight gain is slow at early staging sites, delaying departure, it could reduce fattening rates at later sites if food supplies have already been depleted there (the domino effect of Piersma 1987). On this basis, and in the absence of weather-induced delays, differences between individuals in start dates could be magnified during the course of a journey. Some studies at particular stopover sites have indicated a progressive lowering of body mass and fat levels among successive samples of birds caught during the course of the spring migration. Individuals that passed through earlier in the season were heaviest, and those that passed through latest were lightest: as for example, in three out of four thrush species caught in spring at a stopover site in coastal Louisiana (Yong & Moore 1997), and among Steppe Buzzards *Buteo b. vulpinus* and Levant Sparrowhawks *Accipiter brevipes* caught in spring at a stopover site in Israel (Yosef *et al.* 2002, 2003).

The same has been found for birds arriving in their breeding areas, with the earliest arrivals being the heaviest, as in Barn Swallows *Hirundo rustica* (Møller 1994), Garden Warblers *Sylvia communis* (Widmer & Biebach 2001), American Redstarts (Smith & Moore 2005), and Common Terns *Sterna hirundo* (Dittman & Becker 2003). Such trends could result from the effects of competition relegating lighter birds to the later part of the spring migration period. However, not all species would be expected to show decline in the mean weights of samples taken through the spring migration season: at some stopover sites food supplies increase through spring (through growth or reproduction of prey items), and bird densities decline, as passage comes to an end. Both these changes could provide enhanced feeding conditions for the later migrants (for greater fattening rates in the latest of two godwit populations to pass through the Netherlands in spring, see Drent & Piersma 1990).

Social interactions

Dominance relationships influence the outcome of competitive interactions at stopover areas, just as in breeding or wintering areas, even though the individuals concerned may be present for only short periods (Pienkowski & Evans 1985). In some species, dominant individuals hold short-term territories at stopover sites, and subordinate individuals unable to obtain territories must feed in less good places or leave (Rappole & Warner 1976, Kodric-Brown & Brown 1978, Bibby & Green 1980, Sutherland *et al.* 1982, Hixon *et al.* 1983, Veiga 1986, Sealy 1989, Carpenter *et al.* 1993). Territorial Northern Waterthrushes *Seiurus noveboracensis* were found to accumulate body reserves at a rapid rate, whereas non-territorial ones gained no weight until after they had acquired a feeding territory (Rappole & Warner 1976). The same was found among European Robins *Erithaca rubecula* (Mehlum 1983). Similarly, in non-territorial species, dominants had higher feeding and fuel deposition rates than subordinate conspecifics (for Bluethroat *Luscinia svecica* see Lindström *et al.* 1990, Ellegren 1991, for Brent Goose *Branta bernicla* see Prop & Deerenberg 1991). At similar environmental food levels, average fattening rates were lower at high than low densities of individuals (Moore & Yong 1991, Kelly *et al.* 2002). There is thus no doubt that birds compete for food at stopover sites, and that the fuelling rates of some individuals are thereby lowered. Competition may also help to regulate local densities in relation to resource levels, and result in some birds making longer stopovers, or leaving with lower fat reserves than others, reducing the distance they could travel before the next stop.

Age and sex effects

In many species, adults trapped at stopover sites put on weight more rapidly, achieved greater body reserves, and stayed for shorter periods than first-year birds; and within age groups, males put on weight more rapidly, achieved greater body reserves, and stayed for shorter periods than females.[1] These differences may have arisen because young birds were at a competitive disadvantage in the presence of older ones, and females in the presence of males. Juveniles may also have foraged less efficiently than adults, resulting in generally slower fuel accumulation and migratory progress, with longer or more frequent stopovers (for Dunlin *Calidris alpina* see Rösner 1990, for cranes and raptors see Ueta & Higuchi 2002). In some species in which females are bigger than males, the sex-related difference in fuelling rates was reversed, as females at particular sites tended to have shorter stopovers (Figuerola & Bertolero 1998, Butler *et al.* 1987).

Such age and sex differences could influence the travel speeds, departure and arrival dates, and survival chances of individuals during migration. It is hard to tell how much they result from competition acting here and now at stopover sites, or from genetic age-linked and sex-linked differences in migration strategies that have evolved as a result of competition and other differential selection pressures in the past. However, age and sex differences are not apparent in all species, or at all stopover sites (Morris *et al.* 1996, Maitav & Izhaki 1994). Occasionally, juveniles are found to be heavier than adults on average (Butler *et al.* 1987, Alerstam & Lindström 1990, Woodrey 2000), possibly because juveniles migrate at a more favourable time of year (earlier in autumn or later in spring), so may have the advantage of a better food supply than adults.

In conclusion, increasing bird densities at stopover sites have been shown to intensify competition, reducing food availability through depletion or interference, lowering food intake and fat deposition rates, and thereby slowing migration. Some individuals are affected more than others. Interference competition may also redistribute birds among habitats, with younger less experienced migrants forced into sites where feeding rates are lower. While competition for food may limit the number of birds that can fatten simultaneously at particular sites, it may have influence on the overall population level only if displaced birds find no alternative places to fatten, or migrate more slowly with adverse consequences.

Poor condition and mortality at stopover sites

Birds caught soon after their arrival at stopover sites often show great variation in body condition. At some sites, it is not uncommon to find a two-fold variation among the body weights of individuals of similar size (e.g. Moreau & Dolp 1970). Some individuals seem on the point of starvation, devoid of body reserves, while others have enough residual fuel for several hundred kilometres of further flight. Over three different years, some 39% of 1903 thrushes (*Catharus, Hylocichla*) caught on arrival in a Louisianan coastal woodland after crossing the Gulf of Mexico carried no obvious body fat. The body weights of 23% of these thrushes were at or below the estimated fat-free body mass; while only 5% had estimated

---

[1]*For age differences see Veiga (1986), Serie & Sharp (1989), Ellegren (1991), Carpenter* et al. *(1993), Holmgren* et al. *(1993), Gorney & Yom-Tov (1994), Morris* et al. *(1996), Woodrey & Moore (1997), Lindström* et al. *(1990), Ellegren (1991), Yong* et al. *(1998), Woodrey (2000), Heise & Moore (2003); for sex differences see Morris* et al. *(1994), Otahal (1995).*

fat stores that exceeded 20% of lean body mass (Yong & Moore 1997). Between 21 March and 9 May 1987, 13 individuals of five migrating species were found dead in wooded feeding habitat, apparently having starved, and another 66 corpses of 18 species were found on daily walks on a 2 km stretch of beach (Moore *et al.* 1990). Other instances of birds arriving in Panama in autumn in very poor condition (below fat-free mass) were given by Rogers & Odum (1966), and other severely emaciated birds were collected from oil drilling platforms in the Gulf of Mexico (Gauthreaux (1971, 1999), from ships at sea (Serle 1956, McClintock *et al.* 1978), and from an island of southern Italy (Baccetti *et al.* 1981, cited in Pilastro & Spina 1997).

Birds weakened by starvation are more easily caught by predators, and may account for some of the raptor kills found at stopover sites, among which young birds often predominate (Bijlsma 1990). Many other records of exhausted and starving migrants are mentioned incidentally in the ornithological literature, often in association with sea- or desert-crossings (for hirundines and others see Paynter 1953, Spendelow 1985; for other passerines see Pilastro & Spina 1997; for shorebirds see Dick & Pienkowski 1979, Bijlsma 1990; for raptors see Smith *et al.* 1986). They confirm that some birds die of starvation during the course of their journeys.

A striking example was provided by the 200–300 Chimney Swifts *Chaetura pelagica* which in October 1979 stopped and starved to death on the Caribbean Islas del Cisna (Spendelow 1985). All collected specimens were emaciated, with no visible body fat, and weighed around 40% less than healthy Chimney Swifts in spring and summer. Totalling less than 4 km^2 in area, and lying about 180 km north of Honduras, these two islands were judged too small and isolated to provide sufficient food, and 21 migrants of various other species were also found dead there in the spring and autumn that year, including Blue-winged Teal *Anas discors* and Cattle Egret *Bubulcus ibis*. Such events may occur frequently on small islands and other potential stopping places offering insufficient food. Spendelow (1985) cites B. Robertson that larger birds, such as Cattle Egrets, may survive for weeks on the Dry Tortugas on inadequate food, with as many as 100 dying slowly through the summer on an island group of only 40 ha. Such instances provide examples of birds unable to continue migration, and eventually dying, because of inadequate food at stopover sites.

It is common for newly arrived birds, especially of long-distance migrants, to lose mass for another day or two before they begin to regain it (Rappole & Warner 1976, Moore & Kerlinger 1987). Explanations for this initial weight loss include: (a) effects of capture and handling (Nisbet & Medway 1972), (b) inefficient foraging because of unfamiliarity with the stopping site (Yong & Moore 1997), (c) rebuilding of the digestive system in birds that have undergone long non-stop flights (Chapter 5; Piersma & Gill 1998), and (d) competition among conspecifics (Rappole & Warner 1976, Moore & Yong 1991).

BODY CONDITION AND SUBSEQUENT PERFORMANCE

Many birds have been trapped, ringed and weighed at migration sites, enabling their subsequent survival rates to be assessed from later surveys of the population. Several studies have revealed a link between the body condition of

Table 27.1 Examples of associations between body weights at migration time and subsequent re-sighting probability (reflecting survival)

	Evidence	Reference
Autumn		
Semi-palmated Sandpiper *Calidris pusilla*	Departure weight from a stopover site estimated from a previously established relationship between stopover duration and departure weight. Estimated heaviest birds were more likely to be seen at the same site in a later year	Pfister *et al.* (1998)
Mallard *Anas platyrhynchos*	Birds trapped, weighed, ringed and released soon after arrival at a wintering site. Lightest birds were more likely to be shot in the ensuing months	Dufour *et al.* (1993)
Lesser Scaup *Aythya affinis*	Birds trapped, weighed, ringed and released soon after arrival at a wintering site. Lightest birds were more likely to be shot in the ensuing months	Pace & Afton (1999)
American Redhead *Aythya americana*	Birds trapped, weighed, ringed and released soon after arrival at a wintering site. Lightest birds were more likely to be shot in the ensuing months	Bain (1980)
Canvasback *Aythya valisineria*	Birds trapped, weighted, ringed, and released soon after arrival at wintering site. Lightest birds were more likely to be found dead in the ensuing months	Haramis *et al.* (1986)
Spring		
Great Reed Warbler *Acrocephalus arundinaceus*	Birds trapped, weighed, ringed and released at a migration/wintering site. Heaviest birds were more likely to be recaught at the site in a later year	Nisbet & Medway (1972)
Red Knot *Calidris canutus*	Birds trapped, weighed, ringed and released at a spring stopover site. Those that were above average weight were most likely to survive poor spring weather in breeding areas	Wilson (1988)
Red Knot *Calidris canutus*	Birds trapped, weighed, ringed and released at a spring stopover site. Heavy individuals were more likely to be re-sighted in subsequent years than lighter ones. In years when many birds failed to achieve normal departure weight, mean survival of ringed birds declined (by 37%)	Baker *et al.* (2004)
Bar-tailed Godwit *Limosa lapponica*	Birds trapped, weighed, ringed and released at a stopover site. The heaviest individuals (with the most advanced plumage development) were most likely to be recaught at the site in a subsequent season	Drent *et al.* (2003)
Pink-footed Goose *Anser brachyrhynchus*	Birds at a spring stopover site judged to be in good condition (from the abdominal profile) were more likely than poor condition birds to be later re-sighted in a wintering or migration area	Madsen (1995), Drent *et al.* (2003)
Barnacle Goose *Branta leucopsis*	Birds at a spring stopover site judged to be in good condition (from the abdominal profile) were more likely than poor condition birds to be later re-sighted in a wintering or migration area	Prop & Deerenburg (1991)

individual birds at migration sites, whether autumn or spring, and their subsequent survival or, more strictly, their re-sighting rates (**Table 27.1**). In some such studies, searches for marked birds were not restricted to the capture site in subsequent migration seasons, but were made at several different points along the migration route, allowing the whole population to be surveyed thoroughly on repeated occasions in subsequent years. In these studies, therefore, we can be confident that lower re-sighting rates reflected poorer survival (as concluded by the authors).

Among various species of ducks, individuals which were light in weight when banded at autumn stopover sites, or soon after arrival in wintering areas, survived less well than individuals in good condition. After allowing for age and sex differences in body weight, such variation was evident in Mallards *Anas platyrhynchos* (Hepp *et al.* 1986, Dufour *et al.*1993), Redheads *Aythya americana* (Bain 1980), Canvasbacks *A. valisineria* (Bain 1980, Haramis *et al.* 1986) and Lesser Scaups *Aythya affinis* (Pace & Afton 1999). In adult male Canvasbacks, lower survival of poor condition birds was due to natural factors (because the birds were not hunted at the time, Haramis *et al.* 1986), but in the other species it mainly resulted from greater proportions of poor condition birds being shot within the next few months. Relationships between body weight at migration time and subsequent survival were also noted among passerines, shorebirds and geese (**Table 27.1**).

Among geese and other waterfowl, food availability at wintering and spring migration sites has long been known to affect body condition and subsequent breeding success (**Table 27.2**). In a few species, the mechanisms have been studied. For example, among Brent Geese *Bernicla branta* in the Netherlands, females that had accumulated the greatest body reserves at a spring stopover site were more likely to return with young in the following autumn than were females that had accumulated smaller reserves, whereas males, which accumulated generally smaller reserves than females, showed no such relationship (Ebbinge & Spaans 1995). The favoured spring staging habitat was saltmarsh where nutrient-rich plants allowed the geese to fatten rapidly. However, the number of geese that could feed in saltmarsh was limited, so as the population grew over a period of years, increasing proportions of birds were relegated to less nutritious agricultural grassland. The geese used body reserves accumulated in spring for migration and reproduction, and individuals that had fed on saltmarsh showed better breeding success than those that had fed on grassland (Ebbinge 1992).

In some other species of geese, smaller proportions of females laid, and clutches were smaller, in years when feeding conditions in staging areas were poor than in years when they were good (for Barnacle Geese *Branta leucopsis* see Cabot & West 1973, for Lesser Snow Geese *Chen c. caerulescens* see Davies & Cooke 1983). These various studies show that the breeding success of some migratory geese depends partly on body reserves accumulated in wintering and spring staging areas, and that the effects of such reserves are evident both at the level of the individual and at the level of the population. Breeding success in some ducks is also influenced by body reserves, accumulated partly before and partly after arrival in breeding areas (e.g. Mosbech *et al.* 2006). The females of some species at high latitudes, such as Northern Pintail *Anas acuta* and Steller's Eider *Polisticta stelleri*, start laying within days after arrival in breeding areas, and so may be assumed to have accumulated the necessary body reserves elsewhere (Arzel *et al.* 2006).

Table 27.2 Examples of associations between body weight at migration time and subsequent breeding

	Evidence	Reference
American Redstart *Setophaga ruticilla*	Birds weighed soon after arrival in breeding areas. Heavier individuals laid earlier, and had larger clutches, egg volumes and chick weights	Smith & Moore (2003)
Dusky Warbler *Phylloscopus fuscatus*	Birds weighed soon after arrival in breeding areas. Old males and males with high body mass had a greater chance of mating polygynously, while first-year males and males of low body mass often remained unmated	Forstmeier (2002a)
European Pied Flycatcher *Ficedula hypoleuca*	Birds weighed soon after arrival in breeding areas. Early-arriving females were heaviest, with greater body reserves, laid earlier and had high hatching success	Lundberg & Alatalo (1992)
Red Knot *Calidris canutus*	Birds at stopover sites in best condition produced most young	Morrison (2006)
Pectoral Sandpiper *Calidris melanotos*	Birds weighed after arrival in breeding areas. Females that arrived early, with the largest body reserves, laid the largest eggs	Farmer & Wiens (1999)
Mallard *Anas platyrhynchos*	Birds caught soon after arrival in breeding areas. Females in best body condition laid earlier, larger clutches	Krapu (1981), Pattenden & Boag (1989)
Lesser Snow Goose *Chen c. caerulescens*	Birds shot after arrival in breeding areas. Heavier females, with greatest body reserves, had more developing egg follicles (taken to indicate larger clutches). Average spring body condition at a stopover site from year to year predicted autumn young; old ratios	Ankney & MacInnes (1978), Alisauskas (2002)
Greater Snow Goose *Chen caerulescens atlantica*	Birds at spring stopover site judged to be in good condition (from abdominal profile) laid earlier and larger clutches	Béty et al. (2003), Reed et al. (2004)
Pink-footed Goose *Anser brachyrhynchus*	Birds at spring stopover site judged to be in good condition (from abdominal profile) were most likely to produce young	Madsen (1995, 2001), Drent et al. (2003)
Barnacle Goose *Branta leucopsis*	Birds at spring stopover site judged to be in good condition (from abdominal profile) were most likely to produce young	Prop & Black (1998)
Brent Goose *Branta bernicla*	Birds at spring stopover site judged to be in good condition (from abdominal profile) were most likely to produce young	Ebbinge & Spaans (1995)
Tundra Swan *Cygnus columbianus*	Birds assessed at spring stopover site. Early individuals with high feeding rates were most likely to produce young	Nolet & Drent (1998), Beekman (2005)
Black Kite *Milvus migrans*	Birds weighed soon after arrival in breeding areas. Individuals in best condition occupied the best territories, laid earlier, and raised most young	Sergio et al. (2007)

Among shorebirds nesting in the High Arctic, the extent to which body reserves are available for egg production may depend on climatic and other conditions in particular years, and may differ between protein and lipid. In one study, isotope analysis revealed that in several species egg protein was formed from terrestrial rather than marine foods, and hence was influenced by food eaten after arrival in breeding areas (Klaassen *et al.* 2001). On the other hand, in another study, eggs in the earlier clutches of Red Knots *Calidris canutus* and Ruddy Turnstones *Arenaria interpres* in the northeastern Canadian High Arctic were rich in ^{13}C and ^{15}N, which suggested that some residual marine nutrients were used in their production (Morrison & Hobson 2004).

Some studies of passerines have also suggested that conditions in wintering areas can influence both the timing of spring migration, and also the timing and success of reproduction (Chapter 26; for American Redstart *Setophaga ruticilla* see Marra *et al.* 1998, for Black-throated Blue Warbler *Dendroica caerulescens* see Bearhop *et al.* 2004, Norris *et al.* 2004). In such small species, the weight of a clutch is so large, relative to body weight, that body reserves could provide at most a tiny proportion of the material necessary for egg formation. Presumably they serve some other function, such as maintenance, and the nutrients required for eggs must come mainly from food eaten at the time (Sandberg & Moore 1996). However, both sexes of American Redstarts arrived to breed in northern Michigan with surplus fat, but females arrived with more than males in two out of three years (Smith & Moore 2003). Individuals of both sexes, but especially females, that had more body fat on arrival also showed higher reproductive success, with larger clutch sizes, egg volumes and nestling weights than birds arriving with little or no fat. In cases like this, where the reserve is small, it is hard to tell whether the association with subsequent breeding success reflects a cause–effect relationship, or merely a correlation, with birds that achieve good body condition on migration also achieving good body condition at other times, and hence showing high survival and reproductive success.

All the studies known to me that have related body condition at migration times to subsequent survival or breeding success have shown clear, positive relationships, although their numbers are still small. Possibly other species showed no such relationships, and have therefore less often been reported in the ornithological literature. However, if we are to assess their frequency, it is important that studies showing no relationships are published, so that they can be assessed along with those that do.

INFLUENCE OF PREDATION, DISTURBANCE AND PARASITISM

Sometimes large numbers of predators accumulate at stopover sites, relative to the numbers of potential prey. This can make individual birds more vulnerable than usual. In addition, when accumulating body reserves, migrants often feed intensively, reducing vigilance and spending more time in places where the danger is greater. Accipiters and falcons are often seen catching birds at stopover sites (Rudebeck 1950, Kerlinger 1989, Lindström 1989, Moore *et al.* 1990). On the basis of studies at Falsterbo in south Sweden, Lindström (1989) estimated that raptors (mainly Sparrowhawks *Accipiter nisus*) removed 10% of all Chaffinches *Fringilla coelebs* and Bramblings *F. montifringilla* during the six-week period of autumn migration. This rate of predation was much greater than expected from the annual mortality rate in these species, if losses had been evenly spread through the year.

Although prey-birds can respond to the presence of predators through greater vigilance and selection of safer habitats, both these measures may reduce feeding rates (Metcalfe & Furness 1984, Lindström 1990). This is because continual scanning for predators takes time which could otherwise be used for feeding, while secondary habitats usually offer less food. Moreover, the weight increase associated with fuel deposition reduces lift-off speed and agility, supposedly making prey easier for a predator to catch (Chapter 5). Little wonder that some birds increase their vigilance and feed in safer places as they become heavier (Burns & Ydenberg 2002).

Disturbance at stopover sites, caused by natural predators or people, can have marked effects on the rates and extent of weight gain by migrants, and hence on subsequent survival or breeding success. Pink-footed Geese *Anser brachyrhychus* that migrate from the Netherlands to breed in Spitzbergen make a major feeding stop at Vesterälen in north Norway. For many years, this population had been increasing. From about 1993 on, however, local farmers began systematically to disturb geese from their grasslands. This continual harassment prevented most of the geese from accumulating adequate body reserves, as was apparent from their body shapes ('abdominal profiles'). In contrast, geese using small areas of undisturbed habitat still achieved large reserves. Progressively, the geese began to abandon this site in favour of a less-good site further south, their spring–summer mortality increased, and the mean production of young per pair declined. The population stopped increasing (Madsen 1995, Drent *et al.* 2003).

A second example of the influence of disturbance came from an experiment in Canada, designed to curb the population growth of the Greater Snow Goose *Chen caerulescens atlantica*. On the main spring staging areas beside the St Lawrence River, the geese had been protected since 1917, and their numbers had increased greatly, but in the 1990s a spring hunt was reinstated. The main effects of this hunting came not from the numbers killed, but from the effects of disturbance on the accumulation of body reserves and subsequent nest success. The migration and breeding behaviour of large samples of radio-tagged geese were compared between two non-hunting years and the first two years with spring hunting (Mainguy *et al.* 2002). In the non-hunting years, 85% of the 80 radio-tagged females identified on the spring staging areas in the St Lawrence River valley were subsequently found in breeding areas on Bylot Island, where 56% were known to have nested. By contrast, in hunting years, only 28% of 80 radio-tagged females identified in the St Lawrence valley reached the nesting areas on Bylot Island, and a mere 9% nested. The differences between years in these proportions were statistically significant (proportion present, $\chi^2 = 57.6$, $P < 0.001$; proportion nesting, $\chi^2 = 16.9$, $P < 0.001$). They were not due to loss of birds to spring hunting, because most of the missing radio-tagged birds turned up again in the autumn. Rather, many radio-marked geese had shortened their migration and stopped on Baffin Island, where they did not breed. Moreover, females shot near the nest in hunting years were lighter in weight ($F = 12.7$, df $= 37$, $P < 0.001$), and contained less breast muscle ($F = 12.3$, df $= 36$, $P < 0.001$)

and abdominal fat ($F = 6.1$, df $= 34$, $P < 0.02$) than some shot near the nest in earlier years with no spring hunt. Egg-laying in the colony was delayed about a week in both hunting years, and clutches were significantly smaller, compared with four previous years (Bêty *et al.* 2003). The implication was that spring disturbance at the St Lawrence staging sites reduced the feeding rates and body condition of geese which survived the hunt, adversely affecting their subsequent migration and reproduction.

In another study, the disturbance effects of autumn shooting of waterfowl were tested by setting up experimental reserves in two Danish coastal wetlands (Madsen 1995). Over a five-year period, these undisturbed reserves became important staging areas for waterfowl, increasing the national totals of several species. Hunted species increased the most in these reserves, some four- to 20-fold, while non-hunted species increased two- to five-fold. Furthermore, most quarry species stayed in the area for up to several months longer each winter than in earlier years. No declines in bird use were noted at other Danish sites still open to hunting, so the accumulation of birds in the reserves was attributed to the short-stopping of birds that would otherwise have migrated further south. In this and other studies, hunting disturbance emerged as a major factor influencing the migration and winter distribution of waterfowl, on both local and regional scales. But it is not just waterfowl that are susceptible to human disturbance. The mean body mass of Sandhill Cranes *Grus canadensis* dropped by more than 7% at a staging area in North Dakota after the hunting season was brought forward from November (when most of the cranes had left) to September (Krapu & Johnson 1990).

Other changes in the behaviour of migrant birds towards shorter stopovers at certain sites have been attributed to enhanced predation risk associated with the increasing numbers of Peregrine Falcons *Falco peregrinus* and other raptors as they recovered from the organochlorine pesticide impacts of earlier years. For example, the Strait of Georgia in British Columbia is a major autumn staging site for Western Sandpipers *Calidris mauri* on southward migration. Birds trapped on the extensive mudflats of the Fraser estuary were significantly heaver (by 10%) than others caught on the small mudflat of Sidney Island (Ydenberg *et al.* 2002). The weight difference could not be attributed to seasonal timing, age or sex effects, but was linked with vulnerability to predation. The open expanse of the Fraser estuary offered safety from avian predators, but a lower fattening rate, while the small Sidney Island was more dangerous, but offered a higher fattening rate. The inference was that sandpipers arriving in the Strait with little fat (and hence more rapid escape responses) chose to take advantage of the high feeding rate at small dangerous sites like Sidney Island, whereas individuals encumbered by higher fat reserves elected to feed in larger but safer sites such as the Fraser estuary. Large, open sites are safer because they make it difficult for raptors to approach undetected, giving the shorebird prey earlier warning and longer escape times. From 1985, as Peregrine numbers increased, average migratory body mass and stopover durations of Western Sandpipers *Calidris mauri* at Sidney Island fell steadily (Ydenberg 2004). An accompanying steep decline in sandpiper numbers at Sidney Island was accounted for by shortening stopovers (mean 8.4 days, falling to 2.7 days), rather than by fewer individuals using the site. Under greater danger from predation, these birds seem to have switched from a long stay/high fuelling strategy at this site to a short stay/low fuelling strategy, using only safer sites for further weight gain. The authors suggested that such behavioural adjustments could be widespread among shorebird species, and that predation could be a major factor shaping the migratory routes, timing and behaviour of shorebirds (Lank *et al.* 2003a).

Predation at stopover sites is not the only form of predation endured by migrants. Because many falcons hunt high above the ground, beyond the limits of human vision, predation on high-flying diurnal migrants could easily be underestimated. In many regions, Peregrines *Falco peregrinus* and other bird-eating falcons are so widespread that day-flying migrants could be under continual risk of attack. In addition, falcons and accipiters are frequently seen hunting from oil rigs and ships at sea, taking passing migrants (e.g. Ellis *et al.*1994), and large gulls are often seen chasing and catching tired migrants arriving low over water (Macdonald & Mason 1973, Riddiford 1978). Moreover, in the Mediterranean–North African region, Eleonora's Falcons *Falco eleonorae* breed during the autumn migration season and specialise on passage birds caught by day while on migration (Chapter 24). In the deserts of the Middle East, Sooty Falcons *Falco concolor* fill a similar role but, being smaller than Eleonora's Falcon, they take a narrower range of prey species. By flying at night, migrants can avoid this onslaught for part of their journey (owls not being known to hunt high-flying migrants), although in the Mediterranean region some birds may be taken by bats (Chapter 24).

Parasites are also likely to affect the migratory performance of birds, not only because their effects can be debilitating, but also because, when abundant (especially gut parasites), they can absorb a substantial part of the host's food intake. Their effects are tantamount to lowering the feeding rates of their hosts. Birds might pick up parasites at stopover sites that affect them later in the year. Migratory birds have sometimes been found to contain a greater range of parasites, such as haematozoa, than closely related resident species, a difference attributed to the exposure of migrants, as they pass through different regions, to a wider range of parasite species and their vectors (Bennett & Fallis 1960, Figuerola & Green 2000). Correspondingly, organs concerned with immune defence were found to be larger in migratory than in closely related non-migratory species (Møller & Eritzoe 1998). Many birds caught on migration have been found to harbour ticks, which can pass readily from bird to bird, and act as both reservoirs and vectors of pathogens. Interspecies transfer of avian haemosporidian parasites (*Haemoproteus* and *Plasmodium*) has been found to occur between resident and migratory species wintering in Africa (Waldenström *et al.* 2002), and the reactivation of latent *Borrelia* infections among Redwings *Turdus iliaca* was attributed to the stress of autumn migration (Gylfe *et al.* 2000). Similarly, in some North American species sampled on migration, individuals with blood parasites (mainly *Haemoproteus* and *Plasmodium*) had lower body weights or fat levels than uninfected ones (Garvin *et al.* 2006). These various parasites do not necessarily kill their hosts, but may do so in particular circumstances, or reduce their breeding success, and some of their effects could therefore be regarded as additional costs of migration. Migratory birds may also help to disperse disease organisms, transferring them from one region to another to infect different populations (Ricklefs *et al.* 2005). Avian flu (strain H5N1) is a topical example (Olsen *et al.* 2006), but others have included Lyme disease (Scott *et al.* 2001) and West Nile virus (Rappole *et al.* 2000, Owen *et al.* 2006).

Another known source of mass mortality among migratory waterbirds is botulism, caused by a neurotoxin produced by the bacillus *Clostridium botulinum*. This anaerobic bacterium grows well on rotting organic matter in shallow stagnant waters or mud during warm weather, and occurs in most parts of the world. Affected birds become paralysed and limp, and die from respiratory failure or drowning, but may otherwise appear in good condition (Locke & Friend 1987). Year-to-year losses from botulism are highly variable, but can be spectacular, affecting migratory waterfowl at moulting and stopover sites, especially in western North America. An estimated one million birds died from botulism at a lake in Oregon in 1925, 1–3 million at Great Salt Lake in Utah in 1929, and 250 000 at the northern end of Great Salt Lake in 1932 (Jensen & Williams 1964). Many other outbreaks have involved smaller numbers of birds, but totalled over wide areas or over a period of years they can be substantial, for example the 4–5 million waterfowl deaths attributed to botulism in the western USA in 1952 (Smith 1975, Locke & Friend 1987), or the 1.5 million in California alone during 1954–1970 (Hunter *et al.* 1970).

Although botulism is not infectious, its effects are exacerbated by droughts which cause waterbirds to concentrate in larger than usual numbers on remaining shallow wetlands. The birds become poisoned when they ingest toxin-producing bacteria in the bodies of invertebrates that form their food, including the maggots living in rotting vegetation and carcasses. Dead waterfowl themselves become host to more maggots and perpetuate the outbreak. Mortality ends when the weather cools, when the birds leave the site or switch to other foods, when flies stop breeding, or when water levels rise (Wobeser 1981). Although botulism hits waterfowl seriously only in relatively arid areas, and only in occasional years, its impacts can be so substantial that affected populations could take several years to recover from each serious outbreak, constituting 'the greatest drain upon western waterfowl due to any single natural agency' (Kalmbach & Gunderson 1934).

Various other bacterial and other disease agents and neurotoxins have also caused occasional mass mortality incidents among migrating waterbirds concentrated at stopover sites (for review, see Newton 1998b). Examples include the erysipelas that killed 5000 Eared (Black-necked) Grebes *Podiceps nigricollis* on Great Salt Lake in 1975, and the streptococcal infection that killed another 7500 grebes there in 1977 (Jensen & Cotter 1976, Jensen 1979). Another more recent example was an outbreak of the H5N1 strain of avian influenza that killed many waterfowl, including rare Bar-headed Geese *Anser indicus*, on Qinghaihu Lake in China in 2005 (Liu *et al.* 2005). Influenza viruses are found in a wide range of bird species, especially aquatic ones, including ducks, geese and swans, and gulls, terns and waders. The viruses are excreted into the water, so can easily pass from bird to bird. Some bird species are likely to be little affected, and so act as reservoirs and transmitters of the virus, while others are extremely vulnerable and quickly die.

As with predation, parasitism has been suggested as a selective factor influencing the timing of migration, the habitats used, and the routes taken. In particular, by migrating long distances, many wader species are able to remain year-round in habitats (such as arctic tundra and lower latitude coastlines) which have relatively few parasites (Piersma 1997). As yet, however, the evidence for migration-related effects of either predation or parasitism on the migratory timing, routes and strategies of birds is little more than suggestive.

EFFECTS OF STOPOVER EVENTS ON POPULATIONS

The above sections illustrate different types of evidence indicating that events at stopover sites can influence migratory bird populations. They show that food supplies at staging sites can be heavily depleted, slowing rates of fattening, which in turn can delay spring migration so much that it reduces breeding success, or prevents breeding altogether. Moreover, the intraspecific crowding, or the disturbance caused by predators or human hunters, can affect the behaviour and fattening rates of migrants, causing them to leave stopover sites prematurely and with lower fuel reserves than otherwise, with affects on their subsequent survival and breeding success. The most crucial question, however, is whether these effects on individuals are sufficient to reduce population sizes below what they would otherwise achieve. The mortality from botulism, although sometimes large scale, occurs mainly in late summer or autumn, so could be compensated to some extent by greater overwinter survival of unaffected birds.

In the Greater Snow Geese *Chen caerulescens atlantica* and Pink-footed Geese *Anser brachyrhychus* mentioned above, increased disturbance at a major spring stopover site led to reduced breeding output, and at least in Pink-footed Geese also to reduced survival. This would be expected to influence the subsequent population trend, and the Pink-footed Goose population has now ceased its long-term increase. In the Red Knot *Calidris canutus rufa* population that breeds in arctic Canada and winters in Tierra del Fuego, marked decline has also been tied to changed conditions at Delaware Bay, the last major spring staging site of knots en route to arctic breeding areas. Numbers staging there in spring fell from 51 000 to 27 000 individuals between 2000 and 2002. Decline coincided with collapse (through human overfishing) of the Horseshoe Crab *Limulus polyphemus* population, the eggs of which form the main food of knots at this site (Baker *et al.* 2004). From 1997 to 2002, increasing proportions of knots studied in the Bay failed to reach the usual departure mass of 180–200 g. Annual survival of adults fell by 37%, and the proportions of immature birds in wintering flocks fell by 47%. Of birds caught in the Bay, known survivors were heavier at initial capture than were birds not seen again. By 2004, the numbers of knots had fallen even further, along with those of Turnstone *Arenaria interpres* and Sanderling *Calidris alba*. However, a moratorium was introduced on crab-fishing during May–June; and enough crab eggs were produced that year to feed the reduced numbers of shorebirds, enabling them to fatten at the normal rate. Because Horseshoe Crabs do not breed until they are about 10 years old, the situation could remain precarious for several further years.

Recent problems at stopover sites may have influenced the population levels of some other species too. Widespread decline in the numbers of Lesser Scaups *Aythya affinis* in North America has been linked with females arriving on breeding areas in poorer condition than before (Anteau & Afton 2004). In breeding areas in Minnesota and Manitoba, mean body mass of females was about 8% lower in the years after 2000 than in the 1980s, and lipid reserves were 30% lower. Mineral reserves were also lower in the Manitoba females, and the mean body mass of males was 41% lower in Minnesota. All these downward trends were statistically significant. With lower body reserves, scaup are unable to breed, or must wait until they have replenished reserves on breeding areas, a delay which reduces breeding success.

Among White Storks *Ciconia ciconia* migrating between eastern Europe and Africa, annual survival and population change were found to correlate with autumn conditions on a major staging area in the eastern Sahel zone (Schaub *et al*, 2005). The birds remained in this region for up to two months during their southward migration, and again more briefly in spring. Annual variation in the rainfall-dependent condition of the vegetation in this region (assessed from satellite images) accounted for 88% of the annual variation in stork survival. Survival was high in years when primary production in the eastern Sahel was high. It led to synchrony in the annual survival of adult and juvenile storks, and their population changes, across a large part of eastern Europe over the 19-year study period. In contrast, vegetation conditions further south in Africa, where some of the storks spend the winter, made no significant contribution to variation in annual survival. Similarly, in British Barn Swallows *Hirundo rustica*, breeding numbers from year to year were correlated with the preceding rainfall in the western Sahel zone, at its driest during spring migration, and not with rainfall in the South African wintering areas (Robinson *et al.* 2003).

These various observations imply that conditions at stopover sites, including competition and predation/disturbance, can influence the migration speeds, reproduction and survival chances of individual migrants, and in extreme cases can affect their breeding numbers. Such processes can act in a density-dependent manner, at least among the individuals that are present together at particular sites. Where most individuals depend on the same small number of stopover sites (as in Pink-footed Goose *Anser brachyrhychus* and Red Knot *Calidris canutus*), and birds are competing largely at the same time in the same area, such processes could have density-dependent effects at the level of the entire population. It remains to be seen whether the same holds in species in which individuals are spread over a large number of stopping sites at all stages of the journey, each individual competing with whichever other individuals happen to be at the same sites at the same times, as in many passerines that migrate over land.

RELATIVE IMPORTANCE OF MIGRATION-RELATED MORTALITY

Migration mortality has proved difficult to measure as a distinct component of the overall annual mortality. This is because the chance of getting ring recoveries varies along a migration route so does not reflect the scale of mortality in different places, and with radio-marked birds, it is not usually possible to distinguish death from radio failure. However, in a study of Black-throated Blue Warblers *Dendroica caerulescens*, Sillett & Holmes (2002) were able to assess survival rates during the summer breeding period in New Hampshire, during the winter period in Jamaica, and over the year as a whole, respectively. They concluded that more than 85% of apparent annual mortality occurred during migration, giving a rate which was at least 15 times higher than at other times of year. The relative importance of different types of mortality to the overall migration total remained unknown.

Among Barnacle Geese *Branta leucopsis* travelling 3200 km between Svalbard via Bear Island to Scotland, it proved possible in 1986 to check for colour rings in the same group of birds just before and after this migration (Owen & Black 1989). About 35% of the juveniles were found to have disappeared (presumed dead) on this one journey, compared with about 5% of older birds (about half the annual total). The losses were greatest among young hatched latest in the season, which were lightest in weight at the date of departure. This amount of mortality was deemed exceptional, however, because severe weather in the breeding area forced the birds to leave earlier than usual and also stopped some from staging on Bear Island. Once juveniles reached their wintering areas, mortality dropped to a level equivalent to 10% per year, the same as adults. Evidence of similar mortality during the autumn journey of Light-bellied Brent Geese *Branta bernicla hrota* migrating from northern Canada to Ireland was provided by O'Briain (1987). Their migration covers at least 2500 km, part of which crosses the Greenland ice cap. In each of two years, loss of young averaged 33%, compared with 5% for adults.

As mentioned above, Greater Snow Geese *Chen caerulescens atlantica* migrate from breeding areas on Bylot Island in the Canadian Arctic to staging areas on the St Lawrence River in Quebec. Mortality on this autumn journey was calculated in five successive years, both from banding studies and from comparison of the brood sizes of neck-banded females before and after migration (Menu *et al.* 2005). The two approaches yielded similar mortality estimates, and the same pattern of year-to-year variation. The average monthly mortality of juveniles over the autumn journey was 34%, and that of adults 1%. However, after this migration, juveniles survived as well as adults (both 97% per month). The loss of juveniles over the migration period also varied greatly between years (range 29–88% over five years), and most mortality appeared to be natural (rather than due to shooting). Juvenile mortality was highest in years when: (1) temperatures at the time of fledging and migration were low (at or below freezing), (2) the mean body mass of goslings at fledging was low, and (3) the mean fledging date was late. These studies indicate how conditions on breeding areas can influence subsequent mortality on migration.

In some other species, too, most of the difference in annual mortality between juveniles and adults was attributed to greater losses among juveniles before they reached their winter quarters (Cavé 1983, Gromadzka 1989, Pienkowski & Evans 1985). In other species, mortality on the following spring journey was also heavier in juveniles than in adults. This was deduced, for example, from comparison of age ratios among museum skins of Pacific Slope Flycatchers *Empidonax d. difficilis* obtained at the start and end of each migration (Johnson 1973). In some migratory species, notably raptors and shorebirds, the annual survival rates among adults are so high (>90%) that working out how much of the annual mortality occurs on migration is seldom going to be easy.

CONCLUDING REMARKS

Limitation of breeding populations through migration events has been viewed here mainly as a three-step process, involving: (1) food availability at stopover sites, which influences the migratory performance of individuals, as reflected in rates of weight gain, departure weights, frequency and duration of stopovers, and overall migration speed; (2) carry-over effects of migration performance on subsequent survival or breeding success; (3) which in turn influence population trend (**Figure 27.2**). Many studies have been concerned with the first aspect, providing evidence from stopover sites of interference and depletion competition, and of effects of disturbance and predation on the fuelling rates of individuals (involving some age and sex effects). Relatively few studies have provided evidence of migration conditions influencing subsequent breeding and survival, and even fewer of effects on subsequent breeding numbers. The paucity of examples of migration effects on breeding numbers may reflect the difficulties of study rather than the rarity of the phenomenon. Moreover, on all aspects, the evidence is based primarily on correlations, giving no direct evidence for causal relationships, although the provision of extra food and contrived disturbance to migrants could be classed as experimental. The main challenge for future research, therefore, is to test whether the main types of correlation discussed above reflect causal relationships.

All the various processes envisaged involve competition for food, and its effects on feeding rates (along with predation and parasitism). All these factors could act in a density-dependent manner to regulate local and overall populations partly in relation to the availability and quality of stopover habitat. The second type of mortality affecting migratory bird populations, caused by storms or other adverse weather and discussed in the next chapter, is likely to act in a density-independent manner, in that the proportion of birds removed each year bears no consistent relationship to overall population size. However, the likelihood of any one individual succumbing in an extreme weather event may depend partly on its body condition at the time, and in turn on conditions (including competition) at previously attended stopover sites.

While we yet have few examples of bird populations in which changes in breeding numbers have been unequivocally linked to events at a stopover site, in many species changes in breeding numbers are known to be influenced by conditions in wintering areas (Chapter 26; Newton 2004a). Many of these examples of apparent winter limitation could operate through effects on spring migration, which often occurs when food sources reach their lowest level of the year. For example, many Eurasian migrants have declined in numbers following years of drought in the Sahel zone of Africa (Chapter 24). Most mortality is likely to have occurred towards the end of their stay in the Sahel, at the time of migratory fat deposition, or in the Sahara as they attempt to migrate on inadequate body reserves. Some of the species that have shown declines spend the winter mainly south of the Sahel in less arid habitat, but they still have to migrate through the Sahel in spring. More careful investigation of migration in this region may reveal that much of the apparent overwinter mortality is related to spring migration, as indicated by the study on Barn Swallows *Hirundo rustica* discussed above.

Competitive interactions and food shortage

Although few would question the importance of food supply to successful migration, its precise effects are not always easy to quantify, for they do not always result in direct starvation (loss of body condition to the point of death).

At stopover sites, territorial and other interactions between individuals can operate to adjust densities to local food supplies at the time, causing hungry birds to move elsewhere, to places where they may survive or die of starvation or something else. Secondly, food shortage at stopover sites may reduce population size through lowering breeding rates (as in some geese), not necessarily entailing the starvation of full-grown birds. Without special study, this type of effect may be hard to detect because of the time lag between the food shortage and the resulting decline in breeding numbers. In some long-lived species, individuals do not normally breed until they are several years old, so it may take several years before the effects of poor breeding are reflected in decline in the numbers of adult breeders. As a further complication, the effects of food shortage on any population may be accentuated by the presence of people or natural predators, which limit feeding opportunities, and by parasites which directly or indirectly take part of the host's nutrient intake for themselves.

Residual body reserves

Residual body reserves could be advantageous at any stage of migration, cushioning the bird against adverse weather or other unexpected mishaps during a flight. They give a margin of safety against bad weather immediately after arrival, and allow the bird time to establish itself in a new area. In many species, it is after arrival in breeding areas that residual body reserves are most useful. It is now well established that such reserves support the breeding of arctic-nesting geese (Chapter 5); but it is still uncertain what benefits accrue from residual reserves in smaller birds, although feeding experiments on several species have shown the value of supplementary food in influencing both laying date and clutch size (Newton 1998b). Four types of benefit have been suggested, which are not mutually exclusive, namely that reserves could: (a) increase survival chances if weather conditions deteriorate; (b) allow more time to be spent on other activities important to reproduction, such as territorial defence, song and mate selection; (c) relieve food demands in the early stages of breeding, allowing an earlier start; and (d) allow females to forage selectively for nutrients important to reproduction, such as calcium, while living mainly off their fat (Sandberg & Moore 1996). These benefits must presumably be set against the costs of longer stopovers needed to accumulate the extra reserves, the energy to transport them, and any associated predation risks incurred.

Conservation considerations

The most crucial habitat for travelling migrants is presumably that which lies adjacent to an ocean or other barrier, and forms the last possible feeding place before the barrier, or the first encountered after it. Examples include the remaining woodland and scrub patches on the coastlines of the Mediterranean Sea in Europe or the Gulf of Mexico in North America. Elsewhere, constraints on migratory fuelling are likely to be most obvious in landscapes where patches of suitable habitat are few and widely scattered, because such patches often attract large numbers of passing birds. Such conditions may be encountered, for example, by forest species migrating through essentially open landscapes with few trees, or by wetland species through essentially arid landscapes with few lakes or rivers. These are the conditions that, through their effects on individual birds, are likely to lead in the long term to the 'high-fuel/long-stopover/long-flight' strategy, rather than the 'low-fuel/short-stopover/short-flight' strategy appropriate in more continuous habitat. Because of human effects on landscapes, the situation encountered by migrants has altered greatly in recent centuries. The migrations of many forest birds, for example, evolved in landscapes very different from those of today, as once-continuous forest has been greatly reduced and fragmented. There must presumably come a point in the process of habitat fragmentation when such landscapes become 'ecological barriers' that are best crossed by long, non-stop flights requiring large fuel loads.

Interactions between populations

Another neglected aspect concerns the interactions that occur between different populations of a species. If individuals from two or more breeding populations occur together in the same staging or wintering areas (like many shorebirds), and feed on the same limited prey supply, the dynamics of the separate breeding populations could be interlinked, because the mean survival rate of individuals from one population is likely to depend on the overall size of both populations (Dolman & Sutherland 1995).

Despite so many travel lanes in bird migration, populations from different breeding or wintering areas may often use the same staging sites. This gives great potential for intra- and inter-specific competition: if one population passes first, it may deplete the food stocks for a later one, and if different populations are present at the same time, the individuals in both may suffer from depletion and interference. In other words, although living apart for most of their lives, the annual few weeks of contact on stopover sites could ultimately influence the size of one or both of any two competing populations. This is an aspect of stopover ecology that has so far received little attention, but it could have great repercussions for the numbers and behaviour of individuals in competing populations.

SUMMARY

Some bird populations might be limited by conditions encountered on migration. This could occur at stopover sites where competition for restricted food supplies can reduce fuelling rates, migration speed, and subsequent survival or breeding success.

When preparing for migration, birds must normally obtain more food per day than usual, in order to accumulate the body reserves that fuel their flights. Birds often concentrate in large numbers at particular stopover sites, where food can become scarce, thus affecting migratory performance. Rates of weight gain, departure weights and stopover durations often correlate with food supplies at stopover sites, sometimes influencing the subsequent survival and reproductive success of individuals, which can in turn affect their breeding numbers. Many studies have provided evidence for food depletion and interference competition at stopover sites, but relatively few for migration conditions influencing the subsequent breeding or survival of individuals, and even fewer for effects on subsequent breeding numbers. The scarcity of studies showing effects on breeding numbers is probably due mainly to the difficulty of study, rather than the rarity of the phenomenon.

In addition, disturbance caused by natural predators or people at stopover sites can lower the food intake of birds, and in some populations of geese has proved sufficient to reduce subsequent breeding rates. In addition to disturbance and other food-related effects, birds may be exposed to greater rates of predation at stopover sites, and greater risks of picking up parasites and pathogens. Several mass-mortality events attributed to disease organisms or botulism have been described in waterbirds concentrated at stopover sites. Mortality on migration is hard to estimate as a separate component of the overall annual mortality in birds. However, in one small warbler species, 85% of the total annual mortality was estimated to occur during the two migration periods, and among arctic-nesting geese major losses of juveniles have been recorded over the first autumn migration.

APPENDIX 27.1

Two other types of findings have been cited as evidence that food supplies at stopover sites can be limiting for migrants. Because these types of evidence are open to other interpretations, and are not amenable to experimental testing, they are discussed separately below.

Ecological segregation

Nearly all bird species show some degree of ecological separation during migration stopovers, whether in habitat, foraging sites, foraging times or diet (Bairlein 1981, Spina et al. 1985, Berthold 1988, Fasola & Fraticelli 1990, Moore et al. 1990, Streif 1991). It has been argued that such segregation helps to reduce competition between species at stopover sites, as well as in breeding or wintering areas.

While this may be true, the demonstration of ecological differences between species at a stopover site does not necessarily imply that food is limiting there. Such differences, which depend on the structure and behaviour of the species themselves, may result from food-based competition in the past, or in areas other than stopover sites, or they could result from causes other than food-based competition. Such ecological differences are thus consistent with the idea that interspecific competition for food is limiting for individual migrant performance, but cannot prove it – nor can the idea be tested satisfactorily.

Temporal segregation of migration seasons

Closely related species with similar ecology often pass through particular sites at somewhat different dates during the migration seasons (for shorebirds see Recher 1966, for warblers see Howlett et al. 2000). The same is also true for different populations of the same species. In general, different populations pass north in spring in the sequence in which their breeding areas become habitable, and south in autumn in the sequence in which their breeding areas become unsuitable (Chapter 12). Although such temporal segregation may reduce the opportunity for competition between the individuals in different populations, it might not have evolved for that reason, but have some quite different basis, related to the dates and periods that the nesting areas of different populations are suitable for occupation (Chapter 14). Also, the early migratory populations might, in some situations, deplete the food for later ones, in which case competition would not be eliminated, but its effects could fall especially heavily on the later migrating populations.

Redwings *Turdus iliacus* migrating through a rainstorm at sea

Chapter 28
Mass mortality of migrants

Migration is a season full of peril for great numbers of winged travellers. (W. W. Cooke 1915.)

Great numbers of migrating birds, chiefly warblers, had accomplished nearly 95% of their long flight and were nearing land when, caught by a norther, against which they were unable to contend, hundreds were forced onto the waters and drowned. . . . During the fall migration of 1906, when thousands of birds were crossing Lake Huron, a sudden drop in temperature, accompanied by a heavy snowfall, resulted in the death of incredible numbers. Literally thousands were forced into the water and subsequently cast up along the beaches, where in places their bodies were piled in windrows. On one section of the beach their numbers were estimated at 1,000 per mile, and at another point at five times that number. (Frederick Lincoln 1935.)

Despite its overall benefits, migration is often perceived as hazardous. During their seasonal journeys, migrating birds must travel through unfamiliar areas, and often through alien habitat, making it more difficult than usual for them to find food and avoid predators. They may run out of fuel, suffer from exhaustion, or encounter storms which could kill them. One might imagine, therefore, that the mean daily death rate would be greater during migration than at other times but, for understandable reasons, little coherent information is available on the mortality costs of migration (but see Chapter 27; O'Briain 1987, Owen & Black 1989, Sillett & Holmes 2002, Menu *et al.* 2005).

While the previous chapter alluded to the ongoing mortality resulting from food shortage, predation and other factors operating at stopover sites, this chapter reviews some additional major mortality incidents described in the ornithological literature. They establish the importance of such additional mortality as a frequent hazard for migratory birds, even though their effects on populations are hard to estimate. Almost all such incidents occurred during adverse weather, either during the journey itself, soon after arrival in breeding areas in spring, or just before departure from breeding areas in late summer or autumn. Typically, those incidents that occurred in breeding areas were not matched by concurrent mortality among local resident species, and could have been avoided if the migrants had arrived some days later or left some days earlier. They are therefore classed here as migration-related.

The risks of migration itself vary with the body size and other features of the birds themselves, with the length of the journey and the terrain to be crossed, and with weather at the time. In general, the risks would seem to be greater for small birds than for large ones, over long journeys than short ones, in adverse than favourable weather, and over hostile than favourable terrain. These generalisations are borne out by the events discussed in the following paragraphs (Newton 2007).

WEATHER AND IN-FLIGHT MORTALITY

Migrants often encounter bad weather en route, whether rain, mist or adverse winds. Over land, birds can usually seek shelter from storms, but this option is not available over water. Heavy rain can saturate the plumage, increase wing-loading (already high in many migrants through fuel deposition), and cause the loss of body heat. These stresses, often coupled with disorientation, sometimes force migrants down where they may be killed by collision, drowning or chilling (Frazar 1881, Saunders 1907, Cottam 1929, Williams 1950, Woodford 1963, Kennedy 1970). Flying birds are also sometimes killed by hail, or by electrocution in lightning storms (Hochbaum 1955, Roth 1976, Glasrud 1976). Clearly, it is important for birds to avoid flight at times of adverse weather, and most of the recorded mortality incidents listed in **Table 28.1** refer to landbirds that encountered storms over water or other terrain where they could not take shelter.

Migrant landbirds caught by mist or storms over water must often be lost without trace, being consumed by gulls and other predators, or washed on to remote shorelines. Nevertheless, reports of mass deaths of migrants caught in bad weather (**Table 28.1**), and of birds arriving exhausted and emaciated, are fairly frequent (Morse 1980, Dick & Pienkowski 1979, Evans & Pienkowski 1984, Pienkowski & Evans 1985, Spendelow 1985). Examples include the large numbers of Quail *Coturnix coturnix* found drowned during a spell of sea fog off North Africa (Moreau 1927), the large numbers of dead Swifts *Apus apus* and martins

Table 28.1 Some large-scale mortality incidents associated with migration

Species	Date	Location	Conditions	Numbers	Source
A. Mortality during spring migration					
Various species (>23 species)	April 1881	Off Louisiana coast[a]	Gale	'Many thousands'	Frazar (1881)
Lapland Longspurs *Calcarius lapponicus*	March 1904	Minnesota–Iowa	Snowstorm	1.5 million	Roberts (1907a, 1907b)
Mainly Lapland Longspurs *Calcarius lapponicus*	February 1922	Nebraska	Snowstorm	'Thousands'	Reed (1922), Swenk (1922)
Magnolia Warblers *Dendroica magnolia* and others (39 species)	May 1951	Off Texas coast[a]	Rainstorm	>10000	James (1956)
Ducks, geese and swans	April 1954	Wisconsin	Hailstorm	'Many'	Hochbaum (1955)
Various (>14 species)	May 1954	Minnesota	Snowstorm	>175	Frenzel & Marshall (1954)
Hirundines (4 species)	May 1956	Saskatchewan	Snow and cold	74	Sealy (1966)
Snow Geese *Chen caerulescens*	May 1959	North Dakota	Hail	100	Krause (1959)
Various (43 species)	May 1962	Minnesota[a]	Mist and rain	5500	Green (1962)
Various (56 species)	April 1960	Michigan[a]	Gale, hail	3636	Segal (1960)
White-throated Sparrows *Zonotrichia albicollis* and many others (>9 species)	April 1963	Georgian Bay, Ontario[a]	Rainstorm	'Thousands'	Woodford (1963)
Barn Swallows *Hirundo rustica* and others	April 1965	Morocco	Cold and rain	>30	Ash (1969)
Various (>32 species)	May 1974	Off Texas coast[a]	Rainstorm	5000	Webster (1974), King (1976)
Various (32 species)	April–May 1975	Utah	Cold and snow	569	Whitmore et al. (1977)
Warblers	May 1974	Lake Manitoba[a]	Cold	'Thousands'	Houston & Shadick (1974)

(continued)

Table 28.1 (Continued)

Species	Date	Location	Conditions	Numbers	Source
Scarlet Tanagers *Piranga olivacea* and others	May 1974	NE Maritime Region	Cold and wet	'Many thousands'	Finch (1975)
Turnstone *Arenaria interpres* and Knot *Calidris alpina*	June 1974	Ellesmere Island, Canada	'Bad weather'	24	Morrison (1975)
Jays, thrushes, warblers	May 1976	Lake Huron, Michigan[a]	Rainstorm	200000	Jansson (1976)
Swifts *Apus apus* and martin species	April 1982	Off Tunisia coast[a]	Rainstorm	Large numbers	Perrins et al. (1985)
Raptors, and others (>12 species)	April 1980	Off Israel coast[a]	Wind	>1300	Zu-Aretz & Leshem (1983)
Mainly Rooks *Corvus frugilegus* and many others (20 species)	April 1985	Off Swedish coast[a]	Dense fog	>20000	Alerstam (1988)
Warblers and others	April 1991	Off Spanish coast[a]	Rainstorm	>677	Mead (1991)
Various (45 species)	April 1993	Off Louisiana coast[a]	Tornado	40000	Wiedenfeld & Wiedenfeld (1995)
B. Mortality during autumn migration					
Various (26 species)	October 1906	Lake Huron, Ontario[a]	Snowstorm	>10000	Saunders (1907)
Eared (Black-necked) Grebe *Podiceps nigricollis*	December 1928	Nevada	Snowstorm	'Thousands'	Cottam (1929)
Swifts *Apus apus*	July 1930	England	Rain and wind	30	Watson (1930)
Various ducks (5 species)	October 1951	South Dakota	Fog, rain, snow	ca 500	Schorger (1952)
Various, mainly warblers (37 species)	October 1964	Florida	Low cloud, heavy rain	>4707	White (1965)
Various ducks (8 species)	November 1973	Arkansas	Thunderstorm	>76	Roth (1976)
Chimney Swift *Chaetura pelagica*	October 1979	Islas del Cisne	Sunny	200–300	Spendelow (1985)
Mainly Song Thrush *Turdus philomelos* Blackbird *T. merula* and others (12 species)	October 1988	Off Swedish coast[a]	Dense fog	>20000	Alerstam (1990a)
Eared (Black-necked) Grebe *Podiceps nigricollis*	January 1997	Utah	Snowstorm	35000	Jehl et al. (1999)

C. Mortality after arrival from spring migration[c]

Species	Date	Location	Weather	Number	Reference
Mainly warblers (22 species)	May 1888	Wisconsin	Cold	>645	Deane (1914)
Tree Swallow *Tachycineta bicolor*	May 1906	Maine	Cold	'A number'	Deane (1923)
Least Flycatcher *Empidonax minimus*, various warblers and others	May 1907	Minnesota	Cold and snow	'Hundreds'	Wood (1908)
Warblers	May 1907	Michigan	Snowstorm	'A number'	Wood (1908)
Tree Swallow *Tachycineta bicolor*	Spring 1945	New York	Cold and wet	'Many'	Dence (1946)
Eurasian Chiffchaff *Phylloscopus collybita*	April 1950	Wales	Cold and snow	13	Kent (1951)
American Coots *Fulica americana*	March 1964	Iowa	Re-freezing	'Hundreds'	Fredrickson (1969)
King Eider *Somateria spectabilis*	May–June 1964	Beaufort Sea	Re-freezing	100000	Barry (1968)
Northern Lapwings *Vanellus vanellus*	April 1966	Finland-Sweden	Cold and snow	'Many thousands'	Vepsäläinen 1968), (Marcström & Mascher (1979)
Hirundines and others	May 1965	Minnesota	Cold and snow	'Hundreds'	Anderson (1965)
Various (42 species)	May 1968	Finland	Cold and snow	>3000	Ojanen (1979)
American Coots *Fulica Americana*	May 1967	North Dakota	Cold and snow	>387	Dane & Pearson (1971)
Scarlet Tanagers *Piranga olivacea* and others	May 1974	New Hampshire	Cold and rain	28 per hour of road travel	Zumeta & Holmes (1978)
Dusky Flycatcher *Empidonax oberholseri* and others (10 species)	June 1974	Colorado	Rain and snow	>49	Eckhardt (1977)
Various (32 species)	April–May 1975	Utah	Snowstorms	569	Whitmore et al. (1977)
Eurasian Oystercatcher *Haematopus ostralegus*	March 1979	Scotland	Cold and snow	33	Watson (1980)

(continued)

Table 28.1 (Continued)

Species	Date	Location	Conditions	Numbers	Source
Hirundines (4 species)	April 1982	California	Cold and wet	>100	DuBowy & Moore (1985)
Cliff Swallow Petrochelidon pyrrhonota	May 1996	Nebraska	Cold and rain	'Thousands'	Brown & Brown (1998)
D. Mortality before departure from breeding areas[c]					
Brent Geese Branta bernicla	August 1956	Southampton Is., Canada	Water freezing	21	Barry (1962)
Barn Swallow Hirundo rustica	September 1931	Germany	Cold and snow	'Millions'	Alexander (1933)
Northern House Martin Delichon urbica, Sand Martin Riparia riparia, Barn Swallow Hirundo rustica, Northern House Martin Delichon urbica	October 1974	Germany and Switzerland	Cold and snow	'Many thousands'	Ruge (1974), Bruderer & Muff (1979), Reid (1981)
King Eider Somateria spectabilis	Autumn 1960	Banks Is., Canada	Water re-freezing	50 000[b]	Barry (1968)
Barn Swallow Hirundo rustica	October–November 2000	Kazakhstan	Cold and snow	'Several thousands'	Berezovikov & Anisimov 2002
Sand Martin Riparia riparia	October–November 2000	Kazakhstan	Cold and snow	>1000	Berezovikov & Arisimov 2002
Common Swift Apus apus	August–November 1986	Finland	Cold and rain	2000	Kolunen & Peiponen (1991)
Common Swift Apus apus	July 1930	England	Cold and rain	>30	Watson (1930)

Within categories, incidents are listed in date order.

For other examples of heavy mortality in migrants after arrival, see Smith (1929), Brown & Brown (2000).

For examples of hold-ups in migration through bad weather, see Wood (1908), Williams (1950), Berthold et al. (2002).

[a] Much or all of the recorded mortality occurred over water.

[b] Females and young, the males having already migrated.

[c] Birds classed as killed after arrival in breeding areas may have included some migrants with further to go, and those classed as killed in breeding areas before departure may have included some migrants already en route from higher latitudes (see especially the Swifts discussed by Kolunen & Peiponen 1991).

washed up on a beach in North Africa (Perrins *et al.* 1985), or the hundreds of Garden Warblers *Sylvia borin* and other species washed ashore in Spain following a night of heavy rainstorms (Mead 1991).

Even greater overwater losses have been recorded in the New World (**Table 28.1**). A massive kill, estimated at 40 000 migrants of 45 species, occurred during a tornado and storm on 8 April 1993 off Louisiana. The storm occurred when large numbers of birds were arriving at the coast after an overnight sea-crossing (Wiedenfeld & Wiedenfeld 1995). Other mortality events in the Gulf of Mexico included one of 10 000 birds (half of which were Magnolia Warblers *Dendroica magnolia*) washed up on Padre Island, Texas, in May 1951 (James 1956), and another of 5000 birds on Galveston Island, Texas, in May 1974 (Webster 1974, King 1976). In another incident, over the Great Lakes in May 1976, an estimated 200 000 birds were washed up on one stretch of shore on Lake Huron alone (Janssen 1976).

Lapland Longspurs *Calcarius lapponicus* seem to be frequent victims of spring storms. While migrating northward at night, large numbers have sometimes been grounded by wet, clinging snow. In March 1904, an estimated 750 000 were found dead on the ice of two lakes in Minnesota, each about one square mile (2.6 km^2) in extent. However, carcasses were reported from a much wider area, and it was estimated that at least twice that number (1.5 million) may have been killed. Some had been attracted to town lights, and collided with buildings and other obstacles (Roberts 1907a, 1907b, Lincoln 1935a).

Waterbirds probably suffer fewer losses on overseas flights than do landbirds, being more robust and able to rest on water when navigation or flying conditions become difficult (as shown for radio-tagged Whooper Swans *Cygnus cygnus* on their migration between Iceland and Britain, Pennycuick *et al.* 1996). However, for some waterbirds on migration, problems arise from their being forced down onto dry land. In western North America, Eared (Black-necked) Grebes *Podiceps nigricollis* cross hundreds of kilometres of desert with few places for a waterbird to land in emergency. Snow storms have occasionally brought hundreds or thousands of birds crashing to the ground where they may die:

> 'During an early morning hour (about 2 am) of December 13, 1928, the residents of Caliente, Nevada, were awakened by a heavy thumping of something falling on the roofs of their houses. Those who were curious enough to step outside and investigate the unusual occurrence found scores of waterbirds in the new fallen snow. The next morning several thousand Eared Grebes were found on the ground and flat roofs of business houses throughout the city.'

Most were dead, having struck trees or buildings, but others were busy working themselves out of the snow (Cottam 1929). In January 1997, bad weather in southern Utah several times brought grebes to ground, killing an estimated total of 35 000, forming about 3% of the population that stages at the Great Salt Lake (Jehl *et al.* 1999).

Headwinds provide another hazard for migrants, because in effect they force the birds to fly for longer, depleting their energy reserves. This is especially important for species that cross large expanses of inhospitable habitat, such as oceans or deserts. In strong headwinds, birds exhausted over water sometimes settle on the surface, becoming soaked and unable to take off again. In addition, birds forced by headwinds to fly close to the waves become vulnerable to attacks by gulls, which have no difficulty in forcing small birds into the water, from which they can be snatched and swallowed (Hobbs 1959). However, not all land-birds that settle on water are doomed. Gätke (1895) mentioned coming across, on different occasions, a Song Thrush *Turdus philomelos*, Snow Bunting *Plectrophenax nivalis* and Brambling *Fringilla montifringilla* sitting on the sea surface, all of which flew when approached by a boat. Probably many birds are lost at sea while fighting headwinds, while others die or suffer reduced breeding success after arrival (for Brent Geese *Branta bernicla* see Ebbinge 1989).

Dead landbirds are often found washed up on beaches, and their remains show that many have been eaten at sea, presumably mainly by gulls (e.g. Alerstam 1988). Other individuals arrive on coastlines in an apparently exhausted state, as often witnessed by bird-watchers. Evidently, overwater migration inflicts continual, and occasionally heavy, losses on many landbird species.

Storm-induced deaths over water mostly involved small birds, but have included such robust species as Rook *Corvus frugilegus*, of which 4600 carcasses were found in one incident in southern Sweden (Alerstam 1988). At least 106 species, including passerines, cuckoos, nighthawks, rails, gulls, terns, skimmers, shorebirds and waterfowl, have been washed up following storms over the Gulf of Mexico. In addition, one large overwater incident off Israel in April 1980 involved at least 1300 birds of prey (including eagles), presumably blown over water from their usual landward route, and other incidents involving raptors are mentioned by Kerlinger (1989). One overland hailstorm killed many ducks and geese, and at least 35 Tundra Swans *Cygnus columbianus* (Hochbaum 1955), while another killed 100 Snow Geese *Chen caerulescens* (Krause 1959). It seems, then, that all sizes of birds are vulnerable to in-flight mortality from adverse weather of one type or another.

Sandstorms can be a particular hazard to birds moving through the Sahara and neighbouring deserts of the Middle East, and can be fatal to any grounded migrant. Incidents involving Northern Wheatears *Oenanthe oenanthe*, Barn Swallows *Hirundo rustica*, Quail *Coturnix coturnix* and White Storks *Ciconia ciconia* have been documented (Moreau 1928, Schüz et al. 1971). 'In the hollows between the dunes there were a few small bushes, and under each one of them the carcasses of Quail in various stages of desiccation were huddled.' (Moreau 1928).

UNSEASONABLE COLD SOON AFTER ARRIVAL IN BREEDING AREAS

Another source of loss among migrants is the unseasonable cold which may occur in some years soon after migrants have arrived in their breeding areas. Typically, such losses are restricted to migrants, local resident species being better able to withstand the cold. In each documented example, as mentioned above, the migrants could have avoided the cold spell if they had arrived in breeding areas some days later than they did. Such losses could often have included passage migrants en route to other breeding areas, as well as migrants breeding locally.

Insectivorous migrants seemed especially vulnerable at such times, because cold and snow greatly reduced their food supplies, but large-scale losses also occurred among newly arrived waterfowl and waders. Ground-feeding birds may starve when fresh snow and ice makes their food unavailable (King 1974, Roseberry 1962, Vepsäläinen 1968, Bull & Dawson 1969), and aquatic feeders are susceptible to delayed melting or re-freezing of water surfaces (Smith 1964, Barry 1968, Fredrickson 1969).

One such catastrophe affected newly arrived Cliff Swallows *Petrochelidon pyrrhonota* in Wisconsin, when a sudden drop in temperature caused insects to become dormant. The swallows did not fly during this period, but clung to their old nests, with one observer collecting 'a milk pail full' of birds that had died (Buss 1942). Another unusual period of cold caused heavy mortality of Cliff Swallows across the north-central Great Plains in 1996, reducing a study population by about 53% (Brown & Brown 1998, 2000). When short of food in cold weather, swallows and swifts often seek shelter in buildings, huddle together for warmth, and may suffer from hypothermia and starvation (Koskimies 1950, Smith 1968, Keskpaik 1972, Lyuleeva 1973, Brown & Brown 2000). Other migratory insectivores also die in such conditions, but less conspicuously (**Table 28.1**; Ligon 1968, Eckhardt 1977, Tramer & Tramer 1977, Whitmore *et al.* 1977, Zumeta & Holmes 1978, Mead 1979a).

In spring 1964, an estimated 100 000 King Eiders *Somateria spectabilis* – about one-tenth of the whole west Canadian population – died at migration staging areas in the Beaufort Sea. This mortality occurred when newly opened leads among sea-ice re-froze, preventing the birds from feeding (Barry 1968). In a more recent incident, birds found dead had lost about half of their body weight, and many survivors had insufficient body reserves for breeding, so that the effects of the freeze extended beyond the immediate mortality (Fournier & Hines 1994). Such mass starvation events in spring were documented in this species several times during the twentieth century (Barry 1968, Fournier & Hines 1994), and occasionally in autumn (see later). Similar events, involving the re-freezing of open water, have also killed hundreds of American Coots *Fulica americana* and other newly arrived waterfowl (Fredrickson 1969).

In the spring of 1966, frost and snow on 11–17 April caused massive mortality of newly arrived Lapwings *Vanellus vanellus* across southern Sweden and Finland, with reductions in local breeding populations the same year of 30–90% depending on region (Vepsäläinen 1968, Marcström & Mascher 1979). In such conditions, some birds retreated southward (on 'reverse migration'), others abandoned their territories and re-assembled in flocks, moved from mountains to valleys or from inland to sea coasts, and some females about to lay resorbed their eggs. A similar event occurred in the same region in 1927 (Kalela 1955), and a smaller event affected Eurasian Oystercatchers *Haematopus ostralegus* in Scotland in 1979 (Watson 1980).

Migrants that had died during unusual cold snaps in breeding areas had simply starved to death: carcasses were light in weight and practically devoid of fat reserves (e.g. Ligon 1968, Vepsäläinen 1968, Fredrickson 1969, Morrison 1975, Whitmore *et al.* 1977, Marcström & Mascher 1979, Watson 1980). At such times, birds were often reported in unusual places, such as roadsides or human

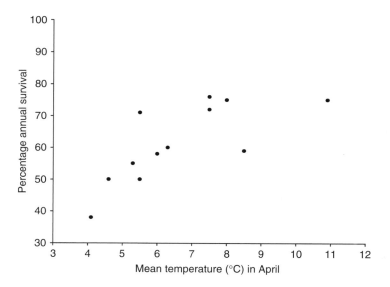

Figure 28.1 Relationship between annual survival of Common Sandpipers *Actitis hypoleucos* and mean temperature in the latter half of April, soon after arrival in breeding areas, England, 1977–1989 ($y = 0.074x + 0.40$, $r = 0.82$, $P < 0.001$). Birds suffered higher mortality than usual in years when snowstorms occurred in the days following their arrival, so that breeding densities in these years were reduced. From Holland and Yalden (1991).

habitation, where food of some sort was available (for Scarlet Tanager *Piranga olivacea* see Manville 1957, Zumeta & Holmes 1978, for Oystercatcher see Watson 1980).

Again, little information is available on the impact of occasional heavy spring losses on the overall annual mortality. However, among Common Sandpipers *Actitis hypoleucos*, which migrate from Africa to breed in Britain, annual survival fluctuated in one area according to the weather in April, when they arrived (**Figure 28.1**). The mean survival over 13 years was 79%, but following late snowstorms in 1981 and 1989 survival fell to 39% and 50% respectively, and breeding pairs from 21 to 14 and from 20 to 12 (equivalent to 33% and 40% reductions). After both these years, recovery in population level was slow, with annual increments of only 1–2 pairs (Holland & Yalden 1991). Similar reductions in breeding densities in cold springs were estimated for Scarlet Tanager *Piranga olivacea* at 33% in a study area in New Hampshire (Zumeta & Holmes 1978), at 30% for New Hampshire as a whole, and 50% for Maine; for various hirundines (*Tachycineta bicolor, Riparia riparia, Hirundo rustica, H. pyrrhonota*) declines were estimated at 30% for Nova Scotia, Maine and New Hampshire, together with an average of 25% in eight species of warblers in New Brunswick, compared to numbers in the previous year (Robbins & Erskine 1975). Hence, all these estimates of local or regional population declines, attributed to cold weather after arrival in breeding areas, fall within the range 25–90%, depending on species and area.

UNSEASONABLE COLD BEFORE DEPARTURE
FROM BREEDING AREAS

Unseasonable weather can also affect birds about to leave their breeding areas in autumn (**Table 28.1**). In central Europe in September 1931, during an early cold snap, 'immense numbers of Swallows *Hirundo rustica* and House Martins *Delichon urbica*, and some Sand Martins *Riparia riparia*, numbed by the cold and unable to find food, sought shelter in stables, barns and houses where, in some cases, they congregated in great masses and were so inert that they could be picked up by hand' (Alexander 1933). The 89000 individuals that were collected and transported by air for release in Italy 'were probably a small fraction of the total lost'. Another event in central Europe in 1974 had similar effects, with hundreds of thousands of hirundines reported dead in the Alps region of Switzerland, Austria and southern Germany, and two million found alive flown by planes further south (Bruderer & Muff 1979, Reid 1981). In the next spring (1975), Northern House Martin *Delichon urbica* populations across Switzerland were reduced by 25–30%, but no overall effect was noted in Barn Swallows *Hirundo rustica* (despite some local reductions) (Ruge 1974, Bruderer & Muff 1979). However, many ringed Barn Swallows from Denmark were recovered in central Europe during this event (presumably on migration), and the following spring breeding numbers in a study area were reduced by more than one half (Moller 1994b).

Occasionally, Swifts *Apus apus* are unable to leave their north European breeding areas because of adverse weather. They usually leave south Finland by the end of August, but in 1986 when late summer was cold and wet, around 2000 remained into November before dying (Kolunen & Peiponen 1991). Seven were found dead in their roost sites, and others still alive were clearly starving. Scarcity of aerial insects prevented the birds from accumulating fat for migration and they remained in southern Finland until their death. Similar events occurred in 1918 and 1957, but in these years delay was caused by late breeding resulting from cold weather in spring.

Autumn mortality events have also occurred among King Eiders *Somateria spectabilis*, following the early freezing of sea-water (Barry 1968). One in 1961 near Banks Island (Canada) involved an estimated 50000 females and young. The males had already migrated and escaped the freeze. This event supposedly killed almost the entire Banks Island population of breeding females. Similarly, 21 Brent Geese *Branta bernicla*, hatched on Southampton Island (Canada) in the relatively late spring of 1956, were found frozen in shoreline ice the following spring. 'They were well preserved, and nothing could be found wrong with them except that their feather development was 4 to 5 days short of allowing them to fly' (Barry 1962).

Other mass-mortality events, which sometimes occur during extreme weather in summer or winter, are not directly associated with migration, and can affect resident as well as migratory species (Newton 1998b). However, it is sometimes difficult to separate migration-associated losses from other weather-induced mortality. For example, an estimated 27000–62000 wintering diving ducks (mainly Tufted Ducks *Aythya fuligula* and Pochards *Aythya ferina*) starved to death in Switzerland and Holland during a cold spell in March 1986, at a time when these birds would normally set off on spring migration (Suter & van Eerden 1992). This incident is not listed in **Table 28.1** as migration mortality. But the authors argued that, if such weather had occurred earlier in winter, the birds would have moved further south and escaped starvation. As it was, they were programmed to move, if at all, towards their breeding areas, which prevented their usual hard-weather response.

Similarly, many large-scale mortality incidents have been recorded from time to time in seabirds (so-called wrecks), when large numbers are blown ashore, sometimes far inland (for review see Newton 1998b). It is usually unknown whether these birds were on migration at the time, or in their wintering areas. However, because some seabird species seem to remain on the move for most of the time between leaving their breeding places in one year and returning there the next, migration periods cannot always be separated from other parts of the annual cycle. The same could be said for some landbird species, which seem able to move long distances at almost any time during the course of a winter to avoid food shortages (Chapter 16).

HUMAN-INDUCED LOSSES

In some of the in-flight incidents mentioned above, human artefacts may have increased the losses at night, because on dark or misty nights birds were attracted to lights, resulting in collisions (Swenk 1922, Roth 1976). Such casualties have long been known from coastal and offshore lighthouses on nights of poor visibility, with low cloud or rain (e.g. Gätke 1895). In the years 1886–1939, some 500–8000 birds per year were reported as killed at lighthouses around Denmark, a total of 33 800 in autumn and 20 700 in spring (Hansen 1954). Relative to their numbers, some species were killed more often than others: for example Brambling *Fringilla montifringilla* more than Chaffinch *F. coelebs*, and Jack Snipe *Lymnocryptes minimus* more than Common Snipe *Gallinago gallinago*. At lighthouses and other structures, rails seemed especially vulnerable. Such losses almost ceased where continuous beam lights were replaced by flashing on–off lights in the late twentieth century.

Even greater losses have been associated with the tall, illuminated masts used for radio, television and mobile phone transmission, especially in North America. Along with tall buildings and ceilometers (light beams for measuring cloud height that attract birds which then collide with nearby buildings), these masts kill many migrant birds (mainly by collision), especially those flying at night (e.g. Brewer & Ellis 1958, Tordoff & Mengel 1956, Taylor & Anderson 1973, Weir 1976, Avery *et al.* 1977, 1978, Lid 1977, Crawford 1978, 1981, Kemper 1996, Kerlinger 2000). In North America in the 1970s, an estimated 1.3 million migrants were killed in this way each year (Banks 1979). By the year 2000, tower numbers had increased roughly four-fold, as had the associated death toll, reaching an estimated 4–5 million birds per year (US FWS 2002). This is likely to be an underestimate, because many birds probably die away from the point of collision, or are removed by nocturnal scavengers before they can be found by people in daylight (for effects of scavenger control on numbers of carcasses found, see Crawford & Engstrom 2001). About 350 species have been recorded as casualties, the vast majority being Nearctic–Neotropical migrants which fly at night, such as Ovenbird *Seiurus aurocapillus*, Tennessee Warbler *Vermivora peregrina*, Black-and-white Warbler *Mniotilta varia*, Blackpoll Warbler *Dendroica striata*, Prairie Warbler *Dendroica discolor* and Magnolia Warbler *Dendroica magnolia*. In one of the most detailed studies, at a television tower in Florida, 44 007 victims of 186 species were found over a 29-year period, an average of 1517 per year. More than 94% were Neotropical migrants, with Red-eyed Vireos *Vireo olivaceous* the most frequent (Crawford & Engstrom 2001). Higher totals in late summer/autumn than in spring could be attributed to the greater numbers of birds migrating in autumn.

Mortality has sometimes been heavy, as exemplified by the 50 000 birds of 53 species killed in one night at a ceilometer in Georgia (Johnston & Haines 1957). In general, towers taller than about 150 m kill the largest numbers of birds, and shorter towers relatively few (apart from one anomalous incident involving the deaths of 5000–10 000 Lapland Longspurs *Calcarius lapponicus* on the snowy night of 22 January 1998 at three 130-m towers in Western Kansas, Kerlinger 2000). Reducing the height of some towers has greatly reduced the fatalities (Crawford & Engstrom 2001), while increasing the height of other towers has greatly increased the fatalities (Kemper 1996). As at lighthouses, mass mortalities are most frequent on nights of low cloud and fog or rain, when birds fly lower than normal. At most other times, birds fly too high for masts and tall buildings to represent a hazard, and I know of no evidence that windows of low buildings kill more birds at migration times than at other times of year.

Gas flares on oilrigs also attract birds on dark foggy nights; up to several thousand per night having been killed at individual flares (Lid 1977). The usual numbers are much lower, however, estimated by Bourne (1979) at a few hundreds of birds per rig per year, a small proportion of the numbers passing. Modern wind turbines are known to kill migrants by night or by day, but information is only just beginning to emerge on the scale of these losses (which generally seem small, being estimated at a total of 33 000 birds per year in the USA, US FWS 2002). The greatest losses seem to occur at windfarms situated on narrow migration routes (with many raptors killed in southwest Spain), or near wetlands, which attract large numbers of gulls and other large birds. As these turbines continue to multiply in the years ahead, they could collectively begin to impose a heavier drain on migratory bird populations, perhaps achieving population levels effects, especially on large species with low reproductive rates. Some forms of mortality are so long established, and affect birds year-round, that it is hard to separate mortality at migration times from that at other times. This holds for mortality caused by collisions with power lines (Ferrer & Janss 1991), and for mortality inflicted by oil pollution on seabirds, although some species are clearly affected mainly during migration. An example is the Magellanic Penguin *Spheniscus magellanicus*, of which an estimated 41 000 are killed each year, mainly when on return migration off the Argentinian coast (Gandini *et al.* 1994). Another form of human-induced loss is the hunting inflicted on many species while on migration, and discussed for Palaearctic–Afrotropical migrants in Chapter 24.

DISCUSSION

Because mortality incidents that occur over sea or desert are usually impossible to detect, it is hard to tell whether they are of more than sporadic occurrence, let alone what proportion they form of the overall annual mortality. Losses that occur in particular storms may involve birds from wide areas of the breeding range, so have less impact on local breeding densities than the number of casualties might suggest. However, losses that occur in spring, when numbers are near their seasonal low, are more likely to affect subsequent breeding densities than are those that occur in autumn when numbers are near their highest. In populations in which overwinter mortality is density dependent (Newton 1998b, 2004b), losses on autumn migration could be largely or entirely offset by reduced overwinter losses, but by spring there is much less scope for such compensation before breeding begins. Most of the documented mortality events in **Table 28.1** were in spring which in many northern regions tends to have more severe weather events than autumn, affecting birds on passage, or soon after arrival in breeding areas.

Of the 23 spring in-flight mortality incidents listed in **Table 28.1**, eight occurred during heavy rain, five during snow, two in hail, two in dense mist, four in strong winds, and two in unspecified 'bad weather'. Of eight such incidents in autumn, three occurred in heavy rain, three in snow, and two in dense mist. Incidents were categorised according to the main weather factor, as mentioned by the author, but in some a combination of factors was involved: for example, rain and mist, or rain and wind. Clearly, mist, rain and snow (or rain and snow associated with strong winds) were important causes of mass mortality among birds in flight. All 18 incidents that occurred soon after arrival in breeding areas were associated with a marked drop in temperature; ten also included snow, four included rain, and two involved the re-freezing of lake or sea-water. All eight incidents that occurred in late summer–autumn before departure were also associated with exceptional cold, including four with snow, two with rain, and two with freezing lake or sea-water. In none of these incidents were resident bird species mentioned as casualties.

In general, the survival of any bird through a difficult journey may depend not only on the weather encountered, but also on the bird's weight (and fat content) on departure, in turn influenced by the prevailing food supply, age and dominance, and levels of competition in the population (Chapter 27). In addition, some juvenile birds may die through their inexperience, making them more vulnerable than adults to various kinds of hazard, while others could die through directional or navigational errors. Vagrants are almost all first-year birds, as are individuals that in spring 'overshoot' their normal breeding range (Chapter 10).

Because the records in **Table 28.1** result from incidental observations, and not from systematic study, it is impossible to estimate how often particular populations experience different types of mass-mortality events. However, in the twentieth century, Northern Lapwings *Vanellus vanellus* in Finland experienced at least two mass starvation events just after arrival, Common Swifts *Apus apus* in Finland experienced at least three such events before departure, hirundines in central Europe experienced at least two such events just before departure, and King Eiders *Somateria spectabilis* in the Beaufort Sea experienced at least four such events at stopover sites in spring. Cliff Swallows *Hirundo pyrrhonota* in Nebraska experienced at least 11 spring mortality events in 123 years (including two of extreme severity), and Lapland Longspurs *Calcarius lapponicus* in Minnesota experienced three such events in 25 years. These figures give some indication of the likely frequency of major mortality incidents, while smaller events in other years may pass without being noticed or documented. If such big events were to increase in frequency, they would presumably lead to changed migration timing or route (through selection), or eliminate altogether those populations occupying particular parts of the breeding range.

In addition to the major weather-induced catastrophes, other birds are presumed to die because they get drifted too far off course, or run out of fuel in places where they cannot feed. Many records exist of emaciated or dehydrated birds that have crossed seas or deserts (for records from a ship off West Africa, see Serle 1956, and from a ship off northeast Cuba, see Johnston 1968), and small islands may often attract weakened migrants from the passing stream (Spendelow 1985).

There can be no doubt, therefore, that many birds die on migration, and that in some weather-induced incidents, the numbers can be very large. In some of the examples mentioned above, the daily mortality during migration, or soon after arrival in breeding areas, was almost certainly much greater than the mean daily rate at other times of year. Estimates from the Black-throated Blue Warbler *Dendroica caerulescens*, Barnacle Goose *Branta leucopsis*, Brent Goose *B. bernicla* and Greater Snow Goose *Chen caerulescens* in the previous chapter provide striking examples, and in other species (such as Northern Lapwing *Vanellus vanellus* and King Eider *Somateria spectabilis*), the losses immediately after arrival in breeding areas in unusually cold springs exceeded the normal total annual mortality expected in these species. Following post-arrival losses in spring, local breeding populations of various species fell by 25–90%, and following heavy pre-departure losses in autumn, subsequent breeding numbers of House Martins *Delichon urbica* in Switzerland fell by 25–30%, and Barn Swallows in part of Denmark by more than 50% (Moller 1994b). Apart from these extreme examples, it is hard to judge the importance of such major weather events in relation to the ongoing but more diffuse losses due to food shortage, predation and other regular mortality agents. Many severe weather events are unlikely to cause mass mortalities, because birds do not normally depart on migration when conditions look bad (Chapter 4). Most losses occur when birds already aloft over water encounter storms en route. Whether the bodies of casualties reach shore, where they might be recorded, depends on many factors, such as how long they remain afloat, the direction and strength of winds and currents, proximity to land, and actions of scavengers (Bibby & Lloyd 1977).

Regarding possible effects of overwater storms on population levels, Butler (2000) pointed out that, among Nearctic–Neotropical migrants, 25 species had declined in eastern North America during the period 1966–1996, compared with only three in western North America over the same period. The eastern species migrate partly over water (western Atlantic or Gulf of Mexico), while the western ones migrate entirely over land. Moreover, among the water-crossers, declines were significantly more frequent among 13 long-distance migrants, wintering in South America, than in shorter distance migrants, wintering in Central America or on Caribbean Islands. In two of the 25 eastern species (Rose-breasted Grosbeak *Pheuticus ludovicianus* and Mourning Warbler *Oporornis philadelphia*), annual population levels (as measured by the Breeding Bird Survey) were related to the number of stormy days during the previous autumn migration (mean 39 stormy days, range 18–59 days in different years). Years with the lowest populations had followed autumns with the most storms. These findings do not necessarily imply causal relationships, but they do suggest that further investigation of weather effects on the population levels of overwater landbird migrants would be worthwhile. The autumn migration of many species between North and South America coincides with the time of hurricanes, the frequency and severity of which have increased in recent years in association with climate change.

Although such losses represent a major cost of migration, for the migratory habit to persist, they are presumably less in the long run than any losses that would be experienced if the birds stayed on their breeding areas year-round. For many species, high-latitude breeding areas are completely uninhabitable in winter. For others, the costs of reaching distant wintering areas may be offset by improved survival there, promoted by milder weather or greater food supplies (Chapter 23).

Migration among birds thus involves a trade-off: the fitness benefits of breeding and wintering in separate regions, set against the fitness costs of the journeys themselves. Mortality that occurs through adverse weather en route presumably has an important selective influence on the birds' behaviour with respect to weather, including the routes taken, while mortality that occurs soon after arrival in breeding areas, or just before departure from breeding areas, has a selective influence on the timing of migration within the annual cycle (for evidence of genetic influence on migration dates see Chapter 20; Berthold 1993, Møller 1994, Brown & Brown 2000). That mortality events associated with migration are not more frequent is testimony to the adaptive behaviour of birds in avoiding bad weather, either by not flying then, or by circumventing it.

In conclusion, although many mass-mortality events among migratory birds almost certainly go unrecorded, and most documented ones cannot be translated into population-level impacts, the sheer scale of some events must inevitably result in temporary reductions in breeding numbers over local or wider areas (as so far documented in a few studies). Almost certainly such mortality represents in some bird species a major cost of migration.

SUMMARY

A major perceived cost of migration in birds is the associated mortality. This mortality has proved difficult to measure, and separate from mortality during stationary periods of the annual cycle. This chapter summarises some major recorded mortality incidents attributed to inclement weather and other factors, including: (a) in-flight losses, caused by storms and other adverse weather en route, usually over water; (b) unseasonable cold weather soon after arrival in breeding areas; and (c) unseasonable cold weather before departure from breeding areas. Cold weather often kills migrants in their breeding areas, but not the local resident species which can better withstand it. For migrants, cold and snow act to cut off the food supply, and could have a major selective effect on the seasonal timing of migration (as demonstrated in a few studies).

Records of in-flight weather-induced mortality, involving up to hundreds or thousands of birds at a time, have affected mainly small passerines, but also larger birds, including eagles and swans. Most occurred in conditions of mist, rain or snow storms, but some also involved nocturnal collisions with illuminated masts and other structures.

Records of post-arrival mortality on breeding areas have involved mainly small insectivores (especially hirundines), but also waders and waterfowl. Such incidents, associated with cold and snow, have reduced regional breeding densities from the previous year by 30–90%, depending on species and area, with up to several years required for recovery.

Records of pre-departure mortality on breeding areas have mainly affected hirundines. Two major incidents in central Europe in September 1931 and 1974 killed hundreds of thousands, or even millions, of swallows and martins. After the latter incident, Northern House Martin *Delichon urbica* populations in Switzerland the following year were reduced by an estimated 25–30%. Such climatic extremes that occurred in spring or late summer in particular parts of the breeding range have been recorded at approximate mean frequencies of 2–10 per century.

In addition to weather effects, many migrants die each year from collisions with tall masts and buildings. Average daily mortality in many bird species can clearly be much greater during migration periods than during stationary periods. Despite the heavy losses of birds on migration, it may be assumed that migration persists in the long term because the costs (in terms of associated mortality) are more than offset by the benefits (in terms of improved overall survival and breeding success).

White Storks *Ciconia ciconia* thermal-soaring on migration

Glossary

German terms that are well established in the migration literature are shown in italics. They date from the time before the Second World War, when German research was pre-eminent in migration studies, and when many underlying concepts were established. Not all the terms defined here are used in this book, but can be found elsewhere in the migration literature.

Abmigration. The movement of a bird from one flyway to another, resulting in long breeding or wintering dispersal.

Accidental. A vagrant or stray species, not normally found in the region concerned.

Adaptation. Any feature of an organism that improves its ability to survive or reproduce, compared with ancestral forms or with other coexisting individuals.

Adult plumage. The final breeding plumage of a bird; also called definitive or basic plumage.

Afrotropical Region. The biogeographical region embracing the portion of Africa south of the Sahara Desert, including Madagascar and other nearby islands.

Alloheimy. Applied to populations which occur in different (usually adjacent) geographical wintering areas from one another.

Allopatry. Applied to populations that occur in geographically different areas from one another, with little or no distributional overlap.

Allozyme. A form of protein, detectable by electrophoresis, produced by a particular allele at a single gene locus. Can reflect relationships between species and populations.

Altitudinal migration. Seasonal return movement between higher and lower elevations in the same region.

Annual cycle. The yearly cycle of activities, made up mainly of breeding, moult and migration, which in most bird species occur in a fixed order at about the same times each year.

Anti-trade winds. Upper winds at low latitudes that flow counter to the tradewinds below them.

Arched migration. A curved, rather than straight, migration route.

Aspect ratio. A measure of the shape of the wing, defined as wingspan squared divided by wing area. It is thus dependent on two other measurements, and varies from about 5 in gallinaceous birds to more than 20 in albatrosses.

Arrested (suspended) moult. A moult which is stopped part way through, while the bird completes some other activity such as migration, after which the process of feather replacement is resumed where it left off.

Austral migration. Migration of southern hemisphere breeding birds.

Australasia. The biogeographical region embracing Australia, New Zealand, New Guinea, Tasmania, Timor, New Caledonia and several smaller islands.

Azimuth. A compass bearing towards the sun in its present position (only the direction, not the vertical angle).

Barrier. Any physical feature that would usually prevent or restrict the movement of organisms from one side to the other.

Basal metabolic rate (BMR). The lowest rate at which an inactive bird consumes energy. See also **Metabolic rate**.

Beaconing. Goal-oriented movement based on following a stimulus emanating from the goal.

Bergman's Rule. The tendency for individuals of a warm-blooded species to increase in body size towards the cooler parts of the range.

Bi-coordinate navigation. Navigation based on a virtual grid map involving gradients in two variables running in different directions from one another, enabling particular locations to be fixed. Indicates a 'map-sense'.

Bottleneck. In migration studies, a narrow corridor through which migrants must pass (e.g. a mountain pass or land bridge).

Botulism. A paralysis caused by a neurotoxin produced by *Clostridium botulinum*, an anaerobic bacterium that thrives in rotting organic matter in shallow stagnant waters or mud during warm weather. It is a major cause of waterbird mortality in late summer, especially in western North America.

Breeding area/range. In migratory birds, the area in which populations reproduce.

Breeding dispersal. The distance between a breeding site in one year and a different breeding site in another year.

Broad-front migration. Migration across a region, with no apparent streaming by topographic or other features.

Calendar effect. The tendency of birds to alter a physiological process with date; for example, moulting or migrating more rapidly, or at an earlier age, as the season advances.

Calendar migrants. An earlier term, used for obligate migrants whose autumn departure is triggered primarily by date, and little influenced by other factors such as prevailing food supply.

Chain migration. The situation in which migratory populations that breed furthest north also winter furthest north, while populations that breed furthest south also winter furthest south, thus retaining the same latitudinal sequence summer and winter. Sometimes one population may occupy an area in summer, and another from higher latitude in winter.

Circadian. From the Latin *circa* meaning 'about', and *dies* meaning 'day'. Literally 'about a day'. Usually applied to patterns of activity, such as sleep and wakefulness, which comprise the daily cycle of birds, and which deviate from a strictly diurnal schedule in individuals kept under constant light or darkness.

Circadian clock. An internal clock that exists in nearly all organisms to time physiological and behavioural processes during the diurnal cycle. It can operate at the molecular level.

Circadian cycle/rhythm. A regular cycle of bodily activities and functions which, in the absence of external cues, repeats with a periodicity of about 24 hours, and is driven by the circadian clock (see above).

Circannual cycle/rhythm. From the Latin *circa* meaning 'about' and *annus* meaning 'year'. Literally 'about a year'. Applied to processes, such as breeding, moult and migration, which comprise the annual cycle of a bird, and are controlled partly by an endogenous rhythm. Circannual organisation becomes apparent when seasonal processes deviate from a strictly annual schedule in individuals kept under a limited range of experimentally constant photoperiodic conditions. Species differ in the degree to which circannual cycles continue without external cues.

Coexistence. Living together in the same local community.

Community. An assemblage of organisms that live in a particular habitat and may interact with one another.

Compass orientation. Orientation in a compass direction (either geographic or magnetic). Keeping a constant angle towards an external reference system to give a straight migration direction.

Competition. Any interaction that is detrimental to one or both participants, excluding parasitism and predation. Inter-specific competition occurs among species that share limited resources. It may affect the survival and breeding success of individuals, or the size and distribution of populations. Intra-specific competition occurs among individuals in the same population, and mainly affects their survival or breeding success. Competition may involve depletion where some individuals reduce the amount of a resource available to others, or interference where some individuals, by aggression or other means, reduce access to a resource by other individuals.

Competitive exclusion. The principle that when two species limited by the same resource live together in the same area, one eventually outcompetes and eliminates the other.

Complete migrants. Obligate migrants in which the entire population of a region migrates every year.

Conditional response. Behaviour that is dependent on specific conditions.

Connectivity. In migration studies, the link between breeding and non-breeding areas. More specifically, the extent to which birds from a particular breeding area use the same non-breeding (wintering) area. Migratory connectivity can be weak (diffuse) or strong, depending on the degree to which individuals from different breeding areas mix in non-breeding areas and vice versa.

Contranuptial area. Non-breeding (or wintering) area, usually reached after a migration.

Coriolis force. An apparent force resulting from the earth's daily rotation which causes objects in motion, including wind and ocean currents, to be deflected to the right in the northern hemisphere and to the left in the southern hemisphere. The force varies from zero at the equator to a maximum at the poles.

Crosswind. A wind blowing more or less perpendicular to the travel direction of a bird or other moving object.

Cue conflict. The situation in which different environmental cues, which normally give the same message, give different messages. More specifically, when under experimental conditions, celestial and magnetic compasses indicate different directions.

Deferred maturity. Breeding for the first time, not in the first year of life, but in a later year, varying between species, and to some extent between the individuals in a population.

Deferred (return) migration. The situation in which young birds, having reaching their non-breeding areas, remain there for one or more years before returning to the breeding range as they approach reproductive age. See also **Graded migration**.

Deflection upcurrent. A rising air current created when horizontally blowing wind strikes a slope, cliff or building, and is deflected upwards.

Delayed maturity. See **Deferred maturity**.

Delayed migration. See **Deferred migration**.

Density dependent. Applied to a population regulating factor which allows numbers to increase when they are low, and causes numbers to decrease when they are high, resulting in density-related changes in reproduction or mortality, immigration or emigration. Potential regulating factors include competition for food and other resources, parasitism (or infectious disease), and predation, all of which can affect a greater proportion of individuals as their numbers rise.

Density independent. Applied to a limiting factor, such as severe weather or other natural disasters, which affects proportions of individuals which bear no consistent relationship to population size. Such factors tend to cause fluctuations in populations, rather than regulate them within restricted limits.

Differential migration. The situation in which some distinguishable classes of individuals (ages, sexes, and sometimes races) differ in one or more aspects of migration, notably timing and distance.

Digiscope. In ornithology, a digital camera attached to a telescope, often used to photograph birds.

Disorientation. Inability to follow a consistent direction on migration.

Dispersal. In ecology, the movement of an individual from its current place of residence, usually of no fixed direction or distance (see also **Natal dispersal, Breeding dispersal** and **Non-breeding dispersal**). In biogeography, range extension over a pre-existing barrier.

Dispersion. The spatial distribution of individuals within a local population.

Displacement experiments. The capture of wild birds, their transport to a different site, release and follow-up. Commonly used to study orientation and homing behaviour.

Diurnal. Active during daylight hours.

Diurnal cycle. The daily cycle of activities, including feeding, rest, sleep and various bodily functions, which affect metabolic rate, body temperature, and hormonal activity.

Diversion line. See **Leading line**.

DNA. Deoxyribonucleic acid, the double-stranded molecule that encodes and transmits genetic inheritance. It is present in the nucleus of every living cell. See also **Mitochondrial DNA**.

Doldrums. A moveable zone of calm or light variable winds in the lower atmospheric layers, over seas situated near the equator, reflecting the position of the Inter-Tropical Convergence Zone.

Doppler effect. An apparent change in the frequency of waves due to a relative motion between transmitter and receiver. It occurs when the transmitting source is moving towards or away from the reception point, or whose source is fixed while the reception point is moving (as in satellite-based radio-tracking).

Drag. A force that resists the motion of a body through a gas or liquid. For birds flying through the atmosphere, three types of drag are experienced: (1) induced drag created by trailing vortices; (2) parasite drag created by the outline of the bird's body; and (3) profile drag created by the bird's flapping wings.

Drift. The extent to which a migrating bird is blown off course by wind.

Dry weight. The weight of a carcass minus the water fraction.

Dynamic soaring. A form of soaring, mainly over water, dependent on a gradient of increasing wind speed upwards from the surface. Typically, a bird heads into the wind to gain height, then turns to glide with the wind, gradually losing height, and then repeats the process, again and again, thereby travelling over long distances.

Eclipse plumage. Dull plumage worn by male ducks between successive bright plumages. Equivalent to 'winter plumage' or 'alternate plumage' in some other species, but worn for a shorter period post-breeding.

El Niño-Southern Oscillation (ENSO). An oceanic–atmospheric phenomenon, centred on the Pacific Ocean, that affects climate over large parts of the world.

Over a period of years, sea temperature and climate fluctuate between El Niño and La Niña conditions. In regions where El Niño brings more rain, La Niña brings less, and vice versa. Through this phenomenon, periodic climatic shifts can thereby be linked, fluctuating in phase, in widely separated regions. Its state can conveniently be measured as the difference in sea temperature between Darwin (Australia) and Tahiti. Low (negative) values reflect El Niño conditions and high values reflect La Niña conditions.

Elliptical migration. See **Loop migration**.

Emigration. Dispersal or migration of organisms away from an area.

Endogenous (rhythm). Internally generated and not dependent on (but may be modified by) an external stimulus. Usually applied to seasonal and diurnal processes, such as gonad growth and moult, which can occur autonomously in birds kept under a limited range of photoperiodic conditions.

ENSO. See **El Niño**.

Equinox. The twice-yearly date when, owing to the changing tilt of the earth, the sun appears to cross the equator and when, throughout the world, day and night are of equal length, usually about 21 March (spring equinox) and 21 September (autumn equinox).

Eruption. In migration studies, a massive emigration from a particular region. See also **Irruption**.

Escape movements. Mass movements from areas where conditions suddenly become hazardous for survival, as after deep snowfall.

Evolutionarily stable strategy. Applied to situations in which different types of individuals in a population (say, males and females, or migrants and non-migrants) occur in a specific ratio that, in the conditions prevailing, cannot be bettered. This is because the advantage of being any one type depends on its proportion with respect to the other type, so that, through natural selection, any deviation from the optimal ratio will be quickly rectified.

Facultative. Applies to organisms that can adopt alternative modes of living. They are able, but not obliged, to behave in any of the ways specified. More specifically, individual facultative migrants have the choice of whether to migrate or not.

Fall. The sudden descent from aloft and settling of large numbers of migrants, as they encounter adverse weather or seek refuge after a long flight.

Fan/funnel migration. Movement through a narrow bottleneck expanding to a broad front, or the reverse.

Fat-free dry weight. The weight of a carcass, minus water and fat (lipid). The same as lean dry weight.

Fat-free weight. The weight of a carcass, minus the entire fat (lipid) content. The same as lean weight.

Fidelity. Faithfulness, usually applied to a locality or mate; see **Mate-fidelity**, **Site-fidelity**.

Flight feathers. The large feathers of the wings, including primaries (attached to the bones of the 'hand') and secondaries (attached to the ulna, the inner bone of the 'forearm').

Flight line. The flight track of an individual bird.

Flight muscles. The 'breast' muscles, including the pectoralis and supracoracoideus muscles of both sides, concerned respectively with downward and upward strokes of the wings.

Flight style. Two main styles involve flapping flight (powered by muscles) and soaring–gliding flight (powered by air currents). The former can be divided into: continuous flapping, flap–gliding and bounding, a distinction useful in calculating the energy needs and flight ranges of migrating birds.

Flyway. An established air route used year after year by large numbers of migratory birds, as illustrated by the twice-yearly movement of large numbers of soaring birds through Panama.

Fresh weight. The weight of a carcass preserved without any evaporative water loss, equivalent to live weight.

Fuel energy height. The height which a bird would reach if all of its stored fuel energy was converted into work by the flight muscles, and used to lift the bird against gravity.

Gause's Principle. The principle stating that no two species can coexist at the same locality if they are limited by resources and have identical ecological requirements. See **Competitive exclusion**.

Gene flow. The movement of genetic material within a population or between populations caused by the dispersal of individuals (or gametes in some animals, pollen or seeds in plants), or the hybridisation of individuals from different taxa.

Genotype. The totality of genetic factors that make up the genetic constitution of an individual. Alternatively, those components of an individual's phenotype that result from genetic (as opposed to environmental) influences. See also **Phenotype**.

Glide coefficient. The ratio between the horizontal distance covered by a bird and its altitude loss in a glide.

Glide polar. The ratio of sinking speed against forward speed in a glide.

Gliding. Coasting downwind on outstretched wings, usually with closed tail, losing height.

Goal orientation. Direct movement towards a known, desired destination.

Graded migration. The situation in which the younger age classes of a species do not return to the breeding range in their first year, but return progressively nearer to the breeding range on each successive spring migration until they reach breeding age.

Gradient map navigation. Goal orientation based on values of at least two gradients of any physical variables that vary systematically over a large region and run in different directions from one another. See also **Bi-coordinate navigation**.

Great circle route. The shortest route between two points on the earth's surface. Because the earth is a sphere, a great circle route requires progressive change in direction along the route. See **Orthodrome** as opposed to **Loxodrome** or **Rhumbline route**.

Heading. The direction of flight, as opposed to the track which results from the net effect of bird speed and direction, and wind speed and direction.

Headwind. A wind aligned against the preferred direction of travel, slowing the bird with respect to the ground.

Holarctic Region. The extra-tropical zone of the northern hemisphere, which includes both the Nearctic and Palaearctic regions.

Hyperphagia. The act of eating more per day than usual, in order to put on weight.

Immature. Non-specific term for 'pre-adult'. Often the stage, with distinct plumage(s), between juvenile and adult plumages.

Immigration. The arrival of new individuals from elsewhere; in biogeography, often to a previously unoccupied area.

Indomalayan Region. The biogeographical region embracing tropical and subtropical parts of Southeast Asia and nearby islands.

Intermittent migration. See **Split migration**.

Internal clock. An endogenous timekeeper, which leads an animal to perform the same activities at approximately the same times of day, or the same times of year. See **Circadian clock**.

Inter-Tropical Convergence Zone (ITCZ). The meteorological equator, a zone of low pressure circling the earth, sandwiched between the northwest and southeast tradewinds, and where rain falls. During the course of each year, the zone moves north then south of the equator, roughly between 5°N and 5°S, and brings distinct wet and dry seasons to large parts of the tropics, but six months out of step between the two hemispheres. Rain falls heaviest and longest on the equator, and declines in duration and amount with increasing distance to the north and south. In sea areas, the position of the zone is marked by almost windless conditions – the doldrums.

Intra-tropical migration. Migration entirely within the tropics.

Invasion. In migration studies, the same as irruption. More generally, the colonisation of an area by a species formerly absent there.

Irruption. In migration studies, a massive immigration to a particular region. More generally, a form of migration in which the proportion of individuals that participate, and the distances they travel, varies greatly from year to year.

Isotherm. Line of equal temperature.

Isotope. A distinct form of an element distinguished by the nuclear mass of its atoms.

ITCZ. See **Inter-Tropical Convergence Zone**.

Jet stream. A stream of fast flowing air at high altitude, usually travelling at more than 30 m per second (108 km per hour), and frequently above 50 m per second (180 km per hour).

Juvenile. Bird in its first feathered plumage, following the down of the nestling, and distinct from later plumages. Sometimes also used (as here) for birds in their first year of life.

Kalendervögel. Literally 'calendar bird'. An old term, applied to populations which migrate at about the same dates each year, to some extent regardless of external conditions and hence under strong endogenous control. Now associated with obligate migration. See **Calendar migrants**.

Kettle. A North American term for a group of soaring raptors, circling upwards in a thermal.

La Niña. See **El Niño-Southern Oscillation**.

Leading line. A topographical feature, such as a coastline, escarpment or river valley, along which migrating birds tend to fly. Usually runs approximately in the preferred direction of migration, leading to concentrated streaming for part of the journey.

Lean dry weight. The dry weight of a carcass, minus the entire fat (lipid) and water contents.

Lean weight. The weight of the bird, minus the entire fat (lipid) content.

Leapfrog migration. The situation in which higher latitude breeding populations over-fly lower latitude breeding populations to winter beyond them, thus reversing their latitudinal sequence of distribution between summer and winter.

Leewave. An oscillating airstream of varying amplitude that sometimes forms on the lee or downward sides of mountain ranges under specific wind and temperature conditions, most readily when horizontal wind is deflected over the mountains. Leewaves commonly reach 2–3 km, and occasionally up to 10 km, above ground. Also called mountain wave or standing wave.

Leitlinie. Leading line.

Limiting factor. The resource or environmental parameter that most limits the size or distribution of a population.

Lipid weight. The weight of lipid (fat) in a carcass.

Live weight. The weight of the living bird.

Long-distance migrant. A loose term, usually applied to species that migrate thousands of kilometres between one continent to another.

Loop migration. The situation in which birds take markedly different routes on their outward and return journeys. Also called **Ellyptical migration.**

Loxodrome. A route between two points on the earth's surface which requires a constant direction throughout. Same as **Rhumbline route**.

Magnetic compass. A compass based on the earth's magnetic field.

Magnetic declination. The deviation of magnetic north from true north.

Magnetic inclination. The angle of the earth's magnetic force lines, which varies from horizontal at the magnetic equator to vertical at the magnetic poles.

Magnetic poles. The two points on the earth's surface between which compass needles line up. Unlike the geographical poles which are in constant position at each end of the earth's axis, the magnetic poles change position somewhat over time.

Map-based navigation. Ability to determine spatial position with respect to home, or at least the direction toward home, based solely on information at the current distant position. Equivalent to map-and-compass navigation, or 'map sense'.

Mass. Frequently confused with weight (W), which is the product of mass and gravity. Mass is reported in grams or kilograms, whereas weight is reported by physicists in Newtons (N), a unit of force. In most of the ornithological literature (as here), the terms weight and mass are used interchangeably.

Mate-fidelity. Tendency to remain with, or re-pair with, the same breeding partner in different years.

Maximum range speed. See **Vmr**.

Metabolic rate. The rate at which food and body reserves are converted to energy, including heat production, and usually expressed as kcal per bird per day. Basal metabolic rate (BMR) is the minimal metabolic rate possible, when the animal is inactive, in a post-absorptive state, in the thermoneutral zone, and not engaged in any physiological activity (such as moulting) that would require additional energy. Standard metabolic rate is the same as BMR, except that the animal is below the thermoneutral zone, usually at 0°C.

Microsatellite DNA. A short region of nuclear DNA in which a motif of a small number (often 2–5) bases is repeated, head to tail, a number of times at a particular locus.

Migrant. A bird that undertakes special movements between widely separated breeding and non-breeding areas.

Migration. Applied to birds, a regular seasonal movement between separate breeding and non-breeding (wintering) areas. Applied to other animals, sometimes used more broadly for travel that occurs in a periodically and geographically predictable way, whether just once or many times. In other branches of ecology, also used for range extension or shift, as in post-glacial times.

Migratory disposition. A state of readiness to migrate. See *Zugdisposition*.

Migratory divide. An imaginary line on a map which separates breeding populations migrating in different directions. On one side of the line birds migrate in one direction (say southwest), and on the other of the line in another direction (say southeast).

Migratory fat. The fat laid down as fuel (usually along with protein) in preparation for migration.

Migratory restlessness. A state of high activity in captive birds, thought to indicate their readiness to migrate. Usually occurs by day in diurnal migrants and by night in nocturnal migrants.

Migratory state/disposition. A state of readiness for migration, usually with appropriate body reserves, directional preferences, and weather responses.

Migratory stream. A stream of migrants concentrated by topographic features, such as a coastline, escarpment or valley.

Minimum power speed. See **Vmp**.

Mirror-image migration. Migration along the correct angle to the north–south axis, but on the wrong side of that axis; for example, southeast rather than southwest.

Misorientation. Taking the wrong migratory direction, but a consistent direction.

Mitochondrial DNA (mt DNA). A small haploid extranuclear DNA molecule, housed in the mitochondria in the cytoplasm of cells. It does not recombine at fertilisation, but is clonally and maternally inherited, transmitted from mother to offspring of both sexes and passed unchanged (except for mutation) down the generations.

Montane. Pertaining to mountains.

Mosaic map navigation. Goal orientation based on learned spatial relationships among features of the landscape and their relation to home. Equivalent to 'topographic map'.

Moult. Scheduled plumage renewal, or partial renewal, but excluding the unscheduled replacement of feathers lost accidentally.

Moult migration. Movement to a special area for moult. Typical participants include some ducks and other waterbirds which, after arrival at the moulting site, shed all their flight feathers simultaneously and become temporarily flightless until the new ones are grown.

Mutation. An error in the replication of a nucleotide sequence or any other alteration in the genome that is not due to recombination.

Narrow front migration. Occurs when migrants from a wide area are concentrated to pass along coastlines, peninsulas or through narrow valleys, or other topographic situations which channel them.

Natal dispersal. The movement of an individual from birthplace to breeding place.

Natural selection. The elimination from a population, through differential survival and reproduction, of those individuals of inferior fitness.

Navigation. Following a specific course to a distant goal. Contrast with **Orientation**, which is compass-directed only.

NDVI (Normalised Difference Vegetation Index). A measure based on satellite images that reflects the primary productivity (or the amount of vegetation related to the level of photosynthetic activity) of particular geographical regions. It is obtained via the red and infrared channels of the Advanced Very High Resolution (AVHRR) sensor of U.S. National Oceanic and Atmosphere Administration (NOAA) satellites.

Nearctic Region. The extra-tropical biogeographical region of North America.

Neotropical Region. The biogeographical region embracing southern Mexico, the West Indies and South America.

Niche. The total requirements of a population or species for resources and physical conditions.

NOA. See **North Atlantic Oscillation**.

Nocturnal. Active during the hours of darkness. Nocturnal migrants include normally diurnal species which migrate at night, as well as normally nocturnally active species, such as owls.

Nomadic. Having no fixed spatial pattern of migration, and no fixed directional preferences.

Nomadism. Movements which lead to irregular changes in distribution.

Non-breeding area. Area occupied between breeding seasons, or by immature birds at any season, usually after a migration. Also called wintering or contranuptial area.

Non-breeding dispersal. The distance between the non-breeding site of an individual in one year and a different non-breeding site of the same individual in another year. See also **Wintering dispersal**.

Non-compass orientation. Oriented movement based on external stimuli, such as topographic features, that do not provide compass directions.

North Atlantic Oscillation (NAO). A large-scale climate phenomenon influencing weather in this region. Positive values indicate warmer and wetter winter–spring weather in northwest Europe, and the opposite conditions in southern Europe.

Obligate. Applies to organisms that have to behave in a particular way to survive and whose behaviour is innate. More specifically, individual obligate migrants migrate every year, and do not have the option of migrating or not, their behaviour being genetically fixed.

Opportunist breeding. The ability to breed at any time of year, whenever food is sufficiently plentiful, unconstrained by season.

Oriental Region. See **Indomalayan Region**.

Orientation. The direction in which a migrant bird heads to a non-specific goal. See also **Compass orientation**.

Orographic lift. Lift gained by soaring birds when they exploit the upward-directed air current caused by a horizontal wind striking a cliff or slope.

Orthodrome. The shortest (great circle) route between two points on the earth's surface, and which requires continual changes in direction. See **Great circle route**.

Ostreue. Site-fidelity (or philopatry), usually in the context of homing after migration.

Outward migration. Movement from breeding to non-breeding areas.

Overshoot. Migration in the customary direction, but further than usual, leading to the appearance of birds in spring or autumn beyond the usual breeding or wintering areas.

Palaearctic Region. The biogeographical region of extra-tropical climates in Eurasia and northernmost Africa.

Parapatry. Applied to different populations occurring in adjacent areas to one another. See also **Allopatry**.

Partial migration. A situation in which some birds from a given breeding area migrate away for the non-breeding season, while others remain in the breeding area year-round. See also **Differential migration**.

Path-integration. Goal orientation based on unspecified information gained on the outward journey that is continually updated relative to the point of departure and thus may make more direct return routes possible.

Pelagic. Ocean-dwelling, usually far from land.

Peninsula effect. The tendency of landbird migrants to travel along peninsulas, thereby reducing the time they must fly over water.

Phenotype. The totality of the characteristics of an individual that results from genetic and environmental influences. See also **Genotype**.

Philopatry. The tendency of an individual to stay in, or return to, its home area (or birthplace).

Photoperiod. The light part of the daily cycle of day–night. The time between sunrise and sunset, or sometimes including civil twilight, and hence the period between dawn and dusk.

Piloting. Goal-directed orientation based on following a sequence of familiar landmarks. See also **Beaconing** and **Navigation**.

Post-fledging dispersal. The movements in any direction from the natal site of newly independent young. In migratory species, such movements usually precede the first migration.

Power–weight ratio. The weight of the wing muscles expressed as a percentage of total body weight.

Pre-alternate moult. The pre-breeding moult into breeding (alternate) plumage.

Pre-basic moult. The post-breeding moult into winter (basic) plumage.

Primaries. The outer large flight feathers, grown from the 'hand' of a bird. Along with the secondaries, these feathers comprise the main 'flight feathers'.

Proximate factors. External stimuli (such as specific daylengths) which are used as cues by a bird to trigger preparation for breeding, migration or other events, or

as time-keepers (*Zeitgebers*) to set their endogenous time programmes at appropriate times of year.

Quiescent period. A part of the annual cycle when a bird is not breeding, moulting or migrating.

Radio-telemetry. The transfer of information collected from a remote source (such as a bird) to a receiver.

Radio-tracking. The location and tracking of a radio-marked individual from a signal emitted frequently by the radio.

Rains migration. Movements in response to rains and rainy seasons, usually in tropical and subtropical areas.

Range. Large-scale geographical distribution.

Rectrices. Large quill feathers of the tail.

Refractory period. A part of the annual cycle when an animal will not respond to otherwise stimulatory environmental cues (such as daylength) and reach a condition that allows breeding, moult or migration. After spontaneous or environmentally induced breaking of the refractory period, the animal once again responds to stimulation.

Refuge. In ecology, an area in which climate and vegetation type have remained relatively unchanged, while adjacent areas have changed markedly. In wildlife management, a protected area of suitable habitat.

Remiges. Large flight feathers of the wing.

Remote sensing. The technology of acquiring, processing and interpreting information by remote means. A common application of remote sensing is the use of satellite-borne instruments to obtain images of the earth's surface, or to determine the location of objects on the earth's surface.

Reproductive window. The period each year when the gonads are in reproductive state, so as to allow breeding if environmental conditions are favourable. Some birds may breed over only a small part of their reproductive window, owing to inadequate food or other external influences.

Resident. Remaining year-round in the same area; applied to individuals and to populations.

Return migration. Movement from non-breeding to breeding areas.

Reversed migration. Migration in a direction opposite to that expected for the time of year. Also known as reversed-direction migration, when applied to the whole journey.

Rhumbline route. A route between two points on the earth's surface that can be followed by maintaining a constant heading (direction) throughout.

Route-based navigation. Homing based on unspecified information perceived during the outward journey; can result in learned routes used repeatedly.

Rückzug. Reversed migration. See also *Umkehrzug*.

Sabbatical. A year-off breeding by individual established breeders, usually applied to albatrosses and other seabirds.

Satellite-telemetry/tracking. A method of determining the map coordinates or other information from a bird attached with a small radio-transmitter. The transmitter (known as a platform terminal transmitter, PTT) emits a signal that is picked up by the Argos array of satellites orbiting the earth. By measuring the Doppler shift of the signal, the system can determine the exact distance between the earth and the satellite, and then, knowing the parameters of the orbit, calculate the coordinates of the transmitter.

Sea thermals. Updrafts over tropical and subtropical seas, some 5–30°N and S, in regions dominated by tradewinds. The warmer surface waters heat the cooler air which therefore rises, producing thermals. Because the temperature differential persists day and night, so do the thermals, unlike the situation on land.

Seasonal. Applied to areas experiencing marked and regular changes in climate (and biological productivity) through the year.

Sedentary. Remaining in the same limited area; non-migratory. See **Resident**.

Serial moult. A pattern of wing-moult in which the bird has more than one locus of feather growth in the primaries or secondaries at once. Feather replacement progresses from feather to feather in wave-like manner from each locus. Found mainly in larger birds, such as albatrosses, it contrasts with the usual situation, in which the bird has only one moult locus in the primaries and one in the secondaries at one time. Also known as step moult, or *Staffelmauser* in German.

Short-distance migrant. A loose term, usually applied to species that migrate up to hundreds of kilometres within a continent.

Short-stopping. Curtailment of an outward journey, usually in the presence of abundant food. If this situation persists, it can give rise to a new wintering area.

Site-fidelity. The tendency to remain in, or return to, the same localities in different years.

Slope-soaring. Soaring in the updrafts produced when a horizontal wind is deflected upwards when it strikes a slope.

Soaring. Gliding in circles with spread wings and tail, while gaining altitude in a rising air current.

Solstice. The dates each year when the tilt of the earth ensures that the sun reaches at each latitude its highest (summer solstice) and lowest (winter solstice) points in the sky, and when days reach their longest and shortest respectively, around 21 June and 21 December. The summer solstice in the northern hemisphere corresponds with the winter solstice in the southern hemisphere.

Speed for best glide. The forward speed for which the glide ratio (ratio of forward speed to sink) is at a maximum.

Split migration. A migration that is separated into two or more main parts by breaks of weeks or months, much longer than needed merely to replenish spent body reserves. Each part of the migration may be associated with a separate period of migratory fat deposition. Also sometimes known as step migration or intermittent migration.

Split moult. A moult that is separated into two or more main parts by periods when moult is arrested. Different feathers are replaced in each part.

Staffelmauser. See **Serial moult**.

Staging area. A place where large numbers of birds break their migratory journeys, usually to refuel but sometimes also to moult.

Standard direction. The regular direction followed by a population on migration.

Star compass. A compass based on steller cues, usually indicating geographical north or south.

Step migration. See **Split migration**.

Step moult. See **Serial moult**.

Stopover. A pause in a migratory journey.

Stopover ecology. The study of the ecology and behaviour of migrants at stopover or staging sites.

Stopover site. A place where a bird stops on migration, including staging areas (see above).

Sub-adult. An immature in pre-adult plumage.

Sun compass. A compass based on the position of the sun in the sky. As this position changes during the day, an animal must take account of this, by way of an internal clock.

Sympatric. In the strictest sense, living in the same local community; in the more general sense, having broadly overlapping geographical ranges.

Synheimy. Applied to different populations occurring together in the same geographical wintering area.

Tailwind. Wind aligned more or less in the direction of travel, and which therefore increases a bird's speed with respect to the ground.

Telescopic migration. The situation in which birds breeding in a wide span of latitude concentrate to spend the non-breeding season together in a narrower span of latitude.

Thermal. A column of rising air, within which birds circling on outstretched wings can gain height.

Thermal corridor. A migration route that provides frequent and predictable thermals, thereby providing ideal travel conditions for soaring birds.

Thermal streets. Lines of thermals, along which birds can glide without losing height. The individual thermals are often topped with cumulus clouds, giving rise to 'cloud streets'.

Time-keeper. Usually applied to environmental cues, such as daylength, to which birds respond when preparing to migrate, breed or moult. See also **Proximate factors** and *Zeitgeber*.

Track. The course resulting from the net effect of the speed and flight direction of the bird and the speed and direction of the wind.

Tradewinds. An old maritime term for the prevailing tropical easterly winds which blow from subtropical high pressure areas in latitudes 30–40°N and S, generally from the northeast in the northern hemisphere and from the southeast in the southern hemisphere.

Tropopause. The boundary between the troposphere below and the stratosphere above, around 11 km above sea level.

Troposphere. The lower part of the atmosphere, up to about 11 km above sea level, and below the stratosphere. Almost all bird migration occurs within the lower half of the tropospere.

Ultimate factors. Factors that determine the value of a behaviour in the evolutionary sense, and lead birds to migrate, breed and moult, for example, at appropriate times of year.

Umkehrzug. Reversed migration. Also *Rückzug*.

Vagrancy. The uncommon appearance of birds outside their regular ranges, and off their usual migration routes.

Vagrant. A bird seen outside its regular range and migration route. See also **Accidental**.

Vector navigation. Migration on an innate predetermined direction (requiring compass information) for a set period (equivalent to distance); also called clock-and-compass, time-and-direction, or bearing-and-distance navigation. Usually applied to young birds migrating to unfamiliar areas.

Vmp. Minimum power speed: the speed at which the bird consumes the least energy per unit time, and which therefore gives the maximum time airborne.

Vmr. Maximum range speed: the speed at which the bird achieves the maximum flight range on the energy available. Faster than **Vmp**.

Wettervögel. Literally 'weatherbirds', that leave their breeding or staging areas in autumn in response to declining food-supplies or adverse weather, such as the onset of cold or snow, and hence at widely different dates in different years. Now associated with facultative migration.

Wind drift. Applied to migrants blown off course, usually by crosswinds.

Wind tunnel. A tunnel, equipped with a powerful fan, for driving through air at speed, facilitating the study of bird flight. The air speed can be adjusted so that the bird flies against it, thereby maintaining a constant position in a short section of the tunnel with a window, so that the bird can be kept under observation.

Wing area. The projected surface area of the wings, including the intervening part of the body, usually expressed in square metres.

Wing-loading. Body weight divided by wing area, measured with the wings outstretched. The lower the wing-loading of a species, the more buoyant the flight.

Wingspan. The distance from wing-tip to wing-tip at full extent.

Wintering area. In migratory birds, the area where populations spend the non-breeding season, usually at lower latitudes.

Wintering dispersal. The distance between the wintering site of an individual in one year and a different wintering site of the same individual in another year.

Zeitgeber. Time-keeper, usually applied to the environmental cues, such as daylength, to which birds respond when preparing to migrate, breed or moult.

Zugdisposition. The state of readiness to migrate, usually with appropriate body reserves, directional preferences and weather responses. See also **Migratory state/disposition**. The older German literature distinguishes between migratory disposition (*Zugdisposition*), meaning readiness to migrate in general terms, and actual preparedness for take-off (*Zugstimmung*).

Zugstau. The accumulation of birds ready for take-off.

Zugstimmung. Readiness for take-off.

Zugunruhe. Migratory restlessness, a state of high-activity in captive birds thought to indicate their readiness to migrate. Usually occurs by day in diurnal migrants, and by night in nocturnal migrants. It is not necessarily absent in birds from resident populations, but is usually much less marked than in birds from migratory ones.

Zugwege. Flight line.

Zwischenzug. Major movements between the main migration seasons.

References

Able, K. P. (1977). The orientation of passerine nocturnal migrants following offshore drift. *Auk* 94: 320–330.

Able, K. P. (1980). Mechanisms of orientation, navigation and homing. pp. 283–373 in *Animal migration, orientation and navigation* (ed. S. A. Gauthreaux). London, Academic Press.

Able, K. P. (1982a). Field studies of avian nocturnal migratory orientation 1. Interaction of sun, wind and stars as directional cues. *Anim. Behav.* 30: 761–767.

Able, K. P. (1982b). Skylight polarization patterns at dusk influence migratory orientation in birds. *Nature* 299: 550–551.

Able, K. P. (1989). Skylight polarization patterns and the orientation of migratory birds. *J. Exp. Biol.* 141: 241–256.

Able, K. P. (1993). Orientation cues by migratory birds: a review of cue-conflict experiments. *Trends Ecol. Evol.* 8: 367–371.

Able, K. P, ed. (1999). *Gatherings of angels*. Ithaca, Cornell University Press.

Able, K. P. & Able, M. A. (1995). Interactions in the flexible orientation system of a migratory bird. *Nature* 375: 230–232.

Able, K. P. & Belthoff, J. R. (1998). Rapid 'evolution' of migratory behaviour in the introduced House Finch of eastern North America. *Proc. R. Soc. Lond. B* 265: 2063–2071.

Able, K. P. & Cherry, J. D. (1986). Mechanisms of dusk orientation in White-throated Sparrows (*Zonotrichia albicollis*) – clock-shift experiments. *J. Comp. Physiol.* A 159: 107–113.

Adkisson, C. S. (1996). Red Crossbill. pp. 1–24 in *The birds of North America*, No. 256 (eds A. Poole & F. Gill). Philadelphia, Academy of Natural Sciences and Washington, D.C., American Ornithologists' Union.

Adamcik, R. S., Todd, A. W. & Keith, L. B. (1978). Demographic and dietary responses of Great Horned Owls during a Snowshoe Hare cycle. *Can. Field Nat.* 92: 156–166.

Adriaensen, F. & Dhondt, A. A. (1990). Population dynamics and partial migration of the European Robin (*Erithacus rubecula*) in different habitats. *J. Anim. Ecol.* 59: 1077–1090.

Adriaensen, F., Ulenaers, P. & Dhondt, A. A. (1993). Ringing recoveries and the increase in numbers of European Great Crested Grebes *Podiceps cristatus*. *Ardea* 81: 59–70.

Adriaensen, F., Verwimp, N. & Dhondt, A. A. (1998). Between cohort variation in dispersal distance in the European Kestrel *Falco tinnunculus* as shown by ringing recoveries. *Ardea* 86: 147–152.

Aebischer, A., Perrin, M., Krieg, M., Struder, J. & Meyer, D. R. (1996). The role of territory choice and arrival date on breeding success in the Savi's Warbler *Locustella luscinioides*. *J. Avian Biol.* 27: 143–152.

Aebischer, N. G., Green, R. E. & Evans, A. D. (2000). From science to recovery: four case studies of how research has been translated into conservation action in the U.K. pp. 43–54 in *Ecology and conservation of lowland farmland birds* (eds N. J. Aebischer, A. D Evans, P. V. Grice, & J. A. Vickery). Tring, British Ornithologists' Union.

Aebischer, N. J. (1995). Philopatry and colony fidelity of Shags *Phalacrocorax aristotelis* on the east coast of Britain. *Ibis* 137: 11–18.

Aebischer, N. J. & Potts, G. R. (1994). Quail *Coturnix coturnix*. pp. 222–223 in *Birds in Europe. Their conservation status* (eds G. M. Tucker & M. F. Heath). Cambridge, UK, Birdlife International.

Afik, D. & Karasov, W. H. (1995). The trade-offs between digestion rate and efficiency in warblers and their ecological implications. *Ecology* 76: 2247–2257.

Agostini, N., Baghino, L., Coleiro, C., Corbi, F. & Premuda, G. (2002). Circuitous autumn migration in the Short-toed Eagle (*Circaetus gallicus*). *J. Raptor Res.* 36: 111–114.

Aidley, D. J. & Wilkinson, R. (1987). The annual cycle of six *Acrocephalus* warblers in a Nigerian reed bed. *Bird Study* 34: 226–234.

Åkesson, S. (2003). Avian long-distance navigation: experiments with migratory birds. pp. 471–492 in *Avian Migration* (eds P. Berthold, E. Gwinner & E. Sonnenschein). Berlin, Springer-Verlag.

Åkesson, S. & Hedenström, A. (2000). Wind selectivity of migratory flight departures in birds. *Behav. Ecol. Sociobiol.* 47: 140–144.

Åkesson, S. & Sandberg, R. (1994). Migratory orientation of passerines at dusk, night and dawn. *Ethology* 98: 177–191.

Åkesson, S. & Weimerskirch, H. (2005). Albatross long-distance navigation: comparing adults and juveniles. *J. Navigation* 58: 365–373.

Åkesson, S., Hedenström, A. & Hasselquist, D. (1995). Stopover and fat accumulation in passerine birds in autumn at Ottenby, southeastern Sweden. *Ornis Svecica* 5: 81–91.

Åkesson, S., Karlsson, L., Walinder, G. & Alerstam, A. (1996). Bimodel orientation and the occurrence of temporary reverse bird migration during autumn in south Scandinavia. *Behav. Ecol. Sociobiol.* 38: 293–302.

Åkesson, S., Morin, J., Muheim, R. & Ottoson, U. (2001). Avian orientation at steep angles of inclination: experiments with migratory White-crowned Sparrows at the magnetic North Pole. *Proc. R. Soc. Lond. B* 268: 1907–1913.

Åkesson, S., Morin, J., Muheim, R. & Ottoson, U. (2005). Dramatic orientation shift of White-crowned Sparrows displaced across longitudes in the high arctic. *Curr. Biol.* 15: 1591–1597.

Akriotis, T. (1998). Post-breeding migration of Reed and Great Reed Warblers breeding in southeast Greece. *Bird Study* 45: 344–352.

Alerstam, T. (1978a). Analysis and a theory of visible bird migration. *Oikos* 30: 273–349.

Alerstam, T. (1978b). Reoriented bird migration in coastal areas: dispersal to suitable resting grounds? *Oikos* 30: 405–408.

Alerstam, T. (1979). Wind as a selective agent in bird migration. *Ornis Scand.* 10: 76–93.

Alerstam, T. (1981). The course and timing of bird migration. pp. 9–54 in *Animal migration* (ed. D. J. Aidley). New York, Cambridge University Press.

Alerstam, T. (1985) Stratregies of migratory flight, illustrated by Arctic and Common Terns, *Sterna paradisaea* and *Sterna hirundo*. pp. 580–603 in *Migration: mechanisms and adaptive significance* (ed. M. A. Rankin). *Contr. Mar. Sci. Suppl.* 27.

Alerstam, T. (1988). Findings of dead birds drifted ashore reveal catastrophic mortality among early spring migrants, especially Rooks *Corvus frugilegus*, over the southern Baltic Sea. *Anser* 27: 181–218.

Alerstam, T. (1990a). *Bird migration*. Cambridge, University Press.

Alerstam, T. (1990b). Ecological causes and consequences of bird orientation. *Experimentia* 46: 405–415.

Alerstam, T. (1996). The geographical scale factor in orientation of migrating birds. *J. Exp. Biol.* 199: 9–19.

Alerstam, T. (2001). Detours in bird migration. *J. Theor. Biol.* 194: 1–13.

Alerstam, T. (2003). Bird migration speed. pp. 253–267 in *Avian migration* (eds P. Berthold, E. Gwinner & E. Sonnenschein). Berlin, Springer-Verlag.

Alerstam, T. & Gudmundsson, G. A. (1999). Bird orientation at high latitudes: flight routes between Siberia and North America across the Arctic Ocean. *Proc. R. Soc. Lond. B* 266: 2499–2505.

Alerstam, T. & Hedenström, A. (1998a). The development of bird migration theory. *J. Avian Biol.* 29: 343–369.

Alerstam, T. & Hedenström, A. (1998b). Optimal migration. *J Avian Biol.* 29: 337–636.

Alerstam, T. & Högstedt, G. (1980). Spring predictability and leapfrog migration. *Ornis Scand.* 11: 196–200.

Alerstam, T. & Högstedt, G. (1982). Bird migration and reproduction in relation to habitats for survival and breeding. *Ornis Scand.* 13: 25–37.

Alerstam, T. & Lindström, Å. (1990). Optimal bird migration: the relative importance of time, energy and safety. pp. 331–351 in *Bird migration: physiology and ecophysiology* (ed. E. Gwinner). Berlin, Springer Verlag.

Alerstam, T. & Pettersson, S. G. (1991). Orientation along great circles by migrating birds using a sun compass. *J. Theor. Biol.* 152: 191–202.

Alerstam, T., Hedenström, A. & Åkesson, S. (2003). Long distance migration: evolution and determinants. *Oikos* 103: 247–260.

Alexander, W. B. (1933). The Swallow mortality in central Europe in September 1931. *J. Anim. Ecol.* 2: 116–118.

Alexander, W. C. (1983). Differential distributions of wintering diving ducks (Aythyini) in North America. *American Birds* 37: 26–29.

Alisauskas, R. T. (2002). Arctic climate, spring nutrition, and recruitment in midcontinent Lesser Snow Geese. *J. Wildl. Manage.* 66: 181–193.

Alisauskas, R. T. & Arnold, T. W. (1994). American Coot. pp. 127–143 in *Migratory shore and upland game bird management in North America* (eds T. C. Tacha & C. E. Braun). Lawrence, Kansas, Allen Press.

Allen, P. E., Goodrich, L. J. & Bildstein, K. L. (1996). Within- and among-year effects of cold fronts on migrating raptors at Hawk Mountain, Pennsylvania, 1934–1991. *Auk* 113: 329–338.

Alon, A., Granit, B., Shamoun-Baranes, J., Leshem, Y., Kirwan, G. M. & Shirihai, H. (2004). Soaring bird migration over northern Israel in autumn. *Br. Birds* 97: 160–182.

Alonso, J. C. & Alonso, J. A. (1992). Male-biased dispersal in the Great Bustard *Otis tarda*. *Ornis Scand.* 23: 81–88.

Alonso, J. C., Alonso, J. A. & Bautista, L. M. (1991). Carrying capacity of staging areas and facultative migration extension in Common Cranes. *J. Appl. Ecol.* 31: 212–222.

Alonso, J. C., Bautista, L. M. & Alonso, J. A. (2004). Family-based territoriality vs flocking in wintering Common Cranes *Grus grus*. *J. Avian Biol.* 35: 434–444.

Altwegg, R., Ringsby, H. & Saether, B. E. (2000). Phenotypic correlates and consequences of dispersal in a metapopulation of House Sparrows *Passer domesticus*. *J. Anim. Ecol.* 69: 762–770.

Ambuel, B. & Temple, S. A. (1983). Area-dependent changes in the bird communities and vegetation of southern Wisconsin forests. *Ecology* 64: 1057–1068.

Ancel, A., Kooyman, G. L., Ponganis, P. J., Gendner, J. P., Lignon, J., Mestre, X., Huin, N., Thorson, P. H., Robisson, P. & Lemaho, Y. (1992). Foraging behavior of Emperor Penguins as a resource detector in winter and summer. *Nature* 360: 336–339.

Anderson, A. (1990). Checklist and status of North Sea birds. pp. 89–151 in *Birds and the North Sea. 10th Anniversary Publication of the North Sea Bird Club* (ed. S. M. D. Alexander). Aberdeen, North Sea Bird Club.

Anderson, D. W. (1965). Spring mortality in insectivorous birds. *Loon* 37: 134–135.

Anderson, M. G., Rhymer, J. M. & Rohwer, F. C. (1992). Philopatry, dispersal, and the genetic structure of waterfowl populations. pp. 365–395 in *Ecology and management of breeding waterfowl* (ed. B.D. J. Batt *et al.*). Minneapolis, University of Minnesota Press.

Andersson, M. (1976). Population ecology of the Long-tailed Skua (*Stercorarius longicaudus*). *J. Anim. Ecol.* 45: 537–559.

Andersson, M. (1980). Nomadism and site tenacity as alternative reproductive tactics in birds. *J. Anim. Ecol.* 49: 175–184.

Andersson, R. (1995). Pattern of territory establishment in males, territory quality and floaters in a marginal population of Black Redstart *Phoenicurus obscurus*. *Ornis Svecica* 5: 143–159.

Andrén, H. (1992). Corvid density and nest predation in relation to forest fragmentation: a landscape perspective. *Ecology* 73: 794–804.

Andrén, H. & Angelstam, P. (1988). Elevated predation rates as an edge effect in habitat islands: experimental evidence. *Ecology* 69: 544–547.

Ankney, C. D. & MacInnes, C. D. (1978). Nutrient reserves and reproductive performance of female Lesser Snow Geese. *Auk* 95: 459–471.

Anteau, M. J. & Afton, A. D. (2004). Nutrient reserves of Lesser Scaup (*Aythya affinis*) during spring migration in the Mississippi Flyway: a test of the spring condition hypothesis. *Auk* 121: 917–929.

Antikainen, E., Skarén, U., Toivanen, J. & Ukkonen, M. (1980). The nomadic breeding of the Redpoll *Acanthis flammea* in 1979 in North Savo, Finland. *Ornis Fenn.* 57: 124–131.

Arcese, P. (1989). Intrasexual competition, mating system and natal dispersal in Song Sparrows. *Anim. Behav.* 38: 958–979.

Ardia, D. R. & Bildstein, K. L. (1997). Sex-related differences in habitat selection in wintering American Kestrels, *Falco sparverius*. *Anim. Behav.* 53: 1305–1311.

Arendt, W. J., Vargas Mora, T. A. & Wiley, J. W. (1979). White-crowned Pigeon: status rangewide and in the Dominican Republic. *Proc. Conf. S.E. Assoc. Fish Wildl. Agencies* 33: 111–122.

Arnold, T. W. (1991). Geographical variation in sex ratios of wintering American Kestrels (*Falco sparverius*). *Ornis Scand.* 22: 20–26.

Arroyo, B. E., Garcia, J. T. & Bretagnolle, V. (2002). Conservation of Montagu's Harrier *Circus pygargus* in agricultural areas. *Anim. Conserv.* 5: 283–290.

Arzel, C., Elmberg, J. & Guillemain, M. (2006). Ecology of spring-migrating Anatidae: a review. *J. Ornithol.* 147: 167–184.

Aschoff, J. (1955). Jahresperiodik der Fortpflanzung bei Warmblütern. *Stud. Gen.* 8: 742–776.

Ash, J. S. (1969). Spring weights of trans-Saharan migrants in Morocco. *Ibis* 111: 1–10.

Ash, J. S. (1980). Migrational status of Palaearctic birds in Ethiopia. *Proc. Pan-Afr. Orn. Congr.* 4: 199–208.

Ashmole, N. P. (1963). The biology of the Wideawake or Sooty Tern *Sterna fuscata* on Ascension Island. *Ibis* 103b: 297–364.

Askenmo, C. (1979). Reproductive effort and return rate of male Pied Flycatchers. *Am. Nat.* 114: 748–753.

Askins, R. A., Lynch, J. F. & Greenberg, R. (1990). Population declines in migratory birds in eastern North America. *Curr. Ornithol.* 7: 1–57.

Atkinson, P. W. (1998). The wintering ecology of the Twite *Carduelis flavirostris* and the consequences of habitat loss. PhD Thesis. University of East Anglia, United Kingdom.

Atwell, I. (2000). Using stable carbon and hydrogen isotopes to examine nutrient allotment by breeding Redhead ducks. MSc thesis, University of Sasketchewan, Saskatoon.

Austin, O. L. (1949). Site tenacity, a behaviour trait of the Common Tern. *Bird-Banding* 20: 1–39.

Avery, M., Springer, P. F. & Cassel, J. F. (1977). Weather influences on nocturnal bird mortality at a North Dakota tower. *Wilson Bull.* 89: 291–299.

Avery, M. C., Springer, P. F. & Dailey, N. S. (1978). *Avian mortality at man-made structures: an annotated bibliography*. Washington, D.C., U.S. Fish and Wildlife Service.

Axell, H. E. (1966). Eruptions of Bearded Tits during 1959–65. *Br. Birds* 59: 513–543.

Baccetti, N., Serra, L., Cherubini, G. & Magnani, A. (1999). Timing of attachment to wintering site as revealed by experimental displacements of Dunlins (*Calidris alpina*). *J. Ornithol.* 140: 309–317.

Bäckman, J. J., Petterson, J. & Sandberg, R. (1997). The influence of fat stores on magnetic orientation in day migrating Chaffinch, *Fringilla coelebs*. *Ethology* 103: 247–256.

Baggott, G. K. (1975). Moult, flight muscle hypertrophy and pre-migratory lipid deposition of the juvenile Willow Warbler, *Phylloscopus trochilus*. *J. Zool., Lond.* 175: 299–314.

Baha el Din, S. M., Salama, W., Greene, A. & Green, R. E. (1996). Trapping and shooting of Corncrakes *Crex crex* on the Mediterranean coast of Egypt. *Bird Conserv. Int.* 6: 213–227.

Bailey, E. P. & Davenport, G. H. (1972). Die-off of Common Murres on the Alaska peninsula and Umimak Island. *Condor* 74: 215–219.

Baillie, S. R. & Peach, W. J. (1992). Population limitation in Palaearctic-African migrant passerines. *Ibis* 134: 120–132.

Bain, G. A. (1980). The relationship between preferred habitat, physical condition and hunting mortality in Canvasbacks (*Aythya valisineria*) and Redheads (*A. americana*) at Long Point, Ontario. MSc thesis, London, Ontario, Univ. Western Ontario.

Bainbridge, I. & Minton, C. D. (1978). The migration and mortality of the Curlew in Britain and Ireland. *Bird Study* 25: 39–50.

Baird, J. & Nisbet, I. C. T. (1960). Northward fall migration on the Atlantic coast and its relation to offshire drift. *Auk* 77: 119–149.

Bairlein, F. (1981). Ökosystemanalyse der Rastplätze von Zugvögeln: Beschreibung und Deutung der Verteilungsmuster von ziehenden Kleinvögeln in verschiedenen Biotopen der Stationen des 'Mettnau-Reit-Illmitz-Programmes'. *Ökol Vögel* 3: 7–137.

Bairlein, F. (1985a). Body weight and fat deposition of Palaearctic passerine migrants in the central Sahara. *Oecologia* 66: 141–146.

Bairlein, F. (1985b). Dismigration und Sterblichkeit in Süddeutschland beringter Schleiereulen (*Tyto alba*). *Vogelwarte* 33: 81–108.

Bairlein, F. (1985c). Efficiency of food utilisation during fat deposition in the long-distance migratory Garden Warbler, *Sylvia borin*. *Oecologia (Berl.)* 68: 118–125.

Bairlein, F. (1985d). Offene Fragen der Erforschung des Zuges paläarktischer Vogelarten in Afrika. *Vogelwarte* 33: 144–155.

Bairlein, F. (1987). Nutritional requirements for maintenance of body weight and fat deposition in the long-distance migratory Garden Warbler, *Sylvia borin* (Boddaert). *Comp. Biochem. Physiol. A Comp. Physiol* 86: 337–347.

Bairlein, F. (1988). Herbstlichter Durchzug, Korpergewichte und Fettedeposition von zugvögeln in einen Rastgebiet in N.-Algerien. *Vogelwarte* 34: 237–248.

Bairlein, F. (1990). Zur Nahrungswahl der Gartengrasmücke *Sylvia borin*: Ein Beitrag zur Bedeutung der Frugivorie bei omnivoren Singvögeln. *Current Topics in Avian Biology, Proc. DOG Meeting Bonn* 100: 103–110.

Bairlein, F. (1991a). Nutritional adaptations in the long distance migratory Garden Warbler *Sylvia borin*. *Proc. Int. Ornithol. Congr.* 20: 2149–2158.

Bairlein, F. (1991b). Body mass of Garden Warbler (*Sylvia borin*) on migration: a review of field data. *Vogelwarte* 36: 48–61.

Bairlein, F. (1992). Recent prospects on trans-Saharan migration of songbirds. *Ibis* 134: 41–46.

Bairlein, F. (1996a). Fruit eating in birds and its nutritional consequences. *Comp. Biochem. Physiol.* 113A: 215–224.

Bairlein, F. (1996b). Long-term ecological studies on birds. *Verh. Dtsch. Zool. Ges.* 89: 165–179.

Bairlein, F. (1998). The effect of diet composition on migratory fuelling in Garden Warbler *Sylvia borin*. *J. Avian Biol.* 29: 546–551.

Bairlein, F. (2001). Results of bird ringing in the study of migration routes. *Ardea* 89 (special issue): 7–19.

Bairlein, F. (2003). Nutritional strategies in migratory birds. pp. 321–332 in *Avian migration* (eds P. Berthold, E. Gwinner & E. Sonnenschein). Berlin, Springer-Verlag.

Bairlein, F. & Henneberg, H. R. (2000). *Der Weissstorch (Ciconia ciconia) im Oldenburger Land.* Oldenburg, Isensee.

Bairlein, F. & Simons, D. (1995). Nutritional adaptations in migrating birds. *Israel J. Zool.* 41: 357–367.

Bairlein, F. & Totzke, U. (1992). New aspects on migratory physiology of trans-Saharan passerine migrants. *Ornis Scand.* 23: 244–250.

Bairlein, F. & Winkel, W. (2001). Birds and climate change. pp. 278–282 in *Climate of the 21st century: changes and risks* (eds J. L. Lozan, H. Grasse & P. Hupfer). Hamburg, Wissenschaftliche Auswertungen.

Baker, A. J. (2002). The deep roots of bird migration: Inferences from the historical record preserved in DNA. *Ardea* 90, special issue: 503–513.

Baker, A. J., Gonzalez, P. M., Piersma, T., Niles, L. J., de Lima Serrano do Nascimento, I., Atkinson, P. W., Clark, N. A., Minton, C. D. T., Peck, M. K. & Aarts, G. (2004). Rapid population decline in Red Knots: fitness consequences of decreased refuelling rates and late arrival in Delaware Bay. *Proc. R. Soc. Lond. B* 271: 875–882.

Baker, J. R. (1938).The evolution of breeding seasons. pp. 161–177 in *Evolution, essays presented to E. S. Goodrich*. Oxford, Oxford University Press.

Baker, J. R. & Ranson, R. M. (1938). The breeding seasons of southern hemisphere birds in the northern hemisphere. *Proc. Zool. Soc. Lond. A* 108: 101–141.

Baker, K. (1977). Westward vagrancy of Siberian passerines in autumn 1975. *Bird Study* 24: 233–242.

Baker, R. R. (1978). *The evolutionary ecology of animal migration*. London, Hodder & Stoughton.

Bakken, V., Runde, O. & Tjørve, E. (2003). *Norsk ringmerkinsatlas*. Vol. 1. Stavanger, Stavanger Museum.

Balança, G. & Schaub, M. (2005). Post-breeding migration ecology of Reed *Acrocephalus scirpaceus*, Moustached *A. melanopogon* and Cetti's Warblers *Cettia cetti* at a Mediterranean stopover site. *Ardea* 93: 245–257.

Ball, G. & Balthazart, D. (2003). Birds return every spring like clockwork, but where is the clock? *Endocrinology* 144: 3738–3741.

Ball, R. M., Freeman, S., James, F. C., Bermingham, E. & Avise, J. C. (1988). Phylogeographic population structure of Red-winged Blackbirds assessed by mitochrondrial DNA. *Proc. Natl Acad. Sci. USA* 85: 1558–1562.

Ball, T. (1983). The migration of geese as an indicator of climate change in the southern Hudson Bay region between 1715 and 1851. *Clim. Change* 5: 85–93.

Banks, R. C. (1979). Human related mortality of birds in the United States. *U.S. Fish & Wildl. Serv. Spec. Sci. Rep.* 215: 1–16.

Barbraud, C., Barbraud, J.-C. & Barbraud, M. (1999). Population dynamics of the White Stork *Ciconia ciconia* in western France. *Ibis* 141: 469–479.

Barbraud, C., Johnson, A. R. & Bertault, G. (2003). Phenotypic correlates of post-fledging dispersal in a population of Greater Flamingos: the importance of body condition. *J. Anim. Ecol.* 72: 246–257.

Barrowclough, G. F. (1978). Sampling bias in dispersal studies based in finite area. *Bird-Banding* 49: 333–341.

Barry, T. W. (1962). Effects of late seasons on Atlantic Brant reproduction. *J. Wildl. Manage.* 26: 19–26.

Barry, T. W. (1968). Observations on natural mortality and native use of Eider Ducks along the Beaufort Sea Coast. *Can. Field-Nat.* 82: 140–144.

Bastian, H. V. (1992). Breeding and natal dispersal of Whinchats *Saxicola rubetra*. *Ringing & Migration* 13: 13–19.

Batschelet, E. (1981). *Circular statistics in biology*. New York, Academic Press.

Battley, P. F. & Piersma, T. (1997). Body composition of Lesser Knots (*Calidris canutus rogersi*) preparing for take-off on migration from northern New Zealand. *Notornis* 44: 137–150.

Battley, P. F., Dekinga, A., Dietz, M. W., Piersma, T., Tang, S. & Hulsman, K. (2001). Basal metabolic rate declines during long distance migratory flight in Great Knots. *Condor* 103: 838–845.

Battley, P. F., Piersma, T., Dietz, M., Tang, S., Dekinga, P. & Hulsman, K. (2000). Empirical evidence for differential organ reductions during trans-oceanic bird flight. *Proc. R. Soc. Lond. B* 267: 191–195.

Bauchinger, U. (2002). Phenotypic flexibility of organs during long distance migration in Garden Warblers: implications for migratory and reproductive periods. PhD thesis, Munich, Technical University.

Bauchinger, U. & Biebach, H. (2005). Phenotypic flexibility of skeletal muscles during long-distance migration of Garden Warblers: muscle changes are differentially related to body mass. *Ann. N.Y. Acad. Sci* 1046: 271–281.

Bauchinger, U. & Klaassen, M. (2005). Longer days in spring than in autumn accelerate migration speed of passerine birds. *J. Avian Biol.* 36: 3–5.

Bauchinger, U., Goymann, W. & Jenni-Eiermann, S. (2005). *Bird hormones and bird migrations*. New York, New York Academy of Sciences.

Baudinette, R. V. & Schmidt-Nielsen, K. (1974). Energy-cost of gliding flight in Herring Gulls. *Nature* 248: 83–84.

Baumann, S. (2001). Observations on the coexistence of Palearctic and African Orioles *Oriolus* sp. in Zimbabwe. *Vogelwarte* 122: 67–69.

Baumgartner, F. M. & Baumgartner, A. M. (1992). *Oklahoma bird life*. Norman, Oklahoma, University Press.

Bearhop, S., Hilton, G. M., Votier, S. C. & Waldron, S. (2004). Stable isotope ratios indicate that body condition in migrating passerines is influenced by winter habitat. *Proc. R. Roc. Lond. B* (Suppl. 4) 271: 215–218.

Bearhop, S., Fiedler, W., Furness, R. W., Votier, S. C., Waldron, S., Newton, J., Bowen, G. J., Berthold, P. & Farnsworth, K. (2005). Assortative mating as a mechanism for rapid evolution of a migratory divide. *Science* 310: 502–503.

Beason, R. C. (1987). Interaction of visual and nonvisual cues during migratory orientation by the Bobolink (*Dolichonyx orizivorus*). *J. Ornithol.* 128: 317–324.

Beck, W. & Wiltschko, W. (1982). The magnetic field as a reference system for the genetically encoded migratory direction in Pied Flycatchers. *Z. Tierpsychol.* 60: 41–46.

Beck, W. & Wiltschko, W. (1988). Magnetic factors control the migratory direction of Pied Flycatcher (*Ficedula hypoleuca* Pallas). *Proc Int. Orn. Congr.* 19: 1955–1962.

Becker, P. H., Wendeln, H. & Gonzàlez-Solís, J. (2001). Population dynamics, recruitment, individual quality and reproductive strategies in Common Terns *Sterna hirundo* marked with transponders. *Ardea* 89 (special issue): 241–252.

Bedard, J. & LaPoint, G. (1984). Banding returns, arrival times, and site fidelity in the Savannah Sparrow. *Wilson Bull.* 96: 196–205.

Bedell, P. A. (1996). Evidence of dual breeding ranges for the Sedge Wren in the Central Great Plains. *Wilson Bull.* 108: 115–122.

Bednarz, J. C. & Kerlinger, P. (1989). Monitoring hawk populations by counting migrants. pp. 328–342 in *Proceedings of the northeast raptor management symposium and workshop* (ed. B. G. Pendleton). Washington, D.C., National Wildlife Federation.

Bednarz, J. C., Klein, D., Goodrich, L. J. & Senner, S. E. (1990). Migration counts of raptors at Hawk Mountain, Pennsylvania, as indicators of population trends, 1934–1986. *Auk* 107: 96–109.

Beekman, J. (2005). Small swans, vast journeys. pp. 44–52 in *Seeking nature's limits, ecologists in the field* (eds R. H. Drent, I. J. Bakker, T. Piersma & J. M. Tinbergen). Utrecht, KNNV Publishing.

Beekman, J. H., Nolet, B. A. & Klaassen, M. (2002). Skipping swans: fuelling rates and wind conditions determine differential use of migratory stopover sites of Bewick's Swans *Cygnus bewickii*. *Ardea* 90: 437–460.

Beheler, A., Rhodes, O. E. & Weeks, H. P. (2003). Breeding site and mate fidelity in Eastern Phoebes (*Sayornis phoebe*) in Indiana. *Auk* 120: 990–999.

Beletsky, L. D. & Orians, G. H. (1991). Effects of breeding experience and familiarity on site fidelity in female Red-winged Blackbirds. *Ecology* 72: 787–796.

Bell, C. P. (1996). Seasonality and time allocation as causes of leap-frog migration in the Yellow Wagtail *Motacilla flava*. *J. Avian Biol.* 27: 334–342.

Bell, C. P. (1997). Leap-frog migration in the Fox Sparrow: minimising the cost of spring migration. *Condor* 99: 470–477.

Bell, C. P. (2000). Process in the evolution of bird migration and pattern in avian ecogeography. *J. Avian Biol.* 31: 258–265.

Bell, D. B. (1991). Recent avifauna changes and the history of ornithology in New Zealand. *Proc. Int. Ornithol. Congr.* 20: 195–230.

Belliure, J., Sorci, G., Møller, A. P. & Clobert, J. (2000). Dispersal distances predict subspecies richness in birds. *J. Evol. Biol.* 13: 480–487.

Bellrose, F. C. (1958). The orientation of displaced waterfowl in migration. *Wilson Bull.* 70: 20–40.

Bellrose, F. C. (1967). Radar in orientation research. *Proc Int. Orn. Congr.* 14: 281–309.

Bellrose, F. C. (1968). Waterfowl migration corridors. *Biol. Notes* 61: 1–24. Illinois Nat. Hist. Survey.

Bellrose, F. C. & Graber, R. R. (1963). A radar study of the flight directions of nocturnal migrants. *Proc Int. Orn. Congr.* 13: 362–389.

Bellrose, F. C., Scott, T. G., Hawkins, A. S. & Low, J. B. (1961). Sex ratios and age ratios in North American ducks. *Illinois Nat. Hist. Surv. Bull.* 27: 391–474.

Belopolskij, L. O. (1971). Migration of Sparrowhawk on Courland Spit. *Notatki Ornitologiczne* 12: 1–12.

Belthoff, J. R. & Gauthreaux, S. A. (1991). Partial migration and differential winter distribution of House Finches in the eastern United States. *Condor* 93: 374–382.

Benhamou, S., Bonnadonna, F. & Jouventin, P. (2003). Successful homing of magnet-carrying White-chinned Petrels released in the open sea. *Anim. Behav.* 65: 729–734.

Benkman, C. W. (1987). Food profitability and the foraging ecology of crossbills. *Ecol. Monogr.* 57: 251–267.

Benkman, C, W. (1992). White-winged Crossbill. *The birds of North America*, No. 27 (eds A. Poole & F. Gill). Philadelphia, Academy of Natural Sciences and Washington, D. C., American Ornithologists' Union.

Benkman, C. W. (1993). Adaptation to single resources and the evolution of Crossbill (*Loxia*) diversity. *Ecol. Monogr.* 63: 305–325.

Benkman, C. W. (1997). Feeding behaviour, flock-size dynamics, and variation in sexual selection in crossbill. *Auk* 114: 163–178.

Bennett, G. M. & Fallis, A. M. (1960). Blood parasites of birds in Algonquin Park, Canada, and a discussion of their transmission. *Can. J. Zool.* 38: 261–273.

Bensch, S. (1999). Is the range size of migratory birds constrained by their migratory programme? *J. Biogeog.* 26: 1225–1235.

Bensch, S. & Hasselquist, D. (1991). Territory infidelity in the polygynous Great Reed Warbler *Acrocephalus arundinaceus*: the effect of variation in territory attractiveness. *J. Anim. Ecol.* 60: 857–871.

Bensch, S. & Hasselquist, D. (1999). Phylogeographic population structure of Great Reed Warblers: an analysis of mt DNA control region sequences. *Biol. J. Linn. Soc.* 66: 171–185.

Bensch, S. & Nielsen, B. (1999). Autumn migration speed of juvenile Reed and Sedge Warblers in relation to date and fat loads. *Condor* 101: 153–156.

Bensch, S., Åkesson, S. & Irwin, D, E. (2002). The use of AFLP to find an informative SNP: genetic differences across a migratory divide in Willow Warblers. *Mol. Ecol.* 11: 2359–2366.

Bensch, S., Andersson, T. & Åkesson, S. (1999). Morphological and molecular variation across a migratory divide in Willow Warblers *Phylloscopus trochilus*. *Evolution* 53: 1925–1935.

Bensch, S., Hasselquist, D., Hedenström, A. & Ottosson, U. (1991). Rapid moult among Palaearctic passerines in West Africa – an adaptation to the oncoming dry season? *Ibis* 133: 47–52.

Benson, A. M. & Winker, K. (2001). Timing of breeding range occupancy among high-latitude passerine migrants. *Auk* 118: 513–519.

Bent, A. C. (1938). *Life histories of North American birds of prey*. New York, Dover.

Bentley, G. E., Spar, B. D., MacDougall-Shackleton, S. A., Hahn, T. P. & Ball, G. F. (2000). Photoperiodic regulation of the reproductive axis in male zebra finches, *Taeniopygia guttata*. *Gen. Comp. Endocrinol.* 117: 449–455.

Benvenuti, S. & Ioalé, P. (1980a). Fedelta al luogo di svernamento, in anni successivi, in alcune species d'uccelli. *Avosetta* 4: 133–139.

Benvenuti, S. & Ioalé, P. (1980b). Homing experiments with birds displaced from their wintering grounds. *J. Ornithol.* 121: 281–286.

Berezovikov, N. N. & Anisimov, E. I. (2002). (Mass mortality of Barn Swallows *Hirundo rustica* in the autumn of 2000 in the Alakol Depression, south-eastern Kazakhstan). *Russ. J. Ornithol.* 11: 258–261.

Berger, A. J. & Radabaugh, B. E. (1968). Returns of Kirtland's Warblers to the breeding grounds. *Bird-banding* 39: 161–186.

Bergmann, G. (1977). Finnish radar investigations on migration of waterfowl between the Baltic and the White Sea. *Congresso per l'Elettronica* 24: unpaginated.

Bermejo, A. & de la Puente (2004). Wintering and migration of Bluethroat *Luscinia svecica* in central Spain. *Ardeola* 51: 285–296.

Berndt, R. & Dancker, P. (1960). Der Kleiber, *Sitta europaea*, als Invasionsvogel. *Vogelwarte* 20: 193–198.

Berndt, R. & Henss, M. (1967). Die Kohlmeise, *Parus major*, als Invasionsvogel. *Vogelwarte* 24: 17–37.

Berndt, R. & Sternberg, H. (1968). Terns, studies and experiments on the problems of bird dispersion. *Ibis* 110: 256–269.

Berndt, R. & Winkel, W. (1979). Verfrachtungs – Experimente zur Frage der Geburtsortsprägung beim Trauerschnapper (*Ficedula hypoleuca*). *J. Ornithol.* 121: 281–286.

Bernis, F. (1975). Migration of Falconiformes and *Ciconia* spp. through the Straits of Gibraltar. Part 2. *Ardeola* 21: 489–580.

Berthold, P. (1973). Relationships between migratory restlessness and migration distance in *Sylvia* species. *Ibis* 115: 594–599.

Berthold, P. (1974a). Circannual Periodik bei Grasmücken (*Sylvia*) III. Periodik der Mauser, der Nachtunruhe und des Körpergewichts bei mediterranean Arten mit unterschiedlivhem Zugverhalten. *J. Ornithol.* 115: 251–272.

Berthold, P. (1974b). Circannual rhythms in birds with different migratory habits. pp. 44–94 in *Circannual clocks* (ed. E. T. Pengelly). New York, Academic Press.

Berthold, P. (1975). Migration: control and metabolic physiology. pp. 77–128 in *Avian biology*, Vol. 5 (eds D. S. Farner & J. R. King). New York, Academic Press.

Berthold, P. (1976). Animalische und vegetabilische Ernährung omnivorer Singvogelarten: Nahrungsbevorzugung Jahresperiodik der Nahrungswahl, physiologische und ökologische Bedeutung. *J. Ornithol.* 117: 145–209.

Berthold, P. (1977). Über die Entwicklung von Zugunruhe bei der Gartengrasmücke (*Sylvia borin*) bei verhinderter Fettdeposition. *Vogelwarte* 29: 113–116.

Berthold, P. (1978a). Circannuale Rhythmik: freilaufende selbsterregte Periodik mit lebenslanger Wirksamkeit bei Vögeln. *Naturwissenschaften* 65: 546.

Berthold, P. (1978b). Die quantitative Erfassung der Zugunruhe bei Tagziehern: Eine Pilotstudie an Ammiern (*Emberiza*). *J. Ornithol.* 119: 334–336.

Berthold, P. (1978c). Endogenous control as a possible basis for varying migratory habits in different bird populations. *Experimentia* 34: 1451.

Berthold, P. (1984a). The control of partial migration in birds: a review. *Ring* 10: 253–265.

Berthold, P. (1984b). The endogenous control of bird migration: a survey of experimental evidence. *Bird Study* 31: 19–27.

Berthold, P. (1986). Wintering in a Mediterranean Blackcap (*Sylvia atricapilla*) population: strategy, control, and unanswered questions. *Proc. First Conf. Birds wintering Mediterranean Region, Aulla* 1885: 261–272.

Berthold, P. (1988). The control of migration in European warblers. *Proc Int. Orn. Congr.* 19: 215–249.

Berthold, P. (1989). Zur Bestandsentwicklung mitteleuropäischer Vogelarten-Ergebnisse aus dem MRI-Programm. *Schriftenr. Bayer. Landesamt Umweltschutz, H.* 92: 71–76.

Berthold, P. (1990a). Die Vogelwelt Mitteleuropas: Entstehung der Diversität, gegenwärtige Veränderungen und Aspekte der Zukunftigen Entwicklung. *Verh. Dtsch. Zool. Ges.* 83: 227–244.

Berthold, P. (1990b). Genetics of migration. pp. 269–280 in *Bird migration. Physiology and ecophysiology* (ed. E. Gwinner). Berlin, Springer-Verlag.

Berthold, P. (1990c). Wegzugbeginn und Einsetzen der Zugunruhe bei 19 Vogelpopulationen – eine vergleichende Untersuchung. *J. Ornithol.* 131: 217–222.

Berthold, P. (1993). *Bird migration. A general survey.* Oxford, Oxford University Press.

Berthold, P. (1995). Microevolution of migratory behaviour illustrated by the Blackcap *Sylvia atricapilla* – 1993 Witherby Lecture. *Bird Study* 42: 89–100.

Berthold, P. (1996). *Control of bird migration.* London, Chapman & Hall.

Berthold, P. (1998). Bird migration: genetic programs with high adaptability. *Zool. Anal. Complex Syst.* 101: 235–245.

Berthold, P. (1999). A comprehensive theory for the evolution, control and adaptability of avian migration. *Ostrich* 70: 1–12.

Berthold, P. (2001). *Bird migration: a general survey* (2nd edn). Oxford, Oxford University Press.

Berthold, P. & Gwinner, E. (1972). Frühe Geschlechtsreife beim Fichtenkreuzschnabel (*Loxia curvirostra*). *Vogelwarte* 26: 356–357.

Berthold, P. & Gwinner, E. (1978). Jahresperiodik der Gonadengrösse beim Fichtenkreuzschnabel (*Loxia curvirostra*). *J. Ornithol.* 119: 338–339.

Berthold, P. & Helbig, A. J. (1992). The genetics of bird migration: stimulus, timing and direction. *Ibis* 134 (suppl): 35–40.

Berthold, P. & Leisler, B. (1981). Migratory restlessness of the Marsh Warbler *Acrocephalus palustris.* *Natuurwis.* 67: 472.

Berthold, P. & Pulido, F. (1994). Heritability of migratory activity in a natural bird population. *Proc. R. Soc. Lond. B* 257: 311–315.

Berthold, P. & Querner, U. (1981). Genetic basis of migrating behaviour in European warblers. *Science* 212: 77–79.

Berthold, P. & Querner, U. (1982a). Genetic basis of moult, wing length, and body weight in a migratory bird species *Sylvia atricapilla.* *Experimentia (Basel)* 38: 801–802.

Berthold, P. & Querner, U. (1982b). Partial migration in birds: experimental proof of polymorphism as a controlling system. *Experimentia (Basel)* 38: 805–806.

Berthold, P. & Terrill, S. B. (1991). Recent advances in studies of bird migration. *Annu. Rev. Ecol. Syst.* 22: 357–378.

Berthold, P., Gwinner, E. & Klein, H. (1970). Vergleichende Untersuchung der Jugendentwicklung eines ausgeprägten Zugvogels, *Sylvia borin*, und eines weniger ausgeprägten Zugvogels, *S. atricapilla.* *Vogelwarte* 25: 297–331.

Berthold, P., Gwinner, E. & Klein, H. (1971). Circannual Periodik bei Grasmücken (*Sylvia*). *Experimentia (Basel)* 27: 399.

Berthold, P., Gwinner, E. & Klein, H. (1972). Circannuale Periodik bei Grasmückken I. Periodic des Körpergewiches, der Mauser und der Nachtunruhe bei *Sylia atricapilla* und *S. borin* unter verschiedenen konstanten Bedingungen. *J. Ornithol.* 113: 170–190.

Berthold, P., Querner, U. & Winkler, H. (1988). Vogelschutz: 100 Jahr lang bis in die 'roten Zahlen' – ein neues Konzept ist unerlasslich. *Natur. Landsch.* 63: 5–8.

Berthold, P., Mohr, G. & Querner, U. (1990a). Steuerung und potentielle Evolutionsgeschwindigkeit des obligaten Teilzieherverhaltens: Ergebnisse eines Zweiweg-Selektionsexperiments mit der Mönchsgrasmücke (*Sylvia atricapilla*). *J. Ornithol.* 131: 33–45.

Berthold, P., Wiltschko, W., Miltenberger, H. & Querner, U. (1990b). Genetic transmission of migratory behaviour into a nonmigratory bird population. *Experimentia* 46: 107–108.

Berthold, P., Helbig, A. H., Mohr, G. & Querner, U. (1992a). Rapid microevolution of migratory behaviour in a wild bird species. *Nature* 360: 668–670.

Berthold, P., Nowak, E. & Querner, U. (1992b). Satelliten-Telemetrie beim Weiss-storch (*Ciconia ciconia*) auf dem Wigzug – eine Pilotstudie. *J. Ornithol.* 136: 73–76.

Berthold, P., Iovchenko, N. P., Mohr, G., Querner, U. & Fertikova, K. P. (2001a). Circannual rhythms in the Whitethroat, *Sylvia communis*. *Zool. Zhurnal* 80: 1387–1394.

Berthold, P., van den Bossch, W., Fieldler, W., Gorney, E., Kaatz, M., Leshem, Y., Nowak, E. & Querner, U. (2001b). Der Zug des Weissstorchs (*Ciconia ciconia*): eine besondere Zugform auf Grundneuer Ergebnisse. *J. Ornithol.* 142: 73–92.

Berthold, P., van den Bossch, W., Fieldler, W., Gorney, E., Kaatz, C., Kaatz, M., Leshem, Y., Novak, E. & Querner, U. (2001c). Detection of a new important staging and wintering area of the White Stork *Ciconia ciconia* by satellite tracking. *Ibis* 143: 450–455.

Berthold, P., van den Bosch, W., Jakubiec, Z., Kaatz, C., Kaatz, M. & Querner, U. (2002). Long-term satellite tracking sheds light on variable migration strategies of White Storks (*Ciconia ciconia*). *J. Ornithol.* 143: 489–495.

Berthold, P., Kaatz, M. & Querner, U. (2004). Long-term satellite tracking of White Stork (*Ciconia ciconia*) migration: constancy versus variability. *J. Ornithol.* 145: 356–359.

Bertin, R. I. (1982). The Ruby-throated Hummingbird and its major food plants – ranges, flowering phenology, and migration. *Can. J. Zool.* 60: 210–219.

Bêty, J., Gauthier, G. & Giroux, J.-F. (2003). Body condition, migration and timing of reproduction in Snow Geese: a test of the condition-dependent model of optimal clutch size. *Am. Nat.* 162: 110–121.

Bibby, C. J. & Green, R. E. (1980). Foraging behaviour of migrant Pied Flycatchers, *Ficedula hypoleuca*, on temporary territories. *J. Anim. Ecol.* 49: 507–521.

Bibby, C. J. & Green, R. E. (1981). Autumn migration strategies of Reed and Sedge Warblers. *Ornis Scand.* 12: 1–12.

Bibby, C. J. & Lloyd, C. S. (1977). Experiments to determine the fate of dead birds at sea. *Biol. Conserv.* 12: 295–309.

Bibby, C. J. & Thomas, D. K. (1984). Sexual dimorphism in size, moult and measurement of Cetti's Warbler *Cettia cetti*. *Bird Study* 31: 25–34.

Bibby, C. J., Green, R. E., Peplar, G. R. M. & Peplar, P. A. (1976). Sedge Warbler migration and reed aphids. *Br. Birds* 69: 384–399.

Biebach, H. (1983). Genetic determination of partial migration in the European Robin (*Erithacus rubecula*). *Auk* 100: 601–606.

Biebach, H. (1985). Sahara stopover in migratory flycatchers: fat and food affect the time program. *Experimentia* 41: 695–697.

Biebach, H. (1990). Strategies of trans-desert migrants. pp. 352–367 in *Bird migration. Physiology and ecophysiology* (ed. E. Gwinner). Berlin, Springer-Verlag.

Biebach, H. (1991). Is water or energy crucial for trans-Saharan migrants? *Proc Int. Orn. Congr.* 20: 773–779.

Biebach, H. (1992). Flight-range estimates for small trans-Sahara migrants. *Ibis* 134: 47–54.

Biebach, H. (1996). Energetics of winter and migratory fattening. pp. 280–323 in *Avian energetics and nutritional ecology* (ed. C. Carey). New York, Chapman & Hall.

Biebach, H. (1998). Phenotypic organ flexibility in Garden Warblers *Sylvia borin* during long distance migration. *J. Avian Biol.* 29: 529–535.

Biebach, H. & Bauchinger, U. (2003). Energetic savings in organ adjustment during long migratory flights in Garden Warblers (*Sylvia borin*). pp. 269–280 in *Avian migration* (eds P. Berthold, E. Gwinner & E. Sonnenschein). Berlin, Springer-Verlag.

Biebach, H., Friedrich, W. & Heine, G. (1986). Interaction of body mass, fat, foraging and stopover period in trans-Sahara migrating passerine birds. *Oecologia* 69: 370–379.

Biebach, H., Biebach, I., Friedrich, W., Heine, G., Partecke, J. & Schmidt, D. (2000). Strategies of passerine migration across the Mediterranean Sea and the Sahara Desert: a radar study. *Ibis* 142: 623–634.

Bijlsma, R. G. (1990). Predation by large falcon species on wintering waders on the Banc d'Arguin, Mauritania. *Ardea* 78: 75–82.

Bildstein, K. L. (2004). Raptor migration in the neotropics: Patterns, processes, and consequences. *Ornitol. Neotropical* 15: 83–99.

Bildstein, K. L. (2006). *Migratory raptors of the world. Their ecology and conservation*. Ithaca, New York, Comstock Publishing Associates.

Bingham, V. P., Budzynski, C. A. & Voggenhuber, A. (2003). Migratory systems as adaptive responses to spatial and temporal variability in orientation stimuli. pp. 457–470 in *Avian migration* (eds P. Berthold, E. Gwinner & E. Sonnenschein). Berlin, Springer-Verlag.

Bingman, V. P. (1987). Earth's magnetism and the nocturnal orientation of migratory European Robins. *Auk* 104: 523–525.

Bingman, V. P. & Wiltschko, W. (1988). Orientation of Dunnocks (*Prunella modularis*) at sunset. *Ethology* 77: 1–9.

Birkhead, T. R. (1974). Movement and mortality rates in British Guillemots. *Bird Study* 21: 241–253.

Bishop, M. A., Warnock, N. & Takekawa, J. Y. (2004). Differential spring migration by male and female Western Sandpipers at interior and coastal stopover sites. *Ardea* 92: 185–196.

Bjerke, T. & Espmark, Y. (1988). Breeding success and breeding dispersal in recovered Redwings *Turdus iliacus*. *Fauna Norveg* 11: 45–46.

Bjorkman, G. & Tyrberg, T. (1982). The Bearded Tit *Panurus biarmicus* in Sweden in 1965–1979. *Vår Fågelvärld* 41: 73–93.

Black, C. P. & Tenney, S. M. (1980). Oxygen transport during progressive hypoxia in high altitude and sea-level waterfowl. *Respir. Physiol.* 39: 217–239.

Black, J. M. (ed.) (1996). *Partnerships in birds. The study of monogamy*. Oxford, University Press.

Black, J. M. & Rees, E. C. (1984). The structure and behaviour of the Whooper Swan population wintering at Caerlaverock, Dumfries and Galloway, Scotland: an introductory study. *Wildfowl* 35: 21–36.

Blackwell, J. A. & Dowdeswell, W. H. (1951). Local movement in the Blue Tit. *Br. Birds* 44: 397–403.

Blake, C. H. (1967). Purple Finches at Hillsborough, N.C., 1961–65. *Bird-Banding* 38: 1–17.

Blake, J. G. & Karr, J. R. (1987). Breeding birds of isolated woodlots: area and habitat relationships. *Ecology* 68: 1724–1734.

Blake, J. G., Niemi, G. J. & Hanowski, J. M. (1992). Drought and annual variation in bird populations. pp. 419–430 in *Ecology and conservation of Neotropical migrant landbirds* (eds J. M. Hagan & D. W. Johnson). Washington, D.C., Smithsonian Press.

Blanchard, B. D. (1941). The White-crowned Sparrows *Zonotrichia leucophrys* of the Pacific seaboard: environment and annual cycle. *Univ. Calif. Publ. Zool.* 46: 1–178.

Blem, C. R. (1990). Avian energy storage. *Curr. Ornithol.* 7: 59–113.

Bloesch, M. (1956). Algerische Störche für den Storchansiedlungsversuch der Vogelwarte Sempach. *Orn. Beob.* 53: 97–104.

Bloesch, M. (1960). Zweiter Bericht über den Einsatz algerischer Störche für den Storchansiedlungversuch in der Schweiz. *Orn. Beob.* 57: 214–223.

Blomqvest, S., Holmgren, N., Åkesson, S., Hedenström, A. & Petterson, J. (2002). Indirect effects of lemming cycles on sandpiper dynamics: 50 years of counts from southern Sweden. *Oecologia* 133: 146–158.

Bluhm, C. (1988). Temporal patterns of pair formation and reproduction in annual cycles and associated endocrinology in waterfowl. *Curr. Ornithol.* 5: 123–185.

Blums, P., Nichols, J. D., Hines, J. E. & Mednis, A. (2002). Sources of variation in survival and breeding site fidelity in three species of European ducks. *J. Anim. Ecol.* 71: 438–450.

Boano, G., Bonardi, A. & Silvano, F. (2004). Nightingale *Luscinia megarhynchos* survival rates in relation to Sahel rainfall. *Avocetta* 28: 77–85.

Bock, C. E. & Lepthien, L. W. (1976). Synchronous eruptions of boreal seed-eating birds. *Am. Nat.* 110: 559–571.

Boddy, M. (1983). Factors influencing timing of autumn dispersal or migration in first-year Dunnocks and Whitethroats. *Bird Study* 30: 39–46.

Boddy, M. & Sellers, R. M. (1983). Orientated movements by Greenfinches in southern Britain. *Ringing & Migration* 4: 129–138.

Böhning-Gaese, K. & Oberrath, R. (2003). Macroecology of habitat choice in long-distance migratory birds. *Oecologia* 137: 296–303.

Böhning-Gaese, K., Taper, M. I. & Brown, J. H. (1993). Are declines in North American insectivorous songbirds due to causes on the breeding range? *Conserv. Biol.* 7: 76–86.

Böhning-Gaese, K., González-Guzmán, L. I. & Brown, J. H. (1998). Constraints on dispersal and the evolution of the avifauna of the northern hemisphere. *Evol. Ecol.* 12: 767–783.

Bojarinova, J. G., Lehikoinen, E. & Eeva, T. (1999). Dependence of postjuvenile moult on hatching date, condition and sex in the Great Tit. *J. Avian Biol.* 30: 437–446.

Bojarinova, J. G., Rymkevich, T. A. & Smirnov, O. P. (2002). Timing of autumn migration of early and late-hatched Great Tits *Parus major* in NW Russia. *Ardea* 90: 401–409.

Boland, J. M. (1990). Leapfrog migration in North American shorebirds: intra- and inter-specific examples. *Condor* 92: 284–290.

Bollinger, E. K. & Gavin, T. A. (1992). Eastern Bobolink populations: ecology and conservation in an agricultural landscape. pp. 497–506 in *Ecology and conservation of Neotropical migrant landbirds* (ed. D. W. J. J.M. Hagan). Washington, D.C., Smithsonian Institution Press.

Bolshakov, C. V. (2001). Specific composition and spatial distribution of passerine nocturnal migrants in the arid and highland zone of western central Asia: the results of large-scale trapping at daytime stopovers. *Avian Ecol. Behav.* 6: 15–16.

Bolshakov, C. (2003). Nocturnal migration of passerines in the desert-highland zone of western central Asia: selected aspects. pp. 225–236 in *Avian mMigration* (eds P. Berthold, E. Gwinner & E. Sonnenschein). Berlin, Springer-Verlag.

Bolshakov, C. V. & Bulyuk, V. N. (1999). Time of nocturnal flight initiation (take-off activity) in the European Robin *Erithacus rubecula* during spring migration: direct observations between sunset and sunrise. *Avian Ecol. Behav.* 2: 19–50.

Bond, R. R. (1957). Ecological distribution of breeding birds in the upland forest communities of southern Wisconsin. *Ecol. Monogr.* 27: 357–384.

Bonadonna, F., Bajzak, C., Benhamou, S., Igloi, K., Jouventin, P., Lipp, H. P. & Dell'Omo, G. (2005). Orientation in the Wandering Albatross: interfering with magnetic perception does not affect orientation performance. *Proc. R. Soc. Lond. B* 272: 489–495.

Borowitz, V. A. (1988). Fruit consumption by birds in relation to fat content of pulp. *Am. Midl. Nat.* 119: 121–127.

Bost, C. A., Charrassin, J. B., Clerquin, Y., Ropert-Coudert, Y. & Le Mayo, Y. (2004). Exploitation of distant marginal ice zones by King Penguins during winter. *Marine Ecology – Progress Series* 283: 293–297.

Both, C. & Visser, M. E. (2001). Adjustment to climatic change is constrained by arrival date in a long-distance migratory bird. *Nature* 411: 296–298.

Both, C., Bouwhuis, S., Lessells, C. M. & Visser, M. E. (2006a). Climate change and population declines in a long-distance migratory bird. *Nature* 441: 81–82.

Both, C., Sanz, J. J., Artemyev, A. V., Blaauw, V., Cowie, R. J., Dekhuizen, A. J., Enemar, A., Järvinen, A., Nyholm, N. E. I., Potti, J., Ravussin, P.-A., Silverin, B., Slater, F. M., Sokolov, L. B., Visser, M. E., Winkel, W., Wright, J. & Zang, H. (2006b). Pied Flycatchers travelling from Africa to breed in Europe: differential effects of winter and migration conditions on breeding date. *Ardea* 94: 511–525.

Boulet, M. & Norris, D. R. (2006). The past and present of migratory connectivity. *Orn. Monog.* 61: 1–13.

Bourne, W. R. P. (1959). Notes on autumn migration in the Middle East. *Ibis* 101: 170–176.

Bourne, W. R. P. (1967). Long distance vagancy in the petrels. *Ibis* 109: 141–167.

Bourne, W. R. P. (1979). Birds and gas flares. *Marine Pollut. Bull.* 10: 124–125.

Bourne, W. R. P. (1980). The midnight descent, dawn ascent and re-orientation of land birds migrating across the North Sea in autumn. *Ibis* 122: 536–540.

Bourne, W. R. P. (1981). A radar film of migration in north-east Scotland. *Ibis* 123: 581.

Bourne, W. R. P. (1983). A satellite view of bird migration between Iceland and Scotland. *Sea Swallow* 32: 80–82.

Bourne, W. R. P. & Casement, M. B. (1996). The migrations of the Arctic Tern. *Bull. Br. Ornithol. Club* 116: 117–123.

Bourne, W. R. P., Knox, A. G., Merrie, T. D. H. & Morley, A. H. (1979). The birds of the Forties oilfield 1975–1978. *North-east Scotland Bird Report 1978*: 47–52.

Brackhill, H. (1956). Unstable migratory behaviour in a Mockingbird. *Bird-Banding* 27: 128.

Bradley, J. S., Wooller, R. D., Skira, R. J. & Serventy, D. L. (1990). The influence of mate retention and divorce upon reproductive success in Short-tailed Shearwater *Puffinus tenuirostris*. *J. Anim. Ecol.* 59: 487–496.

Bradley, N. L., Leopold, A. C., Ross, J. & Huffaker, W. (1999). Phenological changes reflect climate change in Wisconsin. *Proc. Natl Acad. Sci.USA* 96: 9701–9704.

Brandstätteer, R. (2003). Encoding time of day and time of year by the avian circadian system. *J. Neuroendocrinol.* 15: 1–7.

Branson, N. J. B. A., Ponting, E. D. & Minton, C. D. T. (1979). Turnstone populations on the Wash. *Bird Study* 26: 47–54.

Brawn, J. D. & Robinson, S. K. (1996). Source-sink population dynamics may complicate the interpretation of long-term census data. *Ecology* 77: 3–12.

Bray, O. E., Royall, W. C., Guarino, J. L. & Johnson, R. E. (1979). Activities of radio-equipped Common Grackles during fall migration. *Wilson Bull.* 91: 78–87.

Bregnballe, T., Frederiksen, M. & Gregersen, J. (1997). Seasonal distribution and timing of migration of Cormorants breeding in Denmark. *Bird Study* 44: 257–276.

Brensing, D. (1989). Ökophysiologische Untersuchungen der Tagesperiodik von Kleinvögeln. *Oköl. Vögel* 11: 1–148.

Brewer, D., Diamond, A., Woodsworth, E. J., Collins, B. T. & Dunn, E. H. (2000). *Canadian atlas of bird banding*, Vol. 1. Ottawa, Canadian Wildlife Service.

Brewer, R. & Ellis, J. A. (1958). An analysis of migrating birds killed at a television tower in east-central Illinois, September 1955–May 1957. *Auk* 75: 400–415.

Bried, J. & Jouventin, P. (1998). Why do Lesser Sheathbills *Chionis minor* switch territory? *J. Avian Biol.* 29: 257–265.

Brittingham, M. C. & Temple, S. A. (1983). Have cowbirds caused forest song birds to decline? *Bioscience* 33: 31–35.

Brodeur, S., Décarie, R., Bird, D. M. & Fuller, M. (1996). Complete migration cycle of Golden Eagles breeding in northern Quebec. *Condor* 98: 293–299.

Brodeur, S., Savard, J.-P. L., Robert, M., Laporte, P., Lamothe, P., Titman, R. D., Marchand, S., Gilliland, S. & Fitzgerald, G. (2002). Harlequin Duck *Histrionicus histrionicus* population structure in eastern Nearctic. *J. Avian Biol.* 33: 127–137.

Broley, C. L. (1947). Migration and nesting of Florida Bald Eagles. *Wilson Bull.* 59: 3–20.

Broekhuysen, G. J., Schmidt, R. & Martin, J. (1963). Breeding of the Wattled Starling *Creatophora cinereus* in the Southern Cape Province. *Ostrich* 34: 173–174.

Bromley, R. G. & Jarvis, R. L. (1993). The energetics of migration and reproduction of Dusky Canada Geese. *Condor* 95: 193–210.

Brooke M. de L. (1978). The dispersal of female Manx Shearwaters *Puffinus puffinus*. *Ibis* 120: 545–551.

Brooke M. de L. (1979). Differences in the quality of territories held by Wheatears (*Oenanthe oenanthe*). *J. Anim. Ecol.* 48: 21–32.

Brooke, M. (1990). *The Manx Shearwater*. London, T. & A. D. Poyser.

Brooke, R. K., Grobler, J. H. & Irwin, M. P. S. (1972). A study of the migratory eagles *Aquila nipalensis* and *A. pomarina* (Aves: Accipitridae) in southern Africa, with comparative notes on other large raptors. *Occ. Pap. Natn. Mus. Rhod.* 1972 B5 (2): 61–114.

Brooks, B. L. & Temple, S. A. (1990). Dynamics of Loggerhead Shrike population in Minnesota. *Wilson Bull.* 102: 441–450.

Brosselin, M. (1974). *Statut 1974 des hérons arborécoles en France*. Paris, Société Nationale de Protection de la Nature.

Brosset, A. (1984). Oiseaux migrateurs européens hivernant dans la partie guinéenne du Mont Nimba. *Alauda* 52: 81–101.

Brown, C. R. & Brown, M. B. (1992). Ectoparasitism as a cause of natal dispersal in Cliff Swallows. *Ecology* 73: 1718–1723.

Brown, C. R. & Brown, M. B. (1998). Intense natural selection on body size and wing and tail asymmetry in Cliff Swallows during severe weather. *Evolution* 52: 1461–2475.

Brown, C. R. & Brown, M. B. (2000). Weather-mediated natural selection on arrival time in Cliff Swallows (*Petrochelidon pyrrhonata*). *Behav. Ecol. Sociobiol.* 47: 339–345.

Brown, J. L. (1969). Territorial behaviour and population regulation in birds. *Wilson Bull.* 81: 293–329.

Brown, L. & Amadon, D. (1968). *Eagles, hawks and falcons of the world*. London, Country Life Books.

Brown, L. H., Urban, E. K. & Newman, K. (1982). *The birds of Africa*. London, Academic Press.

Brown, R. G. B. (1985). The Atlantic Alcidae at sea. pp. 383–426 in *The Atlantic Alcidae* (eds D. N. Nettleship & T. R. Birkhead). London, Academic Press.

Browne, S. J. & Aebischer, N. J. (2001). The role of agricultural intensification in the decline of the Turtle Dove *Streptopelia turtur*. English Nature Research Report No 421. English Nature, Peterborough.

Browne, S. J. & Mead, C. J. (2003). Age and sex composition, biometrics, site fidelity and origin of Brambling *Fringilla montifringilla* wintering in Norfolk, England. *Ringing & Migration* 21: 145–153.

Bruderer, B. (1971). Radarbeobachtungen über den Frühlingszug in Schweizerischen Mittelland. *Ornithol. Beob.* 68: 89–158.

Bruderer, B. (1994). Nocturnal migration in the Negev (Israel) – a tracking radar study. *Ostrich* 65: 204–212.

Bruderer, B. (1997a). The study of bird migration by radar. Part 1. The technical bases. *Naturwissenschaften* 84: 1–8.

Bruderer, B. (1997b). The study of bird migration by radar. Part 2. Major achievements. *Natuurwissenschaften* 84: 45–54.

Bruderer, B. (1999). Three decades of tracking radar-studies on bird migration in Europe and the Middle East. pp. 107–141 in *Migrating birds know no boundaries* (eds Y. Leshem, Y. Mandelik & J. Shamoun-Baranes). Latron, Israel, International Centre for the study of Bird Migration.

Bruderer, B. (2001). Recent studies modifying current views of nocturnal bird migration in the Mediterranean. *Avian Ecol. Behav.* 7: 11–25.

Bruderer, B. & Boldt, A. (2001). Flight characteristics of birds: 1. Radar measurements of speeds. *Ibis* 143: 178–204.

Bruderer, B. & Jenni, L. (1990). Migration across the Alps. pp. 60–77 in *Bird migration. Physiology and ecophysiology* (ed. E. Gwinner). Berlin, Springer-Verlag.

Bruderer, B. & Liechti, F. (1998). Flight behaviour of nocturnally migrating birds in coastal areas – crossing or coasting. *J. Avian Biol.* 29: 499–507.

Bruderer, B. & Liechti, F. (1999). Bird migration across the Mediterranean. *Proc. Int. Ornithol. Congr.* 22: 1983–1999.

Bruderer, B. & Muff, J. (1979). Bestandesschwankungen schweizerischer Rauch – und Mehlserchwalben, insbesondere im Zusammenhang mit der Schwalbenkatastrophe in Herbst 1974. *Ornithol. Beob.* 76: 229–234.

Bruderer, B., Steuri, T. & Baumgartner, M. (1995). Short-range high precision surveillance of nocturnal migration and tracking of single targets. *Isr. J. Zool.* 41: 207–220.

Brugger, K. E., Arkin, L. N. & Gramlich, J. M. (1994). Migration patterns of Cedar Waxwings in the Eastern United-States. *J. Field Ornithol.* 65: 381–387.

Bryant, D. M. (1975). Breeding biology of the House Martin *Delichon urbica* in relation to aerial insect abundance. *Ibis* 117: 180–215.

Bryant, D. M. & Jones, G. (1995). Morphological changes in a population of Sand Martins *Riparia riparia* associated with fluctations in population size. *Bird Study* 42: 57–65.

Bucher, E. H. (1982). Colonial breeding of the Eared Dove (*Zenaida auriculata*) in northeastern Brazil. *Biotropica* 14: 255–261.

Budnik, J. M., Ryan, M. R. & Thompson, F. R. (2000). Demography of Bell's Vireos in Missouri grassland-shrub habitats. *Auk* 117: 925–935.

Buerckle, C. A. (1999). The historical pattern of gene flow among migratory and non-migratory populations of Prairie Warblers (Aves: Parulinae). *Evolution* 53: 1915–1924.

Bull, E. L. & Duncan, J. R. (1993). Great Grey Owl. *The birds of North America*, No 41. (eds A. Poole, P. Stettenheim & F. Gill). Philadelphia, Academy of Natural Sciences and Washington, D.C., American Ornithologists Union.

Bull, P. C. & Dawson, P. G. (1969). Mortality and survival of birds during an unseasonable snow storm in South Canterbury, November 1967. *Notornis* 14: 172–179.

Bulyuk, V. N. & Mukhin, A. (1999). Commencement of nocturnal restlessness in the European Robin *Erithacus rubecula* during migration. *Avian Ecol. Behav.* 3: 79–90.

Bulyuk, V. N., Mukhin, A., Fedorov, V. A., Tsvey, A. & Kishkinev, D. (2000). Juvenile dispersal in Reed Warblers *Acrocephalus scirpaceous* at night. *Avian Ecol. Behav.* 5: 45–61.

Burbidge, A. A. & Fuller, P. J. (1982). Banded Stilt breeding at Lake Barlee, Western Australia. *Emu* 82: 212–216.

Burgess, J. P. C. (1982). Sexual differences and dispersal in the Blue Tit *Parus caeruleus. Ringing & Migration* 4: 25–32.

Burke, D. M. & Nol, E. (1998). Influence of food abundance, nest site habitat, and forest fragmentation on breeding Ovenbirds. *Auk* 115: 96–104.

Burns, J. G. (2003). Relationship of *Calidris* sandpiper wingshape with relative fuel load and total migration distance. *Auk* 120: 827–835.

Burns, J. G. & Ydenberg, R. C. (2002). The effects of wing loading and gender on the escape flights of Least Sandpipers (*Calidris minutilla*) and Western Sandpipers (*Calidris mauri*). *Behav. Ecol. Sociobiol.* 52: 128–136.

Burton, N. H. K. (2000). Winter site-fidelity and survival of Redshank *Tringa totanus* at Cardiff, south Wales. *Bird Study* 47: 102–112.

Burton, N. H. K. & Evans, P. R. (1997). Survival and winter site fidelity of Turnstones *Arenaria interpres* and Purple Sandpipers *Calidris maritima* in northeast England. *Bird Study* 44: 35–44.

Buss, I. O. (1942). A managed Cliff Swallow colony in southern Wisconsin. *Wilson Bull.* 54: 153–161.

Busse, P. (1992). Migratory behaviour of Blackcaps (*Sylvia atricapilla*) wintering in Britain and Ireland: contradictory hypotheses. *Ring* 14, 1–2: 51–75.

Busse, P. (1995). New technique of a field study of directional preferences of night passerine migrants. *Ring* 17: 97–116.

Busse, P. (2000). *Bird station manual.* Gdansk, Poland, SE European Bird Migration network, University of Gdansk,

Busse, P. (2001). European passerine migration system – what is known and what is lacking. *Ring* 23: 3–36.

Busse, P. & Trocinska, A. (1999). Evaluation of orientation experiment data using circular statistics – doubts and pitfalls in assumptions. *Ring* 21: 107–130.

Butler, C. J. (2003). The disproportionate effect of global warming on arrival dates of short-distance migratory birds in North America. *Ibis* 145: 484–495.

Butler, L. K., Donahue, M. G. & Rohwer, S. (2002). Molt migration in Western Tanagers (*Piranga ludoviciana*): age effects, aerodynamics, and conservation implications. *Auk* 119: 1010–1023.

Butler, P. J. & Woakes, A. J. (1990). The physiology of bird flight. pp. 300–318 in *Bird migration. Physiology and ecophysiology* (ed. E. Gwinner). Berlin, Springer-Verlag.

Butler, P. J., Woakes, A. J. & Bishop, C. M. (1998). Behaviour and physiology of Svalbard Barnacle Geese *Branta leucopsis* during their autumn migration. *J. Avian Biol.* 29: 536–545.

Butler, P. J., Woakes, A. J., Bevan, R. M. & Stevenson, R. (2000). Heart rate and rate of oxygen consumption during flight of the Barnacle Goose, *Branta leucopsis. Comp. Biochem. Physiol. A.* 216: 379–385.

Butler, P. J., Bishop, C. M. & Woakes, A. J. (2003). Chasing a wild goose: post-hatch growth of locomotor muscles and behavioural physiology of migration of an Arctic goose. pp. 527–542 in *Avian migration* (eds P. Berthold, E. Gwinner & E. Sonnenschein). Berlin, Springer-Verlag.

Butler, R. W. (2000). Stormy seas for some North American songbirds: are declines related to severe storms during migration? *Auk* 117: 518–522.

Butler, R. W., Kaiser, G. W. & Smith, G. E. J. (1987). Migration chronology, length of stay, sex-ratio, and weight of Western Sandpipers (*Calidris mauri*) on the south coast of British Columbia. *J. Field Ornithol.* 58: 103–111.

Butler, R. W., Williams, T. D., Warnock, N. & Bishop, M. A. (1997). Wind assistance: a requirement for migration of shorebirds? *Auk* 114: 456–466.

Byholm, P., Saurola, P., Lindén, H. & Wikman, M. (2003). Causes of dispersal in Northern Goshawks (*Accipiter gentilis*) in Finland. *Auk* 120: 706–716.

Byrkjedal, I. & Langhelle, G. (1986). Sex and age biased mobility in Hawk Owls *Surnia ulula. Ornis Scand.* 17: 306–308.

Byrkjedal, I. & Thompson, D. (1998). *Tundra plovers: the Eurasian, Pacific and American Golden Plovers and Grey Plover.* London, T. & A. D. Poyser.

Cabot, D. & West, B. (1973). Population dynamics of Barnacle Geese in Ireland. *Proc. R. Irish Acad.* 73: 415–443.

Cade, B. S. & Hoffman, R. W. (1993). Differential migration of Blue Grouse in Colorado. *Auk* 110: 70–77.

Cade, T. J. (2000). Progress in translocation of diurnal raptors. pp. 343–372 in *Raptors at risk* (eds R. D. Chancellor & B.-U. Meyburg). Washington, D.C., Hancock House Publishers.

Cade, T. J., Enderson, J. H., Thelander, C. G. & White, C. M. (1988). *Peregrine Falcon populations: their management and recovery*. Boise, The Peregrine Fund Inc.

Calmaestra, R. G. & Moreno, E. (2000). Ecomorphological patterns related to migration: a comparative osteological study with passerines. *J. Zool. Lond.* 252: 495–501.

Campbell, B. (1959). Attachment of Pied Flycatchers *Muscicapa hypoleuca* to nest sites. *Ibis* 101: 445–448.

Campredon, P. (1983). Sex-ratios and age-ratios of Wigeon *Anas penelope* during winter in Western Europe. *La Terre et la Vie* 37: 117–128.

Cannell, P. F., Cherry, J. D. & Parkes, K. C. (1983). Variation and migration overlap in flight feather moult of the Rose-breasted Grosbeak. *Wilson Bull.* 95: 621–627.

Cantos, F. J. & Telleria, J. L. (1994). Stopover site fidelity of four migrant warblers in the Iberian Peninsula. *J. Avian Biol.* 25: 131–134.

Carbone, C. & Owen, M. (1995). Differential migration of the sexes of Pochard *Aythya ferina*: results from a European survey. *Wildfowl* 46: 99–108.

Carlisle, J. D., Kaltenecker, G. S. & Swanson, D. L. (2005). Molt strategies and age differences in migration timing among autumn landbird migrants in southwestern Idaho. *Auk* 122: 1070–1085.

Carmi, N. & Pinshow, B. (1995). Water as a physiological limitation to flight duration in migrating birds: the importance of exhaled air temperature and oxygen extraction. *Israel J. Zool.* 41: 369–375.

Carmi, N., Pinshow, B., Porter, W. P. & Jaeger, J. (1992). Water and energy limitations on flight duration in small migrating birds. *Auk* 109: 268–276.

Carpenter, F. L. & Hixon, M. A. (1988). A new function for torpor: fat conservation in a wild migrant hummingbird. *Condor* 90: 373–378.

Carpenter, F. L., Hixon, M. A., Russell, R. W., Paton, D. C. & Temeles, E. J. (1993). Interference asymmetries among age-classes of Rufous Hummingbirds during migratory stopover. *Behav. Ecol Sociobiol.* 33: 297–304.

Case, R. A. & Gerrish, H. P. (1988). Atlantic hurricane season of 1987. *Monthly Weather Review* 116: 939–949.

Casement, M. B. (1966). Migration across the Mediterranean observed by radar. *Ibis* 108: 461–491.

Castro, G. & Myers, J. P. (1989). Flight range estimates for shorebirds. *Auk* 106: 474–476.

Castro, G., Myers, J. P. & Ricklefs, R. E. (1992). Ecology and energetics of Sanderlings migrating to four latitudes. *Ecology* 73: 833–844.

Catley, G. P. & Hursthouse, D. (1985). Parrot Crossbills in Britain. *Br. Birds* 78: 482–505.

Catry, P., Catry, I., Catry, T. & Martins, T. (2003). Within and between-year winter-site fidelity of Chiffchaffs *Phylloscopus collybita*. *Ardea* 9: 213–220.

Catry, P., Campos, A., Almada, V. & Cresswell, W. (2004a). Winter segregation of migrant European robins *Erithacus rubecula* in relation to sex, age and size. *J. Avian Biol.* 35: 204–209.

Catry, P., Encarnacao, V., Arauja, A., Fearon, P., Fearon, A., Armelin, M. & Delaloye, P. (2004b). Are long-distance migrant passerines faithful to their stopover sites? *J. Avian Biol.* 35: 170–181.

Catry, P., Lecoq, M., Araujo, A., Conway, G., Felgueiras, M., King, J. M. B., Rumsey, S., Salima, H. & Tenreiro, P. (2005). Differential migration of Chiffchaffs *Phylloscopus collybita* and *P. ibericus* in Europe and Africa. *J. Avian Biol.* 36: 184–190.

Cavé, A. J. (1968). The breeding of the Kestrel, *Falco tinnunculus* L., in the reclaimed area Oostelijk Flevoland. *Netherlands J. Zool., Lond.* 18: 313–407.

Cavé, A. J. (1983). Purple Heron survival and drought in tropical West Africa. *Ardea* 71: 217–224.

Cawthorne, R. A. & Marchant, J. H. (1980). Effects of the 1978/79 winter on British bird populations. *Bird Study* 27: 163–172.

Chamberlain, C. P., Blum, J. D., Holmes, R. T., Feng, X., Sherry, T. W. & Graves, G. R. (1997). The use of isotope tracers for identifying populations of migratory birds. *Oecologia* 109: 132–141.

Chamberlain, C. P., Bensch, S., Feng, X., Åkesson, S. & Anderson, T. (2000). Stable isotopes examined across a migratory divide in Scandinavian Willow Warblers *Phylloscopus trochilus trochilus* and *Phylloscopus trochilus acredula* reflect their African winter quarters. *Proc. R. Soc. Lond. B* 267: 43–48.

Chan, K. (1994). Nocturnal activity of caged resident and migrant Silvereyes (Zosteropidae: Aves). *Ethology* 96: 313–321.

Chan, K. (2001). Partial migration of Australian landbirds: a review. *Emu* 100: 281–292.

Chase, M. K., Nur, N. & Geupel, G. R. (1997). Survival, productivity, and abundance in a Wilson's Warbler population. *Auk* 114: 354–366.

Cherel, Y. (1995). Nutrient reserve storage, energetics, and food-consumption during the prebreeding and premoulting foraging periods of King Penguins. *Polar Biol.* 15: 209–214.

Cherry, J. D. (1982). Fat deposition and length of stopover of migrant White-crowned Sparrows. *Auk* 99: 725–732.

Cherubini, G., Serra, L. & Bacceti, N. (1996). Primary moult, body mass and moult migration of Little Tern *Sterna albifrons* in NE Italy. *Ardea* 84: 99–114.

Chesser, R. T. (1994). Migration in South America: an overview of the austral system. *Bird Conserv. Int.* 4: 91–107.

Chesser, R. T. (1998). Further perspectives on the breeding distribution of migratory birds: South American austral migrant flycatchers. *J. Anim. Ecol.* 67: 69–77.

Chitty, H. (1950). Canadian arctic wildlife enquiry, 1943–49, with a summary of results since 1933. *J. Anim. Ecol.* 19: 180–193.

Choinière, L. & Gauthier, G. (1995). Energetics of reproduction in female and male Greater Snow Geese. *Oecologia* 103: 3379–389.

Cichoń, M., Olejniczak, P. & Gustafsson, L. (1998). The effect of body condition on the cost of reproduction in female Collared Fycatchers *Ficedula albicollis*. *Ibis* 140: 128–130.

Clapham, C. (1979). The Turnstone populations of Morecambe Bay. *Ringing & Migration* 2: 144–150.

Clark, J. A., Wernham, C. V., Balmer, D. E., Adams, S. Y., Blackburn, J. R., Griffin, B. M. & King, J. (2000). Bird ringing in Britain and Ireland in 1998. *Ringing & Migration* 20: 39–93.

Clark, J. A., Robinson, R. A., Balmer, D. E., Adams, S. Y., Collier, M. P., Grantham, M. J., Blackburn, J. R. & Griffin, B. M. (2004). Bird ringing in Britain and Ireland in 2003. *Ringing & Migration* 222: 85–127.

Clark, W. (1985). Migration of the Merlin along the coast of New Jersey. *Raptor Res.* 19: 85–93.

Clark, W. S. (1999). *A field guide to the raptors of Europe, the Middle East, and North Africa.* Oxford, Oxford University Press.

Clark, W. S. & Wheeler, B. K. (2001). *A field guide to hawks of North America*, 2nd edn. New York, Houghton Mifflin Co.

Clarke, A. L., Saether, B.-E. & Røskaft, E. (1997). Sex biases in avian dispersal: a reappraisal. *Oikos* 79: 429–438.

Clarke, M. F., Griffioen, P. & Loyn, R. H. (1999). Where do all the bush birds go? *Wingspan* 9: 1-XV1.

Clarke, W. E. (1912). *Studies in bird migration*, Vols. 1 & 2. Edinburgh, London, Oliver & Boyd, Gurney & Jackson.

Clausen, P., Green, M. & Alerstam, T. (2003). Energy limitations for spring migration and breeding: the case of Brent Geese *Branta bernicla* tracked by satellite telemetry to Svalbard and Greenland. *Oikos* 103: 426–445.

Clegg, S. M., Kelly, J. F., Kimura, M. & Smith, T. B. (2003). Combining genetic markers and stable isotopes to reveal population connectivity and migration patterns in a neotropical migrant, Wilson's Warbler (*Wilsonia pusilla*). *Mol. Ecol.* 12: 819–830.

Cochran, W. W. & Kjos, C. G. (1985). Wind drift and migration of thrushes: a telemetry study. *Illinois Nat. Hist. Survey Bull.* 33: 297–330.

Cochran, W. W. & Wikelski, M. (2005). Individual migratory tactics of New World *Catharus* thrushes. pp. 274–289 in *Birds of Two Worlds*. (eds Greenberg, R. & Marra, P. P.). Washington, D. C., Smithsonian Institution.

Cochran, W. W., Montgomery, G. G. & Graber, R. R. (1967). Migratory flights of *Hylocichla* thrushes in spring: a radiotelemetry study. *Living Bird* 6: 213–225.

Cochran, W. W., Mouritsen, H. & Wikelski, M. (2004). Migrating songbirds recalibrate their magnetic compass daily from twilight cues. *Science* 304: 405–408.

Cockburn, A., Osmond, H. L., Mulder, R. A., Green, D. J. & Double, M. (2003). Divorce, dispersal and incest avoidance in the cooperatively breeding Superb Fairy-wren *Malurus cyaneus. J. Anim. Ecol.* 72: 189–202.

Cohen, D. (1967). Optimization of seasonal migratory behaviour. *Am. Nat.* 101: 5–17.

Collister, D. M. (1997). Seasonal distribution of the Great Grey Owl (*Strix nebulosa*) in southwestern Alberta. pp. 119–122 in *Biology and conservation of owls of the northern hemisphere* (eds J. R. Duncan, D. H. Johnson & T. H. Nicholls). Second International Symposium, February 5–9, 1997. Winnipeg, Manitoba, Canada. United States Department of Agriculture.

Collister, D. M. & de Smet, K. (1997). Breeding and natal dispersal in the Loggerhead Shrike. *J. Field Ornithol.* 68: 273–282.

Colwell, M. A., Reynolds, J. D., Gratto, C. L., Schamel, D. & Tracy, D. M. (1988). Phalarope philopatry. *Proc. Int. Ornithol. Congr.* 19: 585–593.

Combreau, O., Launay, F. & Bowardi, M. A. (1999). Outward migration of Houbara Bustards from two breeding areas in Kazakhstan. *Condor* 101: 159–164.

Constant, P. & Ebert, M. C. (1995). Population structure of Bluethroats during a phase of recovery. In *Functioning and dynamics of natural and perturbed ecosystems*. (eds D. Bellan, G. Bonin & C. Eming). Lavoisier, Intercept Ltd.

Cooke, F., Rockwell, R. F. & Lank, D. B. (1995). *The Snow Geese of La Pérouse Bay*. Oxford, Oxford University Press.

Cooke, F., Robertson, G. J. & Smith, C. M. (2000). Survival, emigration, and winter population structure of Harlequin Ducks. *Condor* 102: 137–144.

Coppack, T. & Both, C. (2002). Predicting life-cycle adaptation of migratory birds to global climate change. *Ardea* 90: 369–378.

Coppack, T. & Pulido, F. (2004). Photoperiodic response and the adaptability of avian life cycles to environmental change. *Adv. Ecol. Res.* 35: 131–150.

Coppack, T., Pulido, F. & Berthold, P. (2001). Photoperiodic response to early hatching in a migratory bird species. *Oecologia* 128: 181–186.

Coppack, T., Tøttrup, A. P. & Spottiswoode, C. (2002). Degree of protandry reflects levels of extra pair paternity in migratory songbirds. *J. Ornithol.* 147: 260–265.

Coppack, T., Pulido, F., Czish, M., Auer, D. P. & Berthold, P. (2003). Photoperiodic advance may facilitate adaptation to climate change in long-distance migratory birds. *Proc. R. Soc. Lond. B* 270 (suppl): S43–S46.

Coppack, T., Tottrup, A. P. & Spottiswoode, C. (2006). Degree of protandry reflects level of extrapair paternity in migratory songbirds. *J. Ornithol.* 147: 260–265.

Cornwallis, R. K. (1961). Four invasions of Waxwings during 1956–60. *Br. Birds* 54: 1–30.

Cornwallis, R. K. & Townsend, A. D. (1968). Waxwings in Britain and Europe during 1965/66. *Br. Birds* 61: 97–118.

Corso, A. (2001). Raptor migration across the Strait of Messina, southern Italy. *Br. Birds* 94: 196–202.

Cottam, C. (1929). A shower of grebes. *Condor* 31: 80–81.

Cottridge, D. M. & Vinicombe, K. (1996). *Rare birds in Britain and Ireland. A photographic record*. London, Collins.

Coulson, J. C. (1961). Movements and seasonal variation in mortality of Shags and Cormorants ringed on the Farne Islands, Northumberland. *Br. Birds* 54: 225–235.

Coulson, J. C. & Butterfield, J. (1985). Movements of British Herring Gulls. *Bird Study* 32: 91–103.

Coulson, J. C. & Mévergnies, G. N. (1992). Where do young Kittiwakes *Rissa tridactyla* breed, philopatry or dispersal? *Ardea* 80: 187–197.

Coulson, J. C., Butterfield, J., Duncan, N., Kearsey, S., Monaghan, P. & Thomas, C. (1984). Origin and behaviour of Great Black-backed Gulls wintering in northeast England. *Br. Birds* 77: 1–11.

Court, G. S., Bradley, D. M., Gates, C. C. & Boag, C. C. (1988). The population biology of Peregrine Falcons in the Keewater District of the Northwest Territories, Canada. pp. 729–739 in *Peregrine Falcon populations. Their management and recovery* (ed. T. J. Cade, J. H. Enderson, G. J. Thelander & C. M. White). Boise, The Peregrine Fund.

Cowley, E. (1979). Sand Martin population trends in Britain, 1965–1975. *Bird Study* 26: 113–116.

Cowley, E. & Siriwardena, G. M. (2005). Long-term variation in survival rates of Sand Martins *Riparia riparia*: dependence on breeding and wintering ground weather, age and sex, and their population consequences. *Bird Study* 52: 237–251.

Cox, G. W. (1968). The role of competition in the evolution of migration. *Evolution* 22: 180–192.

Cox, G. W. (1985). The evolution of avian migration systems between temperate and tropical regions of the New World. *Am. Nat.* 126: 451–474.

Cramp, S. (1985). *Handbook of the birds of Europe, the Middle East and North Africa*, Vol 4. Oxford, Oxford University Press.

Cramp, S. (1988). *Handbook of the birds of Europe, the Middle East and North Africa*, Vol. 5. Oxford, Oxford University Press.

Cramp, S. (1992). *Handbook of the birds of Europe, the Middle East and North Africa*, Vol. 6. Oxford, Oxford University Press.

Cramp, S. & Perrins, C. M. (1993). *Handbook of the birds of Europe, the Middle East and North Africa*, Vol. 7. Oxford, Oxford University Press.

Cramp, S. & Perrins, C. M. (1994). *Handbook of the birds of Europe, the Middle East and North Africa*, Vols. 8–9. Oxford, Oxford University Press.

Cramp, S. & Simmons, K. E. L. (1977). *Handbook of the birds of Europe, the Middle East and North Africa*, Vol. 1. Oxford, Oxford University Press.

Cramp, S. & Simmons, K. E. L. (1980). *Handbook of the birds of Europe, the Middle East and North Africa*, Vol. 2. Oxford, Oxford University Press.

Cramp, S. & Simmons, K. E. L. (1983). *Handbook of the birds of Europe, the Middle East and North Africa*, Vol. 3. Oxford, Oxford University Press.

Crawford, H. S. & Jennings, D. T. (1989). Predation by birds on Spruce Budworm *Choristoneura fumiferara*: functional, numerical and total responses. *Ecology* 70: 152–163.

Crawford, R. J. M., Dyer, B. M. & Brooke, R. K. (1994). Breeding nomadism in southern African seabirds – constraints, causes and conservation. *Ostrich* 65: 231–246.

Crawford, R. L. (1978). Autumn bird casualties at a northwest Florida TV tower: 1973–1975. *Wilson Bull.* 90: 335–478.

Crawford, R. L. (1981). Bird casualties at Leon County, Florida TV tower: a 25-year migration study. *Bull. Tall Timbers Res. Stn* 22.

Crawford, R. L. & Engstrom, R. T. (2001). Characteristics of avian mortality at a north Florida television tower: A 29-year study. *J. Field Ornithol.* 72: 380–388.

Cresswell, W. (1994) Age-dependent choice of Redshank (*Tringa totanus*) feeding location – profitability or risk. *J. Anim. Ecol.* 63:589–600.

Creutz, G. (1949). Untersuchungen zur Brutbiologie des Feldsperlings (*Passer m. montanus* L.). *Zool. Jahrb.* 78: 133–172.

Creutz, G. (1955). Der Trauerschnapper (*Muscicapa hypoleuca* Pall.), eine Populationsstudie. *J. Ornithol.* 96: 241–326.

Crick, H. Q. P. & Sparks, T. H. (1999). Climate change related to egg-laying trends. *Nature* 399: 423–424.

Cristol, D. A. (1995). Early arrival, initiation of nesting, and social status: an experimental study of breeding female Red-winged Blackbirds. *Behav. Ecol.* 6: 87–93.

Cristol, D. A., Reynolds, E. B., Leclerc, J. E., Donner, A. H., Farabaugh, C. S. & Ziegenfus, C. W. (2003). Migratory Dark-eyed Juncos, *Junco hyemalis*, have better spatial memory and denser hippocampal neurons than non-migratory conspecifics. *Anim. Behav.* 66: 317–328.

Cristol, D. C., Baker, M. B. & Carbone, C. (1999). Differential migration revisited: latitudinal segregation by age and sex class. *Curr. Ornithol.* 15: 33–87.

Crivelli, A. J., Leshem, Y., Mitchev, T. & Jerrentrup, H. (1991). Where do Palearctic Great White Pelicans (*Pelecanus onocrotalus*) presently overwinter? *Rev. Ecol.-Terre Vie* 46: 145–171.

Croxall, J. P., Silk, J. R. D., Phillips, R. A., Afanasyev, V. & Briggs, D. R. (2005). Global circumnavigations: tracking year-round ranges of nonbreeding albatrosses. *Science* 307: 249–250.

Cuadrado, M. (1992). Year to year recurrence and site-fidelity of Blackcaps *Sylvia atricapilla* and Robins *Erathicus rubecula* in a Mediterranean wintering area. *Ringing & Migration* 13: 36–42.

Cuadrado, M. (1995). Winter territoriality in migrant Black Redstarts *Phoenicurus ochruros* in the Mediterranean area. *Bird Study* 42: 232–239.

Cuadrado, M., Senar, J. C. & Copeti, J. C. (1995). Do all Blackcaps *Sylvia atricapilla* show winter site fidelity? *Ibis* 137: 70–75.

Curio, E. (1958). Geburtsortstreue und Lebenser-wartung juger Traverschnapper (*Muscicapa hypoleuca* Pall.). *Vogelwelt* 79: 135–149.

Currie, F. A. & Bamford, R. (1982). Songbird nest-box studies in forests in north Wales. *Q. J. For.* 76: 250–255.

Currie, D., Thompson, D. B. & Burke, T. (2000). Patterns of territory settlement and consequences for breeding success in the Northern Wheatear *Oenanthe oenanthe*. *Ibis* 142: 389–398.

Curry-Lindahl, K. (1958). Internal timer and spring migration in an equatorial migrant, the Yellow Wagtail (*Motacilla flava*). *Ark. Zool.* 11: 541–557.

Curry-Lindahl, K. (1963). Molt, body weights, gonadal development and migration in *Motacilla flava*. *Proc. Int. Ornithol. Congr.* 13: 960–973.

Curry-Lindahl, K. (1981). *Bird migration in Africa. Movements between six continents.* London, Academic Press.

Curson, D. R., Goguen, C. B. & Mathews, N. E. (2000). Long distance commuting by Brown-headed Cowbirds in New Mexico. *Auk* 117: 795–799.

Dalhaug, L. Tombre, I. M. & Erikstad, K. E. (1996). Seasonal decline in clutch size of the Barnacle Goose in Svalbard. *Condor* 98: 42–47.

Dall'Antonia, L., Dall'Antonia, P., Benvenuti, S., Ioalè, P., Massa, B. & Bonadonna, F. (1995). The homing behaviour of Cory's Shearwaters (*Calonectris diomedea*) studied by means of a direction recorder. *J. Exp. Zool.* 198: 359–362.

Dallinga, J. H. & Schoenmakers, S. (1989). Population changes of the White Stork *Ciconia ciconia* since the 1850s in relation to food resources. pp. 231–262 in *White Stork. Status and conservation* (eds G. Rheinwald, J. Ogden & H. Schulz). Bonn, Dachverband Deutscher Avifaunisten, International Council for Bird Preservation.

Dane, C. W. & Pearson, G. L. (1971). Effect of spring storm on waterfowl mortality and breeding activity. pp. 258–266 in *Snow and ice in relation to wildlife and recreation*. Ames: Iowa Cooperative Wildlife Research Unit, Iowa State University.

Danhardt, J. & Lindström, A. (2001). Optimal departure decisions of songbirds from an experimental stopover site and the significance of weather. *Anim Behav.* 62: 235–243.

Daniels, S. J. & J.R. Walters. (2000). Inbreeding depression and its effects on natal dispersal in Red-cockaded Woodpeckers. *Condor* 102: 482–491.

Dau, C. P. (1992). The fall migration of Pacific flyway Brent *Branta bernicla* in relation to climatic conditions. *Wildfowl* 43: 80–95.

Davidson, N. C. (1984). How valid are flight range estimates for waders? *Ringing & Migration* 5: 49–64.

Davidson, N. C. & Evans, P. R. (1988). Prebreeding accumulation of fat and muscle protein by arctic-breeding shorebirds. *Proc. Int. Ornithol. Congr.* 19: 342–352.

Davies, C. & Sharrock, J. T. R. (2000). The European Bird Report. *Br. Birds* 93: 415–427.

Davies, J. C. & Cooke, F. (1983). Annual nesting production in Snow Geese: prairie droughts and arctic springs. *J. Wildl. Manage.* 47: 291–296.

Davies, N. B. (1978). Ecological questions about territorial behaviour. pp. 317–350 in *Behavioural ecology. An evolutionary approach* (eds J. R. Krebs & N. B. Davies). Oxford, Blackwell Scientific Publications.

Davies, S. (1984). Nomadism as a Response to desert conditions in Australia. *J. Arid Environ.* 7: 183–195.

Davies, S. J. J. F. (1976). The natural history of the Emu in comparison with that of other ratites. *Proc. Int. Ornithol. Congr.* 16: 109–120.

Davies, S. J. J. F. (1988). Nomadism in the Australian Gull-billed Tern *Sterna nilotica*. *Proc. Int. Ornithol. Congr.* 19: 744–753.

Davis, D. E. (1937). A cycle in Northern Shrike emigrations. *Auk* 54: 43–49.

Davis, D. E. (1949). Recent emigrations of Northern Shrikes. *Auk* 66: 293.

Davis, D. E. & Morrison, M. L. (1987). Changes in cyclic patterns of abundance in four avian species. *Am. Birds* 4: 1341–1347.

Davis, J. & Williams, L. (1957). Irruptions of the Clark's Nutcracker in California. *Condor* 59: 297–307.

Davis, J. & Williams, L. (1964). The 1961 irruption of the Clark's Nutcracker in California. *Wilson Bull.* 76: 10–18.

Davis, L. S., Boersma, P. D. & Court., G. S. (1996). Satellite telemetry of the winter migration of Adélie Penguins (*Pygoscelis adeliae*). *Polar Biol.* 16: 221–225.

Davis, P. (1966). The great immigration of early September. *Br. Birds* 59: 353–376.

Davis, P. E. (1964). Crossbills in Britain and Ireland in 1963. *Br. Birds* 57: 477–501.

Dawson, A. (1998). Photoperiodic control of the termination of breeding and the induction of moult in House Sparrows *Passer domesticus*. *Ibis* 140: 35–40.

Dawson, A. (2007). Seasonality in a temperate zone bird can be entrained by near equatorial photoperiods. *Proc. R. Soc. Lond. B* in press.

Dawson, A., Hinsley, S. A., Ferns, P. N., Bosser, R. H. C. & Eccleston, L. (2000). Rate of moult affects feather quality: a mechanism linking current reproductive effort to future survival. *Proc R. Soc Lond. B* 267: 2093–2098.

De Groot, K. L., Smith, J. N. M. & Tait, M. J. (1999). Cowbird removal programs as ecological experiments: measuring community-wide impacts of parasitism and predation. *Stud. Avian Biol.* 18: 299–234.

De la Puente, J., Seoane, J. & Bermejo, A. (1997). Nueva cita de invernada de carricero tordal, *Acrocephalus arundinaceus*, en Europa. *Apus* 10: 4–6.

De Sante, D. F. (1973). An analysis of the fall occurrences and nocturnal orientations of vagrant wood warblers (Parulidae) in California. Unpub. PhD dissertation. California, Stanford University.

De Sante, D. F. (1983a). Annual variability in the abundance of migrant landbirds on southeast Farallon Island, California. *Auk* 100: 826–852.

De Sante, D. F. (1983b). Vagrants: when orientation or navigation goes wrong. *Point Reyes Bird Observatory Newsletter* 61: 12–16.

De Sante, D. F. & Ainley, D. G. (1980). The avifauna of the South Farallon Islands, California. *Stud. Avian Biol.* 4: 1–104.

De Smet, K. D. (1997). Burrowing Owl (*Speotyto cunicularia*) monitoring and management activities in Manitoba 1987–1996. pp. 123–130 in *Biology and conservation of owls of the northern hemisphere* (ed. J. R. Duncan, D. H. Johnson & T. H. Nicholls). Second International Symposium, February 5–9, 1997. Winnipeg, Manitoba, Canada. United States Department of Agriculture.

Dean, W. R. J. (2004). *Nomadic desert birds*. London, Springer.

Deane, R. (1914). Serious loss of bird life during spring migration. *Auk* 31: 548–549.

Deane, R. (1923). The Tree Swallow (*Iridoprocne bicolor*) affected by sudden cold. *Auk* 40: 332–333.

DeBenedictis, P. (1971). Wood warblers and vireos in California: the nature of the accidental. *California Birds* 2: 111–128.

DeCandido, R., Nualsri, C., Allen, D. S. & Bildstein, K. L. (2004). Autumn 2003 raptor migration at Chumphon, Thailand: a globally significant raptor migration watch site. *Forktail* 20: 49–54.

DeCandido, R., Bierregaard, R. O., Martell, M. S. & Bildstein, K. L. (2006). Evidence of nocturnal migration by Osprey (*Pandion haliaetus*) in North America and western Europe. *J. Raptor Res.* 40: 156–158.

DeCapita, M. E. (2000). Brown-headed Cowbird control on Kirtland's Warbler nesting areas in Michigan, 1972–95. pp. 333–341 in *Ecology and management of Cowbirds and their hosts* (eds J. N. M. Smith, T. L. Cook, S. I., Rothstein, S. K. Robinson, & S. G. Sealy). Austin, Texas, University of Texas Press.

Dejaifve, P. A. (1994). Ecology and behavior of a Palearctic migrant in Africa – the Wintering of the Whinchat *Saxicola rubetra* (L) in the Zaire and its winter distribution in Africa. *Rev. Ecol.* 49: 35–52.

Delingat, J. & Dierschke, V. (2000). Habitat utilisation by Northern Wheatears (*Oenanthe oenanthe*) stopping over on an offshore island during migration. *Vogelwarte* 40: 271–278.

DeLong, J. & Hoffman, S. W. (1999). Differential autumn migration of Sharp-shinned and Cooper's Hawks in western North America. *Condor* 101: 674–678.

DeLong, J. P. & Hoffman, S. W. (2004). Fat stores of migrant Sharp-shinned and Cooper's Hawks in New Mexico. *Raptor Res.* 38: 163–168.

Dementiev, G. P. & Gladkov, N. A. (1954). *Birds of the Soviet Union*, Vol. 1. Moscow, State Publishing House.

Demey, R. & Fishpool, L. D. C. (1994). The birds of Yapo Forest, Ivory Coast. *Malimbus* 16: 100–122.

den Held, J. D. (1981). Population changes of the Purple Heron in relation to drought in the wintering area. *Ardea* 69: 185–191.

Dence, W. A. (1946). Tree Swallow mortality from exposure during unseasonal weather. *Auk* 63: 440.

Dennis, J. V. (1981). A summary of banded North American birds encountered in Europe. *N. Am. Bird Bander* 6: 88–96.

Dennis, J. V. (1987). Additional records of banded North American birds in Europe. *N. Am. Bird Bander* 12: 11–12.

Dennis, J. V. (1990). Banded North American birds encountered in Europe: an update. *N. Am. Bird Bander* 15: 130–133.

Derégnaucourt, S., Guyomarc'h, J.-C. & Belhamra, M. (2005). Comparison of migratory tendency in European Quail (*Coturnix c. coturnix*), domestic Japanese Quail (*Coturnix c. japonica*) and their hybrids. *Ibis* 147: 25–36.

Desholm, M. (2003). How much do small-scale changes in flight direction increase overall migration distance? *J. Avian Biol.* 34: 155–158.

Dhondt, A. A. (1979). Summer dispersal and survival of juvenile Great Tits in southern Sweden. *Oecologia* 42: 139–157.

Dhondt, A. A. (1983). Variations in the numbers of overwintering Stonechats possibly caused by natural selection. *Ringing & Migration* 4: 155–158.

Dhondt, A. & Hublé, J. (1968). Fledging date and sex in relation to dispersal in young tits. *Bird Study* 15: 127–134.

Diamond, J. M. (1982). Mirror-image navigational errors in migrating birds. *Nature* 295: 277–278.

Diamond, J. M., Karasov, W. H., Phan, D. & Carpenter, F. C. (1986). Digestive physiology is a determinant of foraging bout frequency in hummingbirds. *Nature* 320: 62–63.

Dick, W. J. A. & Pienkowski, M. W. (1979). Autumn and early winter weights of waders in north-west Africa. *Ornis Scand.* 10: 117–123.

Diefenbach, D. R., Nichols, J. D. & Hines, J. E. (1988). Distribution patterns during winter and fidelity to wintering areas of American Black Ducks. *Can. J. Zool.* 66: 1506–1513.

Diefenbach, D. R., Derleth, E. L., Haegen, W. M. V., Nichols, J. D. & Hines, J. E. (1990). American Woodcock winter distribution and fidelity to wintering areas. *Auk* 107: 745–749.

Diehl, R. H., Larkin, R. P. & Black, J. E. (2003). Radar observations of bird migration over the Great Lakes. *Auk* 120: 278–290.

Dierschke, J. (2002). Food preferences of Shorelarks *Eremophila alpestris*, Snow Buntings *Plectrophenax nivalis* and Twites *Carduelis flavirostris* wintering in the Wadden Sea. *Bird Study* 49: 263–269.

Dierschke, V. (1998). Site fidelity and survival of Purple Sandpipers *Calidris maritima* at Helgoland (SE North Sea). *Ringing & Migration* 19: 41–48.

Dierschke, V., Delingat, J. & Schmaljohann, H. (2003). Time allocation in migrating Northern Wheatears (*Oenanthe oenanthe*) during stopover: is refueling limited by food availability or metabolically? *J. Ornithol.* 144: 33–44.

Dijk, A. J., van, Roder, F. E., de, Marteijn, E. C. L. & Spickman, H. (1990). Summering waders on the Banc d'Arguin, Mauritania: a census in June 1988. *Ardea* 78: 145–156.

Dijkstra, C., Bult, A., Bijlsma, S., Daan, S., Meijer, T. & Zijlstra, M. (1990). Brood size manipulations in the Kestrel (*Falco tinnunculus*): effects on offspring and parent survival. *J. Anim. Ecol.* 59: 269–285.

Dingemanse, N. J., Both, C., Noordwijk, A. J., Rutten, A. L. & Drent, P. J. (2003). Natal dispersal and personalities in Great Tits (*Parus major*). *Proc. R. Soc. Lond. B.* 270: 741–747.

Dingle, H. (1996). *Migration: the biology of life on the move.* Oxford, Oxford University Press.

Dingle, H. (2004). The Australo-Papuan bird migration system: another consequence of Wallace's Line. *Emu* 104: 95–108.

Dingle, H. (2005) Animal migration: is there a common migratory syndrome? *J. Ornithol.* 147: 212–220.

Dittman, T. & Becker, P. H. (2003). Sex, age, experience and condition as factors affecting arrival date in prospecting Common Terns, *Sterna hirundo*. *Anim. Behav.* 65: 981–986.

Dobinson, H. M. & Richards, A. J. (1964). The effects of the severe winter of 1962/63 on birds in Britain. *Br. Birds* 59: 373–434.

Dobrynina, I. N. (1990). Bird migration studies in the Soviet Union. *Sitta* 4: 1–26.

Dohn, G. L. (1986). Protein as fuel for endurance exercise. *Exercise Sport Sci. Rev.* 14: 143–173.

Dolbeer, R. A. (1982). Migration patterns for age and sex classes of Blackbirds and Starlings. *J. Field Ornithol.* 53: 28–46.

Dolbeer, R. A. (1991). Migration patterns of Double-crested Cormorants east of the Rocky Mountains. *J. Field Ornithol.* 62: 83–93.

Dolman, P. M. & Sutherland, W. J. (1995). The response of bird populations to habitat loss. *Ibis* 137: 38–46.

Dolnik, V. R. (1963). A quantitative study of vernal testicular growth in several species of finches (Fringillidae). *Dokl. Akad. Nauk USSR* 149: 370–372.

Dolnik, V. R. (1990). Bird migration across arid and mountainous regions of Middle Asia and Kazakhstan. pp. 368–386 in *Bird migration. Physiology and ecophysiology* (ed. E. Gwinner). Berlin, Springer-Verlag.

Dolnik, V. R. & Blyumental, T. I. (1967). Autumnal premigratory and migratory periods in the Chaffinch *Fringilla coelebs* and some other temperate zone passerine birds. *Condor* 69: 435–468.

Dolnik, V. R. & Bolshakov, L. V. (1985). Preliminary results of vernal nocturnal bird passage study over arid and mountain areas of central Asia: latitudinal crossing. pp. 292–296 in *Spring nocturnal bird passage over arid and mountain areas of Asia middle and Kazakhstan* (ed. V. R. Dolnik). Moscow, USSR, Academy of Sciences.

Dolnik, V. R. & Gavrilov, V. M. (1980). Photoperiodic control of the molt cycle in the Chaffinch (*Fringilla coelebs*). *Auk* 97: 50–62.

Donázar, J. A., Negro, J. J. & Hiraldo, F. (1993). Foraging habitat selection, land-use changes and population decline in the Lesser Kestrel *Falco naumanni*. *J. Appl. Ecol.* 30: 515–522.

Dorka, V. (1966). Das jahres- und tageszeitliche Zugmuster von Kurz- und Langstreckenziehern nach Beobachtungen auf den Alpenpässen Cou/Bretolet (Wallis). *Orn. Beob.* 63: 165–223.

Dorst, J. (1961). *The migrations of birds*. London, Heinemann.

Dougall, T. W. (1992). Post-fledging dispersal of British Pied Wagtails *Motacilla alba yarrellii*. *Ringing & Migration* 13: 21–26.

Dow, H. & Fredga, S. (1983). Breeding and natal dispersal of the Goldeneye, *Bucephala clangula*. *J. Anim. Ecol.* 52: 681–695.

Dowsett, R. J. & Fry, C. H. (1971). Weight losses of trans-Saharan migrants. *Ibis* 113: 531–533.

Dowsett-Lemaire, F. & Dowsett, R. J. (1987). European Reed and Marsh Warblers in Africa: migration patterns, moult and habitat. *Ostrich* 58: 65–85.

Doyle, F. I. & Smith, J. M. N. (1994). Population responses of Northern Goshawks to the 10-year cycle in numbers of Snowshoe Hares. *Stud. Avian Biol.* 16: 122–129.

Drent, R. & Piersma, T. (1990). An exploration of the energetics of leap-frog migration in arctic breeding waders. pp. 399–412 in *Bird migration. Physiology and ecophysiology* (ed. E. Gwinner). Berlin, Springer-Verlag.

Drent, R., Both, C., Green, M., Madsen, J. & Piersma, T. (2003). Pay-offs and penalties of competing migratory schedules. *Oikos* 103: 274–292.

Drent, R. H., Fox, A. D. & Stahl, J. (2006). Travelling to breed. *J. Ornithol.* 147: 122–134.

Driedzic, W. R., Crowe, K. L., Hicklin, P. W. & Sephton, D. H. (1993). Adaptations in pectoralis muscle, heart mass, and energy metabolism during premigratory fattening in Semipalmated Sandpipers (*Calidris pusilla*). *Can. J. Zool.* 71: 1602–1608.

Drilling, N. E. & Thompson, L. F. (1988). Natal and breeding dispersal in House Wrens (*Troglodytes aedon*). *Auk* 105: 480–491.

Drost, R. (1930). Über die Tagesaufbruchszeit der Zugvogel und ihre Abhängigkeit vom Licht. *Vogelzug* 1: 117–119.

Drost, R. (1938). Uber den Einflus von Verfrachtungen zur Herbstzugzeit auf den Sperber *Accipiter nisus* (L.). *Proc. Int. Ornithol. Congr.* 9: 502–521.

Drost, R. (1955). Wo verbleiben im Binnenland frei aufgezogne Nordsee-Silbermöwen. *Vogelwarte* 18: 85–93.

Drost, R. (1958). Über die Ansiedbing von jung ins Binnerland verfrachteten Silbermöwen (*Larus argentatus*). *Vogelwarte* 19: 169–173.

Duebbert, H. F. & Kantrud, H. A. (1974) Upland duck nesting related to land use and predator reduction. *J. Wildl. Manage.* 38: 257–265.

Duebbert, H. F. & Lokemoen, J. T. (1980). High duck nesting success in a predator-reduced environment. *J. Wildl. Manage.* 44: 428–437.

Dufour, K. W., Ankney, C. D. & Weatherhead, P. J. (1993). Condition and vulnerability to hunting among Mallards staging at Lake St-Clair, Ontario. *J. Wildl. Manage.* 57: 209–215.

Dugger, B. D. (1997). Factors influencing the onset of spring migration in Mallards. *J. Field Ornithol.* 68: 331–337.

Dugger, K. M., Arendt, W. J. & Hobson, K. A. (2004). Understanding survival and abundance of overwintering warblers: does rainfall matter? *Condor* 106: 744–760.

Duncan, J. R. (1992). *Influence of prey abundance and snow cover on Great Grey Owl breeding dispersal*. M. Sc. thesis. Winnipeg, University of Manitoba.

Duncan, J. R. (1997). Great Grey Owls (*Strix nebulosa nebulosa*) and forest management in North America: a review and recommendations. *J. Raptor Res.* 31: 160–166.

Duncan, J. R. & Duncan, P. A. (1998). Northern Hawk Owl. *The birds of North America*, No. 356. (eds A. Poole & F. Gill). New York, Cornell Laboratory of Ornithology and Philadelphia, Academy of Natural Sciences.

Duncan, W. N. M. & Monaghan, P. (1977). Infidelity to the natal colony by breeding Herring Gulls. *Ringing & Migration* 1: 166–172.

Dunn, P. O., May, T. A., McCollough, M. A. & Howe, M. A. (1988). Length of stay and fat content of migrant Semipalmated Sandpipers in eastern Maine. *Condor* 90: 824–835.

Dunne, P. J. & Clark, W. S. (1977). Fall hawk movement at Cape May Point, N. J. – 1976. *New Jersey Audubon* 3: 114–124.

Dunne, P. J. & Sutton, C. C. (1986). Population trends in coastal raptor migrants over ten years of Cape May Point autumn counts. *Rec. N.J. Birds* 12: 39–43.

Dunning, J. B. & Brown, J. H. (1982). Summer rainfall and winter sparrow densities – a test of the food limitation hypothesis. *Auk* 99: 123–129.

Durand, A. L. (1963). A remarkable fall of American landbirds on the 'Mauritania', New York to Southampton, October 1962. *Br. Birds* 56: 157–164.

Dymond, J. N., Fraser, P. A. & Gantlet, S. J. M. (1989). *Rare birds in Britain and Ireland*. Calton, T. & A.D. Poyser.

Eastwood, E. (1967). *Radar ornithology*. London, Methuen.

Eastwood, E. & Rider, G. C. (1965). Some radar measurements of the altitude of bird flight. *Br. Birds* 58: 393–426.

Ebbinge, B. S. (1985). Factors determining the population size of arctic-breeding geese wintering in western Europe. *Ardea* 73: 121–128.

Ebbinge, B. S. (1989). A multifactorial explanation for variation in breeding performance of Brent Geese *Branta bernicla*. *Ibis* 131: 196–204.

Ebbinge, B. S. (1991). The impact of hunting on mortality rates and spatial distribution of geese wintering in the western Palearctic. *Ardea* 79: 197–210.

Ebbinge, B. S. (1992). Regulation of numbers of Dark-bellied Brent Geese *Branta bernicla* on spring staging sites. *Ardea* 80: 203–228.

Ebbinge, B. S. & Spaans, B. N. (1995). The importance of body reserves accummulated in spring staging areas in the temperate zone for breeding in Dark-bellied Brent Geese *Branta b. bernicla* in the high arctic. *J. Avian Biol.* 26: 105–113.

Eckhardt, R. C. (1977). Effects of a late spring storm on a Dusky Flycatcher population. *Auk* 94: 362.

Edelstam, C. (ed.) (1972). *The visible migration of birds at Ottenby, Sweden*. Sweden, Vår Fagelvärld Sonderh.

Eden, S. F. (1987). Natal philopatry of the Magpie *Pica pica*. *Ibis* 129: 477–490.

Egeler, O., Seaman, D. & Williams, T. D. (2003). Influence of diet on fatty-acid composition of depot fat in Western Sandpipers (*Calidris mauri*). *Auk* 120: 337–345.

Ekman, J., Sklepkovych, B. & Tegelström, H. (1994). Offspring retention in the Siberian Jay (*Perisoreus infaustus*): the prolonged brood care hypotheses. *Behav. Ecol.* 5: 245–253.

Elgood, J. H., Fry, C. H. & Dowsett, R. J. (1973). African migrants in Nigeria. *Ibis* 115: 1–45.

Elkins, N. (1988). *Weather and bird behaviour*. Calton, Poyser.

Elkins, N. (2005). Weather and bird migration. *Br. Birds* 98: 238–256.

Elkins, N. & Etheridge, B. (1977). Further studies of wintering Crag Martins. *Ringing & Migration* 1: 158–165.

Ellegren, H. (1990a). Autumn migration speeds in Scandinavian Bluethroats *Luscinia s. svecica*. *Ringing & Migration* 11: 121–131.

Ellegren, H. (1990b). Timing of autumn migration in Bluethroats *Luscinia svecica* depends on timing of breeding. *Ornis Fenn.* 67: 13–17.

Ellegren, H. (1991). Stopover ecology of autumn migrating Bluethroats *Luscinia s. svecica* in relation to age and sex. *Ornis Scand.* 22: 340–348.

Ellegren, H. (1993). Speed of migration and migratory flight lengths of passerine birds ringed during autumn migration in Sweden. *Ornis Scand.* 24: 220–228.

Ellegren, H. & Fransson, T. (1992). Fat loads and estimated flight-ranges in four *Sylvia* species analysed during autumn migration at Gotland, south-east Sweden. *Ringing & Migration* 13: 1–12.

Elliott, C. C. H., Watter, M., Underhill, L. G., Pringle, J. S. & Dick, W. J. A. (1976). The migration system of the Curlew Sandpiper *Calidris ferruginea* in Africa. *Ostrich* 47: 191–213.

Ellis, D. H., Kepler, A. K. & Kepler, C. B. (1994). A fall land bird migration across the South China Sea from Indo-China to the Greater Sunda Islands. *J. Bombay Nat. Hist. Soc.* 91: 427–434.

Ellis, D. H., Moon, S. L. & Robinson, J. W. (2001). Annual movements of a Steppe Eagle (*Aquila nipalensis*) summering in Mongolia and wintering in Tibet. *J. Bombay Nat. Hist. Soc.* 98: 335–340.

Ellis, D. H., Sladen, W. J. L., Lishman, W. A., Clegg, K. R., Duff, J. W., Gee, G. F. & Lewis, J. L. (2003). Motorised migrations: The future or mere fantasy? *Bioscience* 53: 260–264.

Elmberg, J. & Mo, A. (1984). Smalnäbbad nötkråka *Nucifraga caryocatactes macrorhynchos* – nyeteblerad häckfågel i Västerbotten. *Vår Fågelvärld* 43: 193–197.

Elton, C. S. (1942). *Voles, mice and lemmings*. Oxford, Oxford University Press.

Ely, C. A. (1970). Migration of Least and Traill's Flycatchers in west-central Kansas. *Bird-Banding* 61: 198–204.

Ely, C. A., Latas, P. J. & Lohoefener, R. R. (1977). Additional returns and recoveries of North American birds banded in Southern Mexico. *Bird-Banding* 48: 275–276.

Emlen, S. T. (1967a). Migratory orientation in the Indigo Bunting, *Passerina cyanea*. Part 1: Evidence for use of celestial cues. *Auk* 84: 309–342.

Emlen, S. T. (1967b). Migratory orientation in the Indigo Bunting, *Passerina cyanea*. Part 2: Mechanism of celestial orientation. *Auk* 84: 463–489.

Emlen, S. T. (1969). Bird migration: influence of physiological state upon celestial orientation. *Science* 165: 716–718.

Emlen, S. T. (1975). Migration: orientation and navigation. pp. 129–219 in *Avian biology*, Vol. 5 (eds D. S. Farner, J. R. King & K. C. Parkes). London, Academic Press.

Emlen, S. T. & Emlen, J. T. (1966). A technique for recording migratory orientation of captive birds. *Auk* 83: 361–367.

Enemar, A., Nilsson, L. & Sjöstrand, B. (1984). The composition and dynamics of the passerine bird community in a subalpine birch forest, Swedish Lapland. A 20-year study. *Ann. Zool. Fenn.* 21: 321–338.

Enoksson, B. & Nilsson, S. G. (1983). Territory size and population density in relation to food supply in the Nuthatch *Sitta europea*. *J. Anim. Ecol.* 52: 927–935.

Enquist, M. & Pettersson, J. (1986). The timing of migration in 104 bird species at Ottenby – an analysis based on 39 years trapping data. Special Report from Ottenby Bird Observatory No. 8., Sweden,

Enticott, J. W. (1999). Britain and Ireland's first 'soft-plumaged petrel' – an historical and personal perspective. *Br. Birds* 92: 504–518.

Eriksson, K. (1970a). Wintering and autumn migration ecology of the Brambling *Fringilla montifringilla*. *Sterna* 9: 77–90.

Eriksson, K. (1970b). The autumn migration and wintering ecology of the Siskin *Carduelis spinus*. *Ornis Fenn.* 47: 52–68.

Eriksson, K. (1970c). Ecology of the irruption and wintering of Fennoscandian Redpolls (*Carduelis flammea* coll.). *Ann. Zool. Fenn.* 7: 273–282.

Eriksson, K. (1970d). The invasion of *Sitta europaea asiatica* Gould into Fennoscandia in the winter of 1962/63 and 1963/64. *Ann. Zool. Fenn.* 7: 121–140.

Eriksson, K. (1971). Irruption and wintering ecology of the Great Spotted Woodpecker. *Ornis Fenn.* 48: 69–76.

Erni, B., Liechti, F., Underhill, L. G. & Bruderer, B. (2002). Wind and rain govern the intensity of nocturnal bird migration in central Europe – a log-linear regression analysis. *Ardea* 90: 153–166.

Erskine, A. J. (1979). Man's influence on potential nesting sites and populations of swallows in Canada. *Can. Field Nat.* 93: 371–377.

Esler, D. & Grand, J. B. (1994). The role of nutrient reserves for clutch formation by Northern Pintails in Alaska. *Condor* 96: 422–432.

Evans, I. M., Summers, R. W., O'Toole, L., Orr-Ewing, D. C., Evans, R. D., Snell, N. & Smith, J. (1999). Evaluating the success of translocating Red Kites *Milvus milvus* to the UK. *Bird Study* 46: 129–144.

Evans, K. L., Waldron, S. & Bradbury, R. B. (2003a). Segregation in the African wintering ranges of English and Swiss Swallow *Hirundo rustica* populations: a stable isotope study. *Bird Study* 50: 294–299.

Evans, K. L., Wilson, J. D. & Bradbury, R. B. (2003b). Swallow *Hirundo rustica* population trends in England: data from repeated historical surveys. *Bird Study* 50: 178–181.

Evans, P. R. (1966a). Autumn movements, moult and measurements of the Lesser Redpoll *Carduelis flammea cabaret*. *Ibis* 108: 183–216.

Evans, P. R. (1966b). Migration and orientation of passerine night migrants in northeast England. *J. Zool. (Lond.)* 150: 319–369.

Evans, P. R. (1968). Reorientation of passerine night migrants after displacement by the wind. *Br. Birds* 61: 281–303.

Evans, P. R. & Davidson, N. C. (1990). Migration strategies and tactics of waders breeding in arctic and north temperate latitudes. pp. 387–398 in *Bird migration. Physiology and ecophysiology* (ed. E. Gwinner). Berlin, Springer-Verlag.

Evans, P. R. & Lathbury, G. W. (1973). Raptor migration across the Straits of Gibralter. *Ibis* 115: 572–585.

Evans, P. R. & Pienkowski, M. W. (1984). Population dynamics of shorebirds. pp. 83–123 in *Behaviour of marine animals*, Vol. 5, *Shorebirds, breeding behaviour and populations* (eds J. Burger & B. L. Olla). New York.

Evans, P. R. & Townsend, D. J. (1988). Site faithfulness of waders away from the breeding grounds: how individual migration patterns are established. *Proc. Int. Ornithol. Congr.* 19: 594–603.

Evans, P. R., Davidson, N. C., Piersma, T. & Pienkowski, M. W. (1991). Implications of habitat loss at migration staging posts for shorebird populations. *Proc. Int. Ornithol. Congr.* 20: 2228–2235.

Evans, P. R., Davidson, N. C., Uttley, J. D. & Evans, R. D. (1992). Premigratory hypertrophy of flight muscles: an ultrastructural study. *Ornis Scand.* 23: 238–243.

Eyster, M. B. (1954). Quantitative measurement of the influence of photoperiod, temperature and season on the activity of captive songbirds. *Ecol. Monogr.* 24: 1–28.

Faaborg, J. & Arendt, W. J. (1992). Long-term declines of winter resident warblers in a Puerto Rican dry forest: which species are in trouble? pp. 57–63 in *Ecology and conservation of neotropical migrant landbirds* (eds J. M. Hagan & D. W. Johnston). Washington, D.C., Smithsonian Institute Press.

Faaborg, J. R., Arendt, W. J. & Kaiser, M. S. (1984). Rainfall correlates of bird population fluctuations in a Puerto Rican dry forest. A nine year study. *Wilson Bull.* 96: 557–595.

Farmer, A. H. & Wiens, J. A. (1999). Models and reality: time-energy trade-offs in Pectoral Sandpiper (*Calidris melanotus*) migration. *Ecology* 80: 2566–2580.

Farner, D. S. (1955). The annual stimulus for migration: experimental and physiologic aspects. *Recent studies in avian biology* (ed. A. Wolfson). Urbana, University Illinois Press.

Farner, D. S. (1960). Metabolic adaptation in migration. *Proc Int. Orn. Congr.* 12: 197–208.

Farner, D. S. & Follett, B. K. (1966). Light and other environmental factors affecting avian reproduction. *J. Anim. Sci.* 25: 90–105.

Farner, D. S., Donham, R. S., Moore, M. C. & Lewis, R. A. (1980). The temporal relationship between the cycle of testicular development and moult in the White-crowned Sparrow, *Zonotrichia leucophrys gambelii. Auk* 97: 63–75.

Farnsworth, A., Gauthreaux, S. A. & van Blaricom, D. (2004). A comparison of nocturnal call counts of migrating birds and reflectivity measurements on Doppler radar. *J. Avian Biol.* 35: 365–369.

Fasola, M., Hafner, H., Prosper, J., van der Kooij, H. & Schogolev, I. V. (2000). Population changes in European herons in relation to African climate. *Ostrich* 71: 52–55.

Fauth, P. T. (2000). Reproductive success of Wood Thrushes in forest fragments in northern Indiana. *Auk* 177: 194–204.

Feare, C. J. (1976). Desertion and abnormal development in a colony of Sooty Terns *Sterna fuscata* infested with virus-infected ticks. *Ibis* 118: 112–115.

Feare, C. J. & Burham, S. (1978). Lack of nest site tenacity and mate fidelity in the Starling. *Bird Study* 25: 189–191.

Feare, C. J., Gill, E. L., McKay, H. V. & Bishop, J. D. (1995). Is the distribution of Starlings *Sturnus vulgaris* within roosts determined by competition? *Ibis* 137: 379–382.

Ferns, P. N. (1975). Feeding behaviour of autumn passage migrants in northeast Portugal. *Ringing & Migration* 1: 3–11.

Ferrer, M. (1993). Juvenile dispersal behavior and natal philopatry of a long-lived raptor, the Spanish Imperial Eagle *Aquila adalberti. Ibis* 135: 132–138.

Ferrer, M. & Janss, G. F. E. (eds.) (1999). *Birds and power lines. Collision, electrocution and breeding.* Madrid, Quercus.

Fiedler, W. (2001). Large scale ringing recovery analysis of European White Storks (*Ciconia ciconia*). *The Ring* 23: 73–79.

Fiedler, W. (2003). Recent changes in migratory behaviour of birds: a compilation of field observations and ringing data. pp. 21–38 in *Avian migration* (eds P. Berthold, E. Gwinner & E. Sonnenschein). Berlin, Springer-Verlag.

Fiedler, W. (2005). Ecomorphology of the external flight apparatus of Blackcaps (*Sylvia atricapilla*) with different migration behaviour. *Ann. N.Y. Acad. Sci* 1046: 253–263.

Fiedler, W., Bairlein, F. & Koppen, U. (2004). Using large-scale data from ringed birds for the investigation of effects of climate change on migrating birds: pitfalls and prospects. pp. 35–49 in *Birds and climate change. Advances in Ecological Research*, Vol. 35. London: Elsevier.

Figuerola, J. & Bertolero, A. (1998). Sex differences in the stopover ecology of Curlew Sandpipers *Calidris ferruginea* at a refuelling area during autumn migration. *Bird Study* 45: 313–319.

Figuerola, J. & Green, A. J. (2000). Haematozoan parasites and migratory behaviour in waterfowl. *Evol. Ecol.* 14: 143–153.

Finch, D. W. (1975). Northeastern Maritime Region. *Am. Birds* 29: 125–129.

Finlayson, J. C. (1980). The recurrence in winter quarters at Gibraltar of some scrub passerines. *Ringing & Migration* 3: 32–36.

Fischer, J. H., Munro, U. & Phillips, J. B. (2003). Magnetic navigation by an avian migrant? pp. 423–432 in *Avian migration* (eds P. Berthold, E. Gwinner & E. Sonnenschein). Berlin, Springer-Verlag.

Fischer, S., Mauersberger, G., Schielzeth, H. & Witt, K. (1992). First breeding record of Two-barred Crossbill (*Loxia leucoptera*) in Central-Europe. *J. Ornithol.* 133: 197–202.

Fisher, H. I. (1971). Experiments on homing in Laysan Albatrosses, *Diomedea immutabilis*. *Condor* 73: 389–400.

Fisher, R. M. & Myres, M. T. (1980). A review of factors influencing extralimital occurrences of Clark's Nutcracker in Canada. *Can. Field Nat.* 94: 43–51.

Flegg, J. (2004). *Time to fly. Exploring bird migration.* Thetford, British Trust for Ornithology.

Flynn, L., Nol, E. & Zharikov, Y. (1999). Philopatry, nest-site tenacity, and mate fidelity of Semipalmated Plovers. *J. Avian Biol.* 30: 47–55.

Fogden, M. P. L. (1972a). Premigratory dehydration in the Reed Warbler *Acrocephalus scirpaceus* and water as a factor limiting migratory range. *Ibis* 114: 548–552.

Fogden, M. P. (1972b). The seasonality and population dynamics of equatorial forest birds in Sarawak. *Ibis* 114: 307–343.

Foppen, R., Braak, C. J. T. T., Verbook, J. & Reijnen, R. (1999). Dutch Sedge Warblers *Acrocephalus schoenobaenus* and West-African rainfall: empirical data and simulation modelling show low population resilience in fragmented marshlands. *Ardea* 87: 113–125.

Ford, H. A. (1978). The Black Honeyeater: nomad or migrant? *S. Aust. Ornithol.* 27: 263–269.

Forero, M. G., Donázar, J. A., Blas, J. & Hiraldo, F. (1999). Causes and consequences of territory change and breeding dispersal distance in the Black Kite. *Ecology* 80: 1298–1310.

Forero, M. G., Donázar, J. A. & Hiraldo, F. (2002). Causes and fitness consequences of natal dispersal in a population of Black Kites. *Ecology* 83: 858–872.

Formosov, A. N. (1933). The crop of cedar nuts, invasions into Europe of the Siberian Nutcracker (*Nucifraga caryocatactes macrorhynchos* Brehm) and fluctuations in the numbers of the Squirrel (*Sciurus vulgaris* L.). *J. Anim. Ecol.* 2: 70–81.

Formosov, A. N. (1960). La production de graines dans les forêts de conifères de la taiga de l'USSR et l'envahissement de l'Europe occidentale par certaines espèces d'oiseaux. *Proc. Int. Orn Congr.* 12: 216–229.

Formosov, A. N. (1965). Irregularities in the mass autumn migration of the Coal Tit. *Communs. Baltic Comm. Study Bird Migr.* 3: 82–90. In Russian with English summary.

Förschler, M. I., Förschler, L. & Dorka, U. (2006). Population fluctuations of Siskins *Carduelis spinus*, Common Crossbills *Loxia curvirostra*, and Citril Finches *Carduelis citrinella* in relationship to flowering intensity of Spruce *Picea abies*. *Ornis Fenn.* 83: 91–96.

Forslund, P. & Larsson, K. (1991). Breeding range expansion of the Barnacle Goose in the Baltic area. *Ardea* 79: 343–346.

Forsman, D. (1999). *The raptors of Europe and the Middle East: a handbook of identification.* London, T. & A.D. Poyser.

Forstmeier, W. (2002a). Benefits of early arrival at breeding grounds vary between males. *J. Anim. Ecol.* 7: 1–9.

Forstmeier, W. (2002b). Factors contributing to male mating success in the polygynous Dusky Warbler (*Phylloscopus fuscatus*). *Behaviour* 139: 1361–1381.

Forstmeier, M. C., Post, E. & Stenseth, N. C. (2002). North Atlantic Oscillation timing of long- and short-distance migration. *J. Anim. Ecol.* 71: 1002–1014.

Foster, M. S. (1975). The overlap of molting and breeding in some tropical birds. *Condor* 77: 304–314.

Fournier, M. A. & Hines, J. E. (1994). Effects of starvation on muscle and organ mass of King Eiders *Somateria spectabilis* and the ecological and management implications. *Wildfowl* 45: 188–197.

Fowler, J. A., O'Kill, J. D. & Marshall, B. (1982). A retrap analysis of Storm Petrels tape-lured in Shetland. *Ringing & Migration* 4: 1–7.

Fox, A. D., Francis, I. S., Madsen, J. & Stroud, J. M. (1987). The breeding biology of the Lapland Bunting *Calcarius lapponicus* in West Greenland during two contrasting years. *Ibis* 129: 541–552.

Fox, A. D., Hilmarsson, J. Ó., Einarsson, O., Walsh, A. J., Boyd, H. & Kristiansen, J. N. (2002). Staging site fidelity of Greenland White-fronted Geese *Anser albifrons flavirostris* in Iceland. *Bird Study* 49: 42–49.

Fox, A. D., Glahder, C. M. & Walsh, A. J. (2003). Spring migration routes and timing of Greenland White-fronted Geese – results from satellite telemetry. *Oikos* 103: 415–425.

Francis, C. M. & Cooke, F. (1986). Differential timing of spring migration in Wood Warblers (Parulinae). *Auk* 103: 548–556.

Francis, C. M., Richards, M. H., Cooke, F. & Rockwell, R. F. (1992). Long-term changes in survival rates of Lesser Snow Geese. *Ecology* 73: 1346–1362.

Franklin, D. C., Smales, I. J., Quin, B. R. & Menkhorst, P. W. (1999). Annual cycle in the Helmeted Honeyeater *Lichenostomus flavicollis cassidix*, a sedentary inhabitant of a predictable environment. *Ibis* 141: 256–268.

Fransson, T. (1986). The migration and wintering area of Nordic Spotted Flycatcher *Muscicapa striata*. *Vår Fågelvärld* 45: 5–18.

Fransson, T. (1995). Timing and speed of migration in North and West European populations of *Sylvia* warblers. *J. Avian Biol.* 26: 39–48.

Fransson, T. (1998). A feeding experiment on migratory fuelling in Whitethroats *Sylvia communis*. *Anim. Behav.* 55: 153–162.

Fransson, T. & Jakobsson, S. (1998). Fat storage in male Willow Warblers in spring: do residents arrive lean or fat? *Auk* 115: 759–763.

Fransson, T. & Stolt, B.-O. (2005). Migration routes of North European Reed Warblers *Acrocephalus scirpaceus*. *Ornis Svecica* 15: 153–160.

Fransson, T., Jakobsson, S., Johansson, U. S., Kullberg, C., Lind, J. & Vallin, A. (2001). Bird migration – magnetic cues trigger extensive refuelling. *Nature* 414: 35–36.

Fransson, T., Jakobsson, S. & Kullberg, C. (2005). Non-random distribution of ring recoveries from trans-Saharan migrants indicates species-specific stopover areas. *J Avian Biol.* 36: 6–11.

Franz, D. (1988). Migration of the Penduline Tit (*Remiz pendulinus*) during the breeding period – range, frequency and ecological importance. *Vogelwelt* 109: 188–206.

Fraser, P. A. (1997). How many rarities are we missing? Weekend bias and length of stay revisited. *Br. Birds* 90: 94–101.

Fraser, P. A. & Rogers, M. J. (2002). Report on scarce migrant birds in Britain in 2000. *Br. Birds* 95: 606–630.

Fraser, P. A. & Rogers, M. J. (2003). Report on scarce migrant birds in Britain in 2001. *Br. Birds* 96: 626–649.

Fraser, P. A. & Rogers, M. J. (2004). Report on scarce migrant birds in Britain in 2002. *Br. Birds* 97: 647–664.

Fraser, P. A. & Rogers, M. J. (2005). Report on scarce migrant birds in Britain in 2002. *Br. Birds* 98: 73–88.

Fraser, P. A., Lansdown, P. G. & Rogers, M. J. (1999). Report on scarce migrant birds in Britain in 1997. *Br. Birds* 92: 618–658.

Frazar, A. M. (1881). Destruction of birds by a storm while migrating. *Bull. Nuttall Ornithol. Club* 6: 250–252.

Fredrickson, L. H. (1969). Mortality of Coots during severe spring weather. *Wilson Bull.* 81: 450–453.

Freemark, K. & Collins, B. (1992). Landscape ecology of birds breeding in temperate forest fragments. pp. 443–454 in *Ecology and conservation of neotropical migrant landbirds* (eds J. M. Hagan & D. W. Johnston). Washington, D.C., Smithsonian Institution Press.

Freemark, K. E. & Merriam, H. G. (1986). Importance of area and habitat heterogeneity to bird assemblages in temperate forest fragments. *Biol. Conserv.* 36: 115–141.

Frenzel, L. D. & Marshall, W. H. (1954). Observations on the effect of the May 1954 storm on birds in northern Minnesota. *Flicker* 26: 126–130.

Fretwell, S. D. (1972). *Populations in a seasonal environment*. Princeton, Princeton University Press.

Fretwell, S. (1980). Evolution of migration in relation to factors regulating bird numbers. pp. 517–527 in *Migrant birds in the Neotropics* (eds A. Keast & E. Morton). Washington, D.C., Smithsonian Institution Press.

Frith, H. J. (1967). *Waterfowl in Australia*. Sydney, Angus & Robertson.

Fry, C. H., Ash, J. S. & Ferguson-Lees, I. J. (1970). Spring weights of Palaearctic migrants at Lake Chad. *Ibis* 112: 58–82.

Fry, C. H., Ferguson-Lees, I. J. & Dowsett, R. J. (1972). Flight muscle hypertrophy and eco-physiological variation of Yellow Wagtail *Motacilla flava* races at Lake Chad. *J. Zool.* 167: 293–306.

Fujita, G., Hong-Liang, G., Ueta, M., Goroshko, O., Krever, V., Ozaki, K., Mita, N. & Higuchi, H. (2004). Comparing areas of suitable habitats along travelled and possible shortest routes in migration of White-naped Cranes *Grus vipio* in East Asia. *Ibis* 146: 461–474.

Fuller, M. R., Seegar, W. S. & Schueck, L. S. (1998). Routes and travel rates of migrating Peregrine Falcons *Falco peregrinus* and Swainson's Hawks *Buteo swainsoni* in the western hemisphere. *J. Avian Biol.* 29: 433–440.

Fuller, M., Holt, D. & Schueck, L. (2003). Snowy Owl movements: variation on a migration theme. pp. 359–366 in *Avian migration* (eds P. Berthold, E. Gwinner & E. Sonnenschein). Berlin, Springer-Verlag.

Furness, R. W. (1987). *The skuas*. Calton, Poyser.

Fusani, L. & Gwinner, E. (2005). Melatonin and nocturnal migration. *Ann. N.Y. Acad. Sci.* 1046: 264–270.

Galushin, V. M. (1974). Synchronous fluctuations in populations of some raptors and their prey. *Ibis* 116: 127–134.

Gandini, P., Boersma, P. D., Frere, E., Gandini, M., Holik, T. & Lichtschein, V. (1994). Magellanic Penguins (*Spheniscus magellanicus*) affected by chronic petroleum pollution along the coast of Chubut, Argentina. *Auk* 111: 20–27.

Gannes, L. Z. (2001). Comparative fuel use of migrating passerines: effects of fat stores, migration distance, and diet. *Auk* 118: 665–677.

Gantlet, S. J. M. (1991). *The birds of the Isles of Scilly*, 2nd edn. Privately published.

Ganusevich, S. A., Maechtle, T. L., Seegar, W. S., Yates, M. A., McGrady, M. J., Fuller, M., Schueck, L., Dayton, J. & Henny, C. J. (2004). Autumn migration and wintering areas of Peregrine Falcons *Falco peregrinus* nesting on the Kola Peninsula, northern Russia. *Ibis* 146: 291–297.

Garant, D., Kruuk, L. E. B., Wilkin, T. A., McCleery, R. H. & Sheldon, B. C. (2005). Evolution driven by differential dispersal within a wild bird population. *Nature* 433: 60–65.

Gardarsson, A. & Einarsson, A. (1994). Responses of breeding duck populations to changes in food supply. *Hydrobiologia*: 279–280; 15–27.

Garvin, M. C., Szell, C. C., Moore, F. R. (2006). Blood parasites of Nearctic-Neotropical passerine birds during spring trans-gulf migration: impact on host body condition. *J. Parasitology* 92: 990–996.

Gaston, A. J. (1970). Birds in the central Sahara in winter. *Bull. Br. Ornithol. Club* 90: 53–60.

Gaston, A. J. (1983). Migration of juvenile Thick-billed Murres through Hudson Strait in 1980. *Canad. Field Nat.* 96: 30–34.

Gaston, A. J. (2003). Synchronous fluctuations of Thick-billed Murre (*Uria lomvia*) colonies in the eastern Canadian Arctic suggest population regulation in winter. *Auk* 120: 362–370.

Gaston, A. J., Deforest, C. N., Donaldson, L. & Noble, D. G. (1994). Population parameters of Thick-billed Murres at Coats Island, northwest Territories, Canada. *Condor* 96: 935–948.

Gates, J. E. & Gysel, L. W. (1978). Avian nest dispersion and fledging success in field-forest ecotones. *Ecology* 59: 871–883.

Gätke, H. (1895). *Heligoland as an ornithological observatory. The result of fifty years experience.* Edinburgh, David Douglas.

Gatter, W. (1974). Analyse einer Invasion des Eichelhähers (*Garrulus glandarius*) 1972/73 am Randecker Maar (Schäbische Alb). *Vogelwarte* 27: 278–289.

Gatter, W. (1976). Über den Wegzug des Gimpels *P. pyrrhula*: Geschlechterverhältris end Einfluss von Witterungsfaktoren. *Vogelwarte* 28: 165–170.

Gatter, W. (1993). Explorationsverhalten, Zug und Migrationsevolution beim Fichtenkreuzschnabel *Loxia curvirostra*. *Die Vogelwelt* 114: 38–54.

Gatter, W. (2000). *Vogelzug und Vogelbestände in Mitteleuropa.* Wiebesteim, AULA-Verlag.

Gaunt, A. S., Hikida, R. S., Jehl, J. R. & Fenbert, L. (1990). Rapid atrophy and hypertrophy of an avian flight muscle. *Auk* 107: 649–659.

Gause, G. F. (1934). *The struggle for existence.* Baltimore, Williams & Wilkins.

Gauthier, G., Bédard, J., Huot, J. & Bédard, Y. (1984a). Spring accumulation of fat by Greater Snow Geese in two staging habitats. *Condor* 86: 192–199.

Gauthier, G., J. Bédard, J. Huot & Y. Bédard. (1984b). Protein reserves during spring staging in Greater Snow Geese. *Condor* 86: 210–212.

Gauthreaux, S. A. (1971). A radar and direct visual study of passerine spring migration in southern Louisiana. *Auk* 88: 343–365.

Gauthreaux, S. A. (1972). Behavioural responses of migrating birds to daylight and darkness: a direct visual study. *Wilson Bull.* 84: 136–148.

Gauthreaux, S. A. (1978a). The ecological significance of behavioural dominance. pp. 17–54 in *Perspectives in Ornithology*, Vol. 3. (eds P. P. G. Bateson & P. H. Klopfer). New York, Plenum Press.

Gauthreaux, S. A. (1978b). Importance of daytime flights of nocturnal migrants: redetermined migration following displacement. pp. 219–227 in *Animal migration, navigation and homing* (eds K. Schmidt-Koenig & W. T. Keeton). Berlin, Springer-Verlag.

Gauthreaux, S. A. (1982a). The ecology and evolution of avian migration systems. pp. 93–167 in *Avian Biology*, Vol. 6. (eds D. S. Farner & J. R. King). New York, Academic Press.

Gauthreaux, S. A. (1982b). Age-dependent orientation in migratory birds. pp. 68–74 in *Avian navigation* (eds F. Papi & H. G. Wallraff). Berlin, Springer-Verlag.

Gauthreaux, S. A. (1985). Differential migration of raptors: the importance of age and sex. *Proc. Hawk Migration Conf.* 4: 99–106.

Gauthreaux, S. A. (1991). The flight behaviour in migrating birds in changing wind fields: radar and visual analyses. *Am. Zool.* 31: 187–204.

Gauthreaux, S. A. (1992). The use of weather radar to monitor long-term patterns of trans-Gulf migration in spring. pp. 96–100 in *Migrant birds in the Neotropics: ecology, behaviour, distribution and conservation* (eds A. Keast & E. S. Morton). Washington, D.C., Smithsonian Institution Press.

Gauthreaux, S. A. (1999). Neotropical migrants and the Gulf of Mexico: the view from aloft. pp. 27–47 in *Gatherings of angels. Migratory birds and their ecology* (ed. K. P. Able). Ithaca, Cornell University Press.

Gauthreaux, S. A. & LeGrand, H. E. (1975). The changing seasons. *Am. Birds* 29: 820–826.

Gauthreaux, S. A., Belser, C. G. & van Blaricom, D. (2003). Using a network of WSR-88D weather surveillance radars to define patterns of bird migration at large spatial scales. pp. 335–346 in *Avian migration* (eds P. Berthold, E. Gwinner & E. Sonnenschein). Berlin, Springer.

Gavin, T. A. & Bollinger, E. K. (1988). Reproductive correlates of breeding site fidelity in Bobolinks (*Dolichonyx oryzivorus*). *Ecology* 69: 96–103.

Geering, D. & French, K. (1998). Breeding biology of the Regent Honeyeater *Xanthomyza phrygia* in the Capertee Valley, New South Wales. *Emu* 98: 104–116.

Geller, G. A. & Temple, S. A. (1983). Seasonal trends in body composition of juvenile Red-tailed Hawks during autumn migration. *Wilson Bull.* 95: 492–495.

Germi, F. & Waluyo, D. (2006). Additional information on the autumn migration of raptors in east Bali, Indonesia. *Forktail* 22: 71–76.

Gessaman, J. A. (1979). Premigratory fat in the American Kestrel. *Wilson Bull.* 91: 625–626.

Gessaman, J. A. (1990). Body temperatures of migrant accipiter hawks just after flight. *Wilson Bull.* 102: 133–137.

Gessaman, J. A. & Nagy, K. A. (1988). Transmitter loads affect the flight speed and metabolism of homing pigeons. *Condor* 90: 662–668.

Gibbons, D. W., Reid, J. B. & Chapman, R. A. (1993). *The new atlas of breeding birds in Britain and Ireland: 1988–1991.* London, Poyser.

Gibbs, H. L. & Grant, P. R. (1987). Ecological consequences of an exceptionally strong El Niño event on Darwins Finches. *Ecology* 68: 1735–1746.

Gibson, D. D. (1981). Migrants at Shemya Island, Aleutian Islands, Alaska. *Condor* 83: 65–77.

Gifford, C. E. & Odum, E. P. (1965). Bioenergetics of lipid deposition in the Bobolink, a trans-equatorial migrant. *Condor* 67: 383–403.

Gilg, O., Sittler, B., Sabard, B., Hirstel, A., Sané, R., Delattre, P. & Hanski, I. (2006). Functional and numerical responses of four lemming predators in high arctic Greenland. *Oikos* 113: 193–216.

Gill, F. B., Mostrom, A. M. & Mack, A. L. (1993). Speciation in North American chickadees. I. Patterns of mtDNA genetic divergence. *Evolution* 47: 195–212.

Gill, J. A., Clark, J., Clark, N. & Sutherland, W. (1995). Sex differences in the migration, moult and wintering areas of British-ringed Ruff. *Ringing & Migration* 16: 159–167.

Gill, J. A., Norris, K., Potts, P. M., Gunnarsson, T. G., Atkinson, P. W. & Sutherland, W. (2001). The buffer effect and large scale regulation in migratory birds. *Nature* 412: 436–438.

Gill, R. E., Babcock, C. A., Handel, C. M., Butler, W. R. & Raveling, D. G. (1997). Migration, fidelity and use of autumn staging grounds in Alaska by Cackling Canada Geese *Branta canadensis minima. Wildfowl* 47: 42–61.

Gill, R. E., Piersma, T., Hufford, G., Servrancky, R. & Riegen, A. (2005). Crossing the ultimate ecological barrier: evidence for an 11,000 km-long nonstop flight from Alaska to New Zealand and eastern Australia by Bar-tailed Godwits. *Condor* 107: 1–20.

Gilroy, J. J. & Lees, A. C. (2003). Vagrancy theories: are autumn vagrants really reverse migrants? *Br. Birds* 96: 427–438.

Gilyazov, A. & Sparks, T. (2002). Change in the timing of migration of common birds at the Lapland nature reserve (Kola Peninsula, Russia) during 1931–1999. *Avian Ecol. Behav.* 8: 35–47.

Ginn, H. B. & Melville, D. S. (1983). *Moult in birds.* BTO Guide 19. Thetford, British Trust for Ornithology.

Gladwin, T. W. (1963). Increase in the weights of Acrocephali. *Bird Migration* 2: 319–324.

Glasrud, R. D. (1976). Canada Geese killed during lightning storm. *Can. Field Nat.* 90: 503.

Glutz von Blotzheim, U. N. (1966). Das Auftreten des Seidenschwanzes *Bombycilla garrulus* in der Schweiz und die von 1901 bis 1965/66 West und Mitteleuropa erreichenden Invasionen. *Orn. Beob.* 63: 93–146.

Glutz von Blotzheim, U. N. (1997). *Handbuch der Vögel Mitteleuropas,* Bd 14, II Passeriformes (5, Teil). Wiesbaden, AULA-Verlag.

Glutz von Blotzheim, U. N. & Bauer, K. M. (1985). *Handbuch der Vögel-Mitteleuropas,* Bd 10, II Passeriformes (1, Teil). Wiesbaden, Akademische Verlagsgesellschaft.

Glutz von Blotzheim, U. N., & Bauer, K. M. (1997). *Handbuch der Vögel Mitteleuropas,* Bd 14, 2 (Passeriformes, Fringillidae). Wiesbaden, Akademische Verlagsgesellschaft.

Glutz von Blotzheim, U. N., Bauer, K. M. & Bezzel, E. (1975). *Handbuch der Vogel Mitteleuropas,* Bd 6, Charadriiformes. Wiesbaden, Akademische Verlagsgesellschaft.

Glutz von Blotzheim, U. N., Bauer, K. M. & Bezzel, E. (1977). *Handbuch der Vogel Mitteleuropas*, Bd 7, Charadriiformes. Wiesbaden, Akademische Verlagsgesellschaft.

Goldstein, M. I., Woodbridge, B., Zaccagnini, M. E. & Canavelli, S. B. (1996). An assessment of mortality of Swainson's Hawks on wintering grounds in Argentina. *J. Rap. Res* 30: 106–107.

Gordo, O., Brotons, L., Ferrer, X. & Comas, P. (2005). Do changes in climate patterns in wintering areas affect the timing of the spring arrival of trans-Saharan migrant birds? *Global Change Biol.* 11: 12–21.

Gorney, E. & Yom-Tov, Y. (1994). Fat, hydration condition, and moult of Steppe Buzzards *Buteo buteo vulpinus* on spring migration. *Ibis* 136: 185–192.

Gorney, E., Clark, W. S. & Yom-Tov, Y. (1999). A test of the condition-bias hypothesis yields different results for two species of sparrowhawks (*Accipiter*). *Wilson Bull.* 111: 181–187.

Goss-Custard, J. D., Le V. dit. Durell, S. E. A., Sitters, H. P. & Swinfen, R. (1982). Age-structure and survival of a wintering population of Oystercatchers. *Bird Study* 29: 83–98.

Goss-Custard, J. D., Clarke, R. T. & Durell, S. E. A. (1984). Rates of food intake and aggression of Oystercatchers *Haematopus ostralegus* on the most and least preferred mussel *Mytilus edulis* beds of the Exe Estuary. *J. Anim. Ecol.* 53: 233–245.

Goss-Custard, J. D., Clarke, R. T., Durell, S. E. A. leV. dit., Caldow, R. W. G. & Ens, B. J. (1995). Population consequences of winter habitat loss in a migratory shorebird. II. Model predictions. *J. Appl. Ecol.* 32: 337–351.

Götmark, F. (1982). Irruptive breeding of the Redpoll *Carduelis flammea*, in South Sweden in 1975. *Vår Fågelvärld* 41: 315–322.

Grainger, A. (1990). *The threatening desert. Controlling desertification.* London, Earthscan Publications.

Graber, J. W. & Graber, R. R. (1979). Severe winter weather and bird populations in southern Illinois. *Wilson Bull.* 91: 88–103.

Graves, G. R. (1997). Geographic clines of age ratios of Black-throated Blue Warblers (*Dendroica caerulescens*). *Ecology* 78: 2524–2531.

Green, G. H., Greenwood, J. J. D. & Lloyd, L. S. (1977). The influence of snow conditions on the date of breeding of wading birds in north-east Greenland. *J. Zool. Lond.* 183: 311–328.

Green, J. C. (1962). Arrested passerine migration and kill at Lake Superior. *Flicker* 34: 110–112.

Green, M. (2004). Flying with the wind – spring migration of Arctic-breeding waders and geese over South Sweden. *Ardea* 92: 145–159.

Green, M. & Alerstam, T. (2000). Flight speeds and climb rates of Brent Geese: mass-dependent differences between spring and autumn migration. *J. Avian Biol.* 31: 215–225.

Green, M., Alerstam, T., Clausen, P., Drent, R. & Ebbinge, B. S. (2002). Dark-bellied Brent *Branta bernicla bernicla*, as recorded by satellite telemetry, do not minimise flight distance during spring migration. *Ibis* 144: 106–121.

Green, R. E. (1999). Survival and dispersal of male Corncrakes in a threatened population. *Bird Study* 46 (Suppl.): 218–229.

Green, R. E. & Stowe, T. J. (1993). The decline of the Corncrake *Crex crex* in Britain and Ireland in relation to habitat change. *J. Appl. Ecol.* 30: 689–695.

Green, R. E., Tyler, G. A., Stowe, T. J. & Newton, A. V. (1997). A simulation model of the effect of mowing of agricultural grassland on the breeding success of the Corncrake *Crex crex*. *J. Zool. (Lond).* 243: 81–115.

Greenberg, R. S. (1980). Demographic aspects of long-distance migration. pp. 493–504 in *Migrant birds in the Neotropics* (eds A. Keast & E. Morton). Washington, DC, Smithsonian Institution Press.

Greenberg, R. (1983). Competition in migrant birds in the nonbreeding season. *Curr. Ornithol.* 3: 281–307.

Greenberg, R. (1985). The social behaviour and feeding ecology of neotropical migrants in the non-breeding season. *Proc. Int. Ornithol. Congr.* 18: 769–775.

Greenberg, R. (1986). Competition in migrant birds in the non-breeding season. *Curr. Ornithol.* 3: 281–307.

Greenberg, R. S. & Gradwohl, J. A. (1980). Observations of paired Canada Warblers *Wilsonia canadensis* during migration in Panama. *Ibis* 122: 509–512.

Greenberg, R. & Marra, P. P. (2005). *Birds of Two Worlds*. Washington, D. C., Smithsonian Institution.

Greenberg, R., Cordero, P. J., Droege, S & Fleischer, R. S. (1998). Morphological adaptation with no mitochondrial DNA differentiation in the coastal plain Swamp Sparrow. *Auk* 115: 706–712.

Greenwood, H., Clark, R. G. & Weatherhead, P. J. (1986). Condition bias of hunter-shot Mallards (*Anas platyrhynchos*). *Can. J. Zool.* 64: 599–601.

Greenwood, P. J. (1980). Mating systems, philopatry and dispersal in birds and mammals. *Anim. Behav.* 28: 1140–1162.

Greenwood, P. J. & Harvey, P. H. (1976a). The adaptive significance of variance in breeding area fidelity of the Blackbird (*Turdus merula L.*). *J. Anim. Ecol.* 45: 887–98.

Greenwood, P. J. & Harvey, P. H. (1976b). Differential mortality and dispersal of male Blackbirds. *Ringing & Migration* 1: 75–77.

Greenwood, P. J. & Harvey, P. H. (1982). The natal and breeding dispersal of birds. *Annu. Rev. Ecol. Syst.* 13: 1–21.

Greenwood, P. J., Harvey, P. H. & Perrins, C. M. (1978). Inbreeding and dispersal in the Great Tit. *Nature* 271: 52–54.

Greenwood, P. J., Harvey, P. H. & Perrins, C. M. (1979). The role of dispersal in the Great Tit (*Parus major*): the causes, consequences and heritability of natal dispersal. *J. Anim. Ecol.* 48: 123–142.

Gregory, R. D., Wilkinson, N. I., Noble, D. G., Robinson, J. A., Brown, A. F., Hughes, J., Proctor, D., Gibbons, D. W. & Galbraith, C. A. (2002). The population status of birds in the United Kingdom, Channel Islands and Isle of Man: an analysis of conservation concern 2002–2007. *Br. Birds* 95: 410–448.

Gregory, R. D., Noble, D. G. & Custance, J. (2004). The state of play of farmland birds – population trends and conservation status of lowland farmland birds in the United Kingdom. *Ibis* 146 (suppl 2): 1–13.

Grenquist, P. (1947). Über die Biologie des Hakengimpels. *Orn. Fenn;* 24: 1–10.

Griffioen, P. A. & Clarke, M. F. (2002). Large-scale bird-movement patterns evident in eastern Australian atlas data. *Emu* 102: 97–125.

Griffith, J. T. & Griffith, J. C. (2000). Cowbird control and the endangered Least Bell's Vireo: a management success story. pp. 342–356 in *Ecology and management of Cowbirds and their hosts* (eds J. J. M. Smith, T. L. Cook, S. I. Rothstein, S. K. Robinson & S. G. Sealy). Austin, University of Texas Press.

Griscom, L. (1937). A monographic study of the Red Crossbill. *Proc. Boston Soc. Nat. Hist.* 41: 77–210.

Gromadzka, J. (1989). Breeding and wintering areas of Dunlin migrating through southern Baltic. *Ornis Scand.* 20: 132–144.

Grote, H. (1937). Der sibirische Hakengimpel (*Pinicola enucleator stschur* Port.) in Deutschland. *Orn. Monatsber.* 45: 83–85.

Groth, J. G. (1988). Resolution of cryptic species in Appalachian Red Crossbills. *Condor* 90: 745–760.

Groth, J. G. (1991). Cryptic species of nomadic birds in the Red Crossbill (*Loxia curvirostra*) complex of North America. Berkeley, California, University of California.

Groth, J. G. (1993a). Evolutionary differentiation in morphology, vocalisations, and allozymes among nomadic sibling species in the North American Red Crossbill *Loxia curvirostra* complex. *Univ. Calif. Publ. Zool.* 127: 1–143.

Groth, J. G. (1993b). Call matching and positive assortative mating in Red Crossbills. *Auk* 110: 398–401.

Grubb, T. G., Bowerman, W. W. & Howey, P. W. (1994). Tracking local and seasonal movements of wintering Bald Eagles *Haliaeetus leucocephalus*. pp. 347–358 in *Raptor conservation today* (eds B.-U. Meyburg & R. D. Chancellor). London, Pica Press.

Gruys, R. C. (1993). Autumn and winter movements and sexual segregation of Willow Ptarmigan. *Arctic* 46: 228–239.

Grzybowski, J. A., Clapp, R. B. & Marshall, J. T. (1986). History and current population status of the Black-capped Vireo in Oklahoma. *Am. Birds* 40: 1151–1161.

Gudmundsson, A. G. & Alerstam, T. (1998). Why is there no transpolar bird migration. *J. Avian Biol.* 29: 93–96.

Gudmundsson, A. G. & Sandberg, R. (2000). Sanderlings (*Calidris alba*) have a magnetic compass: orientation experiments during spring migration in Iceland. *J. Exp. Biol.* 203: 3137–3144.

Gudmundsson, G. A. & Lindström, A. (1992). Spring migration of Sanderlings *Calidris alba* through SW Iceland – where from and where top. *Ardea* 80: 315–326,

Gudmundsson, G. A., Lindström, A. & Alerstam, T. (1991). Optional fat loads and long distance flights by migrating Knots *Calidris canutus*, Sanderlings *C. alba* and Turnstones *A. interpres. Ibis* 133: 140–152.

Gudmundsson, G. A., Benvenuti, S., Alerstam, T., Papi, F., Lilliendahl, K. & Åkesson, S. (1995). Examining the limits of flight and orientation performance: satellite tracking of Brent Geese migrating across the Greenland ice cap. *Proc. R. Soc. Lond. B* 261: 73–79.

Gullestad, N., Owen, M. & Nugent, M. J. (1984). Numbers and distribution of Barnacle Geese *Branta leucopsis* on Norwegian staging islands and the importance of the staging area to the Svalbard population. *Nor. Polarinst. Skr.* 181: 57–65.

Gunnarsson, T. G., Gill, J. A., Sigurbjornsson, T. & Sutherland, W. J. (2004). Pair bonds: Arrival synchrony in migratory birds. *Nature* 431: 646.

Gunnarsson, T. G., Gill, J. A., Newton, J., Potts, P. M. & Sutherland, W. J. (2005). Seasonal matching of habitat quality and fitness in migratory birds. *Proc R. Soc. Lond. B* 272: 2319–2323.

Gustafsson, L. (1988). Inter- and intra-specific competition for nest-holes in a population of Collared Flycatcher *Ficedula albicollis. Ibis* 130: 11–15.

Gwinner, E. (1967). Circanuale Periodik der Mauser und der Zugunruhe bei einem Vogel. *Naturwissenschaften* 54: 447.

Gwinner, E. (1968). Circannuale Periodik als Grundlage des jahreszeitlichen Funktionswandels bei Zugvögeln. Untersuchungen am Fitis (*Phylloscopus trochilus*) und am Waldlaubsänger (*P. sibilatrix*). *J. Ornithol.* 109: 70–95.

Gwinner, E. (1971). A comparative study of circannual rhythms in warblers. pp. 405–427 in *Biochronometry* (ed. M. Menaker.). Washington, D.C., National Academy of Sciences.

Gwinner, E. (1972). Adaptive functions of circannual rhythms in warblers. *Proc. Int. Ornithol. Congr.* 15: 218–236.

Gwinner, E. (1981). Circannual rhythms: Their dependence on the circadian system. pp. 153–169 in *Biological clocks in seasonal reproductive cycles* (eds B. K. Follett & D. E. Follett). Bristol, Wright Publishers.

Gwinner, E. (1986). *Circannual rhythms*. Berlin, Springer.

Gwinner, E. (1987). Annual rhythms of gonadal size, migratory disposition and moult in Garden Warblers *Sylvia borin* exposed in winter to an equatorial or a southern hemisphere photoperiod. *Ornis Scand.* 18: 251–256.

Gwinner, E. (1990). Circannual rhythms in bird migration: control of temporal patterns and interactions with photoperiod. pp. 257–267 in *Bird migration. Physiology and ecophysiology* (ed. E. Gwinner). Berlin, Springer-Verlag.

Gwinner, E. (1991). Circannual rhythms in tropical and temperate zone Stonechats: A comparison of properties under constant conditions. *Ökol. Vögel* 13: 5–14.

Gwinner, E. (1996a). Circadian and circannual programmes in avian migration. *J. Exp. Biol.* 199: 39–48.

Gwinner, E. (1996b). Circannual clocks in avian reproduction and migration. *Ibis* 138: 47–63.

Gwinner, E. & Helm, B. (2003). Circannual and circadian contributions to the timing of avian migration. pp. 81–95 in *Avian migration* (eds P. Berthold, E. Gwinner & E. Sonnenschein). Berlin, Springer-Verlag.

Gwinner, E. & Neusser, V. (1985). Die Jugendmauser europäischer und afrikanischer Schwarzkehlchen (*Saxicola torquata rubicola* und *axillaris*) sowie von F1-Hybriden. *J. Ornithol.* 126: 219–220.

Gwinner, E. & Scheuerlein, A. (1999). Photoperiodic responsiveness of equatorial and temperate zone Stonechats. *Condor* 101: 347–359.

Gwinner, E. & Schwabl-Benzinger, I. (1982). Adaptive temporal programming of moult and migratory disposition in two closely related long-distance migrants, the Pied Flycatcher (*Ficedula hypoleuca*) and the Collared Flycatcher (*F. albicollis*). pp. 75–89 in *Avian navigation* (eds F. Papi & H. Wallraff). Berlin, Springer.

Gwinner, E. & Wiltschko, W. (1978). Endogenously controlled changes in migratory direction of the Garden Warbler *Sylvia borin*. *J. Comp. Physiol.* 125: 267–273.

Gwinner, E. & Wiltschko, W. (1980). Circannual changes in migratory orientation of the Garden Warbler, *Sylvia borin*. *Behav. Ecol.* 7: 73–78.

Gwinner, E., Dittami, J. P., Gänshirt, G., Hall, M. & Wozniak, J. (1985a). Endogenous and exogenous components in the control of the annual reproductive cycle of the European Starling. *Proc. Int. Ornithol. Congr.* 18: 501–515.

Gwinner, E., Biebach, H. & Kries, I. V. (1985b). Food availability affects migratory restlessness in caged Garden Warblers (*Sylvia borin*). *Naturwissenschaften* 72: 51–52.

Gwinner, E., Schwabl, H. & Schwabl-Benzinger, I. (1988). Effects of food-deprivation on migratory restlessness and diurnal activity in the Garden Warbler *Sylvia borin*. *Oecologia* 77: 321–326.

Gwinner, E., Schwabl, H. & Schwabl-Benzinger, I. (1992). The migratory time program of the Garden Warbler: is there compensation for interruptions? *Ornis Scand.* 23: 264–270.

Gwinner, H., Van't Hof, T. & Zeman, M. (2002). Hormonal and behavioural responses of Starlings during a confrontation with males or females at nest boxes during the reproductive season. *Hormones Behav.* 42: 21–31.

Gylfe, A., Bergström, S., Lunström, J. & Olsen, B. (2000). Epidemiology – Reactivation of *Borrelia* infection in birds. *Nature* 403: 724–725.

Haapanen, A. (1966). Bird fauna of Finnish forests in relation to forest succession. *Ann. Zool. Fenn.* 3: 176–200.

Haas, C. A. (1998). Effects of prior nesting success on site fidelity and breeding dispersal: an experimental approach. *Auk* 115: 929–936.

Haas, W. & Beck, P. (1979). Zum Frühjahrszug paläarktischer Vögel über die westliche Sahara. *J. Ornithol.* 120: 237–246.

Hafner, H. (2000). Herons in the Mediterranean. pp. 33–54 in *Heron conservation* (eds J. A. Kushlan & H. Hafner). London, Academic Press.

Hagan, J. M. & Johnston, D. W. (eds) (1992). Ecology and conservation of neotropical migrant landbirds. Washington, D.C., Smithsonian Institution Press.

Hagan, J. M., Lloyd-Evans, L. & Alwood, J. L. (1991). The relationship between latitude and the timing of spring migration of North American landbirds. *Ornis Scand.* 22: 129–136.

Hagan, J. M., Lloyd-Evans, T. L., Alwood, J. L. & Wood, D. S. (1992). Long-term changes in migratory landbirds in the northeastern United States: evidence from migration capture data. pp. 115–130 in *Ecology and conservation of neotropical migrant landbirds* (eds J. M. Hagan & D. W. Johnston). Washington, D.C., Smithsonian Press.

Hagemeijer, W. J. M. & Blair, M. J. (1997). *The EBCC atlas of European breeding birds*. London, T. & A. D. Poyser.

Hagen, Y. (1969). Norwegian studies on the reproduction of birds of prey and owls in relation to micro-rodent population fluctuations. *Fauna* 22: 73–126.

Hagner, S. (1965). Cone crop fluctuations in Scots Pine and Norway Spruce. *Studia Forestalia Svecica* No 33.

Hahn, T. P. (1998). Reproductive seasonality in an opportunistic breeder, the Red Crossbill, *Loxia curvirostra*. *Ecology* 29: 2365–2375.

Hahn, T. P., Swingle, J., Wingfield, J. C. & Ramenofsky, M. (1992). Adjustments of the prebasic moult schedule in birds. *Ornis Scand.* 41: 393–406.

Hahn, T. P., Boswell, T., Wingfield, J. C. & Ball, G. F. (1997). Temporal flexibility in avian reproduction: patterns and mechanisms. *Current Ornithol.* 14: 39–80.

Haila, Y., Tiainen, J. & Vepsäläinen, K. (1986). Delayed autumn migration as an adaptive strategy of birds in northern Europe: evidence from Finland. *Ornis Fenn.* 63: 1–9.

Haines, A. M., McGrady, M. J., Martell, M. S., Dayton, B. J., Henke, M. B. & Seegar, W. S. (2003). Migration routes and wintering locations of Broad-winged Hawks tracked by satellite telemetry. *Wilson Bull.* 115: 166–169.

Håke, M., Kjellén, N. & Alerstam, T. (2001). Satellite tracking of Swedish Ospreys *Pandion haliaetus* autumn migration routes and orientation. *J Avian Biol.* 32: 47–56.

Håke, M., Kjellén, N. & Alerstam, T. (2003). Age-dependent migration strategy in Honey Buzzards *Pernis apivorus* tracked by satellite. *Oikos* 103: 385–396.

Hall, K. S. S. & Fransson, T. (2000). Lesser Whitethroats under time constraint moult more rapidly and grow shorter wing feathers. *J Avian Biol.* 31: 583–587.

Hall, K. S. S. & Fransson, T. (2001). Wing moult in relation to autumn migration in adult Common Whitethroats *Sylvia communis communis*. *Ibis* 143: 580–586.

Halley, D. J., Harris, M. P. & Wanless, S. (1995). Colony attendance patterns and recruitment in immature Common Murres (*Uria aalge*). *Auk* 112: 947–957.

Hamerström, F. (1969). A harrier population study. pp. 367–383 in *Peregrine Falcon populations: their biology and decline* (ed. J. J. Hickey). Madison, University of Wisconsin Press.

Hamilton, W. J. (1962a). Bobolink migratory pathways and their experimental analysis under night skies. *Auk* 79: 208–233.

Hamilton, W. J. (1962b). Evidence concerning the function of nocturnal call notes of migratory birds. *Condor* 64: 390–401.

Hamilton, W. J. (1998). Tricolored Blackbird itinerant breeding in California. *Condor* 100: 218–226.

Hansen, L. (1954). Birds killed at lights in Denmark 1886–1939. *Vidensk. Medd. fra Dansk naturh. Foren.* 116: 269–368.

Hanson, H. C. & Jones, R. L. (1976). *The biogeochemistry of Blue, Snow and Ross' Geese.* Urbana, Illinois, Southern Illinois Press.

Hanski, I., Hansson, L. & Henttonen, H. (1991). Specialist predators, generalist predators, and the microtine rodent cycle. *J. Anim. Ecol.* 60: 353–367.

Haramis, G. M., Nichols, J. B., Pollock, K. H. & Hines, J. E. (1986). The relationship between body mass and survival of wintering Canvasbacks. *Auk* 103: 506–514.

Hario, M., Lammi, E., Mikkola, M. & Södersved, J. (1993). Annual fluctuations in numbers of wintering waterfowl in the Åland Islands in 1968–72. *Suomen Riista* 39: 21–32.

Harmata, A. R. (2002). Vernal migration of Bald Eagles from a southern Colorado wintering area. *J. Raptor Res.* 36: 256–264.

Harrap, S. & Quinn, P. (1996). *Tits, nuthatches and treecreepers.* London, A. & C. Black.

Harrington, B. A. & Haase, B. (1994). Latitudinal differences in sex-ratios among nonbreeding Western Sandpipers in Puerto Rico and Ecuador. *Southwest Nat.* 39: 188–189.

Harrington, B. A., Hagan, J. A. & Leddy, L. E. (1988). Site fidelity and survival differences between two groups of New World Knots *Calidris canutus*. *Auk* 105: 439–445.

Harris, M. P. (1966). Breeding biology of the Manx shearwater *Puffinus puffinus*. *Ibis* 108: 17–33.

Harris, M. P. (1970). Abnormal migration and hybridisation of *Larus argentatus* and *L. fuscus* after interspecies fostering experiments. *Ibis* 112: 488–498.

Harris, M. P. (1984). *The puffin.* Calton, T. & A. D. Poyser.

Harris, M. P. & Wanless, S. (1991). Population studies and conservation of Puffins *Fratercula arctica.* pp. 230–248 in *Bird population studies* (eds C. M. Perrins, J.-D. Lebreton & G. J. M. Hirons). Oxford, Oxford University Press.

Harris, M. P., Heubeck, M., Shaw, D. N. & Okill, J. D. (2006). Dramatic changes in the return date of Common Guillemots *Uria aalge* to colonies in Shetland, 1962–2005. *Bird Study* 53: 247–252.

Harrison, J. A., Allan, D. G., Underhill, L. G., Herremans, M., Tree, A. J., Parker, V. & Brown, C. J. (1997). *The atlas of Southern African birds.* Vols 1 & 2. Johannesburg, Birdlife South Africa.

Harrison, J. L. (1962). Distribution of feeding habits among animals in a tropical forest. *J. Anim. Ecol.* 31: 53–63.

Harrison, P. (1983). *Seabirds. An identification guide.* Beckenham, Kent, Croom Helm.

Hart, J. S. & Berger, M. (1972). Energetics, water economy and temperature regulation during flight. *Proc. Int. Ornithol. Congr.* 15: 189–199.

Hartley, R. & Hustler, K. (1993). A less-than-annual breeding cycle in a pair of African Bat Hawks *Machaeramphus alcinus. Ibis* 135: 456–458.

Harvey, P. H., Greenwood, P. J., Campbell, B. & Stenning, M. J. (1984). Breeding dispersal of the Pied Flycatcher (*Ficedula hypoleuca*). *J. Anim. Ecol.* 53: 727–736.

Harvey, P. V. & Riddiford, N. (1996). An uneven sex ratio of migrant Long-eared Owls. *Ringing & Migration* 11: 132–135.

Hau, M., Wikelski, M. & Wingfield, J. C. (1998). A neotropical forest bird can measure the slight changes in tropical photoperiod. *Proc. R. Soc. Lond. B* 265: 89–95.

Hayden, T. J., Faaborg, J. & Clawson, R. L. (1985). Estimates of minimum area requirements for Missouri forest birds. *Trans. Mo. Acad. Sci.* 19: 11–22.

Hayden, T. J., Tazik, D. J., Melton, R. H. & Cornelius, J. D. (2000). Cowbird control program at Fort Hood, Texas: lessons for mitigation of Cowbird parasitism on a landscape scale. pp. 357–370 in *Ecology and management of Cowbirds and their hosts* (eds J. N. M. Smith, T. L. Cook, S. I. Rothstein, S. K. Robinson & S. G. Sealy). Austin, University of Texas Press.

Hays, H., DiCostanzo, J., Cormons G.de T.Z., Antas P. de T.Z., do Nascimento J.L.X., do Nascimento I. de L.S. & Bremer, R. E. (1997). Recovery of Roseate and Common Terns in South America. *J. Field Ornithol.* 68: 79–90.

Hebrard, J. J. (1971). The nightly initiation of passerine migration in spring, a direct visual study. *Ibis* 113: 8–18.

Hedenström, A. (1993). Migration by soaring or flapping flight in birds: the relative importance of energy cost and speed. *Phil. Trans. R. Soc. Lond. B* 342: 353–361.

Hedenström, A. (2004). Migration and morphometrics of Temminck's Stint *Calidris temminckii* at Ottenby, southern Sweden. *Ringing & Migration* 22: 51–58.

Hedenström, A. & Alerstam, T. (1992). Climbing performance of migrating birds as a basis for estimating limits for fuel-carrying capacity and muscle work. *J. Exp. Biol.* 164: 19–38.

Hedenström, A. & Alerstam, T. (1998). How fast can birds migrate? *J. Avian Biol.* 29: 424–432.

Hedenström, A. & Pettersson, J. (1987). Migration routes and wintering areas of Willow Warblers *Phylloscopus trochilus* ringed in Fennoscandia. *Ornis Scand.* 64: 137–143.

Hedenström, A., Alerstam, T., Green, M. & Gudmundsson, G. A. (2005). Adaptive variation of airspeed in relation to wind, altitude and climb rate by migrating birds in the Arctic. *Behav. Ecol. Sociobiol.* 52: 308–317.

Heibl, I. & Braunitzer, G. (1988). Anpassungen der Hämoglobine von Streifengans (*Anser indicus*), Andengans (*Chloephaga melanoptera*) und Sperbergeier (*Gyps rueppellii*) an hypoxische Bedingungen. *J. Ornithol.* 29: 217–226.

Heise, C. W. & Moore, F. R. (2003). Age-related differences in foraging efficiency, moult and fat deposition of Gray Catbirds prior to autumn migration. *Condor* 105: 496–504.

Heitmeyer, M. E., Fredrickson, L. H. & Humburg, D. D. (1993). Further evidence of biases associated with hunter-killed Mallards. *J. Wildl. Manage.* 57: 733–740.

Helander, W. (1985). Winter feeding as a management tool for White-tailed Sea Eagles in Sweden. *ICBP Techn. Publ.* 5: 401–407.

Helbig, A. J. (1990). Depolarization of natural skylight disrupts orientation of an avian nocturnal migrant. *Experientia* 46: 755–758.

Helbig, A. J. (1991a). Dusk orientation of migratory European Robins, *Erithacus rubecula* – the role of sun-related directional information. *Anim. Behav.* 41: 313–322.

Helbig, A. J. (1991b). Inheritance of migratory direction in a bird species: a cross-breeding experiment with SE- and SW-migrating Blackcaps (*Sylvia atricapilla*). *Behav. Ecol. Sociobiol.* 28: 9–12.

Helbig, A. J. (1991c). SE migrating and SW migrating Blackcap (*Sylvia atricapilla*) populations in Central Europe: orientation of birds in the contact zone. *J. Evol. Biol.* 4: 657–670.

Helbig, A. J. (1994). Genetic basis and evolutionary change of migratory directions in a European passerine migrant *Sylvia atricapilla*. *Ostrich* 65: 151–159.

Helbig , A. (1996). Genetic basis, mode of inheritance and evolutionary changes of migratory direction in Palearctic warblers (Aves: Sylviidae). *J. Exp. Biol.* 199: 49–55.

Helbig, A. (2003). Evolution of bird migration: a phylogenetic and biogeographic perspective. pp. 3–20 in *Avian migration* (eds P. Berthold, E. Gwinner & E. Sonnenschein). Berlin, Springer-Verlag.

Helbig, A. J. & Wiltschko, W. (1989). The skylight polarization patterns at dusk affect the orientation behavior of Blackcaps, *Sylvia atricapilla*. *Naturwissenschaften* 76: 227–229.

Helbig, A. J., Berthold, P. & Wiltschko, W. (1989). Migratory orientation of Blackcaps (*Sylvia atricapilla*): population-specific shifts of direction during the autumn. *Ethology* 82: 307–315.

Helm, B. (2003). Seasonal timing in different environments: comparative studies in Stonechats. PhD thesis, Munich, Ludwig-Maximilian University.

Helm, B. & Gwinner, E. (2005). Carry-over effects of day length during spring migration. *J. Ornithol.* 146: 348–354.

Helm, B. & Gwinner, E. (2006). Migratory restlessness in an equatorial non-migratory bird. *PLOS Biol.* 4: 611–614.

Helm, B., Gwinner, E. & Trost, L. (2005). Flexible seasonal timing and migratory behaviour. Results from Stonechat breeding programmes. *Ann. N.Y. Acad. Sci* 1046: 216–227.

Helms, C. W. (1963). The annual cycle and Zugunruhe in birds. *Proc. Int. Ornithol. Congr.* 13: 925–939.

Helms, C. W. & Smythe, R. B. (1969). Variation in major body components of the Tree Sparrow (*Spizella arborea*) sampled within the winter range. *Wilson Bull.* 81: 280–292.

Hengeveld, R. (1993). What to do about the North American invasion by the Collared Dove? *J. Field Ornithol.* 64: 477–489.

Henningsson, S. S. & Alerstam, T. (2005). Barriers and distances as determinants for the evolution of bird migration links: the arctic shorebird system. *Proc. R. Roc. Lond. B* 272: 2251–2258.

Henny, C. J. (1973). Drought displaced movement of North American Pintails into Siberia. *J. Wildl. Manage.* 37: 23–29.

Henny, C. J. & van Velzen, W. T. (1972). Migration patterns and wintering localities of American Ospreys. *J. Wildl. Manage.* 36: 1133–41.

Hepp, G. R. & Hair, J. D. (1984). Dominance in wintering waterfowl (Anatini): effects on distribution of sexes. *Condor* 86: 251–257.

Hepp, G. R. & Hines, J. E. (1991). Factors affecting winter distribution and migration distance of Wood Ducks from southern breeding populations. *Condor* 93: 884–891.

Hepp, G. R. & Kennamer, R. A. (1992). Characteristics and consequences of nest-site fidelity in Wood Ducks. *Auk* 109: 812–818.

Hepp, G. R., Blohm, R. J., Reynolds, R. E., Hines, J. E. & Nichols, J. D. (1986). Physiological condition of autumn-banded Mallards and its relationship to hunting vulnerability. *J. Wildl. Manage.* 50: 177–183.

Herremans, M. (1988). Measurements of moult of irruptive Common Crossbills (*Loxia curvirostra curvirostra*) in central Belgium. *Le Gerfaut* 78: 243–260.

Herremans, M. (1993). Seasonal dynamics in sub-Kalahari bird communities with emphasis on migrants. *Proc. Pan-Afr. Ornithol. Congr.* 7: 555–664.

Herremans, M. (1994). Fifteen years of migrant phenology records in Botswana. A summary and prospects. *Babbler* 28: 47–68.

Herremans, M. (1997). Habitat segregation of male and female Red-backed Shrikes *Lanius collurio* and Lesser Grey Shrikes *Lanius minor* in the Kalahari basin. Botswana. *J. Avian Biol.* 28: 240–248.

Herremans, M. (1998). Strategies, punctuality of arrival and ranges of migrants in the Kalahari basin, Botswana. *Ibis* 140: 585–590.

Herrera, C. M. (1978). Ecological correlates of residence and non-residence in a Mediterranean passerine bird community. *J. Anim. Ecol.* 47: 871–890.

Herrera, C. M. (1981). Fruit food of Robins wintering in southern Spanish Mediterranean scrubland. *Bird Study* 28: 115–122.

Herrera, C. M. & Rodriguez, M. (1979). Year-to-year site constancy among three passerine species wintering at a southern Spanish locality. *Ringing & Migration* 2: 160.

Herzog, P. & Keppie, D. M. (1980). Migration in a local population of Spruce Grouse. *Condor* 82: 366–372.

Hestbeck, J. B., Nichols, J. D. & Malecki, R. A. (1991). Estimates of movement and site fidelity using mark-resight data of wintering Canada Geese. *Ecology* 72: 523–533.

Hestbeck, J. B., Nichols, J. D. & Hines, J. E. (1992). The relationship between annual survival rate and migration distance in Mallards: An examination of the time-allocation hypothesis for the evolution of migration. *Can. J. Zool.* 70: 2021–2027.

Hewitt, G. M. (2000). The genetic legacy of the Quaternary ice ages. *Nature* 405: 907–913.

Hickling, R. A. O. (1984). Lesser Black-backed Gull numbers at British inland roosts in 1979–80. *Bird Study* 31: 157–160.

Hicks, D. L. (1967). Adipose tissue composition and cell size in fall migratory thrushes (Turdidae). *Condor* 69: 387–399.

Higgins, P. J. & Davies, S. J. J. F. (1996). *Handbook of Australian, New Zealand and Antarctic birds*, Vol. 3. Melbourne & Oxford, Oxford University Press.

Higuchi, H., Sato, F., Matrui, S., Soma, M. & Kanmuri, N. (1991). Satellite-tracking of the migration routes of Whistling Swans *Cygnus columbianus*. *Yamashina Inst. Ornithol.* 23: 6–12.

Higuchi, H., Ozaki, K., Fujita, G., Soma, M., Kanmuni, N. & Ueta, M. (1992). Satellite tracking of the migration routes of cranes from southern Japan. *Strix* 11: 1–20.

Higuchi, H., Ozaki, K., Fujita, G., Minton, J., Ueta, M., Soma, M. & Mita, N. (1996). Satellite tracking of White-naped Crane. Migration and the importance of the Korean demilitarized zone. *Conserv. Biol.* 10: 806–812.

Higuchi, H., Shibaev, Y., Minton, J., Okaki, K., Surmach, S., Fugita, G., Momose, K., Momose, Y., Ueta, M., Andronov, V., Mita, N. & Kanai, Y. (1998). Satellite tracking of the migration of the Red-crowned Crane *Grus japonensis*. *Ecol. Res.* 13: 273–282.

Higuchi, H., Nagendram, M., Darman, Y., Tamura, M., Andronov, V., Parilov, M., Shimazaki, H. & Morishita, E. (2000). Migration and habitat use of Oriental White Storks from satellite tracking studies. *Global Environ. Res.* 2: 169–182.

Higuchi, H., Pierre, J. P., Krever, V., Andronov, V., Fugita, G., Ozaki, K., Goroshko, O., Ueta, M., Smirensky, S. & Mita, N. (2004). Using a remote technology in conservation: satellite tracking of White-naped Cranes in Russia and Asia. *Conserv. Biol.* 18: 136–147.

Higuchi, H., Shiu, H.-J., Nakamura, H., Uematsu, A., Kuno, K., Saeki, M., Hotta, M., Tokita, K.-I., Moriya, E., Morishita, E. & Tamura, M. (2005). Migration of Honey Buzzards *Pernis apivorus* based on satellite tracking. *Ornithol. Sci.* 4: 109–115.

Hildén, O. (1965). Habitat selection in birds. A review. *Ann. Zool. Fenn.* 2: 53–70.

Hildén, O. (1969). Über Vorkommen und Brutbiologie des Birkenzeisigs (*Carduelis flammea*) in Finnisch-Lapland in Sommer 1968. *Ornis Fenn.* 46: 93–112.

Hildén, O. (1974). Finnish bird stations, their activities and aims. *Ornis Fenn.* 51: 10–35.

Hildén, O. (1978). Population dynamics in Temminck's Stint *Calidris temminckii. Oikos* 30: 17–28.

Hildén, O. (1979). Territoriality and site tenacity of Temminck's Stint *Calidris temminckii. Ornis Fenn.* 56: 56–74.

Hildén, O. & Helo, P. (1981). The Great Grey Owl *Strix nebulosa* – a bird of the northern taiga. *Ornis Fenn.* 58: 159–166.

Hildén, O. & Saurola, P. (1982). Speed of autumn migration of birds ringed in Finland. *Ornis Fenn.* 59: 140–143.

Hildén, O. & Solonen, T (1987). Status of the Great Grey Owl in Finland. pp. 116–120 in *Biology and conservation of northern forest owls*, eds. Nero, R. W., Clark, R. J., Knapton, R. J. & Hamre, R. H. Fort Collins, USDA For. Gen. Tech. Rep. RM-142. Co., USA.

Hill, D. A. (1988). Population dynamics of Avocets (*Recurvirostra avosetta* L.) breeding in Britain. *J. Anim. Ecol.* 57: 669–683.

Hill, L. A. (1997). Trans-Saharan recoveries of House Martins *Delichon urbica*, with discussion on ringing, roosting and sightings in Africa. *Safring News* 26: 7–12.

Hiraldo, F., Negro, J. H., Donázar, J. A. & Gaona, P. (1996). A demographic model for a population of the endangered Lesser Kestrel in southern Spain. *J. Appl. Ecol.* 33: 1085–1093.

Hixon, M. A., Carpenter, F. L. & Paton, D. C. (1983). Territory area, flower density and budgeting in hummingbirds: an experimental and theoretical analysis. *Am. Nat.* 122: 366–391.

Hjertaas, D. G., Ellis, D. H., Johns, D. W. & Moon, S. L. (2001). Tracking Sandhill Crane migration from Saskatchewan to the Gulf coast. *Proc. N. Am. Crane Workshop* 8: 57–61.

Hjort, C. & Lindholm, C.-G. (1978). Annual bird ringing totals and population fluctuations. *Oikos* 30: 387–392.

Hjort, C., Pettersson, J., Lindström, Å. & King, J. M. (1996). Fuel deposition and potential flight ranges of Blackcaps *Sylvia atricapilla* and Whitethroats *S. communis* on spring migration in the Gambia. *Ornis Svecica* 6: 137–144.

Hobbs, J. N. (1959). Gulls attacking migrant thrushes. *Br. Birds* 52: 313.

Hobson, K. A. (1999). Tracing origins and migration of wildlife using stable isotopes: a review. *Oecologia* 120: 314–326.

Hobson, K. A. (2003). Making migratory connections with stable isotopes. pp. 379–392 in *Avian migration* (eds P. Berthold, E. Gwinner & E. Sonnenschein). Berlin, Springer-Verlag.

Hobson, K. A. (2005). Stable isotopes and the determination of avian migratory connectivity and seasonal interactions. *Auk* 122: 1037–1048.

Hobson, K. A. & Wassenaar, L. I. (1997). Linking breeding and wintering grounds of neotropical migrant songbirds using stable hydrogen isotopic analysis of feathers. *Oecologia* 109: 142–148.

Hobson, K. A. & Wassenaar, L. I. (2001). Isotopic delineation of North American migratory wildlife populations: Loggerhead shrikes. *Ecol. Appl.* 11: 1545–1553.

Hobson, K. A., Sirois, J. & Gloutney, M. L. (2000). Tracing nutrient allocations to reproduction using stable isotopes: a preliminary investigation using the colonial waterbirds of Great Slave Lake. *Auk* 117: 760–774.

Hobson, K. A., McFarland, K. P., Wassenaar, L. I., Rimmer, C. C. & Goetz, J. E. (2001). Linking breeding and wintering grounds of Bicknell's Thrushes using stable isotope analyses of feathers. *Auk* 118: 16–23.

Hochachka, W. M., Wells, J. V., Rosenberg, K. V., Tessaglia-Hymes, D. L. & Dhondt, A. A. (1999). Irruptive migration of Common Redpolls. *Condor* 101: 195–204.

Hochbaum, H. A. (1955). *Travels and traditions of waterfowl*. Minneapolis, University Minnesota Press.

Hockey, P. A. K. (2000). Patterns and correlates of bird migrations in sub-Saharan Africa. *Emu* 100: 401–417.

Hockey, P. A. R., Turpie, J. K. & Velásquez, C. R. (1998). What selective pressures have driven the evolution of deferred northward migration by juvenile waders? *J. Avian Biol.* 29: 325–330.

Hoffman, R. W. & Braun, C. E. (1977). Characteristics of a wintering population of White-tailed Ptarmigan in Colorado. *Wilson Bull.* 89: 107–115.

Hoffman, W. & Darrow, H. (1992). Migration of diurnal raptors from the Florida Keys into the West Indies. *HMANA Hawk Migration Stud.* 17: 7–14.

Hoffman, S. W. & Smith, J. P. (2003). Population trends of migratory raptors in western North America., 1977–2001. *Condor* 105: 397–419.

Hoffman, S. W., Smith, J. P. & Meehan, T. D. (2002). Feeding grounds, winter ranges, and migratory routes of raptors in the mountain west. *J. Raptor Res.* 36: 97–110.

Höglund, N. (1964). Der Habicht *Accipiter gentilis* Linné in Fennoscadia. *Viltrevy* 2: 195–270.

Hogstad, O. (1967). Density fluctuations of *Carduelis spinus* in relation to the cone crops of Norway Spruce. *Sterna* 7: 255–259.

Hogstad, O. (2000). Fluctuation of a breeding population of Brambling *Fringilla montifringilla* during 33 years in a subalpine birch forest. *Ornis Fenn.* 77: 97–103.

Holberton, R. L. (1993). An endogenous basis for differential migration in the Dark-Eyed Junco. *Condor* 95: 580–587.

Holberton, R. L. (1999). Changes in patterns of corticosterone secretion concurrent with migratory fattening in a neotropical migratory bird. *Gen. Comp. Endocrinol.* 11: 49–58.

Holberton, R. L. & Able, K. P. (1992). Persistence of circannual cycles in a migratory bird held in constant dim light. *J. Comp. Physiol.* 171: 477–481.

Holland, P. K. & Yalden, D. W. (1991). Population dynamics of Common Sandpipers *Actitis hypoleucos* breeding along an upland river system. *Bird Study* 38: 151–159.

Holliday, H. O., Pennycuick, C. J. & Fuller, M. (1988). Wind tunnel experiments to assess the effect of back-mounted radio transmitters on bird body drag. *J. Exp. Biol.* 135: 265–273.

Hollyer, J. N. (1970). The invasion of Nutcrackers in autumn 1968. *Br. Birds* 63: 353–373.

Holmes, H. T. (1990). The structure of a temperate deciduous forest bird community: variability in time and space. pp. 121–139 in *Biogeography and ecology of forest bird communities* (ed. A. Keast). The Hague, Netherlands, SPB Academic Publishing.

Holmes, R. T. & Sherry, T. W. (1992). Site fidelity of migratory warblers in temperate breeding and neotropical wintering areas: applications for population dynamics, habitat selection, and conservation. pp. 563–575 in *Ecology and conservation of neotropical migrant landbirds* (eds J. M. Hagan & D. W. Johnston). Washington, D.C., Smithsonian Institution Press.

Holmes, R. T. & Sherry, T. W. (2001). Thirty-year bird population trends in an unfragmented temperate deciduous forest: importance of habitat change. *Auk* 118: 589–609.

Holmes, R. T., Sherry, T. W. & Sturges, F. W. (1991). Numerical and demographic responses of temperate forest birds to annual fluctuations in their food resources. *Proc. Int. Ornithol. Congr.* 20: 1559–1567.

Holmes, R. T., Marra, P. P. & Sherry, T. W. (1996). Habitat-specific demography of breeding Black-throated Blue Warblers (*Dendroica caerulescens*) – implications for population dynamics. *J. Anim. Ecol.* 65: 183–195.

Holmgren, N. & Hedenström, A. (1995). The scheduling of molt in migratory birds. *Evol. Ecol.* 9: 354–368.

Holmgren, N. & Lundberg, S. (1993). Despotic behaviour and the evolution of migration patterns in birds. *Ornis Scand.* 24: 103–109.

Holmgren, N., Ellegren, H. & Petterson, J. (1993). Stopover length, body mass and fuel deposition rate in autumn migrating adult Dunlins *Calidris alpina* – evaluating the effects of molting status and age. *Ardea* 81: 9–20.

Hoover, J. P., Brittingham, M. C. & Goodrich, L. J. (1995). Effects of forest patch size on nesting success of Wood Thrushes. *Auk* 112: 146–155.

Hopp, S. L., Kirby, A. & Boone, C. A. (1999). Banding returns, arrival pattern and site fidelity of White-eyed Vireos. *Wilson Bull.* 111: 46–55.

Hörnfeldt, B. (1978). Synchronous population fluctuations in voles, small game, owls and tularemia in northern Sweden. *Oecologia* 32: 141–152.

Hötker, H. (2002). Arrival of Pied Avocets *Recurvirostra avosetta* at the breeding site: effects of winter quarters and consequences for reproductive success. *Ardea* 90: 379–387.

Houlihan, P. F. (1986). The birds of Ancient Egypt. Warminster, Aris & Phillips.

Houston, C. S. (1978). Recoveries of Saskatchewan-banded Great Horned Owls. *Can. Field Nat.* 92: 61–66.

Houston, C. S. (1999). Dispersal of Great Horned Owls banded in Saskatchewan and Alberta. *J. Field Ornithol.* 70: 343–350.

Houston, C. S. & Francis, C. M. (1995). Survival of Great Horned Owls in relation to the Snowshoe Hare cycle. *Auk* 112: 44–59.

Houston, C. S. & Houston, M. I. (1998). LeRoy and Myrtle Simmons, record-breaking Winnipeg bird-banders. *Blue Jay* 56: 75–81.

Houston, C. S. & Shadick, S. J. (1974). The spring migration, April 1-May 31, 1974, Northern Great Plains. *Am. Birds* 28: 814–817.

Howe, R. W. (1984). Local dynamics of bird assemblages in small forest habitat islands in Australia and North America. *Ecology* 65: 1585–1601.

Howey, D. H. & Bell, M. (1985). Pallas's Warblers and other migrants in Britain and Ireland in October 1982. *Br. Birds* 78: 381–392.

Howlett, J. S. & Stutchbury, B. J. M. (2003). Determinants of between-season site, territory, and mate fidelity in Hooded Warblers (*Wilsonia citrina*). *Auk* 120: 457–465.

Howlett, P., Jüttner, I. & Ormerod, S. J. (2000). Migration strategies of sylviid warblers: chance patterns or community dynamics? *J. Avian Biol.* 31: 20–30.

Hübner, C. E. (2006). The importance of pre-breeding sites in the arctic Barnacle Goose *Branta leucopsis*. *Ardea* 94: 701–713.

Huin, N. & Sparks, T. H. (1998). Arrival and progression of the Swallow *Hirundo rustica* through Britain. *Bird Study* 45: 361–370.

Huin, N. & Sparks, T. H. (2000). Spring arrival patterns of the Cuckoo *Cuculus canorus*, Nightingale *Luscinia megarhynchos* and Spotted Flycatcher *Muscicapa striata*. *Bird Study* 47: 22–31.

Hume, I. D. & Biebach, H. (1996). Digestive tract function in the long-distance migratory Garden Warbler *Sylvia borin*. *J. Comp. Physiol. B* 166: 388–395.

Hummel, D. (1973). Die Leistungsersparnis beim Verbandsflug. *J. Ornithol.* 114: 259–282.

Hummel, D. & Beukenberg, M. (1989). Aerodynamische Interferenzeffekte beim Formationsflug von Vögeln. *J. Ornithol.* 130: 15–24.

Hunt, W. G., Rogers, R. R. & Stowe, D. J. (1975). Migratory and foraging behaviour of Peregrine Falcons on the Texas coast. *Can. Field Nat.* 89: 111–123.

Hunt, W. J., Jackman, R. E., Jenkins, J. M., Thelander, C. G. & Lehman, R. (1992). Northward post-fledging migration of Californian Bald Eagles. *J. Raptor Res.* 26: 19–23.

Hunter, B. F., Clark, W., Perkins, P. & Coleman, P. (1970). Applied botulism research, including management recommendations, Progress Report. Fish and Game, California Dept. Sacramento.

Hupp, J. W., Schmutz, J. A. & Ely, C. R. (2006). The prelaying interval of Emperor Geese on the Yukon-Kuskokwim Delta, Alaska. *Condor* 108: 912–924.

Hüppop, O. & Hüppop, K. (2003). North Atlantic Oscillation and timing of spring migration in birds. *Proc. R. Soc. Lond. B* 270: 233–240.

Hüppop, O. & Winkel, W. (2006). Climate change and timing of spring migration in the long-distance migrant *Ficedula hypoleuca* in central Europe: the role of spatially different temperature changes along migration routes. *J. Ornithol.* 147: 344–353.

Hussell, D. J. T., Davis, T. & Montgomerie, R. D. (1967). Differential fall migration of adult and immature Least Flycatchers. *Bird-Banding* 68: 61–66.

Hussell, D. J. T., Mather, M. H. & Sinclair, P. H. (1992). Trends in numbers of tropical- and temperate-wintering landbirds in migration at Long Point Ontario, 1961–88. pp. 101–114 in *Ecology & conservation of Neotropical migrant landbirds* (eds J. M. Hagan & D. W. Johnson). Washington, D.C., Smithsonian Institution Press.

Hutto, R. L. (1985). Seasonal changes in the habitat distribution of transient insectivorous birds in southeastern Arizona: competition mediated? *Auk* 102: 120–132.

Hutto, R. L. (1986). Migratory landboirds in western Mexico: a vanishing habitat. *Western Wildlands* 11: 12–16.

Huyvaert, K. P. & Anderson, D. J. (2004). Limited dispersal by Nazca Boobies *Sula granti*. *J. Avian Biol.* 35: 46–53.

Igual, J. M., Forero, M. G., Tavecchia, G., Gonzalez-Solis, J., Martinez-Abrain, A., Hobson, K. A., Ruiz, X. & Oro, D. (2005). Short-term effects of data loggers on Cory's Shearwater (*Calonectris diomedia*). *Marine Biology* 146: 619–624.

Imber, M. J. & Lovegrove, T. G. (1982). Leach's Storm Petrels (*Oceanodroma l. leucorhoa*) prospecting for nest sites at the Chatham Islands. *Notornis* 29: 101–108.

Imboden, C. (1974). Zum Fremdansiedlung und Brutperiode des Kiebitz *Vanellus vanellus* in Europa. *Orn. Beob.* 71: 5–13.

Immelmann, K. (1963) Drought adaptations in Australian desert birds. *Proc. Int. Ornithol. Congr.* 13: 649–657.

Insley, H. & Boswell, R. C. (1978). The timing of arrivals of Reed and Sedge Warblers at south coast ringing sites during autumn passage. *Ringing & Migration* 2: 1–9.

Inouye, D. W., Barr, B., Armitage, K. B. & Inouye, B. D. (2000). Climate change is affecting alititudinal migrants and hibernating species. *Proc. Natl Acad. Sci. USA* 97: 1630–1633.

Iolé, P. & Benvenuti, S. (1983). Site attachment and homing ability in passerine birds. *Monit. Zool. Ital.* 17: 279–294.

Irwin, D. E. & Irwin, J. H (2005). Siberian migratory divides. pp. 27–40 in *Birds of Two Worlds.* (eds Greenberg, R. & Marra, P. P.). Washington, D. C., Smithsonian Institution.

Irwin, D. E., Alstrom, P., Olsson, U. & Benowitz-Friederichs, Z. M. (2001). Cryptospecies in the genus *Phylloscopus* (Old World leaf warblers). *Ibis* 143: 233–247.

Iverson, S. A., Esler, D. & Rizzolo, D. J. (2003). Winter philopatry of Harlequin Ducks in Prince William Sound, Alaska. *Condor* 106: 711–715.

Izhaki, I. & Maitav, A. (1998a). Blackcaps *Sylvia atricapilla* stopping over at the desert edge; physiological state and flight-range estimates. *Ibis* 140: 223–233.

Izhaki, I. & Maitav, A. (1998b). Blackcaps *Sylvia atricapilla* stopping over at the desert edge; inter- and intra-sexual differences in spring and autumn migration. *Ibis* 140: 234–243.

Izhaki, I. & Safriel, U. N. (1989). Why are there so few exclusively frugivorous birds? Experiments on fruit digestibility. *Oikos* 54: 23–32.

Jackson, D. B. (1994). Breeding dispersal and site-fidelity in three monogamous wader species in the Western Isles, U.K. *Ibis* 136: 463–473.

Jaeger, E. C. (1949). Further observations on the hibernation of the Poorwill. *Condor* 51: 105–109.

Jaeger, M. M., Bruggers, R. L., Johns, B. E. & Erickson, W. A. (1986). Evidence of itinerant breeding of the Red-billed Quelea *Quelea quelea* in the Ethiopian Rift Valley. *Ibis* 128: 469–482.

Jahn, A. E., Levey, D. J. & Smith, K. G. (2004). Reflections across hemispheres: a system-wide approach to New World bird migration. *Auk* 121: 1005–1013.

Jakober, H. & Stauber, W. (1983). Phenology of a population of Red-backed Shrike (*Lanius collurio*). *J. Ornithol.* 124: 29–46.

James, F. C., McCulloch, C. E. & Wiedenfield, D. A. (1996). New approaches to the analysis of population trends in land birds. *Ecology* 77: 13–25.

James, P. (1956). Destruction of warblers on Padre Island, Texas, in May 1951. *Wilson Bull.* 68: 224–227.

Jansson, R. B. (1976). The spring migration, April 1-May 31, 1976, Western Great Lakes Region. *Am. Birds* 30: 844–846.

Jardine, D. C. (1994). Brood patch on a Common Crossbill *loxia curvirostra* still in juvenile plumage. *Bird Study* 41: 155–156.

Järvinen, A. (1987). Key-factor analyses of two Finnish hole-nesting passerines: comparisons between species and regions. *Ann. Zool. Fenn.* 24: 275–280.

Javed, S., Takekawa, J. Y., Douglas, D. C., Rahmani, A. R., Kanai, Y., Nagendram, M., Choudhury, B. C. & Sharma, S. (2000). Tracking the spring migration of a Bar-headed Goose (*Anser indicus*) across the Himalaya with satellite telemetry. *Global Environ. Res.* 2: 195–205.

Jehl, J. R. (1990). Aspects of the molt migration. pp. 102–113 in *Bird migration: physiology and ecophysiology* (ed. E. Gwinner). Berlin, Springer.

Jehl, J. R. (1994). Field estimates of energetics in migrating and downed Black-necked Grebes. *J. Avian Biol.* 25: 63–68.

Jehl, J. R. (1996). Mass mortality events of Eared Grebes in North America. *J. Field Ornithol.* 67: 471–476.

Jehl, J. R. (1997). Cyclical changes in body composition in the annual cycle and migration of the Eared Grebe *Podiceps nigricollis*. *J. Avian Biol.* 28: 132–142.

Jehl, J. R., Henry, A. E. & Bond, S. I. (1999). Flying the gauntlet: population characteristics, sampling bias, and migrating routes of Eared Grebes downed in the Utah desert. *Auk* 116: 178–183.

Jehl, J., Henry, A. E. & Ellis, H. I. (2003). Optimizing migration in a reluctant and inefficient flyer: the Eared Grebe. pp. 199–209 in *Avian migration* (eds P. Berthold, E. Gwinner & E. Sonnenschein). Berlin, Springer-Verlag.

Jenkins, D. & Watson, A. (2000). Dates of first arrival and song of birds during 1974–99 in mid-Deeside, Scotland. *Bird Study* 47: 249–251.

Jenkins, K. D. & Cristol, D. A. (2002). Evidence of differential migration by sex in White-throated Sparrows (*Zonotrichia albicollis*). *Auk* 119: 539–543.

Jenni, L. (1982). Schweizerische Ringfunde von Bergfinken *Fringilla montifringilla*: ein Beitrag zum Problem der Masseneinflüge. *Der Orn. Beob.* 79: 265–272.

Jenni, L. (1987). Mass concentrations of Bramblings *Fringilla montifringilla* in Europe 1900–1983: their dependence upon beech mast and the effect of snow cover. *Ornis Scand.* 18: 84–94.

Jenni, L. (1993). Structure of a Brambling *Fringilla montifringilla* roost according to sex, age and body mass. *Ibis* 135: 85–90.

Jenni, L. & Jenni-Eiermann, S. (1998). Fuel supply and metabolic constraints in migrating birds. *J. Avian Biol.* 29: 521–528.

Jenni, L. & Jenni-Eiermann, S. (1999). Fat and protein utilisation during migratory flight. *Proc. Int. Ornithol. Congr.* 22: 1437–1449.

Jenni, L. & Kéri, M. (2003). Timing of autumn bird migration under climate change: advances in long distance migrants, delays in short distance migrants. *Proc. R. Soc. Lond. B* 270: 1467–1472.

Jenni, L. & Neuschulz, F. (1985). Schweizerische Ringfunds von Bergfinken *Fringilla montifringilla*: Ein Beitrag zum Problem der Masseneinfluge. *Orn. Beob.* 82: 85–106.

Jenni, L. & Schaub, M. (2003). Behavioural and physiological reactions to environmental variation in bird migration: a review. pp. 155–171 in *Avian migration* (eds P. Berthold, E. Gwinner & E. Sonnenschein). Berlin, Springer.

Jenni, L. & Winkler, R. (1994). *Moult and ageing of European passerines*. London, Academic Press.

Jenni, L., Jenni-Eiermann, S., Spina, F. & Schwabl, H. (2000). Regulation of protein breakdown and adrenocortical response to stress in birds during migratory flight. *Am. J. Physiol. (Regul. Integr. Comp. Physiol.)* 278B: R1182–R1189.

Jenni-Eiermann, S. & Jenni, L. (1991). Metabolic responses to flight and fasting in night migratory passerines. *J. Comp. Physiol. B* 161: 465–474.

Jenni-Eiermann, S. & Jenni, L. (1996). Metabolic differences between the postbreeding, moulting and migratory periods in feeding and fasting passerine birds. *Funct. Ecol.* 10: 62–72.

Jenni-Eiermann, S. & Jenni, L. (2001). Postexercise ketosis in night-migrating passerine birds. *Physiol. Biochem. Zool.* 74: 90–101.

Jenni-Eiermann, S. & Jenni, L. (2003). Interdependence of flight and stopover in migrating birds: possible effects of metabolic constraints during refuelling on flight metabolism. pp. 2933–3006 in *Avian migration* (eds P. Berthold, E. Gwinner & E. Sonnenschein). Berlin, Springer.

Jensen, W. I. (1979). An outbreak of streptococcus in Eared Grebes *Podiceps nigricollis*. *Avian Dis.* 23: 543–546.

Jensen, W. I. & Cotter, S. E. (1976). An outbreak of erysipelas in Eared Grebes (*Podiceps nigricollis*). *J. Wildl. Dis.* 12: 583–586.

Jensen, W. I. & Williams, C. (1964). Botulism and waterfowl. pp. 333–341 in *Waterfowl tomorrow* (ed. J. P. Linduska). Washington, D.C., Government Printing Office.

John, A. W. G. & Roskell, J. (1985). Jay movements in autumn 1983. *Br. Birds* 78: 611–637.

Johnson, C. & Minton, C. D. T. (1980). The primary moult of the Dunlin *Calidris alpina* at the Wash. *Ornis Scand.* 11: 190–195.

Johnson, D. H. (1979). Effects of a summer storm on bird populations. *The Prairie Naturalist* 11: 78–82.

Johnson, D. H. & Grier, J. W. (1988). Determinants of breeding distributions of ducks. *Wildl. Monogr.* 100: 1–37.

Johnson, D. N. & MacLean, G. L. (1994). Altitudinal migration in Natal. *Ostrich* 65: 86–94.

Johnson, D. W. & Downer, A. C. (1968). Migratory features of the Indigo Bunting in Jamaica and Florida. *Bird-Banding* 39: 277–293.

Johnson, M. D. & Geupel, G. R. (1996). The importance of productivity to the dynamics of a Swainson's Thrush population. *Condor* 98: 133–141.

Johnson, N. K. (1965). Differential timing and routes of spring migration in the Hammond's Flycatcher. *Condor* 67: 423–437.

Johnson, N. K. (1973). Spring migration of the Western Flycatcher with notes on seasonal changes in sex and age ratios. *Bird-Banding* 44: 205–220.

Johnson, O. W., Morton, M. L., Bruner, P. L. & Johnson, P. M. (1989). Fat cyclicity, predicted migratory flight ranges, and features of wintering behaviour in Pacific Golden Plovers. *Condor* 91: 156–177.

Johnson, O. W., Warnock, N., Bishop, M. A., Bennett, A. J., Johnson, P. M. & Kienholz, R. J. (1997). Migration by radio-tagged Pacific Golden Plovers from Hawaii to Alaska, and their subsequent survival. *Auk* 114: 4521–4524.

Johnson, O. W., Bruner, P. L., Rotella, J. J., Johnson, P. M. & Bruner, A. E. (2001). Long-term study of apparent survival in Pacific Golden Plovers at a wintering ground on Oahu, Hawaiian Islands. *Auk* 118: 342–351.

Johnson, S. R. (1995). Immigration in a small population of Snow Geese. *Auk* 112: 731–736.

Johnson, S. R. & Herter, D. R. (1990). Bird migration in the arctic: a review. pp. 33–46 in *Bird migration: physiology and ecophysiology* (ed. E. Gwinner). Berlin, Springer-Verlag.

Johnston, D. W. (1966). A review of the vernal fat deposition picture in overland migrant birds. *Bird-Banding* 37: 172–183.

Johnston, D. W. (1968). Body characteristics of Palm Warblers following an overwater flight. *Auk* 85: 13–18.

Johnston, D. W. & Haines, T. P. (1957). Analysis of mass bird mortality in October, 1954. *Auk* 74: 447–458.

Johnston, V. H. & Ryder, J. P. (1987). Divorce in larids: a review. *Colonial Waterbirds* 10: 16–26.

Jones, C. D. (2005). The Ontario Great Grey Owl irruption of 2004–2005: numbers, dates and distribution. *Ontario Birds* 23: 106–121.

Jones, P. (1985). The migration strategies of Palaearctic passerines in West Africa. pp. 9–21 in *Migratory birds: problems and prospects in Africa* (eds A. MacDonald & P. Goriup). Cambridge, International Council for Bird Preservation.

Jones, P. (1995). Migration strategies of Palaearctic passerines in Africa: an overview. *Israel J. Zool.* 41: 393–406.

Jones, P. (1999). Community dynamics of arboreal insectivorous birds in African savannas in relation to seasonal rainfall patterns and habitat change. pp. 421–447 in *Dynamics of tropical communities* (eds D. M. Newberry, H. H. T. Prins & N. D. Brown). Oxford, Blackwells.

Jones, P., Vickery, J., Holt, S. & Cresswell, W. (1996). A preliminary assessment of some factors influencing the density and distribution of Palaearctic passerine migrants in the Sahel zone of West Africa. *Bird Study* 43: 73–84.

Jönsson, P. E. (1991). Reproduction and survival in a declining population of the Southern Dunlin *Calidris alpina schinzii*. *Wader Study Group Bull.* 61, Supplement: 56–68.

Jordano, P. (1987). Frugivory, external morphology and digestive system in Mediterranean sylviid warblers, *Sylvia* spp. *Ibis* 129: 175–189.

Joseph, I., Lessa, E. P. & Christidis, L. (1999). Phylogeny and biogeography in the evolution of migration: shorebirds of the *Charadrius* complex. *J. Biogeog.* 26: 329–342.

Joseph, L. (1997). Towards a broader view of neotropical migrants: consequences of a re-examination of austral migration. *Ornitol. Neotropical* 8: 31–36.

Joseph, L. & Stockwell, D. R. B. (2000). Temperature-based models of the migration of Swainson's Flycatcher (*Myiarchus swainsoni*) across South America: a new use for museum specimens of migratory birds. *Proc. Acad. Natl Sci. Phil.* 150: 293–300.

Jouventin, P. & Weimerskirch, H. (1990). Satellite tracking of Wandering Albatrosses. *Nature* 343: 746–748.

Jukema, J. & Piersma, T. (2000). Contour feather moult of Ruffs *Philomachus pugnax* during northward migration, with notes on homology of nuptial plumages in scolopacid waders. *Ibis* 142: 289–296.

Kaiser, A. (1999). Stopover strategies in birds: a review of methods for estimating stopover length. *Bird Study* 46(suppl.): 299–308.

Kaitala, A., Kaitala, V. & Lundberg, P. (1993). A theory of partial migration. *Am. Nat.* 142: 59–81.

Kalela, O. (1954). Über den Revierbesitz bei Vögeln und Saügetieren als populalionsökologischer Faktor. *Ann. Zool. Soc. Vanamo* 16: 1–48.

Kalela, O. (1955). Die neuzeitliche Ausbreitung des Kiebitzes, *Vanellus vanellus* (L.), in Finnland. *Ann. Zool. Soc. Vanamo* 16,11: 1–80.

Kalmbach, E. R. & Gunderson, M. F. (1934). Western duck sickness, a form of botulism. *U.S. Dept. Agr. Tech. Bull.* 411: 1–81.

Kaminski, R. M. & Gluesing, E. A. (1987). Density- and habitat-related recruitment in Mallards. *J. Wildl. Manage.* 51: 141–148.

Kanai, Y., Sato, F., Ueta, M., Minton, J., Higushi, H., Soma, M., Mita, N. & Matsui, S. (1997). The migration routes and important rest sites of Whooper Swans satellite-tracked from northern Japan. *Strix* 15: 1–13.

Kanai, Y., Minton, J., Nagendran, M., Ueta, M., Aurysana, B., Goroshko, O., Kovhsar, A. F., Mita, N., Suwal, R. N., Uzawa, K., Krever, V. & Higuchi, H. (2000). Migration of Demoiselle Cranes in Asia based on satellite tracking and fieldwork. *Global Environ. Res.* 2: 143–153.

Kanai, Y., Ueta, M., Gerogenov, N., Nagendran, M., Mita, N. & Higuchi, H. (2002). Migration routes and important resting areas of Siberian Cranes (*Grus leucogeranus*) between northeastern Siberia and China as revealed by satellite tracking. *Biol. Conserv.* 106: 339–346.

Kanyamibwa, S., Bairlein, F. & Schierer, A. (1993). Comparison of survival rates between populations of the White Stork *Ciconia ciconia* in central Europe. *Ornis Scand.* 24: 297–302.

Karasov, W. H. (1996). Digestive plasticity in avian energetics and feeding ecology. pp. 61–84 in *Avian energetics and nutritional ecology* (ed. C. Carey). New York, Chapman & Hall.

Karasov, W. H. & Pinshow, B. (2000). Test for physiological limitation to nutrient assimilation in a long-distance passerine migrant at a springtime stopover site. *Physiol. Biochem. Zool.* 73: 335–343.

Karr, J. R. (1976). On the relative abundance of migrants from the north temperate zone in tropical habitats. *Wilson Bull.* 88: 433–458.

Karr, J. R. (1980). Patterns in the migration system between the north temperate zone and the tropics. pp. 529–543 in *Migrant birds in the Neotropics* (eds A. Keast & E. Morton). Washington, D.C., Smithsonian Institution Press.

Kaufman, K. (1977). The changing seasons. *Am. Birds* 31: 142–152.

Kaufman, K. (1984). The changing seasons. *Am. Birds* 38: 992–996.

Keast, A. (1959). Australian birds: their zoogeography and adaptations to an arid continent. pp. 89–114 in *Biogeography and ecology in Australia* (eds A. Keast, R. L. Crocker & C. S. Christian). The Hague, W. Junk.

Keast, A. (1968a). Moult in birds of the Australian dry country relative to rainfall and breeding. *J. Zool. (Lond.)* 155: 185–200.

Keast, A. (1968b). Seasonal movements in the Australian honeyeaters (*Melaphagidae*) and their ecological significance. *Emu* 67: 159–209.

Keast, A. (1980). Spatial relationships between migratory parulid warblers and their ecological counterparts in the neotropics. pp. 109–130 in *Migrant birds in the Neotropics: ecology, behaviour, distribution and conservation* (eds A. Keast & E. S. Morton). Washington, D.C., Smithsonian Institution Press.

Keast, A. (1995). The Nearctic–Neotropical bird migration system. *Israel J. Zool.* 41: 455–476.

Keast, A. & Morton, E. S. (eds) (1980). *Migrant birds in the Neotropics: ecology, behaviour, distribution and conservation*. Washingon, D.C., Smithsonian Institution Press.

Keeton, W. T. (1980). Avian orientation and navigation: new developments in an old mystery. *Proc. Int. Ornithol. Congr.* 17: 137–157.

Keith, L. B. (1963). Wildlife's ten-year cycle. Madison, University Wisconsin Press.

Keith, L. B. & Rusch, D. H. (1988). Predation's role in the cyclic fluctuations of Ruffed Grouse. *Proc. Int. Ornithol. Congr.* 19: 699–732.

Keller, L. F., Arcese, P., Smith, J. N. M., Hochachka, W. M. & Stearns, S. (1994). Selection against inbred Song Sparrows during a natural bottleneck. *Nature* 372: 356–357.

Keller, V. (2000). Winter distribution and population change of Red-crested Pochard *Netta rufina* in southwestern and central Europe. *Bird Study* 47: 176–185.

Kelly, J. F. (2006). Stable isotope evidence links breeding geography and migration timing in wood warblers (Parulidae). *Auk* 123: 431–437.

Kelly, J. F., Atudorei, V., Sharp, Z. D. & Finch, D. M. (2002). Insights into Wilson's Warbler migration from analyses of hydrogen stable-isotope ratios. *Oecologia* 130: 216–231.

Kelly, J. F., Delay, L. S. & Finch, D. M. (2002). Density-dependent mass gain by Wilson's Warblers during stopover. *Auk* 119: 210–213.

Kelly, S. T. & DeCapita, M. E. (1982). Cowbird control and its effect on Kirtland's Warbler reproductive success. *Wilson Bull.* 94: 363–365.

Kelsey, M. G. (1989). A comparison of the song and territorial behaviour of a long-distance migrant, the Marsh Warbler *Acrocephalus palustris*, in summer and winter. *Ibis* 131: 403–414.

Kemp, A. (2001). Concentration of non-breeding Lesser Spotted Eagles *Aquila pomarina* at abundant food: a breeding colony of Red-billed *Queleas Quelea quelea* in the Kruger National Park, South Africa. *Acta Ornithol. Jena* 4: 325–329.

Kemper, C. (1996). A study of bird mortality at a west central Wisconsin TV tower from 1957–1995. *Passenger Pigeon* 58: 219–235.

Kendeigh, S. C. (1941). Territorial and mating behaviour of the House Wren. *Illinois Biol. Monogr.* 18: 1–120.

Kendeigh, S. C. (1947). Bird population studies in the coniferous biome during a Spruce Budworm outbreak. *Biol. Bull. (Dept. Lands and Forest, Ontario)* 1: 1–100.

Kennard, J. H. (1976). A biennial pattern in the winter distribution of the Common Redpoll. *Bird-Banding* 47: 232–237.

Kennedy, R. J. (1970). Direct effects of rain on birds: a review. *Br. Birds* 63: 401–414.

Kent, A. K. (1951). Mortality of Chiffchaffs in a snow-storm in Pembrokeshire. *Br. Birds* 44: 64.

Kenward, R. E. (2001). *A manual for wildlife radio tagging*. London, Academic Press.

Kenward, R. E., Marcström, V. & Karlbom, M. (1993a). Post-nestling behaviour in Goshawks, *Accipiter gentilis*: I. The causes of dispersal. *Anim. Behav.* 46: 365–370.

Kenyon, K. W. & Rice, D. W. (1958). Homing of Laysan Albatrosses. *Condor* 60: 3–6.

Kerlinger, P. (1989). *Flight strategies of migrating hawks*. Chicago, University Press.

Kerlinger, P. (2000). Avian mortality at communication towers: a review of recent literature, research and methodology. http://migratorybirds.fws.gov/issues/towers/review.pdf

Kerlinger, P. & Lein, M. R. (1986). Differences in winter range among age-sex classes of Snowy Owls *Nyctea scandiaca* in North America. *Ornis Scand.* 17: 1–7.

Kerlinger, P. & Moore, F. R. (1989). Atmospheric structure and avian migration. *Curr. Ornithol.* 6: 109–142.

Kerlinger, P., Lein, M. R. & Sevick, B. J. (1985). Distribution and population fluctuations of wintering Snowy Owls (*Nyctea scandiaca*) in North America. *Ecology* 63: 1829–1834.

Keskpaik, J. (1972). (Temporary hypothermia in Sand Martins *Riparia riparia* in natural conditions). *Communs. Baltic Comm. Study Bird Migr.* 7: 176–183.

Kessel, B. (1953). Distribution and migration of the European Starling in North America. *Condor* 55: 49–67.

Ketterson, E. D. & Nolan, V. J. (1976). Geographic variation and its climatic correlates in the sex ratio of eastern-wintering Dark-eyed Juncos (*Junco hyemalis hyemalis*). *Ecology* 57: 679–693.

Ketterson, E. D. & Nolan, V. (1979). Seasonal, annual and geographic variation in sex ratio of wintering Dark-eyed Juncos (*Junco hyemalis*). *Auk* 96: 532–536.

Ketterson, E. D. & Nolan, V. (1982). The role of migration and winter mortality in the life history of a temperate-zone migrant, the Dark-eyed Junco, as determined from demographic analyses of winter populations. *Auk* 99: 243–259.

Ketterson, E. D. & Nolan, V. (1983). The evolution of differential bird migration. *Curr. Ornithol.* 1: 357–402.

Ketterson, E. D. & Nolan, V. (1985). Intraspecific variation in avian migration: evolutionary and regulatory aspects. pp. 553–579 in *Migration: mechanism and adaptive significance. Contributions to Marine Science,* Supplement, Vol. 27. Austin, University of Texas.

Ketterson, E. D. & Nolan, V. (1986). A possible role for experience in the regulation of the timing of bird migration. *Proc. Int. Ornithol. Congr.* 19: 2169–2179.

Ketterson, E. D. & Nolan, V. (1990). Site attachment and site fidelity in migratory birds: experimental evidence from the field and analogies from neurobiology. pp. 117–129 in *Bird migration. Physiology and ecophysiology* (ed. E. Gwinner). Berlin, Springer-Verlag.

Kikkawa, J. (1980). Winter survival in relation to dominance classes among Silvereyes *Zosterops lateralis chlorocephala* of Heron Island, Great Barrier Reef. *Ibis* 122: 437–446.

Kilpi, M. & Saurola, P. (1984). Migration and wintering strategies of juvenile and adult *Larus marinus, L. argentatus* and *L. fuscus. Ornis Fenn.* 61: 1–8.

Kilpi, M. & Saurola, P. (1985). Movements and survival areas of Finnish Common Gulls *Larus canus*. *Ann. Zool. Fenn.* 22: 157–168.

King, J. R. (1961). The bioenergetics of vernal premigratory fat deposition in the White-crowned Sparrow. *Condor* 63: 128–142.

King, J. R. (1968). Cycles of fat deposition and moult in White-crowned Sparrows in constant environmental conditions. *Comp. Biochem. Physiol.* 24: 827–837.

King, J. R. (1972). Adaptive periodic fat storage in birds. *Proc Int. Ornithol. Congr.* 15: 200–217.

King, J. R. (1974). Seasonal allocation of time and energy resources in birds. *Publ. Nuttall Ornithol. Club* 15: 4–85.

King, J. R. & Farner, D. S. (1965). Biochronometry and bird migration. pp. 625–630 in *Chronobiology* (eds L. E. Scheving, F. Halberg & J. E. Pauly). Tokyo, Igaku Shoin.

King, J., R. & Mewaldt, L. R. (1981). Variation of body weight in Gambel's White-crowned Sparrows in winter and spring: latitudinal and photoperiodic correlates. *Auk* 98: 752–764.

King, J. R. & Wales, E. E. (1964). Observations of migration, ecology and population flux of wintering Rosy Finches. *Condor* 66: 24–31.

King, J. R. & Wales, E. E. (1965). Photoperiodic regulation of testicular metamorphosis and fat deposition in three taxa of rosy finches. *Physiol. Zool.* 38: 49–68.

King, J. R., Barker, S. & Farner, D. S. (1963). A comparison of energy reserves during the autumnal and vernal migratory periods in the White-crowned Sparrow *Zonotrichia leucophys gambelii. Ecology* 44: 513–521.

King, J. R., Farner, D. S. & Mewaldt, L. R. (1965). Seasonal sex and age ratios in populations of White-crowned Sparrows of the race *gambelii. Condor* 67: 489–504.

King, K. A. (1976). Bird mortality, Galveston Island, Texas. *Southwest. Natur.* 21: 414.

Kirkwood, J. K. (1983). A limit to metabolisable energy intake in mammals and birds. *Comp. Biochem. Physiol.* 75A: 1–3.

Kirschvink, J. & Gould, J. (1981). Biogenic magnetite as a basis for magnetic field detection in animals. *Biosystems* 13: 181–201.

Kjellén, N. (1992). Differential timing of autumn migration between sex and age groups in raptors at Falsterbo, Sweden. *Ornis Scand.* 23: 420–434.

Kjellén, N. (1994a). Differences in age and sex-ratio among migrating and wintering raptors in southern Sweden. *Auk* 111: 274–284.

Kjellén, N. (1994b). Moult in relation to migration in birds – a review. *Ornis Svecica* 4: 1–24.

Kjellén, N. & Roos, G. (2000). Population trends in Swedish raptors demonstrated by migration counts at Falsterbo, Sweden 1942–97. *Bird Study* 47: 195–211.

Kjellén, N., Håke, M. & Alerstam, T. (1997). Strategies of two Ospreys *Pandion haliaetus* migrating between Sweden and tropical Africa as revealed by satellite tracking. *J. Avian Biol.* 28: 15–23.

Kjellén, N., Håke, M. & Alerstam, T. (2001). Timing and speed of migration in male, female and juvenile Ospreys *Pandion haliaetus* between Sweden and Africa as revealed by field observations, radar and satellite tracking. *J. Avian Biol.* 32: 57–67.

Klaassen, M. (1995). Water and energy limitations on flight range. *Auk* 112: 260–262.

Klaassen, M. (1996). Metabolic constraints on long-distance migration in birds. *J. Exp. Biol.* 199: 57–64.

Klaassen, M. (2004). May dehydration risk govern long-distance migratory behaviour? *J. Avian Biol.* 35: 4–6.

Klaassen, M. & H. Biebach. (2000). Flight altitude of trans-Sahara migrants in autumn: a comparison of radar observations with predictions from meteorological conditions and water and energy balance models. *J. Avian Biol.* 31: 47–55.

Klaassen, M., Lindström, Å. & Zijlatra, R. (1997). Composition of fuel stores and digestive limitations to fuel deposition rate in the long-distance migratory Thrush-Nightingale *Luscinia luscinia. Physiol. Zool.* 70: 125–133.

Klaassen, M., Kvist, A. & Lindström, A. (1999). How body water and fuel stores affect long distance flight in migrating birds. *Proc. Int. Ornithol. Congr.* 22: 1450–1467.

Klaassen, M., Kvist, A. & Lindström, A. (2000). Flight costs and fuel composition of a bird migrating in a wind tunnel. *Condor* 102: 444–451.

Klaassen, M., Lindström, A., Meltofte, H. & Piersma, T. (2001). Arctic waders are not capital breeders. *Nature* 413: 794.

Klaassen, M., Beekman, J. H., Kontiokorpi, J., Mulder, R. J. W. & Nolet, B. A. (2004). Migrating swans profit from favourable changes in wind conditions at low altitude. *J. Ornithol.* 145: 142–151.

Klein, H., Berthold, P. & Gwinner, E. (1973). Der Zug europäischer Garten und Mönchsgrasmücken (*Sylvia borin* und *S. atricapilla*). *Vogelwarte* 27: 73–134.

Klein, N. & Brown, W. M. (1994). Intraspecific molecular phylogeny in the Yellow Warbler and implications for avian biogeography in the West Indies. *Evolution* 48: 1914–1932.

Klomp, A. (1972). Regulation of the size of bird populations by means of territorial behaviour. *Neth. J. Zool.* 22: 456–488.

Klomp, N. I. & Furness, R. W. (1992). The dispersal and philopatry of Great Skuas from Foula, Shetland. *Ringing & Migration* 13: 73–82.

Kluijver, H. N. (1971). Regulation of numbers in populations of Great Tits (*Parus m. major* L.). pp. 507–524 in *Dynamics of populations* (eds P. J. den Boer & G. R. Gradwell). Wageningen, Centre for Agricultural Publishing and Documentation.

Kluijver, H. N. & Tinbergen, L (1953). Territory and the regulation of density in titmice. *Arch. Neerlandaises Zool.* 10: 265–289.

Knox, A. G. (1992). Species and pseudospecies: the structure of crossbill populations. *Biol. J. Linn. Soc.* 47: 325–335.

Kodric-Brown, A. & Brown, J. H. (1978). Influence of economics, interspecific competition and sexual dimorphism on territoriality of migrant Rufous Hummingbirds. *Ecology* 59: 285–296.

Köenig, W. D. & Knops, J. M. H. (1998). Scale of mast seeding and tree-ring growth. *Nature* 396: 225–226.

Köenig, W. D. & Knops, J. M. H. (2000). Patterns of annual seed production by northern hemisphere trees: a global perspective. *Am. Nat.* 155: 59–69.

Köenig, W. D. & Knops, J. M. H. (2001). Seed crop size and eruptions of North American boreal seed-eating birds. *J. Anim. Ecol.* 70: 609–620.

Köenig, W. D. & Mumme, R. L. (1987). *Population ecology of the cooperatively breeding.* Acorn Woodpecker. Princeton, University Press.

Köenig, W. D., Hooge, P. N., Stanback, M. T. & J. Haydock. (2000). Inbreeding depression and its effects on natal dispersal in Red-cockaded Woodpeckers. *Condor* 102: 492–502.

Koerner, J. W., Bookhout, T. A. & Bednarik, K. E. (1974). Movements of Canada Geese color-marked near southwestern Lake Erie. *J. Wildl. Manage.* 38: 27–289.

Kok, O. B., Ee, C. A., van & Nel, D. G. (1991). Daylength determines departure date of the Spotted Flycatcher *Muscicapa striata* from its winter quarters. *Ardea* 79: 63–66.

Kokko, H. (1999). Competition for early arrival in migratory birds. *J. Anim. Ecol.* 68: 940–950.

Kokko, H. & Ekman, J. (2002). Delayed dispersal as a route to breeding: territorial inheritance, safe havens, and ecological constraints. *Am. Nat.* 160: 468–484.

Kolunen & Peiponen. (1991). Delayed autumn migration of the Swift *Apus apus* from Finland in 1986. *Ornis Fenn.* 68: 81–92.

Komar, O., O'Shea, B. J., Peterson, A. T. & Navarro-Siguenza, A. G. (2005). Evidence of latitudinal sexual segregation among migratory birds wintering in Mexico. *Auk* 122: 938–948.

Komdeur, J., Huffstadt, A., Prast, W., Castle, G., Mileto, R. & Wattel, J. (1995). Transfer experiments of Seychelles Warblers to new islands: changes in dispersal and helping behaviour. *Anim. Behav.* 49: 695–708.

Korpimäki, E. (1986). Gradients in population fluctuations of Tengmalm's Owl *Aegolius funereus* in Europe. *Oecologia* 69: 195–201.

Korpimäki, E. (1992). Population dynamics of Fennoscandian Owls in relation to wintering conditions and between-year fluctuations of food. pp. 1–10 in *The ecology and conservation of European Owls* (eds C. A. Galbraith, I. R. Taylor & S. Percival). Edinburgh, Joint Nature Conservation Committee.

Korpimäki, E. (1994). Rapid or delayed tracking of multi-annual vole cycles by avian predators? *J. Anim. Ecol.* 63: 619–628.

Korpimäki, E. & Norrdahl, K. (1989). Predation of Tengmalm's Owls: numerical responses, functional responses and dampening impact on population fluctuations of voles. *Oikos* 54: 154–164.

Korpimäki, E. & Norrdahl, K. (1991). Numerical and functional responses of Kestrels, Short-eared Owls and Long-eared Owls to vole densities. *Ecology* 72: 814–825.

Korpimäki, E., Lagerström, M. & Saurola, P. (1987). Field evidence for nomadism in Tengmalm's Owl *Aegolius funereus*. *Ornis Scand.* 18: 1–4.

Kosarev, V. (1999). Summer movements of Starlings *Sturnus vulgaris* on the Courish Spit of the Baltic Sea. *Avian Ecol. Behav.* 30: 99–109.

Koskimies, J. (1950). The life of the Swift *Micropus apus* (L.) in relation to the weather. *Suom. Tied. Toim. Ann. Acad. Sci. Fenn.* 15: 1–151.

Kramer, G. (1951). Eine neue Methode zur Erforschung der Zugorientierung und die bisher damit erzeilten Ergebnisse. *Proc. Int. Ornithol. Congr.* 10: 269–280.

Kramer, G. (1952). Experiments on bird orientation. *Ibis* 94: 265–285.

Kramer, G. (1957). Experiments on bird orientation and their interpretation. *Ibis* 99: 196–227.

Krapu, G. (1981). The role of nutrient reserves in Mallard reproduction. *Auk* 98: 29–38.

Krapu, G. L. & Johnson, D. H. (1990). Conditioning of Sandhill Cranes during fall migration. *J. Wildl. Manage.* 54: 234–238.

Krapu, G. L., Iverson, G. C., Reinecke, K. J. & Boise, C. M. (1985). Fat deposition and usage by arctic-nesting Sandhill Cranes during spring. *Auk* 102: 362–368.

Krapu, G. L., Eldridge, J. L., Gratto-Trevor, C. L. & Buhl, D. A. (2006). Fat dynamics of arctic-nesting sandpipers during spring in mid-continental North America. *Auk* 123: 323–334.

Krause, H. (1959). Northern Great Plains Region. *Audubon Field Notes* 13: 380–381.

Krebs, C. S., Boutin, S. & Boonstra, R (eds) (2001). *Ecosystem dynamics of the boreal forest: The Kluane Project*. New York, Oxford University Press.

Kress, S. W. & Nettleship, D. N. (1988). Re-establishment of Atlantic Puffins (*Fratercula arctica*) at a former breeding site in the Gulf of Maine. *J. Field Ornithol.* 59: 161–170.

Kuenzi, A. J., Moore, F. R. & Simons, T. R. (1991). Stopover of neotropical landbird migrants on East Ship Island following trans-Gulf migration. *Condor* 93: 869–883.

Kuhnen, K. (1975). Bestandsentwicklung, Verbreitung, Biotop und Siedlungsdichte der Uferschwalbe (*Riparia riparia*) 1966–1973 am Niederrhein. *Charadrius* 11: 1–24.

Kullberg, C., Fransson, T. & Jakobsson, S. (1996). Impaired predator evasion in fat Blackcaps (*Sylvia atricapilla*). *Proc R. Soc Lond. B* 263: 1671–1675.

Kullberg, C., Jakobsson, S. & Fransson, T. (2000). High migratory fuel loads impair predator evasion in Sedge Warblers. *Auk* 117: 1034–1038.

Kullberg, C., Lind, J., Fransson, T., Jakobsson, S. & Vallin, A. (2003). Magnetic cues and time of season affect fuel deposition in migratory Thrush Nightingales (*Luscinia luscinia*). *Proc. R. Soc. Lond. B* 270: 373–378.

Kuroda, N. (1964). Analysis of variations by sex, age, and season of body weight, fat and some body parts in the Dusky Thrush, wintering in Japan: a preliminary study. *Misc. Rep. Yamashina Inst. Ornithol.* 4: 91–104.

Kuyt, E. (1992). Aerial radio-tracking of Whooping Cranes migrating between Wood Buffalo National Park and Aransas National Wildlife Refuge, 1981–84. *Canadian Wildlife Service, occasional paper no.* 74: 1–53.

Kvist, A. & Lindström, Å. (2000). Maximum daily energy intake: it takes time to lift the metabolic ceiling. *Physiol. Biochem. Zool.* 73: 30–36.

Kvist, A. & Lindström, A. (2003). Gluttony in migratory waders – unprecedented energy assimilation rates in vertebrates. *Oikos* 103: 397–402.

Kvist, A., Lindström, A., Green, M., Piersma, T. & Visser, G. H. (2001). Carrying large fuel loads during sustained bird flight is cheaper than expected. *Nature* 413: 730–732.

Lack, D. (1944). Ecological aspects of species formation in passerine birds. *Ibis* 86: 260–286.

Lack, D. (1954). *The natural regulation of animal numbers*. Oxford, Oxford University Press.

Lack, D. (1956). *Swifts in a tower*. London, Methuen.

Lack, D. (1960a). Autumn drift migration onto the English east coast. *Br. Birds* 53: 325–352, 379–397.

Lack, D. (1960b). The influence of weather on passerine migration. A review. *Auk* 77: 171–209.

Lack, D. (1963). Migration across the southern North Sea studied by radar. Part 5. Movements in August, winter and spring, and conclusion. *Ibis* 105: 461–492.

Lack, D. (1971). *Ecological isolation in birds*. Oxford, Blackwells.

Lack, D. & Lack, P. C. (1972). Wintering warblers in Jamaica. *Living Bird* 11: 129–153.

Lack, P. C. (1983). The movements of Palaearctic landbird migrants in Tsavo East National Park, Kenya. *J. Anim. Ecol.* 52: 513–524.

Lack, P. C. (1986a). *The atlas of wintering birds in Britain and Ireland*. Calton, Poyser.

Lack, P. C. (1986b). Ecological correlates of migrants and residents in a tropical African savanna. *Ardea* 74: 111–119.

Lack, P. C. (1987). The structural and seasonal dynamics of the bird community in Tsavo East National Park. *Ostrich* 58: 9–23.

Lack, P. C. (1990). Palaearctic-African systems. pp. 345–356 in *Biogeography and ecology of forest bird communities* (ed. A. Keast). The Hague, Netherlands, SPB Academic Publishing.

Laing, D. K., Bird, D. M. & Chubbs, T. E. (2005). First complete migration cycles for juvenile Bald Eagles (*Haliaeetus leucocephalus*) from Labrador. *J. Raptor Res.* 39: 11–18.

Landmann, A. & Kollinsky, C. (1995). Age and plumage related territory differences in male Black Redstarts *Phoenicurus ochropus* – the (non)-adaptive significance of delayed plumage maturation. *Ethol. Ecol. Evol.* 7: 147–167.

Landys, M. M., Piersma, T., Visser, G. H., Jukema, J. & Wijker, A. (2000). Water balance during real and simulated long-distance migratory flight in the Bar-tailed Godwit. *Condor* 102: 645–652.

Landys-Ciannelli, M., Ramenofsky, M., Piersma, T., Jukema, J. & Wingfield, J. C. (2002). Baseline and stress-induced plasma corticosterone during long-distance migration in the Bar-tailed Godwit, *Limosa lapponica*. *Physiol. Biochem. Zool.* 75: 101–110.

Lane, B. A. & Parish, D. (1991). A review of the Asian-Australasian bird migration system. *ICBP Tech. Bull.* 12: 291–312.

Lank, D. B., Butler, R. W., Ireland, J. & Ydenberg, R. C. (2003a). Effects of predation danger on migration strategies of sandpipers. *Oikos* 103: 303–319.

Lank, D. B. & Ydenberg, R. C. (2003b). Death and danger at migratory stopovers: problems with 'predation risk'. *J. Avian Biol.* 34: 225–228.

Lanner, R. M. (1996). *Made for each other. A symbiosis of birds and pines*. Oxford, Oxford University Press.

Lanyon, S. M. & Thompson, C. F. (1986). Site fidelity and habitat quality as determinants of settlement pattern in male Painted Buntings. *Condor* 88: 206–210.

Larson, D. L. & Bock, C. E. (1986). Eruptions of some North American boreal seed-eating birds, 1901–1980. *Ibis* 128: 137–140.

Larson, T. & Tombre, I. (1989). Cyclic irruptions of Two-barred Crossbills in Scandinavia. *Fauna Norv. Ser. C, Cinclus* 12: 3–10.

Larsson, K. (1996). Genetic and environmental effects on the timing of wing moult in the Barnacle Goose. *Heredity* 76: 100–107.

Larsson, K., Forslund, P., Gustafsson, L. & Ebbinge, B. S. (1988). From the high arctic to the Baltic: the successful establishment of a Barnacle Goose *Branta leucopsis* population on Gotland, Sweden. *Ornis Scand.* 19: 182–189.

Lasiewski, R. C. & Dawson, W. R. (1967). A re-examination of the relation between standard metabolic rate and body weight in birds. *Condor* 69: 13–23.

Latta, S. C. & Baltz, M. E. (1997). Population limitation in Neotropical migratory birds: comments on Rappole & McDonald (1994). *Auk* 114: 754–762.

Latta, S. C. & Faaborg, J. (2001). Winter site fidelity of Prairie Warblers in the Dominican Republic. *Condor* 103: 455–468.

Latta, S. C. & Faaborg, J. (2002). Demographic and population responses of Cape May Warblers wintering in multiple habitats. *Ecology* 83: 2502–2515.

Lavée, D., Safriel, U. N. & Meilijson, I. (1991). For how long do trans-Saharan migrants stop at an oasis? *Ornis Scand.* 22: 33–44.

Lawn, M. R. (1982). Pairing systems and site tenacity of the Willow Warbler *Phylloscopus trochilus* in southern England. *Ornis Scand.* 13: 193–199.

Le Houerou, H. N. & Gillet, H. (1986). Conservation versus desertisation in African arid lands. pp. 444–461 in *Conservation biology* (ed. M. Soule). Sunderland, Ma, Sinauer Associates.

Lee, S. J., Witter, M. S., Cuthill, I. C. & Goldsmith, A. R. (1996). Reduction in escape performance as a cost of reproduction in gravid Starlings, *Sturnus vulgaris. Proc. R. Soc. Lond. B* 263: 619–624.

Lehikoinen, E., Sparks, T. H. & Zalakevicius, M. (2004). Arrival and departure dates. pp. 1–28 in *Birds and climate change. Advances in Ecological Research*, Vol. 35. London: Elsevier.

Leisler, B. (1992). Habitat selection and coexistence of migrants and Afrotropical residents. *Ibis* 134, suppl. 1: 77–82.

Leisler, B. & Winkler, H. (2003). Morphological consequences of migration in passerines. pp. 175–186 in *Avian migration* (eds P. Berthold, E. Gwinner & E. Sonnenschein). Berlin, Springer-Verlag.

Leitner, S., Van't Hof, T. & Gahr, M. (2003). Flexible reproduction in wild canaries is independent of photoperiod. *Gen. Comp. Endocrinol.* 130: 102–108.

Lemon, R. E., Perrault, S. & Lozano, G. A. (1996). Breeding dispersion and site fidelity of American Redstarts (*Setophaga ruticilla*). *Can. J. Zool* 74: 2238–2247.

Lensink, R., van der Bijtel, H. J. V. & Schols, R. M. (1986). Invasion of Redpolls *Carduelis flammea* in the Netherlands in 1986. *Limosa* 62: 1–10.

Lepczyk, C. A., Murray, K. G., Winnett-Murray, K., Bartell, P., Geyer, E. & Work, T. (2000). Seasonal fruit preferences for lipids and sugars by American Robins. *Auk* 117: 709–717.

Leshem, Y. & Bahat, O. (1999). *Flying with the birds*. Israel, Chemed Books.

Leshem, Y. & Yom-Tov, Y. (1996a). The magnitude and timing of migration by soaring raptors, pelicans and storks over Israel. *Ibis* 138: 188–203.

Leshem, Y. & Yom-Tov, Y. (1996b). The use of thermals by soaring migrants in Israel. *Ibis* 138: 667–674.

Leshem, Y. & Yom-Tov, Y. (1998). Routes of migrating soaring birds. *Ibis* 149: 41–52.

Levey, D. J. & Stiles, F. G. (1992). Evolutionary precursors of long-distance migration: resource availability and movement patterns in neotropical landbirds. *Am. Nat.* 140: 447–476.

Lewis, R. A. & Farner, D. S. (1973). Temperature modulation of photoperiodically induced vernal phenomena in White-crowned Sparrows (*Zonotrichia leucophrys*). *Condor* 75: 279–286.

Lewis, R. A. & Orcutt, F. S. (1971). Social behaviour and avian sexual cycles. *Scientia (Rivista di Scienzia, Milano, Italy)*, 5/6: 1–24.

Lid, G. (1977). Fugler brennes ihjel av gassflaminer i Nordsjøen. *Fauna* 30: 185–190.

Liechti, F. (2006). Birds: blowin' by the wind? *J. Ornithol.* 147: 202–211.

Liechti, F. & Bruderer, B. (1998). The relevance of wind for optimal migration theory. *J. Avian Biol.* 29: 561–568.

Liechti, F. & Shaller, E. (1999). The use of low-level jets by birds. *Naturwissenschaften* 86: 549–551.

Liechti, F., Bruderer, B. & Paproth, H. (1995). Quantification of nocturnal bird migration by moonwatching: comparison with radar and infrared observations. *J. Field Ornithol.* 66: 457–468.

Liechti, F., Klaassen, M. & Bruderer, B. (2000). Predicting migratory flight altitudes by physiological migration models. *Auk* 117: 205–214.

Ligon, J. D. (1968). Starvation of spring migrants in the Chiricahua Mountains, Arizona. *Condor* 70: 387–388.

Ligon, J. D. (1971). Late summer-autumnal breeding of the Pinyon Jay in New Mexico. *Condor* 73: 147–153.

Limpert, R. J. (1980). Homing success of adult Buffleheads to a Maryland wintering site. *J. Wildl. Manage.* 44: 905–908.

Lin, W. H. & Severinghaus, L. L. (1998). Raptor migration and conservation in Taiwan. pp. 631–640 in *Holarctic birds of prey* (eds B. U. Meyburg, R. D. Chancellor & J. J. Ferrero). Berlin, WWGBP & ADENEX.

Lincoln, F. C. (1935a). The migration of North American birds. *U.S. Dept. Agric., Washington D.C. Circular No. 363.*: 1–72.

Lincoln, F. C. (1935b). The waterfowl flyways of North America. *U.S. Dept. Agric., Washington D.C. Circular No. 342.*: 1–12.

Lind, J., Fransson, T., Jacobsson, S. & Kullberg, C. (1999). Reduced take off ability in Robins due to migratory fuel load. *Behav. Ecol. Sociobiol.* 46: 65–70.

Lindström, A. (1989). Finch flock size and risk of hawk predation at a migratory stopover site. *Auk* 106: 225–232.

Lindström, A. (1990). The role of predation risk in stopover habitat selection in migrating Bramblings *Fringilla montifringilla*. *Behav. Ecol.* 1: 102–106.

Lindström, A. (1991). Maximum fat deposition rates in migrating birds. *Ornis Scand.* 22: 12–19.

Lindström, Å. (1987). Breeding nomadism and site tenacity in the Brambling *Fringilla montifringilla*. *Ornis Fenn.* 64: 50–56.

Lindström, Å. (2003). Fuel deposition rates in migrating birds: causes, constraints and consequences. pp. 307–320 in *Avian Migration* (eds P. Berthold, E. Gwinner & E. Sonnenschein). Berlin, Springer-Verlag.

Lindström, Å. & Alerstam, T. (1986). The adaptive significance of reoriented migration of Chaffinches *Fringilla coelebs* and Bramblings *F. montifringilla* during autumn in southern Sweden. *Behav. Ecol. Sociobiol.* 19: 417–424.

Lindström, Å. & Alerstam, T. (1992). Optimal fat loads in migrating birds: a test of time minimisation hypothesis. *Am. Nat.* 140: 477–491.

Lindström, Å. & Kvist, A. (1995). Maximum energy intake rate is proportional to basal metabolic rate in passerine birds. *Proc. Zool. Soc. Lond.* 261: 337–343.

Lindström, Å., Hasselquist, D., Bensch, S. & Grahn, M. (1990). Asymmetric contests over resources for survival and migration: a field experiment with Bluethroats. *Anim. Behav.* 40: 453–461.

Lindström, Å., Klaassen, M. & Kvist, A. (1999). Variation in energy intake and basal metabolic rate of a bird migrating in a wind tunnel. *Funct. Ecol.* 13: 352–359.

Lindström, Å., Daan, S. & Visser, G. H. (1994). The conflict between moult and migratory fat deposition: a photoperiodic experiment with Bluethroats. *Anim. Behav.* 48: 1173–1181.

Lindström, Å., Enemar, A., Andersson, G., von Proschwitz, T. & Nyholm, E. (2005). Density-independent reproductive output in relation to a drastically varying food supply: getting the density measure right. *Oikos* 110: 155–163.

Littlefield, C. D. (1995). Demographics of a declining flock of Greater Sandhill Cranes in Oregon. *Wilson Bull.* 107: 667–674.

Litwin, T. S. & Smith, C. R. (1992). Factors influencing the decline of neotropical migrants in a northeastern forest fragment: isolation, fragmentation, or mosaic effects. pp. 483–496 in *Ecology and conservation of neotropical migrant landbirds* (eds J. M. Hagan & D. W. Johnston). Washington, D.C., Smithsonian Institution Press.

Liu, J., Xiao, H., Lei, F., Zhu, Q., Qin, K., Zhang, X.-W., Zhang, X.-I., Zhao, D., Wang, G., Feng, Y., Ma, J., Liu, W., Wang, J. & Gao, G. F. (2005). Highly pathogenic H5N1 influenza virus infection in migratory birds. *Science* 309: 1206.

Liversidge, R. (1989). Factors influencing migration of 'wintering' raptors in southern Africa. pp. 151–158 in *Raptors in the modern world* (eds B.-U. Meyburg & R. D. Chancellor). London, WWGBP.

Lloyd, C. S. & Perrins, C. M. (1977). Survival and age at first breeding in the Razorbill (*Alca torda*). *Bird-Banding* 48: 239–252.

Locke, L. N. & Friend, M. (1987). Avian botulism. pp. 83–94 in *Field guide to wildlife diseases*, Vol. 1. *General field procedures and diseases of migratory birds* (ed. M. Friend). Washington, D.C., U.S. Dept. Interior, Fish & Wildlife Service Res. Publ. 167.

Lockie, J. D. (1955). The breeding habits and food of Short-eared Owls after a vole plague. *Bird Study* 2: 53–69.

Lockwood, R., Swaddle, J. P. & Rayner, J. M. V. (1998). Avian wingtip shape reconsidered: wingtip shape indices and morphological adaptations to migration. *J. Avian Biol.* 29: 273–292.

Löfgren, O., Hörnfeldt, B. & Carlsson, B.-G. (1986). Site tenacity and nomadism in Tengmalm's Owl (*Aegolius funereus* (L)) in relation to cyclic food production. *Oecologia* 69: 321–326.

Loftin, H. (1962). A study of boreal shorebirds summering on Apalachee Bay, Florida. *Bird-Banding* 33: 21–41.

Lofts, B. (1962). Cyclical changes in the interstitial and spermatogenic tissue of migratory waders 'wintering' in Africa. *Proc. Zool. Soc. Lond.* 138: 405–413.

Lofts, B. (1964). Evidence of an autonomous reproductive rhythm in an equatorial bird (*Quelea quelea*). *Nature* 201: 523–524.

Lofts, B. & Marshall, A. J. (1960). The experimental regulation of *Zugunruhe* and the sexual cycle in the Brambling *Fringilla montifringilla*. *Ibis* 102: 209–214.

Lofts, B. & Marshall, A. J. (1961). *Zugunruhe* activity in castrated Bramblings *Fringilla montifringilla*. *Ibis* 103a: 189–194.

Lofts, B. & Murton, R. K. (1968). Photoperiodic and physiological adaptations regulating avian breeding cycles and their ecological significance. *J. Zool. (Lond.)* 155: 327–394.

Lofts, B., Marshall, A. J. & Wolfson, A. (1963). The experimental demonstration of pre-migratory activity in the absence of fat deposition in birds. *Ibis* 105: 99–105.

Lofts, B., Murton, R. K. & Westwood, N. J. (1967). Interspecific differences in photosensitivity between three closely related species of pigeons. *J. Zool. Lond.* 151: 17–25.

Lõhmus, A. (1999). Vole-induced regular fluctuations in the Estonian owl populations. *Ann. Zool. Fenn.* 36: 167–178.

Lõhmus, A. (2001). Habitat selection in a recovering Osprey *Pandion haliaetus* population. *Ibis* 143: 651–657.

Lõhmus, M., Sandberg, R. Holberton, R. L. & Moore, F. R. (2003). Corticosterone levels in relation to migratory readiness in Red-eyed Vireos (*Vireo olivaceus*). *Behav. Ecol. Sociobiol.* 54: 233–239.

Löhrl, H. (1959). Zur Frage des Zeitpunkts einer Prägung auf die Heimatregion beim Halbandschnapper (*Ficedula albicollis*). *J. Ornithol.* 100: 132–140.

Long, J. A. & Holberton, R. L. (2004). Corticosterone secretion, energetic condition, and a test of the migration modulation hypothesis in the Hermit Thrush (*Catharus guttatus*), a short-distance migrant. *Auk* 121: 1094–1102.

Long, J. A. & Stouffer, P. C. (2003). Diet and preparation for spring migration in captive Hermit Thrushes (*Catharus guttatus*). *Auk* 129: 323–330.

Lopez, L. O. & Greenberg, R. (1990). Sexual selection by habitat in migratory warblers in Quintane Roo, Mexico. *Auk* 107: 539–543.

Lorenzana, J. C. & Sealy, S. G. (1999). A meta-analysis of the impact of parasitism by the Brown-headed Cowbird on its hosts. *Stud. Avian Biol.* 18: 241–253.

Loria, D. E. & Moore, F. R. (1990). Energy demands of migration on Red-eyed Vireos, *Vireo olivaceous*. *Behav. Ecol.* 1: 24–35.

Lowery, G. H. & Newman, R. J. (1966). A continent-wide view of bird migration on four nights in October. *Auk* 83: 547–586.

Loxton, R. G. & Sparks, T. H. (1999). Arrival of spring migrants at Portland, Stokholm, Bardsey and Calf of Man. *Bardsey Observatory Report* 42: 105–143.

Loxton, R. G., Sparks, T. H. & Newnham, J. A. (1998). Spring arrival dates of migrants in Sussex and Leicestershire (1966–1996). *The Sussex Bird Report* 50: 182–196.

Lozano, G. A., Perrault, S. & Lemon, R. E. (1996). Age, arrival date and reproductive success of male American Redstarts *Setophaga ruticilla*. *J. Avian Biol.* 27: 164–170.

Ludwigs, J.-D. & Becker, P. H. (2002). The hurdle of recruitment: influences of arrival date, colony experience and sex in the Common Tern *Sterna hirundo*. *Ardea* 90: 389–399.

Lundberg, A. (1979). Residency, migration and compromise: adaptors to nest-site scarcity and food specialisation in three Fennoscandian owl species. *Oecologia* 41: 273–281.

Lundberg, A. & Alatalo, R. (1992). *The Pied Flycatcher*. London, Poyser.

Lundberg, P. & Eriksson, L.-O. (1984). Post-juvenile moult in two northern Scandinavian Starling *Sturnus vulgaris* populations – evidence for difference in the circannual time program. *Orn. Scand.* 15: 105–109.

Lundberg, A., Alatalo, R. V., Carlson, A. & Ulfstrand, S. (1981). Biometry, habitat distribution and breeding success in the Pied Flycatcher *Ficedula hypoleuca*. *Ornis Scand.* 12: 68–79.

Lundberg, P. (1987). Partial bird migration and evolutionary stable strategies. *J. Theor. Biol.* 125: 351–360.

Lundberg, P. (1988). The evolution of partial migration in birds. *TREE* 3: 172–175.

Lundberg, S. & Alerstam, T. (1986). Bird migration patterns: conditions for stable geographical population segregation. *J. Theor. Biol.* 123: 403–414.

Lundgren, B. O. & Kiessling, K.-H. (1988). Comparative aspects of fibre types, areas, and capillary supply in the pectoralis muscle of some passerine birds with differing migratory behaviour. *J. Comp. Physiol. B Biochem. Syst. Environ. Physiol* 158: 165–173.

Lynch, J. F. (1987). Responses of breeding bird communities to forest fragmentation. *Nature conservation* (eds D. A. Saunders, G. W. Arnold, A. A. Burbidge & A. J. M. Hopkins). Chipping Norton, Australia, Surrey Heatly.

Lynch, J. F. & Whigham, D. F. (1984). Effects of forest fragmentation on breeding bird communities in Maryland, USA. *Biol. Conserv.* 28: 287–324.

Lynch, J. F., Morton, E. S. & van der Voort, M. E. (1985). Habitat segregation between the sexes of wintering Hooded Warblers *Wilsonia citrina*. *Auk* 102: 714–721.

Lyngs, P. (2003). Migration and winter ranges of birds in Greenland. An analysis of ring recoveries. *Dansk Orn. Foren Tidsskr.* 97: 1–168.

Lyons, J. E. & Haig, S. M. (1995). Fat content and stopover ecology in spring migrant Semipalmated Sandpipers in South Carolina. *Condor* 97: 427–437.

Lyuleeva, D. S. (1973). Features of Swallow biology during migration. pp. 219–272 in *Bird migrations. Ecological and physiological factors* (ed. B. E. Bykhovskii). New York, Halsted-Wiley.

Mabie, D. W., Meredino, M. T. & Reid, D. H. (1994). Dispersal of Bald Eagles fledged in Texas. *J. Raptor Res.* 28: 2113–219.

MacArthur, R. H. (1972). *Geographical ecology. Patterns in the distribution of species*. New York, Harper & Row.

Macdonald, M. A. (1977). Adult mortality and fidelity to mate and nest-site in a group of marked Fulmars. *Bird Study* 24: 165–168.

Macdonald, S. M. & Mason, C. F. (1973). Predation of migrant birds by gulls. *Br. Birds* 66: 361–363.

MacInnes, C. D. (1966). Population behaviour of eastern Arctic Canada Geese. *J. Wildl. Manage.* 30: 536–535.

MacMillan, J. P., Gauthreaux, S. A. & Helms, C. W. (1970). Spring migratory restlessness in caged birds: a circadian rhythm. *Bioscience* 20: 1259–1260.

Maddock, M. (2000). Herons in Australasia and Oceania. pp. 123–149 in *Heron conservation* (eds J. A. Kushlan & H. Hafner). London, Academic Press.

Maddock, M. & Geering, D. (1994). Range expansion and migration of the Cattle Egret. *Ostrich* 65: 191–203.

Madsen, J. (1985). Relations between spring habitat selection and daily energetics of Pink-footed Geese *Anser brachyrhynchus*. *Ornis Scand.* 16: 222–228.

Madsen, J. (1995). Impacts of disturbance on migratory waterfowl. *Ibis* 137 (suppl.): 67–74.

Madsen, J. (2001). Spring migration strategies in Pink-footed Geese *Anser brachyrhynchus* and consequences for spring fattening and fecundity. *Ardea* 89: 43–55.

Madsen, J., Cracknell, G. & Fox, T. (1999). *Goose populations of the western Palaearctic.* Rönde, National Environment Research Institute, Denmark.

Madsen, J., Frederiksen, M. & Ganter, B. (2002). Trends in annual and seasonal survival of Pink-footed Geese *Anser brachyrhynchus*. *Ibis* 144: 218–226.

Magnin, G. (1991). Hunting and persecution of migratory birds in the Mediterranean region. pp. 59–71 in *Conserving migratory birds* (ed. T. Salathé). Cambridge, International Council for Bird Preservation.

Maher, W. J. (1970). The Pomarine Jaeger as a Brown Lemming predator in northern Alaska. *Wilson Bull.* 82: 130–157.

Main, I. G. (2000). Obligate and facultative partial migration in the Blackbird (*Turdus merula*) and the Greenfinch (*Carduelis chloris*): uses and limitations of ringing data. *Vogelwarte* 40: 286–291.

Main, I. G. (2002). Seasonal movements of Fennoscandian Blackbirds *Turdus merula*. *Ringing & Migration* 21: 65–74.

Mainguy, J., Bêty, J., Gauthier, G. & Giroux, J. F. (2002). Are body condition and reproductive effort of laying Greater Snow Geese affected by the spring hunt? *Condor* 104: 156–161.

Mainwood, A. R. (1976). The movements of Storm Petrels as shown by ringing. *Ringing & Migration* 1: 98–104.

Maisonneuve, C. & Bédard, J. (1992). Chronology of autumn migration by Greater Snow Geese. *J. Wildl. Manage.* 56: 55–62.

Maitav, A. & Izhaki, I. (1994). Stopover and fat deposition by Blackcaps *Sylvia atricapilla* following spring migration over the Sahara. *Ostrich* 65: 160–166.

Malherbe, A. P. (1963). Notes on the birds of prey and some others at Boshoek north of Rustenburg during a rodent plague. *Ostrich* 34: 95–96.

Manolis, J. C., Andersen, D. E. & Cuthbert, F. J. (2002). Edge effect on nesting success of ground nesting birds near regenerating clearcuts in a forest-dominated landscape. *Auk* 119: 955–970.

Manville, R. H. (1957). Effects of unusual spring weather on Scarlet Tanagers. *Wilson Bull.* 69: 111–112.

Marchant, J. H. (1992). Recent trends in breeding populations of some common trans-Saharan migrant birds in northern Europe. *Ibis* 134: 113–119.

Marchant, J. H., Hudson, R., Carter, S. P. & Whittington, P. (1990). *Population trends in British breeding birds.* Tring, British Trust for Ornithology.

Marchant, S. & Higgins, P. J. (1990). *Handbook of Australian, New Zealand and Antarctic birds*, Vol. 1. Melbourne, Oxford University Press.

Marchette, K., Price, T. & Richman, A. (1995). Correlates of wing morphology with foraging behaviour and migration distance in the genus *Phylloscopus*. *J. Avian Biol.* 26: 177–181.

Marcström, V. & Mascher, J. W. (1979). Weights and fat in Lapwings *Vanellus vanellus* and Oystercatchers *Haematopus ostralegus* starved to death during a cold spell in spring. *Ornis Scand.* 10: 235–240.

Marion, L., Ulenaers, P. & van Vessen, J. (2000). Herons in Europe. pp. 1–32 in *Heron Conservation* (eds J. A. Kushlan & H. Hafner). London, Academic Press.

Marjakangas, A. & Kiviniemi, S. (2005). Dispersal and migration of female Black Grouse *Tetrao tetrix* in eastern central Finland. *Ornis Fenn.* 82: 107–116.

Markgren, G. & Lundberg, S. (1959). Irruptions of Pine Grosbeak (*Pinicola enucleator*) and Two-barred Crossbill (*Loxia leucoptera*) in Sweden 1956–57. *Vår Fagelvarld* 18: 185–205.

Markovets, M. Y. & Sokolov, L. V. (2002). Spring ambient temperature and movements of Coal Tits. *Avian Ecol. Behav.* 9: 55–62.

Marks, J. S. & Doremus, J. H. (2000). Are Northern Saw-Whet Owls nomadic? *J. Raptor Res.* 34: 299–304.

Marks, J. S. & Redmond, R. L. (1994). Migration of Bristle-thighed Curlews on Laysan Island: timing, behaviour and estimated flight range. *Condor* 96: 316–330.

Marks, J. S. & Redmond, R. L. (1996). Demography of Bristle-thighed Curlews *Numenius tahitiensis* wintering on Laysan Island. *Ibis* 138: 438–447.

Marks, J. S., Evans, D. L. & Holt, D. W. (1994). Long-eared Owl. *The birds of North America*, No. 133 (eds A. Poole, P. Stettenheim & F. Gill). Philadelphia: Academy of Natural Sciences and Washington, D.C., American Ornithologists' Union.

Marks, J. S., Redmond, R. L., Hendricks, P., Clapp, R. B. & Gill, R. E. (1990). Notes on longevity and flightlessness in Bristle-thighed Curlews. *Auk* 107: 779–781.

Marquiss, M. & Rae, R. (1994). Seasonal trends in abundance, diet and breeding of Common Crossbill (*Loxia curvirostra*) in an area of mixed species conifer plantation following the 1990 crossbill 'irruption'. *Forestry* 67: 31–47.

Marquiss, M. & Rae, R. (2002). Ecological differentiation in relation to bill size amongst sympatric, genetically undifferentiated crossbills (*Loxia* spp.). *Ibis* 144: 494–508.

Marra, P. P. & Holmes, R. T. (1997). Avian removal experiments: do they test for habitat saturation or female availability? *Ecology* 78: 947–952.

Marra, P. P. & Holmes, R. T. (2001). Consequence of dominance-related habitat segregation in a migrant passerine bird during the non-breeding season. *Auk* 118: 92–104.

Marra, P. P., Hobson, K. A. & Holmes, R. T. (1998). Linking winter and summer events in a migratory bird by using stable-carbon isotopes. *Science* 282: 1884–1886.

Marra, P. P., Sherry, T. W. & Holmes, R. T. (1993). Territorial exclusion by a long-distance migrant warbler in Jamaica: a removal experiment with American Redstarts (*Setophaga ruticilla*). *Auk* 110: 565–572.

Marra, P. P., Francis, C. M., Mulvihill, R. S. & Moore, F. R. (2005). The influence of climate on the timing and rate of spring bird migration. *Oecologia* 142: 307–315.

Marsh, R. L. (1983). Adaptations of the Gray Catbird *Dumetella carolinensis* to long-distance migration energy stores and substrate concentrations in plasma. *Auk* 100: 170–179.

Marsh, R. L. (1984). Adaptations of the Gray Catbird *Dumetella carolinensis* to long-distance migration: flight muscle hypertrophy associated with elevated body mass. *Physiol. Zool.* 57: 105–117.

Marshall, A. J. (1952). The condition of the interstitial and spermatogenetic tissue of migratory birds on arriving in England in April and May. *Proc. Zool. Soc. Lond.* 122: 287–295.

Marshall, A. J. (1960). Annual periodicity in the migration and reproduction of birds. *Cold Spring Harb. Symp. Quant. Biol.* 25: 499–505.

Marshall, H. D. & Baker, A. J. (1999). Colonisation history of Atlantic Island Chaffinches (*Fringilla coelebs*) revealed by mitochondrial DNA. *Mol. Phyl. Ecol.* 11: 201–212.

Martell, M. S., Henny, C. J., Nye, P. E. & Solensky, M. J. (2001). Fall migration routes, and wintering sites of North American Ospreys as determined by satellite telemetry. *Condor* 103: 715–724.

Martell, M. S., McMillian, M. A., Solensk, M. J. & Mealey, B. K. (2004). Partial migration and wintering use of Florida by Ospreys. *J. Raptor Res.* 38: 55–61.

Marti, C. D. (1999). Natal and breeding dispersal in Barn Owls. *J. Raptor Res.* 33: 181–189.

Martin, D. D. & Meier, A. H. (1973). Temporal synergism of corticosterone and prolactin in regulating orientation in the migratory White-throated Sparrows (*Zonotrichia albicollis*). *Condor* 75: 369–374.

Martin, G. R. (1990). *Birds by night*. London, Poyser.

Martin, T. E. & Finch, D. M., eds. (1995). *Ecology and management of neotropical migratory birds*. Oxford, Oxford University Press.

Martin, T. E. & Karr, J. R. (1986). Patch utilisation by migrating birds: resource oriented? *Ornis Scand.* 17: 165–174.

Martinez, M. M. (1983). Nidificacion de *Hirundo rustica erythrogaster* (Boddaert) en la Argentina (Aves, Hirundinidae). *Neotropica* 29: 83–86.

Martínez-Meyer, E., Peterson, A. T. & Navarro-Siguenza, A. G. (2004). Evolution of seasonal ecological niches in the *Passerina* buntings (Aves: Cardinalidae). *Proc. R. Soc. Lond. B.* 271: 1151–1157.

Massa, B., Benvenuti, S., Ioalé, P., Lovalvo, M. & Papi, F. (1991). Homing of Cory's Shearwaters (*Calonectris diomedea*) carrying magnets. *Boll. Zool.* 58: 245–247.

Mattes, H. & Jenni, L. (1984). Ortstreue und Zugbewegungen des Tannenhähers *Nucifraga caryocatactes* im Alpenraum und am Randecher Maar/Schwäbische Alb. *Orn. Beob.* 81: 303–315.

Matthews, G. V. T. (1968). *Bird Navigation*, 2nd edn. Cambridge, Cambridge University Press.

Matthysen, E. (1990). Behavioural and ecological correlates of territory quality in the Eurasian Nuthatch (*Sitta europaea*). *Auk* 107: 86–95.

Matthysen, E., Adriaensen, F. & Dhondt, A. A. (2001). Local recruitment of Great and Blue Tits (*Parus major, P. caeruleus*) in relation to study plot size and degree of isolation. *Ecography* 24: 33–42.

Mattocks, P. W. (1976). The role of gonadal hormones in the regulation of the premigratory fat deposition in the White-crowned Sparrow, *Zonotrichia leucophrys gambelii*. MS thesis, University of Washington, Seattle.

May, R. M. & Robinson, S. K. (1985). Population dynamics of avian brood parasitism. *Am. Nat.* 126: 475–494.

Mayaud, N. (1947). Les migrations de Casse-noix mouchetés â travers la France. *Alauda* 15: 34–48.

Mayfield, H. (1977). Brown-headed Cowbird: agent of extermination? *Am. Birds* 31: 107–113.

Mayfield, H. F. (1983). Kirtland's Warbler, victim of its own rarity? *Auk* 100: 974–976.

Mayr, E. (1969). On the origin of bird migration in the Pacific. *Proc. Pacific Sci. Congr.* 7: 387–394.

Mazerolle, D. F., Dufour, K. W., Hobson, K. A. & den Haan, H. E. (2005). Effects of large-scale climatic fluctuations on survival and production of young in a Neotropical migrant songbird, the Yellow Warbler *Dendroica petechia*. *J. Avian Biol.* 36: 155–163.

McCaskie, G. (1970). Occurrence of the eastern species of *Oporornis* and *Wilsonia* in California. *Condor* 72: 373–374.

McClelland, B. R., McClelland, P. T., Tates, R. E., Caton, E. L. & McFadzen, M. E. (1996). Fledging and migration of juvenile Bald Eagles from Glacier National Park, Montana. *J. Raptor Res.* 30: 79–89.

McClintock, C. P., Williams, T. C. & Teal, J. M. (1978). Autumnal bird migration observed from ships in the western North Atlantic Ocean. *Wilson Bull.* 49: 262–277.

McClure, H. E. (1974). *Migration and survival of the birds of Asia*. U.S. Army Medical Component, SEATO Medical Project, Bangkok, Thailand.

McCulloch, M. N., Tucker, G. M. & Baillie, S. R. (1992). The hunting of migratory birds in Europe: a ringing recovery analysis. *Ibis* 134 suppl.1: 55–65.

McGrady, M. J., Maechtle, T. L., Vargas, J. J., Seegar, W. S. & Pena, M. C. P. (2002). Migration and ranging of Peregrine Falcons wintering on the Gulf of Mexico coast, Tamaulipas, Mexico. *Condor* 104: 39–48.

McGrady, M. J., Ueta, M., Potapov, E. R., Utekhina, I., Masterov, V., Ladyguine, A., Zykov, V., Cibor, J., Fuller, M. & Seegar, W. S. (2003). Movements by juvenile and immature Steller's Sea Eagles *Haliaeetus pelagicus* tracked by satellite. *Ibis* 145: 318–328.

McGrady, M. J., Young, G. S. & Seegar, W. S. (2006). Migration of a Peregrine Falcon *Falco peregrinus* over water in the vicinity of a hurricane. *Ringing & Migration* 23: 80–84.

McIlhenny, E. A. (1934). Twenty-two years of banding migratory waterfowl at Avery Island, Louisiana. *Auk* 51: 328–337.

McIlhenny, E. A. (1940). An early experiment in the homing ability of wildfowl *Bird Banding* 11: 58.

McLachlan, G. R. & Liversidge, R. (1978). *Roberts birds of South Africa*. Cape Town, The John Voelcker Bird Book Fund.

McLandress, M. R. & Raveling, D. G. (1981). Changes in diet and body composition of Canada Geese before spring migration. *Auk* 98: 65–79.

McLandress, M. R. & Raveling, R. G. (1983). Hyperphagia and social behaviour of Giant Canada Geese prior to spring migration. *Wilson Bull.* 93: 310–324.

McLaren, I. A. (1981). The incidence of vagrant landbirds on Nova Scotian Islands. *Auk* 98: 243–257.

McMillan, J. P. S., Gauthreaux, S. A. & Helms, C. W. (1970). Spring migratory restlessness in caged birds: a circadian rhythm. *Bioscience* 20: 1259–1260.

McNeil, R. & Cadieux, F. (1972). Fat content and flight range capabilities of some adult spring and fall migrant North American shorebirds in relation to migration routes on the Atlantic coast. *Can. Field Nat.* 99: 589–605.

McNeil, R., Diaz, M. T. & Villeneuve, A. (1994). The mystery of shorebird oversummering: a new hypothesis. *Ardea* 82: 143–152.

McWilliams, S. R. & Karasov, W. H. (2005). Migration takes guts. pp. 67–78 in *Birds of Two Worlds*. (eds Greenberg, R. & Marra, P. P.). Washington, D. C., Smithsonian Institution.

McWilliams, S. R., Guglielmo, C., Pierce, B. & Klaassen, M. (2004). Flying, fasting, and feeding in birds during migration: a nutritional and physiological ecology perspective. *J. Avian Biol.* 35: 377–393.

Mead, C. J. (1970). The winter quarters of British Swallows. *Bird Study* 17: 229–240.

Mead, C. J. (1979a). Mortality and causes of death in British Sand Martins. *Bird Study* 26: 107–112.

Mead, C. J. (1979b). Colony fidelity and interchange in the Sand Martin. *Bird Study* 26: 99–106.

Mead, C. J. (1983). *Bird migration*. Feltham, Middlesex, Country Life Books.

Mead, C. J. (1991). The missing migrants. *BTO News* 176: 1.

Mead, C. J. & Clark, J. A. (1988). Report on bird ringing in Britain and Ireland for 1987. *Ringing & Migration* 9: 169–204.

Mead, C. J. & Watmough, B. R. (1976). Suspended moult of trans-Saharan migrants in Iberia. *Bird Study* 23: 187–196.

Mearns, R. & Newton, I. (1988). Factors affecting breeding success of Peregrines in south Scotland. *J. Anim. Ecol.* 57: 903–916.

Medway Lord & Wells, D. R. (1976). *Birds of the Malay Peninsula*, Vol. 5. London, Witherby.

Mehl, K., Alisauskas, R. T., Hobson, K. A. & Kellett, D. K. (2004). To winter east or west? Heterogeneity in winter philopatry in a central arctic population of King Eiders. *Condor* 106: 241–251.

Mehlum, F. (1983). Resting time in migrating Robins *Erithacus rubecula* at Støre Faerder, Outer Oslofiord, Norway. *Fauna Norv. Ser C, Cinclus* 6: 62–72.

Meltofte, H. (1975). Ornithological observations in northeast Greenland between 76° 00' and 78° 00' N. Lat., 1969–71. *Meddr. Grønland* 191, No. 9.

Meltofte, H. (1985). Population and breeding schedules of waders Charadrii, in high arctic Greenland. *Meddelelser Om. Grønland, Biosci.* 16: 1–43.

Mendelsohn, J. M. (1983). Social behaviour and dispersion of the Black-shouldered Kite. *Ostrich* 54: 1–18.

Mendelssohn, H. & Paz, U. (1977). Mass mortality of birds of prey caused by azodrin, an organophosphorus insecticide. *Biol. Conserv.* 11: 163–70.

Menu, S., Gauthier, G. & Reed, A. (2005). Survival of young Greater Snow Geese (*Chen caerulescens atlantica*) during fall migration. *Auk* 122: 479–496.

Merilä, J., Björkland, M. & Baker, A. J. (1997). Historical demography and present day population structure of the Greenfinch *Carduelis chloris* – an analysis of mtDNA control-region sequences. *Evolution* 51: 946–956.

Merkel, F. W. (1956). Untersuchungen über tages- und jahresperiodische Aktivitätsanderungen bei gehäfigter Zugvögeln. *Z. Tierpsychol.* 13: 278–301.

Merkel, F. W. (1963). Long-term effects of constant photoperiods on European Robins and Whitethroats. *Proc. Int. Ornithol. Congr.* 13: 950–959.

Merkel, F. W. (1966). The sequence of events leading to migratory restlessness. *Ostrich* (suppl. 6): 239–248.

Merom, K., Yom-Tov, Y. & McCleery, R. (2000). Philopatry to stopover site and body condition of transient Reed Warblers during autumn migration through Israel. *Condor* 102: 441–444.

Metcalfe, N. B. & Furness, R. W. (1984). Changing priorities: the effect of pre-migratory fattening on the trade-off between foraging and vigilance. *Behav. Ecol Sociobiol.* 15: 203–206.

Metcalfe, N. B. & Furness, R. W. (1985). Survival, winter population stability and site-fidelity in the Turnstone *Arenaria interpres* . *Bird Study* 32: 207–214.

Metcalfe, N. B. & Ure, S. E. (1995). Diurnal variation in flight performance and hence potential predation risk in small birds. *Proc. R. Soc. Lond. B* 261: 395–400.

Meury, R. (1989). Brutbiologie und Ortstreue einer Baumpieper population *Anthus trivialis* in einem inselartig verteilten Habitat des schweizerischen Mittellandes. *Orn. Beob.* 86: 219–233.

Mewaldt, L. R. & King, J. R. (1978). Latitudinal variation in prenuptial moult in wintering Gambel's White-crowned Sparrows. *North Am. Bird Bander* 3: 138–144.

Mewaldt, R. (1964). California sparrows return from displacement to Maryland. *Science* 146: 941–942.

Meyburg, B.-U. & Meyburg, C. (1998). Satellite tracking of Eurasian raptors. *Torgos* 28: 33–48.

Meyburg, B.-U. & Meyburg, C. (1999). The study of raptor migration in the Old World using satellite telemetry. *Proc. Int. Ornithol. Congr.* 22: 2992–3006.

Meyburg, B.-U., Matthes, J. & Meyburg, C. (2002). Satellite-tracked Lesser Spotted Eagle avoids crossing water at the Gulf of Suez. *Br. Birds* 95: 372–376.

Meyburg, B.-U., Meyburg, C. & Barbraud, J.-C. (1998). Migration strategies of an adult Short-toed Eagle *Circaetus gallicus* tracked by satellite. *Alauda* 66: 39–48.

Meyburg, B.-U., Paillat, P. & Meyburg, C. (2003). Migration routes of Steppe Eagles between Asia and Africa: a study by means of satellite telemetry. *Condor* 105: 219–227.

Meyburg, B.-U., Scheller, W. & Meyburg, C. (1995c). Migration and wintering of the Lesser Spotted Eagle (*Aquila pomarina*) – a study by means of satellite telemetry. *J. Ornithol.* 136: 401–422.

Meyburg, B.-U., Scheller, W. & Meyburg, C. (2000). Migration and wintering of the Lesser Spotted Eagle *Aquila pomarina*: a study by means of satellite telemetry. *Global. Environ. Res.* 4: 401–422.

Meyburg, B.-U., Eichaker, X., Meyburg, C. & Paillat, P. (1995a). Migrations of an adult Spotted Eagle tracked by satellite. *Br. Birds* 88: 357–361.

Meyburg, B.-U., Gallardo, M., Meyburg, C. & Dimitrova, E. (2004a). Migrations and sojourn in Africa of Egyptian Vultures (*Neophron percnopterus*). *J. Ornithol.* 145: 273–280.

Meyburg, B.-U., Ellis, D. H., Meyburg, C., Mendelsohn, J. M. & Scheller, W. (2001). Satellite tracking of two Lesser Spotted Eagles, *Aquila pomarina*, migrating from Namibia. *Ostrich* 72: 35–40.

Meyburg, B.-U., Meyburg, C., Belka, T., Šreibr, O. & Vrana, J. (2004b). Migration, wintering and breeding of a Lesser Spotted Eagle (*Aquila pomarina*) from Slovakia tracked by satellite. *J. Ornithol.* 145: 1–7.

Meyburg, B.-U., Mendelsohn, J. M., Ellis, D. H., Smith, D. G., Meyburg, C. & Kemp, A. (1995b). Year-round movements of a Wahlberg's Eagle *Aquila wahlbergi* tracked by satellite. *Ostrich* 66: 135–140.

Michener, H. & Michener, J. R. (1935). Mockingbirds, their territories and individualities. *Condor* 37: 91–140.

Michener, M. C. & Walcott, C. (1966) Navigation of single homing pigeons: airplane observations by radio-tracking. *Science* 154: 410–413.

Mihelsons, H. A., Mednis, A. A. & Blums, P. B. (1985). Regulatory mechanisms of numbers in breeding populations of migratory ducks. *Proc. Int. Ornithol. Congr. Moscow* 18: 797–802.

Mihelsons, H., Mednis, A. & Blums, P. (1986). *Population ecology of migratory ducks in Latvia.* Riga, Zinatne. In Russian.

Mikkonen, A. (1981). Factors influencing spring arrival of the Brambling *Fringilla montifringilla* in northern Finland. *Ornis Fenn.* 58: 78–82.

Mikkonen, A. V. (1983). Breeding site tenacity of the Chaffinch *Fringilla coelebs* and the Brambling *F. montifringilla* in northern Finland. *Ornis Scand.* 14: 36–47.

Milá, B., Girman, D. J., Kimura, M. & Smith, T. B. (2000). Genetic evidence for the effect of a postglacial population expansion on the phylogeography of a North American songbird. *Proc. R. Soc. Lond. B* 267: 1033–1040.

Miller, A. H. (1954). Breeding cycles in a constant equatorial environment in Columbia, South America. *Proc Int. Ornithol. Congr.* 11: 495–503.

Miller, A. H. (1959). Reproductive cycles in an equatorial sparrow. *Proc. Natl Acad. Sci. USA* 45: 1075–1100.

Mills, A. M. (2005). Changes in the timing of spring and autumn migration in North American migrant passerines during a period of global warming. *Ibis* 147: 259–269.

Mills, J. A. (1973). The influence of age and pair-bond on the breeding biology of the Red-billed Gull *Larus novaehollandiae scopulinus*. *J. Anim. Ecol.* 42: 147–163.

Milne, H. & Robertson, F. W. (1965). Polymorphism in egg albumen protein and behaviour in the Eider Duck. *Nature* 205: 367–369.

Milot, E., Gibbs, H. L. & Hobson, K. A. (2000). Phylogeography and genetic structure of northern populations of the Yellow Warbler (*Dendroica petechia*). *Mol. Ecol.* 9: 667–681.

Milwright, D. (1994). Fieldfare *Turdus pilaris* ringing recoveries during autumn, winter and spring analysed in relation to river basins and watersheds in Europe and the Near East. *Ringing & Migration* 15: 129–189.

Milwright, R. D. P. (2002). Redwing *Turdus iliacus* migration and wintering areas shown by recoveries of birds ringed in the breeding season in Fennoscandia, Poland, the Baltic Republics, Russia, Siberia and Iceland. *Ringing & Migration* 21: 5–15.

Minami, H., Minagawa, M. & Ogi, H. (1995). Changes in stable carbon and nitrogen iso-tope ratios in Sooty and Short-tailed Shearwaters during their northward migration. *Condor* 97: 565–574.

Mindell, D. P., Albuquerque, J. L. B. & White, C. M. (1987). Breeding population fluctua-tions in some raptors. *Oecologia* 72: 382–388.

Minton, C. (2003). The importance of long-term monitoring of reproduction rates in wad-ers. *Wader Study Group Bull.* 199: 178–182.

Møller, A. P. (1978). Sex ratios among migrating and wintering Bullfinches *Pyrrhula pyrrhula* in Northern Jutland. *Dansk Orn. Foren. Tidsskr.* 72: 61–63.

Møller, A. P. (1979). Irruptions of Parrot Crossbills *Loxia pytyopsittacus* in Denmark 1960–1976. *Dansk Orn. Foren Tidsskr.* 73: 303–309.

Møller, A. P. (1989). Population dynamics of a declining Swallow *Hirundo rustica* popula-tion. *J. Anim. Ecol.* 58: 1051–1063.

Møller, A. P. (1994). *Sexual selection and the Barn Swallow.* Oxford, University Press.

Møller, A. P. (2001). Heritability of arrival date in a migratory bird. *Proc. R. Soc. Lond.* 268: 203–206.

Møller, A. P. (2004). Protandry, sexual selection and climate change. *Global Change Biol.* 10: 2028–2035.

Møller, A. P. & Erritzoe, J. (1998). Host immune defence and migration in birds. *Evol. Ecol.* 12: 945–953.

Møller, A. P. & Hobson, K. A. (2004). Heterogeneity in stable isotope profiles predicts coex-istence of populations of Barn Swallows *Hirundo rustica* differing in morphology and reproductive performance. *Proc. R. Soc. Lond. B* 271: 1355–1362.

Monaghan, P. (1980). Dominance and dispersal between feeding sites in the Herring Gull (*Larus argentatus*). *Anim. Behav.* 28: 521–527.

Mönkkönen, M. (1992). Life history traits of Palearctic and Nearctic migrant passerines. *Ornis Fenn.* 69: 161–172.

Mönkkönen, M. (1995). Do migratory birds have more pointed wings? A comparative study. *Evol. Ecol.* 9: 520–528.

Mönkkönen, M., Helle, P. & Welsh, D. (1992). Perspectives on Palaearctic and Nearctic bird migration; comparisons and overview of life-history and ecology of migrant passer-ines. *Ibis* 134: 7–13.

Montalvo, S. & Potti, J. (1992). Breeding dispersal in Spanish Pied Flycatchers *Ficedula hypoleuca*. *Ornis Scand.* 23: 491–498.

Moore, F. R. (1976). The dynamics of seasonal distribution of Great Lakes Herring Gulls. *Bird-Banding* 47: 141–159.

Moore, F. R. (1982). Sunset and the orientation of a nocturnal bird migrant – a mirror experiment. *Behav. Ecol Sociobiol.* 10: 153–155.

Moore, F. R. (1984). Age-dependent variability in the migratory orientation of the Savannah Sparrow (*Passerculus sandwichensis*). *Auk* 101: 875–880.

Moore, F. R. (1985). Integration of environmental stimuli in the migratory orientation of the Savannah Sparrow (*Passerculus sandwichensis*). *Anim. Behav.* 33: 657–663.

Moore, F. R. (1986). Sunrise, skylight polarization, and the early morning orientation of night-migrating warblers. *Condor* 88: 493–498.

Moore, F. R. (1994). Resumption of feeding under risk of predation – effect of migratory condition. *Anim. Behav.* 48: 975–977.

Moore, F. R. & Aborn, D. (2000). Mechanisms of en route habitat selection: how do migrants make habitat decisions during stopovers? *Stud. Avian Biol.* 20: 820–826.

Moore, F. & Kerlinger, P. (1987). Stopover and fat deposition by North American wood-warblers (Parulinae) following spring migration over the Gulf of Mexico. *Oecologia* 74: 47–54.

Moore, F. R. & McDonald, M. V. (1993). On the possibility that inter-continental landbird migrants copulate en route. *Auk* 110: 157–160.

Moore, F. R. & Phillips, J. B. (1988). Sunset, skylight polarization and the migratory orientation of Yellow-rumped Warblers, *Dendroica coronata. Anim. Behav.* 36: 1770–1778.

Moore, F. R. & Simons, T. R. (1992). Habitat suitability and stopover ecology of neotropical landbird migrants. pp. 345–355 in *Ecology and conservation of neotropical migrant landbirds* (eds J. M. Hagen & D. W. Johnson). Washington, D.C., Smithsonian Institution Press.

Moore, F. R. & Yong, W. (1991). Evidence of food-based competition among passerine migrants during stopover. *Behav. Ecol. Sociobiol.* 28: 85–90.

Moore, F. R., Kerlinger, P. & Simons, T. R. (1990). Stopover on a Gulf coast barrier island by spring trans-Gulf migrants. *Wilson Bull.* 102: 487–500.

Moore, F. R., Smith, R. J. & Sandberg, R. (2005). Stopover ecology of intercontinental migrants. pp. 251–261 in *Birds of Two Worlds.* (eds Greenberg, R. & Marra, P. P.). Washington, D. C., Smithsonian Institution.

Moore, F. R., Gauthreaux, S. A., Kerlinger, P. & Simons, T. R. (1995). Habitat requirements during migration: important link in conservation. pp. 121–144 in *Ecology and management of neotropical birds* (eds T. E. Martin & D. M. Finch). Oxford, Oxford University Press.

Moore, F., Mabey, S. & Woodrey, M. (2003). Priority access to food in migratory birds: age, sex and motivational asymmetries. pp. 281–292 in *Avian migration* (eds P. Berthold, E. Gwinner & E. Schonnenschein). Berlin, Springer-Verlag.

Moore, M. C., Donham, R. S. & Farner, D. S. (1982). Physiological preparation for autumnal migration in White-crowned Sparrows. *Condor* 84: 410–419.

Morales, M. B., Alonso, J. C., Alonso, J. A. & Martin, E. (2000). Migration patterns in male Great Bustards (*Otis tarda*). *Auk* 117: 493–498.

Morbey, Y. E. & Ydenberg, R. C. (2001). Protandrous arrival timing to breeding areas: a review. *Ecol. Lett.* 4: 663–673.

Moreau, R. E. (1927). Quail. *Bull. Zool. Soc. Egypt* 1: 6–13.

Moreau, R. E. (1928). Some further notes from the Egyptian deserts. *Ibis* 12, Ser. 4: 453–475.

Moreau, R. E. (1951). The British status of the Quail and some problems of its biology. *Br. Birds* 44: 257–275.

Moreau, R. E. (1961). Problems of Mediterranean-Sahara migration. *Ibis* 103: 373–421, 580–623.

Moreau, R. E. (1969). The recurrence in winter quarters (*Ortstreue*) of trans-Saharan migrants. *Bird Study* 16: 108–110.

Moreau, R. E. (1972). *The Palaearctic-African bird migration system.* London, Academic Press.

Moreau, R. E. & Dolp, R. M. (1970). Fat, water, weights and wing-lengths of autumn migrants in transit on the northwest coast of Egypt. *Ibis* 112: 209–228.

Morel, G. (1973). The Sahel zone as an environment for Palearctic migration. *Ibis* 115: 413–417.

Morel, G. J. & Morel, M.-Y. (1992). Habitat use by Palaearctic migrant passerine birds in West Africa. *Ibis* 134, suppl. 1: 83–88.

Morgan, J. H. & Shirihai, H. (1997). Passerines and passerine migration in Eilat 1984–1993. *Int. Birdw. Cent. Eilat Tech. Publ.* 6: 1–50.

Moritz, D. (1993). Long-term monitoring of Palaearctic-African migrants at Helgoland/ German Bight, North Sea. *Proc. Pan-African Ornithol. Congr.* 8: 579–586.

Morris, R. F., Cheshire, W. F., Miller, C. A. & Mott, D. G. (1958). The numerical response of avian and mammalian predators during a gradation of the spruce budworm. *Ecology* 39: 487–494.

Morris, S., Richmond, M. E. & Holmes, D. W. (1994). Patterns of stopover by warblers during spring and fall migration on Appledore Island, Maine. *Wilson Bull.* 106: 703–718.

Morris, S. R., Holmes, D. W. & Richmond, M. E. (1996). A 10-year study of the stopover patterns of migratory passerines during fall migration on Appledore Island, Maine. *Condor* 98: 395–409.

Morrison, R. I. G. (1975). Migration and morphometrics of European Knot and Turnstone on Ellesmere Island, Canada. *Bird-Banding* 46: 290–301.

Morrison, R. I. G. (2006). Body transformations, condition, and survival in Red Knots *Calidris canutus* travelling to breed at Alert, Ellesmere Island, Canada. *Ardea* 94: 607–618.

Morrison, R. I. G. & Hobson, K. A. (2004). Use of body stores in shorebirds after arrival on High-Arctic breeding grounds. *Auk* 121: 333–344.

Morse, A. H. (1989). American Warblers. Cambridge, MA, Harvard University Press.

Morse, D. H. (1980). Population limitation: breeding or wintering grounds? pp. 505–516 in *Migrant birds in the Neotropics* (eds A. Keast & E. Morton). Washington, D.C., Smithsonian Institution Press.

Morton, E. S. (1980). Adaptations to seasonal changes by migrant land-birds in the Panama Canal Zone. pp. 437–453 in *Migrant birds in the Neotropics* (eds A. Keast & E. S. Morton). Washington, D.C., Smithsonian Institution Press.

Morton, E. S. (1990). Habitat segregation by sex in the Hooded Warbler – experiments on proximate causation and discussion of its evolution. *Am. Nat.* 135: 319–333.

Morton, E. S., Lynch, J. F., Young, K. & Mehlhop, P. (1987). Do male Hooded Warblers exclude females from non-breeding territories in tropical forests? *Auk* 104: 133–135.

Morton, G. A. & Morton, M. L. (1990). Dynamics of postnuptial molt in free-living Mountain White-crowned Sparrows. *Condor* 92: 813–828.

Morton, M. L. (1967). Diurnal feeding patterns in White-crowned Sparrows *Zonotrichia leucophrys gambelii*. *Condor* 69: 491–512.

Morton, M. L. (1984). Sex and age ratios of wintering White-crowned Sparrows. *Condor* 86: 85–87.

Morton, M. L. (1997). Natal and breeding dispersal in the Mountain White-crowned Sparrow *Zonotrichia leucophrys oriantha*. *Ardea* 85: 145–154.

Morton, M. L. (2002). The Mountain White-crowned Sparrow: migration and reproduction at high altitude. *Stud. Avian Biol.* 24: 1–236.

Morton, M. L. & Mewaldt, L. R. (1962). Some effects of castration on a migratory sparrow (*Zonotrichia atricapilla*). *Physiol. Zool.* 35: 237–247.

Morton, R. (1986). Rapid territory establishment by a Wood Warbler. *Ringing & Migration* 7: 56.

Mosbech, A., Gilchrist, G., Merkel, F., Sonne, C., Flagstad, A. & Nyegaard, H. (2006). Year-round migration of Northern Common Eiders breeding in Arctic Canada and west Greenland based on satellite telemetry. *Ardea* 94: 651–655.

Moser, M. E. (1988). Limits to the numbers of Grey Plover *Pluvialis squatarola* wintering on British estuaries: an analyses of long-term population trends. *J. Appl. Ecol.* 25: 473–486.

Moss, R. (1972). Social organisation of Willow Ptarmigan on the breeding grounds in interior Alaska. *Condor* 74: 144–151.

Mouritsen, H. (2001). Ringing recoveries contain hidden information about orientation mechanisms. *Ardea* 89: 31–42.

Mouritsen, H. & Larsen, O. N. (1998). Migrating young Pied Flycatchers *Ficedula hypoleuca* do not compensate for geographical displacements. *J. Exp. Biol.* 201: 2927–2934.

Mouritsen, H. & Mouritsen, O. (2000). A mathematical expectation model for bird navigation based on the clock-and-compass strategy. *J. Theor. Biol.* 207: 283–291.

Mueller, H. C. & Berger, D. D. (1967). Some observations and comments on the periodic invasions of Goshawks. *Auk* 84: 183–191.

Mueller, H. C., Berger, D. D. & Allez, G. (1977). The periodic invasions of Goshawks. *Auk* 94: 652–63.

Mueller, H. C., Berger, D. D. & Mueller, N. S. (2003). Age and sex differences in the timing of spring migration of hawks and falcons. *Wilson Bull.* 115: 321–324.

Mueller, H. C., Mueller, N. S., Berger, D. D., Allez, G., Robichaud, W. & Kaspar, J. L. (2000). Age and sex differences in the timing of fall migration of hawks and falcons. *Wilson Bull.* 112: 214–224.

Muheim, R., Åkesson, S. & Alerstam, T. (2003). Compass orientation and possible migration routes of passerine birds at high arctic latitudes. *Oikos* 103: 341–349.

Muheim, R., Phillips, J. B. & Åkesson, S. (2006). Polarized light cues underlie compass calibration in migratory songbirds. *Science* 313: 837–839.

Mukhin, A. (2002). Post-juvenile moult and change in locomotry activity of caged Reed Warblers *Acrocephalus scirpaceus* from the early and the late broods prior to migration. *Avian Ecol. Behav.* 8: 71–78.

Mukhin, A. (2004). Night movements of young Reed Warblers (*Acrocephalus scirpaceus*) in summer: is it postfledging dispersal? *Auk* 121: 203–209.

Mukhin, A. & Bulyuk, V. (2001). Juvenile dispersal of the Reed Warbler *Acrocephalus scirpaceus*: new interesting facts. *Avian Ecol. Behav.* 6: 42–43.

Mullié, W. C. & Keith, J. O. (1993). The effects of aerially applied fenithrothion and chlorpyrifos on birds in the savannah of northern Senegal. *J. Appl. Ecol.* 30: 536–550.

Munro, U. (2003). Life history and ecophysiological adaptations to migration in Australian birds. pp. 141–154 in *Avian migration* (eds P. Berthold, E. Gwinner & E. Sonnenschein). Berlin, Springer-Verlag.

Munro, U. & Wiltschko, W. (1993). Magnetic compass orientation in the Yellow-faced Honeyeater, *Lichenostomus chrysops*, a day migrating bird from Australia. *Behav. Ecol. Sociobiol.* 32: 141–145.

Murphy, M. T. (1996). Survivorship, breeding dispersal and mate fidelity in Eastern Kingbirds. *Condor* 98: 82–92.

Murphy, R. C. (1936). *Oceanic birds of South America*. New York, American Museum of Natural History.

Murphy-Klaassen, H. M., Underwood, T. J., Sealy, S. G. & Czyrnyj, A. A. (2005). Long-term trends in spring arrival dates of migrant birds at Delta Marsh, Manitoba, in relation to climate change. *Auk* 122: 1130–1148.

Murray, B. G. (1966). Migration of age and sex classes of passerines on the Atlantic coast in autumn. *Auk* 83: 352–360.

Murton, R. K. & Westwood, N. J. (1977). *Avian breeding cycles*. Oxford, Clarendon Press.

Myers, J. P. (1981). A test of three hypotheses for latitudinal segregation of the sexes in wintering birds. *Can. J. Zool.* 59: 1527–1534.

Myers, J. P., Schick, C. T. & Castro, G. (1988). Structure in Sanderling (*Calidris alba*) populations: the magnitude of intra- and inter-year dispersal during the nonbreeding season. *Proc. Int. Ornithol. Congr.* 19: 604–615.

Myers, N. (1980). *Conversion of tropical moist forests*. Washington, D.C., National Academy of Sciences.

Myres, M. T. (1964). Dawn ascent and reorientation of Scandinavian thrushes (*Turdus* spp.) migrating at night over the northeastern Atlantic Ocean in autumn. *Ibis* 106: 7–51.

Nachtigall, W. (1990). Wind tunnel measurements of long-time flights in relation to the energetics and water economy of migrating birds. pp. 319–327 in *Bird migration. Physiology and ecophysiology* (ed. E. Gwinner). Berlin, Springer-Verlag.

Nachtigall, W. (1993). Biophysics of bird flight – aspects of energy-expenditure during sustained flight. *Z. Angew. Math. Mech.* 73: 191–202.

Nager, R. G., Johnson, A. R., Boy, V., Rendon-Martos, M., Calderon, J. & Cézilly, F. (1996). Temporal and spatial variation in dispersal in the Greater Flamingo (*Phoenicopterus ruber roseus*). *Oecologia* 107: 204–211.

Nakazawa, Y., Peterson, A. T., Martinez-Meyer, E. & Navarro-Sigüenza, A. G. (2004). Seasonal niches of Nearctic-Neotropical migratory birds: implications for the evolution of migration. *Auk* 121: 610–618.

Navarro, R. A. (1992). Body composition, fat reserves, and fasting capability of Cape Gannet chicks. *Wilson Bull.* 104: 644–655.

Nebel, S., Lank, D. B., O'Hara, P. D., Fernandez, G., Haare, B., Delgado, F. S., Estela, F. A., Ogden, L. J. E., Harrington, B., Kus, B. E., Lyons, J. E., Mercier, F., Ortego, B., Takewawa, J. Y., Warnock, N. & Warnock, S. E. (2002). Western Sandpipers (*Calidris mauri*) during the nonbreeding season: spatial segregation on a hemispherical scale. *Auk* 119: 922–928.

Negro, J. J., Hiraldo, F. & Donázar, J. A. (1997). Causes of natal dispersal in the Lesser Kestrel: inbreeding avoidance or resource competition? *J. Anim. Ecol.* 66: 640–648.

Nelson, B. (1973). *Azraq: desert oasis*. London, Allen Lane.

Nelson, J. B. (1978). *The Gannet*. Berkhamsted, Herts, Poyser.

Nero, R. W., Copland, H. W. R. & Mezibroski, J. (1984). The Great Grey Owl in Manitoba, 1968–83. *Blue Jay* 42: 129–190.

Nero, R. W. & Copland, H. W. R. (1997). Sex and age composition of Great Grey Owls (*Strix nebulosa*) winter 1995/1996. pp. 587–590 in *Biology and conservation of owls of the northern hemisphere* (ed. J. R. Duncan, D. H. Johnson & T. H. Nicholls). Second International Symposium, February 5–9, 1997. Winnipeg, Manitoba, Canada. United States Department of Agriculture.

Newton, I. (1966). The moult of the Bullfinch *Pyrrhula pyrrhula*. *Ibis* 108: 41–67.

Newton, I. (1967). Feather growth and moult in some captive finches. *Bird Study* 14: 10–24.

Newton, I. (1970). Irruptions of Crossbills in Europe. pp. 337–357 in *Animal populations in relation to their food resources* (ed. A. Watson). Oxford, Blackwell Scientific Publications.

Newton, I. (1972). *Finches*. London, Collins.

Newton, I. (1975). Movements and mortality of British Sparrowhawks. *Bird Study* 22: 35–43.

Newton, I. (1977). Timing and success of breeding in tundra-nesting geese. pp. 113–126 in *Evolutionary ecology* (eds B. Stonehouse & C. M. Perrins). London, Macmillan.

Newton, I. (1979). *Population ecology of raptors*. Berkhamsted, Poyser.

Newton, I. (1986). *The Sparrowhawk*. Calton, T. & A.D. Poyser.

Newton, I. (1993). Age and site fidelity in female Sparrowhawks *Accipiter nisus*. *Anim. Behav.* 46: 161–168.

Newton, I. (1994). The role of nest-sites in limiting the numbers of hole-nesting birds: a review. *Biol. Conserv.* 70: 265–276.

Newton, I. (1995a). Relationship between breeding and wintering ranges in Palaearctic-African migrants. *Ibis* 137: 241–249.

Newton, I. (1995b). The contribution of some recent research on birds to ecological understanding. *J. Anim. Ecol.* 64: 675–696.

Newton, I. (1998a). Migration patterns in West Palaearctic Raptors. pp. 603–612 in *Holarctic birds of prey* (eds R. D. Chancellor, B.-U. Meyburg & J. J. Ferrero). Calamonte, Spain, ADENEX-WWGBP.

Newton, I. (1998b). *Population limitation in birds*. London, Academic Press.

Newton, I. (2000). Movements of Bullfinches *Pyrrhula pyrrhula* within the breeding season. *Bird Study* 47: 372–376.

Newton, I. (2001). Causes and consequences of breeding dispersal in the Sparrowhawk. *Ardea* 89: 143–154.

Newton, I. (2002). Population limitation in Holarctic owls. pp. 3–29 in *Ecology and conservation of owls* (eds I. Newton, R. Kavanagh, J. Olson & I. R. Taylor). Collingwood, Victoria, Australia, CSIRO Publishing.

Newton, I. (2003). *Speciation and biogeography of birds*. London, Academic Press.

Newton, I. (2004a). Population limitation in migrants. *Ibis* 146: 197–226.

Newton, I. (2004b). The recent declines of farmland bird populations in Britain: an appraisal of causal factors and conservation actions. *Ibis* 146: 579–600.

Newton, I. (2006). Advances in the study of irruptive migration. *Ardea* 94: 433–460.

Newton, I. (2006a). Can conditions experienced during migration limit the population levels of birds? *J. Ornithol.* 147: 146–166.

Newton, I. (2006b). Movement patterns of Common Crossbills *Loxia curvirostra* in Europe. *Ibis* 148: 782–788.

Newton, I. & Campbell, C. R. G. (1970). Goose studies at Loch leven in 1967/68. *Scott. Birds* 6: 5–18.

Newton, I. & Dale, L. (1996a). Relationship between migration and latitude among west European birds. *J. Anim. Ecol.* 65: 137–146.

Newton, I. & Dale, L. C. (1996b). Bird migration at different latitudes in eastern North America. *Auk* 113: 626–635.

Newton, I. & Dale, L. C. (1997). Effects of seasonal migration on the latitudinal distribution of west Palaearctic bird species. *J. Biogeog.* 24: 781–789.

Newton, I. & Marquiss, M. (1982). Fidelity to breeding area and mate in Sparrowhawks *Accipiter nisus. J. Anim. Ecol.* 51: 327–341.

Newton, I. & Marquiss, M. (1983). Dispersal of Sparrowhawks between birthplace and breeding place. *J. Anim. Ecol.* 52: 463–477.

Newton, I. & Rothery, P. (2000). Post-fledging recovery and dispersal of ringed Eurasian Sparrowhawks *Accipiter nisus. J. Avian Biol.* 31: 226–236.

Newton, I., Hobson, K. A., Fox, A. D. & Marquiss, M. (2006). An investigation into the provenance of Northern Bullfinches *Pyrrhula p. pyrrhula* found in winter in Scotland and Denmark. *J. Avian Biol.* 37: 431–435.

Newton, S. F. (1996). Wintering range of Palaearctic-African migrants includes southwest Arabia. *Ibis* 138: 335–350.

Nice, M. M. (1933). Zur Naturgeschichte des Singammers. *J. Orn.* 81: 552–595.

Nice, M. M. (1937). Studies in the life history of the Song Sparrow. Vol. 1. *Trans. Linn. Soc. New York* 4: 1–247.

Nichols, J. D. & Haramis, G. M. (1980). Sex-specific differences in winter distribution patterns of Canvasbacks. *Condor* 82: 406–416.

Nichols, J. D. & Hines, J. E. (1987). *Population ecology of the Mallard. VIII. Winter distribution patterns and survival rates of winter-banded Mallards.* Washington, D.C., U.S. Fish and Wildlife Service, Resource Publication 162.

Nichols, J. D. & Johnson, F. A. (1990). Wood Duck population dynamics: a review. pp. 83–105 in *Proc. 1988 North American Wood Duck Symposium* (ed. L. H. Frederickson, G. V. Burger, S. P. Havera, D. A. Graber, R. E. Kirby & T. S. Taylor). St. Louis, MO, 1988. North American Wood Duck Symposium.

Nichols, J. D. & Kaiser, A. (1999). Quantitative studies of bird movement: a methodological review. *Bird Study* 46 (suppl): S289–S298.

Nichols, J. D., Reinecke, K. J. & Hines, J. E. (1983). Factors affecting the distribution of Mallards wintering in the Mississippi Alluvial Valley. *Auk* 100: 932–46.

Niles, D. M., Rohwer, S. A. & Robins, R. D. (1969). An observation of midwinter nocturnal tower mortality in Tree Sparrows. *Bird-Banding* 40: 322–323.

Nilsson, J.-A. (1989). Causes and consequences of natal dispersal in the Marsh Tit, *Parus palustris. J. Anim. Ecol.* 58: 619–636.

Nilsson, L. (1969). The migration of the Goldeneye in northwest Europe. *Wildfowl* 20: 112–118.

Nilsson, L. (1976). Sex ratios of Swedish Mallard during the non-breeding season. *Wildfowl* 27: 91–94.

Nilsson, L. (1979). Variation in the production of young of swans wintering in Sweden. *Wildfowl* 30: 129–134.

Nilsson, L. & Persson, H. (1993). Variation in survival in an increasing population of the Greylag Goose *Anser anser* in Scania, southern Sweden. *Ornis Svecica* 3: 137–146.

Nilsson, L. & Persson, H. (2001). Natal and breeding dispersal in the Baltic Greylag Goose *Anser anser. Wildfowl* 52: 21–30.

Nilsson, S. G. (1984). The relation between the beech mast crop and the wintering of Brambling, *Fringilla montifringilla*, and Woodpigeon, *Columba palumbus*, in south Sweden. *Vår Fågelvarld* 43: 135–136.

Nisbet, I. C. T. (1957). Wader migration at Cambridge sewage farm. *Bird Study* 4: 131–148.

Nisbet, I. C. T. (1959a). Calculation of flight directions of birds observed crossing the face of the moon. *Wilson Bull.* 71.

Nisbet, I. C. T. (1959b). Wader migration in North America and its relation to transatlantic crossings. *Br. Birds* 52: 205–215.

Nisbet, I. C. T. (1962). South-eastern rarities at Fair Isle. *Br. Birds* 55: 74–85.

Nisbet, I. C. T. (1963a). American passerines in western Europe, 1951–62. *Br. Birds* 56: 204–217.

Nisbet, I. C. T. (1963b). Weight-loss during migration. Part II. Review of other estimates. *Bird-Banding* 34: 107–138.

Nisbet, I. C. T. (1967). Migration and moult in Pallas's Grasshopper Warbler *Locustella certhiola*. *Bird Study* 14: 96–103.

Nisbet, I. C. T. (1969). Returns of transients: results of an inquiry. *Ebba News* 32: 269–274.

Nisbet, I. C. T. (1970). Autumn migration of the Blackpoll Warbler: evidence for a long flight provided by regional survey. *Bird-Banding* 41: 207–240.

Nisbet, I. C. T. & Drury, W. H. (1968). Short-term effects of weather on bird migration: a field study using multivariate statistics. *Anim. Behav.* 16: 496–530.

Nisbet, I. C. T. & Medway, L. (1972). Dispersion, population ecology and migration of Eastern Great Reed Warblers *Acrocephalus orientalis* wintering in Malaysia. *Ibis* 114: 451–494.

Nisbet, I. C. T. & Safina, C. (1996). Transatlantic recoveries of ringed Common Terns *Sterna hirundo*. *Ringing & Migration* 17: 28–30.

Nisbet, I. C. T., Drury, W. H. & Baird, J. (1963). Weight-loss during migration. Part 1: Deposition and consumption of fat by the Black-poll Warbler *Dendroica striata*. *Bird-Banding* 34: 107–138.

Nix, H. A. (1976). Environmental control of breeding, post-breeding dispersal and migration of birds in the Australian region. *Proc. Int. Ornithol. Congr.* 16: 272–305.

Nolan, V. (1978). The ecology and behaviour of the Prairie Warbler (*Dendroica discolor*). *Ornithol. Monogr.* 26: 1–595.

Nolan, V. & Ketterson, E. D. (1991). Experiments on winter site attachment in young Dark-eyed Juncos. *Ethology* 87: 123–133.

Nolet, B. A. & Drent, R. H. (1998). Bewick's Swans refuelling on pondweed tubers in the Dvina Bay (White Sea) during their spring migration: first come, first served. *Avian Biol.* 29: 574–581.

Norberg, U. M. (1996). Energetics of flight. pp. 199–249 in *Avian energetics and nutritional ecology* (ed. C. Carey). London: Chapman & Hall.

Norman, S. C. (1990). Factors influencing the onset of post-nuptial moult in Willow Warblers *Phylloscopus trochilus*. *Ringing & Migration* 11: 90–100.

Norman, S. C. (1994). Dispersal and return rates of Willow Warblers *Phylloscopus trochilus* in relation to age, sex and season. *Ringing & Migration* 15: 8–16.

Norman, S. C. & Norman, W. (1985). Autumn movements of Willow Warblers ringed in the British Isles. *Ringing & Migration* 6: 17–18.

Norris, C. A. (1947). Report on the distribution and status of the Corncrake. *Br. Birds* 40: 226–244.

Norris, D. R. (2005). Carry over effects and habitat quality in migratory populations. *Oikos* 109: 178–186.

Norris, D. R., Marra, P. P., Kyser, T. K., Sherry, T. W. & Ratcliffe, L. M. (2004). Tropical winter habitat limits reproductive success on the temperate breeding grounds in a migratory bird. *Proc. R. Soc. Lond. B* 271: 59–64.

Norris, D. R., Marra, P. P., Kyser, T. K., Bowen, G. J., Ratcliffe, L. M., Royle, J. A. & Kyser, T. K. (2006). Migratory connectivity in a widely distributed songbird, the American Redstart (*Setophaga ruticilla*). *Ornithol. Monogr.* 61: 14–28.

Noskov, G. A., Rymkevich, T. A. & Iovchenko, N. P. (1999). Intraspecific variation of moult: adaptive significance and ways of realisation. *Proc. Int. Ornithol. Congr.* 22: 544–563.

Noskov, G. A., Rezvyi, S. P., Rychkova, A. L. & Smirnov, O. P. (2005). On migrations of the slender billed race of the Nutcracker *Nucifraga caryocatactes macrorhychos* L. in north-west Russia and adjacent areas. pp. 61–80 in *Ornithological studies in the Ladoga area* (ed. N. P. Iovchenko). St Petersburg, University Press.

Nott, M. P., De Sante, D. F., Siegel, R. B. & Pyle, P. (2002). Influences of the El Nino/Southern Oscillation and the North Atlantic Oscillation on avian productivity in forests of the Pacific Northwest of North America. *Global Ecol. Biogeog.* 11: 333–342.

Novoa, F. F., Rosenmann, M. & Bozinovic, F. (1991). Physiological responses of four passerine species to simulated altitudes. *Comp. Biochem. Physiol.* 99A: 179–183.

Nowakowski, J. K. (2001). Speed and synchronisation of autumn migration of the Great Tit (*Parus major*) along the eastern and the southern Baltic coast. *Ring* 23: 55–71.

Nowakowski, J. K. & Chruściel, J. (2004). Speed of autumn migration of the Blue Tit (*Parus caeruleus*) along the eastern and southern Baltic coast. *Ring* 26: 3–12.

Nuorteva, P. (1952). Havaintoja pikkukäpylinnun (*Loxia curvirostra* L.) vaellussuunnan suhteesta vesistöön. *Ornis Fenn.* 29: 116–117.

Nyström, B. & Nyström, H. (1991). Effects of bad weather on the breeding of the Redpoll *Carduelis flammea* in a year with a poor birch seed crop in southern Lapland. *Ornis Svecica* 65–68.

Oatley, T. B. (1988). Change in winter movements of Cape Gannets. pp. 41–43 in *Long-term data series relating to Southern Africa's renewable natural resources* (eds I. A. W. Macdonald & K. J. M. Crawford). South African Natural Science Progress Report 157.

Oatley, T. B. (2000). Migrant European Swallows *Hirundo rustica* in southern Africa – a southern perspective. *Ostrich* 71: 205–209.

Obmascik, M. (2004). *The big year*. London, Transworld Publishers.

O'Briain, M. (1987). Families and other social groups of Brent Geese in winter. Unpub. Report, University College, Dublin.

Obst, B. S. & Nagy, K. A. (1993). Stomach oil and the energy budget of Wilson's Storm Petrel nestlings. *Condor* 95: 792–805.

O'Connor, R. J. (1981). Comparisons between migrant and non-migrant birds in Britain. pp. 167–195 in *Animal migration* (ed. D. J. Aidley). Cambridge, Cambridge University Press.

O'Connor, R. J. (1985). Behavioural regulation of bird populations: a review of habitat use in relation to migration and residence. pp. 105–142 in *Behavioural ecology. Ecological consequences and adaptive behaviour* (eds R. M. Sibly & R. H. Smith). Oxford, Blackwells.

O'Connor, R. J. (1986). Biological characteristics of invaders among bird species in Britain. *Phil. Trans. R. Soc. Lond. B.* 314: 583–598.

Odum, E. P. (1960). Lipid deposition in nocturnal migrant birds. *Proc. Int. Ornithol. Congr.* 13: 563–576.

Odum, E. P., Connell, C. E. & Stoddard, H. L. (1961). Flight energy and estimated flight ranges of some migratory birds. *Auk* 78: 515–527.

Oeming, A. F. (1957). Notes on the Barred Owl and the Snowy Owl in Alberta. *Blue Jay* 15: 153–156.

O'Hara, P. D., Lank, D. B. & Delgado, F. S. (2002). Is the timing of moult altered by migration? Evidence from a comparison of age and residency classes of Western Sandpipers *Calidris mouri* in Panama. *Ardea* 90: 61–70.

O'Hara, P. D., Fernandez, G., Becerril, F., de la Cueva, H. & Lank, D. B. (2005). Life history varies with migratory distance in Western Sandpipers *Calidris mauri*. *J. Avian Biol.* 36: 191–202.

Ohmart, R. D. (1973). Observations on the breeding adaptations of the Roadrunner. *Condor* 75: 140–149.

Ojanen, M. (1979). Effects of a cold spell on birds in northern Finland in May 1968. *Ornis Scand.* 56: 148–155.

Ojanen, M. (1984). The relation between spring migration and onset of breeding in the Pied Flycatcher *Ficedula hypoleuca* in northern Finland. *Ann. Zool. Fenn.* 21: 205–208.

Olea, P. P. (2001). Post-fledging dispersal in the endangered Lesser Kestrel *Falco naumanni*. *Bird Study* 48: 110–115.

Olsen, B., Munster, V. J., Wallensten, A., Waldenström, J., Osterhaus, A. D. M. E. & Fouchier, R. A. M. (2006). Global patterns of influenza A virus in wild birds. *Science* 312: 384–312.

Olssen, O. (1958). Dispersal, migration, longevity and death causes of *Strix aluco*, *Buteo buteo*, *Ardea cinerea* and *Larus argentatus*. *Acta Vertebratica* 1: 91–189.

Orell, M. & Ojanen, M. (1979). Mortality rates of the Great Tit *Parus major* in a northern population. *Ardea* 67: 130–139.

Oring, L. W. (1964). Behaviour and ecology of certain ducks during the post-breeding period. *J. Wildl. Manage.* 28: 223–233.

Oring, L. W. & Lank, D. B. (1982). Sexual selection, arrival times, philopatry and site fidelity in the polyandrous Spotted Sandpiper. *Behav. Ecol. Sociobiol.* 10: 185–191.

Oring, L. W., Gray, E. M. & Reed, J. (1997). Spotted Sandpiper (*Actitis macularia*). in *The Birds of North America*, No. 285 (eds A. Poole & F. Gill). Philadelphia, Academy of Natural Sciences and Washington, D. C., American Ornithologists Union.

Ormerod, S. J. (1989). The influence of weather on the body mass of migrating Swallows *Hirundo rustica* in South Wales. *Ringing & Migration* 10: 65–74.

Ormerod, S. J. (1991). Pre-migratory and migratory movements of Swallows *Hirundo rustica* in Britain and Ireland. *Bird Study* 38: 170–178.

Osborne, P. E., Bowardi, M. A. & Bailey, T. A. (1997). Migration of the Houbara Bustard *Chlamydotis undulata* from Abu Dhabi to Turkmenistan: the first results from satellite tracking studies. *Ibis* 139: 192–196.

Osorio-Beristain, M. & Drummond, H. (1993). Natal dispersal and deferred breeding in the Blue-footed Booby. *Auk* 110: 234–239.

Otahal, C. D. (1995). Sexual differences in Wilson Warbler migration. *J. Field Ornithol.* 66: 60–69.

Ottich, I & Dierschke, V. (2003). Exploitation of resources modulates stopover behaviour of passerine migrants. *J. Ornithol.* 144: 307–316.

Ottosson, U., Sandberg, R. & Tetterson, J. (1990). Orientation cage and release experiments with migratory Wheatears (*Oenanthe oenanthe*) in Scandinavia and Greenland: the importance of visual cues. *Ethology* 86: 57–70.

Ottosson, U., Waldenstrom, J., Hjort, C. & McGregor, R. (2005). Garden Warbler *Sylvia borin* migration in sub-Saharan West Africa: phenology and body mass changes. *Ibis* 147: 750–757.

Outlaw, D. C., Voekler, G., Mila, B. & Girman, D. J. (2003). Evolution of long-distance migration in, and historical biogeography of, *Catharus* thrushes: a molecular phylogenetic approach. *Auk* 120: 299–310.

Overskaug, K., Sunde, P. & Kristiansen, E. (1997). Subcutaneous fat accumulation in Norwegian owls and raptors. *Ornis Fenn.* 74: 29–38.

Owen, J., Moore, F., Panella, N., Edwards, E., Bru, R., Hughes, M. & Komar, N. (2006). Migrating birds as dispersal vehicles for West Nile virus. *Ecohealth* 3: 79–85.

Owen, M. (1980). Wild geese of the world. London, B.T. Batsford Ltd.

Owen, M. (1984). Dynamics and age structure of an increasing goose population – the Svalbard Barnacle Goose *Branta leucopsis. Norsk Polarinstitutt Skrifter* 181: 37–47.

Owen, M. & Dix, M. (1986). Sex ratios in some common British wintering ducks. *Wildfowl* 37: 104–112.

Owen, M. & Black, J. M. (1989). Factors affecting the survival of Barnacle Geese on migration from the breeding grounds. *J. Anim. Ecol.* 56: 603–617.

Owen, M. & Black, J. M. (1991). The importance of migration mortality in non-passerine birds. pp. 360–372 in *Bird population studies. Relevance to conservation and management* (eds C. M. Perrins, J.-D. Lebreton & G. J. M. Hirons). Oxford, Oxford University Press.

Owen, M., Atkinson-Willes, G. L. & Salmon, D. G. (1986). *Wildfowl in Great Britain*, 2nd edn. Cambridge, Cambridge University Press.

Owre, O. T. (1967). Hurricanes and birds. *Sea Frontiers* 13: 14–21.

Ozarowska, A., Yosef, R. & Busse, P. (2004). Orientation of Chiffchaff (*Phylloscopus collybita*), Blackcap (*Sylvia atricapilla*) and Lesser Whitethroat (*S. curruca*) on spring migration at Eilat, Israel. *Avian Ecol. Behav.* 12: 1–10.

Pace, R. M. & Afton, A. D. (1999). Direct recovery rates of Lesser Scaup banded in northwest Minnesota: sources of heterogeneity. *J. Wildl. Manage.* 63: 389–395.

Page, G. & Middleton, A. L. A. (1972). Fat deposition in autumn migration in the Semipalmated Sandpiper. *Bird-Banding* 43: 85–96.

Palmer, W. L. (1962). Ruffed Grouse flight capability over water. *J. Wildl. Manage.* 26: 338–339.

Papi, F. (1989). Pigeons use olfactory cues to navigate. *Ethol. Ecol. Evol.* 1: 219–231.

Paradis, E., Baillie, S. R., Sutherland, W. J. & Gregory, R. D. (1998). Patterns of natal and breeding dispersal in birds. *J. Anim. Ecol.* 67: 518–536.

Parks, G. H. & Parks, H. C. (1965). Supplementary notes on the Evening Grosbeak nesting area study. *Bird-Banding* 36: 113–115.

Parmelee, D. F. (1992). Snowy Owl. *The Birds of North America No. 19* (eds A. Poole, P. Stettenheim & F. Gill). Philadelphia: Academy of Natural Sciences & Washington, D.C., American Ornithologists Union.

Parr, R. (1992). The decline to extinction of a population of Golden Plover in north-east Scotland. *Ornis Scand.* 23: 152–158.

Parrish, J. D. (2000). Behavioral, energetic and conservation implications of foraging plasticity during migration. *Stud. Avian Biol.* 20: 52–70.

Parrish, J. R., Rogers, D. T. & Ward, F. P. (1983). Identification of natal locales of Peregrine Falcons (*Falco peregrinus*) by trace-element analysis of feathers. *Auk* 100: 560–567.

Pärt, T. (1990). Natal dispersal in the Collared Flycatcher: possible causes and reproductive consequences. *Ornis Scand.* 21: 83–88.

Pärt, T. (1996). Problems with testing inbreeding avoidance: the case of the Collared Flycatcher. *Evolution* 50: 1625–1630.

Paton, P. W. C. (1994). The effect of edge on avian nest success: how strong is the evidence? *Conserv. Biol.* 8: 17–26.

Paton, P. W. C. & Edwards, T. C. (1996). Factors affecting interannual movements of Snowy Plovers. *Auk* 113: 534–543.

Patten, M. A. & Marantz, C. A. (1996). Implications of vagrant southeastern vireos and warblers in California. *Auk* 113: 911–923.

Pattenden, R. K. & Boag, D. A. (1989). Effects of body mass on courtship, pairing and reproduction in captive Mallards. *Can. J. Zool.* 67: 495–501.

Patterson, I. J. (1982). *The Shelduck. A study in behavioural ecology.* Cambridge, Cambridge University Press.

Payevski, V. A. (1971). Atlas of bird migrations according to banding data at the Courland Spit. pp. 1–124 in *Bird migrations. Ecological and physiological factors* (ed. B. E. Bykhovskii). Leningrad.

Payevski, V. A. (1994). Age and sex structure, mortality and spatial winter distribution of Siskins (*Carduelis spinus*) migrating through eastern Baltic area. *Vogelwarte* 37: 190–198.

Payevski, V. A. (1998). Bird trapping and ringing as an inexhaustible source of most valuable data for demographic investigators (with special reference to the work of the Biological Station Rybachy). *Avian Ecol. Behav.* 1: 76–86.

Payne, R. B. (1969). Breeding seasons and reproductive physiology of Tricolored Blackbirds and Red-winged Blackbirds. *Univ. Calif. Publ. Zool.* 90: 1–115.

Payne, R. B. (1972). Mechanisms and control of molt. pp. 103–155 in *Avian biology* (eds D. S. Farner, J. R. King & K. C. Parkes.). New York, Academic Press.

Payne, R. B. (1997). Avian brood parasitism. pp. 338–369 in *Host–parasite evolution* (eds D. H. Clayton & J. Moore). Oxford, Oxford University Press.

Payne, R. B. & Payne, L. L. (1993). Breeding dispersal in Indigo Buntings: circumstances and consequences for breeding success and population structure. *Condor* 95: 1–24.

Paynter, R. A. (1953). Autumnal migrants on the Campeche Bank. *Auk* 70: 338–349.

Peach, W. J., Baillie, S. R. & Underhill, L. (1991). Survival of British Sedge Warblers *Acrocephalus schoenobaenus* in relation to west Africa rainfall. *Ibis* 133: 300–305.

Peach, W. J., Thompson, P. S. & Coulson, J. C. (1994). Annual and long-term variation in the survival rates of British Lapwings *Vanellus vanellus*. *J. Anim. Ecol.* 63: 60–70.

Peach, W. J., Crick, H. Q. P. & Marchant, J. H. (1995). The demography of the decline in the British Willow Warbler population. *J. Appl. Stat.* 22: 905–922.

Peach, W. J., Baillie, S. R. & Balmer, D. E. (1998). Long-term changes in the abundance of passerines in Britain and Ireland as measured by constant effort mist-netting. *Bird Study* 45: 257–275.

Pearson, D. (1972). The wintering and migration of Palaearctic passerines at Kampala, southern Uganda. *Ibis* 114: 43–60.

Pearson, D. J. (1973). Moult of some Palaearctic warblers wintering in Uganda. *Bird Study* 20: 24–36.

Pearson, D. J. (1975). The timing of complete moult in the Great Reed Warbler *Acrocephalus arundinaceus*. *Ibis* 117: 506–509.

Pearson, D. J. (1990). Palearctic passerine migrants in Kenya and Uganda: temporal and spatial patterns of their movements. pp. 44–59 in *Bird migration. Physiology and ecophysiology* (ed. E. Gwinner). Berlin, Springer-Verlag.

Pearson, D. J. & Backhouse, G. C. (1976). The southward migration of Palaearctic birds over Ngulia, Kenya. *Ibis* 118: 78–105.

Pearson, D. J. & Lack, P. C. (1992). Migration patterns and habitat use by passerine and near-passerine migrant birds in eastern Africa. *Ibis* 134 (suppl 1): 89–98.

Pearson, D. J., Nokolans, G. & Ash, J. S. (1980). The southward migration of Palaearctic passerines through northeast and east tropical Africa: a review. *Proc. Pan-Afr. Ornithol. Congr.* 4: 243–262.

Peiponen, V. A. (1967). Südliche Fortpflanzung und Zug von *Carduelis flammea* (L.) in Jahre 1965. *Ann. Zool. Fenn.* 4: 547–549.

Pennycuick, C. J. (1969). The mechanics of bird migration. *Ibis* 111: 525–556.

Pennycuick, C. J. (1972). Soaring behaviour and performance of some East African birds observed from a motor-glider. *Ibis* 114: 178–218.

Pennycuick, C. J. (1975). Mechanics of flight. pp. 1–75 in *Avian biology*, Vol. 5. (eds D. S. Farner & J. R. King). London, Academic Press.

Pennycuick, C. J. (1989). *Bird flight performance: a practical calculation manual*. Oxford, Oxford University Press.

Pennycuick, C. J. (1998). Field observations of thermals and thermal streets, and the theory of cross-country soaring flight. *J. Avian Biol.* 29: 33–43.

Pennycuick, C. J. (2002). Gust soaring as a basis for the flight of petrels and albatrosses. *Avian Science* 2: 1–12.

Pennycuick, C. J. (2003). The concept of energy height in animal locomotion: separating mechanics from physiology. *J. Theoret. Biol.* 224: 189–203.

Pennycuick, C. J. (2006). Flight 1.16 and 1.17. www.bio.bristol.ac.uk/people/pennycuick.htm.

Pennycuick, C. J. & Battley, P. F. (2003). Burning the engine: a time-marching computation of fat and protein consumption in a 5420 km non-stop flight by Great Knots *Calidris tenuirostris*. *Oikos* 103: 323–332.

Pennycuick, C. J., Einarsson, O., Bradbury, T. A. M. & Owen, M. (1996). Migrating Whooper Swans *Cygnus cygnus*: satellite tracks and flight performance calculations. *J. Avian Biol.* 27: 118–134.

Pennycuick, C. J., Alerstam, T. & Hedenström, A. (1997). A new low-turbulence wind tunnel for flight experiments at Lund University, Sweden. *J. Exp. Biol.* 200: 1441–1449.

Pennycuick, C. J., Bradbury, T. A. M., Einarsson, O. & Owen, M. (1999). Response to weather and light conditions of migrating Whooper Swans *Cygnus cygnus* and flying height profiles, observed with the Argos satellite system. *Ibis* 141: 434–443.

Percival, S. (1991). Population trends in British Barn Owls. *Br. Wildl.* 2: 131–140.

Perdeck, A. C. (1958). Two types of orientation in migrating Starlings, *Sturnus vulgaris* L., and Chaffinches, *Fringilla coelebs* L., as revealed by displacement experiments. *Ardea* 46: 1–37.

Perdeck, A. C. (1964). An experiment on the ending of autumn migration in Starlings. *Ardea* 52: 133–139.

Perdeck, A. C. (1967). Orientation of Starlings after displacement to Spain. *Ardea* 55: 93–104.

Perdeck, A. C. & Clason, C. (1983). Sexual differences in migration and winter quarters of ducks ringed in the Netherlands. *Wildfowl* 34: 137–143.

Pérez-Tris, J. & Tellería, J. L. (2002). Migratory and sedentary Blackcaps in sympatric non-breeding grounds: implications for the evolution of avian migration. *J. Anim. Ecol.* 71: 211–224.

Pérez-Tris, J., Bensch, S., Carbonell, R., Helbig, A. J. & Tellería, J. L. (2004). Historical diversification of migration patterns in a passerine bird. *Evolution* 58: 1814–1832.

Perrins, C. M. (1966). The effect of beech crops on Great Tit populations and movements. *Br. Birds* 59: 419–432.

Perrins, C. M. (2002). Common Swift. pp. 443–445 in *Migration atlas* (eds C. Wernham, M. Toms, J. Marchant, J. Clark, G. Siriwardena & S. Baillie). London, T. & A. D. Poyser.

Perrins, C. M. & Brooke M. de L. (1976). Manx Shearwaters in the Bay of Biscay. *Bird Study* 23: 295–300.

Perrins, C. M., Harris, M. P. & Britton, C. K. (1973). Survival of Manx Shearwaters *Puffinus puffinus*. *Ibis* 115: 535–548.

Perrins, C. M., Richardson, D. & Stevenson, M. (1985). Deaths of swifts and hirundines. *Bird Study* 32: 150.

Peterjohn, B. G., Sauer, J. R. & Robbins, C. S. (1995). Population trends from the North American breeding bird survey. pp. 3–39 in *Ecology and management of neotropical migratory birds* (ed. T. E. Martin & D.M. Finch). Oxford, Oxford University Press.

Petersen, B. S. (1984). The origins of Herring Gulls *Larus argentatus* occurring in the Copenhagen area with special reference to recent changes. *Dansk. Orn. Foren. Tidskkr.* 78: 15–24.

Petrie, S. A. & Rogers, K. H. (1997). Nutrient reserves dynamics of semi-arid breeding White-faced Whistling Ducks: a north temperate contrast. *Can. J. Zool.* 82: 1082–1090.

Petty, S. J. (1992). Ecology of the Tawny Owl *Strix aluco* in the spruce forests of Northumberland and Argyll. PhD thesis, Open University, Milton Keynes.

Petty, S. J. (1999). Diet of Tawny Owls (*Strix aluco*) in relation to Field Vole (*Microtus agrestis*) abundance in a conifer forest in northern England. *J. Zool. Lond.* 248: 451–465.

Petty, S. J., Patterson, I. J., Anderson, D. I. K., Little, B. & Davison, M. (1995). Numbers, breeding performance, and diet of the Sparrowhawk *Accipiter nisus* and Merlin *Falco columbarius* in relation to cone crops and seed-eating finches. *For. Ecol. Manage.* 79: 133–146.

Pfister, C., Kasprzyk, M. J. & Harrington, B. A. (1998). Body fat levels and annual return in migrating Semipalmated Sandpipers. *Auk* 115: 904–915.

Phillips, A. R. (1951). Complexities of migration: a review. *Wilson Bull.* 63: 129–136.

Phillips, A. R. (1975). The migrations of Allen's and other hummingbirds. *Condor* 77: 196–205.

Phillips, J. B. & Moore, F. R. (1992). Calibration of the sun compass by sunset polarized light patterns in a migratory bird. *Behav. Ecol Sociobiol.* 31: 189–193.

Phillips, R. A. & Hamer, K. C. (1999). Lipid reserves, fasting capability and the evolution of nestling obesity in procellariiform seabirds. *Proc. R. Roc. Lond. B* 266: 1329–1334.

Phillips, R. A., Bearhop, S., Hamer, K. C. & Thompson, D. R. (1999). Rapid population growth of Great Skuas *Catharacta skua* at St Kilda: implications for management and conservation. *Bird Study* 46: 174–183.

Phillips, R. A., Xavier, J. C. & Croxall, J. P. (2003). Effects of satellite transmitters on albatrosses and petrels. *Auk* 120: 1082–1090.

Phillips, R. A., Silk, J. R. D., Croxall, J. P., Afanasyev, V. & Bennett, V. J. (2005). Summer distribution and migration of nonbreeding albatrosses: individual consistencies and implications for conservation. *Ecology* 86: 2386–2396.

Phillips, R. A., Silk, J. R. D., Croxall, J. P. & Afanasyev, V. (2006). Year-round distribution of White-chinned Petrels from South Georgia: relationships with oceanography and fisheries. *Biol. Conserv.* 129: 336–347.

Pickering, S. P. C. (1989). Attendance patterns and behavior in relation to experience and pair-bond formation in the Wandering Albatross *Diomedea exulans* at South Georgia. *Ibis* 131: 183–195.

Pienaar, U. d. V. (1969). Observations on the nesting habits and predators of breeding colonies of Red-billed Queleas *Quelea quelea lathami* (A. Smith) in the Kruger National Park. *Bokmakierie* 21 I(suppl): xi–xv.

Pienkowski, M. W. (1976). Recurrence of waders on autumn migration at sites in Morocco. *Vogelwarte* 28: 293–297.

Pienkowski, M. W. (1984). Breeding biology and population dynamics of Ringed Plovers *Charadrius hiaticula* in Britain and Greenland: nest predation as a possible factor limiting distribution and timing of breeding. *J. Zool.* 202: 83–414.

Pienkowski, M. W. & Evans, P. R. (1985). The role of migration in the population dynamics of birds. *Behavioural ecology: the ecological consequences of adaptive behaviour* (eds R. M. Sibley & R. H. Smith). Oxford, Blackwell Scientific Publications.

Pienkowski, M. W., Knight, P. A., Stanyard, D. J. & Argyle, F. B. (1976). The primary moult of waders on the Atlantic coast of Morocco. *Ibis* 118: 347–365.

Pierce, B. J. & McWilliams, S. R. (2005). Seasonal changes in composition of lipid stores in migratory birds: causes and consequences. *Condor* 107: 269–279.

Piersma, T. (1987). Hop, skip or jump? Constraints in migration of arctic waders by feeding, fattening and flight speed. *Limosa* 60: 185–194.

Piersma, T. (1990). Pre-migratory fattening usually involves more than the deposition of fat alone. *Ringing & Migration* 11: 113–115.

Piersma, T. (1997). Do global patterns of habitat use and migration strategies co-evolve with relative investments in immunocompetence due to spatial variation in parasite pressure? *Oikos* 80: 623–631.

Piersma, T. (1998). Phenotypic flexibility during migration: optimisation of organ size contingent on the risks and rewards of fueling and flight? *J. Avian Biol.* 29: 511–520.

Piersma, T. (2002). Energetic bottlenecks and other design constraints in avian annual cycles. *Integr. Comp. Biol.* 442: 51–67.

Piersma, T. & Davidson, N. C. (1992). The migrations and annual cycles of five subspecies of Knots in perspective. *Wader Study Group Bull.* 64, Suppl.: 187–197.

Piersma, T. & Gill, R. E. (1998). Guts don't fly: small digestive organs in obese Bar-tailed Godwits. *Auk* 115: 196–203.

Piersma, T. & Jukema, J. (1993). Red breasts as honest signals of migratory quality in a long-distance migrant, the Bar-tailed Godwit. *Condor* 95: 163–177.

Piersma, T. & Jukema, J. (2002). Contrast in adaptive mass gains: Eurasian Golden Plovers store fat before midwinter and protein before prebreeding flight. *Proc. R. Soc. Lond. B* 269: 1101–1105.

Piersma, T. & Lindström, A. (1997). Rapid reversible changes in organ size as a component of adaptive behaviour. *Trends Ecol. Evol.* 12: 134–138.

Piersma, T., Drent, R. & Wiersma, P. (1991). Temperate versus tropical wintering in the world's northernmost breeder, the Knot: metabolic scope and resource levels restrict subspecific options. *Proc. Int. Ornithol. Congr.* 20: 761–772.

Piersma, T., Gudmundsson, G. A. & Lilliendahl, K. (1999). Rapid changes in size of different functional organ and muscle groups during refuelling in a long-distance migrating shorebird. *Physiol. Biochem. Zool.* 72: 405–416.

Piersma, T., Koolhaas, A. & Dekinga, A. (1993). Interactions between stomach structure and diet choice in shorebirds. *Auk* 110: 552–564.

Piersma, T., Reneerkens & Ramenofski, M. (2000). Baseline corticosterone peaks in shorebirds with maximal energy stores for migration: a general preparatory mechanism for rapid behavioural and metabolic transitions? *Gen. Comp. Endocrinol.* 120: 118–126.

Piersma, T. B., Spaans, A. L. & Dekinga, S. (2002). Are shorebirds sometimes forced to roost on water in thick fog? *Wader Study Group Bull.* 97: 42–44.

Piersma, T., Zwarts, L. & Bruggemann, J. H. (1990b). Behavioural aspects of the departure of waders before long-distance flights: flocking, vocalisations, flight paths and diurnal timing. *Ardea* 78: 157–184.

Piersma, T., Klaassen, M., Bruggemann, J. H., Blomert, A.-M., Gueye, A., Ntiamoa-Baidu, Y. & Brederode, N. E., van. (1990a). Seasonal timing of the spring departure of waders from the Banc-D'Arguin, Mauritania: a census in June 1988. *Ardea* 78: 123–134.

Piersma, T., Rogers, D. I., González, P. M., Zwarts, L., Niles, L. J., De Lima, I. Serrano do Nascimento., Minton, C. D. T. & Baker, A. J. (2005). Fuel storage rates before northward flights in Red Knots worldwide. pp. 262–273 in *Birds of Two Worlds*. (eds Greenberg, R. & Marra, P. P.). Washington, D. C., Smithsonian Institution.

Piertney, S. B., Summers, R. & Marquiss, M. (2001). Microsatellite and mitochondrial DNA homogeneity among phenotypically diverse crossbill taxa in the UK. *Proc. R. Soc. Lond. B* 268: 1511–1517.

Pilastro, A. & Spina, F. (1997). Ecological and morphological correlates of residual fat reserves in passerine migrants at their spring arrival in southern Europe. *J. Avian Biol.* 28: 309–318.

Pineau, O. (2000). Conservation of wintering and migratory habitats. pp. 237–250 in *Heron conservation* (eds J. A. Kushlan & H. Hafner). London, Academic Press.

Pinowski, J. (1965). Overcrowding as one of the causes of dispersal of young Tree Sparrows. *Bird Study* 12: 27–33.

Pitches, A. & Cleeves, T. (2005). *Birds new to Britain, 1980–2004*. London, T. & A.D. Poyser.

Pitelka, F. A., Tomich, P. Q. & Treichel, G. W. (1955). Breeding behaviour of jaegers and owls near Barrow, Alaska. *Condor* 57: 3–18.

Pitra, C., Lieckfeldt, D. & Alonso, J. C. (2000). Population subdivision in Europe's Great Bustard inferred from mitochondrial and nuclear DNA sequence variation. *Mol. Ecol.* 9: 1165–1170.

Ploeger, P. L. (1968). Geographical differentiation in arctic Anatidae as a result of isolation during the last glacial. *Ardea* 56: 1–159.

Poole, A. F. (1989). *Ospreys. A natural and unnatural history*. Cambridge, University Press.

Poole, A. & Agler, B. (1987). Recoveries of Ospreys banded in the United States 1914–84. *J. Wildl. Manage.* 51: 148–155.

Popa-Lisseanu, A. G., Delgado-Huertas, A., Forero, M. G., Rodriguez, A., Arlettaz, R. & Ibanez, C. (2007). Bat's conquest of a formidable foraging niche: the myriads of nocturnally migrating songbirds. *PloS ONE* 2(2): e205. doi: 10.1371/journal/pone.0000205.

Porter, J. M. & Coulson, J. C. (1987). Long-term changes in recruitment to the breeding group, and the quality of recruits at a Kittiwake *Rissa tridactyla* colony. *J. Anim. Ecol.* 56: 675–689.

Porter, R. K., Willis, I., Christensen, S. & Nielson, R. P. (1974). *Flight identification of European raptors*. Berkhamsted, T. & A.D. Poyser.

Pospahala, R. S., Anderson, D. R. & Henny, C. J. (1974). *Population ecology of the Mallard. II Breeding habitat conditions, size of breeding populations, and production indices*. Washington, D.C., U.S. Fish & Wildlife Service, Resource Publication 115.

Post, W., Cruz, A. & McNair, D. B. (1993). The North American invasion pattern of the Shiny Cowbird. *J. Field Ornithol.* 64: 32–41.

Postma, E. & van Noordwijk, A. J. (2005). Gene flow maintains a large genetic difference in clutch size at a small spatial scale. *Nature* 433: 65–68.

Potapov, E. R. (1997). What determines the population density and reproductive success of Rough-legged Buzzards, *Buteo lagopus*, in the Siberian tundra? *Oikos* 78: 362–376.

Potti, J. (1998). Arrival time from spring migration in male Pied Flycatchers: individual consistency and familial resemblance. *Condor* 100: 702–708.

Pratt, A. & Peach, W. (1991). Site tenacity and annual survival of a Willow Warbler *Phylloscopus trochilus* population in southern England. *Ringing & Migration* 12: 128–134.

Prescott, D. R. C. (1991). Winter distribution of age and sex classes in an irruptive migrant, the Evening Grosbeak (*Coccothraustes vespertinus*). *Condor* 93: 694–700.

Prescott, D. R. C. & Middleton, A. L. A. (1990). Age and sex differences in winter distribution of American Goldfinches in eastern North America. *Ornis Scand.* 21: 99–104.

Preston, F. W. (1966). The mathematical representation of migration. *Ecology* 47: 375–392.

Prevett, J. P. & MacInnes, C. D. (1980). Family and other social groups in Snow Geese. *Wildl. Monogr.* 71: 1–46.

Price, T. (1981). The ecology of the Greenish Warbler *Phylloscopus trochiloides* in its winter quarters. *Ibis* 123: 131–144.

Price, T. & Gross, S. (2005). Correlated evolution of ecological differences among the Old World leaf warblers in the breeding and non-breeding seasons. pp. 359–370 in *Birds of Two Worlds*. (eds Greenberg, R. & Marra, P. P.). Washington, D. C., Smithsonian Institution.

Prill, A. G. (1931). Note on 'a land migration of Coots. *Wilson Bull.* 43: 148–149.

Prince, P. A., Wood, A. G., Barton, T. & Croxall, J. P. (1992). Satellite tracking of Wandering Albatrosses (*Diomedea exulans*) in the South Atlantic. *Antarctic Sci.* 4: 31–36.

Prop. J. (1991). Food exploitation patterns by Brent Geese *Branta bernicla* during spring staging. *Ardea* 79: 331–342.

Prop, J. (2004). Food finding. On the trail to successful reproduction in migrating geese. PhD thesis, University of Groningen.

Prop, J. & Black, J. M. (1998). Food intake, body reserves and reproductive success of Barnacle Geese *Branta leucopsis* staging in different habitats. *Norsk Polarinstitutt Skrifter* 200: 175–193.

Prop, J. & Deerenburg, C. (1991). Spring staging in Brent Geese *Branta bernicla*, feeding constraints and the impact of diet on the accumulation of body reserves. *Oecologia* 87: 19–28.

Prop, J., Black, J. M. & Skimmings, P. (2003). Travel schedules to the high arctic: Barnacle Geese trade-off the timing of migration with accumulation of fat deposits. *Oikos* 103: 403–414.

Pulido, F. (2000). Evolutionary quantitative genetics of migratory restessness in the Blackcap (*Sylvia atricapilla*). Edn. Wissenschaft, Reihe Biologie, Bd 224. Marburg, Tectum Verlag.

Pulido, F. & Berthold, P. (1998). The microevolution of migratory behaviour in the Blackcap: effects of genetic covariances on evolutionary trajectories. *Biol. Cons. Fauna* 102: 206–211.

Pulido, F. & Berthold, P. (2003). Quantitative genetic analyses of migratory behaviour. pp. 53–77 in *Avian migration* (eds P. Berthold, E. Gwinner & E. Sonnenschein). Berlin, Springer-Verlag.

Pulido, F. & Berthold, P. (2004). Microevolutionary response to climate change. pp. 151–183 in *Birds and climate change. Advances in Ecological Research*, Vol. 35. London: Elsevier.

Pulido, F., Berthold, P. & van Noordwijk, A. J. (1996). Frequency of migrants and migratory activity are genetically correlated in a bird population: evolutionary implications. *Proc. Nat. Acad. Sci. USA* 93: 14642–14647.

Pulido, F., Berthold, P., Mohr, G. & Querner, U. (2001). Heritability of the timing of autumn migration in a natural bird population. *Proc. R. Soc. Lond. B* 268: 953–959.

Pulliam, H. R. & Parker, T. A. (1979). Population regulation of sparrows. *Fortschr. Zool.* 25: 137–147.

Putnam, L. S. (1949). The life history of the Cedar Waxwing. *Wilson Bull.* 61: 141–182.

Putz, K., Wilson, R. P., Charrassin, J. B., Raclot, T., Lage, J., Le Maho, Y., Kierspel, M. A. M., Culik, B. M. & Adelung, D. (1998). Foraging strategy of King Penguins (*Aptenodytes patagonicus*) during summer at the Crozet Islands. *Ecology* 79: 1905–1921.

Putz, K., Ingham, R. J. & Smith, J. G. (2000). Satellite tracking of the winter migration of Magellanic Penguins *Spheniscus magellanicus* breeding in the Falkland Islands. *Ibis* 142: 614–622.

Pützig, P. (1938). Über das Zug verhalten umgesiedelter englischer Stockenten. *Vogelzug* 9: 139–145.

Pynnönen, A. (1939). Beitrage zur Kenntis der Biologie finnischer Spechte 1. *Ann. Soc. Zool.-Bot. Fenn. Vanamo* 7: 1–71.

Quay, W. B. (1989). Insemination of Tennessee Warblers during spring migration. *Condor* 91: 660–670.

Questiau, S., Gielly, L., Clouet, M. & Taberlet, P. (1999). Phylogeographical evidence of gene flow among Common Crossbill *Loxia curvirostra* (Aves, Fringillidae) populations at the continental level. *Heredity* 83: 196–205.

Rabenold, K. N. & Rabenold, P. P. (1985). Variation in altitudinal migration, winter segregation, and site-tenacity in two subspecies of Dark-eyed Juncos in the southern Appalachians. *Auk* 102: 805–819.

Rabøl, J. (1969a). Orientation of autumn migrating Garden Warblers *Sylvia borin* after displacement from western Denmark (Blavand) to eastern Sweden (Ottenby). A preliminary experiment. *Dansk Orn. Foren Tidskr.* 63: 93–104.

Rabøl, J. (1969b). Reversed migration as the cause of westward vagrancy by four *Phylloscopus* warblers. *Br. Birds* 62: 89–92.

Rabøl, J. (1976). The orientation of Pallas' Leaf Warbler *Phylloscopus proregulus* in Europe. *Dansk. Orn. Foren. Tidskr.* 70: 6–16.

Rabøl, J. (1987). Coexistence and competition between overwintering Willow Warblers *Phylloscopus trochilus* and local warblers at Lake Naivaska, Kenya. *Ornis Scand.* 18: 101–121.

Rabøl, J. (1992). Star-navigation in night-migrating passerines. *Dansk Orn. Foren Tidskr.* 86: 177–181.

Rabøl, J. (1993). The orientation system of long-distance passerine migrants displaced in autumn from Denmark to Kenya. *Ornis Scand.* 24: 183–196.

Rabøl, J. (1994). Compensatory orientation in Pied Flycatchers, *Ficedula hypoleuca*, following a geographical displacement. *Dansk Orn. Foren Tidskr.* 88: 171–182.

Rabøl, J. & Noer, H. (1973). Spring migration of the Skylark (*Alauda arvensis*) in Denmark. *Vogelwarte* 27: 50–65.

Ralph, C. J. (1971). An age differential of migrants in coastal California. *Condor* 73: 243–246.

Ralph, C. J. & Mewaldt, L. (1975). Timing of site fixation upon the wintering grounds in sparrows. *Auk* 92: 698–705.

Ralph, C. J. & Mewaldt, L. R. (1976). Homing success in wintering sparrows. *Auk* 93: 1–14.

Ramenofsky, M. (1990). Fat storage and fat metabolism in relation to migration. pp. 214–231 in *Bird migration: physiology and ecophysiology* (ed. E. Gwinner). Berlin, Springer-Verlag.

Ramenofsky, M., Agatsuma, R., Barga, M., Cameron, R., Harm, J., Landys, M. & Ramfar, T. (2003). Migratory behaviour: new insights from captive studies. pp. 97–111 in *Avian migration* (eds P. Berthold, E. Gwinner & E. Sonnenschein). Berlin, Springer-Verlag.

Rappole, J. (1995). *The ecology of migrant birds. A neotropical perspective.* Washington, D.C., Smithsonian Institution Press.

Rappole, J. H. (2005). Evolution of Old and New World migration systems: a response to Bell. *Ardea* 93: 125–131.

Rappole, J. H. & Jones, P. (2002). Evolution of Old and New World migration systems. *Ardea* 90: 525–537.

Rappole, J. H. & Tipton, A. R. (1992). The evolution of avian migration in the Neotropics. *Ornitol. Neotropical* 3: 45–55.

Rappole, J. H. & Warner, D. W. (1976). Relationships between behaviour, physiology, and weather in avian transients at a migration stopover. *Oecologia* 26: 193–212.

Rappole, J. H. & Warner, D. (1980). Ecological aspects of migrant bird behaviour in Veracruz, Mexico. *Migrant birds in the Neotropics: ecology, behaviour, distribution and conservation* (eds A. Keast & E. J. Morton). Washington, D.C., Smithsonian Institute Press.

Rappole, J. H., Morton, E. S., Lovejoy, T. E. & Ruos, J. L. (1983). *Nearctic avian migrants in the Neotropics.* Washington, D.C., United States Fish & Wildlife Service.

Rappole, J. H., Derrickson, S. R. & Hubalek, Z. (2000). Migratory birds and spread of West Nile virus in the Western Hemisphere. *Emerg. Infect. Dis.* 6: 319–328.

Raveling, D. G. (1979). The annual cycle of body composition of Canada Geese with special reference to control of reproduction. *Auk* 96: 234–252.

Raveling, D. G. & Heitmeyer, M. E. (1989). Relationships of population size and recruitment of Pintails to habitat condition and harvest. *J. Wildl. Manage.* 53: 1088–1103.

Raveling, D. G. & LeFebvre, E. A. (1967). Energy metabolism and theoretical flight range of birds. *Bird-Banding* 38: 97–113.

Rayner, J. M. V. (1979). A new approach to animal flight mechanics. *J. Exp. Biol.* 80: 17–54.

Rayner, J. M. V. (1985). Bounding and undulating flight. *J. Theor. Biol.* 117: 47–77.

Rayner, J. M. V. (1988). Form and function in avian flight. *Curr. Ornithol.* 5: 1–66.

Rayner, J. M. V. (1990). The mechanics of flight and bird migration performance. pp. 283–299 in *Bird migration. Physiology and ecophysiology* (ed. E. Gwinner). Berlin, Springer-Verlag.

Recher, H. F. (1966). Some aspects of the ecology of migrant shorebirds. *Ecology* 47: 393–407.

Reed, A. (1975). Migration, homing, and mortality of breeding female Eiders *Somateria mollissima dresseri* of the St Lawrence estuary, Quebec. *Ornis Scand.* 6: 41–47.

Reed, B. P. (1922). Bird catastrophe at Gordon, Nebraska. *Auk* 39: 428.

Reed, E. T., Gauthier, G. & Giroux, J.-F. (2004). Effects of spring conditions on breeding propensity of Greater Snow Goose females. *Anim. Biodiv. Conserv.* 27: 35–46.

Reed, J. M. & Oring, L. W. (1993). Philopatry, site fidelity, dispersal and survival of Spotted Sandpipers. *Auk* 110: 541–551.

Rees, E. C. (1982). The effect of photoperiod on the timing of spring migration in the Bewick's Swan. *Wildfowl* 33: 119–132.

Rees, E. C. (1987). Conflict of choice within pairs of Bewick's Swans regarding their migratory movement to and from wintering grounds. *Anim Behav.* 35: 1685–1693.

Rees, E. C. (1989). Consistency in the timing of migration for individual Bewick's Swans. *Anim. Behav.* 38: 384–393.

Rehfisch, M. M., Clark, N. A., Langston, R. H. W. & Greenwood, J. J. D. (1996). A guide to the provision of refuges for waders: an analysis of 30 years of ringing data from the Wash, England. *J. Appl. Ecol.* 33: 673–687.

Rehfisch, M. M., Insley, H. & Swann, B. (2003). Fidelity of overwintering shorebirds to roosts on the Moray Basin, Scotland: Implications for predicting impacts of habitat loss. *Ardea* 91: 53–70.

Reid, J. C. (1981). Die Schwalbenkatastrophe vom Herbst 1974. *Egretta* 24: 76–80.

Reid, W. V. (1987). The cost of reproduction in the Glaucous-winged Gull. *Oecologia* 74: 458–467.

Reinikainen, A. (1937). The irregular migrations of the Crossbill, *Loxia c. curvirostra*, and their relation to the cone-crop of the conifers. *Ornis Fenn.* 14: 55–64.

Remisiewicz, M. & Wennerberg, L. (2006). Differential migration strategies of the Wood Sandpiper (*Tringa glareola*) – genetic analyses reveal sex differences in morphology and spring migration phenology. *Ornis Fenn.* 83: 1–10.

Rensch, B. (1933). Zoologische Systematik und Artbildungsproblem. *Verh. dtsch. zool. Ges.* 1933: 19.

Restani, M. & Maddox, W. G. (2000). Natal dispersal of Peregrine Falcons in Greenland. *Auk* 117: 500–504.

Rey, P. (1995). Spatio-temporal variation in fruit and frugivorous bird abundance in olive orchards. *Ecology* 76: 1625–1635.

Reynolds, R. E. (1987). Breeding duck population, production and habitat surveys, 1979–85. *Trans. N. Am. Nat. Res. Conf.* 52: 186–205.

Reynolds, R. T. & Linkhart, B. D. (1987). Fidelity to territory and mate in Flammulated Owls. pp. 234–238 in *Biology and conservation of northern forest owls* (eds R. W. Nero, R. J. Clark, R. J. Knapton & R. H. Hamre). Fort Collins, Colorado, U.S. Dept. Agric., Forest Service, Rocky Mountain Forest & Range Experimental Station.

Rezvyi, S. P. & Savinich, I. B. (2001). Post-breeding migration of juvenile and adult Robins (*Erithacus rubecula* L.) in the Lake Ladoga region. *Avian Ecol. Behav.* 6: 33–34.

Rheinwald, G. (1975). The pattern of settling distances of House Martins *Delichon urbica*. *Ardea* 63: 136–145.

Rheinwald, G. & Gutscher, H. (1969). Dispersion und Ortstreue der Mehlschwalbe (*Delichon urbica*). *Die Vogelwelt* 90: 121–140.

Richardson, W. J. (1978). Timing and amount of bird migration in relation to weather: a review. *Oikos* 30: 224–272.

Richardson, W. J. (1990). Timing of bird migration in relation to weather: updated review. pp. 78–101 in *Bird migration. Physiology and ecophysiology* (ed. E. Gwinner). Berlin, Springer-Verlag.

Richdale, L. E. (1963). Biology of the Sooty Shearwater *Puffinus griseus. Proc. Zool. Soc. Lond.* 141: 1–117.

Ricklefs, R. E., Fallon, S. M., Latta, S., Swanson, B. & Bermingham, E. (2005). Migrants and their parasites. pp. 210–221 in *Birds of Two Worlds.* (eds Greenberg, R. & Marra, P. P.). Washington, D. C., Smithsonian Institution.

Riddiford, N. (1978). Observations on the fitness of migrant birds predated by gulls at Dungeness, Kent. *Ringing & Migration* 2: 46–47.

Riddington, R. & Ward, N. (1998). The invasion of Northern Bullfinches to Britain in autumn 1994, with particular reference to the Northern Isles. *Ringing & Migration* 9: 48–52.

Ridgely, R. & Gwynne, J. A. (1989). *A guide to the birds of Panama with Costa Rica, Nicaragua, and Honduras*, 2nd edn. Princeton, Princeton University Press.

Ridgill, S. C. & Fox, A. D. (1990). *Cold weather movements of waterfowl in western Europe*. IWRB Special Publication 13. Slimbridge, International Waterfowl and Wetlands Research Bureau.

Rimmer, C. C. & Darmstadt, C. H. (1996). Non-breeding site fidelity in Northern Shrikes. *J. Field. Ornithol.* 67: 360–366.

Rising, J. D. (1988). Geographic variation in sex ratios and body size in wintering flocks of Savannah Sparrows (*Passerculus sandwichensis*). *Wilson Bull.* 100: 183–203.

Ritz, T., Thalau, P., Philips, J. B., Wiltschko, R. & Wiltschko, W. (2004). Resonance effects indicate a radical-pair mechanism for avian magnetic compass. *Nature* 429: 177–180.

Robb, M. S. (2000). Introduction to vocalisations of Crossbills in north-western Europe. *Dutch Birding* 22: 2.

Robbins, C. S. (1979). Effect of forest fragmentation on bird populations. *Workshop Proceedings: Management of North Central and Northeastern Forests for Nongame Birds, coordinated by R.M. DeGraaf.* USDA Forest Service General Technical Report NC-51.

Robbins, C. S. (1980). Effect of forest fragmentation on breeding bird populations in the piedmont of the Mid-Atlantic region. *Atlantic Naturalist* 33: 31–36.

Robbins, C. S. & Erskine, A. J. (1975). Population trends in non-game birds in North America. *Trans. N. Am. Wildl. Nat. Res. Conf.* 40: 288–293.

Robbins, C. S., Dawson, D. K. & Dowell, B. A. (1989). Habitat area requirements of breeding forest birds of the Middle Atlantic states. *Wildl. Monogr.* 103: 1–34.

Robbins, C. S., Fitzpatrick, J. W. & Hamel, P. B. (1992). A warbler in trouble: *Dendroica cerulea.* pp. 549–562 in *Ecological conservation of neotropical migrant landbirds* (eds J. M. Hagen & D. W. Johnston). Washington, D.C., Smithsonian Institution Press.

Robert, M., Benoit, R. & Savard, J.-P. L. (2002). Relationship among breeding, molting and wintering areas of male Barrow's Goldeneyes (*Bucephala islandica*) in eastern North America. *Auk* 119: 676–684.

Roberts, T. S. (1907a). A Lapland Longspur tragedy. *Auk* 24: 369–377.

Roberts, T. (1907b). Supplemental note to 'A Lapland Longspur tragedy'. *Auk* 24: 449–450.

Robertson, D. (1980). *Rare birds of the west coast.* Pacific Grove, California, Woodcock Publications.

Robertson, G. & Cooke, F. (1999). Winter philopatry in migratory waterfowl. *Auk* 116: 20–34.

Robertson, W. B. (1969). Transatlantic migration of juvenile Sooty Terns. *Nature* 122: 632–634.

Robinson, J. A. & Oring, L. W. (1997). Natal and breeding dispersal in American Avocets. *Auk* 114: 416–430.

Robinson, R. A., Crick, H. Q. P. & Peach, W. J. (2003). Population trends of Swallows *Hirundo rustica* breeding in Britain. *Bird Study* 50: 1–7.

Robinson, S. K. (1992). Population dynamics of breeding Neotropical migrants in a fragmented Ilinois landscape. pp. 408–418 in *Ecology and conservation of neotropical migrant landbirds* (eds J. M. Hagan & D. W. Johnston). Washington, D.C., Smithsonian Institution Press.

Robinson, S. K., Rothstein, S. I., Brittingham, M. C., Petit, L. J. & Grybowski, J. A. (1995). Ecology and behaviour of Cowbirds and their impact on host populations. pp. 428–460 in *Ecology and management of neotropical migratory birds* (ed. T. E. Martin & D.M. Finch). Oxford, Oxford University Press.

Robinson, T. & Minton, C. (1989). The enigmatic Banded Stilt. *Birds Int.* 1: 72–85.

Rockenbauch, D. (1968). Zur Brutbiologie des Turmfalken (*Falco tinnunculus* L.). *Anz. orn. Ges. Bayern* 8: 267–276.

Rockwell, R. F. & Cooke, F. (1977). Gene flow and local adaptation in a colonially nesting dimorphic bird: the Lesser Snow Goose (*Anser caerulescens caerulescens*). *Am. Nat.* 11: 91–97.

Rodenhouse, N. L. & Holmes, R. T. (1992). Results of experiments and natural food reductions for breeding Black-throated Blue Warblers. *Ecology* 73: 357–372.

Rodenhouse, N. L., Sillett, T. S., Doran, P. J. & Holmes, R. T. (2003). Multiple density-dependence mechanisms regulate a migratory bird population during the breeding season. *Proc. R. Roc. Lond. B* 270: 2105–2110.

Rödl, T. (1994). The wintering of territorial Stonechat *Saxicola torquata* pairs in Israel. *J. Ornithol.* 136: 423–433.

Rogers, C. M., Theimer, T. L., Nolan, V. & Ketterson, E. D. (1989). Does dominance determine how far Dark-Eyed Juncos, *Junco hyemalis*, migrate into their winter range. *Anim Behav.* 37: 498–506.

Rogers, D. T. & Odum, E. P. (1966). A study of autumnal postmigrant weights and vernal fattening of North American migrants in the tropics. *Wilson Bull.* 78: 415–435.

Rogers, M. J. (1992). Report on rare birds in Great Britain in 1991, Parrot Crossbill. *Br. Birds* 85: 550.

Rogers, M. J. & the Rarities Committee (2002). Report on rare birds in Great Britain in 2002. *Br. Birds* 96: 542–609.

Rohner, C. (1995). Great Horned Owls and Snowshoe Hares – what causes the time-lag in the numerical response of predators to cyclic prey. *Oikos* 74: 61–68.

Rohner, C. (1996). The numerical response of Great Horned Owls to the Snowshoe Hare cycle: consequences of non-territorial 'floaters' on demography. *J. Anim. Ecol.* 65: 359–370.

Rohwer, F. C. & Anderson, M. G. (1988). Female-biased philopatry, monogamy, and the timing of pair formation in migratory waterfowl. *Curr. Ornithol.* 5: 187–221.

Rohwer, S. & Manning, J. (1990). Differences in timing and number of moults for Baltimore and Bullock's Orioles: implications to hybrid fitness and theories of delayed plumage maturation. *Condor* 92: 125–140.

Roos, G. (1978). Counts of migrating birds and environmental monitoring: long-term changes in the volume of autumn migration at Falsterbo 1942–1977. *Anser* 17: 133–138.

Roos, G. (1984). Migration, wintering and longevity of birds ringed at Falsterbo (1947–1980). *Anser* (suppl 13): 1–208.

Roos, G. (1991). Visible bird migration at Falsterbo in autumn 1989, with a summary of the occurrence of six *Carduelis* species in 1973–90. *Anser* 30: 229–253.

Root, T. (1989). Energy constraints on avian distributions: a reply to Castro. *Ecology* 70: 1183–1185.

Root, T. L., Price, J. T., Hall, K. R. Schneider, S. H., Rosenzweig, C. & Pounds, J. A. (2003). Fingerprints of global warming on plants and animals. *Nature* 421: 57–60.

Roseberry, J. L. (1962). Avian mortality in southern Illinois resulting from severe weather conditions. *Ecology* 43: 739–740.

Rosenzweig, M. L. (1992). Species diversity gradients: we know more and less than we thought. *J. Mammal.* 73: 715–730.

Roshier, D. A. & Reid, J. R. W. (2002). Broad scale processes in dynamic landscapes and the paradox of large populations of desert waterbirds. pp. 148–155 in *Avian landscape ecology: pure and applied issues in the landscape ecology of birds* (eds D. Chamberlain & A. Wilson). Proc. 11th IALE (UK) Conference, University of East Anglia.

Roshier, D. A., Klomp, N. I. & Asmus, M. (2006). Movements of a nomadic waterfowl, Grey Teal *Anas gracilis*, across inland Australia – results from satellite telemetry spanning fifteen months. *Ardea* 94: 461–475.

Rösner, H. U. (1990). Are there age-dependent differences in migration patterns and choice of resting sites in Dunlin *Calidris alpina*. *J. Ornithol.* 131: 121–139.

Roth, R. R. (1976). Effects of a severe thunderstorm on airborne ducks. *Wilson Bull.* 88: 654–656.

Roth, R. R. & Johnson, R. K. (1993). Long-term dynamics of a Wood Thrush population breeding in a forest fragment. *Auk* 110: 37–48.

Rothstein, S. I. & Cook, T. L. (2000). Cowbird management, host population limitation and efforts to save endangered species. pp. 323–332 in *Ecology and management of Cowbirds and their hosts* (eds J. J. M. Smith, T. L. Cook, S. I. Rothstein, S. K. Robinson & S. G. Sealy). Austin, University of Texas Press.

Rothstein, S. I. & Robinson, S. K. (1994). Conservation and coevolutionary implications of brood parasitism by cowbirds. *TREE* 9: 162–164.

Rowan, M. K. (1968). The origins of European Swallows 'wintering' in South Africa. *Ostrich* 39: 76–84.

Rowan, W. (1925). Relation of light to bird migration and development of changes. *Nature* 115: 494–495.

Rowan, W. (1926). On photoperiodism, reproductive periodicity, and the annual migrations of birds and certain fishes. *Proc. Boston Soc. Nat. Hist.* 38: 147–189.

Rowan, W. (1932). Experiments in bird migration-III. The effects of artificial light, castration and certain extracts on the autumn movements of the American Crow (*Corvus brachyrhynchos*). *Proc. Natl Acad. Sci. Wash* 18: 639–654.

Rowan, W. (1946). Experiments in bird migration. *Trans. R. Soc. Can.* 40: 123–135.

Rowan, W. & Batrawi, A. M. (1939). Comment on the gonads of some European migrants collected in East Africa immediately before their spring departure. *Ibis* 14: 58–65.

Rowley, I. (1974). *Bird life*. Sydney, Collins.

Rubolini, D., Pastor, A. G., Pilastro, A. & Spina, F. (2002). Ecological barriers shaping fuel stores in Barn Swallows *Hirundo rustica* following the central and western Mediterranean flyways. *J. Avian Biol.* 33: 15–22.

Ruchomaki, K., Tanhuanpää, M., Ayres, M. P., Kaitaniemi, P., Tammaru, T. & Haukioja, E. (2000). Causes of cyclicity of *Epirrita autumnata* (Leipdoptera, Geometridae): grandiose theory and tedious practice. *Population Ecol.* 42: 211–223.

Rudebeck, G. (1950). The choice of prey and modes of hunting of predatory birds with special reference to their selective effect. *Oikos* 2: 65–88.

Ruegg, K. C. & Smith, T. B. (2002). Not as the crow flies: a historical explanation for circuitous migration in Swainson's Thrush (*Catharus ustulatus*). *Proc. R. Soc. Lond. B* 269: 1375–1381.

Ruelas Inunza, E., Hoffman, S. W., Goodrich, L. J. & Tingay, R. (2000). Conservation strategies for the world's largest known raptor migration flyway: Veracruz river of raptors. pp. 591–615 in *Raptors at risk* (eds R. D. Chancellor & B. U. Meyburg). Berlin, WWGBT/Hancock House.

Ruge, K. (1974). Europäische Schwalbenkatastrophe im Oktober 1974: Bitte 1975 auf die Brutbestände achten! *Vogelwarte* 27: 299–300.

Runfeldt, S. & Wingfield, J. C. (1985). Experimentally prolonged sexual activity in female sparrows delays termination of reproductive activity in their untreated mates. *Anim Behav.* 33: 403–410.

Rüppell, W. (1944). Versuche über Hamfinden ziehender Nebelkrähen nach Verfrachtung. *J. Ornithol.* 92: 106–133.

Rüppell, W. & Schüz, E. (1948). Ergebnis der Verfrachtung von Nebel-krähen (*Corvus corone cornix*) wahrend des Weg zuges. *Vogelwarte* 1: 30–36.

Russell, A. F. (1999). Ecological constraints and the cooperative breeding system of the Long-tailed Tit *Aegithalos caudatus*. PhD thesis, University of Sheffield.

Russell, R. W., Carpenter, F. L., Hixon, M. A. & Paton, D. C. (1992). The impact of variation in stopover habitat quality on migrant Rufuous Hummingbirds. *Conserv. Biol.* 8: 483–490.

Rute, J. (1976). On the speed of migration as observed with the tit (genus *Parus*). *Zool. Muz. Raksti* 15: 34–39.

Rutschke, E. (1990). Zur Etho-Ökologie einer Wildpopulation der Graugans (*Anser anser*). Proc. Int. 100. DO-G Meeting, Current Topics Avian Biol. *J. Ornithol. Sonderh.* 365–371.

Ryder, R. A. (1963). Migration and population dynamics of American Coots in Western North America. *Proc. Int. Ornithol. Congr.* 13: 441–453.

Rydzewski, W. (1939). Compte rendu de l'activité de la Station pour l'étude de migrations des oisaux pour l'année 1937. *Act. Orn. Mus. Zool. Pol.* 2: 431–527.

Safriel, U. N. (1995). The evolution of Palearctic migration – the case for southern ancestry. *Israel J. Zool.* 41: 417–431.

Sagar, P. (1986). Wader counts in New Zealand. *Stilt* 9: 32–34.

Saino, N., Szép, T., Romano, M., Rubolini, D., Spina, F. & Møller, A. P. (2004). Ecological conditions during winter predict arrival date at the breeding quarters in a trans-Saharan migratory bird. *Ecol. Lett.* 7: 21–25.

Salewski, V. & Jones, P. (2006). Palearctic passerines in Afrotropical environments: a review. *J. Ornithol.* 147: 192–201.

Salewski, V., Bairlein, F. & Leisler, B. (2000). Site fidelity of Paleartic passerine migrants in the Northern Guinea savanna zone, West Africa. *Vogelwarte* 40: 298–301.

Salewski, V., Bairlein, F. & Leisler, B. (2002). Different wintering strategies of two Palaearctic migrants in West Africa – a consequence of foraging strategies? *Ibis* 144: 85–93.

Salomonsen, F. (1951). The immigration and breeding of the Fieldfare (*Turdus pilaris* L.) in Greenland. *Proc. Int. Ornithol. Congr.* 10: 515–525.

Salomonsen, F. (1955). The evolutionary significance of bird-migration. *Dan. Biol. Medd.* 22: 1062.

Salomonsen, F. (1967a). Fuglene på Grønland. Copenhagen, Rhodos.

Salomonsen, F. (1967b). Migratory movements of the Arctic Tern (*Sterna paradisaea* Pontoppidan in the southern Ocean. *Biol. Medd. Dan. Vid. Selsk* 24: 1–42.

Salomonsen, F. (1968). The moult migration. *Wildfowl* 19: 5–24.

Sammut, M. & Bonavia, E. (2004). Autumn raptor migration over Buskett, Malta. *Br. Birds* 97: 318–322.

Sandberg, R. (1991). Sunset orientation of Robins, *Erithacus rubecula*, with different fields of sky vision. *Behav. Ecol Sociobiol.* 28: 77–83.

Sandberg, R. (1994). Interaction of body condition and magnetic orientation in autumn migrating Robins, *Erithacus rubecula*. *Anim. Behav.* 47: 679–686.

Sandberg, R. (1996). Fat reserves of migrating passerines at arrival on the breeding grounds in Swedish Lapland. *Ibis* 138: 514–524.

Sandberg, R. & Moore, F. R. (1996a). Fat stores and arrival on the breeding grounds: reproductive consequences for passerine migrants. *Oikos* 77: 577–581.

Sandberg, R. & Moore, F. R. (1996b). Migratory orientation of Red-eyed Vireos, *Vireo olivaceus*, in relation to energetic condition and ecological context. *Behav. Ecol. Sociobiol.* 39: 1–10.

Sandberg, R., Bäckman, J. & Ottosson, U. (1998). Orientation of Snow Buntings (*Plectrophenax nivalis*) close to the magnetic North Pole. *J. Exp. Biol.* 201: 1859–1870.

Sandberg, R., Ottosson, U. & Pettersson, J. (1991). Magnetic orientation of migratory Wheatears (*Oenanthe oenanthe*) in Sweden and Greenland. *J. Exp. Biol.* 155: 51–64.

Sandberg, R., Moore, F. R., Bäckman, J. & Lõhmus, M. (2002). Orientation of nocturnally migrating Swainson's Thrush at dawn and dusk: the importance of energetic condition and geomagnetic cues. *Auk* 119: 201–209.

Sandercock, B. K. & Gratto-Trevor, C. L. (1997). Local survival in Semipalmated Sandpipers *Calidris pusilla* breeding at La Perouse Bay, Canada. *Ibis* 139: 305–312.

Sanderson, F. J., Donald, P. F., Pain, D. J., Burfield, I. J. & van Bommel, F. P. J. (2006). Long-term population declines in Afro-Palearctic migrant birds. *Biol. Conserv.* 131: 93–105.

Sandström, U. (1991). Enhanced predation rates on cavity bird nests at deciduous forest edges – an experimental study. *Ornis Fenn.* 68: 93–98.

Sansum, E. L. & King, J. R. (1976). Long-term effects of constant photoperiods on testicular cycles of White-crowned Sparrows (*Zonotrichia leucophrys gambelii*). *Physiol. Zool.* 49: 407–416.

Sapir, N., Tsurim, I., Gal, B. & Abramsky, Z. (2004). The effect of water availability on fuel deposition of two staging *Sylvia* warblers. *J. Avian Biol.* 35: 25–32.

Sappington, J. N. (1977). Breeding biology of House Sparrows in Mississippi. *Wilson Bull.* 89: 300–309.

Sauer, E. G. F. & Sauer, E. M. (1955). Zur Frage der nächtlichen Zugorientierung von Grasmücken. *Rev. Suisse Zool.* 62: 250–259.

Sauer, E. G. F. & Sauer, E. M. (1960). Star navigation of nocturnal migrating birds. The 1958 planetarium experiment – Cold Spring Harbour Symposium. *Quant. Biol.* 25: 463–473.

Sauer, F. (1957). Die Sternenorientierung nächtlich ziehender Grasmücken (*Sylvia atricapilla, borin* und *curruca*). *Z. Tierpsych.* 14: 29–70.

Sauer, J. R. & Droege, S. (1992). Geographic patterns in population trends of Neotropical migrants in North America. pp. 26–42 in *Ecology and conservation of Neotropical migrant landbirds* (eds J. M. Hagan & D. W. Johnston). Washington, D.C., Smithsonian Institution Press.

Saunders, W. E. (1907). A migration disaster in Western Ontario. *Auk* 24: 108–111.

Saurola, P. (1981). Migration of the Sparrowhawk *Accipiter nisus* as revealed by Finnish ringing and migration data. *Lintumies* 16: 10–18.

Saurola, P. (1983). Movements of Short-eared Owl (*Asio flammeus*) and Long-eared Owl (*A. otus*) according to Finnish ring recoveries. *Lintumies* 18: 67–71.

Saurola, P. (1987). Mate and nest-site fidelity in Ural and Tawny Owls. *Biology and conservation of northern forest owls* (eds R. W. Nero, R. J. Clark, R. J. Knapton & R. H. Hamre). Fort Collins, CO, USDA Forestry Service General Technical Report RM-142.

Saurola, P. (1989). Ural Owl. pp. 327–345 in *Lifetime reproduction in birds* (ed. I. Newton). London, Academic Press.

Saurola, P. (1994). African non-breeding areas of Fennoscandian Ospreys *Pandion haliaetus*: a ring recovery analysis. *Ostrich* 65: 127–136.

Saurola, P. (1997). Monitoring Finnish owls 1982–1996: methods and results. pp. 363–380 in *Biology and conservation of owls of the northern hemisphere* (eds J. R. Duncan, D. H. Johnson & T. H. Nicholls). Second International Symposium, February 5–9, 1997. Winnipeg, Manitoba, Canada. United States Department of Agriculture.

Saurola, P. (2002). Natal dispersal distances of Finnish owls: results from ringing. pp. 42–55 in *Ecology and conservation of owls* (eds I. Newton, R. Kavanagh, J. Olsen & I. Taylor). Collingwood, Victoria, Australia, CSIRO Publishing.

Sauvage, A., Rumsey, S. & Rodwell, S. (1998). Recurrence of Palaearctic birds in the lower Senegal river valley. *Malimbus* 20: 33–53.

Savard, J.-P. L. (1985). Evidence of long-term pair bonds in Barrow's Goldeneye (*Bucephala islandica*). *Auk* 102: 389–391.

Savard, J.-P. L. & Eadie, J. M. (1989). Survival and breeding philopatry in Barrow and Common Goldeneyes. *Condor* 91: 198–203.

Schaub, M. & Jenni, L. (2000). Body mass of six long-distance migrant passerine species along the autumn migration route. *J. Ornithol.* 141: 441–460.

Schaub, M. & Jenni, L. (2001a). Stopover durations of three warbler species along their autumn migration route. *Oecologia* 128: 217–227.

Schaub, M. & Jenni, L. (2001b). Variation of fuelling rates among sites, days and individuals in migrating passerine birds. *Funct. Ecol.* 15: 584–594.

Schaub, M., Kania, W. & Koppen, U. (2005). Variation of primary production during winter induces synchrony in survival rates in migratory White Storks *Ciconia ciconia*. *J. Anim. Ecol.* 74: 656–666.

Schaub, M., Pradel, R., Jenni, L. & Lebreton, J. D. (2001). Migrating birds stop over longer than usually thought: an improved capture-recapture analysis. *Ecology* 82: 852–859.

Schenk, J. (1934). Die Brutinvasionen des Rosenstares in Ungarn in den Jahren 1932 und 1933. *Aquila* 38–41: 136–153.

Schields, W. M. (1984). Factors affecting nest and site fidelity in Adirondack Barn Swallows *Hirundo rustica*. *Auk* 101: 780–789.

Schifferli, A. (1965). Vom Zug verhalten in der Schweitz brutenden Turmfalken, *Falco tinnunculus*, nach den Ringfunden. *Orn. Beob.* 62: 1–13.

Schifferli, A. (1967). Vom Zug Schweizerischer und Deutscher Schwarzer Milane nach Ringfunden. *Orn. Beob.* 64: 34–51.

Schindler, J., Berthold, P. & Bairlein, F. (1981). Über den Einfluß simulierter Wetterbedingungen auf das endogene Zugzeitprogramm der Gartengrasmücke *Sylvia borin*. *Vogelwarte* 31: 14–32.

Schloss, W. (1984). Ringfunde des Fichtenkreuzschnabels (*Loxia curvirostra*). *Auspicium* 7: 257–284.

Schmidt, R. (1989). Änderungen im Zugverhalten des Kormorans (*Phalacrocorax carbo*) im Zusammenhang mit seinem Bestandsanstieg. *Beitr. Vogelk.* 35: 219–221.

Schmidt, R. C. & Vauk, G. (1981). Zug Ringfunde auf Helgoland durch ziehender Waldohreulen und Sunpfohreulen (*Asio otus* and *A. flammeus*). *Vogelwelt* 102: 180–189.

Schmidt-Koenig, K., Ganzhorn, J. U. & Ranvaud, R. (1991). The sun compass. pp. 1–15 in *Orientation in birds* (ed. P. Berthold). Basel, Birkhäuser.

Schmitz, P. & Steiner, F. (2006). Autumn migration of Reed Buntings *Emberiza schoeniclus* in Switzerland. *Bird Study* 23: 333–338.

Schmueli, M., Izhaki, I., Arieli, A. & Arad, Z. (2000). Energy requirements of migrating Great White Pelicans *Pelecanus onocrotalus*. *Ibis* 142: 208–216.

Schneider, D. C. & Harrington, B. A. (1981). Timing of shorebird migration in relation to prey depletion. *Auk* 98: 801–811.

Scholander, S. I. (1955). Land birds over the western North Atlantic. *Auk* 72: 225–239.

Schorger, A. W. (1952). Ducks killed during a storm at Hot Springs, South Dakota. *Wilson Bull.* 64: 113.

Schroeder, M. A. (1985). Behavioural differences of female Spruce Grouse undertaking short and long migrations. *Condor* 87: 281–286.

Schroeder, M. A. & Braun, C. E. (1993). Partial migration in a population of Greater Prairie Chickens in northeastern Colorado. *Auk* 110: 21–28.

Schueck, L., Maechtle, T., Fuller, M., Bates, K., Seegar, W. S. & Ward, J. (1998). Movements of Ferruginous Hawks through western North America. *J. Idaho Acad. Sci.* 34: 11–12.

Schulz, H. (1998). World status and conservation of the White Stork. *Torgos* 28: 49–65.

Schüz, E. (1938). Auflassung ostpreussischer Jungstörche in England 1936. *Vogelzug* 9: 65–70.

Schüz, E. (1945). Der europäischer Rauhfussbuzzard, *Buteo l. lagopus* (Brünn.), als Invasionsvogel. *Jahr. Vereins Vaterlandische Natuurkunde Württemburg* 97–101: 125–150.

Schüz, E. (1949). Die Spat-Auflassung ostpreussischer Jungstörche in West-Deutschland durch die Vogelwarte Rossitten 1933. *Vogelwarte* 15: 63–78.

Schüz, E. (1950). Früh-Auflassung ostpreussischer Jungstorche in West-Deutschland durch die Vogelwarte Rossitten 1933–36. *Bonner Zool. Beitr.* 1: 239–253.

Schüz, E. & Weigold, H. (1931). *Atlas des Vogelzugs nach den Beringungsergebnissen bei paläarktischen Vögeln*. Berlin, Frudlander & Sohn.

Schüz, E., Berthold, P., Gwinner, E. & Oelke, H. (1971). *Grundriss der Vogelzugskunde*. Berlin, Parey.

Schwab, R. G. (1971). Circannual testicular periodicity in the European Starling in the absence of photoperiodic change. in *Biochronometry* (ed. M. Menaker). Washington, D. C., National Academy of Sciences.

Schwabl, H. (1983). Ausprägung und Bedeutung des Teilzugverhaltens einer sudwestdeutschen Population der Amsel *Turdus merula*. *J. Ornithol.* 124: 101–115.

Schwabl, H. & Silverin, B. (1990). Control of partial migration and autumnal behaviour. pp. 144–155 in *Bird migration: physiology and ecophysiology* (ed. E. Gwinner). Berlin, Springer-Verlag.

Schwabl, H., Schwabl-Benzinger, I., Goldsmith, A. R. & Farner, D. S. (1988). Effects of ovariectomy on long-day-induced premigratory fat deposition, plasma levels of luteinising hormone and prolactin, and molt in White-crowned Sparrows *Zonotrichia leucophrys gambelii*. *Gen. Comp. Endocrinol.* 71: 398–405.

Schwabl, H., Bairlein, F. & Gwinner, E. (1991a). Basal and stress-induced corticosterone levels of Garden Warblers *Sylvia borin*, during migration. *J. Comp. Physiol. B*: 161–576.

Schwabl, H., Gwinner, E., Benvenuti, S. & Ioalé, P. (1991b). Exposure of Dunnocks (*Prunella modularis*) to their previous wintering site modifies autumnal activity pattern: evidence for site recognition? *Ethology* 88: 35–45.

Schwartz, P. (1964). The Northern Waterthrush in Venezuela. *Living Bird* 3: 169–184.

Schwilch, R. & Jenni, L. (2001). Low initial refueling rate at stopover sites: a methodological effect? *Auk* 118: 698–708.

Schwilch, R., Grattarola, A., Spina, F. & Jenni, L. (2002a). Protein loss during long distance migratory flights in passerine birds: adaptation or constraint. *J. Exp. Biol.* 205: 587–695.

Schwilch, R., Piersma, T., Holmgren, N. M. A. & Jenni, L. (2002b). Do migratory birds need a nap after a long non-stop flight? *Ardea* 90: 149–154.

Scott, D. A. & Rose, P. M. (1996). *Atlas of Anatidae populations in Africa and western Asia.* Wetlands International Publication No. 41. The Netherlands, Wetlands International.

Scott, D. K. (1980). The behaviour of Bewick's Swans at the Welney Wildfowl Refuge, Norfolk, and on surrounding fens: a comparison. *Wildfowl* 31: 5–18. OR

Scott, J. D., Fernando, K., Banerjee, S. N., Durden, L. A., Byrne, S. K., Banerjee, M., Mann, R. B. & Morshed, M. G. (2001). Birds disperse ixodid (Acari: Ixodidae) and *Borrelia burgdorferi*-infected ticks in Canada. *J. Med. Entomol.* 38: 493–500.

Sealy, S. G. (1966). Swallow mortality at Moose Mountain. *Blue Jay* 24: 17–18.

Sealy, S. G. (1989). Defence of nectar resources by migrating Cape May Warblers. *J. Field Ornithol.* 60: 89–93.

Sedgewick, J. A. (2004). Site fidelity, territory fidelity, and natal philopatry in Willow Flycatchers (*Empidonax traillii*). *Auk* 121: 1103–1121.

Seegar, W. S., Henke, M. B., Schor, M. & Stone, M. (1999). Next generation satellite-based technology for conservation and bird strike science. pp. 91–99 in *Migrating birds know no boundaries* (eds Y. Leshem, Y. Mandelik & J. Shamoun-Baranes). Latron, Israel, International Centre for the study of Bird Migration.

Segal, S. (1960). Bird tragedy at the dunes. *Indiana Audubon Q.* 38: 23–35.

Selås, V., Hogstad, O., Andersson, G. & von Proschwitz, T. (2001). Population cycles of Autumnal Moth, *Epirrita autumnata*, in relation to birch mast seeding. *Oecologia* 129: 213–219.

Sellers, R. M. (1984). Movements of Coal, Marsh and Willow Tits in Britain. *Ringing & Migration* 5: 79–89.

Semm, P. & Beason, R. (1990). Responses to small magnetic variations by trigeminal system of the Bobolink. *Brain Res. Bull.* 25: 735–740.

Senar, J. C., Borras, A., Cabrera, T. & Cabrera, J. (1993). Testing for the relationship between coniferous crop stability and Common Crossbill residence. *J. Field Ornithol.* 64: 464–469.

Senar, J. C., Burton, P. J. K. & Metcalfe, N. B. (1992). Variation in the nomadic tendency of a wintering finch *Carduelis spinus* and its relationship with body condition. *Ornis Scand.* 23: 63–72.

Senar, J. C., Copete, J. L. & Metcalfe, N. B. (1990). Dominance relationships between resident and transient wintering Siskins. *Ornis Scand.* 21: 129–132.

Sergio, F., Blas, J., Forrero, M. G., Donazur, J. A. & Hiraldo, F. (2007). Sequential settlement and site dependence in a migratory raptor. *Behav. Ecol.* 18: 811–821.

Serie, J. R. & Sharp, D. E. (1989). Body weight and composition dynamics of fall migrating Canvasbacks. *J. Wildl. Manage.* 53: 431–441.

Serle, W. (1956). Migrant land birds at sea off west Africa. *Ibis* 98: 307–311.

Serra, L. (2000). How do Palaearctic Grey Plovers adapt primary moult to time constraints? An overview across four continents. *Wader Study Group Bull.* 93: 11–12.

Serra, L. & Underhill, L. G. (2004). The regulation of primary molt speed in the Grey Plover *Pluvialis squatarola*.

Serrano, D., Tella, J. L., Forero, R. G. & Donázar, J. A. (2001). Factors affecting breeding dispersal in the facultative Lesser Kestrel: individual experience vs. conspecific cues. *J. Anim. Ecol.* 70: 568–578.

Serventy, D. L. (1971). The biology of desert birds. pp. 287–339 in *Avian biology* Vol. 1. (eds D. S. Farner & J. R. King). London, Academic Press.

Serventy, D. L. & Curry, P. J. (1984). Observations on colony size, breeding success, recruitment and inter-colony dispersal in a Tasmanian colony of Short-tailed Shearwaters *Puffinus tenuirostris* over a 30-year period. *Emu* 84: 71–79.

Serventy, D. L., Gunn, B. M., Skira, I. J., Bradley, J. S. & Wooller, R. D. (1989). Fledgling translocation and philopatry in a seabird. *Oecologia* 81: 428–429.

Shaffer, S. A., Tremblay, Y., Weimerskirch, H., Scott, D., Thompson, D. R., Sagar, P. M., Moller, H., Taylor, G. A., Foley, D. G., Block, B. A. & Costa, D. P. (2006). Migratory shearwaters integrate oceanic resources across the Pacific Ocean in an endless summer. PNAS 103: 2799–12802.

Shaw, G. (1990). Timing and fidelity of breeding for Siskins *Carduelis spinus* in Scottish conifer plantations. *Bird Study* 37: 30–35.

Shelford, V. E. (1945). The relation of Snowy Owl migration to the abundance of the Collared Lemming. *Auk* 62: 592–596.

Sherry, T. W. & Holmes, R. T. (1988). Habitat selection by breeding American Redstarts in response to a dominant competitor, the Least Flycatcher. *Auk* 105: 350–364.

Sherry, T. W. & Holmes, R. T. (1989). Age-specific social dominance affects habitat use by breeding American Redstarts (*Setophaga ruticilla*) – a removal experiment. *Behav. Ecol. Sociobiol.* 25: 327–333.

Sherry, T. W. & Holmes, R. T. (1992). Population fluctuations in a long distance neotropical migrant: demographic evidence for the importance of breeding season events in the American Redstart. pp. 431–442 in *Ecology and conservation of neotropical migrant landbirds* (eds J. M. Hagan & D. W. Johnston). Washington, D.C., Smithsonian Press.

Sherry, T. W. & Holmes, R. T. (1995). Summer versus winter limitation of populations: what are the issues and what is the evidence? pp. 85–120 in *Ecology and management of neotropical migratory birds* (eds T. E. Martin & D. M. Finch). Oxford, Oxford University Press.

Shields, W. M. (1984). Factors affecting nest and site fidelity in Adirondack Barn Swallows (*Hirundo rustica*). *Auk* 101: 780–789.

Shirihai, H. (1996). *The birds of Israel*. London, Academic Press.

Shirihai, H., Yosef, R., Alon, D., Kirwan, G. M. & Spaar, R. (2000). *Raptor migration in Israel and the Middle East. A summary of 30 years of field research*. Israel, International Birding & Research Centre.

Shumakov, M. E. (2001). Development of orientation abilities and migratory direction in nocturnal migrants on the Courish Spit. *Avian Ecol. Behav.* 6: 76–77.

Shumakov, M. E., Virogradova, N. V. & Sukhov, A. V. (2001). Position in the migratory route as a factor controlling the duration of the directed movement in migrating birds. *Avian Ecol. Behav.* 6: 78–79.

Siivonen, L. (1941). Über die Kausatzusammenhage der Wanderungen beim Seidenschwanz *Bombicilla g. garrulus* (L). *Ann. Soc. Zool.-Bot. Fenn. Vanamo* 8: 1–38.

Sillett, T. S. & Holmes, R. T. (2002). Variation in survivorship of a migratory songbird throughout its annual cycle. *J. Anim. Ecol.* 71: 296–308.

Sillet, T. S. & Holmes, R. T. (2005). Long-term demographic trends, limiting factors, and the strength of density dependence in a breeding population of a migratory songbird. pp. 426–436 in *Birds of Two Worlds*. (eds Greenberg, R. & Marra, P. P.). Washington, D. C., Smithsonian Institution.

Sillett, T. S., Holmes, R. T. & Sherry, T. W. (2000). Impacts of a global climate cycle on population dynamics of a migratory songbird. *Science* 288: 2040–2042.

Silverin, B. (2003). Behavioural and hormonal dynamics in a partial migrant – the Willow Tit. pp. 127–140 in *Avian migration* (eds P. Berthold, E. Gwinner & E. Sonnenschein). Berlin, Springer-Verlag.

Silverin, B., Massa, R. & Stokkan, K. A. (1993). Photoperiodic adaptation to breeding at different latitudes in Great Tits. *Gen. Comp. Endocrinol.* 90: 14–22.

Silvola, T. (1967). Changes in the bird populations in Utsjoki, Finnish Lapland in 1964–1966, caused by the mass-occurrence of the caterpillar *Oporinia autumnata*. *Ornis Fenn.* 44: 65–67.

Simek, J. (2001). Patterns of breeding fidelity in the Red-backed Shrike (*Lanius collurio*). *Ornis Fenn.* 78: 61–71.

Simons, A. M. (2004). Many wrongs: the advantage of group navigation. *Trends Ecol. Evol.* 19: 453–455.

Sinclair, A. R. E. (1978). Factors affecting the food-supply and breeding season of resident birds and movements of Palaearctic migrants in a tropical African savanna. *Ibis* 120: 480–497.

Siriwardena, G. & Wernham, C. (2002). Movement patterns of British and Irish birds. pp. 70–102 in *Migration atlas: movements of birds of Britain and Ireland* (eds C. V. Wernham, M. P. Toms, J. H. Marchant, J. A. Clark, J. M. Siriwardena & S. R. Baillie). London, T. & A. D. Poyser.

Sitters, H., Minton, C., Collins, P., Etheridge, B., Hassell, C. & O'Connor, F. (2004). Extraordinary numbers of Oriental Pratincoles in NW Australia. *Wader Study Group Bull.* 103: 1–6.

Skira, I. (1991). The Short-tailed Shearwater: a review of its biology. *Corella* 15: 45–52.

Sladen, W. J. L., Lishman, W., Ellis, D. H., Shire, G. & Rininger, D. L. (2002). Teaching migration routes to Canada Geese and Trumpeter Swans using ultralight aircraft, 1990–2001. *Waterbirds* 25 (Spec. Publ. 1): 132–137.

Slagsvold, T. (1976). Arrival of birds from spring migration in relation to vegetational development. *Norw. J. Zool.* 24: 161–173.

Slud, P. (1976). Geographic and climatic relationships of avifaunas with special reference to comparative distribution in the Neotropics. *Smithsonian Contributions to Zoology* 212: 1–149.

Small, M. F. & Hunter, M. l. (1988). Forest fragmentation and avian nest-predation in forested landscapes. *Oecologia* 76: 62–64.

Smallwood, J. A. (1988). A mechanism of sexual segregation by habitat in American Kestrels (*Falco sparverius*). *Auk* 105: 36–46.

Smith, G. R. (1975). Recent European outbreaks of botulism in waterfowl. *I.W.R.B. Bull.* 39/40: 72–74.

Smith, H. G. & Nilsson, J.-Å. (1987). Intraspecific variation in migratory pattern of a partial migrant, the Blue Tit (*Parus caeruleus*): an evaluation of different hypotheses. *Auk* 104: 109–115.

Smith, J. J. M., Tait, M. J. & Zanette, L. (2002). Removing Brown-headed Cowbirds increases seasonal fecundity and population growth in Song Sparrows. *Ecology* 83: 3037–3047.

Smith, J. N. M., Montgomerie, R. D., Taitt, M. J. & Yom-Tov, Y. (1980). A winter feeding experiment on an island Song Sparrow population. *Oecologia* 47: 164–170.

Smith, K. B. (1968). Spring migration through southeast Morocco. *Ibis* 110: 452–492.

Smith, K. W., Reed, J. M. & Trevis, B. E. (1992). Habitat use and site fidelity of Green Sandpipers *Tringa ochropus* wintering in southern England. *Bird Study* 39: 155–164.

Smith, M. A. (1964). Cohoe-Alaska. *Audubon Field Notes* 18: 478–479.

Smith, N. (1997). Observations of wintering Snowy Owls (*Nyctea scandiaca*) at Logan Airport, East Boston, Massachusetts from 1981–1997. pp. 591–597 in *Biology and conservation of owls of the northern hemisphere* (ed. J. R. Duncan, D. H. Johnson & T. H. Nicholls). Second International Symposium, February 5–9, 1997. Winnipeg, Manitoba, Canada. United States Department of Agriculture.

Smith, N. G. (1980). Hawk and vulture migrations in the Neotropics. pp. 50–61 in *Migrant birds in the Neotropics* (eds A. Keast & E. S. Morton). Washington, D.C., Smithsonian Institution Press.

Smith, N. G. (1985a). Dynamics of the transisthmian migration of raptors between Central and South America. pp. 271–290 in *Conservation studies on raptors* (eds I. Newton & R. D. Chancellor). Cambridge, ICPB Technical Bulletin No. 5.

Smith, N. G., Goldstein, D. L. & Bartholomew, G. A. (1986). Is long-distance migration possible for soaring hawks using only stored fat? *Auk* 103: 607–611.

Smith, P. W. & Houghton, N. T. (1984). Fidelity of Semipalmated Plovers to a migration stopover area. *J. Field Ornithol.* 55: 247–249.

Smith, R. D., Marquiss, M., Rae, R. & Metcalfe, N. B. (1993). Age and sex variation in choice of wintering site by Snow Buntings: the effect of altitude. *Ardea* 81: 47–52.

Smith, R. I. (1970). Response of Pintail breeding populations to drought. *J. Wildl. Manage.* 34: 943–946.

Smith, R. J. & Moore, F. R. (2003). Arrival fat and reproductive performance in a long-distance passerine migrant. *Oecologia* 134: 325–331.

Smith, R. J. & Moore, F. R. (2005). Arrival timing and seasonal reproductive performance in a long-distance migratory landbird. *Behav. Ecol. Sociobiol.* 57: 231–239.

Smith, R. W., Brown, I. L. & Mewaldt, L. R. (1969). Annual activity patterns of caged non-migratory White-crowned Sparrows. *Wilson Bull.* 81: 419–440.

Smith, T. B., Meehan, T. D. & Wolf, B. O. (2003). Assessing migration patterns of Sharp-shinned Hawks *Accipiter striatus* using stable isotope and band encounter analysis. *J. Avian Biol.* 34: 387–392.

Smith, T. B., Clegg, S. M., Kimura, M., Ruegg, K., Milá, B. & Lovette, I. (2005). Molecular genetic approaches to linking breeding and overwintering areas in five neotropical migrant passerines. pp. 222–234 in *Birds of Two Worlds*. (eds Greenberg, R. & Marra, P. P.). Washington, D. C., Smithsonian Institution.

Smith, W. P. (1929). Some observations of the effect of a late snowstorm upon birdlife. *Auk* 46: 557–558.

Sniegowski, P. D., Ketterson, E. D. & Nolan, V. (1988). Can experience alter the avian annual cycle? Results of migration experiments with Indigo Buntings. *Ethology* 79: 333–341.

Snow, B. & Snow, D. (1988). *Birds and berries*. Calton, T & A D Poyser.

Snow, D. W. (1956). The annual mortality of the Blue Tit in different parts of its range. *Br. Birds* 49: 174–177.

Snow, D. W. (1962). A field study of the Black-and-White Manakin, *Manacus manacus*, in Trinidad. *Zoologica* 47: 65–104.

Snow, D. W. (1966). Moult and the breeding cycle in Darwin's Finches. *J. Ornithol.* 107: 283–291.

Snow, D. W. (1968). Movements and mortality of British Kestrels *Falco tinnunculus*. *Bird Study* 15: 65–83.

Snow, D. W. (1978). Relationships between European and African avifaunas. *Bird Study* 25: 134–148.

Soikkeli, M. (1970). Dispersal of Dunlin *Calidris alpina* in relation to sites of birth and breeding. *Ornis Fenn.* 47: 1–9.

Sokolov, L. V. (1997). Philopatry of migratory birds. *Phys. Gen. Biol. Rev.* 11: 1–58.

Sokolov, L. (1999). Population dynamics in 20 sedentary and migratory passerine species of the Courish Spit on the Baltic Sea. *Avian Ecol. Behav.* 3: 23–50.

Sokolov, L. V. (2000a). Spring ambient temperature as an important factor controlling timing of arrival, breeding, post-fledging dispersal and breeding success of Pied Flycatchers *Ficedula hypoleuca*. *Avian Ecol. Behav.* 5: 79–104.

Sokolov, L. (2000b). Philopatry, dispersal and population structure of passerines on the Courish Spit of the Baltic Sea. *Vogelwarte* 40: 302–314.

Sokolov, L. V. (2001). Climate influence on year-to-year variations in timing of migration and breeding phenology in passerines on the Courish Spit. *Ring* 231: 159–166.

Sokolov, L. V. (2006). Influence of the global warming on the timing of migration and breeding of passerines in the 20th century. *Zool. Zh.* 85: 317–342.

Sokolov, L. & Payevsky, V. A. (1998). Spring temperatures influence year-to-year variations in the breeding phenology of passerines on the Courish Spit, eastern Baltic. *Avian Ecol. Behav.* 1: 22–36.

Sokolov, L. V. & Vysotsky, V. G. (1999). Homing ability of the Pied Flycatcher *Ficedula hypoleuca*. *Avian Ecol. Behav.* 3: 51–58.

Sokolov, L. V., Markovets, M. Y. & Morozov, Y. G. (1999). Long-term dynamics of the mean date of autumn migration in passerines on the Courish Spit of the Baltic Sea. *Avian Ecol. Behav.* 2: 1–18.

Sokolov, L., Markovets, M. Y., Shapoval, A. P. & Morozov, Y. G. (1998). Long-term trends in the timing of spring migration of passerines on the Courish Spit of the Baltic Sea. *Avian Ecol. Behav.* 1: 1–21.

Sokolov, L. V., Vysotsky, V. G., Gavrilov, V. M. & Kerimov, A. B. (1994). Experimental test of philopatry: population differences in Pied Flycatcher *Ficedula hypoleuca*. *Dokl. Akad. Sci. ser. Biol.* 335: 538–539. (In Russian).

Sokolov, L. V., Baumanis, J., Leivits, A., Poluda, A. M., Yefremov, V. D., Markovets, M. Y., Morozov, Y. G. & Shapoval, A. P. (2001). Comparative analysis of long-term monitoring data of numbers of passerines in nine European countries in the second half of the 20th century. *Avian Ecol. Behav.* 7: 41–74.

Solonen, T. (1979). Population dynamics of the Garden Warbler *Sylvia borin* in southern Finland. *Ornis Fenn.* 56: 3–12.

Solonen, T. (1986). Breeding of the Great Grey Owl *Strix nebulosa* in Finland. *Lintumies* 21: 11–18.

Sonerud, G. (1984). Effect of snow cover on seasonal changes in diet, habitat and regional distribution of raptors that prey on small mammals in boreal zones of Fennoscandia. *Holarctic Ecol.* 9: 33–47.

Sonerud, G. (1997). Hawk Owls in Fennoscandia: population fluctuations, effects of modern forestry, and recommendations on improving foraging habitats. *J. Raptor Res.* 31: 167–174.

Sonerud, G. A., Solheim, R. & Prestrud, K. (1988). Dispersal of Tengmalm's Owl *Aegolius funereus* in relation to prey availability and nesting success. *Ornis Scand.* 19: 175–181.

Southern, H. N. (1938a). The spring migration of the Swallow over Europe. *Br. Birds* 32: 4–7.

Southern, H. N. (1938b). The spring migration of the Willow Warbler over Europe. *Br. Birds* 32: 202–206.

Southern, H. N. (1939). The spring migration of the Redstart over Europe. *Brit. Birds* 33: 34–38.

Southern, H. N. (1940). The spring migration of the Wood Warbler over Europe. *Brit. Birds* 33: 34–38.

Southern, H. N. (1941). The spring migration of the Red-backed Shrike over Europe. *Brit. Birds* 33: 34–38.

Southern, H. N. (1970). The natural control of a population of Tawny Owls (*Stix aluco*). *J. Zool. Lond.* 162: 197–285.

Southern, W. E. (1959). Homing of Purple Martins. *Wilson Bull.* 71: 254–261.

Spaans, A. L. (1977). Are Starlings faithful to their individual winter quarters? *Ardea* 64: 83–87.

Spaar, R. (1997). Flight strategies of migrating raptors: a comparative study of interspecific variation in flight characteristics. *Ibis* 139: 523–535.

Spaar, R. & Bruderer, B. (1996). Soaring migration of Steppe Eagles *Aquila nipalensis* in southern Israel: flight behaviour under various wind and thermal conditions. *J. Avian Biol.* 27: 289–301.

Spaepen, J. & Cauteren, F. v. (1968). Migration of the Skylark, *Alauda arvensis* L. *Gerfaut* 58: 24–77.

Sparks, T. H. (1999). Phenology and the changing pattern of bird migration in Britain. *Int. J. Biometeorol.* 42: 134–138.

Sparks, T. (2000). The long-term phenology of woodland species in Britain. pp. 98–105 in *Long-term studies in British woodlands* (eds K. S. Kirby & M. D. Morecroft). Peterborough, English Nature Science Series.

Sparks, T. & Mason, C. (2001). Dates of arrivals and departures of spring migrants taken from Essex Bird Reports 1950–1998. pp. 154–164 in *Essex Bird Report 1999* (eds A. Mullins & H. Vaughan). Essex, Essex Birdwatching Society.

Sparks, T., Heyen, H., Braslavska, O. & Lehikoinen, E. (1999). Are European birds migrating earlier. *BTO News* 223: 8–9.

Sparks, T. H., Crick, H. Q. C., Bunn, P. O. & Sokolov, L. V. (2003). Birds. pp. 421–436 in *Phenology: an integrative environmental science* (ed. M. D. Schwartz). Dordrecht, Kluwer Academic.

Sparks, T. H., Bairlein, F., Bojarinova, J. G., Huppop, O., Lehikoinen, E. A., Rainios, K., Sokolov, L. V. & Walker, D. (2005). Examining the total arrival distribution of migratory birds. *Global Change Biol.* 11: 22–30.

Sparks, T. H., Huber, K., Bland, R. L., Crick, H. Q. P., Croxton, P. J., Flood, J., Loxton, R. G., Mason, C. F., Newnham, J. A. & Tryjanowski, P. (2007). How consistent are trends in arrival (and departure) dates of migratory birds in the UK? *J. Ornithol.,* 148: in press.

Spear, L. B. (1988). Dispersal patterns of Western Gulls from Southeast Farallon Island. *Auk* 105: 128–141.

Spear, L. B., Pyle, P. & Nur, N. (1998). Natal dispersal in the Western Gull: proximal factors and fitness consequences. *J. Anim. Ecol.* 67: 165–179.

Speirs, J. M. (1939). Fluctations in numbers of birds in the Toronto Region. *Auk* 56: 411–419.

Spendelow, P. (1985). Starvation of a flock of Chimney Swifts on a very small Caribbean island. *Auk* 102: 387–388.

Spendelow, J. A., Nichols, J. D., Nisbet, I. C. T., Hays, H., Cormons, G. D., Burger, J., Safina, C., Hines, J. E. & Gochfeld, M. (1995). Estimating annual survival and movement rates of adults within a metapopulation of Roseate Terns. *Ecology* 76: 2415–2428.

Spina, F. & Bezzi, E. M. (1990). Autumn migration and orientation of the Sedge Warbler *Acrocephalus schoenobaenus* in northern Italy. *J. Ornithol.* 131: 429–438.

Spina, F., Massi, A. & Montemaggiori, A. (1994). Back from Africa: who's running ahead? Aspects of differential migration of sex and age classes in Palearctic-African spring migrants. *Ostrich* 65: 137–150.

Spina, F., Piacentini, D. & Frugis, S. (1985). Vertical distribution of Blackcap (*Sylvia atricapilla*) and Garden Warbler (*Sylvia borin*) within the vegetation. *J. Ornithol.* 126: 431–434.

Staicer, C. A. (1992). Social behaviour of the Northern Parula, Cape May Warbler, and Prairie Warbler wintering in second-growth forest in southwestern Puerto Rico. pp. 308–320 in *Ecology and conservation of neotropical migrant landbirds* (eds J. M. Hagan & D. W. Johnston). Washington, D.C., Smithsonian Institution Press.

Steenhof, K., Fuller, M. R., Kochert, M. N. & Bates, K. K. (2005). Long range movements and breeding dispersal of Prairie Falcons from southwest Idaho. *Condor* 107: 481–496.

Stenning, M. J., Harvey, P. H. & Campbell, B. (1988). Searching for density-dependent regulation in a population of Pied Flycatchers *Ficedula hypoleuca* Pallas. *J. Anim. Ecol.* 57: 307–317.

Stenseth, N. C. (1999). Population cycles in voles and lemmings: density dependence and phase dependence in a stochastic world. *Oikos* 87: 427–461.

Stenzel, L. E., Warriner, J. C., Warriner, J. S., Wilson, K. S., Bidstrup, F. C. & Page, G. W. (1994). Long-distance dispersal of Snowy Plovers in western North America. *J. Anim. Ecol.* 63: 887–902.

Sterling, T. & Dzubin, A. (1967). Canada Goose molt migrations to the Northwest Territories. *Trans. North Am. Wildl. Nat. Resour. Conf.* 32: 355–373.

Stervander, M., Lindstrom, K., Jonzen, N. & Andersson, A. (2005). Timing of spring migration in birds: long-term trends, North Atlantic Oscillation and the significance of different migration routes. *J. Avian Biol.* 36: 210–221.

Stetson, M. H. & Erickson, J. E. (1971). Neuroendocrine control of photoperiodically induced fat deposition in White-crowned Sparrows. *J. Exp. Zool.* 176: 409–414.

Stevens, G. C. (1992). The elevational gradient in altitudinal range: an extension of Rapoport's latitudinal rule to altitude. *Am. Nat.* 140: 893–911.

Stewart, P. A. & Connor, H. A. (1980). Fixation of wintering Palm Warblers to a specific site. *J. Field Ornithol.* 51: 365–367.

Stewart, R. E. & Aldrich, J. W. (1951). Removal and repopulation of breeding birds in a spruce-fir forest community. *Auk* 68: 471–482.

Stiles, F. G. (1973). Food supply and the annual cycle of the Anna Hummingbird. *Univ. Calif. Publ. Zool.* 97: 1–109.

Stiles, F. G. & Skutch, A. F. (1989). *A guide to the birds of Costa Rica.* London, Christopher Helm.

Stoate, C. & Moreby, S. J. (1995). Premigratory diet of trans-Saharan migrant passerines in the western Sahel. *Bird Study* 42: 101–106.

Stoddard, P. K., Marsden, J. E. & Williams, T. C. (1983). Computer simulation of autumnal bird migration over the western North Atlantic. *Anim. Behav.* 31: 173–180.

Storer, R. W. & Jehl, J. R. (1985). Moult patterns and moult migration in the Black-necked Grebe *Podiceps nigricollis. Ornis Scand.* 16: 253–260.

Storey, A. E. & Lien, J. (1985). Development of the first North American colony of Manx Shearwaters. *Auk* 102: 395–401.

Stouffer, P. C. & Dwyer, G. M. (2003). Sex-biased winter distribution and timing of migration of Hermit Thrushes (*Catharus guttatus*) in eastern North America. *Auk* 120: 836–847.

Stowe, T. J., Newton, A. V., Green, R. E. & Mayes, E. (1993). The decline of the Corncrake *Crex crex* in Britain and Ireland in relation to habitat. *J. Appl. Ecol.* 30: 53–62.

Streif, M. (1991). Analyse der Biotoppräferenzen auf dem Wegzug in Süddeutschland rastender Kleinvögel. *Ornithol. Jahresb. Bad.-Württemb.* 7: 1–132.

Stresemann, E. & Stresemann, V. (1966). Die Mauser der Vogel. *J. Ornithol.* 107: 3–448.

Strickland, D. (1991). Juvenile dispersal in Gray Jays: dominant brood members expel siblings from natal territory. *Can. J. Zool* 69: 2935–2945.

Struwe-Juhl, B. & Schmidt, R. (2003). Zur Mauser des Grossgefieders beim Seeadler (*Haliaeetus albicilla*) in Schleswig-Holstein. *J. Ornithol.* 144: 418–437.

Studds, C. E. & Marra, P. P. (2005). Nonbreeding habitat occupancy and population process: an upgrade experiment with a migratory bird. *Ecology* 86: 2380–2385.

Stutchbury, B. J. M. (1997). Effects of female Cowbird removal on reproductive success of Hooded Warblers. *Wilson Bull.* 109: 74–81.

Suddaby, D. & Ratcliffe, N. (1997). The effects of fluctuating food availability on breeding Arctic Terns (*Sterna paradisaea*). *Auk* 114: 524–530.

Summers, R. W. (1994). The migration patterns of the Purple Sandpiper *Calidris maritima. Ostrich* 65: 167–173.

Summers, R. W. (1999). Numerical responses by Crossbills *Loxia* spp. to annual fluctuations in cone crops. *Ornis Fenn.* 76: 141–144.

Summers, R. W. & Piertney, S. B. (2003). The Scottish Crossbill – what we know and what we don't. *Br. Birds* 96: 100–111.

Summers, R. W. & Underhill, L. G. (1987). Factors related to breeding production of Brent Geese *Branta bernicla* and waders (Charadrii) on the Taimyr Peninsula. *Bird Study* 34: 161–171.

Summers, R. W. & Waltner, M. (1979). Seasonal variations in the mass of waders in southern Africa, with special reference to migration. *Ostrich* 50: 21–37.

Summers, R. W., Underhill, L. G. & Prys-Jones, R. P. (1995). Why do young waders in southern Africa delay their first return migration to the breeding grounds? *Ardea* 83: 351–357.

Summers, R. W., Underhill, L. G. & Syroechkovski, E. E. (1998). The breeding productivity of Dark-bellied Brent Geese and Curlew Sandpipers in relation to changes in the numbers of Arctic Foxes and Lemmings on the Taimyr Peninsula, Siberia. *Ecography* 21: 573–580.

Summers, R. W., Jardine, D. C., Marquiss, M. & Proctor, R. (1996). The biometrics of invading Common Crossbills *Loxia curvirostra* in Britain during 1990–1991. *Ringing & Migration* 17: 1–10.

Summers, R. W., Jardine, D. C., Marquiss, M. & Rae, R. (2002). The distribution and habitats of crossbills *Loxia* spp. In Britain, with special reference to the Scottish Crossbill *Loxia scotica*. *Ibis* 144: 393–410.

Summers, R. W., Nicoll, M., Underhill, L. G. & Petersen, A. (1998). Methods for establishing the proportions of Icelandic and British Redshanks *Tringa totanus* in mixed populations wintering on British coasts. *Bird Study* 35: 169–180.

Suter, W. (1995). Are Cormorants *Phalacrocorax carbo* wintering in Switzerland approaching carrying capacity? An analysis of increase patterns and habitat choice. *Ardea* 83: 255–266.

Suter, W. & van Eerden, M. R. (1992). Simultaneous mass starvation of wintering diving ducks in Switzerland and The Netherlands: a wrong decision in the right strategy? *Ardea* 80: 229–242.

Sutherland, G. D., Gass, C. L., Thompson, P. A. & Certzman, K. P. (1982). Feeding territoriality in migrant Rufous Hummingbirds: defense of Yellow-bellied Sapsucker (*Sphyrapicus varius*) feeding sites. *Can. J. Zool.* 60: 2046–2050.

Sutherland, W. J. (1996). Predicting the consequences of habitat loss for migratory populations. *Proc. R. Soc. B* 263: 1325–1327.

Sutherland, W. J. (1998). Evidence for flexibility and constraint in migration systems. *J. Avian Biol.* 29: 441–446.

Sutherland, W. J. & Crockford, N. J. (1993). Factors affecting the feeding distribution of Red-breasted Geese *Branta ruficollis* wintering in Romania. *Biol. Conserv.* 63: 61–65.

Sutherland, W. J. & Dolman, P. M. (1994). Combining behaviour and population dynamics with applications for predicting the consequences of habitat loss. *Proc. R. Soc. Lond. B* 255: 133–138.

Švažas, S., Meisner, W., Serebryakov, V., Kozulin, A & Grishanov, G. (2001). *Changes of wintering sites of waterfowl in central and eastern Europe.* Vilnius, Institute of Ecology.

Svärdson, G. (1953). Visible migration within Fenno-Scandia. *Ibis* 95: 181–211.

Svärdson, G. (1957). The 'invasion' type of bird migration. *Br. Birds* 50: 314–343.

Svensson, E. & Hedenström, A. (1999). A phylogenetic analysis of the evolution of moult strategies in western Palearctic warblers (Aves: Sylviidae). *Biol. J. Linn. Soc.* 67: 263–276.

Svensson, S. E. (1985). Effects of changes in tropical environments on the North European avifauna. *Ornis Fenn.* 62: 56–63.

Swann, R. S. & Baillie, S. R. (1979). The suspension of moult by trans-Saharan migrants in Crête. *Bird Study* 26: 55–58.

Swann, R. L. & Ramsay, A. D. K. (1983). Movements from, and age of return to, an expanding Scottish Guillemot colony. *Bird Study* 30: 207–214.

Swarth, H. S. (1920). Revision of the avian genus *Passerella*, with special reference to the distribution and migration of the races in California, *Univ. Calif. Publ. Zool.* 21: 75–224.

Swartz, L. G., Walker, W., Roseneau, D. G. & Springer, A. M. (1974). Populations of Gyrfalcons on the Seward Peninsula, Alaska, 1968–1972. Raptor Research Foundation. *Raptor Res. Rep.* 3: 71–75.

Swenk, M. H. (1922). An unusual mortality among migrating Lapland Longspurs in north-eastern Nebraska. *Wilson Bull.* 34: 118–119.

Swenson, J. E., Jensen, K. C. & Toepfer, J. E. (1988). Winter movements by Rosy Finches in Montana. *J. Field Ornithol.* 59: 157–160.

Syroechkovski, E. E. & Rogacheva, H. V. (1994). Bird conservation in Siberia: a summary. *Ibis* 137: 191–194.

Szép, T. (1995). Relationship between west African rainfall and the survival of central European Sand Martins *Riparia riparia*. *Ibis* 137: 162–168.

Szép, T., Møller, A. P., Vallner, J., Kovacs, B. & Norman, D. (2003). Use of trace elements in feathers of Sand Martin *Riparia riparia* for identifying moulting areas. *J. Avian Biol.* 34: 307–320.

Takaki, Y., Eguchi, K. & Nagata, H. (2001). The growth bars on tail feathers in the male Styan's Grasshopper Warbler may indicate quality. *J. Avian Biol.* 32: 319–325.

Tamm, S. (1980). Bird orientation – single Homing Pigeons compared with small flocks. *Behav. Ecol. Sociobiol.* 7: 319–322.

Taylor, D., Saksena, P., Sanderson, P. G. & Kucera, K. (1999). Environmental change and rain forests on the Sunda shelf of Southeast Asia: drought, fire and the biological cooling of biodiversity hotspots. *Biodiv. Conserv.* 8: 1159–1177.

Taylor, I. (1994). *Barn Owls. Predator–prey relationships and conservation.* Cambridge, University Press.

Taylor, W. K. & Anderson, B. H. (1973). Nocturnal migrants killed at a central Florida TV tower: autumns 1969–1971. *Wilson Bull.* 85: 42–51.

Tellería, J. L. & Pérez-Tris, J. (2004). Consequences of the settlement of migrant Robins *Erithacus rubecula* in wintering habitats occupied by conspecific residents. *Ibis* 146: 258–268.

Temple, S. A. & Cary, J. R. (1988). Modelling dynamics of habitat-interior bird populations in fragmented landscapes. *Conserv. Biol.* 2: 340–347.

Terborgh, J. W. (1980). The conservation status of Neotropical migrants: present and future. pp. 21–30 in *Migrant birds in the Neotropics* (eds A. Keast & E. Morton). Washington, D.C., Smithsonian Institution Press.

Terborgh, J. (1989). *Where have all the birds gone?* Princeton, University Press.

Terborgh, J. W. & Faaborg, J. (1980). Factors affecting the distribution and abundance of North American migrants in the eastern Caribbean region. pp. 145–55 in *Migrant birds in the Neotropics* (eds A. Keast & E. Morton). Washington, D.C., Smithsonian Institution Press.

Terrill, S. B. (1987). Social dominance and migratory restlessness in the Dark-eyed Junco (*Junco hyemalis*). *Behav. Ecol Sociobiol.* 21: 1–11.

Terrill, S. B. (1990). Ecophysiological aspects of movements by migrants in the wintering quarters. pp. 130–143 in *Bird migration. Physiology and ecophysiology* (ed. E. Gwinner). Berlin, Springer-Verlag.

Terrill, S. B. & Berthold, P. (1989). Experimental evidence for endogenously programmed differential migration in the Blackcap *Sylvia atricapilla*. *Experimentia* 45: 207–209.

Terrill, S. B. & Crawford, R. L. (1988). Additional evidence of nocturnal migration by Yellow-rumped Warblers in winter. *Condor* 90: 261–263.

Terrill, S. B. & Ohmart, R. D. (1984). Facultative extension of fall migration by Yellow-rumped Warblers (*Dendroica coronata*). *Auk* 101: 427–438.

Thapliyal, J. P., Pati, A. K., Singh, V. K. & Lal, P. (1982). Thyroid, gonad and photoperiod in the hemopoesis of the migratory Red-headed Bunting, *Emberiza bruniceps*. *Gen. Comp. Endocrinol.* 46: 327–332.

Thibault, J. C. (1994). Nest-site tenacity and mate fidelity in relation to breeding success in Cory's Shearwater *Calonectris diomedea*. *Bird Study* 41: 25–28.

Thies, H. (1991). Invasionen des Birkenzeisigs (*Carduelis flammea flammea*) nach Norddeutschland und ihre Ursachen – eine Literaturstudie. *Corax* 14: 73–86.

Thies, H. (1996). Zum Vorkommen des Fichtenkreuzschnabels (*Loxia curvirostra*) und anderer *Loxia*-Arten im Segeberger Forst 1970–1995 mit besonderer Erörterung der Zugphänologie. *Corax* 16: 305–334.

Thingstand, G. (1997). Annual and local reproductive variations of a Pied Flycatcher *Ficedula hypoleuca* population near a subalpine lake in central Norway. *Ornis Fenn.* 74: 39–49.

Thiollay, J.-M. (1978). Les migrations des rapaces ens Afrique occidentale: adaptations ecologiques aux fluctuations saisonières de production des ecosystems. *La Terre et la Vie* 32: 89–133.

Thiollay, J. M. (1989). Distribution and ecology of Palaearctic birds of prey wintering in West and Central Africa. pp. 95–107 in *Raptors in the modern world* (eds B.-U. Meyburg & R. D. Chancellor). London, WWGBP.

Thom, V. (1986). *Birds in Scotland*. Calton, Poyser.

Thomas, D. K. (1979). Figs as a food source of migrating Garden Warblers in southern Portugal. *Bird Study* 26: 187–191.

Thompson, C. F. & Nolan, V. (1973). Population biology of the Yellow-breasted Chat (*Icteria virens* L.) in southern Indiana. *Ecol. Monogr.* 43: 145–171.

Thompson, J. D. & Baldasarre, G. A. (1992). Dominance relationships of dabbling ducks wintering in Yucatan, Mexico. *Wilson Bull.* 104: 529–536.

Thompson, M. (1973). Migratory patterns of Ruddy Turnstones in the central Pacific region. *Living Bird* 12: 5–23.

Thompson, P. S., Baines, D., Coulson, J. C. & Longrigg, G. (1994). Age at first breeding, philopatry and breeding site fidelity in the Lapwing *Vanellus vanellus*. *Ibis* 136: 474–484.

Thomson, A. Landsborough (1926). *Problems of bird migration*. London, Houghton Mifflin.

Thomson, A. Landsborough (1936). Recent progress in the study of bird-migration: a review of the literature, 1926–35. *Ibis* 78: 472–530.

Thorup, K. (1998). Vagrancy of Yellow-browed Warbler *Phylloscopus inornatus* and Pallas's Warbler *Ph. proregulus* in north-west Europe: misorientation or great circles? *Ringing & Migration* 19: 7–12.

Thorup, K. (2004). Reverse migration as a cause of vagrancy. *Bird Study* 51: 228–238.

Thorup, K. & Rabøl, J. (2001). The orientation system and migration pattern of long-distance migrants: conflict between model predictions and observed patterns. *J. Avian Biol.* 32: 111–119.

Thorup, K., Alerstam, T., Håke, M. & Kjellen, N. (2003). Can vector summation describe the orientation system of juvenile Ospreys and Honey Buzzards? – An analysis of ring recoveries and satellite tracking. *Oikos* 103: 350–359.

Tiainen, J. (1983). Dynamics of a local population of the Willow Warbler *Phylloscopus trochilus* in southern Finland. *Ornis Scand.* 14: 1–15.

Tiedemann, R. (1999). Seasonal changes in the breeding origin of migratory Dunlins (*Calidris alpina*) as revealed by mitochondrial DNA sequencing. *J. Ornithol.* 140: 319–323.

Tischler, F. (1941). *Die Vogel Ostpreussers*, Vol. 1. Königsberg.

Tomback, D. (1988). Clark's Nutcracker. *The birds of North America, No 331* (eds A. Poole, P. Stettenheim & F. Gill). Philadelphia, Academy of Natural Sciences & Washington, D.C., American Ornithologists Union.

Tomiałojc, L. (1990). *Birds of Poland. Their distribution and abundance*. Warszawa, PWN.

Tomiałojc, L. & Stawarczyk, T. (2003). *Awifauna Polski. Rozmieszczenie, liczebnosc i zmiany*. Wroclaw, PTPP, Pro Natura.

Tomkovich, P. J. & Soloviev, M. Y. (1994). Site fidelity in high arctic breeding waders. *Ostrich* 65: 174–180.

Tordoff, H. G. & Mengel, R. M. (1956). Studies of birds killed in nocturnal migration. *Univ. Kans. Publ. Mus. Nat. Hist.* 10: 1–44.

Totzke, U., Hubinger, A., Dittami, J. & Bairlein, F. (2000). The autumnal fattening of the long-distance migratory Garden Warbler (*Sylvia borin*) is stimulated by intermittent fasting. *J. Comp. Physiol.* 170: 627–631.

Townsend, C. W. (1931). On the post-breeding northern migration of North American herons. *Proc. Int. Ornithol. Congr.* 7: 382.

Townsend, D. J. (1985). Decisions of a lifetime: establishment of spatial defence and movement pattern by juvenile Grey Plovers (*Pluvialis squatarola*). *J. Anim. Ecol.* 54: 267–274.

Trail, P. W. & Baptista, L. F. (1993). The impact of Brown-headed Cowbird parasitism on populations of the Nuttall's White-crowned Sparrow. *Conserv. Biol.* 7: 309–315.

Tramer, E. J. & Tramer, F. E. (1977). Feeding responses of fall migrants to prolonged inclement weather. *Wilson Bull.* 89: 166–167.

Trettau, W. (1952). Planberingung des Trauerfliegenschnappers (*Muscicapa hypoleuca*) in Hessen. *Vogelwarte* 16: 89–95.

Triplet, P. & Yésou, P. (2000). Controlling the flood in the Senegal delta: do waterfowl populations adapt to the new environment? *Ostrich* 71: 106–111.

Troy, D. M. (1983). Recaptures of Redpolls: movements of an irruptive species. *J. Field Ornithol.* 54: 146–151.

Tryjanowski, P. & Sparks, T. H. (2001). Is the detection of the first arrival date of migrating birds influenced by population size? A case study of the Red-backed Shrike *Lanius collurio*. *Int. J. Biometeorol.* 45: 217–219.

Tryjanowski, P. & Yosef, R. (2002). Differences between the spring and autumn migration of the Red-backed Shrike *Lanius collurio*: record from the Eilat stopover (Israel). *Acta Ornithol.* 37: 84–90.

Tryjanowski, P., Kuzniak, S. & Sparks, T. (2002). Earlier arrival of some farmland migrants in western Poland. *Ibis* 144: 62–68.

Tryjanowski, P., Sparks, T. H., Ptaszyk, J. & Kosicki, J. (2004). Do White Storks *Ciconia ciconia* always profit from an early return to their breeding grounds? *Bird Study* 51: 222–227.

Tucker, G. & Heath, M. (1994). *Birds in Europe: their conservation status*. Cambridge, Birdlife International.

Tucker, V. A. (1968). Respiratory exchange and evaporative water loss in the flying budgerigar. *J. Exp. Biol.* 48: 67–87.

Tulp, I., McChesney, S., & de Goeij, P. (1994). Migratory departures of waders from north-western Australia: behaviour, timing and possible migration routes. *Ardea* 82: 201–221.

Turner, A. (2006). *The Barn Swallow*. London, T. & A. D. Poyser.

Turpie, J. K. (1995). Non-breeding territoriality – causes and consequences of seasonal and individual variation in Grey Plover *Pluvialis squatarola* behaviour. *J. Anim. Ecol.* 64: 429–438.

Twedt, D. J. & Crawford, R. D. (1995). Yellow-headed Blackbird (*Xanthocephalus xanthocephalus*). in *The Birds of North America*, No. 192 (eds. A. Poole & F. Gill). Philadelphia, The Academy of Natural Sciences and Washington, D.C., The American Ornithologists' Union.

Tyrväinen, H. (1975). The winter irruption of the Fieldfare *Turdus pilaris* and the supply of rowan-berries. *Ornis Fenn.* 52: 23–31.

Ueta, M. & Higuchi, H. (2002). Difference in migration pattern between adult and immature birds using satellites. *Auk* 119: 832–835.

Ueta, M., Sato, F., Lobkov, E. G. & Naga Lisa, M. (1998). Migration route of White-tailed Sea Eagles *Haliaeetus albicilla* in northeastern Asia. *Ibis* 140: 684–696.

Ueta, M., Sato, F., Nakagawa, H. & Mita, N. (2000). Migration routes and differences of migration schedule between adult and young Steller's Sea Eagles *Haliaeetus pelagicus*. *Ibis* 142: 35–39.

Ueta, M., Melville, D. S., Wang, Y., Ozaki, K., Kanai, Y., Leader, P. J., Wang, L.-C. & Kuo, C.-Y. (2002). Discovery of the breeding sites and migration routes of Black-faced Spoonbills *Platalea minor*. *Ibis* 144: 340–343.

Ulfstrand, S. (1962). On the non-breeding ecology and migratory movements of the Great Tit (*Parus major*) and the Blue Tit (*Parus caeruleus*) in southern Sweden. *Vår Fågelvärld* (suppl 3), Lund.

Ulfstrand, S. & Alerstam, T. (1977). Bird communities in *Brachystegia* and *Acacia* woodlands in Zambia. A quantitative study with special reference to the significance of habitat modification for the Palearctic. *J. Ornithol.* 1118: 156–174.

Ulfstrand, S., Roos, G., Alerstam, T. & Osterdahl, L. (1974). Visible bird migration at Falsterbo, Sweden. *Vår Fågelvärld* (suppl 8): 1–245.

UNEP (United Nations Environment Programme) (2002). *Africa environment outlook: past, present and future perspectives.* Stevenage, England, Earthprint.

Underhill, L. G. (2001). Four of our White Storks vanish into the clouds; one keeps going. *Promerops* 246: 12–14.

Underhill, L. G. & Summers, R. W. (1990). Multivariate analyses of breeding performance in Dark-bellied Brent Geese *Branta b. bernicla.* *Ibis* 132: 477–480.

Underhill, L. G., Prys-Jones, R. P., Dowsett, R. J., Lawn, M. R., Herroelen, P., Johnson, D. N., Norman, S. C., Pearson, D. J. & Tree, A. J. (1992). The biannual primary moult of Willow Warblers, *Phylloscopus trochilus*, in Europe and Africa. *Ibis* 134: 286–297.

Underwood, L. A. & Stowe, T. J. (1984). Massive wreck of seabirds in eastern Britain, 1983. *Bird Study* 31: 79–88.

US FWS. (2002). *Migratory bird mortality.* Arlington, Virginia, US Fish & Wildlife Service (Avian mortality fact sheet: http://birds.FWS.gov/mortality-fact-sheet.pdf).

Vahatalo, A., Rainio, K., Lehikoinen, A. & Lehikoinen, E. (2004). Spring arrival of birds depends on the North Atlantic Oscillation. *J. Avian Biol.* 35: 210–216.

Väisänen, R. A. & Hildén, O. (1993). Siemensoyöjät runsastuvat talviruokinnan ansiosta. *Linnut* 28: 21–24 (In Finnish).

Valikangas, I. (1933). Finnische Zugvögel aus englischer Vögeleiern. *Vogelzug* 4: 159–166.

van Balen, J. H. & Hage, F. (1989). The effects of environmental factors on tit movements. *Ornis Scand.* 20: 99–104.

Van den Brink, B., Bijlsma, R. G. & van der Hawe, T. M. (2000). European Swallow *Hirundo rustica* in Botswana during three non-breeding seasons: the effects of rainfall on moult. *Ostrich* 71: 198–204.

Van der Graaf, A. J., Stahl, J., Klimkowska, A., Bakker, J. P. & Drent, R. H. (2006). Surfing on a green wave – how plant growth drives migration in the Barnacle goose. *Ardea* 94: in press.

Van der Wal, R. J. & Zomerdijk, P. J. (1979). The moulting of Tufted Duck and Pochard on the IJsselmeer in relation to moult concentrations in Europe. *Wildfowl* 30: 99–108.

Van der Winden, J. (2002). The odyssey of the Black Tern *Chlidonias niger* migration ecology in Europe and Africa. *Ardea* 90: 421–435.

van Dyke, A. J., de Roder, F. E., Marteijn, E. C. L. & Spiekman, H. (1990). Summering waders on the Banc d'Arguin, Mauritania: a census in June 1988. *Ardea* 78: 145–156.

van Eerden, M. R. & Munsterman, M. J. (1995). Sex and age dependent distribution in wintering Cormorants *Phalacrocorax carbo sinensis* in western Europe. *Ardea* 83: 285–297.

van Eerden, M. R., Zylstra, M. & Loonen, M. J. J. E. (1991). Individual patterns of staging during autumn migration in relation to body composition in Greylag Geese *Anser anser* in the Netherlands. *Ardea* 79: 261–264.

van Gasteren, H., Monstert, K., Groot, A. & van Ruiten, L. (1992). The irruption of the Coal Tit *Parus ater* in the autumn of 1989 in the Netherlands and Northwest Europe. *Limosa* 65: 57–66.

van Gils, J. A., Dekinga, A., Spaans, B., Vahl, W. K. & Piersma, T. (2005). Digestive bottleneck affects foraging decisions in Red Knots *Calidris canutus*. II Patch choice and length of working day. *J. Anim. Ecol.* 74: 120–130.

van Noordwijk, A. J. (1983). Problems in the analysis of dispersal and a critique on its 'heritability' in the Great Tit. *J. Anim. Ecol.* 53: 533–544.

van Noordwijk, A., van Balen, J. H. & Scharloo, W. (1981). Genetic variation in the timing of reproduction in the Great Tit. *Oecologia* 49: 158–166.

van Zyl, A. J. (1994). Sex-related local movement in adult Rock Kestrels in the eastern Cape Province, South Africa. *Wilson Bull.* 106: 145–158.

Vande Weghe, J.-P. (1979). The wintering and migration of Palearctic passerines in Ruanda. *Gerfaut* 69: 29–43.

Vander Wall, S. B., Hoffman, S. W. & Potts, W. K. (1981). Emigration behaviour of Clark's Nutcracker. *Condor* 83: 162–170.

Vangelewe, P. & Stassin, P. (1991). Hivernage de la Bernache à Cou Roux, *Branta ruficollis*, en Dobroudja Septestriolare Roumanie et revice du Statut hivernal de l'espèce. *Gerfaut* 81: 65–99.

Vaught, R. W. (1964). Results of transplanting flightless young Blue-winged Teal. *J. Wildl. Manage.* 28: 208–212.

Veiga, J. P. (1986). Settlement and fat accumulation by migrant Pied Flycatchers in Spain. *Ringing & Migration* 7: 85–98.

Veit, R. T. (1997). Long distance dispersal and population growth of the Yellow-headed Blackbird *Xanthocephalus xanthocephalus*. *Ardea* 85: 135–143.

Vepsäläinen, K. (1968). The effect of the cold spring 1966 upon the Lapwing *Vanellus vanellus* in Finland. *Ornis Fenn.* 45: 33–47.

Verner, G. & Ritter, L. V. (1983). Current status of the Brown-headed Cowbird in the Sierra National Forest. *Auk* 100: 355–368.

Videler, J. (2005). *Avian flight*. Oxford, Oxford University Press.

Videler, J. J., Weihs, D. & Daan, S. (1983). Intermittent gliding in the hunting flight of the Kestrel, *Falco tinnunculus*. *J. Exp. Biol.* 102: 1–12.

Viehmann, W. (1982). Orientierungsverhalten van Monchsgrasmücken (*Sylvia atricapilla*) im Frühjahr in Abhängigkeit der Wetterlage. *Vogelwarte* 31: 452–457.

Village, A. (1981). The diet and breeding of Long-eared Owls in relation to vole numbers. *Bird Study* 28: 215–224.

Village, A. (1987). Numbers, territory size and turnover of Short-eared Owls *Asio flammeus* in relation to vole abundance. *Ornis Scand.* 18: 198–204.

Village, A. (1990). *The Kestrel*. Calton, T. & A.D. Poyser.

Villarán, A. & Pascual-Para, J. (2003). Biometrics, sex ratio and migration periods of Reed Buntings *Emberiza schoeniclus* wintering in the Tajo Basin, Spain. *Ringing & Migration* 21: 222–226.

Virkkala, R. (1992). Fluctuations of vole-eating birds of prey in northern Finland. *Ornis Fenn.* 69: 97–100.

Virolainen, M. (1984). Breeding biology of the Pied Flycatcher *Ficedula hypoleuca* in relation to population density. *Ann. Zool. Fenn.* 21: 187–197.

Visser, M. E., Both, C. & Lambrechts, M. M. (2004). Global climate change leads to mistimed avian reproduction. pp 89–110 in *Birds and climate change. Advances in Ecological Research*, Vol. 35. London: Elsevier.

Viverette, C. B., Struve, S., Goodrich, L. J. & Bildstein, K. L. (1996). Decreases in migrating Sharp-shinned Hawks (*Accipiter striatus*) at traditional raptor-migration watch sites in eastern North America. *Auk* 113: 32–40.

Vogel, C. & Moritz, D. (1995). Langjährige Ànderungen von Zugzeiten auf Helgoland. *Jber Inst. Vogelforshung* 2: 8–9.

von Haartman, L. (1949). Der Trauerfliegenschnäpper. 1. Ortstreue und Rassenbildung. *Acta Zool. Fenn.* 56: 1–104.

von Haartman, L. (1960). The Ortstreue of the Pied Flycatcher. *Proc. Int. Ornithol. Congr.* 12: 266–273.

von Haartman, L. (1971). Population dynamics. pp. 391–459 in *Avian biology,* Vol. 1 (eds D. S. Farner & J. R. King). London: Academic Press.

Voous, K. H. (1960). *Atlas of European birds.* London, Nelson.

Walcott, C., Gould, J. L. & Kirschvink, J. C. (1979). Pigeons have magnets. *Science* 205: 1027–1029.

Waldenström, J., Bensch, S., Kiboi, S., Hasselquist, D. & Ottoson, U. (2002). Cross species infection of blood parasites between resident and migratory songbirds in Africa. *Mol. Ecol.* 11: 1545–1554.

Walkinshaw, L. H. (1953). Life history of the Prothonotary Warbler. *Wilson Bull.* 65: 152–168.

Walkinshaw, L. H. (1983). *The Kirtland's Warblers.* Broom Field Hills, Michigan, Cranbrook Institute of Science.

Wallin, K., Wallin, M., Levin, M., Järdås, T. & Strandvik, P. (1987). Leap-frog migration in the Swedish Kestrel *Falco tinnunculus* population. *Acta Reg. Soc. Sci. Litt. Gothoburgensis, Zool.* 14: 213–222.

Wallraff, H. G. (1991). Conceptual approaches to avian navigation systems. pp. 128–165 in *Orientation in birds* (ed. P. Berthold). Basel, Birkhäuser.

Wallraff, K. G. (2003). Zur olfakorischen Navigation der Vögel. *J. Ornithol.* 144: 1–31.

Walls, S. S. & Kenward, R. E. (1995). Movements of radio-tagged Common Buzzards *Buteo buteo* in their first year. *Ibis* 137: 177–182.

Walls, S. S. & Kenward, R. E. (1998). Movements of radio-tagged Buzzards *Buteo buteo* in early life. *Ibis* 140: 561–568.

Walsberg, G. E. (1978). Brood size and the use of time and energy by the Phainopepla. *Ecology* 59: 147–153.

Walter, H. (1979). *Eleonora's Falcon. Adaptations to prey and habitat in a social raptor.* Chicago, University of Chicago Press.

Wanink, J. H. & Goudswaard, K. (2000). The impact of Lake Victoria's lakefly abundance on Palaearctic passerines. *Ostrich* 71: 194–197.

Wanless, S. & Okill, J. D. (1994). Body measurements and flight performance of adult and juvenile Gannets *Morus bassanus. Ringing & Migration* 15: 101–103.

Ward, D. & Smith, J. N. M. (2000). Brown-headed Cowbird parasitism results in a sink population in warbling birds. *Auk* 117: 337–344.

Ward, P. (1963). Lipid levels in birds preparing to cross the Sahara. *Ibis* 105: 109–111.

Ward, P. (1971). The migration patterns of *Quelea quelea* in Africa. *Ibis* 113: 275–297.

Ward, P. & Jones, P. J. (1977). Pre-migratory fattening in three races of the Red-billed Quelea *Quelea quelea* (Aves: Ploceidae), an intra-tropical migrant. *J. Zool. Lond.* 181: 43–56.

Warham, J. (1990). *The petrels. Their ecology and breeding systems.* London, Academic Press.

Warham, J. (1996). *The behaviour, population biology and physiology of the petrels.* London, Academic Press.

Wassenaar, L. I. & Hobson, K. A. (2001). A stable-isotope approach to delineate geographical catchment areas of avian migration monitoring stations in North America. *Environ. Sci. Technol.* 35: 1845–1850.

Watson, A. (1980). Starving Oystercatchers in Deeside after severe snowstorm. *Scott. Birds* 11: 55–56.

Watson, J. B. (1930). Mortality amongst swifts caused by cold. *Br. Birds* 39: 107.

Weatherhead, P. J. & Forbes, M. R. L. (1994). Natal philopatry in passerine birds: genetic or ecological influences? *Behav. Ecol.* 5: 426–433.

Weatherhead, P. J. & Hoysak, D. J. (1984). Dominance structuring of a Red-winged Blackbird roost. *Auk* 101: 551–555.

Weber, H. (1972). Über die Fichtenkreuzschnabel invasionen der Jahre 1962 bis 1968 im Naturschutz-gebiet Serrahn. *Der Falke* 19: 16–27.

Weber, T. P. & Houston, A. I. (1997a). Flight costs, flight range and the stopover ecology of migrating birds. *J. Anim. Ecol.* 66: 297–306.

Weber, T. P. & Houston, A. I. (1997b). A general model for time-minimising avian migration. *J. Theor. Biol.* 185: 447–458.

Webster, F. S. (1974). The spring migration, April 1–May 31, 1974, South Texas region. *Am. Birds* 28: 822–825.

Webster, M. S. & Marra, P. P. (2005). The importance of understanding migratory connectivity and seasonal interactions. pp. 199–209 in *Birds of Two Worlds*. (eds Greenberg, R. & Marra, P. P.). Washington, D. C., Smithsonian Institution.

Weeden, R. B. (1964). Spatial separation of sexes in Rock and Willow Ptarmigan in winter. *Auk* 81: 534–541.

Weimerskirch, H. & Wilson, R. P. (2000). Oceanic respite for Wandering Albatrosses. *Nature* 406: 955–956.

Weimerskirch, H., Salamolard, M., Sarrazin, F. & Jouventin, P. (1993). Foraging strategy of Wandering Albatrosses through the breeding season: a study using satellite telemetry. *Auk* 110: 325–342.

Weimerskirch, H., Doncaster, C. P. & Cuenot-Chaillet, F. (1994). Pelagic seabirds and the marine environment: foraging patterns of Wandering Albatrosses in relation to prey availability and distribution. *Proc. R. Soc. Lond. B* 255: 91–97.

Weimerskirch, H., Guionnet, T., Martin, J., Shaffer, S. A. & Costa, D. P. (2000). Fast and fuel efficient? Optimal use of wind by flying albatrosses. *Proc. R. Soc. Lond. B* 267: 1869–1874.

Weimerskirch, H., Chastel, O., Barbraud, C. & Tostain, O. (2003). Frigatebirds ride high on thermals. *Nature* 421: 333–334.

Weinburg, H. J. & Roth, R. R. (1998). Forest area and habitat quality for nesting Wood Thrushes. *Auk* 115: 879–889.

Weindler, P., Wiltschko, R. & Wiltschko, W. (1996). Magnetic information affects the stellar orientation of young bird migrants. *Nature* 383: 158–160.

Weir, R. D. (1976). *Annotated bibliography of bird kills at man-made obstacles: a review of the state of the art and solutions*. Ottawa, Ontario, Department of Fish and Environment, Canadian Wildlife Service.

Weise, C. M. (1962). Migratory and gonadal responses of birds on long-continued short daylengths. *Auk* 79: 161–172.

Weise, C. M. (1963). Annual physiological cycles in captive birds of different migratory habits. *Proc. Int. Ornithol. Congr.* 13: 983–993.

Welch, G. R. & Welch, H. J. (1988). The autumn migration of raptors and other soaring birds across the Bab-el-Mandeb Straits. *Sandgrouse* 10: 26–50.

Welch, G. & Welch, H. (1998). Raptor migration at Bab el Mandeb, Yemen – spring 1998. *The Phoenix* 5: 11–12.

Welham, C. V. J. (1994). Flight speeds of migrating birds: a test of maximum range speed predictions from three aerodynamic equations. *Behav. Ecol.* 5: 1–8.

Wells, D. R. (1990). Migratory birds and tropical forest in the Sunda region. pp. 357–369 in *Biogeography and ecology of forest bird communities* (ed. A. Keast). The Hague, SPB Academic Publishers.

Wenink, P. W., Baker, A. J., Rösner, H.-U. & Tilanus, M. G. J. (1993). Hypervariable-control-region sequences reveal global population structuring in a long-distance migrant shorebird, the Dunlin (*Calidris alpina*). *Proc. Natl Acad. Sci. USA* 90: 94–98.

Wenink, P. W., Baker, A. J. & Tilanus, M. G. J. (1996). Global mitochondrial DNA phylogeography of holarctic breeding Dunlins (*Calidris alpina*). *Evolution* 50: 318–330.

Wennerberg, L. (2001). Breeding origin and migratory pattern of Dunlin (*Calidris alpina*) revealed by mitochondrial DNA analysis. *Mol. Ecol.* 10: 1111–1120.

Wenzel, B. M. (1991). Olfactory abilities of birds. *Proc. Int. Ornithol. Congr.* 20: 1820–1829.

Wernham, C. V., Toms, M. P., Marchant, J. H., Clark, J. A., Siriwardena, G. M. & Baillie, S. R. (2002). *The migration atlas: movements of the birds of Britain and Ireland.* London, T & A. D. Poyser.

Wheelwright, N. T. & Mauck, R. A. (1998). Philopatry, natal dispersal, and inbreeding avoidance in an island population of Savannah Sparrows. *Ecology* 79: 755–767.

Whitcomb, R. F., Robins, C. S., Lynch, J. F., Whitcomb, B. L., Klimkiewicz, M. K. & Bystrak, D. (1981). Effects of forest fragmentation on avifauna of eastern deciduous forest. pp. 125–205 in *Forest island dynamics in man-dominated landscapes* (eds R. G. Burgess & D. M. Sharpe). New York, Springer-Verlag.

White, C. M., Clum, N. J., Cade, T. J. & Hunt, W. J. (2002). Peregrine Falcon *Falco peregrinus*. *The Birds of North America*, No. 660 (eds A. Poole, P. Stettenheim & F. Gill). Philadelphia: Academy of Natural Sciences and Washington, D.C., American Ornithologists Union.

White, G. (1797). *The natural history of Selborne*. London, Bodley Head.

White, W. F. (1965). Weather causes heavy bird mortality. *Florida Naturalist* 38: 29–30.

Whitfield, D. P. (2003). Predation by Eurasian Sparrowhawks produces density dependent mortality of wintering Redshanks. *J. Anim. Ecol.* 72: 27–35.

Whitfield, M. J. (2000). Results of a Brown-headed Cowbird control program for the Southwestern Willow Flycatcher. pp. 371–377 in *Ecology and management of Cowbirds and their hosts* (eds J. N. M. Smith, T. L. Cook, S. I. Rothstein, S. K. Robinson & S. G. Sealy). Austin, University of Texas Press.

Whitmore, R. C., Mosher, J. A. & Frost, H. H. (1977). Spring migrant mortality during unseasonable weather. *Auk* 94: 778–781.

Whittington, P. A., Dyer, B. M., Crawford, R. J. M. & Williams, A. J. (1999). First recorded breeding of Leach's Storm Petrel *Oceanodroma leucorhoa* in the southern hemisphere, at Dyer Island, South Africa. *Ibis* 141: 327–330.

Widlechner, M. P. & Dragula, S. K. (1984). Relation of cone crop size to irruptions of four seed-eating birds in California. *Am. Birds* 38: 840–846.

Widmer, M. (1999). Altitudinal variation of migratory traits in the Garden Warbler *Sylvia borin*. PhD thesis, University of Zurich, Switzerland.

Widmer, M. & Biebach, H. (2001). Changes in body condition from spring migration to reproduction in the Garden Warbler *Sylvia borin*: a comparison of a lowland and a mountain population. *Ardea* 89: 57–68.

Wiedenfeld, D. A. & Wiedenfeld, M. G. (1995). Large kill of neotropical migrants by tornado and storm in Louisiana, April 1993. *J. Field Ornithol.* 66: 70–80.

Weimerskirch, H. & Wilson, R. P. (2000). Oceanic respite for Wandering Albatrosses. *Nature* 406: 955–956.

Wiemerskirch, H., Guionet, T., Martin, J., Shaffer, S. A. & Costa, D. P. (2000). Fast and fuel efficient? Optimal use of wind by flying albatrosses. *Proc. R. Roc. Lond. B* 267: 1869–1874.

Wiens, J. A. & Rotenbury, J. T. (1986). A lesson in the limitations of field experiments: shrubsteppe birds and habitat alteration. *Ecology* 67: 365–376.

Wiggins, D. A., Part, T. & Gustaffson, L. (1994). Seasonal decline in Collared Flycatcher *Ficedula albicollis* reproductive success: an experimental approach. *Oikos* 70: 359–364.

Wijnandts, H. (1984). Ecological energetics of the Long-eared Owl (*Asio otus*). *Ardea* 72: 1–92.

Wikelski, M., Tarlow, E. M., Rain, A., Diehl, R. H., Larkin, R. P. & Visser, G. H. (2003). Costs of migration in free-living songbirds. *Nature* 423: 704.

Wiklund, C. G., Kjellén, N. & Isakson, E. (1998). Mechanisms determining the spatial distribution of microtine predators on the Arctic tundra. *J. Anim. Ecol.* 67: 91–98.

Wilcove, D. S. (1985). Nest predation in forest tracts and the decline of migratory songbirds. *Ecology* 66: 1211–1214.

Wilcove, D. S. & Robinson, S. K. (1990). The impact of forest fragmentation on bird communities in Eastern North America. pp. 319–331 in *Biogeography and ecology of forest bird communities* (ed. A. Keast). The Hague, Netherlands, SPB Academic Publishing.

Wilcove, D. S., McLellan, C. H. & Dobson, A. P. (1986). Habitat fragmentation in the temperate zone. pp. 237–256 in *Conservation biology. The science of scarcity and diversity* (ed. M. E. Soulé). Sunderland, Mass., Sinauer Associates.

Wilcove, D. S. & Whitcomb, R. F. (1983). Gone with the trees. *Nat. Hist.* 92: 82–91.

Williams, C. S. & Kalmbach, E. R. (1943). Migration and fate of transported juvenile waterfowl. *J. Wildl. Manage.* 7: 163–169.

Williams, G. C. (1950). Weather and spring migration. *Auk* 67: 52–65.

Williams, G. R. (1953). The dispersal from New Zealand and Australia of some introduced European passerines. *Ibis* 95: 676–692.

Williams, T. C. & Williams, J. M. (1990). The orientation of transoceanic migrants. pp. 7–21 in *Bird migration. Physiology and ecophysiology* (ed. E. Gwinner). Berlin, Springer-Verlag.

Williams, T. C. & Williams, J. M. (1999). The migration of land birds over the Pacific Ocean. *Proc. Int. Ornithol. Congr.* 22: 1948–1957.

Williams, T. C., Williams, J. M., Ireland, L. C. & Teal, J. M. (1977). Bird migration over the western North Atlantic ocean. *Am. Birds* 31: 251–67.

Williams, T. C., Williams, J. M., Ireland, L. C. & Teal, J. M. (1978). Estimated flight time for transatlantic autumnal migrants. *Am. Birds* 32: 275–280.

Williamson, K. (1955). Migrational drift. *Proc. Int. Ornithol. Congr.* 11: 179–186.

Willis, E. O. (1966). The role of migrant birds at swarms of army ants. *Living Bird* 5: 187–231.

Willis, E. O. (1980). Ecological roles of migratory and resident birds on Barro Colorado Island, Pananma. pp. 205–226 in *Migrant birds in the Neotropics: ecology, behaviour, distribution and conservation* (eds A. Keast & E. S. Morton). Washington, D.C., Smithsonian Institution Press.

Wilson, G. E. (1976). Spotted Sandpipers nesting in Scotland. *Br. Birds* 69: 288–292.

Wilson, H. J., Norriss, D. W., Walsh, A., Fox, A. D. & Stroud, D. A. (1991). Winter site fidelity in Greenland White-fronted Geese *Anser albifrons flavirostris*, implications for conservation and management. *Ardea* 79: 287–294.

Wilson, J. R. (1988). The migratory system of Knots in Iceland. *Wader Study Group Bull.* 45: 8–9.

Wilson, V. J. (1970). Notes on the breeding and feeding habits of a pair of Barn Owls, *Tyto alba* (Scolopi), in Rhodesia. *Arnoldia* 4: 1–8.

Wiltschko, R. (1992). Das Verhalten verfrachteter Vögel. *Vogelwarte* 36: 249–310.

Wiltschko, R. & Wiltschko, W. (1995). *Magnetic orientation in animals*. Berlin, Springer-Verlag.

Wiltschko, R. & Wiltschko, W. (1999). The orientation system of birds – III. Migratory orientation. *J. Ornithol.* 140: 273–308.

Wiltschko, R. & Wiltschko, W. (2003). Mechanisms of orientation and navigation in migratory birds. pp. 433–456 in *Avian migration* (eds P. Berthold, E. Gwinner & E. Sonnenschein). Berlin, Springer-Verlag.

Wiltschko, W. (1968). Uber den Einfluss statischer Magnetfelder auf die Zugorientierung der Rotkehlchen (*Erithacus rubecula*). *Z. Tierpsychol* 25: 537–558.

Wiltschko, W. & Gwinner, E. (1974). Evidence for an innate magnetic compass in Garden Warblers. *Naturwissenschaften* 61: 406.

Wiltschko, W. & Merkel. F. W. (1971). Zugorientierung von Dorngrasmücken (*Sylvia communis*). *Vogelwarte* 26: 245–249.

Wiltschko, W. & Wiltschko, R. (1972). Magnetic compass of European Robins. *Science* 176: 62–64.

Wiltschko, W. & Wiltschko, R. (1975a). The interaction of stars and magnetic field in the orientation system of night-migrating birds. I. Autumn experiments with European warblers (Gen. *Sylvia*). *Z. Tierpsychol.* 37: 265–282.

Wiltschko, W. & Wiltschko, R. (1975b). The interaction of stars and magnetic field in the orientation system of night-migrating birds. II. Spring experiments with European Robins (*Erithacus rubecula*). *Z. Tierpsychol.* 37: 337–355.

Wiltschko, W. & Wiltschko, R. (1978). A theoretical model for migratory orientation and homing in birds. *Oikos* 30: 177–187.

Wiltschko, W., Daum-Benz, P., Fergenbauer-Kimmel, A. & Wiltschko, R. (1987). The development of the star compass in Garden Warblers *Sylvia borin*. *Ethology* 74: 285–292.

Wiltschko, W., Weindler, P. & Wiltschko, R. (1998). Interaction of magnetic and celestial cues in the migratory orientation of passerines. *J. Avian Biol.* 29: 606–617.

Wimpfheimer, D., Bruun, B., Baha el Din, S. M. & Jennings, M. C. (1983). *The migration of birds of prey in the northern Red Sea area*. New York, Holy Land Conservation Fund.

Wingate, D. B. (1973). *A checklist and guide to the birds of Bermuda*. Privately published. 36 pp.

Wingfield, J. C. (1980). Fine temporal adjustment of reproductive functions. pp. 367–389 in *Avian endocrinology* (eds A. Epple & M. H. Stetson). New York, Academic Press.

Wingfield, J. (2003). Avian migration regulation of facultative-type movements. pp. 113–125 in *Avian migration* (eds P. Berthold, E. Gwinner & E. Sonnenschein). Berlin, Springer-Verlag.

Wingfield, J. C. (2005). Flexibility in annual cycles of birds: implications for endocrine control mechanisms. *J. Ornithol.* 146: 291–304.

Wingfield, J. C. & Farner, D. S. (1979). Some endocrine correlates of re-nesting after loss of clutch or brood in the White-crowned Sparrow *Zonotrichia leucophrys gambelii*. *Gen. Comp. Endocrinol.* 38: 322–331.

Wingfield, J. C. & Silverin, B. (2002). Ecophysiological studies of hormone-behaviour relations in birds. *Hormones, Brain & Behaviour* 2: 587–647.

Wingfield, J. C., Schwabl, H. & Mattocks, P. M. (1990). Endocrine mechanisms of migration. pp. 232–256 in *Bird migration: physiology and ecophysiology* (ed. E. Gwinner). Berlin, Springer-Verlag.

Wingfield, J. C., Doak, D. & Hahn, T. P. (1993). Integration of environmental cues regulating transitions of physiological state, morphology and behaviour. pp. 111–122 in *Avian endocrinology* (ed. P. J. Sharp). Bristol, Society for Endocrinology.

Wink, M. (2006). Use of DNA markers to study bird migration. *J. Ornithol.* 147: 234–244.

Winkel, W. & Frantzen, M. (1991). Ringfund-Analyse zum Zug einer niedersächsischen Population des Trauerschnäppers *Ficedula hypoleuca*. *Vogelkdl. Beitr. Nieders.* 23: 90–98.

Winker, K. (1995). Autumn stopover on the isthmus of Tehuantepec by woodland Nearctic-Neotropical migrants. *Auk* 112: 690–700.

Winker, K., Warner, D. W. & Weisbrod, A. R. (1991). Unprecedented stopover site fidelity in a Tennessee Warbler. *Wilson Bull.* 103: 512–514.

Winker, K., Warner, D. W. & Weisbrod, A. R. (1992a). Migration of woodland birds at a fragmented inland stopover site. *Wilson Bull.* 104: 580–598.

Winker, K., Warner, D. W. & Weisbrod, A. R. (1992b). Daily mass gains among woodland migrants at an inland stopover site. *Auk* 109: 853–862.

Winkler, D. W., Wrege, P. H., Allen, P. E., Kast, T. L., Senesac, P., Wasson, M. F., Llambías, P. E., Ferretti, V. & Sullivan, P. J. (2004). Breeding dispersal and philopatry in the Tree Swallow. *Condor* 106: 768–776.

Winkler, H. & Leisler, B. (1992). On the ecomorphology of migrants. *Ibis* 134 (suppl 1): 21–28.

Winstanley, D. R., Spencer, R. & Williamson, K. (1974). Where have all the Whitethroats gone? *Bird Study* 21: 1–14.

Witherby, H. F., Jourdain, F. C. R., Ticehurst, N. F. & Tucker, B. W. (1938). *The Handbook of British birds*, Vol. 1. London, Witherby.

Witter, M. S. & Cuthill, I. C. (1993). The ecological costs of avian fat storage. *Phil. Trans. Roy. Soc., Lond. B.* 340: 73–92.

Witter, M. S., Cuthill, I. C. & Bonser, R. (1994). Experimental investigations of mass-dependent predation risk in the European Starling *Sturnus vulgaris*. *Anim. Behav.* 48: 201–222.

Wobeser, G. A. (1981). *Diseases of wild waterfowl*. New York, Plenum Press.

Woffinden, N. D. & Murphy, J. R. (1977). Population dynamics of the Ferruginous Hawk during a prey decline. *Great Basin Nat.* 37: 411–425.

Wolff, W. J. (1970). Goal orientation versus one-direction orientation in Teal, *Anas c. crecca* during autumn migration. *Ardea* 58: 131–141.

Wolfson, A. (1942). Regulation of spring migration in juncos. *Condor* 44: 237–263.

Wolfson, A. (1945). The role of the pituitary, fat deposition, and body weight in bird migration. *Condor* 47: 95–127.

Wolfson, A. (1952). The occurrence and regulation of the refractory period in the gonadal and fat cycles of the Junco. *J. Exp. Zool.* 121: 311–326.

Wolfson, A. (1953). Gonadal and fat response to a 5:1 ratio of light to darkness in the White-throated Sparrow. *Condor* 55: 187–192.

Wolfson, A. (1954). Production of repeated gonadal, fat and moult cycles within one year in the Junco and White-crowned Sparrow by manipulation of day length. *J. Exp. Zool.* 125: 353–376.

Wolfson, A. (1959). Ecologic and physiologic factors in the regulation of spring migration and reproductive cycles in birds. pp. 38–70 in *Comparative endocrinology*. New York, Wiley.

Wolfson, A. (1970). Light and darkness and circadian rhythms in the regulation of annual reproductive cycles in birds. pp. 93–119 in *Photorégulation de la Reproduction chez les Oiseaux et les Mammiféres* (eds J. Benoit & I. Assenmacher). Paris, Colloques Intern CNRS No. 172.

Wood, B. (1979). Changes in numbers of overwintering Yellow Wagtails *Motacilla flava* and their food supplies in a West African savanna. *Ibis* 121: 228–231.

Wood, B. (1982). The trans-Saharan spring migration of Yellow Wagtails (*Motacilla flava*). *J. Zool. (Lond.)* 197: 267–283.

Wood, B. (1992). Yellow Wagtail *Motacilla flava* migration from West Africa to Europe: pointers towards a conservation strategy for migrants on passage. *Ibis* 134 (suppl 1): 66–76.

Wood, N. A. (1908). Notes on the spring migration (1907) at Ann Arbor, Michigan. *Auk* 25: 10–15.

Woodford, J. (1963). Ontario-Western New York region. *Audubon Field Notes* 17: 399–401.

Woodrey, M. S. (2000). Age-dependent aspects of stopover biology of passerine migrants. *Stud. Avian Biol.* 20: 43–52.

Woodrey, M. S. & Chandler, C. R. (1997). Age-related timing of migration: geographic and interspecific patterns. *Wilson Bull.* 107: 52–67.

Woodrey, M. S. & Moore, F. R. (1997). Age-related differences in the stopover of fall land-bird migrants on the coast of Alabama. *Auk* 114: 695–707.

Woolfenden, G. E. & Fitzpatrick, J. W. (1978). The inheritance of territory in group-breeding birds. *Bioscience* 28: 104–108.

World Resources Institute, United Nations Environment Program, United Nations Development Program (1994). *World resources 1994–94: a guide to the global environment*. New York, Oxford University Press.

Wunderle, J. (1992). Sexual habitat segregation in wintering Black-throated Blue Warblers in Puerto Rico. pp. 299–307 in *Ecology and conservation of neotropical migrant landbirds* (eds J. M. Hagan & D. W. Johnston). Washington, D.C., Smithsonian Press.

Wunderle, J. M. (1995). Population characteristics of Black-throated Blue Warblers wintering in three sites on Puerto Rico. *Auk* 112: 931–946.

Wunderle, J. M. & Latta, S. C. (2000). Winter site fidelity of nearctic migrants in shade coffee plantations of different sizes in the Dominican Republic. *Auk* 117: 596–614.

Wyllie, I. (1981). *The Cuckoo*. London, Batsford.

Wyllie, I. & Newton, I. (1991). Demography of an increasing population of Sparrowhawks. *J. Anim. Ecol.* 60: 749–766.

Wyllie, I., Dale, L. & Newton, I. (1996). Unequal sex-ratio, mortality causes and pollutant residues in Long-eared Owls in Britain. *Br. Birds* 89: 429–436.

Wyndham, E. (1982). Movements and breeding seasons of the Budgerigar. *Emu* 82: 276–282.

Yahner, R. H. & Scott, D. P. (1988). Effects of forest fragmentation on depredation of artificial nests. *J. Wildl. Manage.* 52: 158–161.

Yalden, D. W. & Pearce-Higgins, J. W. (1997). Density dependence and winter weather as factors affecting the size of a population of Golden Plovers *Pluvialis apricaria*. *Bird Study* 44: 227–234.

Ydenberg, R. C. (2004). Western Sandpipers have altered migration tactics as Peregrine Falcon populations have recovered. *Proc. R. Roc. Lond. B* 271: 1263–1269.

Ydenberg, R. C., Butler, R. W., Lank, D. B., Guglielmo, C. G., Lemon, M. & Wolf, N. (2002). Trade-offs, condition dependence and stopover site selection by migrating sandpipers. *J. Avian Biol.* 33: 47–55.

Yésou, P. (1995). Individual migration strategies in Cormorants *Phalacrocorax carbo* passing through or wintering in western France. *Ardea* 83: 267–274.

Yong, W. & Moore, F. R. (1993). Relation between migratory activity and energetic condition among thrushes (Turdinae) following passage across the Gulf of Mexico. *Condor* 95: 934–943.

Yong, W. & Moore, F. R. (1997). Spring stopover of intercontinental migratory thrushes along the northern coast of the Gulf of Mexico. *Auk* 114: 263–278.

Yong, W., Finch, D. M., Moore, F. R. & Kelly, J. F. (1998). Stopover ecology and habitat use of migratory Wilson's Warblers. *Auk* 115: 829–842.

Yosef, R. (2003). Nocturnal arrival at a roost by migrating Levant Sparrowhawks. *J. Raptor Res.* 37: 64–67.

Yosef, R. & Tryjanowski, P. (2002a). Avian species saturation at a long-term ringing station – a never-ending story? *J. Yamashina Inst. Ornithol.* 34: 89–95.

Yosef, R. & Tryjanowski, P. (2002b). Differential spring migration of Ortolan Bunting *Emberiza hortulana* by sex and age at Eilat, Israel. *Ornis Fenn.* 79: 173–180.

Yosef, R., Tryjanowski, P. & Bildstein, K. L. (2002). Spring migration of adult and immature Buzzards (*Buteo buteo*) through Elat, Israel: timing and body size. *J. Raptor Res.* 36: 115–120.

Yosef, R., Fornasari, L., Tryjanowski, P., Bechard, M. J., Kaltenecker, G. S. & Bildstein, K. (2003). Differential spring migration of adult and juvenile Levant Sparrowhawks (*Accipiter brevipes*) through Eilat, Israel. *J. Raptor Res.* 37: 31–36.

Young, B. E. (1991). Annual moult and interruption of the fall migration for moulting in Lazuli Buntings. *Condor* 93: 236–250.

Ytreberg, N.-J. (1972). The stationary bird populations in the breeding season, 1968–1970, in two mountain forest habitats on the west coast of Norway. *Norway J. Zool* 20: 61–89.

Yunick, R. P. (1983). Winter site fidelity of some northern finches (Fringillidae). *J. Field Ornithol.* 54: 254–258.

Yunick, R. P. (1997). Geographical distribution of re-encountered Pine Siskins captured in upstate, eastern New York during the 1989–90 irruption. *N. Am. Bird Bander* 22: 10–15.

Yuri, T. & Rohwer, S. (1997). Molt and migration in the Northern Rough-winged Swallow. *Auk* 114: 249–262.

Zahavi, A. (1971). The social behaviour of the White Wagtail *Motacilla alba alba* wintering in Israel. *Ibis* 113: 203–211.

Zahavi, A. (1989). Arabian Babbler. pp. 253–275 in *Lifetime reproduction in birds* (ed. I. Newton). London, Academic Press.

Žalavicius, M. (2001). A review of the practical problems resulting from the impact of the climate warming on birds. *Acta Orn. Lithuanica* 11: 332–339.

Zalakevicius, M. & Zalakeviciute, R. (2001). Global climate warming and birds: a review of research in Lithuania. *Folia Zool.* 50: 1–17.

Zalakevicius, M., Bartkeviciene, G., Raudonikis, L. & Janultaitis, J. (2006). Spring arrival response to climate change in birds: a case study from eastern Europe. *J. Ornithol.* 147: 326–343.

Zalles, J. I. & Bildstein, K. L. (2000). *Raptor watch. A global directory of raptor migration sites.* Cambridge, U.K., Birdlife International & Hawk Mountain Sanctuary.

Zanette, L., Doyle, P. & Trémont, S. M. (2000). Food shortage in small fragments: evidence from an area-sensitive passerine. *Ecology* 81: 1654–1666.

Zann, R. & Runciman, D. (1994). Survivorship, dispersal and sex-ratios of Zebra Finches *Taeniopygia guttata* in southeast Australia. *Ibis* 136: 136–146.

Zann, R. A., Morton, S. R., Jones, K. R. & Burley, N. T. (1995). The timing of breeding by Zebra Finches in relation to rainfall in central Australia. *Emu* 95: 208–222.

Zehnder, S., Åkesson, S., Liechti, F. & Bruderer, B. (2001). Nocturnal autumn bird migration at Falsterbo, South Sweden. *J. Avian Biol.* 32: 239–248.

Zenatello, M., Serra, L. & Bacceti, N. (2002). Trade-offs among body mass and primary moult patterns in migrating Black Terns *Chlidonias niger. Ardea* 90: 411–420.

Zicus, M. C. (1981). Molt migration of Canada Geese from Crex Meadows, Wisconsin. *J. Wildl. Manage.* 45: 54–63.

Zimin, V. B. (2001). Seasonal changes in the composition of marked population of Robins *Erithacus rubecula* in the Ladoga area. *Avian Ecol. Behav.* 6: 37–38.

Zimin, V. B. (2002). Distribution of birds in the taiga zone of north-western Russia: a review of current data. *Avian Ecol. Behav.* 8: 79–105.

Zimmerman, J. L. (1966). Effects of extended tropical photoperiod and temperature on the Dickcissel. *Condor* 68: 377–387.

Zink, G. (1962). Eine Mönchsgrasmücke (*Sylvia atricapilla*) zicht in Herbst von Oberösterreich nach Irland. *Vogelwarte* 21: 222–223.

Zink, G. (1973–85). *Der Zug europäischer Singvögel.* Möggingen, Vogelzug Verlag.

Zink, R. M. (1994). The geography of mitochondrial DNA variation, population structure, hybridisation, and species limits in the Fox Sparrow (*Passerella iliaca*). *Evolution* 48: 96–111.

Zink, R. M. (1997). Phylogeographic studies of North American birds. pp. 301–324 in *Avian molecular evolution and systematics* (ed D. P. Mindell). London, Academic Press.

Zink, G. & Bairlein, F. (1995). *Der Zug Europäischer Singvögel.* Wiesbaden, AULA-Verlag.

Zink, R. M. & Dittman, D. L. (1993). Population structure and gene flow in the Chipping Sparrow (*Spizella passerina*) and a hypothesis for evolution in the genus *Spizella. Wilson Bull.* 105: 399–413.

Zonfrillo, B. (1983). Isle of May Bird Observatory and Field Station report for 1982. *Scott. Birds* 12: 218–224.

Zu-Aretz, S. & Y. Leshem. (1983). The sea – a trap for gliding birds. *Torgos* 5: 16–17.

Zuckerbrot, Y. D., Safriel, U. N. & Paz, U. (1980). Autumn migration of Quail *Cotunix coturnix* at the north coast of the Sinai peninsula. *Ibis* 122: 1–14.

Zumeta, D. C. & Holmes, R. T. (1978). Habitat shift and roadside mortality of Scarlet Tanagers during a cold wet New England spring. *Wilson Bull.* 90: 575–586.

Zwarts, L. (1990). Increased prey availability drives premigration hyperphagia in Whimbrels and allows them to leave the Banc d'Arguin, Mauritania, in time. *Ardea* 78: 279–300.

Zwarts, L., Blomert, A. M. & Hupkes, R. (1990a). Increase of feeding time in waders preparing for spring migration from the Banc D'Arguin, Mauritania. *Ardea* 78: 237–256.

Index